Peter Aproian
1977

Պետրոս Աբրոյեան
Peter Aproian
Prov R I
1977

AF251648

Peter Aproian
1977

Պետրոս Աբրոյեան
Peter Aproian
Prov R I
1977

The Constitution
of Inorganic Compounds

The Constitution of Inorganic Compounds

Atomic Quantum Mechanics:
Metals and Intermetallic Compounds

John L. T. Waugh

Wiley-Interscience

A Division of John Wiley & Sons, Inc.

New York · London · Sydney · Toronto

In Memory of my Father
JOHN THOMSON WAUGH
of Avonhead, Scotland

Foreword

In *The Constitution of Inorganic Compounds*, Professor John L. T. Waugh discusses in detail the quantum theory of atomic structure and the structure and properties of metals and intermetallic compounds.

His presentation of atomic theory is thorough. The student of inorganic chemistry should find this book especially valuable in that the derivations are for the most part presented in full, without the omission of intermediate mathematical steps.

Professor Waugh has during most of his scientific life been interested in the general problem of the structure and properties of metals and intermetallic compounds, and his treatment of this subject reflects this interest. Adequate discussion is presented both of the simple metals and of many of the complex intermetallic compounds that have been thoroughly investigated during recent years. Professor Waugh has carried out with success his innovation of beginning the discussion of the electronic structure of the chemical bond by treating metals and intermetallic compounds.

I believe that the serious student of inorganic chemistry will find this book to be of great value to him.

LINUS PAULING

Preface

The properties and nature of the great variety of molecules and crystals of chemical compounds are dependent primarily on the nuclear charges and the spatial configuration with respect to one another of their constituent atoms. The nuclear charge on an atom determines the number and energy states of its constituent electrons, which in turn is of principal importance in determining the nature of the interaction between atoms. By the constitution of a compound is meant a knowledge of the type and spatial arrangement of its interatomic bonds, the ground state energy of the compound, and the manner in which it interacts with electromagnetic radiation and other reagents to give rise to excited states or to the formation of other substances. This book is concerned with the constitution of inorganic compounds, interpreted in terms of the generalizing principles of structure theory and quantum mechanics. The other major organizing principle in chemistry, thermodynamics, which is essentially a collection of useful relationships between quantities, each of which is independently measureable (mass, length, time, temperature, quantity of substance, electric current), is incapable of providing any microscopic explanation of macroscopic chemical change or attributing any atomic or molecular significance to its functions. On the other hand, there is reason to believe that all of the facts of chemistry may be deducible as logical consequences of the laws and generalizations of quantum mechanics. However, this state of affairs is still far from being realized, and chemistry remains largely an experimental science; crystal and molecular structures, the numerical values of the physical constants of compounds, have still to be determined experimentally, and theories then designed to organize and account for the observed facts.

Most of the material in this book has already been dealt with, in one manner or another, in various other books, treatises, and reviews. There are available many excellent and detailed accounts of quantum mechanics, descriptive inorganic chemistry, crystallography, valency and bonding, electron theories of metals, coordination compounds, group theory, ligand field theory, crystal lattice dynamics, classical and quantum statistics, and molecular structure. There are also many books in which the principal results of some or other of these disciplines are stated, without any detailed mathematical justification for them being given. It is attempted in this book to discuss the constitution of inorganic compounds, at a level perhaps suitable for senior undergraduate or beginning graduate students in chemistry, without simply stating results. In some places the argument adopted may be more in the direction of an appeal to physical plausibility than a formal or rigorous mathematical deduction, but generally a deductive approach is followed. The necessary mathematics is developed in sufficient detail at the appropriate place in the text where it is first required, rather than placing such material in appendices either at the end of a chapter or of the book. For some compounds or groups of compounds, where neither theoretical nor structural data is available, a more descriptive approach has necessarily had to be used. In most respects the information contained in this book is complementary to the usual textbooks of inorganic chemistry, and a selection of topics has been necessary.

The first part (Part Q) deals with the quantum theory of atomic structure, with particular attention being given to physical interpretations and the calculation of discrete energy values. The second part (Part M) develops and applies electron theories to the interpretation of the properties of metals and intermetallic compounds in perhaps more detail than is usual in chemistry books.

All of the illustrations used in this book have been specially drawn or redrawn to illuminate the specific points concerned. It is difficult to acknowledge all of the sources of information which were used in compiling the original lecture notes on which this book is based, but reference is made throughout the text to papers and books which contain the referred-to results of experiment or calculation, or which contain special arguments and information, and these are necessarily selective.

I wish to thank Dr. J. Adin Mann for his helpful comments on the first part of this book at an early stage in the assembly of the manuscript, and especially Professor Linus Pauling for reading the manuscript and writing the foreword.

John L. T. Waugh

Honolulu, Hawaii

Contents

CONTENTS

PART Q

Quantum Theory
of
Atomic Structure

Introduction

Any contemplated discussion of the structure of atoms necessarily presupposes the existence of such entities. There is now, of course, overwhelming evidence to support the contention that all matter is composed, either of free atoms, as are the rare gases and many metals in the vapor phase, or more generally, of combinations of two or more similar or dissimilar types of atoms, interacting with one another in various ways. Each type is best characterized by a nuclear charge number, Z, equivalent numerically to the number of electrons in the atom, and a mass number, A, as in the symbol $_Z X^A$, where X denotes the chemical name of the type of atom. Atoms, or nuclides,[1] corresponding to some 105 different Z's, including both naturally occurring and synthetic species, and some 1666 different A's, again including the natural and the proportionately larger number of synthetic, radioactive, unstable, types, are now recognized.[2] If a piece of matter is composed entirely of atoms of the same charge number, Z, it is referred to as an element. The major constituents of atoms, from the point of view of the chemical constitution of matter, are the nucleus which accounts for the major part of the mass of the atom and may contain one or more nucleons, and a number of electrons equal in number to the nuclear charge; the nuclide with zero charge is generally regarded as a "fundamental" particle, the neutron.[3] Some of the more important experimental evidence for the existence of atoms includes the observation of Brownian motion in monoatomic liquids, the Cantor experiment, the observation of the radioactive decay of various nuclides, many observations involving X-ray, electron and neutron diffraction

---

[1] See page 11.

[2] "1961 Nuclidic Mass Table," König, L. A., Mattauch, J. H. E., and Wapstra, A. H., *Nucl. Phys.*, 1962, **31**, 18; subsequent "Nuclear Data Sheets" of the Nuclear Data Project of the National Academy of Sciences—National Research Council.

[3] Nucleons, commonly regarded as neutrons or protons, each have a composite structure in which regions of both positive and negative charge may be distinguished.

phenomena, the agreement of the Avogadro number determined by a variety of methods, the syntheses of radioactive, unstable, atoms from other atomic species, and the direct observation of some metal atoms by the technique of phase contrast microscopy.

It is also now well established that matter and energy are interconvertible, and each is associated with four fundamentally important attributes. First, the mass of ordinary matter has been known since antiquity, but the mass equivalent of energy has only been recognized since 1904 (Hasenöhrl, 1904; Einstein, 1905); second, the inertia of matter was apparent by 1560 (Galilei) but the inertia of energy was not discovered until 1923 (Compton); third, the particulate nature of matter inherent in the postulate of the atomic structure of matter was first clearly realized in 1812 (Dalton), but it was not until 1900 that Planck found it necessary to postulate the particulate nature of radiation; fourth, the wave nature of electromagnetic radiation was discussed in 1680 by Huygens (for light in particular) but there was no necessity to be concerned about the wave nature of matter before 1923 (de Broglie).

The present state of knowledge of atomic structure has been arrived at as a result of many fundamental discoveries in both chemistry and physics, but there are three major sources of experimental evidence which have converged with the modern atomic quantum theory. These are: the work extending from the early 1800s to the present concerned initially with the determination of the relative combining masses of atoms of different elements, which changed in emphasis after 1920 with the construction by Aston of the first mass spectrograph and which led by 1869 (Mendelyeev) to the discovery of the Periodic Law and much later to the recognition of nuclear isotopes, isobars, isotones, and isomers;‡ then the work associated with the elucidation of the

---

‡ Isotopes are nuclides with the same charge number but different mass numbers, e.g., $_{10}Ne^{20}$ and $_{10}Ne^{22}$; isobars are nuclear species with the same mass number but different charge numbers, e.g., $_{52}Te^{130}$, $_{54}Xe^{130}$, and $_{56}Ba^{130}$; isotones are nuclides with dissimilar mass and charge numbers, but with the same neutron number, $(A - Z)$, e.g., $_{14}Si^{30}$, $_{15}P^{31}$, and $_{16}S^{32}$, each nucleus containing 16 neutrons; nuclear isomers are different energy states of the same nucleus, e.g., $_{51}Sb^{124}$, $_{51}Sb^{124m_1}$, and $_{51}Sb^{124m_2}$, have half-lives of 60 days, 1.5 minutes, and 21 minutes respectively. The use of the superscript, m (metastable), following the mass number has become the accepted manner of representing isomeric nuclear states (other than the ground state); in the event that there are more than two excited states, the labels m_1, m_2,..., are used in the order of increasing excitation energies.

---

line spectra and the X-ray or high frequency spectra of the elements, extending over the period from about 1880 to 1925; and, finally, work involving the investigation of the scattering of electromagnetic radiation and beams of particles by matter, leading to the birth of wave mechanics. Each of these approaches, necessary for an understanding of atomic quantum mechanics, is briefly discussed in the first two chapters of this section, the remaining chapters being concerned with the development of atomic quantum theory. A knowledge of atomic structure, of the electron configuration and energy states of atoms, is a necessary prerequisite for the discussion of the constitution of chemical compounds.

Q1

Development of Atomic Theory

Q.1.1. The Relative Masses of Atoms

The vague, philosophical idea that matter has a discontinuous structure, formulated by Leucippus and Democritus in the fifth century B.C., later elaborated upon by Epicurus and Lucretius some four hundred years later, was revived very much later during the sixteenth and seventeenth centuries by scientific philosophers such as Bacon, Boyle, Newton, Descartes, and Galilei. Not until the publication of Dalton's *New System of Chemical Philosophy* in 1808, however, was the hypothesis of the atomic structure of matter clearly stated and attempts made to provide experimental confirmation of this concept. There was thus initiated a period during which intensive efforts were made to determine the relative combining or equivalent masses of different elements, resulting in the discovery of many empirical relationships between combining weights, gas densities, combining volumes of gases, heat capacities of solids, and the occurrence of isomorphism between chemically similar compounds. From accumulated analytical data for many simple compounds, there emerged the concept of the chemical atomic weight as the smallest weight of an element which could be contained in any of its compounds, this minimum weight being associated with one atom, relative to some arbitrarily selected reference standard. Actually, many of the early gas density measurements were effectively atomic weight determinations, and since it was convenient to use the least dense known gas, hydrogen, as a basic unit of gas density, in which to express the density of other gases in terms of a convenient numerical scale, this led to the adoption of the first atomic weight scale on which the atomic weight of hydrogen was taken as unity, as suggested by Davy in 1812. Confusion about the atomic weight of an element, representing the smallest conceivable particle of the element which could enter into combination with other elements in compound formation, and the molecular weight of a gas, such as nitrogen (N_2), ozone (O_3), or phosphorus vapor (P_4), where the molecule, representing the smallest unit of the gas which can exist as such, may contain two or more atoms, was not resolved until after the general acceptance of Avogadro's hypothesis. The increasing use of oxidation and reduction methods for the determination of equivalent or combining weights, which were observed to be some simple submultiple of the chemical atomic weight, led to the simultaneous introduction of several oxygen-based atomic weight scales; Berzelius arbitrarily selected 100, Thomson recommended unity, and Wollaston used 10, as the atomic weight of oxygen. On the basis of $H = 1$, the atomic weight of oxygen appeared to be about 16, and Ostwald proposed in 1885 that the adoption of 16.000 as the atomic weight of oxygen would provide a convenient basis for assigning atomic weights to the various elements which would combine the advantages of the $H = 1$ scale, on which all atomic weights are conveniently small numbers greater than unity,[1] and the oxygen-based scales which enabled greater experimental accuracy to be attained in combining weight measurements. In 1889, the International Commission on Atomic Weights was formed, adopting exactly one sixteenth of the mass of the oxygen atoms in atmospheric air as the chemical atomic mass unit; this unit survived as the reference basis for chemical atomic weights until 1961. It is remarkable that Ostwald remained unconvinced of the existence of atoms, even as late as 1904. One of the important consequences of the accumulated work on the determination of atomic weights by chemical methods was the discovery of the Periodic Law, consequent to observations made about the relative magnitudes of the atomic weights of the elements by Döbereiner, Cannizzaro, Chancourtois, Newlands, Lothar Meyer, and Mendelyeev. Unfortunately for the subsequent development of the understanding of the nature of atoms, dominant concern with chemical methods of atomic weight determination, occupying the better part of the life-times of successive generations of chemists such as Berzelius, Stas, Dumas, Richards, Hönigschmidt, and Whytlaw-Gray, for example, delayed recognition of the fact that the atomic masses of the atoms present in a compound determine very few of its characteristic properties. In fact, only density, rate of diffusion, and such other properties as are a function of the momentum or kinetic energy of an atom or molecule are influenced by mass alone. Recognition of the periodic relationship between the physical and chemical properties of the elements and their atomic weights provided the incentive for still more intensive efforts to determine combining and atomic weights with higher accuracy, enabling the existence of then-unknown elements to be predicted

---

[1] Also, as first pointed out by Prout (1816), many of which are close to integers.

and subsequently discovered. If all of the known elements (50 were recognized in 1820; 103 by 1960) are arranged in the order of increasing atomic weight, elements with similar properties are observed to recur after periods of 2, 8, 8, 18, 18, 32, and 19 (incomplete), provided the pairs of elements argon and potassium, cobalt and nickel, tellerium and iodine, thorium and protoactinium, are written in this order, even though the first member of each of these four pairs has a larger atomic weight than the second member. The explanation for the necessity of these reversals did not become apparent until after the discovery of isotopes and precise measurements had been made of the relative abundance of isotopes in naturally occurring species. This sequence of numbers, 2, 8, 18, 32, ..., (neglecting meantime the fact that 8 and 18 appear in the observed series twice), which is equivalent to the series, 2×1^2, 2×2^2, 2×3^2, 2×4^2, 2×5^2, ... appeared again some fifty years later as a result of experimental information gathered from a totally different source, namely from examination of the line spectra of the elements, in terms of the first attempted application of the quantum postulate in this area. The theoretical explanation of this sequence did not emerge until after the development of Schrödinger mechanics. At the time when the Periodic Law first assumed importance (after 1869) and fewer elements were known, the periodic interval, 8, attracted more attention than the other members of the series, 2, 18, and 32. Excessive concern with the number, 8, to the essential exclusion of 2, 18, and 32, undoubtedly delayed the attainment of a better understanding of chemical bonding by many years, and perhaps also is responsible for the fact that stable compounds of the rare gases were not discovered until 1962.

After the discovery of the electron as a result of the work of Crookes, J. J. Thomson, and many others, the pioneering work of Becquerel, the Curies, Rutherford, and Soddy among others on natural radioactivity, the recognition of the nature of the high-frequency or X-ray spectra of the elements by Barkla, van den Broek, and Moseley, it became apparent that the chemical atomic weights of elements were simple numerical averages and of no fundamentally important theoretical consequence. Actually, with the construction of the first mass spectrograph by Aston, it became possible to determine in a few months the atomic masses of a greater number of species and with a greater degree of precision, than had been accomplished by the cumulative efforts of many during the preceding 80 years. The initial discovery of isotopes (atomic species with the same nuclear charge number but different mass numbers) was made among the naturally radioactive elements, then isotopes of lead, the stable end-product of each of the three natural radioactive series, were detected. With his first mass spectrometer, Aston was able to establish that neon, the first known stable material to exhibit isotopy, contained two types of atom with atomic masses almost exactly 20 and 22, compared with the 1961 chemical atomic weight of 20.183. The discovery of stable isotopes of oxygen[1] and of hydrogen[2] however, was of much greater consequence. Initially Giaque and Johnston believed that the presence of isotopes of oxygen would not affect the established basis for reporting chemical atomic weights. However Urey and Greiff[3] a few years later, concluded that the atomic weight of oxygen in atmosphereic carbon dioxide was not identical with that of oxygen in water, and that the two values differed from one another by an amount which was significantly larger than the experimental error involved in mass spectrographic measurements. It is now well established that isotopic abundance ratios differ considerably among naturally occurring compounds. It is therefore now necessary to determine and report separately isotopic abundance ratios and isotopic masses. But an accurate reference scale is still required to report the isotopic masses. Aston and successive mass spectroscopists, before the discovery of oxygen isotopes, adopted the system which was in general use until 1961, of taking the mass of the most abundant isotope of oxygen (which he assumed to be mono-isotopic) to be exactly 16 units. This so-called physical atomic mass unit, exactly one sixteenth of the mass of $_8O^{16}$, thus differs from the previous chemical atomic mass unit, which was taken to be one sixteenth of the mass of the oxygen atoms in atmospheric oxygen.[4] The mass unit is therefore larger on the chemical than on the physical scale, and so the numerical value of any atomic weight is smaller when expressed on the chemical scale. From the relative isotopic abundances of $_8O^{16}$, $_8O^{17}$, and $_8O^{18}$, the conversion factor between the two is:

$$\frac{\text{chemical atomic mass unit}}{\text{physical atomic mass unit}} = 1.000272 \pm 0.000005,$$

where the uncertainty is due to the uncertainty in the isotopic composition of natural oxygen.‡ A complete set of atomic masses on both scales is available.[5]

‡ All of the nine elements in the atmosphere, H, He, Ne, N, O, Ar, C, Kr, and Xe, have isotopes, 32 altogether, and on the basis of natural abundance data and the assumption of random distribution of isotopes in a compound molecule, the isotopic composition of atmospheric air is shown in Table Q.1.1. (Eck, C. F., *J. Chem. Ed.*, 1969, **46**, 706).

[1] Giaque, W. F. and Johnston, H. L., *J. Amer. Chem. Soc.*, 1929, **51**, 1436.
[2] Urey, H. C., Brickwedde, F. G., and Murphy, G. M., *Phys. Rev.*, 1932, **39**, 164.
[3] Urey, H. C. and Greiff, L. J., *J. Amer. Chem. Soc.*, 1935, **57**, 321.
[4] Contains approximately 99.759% O^{16}, 0.0374% O^{17}, and 0.2039% O^{18}.
[5] Everling, F., König, L. A., Mattauch, J. H. E., and Wapstra, A. H., *Nucl. Phys.*, 1960, **18**, 529.

TABLE Q.1.1

Mass Number	Gas	Conc (% by volume)
2	H_2	5.00×10^{-5}
3	HD	1.5×10^{-8}
3	^3He	7.0×10^{-10}
4	^4He	5.24×10^{-4}
20	^{20}Ne	1.645×10^{-3}
21	^{21}Ne	5.0×10^{-6}
22	^{22}Ne	1.68×10^{-4}
28	$^{14}N^{14}N$	77.440
29	$^{14}N^{15}N$	4.674×10^{-1}
30	$^{15}N^{15}N$	1.0×10^{-3}
32	$^{16}O^{16}O$	20.845
33	$^{16}O^{17}O$	1.55×10^{-2}
34	$^{16}O^{18}O$	8.55×10^{-2}
34	$^{17}O^{17}O$	3.0×10^{-6}
35	$^{17}O^{18}O$	1.6×10^{-5}
36	$^{18}O^{18}O$	8.7×10^{-5}
36	^{36}Ar	3.11×10^{-3}
38	^{38}Ar	5.90×10^{-4}
40	^{40}Ar	9.303×10^{-1}
44	$^{12}C^{16}O_2$	3.25×10^{-2}
45	$^{12}C^{16}O^{17}O$	2.6×10^{-5}
45	$^{13}C^{16}O_2$	3.6×10^{-4}
46	$^{12}C^{16}O^{18}O$	1.3×10^{-4}
46	$^{12}C^{17}O_2$	4.0×10^{-9}
46	$^{13}C^{16}O^{17}O$	3.0×10^{-7}
47	$^{13}C^{16}O^{18}O$	1.0×10^{-6}
47	$^{13}C^{17}O_2$	7.0×10^{-11}
47	$^{12}C^{17}O^{18}O$	5.0×10^{-8}
48	$^{13}C^{17}O^{18}O$	5.0×10^{-9}
48	$^{12}C^{18}O_2$	3.0×10^{-7}
49	$^{13}C^{18}O_2$	3.0×10^{-9}
78	^{78}Kr	4.0×10^{-7}
80	^{80}Kr	2.6×10^{-6}
82	^{82}Kr	1.3×10^{-5}
83	^{83}Kr	1.3×10^{-5}
84	^{84}Kr	6.5×10^{-5}
86	^{86}Kr	2.0×10^{-5}
124	^{124}Xe	8.0×10^{-9}
126	^{126}Xe	8.0×10^{-9}
128	^{128}Xe	1.7×10^{-7}
129	^{129}Xe	2.3×10^{-6}
130	^{130}Xe	3.5×10^{-7}
131	^{131}Xe	1.84×10^{-6}
132	^{132}Xe	2.34×10^{-6}
134	^{134}Xe	9.1×10^{-7}
136	^{136}Xe	7.7×10^{-7}

The accurate determination of atomic masses generally involves the measurement of the charge to mass ratio of positive ions from the observed amount of deflection suffered by them in a combination of electric and magnetic fields; ions with a particular charge to mass ratio, ε/M, may be separated by the use of velocity or directional focussing, or both, or in time-of-flight instruments by taking advantage of the fact that ions with the same kinetic energy but differing masses traverse a given path length in different times. It is common practice to refer to instruments which use photographic plates to record mass spectra as mass spectrographs; those which make use of the collection and measurement of ion currents are called mass spectrometers. The most accurate techniques for the exact determination of atomic masses in mass spectrometry relate the measured mass to C^{12}, and the ratio of their masses to this nucleus can be more accurately determined than the ratio of C^{12} to O^{16}; the precision to which a particular mass may be expressed on a C^{12} scale is thus greater than that on the O^{16} scale, which in turn is superior to the precision on the natural oxygen, chemical scale. The preference of C^{12} as a mass reference standard is related to the comparative ease of producing ions containing many carbon atoms in mass spectrometers. Partly in order to eliminate the use of two different scales and the uncertainty in chemical atomic weights caused by the variable isotopic composition of natural compounds, a new scale was recommended by the Council of the International Union of Pure and Applied Chemistry in 1959 and adopted by the International Unions of Pure and Applied Physics and Chemistry in 1960–61; this is based on the mass of an atom of C^{12} being taken as exactly 12.00000 units. On this scale the O^{16} mass is 15.994915 and therefore all masses on the old physical atomic weight scale become reduced by about 0.0318%, whereas the difference between atomic masses reported on the new C^{12} scale only differ by about 0.005% from those on the old chemical atomic weight scale, so that essentially chemical atomic weights remain unchanged. On the new mass scale, as for the older physical and chemical atomic weight scales, it is the usual practice to quote isotopic masses as the mass of the nucleus plus the masses and the binding energies of sufficient electrons to form an electrically neutral atom. The name, isotope, originally proposed by Soddy (1913) to describe one of a group of two or more species with the same Z but dissimilar A numbers, has also been used to signify any particular nuclear species characterized by an A and Z number, but according to a proposal by Kohman (1947), it is now common practice to refer to the latter as a nuclide. Thus, there are ten naturally occurring isotopes of $_{50}Sn^A$, where A has the value, 112, 114, 115, 116, 117, 118, 119, 120, 122, or 124, and there are twenty-two of the known elements composed of only one stable nuclide, namely, $_2He^4$, $_4Be^9$, $_9F^{19}$, $_{11}Na^{23}$, $_{13}Al^{27}$, $_{15}P^{31}$, $_{21}Sc^{45}$, $_{25}Mn^{55}$, $_{27}Co^{59}$, $_{33}As^{75}$, $_{39}Y^{89}$, $_{41}Nb^{93}$, $_{45}Rh^{103}$, $_{53}I^{127}$, $_{55}Cs^{133}$, $_{59}Pr^{141}$, $_{65}Tb^{159}$,

$_{67}$Ho^{165}, $_{69}$Tm^{169}, $_{79}$Au^{197}, $_{83}$Bi^{209}, and $_{90}$Th^{232}. Elements are characterized not only by their mass spectrum but also by their unique X-ray and optical spectra.

On the C^{12} scale of mass, the mass of the electron is 5.48597×10^{-4} mass units, that of the proton is 1.00727663 and of the neutron 1.0086654 mass units respectively;[1] the new mass unit itself, one-twelfth of the mass of C^{12}, is equivalent to 1.6604×10^{-27} kg.‡ Isotopic and nuclidic masses may also be accurately obtained by relating nuclei to one another by means of the energies of nuclear reactions, using the relation between mass and energy which follows as a consequence of Einstein's special theory of relativity, where again it is found that the most satisfactory mass reference is C$^{12} = 12$. From the Hasenöhrl–Einstein equation, the energy, E, equivalent to a mass, M, is given by $E = Mc^2$, where c is the velocity of electromagnetic radiation $(2.997925 \times 10^8 \text{ m s}^{-1})$.[1] Hence one mass unit is equivalent to 1.492×10^{-10} J. In nuclear reactions, it is more convenient to express energies in units of electron volts $(1 \text{ eV} = 1.6021 \times 10^{-19} \text{ J})$, where one electron volt is defined as the energy required to accelerate an electron with charge of 1.60210×10^{-19} C through a potential difference of 1 volt $(1 \text{ V} = 1 \text{ J A}^{-1} \text{s}^{-1})$. Thus, one mass unit on the C^{12} scale is equivalent to 9.315×10^8 eV $= 931.5$ MeV.

‡ In the SI system of units (standard abbreviation in many languages for Système International d'Unités), formally approved by the 11th Conférence Générale des Poids et Mesures in 1960, the six basic units required to express the magnitudes of the numerous properties of substances routinely measured in science and technology are as shown in Tables Q.1.2. to Q.1.5.

TABLE Q.1.2

Physical Quantity	Name of Unit	Symbol for Unit
Length	Meter	m
Mass	Kilogram	kg
Time	Second	s
Electric current	Ampere	A
Thermodynamic temperature	Kelvin	K
Luminous intensity	Candela	cd

With the addition of the dimensionless units of radians and steradians (symbols, rad and sr, respectively) for measurement of plane and solid angles respectively, the comprehensive and rational SI system, now adopted by about thirty different countries but clearly marked to become the universal system of units, make it possible to express the magnitude of all physicochemical properties of substances in terms of these and a relatively

---

[1] All numerical values of constants quoted in this book are those recommended by the National Academy of Sciences—National Research Council Committee on Fundamental Constants in 1963.

small number of derived units. Such a system of units is used throughout this book. The logical superiority and the practical convenience of the SI system make it the preferred system in scientific and commercial practice. Among the major advantages of adopting this system are the elimination of the "electrostatic" and "electromagnetic" units in former use, both being replaced by SI electrical units, the fact that the gravitational constant becomes superfluous since the SI unit of force, the newton ($kg\,m\,s^{-2}$), is independent of the Earth's gravitational field, the unit of energy in all forms is the same, the joule ($N\,m$), and the unit of power is the watt ($J\,s^{-1}$), so that the use of variously defined calories is no longer necessary. Each of the following derived SI units has a special name:

TABLE Q.1.3

Name of unit	Symbol for unit	Definition of unit	Physical quantity
Joule	J	$kg\,m^2\,s^{-2}$	Energy
Newton	N	$kg\,m\,s^{-2} = J\,m^{-1}$	Force
Watt	W	$kg\,m^2\,s^{-3} = J\,s^{-1}$	Power
Coulomb	C	$A\,s$	Electrical charge
Volt	V	$kg\,m^2\,s^{-3}\,A^{-1} = J\,A^{-1}\,s^{-1}$	Potential difference
Ohm	Ω	$kg\,m^2\,s^{-3}\,A^{-2} = V\,A^{-1}$	Electrical resistance
Farad	F	$A^2\,s^4\,kg^{-1}\,m^{-2} = A\,s\,V^{-1}$	Electrical capacitance
Hertz	Hz	s^{-1}	Frequency
Kelvin	K	Celsius $°$ above $-273.15\,K$	Temperature
Weber	Wb	$kg\,m^2\,s^{-2}\,A^{-1} = V\,s$	Magnetic flux
Tesla	T	$kg\,s^{-2}\,A^{-1} = V\,s\,m^{-2}$	Magnetic flux density
Lux	lx	$cd\,sr\,m^{-2}$	Illumination
Lumen	lm	$cd\,sr$	Luminous flux
Henry	H	$kg\,m^2\,s^{-2}\,A^{-2} = V\,s\,A^{-1}$	Inductance

Examples of other derived units, which do not have any special names are:

TABLE Q.1.4

Name of SI unit	Symbol for unit	Physical quantity
Square meter	m^2	Area
Kilogram per cubic meter	$kg\,m^{-3}$	Density
Cubic meter	m^3	Volume
Meter per second	$m\,s^{-1}$	Velocity
Radian per second	$rad\,s^{-1}$	Angular velocity
Meter per second per second	$m\,s^{-2}$	Acceleration
Newton per square meter	$N\,m^{-2}$	Pressure
Square meter per second	$m^2\,s^{-1}$	Diffusion coefficient
Newton second per square meter	$N\,s\,m^{-2}$	Viscosity
Volt per meter	$V\,m^{-1}$	Electrical field strength
Ampere per meter	$A\,m^{-1}$	Magnetic field strength
Candela per square meter	$cd\,m^{-2}$	Luminance

Certain other units, in common use in the older cgs system, which are compatible with the SI system may be redefined as follows:

TABLE Q.1.5

Name of unit	Symbol for unit	Definition of unit	Physical quantity
Barn	b	$10^{-28}\,\mathrm{m}^2$	Nuclear cross section
Liter	l	$10^{-3}\,\mathrm{m}^3$	Volume
Bar	bar	$10^{5}\,\mathrm{N\,m}^{-2}$	Pressure
Stokes	St	$10^{-4}\,\mathrm{m}^2\,\mathrm{s}^{-1}$	Diffusion coefficient
Poise	P	$10^{-1}\,\mathrm{kg\,m}^{-1}\,\mathrm{s}^{-1}$	Viscosity
Gauss	G	$10^{-4}\,\mathrm{T}$	Magnetic induction
Curie	Ci	$3.7 \times 10^{10}\,\mathrm{s}^{-1}$	Radioactivity
Electron volt	eV	$1.6021 \times 10^{-19}\,\mathrm{J}$	Energy

It should be noted that there is no plural form for the symbols of the various units in the SI system, and it is recommended that multiples and submultiples of the basic units should be preferably described by the use of prefixes limited to multiples of $10^{\pm 3n}$, in accordance with the following scheme:

TABLE Q.1.6

Multiple or submultiple	Symbol	Prefix
10^{3}	k	kilo
10^{6}	M	mega
10^{9}	G	giga
10^{12}	T	tera
10^{-3}	m	milli
10^{-6}	μ	micro
10^{-9}	n	nano
10^{-12}	p	pico
10^{-15}	f	femto
10^{-18}	a	atto

The mole (symbol, mol) retains its previous significance. Conversion factors relating the SI units to the units of other systems are listed in various publications.[1-6] Of the four independent base units of the SI system (the units of electric current and light intensity, the ampere and the candela respectively, are not independent of the units of mass, length, time, and temperature), adopted by the 11th General Conference on Weights and Measures in 1960, that of time has been redefined. In 1967, the 13th General Conference defined the second as 9,192,631,770 cycles of the transition between two

<hr>

[1] *Handbook of Mathematical Functions with Formulas, Graphs, and Mathematical Tables*, Edited by Milton Abramowitz and Irene A. Stegun, National Bureau of Standards, U.S. Government Printing Office, Washington, D.C., 20402, 1964.

[2] *Changing to the Metric System*, National Physical Laboratory (Anderton, P. and Bigg, P. H.) H.M.S.O., London, 1969, 3rd ed.

[3] *Standards of Measurement*, Allen V. Astin, *Sci. Am.*, 1968, 50–62.

[4] *Metrication in Scientific Journals*, The Royal Society Conference of Editors, *Am. Sci.*, 1968, 56, 2, 159–164.

[5] McGlashan, M. L., *Physico-Chemical Quantities and Units*, The Royal Institute of Chemistry, London, 1968.

[6] Taylor, B. N., Parker, W. H., and Langenberg, D. N., *The Fundamental Constants and Quantum Electrodynamics*, Academic Press, New York, 1969.

hyperfine energy levels of the fundamental state of the Cs^{133} nuclide. The standard of mass remains, as it has been defined since 1889, as the mass of a particular platinum–iridium cylinder (1 kg) kept at Sèvres; the standard of length is discussed on pages 47–48; the standard of thermodynamic temperature since 1960 is based on six defining points on the International Practical Temperature Scale, the triple point of water at 273.16 K (formally degrees Kelvin), the boiling point of water at 373.15 K, the boiling point of oxygen at 90.18 K, the boiling point of sulfur at 717.75 K, the freezing point of silver at 1233.95 K, and the freezing point of gold at 1336.15 K. It appears quite possible that the basic unit of mass, now referred to as the kilogram will be renamed the "giorgi," symbol G.

The mass of a stable nuclide is generally less than the sum of the masses of its constituent nucleons and electrons, this difference being referred to as the nuclear binding energy.[1] For example, for $_2He^4$, the sum of the masses of the 4 nucleons (2 protons and 2 neutrons, in terms of the neutron–proton hypothesis of nuclear structure) and the 2 electrons is 4.032981 mass units, and the mass determined by mass spectrometric means is found to be 4.002604 mass units. The binding energy is thus 0.030377 mass units or 4.5399 pJ (28.30 MeV). This is equivalent to approximately 1.1375 pJ (7.1 MeV) per nucleon. The average binding energy per nucleon is found to be a rather slowly varying function of the mass number for nuclides of all of the elements, reaching a maximum in the region of $A = 60$ (iron and nickel). If the binding energy of atomic nuclei were known the total mass could be calculated for any nuclide, $_ZX^A$, in a manner similar to that illustrated above for $_2He^4$. Thus if M is the mass of the nuclide, $_ZX^A$, then

$$M = Zm_p + Zm_\varepsilon + (A - Z)m_n - E_b \qquad \text{(q.1.1.)}$$

where m_p, m_ε, m_n represent the proton, electron, and neutron masses, and E_b, Z, and A denote the binding energy, charge number, and mass number respectively of the nuclide. E_b may be calculated, to a fairly good approximation, from the so-called von Weizsäcker,[2] semiempirical mass equation,‡ relating the binding energy of a nuclide to its composition in terms of Z and A,

$$E_b = 14.0A - 13.1A^{2/3} - 18.1(A - 2Z)^2A^{-1} - 0.585Z(Z - 1)A^{-1/3} + \delta A^{-1}$$
$$\text{(q.1.2.)}$$

where δ has the value ± 132, if Z and $N = A - Z =$ number of neutrons in the nucleus, are both even or both odd respectively, and the value, 0, if Z is odd and N is even or if Z is even and N is odd. The nuclidic masses calculated from (q.1.1.) agree reasonably well (within about 1%) with masses measured by mass spectrographic techniques, although the principal values of this

---

[1] This includes the electron binding energy, which is generally a very small fraction of the nucleon binding energy.
[2] Weizsäcker, C. F. von, *Z. Physik*, 1935, **96**, 431; Feenberg, E., *Revs. Mod. Phys.*, 1947, **19**, 239.

expression is associated with its use in predicting nuclear stability and in estimating the radioactive decay energies of nuclides with given mass values; this expression may be used, for example, to calculate the binding energy, which provides an exact measure of the energy liberated in forming an atomic nucleus from its constituent nucleons, even for species which disintegrate spontaneously, under the conditions in which they may be formed in nuclear reactions, so that any experimental determination is quite impossible. The von Weizsäcker, semiempirical mass equation, and the many modifications of it which have been proposed with increasing availability of mass spectroscopic data, was originally derived and the numerical values of the various coefficients evaluated, by fitting to experimental mass data.

‡ Based on the theory of nuclear forces, with the introduction of certain empirical data, this semiempirical equation may be derived as follows. It is assumed that intranuclear forces are wave-mechanical exchange forces, which under the conditions pertaining in nuclei have an extremely minute range of action, leading to the result that the energy liberated from the association of nucleons under the action of these forces is approximately proportional to the mass number, so that if this proportionality constant is denoted by a_1, then $E_b = a_1 A$. The concept that strong exchange forces operate between nucleons is based on the assumption that neutron–proton and proton–neutron states continually alternate with one another. As will be described later in this book, just as the exchange of an electron between two protons gives rise to a chemical bond, so the alternation of the states n–p and p–n results in a binding force of great magnitude in nuclei. Binding forces may also operate between neutrons, but these are considered to be weaker than the force between neutrons and protons. Thus the binding energy of a nucleus containing an excess of neutrons over protons will be smaller than $a_1 A$ by a factor dependent on the neutron excess, given by $(A - 2Z)$. It is observed from calculations based on (q.1.1.) that the binding energies of unstable nuclei compared with those containing equal numbers of protons and neutrons is symmetrical around the region for which $N(= A - Z) = Z$; this indicates a parabolic relationship between the nuclear binding energy and the neutron excess number, that is the reduction in binding energy is proportional to $(A - 2Z)^2$. The binding energy contribution per nucleon pair is proportional to the probability of such a pair being contained within a certain volume, determined by the range of nuclear forces, and this probability in turn is inversely proportional to the nuclear volume, which is proportional to the nuclear mass. Therefore the decreased stability of a nucleus due to the neutron excess is given by $(A - 2Z)^2/A$, or if a_2 is the appropriate proportionality constant, then $E_b = a_1 A - a_2(A - 2Z)^2/A$. The protons in a nucleus, being similarly charged, repel one another, and the work required to overcome this repulsion and build protons into a nucleus amounts to $a_3 Z^2/r$ (Coulomb repulsion between like charges), where a_3 is another proportionality constant and r is the nuclear radius. Since the mass of a nucleus is proportional to its volume or dependent on r^3, r is proportional to $A^{1/3}$. Therefore the binding energy of nuclei must be further reduced by an amount $a_3 Z^2 A^{-1/3}$, or $E_b = a_1 A - a_2(A - 2Z)^2 A^{-1} - a_3 Z^2 A^{-1/3}$. Allowance must also be made for the uncompensated or looser binding of nucleons on the surface of the nucleus than in its interior. The surface of an assumed spherical nucleus is proportional to r^2, or to $A^{2/3}$, so that a further quantity $a_4 A^{2/3}$ must be deducted from the binding energy. So $E_b = a_1 A - a_2(A - 2Z)^2 A^{-1} - a_3 Z^2 A^{-1/3} - a_4 A^{2/3}$. Finally, to take account of the fact that the binding energies calculated by

(q.1.1.) are highest for nuclei containing an even number of both neutrons and protons and that odd-odd nuclei are rare outside the region of the lightest elements (the four stable light nuclei are $_1H^1$, $_3Li^6$, $_5B^{12}$, and $_7N^{14}$), a term δ/A is added to the binding energy expression; the relative stability of even-even, odd-odd, odd-even, and even-odd nuclei can be explained in terms of the tendency of two like particles completing an energy level by pairing opposite spins. So the final semiempirical expression for nuclear binding energies obtained on the basis of the assumptions noted is

$$E_b = a_1A - a_2(A - 2Z)^2A^{-1} - a_3Z^2A^{-1/3} - a_4A^{2/3} + \delta A^{-1},$$

which only differs from (q.1.2.) in that the term containing a_3 is there given as $Z(Z - 1)$ rather than Z^2; the former is preferable since actually each proton of number Z repels each of the other $(Z - 1)$ protons rather than each of the Z protons. The numerical values of the constants a_1, a_2, a_3, a_4, and the parameter δ given in (q.1.2.) are one set of values obtained by best fitting mass spectral data to (q.1.1.) with the aid of the E_b equation.

In most chemical applications of atomic masses, the quantities that are of importance, in the SI system as in the cgs system of units, are the gram atom, the gram molecule or mole, and the gram formula weight. Until such time as a new name may be adopted for the kilogram as the basic SI unit of mass,[1] the gram will undoubtedly be retained in order to avoid the absurdity of mkg. The values of enthalpies, entropies, free energies, bond energies, activation energies, molar refractivities, equivalent conductances, and many other such quantities involve the unit of mass. On the $C^{12} = 12$ scale, for example, a gram atom of diamond is 12 g, a mole of nitrogen is 28.014 g, and a gram formula weight of sodium chloride is 58.4428 g. The important physical constant which relates these quantities to the atomic mass unit is the Avogadro number.

Q.1.2   The Avogadro Number, N_A

The fundamental postulate, originally formulated by Avogadro (1811) and independently by Ampère (1814), but not generally accepted until after 1858 largely at the insistence of Cannizzaro, had profound effects on the development of atomic theory. Proposed initially to circumvent certain difficulties in the interpretation of gas phase reactions, the postulate that equal volumes of all gases under the same conditions of temperature and pressure contained equal numbers of molecules, led after Cannizzaro's time to the clear distinction between atoms and molecules and instigated the tremendous task of determining the magnitude of this number. The fact that this has been

---

[1] The unit of mass has still to be related to some natural constant; one constant that has been considered as a substitute for the present mass standard kilogram is the gyromagnetic ratio of the proton.

accomplished by at least 60 different methods, some of great ingenuity, provides perhaps the most convincing demonstration of the reality of atoms and molecules. Most of these methods are now essentially of historical interest, and it is generally agreed that the X-ray method gives the most precise value for the Avogadro number. However, in view of the great importance of Avogadro's number in atomic theory and its relationship to atomic masses, some of these methods will be discussed briefly.

The first approximate estimate of the magnitude of Avogadro's number was made by Loschmidt in 1865, on the basis of the interpretation in terms of the kinetic theory, of the experimentally measured viscosity and density of a gas, and the density of the liquid that may be formed when the gas is compressed and cooled. Making use of the relatively simple concept that the molecules of a gas are assumed to be in constant motion, frequently colliding with one another and with the walls of any container in which the gas may be confined, Krönig (1856), Clausius (1857), Maxwell and Boltzmann (1860) and others, were able to derive a number of equations relating certain measurable properties of a gas to the characteristics of the individual molecules, such as to afford a simple theoretical explanation of the experimentally discovered Laws of Boyle (1662), Charles (1787), Dalton (1801), Gay-Lussac (1802), and Graham (1829).‡

---

‡ These expressions arise most easily by considering that a gas is composed ideally of a number of molecules, n, each of mass, m, moving with a space velocity, v (with components v_x, v_y, and v_z in the x, y, and z Cartesian coordinate directions) within the confines of a container of volume, $V = xyz$ (for a rectangular container), where V is assumed to be large compared with the total volume of the n molecules and there is assumed to be no interaction between the molecules. Then the kinetic energy of each molecule is $mv^2/2$. If v is common for all molecules (more exact treatments allow for the probability that some molecules have a higher velocity than others), then $v_x = v_y = v_z$, otherwise there would result an accumulation of molecules in the direction of the higher velocity component, which is contrary to experimental observation. So $v^2 = 3v_x^2$. Any particular molecule moving with a mean velocity v_x in the x Cartesian coordinate direction normal to the yz plane will collide with this plane $v_x/2x$ times per second, and suffer a change of momentum of $2mv_x$ (from $+mv_x$ to $-mv_x$) during each collision. The rate of change of momentum, which is equal to the force exerted by the molecule as a result of each collision on the area yz is then $(v_x/2x)(2mv_x) = mv_x^2/x$, and the pressure or force per unit area exerted will be $mv_x^2/xyz = mv_x^2/V$. The pressure exerted by all n molecules will be n times as great, or the total pressure, P, due to the gas will be such that $P = nmv_x^2/V = nmv^2/3V$, or $PV = nmv^2/3 =$ constant, for a definite quantity of gas at a constant temperature, which is Boyle's Law. For constant P, $V = (2n/3P)(mv^2/2)$, or if on the basis of the kinetic theory, the temperature is defined as being proportional to the mean kinetic energy per molecule, this is an expression of the Law of Charles and Gay–Lussac. Now for any number of molecules of several gases, $n_1, n_2, \ldots$, with masses $m_1, m_2, \ldots$, and space velocities $v_1, v_2, \ldots$, from the above expression for the pressure, $p_1 = n_1m_1v_1^2/3V$, $p_2 = n_2m_2v_2^2/3V, \ldots$, and the total pressure, if all the gases 1, 2, $\ldots$ were confined in the same volume V would be $P = p_1 + p_2 + \cdots$, which is Dalton's Law.

Also from the definition of temperature being proportional to the mean kinetic energy of the gas molecules, $mv^2/2$, it follows that $PV \propto (2n/3)(mv^2/2) \propto (2n/3)T = nkT$, where k is a constant for all gases, and for the particular number of molecules contained in 1 mole, $n = N_A$, this becomes $PV_M = N_AkT = RT = N_Amv^2/3$. k is a constant for each gas molecule, referred to as the Boltzmann constant (1.38054×10^{-23} J K^{-1}), R is the corresponding constant for a mole of gas, equivalent to N_Ak, referred to as the universal gas constant (8.3143 J K^{-1} mol^{-1}), and V_M is a constant, known as the molar volume of a gas (at standard conditions of 1 atmosphere, 1.013250×10^5 N m^{-2}, and at a temperature of 273.15 K, the molar volume is 2.24136×10^{-2} m^3 mol^{-1}). Now since Avogadro's number of molecules constitutes a mole, the molecular weight of a gas, M, is given by N_Am, and therefore $RT = Mv^2/3$ or $v = \sqrt{3RT/M}$, which is another statement of Graham's Law, that the mean velocity of a gas molecule is inversely proportional to the square root of its mass. Finally, for any two gases at the same pressure, $P_1V_1 = n_1m_1v_1^2/3$ and $P_1V_2 = n_2m_2v_2^2/3$, if they occupy the same volume ($V_1 = V_2$) at the same temperature ($m_1v_1^2/2 = m_2v_2^2/2$), then $n_1 = n_2$, which is Avogadro's Law.

Viscosity is defined as the force in Newtons that must be exerted between two parallel layers, 1 m^2 in area and 1 m apart, in order to maintain a velocity of streaming of 1 m s^{-1} of one layer past another; the unit of viscosity expressed as dynes cm^{-2} s (1 dyne $\equiv 10^{-5}$ N) is referred to as a poise,[1] after Poiseuille (1844) who derived the equation of viscosity for nonturbulent flow, namely

$$\eta = \frac{\pi r^4 tp}{8VL} \qquad \text{(q.1.3.)}$$

relating the viscosity, η, to the volume, V (ml) of fluid (gas or liquid) which will flow in a narrow tube of length L (cm) and radius r (cm) in the time t (s), under the influence of a driving pressure of p (in units of 10^{-1} N m^{-2}); experimental measurements of viscosity are based on this expression. The expression derived from the kinetic theory for gas viscosity involves the mean free path (mean distance travelled between collisions) of the gas molecules (see pages 423–426),

$$l = \frac{1}{\sqrt{2}\,\pi n\sigma^2} \qquad \text{(q.1.4.)}$$

where l denotes the mean free path, n is the number of molecules per unit volume, and σ is the collision diameter of the molecules (the distance between the centers of the molecules at the point of closest approach). The kinetic relationship between the mean free path of a gas molecule and its viscosity is given by,

$$\eta = 1/3(\bar{v}\rho l) \qquad \text{(q.1.5.)}$$

---

[1] The dynamic unit of viscosity, the poise, p, has the dimensions 10^{-1} kg m^{-1} s^{-1} in the SI system of units.

where ρ is the density of the gas and $\bar{v}$ is the mean velocity of the gas molecules, given in terms of the Maxwell–Boltzmann distribution law as (page 40)

$$\bar{v} = (8RT/\pi M)^{1/2} \qquad \text{(q.1.6.)}$$

M is the molecular weight of the gas, T is the absolute temperature, and R is the gas constant (0.082054 liter atm. deg.$^{-1}$ mole^{-1} = 8.3143 J K mol^{-1}). Since n is the number of molecules in unit volume of the gas, the density ρ may be substituted by nM/N_A,[1] and combining (q.1.4.) and (q.1.6.) with (q.1.5.) there results in

$$\eta = (2/3N_A\pi\sigma^2)(\sqrt{MRT/\pi}) \qquad \text{(q.1.7.)}$$

which relates the Avogadro number to the experimentally measurable viscosity and the collision diameter of the gas molecules, the other quantities being constant for any particular gas at a fixed temperature. Loschmidt pointed out that if molecules could be regarded as spherical in shape, and further if it was supposed that in the condensed phase (liquid or solid) these molecular spheres were packed as closely as possible, another expression could be derived relating the density of the condensed phase to N_A and σ, thus determining each of these quantities. If one mole of the condensed phase is considered, then its density, which is experimentally measurable, is given by

$$\rho \text{ (liquid or solid)} = \frac{3M}{4\pi N_A(\sigma/2)^3}. \qquad \text{(q.1.8.)}$$

Using equations (q.1.7.) and (q.1.8.), Loschmidt found that for nitrogen, oxygen, and carbon dioxide, these molecules gave

$$N_A \simeq 10^{23} \text{ mole}^{-1}, \qquad \sigma \simeq 10^{-10} \text{ m}.$$

Actually Loschmidt's values for N_A and σ are now known to be about six times too small and about five times too large, respectively, partly due to his use of inaccurate viscosity data and partly due to his use of the assumption that a liquid is composed of close-packed spheres. But the real importance of his work is that it represents the first attempt, based on theoretical principles, to estimate the properties of single molecules. Avogadro's hypothesis that the number of molecules in unit volumes of all gases, under the same conditions of temperature and pressure, should be the same is obviously confirmed by Loschmidt's calculations, at least within the limits of their accuracy. It should be noted, incidentally, that one mole of any substance, irrespective of its state of aggregation, gaseous, liquid, or solid must contain

---

[1] Loschmidt originally determined the number of molecules per unit volume of gas but Avogadro's number is defined as the number of molecules per mole of gas.

the same number of molecules. It is actually possible, as pointed out later by Born, to obtain an approximate value of Avogadro's number in terms of the kinetic theory of gases, independently of any knowledge of σ, since the influence of the "actual" volume of gas molecules is exhibited not only in the condensed state but also in the gaseous state, by deviations from the ideal gas laws. The equation of state for the ideal gas of the kinetic theory is

$$PV = nRT$$

for n moles of gas. If the total volume occupied by the gas is reduced to such an extent that it is comparable to the "actual" volume of the gas molecules, $nN_A v_m$, where v_m is the volume of an individual molecule, equal to $\frac{4}{3}\pi(\sigma/2)^3$ for spherical molecules, then the free volume in which the gas molecules are constrained to move will be smaller than V by an amount b, so that the equation of state becomes‡

$$P(V - b) = RT. \tag{q.1.9.}$$

b can be shown to be four times the volume of the gas molecules, independently of their shape,‡‡ so that

$$b = 4nN_A \cdot \tfrac{4}{3}\pi(\sigma/2)^3 = \tfrac{2}{3}\pi nN_A\sigma^3. \tag{q.1.10.}$$

If (q.1.10.) is substituted in (q.1.9.), then another experimentally observable relationship between N_A and σ is available, which in conjunction with (q.1.7.) enables N_A to be determined.

‡ In dense gases there are other effects causing deviation from ideal behavior, apart from the actual volume of the gas molecules. In particular attractive and repulsive forces between molecules have the general effect of reducing the pressure, for a fixed T and V, compared with the ideal gas. Thus a molecule in the interior of a gas sample is surrounded by other molecules equally distributed in all directions and so no resultant force is exerted on the particular molecule under consideration. But for a molecule in imminent contact with the wall of the vessel containing the gas, there is a non-uniform distribution of the other molecules surrounding it, such that the attraction between gas molecules tends to reduce the pressure contribution of this molecule to the total gas pressure, making the actual pressure less than the ideal pressure. The attractive force exerted on a single molecule about to collide with the wall of gas container is proportional to the number of molecules per unit volume in the bulk of the gas sample, and the number of molecules striking the walls is also proportional to this number. So the total attractive force, which is related to the correction term to be added to the measured pressure, is proportional to n^2. If V_M is the volume occupied by one mole of gas, then n is inversely proportional to V_M, and the attractive force will vary as $1/V_M^2$. The correction term may then be written as a/V_M^2, where a is the proportionality constant, and the corrected ideal pressure is given by $P + a/V_M^2$. Combining this pressure correction term, for non-ideal gas behavior, with the volume correction above for the volume occupied by the gas molecules, gives the well-known equation of state of van der Waals (1881),

$$(P + a/V_M^2)(V_M - b) = RT.$$

‡‡ The probability that a certain molecule of a total number, n, will be found in a volume element, v, of a total volume, V, is proportional to the magnitude of v. Thus, if a bird is free to fly at random in an otherwise empty room of volume, xyz (assumed to be rectangular), the probability of finding the bird at any instant in the volume element $dx\,dy\,dz$, is proportional to the size of this volume element; there is a certainty that the bird will be found in the room, the probability that it will be found in one half or the other is 1/2, the probability that it will be found in a space which has a volume of 1/10 of the room will be 1/10, and so on. For ideal gas molecules, then, the probability that any particular molecule will be found in a volume element, v, of the total volume, V, is proportional to v. Since ideally the gas molecules do not occupy any space comparable to the total volume available to them, the probability that a second gas molecule will be found in the volume element, v, is also proportional to v, and the total probability that 2 gas molecules will be found in the volume element v is v^2. Similarly the total probability that n gas molecules will be found in the volume element v will be v^n. For gases at high pressures, however, where the gas molecules are so tightly packed together that their combined volume is comparable to the volume element, this is not so. If v_m is the volume of an individual molecule and the molecules are assumed to be spherical, $v_m = \frac{4}{3}\pi(\sigma/2)^3$, but in terms of the concept of collision diameters, σ, the centers of two molecules cannot approach within a distance equal to σ, so that each molecule has an effective volume of $\frac{4}{3}\pi\sigma^3 = 8v_m$, which is independent of the actual shape of the molecules. For gases at high pressure and most real gases, if one molecule occupies the volume element v, then the space available for a second molecule is $v - 8v_m$; the space available for a third molecule is $v - 2\cdot 8v_m$, and so on. The total probability of finding n molecules in a volume element v, for real molecules is therefore less than v^n, and is proportional to $v(v - 8v_m)(v - 2\cdot 8v_m)\cdots[v - (n - 1)\cdot 8v_m]$. If this product is designated P_n, then v has to be substituted by $\sqrt[n]{P_n}$. If $8v_m/v$ is represented by α, the approximate value of P_n can be obtained by taking logarithms,

$$\ln P_n = \ln v^n(1 - \alpha)(1 - 2\alpha)\cdots[1 - (n - 1)\alpha] = n\ln v + \sum_{\beta=1}^{n-1} \ln(1 - \beta\alpha).$$

If the total volume of the n molecules, nv_m, is still small compared to the total volume available to them, then $n\alpha \ll 1$, and therefore for each value of β in this sum $\beta\alpha \ll 1$, and so $\ln(1 - \beta\alpha) \simeq -\beta\alpha$, and

$$\ln P_n = n\ln v - \alpha\sum_{\beta=1}^{n-1}\beta = n\ln v - \frac{\alpha n(n - 1)}{2}.$$

Since n is assumed large, $(n - 1)$ may be replaced by n, and so $\ln P_n = n\ln v - \alpha n^2/2$. So $P_n = v^n e^{-\alpha n^2/2} = [ve^{-n\alpha/2}]^n$. The nth root of P_n is thus $ve^{-\alpha n/2}$, or since for $\alpha n \ll 1$, $e^{-\alpha n/2} = 1 - \alpha n/2$, therefore, if n molecules each of volume v_m occupy the volume element v of a total volume V, this total volume is reduced to $V(1 - n\alpha/2) = V(1 - n\cdot 4v_m/V) = V - b$. Thus the term b is equivalent to four times the actual volume of the molecules.

Various phenomena associated with the variation in density of a column of gas, the atmosphere in particular, the variation of the density of colloidal suspensions, their rate of sedimentation, the variation of the motion of colloid particles in a liquid due to the Brownian movement, the variation of the refractive index of the atmosphere giving rise to a scattering of the transmitted light, have all been used to estimate the value of Avogadro's number.

Perhaps the most notable of measurements of this type were those of Perrin, beginning about 1908. The discovery by Brown (1828) that microscopic pollen grains suspended in water undergo a continual and random motion, may be regarded as a macroscopic illustration of the type of motion which would be expected of molecules behaving in accordance with the kinetic theory of gases; a similar macroscopic display may be observed in the oscillations of a mirror suspended by a fine wire. The kinetic theory of the Brownian motion has been derived independently and from quite different viewpoints by Einstein,[1] Smoluchowski,[2] and Langevin.[3] Perrin made measurements on various types of suspended particles, and on the assumption that they behaved like molecules obeying the equations derived from the kinetic theory of gases, was able to calculate the Avogadro number from the expression, $N_A = RtT/3r_s\eta\pi\Delta^2$, where R is the gas constant and the particles of radius r_s move in a medium of viscosity η at a temperature T, and are observed to undergo a linear displacement Δ in the time t.‡ Perrin also calculated the Avogadro number by counting the numbers of particles of gamboge and gum mastic, N_1 and N_2, at different heights, H_1 and H_2, in aqueous suspensions, from the expression, $N_A = [RT \ln (N_2/N_1)]/[mg(H_1 - H_2)]$, assuming an effective mass of the particles $(4/3)\pi r^3(\rho - \rho^0)$, where ρ is the particle density and ρ^0 is the density of the suspension medium.‡‡ Others using colloidal gold as the suspended material have repeated Perrin's measurements; all of these methods lead to a value of approximately 6×10^{23}.

‡ Consider firstly a stationary column of gas confined in a cylinder of radius, r. Let the concentration be n moles per liter at some height, H, and $n + dn$ at a greater height $H + dH$. Then the number of moles of gas in the disc of volume $\pi r^2\, dH$ is $n\pi r^2\, dH$. From the gas law equation, the kinetic pressure exerted upwards on the bottom face of the disc is $P = nRT$, and the force exerted must therefore be $\pi r^2 P = \pi r^2 nRT$. Similarly, the kinetic force exerted downward on the upper face of the disc is $\pi r^2 RT(n + dn)$. The net kinetic force on the gas molecules within the disc is then $-\pi r^2 RT\, dn$, acting upward. Therefore the net upward kinetic force acting on one mole is $-[\pi r^2 RT\, dn]/\pi r^2 n\, dH = -RT[d \ln n/dH]$.

If in a liquid, it is assumed that the observed Brownian motion is due to the constant random motion of its constituent molecules, similar to that of gas molecules, and if it is also assumed that liquid molecules tend to move from a region of high to a region of low concentration, then the number of molecules, N, crossing an area, πr^2, per second is proportional to the magnitude of the area and to the concentration gradient normal to this area, or $dN/dt = -D\pi r^2\, dn/dx$ (Fick's Law), where the constant of proportionality is known as the linear diffusion coefficient, D, and is assumed to be positive. Now if Δ is the extent of the linear motion of a molecule of liquid in either the $+x$ or $-x$-direction in a time t, then in an imaginary cylinder of cross-sectional area πr^2 and length Δ in the

[1] Einstein, A., *Ann. der Physik*, 1905, **17**, 549; 1906, **19**, 371.
[2] Smoluchowski, M. von, *Ann. der Physik*, 1906, **21**, 756.
[3] Langevin, P., *Compt. Rend.*, 1908, **146**, 530.

x-direction, the average rate of motion of molecules in either direction through a cross-sectional area of this cylinder is Δ/t. But as both directions of motion are equally probable, the number of molecules crossing the base of the cylinder per second from the $-x$ to the $+x$ direction is $(1/2)\pi r^2(\Delta/t)n_1$, where n_1 is the concentration of molecules in a cylinder with axial length $-\Delta$. Similarly, the number of molecules crossing the yz plane per second from the $+x$ to the $-x$ direction into an identically sized cylinder of liquid is $(1/2)\pi r^2(\Delta/t)n_2$, where n_2 is the concentration of molecules in the cylinder of axial length $+\Delta$. Therefore the net number of molecules crossing the yz plane per second is $dN/dt = (1/2t)\pi r^2\Delta(n_1 - n_2)$. However, the concentration thus defined is $dn/dx = (n_2 - n_1)/\Delta$, and substituting these values for dN/dt and dn/dt in Fick's Law, there results $D = \Delta^2/2t$, which is known as Einstein and Smoluchowski's diffusion law. Considering now the steady velocity of the diffusing molecules, v_x, there must apply, the relationship, $dN/dx = \pi r^2 n v_x$, where n is simply the concentration in molecules per ml at the boundary across which diffusion takes place. It is shown on page 26 that this constant velocity of diffusion is given by the ratio of the driving force to the resisting force, $v_x = F/C$. Thus $dN/dt = \pi r^2 n F/C$. It has also been shown above that the driving force on one mole, due to a concentration gradient is $-RT(d\ln n/dx)$, if the quantity H above is considered to extend in the x-direction. On one molecule the driving force would therefore be, $F = -(RT/N_A)d\ln n/dx$ or $F = -(RT/N_A)(1/n)(dn/dx)$. Therefore

$$dN/dt = \pi r^2 n/C[-(RT/N_A)(1/n)(dn/dx)] = -D\pi r^2\,dn/dx,$$

by Fick's Law. Hence, $D = RT/N_A C$. The resisting force, C, for a sphere of radius r_s moving in a medium of viscosity η is known from Stokes's Law to be $C = 6\pi\eta r_s$, and making use of this fact, it is found that $D = RT/[N_A 6\pi\eta r_s]$, which is known as the Stokes–Einstein diffusion law. Finally eliminating D from the Einstein–Smoluchowski and the Stokes–Einstein equations, $N_A = RTt/3\pi\eta r_s\Delta^2$.

‡‡ In the case of a stationary column of gas, such as in the atmosphere, there is no diffusion or mass movement concerned, and so the net kinetic force upward on a mole of gas must be equally balanced by the gravitational force acting on a mole of gas downward, equivalent to $Mg = N_A mg$. Therefore from the above, $-RT[d\ln n/dH] = N_A mg$. On integration this gives, $\ln(n_2/n_1) = [N_A mg/RT](H_1 - H_2)$. Noting that the concentration at any height in a stationary column of gas must be proportional to the number of molecules, N, present in a plane at that particular height, it follows that

$$N_A = \frac{RT\ln(N_2/N_1)}{mg(H_1 - H_2)}.$$

More accurate methods depend on principles quite independent of the kinetic theory. Among these are measurements based on radioactive decay of natural radioactive species and the method of measuring the elementary electrical charge, ε. The decay of a radionuclide follows the exponential law, $N = N_0 e^{-\lambda t}$, where N_0 is the number of atoms initially present in the radionuclide sample, N is the number remaining after a time t, and λ is the appropriate decay constant. If for example a weight W of a radionuclide of mass number A decays by α-particle emission and the number of α-particles emitted at a distance l from the sample is counted as they pass through a square hole of edge length a to fall on a zinc sulfide screen or with the aid of a Geiger type counter, and the number observed is n, then all the information necessary to determine Avogadro's number is available. Since particles are emitted in all directions from the radio-source, only the fraction $a^2/4\pi l^2$ of the total number emitted is actually counted, so that the total number of disintegrations taking place in the time t is $4\pi n l^2/a^2$. Each α-particle originates from the decay of one atom of the radionuclide, and the weight W of the radionuclide contains $N_A W/A$ atoms initially; after the time t there will have occurred $WN_A(1 - e^{-\lambda t})/A$ disintegrations, which must equal the number of counts recorded. Hence $N_A = [4\pi n l^2 A][Wa^2(1 - e^{-\lambda t})]^{-1}$. The elementary electronic charge is related to Avogadro's number by Faraday's Laws (1831–34), which rank along with the Boltzmann Law, as among the surest and most accurate of physico-chemical generalizations. Faraday discovered that it requires 96487.0 coulombs $= 9.68470 \times 10^3$ electromagnetic units $(\text{cm}^{1/2}\,\text{g}^{1/2}\,\text{mole}^{-1}) = (9.68470 \times 10^3 \times 2.997925 \times 10^{10})$ electrostatic units of electricity to liberate one equivalent weight expressed in grams of any substance during electrolysis of its solutions. Thus the complete electrolytic decomposition of one gram formula weight of sodium chloride, one gram formula weight of barium chloride, one gram formula weight of gallium chloride, from solution requires 1, 2, and 3 Faradays of electricity, respectively, and the atoms of sodium, barium, and gallium which react with 1, 2, and 3 atoms of chlorine also react with 1, 2, and 3 units of electricity, respectively. It was Stoney[1] (1874) who proposed the name, electron, for the unit of negative electricity, of charge $-\varepsilon$, and the smallest unit of positive electricity is of course $+\varepsilon$, so that the Faraday constant, F, is equivalent to $N_A\varepsilon$. The electron is a universal constituent of all matter, as is indicated by numerous experimental observations on the passage of electricity through gases and solutions in various solvents, on radioactive decay, on the photo- and thermo-electric effects, and the scattering of electromagnetic radiation by matter (y-radiation, X-radiation, light, micro- and radiowaves). The magnitude of the electronic charge was first determined by Thomson, later improved values being obtained by C. T. R. Wilson, A. H. Wilson, and Millikan, whose value was regarded as the most precise up until about 1928. Millikan made a very large number of observations of the rate of fall (under the influence of gravity) or ascent (under the influence of an applied electrostatic field) of charged droplets of such dissimilar materials as oil, glycerine, and mercury; the droplets were charged either by being formed from an "atomiser" or by ionization with a beam of X-radiation. In the Millikan experiments, the charged droplets move either upwards or downwards with a uniform velocity in a viscous medium, air. For a particle moving with a uniform velocity, v, under the influence of a constant force, the equation of motion of the particle,

---

[1] The SI charge unit, referred to as the Stoney (S), is a unit of electric charge, such that the force of repulsion between two 1-Stoney charges one meter apart is one Newton.

of mass, m, is $m(dv/dt) = F - Cv$, where F is the force and C is a constant, but if the velocity is uniform then dv/dt is zero and the velocity is simply F/C. With no electrostatic field, the force involved (fall) is the gravitational force and the uniform rate of descent of the particle is $v_1 = mg/C$. In the electrostatic field of strength V, the force on a charge $-\varepsilon$ is $-V\varepsilon$, and if the charged droplet moves upwards against the gravitational force under the influence of the field, V, the resulting steady velocity of ascent is given by $v_2 = (mg - V\varepsilon)/C$. The ratio of the two uniform velocities is independent of C, and consequently ε is obtained as $(mg/V)(1 - v_2/v_1)$. The charge measured in this way was always found, by Millikan, to be either 1, 2, 3, 4, ... times the quantity -1.602×10^{-19} C. In Millikan's method as well as in the two methods adopted by Perrin, for determining the Avogadro number, reliance is placed on Stoke's Law to estimate the radius of the droplets (Millikan) or the colloid particles (Perrin); according to Stokes, the steady velocity of fall of a sphere of density ρ_1 and radius r falling under the influence of gravity in a viscous medium of viscosity η and density ρ_0, is given by

$$v = [2gr^2(\rho_1 - \rho_0)]/9\eta,$$

where, as before, g is the gravitational constant ($980.665 \text{ mm}^{-1}\text{ s}^{-2}$). This expression enables r to be determined from experimentally observed data, from which the mass of the spherical (assumed) droplet or particle may be determined as

$$m = \frac{4\pi\rho_1}{3}\left[\frac{9\eta v}{2g(\rho_1 - \rho_0)}\right]^{3/2}. \tag{q.1.11.}$$

Finally from the value of the electronic charge thus determined, $N_A = F/\varepsilon$.

X-Rays were discovered by Röntgen in 1895, but it was not until 1912 that Friedrich and Knipping at the suggestion of von Laue showed that a crystal of zinc sulfide caused the diffraction of a beam of X-radiation incident upon it. The most accurate measurement of the Avogadro number is based on this phenomenon of X-ray diffraction. However, even before the discovery of the diffraction of X-radiation by crystalline matter, it was possible to determine interatomic and interionic distances in crystals, making use of the approximate values of Avogadro's number, as determined by Millikan and others. For example, if sodium chloride is assumed to be composed of sodium and chloride ions, as is indicated by the behavior on electrolysis of its melt and aqueous solutions and from other evidence, then one gram formula weight must contain, by Avogadro's Law, exactly N_A sodium ions and N_A chloride ions. If no distinction is made between the sodium and chloride ions, then 29.22 g of sodium chloride must contain N_A ions. From the density of sodium chloride (2.165 g ml^{-1}), 0.5 mole would occupy 13.49 ml or if this mass were in the form of a cube, the edge length of the cube would be 2.374 cm and would contain $\sqrt[3]{N_A}$ ions $= 0.8446 \times 10^8$ ions (using Millikan's value of $N_A = 6.024 \times 10^{23}$). The sodium-chlorine interionic distance is therefore deduced to be 2.811×10^{-10} m $= 2.811$ Å. Only after the work of Compton and Doan (1925) and Baecklin and Wadlund (1928), did it become possible to determine accurately the wavelength of X-radiation by the use of ruled gratings, which radiation could then be used to measure the lattice constants of crystals, which information in turn could be applied to the determination of the Avogadro number, reversing the above procedure. The principles of the diffraction of X-radiation by crystalline materials and also the relationships between lattice constants and interatomic distances, are discussed later in this book; meantime to illustrate the principle of this method of determining the Avogadro number, the fundamental Bragg law will be assumed, namely that a crystal scatters an incident beam of monochromatic X-radiation in such a manner that a diffracted beam is only observed at certain specific angles of incidence of the direct beam with respect to a specific orientation of the crystal W. H. and W. L. Bragg (1912) pointed out that when a crystal, regarded as a three-dimensional array of atoms, ions, complex ions or molecules, diffracts an incident, monochromatic beam of X-radiation such that every atom, ion, or molecule scatters in phase with each other, the process is geometrically equivalent to the "reflection" of the incident beam from certain crystal planes. This is usually expressed in the form, $n\lambda = 2d_{hkl}\sin\theta$, where λ is the wavelength of the radiation used, d_{hkl} is the interplanar spacing (page 543) of the set of planes with Miller indices (hkl), and θ is the angle of incidence of the direct beam with respect to these planes, n being an integer. The X-radiation wavelength may be determined independently using a ruled diffraction grating, θ observed experimentally, hence obtaining d_{hkl}, from which the crystal lattice constants and therefore the crystal unit cell volume may be calculated (page 563). Multiplying this volume (in cm^3) by the measured density of the crystal (g cm^{-3}) the mass of material in the crystal unit cell is obtained, which must necessarily be the mass of a small integral number of molecules or formula weights of the material composing the crystal, or $V\rho = ZM$, where V is the volume, ρ is the density, Z is the number of molecules or formula weights of material in the unit cell of molecular weight or formula weight, M. The density then is given by $\rho = ZM/N_A V$, or $N_A = Zm/V\rho$. The precision of this method is limited only by the accuracy of density determinations; interatomic distances in crystalline solids can be measured to an accuracy of better than 1 part in 50,000, but the best density measurements made by the flotation technique are perhaps only one-tenth as accurate. The best value of the Avogadro number obtained by this method is $6.02252 \pm 0.00016 \times 10^{23}$. There is actually quite a disparity between the various values of Avogadro's number determined in this manner, dependent on the crystalline material used. It is now known that many crystalline substances deviate considerably from the crystallographic ideal of a three-dimensional point space lattice, with atoms, ions, or molecules situated exactly at the lattice points; naturally occurring diamond and calcite are relatively free from most structural defects, and the best accepted value of N_A is based on the use of these substances.

From a knowledge of the numerical value of the Avogadro number, absolute atomic masses may be obtained; thus the mass of a hydrogen atom is $1.0080/N_A = 1.673 \times 10^{-24}$ g. The electronic charge, ε, and the Boltzmann constant, k, are as previously noted related to N_A by $\varepsilon = F/N_A$ and $k = R/N_A$. From the electronic charge and a knowledge of its charge-to-mass ratio, obtained from study of its behavior in electric and magnetic fields, the electronic mass is obtained as $m_\varepsilon = 9.1091 \times 10^{-28}$ g. It should be noted that this is the rest mass of the electron; in terms of the special theory of relativity, mass depends on velocity‡ and for an electron moving with a velocity of $0.1\,c$ its mass is $1.005\,m_\varepsilon$ but at $0.99999\,c$, the electron mass becemes $223.6\,m_\varepsilon$.

‡ In accordance with the Einstein Theory of Relativity, to account for the observed facts that there is no absolute time and that the upper physical limit for velocities is the velocity of electromagnetic radiation, c, certain relativistic corrections are necessary in describing particles moving with velocities approaching c. Most of the resultant expressions involve the Fitzgerald factor, $\sqrt{1 - v^2/c^2}$, where v is the velocity of the particle. Thus, for a particle of rest mass, m_0, moving with a velocity, v, its mass, $m = m_0[1 - v^2/c^2]^{-1/2}$, its momentum, $p = mv = m_0v[1 - v^2/c^2]^{-1/2}$, its kinetic energy,

$$K = m_0c^2[(1 - v^2/c^2)^{-1/2} - 1] = mc^2 - m_0c^2,$$

and its total energy, $E = mc^2 = m_0c^2[1 - v^2/c^2]^{-1/2}$. A frequently useful relation between the total energy and the momentum of a particle may be obtained by squaring and rearranging the relativistic mass equation, $m = m_0[1 - v^2/c^2]^{-1/2}$. Thus, $m^2c^2 - m_0^2c^2 = m^2v^2 = p^2$. If both sides of this equation are divided by $m_0^2c^2$, then $[p/m_0c]^2 = [mc/m_0c]^2 - 1 = [mc^2/m_0c^2]^2 - 1 = [E/m_0c^2]^2 - 1$, which, on rearrangement, gives, $E^2 = E_0^2 + p^2c^2$, where $E_0 = m_0c^2$.

For particles with zero rest masses, such as photons, $p = E/c = h\nu/c$, where ν is the frequency and h is Planck's Constant.

Q.1.3.   The Planck Constant, h, and the Quantum Postulate

Another fundamental physical constant, quite independent of those referred to in the previous section, and expressible only in units which have no significance in terms of classical mechanics, classical electromagnetic theory, or classical thermodynamics, the so-called Planck Constant, is of central importance to all of the theories about the behavior and properties of matter and radiation which have been developed since 1900. The apparently simple postulate originally advanced by Planck,[1] as an ingenious solution to what

---

[1] Planck, M., *Ann. Physik*, 1900, **1**, 69; 1901, **4**, 553.

was referred to at that time as the "ultraviolet catastrophe," namely that not only matter but also radiant energy is composed of "atoms" or quanta, has since led to the development of a discipline in terms of which it appears that all of the science of chemistry and much of physics may be deduced and interpreted. The Planck quantum postulate is of statistical origin, devised to account for the type of experimental observations made by Lummer and Pringsheim (1899) with respect to the thermal radiation emitted by heated objects, which are incomprehensible in terms of the natural laws previously established.

All hot objects are observed to emit thermal radiation, just as luminous objects emit visible radiation, and both types of radiation are transmitted through space with a constant velocity ($c = 2.997925 \times 10^8$ m s^{-1} *in vacuo*) which is characteristic of all types of electromagnetic disturbances in space; the wavelength of thermal radiation is approximately ten times longer than that characteristic of visible radiation. It is possible to analyse experimentally the thermal radiation emitted by a hot object into a range of frequencies or wavelengths each associated with a finite energy. It is also observed that all hot objects, if blackened, exhibit a thermal spectrum which is independent of their composition, physical mode of fabrication or construction, and is dependent only on their temperature. A perfectly black body is defined as one which absorbs all of the radiant energy incident on it; the radiation emitted by such an object is called *black body radiation* or isothermal cavity radiation, in allusion to the fact that the practical source of such radiation is commonly an oven, heated electrically, and whose interior walls have been blackened with platinum black or charcoal (Kirchhoff, 1859). A small opening in the wall of such an oven allows purely thermal radiation to escape and any radiation falling on the opening from outside is repeatedly reflected and finally absorbed on the interior walls. The spectral distribution of the radiant energy emitted from such a device is a function only of the interior temperature, quite independent of the type of matter composing it and therefore of fundamental importance.

In order to characterize the electromagnetic energy within the thermal cavity, it is necessary to define the energy density, E, as the amount of radiant energy per unit volume, which at equilibrium is the same at every point in the isothermal cavity. For a continuous distribution of the spectral components of the thermal radiation, the function $E_\nu \, d\nu$ would denote the energy density of all radiation with frequencies, ν, between ν and $\nu + d\nu$, E_ν extending over all frequencies from 0 to ∞. The relationship between the total radiant energy density and the manner in which this total density is spread out over the frequency spectrum may be arrived at by considering that for radiant energy to exist in a cavity of volume, V, from which electromagnetic energy escapes, there can only be a finite number of modes of vibration of the system giving rise to the radiant energy which must be compatible with the frequency concerned and with the total volume of the cavity;‡ this general principle is perhaps more easily appreciated in connection with the propagation of sound waves rather than electromagnetic waves. Thus there are only a finite number of ways in which a block of some elastic material or an organ pipe may be set into vibration with some finite frequency. For electromagnetic waves, the number of modes of vibration of frequency, ν, which can exist in a cavity of volume, V, is

$$Z = \frac{8\pi V \nu^3}{3c^3} = \frac{8\pi V}{3\lambda^3}. \qquad \text{(q.1.12.)}$$

‡ Consider a cubical cavity of edge length, L, and volume, $V = L^3$. In order that electromagnetic radiation should exist in the cavity in equilibrium, the wave motion associated with the radiation must correspond with stationary waves. For stationary waves, the modes of vibration must satisfy special boundary conditions; in particular they must have nodes at the walls, and this condition limits the number of ways in which waves may be formed to a finite value. In terms of classical wave theory, if in Fig. Q.1.1., AA', BB', and CC' are wave fronts, with wavelength λ, moving in the direction indicated by the arrows, and the angles which the normals to the wavefronts make with the x and y Cartesian coordinate axes are α and β, respectively, then along x the effective wavelength λ_x is $\lambda/\cos\alpha$, and along y the corresponding value λ_y is $\lambda/\cos\beta$; in three dimensions, if the normal to the wave fronts makes the angle γ with the z coordinate direction, the effective wavelength λ_z in this direction would similarly be $\lambda/\cos\gamma$. If $1/\lambda_x = k_x$, $1/\lambda_y = k_y$, and $1/\lambda_z = k_z$, then k_x, k_y, and k_z are the components of a vector $\mathbf{k}$ of length

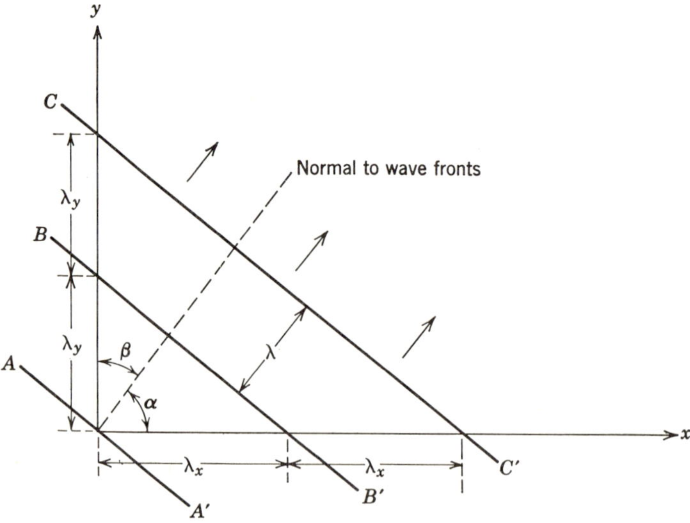

FIG. Q.1.1. The relationship between wave fronts, wave normals, and effective wavelengths.

$1/\lambda$, which is normal to the wave front and is referred to in wave theory as the propagation vector. If the coordinate axes coincide, by choice, with three mutually intersecting edges of the cube, then by analogy with a vibrating string (see page 105), the intercepts of the waves with the cube edge along the x axial direction must be such that there are nodes at $x = 0$ and $x = L$, and so there must be n_x half-wavelengths along the x-axis, where n_x must be a positive integer. Hence $n_x \lambda_x / 2 = L$. Similarly along the y and z axial directions, $n_y \lambda_y / 2 = L$ and $n_z \lambda_z / 2 = L$. Thus the integers n_x, n_y, and n_z satisfy the conditions

$$n_x = 2L/\lambda_x = [2L \cos \alpha]/\lambda = 2Lk_x,$$
$$n_y = 2L/\lambda_y = [2L \cos \beta]/\lambda = 2Lk_y,$$
$$n_z = 2L/\lambda_z = [2L \cos \gamma]/\lambda = 2Lk_z.$$

From the relationship, $k^2 = k_x^2 + k_y^2 + k_z^2$, a necessary consequence is that $n_x^2 + n_y^2 + n_z^2 = 4L^2k^2 = 4L^2\nu^2/c^2$, since $k = 1/\lambda = \nu/c$. This is the equation of a sphere of radius, $2L\nu/c$, but since n_x, n_y, and n_z are restricted to positive integers, this expression describes a region in space which is only the positive octant of a sphere. Also since n_x, n_y, and n_z are limited to positive integral values and L and c are constants, ν must vary discontinuously within the volume described. The volume of this octant of spherical space is $(1/8)(4\pi/3)(2L\nu/c)^3$ or $4\pi V\nu^3/3c^3$. Since the number of stationary waves which may exist in the cube of volume V is determined by the positive, integral values assumed by n_x, n_y, and n_z, and these quantities in turn are such that $n_x^2 + n_y^2 + n_z^2 = 4\pi V\nu^3/3c^3$, the total number of modes of vibration possible corresponding to the frequency ν is $4\pi V\nu^3/3c^3$. However since electromagnetic waves may be polarized in two ways, the total number of vibrational modes must be $Z = 8\pi V\nu^3/3c^3$, which is the expression used in (q.1.12.).

The number of modes of vibration with frequencies or wavelengths lying between ν and $\nu + d\nu$ or λ and $\lambda + d\lambda$, respectively, is obtained by differentiating (q.1.12.) with respect to ν or λ,

$$dZ = \frac{8\pi V\nu^2}{c^3}\, d\nu = -\frac{8\pi V}{\lambda^4}\, d\lambda. \qquad \text{(q.1.13.)}$$

If each vibrational mode corresponds to an average amount of energy, $\bar{E}$, then the amount of electromagnetic energy radiated with frequencies or wavelengths between ν and $\nu + d\nu$ or λ and $\lambda + d\lambda$ is $\bar{E}\, dZ$, or the radiation density within these limits may be defined as $dE = \bar{E}\, dZ/V$. So,

$$dE = E_\nu\, d\nu = \frac{8\pi \bar{E}\nu^2\, d\nu}{c^3} = -E_\lambda\, d\lambda = -\frac{8\pi \bar{E}\, d\lambda}{\lambda^4}. \qquad \text{(q.1.14.)}$$

Therefore,

$$E_\nu = \frac{8\pi\nu^2}{c^3}\, \bar{E}; \qquad E_\lambda = \frac{8\pi}{\lambda^4}\, \bar{E}. \qquad \text{(q.1.15.)}$$

The experimental results of Lummer and Pringsheim, showing E_λ as a function of λ are shown in Fig. Q.1.2.

In the attempt to find an expression for the average energy of an oscillator which would agree with these experimental results, Rayleigh and Jeans

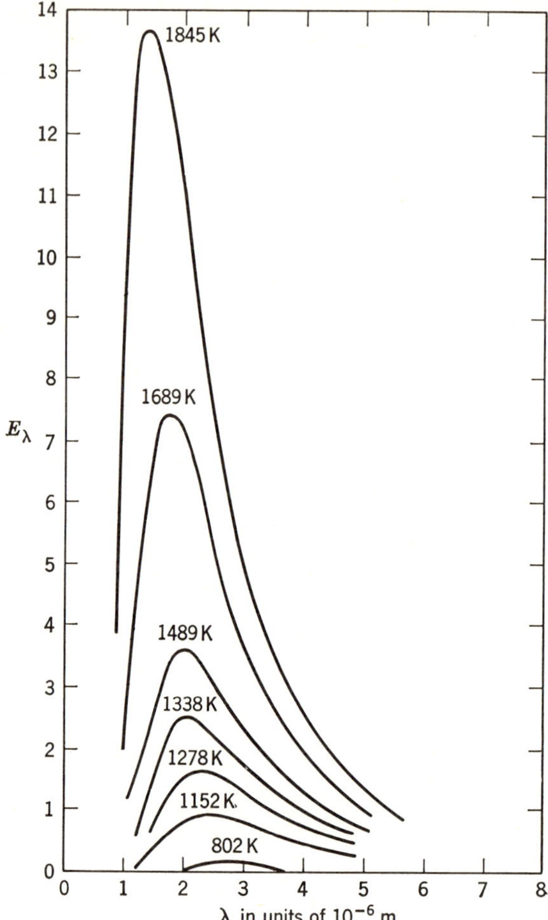

FIG. Q.1.2. Distribution of the intensity of thermal radiation as a function of wavelength, at various temperatures, according to measurements of Lummer and Pringsheim. At higher temperatures, the Planck expression shows that intensity maxima occur at about 1.2×10^{-6} m, 9.1×10^{-7} m, 7.8×10^{-7} m, and 7.0×10^{-7} m, at temperatures of 2500 K, 3000 K, 3500 K, and 4000 K, respectively.

(1900, 1909) suggested that $\bar{E} = kT$, which is the classical, equipartition of energy value (see page 42), so that

$$E_\nu = \frac{8\pi\nu^2 kT}{c^3} \quad \text{and} \quad E_\lambda = \frac{8\pi kT}{\lambda^4}. \qquad (q.1.16.)$$

This is the Rayleigh–Jeans Radiation Law, which fits the experimental data at high temperatures and long wavelengths (or low frequencies), but implies a continual increase of E_ν with increasing values of ν, and affords no indication of the observed maximum in the experimental radiation density curve. This is the "ultraviolet catastrophe" from which there is no escape in terms of arguments based on classical mechanics. Stefan (1879) had earlier observed that the total amount of radiation emitted from unit area per unit time from a black body, S, was proportional to the fourth power of the absolute temperature

$$S = \sigma T^4, \qquad (q.1.17.)$$

where σ is referred to as Stefan's Universal Constant (5.6697×10^{-8} W m^{-2} K^{-4}) and Wien (1893), supplementing Stefan's Law with information deduced on the basis of a classical thermodynamic argument had concluded that the radiation law was

$$E_\lambda = (c_1/\lambda^5)e^{-c_2/\lambda T}, \qquad (q.1.18.)$$

where c_1 and c_2 are constants independent of both temperature and wavelength. The position of the maximum in the $E_\lambda - \lambda$ curve would be given, according to Wien's Law by

$$dE_\lambda/d\lambda = 0 = -(5c_1/\lambda^6)\cdot e^{-c_2/\lambda T} + (c_1/\lambda^5)\cdot(c_2/\lambda^2 T)\cdot e^{-c_2/\lambda T},$$

or

$$\lambda_{\max} = c_2/5T. \qquad (q.1.19.)$$

If this value for $\lambda_{\max}$ is substituted back into (q.1.18.), the corresponding value for $(E_\lambda)_{\max}$ is

$$(E_\lambda)_{\max} = c_1 T^5 (5/c_2 e)^5. \qquad (q.1.20.)$$

Wien's Radiation Law thus accounts for the maximum in the observed data and (q.1.19.) offers a quantitative expression for the well known fact that for a hot object the color of the most intense radiation emitted corresponds to a shorter wavelength as the temperature is raised. From experimentally observed data for the maximum wavelength emitted at various temperatures, the constant c_2 is found to be 1.43879×10^{-2} m K. But the Wien expression does not account for the experimental results observed at low frequencies and high temperatures.

The failure of either classical mechanics (Rayleigh–Jeans) or thermodynamics (Wien) to afford any understanding of the nature of the radiation density of blackbody radiation led Planck to suggest that this phenomenon must indicate the operation of some natural law previously unknown, and his proposed solution for this problem constitutes the birth of the quantum theory. Planck assumed the validity of (q.1.15.), given on the basis of classical electrodynamics; the novelty of his proposals concerned the value to be assigned to the average energy of the oscillators responsible for the emission of the radiation. Specifically, he suggested that the energy of the oscillators cannot be continuously variable but may only differ from one another by finite amounts, so that instead of an infinite number of energy levels only some definite number can exist. The possible energy levels for an oscillator, designated E_1, E_2, E_3, ..., E_n, were assumed by Planck to be such that the smallest unit or quantum of energy was directly proportional to the frequency, $E_0 = h\nu$, and the other energy levels were given by $E_n = nE_0$, where n is an integer, and the proportionality constant, h, relating the energy and frequency of an oscillator is called the Planck Constant, which has proved to be a fundamental natural constant. On the basis of these assumptions, the average energy of an oscillator is,

$$\bar{E} = \sum_0^\infty nE_n \Big/ \sum_0^\infty n. \qquad (q.1.21.)$$

According to classical Boltzmann statistics,‡ the probability that an oscillator

---

‡ The Boltzmann Law, of fundamental importance in both classical and quantum statistical mechanics, may be derived as follows. In a system of total, constant, volume, V, of total energy, E, composed of a total number of particles, N, each of which is identical except for their energy content, if N_1 of the particles possess the energy, E_1, N_2 of the particles possess the energy, E_2, and so on, then the total energy of the system is, $E = N_1E_1 + N_2E_2 + N_3E_3 + \cdots = \sum_i N_iE_i$, and the total number of particles composing the system is, $N = N_1 + N_2 + \cdots = \sum_i N_i$. The probability that any system will exist in any particular state is proportional to the number of distinguishably different ways in which that state may be realized. A great many of the phenomena of physics and chemistry are expressions of the laws of probability. For example, the velocity of certain chemical and nuclear reactions is determined by the probability of molecular or nuclear collisions, the velocity of gas molecules or thermal neutrons from a reactor are distributed in accordance with the laws of probability; similarly with such varied phenomena as the vapor pressure of a liquid, the emission of optical and X-ray spectra of substances, and the radioactive decay of an unstable nuclide, all of which may be interpreted in terms of probability statistics. N particles may be arranged in $N!$ different ways, but if a sub-group, N_1, of these all have the same energy and are thus indistinguishable from one another, then the total number of distinguishable arrangements is only, $N!/N_1!$. Similarly for each of the other sub-groups, N_2, N_3, ..., so that the total statistical probability of the system described above is given by,

$$W = \frac{N!}{N_1!\, N_2!\, N_3! \cdots} = N! \left[\prod_i N_i \right]^{-1}.$$

Also from the laws of thermodynamics, the equilibrium state is statistically the most stable one, and this is also the state of maximum entropy, S. It was Boltzmann who first showed that W is a function of S alone. The probability laws are multiplicative but the entropies of different states are additive, so that $W = W_1 W_2 W_3 \cdots W_n$ but $S = S_1 + S_2 + \cdots + S_n$, and therefore if $W(S)$ is some function of $S(W)$, the functional relationship must be a logarithmic one. Hence the general form of the relationship between entropy and probability must be $S = k \ln W + C$, where k and C are constants. Then, $[S - C]/k = \ln W = \ln N! - \ln N_1! - \ln N_2! - \cdots = \ln N! - \sum_i \ln N_i!$. The factorial of a large number is given by the Stirling approximation (1730), $N! = [\sqrt{2\pi N}][N/e]^N$, or $\ln N! = (\frac{1}{2}) \ln (2\pi N) + N \ln N - N$. Also the logarithm of any large number is negligible in comparison with the number itself, so that $\ln N! = N \ln N - N$. Hence

$$[S - C]/k = N \ln N - N - [N_1 \ln N_1 - N_1] - [N_2 \ln N_2 - N_2] - \cdots$$
$$- [N_n \ln N_n - N_n] - \cdots$$
$$= N \ln N - \sum_i N_i \ln N_i.$$

For systems in equilibrium both the probability of the system existing in any particular state and the entropy of the system are maxima. The condition for a maximum in either the entropy or the probability, since the one is a function of the other, is that $\delta S = \sum_i dS_i = 0$, for all possible variations. To apply this condition it is necessary to know $dS_1, dS_2, \ldots, dS_i$; these quantities may be obtained from $S = C + kN \ln N - k \sum_i N_i \ln N_i$ by differentiation. Differentiating successively with respect to $N_1, N_2, \ldots$, there results $dS = [-k(\ln N_1 + 1)] dN_1$, $dS = [-k(\ln N_2 + 1)] dN_2, \ldots$, and since $-k$ cannot be zero, the condition for a maximum in either the probability or entropy of the system is, $\sum_i dS_i = \sum_i [\ln N_i + 1] dN_i = 0$. The condition that the total number of particles composing the system, and the total energy of the system, should each remain constant, may be similarly expressed as $\delta N = \sum_i dN_i = 0$ and $\delta E = \sum_i E_i dN_i = 0$, respectively. Generally, if any three quantities, $x = y = z = 0$, then $x + Ay + Bz = 0$, where A and B may be any multipliers. Therefore, $\sum_i [\ln N_i + 1 + A + BE_i] dN_i = 0$, and this equation must remain true for any arbitrary variations, dN_i. Therefore each of the factors within the square parentheses in this expression must be zero, or $\ln N_i + 1 + A + BE_i = 0$, or $N_i = e^{-(A+1)}e^{-BE_i}$. The arbitrary constant, $e^{-(A+1)}$, may be eliminated from this expression by making use of the fact that $N = \sum_i N_i$, which must be equivalent to $e^{-(A+1)} \sum_i e^{-BE_i}$, so that $N_i = [Ne^{-BE_i}][\sum_i e^{-BE_i}]^{-1}$. Now taking logarithms of this equation, it is found that $\ln N_i = \ln N - BE_i - \ln [\sum_i e^{-BE_i}]$, which on multiplying both sides by N_i becomes $N_i \ln N_i = N_i \ln N - BN_iE_i - N_i \ln [\sum_i e^{-BE_i}]$. If all of the expressions of this type for each value of i are added, there results, $\sum_i N_i \ln N_i = N \ln N - BE - N \ln [\sum_i e^{-BE_i}]$. Rearranging this latter expression gives, $N \ln N - \sum_i N_i \ln N_i = BE + N \ln [\sum_i e^{-BE_i}]$, which when substituted in the equation previously obtained for the entropy of the system gives,

$$[S - C]/k = BE + N \ln \left[\sum_i e^{-BE_i} \right].$$

Now combining this result with the thermodynamic definition of temperature, namely that $T = [dE/dS]_V$, by differentiating with respect to S, it is found that $[dE/dS]_V = T = 1/kB$, or $B = 1/kT$. Therefore, $N_i = e^{-(A+1)} \cdot e^{-E_i/kT}$, which is the Boltzmann Law, namely that for any system obeying the laws of classical dynamics and thermodynamics, the number of particles, N_i, each possessing the energy, E_i, is proportional to the Boltzmann factor, $e^{-E_i/kT}$, since $e^{-(A+1)}$ is a constant.

The validity of the Boltzmann factor is in no way affected by the Stirling approximation for the factorial of a large number, used in its derivation; the error involved is of the order of $n/12$ of the value of $n!$. Thus for the numerical value of the factorial of the largest number which has been evaluated numerically (R. E. Smith, The Bases of Fortran,

Control Data Institute, Minneapolis, 1967), the Stirling approximation enables the factorial of such a small number as 2206 to be obtained to within $1.79 \times 10^{-6918}\%$.

has the energy nE_1 at some temperature, T, is proportional to the Boltzmann factor, $e^{-nE_1/kT}$, and therefore

$$\bar{E} = \frac{0 \cdot e^{-0/kT} + E_0 \cdot e^{-E_0/kT} + 2E_0 \cdot e^{-2E_0/kT} + \cdots + nE_0 \cdot e^{-nE_0/kT}}{e^{-0/kT} + e^{-E_0/kT} + e^{-2E_0/kT} + \cdots + e^{-nE_0/kT}}$$

$$= \frac{E_0 \sum_0^\infty n e^{-nE_0/kT}}{\sum_0^\infty e^{-nE_0/kT}} = \frac{E_0 \sum_0^\infty n x^n}{\sum_0^\infty x^n}, \tag{q.1.22.}$$

if the substitution, $e^{-E_0/kT} = x$, is made. Then,

$$\bar{E} = \frac{E_0 x \sum_0^\infty n x^{n-1}}{\sum_0^\infty x^n} = E_0 x \frac{d}{dx} \ln \sum_0^\infty x^n, \tag{q.1.23.}$$

since $(d/dx) \ln x^n = n/x$. But for $0 < x < 1$, $(1-x)^{-1} = 1 + x + x^2 + \cdots + x^n = \sum_0^\infty x^n$, and therefore,

$$\bar{E} = E_0 x \frac{d}{dx} \ln \left[\frac{1}{1-x} \right] = E_0 x \cdot \frac{1}{1-x} = \frac{E_0 e^{-nE_0/kT}}{1 - e^{-nE_0/kT}} = \frac{E_0}{e^{E_0/kT} - 1}. \tag{q.1.24.}$$

Substituting this value for $\bar{E}$ in equations (q.1.15.), there results,

$$E_\nu = \frac{8\pi\nu^2}{c^3} \cdot \frac{E_0}{e^{E_0/kT} - 1}; \qquad E_\lambda = \frac{8\pi}{\lambda^4} \cdot \frac{E_0}{e^{E_0/kT} - 1}. \tag{q.1.25.}$$

and if the Planck postulate that $E_0 = h\nu$, is put into (q.1.25.), the two equivalent forms of the Planck radiation law result,

$$E_\nu = \frac{8\pi h\nu^3}{c^3} \cdot \frac{1}{e^{h\nu/kT} - 1}, \qquad \text{and} \qquad E_\lambda = \frac{8\pi hc}{\lambda^5} \cdot \frac{1}{e^{hc/\lambda kT} - 1}. \tag{q.1.26.}$$

Now since

$$e^{hc/\lambda kT} = 1 + \frac{hc}{\lambda kT} + \frac{(hc/\lambda kT)^2}{2!} + \frac{(hc/\lambda kT)^3}{3!} + \cdots,$$

then if λT has a large value,

$$E_\lambda = \frac{8\pi hc}{\lambda^5 (1 + hc/\lambda kT - 1)} = \frac{8\pi kT}{\lambda^4},$$

which is identical with (q.1.16.), the Rayleigh–Jeans expression, valid at high temperatures and long wavelengths. Thus the classical approach based on the assumption that the average energy of an oscillator is given by the equipartition of energy principle, leads to the same result at the Planck quantum postulate at low frequencies or long wavelengths. On the other hand, when the quantity λT is small, then unity is negligible in comparison with $e^{hc/\lambda kT}$, and Planck's expression becomes $E_\lambda = (8\pi hc/\lambda^5)\cdot e^{-hc/\lambda kT}$, which is identical with the Wien Expression of (q.1.18.), provided that $c_1 = 8\pi hc$ and $c_2 = hc/k$, or $h = c_2 k/c$. The magnitude of Planck's constant may therefore be evaluated if the constant c_2 is known, since k and c are the known constants, Boltzmann's constant and the velocity of electromagnetic radiation, respectively. The second radiation constant, c_2, in the Wien radiation law is, as previously pointed out, available from experimental observation (1.43879×10^{-2} m K) and substituting the accepted values of k and c, it is found that

$$h = 6.6256 \times 10^{-34} \text{ J s.}$$

Thus the quantum postulate that $E_0 = h\nu$ successfully accounts for experimentally observed radiation data which are not interpretable in terms of classical ideas, and furthermore the quantum theory of radiation coincides with the classical concepts at low radiation frequencies or long wavelengths; in fact it emerges that quantum theory is the fundamental discipline in which to interpret physical phenomena and that classical mechanical theory is a special limiting case of what has come to be known as quantum mechanics.

For an isolated system in an energy state $E_i = nh\nu$ at a temperature of T K, the average energy of the various oscillators in terms of which the system may be described is

$$\bar{E} = \frac{E}{e^{E/kT} - 1} = \frac{h\nu}{e^{h\nu/kT} - 1},$$

as given by (q.1.24.), where ν is the vibrational frequency characteristic of the normal mode of the oscillators; then the average value of n is, $\bar{n} = 1/(e^{h\nu/kT} - 1)$. In terms of the Planck quantum concept, the energy of an electromagnetic field is said to be quantized; the number n is referred to as a quantum number, so that electromagnetic energy may be regarded as existing as photons of energy $h\nu$. A single wave of the type referred to on page 30, or a single mode of oscillation, is analagous in many ways to a linear oscillator, representing an oscillation of frequency ν. The detailed quantum mechanical description of a linear harmonic oscillator of this type is given in sections Q.2.6. and Q.2.7., where it is shown that the lowest energy possible for such an oscillator is not, however, zero, but $h\nu/2$. Therefore, in quantum mechanics, the energy of a linear oscillator is not simply $nh\nu$ but $[n + (1/2)]h\nu$; a change of the quantum number n by one unit thus corresponds to a change in the energy of the electromagnetic radiation field by $h\nu$, and it is this change in energy which may be interpreted as the addition to or removal of, a photon from the electromagnetic radiation field. The zero point energy of an oscillator in quantum mechanics, the term $h\nu/2$, is simply an additional term in the expression for the total energy of the oscillator, and the quantum number n, with reference to any one oscillator may be regarded as the number of photons of energy $h\nu$, equivalent to that of the particular oscillator concerned. The total energy of a radiation field described in terms of such oscillators is the sum of the energies equivalent to that of the various normal vibrational modes of the oscillators describing the system.

There is another way of evaluating Planck's constant from the above expressions. From the definition of the radiation density, $dE = E_\nu \, d\nu$, the total radiation density, E, may be calculated by substituting for E_ν the value given by (q.1.26.). Then,

$$E = \int dE = \int_0^\infty E_\nu \, d\nu = \frac{8\pi h}{c^3} \int_0^\infty [\nu^3(e^{h\nu/kT} - 1)^{-1}] \, d\nu. \qquad \text{(q.1.27.)}$$

Making the convenient substitution, $x = h\nu/kT$, then $d\nu = (kT/h) \, dx$, and

$$E = \frac{8\pi h}{c^3} \int_0^\infty \frac{[kTx/h]^3[kT/h] \, dx}{e^x - 1} = \frac{8\pi(kT)^4}{(ch)^3} \int_0^\infty [x^3(e^x - 1)^{-1}]dx. \qquad \text{(q.1.28.)}$$

The integral

$$\int_0^\infty [x^3(e^x - 1)^{-1} \, dx = \sum_{n=1}^\infty \int_0^\infty x^3 e^{-nx} \, dx = \sum_{n=1}^\infty [6/n^4] = 6\pi^4/90 = 6.497, \ddagger$$

and therefore

$$E = [8\pi(kT)^4][ch]^{-3} \cdot 6.497. \qquad \text{(q.1.29.)}$$

$\ddagger$ Binomial theorem expansion of $(e^x - 1)^{-1}$ gives,

$$(e^x - 1)^{-1} = (e^x)^{-1} + (-1)(e^x)^{-2}(-1) + \frac{(-1)(-2)(e^x)^{-3}(-1)^2}{2!} + \cdots$$

$$= e^{-x} + (e^{-x})^2 + (e^{-x})^3 + \cdots = \sum_{n=1}^\infty e^{-nx}.$$

Therefore, $\int_0^\infty [x^3(e^x - 1)^{-1}] \, dx = \sum_{n=1}^\infty \int_0^\infty x^3 e^{-nx} \, dx$, and since $\int [x^3 e^{-nx} \, dx = e^{-nx}[-x^3/n - 3x^2/n^2 - 6x/n^3 - 6/n^4]$, it follows that $\sum_{n=1}^\infty \int_0^\infty x^3 e^{-nx} \, dx = \sum_{n=0}^\infty (6/n^4)$, and the evaluation of the integral under consideration depends on being able to find the sum, $\sum_{n=1}^\infty n^{-4}$.

One approach is to make use of the Fourier series representation of functions; thus for a function $f(x)$, continuous in the interval $-\pi \le x \le \pi$, the Fourier series representation is $f(x) = a_0/2 + \sum_{n=1}^\infty [a_n \cos nx + b_n \sin nx]$, where $a_n = \pi^{-1} \int_{-\pi}^\pi f(x) \cos nx \, dx$, and $b_n = \pi^{-1} \int_{-\pi}^\pi f(x) \sin nx \, dx$. If $f(x)$ is assumed to be x^2, then since this is an even function of x, the b_n terms are all zero. The terms containing the coefficients, a, are then given by $a_0 = \pi^{-1} \int_{-\pi}^\pi x^2 \cos (0 \cdot x) \, dx = \pi^{-1} \int_{-\pi}^\pi x^2 \, dx = 2\pi^2/3$ and

$$a_n = \pi^{-1} \int_{-\pi}^\pi x^2 \cos nx \, dx = (4/n^2)(-1)^n,$$

for non-zero values of n, since $\cos n\pi = (-1)^n$. Now substituting these values for a_0 and a_n, together with $b_n = 0$, into the Fourier expression for $f(x) = x^2$, there results, $x^2 = \pi^2/3 + 4 \sum_{n=1}^\infty (-1)^n n^{-2} \cos nx$. Rearranging and squaring both sides of this latter expression gives $[x^2 - \pi^2/3]^2 = 16[\sum_{n=1}^{\infty} (-1)^n n^{-2} \cos nx]^2$, and if both sides of this equation are integrated from $x = -\pi$ to $x = \pi$, making use of the fact that $\int_{-\pi}^{\pi} \cos mx \cos nx \, dx = 0$ if $m \neq n$ or $\int_{-\pi}^{\pi} \cos mx \cos nx \, dx = \pi$ if $m = n$, there is obtained $\int_{-\pi}^{\pi} [x^2 - \pi^2/3]^2 \, dx = 16\pi \sum_{n=1}^{\infty} n^{-4}$. But

$$\int_{-\pi}^{\pi} [x^2 - \pi^2/3]^2 \, dx = 2 \int_{0}^{\pi} [x^2 - \pi^2/3]^2 \, dx = 2 \int_{0}^{\pi} [x^4 - 2\pi^2 x^2/3 + \pi^4/9] \, dx$$

$$= 2[x^5/5 - 2\pi^2 x^3/9 + \pi^4 x/9]_0^{\pi} = 8\pi^5/45.$$

Consequently, $\sum_{n=1}^{\infty} n^{-4} = \pi^4/90$.

An alternative way of finding the sum, $\sum_{n=1}^{\infty} n^{-4}$, makes use of the theory of infinite series, in particular the Bernoulli series, and the expansion of circular functions in terms of such series. The expression $x(e^x - 1)^{-1}$ may be represented as a power series, since

$$\frac{x}{e^x - 1} = \frac{1}{(e^x - 1)/x} = \frac{1}{x^{-1}(1 + x + x^2/2! + x^3/3! + \cdots - 1)}$$

$$= \frac{1}{1 + x/2! + x^2/3! + \cdots} = \frac{1}{\sum_{n=0}^{\infty} x^n[(n + 1)!]^{-1}}.$$

From a theorem in power series, if $\sum_{0}^{\infty} a_\nu x^\nu = g(x)$, then if $a_0 \neq 0$, $1/g(x)$ can also be represented by a power series, such that $1/g(x) = f(x) = \sum_{0}^{\infty} c_n x^n$. By definition, the series expansion above of $x(e^x - 1)^{-1} = \sum_{\nu=0}^{\infty} B_\nu x^\nu/\nu!$, the so-called Bernoulli series. Therefore, $\sum_{n=0}^{\infty} x^n[(n + 1)!]^{-1} \cdot \sum_{\nu=0}^{\infty} B_\nu x^\nu/\nu! = 1$, or writing out the expansion, $[1 + x/2! + x^2/3! + \cdots][B_0 + B_1 x + B_2 x^2/2! + B_3 x^3/3! + \cdots] = 1$. If $x = 0$, then $B_0 = 1$; for any other value of Bernoulli number, B_n, the numerical value may be obtained by multiplying out the above product, collecting the coefficients of the same power of x and applying the identity theorem. Thus, $B_0 = 1$; $B_1 + B_0/2! = 0$ or $B_1 = -1/2$; $B_2/2! + B_1/2! + B_0/3! = 0$ or $B_2 = 1/6$; $B_3/3! + B_2/2! \, 2! + B_1/3! + B_0/4! = 0$ or $B_3 = 0$; similarly for all odd values of ν, $B_{\nu(\text{odd})} = 0$. Proceeding in this way, the general recursion formula enabling any B_ν to be evaluated may be written, $B_n[n!1!]^{-1} + B_{n-1}[(n - 1)!2!]^{-1} + \cdots + B_1[1!n!]^{-1} + B_0[0!(n + 1)!]^{-1} = 0$. Hence, $B_0 = 1$, $B_1 = -1/2$; $B_2 = 1/6$; $B_3 = B_5 = B_7 = \cdots = 0$; $B_4 = -1/30$; $B_6 = 1/42$; $B_8 = -1/30$; $B_{10} = 5/66$; $B_{12} = -691/2730$; $B_{14} = 7/6$; $\cdots$.

Now circular functions may be expanded in terms of the Bernoulli series and they may also be resolved into partial fractions, enabling another power series expansion to be written. Choosing the function $\pi z \cot \pi z$, this may be written in the form $x(e^x - 1)^{-1}$, appropriate to Bernoulli series expansion as follows:

$$\pi z \cot \pi z = \frac{\pi z \cos \pi z}{\sin \pi z} = \pi z \left[\frac{(e^{i\pi z} + e^{-i\pi z})/2}{(e^{i\pi z} - e^{-i\pi z})/2i}\right] = i\pi z \left[\frac{e^{i\pi z} + e^{-i\pi z}}{e^{i\pi z} - e^{-i\pi z}}\right] = i\pi z \left[\frac{e^{2i\pi z} + 1}{e^{2i\pi z} - 1}\right]$$

$$= i\pi z + \frac{2i\pi z}{e^{2i\pi z} - 1},$$

where the second term is now in the form for expansion in a Bernoulli series, with $2i\pi z$ in place of x. Therefore,

$$\pi z \cot \pi z = i\pi z + B_0 + B_1(2i\pi z) + B_2(2i\pi z)^2/2! + \cdots$$

$$= i\pi z + B_0 - 2i\pi z/2 + B_2(2i\pi z)^2/2! + B_4(2i\pi z)^4/4! + \cdots,$$

since $B_1 = -1/2$ and $B_3 = B_5 = \cdots = 0$. Hence,

$$\pi z \cot \pi z = B_0 + B_2(2i\pi z)^2/2! + B_4(2i\pi z)^4/4! + \cdots$$

$$= 1 + \sum_{\nu=1}^{\infty} (i^2)^\nu (2\pi z)^{2\nu} B_{2\nu}/(2\nu)! = \sum_{\nu=0}^{\infty} (-1)^\nu (2\pi z)^{2\nu} B_{2\nu}/(2\nu)!$$

The function $\pi z \cot \pi z$ may alternatively be expressed in terms of a different power series by resolution into partial fractions, using the Mittag–Leffler theorem. (See K. Knopp, *Theory of Functions*, Part II, page 42, Dover Publications, New York, 1947.) Thus $\pi z \cot \pi z = 1 + \sum_{n=1}^{\infty} [2z^2]/[z^2 - n^2]$. Each of the terms in this sum may itself be expressed as a power series, since

$$\frac{2z^2}{z^2 - n^2} = \left[-2\frac{z^2}{n^2} \cdot \frac{1}{1 - z^2/n^2} \right] = -2z^2/n^2 - 2z^4/n^4 - 2z^6/n^6 - \cdots - 2z^{2\nu}/n^{2\nu}.$$

If each of these series is summed for $n = 1, 2, 3, \ldots$, there results, for the sum, $\sum_{n=1}^{\infty} 2z^2/(z^2 - n^2)$,

$$\sum_{n=1}^{\infty} \frac{2z^2}{z^2 - n^2} = -2\left[\sum_{n=1}^{\infty} 1/n^2 \right]z^2 - 2\left[\sum_{n=1}^{\infty} 1/n^4 \right]z^4 - \cdots - 2\left[\sum_{n=1}^{\infty} 1/n^{2\nu} \right]z^{2\nu} - \cdots.$$

This expansion must necessarily give the same result as the Bernoulli Series expansion, so for $\nu = 1, 2, 3, \ldots$, $-2 \sum_{n=1}^{\infty} [1/n^{2\nu}] = (-1)^{\nu}(2\pi)^{2\nu}B_{2\nu}/(2\nu)!$. Therefore for $\nu = 1, 2, \ldots$

$$\sum_{n=1}^{\infty} [1/n^{2\nu}] = \frac{(-1)^{\nu-1}(2\pi)^{2\nu}B_{2\nu}}{2[(2\nu)!]},$$

and for the particular case where $\nu = 2$, the sum

$$\sum_{n=1}^{\infty} [1/n^4] = \frac{-(2\pi)^4 B_4}{2(4!)} = \frac{16 \cdot \pi^4}{30 \cdot 2 \cdot 24} = \pi^4/90,$$

as before.

Now from (q.1.12.), the total number of modes of vibration of frequency ν which can exist in a space of volume V is $Z = [8\pi V\nu^3][3c^3]^{-1}$, and consequently the corresponding number of vibrations per unit volume is $[8\pi\nu^3][3c^3]^{-1}$, and the average energy per vibration is then,

$$\begin{aligned}
\bar{E} = E/Z &= [8\pi(kT)^4][ch]^{-3}[\pi^4/15][3c^3][8\pi\nu^3]^{-1} \\
&= [\pi kT]^4[5(h\nu)^3]^{-1}.
\end{aligned} \tag{q.1.30.}$$

But for electromagnetic radiation in space, the velocity of the radiation is c, and the average value of the component of the velocity in any given direction, which may be assumed to be normal to the direction of the opening in the wall of the isothermal cavity from which the radiation escapes is $c/4$.‡

‡ This result is a consequence of kinetic molecular theory, obtained by applying the Maxwell–Boltzmann distribution law to the light particles in the radiation cavity. Actually the light particles or photons, as will be discussed later in M.3.3., do not obey Maxwell–Boltzmann statistics but Bosé–Einstein quantum statistics, although this distinction does not affect the present argument. In fact, in the derivation of the Boltzmann factor, where it was assumed that different groups of particles had energies, E_1, E_2, $E_3, \ldots E_n$, but no particles had energies intermediate between these values, this characterization of the energies is typical of quantum phenomena. On the other hand, as long as no stipulation is made as to the magnitude of the energy gap between successive energy levels and this energy gap can be assumed to be infinitesimally small, this approach is typical of classical statistics, where a continuous energy distribution rather than a discontinuous one applies. It is therefore equally true for gas molecules or photons, that the

average velocity in three-dimensional space is four times as great as the average velocity in some specific direction. To prove this result, it is necessary firstly to derive the Maxwell–Boltzmann distribution law, which may be done as follows.

In classical mechanics (see Q.2.1.) the energy of a particle is generally given by the sum of its kinetic and potential energies, as $E = K(p_x, p_y, p_z) + V(x, y, z)$, where the kinetic energy, $K = (m/2)(v_x{}^2 + v_y{}^2 + v_z{}^2) = (1/2m)(p_x{}^2 + p_y{}^2 + p_z{}^2)$, is a function of the components of the momentum in the three cartesian coordinate directions, and the potential energy, V, is a function of the position of the particle in cartesian coordinate space. For any specific particle, then, $E = (1/2m)(p_x{}^2 + p_y{}^2 + p_z{}^2) + V(x, y, z)$, and it requires six coordinates, p_x, p_y, p_z, x, y, and z, to specify the total energy of the particle; such a space defined by six coordinates is referred to as molecular phase space, in kinetic molecular theory. The probability that any particle in a collection of N particles shall possess such an energy is proportional to the volume element, $dp_x \, dp_y \, dp_z \, dx \, dy \, dz$, in molecular phase space (see page 22).

Also, if N is very large, the fractional number, dN/N, of particles possessing this specific energy value, is identical to the probability that any one particle selected at random from the total population of particles, shall be found to possess this energy value. Therefore the volume element, $dp_x \, dp_y \, dp_z \, dx \, dy \, dz$, is proportional to the number of particles in the collection, N, which occupy the region of molecular phase space with coordinates between x and $x + dx$, y and $y + dy$, and z and $z + dz$, with momenta coordinates between p_x and $p_x + dp_x$, p_y and $p_y + dp_y$, and p_z and $p_z + dp_z$. The fractional number of particles possessing these energy characteristics must also be proportional to the Boltzmann factor, $e^{-E/kT}$. Hence, the fractional number of particles satisfying these particular conditions must be altogether given by,

$$dN/N = Ce^{-E/kT} \, dx \, dy \, dz \, dp_x \, dp_y \, dp_z,$$

where C is the combined proportionality constant. C may be eliminated from this expression by making use of the fact that $\int_0^\infty dN = N$, so that

$$C = \left[\int\int\int_{-\infty}^{+\infty}\int\int\int e^{-E/kT} \, dp_x \, dp_y \, dp_z \, dx \, dy \, dz \right]^{-1};$$

each integral must be taken over the limits of $-\infty$ to $+\infty$, as shown. Substituting this value for C back into the equation for dN/N gives,

$$dN/N = \frac{e^{-E/kT} \, dp_x \, dp_y \, dp_z \, dx \, dy \, dz}{\int\int\int_{-\infty}^{+\infty}\int\int\int e^{-E/kT} \, dp_x \, dp_y \, dp_z \, dx \, dy \, dz}.$$

This is one form of the distribution law. In terms of the assumptions of the kinetic theory for ideal gases, namely that the molecular energy is entirely kinetic or $V(x, y, z) = 0$, and that the gas molecules do not occupy any significant part of the space available to them, then since $K(p_x, p_y, p_z)$ is independent of the space coordinates of the particles, $\int\int\int_{-\infty}^{\infty} dx \, dy \, dz = V$, the volume available to the gas molecules. The distribution law for ideal gases then becomes,

$$dN/N = \frac{e^{-K/kT} \, dp_x \, dp_y \, dp_z \, dx \, dy \, dz}{V \int\int\int_{-\infty}^{\infty} e^{-K/kT} \, dp_x \, dp_y \, dp_z} = \frac{e^{-K/kT} \, dp_x \, dp_y \, dp_z}{\int\int\int_{-\infty}^{\infty} e^{-K/kT} \, dp_x \, dp_y \, dp_z},$$

since $dx \, dy \, dz = dV$ and $dV/V = 1$ for ideal gases. Now since $K = (1/2m)(p_x{}^2 + p_y{}^2 + p_z{}^2)$, by rearrangement, it follows that $2mK = p_x{}^2 + p_y{}^2 + p_z{}^2 = R^2$, as the sum of the squares of the momenta components may be regarded as the square of the radius, R, of a sphere. It is thus convenient to express the volume element of phase space, $dp_x \, dp_y \, dp_z$, in polar coordinates, when it becomes $4\pi R^2 \, dR$. But from the immediately previous equation, $R = \sqrt{[2mK]}$, and so $dR = [2mK]^{-1/2}m \, dK$, and $dp_x \, dp_y \, dp_z = 4\pi(2mK)(2mK)^{-1/2}m \, dK = 2\pi(2m)^{3/2}K^{1/2} \, dK$. If this result is substituted into the distribution law expression, then

$$dN/N = \frac{e^{-K/kT} \, dp_x \, dp_y \, dp_z}{\iiint\limits_{-\infty}^{\infty} e^{-K/kT} \, dp_x \, dp_y \, dp_z} = \frac{e^{-K/kT} \cdot 2\pi \cdot (2m)^{3/2} \cdot K^{1/2} \, dK}{2\pi(2m)^{3/2} \int_0^\infty e^{-K/kT} K^{1/2} \, dK},$$

where integration over all kinetic energy values only extends over the limits from 0 to ∞, rather than over the limits $-\infty$ to $+\infty$, as the kinetic energy can never be negative although the momenta coordinates can be both negative and positive. It is proved below that (a) $\int_0^\infty e^{-ax}x^n \, dx = [n!]/a^{n+1}$ and that (b) $(1/2)! = \sqrt{\pi}/2$, and making use of these results, the integral in the equation for dN/N above, $\int_0^\infty e^{-K/kT}K^{1/2} \, dK = (1/2)! \, (kT)^{3/2}$ or $[\sqrt{\pi}/2][kT]^{3/2}$, and therefore on substitution into the distribution law equation,

$$dN/N = \frac{2\pi e^{-K/kT}(2m)^{3/2}K^{1/2} \, dK}{2\pi(2m)^{3/2}(\sqrt{\pi}/2)(kT)^{3/2}} = \frac{2K^{1/2}e^{-K/kT} \, dK}{\sqrt{\pi}(kT)^{3/2}},$$

which is the Maxwell–Boltzmann distribution law for the energies of the molecules of an ideal gas.

The average space velocity and the average value of the component of the velocity of an ideal gas molecule may now be found. The average value of any property, denoted by $\bar{P}$, of a number of particles, N, is obviously obtained by, $\bar{P} = [\sum_i P_i n_i]/N$ or by $[\int P \, dn]/N$, where n_i of the particles have the value of the property P_i or a fraction dN of the particles have the property value P, respectively; the first alternative would be applicable to quantum properties where only discrete values P_i are possible for the property concerned, and the second alternative would be applicable to ideal gases, for example, where the velocity of the molecules is continuously variable from 0 to ∞, and the integral concerned is taken over the limits of the appropriate coordinates required to define the particular property. For the average value of the kinetic energy of a gas molecule, $\bar{K} = [\int_0^\infty K \, dN]/N$, and substituting into this expression the quantity dN/N given by the Maxwell–Boltzmann law gives, $\bar{K} = 2(\pi)^{-1/2}(kT)^{-3/2} \int_0^\infty e^{-K/kT}K^{3/2} \, dK = 3kT/2$, making use of the integral, $\int_0^\infty e^{-ax}x^n \, dx = [n!]/a^{n+1}$; this is the classical value for $\bar{E}$, apart from the numerical factor, used in the Rayleigh–Jeans radiation law. If $K = mv^2/2$ is put into the Maxwell–Boltzmann energy distribution equation and the variable of integration changed to velocity using, $dK = mv \, dv$, there results the corresponding distribution of velocity law, $dn/N = [2/\pi]^{1/2}[m/kT]^{3/2}e^{-mv^2/2kT}v^2 \, dv$. Making use of this latter equation in the expression for the average space velocity, $\bar{v} = [\int_0^\infty v \, dN]/N$, gives

$$\bar{v} = [2/\pi]^{1/2}[m/kT]^{3/2} \int_0^\infty e^{-mv^2/2kT}v^3 \, dv = [8kT/\pi m]^{1/2},$$

making use of the integral, $\int_0^\infty e^{-ax^2}x^3 \, dx = [1/2a^2]$, proved in (c) below.

To find the average value of the component of the velocity in one specific direction, it is necessary to return to the Maxwell–Boltzmann distribution law expressed as, $dN/N = e^{-K/kT} \, dp_x \, dp_y \, dp_z \left[\iiint\limits_{-\infty}^{\infty} e^{-K/kT} \, dp_x \, dp_y \, dp_z\right]^{-1}$. The denominator of this expression has already been shown to be $2\pi(2m)^{3/2} \int_0^\infty e^{-K/kT}K^{1/2} \, dK = [2\pi mkT]^{3/2}$, and the numerator expressed in terms of the velocity components becomes

$$m^3 e^{-[m/2kT][v_x^2 + v_y^2 + v_z^2]} \, dv_x \, dv_y \, dv_z.$$

The fraction of the total number of particles which has velocity components resolved along the x axis, for example, which lie between v_x and $v_x + dv_x$, may now be found by integrating the quantity dN/N over all values of v_y and v_z between $-\infty$ and $+\infty$; thus $dN(v_x)/N = [m/2\pi kT]^{3/2}e^{-mv_x^2/2kT}\,dv_x \int_{-\infty}^{\infty} e^{-mv_y^2/2kT}\,dv_y \int_{-\infty}^{\infty} e^{-mv_z^2/2kT}\,dv_z$, and since $\int_{-\infty}^{\infty} e^{-ax^2}\,dx = \sqrt{\pi/a}$ [see below], it follows that

$$dN(v_x)/N = [m/2\pi kT]^{1/2}e^{-mv_x^2/2kT}\,dv_x.$$

The average value of the component of the velocity of an ideal gas molecule in the $+v_x$ direction is now $\bar{v}_x = \int_0^\infty [v_x\,dN(v_x)]/N = [m/2\pi kT]^{1/2}\int_0^\infty v_x e^{-mv_x^2/2kT}\,dv_x = [kT/2\pi m]^{1/2}$; since $\int_0^\infty xe^{-ax^2}\,dx = \dfrac{1}{2a}$. Finally, comparing the values found for the average space velocity, $\bar{v} = [8kT/\pi m]^{1/2}$, and the average component of the velocity in the $+x$ direction, $\bar{v}_x = [kT/2\pi m]^{1/2}$, $\bar{v}_x = \bar{v}/4$, the result quoted on page 40.

(A) PROOF THAT $\int_0^\infty e^{-ax}x^n\,dx = [n!]/a^{n+1}$.

Integrating initially, $\int e^{ax}\,dx = e^{ax}/a + C_1$. Then integrating by parts, $\int e^{ax}\,x\,dx = e^{ax}x/a - \int [e^{ax}/a]\,dx = [xe^{ax}]/a - e^{ax}/a^2 + C_2 = [e^{ax}/a][x - 1/a] + C_2$. Similarly, $\int e^{ax}x^2\,dx = [x^2e^{ax}]/a - 2\int [xe^{ax}/a]\,dx = [2e^{ax}]/a^3[a^2x^2/2 - ax + 1] + C_3$. Generally,

$$\int e^{ax}x^n\,dx = \frac{e^{ax}(n!)}{a^{n+1}}\left[\frac{(ax)^n}{n!} - \frac{(ax)^{n-1}}{(n-1)!} + \frac{(ax)^{n-2}}{(n-2)!} - \cdots \frac{(ax)^{n-m}}{(n-m)!}\cdots + 1\right] + C_n.$$

For negative values of the coefficient a, a similar result pertains,

$$\int e^{-ax}x^n\,dx = -\frac{e^{-ax}(n!)}{a^{n+1}}\left[\frac{(ax)^n}{n!} + \frac{(ax)^{n-1}}{(n-1)!} + \cdots + \frac{(ax)^{n-m}}{(n-m)!} + \cdots + 1\right] + C_m.$$

Thus, $\int_0^\infty e^{-ax}x^n\,dx = n!/a^{n+1}$.

(B) PROOF THAT $(1/2)! = \sqrt{\pi}/2$.

The factorial of a non-integral quantity, like $1/2$, is accessible from the properties of gamma functions. The gamma function, $\Gamma(n)$, is a generalization of $n!$ for such non-integral values of n and is so chosen that if n is an integer, $\Gamma(n) = (n-1)!$. One definition of $\Gamma(n)$ is,

$$\Gamma(n) = \lim_{r\to\infty} \frac{1\cdot 2\cdot 3\cdots(r-1)}{n(n+1)(n+2)\cdots(n+r-1)}\,r^n,$$

from which obviously $\Gamma(1) = \lim r!/r! = 1$. Also from this definition of $\Gamma(n)$,

$$\Gamma(n+1) = \lim_{r\to\infty} \frac{1\cdot 2\cdot 3\cdots(r-1)}{(n+1)(n+2)\cdots(n+r)}\,r^{n+1}$$

$$= \lim_{r\to\infty} \frac{rn}{(n+r)}\frac{1\cdot 2\cdot 3\cdots(r-1)}{n(n+1)(n+2)\cdots(n+r-1)}\,r^n = n\Gamma(n).$$

Thus, if n is a positive integer, $\Gamma(n) = (n-1)!$. The gamma function is alternatively defined by the integral, $\int_0^\infty x^{n-1}e^{-x}\,dx = \Gamma(n)$, where $n > 0$; the first definition is due to Euler, the second to Legendre. Since on integrating by parts, $\int_0^\infty e^{-x}x^n\,dx = n\int_0^\infty e^{-x}x^{n-1}\,dx - e^{-x}x^n|_0^\infty$, and the last term vanishes at the limits $x = 0$ and $x = \infty$, $\int_0^\infty e^{-x}x^n\,dx = n\int_0^\infty e^{-x}x^{n-1}\,dx$, or $\Gamma(n+1) = n\Gamma(n)$, as before. If n is an integer, also as before, $\Gamma(n+1) = n!$, although this important relation is true for any function of n. Thus $\Gamma(1/2) = \int_0^\infty e^{-x} x^{-1/2}\, dx$. If for convenience, the substitution is made, $x = t^2$, then $x^{-1/2} = 1/t$ and $dx = 2t\, dt$. Thus

$$\Gamma(1/2) = \int_0^\infty e^{-t^2}(1/t)2t\, dt = 2\int_0^\infty e^{-t^2}\, dt = 2\int_0^\infty e^{-x^2}\, dx,$$

since a definite integral is a function only of its limits and not of the variable of integration. But $\int_0^\infty e^{-t^2}\, dt \int_0^\infty e^{-x^2}\, dx = \int_0^\infty \int_0^\infty e^{-(x^2+t^2)}\, dx\, dt = [\int_0^\infty e^{-x^2}\, dx]^2$. If $t = vx$ is substituted into this equation, then $dt = x\, dv$, and

$$\int_0^\infty \int_0^\infty e^{-(x^2+t^2)}\, dx\, dt = \int_0^\infty \int_0^\infty x e^{-(1+v^2)x^2}\, dx\, dv.$$

But

$$\int_0^\infty x e^{-x^2(1+v^2)}\, dx = \left[-\frac{e^{-x^2(1+v^2)}}{2(1+v^2)} \right]_0^\infty,$$

which evaluated over the limits from 0 to ∞ is $[2(1+v^2)]^{-1}$. Therefore,

$$\int_0^\infty \int_0^\infty x e^{-(1+v^2)x^2}\, dx\, dv = \int_0^\infty [2(1+v^2)]^{-1}\, dv = [(1/2)\tan^{-1} v]_0^\infty = \pi/4 = [\int_0^\infty e^{-t^2}\, dt]^2.$$

Thus, $\int_0^\infty e^{-t^2}\, dt = \sqrt{\pi}/2$, and since $\Gamma(1/2) = 2\int_0^\infty e^{-t^2}\, dt$ then $\Gamma(1/2) = \sqrt{\pi}$. But from the definition of a gamma function, $(1/2)! = (1/2)\Gamma(1/2)$, and hence $(1/2)! = \sqrt{\pi}/2$.

(C) INTEGRALS OF THE TYPE, $\int_0^\infty x^n e^{-ax^2}\, dx = I_n$.

This type of integral occurs in many applications of the kinetic theory of gases and in the Debye theory of the heat capacity of solids. In kinetic theory applications, $a = m/2kT$. All of these integrals may be related to the first two members, $I_0 = \int_0^\infty e^{-ax^2}\, dx$ and $I_1 = \int_0^\infty x e^{-ax^2}\, dx$, by differentiations with respect to a. For example, $I_2 = -dI_0/da$, $I_3 = -dI_1/da$, $I_4 = d^2I_0/da^2$, $I_5 = d^2I_1/da^2$. It is proved above that $\int_0^\infty e^{-x^2}\, dx = \int_0^\infty e^{-t^2}\, dt = \sqrt{\pi}/2$, and by a similar method it may be shown that $\int_0^\infty e^{-ax^2}\, dx = (1/2)\sqrt{\pi/a} = I_0$; this is the Gauss probability integral. Also from the above it is evident that $I_1 = 1/2a$, by simply making the substitution of a for $(1+v^2)$. Then, by differentiation, $I_2 = -d/da[\int_0^\infty e^{-ax^2}\, dx] = (1/4)\sqrt{\pi/a^3}$, and $I_3 = -d/da[\int_0^\infty x e^{-ax^2}\, dx] = (1/2a^2)$, and generally for even values of n, $\int_0^\infty e^{-ax^2}x^n\, dx = 1\cdot3\cdot5\cdots(n-1)(\pi a)^{1/2}[2a]^{-n/2-1}$, and when n is odd, $\int_0^\infty e^{-ax^2}x^n\, dx = [(1/2)(n-1)]!\,[2\{a^{1/2(n+1)}\}]^{-1}$.

Hence the rate of escape of radiation per unit area, from the cavity, is

$$S = cE/4 = \sigma T^4, \tag{q.1.31.}$$

from Stefan's Law, given in (q.1.17.). If (q.1.29.) is now substituted in (q.1.31.), it is possible to evaluate the Planck constant, if the Stefan–Boltzmann constant, σ, is known; the latter can be accurately determined by experimental measurement. Thus

$$\sigma T^4 = \frac{c}{4}\cdot\frac{8\pi(kT)^4}{(ch)^3}\cdot\frac{\pi^4}{15},$$

or

$$\sigma = \frac{2\pi^5 k^4}{15 c^2 h^3} = 5.6697 \times 10^{-8}\ \mathrm{W\,m^{-2}\,K^{-4}},$$

and substituting for c, k, π, all of which may be determined independently,

$$h = 6.6256 \times 10^{-34} \text{ J s.}$$

The dimension of the Planck constant, energy $\times$ time, is equivalent to ml^2t^{-1}, which in turn is equivalent to mass $\times$ velocity $\times$ distance, the dimension of angular momentum. As will be discussed in the next section, this observation was applied later by Bohr in formulating the first quantum description of the hydrogen atom. The magnitude of h given above has been confirmed as the result of numerous observations involving phenomena in which energy, especially electrical energy, is used to excite radiation not only in the infrared, thermal region of the spectrum but also in the visible, ultraviolet, and X-ray regions.

The concept of energy quanta soon found application, by Einstein, to account for two other types of experimental observations which could not be interpreted in terms of classical ideas, namely the heat capacities of solids at low temperatures and the photoelectric effect. The heat capacities of solids are discussed in some detail in M.1.2.; due to the successive efforts of Einstein, Debye, Blackman, Born, and many others, continued interest in this area has led to the discovery of many important principles of the physics and chemistry of the solid state and to the development of crystal lattice dynamics.

The photoelectric effect refers to the fact that when light of some particular frequency falls on the clean surface of a metal in an evacuated container, electrons are emitted. The velocity of the emitted electrons may be determined by the application of an external electric field of sufficient strength, V, to stop the photoemission process. Then $V\varepsilon = mv^2/2$. The experimentally observed facts are that the number of electrons emitted per second is proportional to the intensity of the incident radiation, the energy of the photoelectrons is independent of the intensity but proportional to the frequency of the incident radiation, and that no electrons are emitted if the frequency, v, involved is below a certain critical value, v_c. The relation between the maximum energy of the emitted photoelectrons and these frequencies can be described by $mv^2/2 = C(v - v_c)$. Classical electromagnetic theory would anticipate that the energy of the photoelectrons would vary with the intensity but would be independent of the frequency of the incident radiation. Actually the photoelectric effect was the first event involving electrons to be discovered which disagreed with classical electromagnetic theory. Einstein showed that the difficulties involved could be resolved if instead of regarding the incident radiation in terms of electromagnetic theory as a wave motion of frequency v, it be considered as a stream of particles or photons each possessing the energy hv, such that when a photon collides with the metal it transfers its energy to an electron which is at or near the surface. Part of the photon energy is required to liberate the electron from the attraction of the metal, the remainder appearing as kinetic energy of the photoelectron, so that $mv^2/2 = h\nu - W = h(\nu - \nu_c)$, where W, the energy required to just remove an electron from the metal surface, is referred to as the work function of the metal, and is quite characteristic of the metal concerned. Thus $W = h\nu_c$, and $h\nu = W + mv^2/2 = W + V\varepsilon = h\nu_c + V\varepsilon$, which is Einstein's photoelectric law. Expressed in the form, $\nu = \nu_c + (\varepsilon/h)V$, this is obviously the equation of a straight line where ν_c represents the intercept of the line formed by plotting frequency as a function of the retarding voltage V, and the slope of the line should be a universal constant for all metals, ε/h; these deductions have been conclusively confirmed. As will be discussed in section M, the work function of most metals is of the order of a few electron volts, reaching a maximum for platinum (1.008×10^{-18} J), while cesium has the lowest work function (2.998×10^{-19} J). Also the photoelectric work functions of metals displays a periodic variation with the nuclear charge of the metal, similar to many other physicochemical properties of the elements. It is also perhaps worth noting here that the work functions of metals correspond to lower energy values than the ionization constants of individual atoms of the same metals; the electrons in metals in bulk are not so firmly bound to the atomic nuclei as in free atoms and in fact this characteristic of metals is responsible for many of their typically metallic properties such as their generally high electrical and thermal conductivities and their relatively uniform optical properties (Chapter M.4.).

The first application of the quantum postulate to phenomena involving individual atoms was made by Bohr[1] to describe the line spectrum of hydrogen.

Q.1.4.   Line Spectra; The First Quantum Theory of Atomic Structure

According to the conditions of experiment a material body may either emit or absorb radiation incident upon it; the observed effects are described in terms of the two principal classes of spectra, emission or absorption spectra. Emission spectra may be of various types ranging from continuous to band type to discontinuous or line spectra. Thermal, soft X-ray spectra, and tesla-luminescence spectra are continuous; raman, rotational, vibrational, and rotational-vibrational spectra are of the band type, while the emission spectra of ionized gases are observed to be discontinuous, consisting of a number of discrete lines. Many other types of spectra have attracted attention more recently, such as electron and proton spin resonance spectra, nuclear magnetic resonance spectra, microwave, and charge-transfer spectra; the discussion in

---

[1] Bohr, N., *Phil. Mag.*, 1913, **26**, 476.

this section will be limited to the line spectra of the elements and of hydrogen in particular. A typical line spectrum is produced by passing an electrical discharge through a gas at a low pressure and observing the radiation emitted after it has been resolved by passage through a suitable prism system, thus giving rise to a number of isolated bright lines. To describe spectra reference is made to the wavelength or frequency of these characteristic lines. Wavelengths may be given in Ångström units‡ for visible and ultraviolet spectra, but infrared and longer wavelengths may be conveniently expressed in microns or millimicrons ($1\ \mu = 10^{-6}\ m = 10^{-4}\ cm = 10^4\ Å = 10^3\ m\mu$).

‡ The first systematic measurements of wavelengths, those of the solar spectrum, using a diffraction grating, were made by Ångström, who published a "normal" scale in 1869. His unit, equalling one ten-millionth of a millimeter (10^{-8} cm) has come to be known as the Ångström unit (Å), and is based on the distance between two lines on a certain platinum bar at 0°C. In 1907 the International Solar Union adopted a new "International Scale" based essentially on Fabry's measurements of the wavelengths of the red line in the cadmium spectrum and, on subsequent comparative interferometric measurements of other selected lines. This unit was defined as unity (one) in terms of the wavelength of the cadmium red line in dry air at 288 K and at a pressure of 760 mm. of mercury (gravitational constant taken as $g = 980.67$ cgs units) taken to be 6438.4696, and it was called the International Ångström unit (I.A.) and later in 1938 the "angstrom" (A). The "angstrom" and the Ångström Unit are equivalent to seven significant figures. Shortly after the first crystal structures were determined, it became apparent that X-ray diffraction measurements were more precise than the measurement of constants such as Avogadro's number, on which the absolute value of crystal spacings were based. To avoid changes in crystal spacings with changing values of the fundamental constants, Siegbahn in 1925 proposed to express X-ray wavelengths in terms of a new unit of measurement, the x unit (xu), defined in terms of the {100} spacing of sodium chloride, taken as 2814.00 xu. Shortly afterwards it was found that natural crystals of the mineral calcite ($CaCO_3$) gave much sharper reflections than sodium chloride (more perfect crystal lattice) and it came to be the standard crystal for measuring X-ray wavelengths, taking the effective spacing of the first order of the calcite cleavage as $d_1 = 3029.04$ xu. In terms of the accepted value of Avogadro's number, the x unit was equal to 10^{-11} cm, within the limits of the experimental error in N_A, which was assumed to be about ± 2 in 3000. Later measurements of X-ray wavelengths after 1928, by a ruled grating in optical wavelength units (A) revealed a discrepancy of about 0.2% in the accepted value of the Avogadro number, which was subsequently confirmed by a redetermination of the electronic charge, ε. After international discussion the conversion factor was agreed upon as 1.00202 (1 kx = 1000 xu = 1.00202 Å). Whether or not the "converted kx" values should be regarded as Å or A is of no consequence, since it is hardly possible to achieve seven-figure accuracy in X-ray wavelength measurements, nor is there any possibility that the calcite spacing standard can compare in accuracy with the cadmium redline standard. Since 1960, the cadmium standard has been changed to a krypton standard, adopting the vacuum wavelength of radiation corresponding to the transition $^2P_{10} - {}^5D_5$ of krypton 86 as an absolute standard; with reference to this Kr^{86} wavelength, λ_{Kr}, the standard unit of length, the meter is given by, 1 meter = 1650763.73 λ_{Kr}. The krypton standard still suffers from the defect that it cannot be used to measure distances as long as a meter in one operation. The development of lasers in the period since 1960, however, provided the

means of measuring optical path differences of hundreds of meters in one step (see page 51).

Frequencies may be expressed as wave numbers or oscillation frequencies. Wave numbers are given by, $\nu = 10^8/\lambda$ in Å, are independent of c and of the order of 10^5, whereas oscillation frequencies are given by c/λ and are of the order of 10^{15-16}. It should be noted, incidentally, that the atoms in a gas at low pressure emitting light under the influence of an electrical discharge, in the production of line spectra, are not necessarily in the same state as the atoms in normal chemical compounds; since 1960 a variety of techniques have been discovered for inverting the normally stable electronic states of both gases and solids to produce strictly monochromatic, coherent, light emission under the stimulation of an external, electrical power source.‡

‡ Devices using this principle, referred to as lasers (Light Amplification by the Stimulated Emission of Radiation), have given rise to an extensive technology since 1960. Lasers emit monochromatic radiant energy of the highest purity at very coherent wavelengths. The emitted beam possesses quite considerable power and the ability to travel great distances with almost negligible divergence. Dependent on the type of ion source used or whether the gas phase or solid state type is employed, lasers may emit in the visible, ultraviolet, or infrared regions, of the electromagnetic spectrum. The argon laser, for example, making use of the electronic transitions between inverted energy states in singly-ionized argon gas, emits energy at six discrete wavelengths, 4.579×10^{-7} m, 4.765×10^{-7} m, 4.880×10^{-7} m, 4.956×10^{-7} m, 5.017×10^{-7} m, and 5.145×10^{-7} m, in the blue-green region of the visible spectrum. Laser technology includes the application of these devices to long-distance communication, spectroscopy, chemistry, biochemistry, surgery, and high-speed evaporation processes in the production of electronic components.

In order to understand how laser action arises, it is necessary to examine in greater detail the nature of the interaction between electromagnetic radiation and matter and also to inquire under what circumstances electromagnetic radiation may be described in terms of classical theory or in terms of quantum theory. The most remarkable features of a laser beam are the coherence of the radiation and the energy density or intensity of the emitted radiant energy. The coherence of a radiation field refers to the fact that the phase of the radiation is definitely determined, or to the fact that there is little out-of-phase interference. From the definitions of E_ν and $E_\nu\, d\nu$ given on page 29, and the Planck radiation law of (q.1.26.), it is apparent that if $E_\nu\, d\nu$ is large for some particular emitted frequency ν, then the quantum number n (page 37) must be large. The phase of electromagnetic radiation (page 112), in terms of the quantum mechanical description, is a quantity which cannot be uniquely determined, but observable information about the phase of radiation is related to its wavelength or wave vector (page 30) and the magnitude of Planck's constant by the Heisenberg Uncertainty Principle (sections Q.2.5. and Q.2.6., also pp. 157–159). The Uncertainty Principle leads to the result that a coherent radiation field of definitely determined phase, is only possible for the situation where the quantum number n for each of the various normal modes of vibration of the oscillators in terms of which the system may be described is large compared with unity. Thus, the average value of n, given on page 37, is large (small) compared with unity if $h\nu/kT$ is small (large) compared with unity. As a consequence, radiation of the intensity normally encountered, for example, in the microwave and radiofrequency parts of the electromagnetic spectrum, may be conveniently described in terms of classical wave theory, without taking account of the Uncertainty Principle or quantum-type behavior. However, as has already been discussed for black body or isothermal cavity radiation, and more generally in the infrared and visible regions of the electromagnetic spectrum, where conventional light sources emit at relatively low intensity, n is small and the radiation is not coherent, it is necessary to describe such radiation in terms of quantum theory. The very intense radiation produced by lasers, even in the visible region of the electromagnetic spectrum, corresponds to large values of the quantum number n and coherent radiation, which again may be described with reasonable accuracy in terms of classical electromagnetic theory, without any reference to the quantum concept.

The average value of the quantum number n for isothermal cavity radiation is given by, $\bar{n} = [e^{h\nu/kT} - 1]^{-1}$, as on page 37. For any frequency in the visible region of the spectrum, where $\nu \simeq 6 \times 10^{15} \, \text{s}^{-1}$, the quantity $h\nu/kT$ is large compared with unity, and the Planck radiation law is required to provide an adequate description of the energy density–frequency distribution. It is only for very large values of the temperature, of the order of thousands of degrees, that $h\nu/kT$ becomes comparable with unity; at the surface temperature of the sun, about 6000 K, the maximum of the Planck distribution curve actually lies in the middle of the visible spectrum, but for all energy densities of radiation in the visible region of the electromagnetic spectrum which are less than that present at the surface of the sun in this frequency range, the Planck radiation law must be used to describe the situation. On the other hand for low frequencies, long wavelength radiation, especially at high temperatures, the Planck radiation law reduces to the classical Rayleigh–Jeans law (page 33), and the quantum description is unnecessary. In lasers, optical energy densities may be built up which would correspond to temperatures very much in excess of the sun's surface temperature, if their radiant energy were emitted as part of an isothermal cavity spectrum. At these very high temperatures, even although the laser energy is emitted at optical frequencies, the Rayleigh–Jeans radiation law is applicable. Radiation in the microwave region of the electromagnetic spectrum may also be quite adequately described in terms of classical theory, but for a different reason. Thus, the quantity $h\nu/kT = 1$ at 300 K for a frequency of about $6 \times 10^{12} \, \text{s}^{-1}$ or a wavelength of about 5×10^{-5} m, which is small compared with all microwave wavelengths; for a typical microwave wavelength of 1 mm, $h\nu/kT \simeq 1/20$ at 300 K, which corresponds to a quantum number of about 20 for thermal radiation due to oscillators of this frequency. Actually, this type of thermal radiation is that responsible for the thermal or Johnson noise in electrical systems, but microwave frequencies are usually generated at power levels well above that of thermal noise, and so classical electromagnetic theory provides an adequate description of events at microwave frequencies, except perhaps very short microwave wavelengths at temperatures near 0 K at very low power levels.

The induced or stimulated emission of radiation by a substance in a particular energy state was actually discovered in the microwave region of the spectrum before such a possibility was realized in the visible region; the former type of behavior is referred to as maser (Microwave Amplification by the Stimulated Emission of Radiation) action and the latter as laser action. Any material in thermal equilibrium with its surroundings has a distribution of atoms or molecules over the various allowed energy states given by the Maxwell–Boltzmann distribution; for example,

$$N_i = N_0 e^{-(W_i - W_0)/kT},$$

where N_i is the population of the state of energy W_i, N_0 is the ground state population of energy W_0, T is the absolute temperature, and k is the Boltzmann constant. This implies that excited states are always less densely populated than the ground state when the system is in thermal equilibrium with its surroundings. When a photon is absorbed by an atom or molecule, to produce an excited state, the atom or molecule, in the absence of any external stimulus, will spontaneously decay to a lower energy state, emitting a photon. If, instead of allowing the excited state to decay naturally, it is irradiated with another photon that has the same energy as the one spontaneously emitted, then the irradiation triggers the release of the photon, providing two photons where there was only one before, thus amplifying the incident radiation. The spontaneous emission lifetime depends on the frequency of the incident radiation, such that in the microwave region spontaneous emission has a lower probability than induced emission, while in the visible region spontaneous emission has a high probability than induced emission (see page 55). The problem in inducing both maser and laser action is to make induced or stimulated emission predominate over spontaneous emission and absorption. For this to occur there must be more atoms or molecules in an excited upper energy state than a lower unexcited state, that is population inversion has to be brought about. This is not possible for any system in thermal equilibrium with its surroundings, but if this equilibrium can be prevented from being established, then population inversion is possible.

Such population inversion was first accomplished for ammonia, in the microwave region of the electromagnetic spectrum: the ammonia maser. Ammonia molecules have two closely spaced energy levels in the microwave region which, in the presence of an electric field, have different induced dipole moments, so that separation of the excited and normal molecules can be accomplished by the use of an inhomogeneous electric field. The excited molecules when passed into a microwave cavity and stimulated with weak microwave power, oscillate with a very stable frequency of $2.387 \times 10^{10}\,\mathrm{s}^{-1}$ determined by the natural frequency of vibration of the nitrogen atom through the plane of hydrogen atoms in the ammonia molecule. A similar type of separation maser has been made with atomic hydrogen as the active material.

Another manner of accomplishing population inversion, which has been used in producing both masers and lasers, is to make use of a substance which has three closely spaced energy levels. If W_1, W_2, and W_3, are three such energy states and $W_1 < W_2 < W_3$, then if a microwave signal of frequency $\nu_{13} = (W_3 - W_1)/h$ is fed into the system to saturation, such that $N_1 = N_3$, and if the populations of states 2 and 3 remain in thermal equilibrium, then $N_3 < N_2$ and $N_1 = N_3$, so that $N_1 < N_2$ and a population inversion between energy levels W_1 and W_2 may be obtained. Amplification by stimulated emission may then be obtained with radiation of frequency $\nu_{21} = (W_2 - W_1)/h$. The frequency ν_{31} is called the pumping frequency and the frequency ν_{21} the signal frequency. Successful masers and lasers using this principle were first produced using the energy states of the tripositive chromium ions in ruby crystals (Cr^{3+} ions in alumina).

Laser action has been accomplished in several types of material. Solid state lasers using ruby, neodymium in calcium tungstate, uranium in calcium fluoride, and other materials have been operated in pulses and continuously. Organic liquids such as benzene, toluene, and nitrobenzene, are capable of laser action in the visible and infrared regions of the spectrum. Rare gas lasers using mixtures of helium and neon, argon, krypton, and xenon, as the active material, which are more efficient than solid state lasers and operate in the infrared and visible regions of the spectrum, are widely used. The helium gas serves as an intermediary to provide the pumping frequency. Semiconductor lasers, using such materials as gallium arsenide, are more efficient than either the solid state or rare gas types, converting electrical energy directly into optical radiation. In the gallium arsenide laser, nearly all of the input electrical energy is converted into light, of which at least 50% is coherent.

The problem of ensuring that all of the individual active atoms in these materials radiate in phase is resolved by using two accurately parallel, highly reflecting plates between which the active material is placed. The active material is then pumped until population inversion is accomplished and then light of the appropriate signal frequency is applied through the material, which is amplified by the induced emission, but only the radiation traveling at right angles to the plates is reflected many times through the active material. If the amplification is sufficient to offset the reflection losses, an intense stationary wave is set up between the reflecting plates, and if one plate is partially transparent an accurately parallel beam of radiation emerges, which may be focused to a spot of the order of magnitude of 10^{-4} m.

Balmer originally identified 14 lines in the visible region of the hydrogen spectrum (there are now known to be 33 identifiable lines in the Balmer series). He found that the wavelengths of these lines could be described by, $\lambda = C_1 m^2 (m^2 - 2^2)^{-1}$, where C_1 is a constant and m is an integer greater than 2. Inspection of the wavelengths of the lines in the hydrogen spectrum showed that as m increased, the difference between the wavelength of adjacent lines became less, tending toward a limit for shorter wavelengths; when m^2 is very large, Balmer's limit was 3.6456×10^{-7} m. Rydberg later plotted the wave number of the lines as a function of m and arrived at the expression, $v = v_\infty - C_2 (m + C_3)^{-1}$, where C_2 and C_3 are constants and $m = 3, 4, 5, \ldots$; he later modified this expression to $v = v_\infty - R_\infty (m + C_3)^{-2}$, where R_∞ now appeared to be a constant for all series of lines in a spectrum, unlike Balmer's Series, which only described the visible region. If $v = 1/\lambda$ and $v_\infty = 1/C_1$ are substituted in Balmer's series, then $v = v_\infty - 4v_\infty/m^2 = v_\infty - R_\infty/m^2$, where $R_\infty = 4v_\infty$ and is defined as the Rydberg Constant, and this expression is identical to Rydberg's when $C_3 = 0$. Thus Balmer's series is a special case of Rydberg's expression; also R_∞ must be equal to $4m^2 v (m^2 - 4)^{-1}$, where m and v for any given line in the hydrogen spectrum may be substituted. The Rydberg Constant has the value, 10973731 m^{-1}, and it is not a fundamental constant, as it may be resolved into simpler components; it will be shown below that $R_\infty = [2\pi^2 m e^4][ch^3]^{-1}$. For the hydrogen spectrum, Rydberg's formula may be written as, $v = v_\infty[1 - 4(m + C_3)^{-2}]$, substituting $R_\infty = 4v_\infty$. Therefore, $v_\infty = [v(m + C_3)^2][(m + C_3)^2 - 4]^{-1}$ and $R_\infty = 4v(m + C_3)^2[(m + C_3)^2 - 4]^{-1}$. Hence,

$$v = v_\infty - R_\infty (m + C_3)^{-2} = R_\infty/4 - R_\infty[m + C_3]^{-2},$$

or more generally,

$$v = \frac{R_\infty}{n^2} - \frac{R_\infty}{m^2} \qquad (q.1.32.)$$

where for the Balmer series, $n = 2$, and $m = 3, 4, 5, \ldots$. This expression is what is referred to as the Ritz Combination Principle, namely, that all the lines in a given spectrum are connected with a number of inferred wave numbers, or "terms," such that every line in the spectrum has a wave number which is the difference of two terms. The advantage of this principle is that the complete description of a spectrum requires fewer terms than the number of lines observed, so that the coordination of observed spectral data is much simplified. If $n = 3$ and $m = 4$ in (q.1.32.), v corresponds to the first line in the infrared spectrum of hydrogen, originally identified by Paschen. Other lines in the Paschen series result if $n = 3$ and $m = 4, 5, 6, \ldots$. Similarly if $n = 1$ and $m = 2, 3, \ldots$, line frequencies are obtained which correspond to a series of lines of short wavelength, first observed and photographed by Lyman. Other lines in the far infrared and the extreme infrared region of the hydrogen spectrum were observed to be given by Brackett and Pfund, respectively, when $n = 4$ and $n = 5$, and when $m = 5, 6, \ldots$ and $m = 6, 7, \ldots$, so that the whole line spectrum of hydrogen may be represented as shown in Table Q.1.7.

TABLE Q.1.7

$v = R_\infty/1^2 - R_\infty/m^2$, where $m = 2, 3, 4, \ldots$, Lyman series (ultraviolet)
$v = R_\infty/2^2 - R_\infty/m^2$, where $m = 3, 4, 5, \ldots$, Balmer series (visible)
$v = R_\infty/3^2 - R_\infty/m^2$, where $m = 4, 5, 6, \ldots$, Paschen series (infrared)
$v = R_\infty/4^2 - R_\infty/m^2$, where $m = 5, 6, 7, \ldots$, Brackett series (far infrared)
$v = R_\infty/5^2 - R_\infty/m^2$, where $m = 6, 7, 8, \ldots$, Pfund series (extreme infrared)

By the time the above spectroscopic facts were discovered, it had been thoroughly established that the hydrogen atom was the smallest and so presumably the simplest known unit of neutral matter. It had also been established that hydrogen atoms could be ionized by the removal of an electron leaving the nucleus (a proton, a particle with a charge equal but opposite to that of the electron) which had a mass about 1837 as great as the electron rest mass. In terms of the established composition of a hydrogen atom, one electron and one proton, it is necessary that some part of this atomic system should be in motion during the emission or absorption of energy by the atom, otherwise no mechanism would be present which could account for the stopping and starting of spectral phenomena. Also, since the proton is so much more massive than the electron, it is reasonable to assume that the motion of the electron is responsible in some way for the line spectrum. However, according to classical mechanics, this system consisting of one charged particle moving with respect to another charged particle, would radiate energy continuously and with the continued emission of radiation, the two charged particles would ultimately coalesce. Any such classical mechanism would not provide the stability which is one of the principal characteristics of hydrogen, nor could it account for the discontinuous nature of the line spectrum. So, not only does classical mechanics appear to be quite incapable of accounting for the mechanism of the emission of a discontinuous spectrum, but it rules out any permanent existence for such a mechanism. At the same time, the relatively simple expressions describing the wave numbers of all the conceivable lines in the spectrum of hydrogen suggested some equally simple mechanism as being responsible for their origin.

Bohr applied the Planck quantum postulate to this problem, evolving an explanation of atomic radiation which has had considerable success. In addition to the fundamental assumption that the exchange of energy between radiation and matter takes place discontinuously in units of $E = h\nu$, Bohr made the additional postulates that: the number of modes of motion of an electron around an atomic nucleus is not unlimited, but that only those modes are permissible which conform to certain quantum conditions (he referred to those modes as stationary states); that an electron moving in one of these permitted modes does not radiate energy as it would in classical theory; and that an electron may change from one of these permitted modes to another, but may not exist in any intermediate state. On the basis of these assumptions, if an electron were to change from a mode of motion more distant from the nucleus to one nearer to the nucleus, radiation would be emitted, and vice versa. Bohr initially chose the simplest possible mode of motion of the electron with respect to the proton in a hydrogen atom, namely a circular motion. Simple harmonic motion has to be discarded on the basis that the electron is unlikely to move through the proton. On the assumption, then, of an electron of mass m_ε and charge $-\varepsilon$ moving in a circular path of radius r, with a velocity v, and rotational frequency ν_r, the electronic angular velocity would be, $\omega = 2\pi\nu_r$ or $v = r\omega = 2\pi r\nu_r$, and the angular momentum would be, $m_\varepsilon rv = m_\varepsilon r^2\omega = 2\pi m_\varepsilon r^2\nu_r$. The electrostatic force of attraction between the electron and the proton would be opposed by the centrifugal force in the opposite direction, and for one of Bohr's stationary states, these two forces would be equal, or $\varepsilon^2/r^2 = m_\varepsilon v^2/r = m_\varepsilon r^2\omega^2/r = m_\varepsilon r\omega^2$. So far, this is in accordance with classical mechanics. To introduce the quantum concept into the problem, W. Wilson pointed out that the Planck Constant, h, has the same dimensions as angular momentum (energy $\times$ time $\equiv ml^2t^{-1} \equiv$ angular momentum). Since the velocity, and thus the momentum (presently ignoring relativistic effects), of the electron moving in a Bohr stationary orbit remain constant during a revolution in a circular path, it appears consistent with the previous applications of the quantum postulate to define the angular momentum of the electron, $m_\varepsilon v2\pi_r$, as some multiple of h, or $m_\varepsilon v2\pi r = nh = m_\varepsilon r^2 2\pi\omega$. Squaring this equation gives, $m_\varepsilon^2 r^4\omega^2 = n^2h^2/4\pi^2$, or $r^4 = n^2h^2[4\pi^2 m_\varepsilon^2\omega^2]^{-1}$. But from $\varepsilon^2/r^2 = m_\varepsilon r\omega^2$, there is obtained, $r^3 = \varepsilon^2[m_\varepsilon\omega^2]^{-1}$. Dividing the expression for r^4 by the one for r^3, gives $r = n^2h^2[4\pi^2 m_\varepsilon\varepsilon^2]^{-1}$. By allowing n to adopt the integral values, $1, 2, \ldots$, the radii of Bohr's stationary paths are seen to be in the proportion of 1^2, 2^2, 3^2, ..., and electrons occupying these states are assumed not to radiate energy. Also from $\varepsilon^2/r^2 = m_\varepsilon v^2/r$, $v^2 = \varepsilon^2/m_\varepsilon r = [\varepsilon^2 4\pi^2 m_\varepsilon \varepsilon^2] \times [mn^2h^2]^{-1}$, or $v = 2\pi\varepsilon^2(nh)^{-1}$. Now regarding the emission of radiation from the atom, it is necessary to consider changes in energy in the system, when the electron transfers from one Bohr stationary state to another. If the proton in the hydrogen atom is assumed stationary, then the total energy of the system is simply the sum of the potential $(-\varepsilon^2/r)$ and the kinetic $(m_\varepsilon v^2/2)$ energies of the electron. This energy can be expressed as $W = -\varepsilon^2/r + m_\varepsilon v^2/2$, and if the values found for v and r are substituted, the total energy of the atomic system is

$$W = \frac{m_\varepsilon}{2}\frac{4\pi^2\varepsilon^4}{n^2h^2} - \frac{\varepsilon^2 4\pi^2 m_\varepsilon \varepsilon^2}{n^2h^2} = -\frac{2\pi^2 m_\varepsilon \varepsilon^4}{n^2h^2}. \qquad \text{(q.1.33.)}$$

For an electron changing from one circular path for which $n = m + 1$ to a circular path of smaller radius for which $n = m$, where the energy for the first state is W_{m+1} and for the second is W_m, then the change of energy associated with this transfer is

$$W_{m+1} - W_m = -\frac{2\pi^2 m_\varepsilon \varepsilon^4}{(m+1)^2 h^2} + \frac{2\pi^2 m_\varepsilon \varepsilon^4}{m^2 h^2} = \frac{2\pi^2 m_\varepsilon \varepsilon^4}{h^2}\left[\frac{1}{m^2} - \frac{1}{(m+1)^2}\right].$$

Since this must be a positive quantity, it is apparent that according to Bohr's assumptions, there is a liberation of energy when the electron moves from a circular path more distant to one closer to the nucleus. Now applying the quantum postulate that one quantum of energy of frequency v, is liberated in this process, then since if m is an integer so is $(m + 1)$, designating $m = n_1$ and $(m + 1) = n_2$ and the corresponding energies, W_1 and W_2

$$h\nu = W_2 - W_1 = \frac{2\pi^2 m_\varepsilon \varepsilon^4}{h^2}\left[\frac{1}{n_1^2} - \frac{1}{n_2^2}\right] \qquad \text{or} \qquad \nu = \frac{2\pi^2 m_\varepsilon \varepsilon^4}{h^3}\left[\frac{1}{n_1^2} - \frac{1}{n_2^2}\right].\ddagger$$

The ν in the above expression is c/λ, so converting to wave numbers, $1/\lambda = v/c$, then

$$\text{wave number, } \nu = \frac{2\pi^2 m_\varepsilon \varepsilon^4}{ch^3}\left[\frac{1}{n_1^2} - \frac{1}{n_2^2}\right]. \qquad \text{(q.1.34.)}$$

This is obviously simply Rydberg's expression with $R_\infty = 2\pi^2 m_\varepsilon \varepsilon^4/ch^3$.

$\ddagger$ If $n_1 = 1$ and $n_2 = \infty$, in this expression, then $W_2 - W_1$ would correspond to the energy released when an electron is brought from an infinite distance away to occupy the state where the electron is closest to the nucleus of the hydrogen atom. The opposite of this quantity, the amount of energy which would have to be expended to remove the electron from its most stable state to an infinite distance away from the nucleus, is accessable to experimential measurement. This is the ionization energy of hydrogen determined experimentally as 2.131×10^{-18} J. The value is calculated from the Bohr expression, $W_1 - W_2 = 2\pi^2 m_\varepsilon \varepsilon^4/h^2 = 2.177 \times 10^{-18}$ J.

The Bohr stationary states W_1, W_2, $W_3, \ldots$, of an atomic system are assumed to remain indefinitely as such, in the absence of any interaction with external radiation. However, under any actual conditions, it has to be considered that there is a certain probability of a transition taking place between two different stationary states. Such transitions are generally brought about by interaction with external radiation incident on the system. If the energy of the state W_i is less than W_j, then the transition is accompanied by the loss of energy by the system, which appears as energy in the external radiation field emitted with a frequency given by the Bohr frequency condition, so that $h\nu = W_j - W_i$. This energy, which appears in the external radiation field, gives rise to a photon, which in terms of Einstein's photoelectric effect hypothesis, behaves in many ways like a material particle. If W_j is less than W_i, then the energy of the system is increased during the transition and a photon is absorbed.

By making certain postulates about the probabilities of the above type of transitions taking place, Einstein[1] arrived at an independent derivation of the Planck radiation law, given by (q.1.26.); actually (q.1.26.) was arrived at by a modification of an argument originally given by Debye,[2] but Einstein's derivation leads to quite a different method of regarding the interaction between radiation and matter. If E_ν is defined as on page 29, Einstein assumed that the probability of transition of a system in the state W_i to a lower energy state W_j with the emission of a photon of radiation with frequency given by the Bohr frequency rule is $P + QE_\nu$, where P and Q are constants characteristic of the transition involved. The constant P is referred to as the probability of spontaneous emission, whether or not the system is present in an external radiation field, and the constant Q as the probability for induced emission, proportional to the external radiation density E_ν. In the case of a transition from a state W_i to a higher energy state W_j, with the absorption of a photon of radiation, Einstein assumed that the probability of absorption is QE_ν.

On the basis of these postulates, Einstein considered the case of a system in thermal equilibrium at some temperature T. If, for example, W_i is a higher stationary energy state than W_j, and at equilibrium there are N_i and N_j parts of the system (atoms or molecules), respectively, in the states W_i and W_j, then, at equilibrium the numbers N_i and N_j would be independent of time. The number of transitions down from W_i to W_j per unit time will be $N_i[P + QE_\nu]$, and the number of transitions up from W_j to W_i will be $N_j QE_\nu$, with $h\nu = W_i - W_j$, such that these numbers of transitions are equal at equilibrium. Also, in a state of thermal equilibrium, the ratio of N_i to N_j must be equal to the ratio of the Boltzmann factors (page 34) for these two states, such that

$$N_i/N_j = e^{-W_i/kT}/e^{-W_j/kT} = e^{-h\nu/kT}.$$

Substituting this relationship into the statement of the equilibrium condition applying to the numbers of transitions to and from the states W_i and W_j, then $N_i[P + QE_\nu] = N_j QE_\nu = QE_\nu N_i e^{-h\nu/kT}$, from which $QE_\nu[e^{-h\nu/kT} - 1] = P$, or

$$E_\nu = [P/Q][1/(e^{-h\nu/kT} - 1)],$$

which is similar with the Planck radiation law of (q.1.26.) if

$$P/Q = 8\pi h\nu^3/c^3.$$

The Einstein transition probability, $P + QE_\nu$, may be interpreted in terms of the factors governing the distinction between classical and quantum mechanical descriptions of electromagnetic radiation in a rather interesting manner. From the Einstein expression

---

[1] Einstein, A., *Physik. Z.*, 1917, **18**, 121.
[2] Debye, P. J. W., *Ann. Physik*, 1910, **33**, 1927.

for the Planck radiation law, $E_v = [P/Q][1/(e^{-hv/kT} - 1)]$, where $P/Q = 8\pi hv^3/c^3$, it may be deduced that

$$P + QE_v = P + [Pc^3/8\pi hv^3][P/(e^{hv/kT} - 1)][8\pi hv^3/Pc^3] = P[1 + 1/(e^{hv/kT} - 1)].$$

But $1/(e^{hv/kT} - 1) = \bar{n}$ (page 37), so that the Einstein transition probability, $P + QE_v = P[1 + \bar{n}]$, where $\bar{n}$ is the average quantum number of the oscillators of frequency v, in terms of which the system is described. This statement therefore indicates that the spontaneous radiation probability P is only $1/\bar{n}$ times the induced radiation probability QE_v. But since for ordinary visible radiation, $\bar{n}$ is generally small compared with unity, this means that induced emission in the visible region of the electromagnetic spectrum is small compared with the probability of spontaneous emission. On the other hand, in situations where $\bar{n}$ is large, compared with unity, as for microwave radiation and the radiation emitted by lasers in the visible region, it is deduced that the probability of induced emission is large compared with the probability of spontaneous emission, at least to the same accuracy to which it may be regarded that $\bar{n}$ is large compared with unity. In turn, this leads to the conclusion that induced emission may be regarded as a classical type of phenomenon, corresponding to radiation described by large quantum numbers, while spontaneous radiation is a quantum mechanical event, which cannot be comprehended in terms of classical electromagnetic theory. This interpretation of the Einstein transition probabilities is confirmed by the fact that it is possible to calculate the induced emission probability QE_v, in reasonable agreement with experimental observation, by regarding the radiation field as being described in terms of classical electromagnetic theory but the atomic systems interacting with this radiation field in terms of quantum mechanics. Such a method, however, fails to account for the spontaneous emission probabilities, but this is not so if both the radiation field and the interacting atomic system are described in terms of quantum mechanics.

The Einstein assumptions regarding the probability of emission or absorption of radiation by a system are not limited to events involving thermal equilibrium nor to cases where E_v is temperature radiation. In the more general case of the interaction of radiation and matter where equilibrium has not been attained, the situation is rather similar to that involved in the theory of radioactive decay, where there may be several or many radionuclides decaying successively into each other.[1] If it is assumed, in particular, that the matter in a cavity is composed of atoms which among other energy states, have energy levels W_i and W_j; if $W_i > W_j$, then the atom concerned will emit a photon of frequency $hv = W_i - W_j$ when a transition takes place from the state i to the state j, and a photon of the same frequency will be absorbed when the atom is excited from the state j to the state i.

If there are initially N_i atoms in state i and N_j in state j, and a radiation density E_v at a frequency suitable to cause transitions between the two states, then Einstein's assumptions indicate that $N_i[P + QE_v]$ atoms will make the transition per unit time from state i to state j, emitting photons, while $N_j QE_v$ atoms will absorb photons in being excited from the state j to the state i, per unit time. Focussing attention on these two particular energy states, then

$$dN_i/dt = -dN_j/dt = -N_i[P + QE_v] + N_j QE_v.$$

The equilibrium situation, where these derivatives with respect to time are zero, has already been considered, in arriving at the Einstein formulation of the Planck radiation

---

[1] See, for example, Friedlander, G., Kennedy, J. W., and Miller, J. M., *Nuclear and Radiochemistry*, 2nd edition, 1964, John Wiley & Sons, Inc., New York.

law. However, these expressions may be equally well applied to transient phenomena as well as steady state events, as in the case of radioactive decay. If, initially at $t = 0$, N_i and N_j do not have their equilibrium values, but these equilibrium values are only achieved after the lapse of an infinite time, then

$$N_{i\infty} = N[QE_v/(P + 2QE_v)] \quad \text{and} \quad N_{j\infty} = N[(P + QE_v)/(P + 2QE_v)],$$

where $N_i + N_j = N$ and $N_{i\infty}$ and $N_{j\infty}$ are the equilibrium values of N_i and N_j. Substituting these latter expressions into the time derivative expressions above, gives

$$d(N_i - N_{i\infty})/dt = -[P + 2QE_v][N_i - N_{i\infty}]$$

and

$$d(N_j - N_{j\infty})/dt = -[P + 2QE_v][N_j - N_{j\infty}],$$

which have the solutions,

$$N_i = N_{i\infty} + [N_{i0} - N_{i\infty}]e^{-(P + 2QE_v)t}$$

and

$$N_j = N_{j\infty} + [N_{j0} - N_{j\infty}]e^{-(P + 2QE_v)t}$$

respectively. Obviously, N_i and N_j approach their equilibrium values exponentially, and the time required for the exponential quantities to fall off to $1/e$ of their initial values is $1/(P + 2QE_v)$. This argument may similarly be extended to any number of energy states, $W_1, W_2, W_3, \ldots, W_n$, with the probability of transitions from any one state to any other being possible. For each state there would be an expression similar to $dN_i/dt = -dN_j/dt = -N_i[P + QE_v] + N_jQE_v$, which would give the rate of change of the number of atoms in that state, but which would also contain a linear combination of the N's of all the states, each with coefficients involving $P + QE_v$ or QE_v, with the energy density of the appropriate frequency. This energy density E_v would need not necessarily be a radiation density characteristic of thermal radiation at any particular temperature. Actually, if E_v were sufficiently large, the term QE_v would become so large compared with the term P, that the ratio $N_{i\infty}/N_{j\infty}$ would become comparable to unity, which would only be possible for thermal radiation for a very high temperature.

In lasers, where there is an external means, usually the collision of electrons with atoms in specific energy states, of raising atoms into a higher energy state W_i, coupled with a large probability of emission of radiation which reduces the concentration of atoms in a lower energy state W_j, the steady-state values of the N's are very different from the thermal equilibrium values. For a laser, operated with a sufficiently high incident radiation density, so that the term P is negligibly small compared to $[N_i - N_j]QE_v$, the emitted radiation density becomes proportional to the incident radiation density, and if radiation of frequency v is incident on the atoms concerned, more radiation of the same frequency is emitted, the device acting as an amplifier. In order to realize any output from a laser, it is necessary to provide partial transparency in at least one of the reflecting walls of the cavity or space in which the device is operated. Then, those atoms which are excited to the higher energy state W_i by electronic collisions from lower or ground states, are stimulated to emit radiant energy at the frequency v, ultimately leading to a very high energy density E_v of the emitted radiation. The great coherence of laser radiation is attributable to the fact that stimulated radiation emitted by the active material is strictly in phase with one of the normal modes of the containing system, which in the case of a gas laser is the cavity between two almost completely reflecting mirrors or in solid state devices the cavity between reflecting ends of a crystal.

The Einstein transition probabilities for spontaneous and induced radiation emission, the quantities P and QE_v, respectively, for atoms with electric dipole moment M, and for the transition between two states W_i and W_j, may be calculated by the method of time-dependent perturbation theory;[1-2] this quantum mechanical calculation leads to the ratio of $P/Q = 8\pi h\nu^3/c^3$, as given on page 55.

The first major triumph of the Bohr model is thus the calculation of the frequencies of all of the lines in the hydrogen spectrum, as the difference of the reciprocals of the squares of two integers multiplied by the constant, $[2\pi^2 m_\varepsilon \varepsilon^3][ch^3]^{-1}$; thus, if as in Table Q.1.8

TABLE Q.1.8

Electron initially in state designated by n_2	makes	Transition to state designated by n_1	then	Spectral series produced
2, 3, 4, . . .		1		Lyman (ultraviolet)
3, 4, 5, . . .		2		Balmer (visible)
4, 5, 6, . . .		3		Paschen (infrared)
5, 6, 7, . . .		4		Brackett (far infrared)
6, 7, 8, . . .		5		Pfund (extreme infrared)

The most impressive initial success of the simple Bohr model is associated with the numerical value of the Rydberg constant, since each of the constants, π, m_ε, ε, c, and h is known independently, and when substituted in the expression deduced in terms of the Bohr model enables R_∞ to be calculated; R_∞ can also be obtained directly from experimental measurement. The calculated value (using constants referred to C^{12} mass scale) is 10973731 m^{-1}, which agrees beautifully with the experimentally observed value of 10980000 m^{-1}. Very few theories in their initial stages have achieved such remarkable success!

In the above discussion, it has been assumed that the electron was revolving around a stationary proton. In atoms other than hydrogen the nuclear charge is greater than a single electronic charge so that if the nuclear charge number is Z, the electrostatic attractive force would be $Z\varepsilon^2$ rather than ε^2. Also Bohr has shown that if the more probable assumption is made that both the electron and the nucleus are in motion around their common center of mass, it would be necessary to substitute $m_\varepsilon M_H(m_\varepsilon + M_H)^{-1}$ in place of m_ε (see page 170). This quantity is referred to as the reduced mass of the atom and is

---

[1] J. C. Slater, *Quantum Theory of Atomic Structure*, **Vol. I**, 1960, App. 12, 443–454, McGraw-Hill Book Company, Inc., New York.
[2] J. C. Slater, *Quantum Theory of Matter*, 1968, 265–287, McGraw-Hill Book Company, Inc., New York.

commonly denoted by μ_M. On inserting these corrections into the expression for the Rydberg constant,

$$R_\infty = \frac{2\pi^2 Z \varepsilon^4}{ch^3} \cdot \frac{m M_H}{(m_\varepsilon + M_H)}$$

is obtained, where M is the atomic mass. Obviously as the nuclear mass, M, becomes infinite, the value obtained for the Rydberg constant approaches that obtained using m_ε. This is the quantity designated above as R_∞.

In the expression obtained above for the electronic velocity, $v = 2\pi\varepsilon^2/nh$, would correspond to a velocity of 2.187×10^6 m s^{-1} for hydrogen and to a velocity of 2.01×10^8 m s^{-1} for uranium with $Z = 92$, both for $n = 1$. The ratio of the velocity of the electron for $n = 1$ for hydrogen to the velocity of electromagnetic radiation is $v/c = 2\pi\varepsilon^2/ch = 1/137$, a remarkable quantity which is dimensionless (ε^2 has dimensions of energy × length, h has dimensions of energy × time, and c has dimensions of length/time); it is referred to as the fine structure constant and has the magnitude 7.29720×10^{-3}, referred to the C^{12} mass scale. The velocity of the electron for other values of the quantum number, n, may be referred to conveniently as multiples of $[c/$fine structure constant$] = \alpha$.

The next simplest atom after hydrogen is the hydrogen isotope of mass 2, consisting of a nucleus containing one proton and one neutron, with charge $+\varepsilon$. The theoretical value of the Rydberg constant would then be, $R_D = 2\pi^2\varepsilon^4\mu_D[ch^3]^{-1}$, where $\mu_D = m_\varepsilon M_D[m_\varepsilon + M_D]^{-1}$, which has a numerical value very close to the mass of the hydrogen atom. It would be expected therefore, that the line spectrum of the isotopes of hydrogen would be very similar, apart from a very small displacement of each of the spectral lines. In fact, the hydrogen isotope of mass 2, deuterium, was discovered by Urey, Brickwedde, and Murphy in 1932 by this spectroscopic method.

The neutral helium atom has a charge of $+2\varepsilon$ and therefore two extra-nuclear electrons; the species He^{++}, the α particle, has been well characterized from the study of radioactive nuclides. If, however, one of the electrons is removed from a neutral helium atom by ionization, then the nucleus and one electron remain, He$^+$, quite analogous to the hydrogen atom. If the charge, $+2\varepsilon$, and the reduced mass of the helium atom is substituted in the expression for the Rydberg constant, then $R_{He} = [8\pi^2\varepsilon^4][ch^3]^{-1}[m_\varepsilon M_{He}] \times [m_\varepsilon + M_{He}]^{-1}$. Such an alteration of the value of the Rydberg Constant has been confirmed by experimental observation. Fowler, for example, found that $R_{He}/R_H = 4.001638$, in agreement with $R_{He}/R_H = [4M_{He}(m_\varepsilon + M_H)] \times [M_H(m_\varepsilon + M_{He})]^{-1}$, calculated from Bohr's theory as 4.001626. The similarity between the line spectrum of singly ionized helium and neutral hydrogen is so great that Pickering, investigating a series of lines in the spectrum of the star, ζ PUPPIS initially ascribed them to neutral hydrogen. Pickering found that the wavelength of the lines in both ζ PUPPIS and hydrogen could be related by, $\lambda = 3646 \cdot 1n^2[n^2 - 16]^{-1}$, and that if $n = 6, 8, 10, \ldots$, the lines corresponded to normal hydrogen, and that if $n = 11, 13, \ldots$, the wavelengths corresponded to the observed lines in the star. At the time, these observations were explained by the assumption that under the conditions existing in the star, a new hydrogen spectrum was being produced which had not previously been observed under laboratory conditions. However, further support for the Bohr theory is supplied by the relatively simple solution which it presents for these observations, namely that they are due to ionized helium. By substituting $+2\varepsilon$ for the nuclear charge and μ_{He} for μ_{H} in the Rydberg expression, there is obtained, for the wave number of the various observed lines,

$$
\nu = \frac{8\pi^2 \varepsilon^4 m_\varepsilon M_{\text{He}}}{ch^3(m_\varepsilon + M_{\text{He}})} \left[\frac{1}{n_1{}^2} - \frac{1}{n_2{}^2}\right] = R_{\text{He}}\left[\frac{1}{n_1{}^2} - \frac{1}{n_2{}^2}\right] = 4R_{\text{H}}\left[\frac{1}{n_1{}^2} - \frac{1}{n_2{}^2}\right]
$$

$$
= R_{\text{H}}\left[\frac{1}{(n_1/2)^2} - \frac{1}{(n_2/2)^2}\right],
$$

where, if $n_1 = 4$ and $n_2 = 6, 8, 10, \ldots$, wave numbers almost identical with the Balmer series of hydrogen are obtained, and if $n_2 = 5, 7, 9, \ldots$, the wave numbers of the Pickering lines are obtained, since for He$^+$, n_2 may have all integral values greater than 5. The Bohr explanation also accounts for the fact that the observed lines in the star are four times as numerous as in normal hydrogen.

Q.1.5.   Sommerfeld Extension of Bohr Theory

The explanation of the line spectrum of hydrogen in terms of the quantum postulate, proposed by Bohr, appears to be spectacularly successful. However, on closer examination, each of the lines in the spectrum, so far described as a single line, is found to consist generally of two or more closely spaced lines. Sommerfeld[1] suggested two modifications to the Bohr model, first by allowing the electron to move in an elliptical path around the nucleus situated at one of the foci of the ellipse, and second by using the relativity mass for the electron. It requires two positional coordinates to describe the position of a particle moving in an elliptical path; it is convenient to use the radius vector, r, and the azimuthal angle, ϕ, for this purpose. In the Bohr model, for circular electronic paths, the allowed stationary states were defined by n^2, where n could adopt the integral values, $1, 2, 3, \ldots$, and these integers are referred to as the first quantum number. Since elliptical motion presents

---

[1] Sommerfeld, A., *Ann. Physik.*, 1916, **51**, 1.

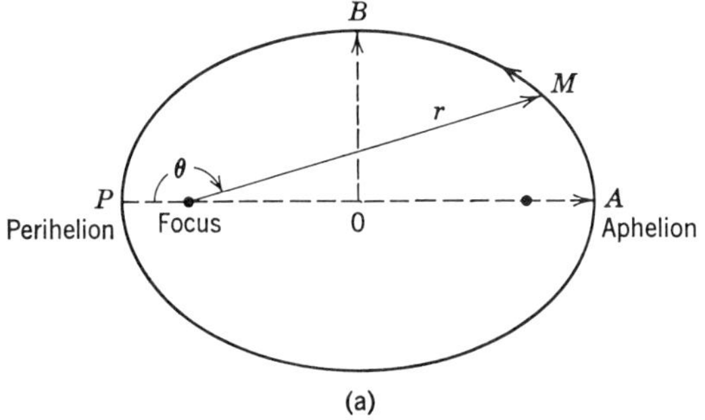

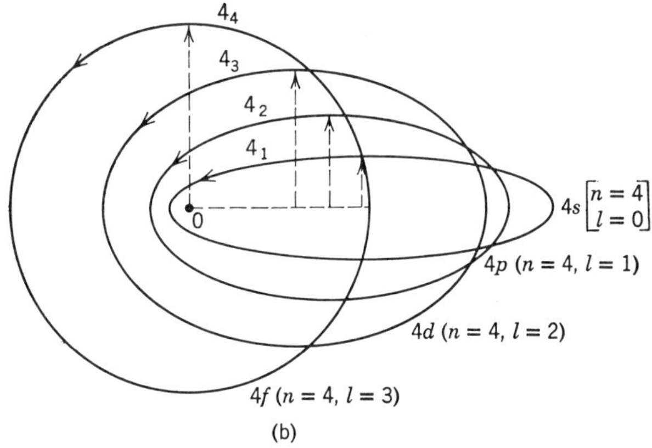

FIG. Q.1.3. Elliptical motion of electron around nucleus of hydrogen atom and hydrogen-like atoms, according to Sommerfeld model. (a) OA, major semiaxis of ellipse, OB, minor semi-axis of ellipse, r, radius vector, θ, azimuthal angle. (b) Permissible elliptical electronic paths corresponding to principal quantum number 4.

two degrees of freedom to the particle concerned, requiring two positional coordinates to define the position of the particle, Sommerfeld proposed that two quantum numbers are required to restrict the motion to ellipses of specific shapes. The shape of an ellipse is defined, in turn, by its eccentricity, defined as $1 - \xi^2 = (b/a)^2$, where ξ is the eccentricity, b and a are the minor and major semi-axes respectively (see Fig. Q.1.3.). In other words, in accordance with Sommerfeld's proposals, in order to quantize elliptical motion, two quantum numbers are required: the radial quantum number, n_r, and the azimuthal quantum number, n_a, such that the sum of the two is equivalent to the total quantum number, n. Transforming the equation of an ellipse to polar coordinates, the result is, $r = a[1 - \xi^2][1 + \xi \cos \phi]^{-1}$, and repeating the Bohr argument as it applies to elliptical rather than circular motion, Sommerfeld arrived at the following expressions:

$$1 - \xi^2 = n_a{}^2[n_a + n_r]^{-2} \qquad \text{(q.1.35.)}$$

$$a = h^2[4\pi^2 m_\varepsilon Z\varepsilon^2]^{-1}[n_a + n_r]^2 \qquad \text{(q.1.36.)}$$

$$W = -[h^2]^{-1}[2\pi^2 m_\varepsilon Z^2\varepsilon^4][n_a + n_r]^{-2}. \qquad \text{(q.1.37.)}$$

The expression (q.1.37.) is the same as Bohr's equation, (q.1.33.), with $(n_a + n_r)$ substituted for n, and the nuclear charge, $+Z\varepsilon$, rather than simply ε. Thus, if $n = n_a + n_r$, the total energy, W, of the electron moving in an elliptical path around a nucleus of charge $Z\varepsilon$, will be identical with the energy for the corresponding Bohr circular path, but this energy will now be dependent upon the total or principal quantum number $(n_a + n_r)$. This result of course, is simply a consequence of the fact that a circle is a special type of ellipse with equal major and minor axes. Also, according to the Sommerfeld model, all elliptical paths having the same principal or total quantum number, have equal major axes, but the eccentricity will depend on the choice of n_a and n_r. For example, for principal quantum number, $4 = n_a + n_r$, the major semi-axes, $a = h^2[4\pi^2 m_\varepsilon Z\varepsilon^2]^{-1} \times 4^2 = 4h^2[\pi^2 m_\varepsilon Z\varepsilon^2]^{-1}$, or all ellipses corresponding to principal quantum number 4 have major axes $8h^2[\pi^2 m_\varepsilon Z\varepsilon^2]^{-1}$. Now, if $4 = n_a + n_r$, and according to the quantum postulate, n_a and n_r may only assume integral values, there are obviously various values which may be assigned to n_a and n_r. The eccentricity of an ellipse, $\xi = [1 - (n_a)^2(n_a + n_r)^{-2}]^{1/2}$, as illustrated below in Table Q.1.9.

TABLE Q.1.9

n_a	n_r	ξ	designated
4	0	0	4_4
3	1	$\sqrt{7}/4$	4_3
2	2	$\sqrt{3}/2$	4_2
1	3	$\sqrt{15}/4$	4_1
0	4	1	4_0

The elliptical designation 4_4 is obviously a circle and the designation 4_0 is a straight line through the nucleus, which has to be excluded as there is no knowledge available of any such type of electronic motion. Accordingly, in the Sommerfeld model, circular motion alone of the electron around the nucleus is not the only possible mode of quantized motion for the electron, but there must also be included 4_3, 4_2, and 4_1, which refer to elliptical paths of successively greater elongation (see Fig. Q.1.3.). The subscripts, 4, 3, 2, 1, in this notation are chiefly referred to as κ in the chemical literature of the period, but another notation, referring to these integers as 3, 2, 1, and 0, usually denoted by l, is also used. Thus, according to Sommerfeld, the meaning of the phrases, "with a principal quantum number 4, four auxiliary or azimuthal quantum numbers (either 4, 3, 2, and 1 in the κ notation, or 3, 2, 1, and 0 in the l notation) may be associated" is the following. For an electron moving in a circular path of radius 4, there may be associated four modes of motion, such that the minor axes of the elliptical paths have the relative lengths 4, 3, 2, and 1. The first of these ellipses is actually a circle (the Bohr case). The most elongated ellipse is designated either 4_1 or 4_0 (n_κ or n_l). And the other ellipses are designated 4_3 or 4_2, and 4_2 or 4_1, respectively. The second quantum number results in the electronic motion being doubly periodic, which gives rise to a rotation of the major axis of the ellipse round the focus in the plane of the ellipse (precession of the perihelion). The Sommerfeld modification of the Bohr theory therefore allows for a larger number of closely spaced energy levels than it is possible to describe by the use of only one quantum number, although as far as determining the position of the spectral lines is concerned, the theory of elliptical electron paths leads to exactly the same result as the theory of circular paths.

In order to appreciate the significance of Sommerfeld's contribution to the theory of atomic spectra, it is necessary to examine such effects as the Zeeman effect, the Stark effect, and the fine structure of the lines in the hydrogen spectrum. According to Sommerfeld, when the hydrogen atom is in its normal state, that is in the absence of any external electric or magnetic fields, the methods generally used for exciting spectra only produce single lines by accidental coincidence. This state of affairs is altered if some external influence or some effect within the atom alters the elliptical paths of the electron(s) in any way. Zeeman discovered that powerful magnetic fields have an influence on the appearance of spectral lines. A line which normally appears single may appear as a doublet when observed in the direction parallel to that of an externally applied magnetic field or as a triplet when observed normal to the direction of the applied field. The light emitted under the influence of an external magnetic field is actually circularly polarized in the doublet and in the triplet two of the lines correspond to plane polarized light in the position of the original lines and one is plane-polarized parallel to the applied field. These phenomena associated with the Zeeman effect could again not be explained in terms of classical dynamics, but, according to the Sommerfeld model, the magnetic field disturbs the various elliptical electron paths leading to a change in the associated energies.

If a strong electric field is applied to a luminescing gas in an evacuated tube, the original spectral lines are replaced by a much more complicated pattern of lines, a phenomenon again for which classical dynamics affords no explanation. It has also been found that if efficient diffraction gratings or interferometers are used to observe the spectral lines of hydrogen, certain lines may be further resolved; for example, all of the lines in the visible or Balmer region have been resolved into doublets.

It has already been mentioned on page 59 that the velocity of the electron in the Bohr circular path for which $n = 1$ may be calculated as 2.187×10^6 m s^{-1}, or $c/137$. According to the theory of relativity, the mass of a particle in motion is a function of its velocity and its rest mass, m_0, and is related to the velocity of electromagnetic radiation by the Lorentz equation, $m = m_0[1 - v^2/c^2]^{-1/2}$.

With $v = c/137$, the mass of the electron is sufficiently altered to significantly affect its energy. Sommerfeld has also shown that by making use of this relativistic correction to the electron mass, a theoretical spectral line separation of 0.365 cm^{-1} would result, which is in agreement with the experimentally observed multiplet separation of the lines in the hydrogen spectrum. These effects are discussed in more detail in Chapter Q.4.

According to the Bohr model hydrogen atom, when an electron makes the transition from one stationary state to another, a definite energy change is involved; this is the so-called Bohr frequency rule, $h\nu = E_2 - E_1$. Each stationary state, described by an integer, $n = 1, 2, \ldots$ may be regarded as an energy level corresponding to a definite amount of energy. The difference between the Bohr and Sommerfeld interpretations of the first four energy levels of the hydrogen atom may be illustrated as in Fig. Q.1.4. If spectral lines are regarded as being produced when an electron makes a transition from an upper to a lower energy state, then in terms of the Sommerfeld model many transitions are possible, especially between the higher energy states and the lower states. For example, using the n_κ notation for the first and second quantum numbers, there would be 12 possible transitions between the fourth and the third energy levels, and therefore twelve possible frequencies to be observed: $4_4 - 3_3$, $4_4 - 3_2$, $4_4 - 3_1$, $4_3 - 3_3$, $4_3 - 3_2$, $4_3 - 3_1$, $4_2 - 3_3$, $4_2 - 3_2$, $4_2 - 3_1$, $4_1 - 3_3$, $4_1 - 3_2$, and $4_1 - 3_1$. However, even under the best conditions of resolution in the presence of an external magnetic field, practically no spectral line has ever been observed to consist of 12 components. This type of difficulty has been overcome by the arbitrary imposition of certain restrictions or selection principles, designed to make the theoretical deductions compatible with experimental observation. For the description of electronic motion by means of a principal quantum number, n, and an azimuthal or subsidiary quantum number, κ, two selection principles are found to be adequate, the first of which applies when the first or principal quantum numbers differ by unity, namely that only transitions between circular paths are allowed; and the second of which applies for all other transitions for which the azimuthal or second quantum numbers must differ

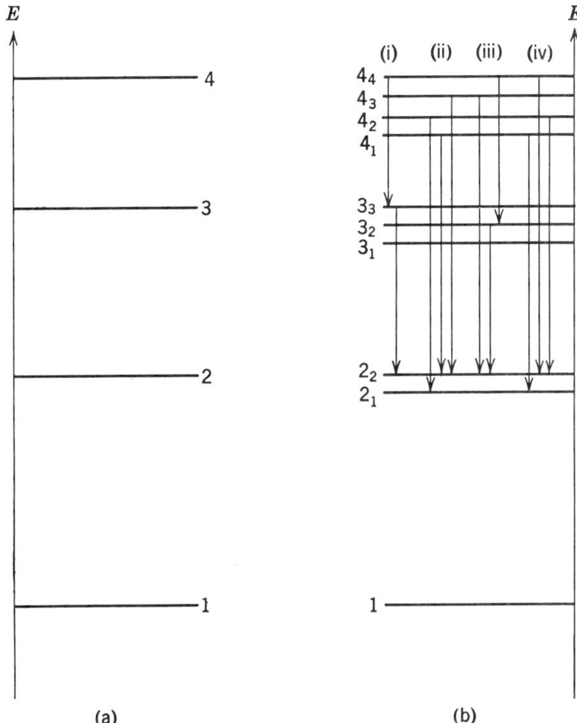

FIG. Q.1.4. First four energy levels of hydrogen atom according to (a) Bohr (b) Sommerfeld. In terms of Sommerfeld model, the energy transitions (i) and (ii) are allowed by the first and second selection principles for n and κ, and (iii) and (iv) are disallowed by the first and second selection principles for n and κ, respectively.

by ± 1. The applicability of these rules is illustrated in Fig. Q.1.4. The application of such arbitrary selection principles of course greatly limits the number of possible electronic transitions between different energy states and therefore the number of possible components of spectral lines, and so helps to fit theory to experimental observation, but offers nothing by way of explanation. The first selection principle described above ensures that only the first line of any series is produced by a circular-circular transition; the second selection principle described above allows one transition from any level to the first level, three transitions from any energy level to the second level, and five transitions to the third level. In the absence of these selection principles, say in the case of the Lyman series of the hydrogen spectrum, where the various lines are produced by transitions to the first level, successive lines of this series would increase in complexity. But the selection principles allow only one transition to this first energy level and therefore each line of the Lyman series should be a single line. It is found that this is actually so. No line in the Lyman series of hydrogen has ever been resolved into further components. Both of these selection principles may be interpreted theoretically, as will be discussed in Chapter Q.3.

Q.1.6.   The Magnetic and Spin Quantum Numbers

The logical necessity of a third quantum number to describe the motion of an electron around the nucleus of an atom is inherent in the fact that this motion takes place in three-dimensional space. The third, or magnetic, quantum number was originally postulated by Landé and justified by the argument that an electron moving in space, which constitutes a moving electric charge, is equivalent to a magnet of strength proportional to the area of the surface enclosed by the path of motion of the electron, thus giving rise to an ellipsoid of revolution around the nucleus rather than simply an ellipse. If this magnet is situated in a magnetic field, then a force will be exerted on it for all positions in which the surface enclosed by the path of motion of the electron is oblique or normal to the direction of the magnetic field, which will tend to rotate the Sommerfeld ellipse.

The Faraday laws of electromagnetism, established experimentally, indicate that a current, i, flowing in a small closed circuit of area, A, is equivalent to a small magnet of magnetic moment M_z placed at its center and acting normal to the plane of the circuit (assuming the area enclosed by the circuit is in the xy Cartesian coordinate plane). In the Bohr model atom, for an electron rotating in a circular path of radius r, around the nucleus of an atom, with an angular frequency, ω, the current involved would be $\omega\varepsilon/2\pi$, since the electron goes around the nucleus $\omega/2\pi$ times per second. The associated magnetic moment would be $M_z = [\omega\varepsilon/2\pi][\pi r^2] = \omega\varepsilon r^2/2$ esu or $\omega\varepsilon r^2/2c$ emu. Also the mechanical angular momentum of the electron of mass m_ε would be, $p_\varepsilon = m_\varepsilon r^2\omega$, and substituting the value for $r^2\omega = p_\varepsilon/m_\varepsilon$ from this latter expression into the equation for the magnetic moment gives, $M_z = [\varepsilon p_\varepsilon][2m_\varepsilon c]^{-1}$. But the Bohr quantum condition, that $mv2\pi r = nh$, implying the quantization of mechanical momentum, leads to $p_\varepsilon = nh/2\pi$. Thus

$$M_z = n\frac{h\varepsilon}{4\pi m_\varepsilon c}, \tag{q.1.38.}$$

or as a consequence of the Bohr quantization of mechanical angular momentum, the magnetic moment associated with a moving electron is also quantized, such that the unit of magnetic moment is $h\varepsilon/4\pi m_\varepsilon c$, the so-called Bohr Magneton; this unit is conventionally designated $\mu_B = \varepsilon h/4\pi m_\varepsilon c =$

9.2732 $\times$ 10^{-24} J T^{-1} (T, tesla) and its magnitude has been confirmed by the direct experimental measurements of Stern and Gerlach of the magnetic moments of hydrogen and other hydrogen-like atoms, such as silver, in their lowest quantum state. To the two coordinates required to describe the motion of the electron around the nucleus of an atom in terms of the Sommerfeld model, a third coordinate must therefore be added. This coordinate may be conveniently taken to be the angle between the direction of the magnetic moment and the direction of an externally applied magnetic field, θ. If p_ϕ is the momentum associated with the Sommerfeld azimuthal coordinate ϕ, then the momentum associated with the third coordinate θ, will be $p_\theta = p_\phi \cos \theta$, and the Bohr quantization condition applied to this quantity may be expressed as $p_\theta = m(h/2\pi)$, where m is another integer, different from κ. Eliminating p_θ from the latter two expressions gives $\cos \theta = mh/2\pi p_\phi = m/\kappa$; on substituting, $p_\phi = \kappa h/2\pi$. But since the cosine of an angle is limited to values ranging between $+1$ and -1, therefore the limiting values of m are $\pm \kappa$, so that m may only adopt values lying between $m = -\kappa$, $m = -\kappa + 1$, $\ldots$, $-1, 0, +1, +2, \ldots, \kappa - 1, \kappa$. Thus, for any value of the azimuthal quantum number κ, there are $(2\kappa + 1)$ possible values for m. The geometrical interpretation of this restriction is illustrated in Fig. Q.1.5. for the case where $\kappa = 4$.

A magnet with a magnetic moment M_z, inclined at an angle θ to a magnetic field of strength H, would have a component in the H direction of $M_z \cos \theta$, and would acquire an energy $HM_z \cos \theta$, by virtue of its situation in the field. Since M_z is given by $\kappa h \varepsilon/4\pi m_\varepsilon c$ and $\cos \theta = m/\kappa$, this magnetic energy is given by,

$$E_M = -\varepsilon Hmh/4\pi m_\varepsilon c. \qquad \text{(q.1.39.)}$$

As m may adopt positive or negative values, the total energy of the electron may be either increased or decreased by the application of a magnetic field. The spectrum of atomic hydrogen or hydrogen-like atoms should therefore, when observed in the presence of a magnetic field, show a number of lines on either side of the line observed in the absence of the field; moreover these lines should be evenly spaced, with a separation proportional to the magnetic field strength, H. This is indeed observed to be so in the normal Zeeman effect.

A fourth quantum number was postulated by Uhlenbeck and Goudsmit, on the basis of consideration of a large number of spectroscopic facts. In particular, the alkali metal atoms, $_3$Li, $_{11}$Na, $_{19}$K, $_{37}$Rb, and $_{55}$Cs, exhibit atomic spectra which are significantly different from the line spectrum of hydrogen. For sodium, for example, the most prominent line in the visible region of the spectrum is a close doublet at (5.896 $\times$ 10^{-7} m and 5.890 $\times$ 10^{-7} m), which must be due to two very similar transitions. Although the

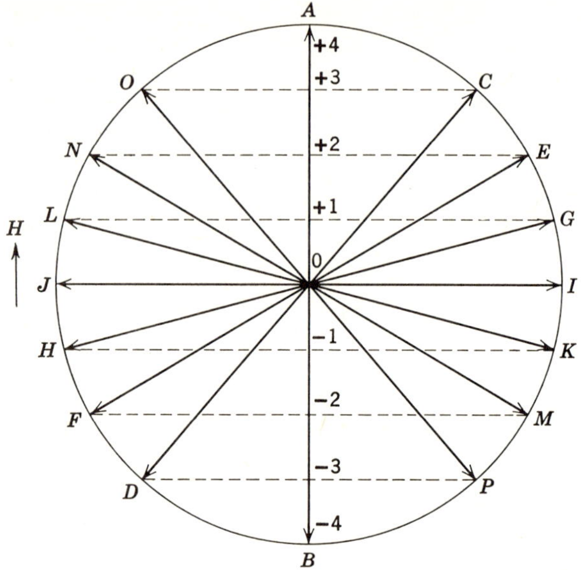

FIG. Q.1.5.   Example of the quantization of the angular momentum of an electron for $\kappa = 4$. The vectors *AB*, *CD*, *EF*, *GH*, *IJ*, *KL*, *MN*, *PQ*, represent the possible settings of the direction of the angular momentum, with respect to the direction, *H*, of an external magnetic field. The projection of each of these vectors on the line *AB* parallel to the magnetic field are such that they correspond to integral numbers (of units $h/2\pi$).

energy separation between such a pair of spectral lines is sufficiently small, amounting for the sodium lines above to only about 2.93×10^{-2} J (g atom)$^{-1}$, to be conceivably due to the Sommerfeld type of relativity correction to the electron mass, the actual explanation of the nature of the lines in the spectrum of the alkali metals, each of which is observed to be a doublet, has been found elsewhere. For the prominent doublet in the visible line spectrum of sodium, it is found that under the influence of a not too strong magnetic field, the doublet is resolved in a remarkable and complicated way. Both of the original lines vanish, one being replaced by four lines symmetrically spaced around the position of the original line, and the second member of the doublet is similarly resolved into six components. The separation in both cases is less than the value $\varepsilon H/4\pi m_\varepsilon c$, found in systems exhibiting the normal Zeeman effect. In order to account for this and for a large residuum of other spectroscopic data, not interpretable in terms of the three previous quantum numbers, Uhlenbeck and Goudsmit proposed that it is necessary to assign a fourth degree of freedom to the electrons in atoms which is independent of the other three quantum numbers, and which may best be described by assigning a proper rotation to the electron itself around its own axis, endowing it therefore with the properties of a small magnet. This fourth quantum number is generally denoted by s, the spin quantum number. By analogy with the angular momentum due to the motion of the electron around the nucleus of an atom, where the azimuthal quantum number κ gives rise to an angular momentum $\kappa(h/2\pi)$ and there are $(2\kappa + 1)$ possible orientations of the motion described by n and κ in a magnetic field. The angular momentum due to the spin is assumed to be similarly quantized to give $s(h/2\pi)$, and $(2s + 1)$ orientations are possible for the spin magnetic moment. Since all the spectroscopic terms of the alkali metals, except those of the subsidiary series are observed to be doublets, it is taken that $(2s + 1) = 2$, or $s = 1/2$. The two different spin states of the electron are assumed to differ only in that the component of the spin angular momentum in any chosen direction is $+(1/2)(h/2\pi)$ in one state and $-(1/2)(h/2\pi)$ in the other. There is now an overwhelming accumulation of evidence, although much of it rather indirect, to support these assumptions. The most direct piece of experimental evidence to support the concept of electron spin comes from the observation that neutral atoms of silver, copper, gold, and hydrogen, for each of which there is good reason to believe that in the lowest or ground energy state there is one electron which is unique in hydrogen and distinguished from the other electrons in the other atoms mentioned, for which there is no angular momentum associated with its motion around the nuclei, but in each case beams of atoms of these species are split into two components only in an inhomogeneous magnetic field. The assumption of electron spin accounts for the "fine structure" of many spectral lines and for the anomalous Zeeman effect. This latter effect and other aspects of angular momentum and electron spin will be discussed in greater detail in Chapter Q.5. It is worth noting at this stage that the concept of electron spin does not emerge from, or play any part, in the subsequent quantum mechanical theories of de Broglie–Schrödinger–Slater–Heitler nor in the matrix mechanics of Born–Heisenberg–Jordan, although a satisfactory mathematical scheme may be devised to graft on this idea to both wave mechanics and matrix mechanics. Only in the much more elaborate relativistic quantum mechanics of Dirac is electron spin included as an integral part of the mathematical theory. The major point to be made here is that even before the formulation of more elaborate quantum mechanical theories, essentially from the attempt to correlate experimentally observed line spectra data, the necessity of four quantum numbers for the complete description of electronic motion was recognized. The first three quantum numbers are dependent on one another in a particular fashion, such that each may only assume integral values, but the fourth or spin quantum number may apparently only adopt two values, $+1/2$ and $-1/2$, independently of the value of the other three quantum numbers.

The arbitrary selection principles governing the operation of the principal quantum number n, and the azimuthal or subsidiary quantum number κ, referred to on page 64, have been extended to include the magnetic and spin quantum numbers, m and s, respectively. These general selection principles, designed to correlate experimental observations regarding the number and frequencies of the lines observed in the atomic spectra of the elements with the theory developed so far, may now be stated. There are five such general selection principles:

(a) The first or principal quantum number, which determines the energy of an electron may have any integral value, $1, 2, 3, \ldots, n$.

(b) The azimuthal or subsidiary quantum number κ, may have any integral value from $n, (n-1), (n-2), \ldots, 1$. Actually, although Sommerfeld referred to this number by the designation κ, it is now universal practice to use the second notation referred to on page 63, and to refer to the second quantum number, which determines the angular momentum of an electron, as l, with allowed values ranging from $(n-1)$ to 0. This notation is in keeping with subsequent developments in quantum theory and with the fifth selection principle below. There is the further convention with respect to the second quantum number, which has been universally adopted, of referring to $l = 0, 1, 2, 3, \ldots$ as $s, p, d, f, g, \ldots$. Thus the designation $4f$ refers to an electronic path around a nucleus describable by principal quantum number 4 and azimuthal quantum number 3. The origin of this convention is a spectroscopic one, namely that since transitions are permissible only in terms of these selection principles, between electron energy states for which l differs by ± 1, the important spectral lines are those corresponding to the transitions n_1 to 1_0, n_0 to 2_1, n_2 to 2_1, and n_3 to 3_2, referred to spectroscopically as giving lines is the principal, sharp, diffuse, and fundamental series, respectively.

(c) The third or magnetic quantum number, m or m_l, to denote its dependence on l, may adopt any integral value from $+l$ to $-l$, so that it may have $2l + 1$ values, denoting the number of ways in which the magnetic moment associated with the angular momentum of the electron may be oriented in space.

(d) The fourth or spin quantum number, s or m_s, may only be either $+1/2$ or $-1/2$ independently of the other three quantum numbers.

(e) *No two electrons in any polyelectronic system may have all four quantum numbers the same.* This principle is perhaps one of the most important in atomic theory. It was originally formulated by Pauli on the basis of the consideration of accumulated spectroscopic data, and still lacks any satisfactory theoretical construction, but must be regarded as a fundamental law of nature.

It should be emphasized that the values of the quantum numbers are not precise constants, expressing any physical conception. Their prime justification is that their use correlates theory and experiment and, as will be seen in the following chapters, they do not have a purely spectroscopic application. The straight-forward application of these selection principles leads to a most interesting result. For $n = 1$, l can only be 0 and therefore m also can only be 0, but s may be either $+1/2$ or $-1/2$, so that there are only two possible, dissimilar sets of four quantum numbers corresponding to the value 1 for the principal quantum number: 1, 0, 0, $+1/2$ and 1, 0, 0, $-1/2$, respectively. For $n = 2$, there are two possible values for l, namely 0 and $+1$; for $l = 0$, m may only be 0, but s may be either $+1/2$ or $-1/2$ as before, and for $l = 1$, m may be either of the three values, $+1$, 0, or -1, each of these m values in turn corresponding to two possible s values, so that for $n = 2$, there are altogether eight possible sets of four quantum numbers such that no two sets are alike, as shown in Table Q.1.10 below.

TABLE Q.1.10

n	2	2	2	2	2	2	2	2
m	0	0	1	1	1	1	1	1
l	0	0	$+1$	$+1$	0	0	-1	-1
s	$+1/2$	$-1/2$	$+1/2$	$-1/2$	$+1/2$	$-1/2$	$+1/2$	$-1/2$

For $n = 3$, l may be 2, 1, or 0, and for each of these l values m may be ± 2, ± 1, or 0, ± 1, or 0 or only 0, respectively, and for each m value s can be $\pm 1/2$, a total of eighteen possible combinations of the four quantum numbers corresponding to $n = 3$, without any violation of the Pauli Exclusion Principle taking place, as illustrated in Table Q.1.11. below.

TABLE Q.1.11

n	3	3	3	3	3	3	3	3	3
l	0	0	1	1	1	1	1	1	2
m	0	0	$+1$	$+1$	0	0	-1	-1	$+2$
s	$+1/2$	$-1/2$	$+1/2$	$-1/2$	$+1/2$	$-1/2$	$+1/2$	$-1/2$	$+1/2$

n	3	3	3	3	3	3	3	3	3
l	2	2	2	2	2	2	2	2	2
m	$+2$	$+1$	$+1$	0	0	-1	-1	-2	-2
s	$-1/2$	$+1/2$	$-1/2$	$+1/2$	$-1/2$	$+1/2$	$-1/2$	$+1/2$	$-1/2$

Similarly for $n = 4$, l may adopt the values, 3, 2, 1, or 0, for each of which m can be ± 3, ± 2, ± 1, or 0 and ± 2, ± 1, or 0, and ± 1, or 0, and only 0, respectively; again for every m value, s can be either $\pm 1/2$, leading to a total number of thirty-two dissimilar sets of four quantum numbers corresponding to $n = 4$, as shown in Table Q.1.12. below.

TABLE Q.1.12

n	l	m	s	Possible sets	Total
	3	± 3, ± 2, ± 1, 0	$\pm 1/2$	14	
	2	± 2, ± 1, 0	$\pm 1/2$	10	
4	1	± 1, 0	$\pm 1/2$	6	32
	0	0	$\pm 1/2$	2	

In the same manner, it can be shown that there are fifty possible sets of four quantum numbers, corresponding to $n = 5$, such that no two sets are the same, as shown in Table Q.1.13. below.

TABLE Q.1.13

n	l	m	s	Possible sets	Total
	4	± 4, ± 3, ± 2, ± 1, 0	$\pm 1/2$	18	
	3	± 3, ± 2, ± 1, 0	$\pm 1/2$	14	
5	2	± 2, ± 1, 0	$\pm 1/2$	10	50
	1	± 1, 0	$\pm 1/2$	6	
	0	0	$\pm 1/2$	2	

It is therefore found that for the various possible integral values for the principal quantum number $n = 1, 2, 3, 4, 5 \ldots$, there are 2, 8, 18, 32, 50, $\ldots$ ways, respectively, in which the four quantum numbers, n, l, m, and s, may be combined subject to the general selection principles and in such a manner that no two sets of combinations for any value of n are the same. This is exactly the same series which was found to describe the periods at which similar chemical behavior was found among the various known elements when they were arrayed in the order of their increasing atomic mass. This series may, as previously pointed out, be written as 2×1^2, 2×2^2, 2×3^2, 2×4^2, 2×5^2, $\ldots$. This series may alternatively be interpreted as denoting the maximum number of electrons which can be accommodated in the quantum groups characterized by the principal quantum numbers 1, 2, 3, $\ldots$; and on the assumption that in atoms of the various elements with nuclear charges ranging from 1 upward the electrons in the ground or lowest energy states in each atom are added one at a time, according to the so-called Aufbau principle, it is possible to immediately write down the ground state electron configuration of any particular atom. However, this process does not always lead to the same result as the ground state configuration deduced from direct spectroscopic observation. The exceptional stability of atoms with nuclear charges 2, 10, 18, 36, 54, and 86, corresponding to the rare gases He, Ne, Ar, Kr, Xe, and Rn, respectively has long been attributed to the fact that these atoms contain just the requisite number of electrons to compose a closed quantum group, although this is actually only true for the first two members. Before proceeding further in the attempt to understand the significance of this series of numbers, which arises from two such apparently dissimilar sources of experimental data, i.e., the determination of atomic masses and the examination of line spectra, it is necessary to consider the question of the ordinal numbers to be assigned to atoms of the various elements, the so-called atomic numbers which are numerically identical to the nuclear charges on atoms, and how these latter numbers are related to the chemical and physical periodicity of properties exhibited by the known elements.

Q.1.7. High Frequency or X-Ray Spectra

From the study of the effects produced by passing electric currents through highly evacuated tubes[1] and the study of the radiations emitted by naturally radioactive materials, it was recognized that in both types of events three different types of radiation were involved. One type of radiation, cathode rays from the discharge tube and β rays from radioactive materials, is identical with a stream of electrons; another type is found to be composed of positively charged particles, variously referred to as positive rays, Goldstein rays, Kanalstrahlen, the magnitude of the charge and the nature of the species being dependent on the type of gas left in the discharge tube, or in the case of radioactive materials as α rays. The latter radiation is composed of a stream of positively charged helium ions, He^{++}. The relative charges and masses, the ratio ε/m, and the relative penetrating powers of each of these types of radiation were quickly established by subjecting them to the influence of electrostatic and magnetic fields and by the interposition of thin sheets of different materials in their paths. In fact, it is on the basis of extensive experimental observations of this type of phenomenon that it is concluded that

---

[1] At 10^{-7} mm pressure, there are still approximately 10^{10} molecules in 1 ml of gas in an evacuated tube, compared with about 2.7×10^{19} molecules/ml of gas at normal atmospheric pressure and about 10^{22} molecules/ml of liquids and solids.

electrons, whether emitted during the passage of an electrical discharge through any known gaseous material, or from natural and synthetic radio-nuclides, or produced by the photoelectric or thermionic effect, are universal constituents of all matter.‡

‡ The number of electrons in a neutral atom of any element is numerically equivalent to the nuclear charge of that atom, Z. The value of Z for each element deduced from the examination of their X-ray spectra is confirmed by direct experimental determination of the charge number, by the Rutherford (1913) scattering technique. In fact, the definite, concrete, quantitative concept of the structure of atoms is due to Rutherford, who systematically investigated the manner in which α-particles from various radionuclides are scattered by metal foils. While Lenard (1903) had observed that atoms are essentially transparent to β-particles, Rutherford found that the more massive α-particles are not appreciably deflected by the electrons in atoms, but do suffer occasional deflections through a large angle, which he interpreted as due to repulsion by the positively charged atomic nucleus, and by measurement of the number of α-particles scattered through a finite angle, he was able to determine the magnitude of the nuclear charge, $Z\varepsilon$.

When an α-particle (He^{++}) is scattered by an atomic nucleus, the path described by the particle before and after the encounter with an atomic nucleus is observed to be a hyperbola, with the nucleus acting as if situated at the focus. In Fig. Q.1.6., if F is the focus of one branch of a hyperbola, at which it is imagined the nucleus of an atom is situated, with the α-particle approaching and being repelled by this nucleus along the path described by the other branch of the hyperbola, then at the point of closest approach

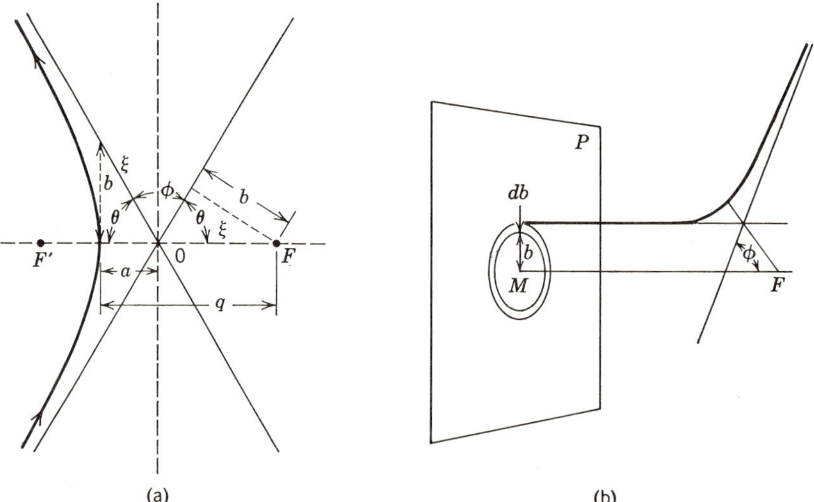

(a)   (b)

FIG. Q.1.6.   The Rutherford scattering of α-particles by atomic nuclei. The hyperbolic path followed by the α-particle before and after the encounter with a nucleus is shown at (a); a and b are the major and minor semi-axes of the hyperbola, ξ is the linear eccentricity, ϕ is the angle of scattering, and θ is the angle between the asymptotes and the line joining the foci. (b) shows the geometrical arrangement for counting the relative number of α-particles scattered through an angle between ϕ and $(\phi + d\phi)$.

between the two positively charged species, the Coulomb repulsive force between them is $Z\varepsilon E/q^2$, where $Z\varepsilon$ is the nuclear charge, E is the α-particle charge ($+2\varepsilon$) and q is the distance from the vertex of one branch to the focus of the other branch of the hyperbola. Let ϕ be the angle of deflection of the α-particle, and θ the angle between the asymptotes and the line joining the foci of the hyperbola, as shown. The other distances noted are determined by the characteristic properties of hyperbolas; ξ, the linear eccentricity or distance from the center to a focus, b, the minor semi-axis or distance along a normal from a focus to an asymptote, a, the major semi-axis or distance from the center to a vertex. Thus, $q = \xi(1 + \cos \theta)$ and $\xi = b/\sin \theta$, and therefore,

$$q = (b/\sin \theta)(1 + \cos \theta) = b \cot (\theta/2).$$

Also from geometry, $\phi = \pi - 2\theta$. In the encounter between the α-particle (of mass M, and moving at very large distances away from the nucleus with a velocity v) and the nucleus (assumed stationary because of its generally large mass compared with the α-particle), energy and momentum must be conserved before and after the encounter, giving rise to the deflection through the angle ϕ. Considering first the conservation of energy, let the velocity of the α-particle at its distance of closest approach, q, to the nucleus be v_c; then $Mv^2/2 = Mv_c{}^2/2 + Z\varepsilon E/q$, or dividing both sides of this equation by $Mv^2/2$, results in

$$(v_c/v)^2 = 1 - (2Z\varepsilon E)/(qMv^2),$$

and substituting the trigonometric expression for q and letting $Z\varepsilon E/Mv^2 = C$, then

$$(v_c/v)^2 = 1 - [2C \sin \theta]/[b(1 + \cos \theta)].$$

From the law of conservation of angular momentum, $Mvb = Mv_c q$ or $v_c/v = b/q$; but $b/q = \sin \theta/(1 + \cos \theta)$ and so,

$$(v_c/v)^2 = \sin^2 \theta/(1 + \cos \theta)^2 = (1 - \cos^2 \theta)/(1 + \cos \theta)^2$$
$$= (1 - \cos \theta)/(1 + \cos \theta).$$

Eliminating $(v_v/v)^2$ from the expressions of the conservation of energy and angular momentum, gives $b/C = \sin \theta/\cos \theta = \tan \theta$, and since $\phi = \pi - 2\theta$, or $\theta = \pi/2 - \phi/2$, and $\tan (\pi/2 - x) = \cot x$, therefore $b = C \cot (\phi/2)$. This gives the angle of deflection of the α-particle as a function of the minor semi-axis of the hyperbola, and of the atomic constants $C = Z\varepsilon E/Mv^2$. The actual number of α-particles from the radionuclide source which are scattered through a specific angle can be determined by experiment. Rutherford counted the number falling on unit area at some constant distance from the source. In Fig. Q.1.6., P is a plane at right angles to the original path of the α-particles and at a large distance from the focus F, denoting the position of the scattering material; M is the foot of the normal from F to this plane P. The number of α-particles which can be detected at the ring formed by two circles with radii b and $(b + db)$ will be those which have been scattered through an angle between ϕ and $(\phi + d\phi)$, and the area of this ring is $2\pi b\, db$. If the number of α-particles arriving at this area, per square centimeter per second is N_α, then the number detected per second over the whole area $2\pi b\, db$, will be $dn = N_\alpha 2\pi b\, db$. But db can be obtained from $b = C \cot (\phi/2)$ as $db = C\, d(\cot \phi/2) = -C[2 \sin^2 (\phi/2)]^{-1}$, and substituting this value for db and for b into the expression for number of α-particles counted per second, it is found that

$$|dn| = N_\alpha \pi C^2[\cot (\phi/2)][\sin^2 (\phi/2)]^{-1}|d\phi| = N_\alpha \pi C^2[\cos (\phi/2)][\sin^3 (\phi/2)]^{-1}|d\phi|.$$

Since the particles are scattered in three dimensions, this is the number which is distributed over the area of the zone of a sphere of radius r, where r is the distance of the detecting plane from the scattering source located at F; the area of this zone of spherical surface is $2\pi r^2 \sin \phi\, d\phi$. Now substituting $\sin \phi = 2 \sin (\phi/2) \cos (\phi/2)$, the number of deflected α-particles detected per unit area is $n = N_\alpha C^2[4r^2 \sin^4 (\phi/2)]^{-1}$ or,

$$n = \frac{N_\alpha}{r^2} \left[\frac{Z\varepsilon E}{2Mv^2}\right]^2 \operatorname{cosec}^4 (\phi/2),$$

which is Rutherford's scattering law. By using differing radionuclide sources of α-particles (and therefore differing v's) and carrying out measurements at differing angles of scattering, ϕ, and measuring N_α and r experimentally, this relation is found to give extremely accurate agreement with experiment. From a knowledge of $M = 4M_H$ and $E = +2\varepsilon$, determined independently, $Z\varepsilon$, the nuclear charge of the scattering atoms can be determined directly. Anomalous results are observed, however, for scattering from light atoms through angles of about 2π, under which conditions the scattering atoms undergo some recoil and the α-particle makes an almost head-on collision with the nucleus concerned.

The Rutherford type scattering experiments have been conducted many times with various particles of ever increasing energy, ranging from the relatively low energy natural α-emitting radionuclides up to 21 GeV electrons provided by the two-mile-long electron accelerator at the Stanford Linear Accelerator Center. At first, the data acquired from such scattering experiments was used to deduce information about the structure of atoms, and then with increasing energy of the incident particle beams to investigate the structure of atomic nuclei, and then with the availability of ultra-high energy electron beams to investigate the structure of the basic constituents of nuclei, protons and neutrons. Using liquid hydrogen as the source of target protons and liquid deuterium as a source of target neutrons, and incident electron energies in the range of 20 GeV, it emerges that nucleons scatter such very high energy electrons in a manner which indicates that they have a complex internal structure consisting of pointlike entities (H. W. Kendall and W. K. H. Panofsky, 1971). These point particle constituents of nucleons, in comparison with the indicated average radius of the proton of 0.8 fermi (8×10^{-16} m) have been referred to as "partons" (R. P. Feynman) and it has been suggested that they may be identical with the hypothetical entities proposed earlier (M. Gell-Mann and G. Zweig, 1964) as quarks, in order to account for the structure of certain types of mesons. The hypothetical quarks are unlike any other known sub-atomic particle in having fractional electric charges of $\pm(2/3)\varepsilon$ and $\mp(1/3)\varepsilon$, for quarks and antiquarks, respectively, so that mesons may be imagined to be assembled from a quark and an antiquark and nucleons (and similar particles like the baryons) would have to be composed of three quarks.

The third type of radiation, X-rays from the discharge tube and γ rays from radionuclides, are quite different in nature from either of the above types. Both X-rays and γ-rays are unaffected by electric and magnetic fields and are found to correspond to electromagnetic radiation of short wavelength. The initial discovery of X-rays by Röntgen (1895) was based on the observation that the radiation produced when a beam of high velocity electrons impinged on matter was capable of inducing intense luminescence in substances like barium platinocyanide, zinc sulfide and zinc silicate. X-rays can be diffracted and polarized like visible light and the subsequent discovery that a beam of monochromatic X-radiation may be coherently scattered by crystalline materials now forms the basis of the principal experimental method for determining the positions of atoms and molecules in solid materials. From the energy associated with γ-rays, $E = h\nu = hc/\lambda$, it is deduced that their wavelength is of the order of $10^{-12} - 10^{-13}$ m, although no diffraction grating is known by means of which such short wavelengths can be measured directly. X-ray wavelengths range from about $(2 \times 10^{-12}$ m to 1.16×10^{-8} m) and although it has been known since 1912 (Friedrich, Knipping, and von Laue) that most crystalline materials behave as natural diffraction gratings for wavelengths of the order of several multiples of 10^{-10} m, it was not until 1927 that Baecklin, Wadlund, and Compton were able to construct synthetic diffraction gratings capable of enabling direct wavelength measurements of this order of magnitude to be made.

The interaction with matter, of X-radiation and electromagnetic radiation generally, may alternatively be interpreted in terms of collision processes between a stream of particles (photons) and the constituent electrons contained in the piece of matter concerned, as in the Compton effect (page 94). In a similar manner, the behavior of a beam of particles such as electrons, may be interpreted as a wave phenomenon, as in electron microscopy, or as a beam of particulate matter interacting with and leaving traces, for example, in a Wilson cloud chamber or in a hydrogen bubble chamber. The scattering of electromagnetic radiation by crystalline material (page 543) is most easily interpreted as a diffraction or wave interference phenomenon; expressed in quantum mechanical language, the observed diffraction effects are a consequence of interference between the system of stationary waves arising from the motion of the particles (atoms, ions, or molecules) composing the crystalline material and so acting as a diffraction grating for the electromagnetic radiation. The possibility of observing the inverse of this process, that is a beam of stationary waves of electromagnetic radiation acting as a diffraction grating for a stream of particles such as electrons, although suggested in 1933 by Kapitza and Dirac[1] was not realized until 1965, several years after the development of lasers[2] (page 48). The coherence and good monochromasy associated with the intense stationary wave system of lasers, enables not only the Kapitza–Dirac effect to be observed in the visible region of the electromagnetic spectrum[3,4] but also makes it possible to observe the "modulation" of electron beams by electromagnetic radiation[5]. This type of modulation of electromagnetic radiation of one frequency by electromagnetic radiation of another frequency has previously been established as common practice in electrical communication techniques as amplitude modulation (AM) and frequency modulation (FM).

---

[1] Kapitza, P. L., and Dirac, P. A. M., *Proc. Camb. Phil. Soc.*, 1933, **29**, 297.

[2] Schwarz, H., Tourtellotte, H. A., and Gaertner, W. W., *Physics Letters*, 1965, **19**, 202.

[3] Bartell, L. S., Roskos, R. R., and Thompson, H. B., *Phys. Rev.*, 1968, **166**, 1494.

[4] Oliver, B. M., and Cutler, L. S., *Phys. Rev. Lett.*, 1970, **25**, 273.

[5] Schwarz, H., *Trans. N.Y. Acad. Sci.*, 1971, **33**, 150.

If a beam of X-radiation produced by accelerating a stream of electrons against a metal target is incident on some other metal surface, the latter metal may be excited to emit a characteristic X-radiation with a specific wavelength. The nature of this emitted radiation is found to be dependent only on the metal from which it is generated, provided only that the parent radiation has a shorter wavelength than the characteristic radiation of the metal used. If the parent radiation is of too long a wavelength, no characteristic radiation is emitted. This is obviously therefore, a quantum phenomenon similar to the photoelectric effect.

Generally the characteristic radiation of any metal or type of atom is capable of exciting the corresponding characteristic X-radiation of atoms of lower mass but not of larger mass. This behavior is similar to optical fluorescence effects as well as the photoelectric effect. It is also established that some elements emit several kinds of characteristic X-radiation, referred to as belonging to the $K, L, M, \ldots$ series, and that for any element this characteristic radiation is independent of the state of chemical combination. Thus, Fe, Fe_2O_3, $FeSO_4$, and $K_4Fe(CN)_6$, all emit the characteristic X-radiation of iron.

The discovery of the fundamental relationship between the X-ray spectra of the various elements is due to Moseley (1913), who systematically investigated the wavelengths of the lines in the X-ray spectrum of a variety of compounds, using a crystal of potassium hexacyanoferrate as analysing grating (of constant 8.408×10^{-10} m). X-ray spectra are much simpler than optical spectra. For all elements the K series consists of two principal strong lines, the stronger line being referred to as the $K\alpha$ line and the weaker as the $K\beta$ line. The excitation of characteristic X-ray spectra is particularly simple compared with optical spectra, as, for example, solid potassium chloride may be used as a source for chlorine. Moseley's original discovery was that the frequency of the characteristic lines in the K and L series of the X-ray spectra for the elements with nuclear charges between 20 and 30 are given by,

$$Q = \left[\frac{\nu}{R[1/n^2 - 1/(n + 1)^2]} \right]^{1/2}, \qquad \text{(q.1.40.)}$$

where n is an integer, R is the Rydberg constant, and Q is a constant for each element. For $n = 1$, the lines of the K series are observed and Moseley deduced from direct measurement of ν that $Q = Z - 1$, where Z is the nuclear charge of the particular element. In other words, for the K series of X-ray spectra, $Z - 1 = \sqrt{4\nu/3R}$, or $\sqrt{\nu} = [\sqrt{3R/4}][Z - 1]$. Therefore, if the nuclear charge is plotted as a function of $\sqrt{\nu}$ a linear relation is obtained. This means that the frequency of the $K\alpha$ line in the X-ray spectrum of any element may be calculated, which cannot be done for optical spectra. This is the origin of the recognition of the fundamental nature of the nuclear charge or atomic number of the elements as the principal factor in determining their chemical properties and behavior rather than their atomic masses. Moseley's law therefore, for the first time, enabled the order of succession of the elements in the direction of increasing atomic number to be established. In this manner, for example it is clearly established that the atomic numbers or nuclear charges of the pairs of elements, Ar–K, Co–Ni, Te–I, and Th–Pa, are 18–19, 27–28, 52–53, and 90–91, respectively, in agreement with their observed chemical properties, although for each of these pairs the first member has a larger atomic mass than the second member. Another important result of Moseley's work was the final determination of the number of elements between hydrogen with nuclear charge at $Z = 1$, and the heaviest known naturally occurring element, uranium, with a nuclear charge of $Z = 92$. Thus in Moseley's time, the elements corresponding to $Z = 43, 61, 72, 75, 85$, and 87 were unknown and this fact could be pinpointed; these elements were subsequently discovered and named technetium ($_{43}Tc^{95}$, 1937), promethium ($_{61}Pm^{147}$, 1945), hafnium ($_{72}Hf$, 1923), rhenium ($_{75}Re$, 1925), astatine ($_{85}At^{210}$, 1940), and francium ($_{87}Fr^{223}$, 1939); protoactinium, $_{91}Pa$, was discovered in 1917. The thirteen synthetic radioactive elements with $Z > 92$, were not predicted in Moseley's time and have all been synthesized since 1940 by various nuclear reactions, as follows: neptunium ($_{93}Np^{239}$, 1940), plutonium ($_{94}Pu^{238}$, 1940), americium ($_{95}Am^{241}$, 1944), curium ($_{96}Cm^{242}$, 1944), berkelium ($_{97}Bk^{243}$, 1949), californium ($_{98}Cf^{249}$, 1950), einsteinium ($_{99}Es^{253}$, 1952), fermium ($_{100}Fm^{255}$, 1953), mendelevium ($_{101}Mv^{256}$, 1955), nobelium ($_{102}Nb^{254}$, 1958), lawrencium ($_{103}Lw^{257}$, 1961), kurchatovium ($_{104}Kh$, 1969), and hahnium ($_{105}Hn$, 1970).

The similarity between the expression, (q.1.40.), for the frequencies of the lines in the X-ray and optical spectra becomes quite apparent if written as,

$$\text{for the } K \text{ series,} \quad \nu = (Z - 1)^2 R[1/1^2 - 1/2^2],$$
$$\text{for the } L \text{ series,} \quad \nu = (Z - 7.4)^2 R[1/2^2 - 1/3^2],$$
$$\text{and generally,} \quad \nu = (Z - b)^2 R[1/n^2 - 1/(n + 1)^2].$$

The term b is a screening or shielding constant, which for the L series lines is found empirically to have the numerical value 7.4. In fact if $Z = 1$ and b is very small, this is identical with the expression for the frequencies of the lines in the hydrogen line spectrum. On the other hand if Z is large, for example 90 for Th, then ν is about 10^4 times larger and therefore the wavelengths are correspondingly shorter. The Ritz Combination Principle obviously applies equally well to both X-ray and optical line spectra. Since X-ray spectral frequencies are so much larger than optical frequencies, it requires higher energy electrons, or radiation from an element with Z of $(Z + 1)$ or greater to excite them. Making use of the Ritz Combination Principle, the various lines in the X-ray spectra of the elements found by experiment, which are denoted by $K_\alpha, K_\beta, \ldots, L_\alpha, L_\beta, \ldots, M_\alpha, M_\beta, \ldots$, can be arranged in a term scheme by

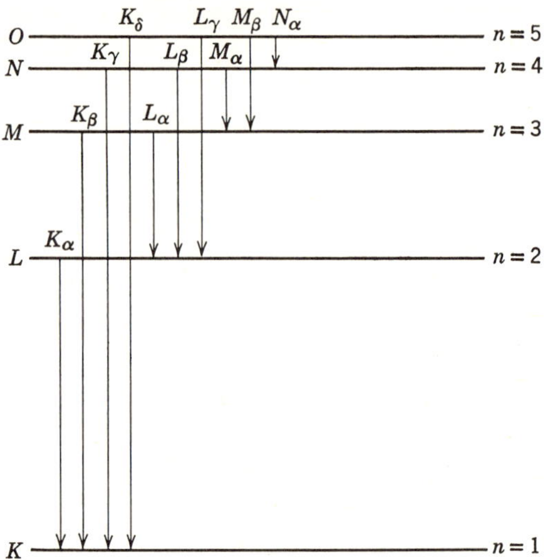

FIG. Q.1.7. Term diagram, showing the origin of the X-ray spectra of the elements; the electronic transitions corresponding to specific X-ray lines are shown.

means of the difference relations: $K_\beta = K_\alpha + L_\alpha, K_\gamma = K_\alpha + L_\beta = K_\beta + M_\alpha$, and so on; these relations are found to agree exactly with experimental observation. Emission or absorption of electromagnetic radiation, whether in the X-ray or optical region of the spectrum, thus coincides with the removal of an electron from one energy level to another. From the similarity and simplicity of the K series of X-ray lines shown by all elements, and the fact that only the heavier elements display L, M, N, ..., spectral lines, it is concluded that the K series corresponds to electron transitions into the first quantum group for which $n = 1$; the L series corresponds to transitions into the second quantum group for which $n = 2$; the M series lines to transitions into the quantum group for which $n = 3$; and so on. Also since in hydrogen there is only one electron, there is no screening constant involved, so that the ultraviolet, Lyman spectrum of hydrogen is identical with its X-ray spectrum. This term scheme is illustrated in Fig. Q.1.7.

The screening constant b, arises because of the effect of the group of electrons closer to the nucleus of an atom reducing the attractive force of the nucleus on electrons further away from the nucleus. Any particular electron is not only attracted to the nucleus but is also repelled by neighboring electrons in the atom. This reduces the nuclear attraction for any electron by the amount calculated by Sommerfeld as $S_n = (1/4) \sum_{\kappa=1}^{\kappa=n-1} \csc \pi\kappa/n$; actually the screening constant cannot be accurately calculated, but is usually deduced from experimental data. Slater has formulated a set of empirical rules for determining the screening constant for any electron in atoms. For this purpose Slater divides the electrons in any atom into groups, each with a different screening constant, as follows,

$$1s \quad 2s \quad 3s \quad 3d \quad 4s \quad 4f \quad 5s \quad 5d \quad 6s \quad 6d \quad 5f,$$
$$2p \quad 3p \quad \quad 4p \quad \quad 5p \quad \quad 6p$$

such that ns and np electrons are grouped together but nd and nf are separate. The screening constant S, is then given by the sum of:

(a)   0 from electrons in any group further from nucleus than the one considered,

(b)   0.35 from each electron in the group considered, except $1s$, where 0.30 is used,

(c)   if an s or p quantum group, 0.85 from each electron in next inner group and 1.00 from all electrons still closer to nucleus, but if group considered is a d or f group, then 1.00 from every electron closer to nucleus.

For example for a potassium atom with a nuclear charge of $Z = 19$, if the 19 electrons are arranged around the nucleus according to the Aufbau Principle, then when each electron is accommodated in the lowest energy state possible consistent with the Pauli Exclusion Principle the electron configuration of the ground state of the potassium atom would be $1s^2 2s^2 2p^6 3s^2 3p^6 4s^1$, and according to Slater's rules, the screening constant for each of the $1s$ electrons would be $(19 - 0.3)$, for the $2s$ and $2p$ electrons, since these are grouped together the screening constant would be $19 - (7 \times 0.35) - (2 \times 0.85)$ or 14.85, for any of the $3s$ or $3p$ electrons, it would be $[19 - 7 \times 0.35 - 8 \times 0.85 - 2 \times 1]$ or 7.75, and for the $4s$ electron the screening constant would be $[19 - 8 \times 0.85 - 10]$ or 2.2.

The most important consequence of work on X-ray spectra is thus the realization that the atomic number or nuclear charge of an element is the factor which determines its chemical identity, and therefore the chemical properties of any element are a function of its atomic number and not of its mass number. The higher frequencies of the X-ray spectra and their regular variation from one element to another such that the square root of the frequency is a linear function of the atomic number, in contrast to the optical spectra which display a periodic variation with atomic number, similar to the periodicity of other physical and chemical properties, is apparently due to the fact that the electrons in the lowest or most stable energy state are involved in the process of exciting or producing the characteristic X-ray spectra. Since all elements with $Z > 2$ contain a K group of electrons, the similarity of the K series of X-ray spectra follows. The optical spectra, on the other hand, of lower frequencies, involve transitions between electronic states in the outer

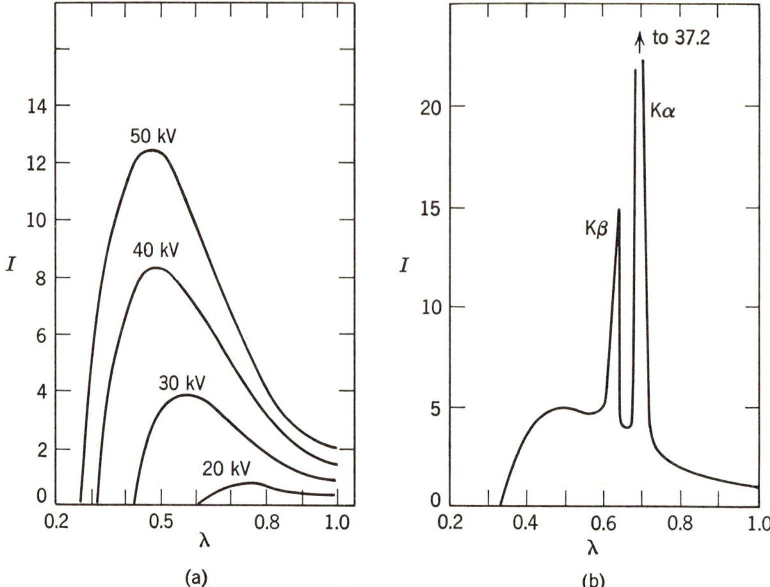

FIG. Q.1.8. The distribution of intensities as a function of wavelength for the X-radiation emitted by metals at different applied voltages. (a) shows the intensity distribution in the continuous or white X-ray spectrum of tungsten for voltages between 20 and 50 kV. (b) shows the superposition of the characteristic X-ray spectrum of molybdenum on the continuous distribution curve which occurs for an applied voltage of 35 kV. (after Ulrey).

region of atoms further away from the nuclei, and follow a periodic pattern. It is therefore deduced that the periodic properties of atoms are related to the behavior of the electrons in the higher energy states and that these must be arranged in some sort of repetitive pattern.

In actual practice, when X-radiation is generated by irradiating or bombarding some metal with high energy electrons, some X-radiation with quite a range of wavelengths is usually produced, apart from the characteristic radiation. This so-called white radiation is not due to electronic transitions in the target metal, but arises from the energy emitted as a result of the deceleration of the high velocity electrons upon collision with the target; this is referred to as bremsstrahlung. The amount of kinetic energy of the incident electron beam converted into bremsstrahlung in this manner increases with the velocity of the electrons used and with the nuclear charge of the target material. So tungsten is generally used as a source of white X-radiation, as for example in the Laue method of X-ray crystallography. The energy distribution of this white radiation as a function of the accelerating voltage applied to the incident electron beam, is identical with the distribution of wavelengths in the thermal radiation emitted by hot objects, provided the applied voltage is below the critical excitation voltage required to excite the characteristic X-radiation of the target; the maximum intensity tends toward the shorter wavelength region with increasing applied voltage, just as with thermal radiation the maximum intensity tends toward the low wavelength region with increasing temperature (see Fig. Q.1.8.). For an X-ray tube, the intensity is given by, $I = Ai(V_t - V_c)^{3/2}$, where V_t is the applied potential difference in volts and V_c is the critical excitation voltage for the target, i is the current conducted by the tube, and A is a constant depending on the geometry and size of the tube and electrodes. In a typical X-ray tube, designed to produce X-radiation of a particular wavelength, applied voltages of the order of $3-7 \times 10^4$ are used, so that if the kinetic energy of an electron after being accelerated through a potential difference $V_t = V_t\varepsilon$, is converted entirely into radiation, the wavelength of this radiation from the Planck relation, $E = h\nu = hc/\lambda = V_t\varepsilon$, is given by $\lambda_{\text{minimum}} = hc/V_t\varepsilon = (1.24/V_t) \times 10^{-6}$ m, if V_t is expressed in volts and the other constants have their usual values, and this is obviously in the X-ray region of the electromagnetic spectrum.

There are several other aspects of X-ray spectra and optical line spectra related to the Bohr-Sommerfeld model atom which are of consequence in the development of atomic theory, namely the Bohr Correspondence Principle, the symbolism which has come into use dealing with spectroscopic terms and term values, and the system of atomic units resulting from the Bohr theory. These are taken up in the following sections.

Q.1.8. Bohr Correspondence Principle

According to classical dynamics, a particle moving in a circular path of radius r, with a velocity v, has an angular velocity $\omega = v/r$, and a frequency of revolution $\nu_r = \omega/2\pi = v/2\pi r$. Also, in classical mechanics, the rotational frequency of a particle is equivalent to the frequency of the radiation which it either emits or absorbs. If the values found for v and r in terms of the Bohr theory, $v = 2\pi\varepsilon^2/nh$ and $r = n^2h^2/4\pi^2m_\varepsilon\varepsilon^2$, respectively, are substituted in the classical expression for rotational frequency, then for an electron, $\nu_r = \omega/2\pi = v/2\pi r$, or $\nu_r = [2\pi\varepsilon^2/nh][2\pi n^2h^2/4\pi^2m_\varepsilon\varepsilon^2]^{-1} = 4\pi^2m_\varepsilon\varepsilon^4/n^3h^3 = 2R/n^3$, since $R = 2\pi^2m_\varepsilon\varepsilon^4/h^3$. But according to the Bohr frequency rule, the frequency of the radiation emitted when an electron makes the transition from the $(n + 1)$th to the nth stationary state is, $\nu = R[1/n^2 - 1/(n + 1)^2] = R(2n + 1)[n^2(n + 1)^2]^{-1} = 2R/n^3$, when n is very large and therefore the quantity unity in both the numerator and denominator of the penultimate expression above may be neglected. In other words, both the classical expression and the Bohr quantum expression for the frequency of the radiation emitted by the particle lead to the same result when n is large. This correspondence between the results of classical electromagnetic theory and quantum theory, which is applicable when the quantum number n is sufficiently large, takes place therefore for low frequencies, since the factor n^3 occurs in the denominator of the frequency expression. Both theories should coincide then, for far infrared radiations, and the deviation from classical laws should be most pronounced when the frequency of motion is high, or when the time taken to perform a complete cycle of changes is small. The classical theory of the motion of gas molecules is therefore obviously quite valid, since the time interval between the occurrence of two identical states when all of the gas molecules have the same positional and momentum coordinates must be large. In the solid state, however, where the distance between adjacent atoms, ions, or molecules is small, of the order of a few units of 10^{-10} m, the time interval between successive identical states is low, and quantum theory is necessary.

Q.1.9. Spectroscopic Terms and Term Symbols

Each spectral line in the line spectrum of an atom is associated with the transition of an electron from one energy state to another, each corresponding to a definite amount of energy, as for example E_m and E_n. The frequency of the emitted spectral line is given by $\nu = (E_m - E_n)/h$. The analysis of the structure of any spectrum consists primarily in determining the energy states of a given atom or ion from the observed frequencies or wavelengths of the spectral lines. The various energy states E_m/h and E_n/h are called terms in spectroscopy and their numerical values are called term values. Spectroscopic data, whether in the infrared, visible, ultraviolet, or X-ray regions, give only differences between various energy states. A suitable zero of energy has therefore to be chosen. The lowest energy state is usually taken as zero, in which event the various term values are positive for all other states, the energy increasing with the height of excitation of the electron. In some cases it is possible to calculate very exactly the energy necessary to remove an electron from the lowest energy state to an infinite distance away from the nucleus of the atom; this is the ionization energy.

As has already been discussed, each electron state may be characterized by describing the electron in terms of four quantum numbers, and order has been brought to the complexity of atomic spectra by the imposition of certain selection principles to govern the various electronic transitions which take place in atoms or ions. Before the advent of wave mechanics, these selection principles had a purely empirical character, but there is a logical mathematical justification for them, which is the subject of the remaining chapters of this section. For a single electron the azimuthal or angular momentum quantum number, l, may only adopt values dependent on the principal quantum number, n, such that $l = 0, 1, \ldots (n - 1)$, and a convenient description of an electron state is with respect to these two quantum numbers. Thus, $3p$, denotes an electron with principal quantum 3 and $l = 1$, and generally for $l = 0, 1, 2, 3, 4, 5, \ldots$, the second member of the symbol nl is denoted by $s, p, d, f, g, h, \ldots$. To conform with the selection principles, if two or more electrons in an atom with $Z > 2$, have the same n and l quantum numbers as for example in the system $\cdots 3d^2$ (this means two electrons with principal quantum 3 and $l = 2$), then either the m or s quantum numbers must differ from one another. Obviously there are a number of ways in which this can occur since m may have any value from $+l \cdots 0 \cdots -l$, and for each l value s may be $\pm 1/2$. In the example above, an atom containing two $3d$ electrons, m may have 5 possible values, $+2, +1, 0, -1$, and -2, and each of these m values may have $s = \pm 1/2$, so that there are ten possible ways in which one electron may be assigned m and s values, and therefore nine ways in which two $3d$ electrons may be assigned m and s values simultaneously. However, there is no way known in which one electron may be distinguished from another in an atom, so that the total of 90 modes of description above must be reduced by half; that is, there are $(10 \times 9)/2 = 45$ different ways of assigning m and s quantum numbers to the two $3d$ electrons. The system of term symbolism which has come into use, involves designing a single compact symbol for an atomic state, where the atom may contain more than one electron. This symbol describes the total angular momentum due to the electronic motion around the nucleus, the total angular momentum due to the electronic spin, the total resultant angular momentum due to both of these, and the multiplicity of the spectral lines which result from electronic transitions between this state and some other state. This symbol is of the form $^{2S+1}L_J$, where S denotes the resultant spin due to the various electrons, L denotes the resultant angular momentum, and J is the total resultant of S and L. Thus for a hydrogen atom in its lowest energy state where $Z = 1$, and therefore since n is 1, l must be 0 and also m, the electron is described by $1s$, and it is assumed that the positive spin state with $s = +1/2$ is most stable; the term symbol then for the ground state of hydrogen is $^2S_{1/2}$. For helium, $1s^2$, since in the ground state if one electron has $s = +1/2$, the other must be $-1/2$, the total S is 0, and the term symbol is $^1S_0$. For Li, Na, K, Rb, Cs, which according to the Aufbau Principle, each contain in their ground states one $2s$, $3s$, $4s$, $5s$, $6s$, electron, respectively, the term symbol is also $^2S_{1/2}$; similarly for Fr with one $7s$ electron, Cu with one $4s$ electron (ground state configuration, $1s^2 2s^2 2p^6 3s^2 3p^6 3d^{10} 4s^1$, compared with potassium, $1s^2 2s^2 2p^6 3s^2 3p^6 4s^1$), silver (ground state configuration, $1s^2 2s^2 2p^6 3s^2 3p^6 3d^{10} 4s^2 4p^6 4d^{10} 5s^1$, compared with rubidium, $1s^2 2s^2 2p^6 3s^2 3p^6 3d^{10} 4s^2 4p^6 5s^1$) with one $5s$ electron, and gold with one $6s$ electron (ground state electron configuration, $1s^2 2s^2 2p^6 3s^2 3p^6 3d^{10} 4s^2 4p^6 4d^{10} 5s^2 5p^6 4f^{14} 5d^{10} 6s^1$, compared with cesium with the configuration, $1s^2 2s^2 2p^6 3s^2 3p^6 3d^{10} 4s^2 4p^6 4d^{10} 5s^2 5p^6 6s^1$), for each of which the term symbol is $^2S_{1/2}$. In fact, at an early stage in the development of spectroscopy, the spectra of hydrogen, of the alkali metals, and of copper, silver, and gold were apparently the only simple ones. All the other atomic spectra showed a bewildering complexity; for example over 5000 lines appear in the line spectrum of iron. It is remarkable that these complex spectra were interpreted and explained even partially in terms of a relatively small number of arbitrary selection principles before the discovery of wave mechanical methods in the period between 1920–25. On the other hand there were many apparent anomalies which were only subsequently explained by the mathematical theory of quantum mechanics. The general concept which enabled these complex spectra to be interpreted was that of angular momentum; by representing angular momentum for each of the electrons in an atom by a vector and combining these vectors for the angular momentum due to the motion of the electrons around the nuclei and for the spin of the electrons separately, and then combining the two resultant angular momenta vectors, there arose the concept of the so-called inner quantum number, $J = L + S$. The apparent simplicity of the spectra of the alkali metals, and of copper, silver, and gold, appeared to be due to the fact that atoms of these elements were similar to hydrogen in that each contained one electron moving in a central field of force provided by the nuclei and a group of electrons which themselves screened the single electron from the attractive force of the nuclei. This suggested that these closed groups of electrons acted on the single electron most distant from the nuclei like spherical force fields. It was a further deduction that only those electrons outside the closed groups had to be considered in discussing the vector model or the interaction of angular momentum vectors. Thus, all atoms with two electrons outside a closed group of electrons displayed similar spectra. These include the alkaline earth metals, Be, Mg, Ca, Sr, Ba, and Ra, as well as Zn, Cd, and Hg. Also, the singly charged positive ions of the alkaline earth metals showed similar spectra to the neutral alkali metals. On this basis the spectra of the alkaline earth metals, were presumed to be the next simplest to the alkali metal spectra, and in fact it was the study of the spectrum of calcium which led Russell and Saunders[1] to propose the symbolism already referred to. A term symbol, or Russell–Saunders symbol, of the type $^2S_{1/2}$, implies a so-called Russell–Saunders coupling between the total angular momentum due to the motion of the various electrons in an atom around the nucleus, $L = \sum_i l_i$, and the total resultant spin momentum of the various electrons, $S = \sum_i s_i$, according to conventional vector addition methods. The superscript in the symbol, represents the multiplicity of the energy levels and corresponds to the number

---

[1] Russell, H. N., and Saunders, F. A., *Astrophys. J.*, 1925, **61**, 38.

of ways in which the quantum number S may be oriented in space. The central capital letter in the symbol denotes the total resultant angular momentum due to the electronic motion around the nucleus, using the same convention as for l, with capitals; thus for $L = 0, 1, 2, 3, 4, 5, 6, 7, 8, 9, 10, \ldots$, the corresponding letters $S, P, D, F, G, H, J, K, L, M, N, \ldots$, are used (omitting J, since this is used to denote the resultant of S and L, and $J = L \pm S$). In order to write a Russell–Saunders symbol for polyelectronic atoms containing two or more p, d, or f, electrons it is necessary to adopt some system to distinguish the difference between states describable by dissimilar m and s quantum numbers. This is usually done by means of a set of empirical rules called Hund's[1] rules, again designed to ensure agreement with the theory thus developed of the experimentally observed spectroscopic data. These rules are as follows:

(a) The most stable Russell–Saunders state is the one corresponding with the largest S value, or the largest multiplicity.

(b) For a group of states with any given value of S, the most stable is that with the largest value of L.

(c) For states with given values of S and L, corresponding to any electron configuration of an atom, if the particular configuration contains less than one half of the number of electrons which can be described by the same n and l quantum numbers (less than 3 for np, less than 5 for nd, and less than 7 for nf), the state with the minimum J value is the most stable. For an atom or ion with an electron configuration, containing more than one half of the number of electrons which can be described by the same n and l quantum numbers, and with any given S and L values, the state with the maximum value of J is the most stable.

The multiplicity of spectral lines, or the number of components into which an otherwise single line is resolved in a magnetic field, is referred to as the multiplet structure of the line. Thus one-electron atoms (H, alkali metals, Cu, Ag, Au) exhibit doublets (two closely spaced levels), the spectra of two-electron atoms show singlets and triplets, three-electron atoms show doublets and quartets in their spectra, four-electron atoms show singlets, triplets, and quintets, in their spectra. Multiplets corresponding to minimum J values and maximum J values, in terms of the third of Hund's rules above, are referred to as normal multiplets and inverted multiplets, respectively.

Making use of Hund's rules, the Russell–Saunders or term symbol of any atom or ion may be written down. Thus for neodymium with $Z = 60$, the ground state configuration, by application of the arithmetic of the four quantum numbers and the Pauli Exclusion Principle is, $1s^2 2s^2 2p^6 3s^2 3p^6 3d^{10} 4s^2 4p^6 4d^{10} 5s^2 5p^6 4f^3 5d^1 6s^2$, and the only electron groups which need be

---

[1] Hund, F., *Z. Physik*, 1925, **33**, 345.

considered as far as their motion around the nucleus and spin are concerned in determining J are $4f^3$ and $6d^1$. Since for 3 electrons in f states ($l = 3$, and m may be $+3$, $+2$, $+1$, 0, -1, -2, or -3) the maximum S would be 3/2, and for the single $6d$ electron with $s = 1/2$, the total maximum S would be 2 and therefore $2S + 1 = 5$. The three $4f$ electron states and the one $6d$ state would together possess a maximum L of $(3 + 2 + 1 + 2) = 8$, corresponding to L as the central part of the composite symbol; also since both the $4f$ and $6d$ states involved in Nd correspond to fewer than one half of the maximum number of electrons which may be described as $4f$ or $6d$ (with different m and s values), $J = L - S = 8 - 2 = 6$. Therefore the complete Russell–Saunders symbol for the lowest energy state of the neutral Nd atom is $^5L_6$. As another example, consider the tripositively charged holmium ion, Ho^{3+}. For a neutral holmium atom $Z = 67$, with ground state configuration, $1s^2 2s^2 2p^6 3s^2 3p^6 3d^{10} 4s^2 4p^6 4d^{10} 5s^2 5p^6 4f^{10} 5d^1 6s^2$, and on the assumption that in the formation of the tricharged ion the three electrons in the three highest states are removed, the only group of electrons which need be considered in the Ho^{3+} species as contributing to the angular momenta is $4f^{10}$. The maximum S is thus $4 \times 1/2 = 2$, and maximum $L = (3 + 2 + 1) = 6$, and $J = L + S = 8$, so the Russell–Saunders symbol would be $^5J_8$. Actually, the Russell–Saunders coupling scheme applies best to light atoms, and although it is common practice to describe the ground states and for that matter various excited states of the heavier atoms in terms of Russell–Saunders symbols, the spectra of heavier atoms is considerably more complicated than is indicated by this system. For heavy atoms, for which the interaction between L and S is large, it appears that the Russell–Saunders concept breaks down, and that the observed spectra appear to follow what is referred to as j–j coupling. This is so when the interaction between the l and s of a single electron is much greater than the interaction between the l's and the s's of different electrons; in this event it is preferable to couple the l and s of a single electron to form a j term, and then combine the j's to form J.

An alternative interpretation (as will become apparent later) of the Hund's rules that the most stable state in polyelectronic systems is the one with maximum S, or if two or more states have the same S, then it is the one with maximum L, is that these rules are a consequence of electronic repulsion. Apart from this electron repulsive effect between pairs of electrons in atoms, another interaction arises from the magnetic moments due to the motion of the electrons around the nuclei and that due to the electron spin itself. This interaction has the effect of separating the energy levels which arise from the electron repulsive influence into a number equivalent to $(2S + 1)$, where S is the total spin. In turn, the effect of a magnetic field on these energy levels is to separate them into $(2J + 1)$ separate levels, so that the J quantum number is most directly obtained from observation of the Zeeman effect, the observed separation being proportional to the applied magnetic field strength. To illustrate how the possible energy states of any given system may be deduced, consider an atom with two $3d$ electrons ($n = 3, l = 2$); the method is applicable to all possible atomic systems.[1] Not all combinations of L (maximum value = 4) and S (maximum value = 1) are allowed, becuase of the restriction imposed by the Pauli Exclusion Principle, and as already noted on page 85, there are 45 different ways of assigning m and s to the two $3d$ electrons so that no two electrons are described by the same combination of four quantum numbers. For two electrons, S may be either, $+1$, 0, or -1, and L may be either $+4, +3, +2, +1, 0, -1, -2, -3$, or -4. If all 45 allowed possibilities are written down, denoting $s = +1/2$ by $\uparrow$ and $s = -1/2$ by $\downarrow$, the values obtained are given in Table Q.1.14.

These 45 possibilities describing pairs of $3d$ electrons are derivable from a much smaller number of terms. For example, for the combination $(2\uparrow, 2\downarrow)$, for one electron, $n = 3, l = 2, m = 2, s = +1/2$, and for the other, $n = 3, l = 2, m = 2$, and $s = -1/2$; since $L = 4$ and $S = 0$, this describes a ^1G state, but as M_L can adopt values ranging from $+4$ to -4, this one term must refer to 9 of the 45 states. In the same way, for the combination of two electrons, described as $(2\uparrow, 1\uparrow)$, $L = 3$, $S = 1$, and the corresponding term symbol is ^3F; now M_s may be $+1$, 0, or -1, for each value of M_L, which itself may have 7 possible values ranging from $+3$ to -3, or altogether the term ^3F corresponds to $3 \times 7 = 21$ of the 45 two-electron states. In a similar manner, it is found that 5 of the 45 two-electron states are included in the symbol, ^1D; nine are included in the symbol, ^3P; and one is included in the symbol, ^1S. If there were no interaction between the two $3d$ electrons, there would of course be only one energy level, but the effect of the LS interaction is to split this single energy level into five others, referred to as $^1G_4$, $^3F_2$, $^1D_2$, $^3P_0$, and $^1S_0$. Then as a result of the interaction of the magnetic moments due to the motion of the electrons around the nucleus and due to the electron spin, each of these 5 levels is split into $2S + 1$ other levels, so that the $^1S_0$, $^1G_4$, and $^1D_2$ levels remain unaffected, since in each $S = 0$, but the ^3F and ^3P levels each produce a triplet. Finally, in a magnetic field, each of the latter energy levels separate into $2J + 1$ different levels. The complete energy level diagram for the $3d^2$ is shown in Fig. Q.1.9. Each of the 45 energy levels in thus characterized by the value of the J quantum number, which is deducible from the Zeeman patterns of spectral lines of an atom. It was Landé[2] who showed that complex atomic spectra could be interpreted in this manner, in the period before the development of wave mechanics.

---

[1] Condon, E. U., and Shortley, G. H., *The Theory of Atomic Spectra*, 2nd Ed., Cambridge University Press, London and New York, 1953.
[2] Landé, A., *Z. Physik*, 1921, **5**, 231; **7**, 398; 1922, **11**, 353; 1923, **15**, 189; **19**, 112.

TABLE Q.1.14

M_L	$S = 1$	$S = 0$	$S = -1$
4		$(2\uparrow, 2\downarrow)$	
3	$(2\uparrow, 1\uparrow)$	$(2\uparrow, 1\downarrow)(2\downarrow, 1\uparrow)$	
2	$(2\uparrow, 0\uparrow)$	$(2\uparrow, 0\downarrow)(2\downarrow, 0\uparrow)(1\uparrow, 1\downarrow)$	$(2\downarrow, 1\downarrow)$
1	$(2\uparrow, -1\uparrow)(1\uparrow, 0\uparrow)$	$(2\uparrow, -1\downarrow)(2\downarrow, -1\uparrow)(1\uparrow, 0\downarrow)(1\downarrow, 0\uparrow)$	$(2\downarrow, 0\downarrow)$
0	$(2\uparrow, -2\uparrow)(1\uparrow, -1\uparrow)$	$(2\uparrow, -2\downarrow)(2\downarrow, -2\uparrow)(1\uparrow, -1\downarrow)(1\downarrow, -1\uparrow)(0\downarrow, 0\uparrow)$	$(2\downarrow, -1\downarrow)(1\downarrow, 0\downarrow)$
-1	$(-2\uparrow, 1\uparrow)(-1\uparrow, 0\uparrow)$	$(-2\uparrow, 1\downarrow)(-2\downarrow, 1\uparrow)(-1\uparrow, 0\downarrow)(-1\downarrow, 0\uparrow)$	$(2\downarrow, -2\downarrow)(1\downarrow, -1\downarrow)$
-2	$(-2\uparrow, 0\uparrow)$	$(-2\uparrow, 0\downarrow)(-2\downarrow, 0\uparrow)(-1\uparrow, -1\downarrow)$	$(-2\downarrow, 1\downarrow)(-1\downarrow, 0\downarrow)$
-3	$(-2\uparrow, -1\uparrow)$	$(-2\uparrow, -1\downarrow)(-2\downarrow, -1\uparrow)$	$(-2\downarrow, 0\downarrow)$
-4		$(-2\uparrow, -2\downarrow)$	$(-2\downarrow, -1\downarrow)$

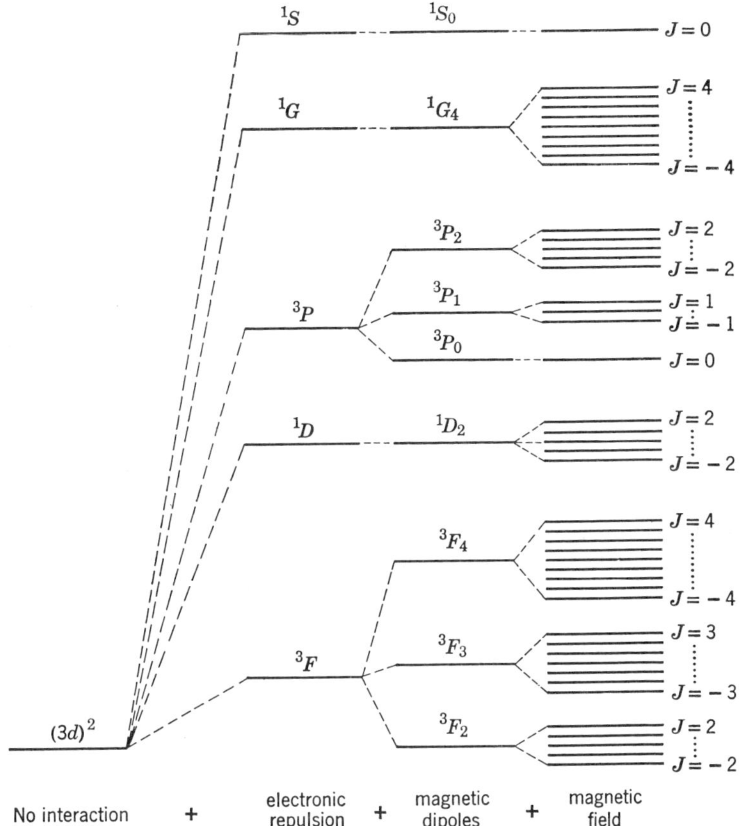

FIG. Q.1.9. Energy level scheme for atomic system $3d^2$ (after Ballhausen). The separation between the various energy levels is not drawn to scale.

Q.1.10. Atomic Units

In the calculation of the energy states of atoms, ions, and molecules, according to the methods of quantum mechanics, it is convenient to adopt a system of units, such that the universal constants, m_ε, c, h, ε, and π, all of which appear in the expressions arising from the Bohr–Sommerfeld theory, may be cancelled from the expressions used. The numerical values of the resultant calculations do not then require the use of large positive or negative exponents of 10.

In this system of atomic units, it is common practice to select the unit of length as the radius of the circular path of the electron around the nucleus of the hydrogen atom, in the Bohr model atom for $n = 1$. This quantity (page 53), is designated the Bohr unit $= h^2/4\pi^2 m_\varepsilon e^2 = 5.29167 \times 10^{-11}$ m $= 0.529167$ Å, is generally written as a_0 and also referred to as the Bohr radius or the first Bohr radius. The unit of mass is taken as the rest mass of the electron, m_ε, and the unit of charge as the electronic charge, ε. There are two atomic units of energy which are in use. The Hartree[1] unit of energy is defined as the mutual potential energy of two unit charges at unit distance apart, $\varepsilon^2/a_0 = 4\pi^2 m_\varepsilon \varepsilon^4/h^2$, which is numerically twice the energy of the ground state of the hydrogen atom or twice the ionization energy of hydrogen $= (2 \times 1.60210 \times 10^{-19})\,\mathrm{J} = 27.21$ e.v. The Slater[2] atomic unit of energy is one half this amount, and defined as the Rydberg energy $= 2\pi^2 m_\varepsilon \varepsilon^4/h^2 = 13.605$ e.v. It has already been seen that the Rydberg number, R_∞, from ν (wave number) $= [2\pi^2 m_\varepsilon \varepsilon^4][ch^3]^{-1}(1/n_1{}^2 - 1/n_2{}^2) = R_\infty(1/n_1{}^2 - 1/n_2{}^2)$, is $109737.31\ \mathrm{cm}^{-1}$, which has to be distinguished from the Rydberg frequency, derived from $E = h\nu = hc/\lambda$, which is $2\pi^2 m_\varepsilon \varepsilon^4/h^3 = 3.290 \times 10^{15}\ \mathrm{s}^{-1}$. The advantage of using such atomic units, apart from the simplification of many formulae, is that the results calculated in these units are not affected by revisions of the numerical values of the fundamental constants.

In atoms of large Z, relativistic effects are more important than in light atoms. The variation of mass with velocity involves the quantity, $\beta = v/c$, which for the Bohr model of the hydrogen atom for $n = 1$ (page 59) has the value, $2\pi\varepsilon^2/ch = 1/137.0388$. As pointed out by Hartree, this means that in a system of units in which a_0, ε, and m_ε are unity, then $c \simeq 137$ and $h/2\pi$ is also unity. The significance of these quantities from the point of view of atomic structure are that, since electronic velocities in atoms of nuclear charge Z are of the order of magnitude Z (in atomic units), then relativistic mass effects are of the order of $(Z/137)^2$, and thus small for low Z; also if the electron mass is assumed to be entirely electromagnetic, this mass is given in classical electrodynamic theory as $m_\varepsilon = C\varepsilon^2/c^2 r$, where the electron is assumed to be a sphere of radius r, and $C = 1$, so that $r = C\varepsilon^2/m_\varepsilon c^2$ and

$$a_0/r = [h^2/4\pi^2 m_\varepsilon \varepsilon^2][C\varepsilon^2/m_\varepsilon c^2] = [ch/2\pi\varepsilon^2]/ \simeq (137)^2.$$

The relativistic electron energy, $m_\varepsilon c^2$, in atomic units, is also $(137)^2$, since in atomic units of energy, $m_\varepsilon c^2 = m_\varepsilon c^2[4\pi^2 m_\varepsilon \varepsilon^4/h^2]^{-1} = [hc/2\pi\varepsilon^2]^2 = (137)^2$. This small value for $r/a_0 = C/(137)^2 \simeq 1/(137)^2$, consequently means that it is a good approximation to regard the electron as a point charge, and so in the calculation of the energy states of atoms it is not necessary to use quantum electrodynamics.

In the event that electron energies are very large and of the order of $m_\varepsilon c^2$, it is sometimes convenient to define a system of units in which $m_\varepsilon = 1$, $c = 1$, and $h/2\pi = 1$, when the unit of length becomes $a_0/137$, the unit of charge $\sqrt{137}\,|\varepsilon|$, and the unit of energy $m_\varepsilon c^2$.

---

[1] Hartree, D. R., *The Calculation of Atomic Structures*, John Wiley & Sons, New York, 1957.
[2] Slater, J. C., *Quantum Theory of Atomic Structure*, **Vol. I**, McGraw-Hill Book Co. Inc., New York, 1960.

Q2

Quantum Mechanical Concepts

Q.2.1. The de Broglie Hypothesis and the Schrödinger Equation

The Bohr–Sommerfeld model provided the first satisfactory, dynamical theory of atoms. Various other attempts had been made from the time of the discovery of the electron to devise a theoretical explanation of the manner in which atoms combine together to form molecules and other compounds, to account for the periodic variation of the valencies of atoms, and to explain directed bond formation in compounds, in terms of static distributions of electrons around nuclei. Beginning with Thomson, the concept that a pair of electrons constitutes a bond between atoms in compounds and that in many of the most stable molecules each atomic nucleus is associated with a group of eight electrons in the highest energy states, has been modified and extended successively by Kossel, Lewis, Langmuir, Bury, and later by Sidgwick, Remy, and many others. Indeed, after the development of wave mechanics, this same concept has continued to play an important part in chemical bond theory, due to the work of Mulliken, Pauling, Coulsen, and has again been revived in modified form by Linnett.[1] The theories, however, of Kossel, Lewis, Langmuir, and Bury, were all fundamentally false, because they presupposed a static distribution of electrons in compounds; they provided a model which aided further theoretical developments, but on the other hand they delayed any understanding of the nature of metallic bonding, and furthermore they violated a fundamental theorem in electrostatics, that no system of charges can be in equilibrium while at rest (Earnshaw's Theorem). The Bohr–Sommerfeld model was a great advance over the static models. It took account of the

---

[1] Linnett, J. W., *The Electronic Structure of Molecules*, John Wiley & Sons, Inc., New York, 1964.

discontinuous nature of light emission and absorption phenomena and its major triumph was the calculation of the frequencies of the lines in the hydrogen spectrum. On the other hand, it made use of a completely unproved assumption in its derivation, and was only able to make qualitative calculations in systems containing more than one electron, and as will be discussed below was inadequate in that it assigned too definite and measurable a character to electrons in atoms by supposing that electrons could be individually recognized and located with a precision which is now known to be quite impossible (see page 132).

The phenomena of thermal radiation emission, the photoelectric effect, and the emission of optical and X-ray line spectra, clearly establish that the energy exchange between electromagnetic radiation and atoms, or between electrons and atoms, takes place discontinuously in units of $h\nu$. The use of the term, photon, for a quantum of electromagnetic energy did not come into general use until after about 1928, although Einstein had proposed much earlier that electromagnetic radiation is not only exchanged with matter in quanta but is also propagated through space in definite quanta or photons. In fact, it was not until the discovery of the so-called Compton[1] effect, that the first definite proof was made available for the particulate nature of radiant energy.‡

‡ The observation of Compton scattering was the first clear indication of the experimental relationship between mass and wavelength of a particle. It represents an apparent wave phenomenon only comprehensible in terms of particles. The Compton effect refers to the observed change in wavelength of X-radiation scattered by electrons, although wave theory predicts no change in frequency or wavelength. Compton investigated the scattering of X-rays by a block of paraffin and found that the radiation scattered at an angle of less than 90° possesses a greater wavelength than the incident radiation. This result can be easily explained as a process involving a simple collision between an electron and a photon, governed by the laws of conservation of energy and of momentum. If the X-ray photon has initially an energy, $E = h\nu$, and travels with velocity c, then its initial momentum is given by $(E/c^2)c = h\nu/c$. If the energy of the photon after the collision with the electron is denoted by $h\nu'$, then its resultant momentum will be $h\nu'/c$. If the electron is regarded as initially at rest, all its energy will be $m_\varepsilon c^2$ and its momentum will be zero. If, as a result of the collision, the electron acquires a velocity, v, then its mass will be $m_\varepsilon[1 - v^2/c^2]^{-1/2}$, its energy, $m_\varepsilon c^2[1 - v^2/c^2]^{-1/2}$, and its momentum, $m_\varepsilon v[1 - v^2/c^2]^{-1/2}$. The amount of energy transferred to the electron as a result of the collision is then, $m_\varepsilon c^2[1 - v^2/c^2]^{-1/2} - m_\varepsilon c^2$, or $m_\varepsilon c^2[(1 - v^2/c^2)^{-1/2} - 1]$, which on expanding in powers of v/c obviously agrees with the formula $m_\varepsilon v^2/2$ for kinetic energy in terms of non-relativistic mechanics for small values of v. If θ and ϕ denote the angles of scattering, as in Fig. Q.2.1., then the law of conservation of energy gives, $h\nu + m_\varepsilon c^2 = h\nu' + m_\varepsilon c^2[(1 - v^2/c^2)^{-1/2} - 1]$. The law of conservation of momentum gives, in the direction parallel with the original direction, $h\nu/c = [h\nu' \cos \phi]/c + m_\varepsilon[1 - v^2/c^2]^{-1/2}v \cos \theta$, and resolved in the direction normal to the original direction of the incident photon, there results

$$0 = [h\nu'/c] \sin \phi - m_\varepsilon[1 - v^2/c^2]^{-1/2}v \sin \theta.$$

---

[1] Compton, A. H., *Phys. Rev.*, 1923, **22**, 409.

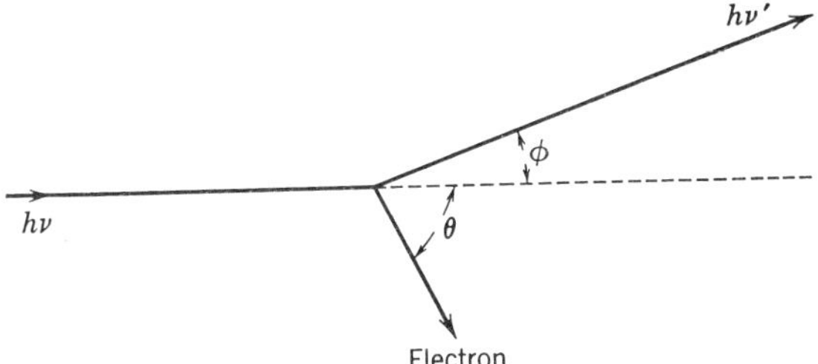

FIG. Q.2.1. Compton scattering. The incident X-ray photon is scattered through the angle, ϕ, and the electron recoils in a direction at an angle, θ, to the incident photon direction.

Writing simply m for the relativity mass of the electron, θ may be eliminated from the two conservation of momentum equations by squaring and adding, arriving at $h^2/c^2[(\nu - \nu'\cos\phi)^2 + (\nu')^2 \sin^2\phi] = m^2\nu^2$, or $m^2\nu^2 c^2 = h^2[\nu^2 + (\nu')^2 - 2\nu\nu'\cos\phi]$. Similarly, by squaring and rearranging the conservation of energy equation, there is obtained,

$$m^2 c^4 = [h(\nu - \nu') + m_\varepsilon c^2]^2 = h^2[\nu^2 + (\nu')^2 - 2\nu\nu'] + m_\varepsilon^2 c^4 + 2hm_\varepsilon c^2(\nu - \nu').$$

Subtracting these two squared equations (the latter from the former) gives,

$$m^2\nu^2 c^2 - m^2 c^4 = -2h^2\nu\nu'\cos\phi + 2h^2\nu\nu' - m_\varepsilon^2 c^4 - 2hm_\varepsilon c^2(\nu - \nu'),$$

or

$$m^2 c^2(c^2 - \nu^2) + 2h^2\nu\nu'(1 - \cos\phi) - 2hm_\varepsilon c^2(\nu - \nu') = m_\varepsilon^2 c^4.$$

But from $m = m_\varepsilon[1 - \nu^2/c^2]^{-1/2}$, $m_\varepsilon^2 = (m^2/c^2)(c^2 - \nu^2)$, and therefore,

$$1 - \cos\phi = (m_\varepsilon c^2/h\nu\nu')(\nu - \nu') = (m_\varepsilon c^2/h)(1/\nu' - 1/\nu),$$

and since $\lambda\nu = c$, then $1/\nu = \lambda/c$ and $1/\nu' = \lambda'/c$, therefore

$$1 - \cos\phi = [m_\varepsilon c^2/hc](\lambda' - \lambda),$$

or

$$\lambda' - \lambda = \Delta\lambda = (h/m_\varepsilon c)(1 - \cos\phi) = (h/m_\varepsilon c)[1 - 1 + 2\sin^2(\phi/2)] = 2\lambda_0 \sin^2(\phi/2),$$

where $\lambda_0 = h/m_\varepsilon c = 2.42621$ pm. This quantity, λ_0, is referred to as the Compton wavelength and is a natural constant; it is the wavelength of a quantum of electromagnetic radiation with the mass of an electron. The increase in wavelength of the photon on Compton scattering is thus independent of the wavelength itself, and depends only on the angle of scattering, ϕ. This conclusion has been thoroughly tested by experiment, first by Compton and later (1936) by Bothe, Jacobsen, and others, who have positively confirmed the simultaneous appearance of the recoiling X-ray photon and electron, when X-radiation is scattered by hydrogen. Incidentally, it is only for short wavelength radiation (X-radiation of high energy, $h\nu$) that the mass of a photon is appreciable compared with the rest mass of an electron, so that this phenomenon is not observed with visible light.

Classical electrodynamics and classical mechanics fail completely to provide any explanation of these phenomena. However, there still remain the phenomena of interference and diffraction, which are only intelligible in terms of the wave theory of electromagnetic radiation, developed successively by

Huygens, Young, Fresnel, and others. The principal method of determining the positions of atoms in molecules and crystals is indeed based on the assumption that an incident beam of monochromatic X-radiation is Bragg scattered or undergoes Bragg reflexion‡ from different crystal lattice planes, which subsequently interfere with one another to produce diffracted beams of varying intensities, without any change in wavelength from the incident beam.

‡ Bragg reflexion is the geometrical equivalent of diffraction by a crystal lattice, and only occurs for some specific angle of incidence, compared with specular reflection (of light) which occurs over a continuous range of angles.

These apparent contradictions in the behavior of electromagnetic radiation led de Broglie in 1923 to propose that a similar type of behavior should be associated with material particles and with electrons in particular. The connection between the two types of behavior was presumed to involve the two different ways of expressing energy. The Planck relation, $E = h\nu$, gives the energy associated with a frequency, and the Hasenöhrl–Einstein relation gives the energy associated with a mass, $E = mc^2$. If therefore, these two expressions are equated, $mc^2 = h\nu = hc/\lambda$, for a particle such as a photon traveling with a velocity c, and thus, $mc = h/\lambda =$ momentum. De Broglie's proposal was that any particle moving with a velocity v, should exhibit a momentum given by a similar relation, namely that p (momentum) $= mv = h/\lambda$, or $\lambda = h/p$.

The novel suggestion, actually made by de Broglie in his 1924 thesis published in 1925,[1,2] was that just as electromagnetic radiation could be regarded as a stream of photons, so could the mechanical motion of particles be referred to a wave motion. He proposed that waves somehow accompany electrons and other particles, in the same manner in which, apparently, photons behave. It was this specific proposal which gave the first plausible explanation as to why quantum conditions exist, and which predicted the wave nature of electrons. De Broglie's argument was actually phrased in relativistic mechanical terms, but it may just as easily be expressed in classical mechanical terms. Thus for an electron moving around the nucleus of an atom in one of Bohr's stationary states, the stability of the stationary state is associated with a stationary wave, containing an integral number of wavelengths in its path length, the Bohr quantum condition being simply a statement of this integral number of wavelengths associated with the moving particle. Thus the various ways of stating the Bohr quantum condition, that the angular momentum of an electron moving in a circular path of radius r is some multiple of Planck's

[1] de Broglie, Louis, *Ann. Phys.*, 1925, **3**, 22.
[2] de Broglie, Louis, *Non-Linear Wave Mechanics, Causal Interpretation*, 1960, Elsevier Publishing Co., Amsterdam.

constant ($m_\varepsilon v 2\pi r = nh$, as on page 53), the Sommerfeld (page 60) statement that the momentum of the electron integrated over its path of motion, $\oint p\, d(2\pi r) = 2\pi m_\varepsilon vr = nh$, where the integral sign $\oint$ indicates that the integral is to be taken over a complete cycle, are simply alternative ways of stating the stationary wave condition. This is so, since if the number of waves in a small interval dq of path length is dq/λ (where λ is the wavelength) then the total number of wavelengths will be $\oint dq/\lambda$, and if $\lambda = h/m_\varepsilon v = h/p$, then $\oint dq/\lambda = \oint p\, dq/h$, and if this is equal to some integer it follows that $\oint p\, dq = \oint p\, d(2\pi r) = nh$.

The relation, $\lambda = h/p$ was verified experimentally for electrons by Davisson and Germer[1] and by Thomson.[2] For electrons accelerated through a potential difference of V volts, the kinetic energy acquired by the electrons is $V\varepsilon = mv^2/2$, where v is the velocity after acceleration. Thus $mv = p = \sqrt{2m\varepsilon V}$ and substituting into the de Broglie equation gives $\lambda = h[2m\varepsilon V]^{-1/2}$. If the appropriate numerical values for the constants are put into this expression, it is found that $\lambda \simeq \sqrt{150/V} \times 10^{-10}$ m; so that electrons accelerated through potential differences ranging from 1.5 to 15,000 volts should be associated with wavelengths ranging from about 10^{-9} m to 10^{-11} m, respectively. Electron diffraction is now a well established technique for the examination of the structure of materials, providing several advantages compared with X-ray diffraction methods. Electron beams are found to produce higher intensities than X-ray beams. Their wavelength may be easily altered by changing the applied voltage, and most valuably, electron beams may be deflected by electric and magnetic fields thus enabling the construction of lenses and microscopes, which due to the short wavelength of the beam (at suitable applied potentials) provides a much higher resolving power than conventional optical devices. Since 1927, it has been amply demonstrated that not only electrons, but beams of neutrons, protons, and molecular beams of hydrogen and helium also exhibit diffraction phenomena when they are scattered by crystalline matter, the associated wavelength being that calculated by the de Broglie expression, $\lambda = h/p$. The realization that both electromagnetic radiation and matter behave, or could be regarded as behaving, as though they both exhibit wave properties and particle properties, on the basis of unequivocal experimental evidence, has led to the development of a system for the description of natural events which basically takes account of the fact that mechanical events are accompanied by a wave process. This is the subject of quantum mechanics, which is essentially statistical in outlook, and in terms of which it appears that the various phenomena of chemistry

---

[1] Davisson, C. J., and Germer, L. H., *Phys. Rev.*, 1927, 30, 705.
[2] Thomson, G. P., *Proc. Roy. Soc.*, 1928, A, 117, 600.

and much of physics can be deduced from a relatively small number of initial postulates. Quantum mechanics, however, is not a completely new or radical way of describing natural phenomena, but retains the conventional concepts of mass, velocity, momentum, angular momentum, energy, time, position and so on, as the centrally important and measurable properties of any system; it constitutes a generalization of classical mechanics, formulated in a mathematical language not previously necessary, such that classical mechanics is a special case of the more generally applicable quantum mechanics. However, there are several alternative sets of initial postulates, from which the subject can be developed, and there are two mathematical formalisms which appear initially to be two different theories. Wave mechanics, due to de Broglie, Schrödinger, Heitler, and others, is essentially a way of rewriting the laws of classical mechanics in terms of wave theory, while matrix mechanics, due to Born, Jordan, and Heisenberg, presents the subject in such a manner that every physical magnitude is associated with a representative matrix, a mathematical representation which is not quite so easily related to physical interpretation. Schrödinger has shown that the two methods are equivalent. Also, there is a wider generalization of matrix mechanics due to Dirac. The choice of describing any given physical situation in terms of wave mechanics or matrix mechanics is a matter of convenience, just as in classical mechanics, the nature of the problem involved decides whether to describe it in terms of Newtonian, Lagrangian, or Hamiltonian mechanics.‡

‡ The Schrödinger equation, which is the centrally important postulate of wave mechanics is closely related to classical mechanical concepts, as successively developed by Newton, Lagrange, Laplace, and Hamilton. It is therefore instructive to review the principles of classical mechanics which led to this development by Schrödinger. Wave mechanics actually makes extensive use of classical mechanical concepts, and furthermore there are still many problems in chemistry and physics, especially those involving radiation, which may be adequately dealt with, in non-wave mechanical terms.

In ordinary dynamics, the motion of a particle of mass, m, acted on by a force F, may be described by the differential equation, $\mathbf{F} = d/dt(m\mathbf{v})$. The position of the particle with respect to some particular reference system can then be determined by solving this equation, usually subject to some boundary conditions. However, the position of the particle may be described in a large number of different ways, and the choice of any particular coordinate system is a matter of convenience. This is particularly so in more complex dynamical problems where difficulties may be associated with the solutions of the problems of motion, dependent on the coordinate system used. One of the advantages of the method due to Lagrange is that, for each individual problem, it uses that coordinate system which best suits the description of the type of motion involved in the given problem. Generally, the complete description of the position of any system requires that one coordinate must be associated with each degree of freedom of the system. For systems whose coordinates are independent of one another, Lagrange's generalized coordinates are denoted, $q_1, q_2, \ldots, q_n$, and these collectively describe a system of n degrees of freedom. The generalized nature of the coordinates signifies that they are not necessarily lengths or angles, but may be any quantity appropriate to the description of the position of the system. For example, to write the kinetic energy K of a particle as a function of generalized coordinates, the procedure is to first express K in terms of Cartesian coordinates and then to transform these coordinates to the $q_1 q_2 q_3$ coordinates of the generalized system. This requires a knowledge of the relations between the two sets of coordinates. Suppose that these relations are

$$x = \phi_1(q_1, q_2, q_3),$$
$$y = \phi_2(q_1, q_2, q_3),$$
$$z = \phi_3(q_1, q_2, q_3).$$

The kinetic energy, K, of a particle in terms of x, y, z, may be written as

$$K = (m/2)[\dot{x}^2 + \dot{y}^2 + \dot{z}^2],$$

where $\dot{x}$, $\dot{y}$, $\dot{z}$ denote differentiation with respect to time, dx/dt, dy/dt, and dz/dt. The velocity components of the particle, in terms of generalized coordinates, may similarly be written in Newtonian notation as

$$dx/dt = \dot{x} = (\partial\phi_1/\partial q_1)(dq_1/dt) + (\partial\phi_1/\partial q_2)(dq_2/dt) + (\partial\phi_1/\partial q_3)(dq_3/dt),$$

with similar expressions for $\dot{y}$ and $\dot{z}$. The kinetic energy then becomes

$$K = (m/2) \sum_{i=1}^{3} [\dot{q}_1(\partial\phi_i/\partial q_1) + \dot{q}_2(\partial\phi_i/\partial q_2) + \dot{q}_3(\partial\phi_i/\partial q_3)]^2.$$

Another form of this expression may be obtained by differentiating it with respect to $\dot{q}_1$ and then multiplying the result obtained by $\dot{q}_1$, resulting in,

$$(\partial K/\partial\dot{q}_1)\dot{q}_1 = m\left\{ \sum_{i=1}^{3} [\dot{q}_1(\partial\phi_i/\partial q_1) + \dot{q}_2(\partial\phi_i/\partial q_2) + \dot{q}_3(\partial\phi_i/\partial q_3)]\dot{q}_1(\partial\phi_i/\partial q_1) \right\}.$$

Similar expressions result by differentiation of K with respect to $\dot{q}_2$ and $\dot{q}_3$,

$$(\partial K/\partial\dot{q}_2)\dot{q}_2 = m\left\{ \sum_{i=1}^{3} [\dot{q}_1(\partial\phi_i/\partial q_1) + \dot{q}_2(\partial\phi_i/\partial q_2) + \dot{q}_3(\partial\phi_i/\partial q_3)]\dot{q}_2(\partial\phi_i/\partial q_2) \right\},$$

and

$$(\partial K/\partial\dot{q}_3)\dot{q}_3 = m\left\{ \sum_{i=1}^{3} [\dot{q}_1(\partial\phi_i/\partial q_1) + \dot{q}_2(\partial\phi_i/\partial q_2) + \dot{q}_3(\partial\phi_i/\partial q_3)]\dot{q}_3(\partial\phi_i/\partial q_3) \right\}.$$

Adding these three expressions gives,

$$\sum_{j=1}^{3} \dot{q}_j(\partial K/\partial\dot{q}_j) = m \sum_{i=1}^{3} [(\partial\phi_i/\partial q_1)\dot{q}_1 + (\partial\phi_i/\partial q_2)\dot{q}_2 + (\partial\phi_i/\partial q_3)\dot{q}_3]^2 = 2K,$$

or

$$K = (1/2) \sum_{j=1}^{3} (\partial K/\partial\dot{q}_j)\dot{q}_j.$$

In this manner, it is always possible to obtain the equations of motion of a particle with respect to any desired coordinate system, first, by writing them in Cartesian coordinates, and then transforming the variables. However, the tedious algebraic manipulations make this procedure rather impractical. It was first pointed out by Lagrange that the equations of motion in any coordinate system can be written directly, if the expression for the kinetic energy in the desired system is known. This requires a knowledge of the relation between work and kinetic energy.

First, in Newton's system of mechanics, force, F, is defined as mass multiplied by acceleration. Thus for a force F moving a particle of mass m through a distance r in a time t to produce an acceleration a,

$$F = ma = m(dv/dt) = m\, d/dt(dr/dt) = d/dt(mv).$$

Also in newtonian mechanics, $F(r)\,dr = dW = -dU(r)$, where W is work performed by the force and $U(r)$ is the potential energy of the particle. Therefore,

$$\int_{r_0}^{r_1} dW = W = \int_{r_0}^{r_1} F(r)\,dr = \int_{t_0}^{t_1} F(r)(dr/dt)\,dt = \int_{t_0}^{t_1} m(d^2r/dt^2)(dr/dt)\,dt$$

$$= \int_{t_0}^{t_1} d[(m/2)(dr/dt)^2],$$

since $d/dt(dr/dt)^2 = 2(d^2r/dt^2)(dr/dt)$. Thus,

$$\int_{r_0}^{r_1} dW = W = (m/2)\int_{t_0}^{t_1} d(v)^2 = mv_1^2/2 - mv_0^2/2 = K_1 - K_0, \qquad \text{or} \qquad dW = dK.$$

Also, from

$$F(r)\,dr = -dU(r), \qquad F(r) = -d/dr[U(r)],$$

or

$$W = \int_{r_0}^{r_1} F(r)\,dr = U(r_0) - U(r_1),$$

and therefore $U(r_0) + K_0 = U(r_1) + K_1$. Now in terms of Lagrange's generalized coordinates, q_1, q_2, q_3, the increment of work done on a particle giving rise to the increment dK in its kinetic energy is given by, $dW = Q_1\,dq_1 + Q_2\,dq_2 + Q_3\,dq_3 = \Sigma_i\,Q_i\,dq_i$, where Q_i is referred to as the generalized force, since generally it has the dimensions of force. Since as has already been deduced, the kinetic energy in Lagrangian mechanics, is a function of the q's and $\dot{q}$'s, or $K = f(q_i, \dot{q}_i)$, it is possible to write $dK = (\partial K/\partial q_i)\,dq_i + (\partial K/\partial \dot{q}_i)\,d\dot{q}_i$, which on differentiation with respect to time gives

$$dK = (dK/dt)\,dt = \sum_{i=1}^{3} [(\partial K/\partial q_i)(dq_i/dt) + (\partial K/\partial \dot{q}_i)(d\dot{q}_i/dt)]\,dt. \qquad \text{(a)}$$

But

$$K = (1/2)\sum_{i=1}^{3} (\partial K/\partial \dot{q}_i)\dot{q}_i,$$

and differentiation of this with respect to time gives

$$dK/dt = (1/2)\sum_{i=1}^{3} [d/dt(\partial K/\partial \dot{q}_i)(\dot{q}_i) + (d\dot{q}_i/dt)(\partial K/\partial \dot{q}_i)],$$

or

$$2\,dK = \sum_{i=1}^{3} [d/dt(\partial K/\partial \dot{q}_i)\,dq_i + (\partial K/\partial \dot{q}_i)\,d\dot{q}_i]. \qquad \text{(b)}$$

Subtraction of equation (a) from equation (b) above gives,

$$dK = \sum_{i=1}^{3} dq_i[d/dt(\partial K/\partial \dot{q}_i) - (\partial K/\partial q_i)].$$

Now equating dK and dW and collecting terms, there is obtained,

$$\sum_{i=1}^{3} [d/dt(\partial K/\partial \dot{q}_i) - (\partial K/\partial q_i) - Q_i]\,dq_i = 0.$$

Since the generalized coordinates are independent of one another in the sense that one may be given an incremental change while the others remain unchanged, the term in parenthesis in the latter equation must be identically zero, or

$$d/dt(\partial K/\partial \dot{q}_i) - (\partial K/\partial q_i) = Q_i,$$

where $i = 1, 2, 3$. These are Lagrange's equations of motion of a particle. If the generalized force, Q, acting on a particle is derivable from a potential function, U, that is if $Q_i = -\partial U/\partial q_i$ (analagous to the Newtonian expression, $F(r) = -dU(r)/dr$), and if the potential energy is independent of the velocity, then it follows that, $\partial U/\partial \dot{q} = 0$ and Lagrange's equations become,

$$d/dt(\partial L/\partial \dot{q}_i) - \partial L/\partial q_i = 0,$$

where $L = K - U$ and L is called the Lagrangian function of the system. This latter relation applies only to conservative systems ($U + K = $ constant), whereas in the form $d/dt(\partial K/\partial \dot{q}_i) - \partial K/\partial q_i = Q_i$, Lagrange's equations apply to non-conservative systems as well.

It is also possible to define generalized momenta compatible with generalized coordinates; the generalized momentum corresponding to a coordinate q_i is defined to be $p_i = \partial L/\partial \dot{q}_i$ (in Newtonian mechanics $\partial/\partial v[mv^2/2 - U] = mv = p$). The differentiation is made with respect to $\dot{q}_i$, the other $\dot{q}$'s being regarded as constant; p_i and q_i are referred to as the conjugated momentum and coordinate. The Lagrangian is generally a function of the generalized coordinates, generalized velocities, and time; that is $L(q, \dot{q}, t)$.

Just as Lagrangian mechanics is more general than Newtonian mechanics, there is an even more general principle from which Lagrange's equations of motion may be deduced. Hamilton,[1] almost a hundred years before Schrödinger, had noted the resemblance between the equations of mechanics and geometrical optics, and indeed Hamilton's Principle is a generalization of the laws of mechanics and optics. The principle known as de Maupertuis Principle of Least Action (about 1730) refers to the fact that in any field of force, a particle will travel from one point A to another point B by following a path such that the integral $\int_A^B 2K\, dt$ has an extremum; thus in the absence of any force field the path will be a straight line, and in a uniform field of force, such as the gravitational field of the earth, a particle will follow a parabolic path. It should be noted that the dimensions of this integral are ml^2t^{-1}, as for Planck's constant; furthermore the value of the integral depends only on the space coordinates of A and B and is independent of time. Similarly, the laws of geometrical optics may be generalized in the form known as Fermat's Principle of Least Time; this refers to the fact that for a beam of light traveling from point A to point B through a medium of variable refractive index, the actual optical path followed is characterized by an extremum for the integral, $\int_A^B ds/v$, where v is the velocity of the beam of light at the position ds and ds is an element of length of the path. Hamilton's Principle may be stated as: the motion of a system from some time t_1 to time t_2 is such that the line integral, $I = \int_{t_1}^{t_2} L\, dt$, where $L = K - U$, is an extremum for the actual path of motion. This implies that the system in its motion from position 1 at time t_1 to position 2 at time t_2 chooses a path such that the integral I is an extremum, whether a maximum or a minimum. This in turn implies that the variation of the integral, δI is identically zero. Hamilton's Principle, and these resemblances between the laws of mechanics and optics were considered to be nothing more than interesting mathematical abstractions, until the work of de Broglie and Schrödinger indicated that a supremely important principle was involved. Because of its variational character, Hamilton's Principle is best interpreted in terms of the calculus of variations.

The technique of the calculus of variations plays an important part in the development of advanced dynamics, and certain features of these techniques have a direct application

---

[1] See, for example, "William Rowan Hamilton—An Appreciation," C. Lanczos, *American Scientist*, 1967, **55**, 2, 129.

to wave mechanics or Schrödinger mechanics.[1] In elementary calculus, a function of several variables, $y = f(x_1, x_2, x_3, \ldots)$, is said to possess extreme values at $x_1, x_2, \ldots$, if $\partial y/\partial x_1 = 0$, $\partial y/\partial x_2 = 0, \ldots$. In the calculus of variations, there is also the problem of finding an extremum, but the type of expression to be investigated depends on the behavior of one or more dependent variables as opposed to a number of independent variables, as in the applications of elementary calculus. The type of expression to be investigated has instead the form, $I = \int_{x_1}^{x_2} f(x, y, dy/dx)\, dx$. This type of problem, in which $y = \xi(x)$, resolves itself into the question of being given the function $f(x, y, dy/dx)$, to determine the function $\xi(x)$, such that the integral I has a stationary value. The simplest example of a problem of this type is to find the plane curve which has the minimum length between some two specified points; that is, if A and B are any two points find the curve $\xi(x)$ which has the minimum length between A and B. From elementary calculus, the length of the curve between two such points as A and B is

$$L = \int_A^B \sqrt{[1 + (dy/dx)^2]}\, dx$$

and this is the integral which has to be minimized. This integral is a special case of the equation, $I = \int_{x_1}^{x_2} f(x, y, dy/dx)\, dx$, so that it may be written as $L = \int_A^B f(dy/dx)\, dx$. If some particular curved line drawn in a plane represents the curve, $y = \xi(x)$, whose length is the shortest between A and B, then any other curve must have a greater length; any such other curve is referred to as the varied path, and may be chosen for simplicity to also pass through A and B. The ordinates of any point on a varied path are given by, $y + \delta y = \xi(x) + \alpha\eta(x)$, where δy is the variation in y, and the small quantity α indicates that the length of the varied path is just slightly greater than the minimum. The value of the integral L, then changes by the amount δL as the path changes from the original to the varied path. This variation in the integral is given by $\delta L = \delta \int_A^B \sqrt{[1 + (dy/dx)^2]}\, dx$. The integral is a function of $dy/dx = y'$ only and thus the interchange of the integral and the variation operations will not affect the integration. Then,

$$\delta L = \int_A^B \delta\sqrt{[1 + (y')^2]}\, dx = \int_A^B \partial/\partial y' \sqrt{[1 + (y')^2]}\, \delta y'\, dx = \int_A^B y'[1 + (y')^2]^{-1/2}\, \delta y'\, dx.$$

Thus the value of the integral depends on the variation $\delta y'$, which is the variation in the slope of the curve; this variation in turn depends on the variation δy. This is so since the slope of the varied path, $d/dx(y + \delta y) = y' + d/dx(\delta y)$; therefore the variation of the slope, $\delta y' = d/dx(\delta y)$, or $\delta(dy/dx) = d/dx(\delta y)$. Commutative operators of this type, here δ and d/dx, are of frequent occurrence in wave mechanics, and since $\delta(dy/dx) = d/dx(\delta y)$, it is found that,

$$\delta L = \int_A^B y'[1 + (y')^2]^{-1/2}\, \delta y'\, dx = \int_A^B y'[1 + (y')^2]^{-1/2}\, d/dx(\delta y)\, dx,$$

which on integration by parts gives,

$$\delta L = y'[1 + (y')^2]^{-1/2}|\delta y|_A^B - \int_A^B d/dx[y'\{1 + (y')^2\}]^{-1/2}\, \delta y\, dx.$$

The first term vanishes at the limits since $\delta y = 0$ at both A and B. Since $y = \xi(x)$ gives the integral L a stationary value, or makes the length represented by the integral a minimum, therefore an infinitesimal variation δy in the curve does not change the value of the integral. Thus, $\int_A^B [d/dx[y'\{1 + (y')^2\}]^{-1/2}]\, \delta y\, dx = 0$, and this equation can only

---

[1] Cf. Margenau, H., and Murphy, G. M., *The Mathematics of Physics and Chemistry*, 1961, 2nd edition, D. Van Nostrand Co. Inc., Princeton, N.J.

be satisfied for an arbitrary variation δy if, $d/dx[y'\{1 + (y')^2\}^{-1/2}] = 0$. On integration of this equation, it is found that $y = a_1 x + a_2$, where a_1 and a_2 are integration constants. But this is the function, $y = \xi(x)$, which makes the length of the curve between A and B a minimum, and is obviously a straight line, as expected. This same type of argument may be extended to a more general case; for example, the integral, $I = \int_{A(x_1)}^{B(x_2)} f(x, y, y') \, dx$, where again $y = \xi(x)$ is the function which gives the integral its extremum value. But the condition for an extremum is as before, that $\delta \int_{A(x_1)}^{B(x_2)} f(x, y, y') \, dx = 0$, which may be written as, $\delta I = \int_A^B [\partial/\partial y f(x, y, y') - d/dx(\partial/\partial y'\{f(x, y, y')\})] \, \delta y \, dx = 0$. Again, since for any arbitrary variation δy, the expression within the square brackets must be zero in order to ensure the vanishing of the integral, therefore,

$$d/dx[\partial/\partial y'\{f(x, y, y')\}] - \partial/\partial y\{f(x, y, y')\} = 0.$$

This is known as Euler's differential equation, and its solution $y = \xi(x)$ gives the integral I an extremum value, where

$$I = \int_A^B f(x, y, y') \, dx.$$

In the same manner, it can be shown that if the integral to be extremized is a function of more than two variables, such as,

$$I = \int_{t_1}^{t_2} f(x, y, z, \dot{x}, \dot{y}, \dot{z}, t) \, dt,$$

where t is an independent variable and x, y, z, are the dependent variables, then the conditions necessary for the vanishing of δI are,

$$\frac{d}{dt}\left[\frac{\partial f}{\partial \dot{x}}\right] - \frac{\partial f}{\partial x} = 0,$$

$$\frac{d}{dt}\left[\frac{\partial f}{\partial \dot{y}}\right] - \frac{\partial f}{\partial y} = 0,$$

$$\frac{d}{dt}\left[\frac{\partial f}{\partial \dot{z}}\right] - \frac{\partial f}{\partial z} = 0.$$

These are the Euler's equations for this problem. The Euler's equations become identically Lagrange's equations of motion for a conservative system, if f is substituted by $L = K - U$. Thus it can be demonstrated in this way, that these equations can be deduced directly from Hamilton's Principle. For a non-conservative system, this Principle may be stated as, $\int_{t_1}^{t_2} (\delta K + \delta W) \, dt = 0$, where δW is the increment of work corresponding to the variation δq_i of the coordinate q_i.

In many dynamical problems it is convenient to deal with a function called the Hamiltonian function rather than the Lagrangian function. This Hamiltonian function is defined in terms of the generalized momenta, p_i, generalized coordinates, q_i, and time as, $H(p, q, t) = \sum_i p_i \dot{q}_i - L(q, \dot{q}, t)$. The differential of H is $dH = \sum_i (\partial H/\partial q_i) \, dq_i + \sum_i (\partial H/\partial p_i) \, dp_i + (\partial H/\partial t) \, dt$. But from the definition of the Hamiltonian function,

$$dH = \sum_i \dot{q}_i \, dp_i + \sum_i p_i \, d\dot{q}_i - \sum_i (\partial L/\partial \dot{q}_i) \, d\dot{q}_i - \sum_i (\partial L/\partial q_i) \, dq_i - (\partial L/\partial t) \, dt.$$

But since the generalized momenta, p_i, are defined as $p_i = \partial L/\partial \dot{q}_i$, the two terms $\sum_i p_i \, d\dot{q}_i$ and $\sum_i (\partial L/\partial \dot{q}_i) \, d\dot{q}_i$ cancel, and also from Lagrange's equation,

$$d/dt(\partial L/\partial \dot{q}_i) - \partial L/\partial q_i = 0,$$

it follows that $\dot{p}_i = \partial L/\partial q_i$. Therefore dH becomes,

$$dH = \sum_i \dot{q}_i \, dp_i - \sum_i \dot{p}_i \, dq_i - (\partial L/\partial t) \, dt.$$

Comparison of the two equations for dH shows that,

$$\partial H/\partial t = -\partial L/\partial t,$$
$$\partial H/\partial q_i = -\dot{p}_i,$$
$$\partial H/\partial p_i = \dot{q}_i.$$

These are Hamilton's canonical equations of motion, which in Hamiltonian mechanics replaces Lagrange's equations. For conservative systems, in which a force is derivable from a potential function, $F = -\nabla U$, U being independent of the velocities. If in addition, the restriction is imposed that the Lagrangian function of the system is not an explicit function of time, then the total derivative of L may be written as,

$$dL/dt = \sum_i (\partial L/\partial q_i)(dq_i/dt) + \sum_i (\partial L/\partial \dot{q}_i)(d\dot{q}_i/dt).$$

But from Lagrange's equations, $\partial L/\partial q_i = d/dt(\partial L/\partial \dot{q}_i)$ and therefore

$$dL/dt = \sum_i d/dt(\partial L/\partial \dot{q}_i)\dot{q}_i + \sum_i (\partial L/\partial \dot{q}_i)(d\dot{q}_i/dt),$$

or

$$dL/dt = \sum_i d/dt[\dot{q}_i(\partial L/\partial \dot{q}_i)],$$

from which

$$d/dt\left(\sum_i \dot{q}_i(\partial L/\partial \dot{q}_i) - L\right) = 0.$$

Substituting $\partial L/\partial \dot{q}_i$ by p_i, this becomes,

$$d/dt\left(\sum_i p_i\dot{q}_i - L\right) = 0,$$

or the expression within the parentheses, which is H, must be a constant. Thus for a conservative system, $H = \sum_i p_i\dot{q}_i - L =$ constant. But $p_i = \partial L/\partial \dot{q}_i = \partial K/\partial \dot{q}_i$, since for conservative systems $L = K - U = K$, and it has already been shown that $K = (1/2) \sum_i (\partial K/\partial \dot{q}_i)\dot{q}_i$. Therefore $H = \sum_i p_i \dot{q}_i - L =$ constant $= \sum_i \dot{q}_i(\partial K/\partial \dot{q}_i) - L = 2K - L$, or $H = 2K - L = K + U(L = K - U)$, which is the total energy of the system. Thus the classical Hamiltonian function of a system is equivalent to the total energy of the system.

There is indeed a close resemblance between the laws governing the trajectories of particles in mechanical problems and the paths of beams of light in optical problems, since in both cases the actual path adopted is characterized by the minimum value of an integral (page 101). This correspondence can be expressed in the form of a differential equation, such that the problem of describing the motion of a particle in a field of force becomes mathematically similar to that of the path of a beam of light moving through a medium of varying refractive index; this equation is the so-called Schrödinger equation, which results if the de Broglie wavelength, $\lambda = h/mv$, is substituted into the classical equations describing wave motion. In the classical electromagnetic theory of light, the propagation of plane, harmonic, waves in one dimension is described by the equation,

$$\psi = Ae^{\pm 2\pi i(kx - vt)},\ddagger \qquad \text{(q.2.1.)}$$

where ψ is a function which varies periodically with the x-coordinate and with time, that is, $\psi(x, t)$, A is the maximum amplitude of the wave motion, ν is the frequency of the wave motion of wavelength λ, propagated with a velocity c, t is the time, and $i = \sqrt{-1}$.

‡ This is actually a more general expression for a wave motion which is periodic in both space (in this case, one-dimensional space) and time, and for which both the phase and the amplitude (page 112) are complex quantities.

All types of wave motion, whether in the case of the wave motion which spreads out over the surface of a pool when a stone is dropped into it, or the motion associated with a vibrating string or diaphragm, or the sound waves in a wind instrument such as an organ or clarinet, or the motion involved in the transmission of light or other electromagnetic waves through space, are characterized by the transmission of energy from one point to another, without the permanent displacement of the intervening medium; all of these types of wave motion have another common characteristic in that they may all be described by a wave equation of a similar form. One of the simplest types of wave motion is that associated with the vibration of a string stretched between two fixed points and set into motion by plucking; in this manner waves travel back and forth along the string but obviously the string does not undergo any permanent translation. Such a type of wave motion is said to involve stationary waves. On the other hand, if one end of a long string is moved up and down, the wave motion produced travels along the length of the string as a traveling or progressive wave, such that the displacement of the string at different positions along its length and at different times is different. If this displacement is represented by ϕ, then ϕ is a function of position and time. If this function is known then all possible modes of vibration of the string are determined, and this function, in turn, may be obtained as a solution of the wave equation of motion of the string. This wave equation may be obtained as follows.

Consider a string maintained under a tension T and stretched between two points in a line parallel with the x Cartesian coordinate direction and set into vibration with small vibrational amplitude in the xz coordinate plane. If the string is assumed to be flexible, then the tension is everywhere directed along the string. Consider μ to be the mass of the string per unit length, which need not necessarily be constant along the string. Applying Newton's Law of motion to a small segment, dx, of the string, $F = ma$. There will generally be a force acting on this segment dx, since the tension acting on the flexible string will be in different directions at x and at $x + dx$; only if the segment happens to be straight will there be no net force on the segment dx. At x (see Fig. Q.2.2b.), the component of the tension, T_1, causing a displacement of the string in the z direction is $T_1 = -T \sin \theta \simeq -T \tan \theta \simeq -T \, \partial\phi/\partial x$; this is so, since for a small displacement $\sin \theta \simeq \tan \theta$, and the sign is negative if $\partial\phi/\partial x$ is positive and the force acts downward. At $x + dx$, the component of the tension, T_2, acting upward is $T_2 = T \sin(\theta + d\theta) \simeq T \tan(\theta + d\theta) \simeq T \, \partial/\partial x(\phi + d\phi)$. The net force acting on the string is therefore,

$$T_1 + T_2 \simeq T \, \partial\phi/\partial x + T(\partial^2\phi/\partial x^2) \, dx - T \, \partial\phi/\partial x \simeq T(\partial^2\phi/\partial x^2) \, dx.$$

According to Newton's Law, this force must be equal to the mass of the element, $\mu \, dx$ multiplied by its acceleration in the z direction, $\partial^2\phi/\partial t^2$. Thus,

$$T(\partial^2\phi/\partial x^2) \, dx = \mu \, dx(\partial^2\phi/\partial t^2) \quad \text{or} \quad T(\partial^2\phi/\partial x^2) = \mu(\partial^2\phi/\partial t^2),$$

which is the wave equation of the vibrating string (one-dimensional wave equation). This equation is not valid if T depends on x but is so if μ depends on x, although generally both μ and T may depend on the time.

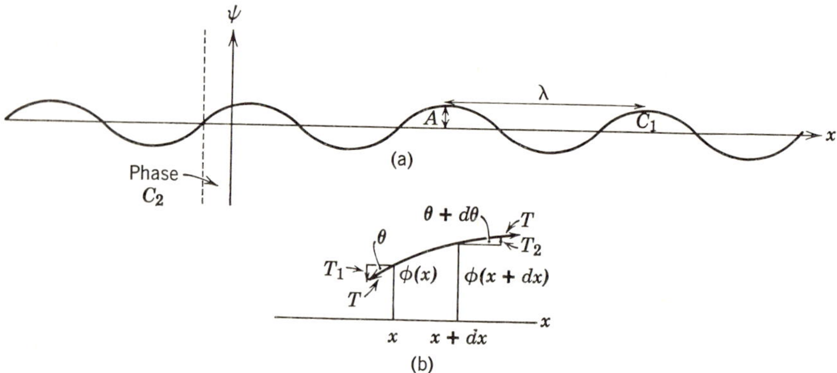

Fig. Q.2.2. (a) The function, $\psi(x) = C_1 \sin[2\pi x/\lambda + C_2]$. (b) The forces acting on a segment dx of a vibrating string.

In this particular example μ and T are constants. Since $\phi(x, t)$ is a function of x and t, solutions of the wave equation must correspond to periodic expressions containing x and t, which are differentiable. The best known periodic function is a sine or cosine function; each of these is characterized by the property that the amplitude returns to similar values at intervals of $n\lambda$, where n is any integer; that is, $\phi(x \pm n\lambda) = \phi(x)$. The most general expression of this kind is $\phi(x) = A \sin[2\pi(c_1 x + c_2)]$ or $\phi(x) = A \cos[2\pi(c_3 x + c_4)]$. These two expressions are equivalent, since the one can be generated from the other by suitable selection of the parameters, c_1, c_2, c_3, and c_4, as $\cos x = \sin(\pi/2 - x) = \sin(\pi/2 + x)$. The maximum value that $\phi(x)$ may adopt is A. If the sin expression is considered, the parameter c_1 is obviously related to the quantity λ, since $c_1\lambda = 2\pi$, and therefore $\phi(x) = A \sin[2\pi x/\lambda + c_2]$. The parameter c_2, referred to as the phase constant, is such that any change in c_2 simply causes the profile of the function $\phi(x)$ to shift in the $+x$ or $-x$ direction without any other change in the profile of the sine curve. In order to examine the behavior of A with respect to time, it is necessary to introduce the idea of the velocity of progression of the sine wave function. If the velocity is v in the $+x$ direction, then the wave profile will move in the $+x$ direction a distance vt in the time t; that is, the wave profile will have exactly the same form after an interval t, if it is imagined that the origin is shifted from $x = 0$ to $x = vt$. Therefore if this quantity vt is subtracted from x, then the motion of the profile of the sine expression may be referred back to the origin $x = 0$, and the equation of the traveling wave becomes, $\phi(x, t) = A \sin[(2\pi/\lambda)(x - vt)]$. But v/λ for the wave motion is equivalent to the frequency ν as $\nu = v/\lambda$. If $1/\lambda$, referred to as the wave number is defined as k, then $\phi(x, t) = A \sin 2\pi[kx - \nu t]$. Similarly, this could be written as $\phi(x, t) = A \cos 2\pi[kx - \nu t]$. The inverse of the frequency ν is referred to as the period, T, of the wave motion, or $T = 1/\nu = \lambda/v$, and it is another characteristic of a wave motion that $\phi(x, t) = \phi(x, t + T)$. With reference to the sine expression, $\partial/\partial x[\phi(x, t)] = \partial/\partial x[A \sin 2\pi(kx - \nu t)]$, which is $2\pi k A[\cos 2\pi(kx - \nu t)]$, and $\partial^2/\partial x^2[\phi(x, t)] = -4\pi^2 k^2 A[\sin 2\pi(kx - \nu t)]$. Similarly, $\partial/\partial t[\phi(x, t)] = -2\pi\nu A[\cos 2\pi(kx - \nu t)]$ and $\partial^2/\partial t^2[\phi(x, t)] = -4\pi^2\nu^2 A[\sin 2\pi(kx - \nu t)]$. If these second derivatives are substituted in the original wave equation, $T(\partial^2\phi/\partial x^2) = \mu(\partial^2\phi/\partial t^2)$, then $-4\pi^2 k^2 \phi(x, t) = (\mu/T)[-4\pi^2\nu^2\phi(x, t)]$.

Therefore, $\mu/T = k^2/\nu^2 = 1/\nu^2\lambda^2 = 1/v^2$, and substituting this quantity into the wave equation of the vibrating string, there results $\partial^2\phi/\partial x^2 = [1/v^2]\,\partial^2\phi/\partial t^2$, which is the more general equation of a one-dimensional wave motion; for electromagnetic radiation, $v = c$. The cosine expression above is also a solution of this same wave equation, and also any linear combination of the sine and cosine functions, both of which are real functions. For the particular linear combinations in which either the amplitude part or the phase factor, or both, are imaginary (page 112), these expressions may be written in a still more general form by noting that $(e^{ix} + e^{-ix}) = 2 \cos x$ and $(e^{ix} - e^{-ix}) = 2i \sin x$. Thus, for $\phi(x, t)$ given by $A \cos 2\pi[kx - vt] \pm iA \sin 2\pi[kx - vt] = Ae^{\pm 2\pi i(kx - vt)}$, for which the amplitude is real and the phase factor is imaginary, and for $\phi(x, t)$ given by the combination $A \sin 2\pi[kx - vt] \pm iA \cos 2\pi[kx - vt] = \pm iAe^{\mp 2\pi i(kx - vt)}$, for which both the amplitude and the phase parts are imaginary.

The motion described by this equation is planar because it describes the dependence of ψ on only the x-coordinate; for various values of A, this equation would describe a whole system of waves characterized by A. If $kx = vt$, then $\psi = A$, since $e^0 = 1$; also the velocity of propagation of the wave, $c = x/t$, so that when $\psi = A$, $c = x/t = v/k = v\lambda$, or $k = 1/\lambda$, the wave number. According to the de Broglie postulate, $\lambda = h/p = h/mv$, or $k = 1/\lambda = mv/h$. In the system of wave mechanics, formulated by Schrödinger, it is postulated that not only for light but for all material systems including, light, electrons, atoms, ions, molecules, and crystals, that their observable and measurable properties are characterized by a periodically varying function ψ, which is referred to as the state function. If this function, given by (q.2.1.) is differentiated twice with respect to x, there results, $\partial^2\psi/\partial x^2 = -4\pi^2 k^2 \psi$, and putting $k = mv/h$ into this equation, it is found that $\partial^2\psi/\partial x^2 = -8\pi^2 mh^{-2}(mv^2/2)\psi$, and making the further substitution for $K = mv^2/2 = E - U$, where E and U are the total and potential energies respectively, associated with the wave number k, this equation is obtained

$$\partial^2\psi/\partial x^2 + 8\pi^2 mh^{-2}(E - U)\psi = 0, \qquad \text{(q.2.2.)}$$

which is the one-dimensional form of the time-independent Schrödinger equation. In three dimensions, ψ becomes

$$\psi = Ae^{2\pi i(k_x x + k_y y + k_z z)},$$

where k_x, k_y, k_z, are the components of the three-dimensional wave vector resolved in the x, y, and z, cartesian coordinate directions respectively. The kinetic energy now becomes

$$K = (m/2)(v_x^2 + v_y^2 + v_z^2) = (h^2/2m)(k_x^2 + k_y^2 + k_z^2),$$

since by the de Broglie relations

$$k_x = 1/\lambda_x = mv_x/h,$$
$$k_y = 1/\lambda_y = mv_y/h,$$
$$k_z = 1/\lambda_z = mv_z/h.$$

Differentiating twice with respect to x, y, and z, and adding,

$$\partial^2\psi/\partial x^2 + \partial^2\psi/\partial y^2 + \partial^2\psi/\partial z^2 = -4\pi^2\psi(k_x{}^2 + k_y{}^2 + k_z{}^2)$$
$$= -8\pi^2 mh^{-2}(E - U)\psi,$$

or writing the Laplacian symbol, $\nabla^2 = \partial^2/\partial x^2 + \partial^2/\partial y^2 + \partial^2/\partial z^2$, the three-dimensional time-independent, Schrödinger Equation is obtained as,

$$\nabla^2\psi + 8\pi^2 mh^{-2}(E - U)\psi = 0. \qquad\qquad (q.2.3.)$$

A principal aim of wave mechanics is concerned with discovering the forms of ψ appropriate to the particular problem under consideration, and then to derive from these forms the values of observable quantities. There is no strictly logical argument by means of which the Schrödinger equation may be derived from more fundamental principles, any more than Newton's equations of motion can be so derived, but there are many points of view which lead to the conclusion that this Schrödinger equation itself represents the statement of a natural law.‡

‡ There is indeed no derivation of the Schrödinger equation. A second order differential equation of this type results whether the de Broglie relationship is substituted into the equations of classical wave theory or into the equations of motion of classical mechanics. In the latter case, if for simplicity, a single particle is imagined to be moving along a certain direction such as the $+x$ direction of the Cartesian Coordinate system in a field of potential energy $U(x)$, then if the particle has mass, m, the Law of Conservation of Energy may be stated as $E = K + U = mv_x{}^2/2 + U(x)$, and integration of this equation gives the classical laws of motion of the particle. If the de Broglie relationship is introduced into this equation, by substituting $\lambda = h/mv_x = h/p_x$, then $E = p_x{}^2/2m + U(x)$ or $E - U(x) = p_x{}^2/2m$, and in three dimensions this would be, $E - U(x, y, z) = (2m)^{-1}[p_x{}^2 + p_y{}^2 + p_z{}^2]$. In Hamiltonian mechanics the same result applies where H, the classical Hamiltonian function is such that $H = K + U$. If now, equation (q.2.3.) is rearranged to read $(E - U)\psi = -h^2(8\pi^2 m)^{-1}\nabla^2\psi$, where U and ψ are each functions of the three space coordinates of the particle, then on comparison of this Schrödinger equation with the classical expression, it may be deduced that the relationship between them is that in the former the classical momenta expressions have been replaced by the differential operators, such that p_x has been substituted by $(h/2\pi i)\,\partial/\partial x$, p_y by $(h/2\pi i)\,\partial/\partial y$, and p_z by $(h/2\pi i)\,\partial/\partial z$. Apparently then, the classical Hamiltonian equation of motion can be transformed into the Schrödinger wave equation by writing the total energy of a particle, or system of particles, as a function of its or their momenta and space coordinates and then making the appropriate operator substitution for the momenta coordinates. As will be demonstrated in the following, not only momentum but other dynamical variables of any particle or system of particles in motion, have a corresponding operator expression in quantum mechanics. Generally, a dynamical variable of a system is a quantity which characterizes the motion of the system, as opposed to those quantities such as charge and mass, which are parameters of the system. This relationship between dynamical variables, operators, and wave or state functions, which have to be regarded as three of the principal concepts of quantum mechanics, may alternatively be illustrated by considering the plane, one-dimensional wave function, (q.2.1.). If this function is differentiated once with respect to x, then $\partial\psi/\partial x = 2\pi i k\psi$, and if both sides of this equation

are multiplied by $h/2\pi i$, there results, $(h/2\pi i)\,\partial\psi/\partial x = hk\psi$, and on substituting the de Broglie wavelength, $k = 1/\lambda = p_x/h$ into this equation it is found that $p_x\psi = [(h/2\pi i)\,\partial/\partial x]\psi$, as before. Similarly, on differentiating (q.2.1.) once with respect to time and multiplying the result obtained by $h/2\pi i$, there results, $(h/2\pi i)\,\partial\psi/\partial t = -h\nu\psi$, and on substituting in the latter expression the Planck expression, $E = h\nu$, it is found that $E\psi = [-(h/2\pi i)\,\partial/\partial t]\psi$. Thus the quantum mechanical equivalent of the classical mechanical total energy is the operator, $-(h/2\pi i)\,\partial/\partial t$. It should be noted that the factors $h/2\pi$ and $h/2\pi i$ are frequently written as $\hbar$ (crossed-h) and $-i\hbar$, respectively. Also, if the operator equations, $p_x\psi = [(h/2\pi i)\,\partial/\partial x]\psi$ and $E\psi = [-(h/2\pi i)\,\partial/\partial t]\psi$, are substituted into the classical Hamiltonian, $H = E = K + U$, there results, for the case of a particle moving in one-dimensional space, where $E = p_x^2/2m + U(x)$, the quantum Hamiltonian equation

$$[-(h^2/8\pi^2 m)\,\partial^2/\partial x^2 + U(x)]\psi = -\frac{h}{2\pi i}\frac{\partial\psi}{\partial t},$$

which is the one-dimensional time dependent Schrödinger equation. In the latter expression, the quantity $U(x)$ is part of the total operator, although it has the simple significance of the operation of multiplication. In three-dimensional space, the time dependent Schrödinger equation becomes, by appropriate substitution of the operator forms for p_y and p_z, $[-(h^2/8\pi^2 m)\nabla^2 + U(x, y, z)]\psi = -(h/2\pi i)\,\partial\psi/\partial t$.

It may seem remarkable that the time dependent Schrödinger equation does not contain the second derivative of the wave function with respect to time, as does the classical equation of wave motion (see page 106), $\partial^2\phi/\partial x^2 = [1/v^2]\,\partial^2\phi/\partial t^2$, but this fact is actually a consequence of the Heisenberg Uncertainty Principle (see page 132). According to this Principle, the position and the velocity of a particle, x and dx/dt, in the one-dimensional case, cannot be simultaneously known with any certainty, and since $x = f(t)$ and $dx/dt = \dot{x} = f'(t)$, then if the Schrödinger equation were to contain $d^2\psi/dt^2$ this would mean that both $\psi(x, t)$ and $d/dt[\psi(x, t)]$ would require to be known in order to obtain a solution of the Schrödinger equation, which would be in violation of the Heisenberg Uncertainty Principle.

The time dependent Schrödinger equation is actually the more general one, the time independent equation corresponding to the special case applicable to conservative systems, or for the description of stationary states, whose dependence on time is trivial. For a non-conservative system, the classical Hamiltonian H and the energy E would both be functions of the time t, corresponding to the energy equation $H(q, p, t) = E(t)$, where $H(q, p, t)$ is as defined on page 103. If another function is defined as $\mathscr{K}(q, p, t, E) = H(q, p, t) - E$, then since $\partial\mathscr{K}/\partial p = \partial H/\partial p = \dot{q}$, and $-\partial\mathscr{K}/\partial q = -\partial H/\partial q = \dot{p}$, it is apparent that Hamilton's equations have the same form whether H or $\mathscr{K}$ is used. However, in the case of the function $\mathscr{K}(q, p, t, E)$, two other equations, $\partial\mathscr{K}/\partial E = -1 = i^2$, and $\partial\mathscr{K}/\partial t = \partial H/\partial t$, may be written; this latter equation may be written as $-\partial\mathscr{K}/\partial(-t) = \dot{E}$, when it becomes apparent that E is conjugate to $-t$. Now, since the time independent Schrödinger equation has been shown above to be obtainable from the classical energy equation by the replacement of the momentum p conjugate to the coordinate q by the operator $(h/2\pi i)\,\partial/\partial q$ (generalizing from the one-dimensional case discussed on page 108), the conjugate relationship between E and $-t$ suggests that E should be replaced by $(h/2\pi i)\,\partial/\partial(-t)$ as its operator equivalent. If these operator forms of p and E are now substituted into the energy equation $H(q, p, t) = E(t)$, and allowed to operate on a function of both the space coordinates and time, $\Psi(q, t)$, there results

$$H[q, (h/2\pi i)\,\partial/\partial q, t]\Psi(q, t) = -(h/2\pi i)\,\partial/\partial t[\Psi(q, t)] = ih/2\pi\,\partial/\partial t[\Psi(q, t)]$$
$$= i\hbar\,\partial/\partial t[\Psi(q, t)],$$

which corresponds to various equivalent ways of writing the time-dependent Schrödinger equation. In the above formulation, it should be noted that ψ is a function of the space coordinates only and Ψ is a function of the space and time coordinates. In three-dimensional Cartesian coordinate space, for a particle of mass m, whose potential energy is a function of time as well as its wave function, this equation becomes

$$[-(h^2/8\pi^2 m)\,\nabla^2 + U(x, y, z, t)]\Psi(x, y, z, t) = (ih/2\pi)\,\partial/\partial t[\Psi(x, y, z, t)].$$

For a conservative system, where H is not explicitly a function of t, the Schrödinger equation is separable (see page 165 and following sections) by writing $\Psi(q, t) = \psi(q)f(t)$, which expression when substituted into the time-dependent Schrödinger equation above and dividing throughout by $\Psi(q, t)$ leads to

$$[1/\psi(q)]H[q, (h/2\pi i)\,\partial/\partial q]\psi(q) = [ih/2\pi f(t)]\,d/dt[f(t)]. \qquad \text{(A)}$$

The left hand side of this equation is a function of q only and the right hand side is a function of t only, which can only be so if both sides are equal to the same constant, the so-called separation constant, λ. Equating both sides of the above equation to λ, there results for the left hand side,

$$[1/\psi(q)]H[q, (h/2\pi i)\,\partial/\partial q]\psi(q) = \lambda \qquad \text{or} \qquad H[q, (h/2\pi i)\,\partial/\partial q]\psi(q) = \lambda\psi(q),$$

which is of the form of the Schrödinger time-independent equation, $\hat{H}\psi = E\psi$ (see page 108), indicating that λ must correspond to some energy value, say E_n, and so $\psi(q)$ to some function giving rise to the energy E_n, say $\psi_n(q)$. Similarly, equating the right hand side of (A) to the same separation constant $\lambda = E_n$, gives

$$[(ih/2\pi)f(t)]\,d/dt[f(t)] = \lambda = E_n \qquad \text{or} \qquad [1/f(t)]\,d/dt[f(t)] = -[2\pi i/h]E_n.$$

This latter equation may be shown to have the general solution

$$f(t) = Ae^{-2\pi i E_n t/h},$$

where A is any arbitrary constant, by direct substitution. Thus the separable solution $\Psi_n(q, t) = \psi_n(q)f(t)$, of the time dependent Schrödinger equation for a conservative system is

$$\Psi_n(q, t) = A\psi_n(q)e^{-2\pi i E_n t/h},$$

and since the equation is a linear one, the general solution may be written as the sum of all such solutions for all the energy states described by the various functions ψ_n. Therefore,

$$\Psi(q, t) = \sum_n A_n\psi_n(q)e^{-2\pi i E_n t/h}, \qquad \text{(B)}$$

where the various coefficients in this sum, A_n, are constants. If $t = 0$, this equation becomes $\Psi(q, 0) = \sum_n A_n\psi_n(q)$. If the various functions ψ_n, satisfy certain conditions, that they constitute an orthonormal set of functions (see section Q.2.4.), the constants may be evaluated as $A_n = \int \psi_n^*(q)\Psi(q, 0)\,dq$, the integration being carried out over all of the coordinates of the system.

It has to be concluded then, that on the basis of observable experimental results, such as the energy distribution of radiant energy, the photoelectric effect, the line spectra and X-ray spectra of atoms, the diffraction of electrons, and the failure of classical mechanics to provide any understanding of these and other phenomena, that the time dependent Schrödinger equation itself constitutes the statement of a natural law describing the behavior of particles and systems of particles. The Schrödinger formulation of quantum mechanics, based on the concepts of operators, dynamical variables, and state or wave functions, may be conveniently placed on a postulatory basis in terms of these and a relatively small number of other laws expressed in terms of these concepts. It is common practice to write the Schrödinger equation in the compact form $\hat{H}\Psi = -(h/2\pi i)\,\partial\Psi/\partial t = \hat{E}\Psi$, as the quantum mechanical equivalent of the classical mechanical energy equation $H = E$. In the quantum mechanical form, the quantities with a circumflex over them are operators, such that $\hat{E} = -(h/2\pi i)\,\partial/\partial t$, $\hat{p}_x = (h/2\pi i)\,\partial/\partial x$, $\hat{H} = [-(h^2/8\pi^2 m)\,\nabla^2 + U(x, y, z)]$, for example for a single particle. For any system the Hamiltonian operator $\hat{H}$, may always be written down by firstly writing the classical expression for the total energy of the system in terms of its momenta, space coordinates, and time, and then making the appropriate operator substitution for the momenta components. Thus, for a system of N particles with a potential energy $U(x, y, z, t)$ the Schrödinger equation is

$$\left[-\sum_{i=1}^{N} (h^2/8\pi^2 m_i)\,\nabla_i^2 + U(x_1, y_1, z_1, \ldots x_N, y_N, z_N, t) \right]\Psi(x_1, \ldots z_N, t)$$
$$= (ih/2\pi)\,\partial/\partial t[\Psi(x_1, \ldots z_N, t)].$$

Equation (q.2.3.) represents a family of surfaces, each corresponding to one value of E; like all differential equations, it has an infinite number of solutions for a given form of the potential energy, $U(x, y, z)$, and the correct solution has to be chosen so that it represents something which is physically possible. Such acceptable solutions, corresponding to particular values of ψ result when certain impositions, compatible with the problem under consideration, are placed on ψ; the values of ψ which comply with the requirements that both ψ and its first derivatives with respect to the positional coordinates $\partial\psi/\partial x$, $\partial\psi/\partial y$, $\partial\psi/\partial z$, are everywhere finite, single-valued and continuous are referred to as proper functions or eigenfunctions of the corresponding Schrödinger equation. The most important reason for making such restrictions is that, if the behavior of material particles is determined by the quantity ψ conforming to these boundary conditions, then the stationary states of the earlier Bohr–Sommerfeld quantum theory are logically comprehensible, each eigenfunction corresponding to a given energy level. In most cases where the solution represents a stationary energy state, the boundary conditions result in the wavelength associated with the particles in any particular problem being simply related to some length which the problem involves. This implies that only certain energies can give rise to acceptable solutions to the wave equation, these energies being interpreted as the stationary states of the older quantum theory, and the whole number relations which the wavelengths obey are similarly interpreted as being connected with the older quantum numbers. In fact it emerges that the quantum numbers and stationary energy states which appear in wave mechanics, are not simply arbitrary postulates imposed on classical mechanics, but result as the necessary and logical consequences of solutions of the wave equation corresponding to acceptable values of ψ.

It has to be concluded, as will become increasingly apparent from the ensuing discussion, that all of the properties and all of the possible modes of motion of a particle of mass m in a potential energy field $U(x, y, z)$ may be deduced by finding the proper solutions, the various eigenfunctions each corresponding to a particular eigenvalue E, of the second order partial differential equation (q.2.3.). From the form of this equation, it is apparently only the mass and in particular the potential energy function which distinguish one system from another. Without being concerned in the meantime about the interpretation to be assigned to the function ψ, or the more formal structure of wave mechanics, it is easily possible to illustrate the manner in which the Schrödinger equation can be formulated and solved for simple systems.

Q.2.2.   Motion of a Single Particle In One Dimension

Consider a particle of mass m moving along the direction of the $+x$ cartesian coordinate direction, under the action of some force F acting in the same direction. For a conservative system, where the total energy of the particle remains constant, the force F and the potential energy of the particle $U(x)$ are related, according to classical mechanics (pages 101 and 104), by $F = -dU(x)/dx$. The Schrödinger wave equation for the particle is then,

$$d^2\psi/dx^2 + 8\pi^2 m h^{-2}[E - U(x)]\psi = 0, \qquad (q.2.4.)$$

where $\psi(x)$ is the wave function for this particular system. If the potential energy $U(x)$ of the particle is constant, then since the total energy E is assumed constant, K must also be constant, and $K = mv^2/2 = E - U$, where $v = \dot{x}$. Substituting the de Broglie relation $\lambda = h/p = h/mv$ into the expression for K gives $E - U = h^2/2m\lambda^2$, which on insertion into (q.2.4.) gives,

$$d^2\psi/dx^2 + 4\pi^2\psi/\lambda^2 = 0, \qquad (q.2.5.)$$

where λ must be constant. The general solution of this equation is

$$\psi(x) = C_1 \sin\,[2\pi x/\lambda + C_2], \qquad (q.2.6.)$$

as may easily be proved by differentiating (q.2.6.) twice with respect to x and substituting in (q.2.5.). C_1 and C_2 are arbitrary constants, which in wave theory are referred to as the amplitude and phase respectively of the wave described by (q.2.6.). If C_1 and C_2 are real‡ then $\psi(x)$ plotted as a function of x presents the sinusoidal wave profile shown in Fig. Q.2.2. If C_1 and C_2 are imaginary, then it is preferable to express (q.2.6.) in the alternative manner,

$$\psi(x) = C_3 e^{2\pi i x/\lambda} + C_4 e^{-2\pi i x/\lambda}, \qquad (q.2.7.)$$

where C_3 and C_4 are constants and both the real and imaginary parts of ψ are sinusoidal type functions.

‡ The totality of integral, fractional, positive, and negative numbers, and zero, which are conventionally used to express the results of physical measurements or observations is called the system of (real) rational numbers. Another category of numbers is necessary

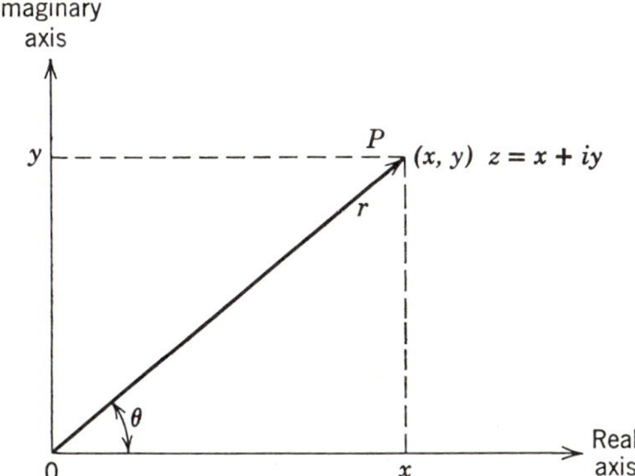

FIG. Q.2.3.   The Argand diagram to represent complex numbers.

to express the roots of algebraic equations with real coefficients such as $x^2 - 2x + 5 = 0$, or $x^2 + \alpha^2 = 0$, which have no real roots, but which are formally satisfied by the solutions, $1 \pm \sqrt{-4}$ or $1 \pm 2i$ and $\pm \alpha \sqrt{-1}$ or $\pm \alpha i$ respectively, where $i = \sqrt{-1}$. It was proved by Gauss (about 1800) that every algebraic equation with real coefficients has complex roots (involving i), real numbers being a special kind of complex numbers for which the coefficient of i is zero. Every number that can be expressed as the product of a real number and i is known as a complex number. The most general complex number is the sum of a real and an imaginary part, as $x + iy$, where x and y are real and their sum z is imaginary; thus $z = x + iy$. Whereas each number in the real number system may be represented by a point on a straight line, it requires two coordinates to designate a complex number; thus $x + iy$ corresponds to only one point of a plane in which a system of rectangular coordinates has been set up. In Fig. Q.2.3., the point P with the Cartesian coordinates (x, y) or the polar coordinates (r, θ) represents the complex number $x + iy$; the abscissa corresponds to the system of all real numbers and the ordinate to all imaginary numbers. Such plane diagrams for the representation of complex numbers are referred to as Argand diagrams; any arbitrary complex number, $x + iy$, has x as its projection on the real axis and y as its projection on the imaginary axis. Since $x = r \cos \theta$ and $y = r \sin \theta$, then $z = x + iy = r(\cos \theta + i \sin \theta)$. The complex conjugate of a complex number is defined as the number written with a minus sign in front of the i; so $x - iy$ is the complex conjugate of $x + iy$. It is conventional to represent the complex conjugate of a complex number by the symbol $*$; thus z^* is the complex conjugate of z. The product $zz^* = (x + iy)(x - iy) = x^2 + y^2$, which is always a positive number. The quantity written usually as $|z| = \sqrt{zz^*}$, is known as the modulus of z or the absolute value of z and is equal to OP. The angle θ is called the amplitude or phase of the complex number z.

The complex number, $z = x + iy = r(\cos \theta + i \sin \theta)$, may be written in another manner by taking advantage of the well known fact that for all values of x, $e^x = 1 + x + x^2/2! + x^3/3! + x^4/4! + \cdots = \sum_{n=0}^{\infty} [x^n/n!]$. By substituting $i\theta$ for x in this exponential series, there is obtained, using $i^2 = -1$, $i^3 = -i$, $i^4 = 1$, etc.:

$$e^{i\theta} = \sum_{n=0}^{\infty} [i^n \theta^n/n!] = 1 + i\theta - \theta^2/2! - i\theta^3/3! + \theta^4/4! + i\theta^5/5! - \theta^6/6! - i\theta^7/7! + \cdots$$
$$= [1 - \theta^2/2! + \theta^4/4! - \theta^6/6! + \cdots] + i[\theta - \theta^3/3! + \theta^5/5! - \theta^7/7! + \cdots]$$
$$= \cos \theta + i \sin \theta.$$

Thus $z = x + iy = r(\cos \theta + i \sin \theta) = re^{i\theta}$. This result is particularly convenient for expressing the product of two complex numbers, such as $z_1 = r_1 e^{i\theta_1}$ and $z_2 = r_2 e^{i\theta_2}$, which becomes $z_1 z_2 = r_1 r_2 e^{\theta_2 + \theta_1}$. θ is usually referred to as the argument of z.

The above power series expressions for e^x, $\sin \theta$ and $\cos \theta$, are all consequences of the assumption that any arbitrary function of a single variable, $u = f(x)$ may be expanded as a power series such that $u = f(x) = \sum_{n=0}^{\infty} A_n x^n$. If this is so, then $du/dx = A_1 + 2A_2 x + 3A_3 x^2 + \cdots = f'(x)$, $d^2 u/dx^2 = 2A_2 + (3!)A_3 x + \cdots = f''(x)$, $d^3 u/dx^3 = (3!)A_3 + (4!)A_4 x + \cdots = f'''(x)$, etc. If $u = \sum_{n=0}^{\infty} A_n x^n$ for all values of x the coefficients A_n have the same values independently of the value of x. So that if $x = 0$, then $f'(0) = A_1$ and $u = A_0$, $f''(0) = 2A_2$, $f'''(0) = (3!)A_3$, or $u = A_0$, $A_1 = f'(0) = [du/dx]_{x=0} = dA_0/dx$, $A_2 = f''(0) = [d^2 u/dx^2]_{x=0} = d^2 A_0/dx^2(1/2!)$, $A_3 = f'''(0)/3! = [d^3 u/dx^3]$, etc. Thus $u = f(x) = A_0 + du/dx(x) + d^2 u/dx^2(x^2/2!) + \cdots$, which is the Maclaurin Series. By now letting $u = f(x) = \sin x$ and noting that $du/dx = d(\sin x)/dx = \cos x$, $d^2 u/dx^2 = d^2(\sin x)/dx^2 = d(\cos x)/dx = -\sin x$, etc., and that $\sin 0 = 0$, $-\sin 0 = 0$, $\cos 0 = 1$, $-\cos 0 = -1$, the above expansion for $\sin x$ is obtained; similarly for $\cos x$. Note also that from the expression $e^{ix} = \cos x + i \sin x$, it follows that $e^{2\pi i} = \cos 2\pi + i \sin 2\pi = 1$. Similarly it can be shown by application of the Maclaurin Series that $e = 1 + 1 + 1/2! + 1/3! + \cdots$, and that $e^x = 1 + x/1 + x^2/2! + x^3/3! + \cdots$.

It should be noted that (q.2.7.) describes a stationary wave (see page 104) for which the amplitude is independent of time. Thus if the potential energy of a moving particle is constant, there are no forces acting on it ($dU/dx = 0$), and the wavelength associated with the freely moving particle is that given by the de Broglie relation. In other words, a free particle may have any energy; whatever the value of E there is a corresponding value of λ given by $E = h^2/2m\lambda^2$; there is no question of discrete energy levels. However, wave mechanics shows that this is not so for a particle acted on by a force which would, in terms of classical mechanics confine the particle to a finite region of motion along the x-axis.

Suppose, for example, that a particle has the total energy E and the potential energy $U(x)$ as shown in Fig. Q.2.4. In classical mechanics the particle could never be found outside the region $a \le x \le b$, since outside this region the kinetic energy $K = E - U(x)$, would be negative. The particle would therefore be confined to motion between a and b, although according to classical mechanics it could have any value of the total energy greater than the minimum denoted on the potential energy curve. Wave mechanics, however, leads to quite different conclusions, and as will be seen the imposition of some natural limitations on the wave function ψ results in a set of discrete energy levels, with energy values intermediate between these not being allowed. The restrictions imposed on ψ, are that it cannot be zero but should have a finite value everywhere, be single valued and continuous everywhere, and so also should its first derivative $d\psi/dx$. These conditions are partly related to the requirements necessary for finding a solution of an equation such as (q.2.4.) and partly related to the physical interpretation of ψ, as will be discussed later.

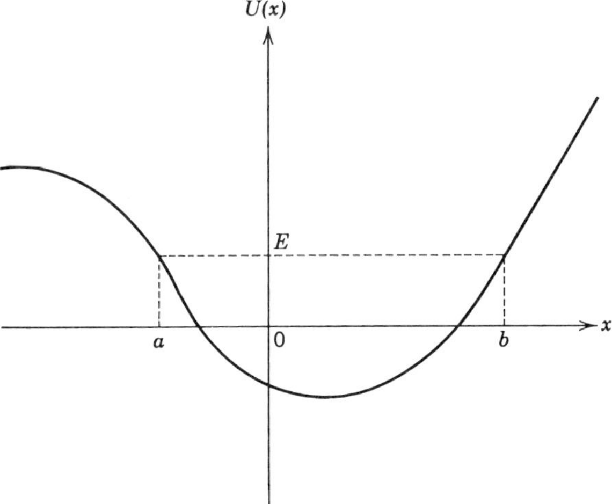

FIG. Q.2.4. A typical potential energy curve of a particle confined to motion between the points a and b along the direction of the x-axis.

The possible energy states for a particle confined to motion along a specific region of the x-axis may be found as follows. Suppose, as in Fig. Q.2.5. that the potential energy $U(x)$ of a particle of mass m has some constant value U_c, when $x < 0$ and when $x > L$, and is zero when $0 \leq x \leq L$. For values of $E < U_c$, the particle is confined to the motion in the region between $x = 0$ and $x = L$. In regions A and B, where $U(x) = U_c$, the Schrödinger equation describing the state of the particle is

$$d^2\psi/dx^2 + (8\pi^2 m/h^2)[E - U_c]\psi = 0, \qquad (\text{q.2.8.})$$

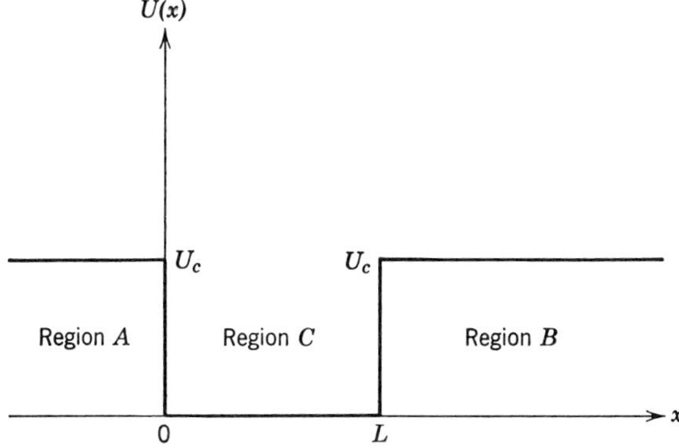

FIG. Q.2.5. The motion of a particle in a rectangular potential well.

which as before has a general solution of form

$$\psi = C_5 e^{\alpha x} + C_6 e^{-\alpha x}, \qquad \text{(q.2.9.)}$$

where C_5 and C_6 are arbitrary constants; by differentiating this expression twice and substituting into (q.2.8.), it may readily be shown that (q.2.9.) is an exact solution of (q.2.8.) if the quantity $\alpha = \sqrt{[8\pi^2 m h^{-2}(U_c - E)]}$. Since $E < U_c$, α is a real quantity. If the positive value of the square root is taken, then $C_6 e^{-\alpha x}$ becomes infinitely large for infinitely large negative values of x, and therefore ψ would become infinite under these conditions; to ensure that ψ is finite everywhere, it is therefore necessary that $C_6 = 0$. In the region A, where $x < 0$, there is no necessity to consider what happens to ψ when x is an infinitely large positive quantity. However, in region B, where $x > L$, for positive values of α, $C_5 e^{\alpha x}$ becomes infinite when x is infinitely large, and therefore so does ψ under these conditions; again to ensure that ψ is finite everywhere, it is necessary to set $C_5 = 0$, in region B; also since x cannot be negative in this region, there is no concern about what happens to ψ for large negative values of x. In region C, the Schrödinger equation for $U(x) = 0$, becomes

$$d^2\psi/dx^2 + 8\pi^2 mE\psi/h^2 = 0,$$

with the general solution, $\psi = C_7 \cos \beta x + C_8 \sin \beta x$, which again may be shown to be an exact solution of the appropriate wave equation if $\beta = \sqrt{[8\pi^2 mE/h^2]}$. The appropriate forms of the solutions of the wave equation in the three regions A, B, and C, are therefore as follows:

$$\text{in region } A, \text{ where } x < 0, \psi = C_5 e^{\alpha x}, \qquad \text{(q.2.10.)}$$

$$\text{in region } B, \text{ where } x > L, \psi = C_6 e^{-\alpha x}, \qquad \text{(q.2.11.)}$$

$$\text{in region } C, \text{ where } 0 \leq x \leq L, \psi = C_8 \sin \beta x + C_7 \cos \beta x. \qquad \text{(q.2.12.)}$$

In order that ψ and $d\psi/dx$ should be continuous, it is necessary that at the boundary points between the three regions, namely at $x = 0$ and at $x = L$, the solutions in adjacent regions should give the same values. At $x = 0$, ψ is continuous if $C_5 e^{\alpha 0} = C_8 \sin 0 + C_7 \cos 0$, or if $C_5 = C_7$, and $d\psi/dx$ is continuous if $C_5 \alpha e^0 = +C_8 \beta \cos 0 - C_7 \beta \sin 0$, or if $C_5 \alpha = C_8 \beta$. Similarly, the condition that ψ and $d\psi/dx$ should be continuous at $x = L$, leads to

$$C_6 e^{-\alpha L} = C_8 \sin \beta L + C_7 \cos \beta L, \qquad \text{(q.2.13.)}$$

and

$$-C_6 \alpha e^{-\alpha L} = C_8 \beta \cos \beta L - C_7 \beta \sin \beta L. \qquad \text{(q.2.14.)}$$

If (q.2.14.) is divided by (q.2.13.), then

$$\frac{\alpha}{\beta} = \frac{C_7 \sin \beta L - C_8 \cos \beta L}{C_8 \sin \beta L + C_7 \cos \beta L} = \frac{C_7 \tan \beta L - C_8}{C_8 \tan \beta L + C_7},$$

or $\tan \beta L = [\alpha C_7 + \beta C_8][\beta C_7 - \alpha C_8]^{-1}$, since $C_5 = C_7$, and $\alpha C_5 = \beta C_8$; therefore, $\tan \beta L = [2\alpha\beta][\beta^2 - \alpha^2]^{-1}$. Now substituting for α and β, it is found that

$$\tan \{L\sqrt{[8\pi^2 mE/h^2]}\} = 2[E(U_c - E)]^{1/2}[2E - U_c]^{-1}. \qquad \text{(q.2.15.)}$$

For any given values of U_c and L, this equation can obviously be solved for E, and it is apparent that E may adopt one or more discrete values but not a continuous range of values. This simple example, usually referred to as an example of a particle confined to a rectangular potential well, shows that the use of the Schrödinger equation, together with a set of simple boundary conditions applied to ψ, leads to a set of discrete energy levels for the particle thus confined. For the idealized case where the potential energy $U_c \to \infty$, the problem is referred to as that of a particle moving in a one-dimensional box with perfectly reflecting walls. If $U_c = \infty$, it is seen from (q.2.15.) that the right-hand side of this equation becomes 0, and thus $\tan \{L\sqrt{[8\pi^2 mE/h^2]}\} = 0$, or $L\sqrt{[8\pi^2 mE/h^2]} = n\pi$, where $n = \pm 1, \pm 2, \pm 3, \ldots$; the possible values for E are then, $E = n^2 h^2/8mL^2$, where $n = 1, 2, 3, \ldots$. Alternatively, this result can be deduced from Fig. Q.2.5. As $U_c \to \infty$, α also tends to infinity, and the solutions in regions A and B given by (q.2.10.) and (q.2.11.) both approach zero. In the limit, the solution (q.2.12.) must be such that the constants C_7 and C_8 are determined by the condition that this solution is zero at $x = 0$ and at $x = L$. In (q.2.12.), if $\psi(0) = 0$ and $\psi(L) = 0$, then $C_8 \sin \beta L = 0$, or since C_8 and β cannot be zero (either of which would make ψ zero, which has to be disallowed) then $\beta = n\pi/L$, where $n = \pm 1, \pm 2, \pm 3, \ldots$; but since $\beta = \sqrt{[8\pi^2 mE/h^2]}$, the allowed energy levels are given by

$$E_n = n^2 h^2/8mL^2, \text{ with } n = 1, 2, 3, \ldots; \qquad \text{(q.2.16.)}$$

the corresponding wave functions valid in region C are given by

$$\psi_n = C_8 \sin (n\pi x/L), \text{ with } n = 1, 2, 3, \ldots, \qquad \text{(q.2.17.)}$$

where C_8 is an arbitrary constant which is not zero. If (q.2.17.) is used to plot ψ as a function of x, assuming some constant real value for C_8, the curves shown in Fig. Q.2.6. result for $n = 1, 2, 3$, and 4, respectively.

The wave function for $n = 1$, corresponding to the lowest or ground state energy is zero at $x = 0$ and at $x = L$. Apart from these zero values at $x = 0$ and at $x = L$, the other wave functions have zero values or nodes at $x = L/2$ (for $n = 2$), and at $x = L/3$ and $2L/3$ (for $n = 3$), and at $x = L/4, L/2, 3L/4$

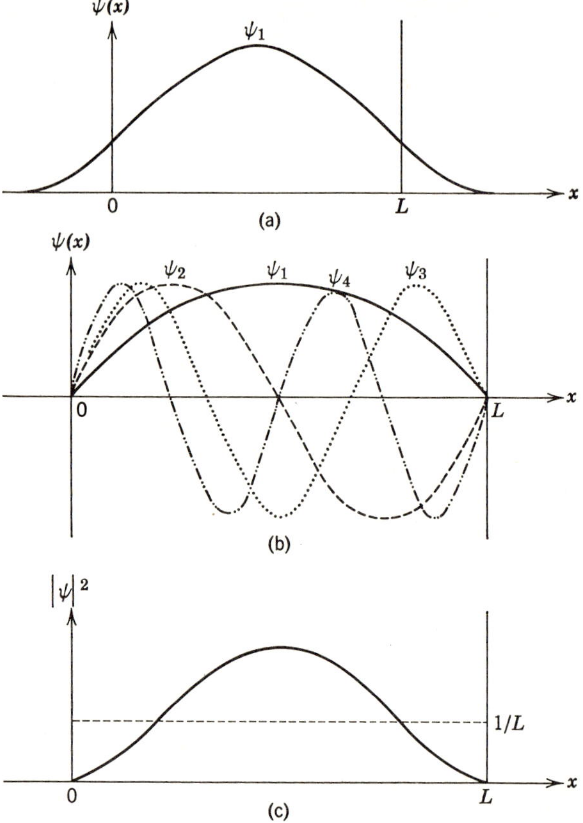

FIG. Q.2.6. The wave functions of a particle moving in a one-dimensional potential box. (a) The ground-state wave function for the particle where the potential energy of the particle is finite; (b) the wave functions corresponding to the four lowest energy states for the particle where the potential energy is infinite; (c) the probability distribution for the particle in the lowest energy state where the potential energy is infinite.

(for $n = 4$); generally ψ_n has $(n - 1)$ nodes, and this criterion provides a useful method of estimating which of two or several wave functions corresponds to the lowest energy, if the value of the energy level cannot otherwise be obtained. The wave functions are all zero outside region C, for $x > L$ and for $x < 0$; there is a discontinuity in slope at these points (due to the limiting process invoked). This is not so for any finite value for U_c, no matter how great it may be; there is no discontinuity in slope but a gradual exponential decrease in ψ in regions A and B and the wave function for the state corresponding to the ground state for a finite U_c is also shown in Fig. Q.2.6. For increasing values of U_c, the exponential decrease in ψ becomes increasingly rapid, and it is only in the limit when $U(x) = \infty$ that the set of curves in the center of Fig. Q.2.6. result. The fact that for finite values of U_c the wave function only falls off to zero in the region outside the one-dimensional box has important consequences, as will be discussed after examining the interpretation to be assigned to the wave function ψ. In the simple application above of the Schrödinger wave equation, it has not been necessary to consider ψ as anything other than a convenient tool to calculate the values of the energy levels. There is of course no way of comparing the calculated energy values in this problem with experimentally observed values, since the problem is purely hypothetical. Metals, however, can be approximately regarded as examples of electrons moving in three-dimensional potential boxes (where $U_c \simeq 1.6 \times 10^{-18}$ J) as in Chapter M.3. It will be seen that exceptional agreement with experiment is obtained by the use of such a model. It is common practice to refer to E_n and ψ_n as eigenvalues and eigenfunctions of the Schrödinger equation, for stationary states.

Q.2.3.   The Interpretation of the Wave Function

The unifying concept relating the description of natural events in terms of wave theory is that of probability, as first pointed out by Max Born. The Schrödinger wave equation makes use of an analytical function ψ to describe the behavior of a particle, which has to be regarded as the most convenient variable for specifying or defining a material system. This function most generally is a complex quantity, or may be written in the form $\psi(x) = a + ib$ (for a one-dimensional function), where a and b are real functions of x and $i = \sqrt{-1}$. The complex conjugate of ψ, denoted by ψ^*, is obtained by changing i to $-i$ wherever it appears in ψ, so that $\psi^* = a - ib$. The product of ψ and ψ^* is real and positive, and its positive square root, denoted by $|\psi|$ is referred to as the modulus of ψ. So, $\psi\psi^* = |\psi|^2 = a^2 + b^2$. In terms of electromagnetic wave theory, the intensity of a beam of light, for example, is proportional to the square of the amplitude of the wave motion; since the function ψ used in the Schrödinger wave equation describing the motion of a particle has the nature of an amplitude function, it is assumed that the probability of finding the particle in a specific region of space is proportional to $\psi\psi^*$ (or simply ψ^2 if ψ is real). For a particle, the probability of finding it in a volume element dv is also proportional to dv (page 22); thus the total probability of finding a particle described by the function ψ in the volume element

dv is $w \propto \psi\psi^* \, dv$. The proportionality constant may be evaluated by making the integral of w over the whole of space equal to unity,

$$\int_0^\infty w \, dv = \int_0^\infty \psi\psi^* \, dv = 1;$$

this procedure is equivalent to making the condition that the particle must be somewhere, that there is a certainty or total probability of unity that the particle is somewhere in the whole of space. This procedure is used in all problems in quantum mechanics and is referred to as the normalization process; by using $\psi\psi^*$ instead of ψ^2, the probability must always be real and positive. Since Schrödinger's equation is linear, if ψ is a solution then so must $C\psi$ be a solution, where C is any constant. Thus there arises the situation that every possible wave function is arbitrary to the extent of a multiplying constant, but it is always possible to choose ψ so that

$$\int_0^\infty \psi\psi^* \, dv = 1, \qquad\qquad (q.2.18.)$$

such that ψ is said to be normalized. There are mathematical difficulties in dealing with the wave function of a free particle, where as already seen the energies of such a free particle belong to a continuous spectrum, and so (q.2.18.) cannot be satisfied; this difficulty vanishes, however, if it simply assumed that every system can be imagined to be confined in a box, and if the box is sufficiently large compared with the actual system it cannot affect any experimental observation but the limits it imposes on the range of motion of any particle makes the normalization procedure universally applicable.

As an example of the normalization procedure, consider the simple wave functions of a particle confined to motion in a one-dimensional box (q.2.17.). In order to normalize these functions, C_8 has to be chosen so that,

$$\int_0^L C_8{}^2 \sin^2 (n\pi x/L) \, dx = 1,$$

since here ψ_n is real and the whole of one-dimensional space for this problem is the region of the x-axis between $x = 0$ and $x = L$. But

$$\int_0^L \sin^2 (n\pi x/L) \, dx = \int_0^L (1/2)[1 - \cos (2n\pi x/L)] \, dx$$

$$= \left.\frac{x}{2}\right|_0^L - |(L/2\pi n) \sin (2\pi nx/L)|_0^L = L/2.$$

Therefore, $C_8 = \sqrt{2/L}$, which is independent of n, so that the normalized wave functions in this case are $\psi_n = [\sqrt{2/L}] \sin (n\pi x/L)$, where $n = 1, 2, 3, \ldots$.

Since ψ is generally complex, it obviously cannot itself have any direct physical meaning; there is no means of ever observing ψ directly. However, when this function is normalized, $|\psi^2|$ may be interpreted as a probability density, so that $|\psi^2|\,dx\,dy\,dz$ represents the probability that a particle describable by the wave function ψ may be found in the small volume element $dx\,dy\,dz$. Alternatively, it may be said that if a very large number of observations of the position of a particle in the state ψ were made at random, then in a fraction $|\psi^2|\,dx\,dy\,dz$ of them the particle would be found in the volume element $dx\,dy\,dz$. Also, more generally, ψ is dependent not only on the positional coordinates of the particle but also on the time. This distinction is generally made by writing for the state function of a particle, $\Psi(x, y, z, t)$ and the stationary state function as $\psi(x, y, z)$. Under those circumstances where the method of the separation of variables is applicable, $\Psi(x, y, z, t) = \psi(x, y, z)e^{2\pi i v t}$, and then $\psi(x, y, z)$ is sometimes referred to as the space factor and $e^{2\pi i v t}$ as the amplitude factor of the state function. For stationary states in which the energy is fixed and equal to $h\nu$, this simply amounts to the assumption that the particle density or probability density is independent of time, since $e^{2\pi i v t} = \cos 2\pi v t + i \sin 2\pi v t$, which is of the form $a + ib$ and so the scalar value of $e^{2\pi i v t}$ is unity. Thus for stationary states Ψ and ψ have the same scalar value and are solutions of the same differential equation and the particle density is $\psi\psi^*$. In fact, any wave function, even a normalized wave function is arbitrary to the extent of a multiplying constant with a scalar value or modulus of unity, such as e^{ib}, where b may be any real number. This again simply emphasizes the fact that ψ itself can have no direct physical meaning, but $|\psi^2|$, which is independent of any factor e^{ib}, may be interpreted physically.

When dealing with three-dimensional wave functions $\psi(x, y, z)$, it is convenient to write $\psi(\mathbf{r})$ instead of $\psi(x, y, z)$, where $\mathbf{r}$ is the position vector of the point (x, y, z); then $d\mathbf{r}$ may be written in place of $dx\,dy\,dz$ and integration over all space may be represented by a single integral as $\int \psi(\mathbf{r})\psi^*(\mathbf{r})\,d\mathbf{r}$, where by analogy to the one-dimensional case $|\psi(\mathbf{r})|^2$ gives the probability density such that $|\psi(\mathbf{r})|^2\,d\mathbf{r}$ is the probability that the particle described by $\psi(\mathbf{r})$ will be found in the small volume element $d\mathbf{r}$ at $\mathbf{r}$. This interpretation of $|\psi(\mathbf{r})|^2$ leads to results which are quite different from those presented by classical mechanics. For example, for the normalized ground state wave function of (q.2.17.), $\psi_1 = [\sqrt{2/L}] \sin (\pi x/L)$, if $|\psi_1|^2$ is plotted as a function of x there is obtained the curve shown in the lower part of Fig. Q.2.6. This shows that according to wave mechanics there is a much greater probability of finding the particle in the middle of the one-dimensional box, whereas in classical mechanics, the particle would simply travel back and forth between the walls with uniform velocity and the probability would be the same that the particle could be anywhere in the box, or its probability distribution would be a straight line at a height of $1/L$ above the x-axis. For finite potentials at the walls, there is an even greater dissimilarity between the wave-mechanical and the classical mechanical results; in this case the probability distribution curve, which has a similar form to the ψ_1 curve, does not become zero at the walls but also falls off exponentially to zero in the region beyond the walls, which means that there is a finite probability of finding the particle outside the box; in classical mechanics, if $E < U_c$ then K would become negative, which is impossible. This quantum mechanical result is known as the "tunnel effect" and is of importance in the theory of nuclear decay and nuclear reactions.

Q.2.4. The Orthogonality of Eigenfunctions

The Schrödinger equation, like all differential equations, is satisfied by an infinite number of solutions. This applies to even the simplest differential equation, $dy/dx = 1$, the solutions of which are $y = x + a$, where a is an arbitrary integration constant which may adopt any value; a graphical solution of the equation, $dy/dx = 1$ would consist of an infinite number of parallel straight lines with a variety of different intercepts, a, on the y co-ordinate axis. For the time-independent Schrödinger equation, $\hat{H}\psi_n = E_n\psi_n$, there are an infinite number of energy values, E_n, and eigenfunctions, ψ_n, which satisfy this equation. If ψ_1 and ψ_2 are any two such eigenfunctions corresponding to the eigenvalues, E_1 and E_2, respectively, and $\psi_1{}^*$ and $\psi_2{}^*$ are the complex conjugates of ψ_1 and ψ_2, then $\psi_1{}^*$ and $\psi_2{}^*$ each satisfy the same Schrödinger equation as ψ_1 and ψ_2 respectively, provided that the potential energy function $U(x)$ for the one-dimensional case, is real. For example, if $\psi_1 = a_1 + ib_1$, where a_1 and b_1 are real functions of x, is a solution of equation (q.2.4.), then a_1 and b_1 must separately satisfy the same equation, and therefore any linear combination of the type $Aa_1 + Bb_1$, where A and B are arbitrary constants must also satisfy it; in particular the linear combination $\psi_1{}^* = a_1 - ib_1$ must be a solution. Similarly for ψ_2 and $\psi_2{}^*$. Therefore $\psi_1{}^*$ and ψ_2 must satisfy the equations

$$d^2\psi_1{}^*/dx^2 + 8\pi^2mh^{-2}[E_1 - U(x)]\psi_1{}^* = 0$$

and

$$d^2\psi_2/dx^2 + 8\pi^2mh^{-2}[E_2 - U(x)]\psi_2 = 0.$$

If the first of these equations is multiplied by ψ_2 and the second equation multiplied by $\psi_1{}^*$ is subtracted from it, it is found that

$$[d^2\psi_1{}^*/dx^2]\psi_2 - [d^2\psi_2/dx^2]\psi_1{}^* = 8\pi^2mh^{-2}[E_2 - E_1]\psi_1{}^*\psi_2. \quad \text{(q.2.19.)}$$

If both sides of (q.2.19.) are integrated over all values of x, then

$$\int_{-\infty}^{\infty} [(d^2\psi_1{}^*/dx^2)\psi_2 - (d^2\psi_2/dx^2)\psi_1{}^*] \, dx = 8\pi^2 mh^{-2}[E_2 - E_1] \int_{-\infty}^{\infty} \psi_1{}^*\psi_2 \, dx.$$

$$\text{(q.2.20.)}$$

But since

$$(d^2\psi_1{}^*/dx^2)\psi_2 - \psi_1{}^*(d^2\psi_2/dx^2) = d/dx[\psi_2(d\psi_1{}^*/dx - \psi_1{}^*(d\psi_2/dx)],$$

therefore

$$[\psi_2(d\psi_1{}^*/dx) - \psi_1{}^*(d\psi_2/dx)] = 8\pi^2 mh^{-2}(E_2 - E_1) \int_{-\infty}^{\infty} \psi_2\psi_1{}^* \, dx.$$

$$\text{(q.2.21.)}$$

But in order that equation (q.2.19.) should be satisfied, both $\psi_1{}^*$ and ψ_2 must approach zero as x approaches ∞ or $-\infty$, and therefore the left hand side of equation (q.2.21.) vanishes at the limits. Also if E_1 and E_2 are different eigenvalues, $(E_2 - E_1)$ cannot be zero and thus $\int_{-\infty}^{\infty} \psi_1{}^*\psi_2 \, dx$ must be identically zero. Any such pair of eigenfunctions corresponding to two different eigenvalues which satisfy the relationship

$$\int_{-\infty}^{\infty} \psi_1{}^*\psi_2 \, dx = 0 \qquad \text{(q.2.22.)}$$

are said to be orthogonal to one another. In certain circumstances, it may happen that $\psi_1{}^*$ and ψ_2 could both be zero at $x = a$ and at $x = b$, in which event $\int_b^a \psi_1{}^*\psi_2 \, dx = 0$ and the pair of functions would be said to be orthogonal in the interval (b, a). Similarly, if $\int_b^a \psi\psi^* \, dx = 1$, the function ψ is said to be normalized in the interval (b, a). In the case of a set of functions ψ_n, where $n = 1, 2, 3, \ldots$, which are all normalized and orthogonal to one another in pairs in the interval (b, a), such a set is considered to be an orthonormal set in this interval. It has been shown on page 120 that the set of functions, $\psi_n = [\sqrt{2}/L] \sin(n\pi x/L)$, where $n = 1, 2, 3, \ldots$, are all normalized in the interval $(0, L)$. These functions are also orthogonal to one another in pairs in the same interval, since for any pair of different integers m and n,

$$\int_0^L \sin(m\pi x/L) \sin(n\pi x/L) \, dx$$

$$= (1/2) \int_0^L \left[\cos \frac{(m-n)\pi x}{L} - \cos \frac{(m+n)\pi x}{L} \right] dx = 0.\ddagger$$

‡ Making use of the exponential expansions given on page 114, it can be shown in a similar manner that $\sin x = [e^{ix} - e^{-ix}](2i)^{-1}$ and that $\cos x = [e^{ix} + e^{-ix}]/2$ and hence generally,

$$\int_{-\pi}^{\pi} \cos mx \cos nx \, dx = \int_{-\pi}^{\pi} \sin mx \sin nx \, dx = \begin{Bmatrix} 0 \text{ if } m \neq n \\ \pi \text{ if } m = n \neq 0 \end{Bmatrix}$$

and that $\int_{-\pi}^{\pi} \cos mx \sin nx \, dx = 0$, for all integral values of m and n.

Thus the functions ψ_n constitute an orthonormal set in the interval $(0, L)$. In this particular case, of course, the functions are all real, so that $\psi_n{}^* \equiv \psi_n$, and also since all of the functions are zero outside the interval $(0, L)$, they must be orthonormal over the whole range of x from ∞ to $-\infty$.

If ψ_1 and ψ_2 are different wave functions corresponding to the same eigenvalue then $E_1 = E_2$ and the above proof of orthogonality which depends on $E_1 \neq E_2$ is not applicable, and the two functions need not be orthogonal. Nevertheless the functions may be orthogonal. In the event that they are not, it is always possible to select some appropriate linear combination of them which are indeed orthogonal. For example, if ψ_1 is normalized, and this can always be done by the selection of a suitable multiplicative constant, then $\int_{-\infty}^{\infty} \psi_1\psi_1{}^* \, dx = 1$, and if ψ_1 and ψ_2 are non-orthogonal, $\int_{-\infty}^{\infty} \psi_1{}^*\psi_2 \, dx \neq 0 = p$, where p is some constant. If now $\psi_3 = p\psi_1 - \psi_2$, then

$$\int_{-\infty}^{\infty} \psi_1{}^*\psi_3 \, dx = p \int_{-\infty}^{\infty} \psi_1\psi_1{}^* \, dx - \int_{-\infty}^{\infty} \psi_1{}^*\psi_2 \, dx = 0.$$

In this manner, the pair of functions ψ_1 and ψ_3 are formed which are independent, degenerate, wave functions, orthogonal to one another. They are degenerate, since if ψ_1 and ψ_2 satisfy the equations $\hat{H}\psi_1 = E_1\psi_1$ and $\hat{H}\psi_2 = E_1\psi_2$, then on subtracting the second of these equations from the first one multiplied by p, it is found that $\hat{H}(p\psi_1 - \psi_2) = E(p\psi_1 - \psi_2)$, indicating that the linear combination function $(p\psi_1 - \psi_2)$ has the same eigenvalue as ψ_1 and ψ_2. The number of independent wave functions, not including those which are merely linear combinations of the others, which correspond to the same eigenvalue, is called the degree of degeneracy of the particular energy value concerned. The above process of forming orthogonal wave functions from a degenerate pair, may be extended to any number of degenerate wave functions, so that it is always possible to form a number of independent, orthogonal, functions, equal in number to the degree of degeneracy of a set of initially non-orthogonal functions. This statement is equally true and may be proved in exactly the same way for three-dimensional wave functions, that is for the wave functions describing the motion of a particle moving in three-dimensional space.

The proof of the orthogonality of two wave functions corresponding to two different, non-degenerate, states is, however, a little more complicated for a function of three space variables, $\psi(x, y, z)$. Applying the same argument as in the one-dimensional case, there is obtained instead of equation (q.2.19.),

$$\int [\psi_2 \nabla^2 \psi_1{}^* - \psi_1{}^* \nabla^2 \psi_2] \, d\mathbf{r} = 8\pi^2 m h^{-2}(E_2 - E_1) \int \psi_1{}^*\psi_2 \, d\mathbf{r}. \quad \text{(q.2.23.)}$$

Now $\int \psi_1{}^*\psi_2 \, d\mathbf{r}$ will only equal zero, the orthogonality condition, if the triple integral on the left hand side of (q.2.23.) is zero. This triple, volume, integral may be converted to a surface integral by the use of Green's theorem (see page 243), to give

$$\int [\psi_2 \nabla^2 \psi_1{}^* - \psi_1{}^* \nabla^2 \psi_2]\, d\mathbf{r} = \int [\psi_2(\partial\psi_1{}^*/\partial n) - \psi_1{}^*(\partial\psi_2/\partial n)]\, dS, \quad \text{(q.2.24.)}$$

where the volume concerned is all of space and so the surface involved is infinitely large, and $\partial/\partial n$ indicates differentiation in the direction of the normal pointing outward to the surface from the origin. For a normalized wave function, which must vanish at infinity, the integrand of the surface integral must be zero, and so the integral is zero, and the left hand side of equation (q.2.23.) must be zero. Thus since $E_2 \neq E_1$, $\int \psi_1{}^* \psi_2\, dr = 0$, as for the one-dimensional case.

For a set of normalized wave functions, ψ_1, ψ_2, ψ_3, ..., all of which are mutually orthogonal to one another in pairs, that is, an orthonormal set, this condition is frequently expressed as,

$$\int_a^b \psi_m{}^*(x)\psi_n(x)\, dx = \delta_{mn}, \quad \text{(q.2.25.)}$$

where ψ_m and ψ_n are functions of the real variable, x, defined in the interval $a \leq x \leq b$, and δ_{mn} is the so-called Kronecker delta (sometimes referred to as the Weierstrass delta). A more compact notation, originally introduced by Dirac,‡ is also commonly used to express this condition, as

$$\int_a^b \psi_m{}^*(x)\psi_n(x)\, dx = \langle \psi_m \mid \psi_n \rangle \quad \text{or} \quad \langle m \mid n \rangle \quad \text{or} \quad (\psi_n, \psi_m).$$
$$\text{(q.2.26.)}$$

‡ Expressions such as (q.2.25.) and (q.2.51.) and many others occurring in quantum mechanics are very conveniently written in the notation introduced by Dirac,[1] which avoids the necessity of writing so many subscripts and superscripts. Generally, in an integral of the type, $\int \phi_n{}^* \hat{A}\phi_m\, d\tau$, where $\hat{A}$ is any operator, real or complex, ϕ_m and ϕ_n are any functions of any number of variables ($\phi_n{}^*$ being the complex conjugate of ϕ_n), again real or complex, and $d\tau$ represents integration over all values of all of the variables involved, the corresponding Dirac notation is written as $\langle n|\hat{A}|m\rangle$, where $\langle n|$ stands for $\phi_n{}^*$ and is called a *bra*; $|m\rangle$ is called a "ket" and stands for ϕ_m. The other variations of this notation are as defined in (q.2.26.). The terms *bra* and *ket* are derived from the word *bracket*. The other notation, in common use to describe such integrals since its introduction by Heisenberg,[2] who developed matrix methods before the advent of wave mechanics, is to refer to $\int \phi_n{}^* \hat{A}\phi_m\, d\tau$ as F_{nm}, otherwise called the matrix component or matrix element of the operator $\hat{A}$. In this notation expressions such as $\int \phi_n{}^* \phi_m\, d\tau$ may be formally regarded as the matrix components of unity.

[1] Dirac, P. A. M., *Quantum Mechanics*, 4th edition, Oxford University Press, New York, 1958.
[2] Heisenberg, W., *Z. Physik*, 1925, **33**, 879.

In the more complete formalism of quantum mechanics,[1] taking into account the fact that if $\psi_1(r_1)$, $\psi_2(r_2)$, $\psi_3(r_3)$, ... are solutions of the Schrödinger equation of a system, $\hat{H}\psi_n = E_n\psi_n$, then so is any linear combination of these functions and also so is the convergent series $\sum_{n=1}^{n} a_n\psi_n$ (provided that the operator $\hat{H}$ can be applied to each member of the series in turn). It is assumed that the wave functions or state functions of a system are related in a more general way to mathematical constructs of a type which when combined lead to new mathematical constructs of the same general type. Such a type of behavior is characteristic of vectors; also most generally such vectors must be defined in a complex infinite dimensional space, referred to as abstract Hilbert space. The wave functions describing certain stationary states, in the Dirac terminology, are represented by ket vectors. The mathematical formalism devised by Hilbert[2] defining a set of vectors having certain properties, requires a second set of dual vectors to exist in the abstract Hilbert space; these are the bra vectors of Dirac. In this terminology, the state of any system may be said to be represented by either a bra vector, $\langle\,|$ or a ket vector, $|\,\rangle$, such that the product of the two is a scalar quantity. The concept of dual sets of vectors is related to the fact that every pair of real numbers, x and y, are associated with two complex numbers, $x + iy$ and $x - iy$, so that the two complex numbers have the same projection on the real axis of the Argand plane (see page 113) and so leading to the concept of dual space. The scalar product of two vectors is defined as the result of the multiplication of the magnitudes of the two vectors by the cosine of the angle between them; the scalar product of a bra vector, $\langle X|$ and a ket vector $|Y\rangle$, is written as $\langle X\,|\,Y\rangle$, and is said to be a linear function of the ket vector and an antilinear function of the bra vector. The normalization of state functions to unity, when the states are described by such bra and ket vectors, means that the corresponding bras and kets have unit moduli.

The Kronecker delta symbol is defined to be equal to unity when $m = n$ and equal to zero when $m \neq n$. The orthonormal set of functions, ψ_n, may consist of either a finite or an infinite number of functions. Thus the set of functions, $\psi_n = \sqrt{2/L} \sin(n\pi x/L)$, composes an infinite orthonormal set, where $n = 1, 2, 3, \ldots$, and if $\langle\psi_m\,|\,\psi_n\rangle = 0$, the functions ψ_m and ψ_n are orthogonal to one another and if $\langle m\,|\,m\rangle = 1$, then ψ_m is normalized. The type of expression given by (q.2.25.) will normally be used in this book unless it is especially convenient to use the Dirac bra-ket notation of (q.2.26.).

A set of functions defined by (q.2.25.) is said to constitute a complete set, for all integral values of m and n, if there is no function, which is not identically zero, which is orthogonal to all the functions ψ_n; otherwise stated, no function $\phi(x)$ exists such that

$$\int_a^b \psi_m{}^*(x)\phi(x)\,dx = 0,$$

for all values of m, except $\phi(x) \equiv 0$. If this is so, then it is possible to expand any arbitrary function $f(x)$, in the interval (a, b), as an infinite series

[1] Margenau, H., and Murphy, G. M., *The Mathematics of Physics and Chemistry*, **Vol. II**, D. Van Nostrand Co. Inc., Princeton, New Jersey, 1964.
[2] Bell, E. T., *The Development of Mathematics*, McGraw-Hill Book Company, Inc., New York, 1945.

of the type

$$f(x) = a_1\psi_1(x) + a_2\psi_2(x) + a_3\psi_3(x) + \cdots = \sum_n a_n\psi_n(x). \quad \text{(q.2.27.)}$$

The constant coefficients, a_n, which may be complex or real, may be determined by multiplying both sides of (q.2.27.) by $\psi_m{}^*(x)$ and integrating over (a, b) to give,

$$\int_a^b \psi_m{}^*(x)f(x)\,dx = \sum_n a_n \int_a^b \psi_m{}^*(x)\psi_n(x)\,dx.$$

But according to (q.2.25.), all the terms on the right hand side of this equation vanish except that for which $m = n$, and for this term the integral is unity. Therefore

$$a_m = \int_a^b \psi_m{}^*(x)f(x)\,dx, \qquad\qquad \text{(q.2.28.)}$$

uniquely determining each of the coefficients. Thus

$$f(x) = \sum_{n=1}^{\infty} \left[\int_a^b \psi_m{}^*(x)f(x)\,dx\right]\psi_n(x)$$

and this expansion is always valid if the functions $\psi_n(x)$ form a complete set. If this were not so then some other function $\chi(x)$ could exist, not identically zero in the interval $a \le x \le b$, such that

$$\chi(x) = f(x) - \sum_{n=1}^{\infty} \left[\int_a^b \psi_m{}^*(x)f(x)\,dx\right]\psi_n(x).$$

If this equation is multiplied by $\psi_m{}^*(x)$ and integrated throughout between the limits (a, b) it is found that

$$\int_a^b \psi_m{}^*(x)\chi(x)\,dx = \int_a^b \psi_m{}^*(x)f(x)\,dx - \int_a^b \psi_m{}^*(x)f(x)\,dx = 0,$$

for all values of m. But this would be incompatible with the definition of a complete set of functions and therefore $\chi(x)$ should be identically zero for $a \le x \le b$.

In the event that the functions ψ_n are orthogonal but not normalized, the right hand side of (q.2.28.) must be divided by $\int_a^b |\psi_m(x)|^2\,dx$. An especially important complete orthonormal set of functions is the set, $[\sqrt{2\pi}]^{-1}$, $[\sqrt{\pi}]^{-1} \cos nx$, $[\sqrt{\pi}]^{-1} \sin nx$, where $n = 1, 2, 3, \ldots$; this set of functions is complete in the interval $(-\pi, \pi)$. The expansion of any arbitrary function in terms of this particular set gives

$$f(x) = a_0/\sqrt{2\pi} + [1/\sqrt{\pi}] \sum_{n=1}^{\infty} a_n \cos nx + [1/\sqrt{\pi}] \sum_{n=1}^{\infty} b_n \sin nx, \quad \text{(q.2.29.)}$$

where

$$a_0 = [\sqrt{2\pi}]^{-1} \int_{-\pi}^{\pi} f(x)\, dx, \quad a_n = [\sqrt{\pi}]^{-1} \int_{-\pi}^{\pi} f(x) \cos nx\, dx,$$

and

$$b_n = [\sqrt{\pi}]^{-1} \int_{-\pi}^{\pi} f(x) \sin nx\, dx,$$

as given by (q.2.28.). In fact, this expansion results in the Fourier Series (see page 38) representing $f(x)$ in the interval $(\pi, -\pi)$. This is an example where all of the functions composing the set do not necessarily appear in the expansion, since some of the coefficients are zero. If, for example $f(x)$ is an odd function,‡ the coefficients a_0 and all the a_n coefficients turn out to be zero and

$$f(x)_{\text{odd}} = \sum_{n=1}^{\infty} (2/n)(-1)^{n+1} \sin nx.$$

‡ An odd function, such as that of a single variable x, changes its magnitude when the sign of the variable is changed; thus x, x^3, $\sin x$, etc. Generally if $f(x)$ is an odd function of x, $f(x) = -f(-x)$. An even function of x is one which retains its value unchanged when the sign of the variable, x, is changed; for example, x, $\cos x$. If $f(x)$ is an even function of x, $f(x) = f(-x)$.

The importance of these properties of orthogonal sets of functions is that the set of eigenfunctions of a system corresponding to specific eigenvalues constitutes a complete orthogonal set, which functions may obviously be normalized. In order to extend the applicability of these properties of orthonormal sets of functions to functions of any number of variables, it is only necessary to carry out the integration implied by (q.2.28.) over the whole of configuration space of the system concerned. The various complete sets of functions of importance in quantum mechanics, besides being orthonormal, usually satisfy certain boundary conditions. These are certain linear relationships between the functions $\psi_n(x)$ and their first derivatives $\psi_n'(x)$ at the points $x = a$ and $x = b$. For example, making use of the relationship between the exponential and circular functions given on page 123, a Fourier Series may be written as an expansion of a function $f(x)$ in terms of the orthonormal set of functions, $\phi_n(x) = [\sqrt{2\pi}]^{-1}e^{inx}$, where $n = 0, \pm 1, \pm 2, \ldots$, so that $f(x) = \sum_{n=-\infty}^{\infty} a_n e^{inx}$, with $a_m = [2\pi]^{-1} \int_{-\pi}^{\pi} e^{-imt}f(t)\, dt$. Here $\phi_n(0) = \phi_n(2\pi) = [\sqrt{2\pi}]^{-1}$ and $\phi_n'(0) = \phi_n'(2\pi)$, for all integral values of n, and the functions $\phi_n(x)$ are defined in the interval $0 \le x \le 2\pi$, since the integral

$$\int_{0}^{2\pi} \phi_n{}^*(x)\phi_m(x)\, dx = [2\pi]^{-1} \int_{0}^{2\pi} e^{i(m-n)x}\, dx.$$

From the Fourier expansion for $f(x)$ above and the orthonormality of $\phi_n(x)$, it may be deduced that

$$\int_0^{2\pi} f(x)^* f(x)\, dx = \sum_{n=1}^{\infty} \sum_{m=1}^{\infty} a_n{}^* a_m \int_0^{2\pi} \phi_n(x)^* \phi_m(x)\, dx = \sum_{n=1}^{\infty} \sum_{m=1}^{\infty} a_n{}^* a_m\, \delta_{nm}$$

$$= \sum_{n=1}^{\infty} a_n a_n{}^*.$$

Each of the terms $a_n a_n{}^*$ (see page 112) must be a positive number, and therefore if the infinite sum given as $\sum_{n=1}^{\infty} a_n a_n{}^*$ is curtailed after some finite number of terms, say N, then,

$$\sum_{n=1}^{\infty} a_n a_n{}^* \leq \int_0^{2\pi} f(x)^* f(x)\, dx, \qquad \text{(q.2.30.)}$$

which is a form of what is known as Bessel's inequality. The useful relationship may be deduced that,

$$\lim_{N \to \infty} \sum_{n=1}^{\infty} a_n a_n{}^* = \int_{-\pi}^{\pi} f(x)^* f(x)\, dx, \qquad \text{(q.2.31.)}$$

which may be used as a criterion for the convergence of expansions in terms of complete sets of functions; if some particular expansion is such that the sum on the left hand side of (q.2.31.) converges, it is said that the expansion $\sum a_n \phi_n(x)$ converges to $f(x)$ in the mean.

The discussion of the properties of waves and of eigenfunction expansions is best managed in terms of Fourier analysis.[1] The Fourier Series

$$f(x) = \sum_{n=-\infty}^{\infty} a_n e^{inx} \qquad \text{with} \qquad a_m = [2\pi]^{-1} \int_{-\pi}^{\pi} e^{-imt} f(t)\, dt \qquad \text{(q.2.32.)}$$

written as an expansion of the function $f(x)$ in terms of the orthonormal set of functions

$$\phi_n(x) = [\sqrt{2\pi}]^{-1} e^{inx} \qquad \text{with} \qquad n = 0, \pm 1, \pm 2, \ldots, \qquad \text{(q.2.33.)}$$

is convergent if $f(x)$ lies within and is continuous in the interval $-\pi \leq x \leq \pi$. The Fourier expansion is also valid outside the interval $-\pi \leq x \leq \pi$, if $f(x)$ is a periodic function satisfying the condition

$$f(x + 2\pi) = f(x). \qquad \text{(q.2.34.)}$$

These ideas may be extended to functions of more than one variable, which may be similarly expanded as multiple Fourier series; for example $F(x, y)$

---

[1] Hameka, H. F., *Advanced Quantum Chemistry*, Addison-Wesley Publishing Co., Reading, Mass., 1965.

can be written as

$$F(x, y) = \sum_{m=-\infty}^{\infty} \sum_{n=-\infty}^{\infty} a_{m,n} e^{i(mx + ny)}$$

where

$$a_{m,n} = [4\pi^2]^{-1} \int_{-\pi}^{\pi} \int_{-\pi}^{\pi} F(x, y) e^{-i(mx + ny)} \, dx \, dy.$$

An especially important use of a triple Fourier series is for the representation of the electron density in crystalline materials, where the function involved is triply periodic, in each of the crystallographic unit cell directions (see section M.5.6.).

The Fourier integral theorem, the essential nature of which may be deduced from the Fourier series expansion, is of wide use in quantum mechanics and in many kinds of crystallographic problems.[1] If $f(y)$ is a function, which need not even be so restricted as to be continuous and bounded in the interval $-\pi \leq x \leq \pi$, but at least satisfies Dirichlet's conditions that it has only a finite number of finite discontinuities and a finite number of maxima or minima in the interval $-\pi \leq x \leq \pi$, and can be expanded as

$$f(y) = \sum_{n=-\infty}^{\infty} a_n e^{iny}$$

where

$$a_n = [2\pi]^{-1} \int_{-\pi}^{\pi} f(s) e^{-ins} \, ds, \tag{q.2.35.}$$

then (q.2.34.) can be modified by changing the variables from y and s to x and t by means of the substitutions $y = \pi x/l$ and $s = \pi t/l$, to give

$$f(x) = \sum_{n=-\infty}^{\infty} a_n e^{i\pi nx/l},$$

where now

$$a_n = [2l]^{-1} \int_{-l}^{l} \{e^{-i\pi nt/l}\} f(t) \, dt. \tag{q.2.36.}$$

The change of variable has the effect of making x valid in the range $-l \leq x \leq l$, and this range of validity may be extended to cover all values of x and t if $l \to \infty$. If a_n from (q.2.36.) is substituted into the expression given there for $f(x)$, $f(x)$ becomes

$$f(x) = [2l]^{-1} \sum_{n=-\infty}^{\infty} \int_{-l}^{l} f(t) e^{in\pi(x-t)/l} \, dt, \tag{q.2.37.}$$

---

[1] Titchmarsh, E. C., *Introduction to the Theory of Fourier Integrals*, Oxford, New York, 1937.

and by defining the small quantity $\delta = \pi/l$, which approaches zero as $l \to \infty$, this becomes

$$f(x) = [2\pi]^{-1} \sum_{n=-\infty}^{\infty} \delta \int_{-l}^{l} f(t)e^{in\delta(x-t)}\, dt. \qquad (q.2.38.)$$

Now when $l \to \infty$, $\delta \to 0$, and the summation in (q.2.38.) may be replaced by an integral over some variable, say u, such that

$$f(x) = [2\pi]^{-1} \int_{-\infty}^{\infty} du \int_{-\infty}^{\infty} f(t)e^{iu(t-x)}\, dt, \qquad (q.2.39.)$$

which is the Fourier integral theorem, and may be alternately expressed as‡

$$f(x) = [\sqrt{2\pi}]^{-1} \int_{-\infty}^{\infty} F(u)e^{iux}\, du$$

where

$$F(u) = [\sqrt{2\pi}]^{-1} \int_{-\infty}^{\infty} f(t)e^{-iut}\, dt. \qquad (q.2.40.)$$

Written in this latter manner, the function $F(u)$ is referred to as the Fourier transform of $f(t)$.

‡ Another way of expressing the Fourier Integral Theorem is to make use of the Dirac δ function. If this is defined as $f(x) = \int f(x)\delta(x - x)\, dx$ (see page 538) and comparison is made with (q.2.39.), then the Fourier Integral Theorem may be written as.

$$\delta(x - t) = (2\pi)^{-1} \int_{-\infty}^{\infty} e^{iu(x-t)}\, du.$$

The particular value of the Fourier transform is that when the transform is known the original function can be determined; if the Fourier transform of a function is taken twice, the original function is reproduced. The Fourier integral theorem is also valid for functions of more than one variable. For example, if $F(h, k, l)$ is the Fourier transform of the function $f(a, b, c)$, such that

$$F(h, k, l) = [\sqrt{2\pi}]^{-3} \int_{-\infty}^{\infty} \int_{-\infty}^{\infty} \int_{-\infty}^{\infty} f(a, b, c)e^{-i(ha+kb+lc)}\, da\, db\, dc,$$

then the function $f(x, y, z)$ is given by

$$f(x, y, z) = [\sqrt{2\pi}]^{-3} \int_{-\infty}^{\infty} \int_{-\infty}^{\infty} \int_{-\infty}^{\infty} F(h, k, l)e^{i(hx+ky+lz)}\, dh\, dk\, dl. \qquad (q.2.41.)$$

These are the expressions which find major use in the calculation of crystal structure factors in X-ray crystallography. The latter two equations may be combined as in (q.2.39.), to give

$$f(x, y, z) = [2\pi]^{-3} \int_{-\infty}^{\infty} \int_{-\infty}^{\infty} \int_{-\infty}^{\infty} \int_{-\infty}^{\infty} \int_{-\infty}^{\infty} \int_{-\infty}^{\infty} f(a, b, c)$$
$$\times \, e^{-(i[h(a-x)+k(b-y)+l(c-z)]} \, da \; db \; dc \; dh \; dk \; dl.$$

Q.2.5.   The Heisenberg Uncertainty Principle and the Uncertainty Relationships

In quantum mechanics there is no such concept as the path of motion of a particle. Even if the wave or state function ψ of a particle is known, it is not possible to predict accurately both the position and linear momentum of the particle simultaneously. This statement constitutes one of the fundamental principles of quantum mechanics, called the Uncertainty Principle, which was first formulated by Heisenberg in 1927. Actually the essentially complete mathematical formalism of quantum mechanics had been outlined by Heisenberg and Schrödinger in 1925–26, before the discovery of the Uncertainty Principle, which provided the physical interpretation of the matrix and wave formalism of Heisenberg and Schrödinger, respectively.

In principle, such a rejection of the ordinary ideas of classical mechanics, as is implied by a negative statement of this type, perhaps obscures the fact that it is impossible to state the basic concepts of quantum mechanics without using classical mechanics. Any attempt to describe the motion of a particle such as an electron requires the presence of some physical object or objects which conform with the classical mechanical description of their behavior with sufficient accuracy, in order to interact with and detect the electron. In fact, generally, all physical observations and measurements, are based on the interactions observed to take place between some form of electromagnetic radiation and matter. Also it is generally meaningless to speak of some physical quantity unless a method can be devised whereby this quantity can be measured. In all ordinary measurements it is usually assumed that no alteration in the quantity being measured is produced by the act of measurement. This, of course, is not always so, even in the macroscopic sense where quantum effects are inconsequential. For example, the attempt to detect the nature and number of the ionic species which are assumed, in order to account for some other type of observed behavior of the system, to coexist in aqueous solutions of many inorganic compounds (borates, phosphates, molybdates, etc.), by pH, cell potential, or some other type of electrometric method perhaps cannot work since the act of inserting an electrode into such solutions immediately disturbs the equilibrium between the ionic species which it is intended to observe. For electrons and other particles whose behavior cannot be adequately described in terms of classical mechanics, the failure of attempts to determine simultaneously and precisely both their position and momentum or velocity or kinetic energy or any other property which is dependent on velocity, is however, independent of the sophistication of the tools, equipment, or instrumentation used in the process. The nature of this problem may be illustrated by the hypothetical attempt to measure simultaneously the position and velocity of an electron.

The velocity of an electron could be measured if it were possible to observe it at two different times during the course of its motion and to determine the time interval concerned. This would require some optical tool such as a microscope. The resolving power of an optical microscope is given by

$$\Delta x = \lambda/2\mu \sin \alpha, \qquad \text{(q.2.42.)}$$

where λ is the wavelength of the radiation used, μ is the refractive index of the medium between the objective and the object being observed, and $\sin \alpha$ is the numerical aperture of the objective used. Thus the greatest precision in observing a small object would result by using an instrument which had the largest numerical aperture, the largest refractive index for the medium surrounding the objective and object, making use of radiation with the shortest possible wavelength. For visible light with $\lambda \simeq 5.3 \times 10^{-7}$ m, using the maximum numerical aperture possible, which is about 1.6, in air, $\Delta x \simeq 1.7 \times 10^{-7}$ m, and since the least angle that can be subtended by the eye at the distance of closest vision (about 25 cm) is of the order of 1.5', the maximum virtual image that can be observed is about $25 \times \tan 1.5' \simeq 1.1 \times 10^{-4}$ m. So the maximum magnification possible with visible light in air is of the order of $(1.1 \times 10^{-4})/(1.7 \times 10^{-7}) \simeq 650$. No greater detail could be observed beyond this magnification; optical microscopes using oil immersion objectives and short wavelength light provide a maximum magnification of about 750–800. Even if it were possible to construct an instrument using ultrashort gamma radiation ($\lambda \simeq 10^{-14}$ m), the other factors being unchanged and to see an electron, the limitation imposed by (q.2.42.) remains.

There is a further difficulty in attempting to precisely locate a particle such as an electron, which is independent of the laws of optics. In order that the interaction between the electron and the radiation used should be detectable, the interacting photon has to be scattered into the aperture of the lens, that is within the cone of angle 2α in Fig. Q.2.7. But in terms of the Compton effect (page 94), when short wavelength radiation is scattered by electrons, it suffers a change in wavelength. If λ for the incident radiation is substituted in the de Broglie relationship, then the momentum of the incident photon is given by

$$h/\lambda = h\nu/c.$$

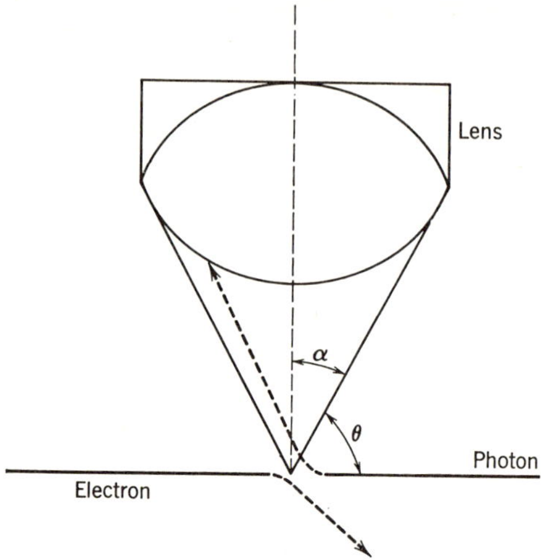

FIG. Q.2.7. Compton scattering of a photon into the aperture of an objective lens after collision with an electron.

If the electron under observation scatters the incident photon through an angle between $\pi/2 \pm \theta$, the effect of the interaction will be detectable, and for this angle of scattering the interacting photon must have momentum between $h\nu/c[1 - \cos(\pi/2 - \alpha)]$ and $h\nu/c[1 + \cos(\pi/2 + \alpha)]$, or between $h\nu/c[1 \pm \sin\alpha]$. If this range of possible values for the scattered photon is denoted by Δp, then

$$\Delta p = [2h\nu/c]\sin\alpha. \qquad (q.2.43.)$$

This uncertainty about the precise momentum of the scattered photon could be reduced by using an objective of smaller aperture, but doing so would increase the uncertainty about the position of the electron, in terms of (q.2.42.). Therefore, in such a hypothetical attempt to measure the position and momentum of the electron simultaneously,

$$\Delta p \times \Delta x = [2h\nu/c]\sin\alpha \times [\lambda/2\mu\sin\alpha] = h/\mu \simeq h, \qquad (q.2.44.)$$

and this is one expression of the Uncertainty Principle. For electronic motion in three-dimensional space, a similar relationship to (q.2.44.) applies for each of the other two Cartesian coordinate directions. For example, if the position of an electron could be determined to within 10^{-14} m, then $\Delta p \simeq 6.6 \times 10^{-34}/10^{-14}$ J s m^{-1}, or 6.6×10^{-20} J s m^{-1}. This uncertainty in momentum is of course negligibly small for macroscopic systems but for electrons with masses of the order of magnitude of 10^{-31} kg, this is certainly not so. It is unavoidable that in the course of attempting to observe a particle such as an electron, its velocity or momentum or energy is altered by an amount which increases with the frequency of the radiation used in the process; in fact the conditions which lead to the most accurate determination of the position lead to the greatest uncertainty in the velocity, and vice versa. It might appear that this difficulty could be overcome by making use of the wave properties of electrons and measuring their wavelength by a crystal diffraction grating and thus obtaining the momentum from $\lambda = h/p$. But electron waves undergo diffraction resulting in a change in their direction and therefore their position is no longer defined. The introduction of probabilities into the description of electronic behavior is a direct result of the operation of the Uncertainty Principle, which refers to the product of two uncertainties and in view of the approximate nature of the above argument, which in any event deals with a purely hypothetical situation, can only refer to the minimum value of this product. The argument that the methods of measurement are at fault and that if improved techniques could be developed it would then be possible to measure real positions, times, and momenta, only leads to philosophical difficulties about the meaning of reality which cannot be resolved since reference is then made to something which cannot be measured. The Uncertainty Principle has to be accepted as a characteristic of natural events involving electronic and nuclear phenomena, where quantities of the order of magnitude of Planck's constant are concerned. As already discussed on page 120, there are however, difficulties inherent in the probability description of free particles; for example if the Uncertainty Principle is applied to the case of an electron confined in a box or any finite space, the conclusion is arrived at that the electron cannot be at rest, for then its velocity would be zero and the uncertainty in its position would be infinite so that it could not be contained in any system of finite dimensions.

There is a similar expression to (q.2.44.) relating the quantities energy and time. For a free particle of mass m, the energy is given by, $E = p^2/2m$ and the uncertainty in the measurement of its energy by $\delta E = (p/m)\,\delta p$, where δp is the indeterminancy in the measurement of p. The velocity of the particle, p/m, is determined by the time, Δt, required for the particle to travel the distance Δx, or $p/m = \Delta x/\Delta t$. But Δx must be the maximum uncertainty in the position of the particle while p is being measured, and so $\delta p \simeq h/\Delta x$. Therefore the minimum uncertainty in the measurement of E must be

$$\delta E \simeq (\Delta x/\Delta t)(h/\Delta x) = h/\Delta t \qquad \text{or} \qquad \delta E \cdot \Delta t \simeq h, \qquad \text{(q.2.45.)}$$

which is another of the Heisenberg Uncertainty Relations of quite general validity. A similar conclusion is arrived at from consideration of the measurement of photon energies. Here, $E = h\nu$, and the indeterminacy in the energy measurement is $\delta E = h\,\delta\nu$, where $\delta\nu$ is the uncertainty in the frequency measurement. Frequency is given by, $\nu = n/\Delta t$, the number of light waves per time interval, Δt. Since n must be integral, the uncertainty in the measurement of ν must be $1/\Delta t$, and therefore $\delta E \simeq h/\Delta t$. This may be expressed alternatively by saying that if τ is the mean lifetime of any particular energy state of a system, then since measurement of the energy of the state must be accomplished within this time, the minimum uncertainty about the energy of that state is, $\delta E \simeq h/\tau$.

One of the logically satisfying features of quantum mechanics is the manner in which it automatically takes account of the uncertainty relationships and the limitations imposed by them in terms of the three types of concepts on which the discipline can be erected. The concept of state functions and some of their more important properties has been discussed above. By combining this concept with the other important quantum mechanical concepts of operators and dynamical variables, it is possible to deduce the uncertainty relationships in a straightforward manner. Before doing so, it is therefore necessary to examine these ideas in some more detail.

Q.2.6. Operators and Their Eigenfunctions

Any expression of the type $d/dx[f(x)]$ may be regarded as being composed of two parts, the operator d/dx and the operand, $f(x)$. Equally, $x f(x)$ may be formally regarded in the same manner, where in this case the operation involved is simply one of multiplication. Operators may be combined into sums and products, the combined operator exhibiting quite distinct properties from its components. For example, $d/dx[e^{-\alpha x}] = -\alpha e^{-\alpha x}$ and operating with x^2 on $e^{-\alpha x}$ gives $x^2 e^{-\alpha x}$, while the sum of these operators operating on the same function, $[d/dx + x^2]e^{-\alpha x} = e^{-\alpha x}[x^2 - \alpha]$; operator sums obey the ordinary algebraic commutative law, the result of operating with the combined operator being independent of the order in which the operators are applied. The product of two operators is also a new operator, but here the order of the operations is non-commutative. Thus,

$$x^2 \, d/dx[e^{-\alpha x}] = -\alpha x^2 e^{-\alpha x},$$

but operating with $(d/dx)x^2$ on $e^{-\alpha x}$ gives $xe^{-\alpha x}[2 - \alpha x]$. Generally, the operator immediately preceding the operand is applied first, then the next operator written to the left of the first one is applied to the resulting function, such that if $\hat{N}\cdots\hat{D}\hat{C}\hat{B}\hat{A}$ is a product of the operators $\hat{N}, \ldots \hat{D}, \hat{C}, \hat{B}$, and $\hat{A}$ operating on some arbitrary function Φ,

$$[\hat{N}\cdots\hat{D}\hat{C}\hat{B}\hat{A}]\Phi = [\hat{N}\cdots\hat{D}\hat{C}\hat{B}]\hat{A}\Phi = [\hat{N}\cdots\hat{D}\hat{C}\hat{B}]\Phi_1 = [\hat{N}\cdots\hat{D}\hat{C}]\hat{B}\Phi_1$$

$$= [\hat{N}\cdots\hat{D}\hat{C}]\Phi_2 = \cdots.$$

Differential operators do not usually obey the commutative laws of multiplication. Any two operators, $\hat{A}$ and $\hat{B}$ are formally defined to be equal if for all operands, Φ, $\hat{A}\Phi = \hat{B}\Phi$; the operators $x^2(d/dx)$ and $(d/dx)x^2$ are not equal. The difference operator, $\hat{A}\hat{B} - \hat{B}\hat{A} = [\hat{A}, \hat{B}]$, is referred to as the commutator of the operators $\hat{A}$ and $\hat{B}$. By ordinary differentiation,

$$\partial/\partial x(x\psi) = \psi + x(\partial\psi/\partial x)$$

and so $[h/2\pi i] \, \partial/\partial x(x\psi) - x(h/2\pi i)(\partial\psi/\partial x) = h\psi/2\pi i$, and substituting $\hat{p}_x$ for $(h/2\pi i) \, \partial/\partial x$ (page 108), this may be written alternatively as $\hat{p}_x x\psi - x\hat{p}_x\psi = h\psi/2\pi i$, where $\hat{p}_x$ is the operator representing the momentum of a particle moving in one dimension. This is called an operator equation; thus

$$\hat{p}_x x - x\hat{p}_x = h/2\pi i, \tag{q.2.46.}$$

from which it follows that the commutator of $\hat{p}_x$ and x,

$$[\hat{p}_x, x] = h/2\pi i. \tag{q.2.47.}$$

For the function $y\psi$, $\partial/\partial x(y\psi) = y(\partial\psi/\partial x)$, as x and y are independent variables, and so

$$(h/2\pi i) \, \partial/\partial x(y\psi) - y(h/2\pi i)(\partial\psi/\partial x) = 0 = \hat{p}_x y - y\hat{p}_x = [\hat{p}_x, y]. \tag{q.2.48.}$$

Two variables, such as these, whose commutator is zero, are said to commute. Also, in this connection, quantities which are related to one another in the manner described by Hamilton's canonical equations of motion (page 104) are said to be canonically conjugate to one another; thus p_x and x are canonically conjugate, and do not commute. In fact, in quantum mechanics, the commutation relationship (q.2.47.) is regarded as the definition of a pair of canonically conjugate coordinates and momenta, instead of the Hamilton equations. This equation is indeed the fundamental equation of matrix mechanics, which was developed by Heisenberg, Born, and Jordan before the advent of wave mechanics. In matrix mechanics, however, both p_x and x would be represented by matrices rather than operators. The fundamental difference between classical and quantum mechanics is also illustrated by this same operator equation; in classical mechanics, the right hand side of (q.2.47.) would be zero.

An alternative definition of eigenfunctions is also based on (q.2.47.). Generally, if an operator, $\hat{A}$ and an operand $\phi(x)$, which must obey certain supplementary conditions, are such that operating by $\hat{A}$ regenerates the function $\phi(x)$ multiplied by some constant, α, so that $\hat{A}[\phi(x)] = \alpha\phi(x)$, then $\phi(x)$ is an eigenfunction of the operator $\hat{A}$ belonging to the eigenvalue, α. The supplementary restrictions which must be imposed on $\phi(x)$ are that it must be finite, single-valued, and continuous throughout the whole range of possible values of x from $-\infty$ to ∞; only those solutions, $\phi(x)$, which conform to these boundary conditions are acceptable eigenfunctions of the operator, $\hat{A}$. It has already been shown (page 108) that substitution of the operator, $[(h/2\pi i)\ \partial/\partial x]$ for p_x into the classical Hamiltonian (total energy) expression leads to the time independent Schrödinger equation, where the classical Hamiltonian is a function of the coordinate and the momentum (in one dimension). This may now be stated in an alternative manner; substitution of p_x by $[(h/2\pi i)\ \partial/\partial x]$ in $H(x, p_x)$ gives $\hat{H}[x, (h/2\pi i)\ \partial/\partial x]\psi(x) = E\psi(x)$, where H is the classical Hamiltonian and $\hat{H}$ is the corresponding quantum Hamiltonian. In this one-dimensional Schrödinger equation, the energy E is constant for any eigenfunction $\psi(x)$; for any such stationary state, the total energy E is said to be a constant of the motion of the system. More generally, any dynamical variable of a system, which may be written as a function $F(q, p)$, where q denotes any generalized coordinate, conjugate to to the momentum, may be transformed into a quantum mechanical operator in the same manner as the Hamiltonian.

If an eigenfunction of the quantum Hamiltonian, $\psi(x)$ in the equation, $\hat{H}[x, (h/2\pi i)\ \partial/\partial x]\psi(x) = E\psi(x)$, also is a solution of an equation such as $\hat{F}[x, (h/2\pi i)\ \partial/\partial x]\psi(x) = \mu\psi(x)$, then $\hat{F}$ and $\hat{H}$ have simultaneous eigenfunctions, and it is said that the dynamical variable represented by $\hat{F}$ is also a constant of the motion of the system and it has the value μ in the state $\psi(x)$. This statement is generally regarded as one of the fundamental postulates of quantum mechanics. An operator such as $\hat{F}$ will, of course, only have simultaneous eigenfunctions with $\hat{H}$ under certain circumstances; the usual result of operating with some other operator than the quantum Hamiltonian, on $\psi(x)$ would be to produce some function quite dissimilar to $\psi(x)$ and not necessarily just $\psi(x)$ multiplied by a constant. For example, $\hat{H}$ and $[(h/2\pi i)\ \partial/\partial x]$ have the simultaneous eigenfunctions, e^{ikx}, in the simple case of a particle moving in one-dimensional space (along the x Cartesian coordinate direction) where, $H(x, p_x) = p_x{}^2/2m$ (if $V(x) = 0$) and $\hat{H} = -[h^2/8\pi^2 m)\ \partial^2/\partial x^2]$, the eigenvalues of $\hat{H}$ corresponding to the energies, $E = [h^2 k^2/8\pi^2 m]$ and the eigenvalues of $\hat{p}_x$ corresponding to $hk/2\pi$, respectively; but obviously, $xe^{ikx} \neq \mu e^{ikx}$ for any constant μ, so that x is not a constant of the motion of the particle. It is worth noting that the condition that any operator, $\hat{F}$, will represent a constant of the motion of a particle (that is, will have simultaneous eigenfunctions with $\hat{H}$) may also be expressed as a commutation relationship; for this to be so $\hat{H}\psi = E\psi$ and $\hat{F}\psi = \mu\psi$, must be satisfied by the same ψ. If now $\hat{H}\psi = E\psi$ is operated on by $\hat{F}$, then $\hat{F}\hat{H}\psi = \hat{F}E\psi = E\hat{F}\psi = E\mu\psi$, and similarly if $\hat{F}\psi = \mu\psi$ is operated on by $\hat{H}$, then $\hat{H}\hat{F}\psi = \hat{H}\mu\psi = \mu\hat{H}\psi = \mu E\psi$, from which $(\hat{F}\hat{H} - \hat{H}\hat{F})\psi = 0$, or the commutator of the operators $\hat{F}$ and $\hat{H}$ is $[\hat{F}, \hat{H}] = 0$. Thus, any operator which corresponds to a constant of the motion of a system commutes with $\hat{H}$, and vice versa.

All of the operators which occur in quantum mechanical expressions are both linear and Hermitean. A linear operator, one like $(x^2 + \partial/\partial x)$, is defined by the property, $\hat{A}(af_1 + bf_2) = a\hat{A}f_1 + b\hat{A}f_2$, where $\hat{A}$ is any operator, a and b are any constants, and f_1 and f_2 are two different functions. A real operator, $\hat{A}$, is said to be Hermitean in the range (a, b) with respect to a certain type of function if

$$\int_a^b \psi\hat{A}\phi\, dx = \int_a^b \phi\hat{A}\psi\, dx, \tag{q.2.49.}$$

where ψ and ϕ are any real functions of the type concerned. If the operator is complex, it is said to be Hermitean if

$$\int_b^b \psi\hat{A}\phi\, dx = \int_b^b \phi\hat{A}^*\psi\, dx, \tag{q.2.50.}$$

where $\hat{A}^*$ is the complex conjugate of $\hat{A}$. For example, $\hat{p}_x$ is Hermitean, since

$$\int_{-\infty}^{\infty} \{\phi(x)(h/2\pi i)[\partial\psi(x)/\partial x]\}\, dx$$
$$= (h/2\pi i)|\phi(x)\psi(x)|\Big|_{-\infty}^{\infty} - \int_{-\infty}^{\infty} \{\psi(x)[h/2\pi i][\partial\phi(x)/\partial x\}\, dx$$
$$= \int_{-\infty}^{\infty} \{\psi(x)[(h/2\pi i)(\partial/\partial x)]^*\phi(x)\}\, dx, \tag{q.2.51.}$$

since both $\psi(x)$ and $\phi(x)$ must vanish at infinity. This expression, (q.2.51.) is much more compactly expressed in the Dirac notation, when it becomes, $\langle\phi|\hat{p}_x|\psi\rangle = \langle\psi|\hat{p}_x^*|\phi\rangle$. Hermitean operators play an important part in quantum mechanics, since the subject can be formally developed on the basic assumption that every observable quantity can be represented by a Hermitean operator. The eigenvalues, h_n, of any Hermitean operator, $\hat{H}$ are all real, since if $\hat{H}\psi_n = h_n\psi_n$, then multiplication throughout by ψ_n^* and integration over all of the variables involved gives, $\langle\psi_n|\hat{H}|\psi_n\rangle = h_n\langle\psi_n\mid\psi_n\rangle$, and the complex conjugate of this equation is $\langle\psi_n|\hat{H}|\psi_n\rangle^* = h_n^*\langle\psi_n\mid\psi_n\rangle^*$; but since $\hat{H}$ is Hermitean by definition, therefore $\langle\psi_n|\hat{H}|\psi_n\rangle = h_n^*\langle\psi_n\mid\psi_n\rangle$, and so $h_n = h_n^*$, or the eigenvalues are real.

As an illustration of some of the techniques required to find the eigenfunctions of an operator, consider the linear operator, $[-d^2/dx^2 + x^2]$,‡ which is representative of an important type of quantum mechanical operator.

‡ The quantum Hamiltonian for the linear harmonic oscillator is of this form. The harmonic oscillator is a system of importance, which has many applications in the whole realm of physics and chemistry.

In classical mechanics, the linear harmonic oscillator is defined as a particle of mass m, moving in a straight line under the influence of a force directed towards a fixed point in the line and of magnitude proportional to the distance of the particle from this point. If the fixed point is taken to be the origin, the line to be the x-axis, the force to be $-kx$ (where k is positive), and the frequency of the oscillation of the particle to be v, then the Newtonian equation of motion is

$$m[d^2x/dt^2] = -kx = m[dv/dt] = m[dv/dx][dx/dt] = mv[dv/dx],$$

where v is the particle's velocity of motion in the x-direction. Therefore, on rearranging this equation, $v\,dv = -(k/m)x\,dx$, which on integration gives $[v^2/2] = [-k/m][x^2/2] + C$, where C is an arbitrary integration constant. If the particle reaches its maximum displacement at $x = A$, then at $x = A$, the kinetic energy of the particle, $mv^2/2$, is zero and thus $v = 0$, which enables C to be evaluated as $C = [k/m][A^2/2]$. Now substituting for C gives,

$$v^2 = [-k/m][x^2 - A^2] = (k/m)[A^2 - x^2], \qquad \text{or} \qquad v = dx/dt = \sqrt{[(k/m)(A^2 - x^2)]}$$

or

$$\frac{dx}{\sqrt{[1 - (x/A)^2]}} = \sqrt{[k/m]}\,dt.$$

This equation may be integrated, making use of the fact that $d/dx(\sin^{-1} x) = [1 - x^2]^{-1/2}$, to give $\sin^{-1}\sqrt{[x/A]} = \sqrt{[k/m]}\,t + C_1$; the integration constant C_1 may be evaluated from the condition that, if the particle starts its motion from $x = 0$ at $t = 0$, then $C_1 = 0$. Therefore, $\sin[\sqrt{(k/m)}\,t] = (x/A)$ or $x = A\sin[\sqrt{(k/m)}\,t]$. In this expression for the position of the particle at some time t, the quantity $\sqrt{[k/m]}$ must define an angle, since with respect to the circle of reference of radius A which may be constructed with the origin as center, the product (angle $\times$ time) defines amplitude. Also, generally, if the particle initially does not start its motion from the origin but from some other position with amplitude δ, then $x = A\sin[\sqrt{(k/m)}\,t \pm \delta]$; alternatively the corresponding cosine expression would be equally valid (see page 106). If $[\sqrt{(k/m)}\,t] = \theta$, then since $\sin\theta = \sin(2\pi + \theta)$, therefore the motion repeats itself when $[\sqrt{(k/m)}\,t] = 2\pi$; if the time required for this to happen is T (the period of the motion), then $1/T$ represents the frequency of oscillation of the particle. From $[\sqrt{(k/m)}\,T] = 2\pi$, it may be deduced that $v = 1/T = (1/2\pi)\sqrt{(k/m)}$, and so $\sqrt{k/m} = 2\pi v$ and $x = A\sin[2\pi vt \pm \delta]$; since the sine or cosine of an angle may only vary between ± 1, x can only vary between $\pm A$. The velocity of motion of the particle, $v = dx/dt$ may now be expressed as

$$v = A2\pi v\cos[2\pi vt + \delta],$$

and the total energy as $E = mv^2/2 + U(x)$, where $U(x)$ is defined (conservative system) by $-dU(x)/dx = -kx$ with $k = 4\pi^2 v^2 m$ (from the expression for the frequency). Thus $U(x) = \int 4\pi^2 v^2 mx\,dx = 2\pi^2 v^2 mx^2 = kx^2/2$, or making use of the relationship between the linear frequency of the motion v, and the angular frequency, $\omega = 2\pi v$, $U(X) = m\omega^2 x^2/2$. There is thus a parabolic relationship between the potential energy of the particle and its displacement, x, as shown in Fig. Q.2.8. The total energy of the particle may now be expressed as

$$\begin{aligned} E = K + U(x) &= (m/2)(dx/dt)^2 + (m/2)(2\pi v)^2[A\sin(2\pi vt + \delta)]^2 = 2m\pi^2 v^2 A^2 \\ &= kA^2/2, \end{aligned}$$

so that the total energy of the particle is independent of the phase angle δ and may adopt any positive value, limited by the value of A.

If the expression for the classical potential energy above, $U(x) = m\omega^2 x^2/2$, is now

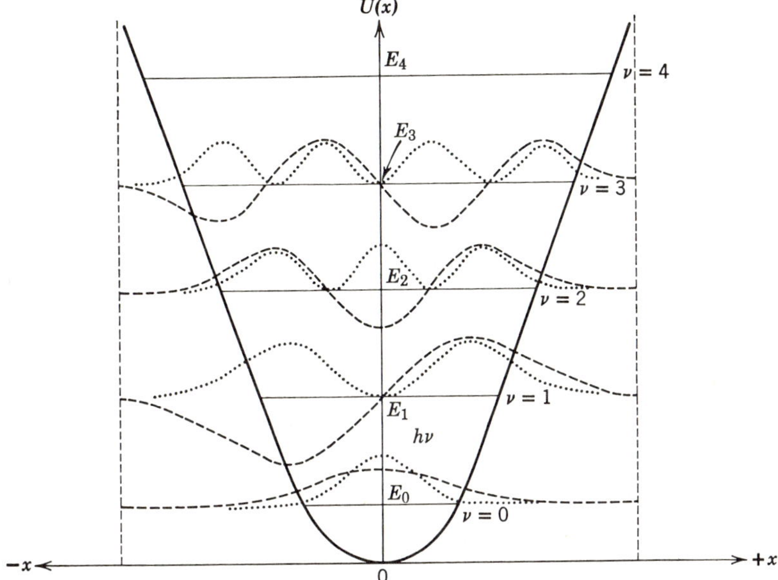

Fig. Q.2.8.   The potential energy, $U(x) = kx^2/2 = 2\pi^2\nu^2 mx^2 = m\omega^2 x^2/2$ (———)
and the energy levels, $E = [n + (1/2)]h\nu$ (———) for the linear harmonic oscillator.
Shown superimposed on each energy level are the wave functions $\psi_n(x/A)$ [------]
and the probability distributions $\psi_n^2(x/A)$ [······] for the ground state and the first
three excited states.

substituted in the Schrödinger equation (q.2.2.), the quantum mechanical description of
the linear harmonic oscillator obtained is

$$d^2/dx^2[\psi(x)] + 8\pi^2 mh^{-2}[E - m\omega^2 x^2/2]\psi(x) = 0.$$

If this equation is multiplied throughout by $[h^2/8\pi^2 m]$ and rearranged, it may be written
in the Hamiltonian form as

$$[-(h^2/8\pi^2 m)\, d^2/dx^2 + m\omega^2 x^2/2]\psi(x) = E\psi(x) = \hat{H}\psi(x),$$

so that apart from the multiplicative constants defining the problem, the quantum
Hamiltonian for the linear harmonic oscillator is of the form $[d^2/dx^2 + x^2]$.

The eigenfunctions of this operator must satisfy the differential equation,
$[-d^2/dx^2 + x^2]f(x) = af(x)$, and be such that they satisfy the boundary
conditions that for all $f(x)$ they are finite, single-valued, and continuous over
the whole range of values of x; the various possible values of a determine the
corresponding eigenvalues. For very large values of x, this equation approximates to

$$-d^2/dx^2[f(x)] + x^2 f(x) = 0, \qquad \text{(q.2.52.)}$$

since under these conditions $af(x)$ may be neglected in comparison with $x^2f(x)$. An approximate solution for (q.2.52.) is $f(x) = e^{\pm x^2/2}$, since

$$d^2/dx^2[e^{\pm x^2/2}] = [x^2 \pm 1]e^{\pm x^2/2}$$

and ± 1 may be neglected also when x is large. The solution with the positive exponent violates the boundary conditions by becoming infinite when $x \to \infty$, but $e^{-x^2/2}$ is satisfactory in this respect. Substituting this approximate solution into the original equation gives

$$[-d^2/dx^2 + x^2]e^{-x^2/2} = e^{-x^2/2}, \tag{q.2.53.}$$

so that the function, $e^{-x^2/2}$, is one possible eigenfunction of the operator $[-d^2/dx^2 + x^2]$ belonging to the eigenvalue, $+1$. If $f(x) = e^{-x^2/2}$ is plotted as a function of x, the resultant curve is similar to Fig. Q.2.6a., except that the maximum occurs for $x = 0$; this function is continuous, finite, and single-valued for all values of x from $-\infty$ to ∞. Similarly, $xe^{-x^2/2}$ is a possible eigenfunction, since

$$[-d^2/dx^2 + x^2]xe^{-x^2/2} = 3xe^{-x^2/2}; \tag{q.2.54.}$$

in this case the corresponding eigenvalue is $+3$. It would be expected that any function of the form $f(x) = g(x)e^{-x^2/2}$, may be a suitable eigenfunction of the operator $[-d^2/dx^2 + x^2]$, where $g(x)$ is some other function of x but must reduce to a constant for certain values, as in (q.2.53.) where it becomes $+1$.

If a function such as

$$f(x) = g(x)e^{-x^2/2} \tag{q.2.55.}$$

is substituted into the original eigenvalue equation,

$$[-d^2/dx^2 + x^2]f(x) = af(x), \tag{q.2.56.}$$

then

$$-e^{-x^2/2}d^2/dx^2[g(x)] + xe^{-x^2/2}d/dx[g(x)] + g(x)e^{-x^2/2} + xe^{-x^2/2}\,d/dx[g(x)]$$
$$- x^2e^{-x^2/2}g(x) + x^2g(x)e^{-x^2/2} = ag(x)e^{-x^2/2},$$

which reduces to

$$d^2/dx^2[g(x)] - 2x\,d/dx[g(x)] + g(x)[a - 1] = 0. \tag{q.2.57.}$$

Since this is the general condition which has to be satisfied, the two special cases of $f(x) = e^{-x^2/2}$ and $f(x) = xe^{-x^2/2}$, must be included in this generalization.

By substitution, when $g(x) = 1$, $a = 1$, and when $g(x) = x$, $a = 3$; therefore (q.2.57.) does represent the general equation of condition which must be satisfied in order that the function given by (q.2.55.) should be a solution of (q.2.56.). Equation (q.2.57.) is of a similar form to a certain classical differential equation known as Hermite's differential equation, which is

$$d^2/dx^2[v(x)] - 2x\, d/dx[v(x)] + 2^n v(x) = 0, \qquad \text{(q.2.58.)}$$

the solutions of which are known to be,‡

$$v(x) = (-1)^n e^{x^2}\, d^n/dx^n[e^{-x^2}] = H_v(x), \qquad \text{(q.2.59.)}$$

for all positive values of n, including 0; $H_v(x)$ is usually referred to as a Hermite polynomial of degree n.

‡ It may be shown that (q.2.59.) is a solution of (q.2.58.) for all values of n, by straightforward differentiation. Thus differentiating (q.2.59.) once, with respect to x, gives, omitting writing the factor $(-1)^n$ which is common to every term,

$$d/dx[v(x)] = e^{x^2}\, d^{n+1}[e^{-x^2}] + 2xe^{x^2}\, d^n/dx^n[e^{-x^2}],$$

and differentiating a second time

$$d^2/dx^2[v(x)] = e^{x^2}\, d^{n+2}/dx^{n+2}[e^{-x^2}] + 4xe^{x^2}\, d^{n+1}/dx^{n+1}[e^{-x^2}]$$
$$+ [2e^{x^2} + 4x^2 e^{x^2}]\, d^n/dx^n[e^{-x^2}].$$

If these first and second derivatives are substituted in (q.2.58.), there results, after dividing throughout by e^{x^2},

$$d^{n+2}/dx^{n+2}[e^{-x^2}] + 2x\, d^{n+1}/dx^{n+1}[e^{-x^2}] + [2 + 2n]\, d^n/dx^n[e^{-x^2}]. \qquad \text{(A)}$$

When $n = 0$, this equation becomes $d^2/dx^2[e^{-x^2}] + 2x\, d/dx[e^{-x^2}] + 2e^{-x^2} = 0$, and performing the indicated differentiations, this reduces to $4x^2 e^{-x^2} - 4x^2 e^{-x^2} - 2e^{-x^2} + 2e^{-x^2} = 0$, so equation A is obviously satisfied when $n = 0$. When $n = 1$, equation A becomes $d^3/dx^3[e^{-x^2}] + 2x\, d^2/dx^2[e^{-x^2}] + 4\, d/dx[e^{-x^2}] = 0$, which is the same equation that results if the $n = 0$ equation is itself differentiated with respect to x. Therefore, differentiating the $n = 0$ equation once with respect to x gives the same result as substituting $n = 1$ directly into equation A. But if equation A is satisfied for $n = 0$, as has been proved, then it must also be satisfied for $n = 1$; similarly by the extension of this procedure, equation A is satisfied for $n = 2$, and so on for all positive integral values of n, including 0.

By examination of equations (q.2.60.), a general algebraic expression may be written for the vth Hermite polynomial as,

$$H_v(x) = (2x)^v - v(v - 1)(2x)^{v-2}(1!)^{-1} + v(v - 1)(v - 2)(v - 3)(2x)^{v-4}(2!)^{-1} - \cdots.$$

If this expression for $H_v(x)$ is differentiated with respect to x, it is found that,

$$\begin{aligned}
d/dx[H_v(x)] &= 2^v v x^{v-1} - v(v - 1)(v - 2)2^{v-2} x^{v-3}(1!)^{-1} \\
&\quad + v(v - 1)(v - 2)(v - 3)(v - 4)2^{v-4} x^{v-5}(2!)^{-1} - \cdots] \\
&= 2v[(2x)^{v-1} - (v - 1)(v - 2)(2x)^{v-3}(1!)^{-1} \\
&\quad + (v - 1)(v - 2)(v - 3)(v - 4)(2x)^{v-5}(2!)^{-1} - \cdots] \\
&= 2v H_{v-1}(x).
\end{aligned}$$

Differentiating a second time,

$$d^2/dx^2[H_v(x)] = 2v\, d/dx[H_{v-1}(x)] = 4v(v - 1)H_{v-2}(x).$$

If these expressions for the first and second derivatives of $H_v(x)$ are substituted into the Hermite equation, (q.2.58.), then

$$4v(v - 1)H_{v-2}(x) - 4xH_{v-1}(x) + 2v H_v(x) = 0,$$

or

$$(v - 1)H_{v-2}(x) - xH_{v-1}(x) + H_v(x)/2 = 0.$$

Now replacing $(v - 1)$ by v, the recursion formula for Hermite polynomials results,

$$xH_v(x) = vH_{v-1}(x) + H_{v+1}(x).$$

By performing the indicated differentiations, the first six Hermite polynomials are found to be,

$$\begin{aligned}
H_0(x) &= 1, \\
H_1(x) &= 2x, \\
H_2(x) &= 4x^2 - 2, \\
H_3(x) &= 8x^3 - 12x, \\
H_4(x) &= 16x^4 - 48x^2 + 12, \text{ and} \\
H_5(x) &= 32x^5 - 160x^3 + 120x.
\end{aligned} \qquad \text{(q.2.60.)}$$

So that by substituting $g(x) = H_0(x)$ and $g(x) = H_1(x)$ into (q.2.55.), then apart from a numerical constant, the first two Hermite polynomials are the factors of $e^{-x^2/2}$ in the first two solutions of (q.2.56.) which have already been found. The most general solution of (q.2.56.), or the complete set of eigenfunctions of the operator $[-d^2/dx^2 + x^2]$ is then

$$f(x) = N_v H_v(x)e^{-x^2/2}, \qquad \text{(q.2.61.)}$$

where N_v is a numerical constant; it is, of course, quite a general condition, that when any eigenfunction of an operator belonging to some particular eigenvalue is multiplied by a numerical constant the product so obtained is still an eigenfunction belonging to the same eigenvalue. The constant N_v has the significance of the normalization constant of the functions, $f(x)$ of (q.2.55.).‡

‡ To find the normalization constant, N_v, in (q.2.61.), it is necessary to solve the equation, $\int_{-\infty}^{\infty} [N_v H_v(x)H_v{}^*(x)e^{-x^2/2}]^2 \, dx = 1$; $H_v(x) = H_v{}^*(x)$, since the functions given by (q.2.59.) and (q.2.61.) are real. Thus, $N_v = [\int_{-\infty}^{\infty} \{H_v(x)\}^2 e^{-x^2} \, dx]^{-1/2}$. For any two different integral values of v, such as m and n,

$$\int_{\infty}^{\infty} H_m(x)H_n(x)e^{-x^2} \, dx = (-1)^n \int_{\infty}^{\infty} H_m(x)[d^n/dx^n(e^{-x^2})] \, dx.$$

Integration by parts of the right-hand member of this equation gives

$$\int_{-\infty}^{\infty} H_m(x)[d^n/dx^n(e^{-x^2})] \, dx$$

$$= H_m(x) \, d^{n-1}/dx^{n-1}(e^{-x^2})\Big|_{-\infty}^{\infty} - \int_{\infty}^{\infty} [d^{n-1}/dx^{n-1}(e^{-x^2}) \, d/dx\{H_m(x)\}] \, dx.$$

The first member of the right-hand side of this equation is zero, since e^{-x^2} and its derivatives vanish at $x = \pm\infty$. If the recursion relationship between Hermite polynomials given on page 143 is now substituted, namely $d/dx[H_m(x)] = 2mH_{m-1}(x)$, then

$$\int_{-\infty}^{\infty} [H_m(x)H_n(x)e^{-x^2}] \, dx = (-1)^{n+1}2m \int_{-\infty}^{\infty} [H_{m-1}(x) \, d^{n-1}/dx^{n-1}(e^{-x^2})] \, dx.$$

Repeated integration by parts and the use of the recursion relationship, until this procedure has been carried out m times gives,

$$\int_{-\infty}^{\infty} [H_m(x)H_n(x)e^{-x^2}]\,dx = (-1)^{n+m}2^m(m!)\int_{-\infty}^{\infty} [H_0(x)\,d^{n-m}/dx^{n-m}(e^{-x^2})]\,dx$$

$$= (-1)^{n+m}2^m(m!)[d^{n-m-1}/dx^{n-m-1}(e^{-x^2})]_{-\infty}^{\infty} = 0.$$

Thus the functions defined by (q.2.61.) are orthogonal. Now when $n = m = \nu$,

$$\int_{-\infty}^{\infty} \{[H_\nu(x)]^2 e^{-x^2}\}\,dx = (-1)^{2\nu}2^\nu(\nu!)\int_{-\infty}^{\infty} e^{-x^2}\,dx,$$

and since $\int_{-\infty}^{\infty} e^{-x^2}\,dx = \sqrt{\pi}$ (see page 43), therefore $\int_{-\infty}^{\infty} \{[H_\nu(x)]^2 e^{-x^2}\}\,dx = 2^\nu(\nu!)\sqrt{\pi}$, and thus $N_\nu = [2^\nu(\nu!)\sqrt{\pi}]^{-1/2}$.

The plot of $f(x)$, given by (q.2.61.) as a function of x for $\nu = 0, 1, 2,$ and 3, is shown in Fig. Q.2.9., where

$$N_\nu = [2^\nu(\nu!)\sqrt{\pi}]^{-1/2}. \tag{q.2.62.}$$

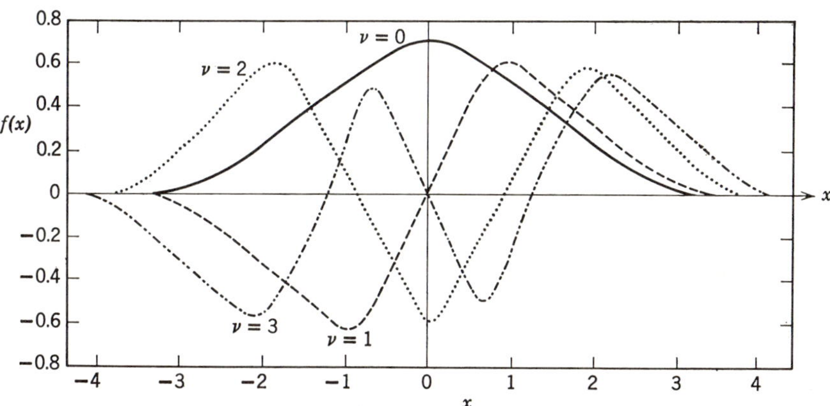

FIG. Q.2.9.   Eigenfunctions $f(x) = N_\nu H_\nu(x)e^{-x^2/2}$ of the operator $[-d^2/dx^2 + x^2]$ for $\nu = 0, 1, 2,$ and 3.

All of these functions satisfy the standard boundary conditions by being everywhere finite, single-valued, and continuous, for all values of x. Comparison of equations (q.2.57.) and (q.2.58.) shows that the constant n of the Hermite equation is related to the eigenvalues, a, by

$$a = 2n + 1, \quad \text{with} \quad n = 0, 1, 2, \ldots, \tag{q.2.63.}$$

so that the eigenvalues of the operator, $[d^2/dx^2 + x^2]$ consist of the set of odd positive integers, $+1, +3, +5, \ldots.\ddagger$

$\ddagger$ Having found the manifold of possible eigenfunctions of the operator $[d^2/dx^2 + x^2]$, given by (q.2.61.), it is a fairly simple matter to find all the possible eigenfunctions and eigenvalues of the simple linear harmonic oscillator. It has been shown on page 141 that the Hamiltonian for this system is related to this operator by the inclusion of certain constants defining the system, such that the actual Hamiltonian for the linear harmonic oscillator is

$$[-(h^2/8\pi^2 m)\, d^2/dx^2 + m\omega^2 x^2/2] = \hat{H},$$

and the appropriate Schrödinger equation is,

$$d^2/dx^2[\psi(x)] + 8\pi^2 mh^{-2}[E - m\omega^2 x^2/2]\psi(x) = 0,$$

or in Hamiltonian form, $[-(h^2/8\pi^2 m)\, d^2/dx^2 + m\omega^2 x^2/2]\psi_n(x) = E_n\psi_n(x)$, where $m\omega^2 x^2/2$ is the potential energy of the system, $U(x)$. It should be noted that $U(x)$ may be alternatively expressed, as shown on page 141, as $U(x) = kx^2/2 = 2m\pi^2\nu^2 x^2$, and the classical energy as $E = kA^2/2 = 2m(\pi\nu A)^2$. An approximate solution to the Schrödinger equation above, the asymptotic solution, may be obtained by noting that when x is very large, as $x \to \infty$, E (if finite) will be negligible compared to $U(x)$ (function of x^2), so that the Schrödinger equation becomes,

$$\begin{aligned}
d^2/dx^2[\psi(x)] &= [16\pi^4 m^2\nu^2 h^{-2}]x^2\psi(x), &&\text{or} \\
&= [4\pi^2 m^2\omega^2 h^{-2}]x^2\psi(x), &&\text{on substitution of } \omega = 2\pi\nu, \text{ or} \\
&= [4\pi^2 mkh^{-2}]x^2\psi(x), &&\text{on substitution of } \nu = [1/2\pi]\sqrt{k/m}, \text{ or} \\
&= \alpha^2 x^2\psi(x), \\
&\quad \text{on substitution of } \alpha = 4\pi^2 m\nu h^{-1} = 2\pi m\omega h^{-1} = 2\pi h^{-1}\sqrt{mk}.
\end{aligned}$$

Comparison of this equation with (q.2.50.) suggests the solution, $\psi(x) = e^{-\alpha x^2/2}$; differentiating this function twice with respect to x gives,

$$\begin{aligned}
d^2/dx^2[e^{-\alpha x^2/2}] &= -[\alpha - \alpha^2 x^2]e^{-\alpha x^2/2}, &&\text{or} \\
&= -[2\pi m\omega h^{-1} - 4\pi^2 m^2\omega^2 x^2 h^{-2}]e^{-x/2}, \\
&\quad \text{on substituting } \alpha = 2\pi m\omega h^{-1}, \text{ or,} \\
&= -8\pi^2 mh^{-2}[(h\omega/4\pi) - m\omega^2 x^2/2]e^{-\alpha x^2/2}, \\
&\quad \text{on factoring out } 8\pi^2 mh^{-2},
\end{aligned}$$

which is identical with the original Schrödinger equation, if $E = h\omega/4\pi = h\nu/2 = \hbar\omega/2$.

So, the function $\psi(x) = e^{-\alpha x^2/2}$, which satisfies all the requirements of a good wave function in that it is smooth, continuous and single valued for all values of x, is an eigenfunction of the linear harmonic oscillator corresponding to the eigenvalue, $E = h\nu/2$; furthermore this function possesses no nodes (see Fig. Q.2.9.), and this property of a wave function (see page 117) is characteristic of the lowest or ground state of a system. If then $e^{-\alpha x^2/2}$ is designated $\psi_0(x)$, then the ground state energy of the linear oscillator is given by $h\nu/2 = E_0$. In a similar empirical manner it may be shown that $\psi_1(x) = xe^{-\alpha x^2/2}$, a function which has one node at the origin, is also an eigenfunction of the linear oscillator, since,

$$\begin{aligned}
d^2/dx^2[xe^{-\alpha x^2/2}] &= xe^{-\alpha x^2/2}[-3\alpha + \alpha^2 x^2] = xe^{-\alpha x^2/2}[-6\pi m\omega h^{-1} + 4\pi^2 m^2\omega^2 x^2 h^{-2}] \\
&= -8\pi^2 mh^{-2}[3h\omega/4\pi - m\omega^2 x^2/2]xe^{-\alpha x^2/2},
\end{aligned}$$

corresponding to the eigenvalue $3h\nu/2 = 3E_0$, and that $\psi_2(x) = (1 - 2\alpha x^2)e^{-\alpha x^2/2}$, a function with two nodes, is also an eigenfunction corresponding to the eigenvalue $5h\nu/2$, since

$$\begin{aligned}
d^2/dx^2[(1 - 2\alpha x^2)e^{-\alpha x^2/2}] &= (1 - 2\alpha x^2)e^{-\alpha x^2/2}[-5\alpha + \alpha^2 x^2] \\
&= (1 - 2\alpha x^2)e^{-\alpha x^2/2}[-10m\pi\omega h^{-1} + 4\pi^2 m^2\omega^2 x^2 h^{-2}] \\
&= -8\pi^2 mh^{-2}[5h\omega/4\pi - m\omega^2 x^2/2](1 - 2\alpha x^2)e^{-\alpha x^2/2}.
\end{aligned}$$

The eigenvalues of the linear oscillator then appear to be the multiples, 1, 3, 5, ..., of $h\nu/2 = E_0$, or $[n + 1/2]h\nu$, where $n = 0, 1, 2, \ldots$. The significant conclusion from the above considerations is that the lowest energy of the harmonic oscillator is not zero, as it would be according to classical mechanics, but $h\nu/2$, which is referred to as the zero point energy of the oscillator.

These results may of course be obtained by the straightforward process of casting the Hamiltonian equation for the oscillator, $[-(h^2/8\pi^2 m)\,d^2/dx^2 + kx^2/2]\psi_n(x) = E_n\psi_n(x)$, into the same form as (q.2.56.) by means of appropriate substitutions. Since $\psi_0(x)$ is $e^{-\alpha x^2/2}$, this may be written as $\psi_0(x) = e^{-2\pi^2 m\nu x^2/h}$ (substituting for α) and $E_0 = 2\pi^2\nu^2 A^2 m = h\nu/2$, therefore the exponent, $-2\pi^2 m\nu x^2/h$ may be written as $-(1/2)(x/A)^2$, so that $\psi_0(x) = e^{-(1/2)(x/A)^2}$. By substituting $(x/A) = \beta$, then $dx = A\,d\beta$, and the Hamiltonian form of the equation of motion of the oscillator becomes

$$[-(h^2/8\pi^2 mA^2)\,d^2/d\beta^2 + kA^2\beta^2/2]A\psi_n(\beta) = E_n A\psi_n(\beta).$$

Since the eigenvalues, $E_n = a(h\nu/2)$, $a = 1, 3, 5, \ldots = 2n + 1$, with $n = 0, 1, 2, \ldots$, therefore $E_n = ah\nu/2 = (ah/4\pi)\sqrt{k/m}$ (substituting $\nu = (1/2\pi)\sqrt{k/m}$) and since $E_0 = h\nu/2 = kA^2/2$, therefore $A = \sqrt{h\nu/k} = [h^2/4\pi^2 km]^{1/4}$. Now substituting this value for A in the oscillator equation, there results

$$[-(h^2/8\pi^2 m)(4\pi^2 km)^{1/2}(h^2)^{-1/2}d^2/d\beta^2 + (kh\beta^2/2)(4\pi^2 km)^{-1/2}]\psi_n(\beta) = (ah/4\pi)\sqrt{k/m}\,\psi_n(\beta),$$

and if this equation is multiplied throughout by $(4\pi/h)\sqrt{m/k}$, it becomes

$$[-d^2/d\beta^2 + \beta^2]\psi_n(\beta) = a\psi_n(\beta),$$

of exactly the same form as (q.2.56.), and therefore with some eigenvalues, $a = 1, 3, 5, \ldots$, with $a = 2n + 1$. Thus the complete set of eigenvalues of the linear oscillator is $E_n = [n + (1/2)](h/2\pi)\sqrt{k/m}$, and the complete set of eigenfunctions is $\psi_n(\beta) = N_\nu H_\nu(\beta)e^{-\beta^2/2}$ [from (q.2.61.)]. From (q.2.63.), the normalization factor N_ν, which is obtained from the condition $N_\nu^2 A\int_{-\infty}^{\infty}[H_\nu(\beta)]^2 e^{-\beta^2}\,d\beta = 1 = N_\nu^2 A2^\nu(\nu!)\sqrt{2\pi}$, or $N_\nu = [A2^\nu(\nu!)\sqrt{2\pi}]^{-1/2}$ [(q.2.62.)].

The displacement of the linear harmonic oscillator is periodic in time. In a fluid medium, such an oscillator would give rise to a disturbance which would travel through the medium giving rise to a wave motion, periodic in time and space; for example a vibrating disc in air gives rise to sound waves and a vibrating or oscillating dipole gives rise to electromagnetic waves. The measure of the periodicity in space is the wavelength and the measure of periodicity in time is the period, $T = 1/\nu$.

While according to classical mechanics the oscillator could have zero energy, in which case the particle concerned would be at rest at the origin, this is not so according to the quantum mechanical description, which indicates that the lowest energy that the oscillator can have is the zero-point energy, $h\nu/2$. This fact is of importance in interpreting the behavior of atomic nuclei, atoms, molecules, and crystals; even at the absolute zero of temperature solids must still vibrate with the zero-point energy of their constituent atoms, molecules, or ions. Also according to quantum mechanics, there is a finite probability of finding a particle vibrating harmonically at large distances from the origin, contrary to the classical mechanical description. The ranges of motion of a particle undergoing simple harmonic motion in terms of classical mechanics are limited to the lengths of the horizontal line segments in Fig. Q.2.8.

Q.2.7. The Average Values of Dynamical Variables

In classical statistical mechanics, the average value of any property, P, of the particles composing a system, where the particular property concerned is a function of the coordinates and momenta of the total number of particles, N, in the system, is obtained by multiplying the number of particles, N_i, in any group by the extent to which each particle in that group possesses the property P_i, by adding all such products together, and by dividing by the total number of particles in the system: $\bar{P} = [\sum_i N_i P_i]/\sum_i N_i$, or where there are a very large number of particles with an essentially continuous range of properties, this average value may be alternatively expressed as

$$\bar{P} = \left[\int_0^N P_i \, dN_i\right] \Big/ \int_0^N dN_i,$$

the summation sign being replaced by an integral.

By analogy to classical probability theory, for a system in a state ψ in quantum mechanics, from the interpretation of $|\psi|^2$ as the probability density of the system, the average value of some property P, which is a function of the coordinates only is given by

$$\bar{p} = \left[\int\int\int_{-\infty}^{\infty} P(x, y, z)\psi\psi^* \, dx \, dy \, dz\right] \Big/ \int\int\int_{-\infty}^{\infty} \psi\psi^* \, dx \, dy \, dz. \quad \text{(q.2.64.)}$$

The expression is similar to (q.1.22.) for the average value of the energy associated with Planck's radiation oscillators, where the probability distribution involved was that of the classical Maxwell–Boltzmann statistics. The denominator of (q.2.64.) is, of course, unity, if the function ψ is normalized. Generally, if a large number of measurements of a dynamical variable is made on a system in the state ψ, or if the measurements are made on a large number of identical systems all in the same state ψ, then the average or expectation value of the dynamical variable may be defined as the mean of the measured values, denoted by a bar over the quantity, as $\bar{P}$ or $\bar{x}$. For example, in the simple case, of a single particle in the state ψ, the mean value of the x coordinate is, $\bar{x} = \int x|\psi|^2 \, dr = \int \psi^* x\psi \, dr$. The second alternative is to be preferred as then the expression is also valid when the property concerned is a function of not only the coordinates of the particles concerned but also of their momenta; in the latter event an operator is involved, which is not simply the operation of multiplication. The same applies to any function of the coordinates, say $f(x, y, z)$, when $\bar{f} = \int \psi^* f\psi \, dr$, the integration being throughout the configuration space of the system concerned. More specifically, the function, $\psi = [\pi r_0^3]^{-1/2} e^{-r/r_0}$, which is actually the ground state wave function for the hydrogen atom, as will be shown later, contains the single variable, r, the distance of the electron from the nucleus, and so the average distance of the electron from the nucleus is,

$$\bar{r} = \int \psi^* r\psi \, dr = [\pi r_0^3]^{-1} \int re^{-2r/r_0} \, dr = [\pi r_0^3]^{-1} \int re^{-2r/r_0} 4\pi r^2 \, dr$$

$$= [4/r_0] \int_0^{\infty} r^3 e^{-2r/r_0} \, dr = [4/r_0][3r_0^4/8] = 3r_0/2. \quad \text{(q.2.65.)}$$

The definition of average values given above may now be generalized to obtain the average value of any property of a system which is a function of both the coordinates and the momenta, by combining it with the concepts that every dynamical variable corresponds to a quantum mechanical operator and that the result of the precise measurement of such a dynamical variable is an eigenvalue of the operator. The Hamiltonian or energy function is an example of this type, where $H(p, q)$ is a function of the coordinates and momenta of a system, the corresponding operator is $\hat{H}$ and the various possible energy values of the system so described are obtained as eigenvalues of the equation $\hat{H}\psi_n = E_n\psi_n$; so the average or expectation value of the classical Hamiltonian for the system in the state ψ is the constant, E. But this result is also a consequence of the application of the mean value definition, $\bar{H} = \int \psi^*\hat{H}\psi \, d\tau$, since if ψ is normalized, $\bar{H} = \int \psi^*\hat{H}\psi \, d\tau = \int \psi^*E\psi \, d\tau = E$; here the quantity $\hat{H}$ in the integrand is a differential operator, and not merely a multiplier, as is the case where the property whose mean value is being sought is only a function of the coordinates. A similar result must pertain for any quantity which is a constant of the motion of the system, other than the energy. This generalization, that the expectation value of a dynamical variable, F, represented in quantum mechanics by the operator, $\hat{F}$, for a state whose normalized wave function is ψ, is given by

$$F = \int \psi^*\hat{F}\psi \, d\tau, \tag{q.2.66.}$$

the integration being taken throughout the configuration space of the system, is normally regarded as another of the fundamental postulates of quantum mechanics, which is justified by the results obtained in its application. As a specific example, the mean value of the x-component of the linear momentum of a particle in the state ψ is $\bar{p}_x = \int \psi^*(h/2\pi i) \, \partial\psi/\partial x \, d\tau$, and if

$$\psi = [\sqrt{2/L}] \sin(\pi x/L),$$

as for the particle in the one-dimensional box with $n = 1$ (page 120), then

$$\bar{p}_x = \int_0^L \{[\sqrt{2/L}] \sin(\pi x/L)(h/2\pi i) \, d/dx[(\sqrt{2/L}) \sin(\pi x/L)]\} \, dx$$

$$= [h/L^2 i] \int_0^L \sin(\pi x/L) \cos(\pi x/L) \, dx = 0 \text{ (page 123)},$$

which is in agreement with the classical mechanical result, where the particle would be expected to move back and forth between the confining walls with constant velocity.

Quantities such as (q.2.66.) are of special importance in matrix mechanics. If a set of functions, ψ_i, are eigenfunctions of the Hamiltonian, $\hat{H}$, such that

$\hat{H}\psi_i = E_i\psi_i$, then $\int \psi_i^*\hat{H}\psi_j\, d\tau = E_j\, \delta_{ij}$, are referred to as the matrix elements of the operator $\hat{H}$ with respect to the set of functions ψ_i and the matrix‡ of $\hat{H}$ is an array of such matrix elements; in Dirac terminology, the matrix of $\hat{H}$ is represented as:

$$\begin{bmatrix} \langle\psi_1|\hat{H}|\psi_1\rangle & \langle\psi_1|\hat{H}|\psi_2\rangle & \langle\psi_1|\hat{H}|\psi_3\rangle & \cdots \\ \langle\psi_2|\hat{H}|\psi_1\rangle & \langle\psi_2|\hat{H}|\psi_2\rangle & \langle\psi_2|\hat{H}|\psi_3\rangle & \cdots \\ \langle\psi_3|\hat{H}|\psi_1\rangle & \langle\psi_3|\hat{H}|\psi_2\rangle & \langle\psi_3|\hat{H}|\psi_3\rangle & \cdots \\ \cdots\cdots\cdots & \cdots\cdots\cdots & \cdots\cdots\cdots & \cdots \end{bmatrix} \qquad \text{(q.2.67.)}$$

‡ The labor involved in merely writing down, as well as the computational effort required to evaluate, many of the involved and tedious expressions and operations encountered in quantum mechanics, X-ray crystallography, and other branches of applied mathematics, may be considerably simplified and shortened, by the use of the abstract constructs of matrices.

A matrix is defined as a two-dimensional array of numbers, functions, or generally elements, subject to certain rules of operation. For example, the arbitrary matrix, $\mathbf{M}$, is represented as

$$\mathbf{M} = \begin{bmatrix} a_{11} & a_{12} & a_{13} & \cdots & a_{1n} \\ a_{21} & a_{22} & a_{23} & \cdots & a_{2n} \\ a_{31} & a_{32} & a_{33} & \cdots & a_{3n} \\ \vdots & \vdots & \vdots & & \vdots \\ a_{m1} & a_{m2} & a_{m3} & \cdots & a_{mn} \end{bmatrix}, \text{ or briefly as } \mathbf{M}_{ij} \text{ or } [\mathbf{M}_{ij}], \text{ or } \mathbf{M},$$

and is composed of m horizontal rows of n elements each and n columns of m elements each, thus describing an $m \times n$ matrix; in this matrix, $\mathbf{M}_{ij}$, $i = 1, 2, 3,\ldots, m$, and $j = 1, 2, 3,\ldots, n$. If $m = n$, the matrix is said to be a square matrix, otherwise referred to as a determinant and written as $|M|$; if $n = 1$ and $m > 1$, $\mathbf{M}$ is a column matrix; and when $n > 1$ and $m = 1$, $\mathbf{M}$ is a row matrix. Two matrices are equal, if all of their corresponding elements are equal; thus, $\mathbf{M} = \mathbf{N}$ if $a_{ij} = b_{ij}$ for all i, j, where a_{ij} and b_{ij} are the corresponding elements of the two matrices $\mathbf{M}$ and $\mathbf{N}$. If every element of the matrix $\mathbf{M}$ is zero, that is $a_{ij} = 0$ for all i, j, then $\mathbf{M}$ is said to be a null matrix. If $\mathbf{M} = 1$ such that $a_{ij} = \delta_{ij}$, $\mathbf{M}$ is a unit matrix. The unit matrix has elements which are unity when $i = j$ and zero when $i \neq j$; the former elements are called the diagonal elements if $\mathbf{M}$ is square and the latter the off-diagonal elements.

The sum of two matrices $\mathbf{M}$ and $\mathbf{N}$ is a new matrix $\mathbf{C}$, whose elements c_{ij} are the sum of the corresponding elements a_{ij} and b_{ij} of $\mathbf{M}$ and $\mathbf{N}$, that is, $a_{ij} + b_{ij} = c_{ij}$. This definition for the sum of two matrices is logical only if the two matrices contain the same number of rows and columns each or are conformable for addition (or subtraction), and is compatible with the ordinary commutative law of algebra. The product of two matrices $\mathbf{MN}$ (regarded as $\mathbf{N}$ multiplied by $\mathbf{M}$, as when dealing with operators) gives a new matrix $\mathbf{U}$, the elements of which are given by $u_{ij} = \sum_k a_{ik} \cdot b_{kj}$; this implies that u_{ij} are obtained by taking the ith row of the first matrix and the jth column of the second matrix, multiplying each of the corresponding elements together and adding them together. For this operation to be logical, it is necessary that the matrix $\mathbf{M}$ has the same number of rows as $\mathbf{N}$ has columns, or $\mathbf{M}$ and $\mathbf{N}$ are only conformable for multiplication if the number of elements in the rows of $\mathbf{M}$ is equal to the number of elements in the columns of $\mathbf{N}$; thus, in general, the multiplication of matrices, like the successive application of operators is non-commutative, or $\mathbf{MN} \neq \mathbf{NM}$. The multiplication of a matrix by a constant λ implies that every element of the matrix is multiplied by λ, or $\lambda\mathbf{M} = \lambda_{ij}$ for all i, j. The expression $\mathbf{M}^q\mathbf{N}$ denotes q successive multiplications of $\mathbf{N}$ by $\mathbf{M}$; this operation is uniquely defined only when q is an integer, positive or negative or $\pm(1/2)$. It should be noted, incidentally, that $\mathbf{MN} = 0$ does not necessarily mean that either $\mathbf{M}$ or $\mathbf{N}$ is a null matrix; for example

$$\begin{bmatrix} -4 & 1 & 7 \\ 2 & -3 & 1 \end{bmatrix} \begin{bmatrix} 8 & 7 \\ -3 & 5 \\ 5 & 1 \end{bmatrix} = 0.$$

Various special names are used for square matrices, the elements of which are related to one another in particular ways. Thus if the elements $a_{ij} = a_{ji}$, the matrix is symmetric; if $a_{ij} = -a_{ji}$, the matrix is skew-symmetric and since also $a_{ii} = -a_{ii}$, all the diagonal elements of a skew-symmetric matrix must be zero; if the elements are all real or all imaginary, the matrix is real or imaginary respectively; if $a_{ij} = a_{ji}^*$, the matrix is Hermitean; if $\sum_k a_{ik}a_{jk}^* = \delta_{ij}$ or $\sum_k a_{ki}a_{kj}^* = \delta_{ij}$, the matrix is unitary and each of the rows and each of the columns must form mutually orthogonal sets of vectors. Rotation of coordinate axes giving rise to coordinate transformations are described by unitary transformations. For two square matrices, $\mathbf{M}$ and $\mathbf{N}$, with elements m_{ij} and n_{ij}, respectively: if $n_{ij} = m_{ji}$, that is the row and column indices are transposed, $\mathbf{N}$ is said to be the transposed matrix to $\mathbf{M}$ and written as $\mathbf{N} = \tilde{\mathbf{M}}$; if $b_{ij} = a_{ij}^*$, $\mathbf{N}$ is the complex conjugate of $\mathbf{M}$ or $\mathbf{N} = \mathbf{M}^*$; if $b_{ij} = a_{ji}^*$, $\mathbf{N}$ is the associate matrix of $\mathbf{M}$ or $\mathbf{N} = \mathbf{M}\dagger = \tilde{\mathbf{M}}^*$; and if $\mathbf{N} = (\mathbf{M})^{-1}$, $\mathbf{M}$ is the inverse matrix of $\mathbf{N}$ or $\mathbf{MN} = \mathbf{I}$, the unit or identity matrix. From these definitions, a unitary matrix $\mathbf{M}$ must satisfy the condition that $\mathbf{MM}\dagger = \mathbf{M}\dagger\mathbf{M} = \mathbf{I}$.

For an $n \times n$ matrix, a square matrix, the sum of the product of its elements obtained by the application of certain rules is called the determinant corresponding to that matrix; a determinant is always a single number even although it may be represented by an array of numbers or functions. The process of evaluating a determinant by taking the sum of the products of its elements in a certain order is referred to as the expansion of the determinant. The older methods of Gauss and Chiò[1] of reducing the order of the determinant successively to $(n - 1), (n - 2), \ldots$, until finally a 2×2 determinant remains, the value of which is then given by the difference between the product of the two diagonal elements, are extremely laborious and time-consuming if computer time is being utilized. The method of Crout,[2] of forming a single auxiliary matrix from the $n \times n$ initial array, and then taking the product of the elements of the principal diagonal, provides a very convenient and fast way of evaluating determinants. The general analytical expression for the value of a determinant is readily obtained by generalization of the process required to evaluate, for example, a 3×3 determinant, which is easily deduced from consideration of the condition which must be satisfied for three straight lines to mutually intersect one another in a point. First, for any two straight lines passing through the origin, $a_1x + b_1y = 0$ and $a_2x + b_2y = 0$, if the first equation is multiplied by b_2 and the second by b_1, or the first by a_2 and the second by a_1, and the second product subtracted from

[1] Whittaker, E., and Robinson, G., *The Calculus of Observations*, Blackie & Son, Glasgow, 4th edition, 1954.

[2] Crout, P. D., *A Short Method for Evaluating Determinants and Solving Systems of Linear Equations*, Paper 41–103, AIEE Summer Convention, Toronto, Canada, 1941.

the first, then in either case, $x(a_1b_2 - a_2b_1) = 0$ and $y(a_2b_1 - a_1b_2) = 0$, and $x = 0$, $y = 0$, or $a_1b_2 - a_2b_1 = 0$ and $a_2b_1 - a_1b_2 = 0$. Or, written as a simple determinant

$$\begin{vmatrix} a_1 & b_1 \\ a_2 & b_2 \end{vmatrix} = 0 = \begin{vmatrix} a_2 & b_2 \\ a_1 & b_1 \end{vmatrix};$$

in other words the value of a 2×2 determinant is simply the difference of the product of the diagonal elements of the determinant. Now, the condition that any three straight lines, $a_1x + b_1y + c_1 = 0$, $a_2x + b_2y + c_2 = 0$, $a_3x + b_3y + c_3 = 0$, should mutually intersect in a point is that the roots of any two of the three equations should satisfy the third, giving three simultaneous equations in X, Y, Z, the coordinates of the point of mutual intersection. In this event, then $a_1 X + b_1 Y + c_1 Z = a_2 X + b_2 Y + c_2 Z = a_3 X + b_3 Y + c_3 Z = 0$, by writing $x = X/Z$ and $y = Y/Z$ in the equations of the three lines. From any two of these equations, it is easily deduced that

$$X : Y : Z = \begin{vmatrix} b_2 & c_2 \\ b_3 & c_3 \end{vmatrix} : \begin{vmatrix} c_2 & a_2 \\ c_3 & a_3 \end{vmatrix} : \begin{vmatrix} a_2 & b_2 \\ a_3 & b_3 \end{vmatrix},$$

which values of x and y must satisfy the third of the line equations, and then by substitution, it is apparent that

$$a_1 \begin{vmatrix} b_2 & c_2 \\ b_3 & c_3 \end{vmatrix} + b_1 \begin{vmatrix} c_2 & a_2 \\ c_3 & a_3 \end{vmatrix} + c_1 \begin{vmatrix} a_2 & b_2 \\ a_3 & b_3 \end{vmatrix} = 0,$$

which on expansion by the rule for 2×2 determinants gives $a_1b_2c_3 + a_3b_1c_2 + a_2b_3c_1 - a_1b_3c_2 - a_2b_1c_3 - a_3b_2c_1 = 0$, and this expression is conveniently written in determinantal form as

$$\begin{vmatrix} a_1 & b_1 & c_1 \\ a_2 & b_2 & c_2 \\ a_3 & b_3 & c_3 \end{vmatrix} = 0.$$

Thus in the 3×3 determinant, the value is given by the sum of a number of terms each of the type $a_ib_jc_k$, where (ijk) is a permutation of the numbers (123), and all possible permutations are included. The permutations (132), (213), which may be obtained by a single interchange of the numbers (123), are referred to as odd permutations or as having odd parity; similarly with permutations such as (321) which requires three successive interchanges from (123), and generally for any odd number of interchanges. All the other permutations may be obtained by an even number of interchanges from (123) are called even permutations. Also the terms for which (ijk) is an odd permutation have negative signs and the others have positive signs. So, if a permutation operator, P, is such that by operating on the subscripts (123) so that $Pa_1b_2c_3 = a_2b_3c_1$, for example, and there are p interchanges involved in arriving at this permutation, then the 3×3 determinant may be conveniently and concisely written as

$$\begin{vmatrix} a_1 & b_1 & c_1 \\ a_2 & b_2 & c_2 \\ a_3 & b_3 & c_3 \end{vmatrix} = \sum_P (-1)^p P a_1 b_2 c_3,$$

the sum being over all six permutation operators, including the identity operator which leaves (123) in their natural order. If the elements of the determinant are functions like

$$\begin{vmatrix} \psi_1(x_1) & \psi_1(x_2) & \psi_1(x_3) \\ \psi_2(x_1) & \psi_2(x_2) & \psi_2(x_3) \\ \psi_3(x_1) & \psi_3(x_2) & \psi_3(x_3) \end{vmatrix},$$

then in a similar manner this determinant may be written as $\sum_P (-1)^p P\psi_1(x_1)\psi_2(x_2)\psi_3(x_3)$, where the operator P now acts on either the subscripts of the ψ_i terms or the x_i terms. A determinant of any order, n, may be expanded, by successive application of this argument to give $\sum_P (-1)^p P\psi_1(x_1)\psi_2(x_2)\cdots\psi_n(x_n)$, the sum being over all of the $n!$ permutations of the subscripts $(123\cdots n)$ of either the ψ_i or the x_i parts, with p representing the number of interchanges in the permutation P. It is also a logical deduction from this definition of the value of a determinant that if any two rows or any two columns of a determinant are changed the sign of the determinant is changed and that if the corresponding elements in any two rows or any two columns are the same, the determinant is zero. Also, for any system of n linear, simultaneous equations, a non-zero solution can only exist if the determinant of the coefficients of the variables in the equations is zero. These properties of determinants are particularly useful in quantum mechanical applications.

Another special type of square matrix may be defined in terms of the properties of determinants known as their complementary minors and cofactors. The complementary minor of any element m_{ij} of a determinant M of order m, is the determinant of order $(m - 1)$ which remains after eliminating the ith row and jth column, and the cofactor of m_{ij} is $(-1)^{i+j}$ times its complementary minor, which may be denoted by M^{ij}. Thus, if

$$|M| = \begin{vmatrix} m_{11} & m_{12} & m_{13} & \cdots & m_{1m} \\ m_{21} & m_{22} & m_{23} & \cdots & m_{2m} \\ m_{31} & m_{32} & m_{33} & \cdots & m_{3m} \\ \vdots & \vdots & \vdots & & \vdots \\ m_{m1} & m_{m2} & m_{m3} & \cdots & m_{mm} \end{vmatrix} = \sum (-1)^p P m_{1r_1} m_{2r_2} m_{3r_3} \cdots m_{mr_m},$$

according to the general expression given for the expansion of a determinant of order m on page 152, where P is the permutation operator which forms $m!$ different permutations of the row indices, $r_1, r_2, r_3, \ldots, r_m$ (or column indices), then the complementary minor of m_{ij} is an $(m-1)$ order determinant such that $|M| = \sum_{j=1}^{m} m_{ij}M^{ij}$; this expression is usually referred to as the Laplace development of a determinant.[1] An adjoint matrix may now be defined as the matrix which results when the cofactor of each element in the original matrix is written down and then this matrix is transposed, such that the adjoint matrix of M, written $\hat{M} = |M^{ji}|$; from the properties of determinants, $M\hat{M} = \hat{M}M = |M|E$, and if $|M| = 0$ (called a singular matrix), then $M\hat{M} = \hat{M}M = 0$ but if M is non-singular then $\hat{M}/|M| = M^{-1}$ or $M^{-1}M = MM^{-1} = E$. These properties of matrices are useful in coordinate transformations. Some authors refer to what is defined above as an adjoint matrix as an adjugate matrix and to an associate matrix, as defined above, by the term adjoint.[2]

Similar matrices may be written for other operators, and matrix mechanics is concerned with the manipulation by addition and multiplication of such quantities, rather than with the operators and wave functions of wave mechanics. Generally, a linear combination wave function expressed as a convergent series, $\sum_i a_i\psi_i$, may be regarded as a vector, which in turn may be

[1] Muir, T., *Theory of Determinants in the Historical Order of Development*, 4 vols., Blackie & Son, Ltd., London, 1930.
[2] Margenau, H., and Murphy, G. M., *The Mathematics of Physics and Chemistry*, 2nd edition, D. Van Nostrand Co., Inc., Princeton, New Jersey, 1961.

represented as a particularly simple kind of matrix, namely a row or a column matrix, such as

$$[x_1, x_2, x_3, \ldots, x_n] \qquad \text{or} \qquad \begin{bmatrix} x_1 \\ x_2 \\ x_3 \\ \vdots \\ x_n \end{bmatrix}. \qquad \text{(q.2.68.)}$$

For an orthonormal set of functions, ψ_i, where (q.2.25.) applies, the Hamiltonian or energy matrix reduces to

$$\begin{bmatrix} E_1 & 0 & 0 & \cdots \\ 0 & E_2 & 0 & \cdots \\ 0 & 0 & E_3 & \cdots \\ \cdots\cdots\cdots\cdots\cdots \end{bmatrix}, \qquad \text{(q.2.69.)}$$

which is called a diagonal matrix and the various energy levels of the system are given by the diagonal elements of the matrix; the process of finding the energy levels and wave functions of the system by this method is thus referred to as the diagonalization of the energy matrix.‡

‡ If Ψ_n represents the wave functions of the stationary states of a system, then any arbitrary wave function may be expanded, according to (q.2.27.), in terms of these functions as $\Psi = \sum_n a_n \Psi_n$, and if this expansion is substituted into (q.2.66), the mean value of some property F of the system is obtained as (see page 149)

$$\bar{F} = \sum_n \sum_m a_n{}^* a_m \int \Psi_n{}^* \hat{F} \Psi_m \, d\tau = \sum_n \sum_m a_n{}^* a_m F_{nm}(t) = \sum_n \sum_m a_n{}^* a_m \langle n|\hat{F}|m\rangle,$$

where in matrix mechanics the quantity $F_{nm}(t)$ with all possible values of m and n is referred to as the matrix of the quantity F, and each of the quantities $F_{nm}(t)$ is referred to as the matrix element corresponding to the transition from the state n to the state m. If the operator $\hat{F}$ is not explicitly dependent on time, then the dependence on time of the matrix elements $F_{nm}(t)$ is determined by the manner in which the functions Ψ_n depend on time, which is given by equation (B) on page 110. If these time-dependent functions are substituted into the expression for $F_{nm}(t)$, there is obtained

$$F_{nm}(t) = \int \Psi_n{}^* \hat{F} \psi_m \, d\tau = e^{2\pi i (E_n - E_m)t/h} \int \psi_n{}^*(q) \hat{F} \psi_m(q) \, d\tau = F_{nm} e^{2\pi i \nu_{nm} t},$$

where $\nu_{nm} = \omega_{nm}/2\pi = (E_n - E_m)/h$ is the transition frequency between the states n and m, and the quantities $F_{nm} = \int \psi_n{}^* \hat{F} \psi_m \, d\tau$ constitute the matrix of the quantity F which is time-independent. It should be noted, that in accordance with the discussion on pages 119–120, relating to the indeterminacy of the phase factor in normalized wave functions, the matrix elements F_{nm} and $F_{nm}(t)$ are also determined only to within a factor of the form $e^{ib} = e^{i(a_m - a_n)}$, although again this does not affect any physical results. Exactly similar relationships apply, therefore, to all relationships between matrix elements whether or not they are time-dependent or time-independent.

For some purposes, the matrix elements of the time derivative of F, of the quantity F are required. Since the mean value $\bar{F}$ is equal to $\dot{\bar{F}}$, or $\dot{\bar{F}} = \sum_n \sum_m a_n^* a_m \dot{F}_{nm}(t)$, the matrix elements of $\dot{F}$ may be obtained as

$$\dot{F}_{nm}(t) = 2\pi i \nu_{nm} F_{nm}(t),$$

or on dividing both sides of this equation by the time factor $e^{2\pi i \nu_{mn} t}$, the time-independent matrix elements of $\dot{F}_{nm}$ are obtained as

$$\dot{F}_{nm} = 2\pi i \nu_{nm} F_{nm} = (2\pi i/h)(E_n - E_m)F_{nm}.$$

Making use of the definition of a Hermitean operator, as given by (q.2.50.), the matrix elements of the complex conjugate F^* of the quantity F are obtained as

$$F_{nm}^* = \int \psi_m \hat{F}^* \psi_n^* \, d\tau = \int \psi_n^* \hat{F} \psi_m \, d\tau, \qquad \text{or} \qquad F_{nm}^* = F_{mn}^*.$$

Therefore, for real physical quantities $F_{nm} = F_{mn}^*$, and such matrices are Hermitean, like the corresponding operators. As in ordinary matrix algebra, the matrix elements with $n = m$ are called diagonal elements; these are real and time-independent, such that the element F_{nn} represents the mean value of the quantity F in the state ψ_n.

The conventional matrix multiplication rules apply equally well to these quantities F_{nm}. Thus, if ψ_n is expanded in terms of the functions ψ_m and the expansion coefficients determined according to equations (q.2.27.) and (q.2.28.), respectively, then

$$\psi_n = \sum_m a_m \psi_m = \sum_m \int \psi_m^* \psi_n \, d\tau \psi_m, \qquad \text{and so} \qquad \hat{F}\psi_n = \sum_m F_{mn}\psi_m.$$

In a similar manner, the product of two operators acting on the function ψ_n may be obtained as

$$\hat{F}\hat{G}\psi_n = \hat{F}(\hat{G}\psi_n) = \hat{F}\sum_k G_{kn}\psi_k = \sum_k G_{km}\hat{F}\psi_k = \sum_k \sum_m G_{km}F_{mk}\psi_m.$$

However, from the expression above for $\hat{F}\psi_n$, the result of the double operation on ψ_n with the operators $\hat{F}\hat{G}$ may also be written as $\hat{F}\hat{G}\psi_n = \sum_m (FG)_{mn}\psi_m$, so that the matrix elements of the product FG are given by $FG_{mn} = \sum_k F_{mk}G_{kn}$, as given on page 151 for ordinary matrix multiplication. In general, then, if the matrix is known, not only is the operator known, but it is possible to determine the eigenvalues of the physical quantity concerned and the corresponding eigenfunctions.

To indicate now how the diagonalization of the Hamiltonian or energy matrix of a system comes about, let it be considered that any arbitrary wave function Ψ is expanded in terms of the eigenfunctions of the Hamiltonian operator $\hat{H}$, or alternatively expressed in terms of the wave functions ψ_m of the stationary states of the system, which are time-independent, then $\Psi = \sum_m a_m \psi_m$. If this expansion is now substituted into the equation $\hat{F}\Psi = F\Psi$, determining the eigenfunctions and eigenvalues of the quantity F (if F is the energy of the system, then this equation is of course $\hat{H}\Psi = H\Psi = E\Psi$), then

$$\sum_m a_m[\hat{F}\psi_m] = F\sum_m a_m\psi_m.$$

Multiplying this equation throughout by ψ_n^* and integrating over all of the coordinate space involved leads to the result that $\sum_m F_{nm}a_m = Fa_n$, since by virtue of (q.2.25), each of the integrals on the left-hand side of this equation $\int \psi_n^* \hat{F} \psi_m \, d\tau$ give the corresponding matrix elements F_{nm}, and on the right-hand side all the integrals $\int \psi_n^* \psi_m \, d\tau$ with $m \neq n$ vanish due to the orthogonality of the functions ψ_m and the integrals $\int \psi_n^* \psi_n \, d\tau = 1$, because of their normalization. This result may be alternatively expressed as

$$\sum_m [F_{nm} - F\,\delta_{nm}]a_m = 0,$$

where δ_{nm}, the Kronecker or Weierstrass delta, is equal to 0 for $n \neq m$ and equal to unity for $n = m$. This series of homogeneous algebraic equations is similar to the system of equations, given by (q.4.94.), obtained by the application of general perturbation theory. As before, such a system of equations have non-zero solutions only if the determinant of their coefficients a_m vanishes, or if $|F_{nm} - F\,\delta_{nm}| = 0$. The roots of this equation correspond to the various possible values of F, and the set of values of a_m satisfying the equations $\sum_m [F_{nm} - F\,\delta_{nm}]a_m = 0$ for each of these values of F determines the corresponding eigenfunction. From the definition of the matrix elements of the quantity F, $F_{nm} = \int \psi_n{}^*\hat{F}\psi_m\,d\tau$, if ψ_n are the eigenfunctions of this quantity, then from $\hat{F}\psi_n = F_n\psi_n$, it is deduced that $F_{nm} = \int \psi_n{}^*\hat{F}\psi_m\,d\tau = F_m \int \psi_n{}^*\psi_m\,d\tau$, and again because of the orthogonality of the functions ψ_m, this leads to the conclusion that $F_{nm} = 0$ for all $n \neq m$ and $F_{mm} = F_m$; therefore, all the non-diagonal matrix elements are zero and each of the diagonal matrix elements is equal to the corresponding eigenvalue of the quantity F. It is such a matrix, where only the diagonal elements have non-zero values, which is said to be diagonalized. Generally, the matrix of any quantity F, defined with respect to the eigenfunctions of some operator $\hat{G}$, is referred to as the matrix of F in a representation in which G is diagonal, and specifically with the wave functions of the stationary states of a system chosen as the functions ψ_n, the energy matrix is diagonal, as given by equation (q.2.69.); the matrices of all other physical quantities having definite values in the stationary states of a system are also diagonal. The sum of the diagonal elements of a matrix is referred to as the trace or spur of the matrix, and is generally denoted by $tr\,F$ or $sp\,F$, so that $tr\,F = \sum_n F_{nn}$; this sum is only defined if the sum over n is convergent. The trace of the product of two matrices is independent of the order of multiplication, since by the rule for matrix multiplication

$$tr\,[FG] = \sum_n \sum_k F_{nk}G_{kn} = \sum_k \sum_n G_{kn}F_{nk} = tr\,[GF].$$

In the same way, it may be shown that for a product of several matrices the trace is unaffected by any cyclic permutation of the products, or $tr\,[FGH] = tr\,[HFG] = tr\,[GHF]$. The trace of a matrix has the further important property that it is independent of the choice of the set of functions with respect to which the matrix elements are defined.

A matrix F_{nm} may be regarded as the operator $\hat{F}$ in the energy representation, since the set of coefficients a_n in the expansion $\sum_n a_n\psi_n$, in terms of the eigenfunctions ψ_n of the Hamiltonian of a system, may be considered to be the wave functions in the energy representation; the variable n gives the number of the stationary state. Thus, the expression $\bar{F} = \sum_n \sum_m a_n{}^*a_m F_{nm}$, for the mean value of a quantity F corresponds to the general quantum mechanical expression for the mean value of a quantity in terms of its operator and the wave function of the state involved.

The commutation properties of operators, discussed on page 138, namely that if two operators commute with one another, they have simultaneous eigenfunctions, may also be proved by use of the matrix representation of operators. Thus, for two operators $\hat{F}$ and $\hat{G}$, which commute with one another, $\hat{F}\hat{G} = \hat{F}\hat{G}$, and applying the matrix multiplication rule, $\sum_k F_{mk}G_{kn} = \sum_k G_{mk}F_{kn}$. If ψ_n are the set of functions, representing the eigenfunctions of the operator $\hat{F}$, with respect to which the matrix elements are calculated, then $F_{mk} = 0$ for $m \neq k$, and therefore the equation above reduces to $F_{mm}G_{mn} = G_{mn}F_{nn}$, or $G_{mn}[F_m - F_n] = 0$. So, if all the eigenvalues F_n of the quantity F are dissimilar, then for all values of $m \neq n$, $F_m - F_n \neq 0$, or G_{mn} must be zero. Therefore, the matrix G_{mn} must also be diagonal, or the functions ψ_n must also be eigenfunctions of the physical quantity G. If there are degeneracies among the values F_n (if there are several different eigenfunctions corresponding to the same eigenvalue), then the matrix elements

G_{mn} corresponding to each such group of functions ψ_n, will generally differ from zero. However, linear combinations of the functions ψ_n which correspond to a single value of the quantity F must also be eigenfunctions of F, and such combinations may always be chosen so that the corresponding nondiagonal matrix elements G_{mn} are zero. Therefore, even in this event, a set of functions which are simultaneously eigenfunctions of $\hat{F}$ and $\hat{G}$ may be obtained.

Since the matrix elements of any particular physical quantity may be defined with respect to various sets of functions, it is sometimes necessary to transform matrices from one representation to another. If, for example $\psi_n(q)$ and $\psi_n'(q)$, where $n = 1, 2, \ldots$, are two complete sets of orthonormal functions which are related to one another by some linear transformation such that $\psi_n' = \sum_m \Omega_{mn}\psi_m$, which amounts to expanding the functions ψ_n' in terms of the complete set of functions ψ_n, then this transformation may be written in operator form as $\psi_n' = \hat{\Omega}\psi_n$. In order that the functions ψ_n' should be orthonormal when the functions ψ_n are orthonormal, the operator $\hat{\Omega}$ must conform to certain conditions. If $\psi_n' = \hat{\Omega}\psi_n$ is substituted into the orthonormality condition, that $\int \psi_m'^*\psi_n' \, d\tau = \delta_{mn}$, and making use of (q.2.50), there results $\int [\hat{\Omega}\psi_n]\hat{\Omega}^*\psi_n^* \, d\tau = \int \psi_m^*\hat{\Omega}^*\hat{\Omega}\psi_n \, d\tau = \delta_{mn}$, and for these equations to be valid for all values of m and n $\hat{\Omega}^*\hat{\Omega}$ must be equal to unity. Therefore $\hat{\Omega}^* = \hat{\Omega}^{-1}$, or the inverse operator must be equal to the Hermitean conjugate operator; such operators are referred to as unitary operators (see page 151), and because of this property, the transformation $\psi_n = \hat{\Omega}^{-1}\psi_n$ which is the inverse of $\psi_n' = \sum_m \Omega_{mn}\psi_m$ must be given by $\psi_n = \sum_m \Omega_{nm}^*\psi_m'$, If now, the relationship $\hat{\Omega}\hat{\Omega}^* = 1$ is written in matrix form, then it becomes $\sum_k \Omega_{km}^*\Omega_{kn} = \delta_{mn}$ or $\sum_k \Omega_{mk}^*\Omega_{nk} = \delta_{mn}$.

By making use of the concept of mean or average or expectation value of a variable, it is now possible to define more exactly what is meant by the uncertainty in a variable, and hence to deduce the Uncertainty Principle. Of the various possible definitions of the uncertainty in a measured quantity, perhaps the most widely used is the root-mean-square deviation from the mean value. Thus, applying the foregoing definition of mean values to the one-dimensional motion of a particle, describable by a normalized wave function, ψ, the quantities Δx and Δp of equations (q.2.42.) and (q.2.43.) may now be written as,

$$[\Delta x]^2 = \overline{[x - \bar{x}]^2} = \overline{[x^2 - 2x\bar{x} + \bar{x}^2]} = \overline{[x^2]} - \bar{x}^2,$$

since $\overline{x\bar{x}} = \bar{x}^2$, and similarly,

$$[\Delta p_x]^2 = \overline{[p_x - \overline{p_x}]^2} = \overline{p_x^2} - \overline{[p_x]}^2, \qquad \text{(q.2.70.)}$$

where

$$\bar{x} = \int_{-\infty}^{\infty} \psi^* x \psi \, dx, \quad \overline{[x^2]} = \int_{-\infty}^{\infty} \psi^* x^2 \psi \, dx, \qquad \overline{p_x} = [h/2\pi i] \int_{-\infty}^{\infty} \psi^* (d\psi/dx) \, dx,$$

and

$$\overline{[p_x^2]} = -[h^2/4\pi^2] \int_{-\infty}^{\infty} \psi^* (d^2\psi/dx^2) \, dx. \qquad \text{(q.2.71.)}$$

Now if A, B, and C, are real constants, then $|(d\psi/dx) + (Ax + B + iC)\psi|^2 \geq 0$, since the squared modulus of a complex number must be positive or zero. This expression, which cannot be negative may be alternatively written as,

$$[(d\psi^*/dx) + (Ax + B - iC)\psi^*][(d\psi/dx) + (Ax + B + iC)\psi]$$
$$= (d\psi^*/dx)(d\psi/dx) + (Ax + B)[\psi(d\psi^*/dx) + \psi^*(d\psi/dx)]$$
$$+ iC[\psi(d\psi^*/dx) - \psi^*(d\psi/dx)] + A^2x^2\psi^*\psi$$
$$+ (B^2 + C^2)\psi^*\psi + 2ABx\psi^*\psi \geq 0. \qquad (q.2.72.)$$

If (q.2.70.) is now integrated throughout, with respect to x, there is obtained,

$$\int_{-\infty}^{\infty} (d\psi^*/dx)(d\psi/dx)\, dx + A \int_{-\infty}^{\infty} x[\psi(d\psi^*/dx) + \psi^*(d\psi/dx)]\, dx$$

$$+ B \int_{-\infty}^{\infty} [\psi(d\psi^*/dx) + \psi^*(d\psi/dx)]\, dx + iC \int_{-\infty}^{\infty} [\psi(d\psi^*/dx) - \psi^*(d\psi/dx)]\, dx$$

$$+ A^2 \int_{-\infty}^{\infty} \psi^*x^2\psi\, dx + [B^2 + C^2] \int_{-\infty}^{\infty} \psi^*\psi\, dx + 2AB \int_{-\infty}^{\infty} \psi^*x\psi\, dx \geq 0.$$
$$(q.2.73).$$

If the various integrals are integrated in the usual manner by parts, then since ψ vanishes at both $+\infty$ and $-\infty$, it is found that,

$$\int_{-\infty}^{\infty} (d\psi^*/dx)(d\psi/dx)\, dx = -\int_{-\infty}^{\infty} \psi^*(d^2\psi/dx^2)\, dx,$$

$$\int_{-\infty}^{\infty} \psi(d\psi^*/dx)\, dx = -\int_{-\infty}^{\infty} \psi^*(d\psi/dx)\, dx,$$

and

$$\int_{-\infty}^{\infty} x[\psi(d\psi^*/dx) + \psi^*(d\psi/dx)]\, dx$$
$$= \int_{-\infty}^{\infty} x[d/dx(\psi^*\psi)]\, dx = -\int_{-\infty}^{\infty} \psi^*\psi\, dx = -1. \quad (q.2.74.)$$

If now (q.2.74.) is substituted in (q.2.73.), making use of the mean value expressions for the various integrals given in (q.2.71.) there results,

$$(4\pi^2/h^2)\overline{[p_x^2]} - A + 2C(2\pi/h)\overline{p_x} + A^2\overline{x^2} + B^2 + C^2 + 2AB\overline{x} \geq 0.$$
$$(q.2.75.)$$

Since this latter expression has been derived from (q.2.72.), it must be valid for any real values of A, B, and C. If therefore, A, B, and C, are set equal to, $A = [1/2(\Delta x)^2]$, $B = -[\bar{x}/2(\Delta x)^2]$, and $C = -\overline{p_x}(2\pi/h)$, respectively, then (q.2.75.) becomes,

$$(4\pi^2/h^2)[\overline{p_x^2} - \overline{(p_x)}^2] - [1/2(\Delta x)^2] + [1/4(\Delta x)^4][\overline{x^2} - (\bar{x})^2] \geq 0,$$

which reduces to

$$[4\pi^2/h^2][\Delta p_x]^2 - [1/4(\Delta x)^2] \geq 0,$$

or

$$\Delta x \cdot \Delta p_x \geq h/4\pi, \qquad\qquad (q.2.76.)$$

which is one of the four Uncertainty Relationships, already discussed in Q.2.5.

Q.2.8. The Laws of Quantum Mechanics

The various topics discussed in the foregoing part of this chapter may be conveniently summarized as a set of four general laws, expressing the relationships between three general concepts. The quantum mechanical concepts of state functions, operators, and dynamical variables, which may be variously described in terms of different mathematical formalisms, provide a particularly convenient method of describing the behavior of electrons, which are a universal constituent of all atoms, molecules, ions (simple and complex), and crystals, in such a manner that the experimentally observed properties of substances may be interpreted and correlated with one another. The four general laws or principles, in terms of which most of the subject matter of chemistry and physics is merely a mathematical consequence may be stated as follows, in the Schrödinger formulation:

I.   The various possible state functions, Ψ, of any system are those which satisfy the second order differential equation, $\hat{H}\Psi = -[h/2\pi i]\,\partial\Psi/\partial t$.

In terms of Schrödinger wave mechanics, $\hat{H}$ is the Hamiltonian operator for the system, which for any number of particles, n, with the ith particle having a mass, m_i, is given by $\hat{H} = -\sum_{i=1}^{n} [h^2/8\pi^2 m_i]\nabla_i^2 + U(x, y, z, \ldots, x_n, y_n, z_n, t)$; Ψ is a function of the time and of all of the positional coordinates of the n particles and $\nabla_i^2 = \partial^2/\partial x_i^2 + \partial^2/\partial y_i^2 + \partial^2/\partial z_i^2$. In the event that the potential energy of the system does not depend explicitly on the time, the above, time dependent Schrödinger equation reduces to $\hat{H}\psi = E\psi$, where $\hat{H}$ is the same operator and ψ and U are functions of the positional coordinates only of the n particles composing the system. This latter, time-independent, steady state (or stationary state) equation constitutes the basis for the discussion of most systems of chemical interest.

In the matrix representation of quantum mechanics, operators are designated by expressions of the type of (q.2.67.) and state functions by vectors such as (q.2.68.), and in fact a much more general discussion of quantum mechanics is in terms of an infinite, linear vector space (Hilbert space) and making use of the concept of density matrices, originally introduced by Dirac,[1] which has been extensively applied in the application of quantum mechanics to many-electron molecular systems.[2,3] The best general understanding of the formal, mathematical, structure of quantum mechanics is in terms of matrix methods.[4,5]

II.   In any assembly of systems, the only possible values which can be observed for a dynamical variable, are the eigenvalues, a, of the operator, $\hat{A}$, operating on the eigenfunctions, ϕ, in the equation, $\hat{A}\phi = a\phi$.

The only restrictions that must be placed on the eigenfunctions of this equation are that the various ϕ's should be finite, single-valued, and quadratically integratable, over the whole range of values of the variable concerned.

III.   If a large number of measurements of the dynamical variable a is made on a system in the state Ψ, or if the measurements are made on a large number of identical systems all in the same state Ψ, the average or expectation value obtained will be $\bar{a} = \int \Psi^* \hat{A} \Psi \, d\tau / \int \Psi^* \Psi \, d\tau$.

In this expression the integrations are carried out over the whole range of values accessable to the various variables used in describing the system.

IV.   For any system of electrons, that is for all atoms, ions (simple and complex), molecules, and crystals, the total state function must be antisymmetric in the electron coordinates.

The above statement is really the quantum mechanical equivalent of the Pauli Exclusion Principle (page 70). The validity of this statement is not restricted to electrons alone; nucleons, protons and neutrons may also only be adequately described in terms of anti-symmetric state functions, whereas photons, for example, are describable by a symmetric wave or state function. It is only possible to understand this interpretation of the Pauli Exclusion

---

[1] Dirac, P. A. M., *Proc. Cambridge Phil. Soc.*, 1930, **26**, 376; **27**, 240.
[2] Löwdin, P.-O., *Phys. Rev.*, 1955, **97**, 1474.
[3] Fraga, S., and Malli, G., *Many Electron Systems: Properties and Interactions*, 1968, W. B. Saunders & Co., Philadelphia.
[4] Dirac, P. A. M., *Quantum Mechanics*, 1958, 4th edition, Oxford University Press, New York.
[5] Neumann, J. von, *Mathematical Foundations of Quantum Mechanics*, 1955, Princeton University Press, Princeton, N.J.

Principle after some discussion of the important concept of angular momentum and electron spin (Chapter Q.5.); in fact, the idea of electron spin and spin quantum numbers does not appear in Schrödinger wave mechanics, but only emerges as a mathematical consequence of Dirac's relativistic quantum mechanics. The quantum statistics of Fermi–Dirac is discussed in Chapter M.3.

Q3

The Motion of a Particle in a Central Field of Force

Q.3.1. Potential Energies in Quantum Mechanics

In the application of quantum mechanical theory to chemical systems, the energy is the predominantly important dynamical variable.‡

‡ It is useful to keep in mind the relative magnitudes of the various energy quantities associated with atoms and molecules. It is common practice, since most such energy quantities are customarily obtained by spectroscopic techniques, to express these atomic and molecular energies in reciprocal centimeters, where

$$1 \text{ cm}^{-1} \simeq 2.998 \times 10^{10} \text{ cm s}^{-1} \simeq 2.859 \text{ cal mole}^{-1} \simeq 11.9626 \text{ J}.$$

Experimental observations carried out in the ultraviolet region of the electromagnetic spectrum provide information about the symmetry of atoms and molecules and the energy levels between which electronic transitions take place; similar information is also obtainable from observations made in the visible and near infrared regions. Infrared,

TABLE Q.3.1.

Spectral region	Energy range in J mole^{-1}
Ultraviolet	2.5×10^6 -2.5×10^5
Visible and near infrared	3.0×10^5 -5.0×10^4
Infrared	1.0×10^5 -5.0×10^3
Raman	5.0×10^4 -1.0×10^3
Microwave	1.5×10^3 -1.0×10^1
Electron spin	1.5×10^1 -1.0×10^{-2}
Nuclear spin	1.5×10^{-1} -1.0×10^{-3}

162

Raman, and microwave spectra, provide information about molecular geometry and the rotational and vibrational energies of molecules. Electron spin resonance spectra enable electron distributions in atoms and molecules to be deduced, and nuclear spin resonance spectra provide information about the environment of particular atoms in molecules. The energy ranges associated with these various types of spectra are approximately as shown in Table Q.3.1.

The excitation of the high frequency or X-ray spectra of atoms requires the expenditure of energies of the order of 10^6–10^{10} J mole^{-1}, and radioactive decay processes in atoms are associated with energies of the order of 10^{-15}–10^{-12} J atom^{-1}. On the still increasing energy scale, nuclear reactions take place at energies up to 10^{-5} J per nuclear event, particle accelerators provide energies up to 10^{-2} J per particle, energetic cosmic ray particles may have energies up to 50 J, and individual particle events may take place in quasers with energies up to 10^3 J.

For all atoms and molecules, in the absence of interaction with some form of electromagnetic radiation, their energies will not depend explicitly on the time, and will be obtained as the eigenvalues of the steady state Schrödinger equation, $\hat{H}\psi = E\psi$, where $\hat{H}$ is the Hamiltonian for the system concerned and the ψ's are the corresponding eigenfunctions. The total energy, the sum of the kinetic and potential energies is actually incorporated in the Schrödinger equation as the kinetic energy, $K = E - U$, rather than as $E = K + U$; the logical necessity for this is easily apparent from the consideration that for a particle not moving in an external potential field, where $U = 0$, the energy involved is simply its kinetic energy, which in this case is the same as its total energy, but if in the presence of an external potential field, E were to be represented by $K + U$, then the corresponding Schrödinger equation with its corresponding eigenvalues and eigenfunctions would remain unaltered, no matter what value of U would be used, which certainly cannot be true. So that, as in (q.2.3.), both E and U appear; this equation only differs in form, from one system to another, by the mass and the nature of the potential energy function. In fact, it is only for certain rather particular forms of this potential energy function in the Schrödinger equation that solutions for the total energy may be obtained in closed terms; this is so for a free particle ($U = 0$), a rigid rotator ($U = $ constant), for the linear harmonic oscillator, the linear anharmonic oscillator, the three-dimensional harmonic oscillator, for linear electronic oscillators in a uniform field and for coupled electronic oscillators, for hydrogen-like atoms and hydrogen-like atoms in a uniform field, and for the hydrogen molecule. For all other systems, including the majority of those of real chemical interest, various approximation methods have to be used to represent their potential energy functions in order to arrive at approximate eigenvalues. Some of these approximation techniques will be discussed in subsequent chapters. In the present chapter, attention will be focused on the detailed solution of the problem of a particle moving in a central field of force. Actually, the knowledge of the detailed solution, in quantum mechanical terms, of the problem of the motion of the electron in the hydrogen atom, provides the model for most other quantum mechanical calculations, and a starting point for devising solutions for other more complicated problems.

In terms of the laws of classical thermodynamics, any system, whether an atom, a molecule, a complex ion, or a crystal, exists as a stable entity if the total energy of the system is less than the sum of the energies of the constituent particles from which that system can be described as being constituted. For most systems of chemical interest, this energy is effectively the electronic energy, which may be conveniently regarded as the sum of the kinetic energies and the mutual Coulomb energies of electrons with electrons, electrons with nuclei, and nuclei with one another. Discrete molecules may in addition have considerable energy of rotation and vibration, but these are generally small compared with the electronic energies. In the case of the simple system of the hydrogen atom, which on the basis of much accumulated information is composed of an electron moving in the spherically symmetrical Coulomb potential field of a proton, as well as for other more complicated systems, a substantial amount of information about the system may be summarized in what is usually referred to as its potential energy curve. This may be constructed if the energy of the system can be calculated for a variety of values of the positional parameters of the two particles with respect to one another. It has also to be kept in mind that in view of the description of the state of motion of a particle, in quantum mechanics, by means of a state function, which for stationary states is a function of the positional coordinates of the particle, and the probability interpretation of the state function, that there are no sharp or definite boundaries to the potential energy of the particle; in other words, there is a finite probability of finding the particle in regions of space where the kinetic energy of the particle may be negative.

Before embarking on the discussion of the problem of the motion of a particle in a central field of force, and the specific case of the hydrogen atom, which is of primary importance in atomic quantum mechanics, it is perhaps instructive to examine a simple three-dimensional problem which introduces some of the techniques required; this is the extension of the problem considered in Q.2.2. to three dimensions and usually referred to as the problem of the particle in a box.

Q.3.2. The Particle in a Box

While apparently a trivial problem, the idealized model of a particle moving in the three-dimensional space inside a cubical potential box of edge length L, is of importance in discussing the behavior of several systems of real physical interest, such as the motion of molecules of an ideal gas, the electron distribution in certain molecules, and the motion of electrons in metals. In fact, with the discovery of such compounds as the cubanes and their substituents,[1,2] the real possibility exists of synthesizing organic metals in which electrons move in the space provided by cubical arrays of carbon atoms.

If it is assumed that the potential energy of the particle is zero inside the box and infinite outside the boundaries of the box, then the wave function describing the state of motion of the particle, $\psi(x, y, z)$, must, by analogy with the one-dimensional case, fall to zero at the surface of the box. If the origin of a Cartesian coordinate reference axial system is taken as one corner of the box and the reference axes as being coincident with three adjacent edges of the box, then $U(x) = U(y) = U(z) = 0$ for

$$0 < \begin{vmatrix} x \\ y \\ z \end{vmatrix} < L.$$

Inside the box, the Schrödinger equation for the particle becomes

$$\partial^2/\partial x^2[\psi(x, y, z)] + \partial^2/\partial y^2[\psi(x, y, z)] + \partial^2/\partial z^2[\psi(x, y, z)]$$
$$+ 8\pi^2 mh^{-2}E(x, y, z) = 0.$$

If this differential equation can be solved by the method of separation of the variables,‡ that is if solutions of the form, $\psi(x, y, z) = X(x)Y(y)Z(z)$, exist, where X, Y, and Z are functions of the coordinates, x, y, z, respectively, then all possible solutions of the original equation can be obtained by constructing linear combinations of the separable solutions.

‡ A *differential equation* is one which contains a dependent variable, y, and one or more independent variables, x_1, x_2, x_3, ..., x_n, and the derivatives of the dependent variable with respect to the independent variables. If the equation contains more than one independent variable and various partial derivatives, then it is a partial differential equation. Those equations containing only one independent variable are called ordinary differential equations. The order of a differential equation is equal to the order of the highest derivative in the equation; the degree of the equation is equal to the power of the highest derivative in the equation after all fractional powers have been eliminated. Thus, $d^2y/dx^2 + l(dy/dx)^2 + \mu x^4 = 0$, is of the second order and first degree. As a consequence, the complete integral of a differential equation of the nth order, must contain n, and no more than n, arbitrary constants. The various phenomena, including quantum mechanical descriptions of natural phenomena, represented by differential equations, must provide on integration a sufficient number of undetermined constants to define the initial conditions of the process symbolized by the differential equation. There are obviously a great many types of differential equations; first order equations can generally

[1] Eaton, P. E., and Cole, T. W., Jr., *J. Amer. Chem. Soc.*, 1964, **86**, 3157.
[2] Fleischer, E. B., *J. Amer. Chem. Soc.*, 1964, **86**, 3889.

be solved in a relatively straightforward manner, but equations of a higher order than second generally cannot be solved analytically in closed terms. Many of the equations which occur in quantum mechanics are second order, homogeneous, differential equations, such as $d^2y/dx^2 + f_1(x)\,dy/dx + f_2(x)y = 0$; if the right hand side of such an equation is equal to some other function of x, say $f_3(x)$, rather than zero, then the equation is inhomogeneous. This equation is a special case of the general class of linear, inhomogeneous, differential equations, represented by,

$$f_n(x)\,d^ny/dx^n + f_{n-1}(x)\,d^{n-1}y/dx^{n-1} + \cdots + f_1(x)\,dy/dx + f_0(x)y = f_m(x).$$

Certain of the differential equations of frequent occurrence in the description of physical problems have been extensively studied, and the functions which satisfy them are referred to by special names, such as gamma functions (page 43) and Hermite functions (page 144); other special functions such as Bessel functions, Legendre functions, and Laguerre functions, are of special importance in quantum mechanics. In most physical problems involving differential equations, there are necessarily certain boundary conditions limiting the number of possible solutions; in quantum mechanics, generally, the solutions must be well behaved (single-valued, continuous, and quadratically integratable), but there are still an infinity of solutions to any differential equation, and these solutions may be derived in either a deductive or an inductive manner. The basic reason why many differential equations are so difficult to solve is that they are formed by the elimination of constants as well as by the omission of some common factor from the original equation, so that they do not actually represent the complete or total differential of the original equation. There are however, certain widely applicable techniques which may provide some indication of the nature and properties of the functions which constitute solutions. The solutions of inhomogeneous equations are customarily derived from the solutions of the corresponding homogeneous equations. It is frequently of value to examine the form which the equation assumes for infinite or zero values of the dependent variables, to attempt to cast the equation under consideration into the form of some classical differential equation, for which the solutions are otherwise known, by some appropriate substitutions, to attempt to find solutions by the use of series expansion or series integration techniques, or to transform the equation into some form in which the variables may be separated.[1-3] Some of these techniques will be illustrated in the following sections.

Separable solutions of the Schrödinger equation do not always exist, dependent on the form of the potential energy function,[4] and such separable solutions are only possible for a few of the general types of equations which arise in quantum mechanics; in some cases the separable solutions are not conformable with the boundary conditions which must be imposed for the particular problem concerned, or the coefficients in the resulting separable solutions contain the variables in such combinations that there is no way of

---

[1] Margenau, H., and Murphy, G. M., *The Mathematics of Physics and Chemistry*, 1961, D. Van Nostrand Co. Inc., Princeton, New Jersey.
[2] Pauling, L., and Wilson, E. B., Jr., *Introduction to Quantum Mechanics*, 1935, McGraw-Hill Book Co., New York.
[3] Morse, P. M., and Feshbach, H., *Methods of Theoretical Physics*, 1953, McGraw-Hill Book Co., New York.
[4] Eisenhart, L. P., *Phys. Rev.*, 1934, **45**, 427; 1948, **74**, 87.

disentangling them, even when the differential equation is transformed by the use of some other coordinate system of reference. For the particle in the box, the method of the separation of variables does work, since on substituting the product function, $\psi(x, y, z) = X(x)Y(y)Z(z)$, into the Schrödinger equation for the particle, the resulting equation is

$$Y(y)Z(z)\, d^2/dx^2[X(x)] + X(x)Z(z)\, d^2/dy^2[Y(y)] + X(x)Y(y)\, d^2/dz^2[Z(z)]$$
$$+ 8\pi^2 m h^{-2} E X(x) Y(y) Z(z) = 0, \quad \text{(q.3.1.)}$$

which on division throughout by $X(x)Y(y)Z(z)$ reduces to,

$$[X(x)]^{-1}\, d^2/dx^2[X(x)] + [Y(y)]^{-1}\, d^2/dy^2[Y(y)] + [Z(z)]^{-1}\, d^2/dz^2[Z(z)]$$
$$+ 8\pi^2 m E h^{-2} = 0.$$

In this equation the first three terms are functions of x only, y only, and z only, respectively, and the sum of these three terms is equal to a constant. This condition can only be satisfied if each of these three terms is separately equal to some constant. Thus, if E is written as $E = E_x + E_y + E_z$, then

$$[X(x)]^{-1}\, d^2/dx^2[X(x)] = -8\pi^2 m h^{-2} E_x,$$
$$[Y(y)]^{-1}\, d^2/dy^2[Y(y)] = -8\pi^2 m h^{-2} E_y, \qquad \text{(q.3.2.)}$$
$$[Z(z)]^{-1}\, d^2/dz^2[Z(z)] = -8\pi^2 m h^{-2} E_z,$$

thereby accomplishing the separation of the original partial differential equation into three ordinary differential equations, each of which is of the same form, although now expressed in a slightly different notation, as the one-dimensional case discussed in Q.2.2. The same boundary conditions apply as in Q.2.2., namely, $X(0) = X(L) = Y(0) = Y(L) = Z(0) = Z(L) = 0$, and so, for example, the eigenvalues E_x are given by $E_x = n_x^2 h^2/8mL^2$, where $n_x = 1, 2, 3, \ldots$, and the corresponding normalized eigenfunctions by $X(x) = [\sqrt{2/L}] \sin (n_x \pi x/L)$. The solutions for the functions, $Y(y)$ and $Z(z)$, are of exactly the same form, so that the total wave functions of the stationary states of the particle in a box are

$$\psi_{n_x n_y n_z} = [\sqrt{8/L^3}] \sin (n_x \pi x/L) \sin (n_y \pi y/L) \sin (n_z \pi z/L), \quad \text{(q.3.3.)}$$

and the corresponding energy value are

$$E_{n_x n_y n_z} = [h^2/8mL^2][n_x^2 + n_y^2 + n_z^2], \qquad \text{(q.3.4.)}$$

where n_x, n_y, and n_z are integers. The wave functions are zero outside the box. The integral values for n_x, n_y, and n_z, arise as a logical consequence of the requirement that the wave functions be well behaved. Any change in the sign of these integral quantities, the quantum numbers in this case, does not change the energy nor does it lead to different wave functions, but simply multiplies (q.3.3.) by -1. All of the stationary states are thus given by the set of positive integral values of n_x, n_y, and n_z, none of which may adopt the value, zero, for this would make the corresponding wave function zero. Equation (q.3.4.) also shows that the allowed energies are inversely proportional to the square of the edge length of the box, which implies that as the edge length L increases, the allowed energy levels become increasingly more closely spaced. If the box is infinitely large, then the problem of the particle in a box becomes simply that of a free particle with nonquantized energies, so closely spaced that they are virtually continuous. There is thus theoretical justification for the classical statistical thermodynamical treatment of ideal gases, where the translational energies of gas molecules may be regarded as continuous functions of their coordinates; for any macroscopic sample of gas, the size of any container in which the gas may be confined is large compared to the size of the gas molecules.

It is easily shown that the functions (q.3.3.) constitute an orthonormal set, since

$$[8/L^3] \int_0^L [\sin (n_x\pi x/L)]^2 \, dx \int_0^L [\sin (n_y\pi y/L)]^2 \, dy \int_0^L [\sin (n_z\pi z/L)]^2 \, dz = 1$$

(page 120),

and

$$\int_0^L \sin (n\pi x/L) \sin (m\pi x/L) \, dx = 0 \text{ (page 123)},$$

for dissimilar values of m and n.

The integral quantum numbers, n_x, n_y, and n_z, completely specify each stationary state of the particle, the energy E depending only on the sum of the squares of these integers. It is obvious that the lowest allowed energy of the particle is $3h^2/8mL^2$, corresponding to $n_x = n_y = n_z = 1$, and no other state has this energy. This is referred to as the zero-point energy of the particle in a box, and generally such zero-point energies are characteristic of systems known as bound systems, where the system is confined to a finite region of space. The particle can never have zero energy, since in that event the wave function would have to be zero everywhere and there would be no particle; in fact, generally, the trivial solution $\psi = 0$, denotes the non-existence of any state. Furthermore, if $E = 0$, then the momenta of the particle resolved in each of the three coordinate directions would be zero, and thus Δp in each of the three coordinate directions would be zero, which would constitute a violation of the Uncertainty Principle. In general, since E depends on the sum of the squares of three integers, there may be several different stationary states, or different wave functions, each corresponding to the same energy. For example, the nine stationary states having values for n_x, n_y, and n_z, of (8, 1, 1), (1, 8, 1), (1, 1, 8), (7, 4, 1), (1, 4, 7), (7, 1, 4), (5, 5, 4), (4, 5, 5), and (5, 4, 5), respectively, all have the energy, $33h^2/4mL^2$; such states are said to be degenerate, and in this particular case, are nine-fold degenerate. The ground state, on the other hand, is said to be non-degenerate. It is easily shown that any linear combination of several degenerate wave functions is also a wave function corresponding to the same energy. Thus, if ψ_1 and ψ_2 are two wave functions corresponding to the same energy, then $\hat{H}\psi_1 = E\psi_1$, and $\hat{H}\psi_2 = E\psi_2$, and if these two equations are multiplied by the arbitrary constants c_1 and c_2, respectively, and added, then $\hat{H}[c_1\psi_2 + c_2\psi_2] = E[c_1\psi_1 + c_2\psi_2]$, denoting that the linear combination $c_1\psi_1 + c_2\psi_2$ is a wave function corresponding to the same energy E. In general, if p linearly independent wave functions (wave functions which are not simple linear combinations of others) lead to the same energy value, the energy level is said to be p-fold degenerate, and the number p is called the degree of degeneracy of the level. The occurrence of degeneracies in quantum mechanical problems is quite common and is in most cases a consequence of certain symmetry properties of the system concerned. In the present problem of a particle confined within a cubic potential box, the degeneracies are due to the cubic symmetry of the box. If all three edge lengths of the box are of dissimilar magnitudes, the degeneracy is removed, and certain energy levels, which for the cubic potential box have the same magnitude become separated or split into several different energy levels. The general theory of the effect of distortions or perturbations on the dimensions of the box or the effect of any disturbances on the state of motion of the particle is referred to as generalized perturbation theory. The degeneracies occurring in any given system may be deduced from the symmetry properties of the system by the application of the formal theory of groups; group theory is a branch of applied mathematics, which has many applications not only in quantum mechanics but in crystallography and in many other areas of physics and chemistry.[1]

Q.3.3. The Hydrogen Atom

The discussion of the nature of the simplest of all atomic types, the hydrogen atom, in terms of quantum mechanics, is not only of importance by itself, but it furnishes a model for the examination of all other atoms as well as forming a starting point for the development of molecular quantum mechanics. Just as Bohr's explanation of the line spectrum of hydrogen was one of the early successes of the application of the quantum postulate to a system of major chemical interest, the problem of the structure of the hydrogen atom was one of the first to be solved by the application of Schrödinger mechanics.

---

[1] McWeeny, R., *Symmetry—An Introduction to Group Theory and Its Applications*, 1963, The Macmillan Co., New York.

The successful solution of this problem, without the introduction of any unproved postulates, leading to the calculation of quantities which may be directly compared with the results of experimental observation, constitutes convincing proof of the correctness of the theory.

In the system which is a hydrogen atom, a single electron moves in the force field provided by the atomic nucleus, which is a single proton of equal but opposite charge. The well understood electrostatic (Coulomb) inverse square law is responsible for the attraction between the electron and the proton. This attractive force is directed towards a fixed point and its magnitude is a function only of the distance between the electron and the proton, meantime regarded as point charges. The potential energy of the electron is then only a function of its distance from the proton and may be written as $U(r)$, where r is the radial distance between the two particles, the electron and the proton. Such a potential function is said to be spherically symmetrical. The important characteristic of central fields of force, in atomic quantum mechanics, is that a central field of force cannot exert any moment of force or torque on a particle moving under its influence; so that in classical mechanics the angular momentum of the particle is constant and there is a simple analogue of this situation in quantum mechanics. This consideration has the consequence that any discussion of the hydrogen atom can be easily generalized to cover all central force fields. In atomic quantum mechanics, this means that the same argument is applicable not only to the special case of the hydrogen atom but to all hydrogenic or hydrogen-like atoms; these are atoms which are composed of a single electron and a nucleus, such as the singly ionized helium atom, the doubly ionized lithium atom, and so on. Hydrogen-like atoms therefore differ from one another in the mass and the magnitude of the charge on their nuclei. If the electronic charge is denoted by $-\varepsilon$ and the proton charge by ε, then if the nuclear charge or atomic number of an atom of an element is Z, the nuclear charge is $Z\varepsilon$, and the attractive force between such a nucleus and an electron is $Z\varepsilon^2/r^2$, where r is the distance separating the two particles. The potential energy of an electron in the central force field of such a nucleus is then

$$U(r) = -Z\varepsilon^2/r, \qquad \text{(q.3.5.)}$$

on the assumption that the zero of potential energy corresponds to the infinite separation of the two particles.

The proton rest mass, m_p, is 1836 times the electron rest mass, m_e; it is found that for any such system of two particles interacting with one another, both in quantum mechanics as well as in classical mechanics, the center of gravity of the system will either remain at rest or move with a uniform velocity in a straight line, each particle rotating about the common center of gravity. The important aspect of such a system is that the relative motion of

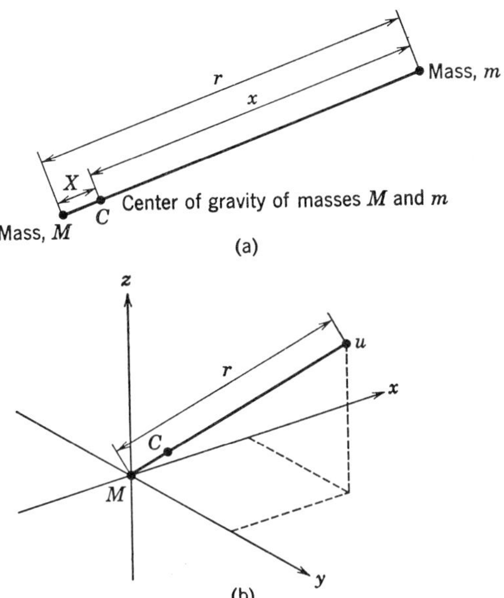

FIG. Q.3.1.   (a) The center of gravity, located at
C, of two masses m and M, distant x and X,
respectively, from C. The two masses are separated
by a distance r. (b) The center of mass coordinate
system. Two masses, m and M, rotating around a
common center of gravity at C, are equivalent to
a reduced mass μ, located at the radial distance r
from the center of mass coordinate system origin.

the two particles is independent of whether the center of mass is stationary or
moving. If, therefore for convenience, it is assumed as in Fig. Q.3.1. that two
masses, m and M, are located at opposite ends of a line passing through the
origin located at the center of gravity of the pair, the distances of the two
masses from the origin being x and X respectively, such that by this definition
of the center of mass, $MX = mx$ and $x + X = r$. Then $M(m + M)^{-1} =
x(x + X)^{-1}$, or $M(m + M)^{-1} = x(x + r - x)^{-1} = x/r$, or

$$x = Mr(m + M)^{-1}.$$

In a similar manner, $m(m + M)^{-1} = X(x + X)^{-1} = X(r - X + X)^{-1} =
X/r$ or $X = mr(m + M)^{-1}$. If the angular velocity of rotation of the line
defined by mM around the common center of mass is ω, then the linear
velocity of motion of mass m is $v = \omega x$ and of mass M is $V = \omega X$. The total
energy of the pair of rotating masses is then the sum of their kinetic energies, or

$$E_{\text{total}} = MV^2/2 + mv^2/2 = M\omega^2 X^2/2 + m\omega^2 x^2/2 = (\omega^2/2)[MX^2 + mx^2],$$

and substituting for x, X,

$$E_{\text{total}} = (\omega^2/2)[Mr^2m^2(m + M)^{-2} + mM^2r^2(m + M)^{-2}]$$
$$= (\omega^2r^2/2)[(Mm^2 + mM^2)(m + M)^{-2}]$$
$$= (\omega^2r^2/2)[mM(m + M)^{-1}] = \mu\omega^2r^2/2, \text{ where } \mu = mM(m + M)^{-1}.$$

$$(q.3.6.)$$

The quantity μ is referred to as the reduced mass of the system, and from (q.3.6.) it is apparent that in this so-called center of mass coordinate system, the two masses behave as a single particle of mass μ, moving at a distance r from the origin. For two masses such as an electron and a proton, the center of gravity obviously subdivides the vector joining the electron and proton in the ratio of $1:1836$, so that the nucleus of the hydrogen atom executes only a very slight motion, and to a very good approximation the proton may be regarded as stationary with the electron moving about this fixed center of attraction at the nucleus. There is, of course, no approximation involved if the reduced mass of the electron–proton system is used rather than m_ε in describing the motion of the electron with respect to the proton. The reduced mass μ is less than m_ε by about 1 part in 1836. While the use of m_ε instead of μ in the equation of motion of the electron with respect to the proton in a hydrogen atom introduces only a small error, which is quite inconsequential in comparison with the other approximations which have to be introduced in dealing with systems other than the hydrogen atom, this would only be strictly correct if the nuclear mass were infinite or if the nucleus were at rest.

The Schrödinger equation of the hydrogen atom is then

$$\nabla^2\psi(x, y, z) + 8\pi^2\mu h^{-2}[E + \varepsilon^2/r]\psi(x, y, z) = 0,$$

where ∇^2 is the Laplacian operator. For any hydrogenic atom, this equation becomes

$$-[(h^2/8\pi^2\mu)\nabla^2 + Z\varepsilon^2/r]\psi(x, y, z) = E\psi(x, y, z), \qquad (q.3.7.)$$

where the potential energy of the electron is a function only of the distance r of the electron from the nucleus of charge $+Z\varepsilon$; the total energy E is the energy of the particle of reduced mass μ relative to the energy of translation of the system. The solutions of equation (q.3.7.) constitute rigorous descriptions of all one-electron atoms, of rigid rotators, and are closely related to the approximate solutions of the Schrödinger equations for many other systems, including many-electron atoms. The problem of finding the solutions of this equation is what is usually referred to. as the central-field problem, and this is one of the most important single problems in the quantum mechanical description of atoms and molecules.

Q.3.4. The Central Field Equation

The equation (q.3.7.) is a second order partial differential equation in three independent variables, which has to be separated in order to find solutions. Such equations written in Cartesian coordinates can very seldom be separated; in fact, separable solutions in the Cartesian coordinate system can only be obtained if the potential function is constant or proportional to r^2, as for the free particle and the three-dimensional harmonic oscillator, respectively. For most other problems, especially those dealing with electrons in atoms, the spherical symmetry of the potential function in a central field of force implies the preferable use of the spherical polar coordinate system rather than Cartesian coordinates. This requires the transformation of $\psi(x, y, z)$ to $\psi(r, \theta, \phi)$; the Laplacian operator in spherical polar coordinates is‡

$$\nabla^2 = \partial^2/\partial x^2 + \partial^2/\partial y^2 + \partial^2/\partial z^2$$
$$= \partial^2/\partial r^2 + (2/r)\,\partial/\partial r + (1/r^2 \sin \theta)\,\partial/\partial \theta[\sin \theta\,\partial/\partial \theta] + (1/r^2 \sin^2 \theta)\,\partial^2/\partial \phi^2.$$
$$= (1/r^2)\,\partial/\partial r[r^2\,\partial/\partial r] + (1/r^2 \sin \theta)\,\partial/\partial \theta[\sin \theta\,\partial/\partial \theta] + (1/r^2 \sin^2 \theta)\,\partial^2/\partial \phi^2.$$

$$(q.3.8.)$$

‡ The relationship between the Cartesian and spherical polar coordinate systems is shown in Fig. Q.3.2., from which it may be seen that:

$$OM = (x^2 + y^2)^{1/2}; \qquad\qquad OP = r = (x^2 + y^2 + z^2)^{1/2};$$
$$\tan \theta = \cot\left(\frac{\pi}{2} - \theta\right) = \frac{(x^2 + y^2)^{1/2}}{z}; \qquad \tan \phi = \frac{y}{x}.$$

From these equations, it is possible to express x, y, and z in terms of r, θ, and ϕ, and conversely as follows:

$$x = r \sin \theta \cos \phi; \qquad r = (x^2 + y^2 + z^2)^{1/2};$$
$$y = r \sin \theta \sin \phi; \qquad \theta = \tan^{-1} \frac{(x^2 + y^2)^{1/2}}{z};$$
$$z = r \cos \theta; \qquad \phi = \tan^{-1} \frac{y}{x}.$$

The first derivatives of r, θ, and ϕ, with respect to x, y, and z, are then:

$$\frac{\partial r}{\partial x} = \frac{1}{2}\,[x^2 + y^2 + z^2]^{-1/2}(2x) = \frac{x}{r} = \sin \theta \cos \phi;$$

$$\frac{\partial \theta}{\partial x} = \frac{\partial}{\partial x} \tan^{-1}\left[\frac{x^2 + y^2}{z^2}\right]^{1/2} = \left[\frac{1}{1 + [(x^2 + y^2)/z^2]}\right] \frac{1}{2}\left[\frac{x^2 + y^2}{z^2}\right]^{-1/2}\left(\frac{2x}{z^2}\right)$$

$$= \frac{1}{1 + \tan^2 \theta}\,\frac{x}{\tan \theta}\,\frac{1}{z^2} = \frac{r \sin \theta \cos \phi \cos^3 \theta}{\sin \theta\, r^2 \cos^2 \theta} = \frac{\cos \theta \cos \phi}{r};$$

$$\frac{\partial \phi}{\partial x} = \frac{1}{1 + (y/x)^2}\,(-y/x^2) = -\frac{r \sin \theta \sin \phi}{\sec^2 \phi\, r^2 \sin^2 \theta \cos^2 \phi} = -\frac{\sin \phi}{r \sin \theta};$$

$$\frac{\partial r}{\partial y} = \frac{1}{2}\,[x^2 + y^2 + z^2]^{-1/2}(2y) = \frac{y}{r} = \sin \theta \sin \phi;$$

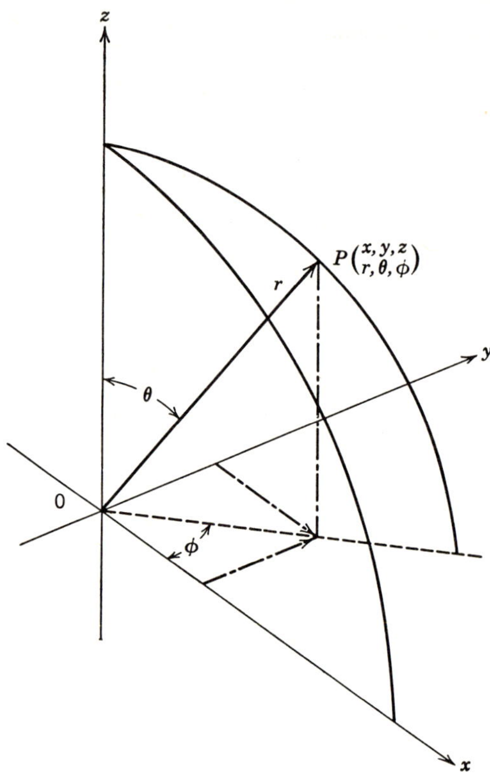

Fig. Q.3.2. Relationship between the Cartesian and the spherical polar-coordinate systems.

$$\frac{\partial\theta}{\partial y} = \frac{1}{1 + [(x^2 + y^2)/z^2]} \frac{1}{2}\left[\frac{x^2 + y^2}{z^2}\right]^{-1/2} \frac{2y}{z^2} = \frac{\cos^2\theta\, r\sin\theta\sin\phi}{\tan\theta\, r^2\cos^2\theta} = \frac{\cos\theta\sin\phi}{r};$$

$$\frac{\partial\phi}{\partial y} = \frac{1}{1 + (y/x)^2}\frac{1}{x} = \frac{\cos^2\theta}{r\sin\theta\cos\phi} = \frac{\cos\phi}{r\sin\theta};$$

$$\frac{\partial r}{\partial z} = \frac{1}{2}[x^2 + y^2 + z^2]^{-1/2}2z = \frac{r\cos\theta}{r} = \cos\theta;$$

$$\frac{\partial\theta}{\partial z} = -\frac{1}{1 + [(x^2 + y^2)/z^2]}\frac{1}{2}\left[\frac{x^2 + y^2}{z^2}\right]^{-1/2}\frac{(x^2 + y^2)2z}{z^4} = -\frac{\cos^2\theta\tan^2\theta\, r\cos\theta}{\tan\theta\, r^2\cos^2\theta}$$

$$= -\frac{\sin\theta}{r};$$

$$\frac{\partial\phi}{\partial z} = 0.$$

Thus, collecting these first derivatives of r, θ, and ϕ, with respect to x, y, and z:

$$\frac{\partial r}{\partial x} = \sin\theta\cos\phi; \qquad \frac{\partial\theta}{\partial x} = \frac{\cos\theta\cos\phi}{r}; \qquad \frac{\partial\phi}{\partial x} = -\frac{\sin\phi}{r\sin\theta};$$

$$\frac{\partial r}{\partial y} = \sin\theta\sin\phi; \qquad \frac{\partial\theta}{\partial y} = \frac{\cos\theta\sin\phi}{r} \qquad \frac{\partial\phi}{\partial y} = \frac{\cos\phi}{r\sin\theta};$$

$$\frac{\partial r}{\partial z} = \cos\theta; \qquad \frac{\partial\theta}{\partial z} = -\frac{\sin\theta}{r}; \qquad \frac{\partial\phi}{\partial z} = 0.$$

If the function which has to be expressed in either set of coordinates is $\psi\begin{bmatrix} x,\ y,\ z \\ r,\ \theta,\ \phi \end{bmatrix}$, then the first partial derivatives of $\psi(r,\ \theta,\ \phi)$ with respect to x, y, and z, are:

$$\frac{\partial\psi}{\partial x}(r,\ \theta,\ \phi) = \frac{\partial\psi}{\partial r}\frac{\partial r}{\partial x} + \frac{\partial\psi}{\partial\theta}\frac{\partial\theta}{\partial x} + \frac{\partial\psi}{\partial\phi}\frac{\partial\phi}{\partial x} = \frac{\partial\psi}{\partial r}\sin\theta\cos\phi + \frac{\partial\psi}{\partial\theta}\frac{\cos\theta\cos\phi}{r} - \frac{\partial\psi}{\partial\phi}\frac{\sin\phi}{r\sin\theta};$$

$$\frac{\partial\psi}{\partial y}(r,\ \theta,\ \phi) = \frac{\partial\psi}{\partial r}\sin\theta\sin\phi + \frac{\partial\psi}{\partial\theta}\frac{\cos\theta\sin\phi}{r} + \frac{\partial\psi}{\partial\phi}\frac{\cos\phi}{r\sin\theta};$$

and,

$$\frac{\partial\psi}{\partial z}(r,\ \theta,\ \phi) = \frac{\partial\psi}{\partial r}\cos\theta - \frac{\partial\psi}{\partial\theta}\frac{\sin\theta}{r}.$$

Now, taking the second derivatives of $\psi(r,\ \theta,\ \phi)$ with respect to x, y, and z, respectively:

$$\frac{\partial^2\psi}{\partial x^2} = \frac{\partial}{\partial x}\left[\sin\theta\cos\phi\,\frac{\partial\psi}{\partial r} + \frac{\cos\theta\cos\phi}{r}\frac{\partial\psi}{\partial\theta} - \frac{\sin\phi}{r\sin\theta}\frac{\partial\psi}{\partial\phi}\right]$$

$$= \sin\theta\cos\phi\,\frac{\partial}{\partial r}\left[\frac{\partial\psi}{\partial x}\right] + \frac{\cos\theta\cos\phi}{r}\frac{\partial}{\partial\theta}\left[\frac{\partial\psi}{\partial x}\right] - \frac{\sin\phi}{r\sin\theta}\frac{\partial}{\partial\phi}\left[\frac{\partial\psi}{\partial x}\right]$$

$$= \sin\theta\cos\phi\left[\sin\theta\cos\phi\,\frac{\partial^2\psi}{\partial r^2} + \frac{\cos\theta\cos\phi}{r}\frac{\partial^2\psi}{\partial r\,\partial\theta} - \frac{\cos\theta\cos\phi}{r^2}\frac{\partial\psi}{\partial\theta}\right.$$
$$\left. - \frac{\sin\phi}{r\sin\theta}\frac{\partial^2\psi}{\partial r\,\partial\phi} + \frac{\sin\phi}{r^2\sin\theta}\frac{\partial\psi}{\partial\phi}\right]$$

$$+ \frac{\cos\theta\cos\phi}{r}\left[\sin\theta\cos\phi\,\frac{\partial^2\psi}{\partial r\,\partial\theta} + \cos\theta\cos\phi\,\frac{\partial\psi}{\partial r} + \frac{\partial^2\psi}{\partial\theta^2}\frac{\cos\theta\cos\phi}{r}\right.$$
$$\left. - \frac{\sin\theta\cos\phi}{r}\frac{\partial\psi}{\partial\theta} - \frac{\partial^2\psi}{\partial\phi\,\partial\theta}\frac{\sin\phi}{r\sin\theta} + \frac{\partial\psi}{\partial\phi}\frac{\sin\phi\cos\theta}{r\sin^2\theta}\right]$$

$$- \frac{\sin\phi}{r\sin\theta}\left[\sin\theta\cos\phi\,\frac{\partial^2\psi}{\partial r\,\partial\phi} - \sin\theta\sin\phi\,\frac{\partial\psi}{\partial r} + \frac{\cos\theta\cos\phi}{r}\frac{\partial^2\psi}{\partial\theta\,\partial\phi}\right.$$
$$\left. - \frac{\cos\theta\sin\phi}{r}\frac{\partial\psi}{\partial\theta} - \frac{\partial\psi^2}{\partial\phi^2}\frac{\sin\varphi}{r\sin\theta} - \frac{\cos\phi}{r\sin\theta}\frac{\partial\psi}{\partial\phi}\right];$$

$$\frac{\partial^2\psi}{\partial y^2} = \frac{\partial}{\partial y}\frac{\partial\psi}{\partial y} = \frac{\partial}{\partial y}\left[\frac{\partial\psi}{\partial r}\sin\theta\sin\phi + \frac{\partial\psi}{\partial\theta}\frac{\cos\theta\sin\phi}{r} + \frac{\partial\psi}{\partial\phi}\frac{\cos\phi}{r\sin\theta}\right]$$

$$= \frac{\partial}{\partial r}\left[\frac{\partial\psi}{\partial y}\sin\theta\sin\phi\right] + \frac{\partial}{\partial\theta}\left[\frac{\partial\psi}{\partial y}\frac{\cos\theta\sin\phi}{r}\right] + \frac{\partial}{\partial\phi}\left[\frac{\partial\psi}{\partial y}\frac{\cos\phi}{r\sin\theta}\right]$$

$$= \sin\theta\sin\phi\left[\frac{\partial^2\psi}{\partial r^2}\sin\theta\sin\phi + \frac{\partial^2\psi}{\partial\theta\,\partial r}\frac{\cos\theta\sin\phi}{r} - \frac{\partial\psi}{\partial\theta}\frac{\cos\theta\sin\phi}{r^2}\right.$$
$$\left. + \frac{\partial^2\psi}{\partial\phi\,\partial r}\frac{\cos\phi}{r\sin\theta} - \frac{\partial\psi}{\partial\phi}\frac{\cos\phi}{r^2\sin\theta}\right]$$

$$+ \frac{\cos\theta\sin\phi}{r}\left[-\frac{\partial^2\psi}{\partial r\,\partial\theta}\sin\theta\sin\phi + \frac{\partial\psi}{\partial r}\cos\theta\sin\phi + \frac{\partial^2\psi}{\partial\theta^2}\frac{\cos\theta\sin\phi}{r}\right.$$
$$\left. - \frac{\partial\psi}{\partial\theta}\frac{\sin\theta\sin\phi}{r} + \frac{\partial^2\psi}{\partial\phi\,\partial\theta}\frac{\cos\phi}{r\sin\theta} - \frac{\partial\psi}{\partial\phi}\frac{\cos\phi\cos\theta}{r\sin^2\theta}\right]$$

$$+ \frac{\cos\phi}{r\sin\theta}\left[\frac{\partial^2\psi}{\partial r\,\partial\phi}\sin\theta\sin\phi + \frac{\partial\psi}{\partial r}\sin\theta\cos\phi + \frac{\partial^2\psi}{\partial\theta\,\partial\phi}\frac{\cos\theta\sin\phi}{r}\right.$$
$$\left. + \frac{\partial\psi}{\partial\theta}\frac{\cos\theta\cos\phi}{r} + \frac{\partial^2\psi}{\partial\phi^2}\frac{\cos\phi}{r\sin\theta} - \frac{\partial\psi}{\partial\phi}\frac{\sin\phi}{r\sin\theta}\right];$$

and,

$$\frac{\partial^2 \psi}{\partial z^2} = \frac{\partial}{\partial z}\frac{\partial \psi}{\partial z} = \frac{\partial}{\partial z}\left[\frac{\partial \psi}{\partial r}\cos\theta - \frac{\partial \psi}{\partial \theta}\frac{\sin\theta}{r}\right] = \frac{\partial}{\partial r}\left[\frac{\partial \psi}{\partial z}\cos\theta\right] - \frac{\partial}{\partial \theta}\left[\frac{\partial \psi}{\partial z}\frac{\sin\theta}{r}\right]$$

$$= \cos\theta\left[\frac{\partial^2 \psi}{\partial r^2}\cos\theta - \frac{\partial^2 \psi}{\partial \theta\,\partial r}\frac{\sin\theta}{r} + \frac{\partial \psi}{\partial \theta}\frac{\sin\theta}{r^2}\right]$$

$$- \frac{\sin\theta}{r}\left[\frac{\partial^2 \psi}{\partial r\,\partial \theta}\cos\theta - \frac{\partial \psi}{\partial r}\sin\theta - \frac{\partial^2 \psi}{\partial \theta^2}\frac{\sin\theta}{r} - \frac{\partial \psi}{\partial \theta}\frac{\cos\theta}{r}\right].$$

Now adding the three second derivatives and collecting terms with similar coefficients:

$$\frac{\partial^2 \psi}{\partial x^2} + \frac{\partial^2 \psi}{\partial y^2} + \frac{\partial^2 \psi}{\partial z^2} = \nabla^2\psi(r,\theta,\phi) = \frac{\partial^2 \psi}{\partial r^2}[\sin^2\theta\cos^2\phi + \sin^2\theta\sin^2\phi + \cos^2\theta]$$

$$+ \frac{\partial^2 \psi}{\partial r\,\partial \theta}\left[\frac{2\sin\theta\cos\theta\cos^2\phi}{r} + \frac{2\sin\theta\cos\theta\sin^2\phi}{r} - \frac{2\sin\theta\cos\theta}{r}\right]$$

$$+ \frac{\partial \psi}{\partial \theta}\left[\frac{-2\sin\theta\cos\theta\cos^2\phi}{r^2} + \frac{\cos\theta\sin^2\phi}{r^2\sin\theta} - \frac{2\sin\theta\cos\theta\sin^2\phi}{r^2}\right.$$
$$\left. + \frac{\cos\theta\cos^2\phi}{r^2\sin\theta} + \frac{2\sin\theta\cos\theta}{r^2}\right]$$

$$+ \frac{\partial^2 \psi}{\partial r\,\partial \phi}\left[\frac{-\sin\phi\cos\phi}{r} - \frac{\cos\phi\sin\phi}{r} + \frac{2\sin\phi\cos\phi}{r}\right]$$

$$+ \frac{\partial \psi}{\partial \phi}\left[\frac{\sin\phi\cos\phi}{r^2} + \frac{\cos^2\theta\sin\phi\cos\phi}{r^2\sin^2\theta} + \frac{\sin\phi\cos\phi}{r^2\sin^2\theta}\right.$$
$$\left. - \frac{\sin\phi\cos\phi}{r^2} - \frac{\cos^2\theta\sin\phi\cos\phi}{r^2\sin^2\theta} - \frac{\sin\phi\cos\phi}{r^2\sin^2\theta}\right]$$

$$+ \frac{\partial \psi}{\partial r}\left[\frac{\cos^2\phi\cos^2\theta}{r} + \frac{\sin^2\phi}{r} + \frac{\cos^2\theta\sin^2\phi}{r} + \frac{\cos^2\phi}{r} + \frac{\sin^2\theta}{r}\right]$$

$$+ \frac{\partial^2 \psi}{\partial \theta^2}\left[\frac{\cos^2\theta\cos^2\phi}{r^2} + \frac{\cos^2\theta\sin^2\phi}{r^2} + \frac{\sin^2\theta}{r^2}\right]$$

$$+ \frac{\partial^2 \psi}{\partial \phi^2}\left[\frac{\sin^2\phi}{r^2\sin^2\theta} + \frac{\cos^2\phi}{r^2\sin^2\theta}\right]$$

$$+ \frac{\partial^2 \psi}{\partial \theta\,\partial \phi}\left[-\frac{2\cos\theta\sin\phi\cos\phi}{r^2\sin\theta} + \frac{2\sin\phi\cos\theta\cos\phi}{r^2\sin\theta}\right].$$

The trigonometric expressions, which are multipliers of the differential coefficients, $\partial^2\psi/\partial r\,\partial\theta$, $\partial^2\psi/\partial r\,\partial\phi$, $\partial^2\psi/\partial\theta\,\partial\phi$, and $\partial\psi/\partial\phi$, all vanish, giving the Laplacian operator in terms of spherical polar coordinates as:

$$\partial^2/\partial x^2 + \partial^2/\partial y^2 + \partial^2/\partial z^2 = \nabla^2 = \partial^2/\partial r^2 + (2/r)(\partial/\partial r) + (1/r^2)(\partial^2/\partial\theta^2)$$
$$+ [(\cot\theta)/r^2](\partial/\partial\theta) + [1/(r^2\sin^2\theta)](\partial^2/\partial\phi^2),$$

which may be alternatively expressed as:

$$\nabla^2 = \frac{\partial^2}{\partial r^2} + \frac{2}{r}\frac{\partial}{\partial r} + \frac{1}{r^2\sin\theta}\left[\sin\theta\frac{\partial^2}{\partial\theta^2} + \cos\theta\frac{\partial}{\partial\theta}\right] + \frac{1}{r^2\sin^2\theta}\frac{\partial^2}{\partial\phi^2},$$

$$= \frac{\partial^2}{\partial r^2} + \frac{2}{r}\frac{\partial}{\partial r} + \frac{1}{r^2\sin\theta}\frac{\partial}{\partial\theta}\left[\sin\theta\frac{\partial}{\partial\theta}\right] + \frac{1}{r^2\sin^2\theta}\frac{\partial^2}{\partial\phi^2},$$

$$= \frac{1}{r^2}\frac{\partial}{\partial r}\left[r^2\frac{\partial}{\partial r}\right] + \frac{1}{r^2\sin\theta}\frac{\partial}{\partial\theta}\left[\sin\theta\frac{\partial}{\partial\theta}\right] + \frac{1}{r^2\sin^2\theta}\frac{\partial^2}{\partial\phi^2},$$

$$= \frac{1}{r}\frac{\partial^2}{\partial r^2}[r\psi] + \frac{1}{r^2\sin\theta}\frac{\partial}{\partial\theta}\left[\sin\theta\frac{\partial}{\partial\theta}\right] + \frac{1}{r^2\sin^2\theta}\frac{\partial^2}{\partial\phi^2}.$$

Alternatively, the foregoing coordinate transformation may be written particularly neatly in matrix notation. From the relationships, $x = r \sin \theta \cos \phi$, $y = r \sin \theta \sin \phi$, and $z = r \cos \theta$, the following matrix may be constructed:

$$
\begin{bmatrix} \partial x/\partial r & \partial x/\partial \theta & \partial x/\partial \phi \\ \partial y/\partial r & \partial y/\partial \theta & \partial y/\partial \phi \\ \partial z/\partial r & \partial z/\partial \theta & \partial z/\partial \phi \end{bmatrix} = \begin{bmatrix} \sin \theta \cos \phi & r \cos \theta \cos \phi & -r \sin \theta \sin \phi \\ \sin \theta \sin \phi & r \cos \theta \sin \phi & r \sin \theta \cos \phi \\ \cos \theta & -r \sin \theta & 0 \end{bmatrix} = J,
$$

where J is said to be the *Jacobian* of the transformation. Generally a matrix, each of the elements of which is a first derivative is referred to as a Jacobian, and if each of the elements is a second derivative, the matrix is said to be a *Hessian*. It is easily shown that $|J| = |\tilde{J}| = r^2 \sin \theta$. If the column matrices,

$$
\begin{bmatrix} \partial/\partial r \\ \partial/\partial \theta \\ \partial/\partial \phi \end{bmatrix} \quad \text{and} \quad \begin{bmatrix} \partial/\partial x \\ \partial/\partial y \\ \partial/\partial z \end{bmatrix},
$$

are referred to by the symbols ∇_r and ∇_x, respectively, then since x, y, and z, are independent variables and therefore their derivatives with respect to one another are zero, it follows that $\nabla_r = J\nabla_x$, or multiplying both sides of this equation by J^{-1}, $\nabla_x = J^{-1}\nabla_r$, since $JJ^{-1} = 1$. Making use of the relationship between adjoint and reciprocal matrices, ∇_x may now be written as

$$
\nabla_x = J^{-1}\nabla_r = \frac{\hat{J}}{|J|}\,\nabla_r,
$$

and therefore the transpose of each side of this equation is

$$
\tilde{\nabla}_x = \frac{1}{|J|}\,\tilde{\nabla}_r[\hat{J}]^\sim = \frac{1}{|J|}\,\tilde{\nabla}_r[|J|J^{-1}]^\sim.
$$

Now since, by the normal rules of matrix multiplication,

$$
\nabla_x{}^2 = \tilde{\nabla}_x\nabla_x = \frac{1}{|J|}\,\tilde{\nabla}_r[|J|J^{-1}]\tilde{J}^{-1}\nabla_r = \frac{1}{|J|}\,\tilde{\nabla}_r|J|[\tilde{J}J]^{-1}\nabla_r,
$$

since $[\tilde{J}]^{-1} = [J^{-1}]^\sim$. Now, by matrix multiplication,

$$
\tilde{J}J = \begin{bmatrix} 1 & 0 & 0 \\ 0 & r^2 & 0 \\ 0 & 0 & r^2 \sin^2 \theta \end{bmatrix},
$$

and therefore

$$
[\tilde{J}J]^{-1} = \frac{1}{|J|\,|J|^\sim}\,[\hat{J}^\sim \hat{J}] = \frac{1}{r^2 \sin^2 \theta} \begin{bmatrix} r^2 \sin^2 \theta & 0 & 0 \\ 0 & \sin^2 \theta & 0 \\ 0 & 0 & 1 \end{bmatrix},
$$

since $|J| = |\tilde{J}| = r^2 \sin \theta$. By now substituting this latter expression into the equation given for $\nabla_x{}^2$ above, the following is obtained,

$$
\nabla_x{}^2 = \frac{1}{|J|}\,\tilde{\nabla}_r|J|[\tilde{J}J]^{-1}\nabla_r = \frac{1}{r^2 \sin^2 \theta}\,\tilde{\nabla}_r r^2 \sin \theta\, \frac{1}{r^2 \sin^2 \theta} \begin{bmatrix} r^2 \sin^2 \theta & 0 & 0 \\ 0 & \sin^2 \theta & 0 \\ 0 & 0 & 1 \end{bmatrix} \begin{bmatrix} \partial/\partial r \\ \partial/\partial \theta \\ \partial/\partial \phi \end{bmatrix}
$$

$$
= \frac{1}{r^2 \sin \theta}\,\hat{\nabla}_r\, \frac{1}{\sin \theta} \begin{bmatrix} r^2 \sin^2 \theta\, \partial/\partial r \\ \sin^2 \theta\, \partial/\partial \theta \\ \partial/\partial \phi \end{bmatrix} = \frac{1}{r^2 \sin \theta}\,[\partial/\partial r \quad \partial/\partial \theta \quad \partial/\partial \phi] \begin{bmatrix} r^2 \sin \theta\, \partial/\partial r \\ \sin \theta\, \partial/\partial \theta \\ \sin^{-1} \theta\, \partial/\partial \phi \end{bmatrix}
$$

$$
= \frac{1}{r^2 \sin \theta}\,\partial/\partial r(r^2 \sin \theta\, \partial/\partial r) + \frac{1}{r^2 \sin \theta}\,\partial/\partial \theta(\sin \theta\, \partial/\partial \theta) + \frac{1}{r^2 \sin \theta}\,\partial/\partial \phi(\sin^{-1} \theta\, \partial/\partial \phi
$$

$$
= \frac{1}{r^2}\frac{\partial}{\partial r}\left(r^2 \frac{\partial}{\partial r}\right) + \frac{1}{r^2 \sin \theta}\frac{\partial}{\partial \theta}\left(\sin \theta \frac{\partial}{\partial \theta}\right) + \frac{1}{r^2 \sin^2 \theta}\frac{\partial^2}{\partial \phi^2}
$$

as before.

The central field equation expressed in spherical polar coordinates then becomes

$$(1/r^2)\,\partial/\partial r[r^2\,\partial\psi(r,\theta,\phi)/\partial r] + (1/r^2\sin\theta)\,\partial/\partial\theta[\sin\theta\,\partial\psi(r,\theta,\phi)/\partial\theta]$$
$$+ (1/r^2\sin^2\theta)\,\partial^2\psi(r,\theta,\phi)/\partial\phi^2 + 8\pi^2\mu h^{-2}[E + Z\varepsilon^2/r]\psi(r,\theta,\phi) = 0.$$
$$(q.3.9.)$$

If separable solutions of this equation exist such that

$$\psi(r,\theta,\phi) = R(r)\Theta(\theta)\Phi(\phi),$$

where R, Θ, and Φ, are functions of the single variables r, θ, and ϕ, respectively, then

$$r^{-2}\,\partial/\partial r[r^2\,\partial R(r)/\partial r]\Theta(\theta)\Phi(\phi) + (r^2\sin\theta)^{-1}\,\partial/\partial\theta[\sin\theta\,\partial\Theta(\theta)/\partial\theta]R(r)\Phi(\phi)$$
$$+ (r^2\sin^2\theta)^{-1}R(r)\Theta(\theta)\,\partial^2\Phi(\phi)/\partial\phi^2 + 8\pi^2\mu h^{-2}[E + Z\varepsilon^2/r]R(r)\Theta(\theta)\Phi(\phi) = 0.$$
$$(q.3.10.)$$

If this equation is multiplied throughout by $r^2\sin^2\theta[R(r)\Theta(\theta)\Phi(\phi)]^{-1}$, there results

$$\sin^2\theta[R(r)]^{-1}\,d/dr[r^2\,dR(r)/dr] + \sin\theta[\Theta(\theta)]^{-1}\,d/d\theta[\sin\theta\,d\Theta(\theta)/d\theta]$$
$$+ 8\pi^2\mu h^{-2}[E + Z\varepsilon^2/r]r^2\sin^2\theta = -[\Phi(\phi)]^{-1}\,d^2\Phi(\phi)/d\phi^2. \quad (q.3.11.)$$

Equation (q.3.11.) is free of partial derivatives, and since the right hand side is a function of the polar coordinate ϕ only, while the left hand side is independent of ϕ, both parts must be equal to some constant, say C, so that

$$d^2\Phi(\phi)/d\phi^2 = -\Phi(\phi)C, \qquad (q.3.12.)$$

and also

$$\sin^2\theta[R(r)]^{-1}\,d/dr[r^2\,dR(r)/dr] + \sin\theta[\Theta(\theta)]^{-1}\,d/d\theta[\sin\theta\,d\Theta(\theta)/d\theta]$$
$$+ 8\pi^2\mu h^{-2}[E + Z\varepsilon^2/r]r^2\sin^2\theta = C. \quad (q.3.13.)$$

Dividing equation (q.3.13.) throughout by $\sin^2\theta$, it is found that

$$[R(r)]^{-1}\,d/dr[r^2\,dR(r)/dr] + 8\pi^2\mu h^{-2}[E + Z\varepsilon^2/r]r^2$$
$$= C[\sin^2\theta]^{-1} - [\Theta(\theta)\sin\theta]^{-1}\,d/d\theta[\sin\theta\,d\Theta(\theta)/d\theta]. \quad (q.3.14.)$$

The right-hand side of equation (q.3.14.) is a function of θ only and the left-hand side contains terms dependent on the radial coordinate r only, so that each side must be equivalent to some constant, β, for such a condition to be satisfied. Therefore, $C[\sin^2\theta]^{-1} - [\Theta(\theta)\sin\theta]^{-1}\,d/d\theta[\sin\theta\,d\Theta(\theta)/d\theta] = \beta$, or multiplying throughout by $-\Theta(\theta)$,

$$[\sin\theta]^{-1}\,d/d\theta[\sin\theta\,d\Theta(\theta)/d\theta] + [\beta - C(\sin^2\theta)^{-1}]\Theta(\theta) = 0. \quad (q.3.15.)$$

Also,

$$[R(r)]^{-1}\,d/dr[r^2\,dR(r)/dr] + 8\pi^2\mu h^{-2}[E + Z\varepsilon^2/r]r^2 = \beta,$$

or multiplying this equation throughout by $R(r)/r^2$,

$$d^2R(r)/dr^2 + (2/r)\,dR(r)/dr + R(r)[8\pi^2\mu h^{-2}E - 8\pi^2\mu h^{-2}Z\varepsilon^2 r^{-1}]$$
$$- \beta R(r)r^{-2} = 0. \quad \text{(q.3.16.)}$$

Thus the equation of motion of a particle of reduced mass μ moving in a spherically symmetrical potential field, when expressed in spherical polar coordinates, is separable into three equations, (q.3.12.), (q.3.15.), and (q.3.16.), each of which is homogeneous in only one of the variables, r, θ, and ϕ, respectively. Only the equation in the radial coordinate, r, contains the potential function, $-Z\varepsilon^2/r$, so that the radial function $R(r)$ is the only one which depends on the particular form of the central field. The angular functions, $\Theta(\theta)$ and $\Phi(\phi)$ are apparently independent of the central field and therefore the same for all central fields; it is essentially for this reason that the widely used polar diagrams showing the product function, $[\Theta(\theta)\Phi(\phi)]^2$, plotted as a function of the polar angles θ and ϕ, to represent the electron distribution in atoms are valid. Also, in the process of separating the original central field equation into the three differential equations, each homogeneous in only one of the polar coordinate functions, it has been necessary to introduce two separation constants, C and β. As usual, in the solution of differential equations, these constants are not completely arbitrary but, as might be expected, they must satisfy specific conditions in order that the product function $\psi(r, \theta, \phi) = R(r)\Theta(\theta)\Phi(\phi)$, should represent the solutions of equation (q.3.9.).

Q.3.5. The Angular Functions

The equation, (q.3.12.), in the polar angular function $\Phi(\phi)$, is by far the simplest. There are obviously two independent solutions of this equation,

$$\Phi(\phi) = e^{i\phi\sqrt{C}} \quad \text{and} \quad \Phi(\phi) = e^{-i\phi\sqrt{C}}, \quad \text{(q.3.17.)}$$

or alternatively by forming linear combinations of this pair of solutions,

$$\Phi(\phi) = \sin\phi\sqrt{C} \quad \text{and} \quad \Phi(\phi) = \cos\phi\sqrt{C}. \quad \text{(q.3.18.)}$$

The solutions given by (q.3.17.) and (q.3.18.) are alternative solutions; they are not linearly independent. In the spherical polar coordinate system, the point designated by the coordinates (r, θ, ϕ) is indistinguishable from the point designated by the coordinates $(r, \theta, \phi + 2\pi)$; so if the function $\psi(r, \theta, \phi)$ is to satisfy the condition that it be single-valued for all values of the variables, the function $\Phi(\phi)$ must not change in value when the angular coordinate ϕ is increased by $\pm 2\pi$ or some multiple of $\pm 2\pi$, or $\Phi(\phi)$ must have the period

2π. For this to be so, the constant $\sqrt{C}$ in equations (q.3.17.) and (q.3.18.) must be zero or a positive or negative integer, such that

$$\sqrt{C} = m_l, \qquad \text{with} \qquad m_l = 0, \pm 1, \pm 2, \ldots \qquad \text{(q.3.19.)}$$

The eigenfunctions of equation (q.3.12.) may therefore be written as,

$$\Phi_{m_l}(\Phi) = e^{im_l\phi}, \qquad \text{with} \qquad m_l = 0, \pm 1, \pm 2, \ldots, \qquad \text{(q.3.20.)}$$

where the positive and negative values correspond to different eigenfunctions. The alternative manner of writing these solutions is,

$$\Phi_{m_l}(\Phi) = \sin m_l\phi, \qquad \text{with} \qquad m_l = 1, 2, 3, \ldots,$$

$$\Phi_{m_l}(\Phi) = \cos m_l\phi, \qquad \text{with} \qquad m_l = 0, -1, -2, \ldots \qquad \text{(q.3.21.)}$$

In order that the functions (q.3.20.) or (q.3.21.) be normalized, it is necessary that $\int_0^{2\pi} \{[N_\phi\Phi(\phi)][N_\phi\Phi^*(\phi)]\}\, d\phi = 1$, where N_ϕ is the appropriate normalizing constant. This is equivalent to the condition that

$$[N_\phi]^2 \int_0^{2\pi} e^{\pm im_l\phi} \cdot e^{\mp im_l\phi}\, d\phi = [N_\phi]^2 \cdot 2\pi = 1,$$

or $N_\phi = \sqrt{1/2\pi}$, so that the normalized functions which are solutions of (q.3.12.) are

$$\Phi(\phi) = [2\pi]^{-1/2}e^{im_l\phi}, \qquad \text{with} \qquad m_l = 0, \pm 1, \pm 2, \ldots \qquad \text{(q.3.22.)}$$

The functions (q.3.22.) are also mutually orthogonal to one another in pairs, by virtue of the relationships given on page 123.

If in Fig. Q.3.2., the particle located at P with coordinates (r, θ, ϕ) is imagined to rotate around the z-Cartesian coordinate axis with r and θ remaining constant, then p_ϕ would describe the angular momentum of the particle about the z direction. The quantum mechanical operator corresponding to this variable is $(h/2\pi i)\, d/d\phi$, and if this operator operates on the function $\psi(r, \theta, \phi) = R(r)\Theta(\theta)\Phi(\phi) = R(r)\Theta(\theta)Ce^{\pm im_l\phi}$, where C is a constant, then

$$p_\phi\psi(r, \theta, \phi) = (h/2\pi i)\, \partial/\partial\phi[\psi(r, \theta, \phi)] = (h/2\pi i)CR(r)\Theta(\theta)\, d/d\phi[e^{\pm im_l\phi}]$$

$$= (h/2\pi i)(\pm im_l)\psi(r, \theta, \phi),$$

or

$$p_\phi = \pm m_l(h/2\pi). \qquad \text{(q.3.23.)}$$

It is thus deduced that the integers m_l in (q.3.20.) denote the number of units of angular momentum of the particle about the z-Cartesian axis; this is the reason for referring to the separation constant C in (q.3.12.) as m_l, in order to indicate that it has the same significance as the magnetic quantum number in the older quantum theory discussed in Chapter Q.1. The only restriction so far imposed on m_l is that it should be integral, but the inclusion of the subscript l, as will be proved in the following pages, denotes that it is not independent of the other separation constant, β in (q.3.15.), but may only adopt values ranging from $+l$ to $-l$.

The solution of the other equation containing the angular function $\Theta(\theta)$, given by (q.3.15.) requires a little more effort. If the value of the first separation constant given by (q.3.19.) is substituted in (q.3.15.), then there is obtained

$$[\sin\theta]^{-1}\, d/d\theta[\sin\theta\, d\Theta(\theta)/d\theta] + [\beta - m_l^2/\sin^2\theta]\Theta(\theta) = 0. \quad \text{(q.3.24.)}$$

This equation is similar in form to a well known classical differential equation known as the associated Legendre equation, usually expressed as

$$[1 - x^2]\, d^2y/dx^2 - 2x\, dy/dx + [l(l+1) - m^2/(1-x^2)]y = 0, \quad \text{(q.3.25.)}$$

which in turn is related to the Legendre differential equation, which is usually given as

$$[1 - x^2]\, d^2y/dx^2 - 2x\, dy/dx + l(l+1)y = 0. \quad \text{(q.3.26.)}$$

The solutions of both the Legendre and the associated Legendre differential equations are known and have been much studied, since both of these equations play an important part in mathematical physics. The equation in the angular function $\Theta(\theta)$ may be transformed into the form of the associated Legendre equation, exactly, by changing variables; if the substitution $\cos\theta = x$ is made, then since the limiting values of the cosine (or sine) of any angle are ± 1, the interval of validity of the new variable x is $-1 < x < 1$, and since $\sin\theta = \sqrt{(1 - \cos^2\theta)} = \sqrt{(1 - x^2)}$, then by making these substitutions into (q.3.24.) the result is

$$\frac{1}{\sin\theta\, d\cos\theta}\frac{d}{d\theta}\frac{d\cos\theta}{d\theta}\left[\sin\theta\,\frac{d\Theta(\theta)}{d\cos\theta}\frac{d\cos\theta}{d\theta}\right] + \left[\beta - \frac{m_l^2}{\sin^2\theta}\right]\Theta(\theta) = 0,$$

or

$$\frac{1}{\sin\theta\, d\cos\theta}\frac{d}{(-\sin\theta)}[\sin\theta\,\frac{d\Theta(\theta)}{d\cos\theta}(-\sin\theta)] + \left[\beta - \frac{m_l^2}{1 - x^2}\right]\Theta(\theta) = 0,$$

or

$$-d/dx[-(1 - x^2)\, d\Theta(\theta)/dx] + [\beta - m_l^2/(1-x^2)]\Theta(\theta) = 0,$$

or

$$(1 - x^2)\, d^2\Theta(\theta)/dx - 2x\, d\Theta(\theta)/dx + [\beta - m_l^2/(1-x^2)]\Theta(\theta) = 0.$$
$$\text{(q.3.27.)}$$

This latter equation becomes identical with the associated Legendre equation if $\Theta(\theta)$ is substituted by y and $\beta = l(l+1)$.

Now the solutions of the associated Legendre differential equation, (q.3.25.) are known to be

$$y = [1 - x^2]^{m/2} P_l^m(x), \qquad \text{(q.3.28.)}$$

where $P_l^m(x)$ is the so-called associated Legendre polynomial, which may be alternatively expressed as a power series in x

$$P_l^m(x) = (2l)! \, [2^l l! \, (l - m)!]^{-1} \{ x^{l-m} - (l - m)(l - m - 1) \\ \times \, [2(2l - 1)]^{-1} x^{l-m-2} + \cdots \}, \qquad \text{(q.3.29.)}$$

or in the derivative form

$$P_l^m(x) = [2^l l!]^{-1} \, d^{l+m}/dx^{l+m} [(x^2 - 1)^l].\ddagger \qquad \text{(q.3.30.)}$$

‡ In order to prove that the functions defined by (q.3.28.), where $P_l^m(x)$ may be expressed either as the polynomial (q.3.29.) or the derivative (q.3.30.) are solutions of the associated Legendre differential equation (q.3.25.), it is convenient to write

$$y = [2^l l!]^{-1} [1 - x^2]^{m/2} v(x), \qquad \text{(q.3.31.)}$$

where

$$v(x) = d^{l+m}/dx^{l+m} [(x^2 - 1)^l] \\ = (2l)! \, [(l - m)!]^{-1} \{ x^{l-m} - (l - m)(l - m - 1)[2(2l - 1)]^{-1} x^{l-m-2} + \cdots \}. \\ \text{(q.3.32.)}$$

Taking the first and second derivatives, respectively, of (q.3.31.) with respect to x,

$$dy/dx = [2^l l!]^{-1} \{ -mx(1 - x^2)^{m/2 - 1} v(x) + (1 - x^2)^{m/2} \, dv(x)/dx \},$$

and

$$d^2y/dx^2 = [2^l l!]^{-1} \{ -mv(x)(1 - x^2)^{m/2 - 1} - 2mx(1 - x^2)^{m/2 - 1} \, dv(x)/dx \\ + \, mx^2(m - x)v(x)(1 - x^2)^{m/2 - 2} + (1 - x^2)^{m/2} \, d^2v(x)/dx^2 \}.$$

If these derivatives are now substituted in equation (q.3.25.), and the resulting expression is divided throughout by $(1 - x^2)^{m/2}$, there results

$$(1 - x^2) \, d^2v(x)/dx^2 - [2mx + 2x] \, dv(x)/dx + v(x)[-m + mx^2(m - 2)(1 - x^2)^{-1} \\ + \, 2mx^2(1 - x^2)^{-1} + l(l + 1) - m^2(1 - x^2)^{-1}] = 0,$$

which may be simplified to

$$(1 - x^2) \, d^2v(x)/dx^2 - 2x(m + 1) \, dv(x)/dx + [l(l + 1) - m(m + 1)]v(x) = 0. \\ \text{(q.3.33.)}$$

The derivative expression for $v(x)$ in (q.3.32.) may be shown to satisfy this equation by firstly letting $l + m = n$, when the derivative form of (q.3.32.) becomes

$$d^n/dx^n [(x^2 - 1)^l],$$

and if this is substituted into (q.3.33.), it is found that

$$(1 - x^2) \, d^{n+2}/dx^{n+2} [x^2 - 1]^l - 2x(n - l + 1) \, d^{n+1}/dx^{n+1} [x^2 - 1]^l \\ + \, (n + 1)(2l - n) \, d^n/dx^n [x^2 - 1]^l = 0. \quad \text{(q.3.34.)}$$

Now since $d/dx[x^2 - 1] = l(x^2 - 1)^{l-1} \cdot 2x$ and

$$d^2/dx^2 [x^2 - 1]^l = 2l(x^2 - 1)^{l-1} + 2xl(l - 1)(x^2 - 1)^{l-2} \cdot 2x,$$

if for the special case where $n = 0$ and these derivatives are substituted in (q.3.34.), this equation becomes

$$(1 - x^2)[2l(x^2 - 1)^{l-1} + 4x^2l(l - 1)(x^2 - 1)^{l-2}]$$
$$- 2(1 - l)xl(x^2 - 1)^{l-1}2x + 2l(x^2 - 1)^l,$$

which when factored by $2l(x^2 - 1)^{l-2}$ and expanded is seen to be identically zero. So for the special case when $(m + l) = n = 0$, the function (q.3.31.) with $v(x)$ expressed in the derivative form given by (q.3.32.), is a solution of (q.3.33.) and therefore also of the associated Legendre equation (q.3.25.). Now if equation (q.3.34.) is differentiated once with respect to x, there is obtained

$$(-2x)\, d^{n+2}/dx^{n+2}[x^2 - 1]^l + (1 - x^2)\, d^{n+3}/dx^{n+3}[x^2 - 1]^l$$
$$- 2(n - l + 1)\, d^{n+1}/dx^{n+1}[x^2 - 1]^l - 2x(n - l + 1)\, d^{n+2}/dx^{n+2}[x^2 - 1]^l$$
$$+ (2l - n)(n + 1)\, d^{n+1}/dx^{n+1}[x^2 - 1]^l = 0.$$

On collecting similar terms and rearranging, this equation is found to be of identical form to (q.3.34.) with $(n + 1)$ substituted for n. Similarly equation (q.3.34.) is true for all integral values of $n = m + l$, and thus the function

$$y(x) = [2^l l!]^{-1}[1 - x^2]^{m/2}\, d^{l+m}/dx^{l+m}[x^2 - 1]^l$$

is a solution of the associated Legendre equation (q.3.25.), with $(l + m) = 0, 1, 2, 3, \ldots$. Now, since the highest power of x in this expression for $y(x)$ is x^{2l} and this expression contains the $(m + l)$th derivative with respect to x, $(m + l)$ cannot exceed $2l$, otherwise the expression vanishes; therefore $(m + l) \leq 2l$ or $m \leq l$ for a non-vanishing solution.

The expression $v(x) = d^{l+m}/dx^{l+m}[x^2 - 1]^l$, in (q.3.32.) may be expressed alternatively as a sum of a series of terms by making use of the binomial expansion of

$$[x^2 - 1]^l = x^{2l} - lx^{2l-2} + l(l - 1)[2!]^{-1}x^{2l-4} - l(l - 1)(l - 2)[3!]^{-1}x^{2l-6} \cdots$$

$$= \sum_{v=0}^{l} (-1)^v(l!)[v!\,(l - v)!]^{-1}x^{2l-2v}, \tag{q.3.35.}$$

with $v = 0, 1, 2, \ldots, l$. If this series is substituted for the derivative form of $v(x)$ in (q.3.32.), then

$$v(x) = \sum_{v=0}^{l} (-1)^v(l!)[(v!)(l - v)!]^{-1}\, d^{l+m}/dx^{l+m}[x^{2l-2v}],$$

and on performing the $(l + m)$ differentiations, this becomes

$$v(x) = \sum_{v=0}^{l} (-1)^v(l!)[(v!)(l - v)!]^{-1}(2l - 2v)(2l - 2v - 1)\cdots(l - 2v - m + 1)x^{l-2v-m}$$

$$= \sum_{v=0}^{l} (-1)^v(l!)[(v!)(l - v)!]^{-1}[(2l - 2v)!][(l - m - 2v)!]^{-1}x^{l-2v-m}. \tag{q.3.36.}$$

If, for convenience, $(l - m - 2v) = v'$, then (q.3.36.) may be written as

$$v(x) = \sum_{v=0}^{l} A_{v'}x^{v'},$$

where

$$A_{v'} = A_{l-m-2v} = (-1)^v(l!)[(v!)(l - v)!]^{-1}[(2l - 2v)!][(l - m - 2v)!]^{-1}.$$

The adjacent term to this in the series, the $(v + 1)$th, would then have the coefficient

$$A_{l-m-2(v+1)} = A_{v'-2} = (-1)^{v+1}(l!)[(2l - 2v - 2)!]$$
$$\times [(v + 1)!\,(l - v - 1)!\,(l - m - 2v + 2)!]^{-1}.$$

The ratio of two adjacent coefficients of powers of x in the series, is then given by

$$\frac{A_{v'}}{A_{v'-2}} = \frac{(-1)(2l-2v)(2l-2v-1)(v+1)}{(l-v)(l-m-2v)(l-m-2v-1)} = \frac{-2(2l-2v-1)(v+1)}{(l-m-2v)(l-m-2v-1)}$$

$$= \frac{(v'+m+l-1)(v'-l+m-2)}{v'(v'-1)} = \frac{(m+v'-1)(m+v'-2)-l(l+1)}{v'(v'-1)}.$$

$$(q.3.37.)$$

Similarly in the power series in x given by (q.3.32.), where only two terms are written down explicitly, it may be deduced by extrapolation that the coefficients of the vth and the $(v+1)$th terms, corresponding to the adjacent coefficients of x with exponents $(l-m-2v)$ and $[l-m-2(v+1)]$, respectively, are

$$A_{l-m-2v} = \frac{(-1)^v(l-m)(l-m-1)\cdots(l-m-2v+1)}{2\cdot4\cdot6\cdots2v(2l-1)(2l-3)\cdots(2l-2v+1)},$$

and

$$A_{l-m-2(v+1)} = \frac{(-1)^{v+1}(l-m)(l-m-1)\cdots(l-m-2v-1)}{2\cdot4\cdot6\cdots\cdot2(v+1)(2l-1)(2l-3)\cdots(2l-2v-1)}.$$

Again, if $(l-m-2) = v'$, then $[l-m-2(v+1)] = v'-2$, and the ratio of the two adjacent coefficients is

$$\frac{A_{v'}}{A_{v'-2}} = \frac{(-1)(2v+2)(2l-2v-1)}{(l-m-2v)(l-m-2v-1)} = \frac{(m+v'+l-1)(m+v'-l-2)}{v'(v'-1)}$$

$$= \frac{(m+v'-1)(m+v'-2)-l(l+1)}{v'(v'-1)},$$

as before. Actually the polynomial form of $v(x)$ in (q.3.32.) and also in $P_l{}^m(x)$ in (q.3.29.) arises from term by term differentiation of the binomial expansion of $(x^2-1)^l$. As pointed out on page 183, $v(x)$ can only have non-zero values if $m \le l$, so that the series is not infinite but only summed over l terms, as in (q.3.36.). Thus, differentiating the first two terms and the general term of the series given by (q.3.35.) $(m+l)$ times with respect to x gives:

$$d^{m+l}/dx^{m+l}[x^{2l}] = 2l(2l-1)(2l-2)\cdots(l-m+1)x^{l-m}$$

$$= [(2l)!][(l-m)!]^{-1}x^{l-m};$$

$$d^{m+l}/dx^{m+l}[-lx^{2l-2}] = (-1)l(2l-2)(2l-3)\cdots(l-m-1)x^{l-m-2}$$

$$= (-1)[(2l)!][2(2l-1)\{(l-m)!\}]^{-1}(l-m)(l-m-1)x^{l-m-2};$$

and for the vth term, which would be $(-1)^v l(l-1)\cdots(l-v+1)[(v)!]^{-1}x^{2l-2v}$,

$$d^{m+l}/dx^{m+l}[(-1)^v l(l-1)\cdots(l-v+1)\{(v)!\}^{-1}x^{2l-2v}]$$

$$= [(v)!]^{-1}(-1)^v l(l-1)(l-2)\cdots(l-v+1)(2l-2v)(2l-2v-1)\cdots$$

$$\times\ [l-2v-(l+m)+1]x^{2l-2v-(m+l)}$$

$$= \frac{(-1)^v[(2l)!](l-m)(l-m-1)\cdots(l-m-2v+1)x^{l-m-2v}}{[(l-m)!]2\cdot4\cdot6\cdots\cdot2v(2l-1)(2l-3)\cdots(2l-2v+1)},$$

after multiplying the numerator and denominator by suitable factors to form complete factorials of $2l$ and $(l-m)$. Thus, it is proved that the two different modes of expressing $v(x)$ in (q.3.32.) are equivalent, and since it has already been proved that the derivative form of (q.3.32.) is a solution of the associated Legendre equation, then the polynomial form must also constitute a solution of the same equation. It should also be noted that the appearance of the factor $2^l l!$ in the denominator of (q.3.30.) and therefore also of (q.3.28.) ensures that the Legendre polynomial in x, $P_l(x) = P_l{}^0(x) = 1$, when $x = \cos\theta = 1$ or when $\theta = 0$, as may be established alternatively as follows: From (q.3.30.), when $m = 0$,

$$P_l{}^0(x) = [2^l l!]^{-1}\ d^l/dx^l(x^2-1)^l$$

$$= [2^l l!]^{-1}\ d^l/dx^l[(x-1)^l(x+1)^l].$$

If this latter expression is differentiated as a product, in the usual manner, then

$$P_l^0(x) = [2^l l!]^{-1}[(x + 1)^l \, d^l/dx^l(x - 1)^l + (x - 1)^l \, d^l/dx^l(x + 1)^l]$$
$$= [2^l l!]^{-1} 2^l l! = 1,$$

since when $x = 1$, the second term in the square brackets above vanishes as $(x - 1)^l = 0$, and $(x + 1)^l = 2^l$ while the lth derivative of $(x - 1)^l$ is

$$d^l/dx^l(x - 1)^l = l(l - 1)(l - 2) \cdots [l - (l - 1)](x - 1)^{l-l} = l!.$$

It is also possible to derive the general nature of the associated Legendre function by assuming that the solutions of (q.3.25.) may be represented by a power series in x, such that

$$y = (1 - x^2)^{m/2}[A_0 + A_1 x + A_2 x^2 + A_3 x^3 + \cdots] = (1 - x^2)^{m/2} \sum_{v=0}^{\infty} A_v x^v,$$

$$\text{(q.3.38.)}$$

where the coefficients A_v have to be determined. If the first and second derivatives of (q.3.38.) are taken and substituted into (q.3.25.), it is found that the condition which must be satisfied for (q.3.38.) to constitute a solution of the associated Legendre equation is,

$$-mx(1 - x^2)^{m/2} \, d/dx \sum_{v=0}^{\infty} A_v x^v - m(1 - x^2)^{m/2} \sum_{v=0}^{\infty} A_v x^v$$

$$+ m(m - 2)x^2(1 - x^2)^{m/2-1} \sum_{v=0}^{\infty} A_v x^v + (1 - x^2)^{m/2+1} \, d^2/dx^2\left[\sum_{v=0}^{\infty} A_v x^v\right]$$

$$- mx(1 - x^2)^{m/2} \, d/dx \sum_{v=0}^{\infty} A_v x^v + 2mx^2(1 - x^2)^{m/2-1} \sum_{v=0}^{\infty} A_v x^v$$

$$- 2x(1 - x^2)^{m/2} \, d/dx \sum_{v=0}^{\infty} A_v x^v + l(l + 1)(1 - x^2)^{m/2} \sum_{v=0}^{\infty} A_v x^v$$

$$- m^2(1 - x^2)^{m/2-1} \sum_{v=0}^{\infty} A_v x^v = 0.$$

If this expression is divided throughout by $(1 - x)^{m/2}$ and the coefficients of similar terms collected, then

$$(1 - x^2) \, d^2/dx^2\left[\sum_{v=0}^{\infty} A_v x^v\right] - 2x(m + 1) \, d/dx\left[\sum_{v=0}^{\infty} A_v x^v\right]$$

$$+ [l(l + 1) - m^2 - m] \sum_{v=0}^{\infty} A_v x^v = 0.$$

Now substituting into this expression the expanded form of the sum in (q.3.38.) and performing the indicated differentiations, this becomes

$$(1 - x^2)[2A_2 + 2 \cdot 3A_3 x + \cdots v(v - 1)A_v x^{v-2} + \cdots]$$
$$- 2x(m + 1)[A_1 + 2A_2 x + \cdots vA_v x^{v-1} + \cdots]$$
$$+ [l(l + 1) - m^2 - m][A_0 + A_1 x + \cdots A_v x^v + \cdots] = 0. \quad \text{(q.3.39.)}$$

This equation can only be true if the sum of the coefficients of each power of the variable x is zero; thus the coefficient of the $(v - 2)$th power of x, from (q.3.39.) is,

$$v(v - 1)A_v - A_{v-2}(v - 2)(v - 3) - 2mA_{v-2}(v - 2) - 2(v - 2)A_{v-2}$$
$$+ l(l + 1)A_{v-2} - m^2 A_{v-2} - mA_{v-2} = 0,$$

or, after some algebraic rearrangement,

$$\frac{A_v}{A_{v-2}} = \frac{v^2 + 2 - 3v + 2mv - 3m + m^2 - l(l + 1)}{v(v + 1)}$$

$$= \frac{(v + m - 1)(v + m - 2) - l(l + 1)}{v(v - 1)}.$$

This expression providing a recursion formula for the A's is such that the numerator must be zero, if l and m are integral for some integral value of v; that is, the series (q.3.38.) cannot be infinite for integral values of m and l. The last non-vanishing term must be $v = l - |m|$; for example, if $v = l - m + 2$, then

$$\frac{A_{l-m}}{A_{l-m-2}} = \frac{(l-1)(l-2) - l(l+1)}{(l-m)(l-m-1)} \neq 0$$

and

$$\frac{A_{l+m+2}}{A_{l-m}} = \frac{(l+1)l - l(l+1)}{(l-m+2)(l-m+1)} = 0,$$

so that A_{l-m} must be the coefficient of the last term of series. This is the reason for writing the series in (q.3.29.) and (q.3.32.) in descending powers of x. In equation (q.3.27.), where $x = \cos\theta$, the function $\Theta(\theta)$ would become infinite in the limit when $\cos\theta = 1$, if the series were not limited to a finite number of terms, and this is the reason why l must be chosen to be an integer; also, in order for the series to be limited to a finite number of terms, $l \geq m$. Also, in order for the above relationships to be valid m must be zero or a positive integer, and in fact the associated Legendre functions are not usually defined for negative values of m, although this is possible.[1] If $l - |m|$ is even, it is only the expansion in even powers which is limited, so that A_1 must be zero, while if $l - |m|$ is odd, only the expansion in odd powers is limited and A_0 must be zero.

The associated Legendre functions of zero order are just the ordinary Legendre functions, and a few of the functions of a higher order are:

$$P_1^1(x) = (1 - x^2)^{1/2}; \ P_1^2(x) = 3x(1 - x^2)^{1/2}; \ P_1^3(x) = (3/2)(5x^2 - 1)(1 - x^2)^{1/2};$$
$$P_2^2(x) = 3(1 - x^2); \ P_2^3(x) = 15x(1 - x^2); \ P_3^3(x) = 15(1 - x^2)^{1/2}.$$

The detailed properties of these functions are discussed in books on differential equations.[2-4]

From the discussion of the properties and nature of the associated Legendre functions, the solutions of the equation in the angular function, $\Theta(\theta)$, are given by

$$\Theta(\theta) = \sin^m \theta P_l^m(\cos\theta), \tag{q.3.40.}$$

where $l = 0, 1, 2, \cdots$, and the separation constant $\beta = l(l+1)$, and

$$P_l^m(\cos\theta) = [2^l l!]^{-1} d^{l+m}/d\cos\theta^{l+m}[\cos^2\theta - 1]^l$$
$$= \frac{(2l)!}{2^l l! (l-m)!}\left[(\cos\theta)^{l-m} - \frac{(l-m)(l-m-1)}{2(2l-1)}(\cos\theta)^{l-m-2}\right.$$
$$+ \frac{(l-m)(l-m-1)(l-m-2)(l-m-3)}{2\cdot4\cdot(2l-1)(2l-3)}(\cos\theta)^{l-m-4}$$
$$+ \frac{(-1)^v(l-m)(l-m-1)\cdots(l-m-2v+1)}{2\cdot4\cdots(2v)(2l-1)(2l-3)\cdots(2l-2v+1)}$$
$$\left. \times (\cos\theta)^{l-m-2v} + \cdots\right], \tag{q.3.41.}$$

[1] Forsyth, A. R., *Differential Equations*, 1914, The Macmillan Company, London.
[2] Rainville, Earl D., *Intermediate Course in Differential Equations*, 1943, John Wiley & Sons, Inc., New York.
[3] Ince, E. L., *Ordinary Differential Equations*, 1944, Dover Publications, New York.
[4] Bateman, H., *Partial Differential Equations of Mathematical Physics*, Chap. VI, 1932, Cambridge University Press.

where m is zero or a positive integer and $|m| \leq l$. The significance of the quantum number l is the same as that in the older quantum theory, although in Schrödinger mechanics it emerges as a necessary, analytical condition in order that $\Theta(\theta)$ should be finite and have non-zero values; it is referred to, as before, as the azimuthal quantum number corresponding to the azimuthal angle θ. From the condition that $|m| \leq l$, and since from the solution of the $\Theta(\theta)$ equation, $m_l = 0, \pm 1, \pm 2, \ldots$, it is apparent that there must be $(2l + 1)$ values of m_l for each value of l, ranging from

$$-l, -(l - 1), \ldots -1, 0, 1, \ldots (l - 1), l. \tag{q.3.42.}$$

As before, in the terminology which was developed originally to describe the application of quantum theory to the description of the line spectra of atoms, if the l quantum number of a system is $0, 1, 2, \ldots$, the system is said to be in a $s, p, d, f, g, h, \ldots$ state. For an electron in a central field in a state described by $l = 0$, $m_l = 0$ and equation (q.3.24.) reduces to $d/d\theta[\sin \theta \, d\Theta(\theta)/d\theta] = 0$, for which the only finite solution is $\Theta(\theta) = $ constant, so that it is said that s-states are spherically symmetrical. This may be easily established by substituting directly $\psi(r, \theta, \phi) = R(r)$ in the equation (q.3.10.) with $\Theta(\theta)$ and $\Phi(\phi)$ constant, when it reduces to

$$d^2R(r)/dr^2 + (2/r) \, dR(r)/dr + 8\pi^2\mu h^{-2}[E + Z\varepsilon^2/r]R(r) = 0$$

so that the wave function for an s state is simply the eigenfunction of this equation, and comparison with (q.3.16.) with $\beta = l(l + 1)$ shows that $l = 0$, or all spherically symmetrical states are s-states.

In order that $\psi(r, \theta, \phi) = R(r)\Theta(\theta)\Phi(\phi)$ should be a solution of (q.3.10.) for a particle moving in a central field, each of the product functions should be separately normalizable. For the $\Theta(\theta)$ function, this means that

$$\int_{-1}^{1} (N^\theta)^2 \Theta_l{}^m(\theta) \Theta_l^{*m}(\theta) \, d\cos \theta = 1 = (N_\theta)^2 \int_0^{2\pi} \Theta_l{}^m(\theta) \Theta_l^{*m}(\theta) \sin \theta \, d\theta$$

$$= (N_\theta)^2 \int_0^{2\pi} [\Theta_l{}^m(\theta)]^2 \sin \theta \, d\theta$$

$$= (N_\theta)^2 \int_{-1}^{1} [P_l{}^m(x)]^2 \, dx, \tag{q.3.43.}$$

where N_θ is the appropriate normalizing constant, which has to be found. Then,

$$N_\theta = \left[\int_{-1}^{1} \{P_l{}^m(x)\}^2 \, dx \right]^{-1/2}. \tag{q.3.44.}$$

To evaluate the integral $\int_{-1}^{1} [P_l{}^m(x)]^2 \, dx$, it is convenient to note that the associated Legendre polynomials are the mth derivatives of the Legendre polynomials, that is $P_l{}^m(x) = d^m/dx^m P_l(x)$, and to find first of all the value of $\int_{-1}^{1} [P_l(\check{x})]^2 \, dx$, where

$$P_l(x) = [2^l l!]^{-1} \, d^l/dx^l (x^2 - 1)^l. \tag{q.3.45.}$$

Therefore,

$$\int_{-1}^{1} [P_l(x)]^2 dx = \int_{-1}^{1} [\{2^l l!\}^{-1} \, d^l/dx^l (x^2 - 1)^l]^2 \, dx$$

$$= [2^{2^l}(l!)^2]^{-1} \int_{-1}^{1} d^l/dx^l (x^2 - 1)^l \, d^l/dx^l (x^2 - 1)^l \, dx.$$

Integrating by parts, in usual manner,

$$\int_{-1}^{1} [d^l/dx^l (x^2 - 1)^l]^2 \, dx = d^l/dx^l (x^2 - 1)^l \, d^{l-1}/dx^{l-1} (x^2 - 1)^l \Big|_{-1}^{+1}$$

$$- \int_{-1}^{1} d^{l-1}/dx^{l-1} (x^2 - 1)^l \, d^{l+1}/dx^{l+1} (x^2 - 1)^l \, dx;$$

The first term on the right hand side of this equation vanishes at the limits, leaving simply the second term. After l successive integrations in this manner, the first term each time (the conventional uv term in the integration by parts) vanishes, leaving

$$\int_{-1}^{1} [d^l/dx^l (x^2 - 1)^l]^2 \, dx = (-1)^l \int_{-1}^{1} (x^2 - 1)^l \, d^{2l}/dx^{2l} (x^2 - 1)^l \, dx. \tag{q.3.46.}$$

Therefore,

$$\int_{-1}^{1} [P_l(x)]^2 \, dx = (-1)^l [2^{2l}(l!)^2]^{-1} \int_{-1}^{1} (x^2 - 1)^l \, d^{2l}/dx^{2l} (x^2 - 1)^l \, dx \tag{q.3.47.}$$

The highest power of x contained in $(x^2 - 1)^l$ is x^{2l}, so that after $2l$ differentiations all the other terms containing x vanish, if power of x is $< 2l$; the coefficient of the term remaining after $2l$ differentiations is

$$2l(2l - 1)(2l - 2) \cdots [2l - (2l - 1)] = (2l)!.$$

Thus,

$$\int_{-1}^{1} [P_l(x)]^2 \, dx = (-1)^l (2l)! \, [2^{2l}(l!)^2]^{-1} \int_{-1}^{1} (x^2 - 1)^l \, dx$$

$$= (2l)! \, [2^{2l}(l!)^2]^{-1} \int_{-1}^{1} (1 - x^2)^l \, dx,$$

since

$$(x^2 - 1) = (-1)(1 - x^2)$$

and

$$(x^2 - 1)^l = (-1)^l(1 - x^2)^l$$

and

$$(-1)^{l+l} = +1.$$

Now

$$\int_{-1}^{1} (1 - x^2)^l \, dx = \int_{-1}^{1} (1 - x)^l(1 + x)^l \, dx$$

$$= [l/(l + 1)] \int_{-1}^{1} (1 - x)^{l-1}(1 + x)^{l+1} \, dx,$$

and after l successive integrations, there is obtained

$$\int_{-1}^{1} (1 - x^2)^l \, dx = l(l - 1)(l - 2) \cdots 1[(l + 1)(l + 2) \cdots 2l \int_{-1}^{1} (1 + x)^{2l} \, dx$$

$$= l! \, l! \, 2^{2l+1}[(2l)! \, (2l + 1)]^{-1}.$$

Thus

$$\int_{-1}^{1} [P_l(x)]^2 \, dx = (2l)! \, [2^{2l}l!]^{-1}l! \, l! \, 2^{2l+1}[(2l)! \, (2l + 1)]^{-1} = 2(2l + 1)^{-1}.\ddagger$$

$$(q.3.48.)$$

$\ddagger$ This may be alternatively established by making use of the properties of beta functions. An alternative definition of the Γ-function (page 43), actually its reciprocal, due to Weierstrass, which may be derived from the Euler definition[1,2] gives

$$2 \int_{0}^{\pi/2} \cos^{m-1} x \sin^{n-1} x \, dx = [\Gamma(m/2)\Gamma(n/2)][\Gamma(\{m + n\}/2)]^{-1}.$$

By making the substitutions, $m = 2r$, $n = 2s$, and $\cos^2 x = u$, then $\cos^{m-1} x = [u^{1/2}]^{2r-1}$, $\sin^{n-1} x = [(1 - u)^{1/2}]^{2s-1}$, $du/dx = -2 \sin x \cos x = -2(1 - u)^{1/2}u^{1/2}$, so that

$$2 \int_{0}^{\pi/2} \cos^{m-1} x \sin^{n-1} x \, dx = 2 \int_{0}^{1} [u^{(2r-1)/2}(1 - u)^{(2s-1)/2}[-2(1 - u)^{1/2}u^{1/2}]^{-1}] \, du$$

$$= \int_{0}^{1} [u^{r-1}(1 - u)^{s-1}] \, du = \Gamma(r)\Gamma(s)[\Gamma(r + s)]^{-1}.$$

Now the integral, $\int_{-1}^{1} [P_l(x)]^2 \, dx$ may be expressed as a β-function by making the substitution, $(1 + x) = 2p$, for then

$$(1 - x^2)^l = [(1 - x)(1 + x)]^l = [2p(1 - 2p + 1)]^l$$
$$= [2^2p(1 - p)]^l = 2^{2l}p^l(1 - p)^l,$$

[1] Margenau, H., and Murphy, G. M., *The Mathematics of Physics and Chemistry*, 1951, D. Van Nostrand Co. Inc., Princeton, New Jersey.
[2] Whittaker, E. T., and Watson, G. N., *A Course of Modern Analysis*, 1940, 4th edition, Cambridge University Press.

and so $\int_{-1}^{1} [P_l(x)]^2\, dx = 2(2l!)[(l!)]^{-2} \int_{0}^{1} p^l(1-p)^l\, dp$, since $dx/dp = 2$, and the integrand is now in the form of a β-function. From the definition of Γ-functions, $\Gamma(n+1) = n\Gamma(n) = n!$, so that

$$\int_{0}^{1} p^l(1-p)^l\, dp = \beta[(l+1), (l+1)] = [\Gamma(l+1)]^2[\Gamma(2l+2)]^{-1}$$
$$= (l!)^2[\Gamma(2l+2)]^{-1} = (l!)^2[(2l+1)!]^{-1}.$$

Therefore, as before, in (q.3.48.),

$$\int_{-1}^{1} [P_l(x)]^2\, dx = 2(2l!)[(l!)]^{-2} \int_{0}^{1} p^l(1-p)^l\, dp$$
$$= 2(2l!)(l!)^{-2}(l!)^2[(2l+1)!]^{-1} = 2(2l+1)^{-1}.$$

By repeating this procedure, writing $[P_l^m(x)]^2$, as a product of the expressions given by (q.3.30), and integrating $(l-m)$ times by parts, the final result is

$$\int_{-1}^{1} [P_l^m(x)]^2\, dx = 2(2l+1)^{-1}(l+m)!\,[(l-m)!]^{-1}. \qquad \text{(q.3.49.)}$$

It is thus found that, from (q.3.44.),

$$N_\theta = \left[\frac{(2l+1)(l-m)!}{2(l+m)!}\right]^{1/2}, \qquad \text{(q.3.50.)}$$

and so the expression for the normalized functions $\Theta(\theta)$ is obtained,

$$\Theta_l^m(\theta) = [(2l+1)(l-m)!\,\{2(l+m)!\}^{-1}]^{1/2} \sin^m \theta P_l^m(\cos\theta), \qquad \text{(q.3.51.)}$$

where

$$l = 0,\, \pm 1,\, \pm 2, \ldots, \qquad \text{and} \qquad m = m_l = 0,\, \pm 1,\, \pm 2, \ldots, \pm l. \qquad \text{(q.3.52.)}$$

It is easily proved that the functions (q.3.51.) are mutually orthogonal to one another in pairs, since from (q.3.49.), with same m but different l,

$$\int_{-1}^{1} P_l^m(x)P_k^m(x)\, dx = 0 \quad (l \neq k). \qquad \text{(q.3.53.)}$$

Or, if $P_l^m(x)$ is substituted directly into (q.3.26.) and the resulting equation multiplied by $P_k^m(x)$, then a similar equation formed by substituting $P_k^m(x)$ into (q.3.26.) and multiplying by $P_l^m(x)$, the result obtained on subtracting the second equation from the first one and integrating over all x is,

$$\int_{-1}^{1} d/dx\{(1-x^2)[P_k^m(x)\, dP_l^m(x)/dx - P_l^m(x)\, dP_k^m(x)/dx]\}\, dx$$
$$= [l(l+1) - k(k+1)] \int_{-1}^{1} P_l^m(x)P_k^m(x)\, dx = 0,$$

and since the first integral is zero, then $\int_{-1}^{1} P_l^m(x)P_k^m(x)\, dx = 0\ (\neq k)$.

In many quantum mechanical problems, functions of the product type

$$Y_{l,m}(\theta, \phi) = P_l^m(\cos\theta)e^{im\phi}, \qquad \text{(q.3.54.)}$$

are encountered. These functions are assumed to be defined on the surface of a sphere of constant radius, such that $0 \leq \phi \leq 2\pi$ and $0 \leq \theta \leq \pi$, and are known as spherical harmonics. If the product of two such functions is integrated over the surface of a sphere,

$$\int_0^{2\pi} \int_0^{\pi} Y^*_{l'm'}(\theta, \phi) Y_{l,m}(\theta, \phi) \sin \theta \, d\theta \, d\phi = \int_{-1}^{1} P_{l'}^{m'}(x) P_l^m(x) \, dx \int_0^{2\pi} e^{i(m-m')\phi} \, d\phi.$$

The integral with respect to the variable ϕ is zero for $m \neq m'$, and if $m = m'$, the integral with respect to θ is zero unless $l = l'$; therefore, from (q.3.49.),

$$\int_0^{2\pi} d\phi \int_0^{\pi} Y^*_{l'm'}(\theta, \phi) Y_{l,m}(\theta, \phi) \sin \theta \, d\theta$$
$$= 4\pi(l + m)! \, [(2l + 1)(l - m)!]^{-1} \, \delta_{ll'} \, \delta_{mm'}.$$

The spherical harmonics defined by

$$Y_{l,m}(\theta, \phi) = \left[\frac{(2l + 1)\{(l - m)!\}}{4\pi\{(l + m)!\}} \right]^{1/2} P_l^m(\cos \theta) e^{im\phi}, \qquad \text{(q.3.55.)}$$

form an orthonormal set of functions on a spherical surface.

For an electron in a central field with l values ranging from 0 to 3, which are the most important cases in atomic quantum mechanics, corresponding to electrons in s, p, d, and f, states, the normalized angular functions, $\Theta(\theta)$ and $\Phi(\phi)$ as well as the orthonormal spherical harmonics, $Y_{l,m}(\theta, \phi) = \Theta(\theta)\Phi(\phi)$, have the following values; these values are obtained by substituting the appropriate values for l and m_l in the expressions given by (q.3.52.), (q.3.41.), (q.3.22.), and (q.3.55.), taking note of the limitations given in (q.3.52.) (see Table Q.3.2.).

It should be noted that the complex part of the $\Phi(\phi)$ and $\Theta(\theta)\Phi(\phi)$ functions may be eliminated by the use of De Moivre's theorem, $e^{im\phi} = \cos m\phi + i \sin m\phi$, and the selection of a suitable linear combination of the imaginary functions; thus, since

$$(1/\sqrt{2})e^{in\phi} = (1/\sqrt{2})[\cos n\phi + i \sin n\phi]$$

and

$$(1/i\sqrt{2})e^{-in\phi} = (1/i\sqrt{2})[\cos n\phi - i \sin n\phi],$$

by taking the sum and difference of these expressions, respectively, it is found that

$$(1/\sqrt{2})e^{in\phi} + (1/\sqrt{2})e^{-in\phi}$$
$$= (1/\sqrt{2})[\cos n\phi + i \sin n\phi] + (1/\sqrt{2})[\cos n\phi - i \sin n\phi],$$

and

$$(1/i\sqrt{2})e^{in\phi} - (1/i\sqrt{2})e^{-in\phi}$$
$$= (1/i\sqrt{2})[(\cos n\phi + i \sin n\phi) - (\cos n\phi - i \sin n\phi)],$$

which may be simplified to give $\sqrt{2} \cos n\phi$ and $\sqrt{2} \sin n\phi$, respectively.

TABLE Q.3.2

State	l	m_l	$\Theta(\theta)$	$\Phi(\phi)$	$\Theta(\theta)\Phi(\phi)$
				Normalized Angular Functions	
s	0	0	$1/\sqrt{2}$	$1/\sqrt{2\pi}$	$1/2\sqrt{\pi}$
p	1	0	$\sqrt{3/2}\,(\cos\theta)$	$1/\sqrt{2\pi}$	$[\sqrt{3/4\pi}]\,(\cos\theta)$
	1	± 1	$[(\sqrt{3})/2]\sin\theta$	$[1/\sqrt{2\pi}]\,e^{\pm i\phi}$	$(1/2)\sqrt{(3/2\pi)}\,\sin\theta e^{\pm i\phi}$
d	2	0	$(\sqrt{5/8})\,(3\cos^2\theta - 1)$	$1/\sqrt{2\pi}$	$(1/4)[\sqrt{5/\pi}]\,(3\cos^2\theta - 1)$
	2	± 1	$(\sqrt{15/4})\sin\theta\cos\theta$	$[1/\sqrt{2\pi}]\,e^{\pm i\phi}$	$[\sqrt{15/8\pi}]\sin\theta\cos\theta e^{\pm i\phi}$
	2	± 2	$(1/4)\sqrt{15}\,\sin^2\theta$	$[1/\sqrt{2\pi}]\,e^{\pm 2i\phi}$	$[\sqrt{15/32\pi}]\sin^2\theta e^{\pm 2i\phi}$
f	3	0	$[(\sqrt{14})/4](5\cos^3\theta - 3\cos\theta)$	$1/\sqrt{2\pi}$	$[(1/4)\sqrt{7/\pi}]\,(5\cos^3\theta - 3\cos\theta)$
	3	± 1	$[(1/4)\sqrt{21/2}]\sin\theta(5\cos^2\theta - 1)$	$e^{\pm i\phi}/\sqrt{2\pi}$	$[(1/8)\sqrt{21/\pi}]\sin\theta(5\cos^2\theta - 1)e^{\pm i\phi}$
	3	± 2	$[(1/4)\sqrt{105}]\sin^2\theta\cos\theta$	$e^{\pm 2i\phi}/\sqrt{2\pi}$	$[(1/4)\sqrt{105/2\pi}]\sin^2\theta\cos\theta\,e^{\pm 2i\phi}$
	3	± 3	$(1/4)[\sqrt{35/2}]\sin^3\theta$	$e^{\pm 3i\phi}/\sqrt{2\pi}$	$[(1/8)\sqrt{35/\pi}]\sin^3\theta e^{\pm 3i\phi}$

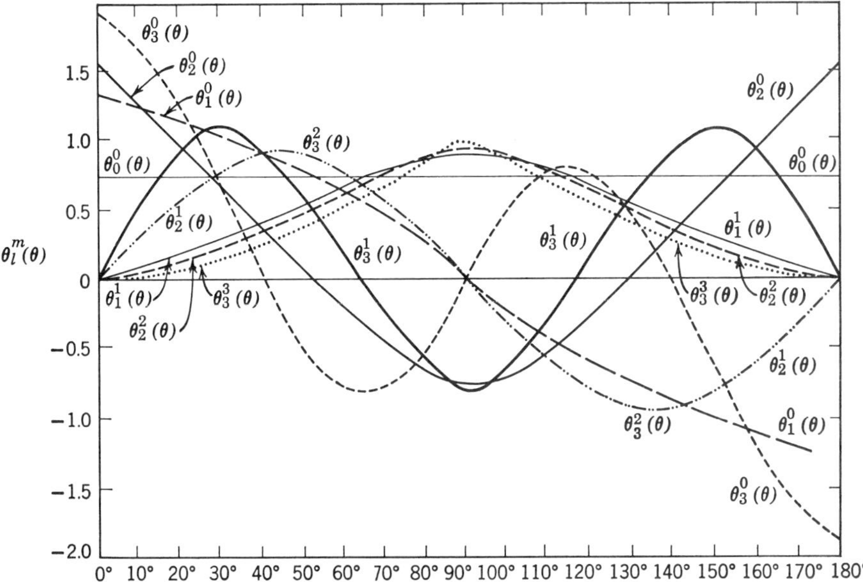

FIG. Q.3.3. The normalized angular function, $\theta_l^m(\theta)$, plotted as a function of the polar angle, θ.

As far as the $\Theta(\theta)$ function is concerned, it is apparent that it has the constant value $1/\sqrt{2} = 0.707$, for $l = 0$, $m_l = 0$, that is for a particle moving in a central field describable by an s-state, its position is independent of the polar angle θ. The other $\Theta(\theta)$ functions referred to above have one or more nodes, and they are shown plotted as a function of the polar angle θ in Fig. Q.3.3. The probability that the particle will be found in the region of space with a polar angle between θ and $\theta + d\theta$ is given by $[\Theta(\theta)]^2 \sin \theta \, d\theta$, and this function is determined completely by the magnitude of the angle θ, and in contrast to the function $\Theta(\theta)$, which as seen in Fig. Q.3.3. may adopt positive and negative values, must always be positive. For a particle in an s-state, $[\Theta(\theta)]^2$ is $1/2$, a constant independent of the angle θ; for $l = 0$, m_l must also be 0, and so $[\Phi(\phi)]^2 = 1/2\pi$, a constant also, independent of the other polar angle ϕ, and therefore $[\Theta(\theta)]^2[\Phi(\phi)]^2$ for a particle in an s-state is independent of both the polar angles θ and ϕ, so that the probability of finding the particle is equal in all directions (for some constant radial distance r). This, of course, is what is meant by saying that an electron in an s-state in an atom has a spherically symmetrical charge cloud density; the motion of an electron, for example, in an atom, describable as being in an s-state, since it is independent of both the polar angles must therefore be entirely a radial motion, away from and towards the nucleus. The three-dimensional representation of a particle in an s-state would then correspond to a sphere concentric with the nucleus (in an atom) and having a constant surface density; the projection of such a sphere on the plane $\phi = 0$ (the xz plane of the Cartesian Coordinate System shown in Fig. Q.3.2.) would be a circle. Similar polar diagrams may be constructed for other values of l and the related $(2l + 1)$ values of m_l. A two-dimensional diagram is sufficient to show how either $[\Theta(\theta)]^2$ or $[\Phi(\phi)]^2$ vary with the polar angles θ and ϕ respectively, and the manner in which the function $[\Theta(\theta)]^2[\Phi(\phi)]^2$ varies with the two angles θ and ϕ simultaneously may be represented in a single three-dimensional diagram. Fig. Q.3.4. shows the probability density distribution function $[\Theta(\theta)]^2$ plotted as a function of the angle θ for $l = 0, 1, 2,$ and 3 and the corresponding values of m_l; if these plane diagrams are imagined to be rotated about the vertical axis ($\theta = 0$ or the z Cartesian coordinate axis of Fig. Q.3.2.), then, since they denote projections on the plane $\phi = 0$, the xz plane of Fig. Q.3.2., the corresponding three-dimensional polar diagrams result. The importance and wide use of these diagrams is due to the fact that they are the same for all central fields. All the states with $m_l = 0$, with the exception of the s-states, have a maximum probability density in the direction of $\theta = 0$ and $\theta = \pi$, and for all states describable by $l = 0$, except the s-states, the maximum value of $[\Theta(\theta)]^2$ is along the equator of the diagrams.

It will be shown later (page 214) that the integral number l determines the number of units of angular momentum of the state to which it refers, such that the total angular momentum of the state concerned is given by

$$L = \pm(h/2\pi)\sqrt{l(l + 1)}, \qquad \text{with} \qquad l = 0, 1, 2, \ldots . \qquad \text{(q.3.56.)}$$

This relationship will be proved after discussion of the nature of the solution of the radial equation, (q.3.16.).

Q.3.6.   The Radial Function

The equation of motion of a particle of reduced mass μ moving in a spherically symmetrical potential field, expressed in spherical polar coordinates, has been separated into three homogeneous equations, the equation in the radial function, $R(r)$, being given by (q.3.16.). This equation contains the potential function, $-Z\varepsilon^2/r$, and so the radial function is dependent on the form of the central field, whereas the equations in the angular functions are not so dependent. Substituting the value found for the separation constant, β, the value $l(l + 1)$ found from the solution of the equation in the angular function $\Theta(\theta)$, gives

$$d^2R(r)/dr^2 + (2/r)\,dR(r)/dr + [8\pi^2\mu E h^{-2} + 8\pi^2\mu Z\varepsilon^2 r h^{-2}$$
$$- l(l + 1)r^{-2}]R(r) = 0. \qquad \text{(q.3.57.)}$$

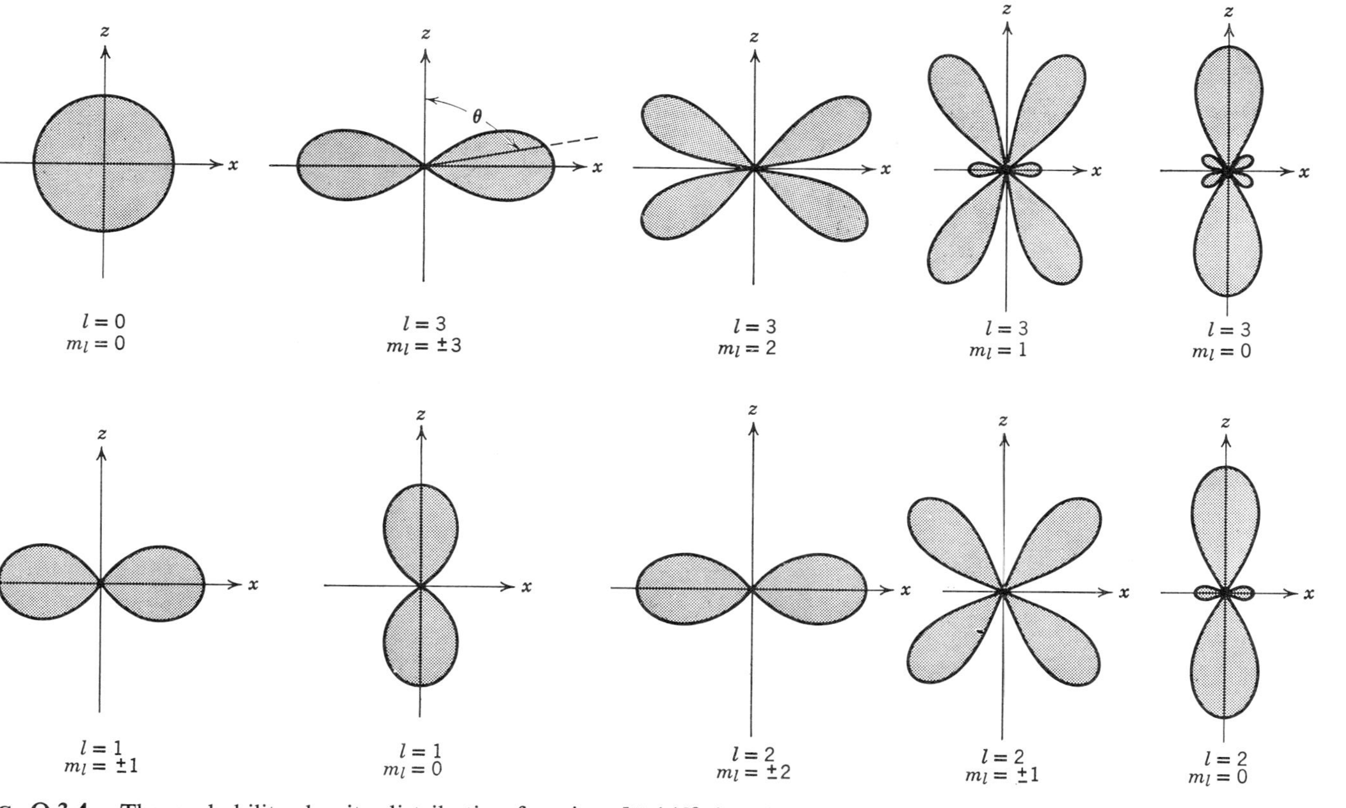

FIG. Q.3.4.   The probability density distribution function, $[\theta^m(\theta)]^2 \sin\theta \, d\theta$, plotted as a function of the polar angle θ, for various quantum states of hydrogen-like atoms. The vertical axis in each diagram corresponds to $\theta = 0°$ and the other polar angle $\phi = 0°$ in the plane of the paper, so that each drawing is a projection on the xz plane of Fig. Q.3.2. A straight line drawn from the origin to the edge of any shaded area has a length proportional to the probability of finding the electron in a hydrogenic atom in the direction of that line. The scale of the diagrams for states with $m = 0$, except for $l = 0$, is approximately $1/(l + 1)$ times that of the other states with the same l value.

For the hydrogen atom, in particular, where $Z = 1$, the total energy E is negative, as giving by (q.1.33.), since the potential energy has a greater negative value than the kinetic energy. It is therefore convenient to let the term containing the total energy in (q.3.57.) be represented by, say, $-\alpha^2$; that is, let

$$8\pi^2\mu Eh^{-2} = -\alpha^2, \qquad (q.3.58.)$$

thus ensuring that α^2 is a positive quantity and so α must be real. It is also convenient, to substitute the quantity $8\pi^2\mu Z\varepsilon^2h^{-2}$ in equation (q.3.57.) by $2z$; that is,

$$8\pi^2\mu Z\varepsilon^2h^{-2} = 2z. \qquad (q.3.59.)$$

The effect of these substitutions is essentially to make use of the atomic system of units (page 91), since $1/z = a_0 = h^2/4\pi^2\mu\varepsilon^2$, the Bohr unit. Making these substitutions in (q.3.57.), the radial equation becomes

$$d^2R(r)/dr^2 + (2/r)\,dR(r)/dr + [-\alpha^2 + 2z/r - l(l+1)r^{-2}]R(r) = 0. \qquad (q.3.60.)$$

Equations of this type are generally easier to solve if the coefficients are constants rather than quantities containing the variable, such as r, in this case. Otherwise, approximate solutions may be obtained by the standard techniques of considering what happens when the variable tends to infinity or zero, and by expressing the function $R(r)$ as a power series in the variable r, differentiating this series to find the first and second derivatives, and comparing coefficients to evaluate the constants in the power series.

For large values of r, as $r \to \infty$, equation (q.3.60.) reduces to $d^2R(r)/dr^2 = \alpha^2R(r)$, which is of the form of an eigenvalue equation, indicating that $R(r)$ is an eigenfunction of the operator d^2/dr^2 corresponding to the eigenvalue α^2. This equation has a simple solution, namely, $R(r) = Ce^{\pm\alpha r}$. For $\psi(r, \theta, \phi) = R(r)\Theta(\theta)\Phi(\phi)$ to be an acceptable solution of the central field equation, (q.3.9.), $R(r)$ must be a well behaved function. In particular, $R(r)$ must be finite everywhere, but the solution $R(r) = Ce^{\alpha r}$ of the above asymptotic equation becomes infinite for large value of r, and therefore has to be rejected. The other solution, $R(r) = Ce^{-\alpha r}$, provides an approximate solution of the equation (q.3.60.), but it is not entirely satisfactory since it would fall off to zero at not too large a value of r. This difficulty may be removed by allowing C to be some slowly varying function of r, the nature of which has still to be deduced. So if

$$R(r) = U(r)e^{-\alpha r}, \qquad (q.3.61.)$$

then

$$dR(r)/dr = d/dr[U(r)e^{-\alpha r}] = dU(r)/dr(e^{-\alpha r}) - \alpha e^{-\alpha r}U(r)$$

$$= R(r)/U(r)[dU(r)/dr - \alpha U(r)],$$

and

$$d^2R(r)/dr^2 = e^{-\alpha r}\, d^2U(r)/dr^2 - \alpha e^{-\alpha r}\, dU(r)/dr - \alpha e^{-\alpha r}\, dU(r)/dr$$
$$+ \alpha^2 e^{-\alpha r} U(r)$$
$$= R(r)/U(r)[\alpha^2 U(r) - 2\alpha\, dU(r)/dr + d^2U(r)/dr^2].$$

Substituting these first and second derivatives into (q.3.60.) there is obtained after factoring out $R(r)/U(r)$ and rearranging,

$$d^2U(r)/dr^2 + 2\, dU(r)/dr[r^{-1} - \alpha] + [2(z - \alpha)r^{-1} - l(l + 1)r^{-2}]U(r) = 0.$$
$$\text{(q.3.62.)}$$

Now, if $U(r)$ is expressed as a power series in r, and substituted into (q.3.62.), it should be possible to discover the relationship between the adjacent coefficients of powers of r. So, if

$$U(r) = A_0 r^0 + A_1 r + A_2 r^2 + \cdots A_n r^n + \cdots = \sum_{n=0}^{\infty} A_n r^n, \quad \text{(q.3.63.)}$$

then in the most general case, n need not necessarily be integral, but may be fractional. A more general representation of $U(r)$ would therefore be

$$U(r) = A_0 r^\mu + A_1 r^{\mu+1} + A_2 r^{\mu+2} + \cdots + A_{\nu-1} r^{\mu+\nu-1} + A_\nu r^{\mu+\nu}$$
$$+ A_{\nu-1} r^{\mu+\nu+1} + r^\mu \sum_{\nu=0}^{\infty} A_\nu r^\nu, \quad \text{(q.3.64.)}$$

where ν is integral and μ is fractional. If the first and second derivatives of $U(r)$ as given by (q.3.64.) are substituted into (q.3.62.), there results

$$\cdots A_{\nu-1}(\mu + \nu - 1)(\mu + \nu - 2)r^{\mu+\nu-3} + A_\nu(\mu + \nu)(\mu + \nu - 1)r^{\mu+\nu-2}$$
$$+ A_{\nu+1}(\mu + \nu + 1)(\mu + \nu)r^{\mu+\nu-1} + \cdots + 2A_{\nu-1}(\mu + \nu - 1)r^{\mu+\nu-3}$$
$$+ 2A_\nu(\mu + \nu)r^{\mu+\nu-2} + A_{\nu+1}(\mu + \nu + 1)r^{\mu+\nu-1} + \cdots$$
$$- 2\alpha A_{\nu-1}(\mu + \nu - 1)r^{\mu+\nu-2} - 2\alpha A_\nu(\mu + \nu)r^{\mu+\nu-1}$$
$$- 2\alpha A_{\nu+1}(\mu + \nu + 1)r^{\mu+\nu} - \cdots + 2z A_{\nu-1}r^{\mu+\nu-2} + 2z A_\nu r^{\mu+\nu-1}$$
$$+ 2z A_{\nu+1}r^{\mu+\nu} + \cdots - 2\alpha A_{\nu-1}r^{\mu+\nu-2} - 2\alpha A_\nu r^{\mu+\nu-1}$$
$$- 2\alpha A_{\nu+1}r^{\mu+\nu} - \cdots - l(l + 1)A_{\nu-1}r^{\mu+\nu-3} - l(l + 1)A_\nu r^{\mu+\nu-2}$$
$$- l(l + 1)A_{\nu+1}r^{\mu+\nu-1} - \cdots = 0. \quad \text{(q.3.65.)}$$

This equation can only be valid if the sum of the coefficients of each power of r is zero; for example, the coefficient of $r^{\mu+\nu-2}$, on being equated to zero, results in the expression

$$A_\nu[(\mu + \nu)(\mu + \nu - 1) + 2(\mu + \nu)] - 2\alpha A_{\nu-1}(\mu + \nu - 1)$$
$$+ 2z A_{\nu-1} - 2\alpha A_{\nu-1} - l(l + 1)A_\nu = 0,$$

which may be rearranged to read

$$A_\nu[(\mu + \nu)(\mu + \nu + 1) - l(l + 1)] = 2A_{\nu-1}[\alpha(\mu + \nu) - z]. \quad \text{(q.3.66.)}$$

The series (q.3.64.) begins with the term A_0, so that A_{-1} must be zero, and if $\nu = 0$ is substituted into the expression (q.3.66.) giving the general recurrence relationship between adjacent coefficients of consecutive powers of r, then $A_{\nu-1} = A_{-1} = 0$, and so

$$A_\nu[\mu(\mu + 1) - l(l + 1)] = 0. \tag{q.3.67.}$$

This equation has two solutions. Either $\mu = l$, for the first solution, or $\mu = -(l + 1)$, for the second solution. In the latter event, the powers of r in (q.3.64.) would be negative and $U(r)$ could not be finite everywhere, nor could $R(r)$ be given by (q.3.61.) and so the second solution has to be rejected. It is thus found that $\mu = l$, and since $l = 0, 1, 2, \ldots$, μ must be integral and the series (q.3.64.) becomes

$$U(r) = A_0 r^l + A_1 r^{l+1} + A_2 r^{l+2} + \cdots + A_\nu r^{l+\nu} + \cdots \tag{q.3.68.}$$

$$= r^l \sum_{\nu=0}^{\infty} A_\nu r^\nu.$$

The recurrence formula (q.3.66.), on the substitution of $\mu = l$ becomes

$$A_\nu[(l + \nu)(l + \nu + 1) - l(l + 1)] = 2A_{\nu-1}[\alpha(l + \nu) - z], \tag{q.3.69.}$$

which enables the coefficient of any power of r in the series representation of $U(r)$ to be expressed in terms of the first coefficient A_0, z, α, and the integers ν and l. Thus, if $\nu = 1, 2, 3, \ldots$, the coefficients $A_1, A_2, A_3, \ldots$ become

$$\begin{aligned}
A_1 &= 2A_0[\alpha(l + 1) - z][(l + 1)(l + 2) - l(l + 1)]^{-1} \\
&= 2A_0[\alpha(l + 1) - z][2l + 2]^{-1};
\end{aligned}$$

$$\begin{aligned}
A_2 &= 2A_1[\alpha(l + 2) - z][(l + 2)(l + 3) - l(l + 1)]^{-1} \\
&= 2A_1[\alpha(l + 2) - z][2(2l + 3)]^{-1} \\
&= 2^2 A_0[\alpha(l + 1) - z][\alpha(l + 2) - z][2(2l + 2)(2l + 3)]^{-1};
\end{aligned}$$

$$\begin{aligned}
A_3 &= 2A_2[\alpha(l + 3) - z][(l + 3)(l + 4) - l(l + 1)]^{-1} \\
&= 2A_2[\alpha(l + 3) - z][3(2l + 4)]^{-1} \\
&= 2^3 A_0[\alpha(l + 1) - z][\alpha(l + 2) - z][\alpha(l + 3) - z] \\
&\quad \times [2(2l + 2)(2l + 3)(2l + 4)]^{-1}.
\end{aligned}$$

Thus the series (q.3.68.) adopts the form

$$\begin{aligned}
U(r) = A_0 r^l\bigg[&1 + 2[\alpha(l + 1) - z][2l + 2]^{-1}r \\
&+ \frac{2^2[\alpha(l + 1) - z][\alpha(l + 2) - z]}{2(2l + 2)(2l + 3)}r^2 + \cdots \\
&+ \frac{2^\nu[\alpha(l + 1) - z][\alpha(l + 2) - z]\cdots[\alpha(l + \nu) - z]}{\nu!\,(2l + 2)(2l + 3)(2l + 4)\cdots(2l + \nu + 1)}r^\nu + \cdots\bigg]
\end{aligned}$$

$$\tag{q.3.70.}$$

If $R(r)$ and therefore $U(r)$ is to be finite, this series cannot be infinite, but must terminate after some finite number of terms. As $v \to \infty$, the recurrence expression (q.3.69.) becomes

$$\frac{A_v}{A_{v-1}} = \frac{2[\alpha(l+v)-z]}{(l+v)(l+v+1)-l(l+1)} = \frac{2\alpha - 2z(l+v)^{-1}}{l+v+1-l(l+1)(l+v)^{-1}}$$
$$\simeq \frac{2\alpha}{l+v+1}.$$

However, this property of the two consecutive coefficients of the series $U(r)$ having ratio 2α to $(l+v+1)$ is also a property of the exponential series, $e^{2\alpha r}$, because

$$e^{2\alpha r} = 1 + (2\alpha r) + (2\alpha r)^2/2! + (2\alpha r)^3/3! + \cdots (2\alpha r)^{l+v}/(l+v)!$$
$$+ (2\alpha r)^{l+v+1}/(l+v+1)! + \cdots,$$

where the ratio of two successive coefficients is

$$\frac{A_v}{A_{v-1}} = \frac{(2\alpha)^{l+v+1}(l+v)!}{(l+v+1)!\,(2\alpha)^{l+v}} = \frac{2\alpha}{l+v+1}. \tag{q.3.71.}$$

Therefore, for large values of v, $U(r) \simeq e^{2\alpha r}$, and so $R(r) = U(r)e^{-\alpha r} \simeq e^{\alpha r}$. But this approximation for $R(r)$ becomes infinite at large values of r and is thus not an acceptable expression for $R(r)$. Therefore the number of terms in the series expression (q.3.70.) for $U(r)$ cannot be large, but must be limited to some finite number. If it is assumed that this finite number of terms is indeed v, then the polynomial given by (q.3.70.) must terminate with the $(v-1)$th term, so that the coefficient $A_v = 0$. This is easily seen for some small finite value of v, such as 4, when (q.3.68.) becomes $U(r) = A_0 r^l + A_1 r^{l+1} + A_2 r^{l+2} + A_3 r^{l+3}$, the coefficient A_4 of r^{l+4} and all the successively higher coefficients vanishing for $v = 4$. The series (q.3.70.) thus becomes

$$U(r) = A_0 r^l \left[1 + \frac{2[\alpha(l+1)-z]}{2l+2} r + \frac{2^2[\alpha(l+1)-z][\alpha(l+2)-z]}{2!\,(2l+2)(2l+3)} r^2 \right.$$
$$+ \frac{2^3[\alpha(l+1)-z][\alpha(l+2)-z][\alpha(l+3)-z]}{3!\,(2l+2)(2l+3)(2l+4)} r^3 + \cdots$$
$$\left. + \frac{(2)^{v-1}[\alpha(l+1)-z][\alpha(l+2)-z]\cdots[\alpha(l+v-1)-z]}{(v-1)!\,(2l+2)(2l+3)\cdots(2l+v)} r^{v-1} \right],$$
$$\tag{q.3.72.}$$

the coefficient A_v of the $(v+1)$th being zero. From the recurrence expression relating two adjacent coefficients in the series, (q.3.69.), it follows then that $A_v[(l+v)(l+v+1)-l(l+1)] = 0 = 2A_{v-1}[\alpha(l+v)-z]$, which can only be true if

$$\alpha(l+v) = z, \tag{q.3.73.}$$

and since $l = 0, 1, 2, \ldots$ (from the Legendre equation) and v is integral from the above argument, being limited to values $1, 2, 3, \ldots$, the quantity $(l + v)$ must also be integral; hence, if $(l + v) = n$, then

$$\alpha n = z, \qquad \text{with} \qquad n = 1, 2, 3, \ldots, \qquad \text{and} \qquad n = (l + v). \qquad \text{(q.3.74.)}$$

Substituting for α and z from (q.3.58.) and (q.3.59), respectively, it is found that

$$\alpha^2 n^2 = -8\pi^2 \mu E h^{-2} n^2 = z^2 = [4\pi^2 \mu Z \varepsilon^2 h^{-2}]^2,$$

or

$$E = -2\pi^2 \mu Z^2 \varepsilon^4 / n^2 h^2, \qquad \text{with} \qquad n = 1, 2, 3, \ldots. \qquad \text{(q.3.75.)}$$

This is exactly the same result as given by (q.1.33.), obtained by Bohr's argument, with $Z = 1$ for hydrogen, but here the necessity for the integral values of the parameter n is determined by the condition of finiteness of the function $U(r)$ and therefore of $R(r)$, for all values of the radial distance r.

If (q.3.74.) is substituted into the recurrence expression (q.3.69.), then

$$\frac{A_v}{A_{v-1}} = \frac{2[\alpha(l + v) - z]}{[(l + v)(l + v + 1) - l(l + 1)]} = -\frac{2\alpha[n - l - v]}{v(v + 2l + 1)}, \qquad \text{(q.3.76.)}$$

and by a similar substitution, (q.3.72.) becomes

$$U(r) = A_0 r^l \left[1 - \frac{2\alpha(n - l - 1)}{1(2l + 2)} r + \frac{(2\alpha)^2(n - l - 1)(n - l - 2)}{2!\,(2l + 2)(2l + 3)} r^2 \right.$$

$$- \frac{(2\alpha)^3(n - l - 1)(n - l - 2)(n - l - 3)}{3!\,(2l + 2)(2l + 3)(2l + 4)} r^3 + \cdots \qquad \text{(q.3.77.)}$$

$$\left. + \frac{(-1)^{2l+v}(2\alpha)^{v-1}(n - l - 1)(n - l - 2)\cdots(n - l - v + 1)}{(v - 1)!\,(2l + 2)(2l + 3)\ldots(2l + v)} r^{v-1} \right].$$

This polynomial expression is a function of the integers n and l, and since $\alpha = z/n = 1/na_0$ (page 196), the radial function given by (q.3.61.) may be now written

$$R(r) = U(r)e^{-\alpha r} = U_{n,l}(r)e^{-r/na_0}, \qquad \text{(q.3.78.)}$$

with $U_{n,l}$ given by (q.3.77.), and a_0 being the Bohr atomic unit of length.

The polynomial expressions in (q.3.77.) and (q.3.78.) are special forms of functions which are well known in mathematical physics as associated Laguerre polynomials, which are solutions of the associated Laguerre differential equation. In fact, as with the equation in the radial function $\Theta(\theta)$, the equation in the radial function $R(r)$ given by (q.3.60.) may be transformed into exactly the form of the classical associated Laguerre differential equation by appropriate substitutions. The relationship between consecutive coefficients of the associated Laguerre polynomials obeys the same equation as that given by (q.3.76.). The manner in which this transformation may be accomplished is as follows:

It is necessary to first define a new function such that

$$rR(r) = Y(r), \qquad (q.3.79.)$$

and to make the change of variable from r to x by means of the substitution,

$$x = 2r/na_0. \qquad (q.3.80.)$$

Now, rewriting equation (q.3.60.) as an equation in $Y(r)$, incorporating the substitutions given by

$$-\alpha^2 = 8\pi^2 \mu E h^{-2} = -(z^2/n^2) = -(1/a_0{}^2 n^2), \qquad (q.3.81.)$$

where $z = 1/a_0 = 4\pi^2 \mu \varepsilon^2 h^{-2}$ (page 196), an equation results which resembles the one-dimensional wave equation, namely,

$$d^2/dr^2[Y(r)/r] + (2/r)\, d/dr[Y(r)/r]$$
$$+ [-a_0{}^{-2}n^{-2} + 2(ra_0)^{-1} - l(l+1)r^{-2}]Y(r)r^{-1} = 0.$$

Performing the indicated differentiations and multiplying throughout by r, this equation simplifies to

$$d^2/dr^2[Y(r)] + [-(1/n^2 a_0{}^2) + (2/ra_0) - l(l+1)r^{-2}]Y(r) = 0. \quad (q.3.82.)$$

On now making the change of variable given by (q.3.80.), this becomes after multiplication throughout by $n^2 a_0{}^2/4$,

$$d^2/dx^2[Y(r)] + [-(1/4) + (n/x) - l(l+1)x^{-2}]Y(r) = 0. \quad (q.3.83.)$$

This equation is one form of the classical associated Laguerre differential equation, which is known to have solutions of the type‡

$$Y_{n,l}(r) = x^{l+1}e^{-x/2}L_{n+l}^{2l+1}(x), \qquad \text{with} \qquad na_0 x/2 = r, \quad (q.3.84.)$$

where $L_{n+l}^{2l+1}(x)$ is the so-called associated Laguerre polynomial, defined as

$$L_{u+l}^{2l+1}(x) = d^{2l+1}/dx^{2l+1}[e^x d^{n+l}/dx^{n+l}(x^{n+l}e^{-x})], \qquad (q.3.85).$$

or in the usual polynomial form

$$L_{n+l}^{2l+1}(x) = A_0 + A_1 x + A_2 x^2 + \cdots A_p x^p + \cdots A_{n-l-1}x^{n-l-1} = \sum_{p=0}^{n-l-1} A_p x^p$$

$$(q.3.86.)$$

where

$$\frac{A_p}{A_{p-1}} = \frac{-(n-l-p)}{p(p+2l+1)}$$

and

$$A_{n-l-1} = (-1)^{n+l}(n+l)!\,[(n-l-1)!]^{-1}. \qquad (q.3.87.)$$

‡ To prove that (q.3.84.) is a solution of (q.3.83.), it is convenient to write $2l + 1 = p$ and $n + l = q$, so that $L_{n+1}^{2l+1}(x)$ may be written as $L_q^p(x)$. Then

$$d/dx[x^{l+1}e^{-x/2}L_q{}^p(x)] = (l+1)x^le^{-x/2}L_q{}^p(x)$$
$$+ x^{l+1}[-(1/2)e^{-x/2}L_q{}^p(x) + e^{-x/2}\,d/dxL_q{}^p(x)]$$
$$= (l+1)x^le^{-x/2}L_q{}^p(x) - (1/2)x^{l+1}e^{-x/2}L_q{}^p(x)$$
$$+ e^{-x/2}x^{l+1}\,d/dxL_q{}^p(x),$$

and

$$d^2/dx^2[x^{l+1}e^{-x/2}L_q{}^p(x)] = l(l+1)x^{l-1}e^{-x/2}L_q{}^p(x)$$
$$+ (l+1)x^le^{-x/2}[-L_q{}^p(x)/2 + d/dxL_q{}^p(x)]$$
$$- (1/2)(l+1)x^le^{-x/2}L_q{}^p(x) + (1/4)x^{l+1}e^{-x/2}L_q{}^p(x)$$
$$- (1/2)x^{l+1}e^{-x/2}\,d/dxL_q{}^p(x) + (l+1)x^le^{-x/2}\,d/dxL_q{}^p(x)$$
$$- (1/2)x^{l+1}e^{-x/2}\,d/dxL_q{}^p(x) + x^{l+1}e^{-x/2}\,d^2/dx^2L_q{}^p(x).$$

Substituting these first and second derivatives into (q.3.83.) and dividing throughout by $e^{-x/2}x^{l-1}$, gives an equation which is easily simplified to

$$x\,d^2/dx^2[L_q{}^p(x)] + d/dxL_q{}^p(x)[p+1-x] + L_q{}^p(x)[q-p] = 0. \qquad \text{(q.3.88.)}$$

This equation is exactly the form of the classical associated Laguerre differential equation Now from the definition of $L_q{}^p(x) = d^p/dx^p[e^x\,d^q/dx^q(x^qe^{-x})]$, as given by (q.3.94.), equation (q.3.85.) may be rewritten as

$$x\,d^{p+2}/dx^{p+2}[L_q] + [p+1-x]\,d^{p+1}/dx^{p+1}[L_q] + [q-p]\,d^p/dx^p[L_q] = 0. \quad \text{(q.3.89.)}$$

If equation (q.3.86.) is differentiated once more with respect to x, then

$$d^{p+2}/dx^{p+2}[L_q] + x\,d^{p+3}/dx^{p+3}[L_q] + (p+1)\,d^{p+2}/dx^{p+2}[L_q] - d^{p+1}/dx^{p+1}[L_q]$$
$$- x\,d^{p+2}/dx^{p+2}[L_q] + (q-p)[d^{p+1}/dx^{p+1}(L_q)] = 0,$$

or

$$x\,d^{p+3}/dx^{p+3}[L_q] + [p+2-x]\,d^{p+2}/dx^{p+2}[L_q] + [q-p-1]\,d^{p+1}/dx^{p+1}[L_q] = 0.$$
$$\text{(q.3.90.)}$$

Equation (q.3.87.) is the same as equation (q.3.86.) with $(p+1)$ substituted for p. Therefore, if (q.3.86.) is valid for any value of p, it is true for $(p+1)$, and thus for all values of p. For the special case where $p = 0$, it becomes

$$x\,d^2/dx^2[L_q] + (1-x)\,d/dx[L_q] + qL_q = 0. \qquad \text{(q.3.91.)}$$

Now L_q, the so-called Laguerre polynomial is defined as

$$L_q = e^x\,d^q/dx^q(x^qe^{-x}), \qquad \text{(q.3.92.)}$$

and the first and second derivatives of this function are, respectively,

$$d/dx[L_q] = e^x\,d^q/dx^q(x^qe^{-x}) + e^x\,d^{q+1}/dx^{q+1}(x^qe^{-x}),$$

and

$$d^2/dx^2[L_q] = e^x\,d^q/dx^q(x^qe^{-x}) + 2e^x\,d^{q+1}/dx^{q+1}(x^qe^{-x}) + e^x\,d^{q+2}/dx^{q+2}(x^qe^{-x}),$$

which when substituted into (q.3.88.) and the resulting equation divided throughout by e^x leads to

$$x\,d^{q+2}/dx^{q+2}(x^qe^{-x}) + [1+x]\,d^{q+1}/dx^{q+1}(x^qe^{-x}) + [q+1]\,d^q/dx^q(x^qe^{-x}) = 0.$$
$$\text{(q.3.93.)}$$

This equation is easily shown to be true for $q = 0$ and $q = 1$, respectively, for then $x\,d^2/dx^2e^{-x} + (1+x)\,d/dxe^{-x} + e^{-x} = xe^{-x} - e^{-x} - xe^{-x} + e^{-x} = 0$, and $x\,d^3/dx^3(xe^{-x}) + (1+x)\,d^2/dx^2(xe^{-x}) + 2\,d/dx(xe^{-x}) = 0$, but more generally, for any value of q, it is necessary to evaluate the derivative, $d^q/dx^q(x^qe^{-x})$. The general expression for this derivative may be deduced by examining the first few derivatives for $q = 1, 2, 3, \ldots$. Thus,

$$d/dx[x^q e^{-x}] = qx^{q-1}e^{-x} - e^{-x}x^q = e^{-x}q! \, [x^{q-1}/(q-1)! - x^q q!];$$

$$d^2/dx^2[x^q e^{-x}] = q(q-1)x^{q-2}e^{-x} - e^{-x}qx^{q-1} + e^{-x}x^q - e^{-x}qx^{q-1} \qquad \text{(q.3.94.)}$$

$$= e^{-x}q! \, [x^{q-2}/(q-2)! - 2x^{q-1}/(q-1)! + x^q/q!];$$

$$d^3/dx^3[x^q e^{-x}] = e^{-x}q! \, [x^{q-3}/(q-3)! - 3x^{q-2}/(q-2)! + 3x^{q-1}/(q-1)! - x^q/q!].$$

It is thus possible to find that for any integer t,

$$d^t/dx^t[x^q e^{-x}] = e^{-x}q! \left[\sum_{s=0}^{t} \frac{(-1)^s t! \, x^{q-t+s}}{s! \, (t-s)! \, (q-t+s)!} \right]. \qquad \text{(q.3.95.)}$$

For $t = 0$, this expression is obviously true, as $d^0/dx^0(x^q e^{-x}) = x^q e^{-x} = e^{-x}q! \, [x^q/q!]$, and by substituting $t = 1, 2, 3$, it is easily shown that equations (q.3.91.) are reproduced. Alternatively, if (q.3.92.) is differentiated once with respect to x, it is found that the same expression is obtained with t replaced by $(t + 1)$, so that (q.3.92.) is valid for all integral values of t. Making use of (q.3.92.), the derivatives required in (q.3.90.) may now be written as

$$d^{q+2}/dx^{q+2}[x^q e^{-x}] = e^{-x}q! \sum_{s=0}^{q+2} \frac{(-1)^s(q+2)! \, x^{s-2}}{s! \, (q+2-s)! \, (s-2)!};$$

$$d^{q+1}/dx^{q+1}[x^q e^{-x}] = e^{-x}q! \sum_{s=0}^{q+1} \frac{(-1)^s(q+1)! \, x^{s-1}}{s! \, (q+1-s)! \, (s-1)!}; \qquad \text{(q.3.96.)}$$

$$d^q/dx^q[x^q e^{-x}] = e^{-x}q! \sum_{s=0}^{q} \frac{(-1)^s q! \, x^s}{s! \, (q-s)! \, s!}.$$

If these three derivatives, which have been shown to be valid for all integral values of q, are substituted into equation (q.3.90.), it is found that this equation is satisfied. Therefore the Laguerre polynomials, as defined by (q.3.89.), for all integral values of q are solutions of equation (q.3.88.), for the case where $p = 0$. But by the previous argument, if (q.3.88.) is true for any value of p, it is true for all integral values of p, so that the associated Laguerre polynomials, defined by (q.3.94.) must constitute solutions of equation (q.3.85.), and also of equation (q.3.83.) from which it was derived. Thus, it is proved that the functions defined by (q.3.84.) are solutions of the (q.3.83.) and therefore of the radial equation (q.3.60.), from which it was in turn derived.

If the quantities defined by (q.3.59.), (q.3.73.), and (q.3.80.), that is $\alpha = z/n = 1/na_0$, $r = na_0 x/2$ or $x = 2r/na_0$, and $v = (n - l)$, are substituted into $U(r)$ by (q.3.77.), then it is seen that,

$$U(na_0 x/2) = A_0 x^l (na_0/2)^l \left[1 - \frac{(n-l-1)}{1! \, (2l+2)} x + \frac{(n-l-1)(n-l-2)}{2! \, (2l+2)(2l+3)} x^2 \right.$$

$$- \frac{(n-l-1)(n-l-2)(n-l-3)}{3! \, (2l+2)(2l+3)(2l+4)} x^3 + \cdots$$

$$\left. + \frac{(-1)^{n+l}}{(2l+2)(2l+3)\cdots(n+l)} x^{n-l-1} \right], \qquad \text{(q.3.97.)}$$

so that for this series the ratio of any two adjacent coefficients is the same as that given by (q.3.96.) for the associated Laguerre polynomial defined as a power series; also the power of x of the last term in both series, the highest power of x which occurs, neglecting the common multiplier in (q.3.97.) is the same. The highest power of x which occurs in the derivative form of the associated Laguerre polynomial given by (q.3.94.) may be deduced by considering the effect of the double differentiations involved. In the differentiation of $(x^{n+l}e^{-x})$, the factor of (-1) is introduced each time differentiation with respect to x is carried out, due to the e^{-x} term, a term of the type $(x^{n+l}e^{-x})$ remaining each time, while all of the other terms resulting have lower powers of x. After $(n+l)$ such differentiations, the result is $d^{n+l}/dx^{n+l}(x^{n+l}e^{-x}) = (-1)^{n+l}(x^{n+l}e^{-x}) +$ all other terms containing lower powers of x. Therefore $L_{n+l}^{2l+1}(x) = d^{2l+1}/dx^{2l+1}[e^x\{(-1)^{n+l}(x^{n+l}e^{-x}) +$ terms of lower powers of $x\}]$

$$= (-1)^{n+l}\frac{(n+l)!}{(n-l-1)!}x^{n-l-1}$$

$+$ all other terms containing lower powers of x,

so that

$$A_{n-l-1} = (-1)^{n+l}(n+l)[(n-l-1)!]^{-1},$$

as given by (q.3.96.).

From comparison of equations (q.3.84.) and (q.3.78.), it is apparent that

$$U_{n,l}(r) = R(r)[e^{-r/na_0}]^{-1} = Y_{n,l}(r)[re^{-r/na_0}]^{-1} = (2/na_0)x^l L_{n+l}^{2l+1}(x),$$
$$(q.3.98.)$$

and that since (q.3.84.) is a solution of (q.3.83.), so also is $L_{n+l}^{2l+1}(x)$ and $U_{n,l}(r)$, as the multipliers of the associated Laguerre polynomial simply cancel out on substitution into the equation in $Y(r)$. Also, from comparison of equation (q.3.95.) using the relations (q.3.96.), with $U(r)$ given by (q.3.97.), A_0 must be unity. This latter fact also follows from the consideration that if two functions, expressed in two different ways satisfy the same differential equations and have coefficients which obey the same recurrence relationships, they must be identical.

The solutions of the radial equation, (q.3.16.), variously expressed as (q.3.57.), (q.3.60.), (q.3.62.), or (q.3.82.) by means of the substitutions noted, are given by $R(r) = Y_{n,l}(r)/r = U_{n,l}(r)e^{-r/na_0}$. To find the normalization constant for the radial function, ultimately requires the evaluation of an integral containing the square of the associated Laguerre polynomial in the integrand. It is proved below that the normalized function $Y_{n,l}(r)$ is given by (q.3.109.)‡

‡ To evaluate the integral, $\int_0^\infty [Y_{n,l}(r)]^2\, dr$, required in calculating the normalization factor for the radial function $R(r)$, it is convenient to write one part of the product which composes the integrand as a polynomial given by (q.3.95.) and the other part in the derivative form given by (q.3.94.), these two expressions being equivalent to one another; $Y_{n,l}(r)$ is given by (q.3.84.). Since $x = 2r/na_0$ by (q.3.80.), $dr = (na_0/2)\, dx$, so

the integration over all values of the variable x is properly taken over the limits 0 to ∞, and not $-\infty$ to ∞. Thus,

$$\int_0^\infty [Y_{n,l}(r)]^2\, dr = \int_0^\infty [x^{l+1}e^{-x/2}L_{n+l}^{2l+1}(x)]^2(na_0/2)\, dx$$

$$= (na_0/2)\int_0^\infty x^{2l+2}e^{-x}[L_{n+l}^{2l+1}(x)]^2\, dx$$

$$= (na_0/2)\sum_{p=0}^{n-l-1} A_p \int_0^\infty x^{2l+2+p}$$

$$\times\, e^{-x}\, d^{2l+1}/dx^{2l+1}[e^x\, d^{n+l}/dx^{n+l}(e^{n+l}e^{-x})]\, dx$$

$$\text{(q.3.99.)}$$

Apart from the multiplier, $(na_0/2)A_p$, if this expression is integrated by parts, in the usual manner, the following is obtained,

$$[[x^{2l+2+p}e^{-x}]\, d^{2l}/dx^{2l}[e^x\, d^{n+l}/dx^{n+l}(x^{n+l}e^{-x})]]_0^\infty$$

$$-\int_0^\infty d^{2l}/dx^{2l}[e^x\, d^{n+l}/dx^{n+l}(x^{n+l}e^{-x})]\, d/dx[x^{2l+2+p}e^{-x}]\, dx. \qquad \text{(q.3.100.)}$$

The first integrated term above vanishes at the limits, and if this integration by parts is repeated $(2l+1)$ times on the second term, the integrated term vanishes each time and the integral finally reduces to,

$$-\int_0^\infty [e^x\, d^{n+l}/dx^{n+l}(x^{n+l}e^{-x})]\, d^{2l+1}/dx^{2l+1}[x^{2l+2+p}e^{-x}]\, dx, \qquad \text{(q.3.101.)}$$

since $(-1)^{2l+1} = -1$. Now making use of the expression (q.3.92.), the factor $d^{2l+1}/dx^{2l+1}[x^{2l+2+p}e^{-x}]$, in the integrand of (q.3.101.) may be rewritten as

$$d^{2l+1}/dx^{2l+1}[x^{2l+2+p}e^{-x}] = (2l+2+p)!\, e^{-x}\sum_{g=0}^{2l+1}\frac{(-1)^g(2l+1)!\, x^{p+1+g}}{g!(2l+1-g)!\,(1+p+g)!}.$$

$$\text{(q.3.102.)}$$

If (q.3.102.) is now substituted in (q.3.99.), it is found that,

$$\int_0^\infty [Y_{n,l}(r)]^2\, dr = (na_0/2)\int_0^\infty x^{2l+2}e^{-x}[L_{n+l}^{2l+2}(x)]^2\, dx$$

$$= (na_0/2)\left[-\sum_{p=0}^{n-l-1} A_p\right]\sum_{g=0}^{2l+1}\frac{(2l+2+p)!\,(2l+1)!\,(-1)^g}{g!\,(2l+1-g)!\,(p+1+g)!}$$

$$\int_0^\infty x^{p+1+g}\, d^{n+l}/dx^{n+l}(x^{n+l}e^{-x})\, dx. \qquad \text{(q.3.103.)}$$

The integral in equation (q.3.103.) may be evaluated as follows:

$$\int_0^\infty x^{p+1+g}\, d^{n+l}/dx^{n+l}(x^{n+l}e^{-x})\, dx = (-1)^{n+1}\int_0^\infty x^{n+l}e^{-x}\, d^{n+l}/dx^{n+l}(x^{p+1+g})\, dx,$$

since after $(n+l)$ integrations by parts, the integrated term vanishes each time and a minus sign is introduced into the result after each of the $(n+l)$ integrations. The term x^{p+1+g}, in this integral can be differentiated $(n+l)$ times to give a non-vanishing result, only if $(p+1+g) \geq (n+l)$, or if

$$p + 1 + g - n - l \geq 0. \qquad \text{(q.3.104.)}$$

If this latter condition is satisfied, then

$$d^{n+l}/dx^{n+l}(x^{p+1+g}) = (p+1+g)!\, x^{p+1+g-n-l},$$

and the integral in (q.3.103.) reduces to,

$$\int_0^\infty x^{p+1+g}\, d^{n+l}/dx^{n+l}(x^{n+l}e^{-x})\, dx = (-1)^{n+1}\int_0^\infty \frac{(p+1+g)!}{(p+1+g-n-l)!}\, x^{p+1+g}e^{-x}\, dx$$

Now, $\int_0^\infty x^{p+1+g} e^{-x}\, dx = (p+1+g)! \int_0^\infty x e^{-x}\, dx$, after $(p+g)$ integrations by parts, and $\int_0^\infty x e^{-x}\, dx = 1$. These integrals may, of course, be alternatively evaluated by making use of the Legendre definition of a gamma function, as on page 43, which gives $\int_0^\infty x^{p-1} e^{-x}\, dx = \Gamma(p) = (p-1)!$. Substituting the values found for these integrals into (q.3.103.) and combining the minus signs into one term, there is finally obtained for $\int_0^\infty x^{2l+2} e^{-x}[L_{n+l}^{2l+1}(x)]^2\, dx$, the expression,

$$\int_0^\infty x^{2l+2} e^{-x}[L_{n+l}^{2l+1}]^2\, dx$$
$$= \sum_{p=0}^{n-l-1} A_p \sum_{g=0}^{2l+1} \frac{(2l+2+p)!\,(2l+1)!\,(p+1+g)!\,(-1)^{n+l+g+1}}{g!\,(2l+1-g)!\,(p+1+g-n-l)!} \cdot \quad\text{(q.3.105.)}$$

It should be noted that if $(p+1+g-n-l) < 0$, the whole integral vanishes and this corresponds to the situation where the $(n+l)$ differentiations of x^{p+1+g} reduce the integrand to zero. Therefore for the integral above to have a non-vanishing value, the condition (q.3.104.) must be satisfied. However, since the double sum, representing the value of this integral as given by (q.3.105.) is summed over the values of p from 0 to $(n-l-1)$ and of g from 0 to $(2l+1)$, it emerges that there are only three possible values for the pair of indices p and g, which enable the integral above to adopt a non-zero value, namely,

(i)   if $p = n-l-1$ and $g = 2l$,   when $(p+1+g-n-l) = 0$,

and

(ii)   if $p = n-l-1$ and $g = 2l+1$,
                        when $(p+1+g-n-l) = 1$,   and

(iii) if $p = n-l-2$ and $g = 2l+1$,   when $(p+1+g-n-l) = 0$.

For all other values of p up to and including $(n-l-3)$ and of g up to and including $(2l-1)$, the quantity $(p+1+g-n-l)$ is negative, since n and l must both be integral, as previously established. The coefficients A_{n-l-1} and A_{n-l-2} may be obtained from (q.3.96.), and by substituting the three pairs of values of p and g given above in (q.3.105.), the result is as follows:

$$\int_0^\infty x^{2l+2} e^{-x}[L_{n+l}^{2l+1}(x)]^2\, dx$$
$$= \left[\frac{(-1)^{n+l}(n+l)!}{(n-l-1)!}\right]\left[\frac{(n+l+1)!\,(2l+1)!\,(n+l)!\,(-1)^{n+3l+1}}{2l!\,1!\,1!}\right]$$
$$+ \left[\frac{(-1)^{n+l}(n+l)!}{(n-l-1)!}\right]\left[\frac{(n+l+1)!\,(2l+1)!\,(n+l+1)!\,(-1)^{n+3l+2}}{(2l+1)!\,1!\,1!}\right]$$
$$- \left[\frac{(-1)^{n+l}(n+l)!}{(n-l-1)!} \frac{[(n-1)n - l(l+1)]}{1}\right]$$
$$\times \left[\frac{(n+l)!\,(2l+1)!\,(n+l)!\,(-1)^{n+3l+2}}{(2l+1)!\,1!\,1!}\right]$$
$$= \left[\frac{(-1)^{n+l}(n+l)!}{(n-l-1)!}\right]\left[\frac{(n+l+1)!\,(2l+1)!\,(n+l)!\,(-1)^{n+3l+1}}{2l!}\right.$$
$$+ [(n+l+1)!]^2(-1)^{n+3l+2} - [n(n-1) - l(l+1)][(n+l)!]^2(-1)^{n+3l+2}\bigg].$$

This expression, on factoring out $[(n+l)!]^2$, reduces to,

$$\int_0^\infty x^{2l+2} e^{-x}[L_{n+l}^{2l+1}(x)]^2\, dx$$
$$= \left[\frac{[(n+l)!]^3}{(n-l-1)!}\right][(-1)^{n+l}\{(n+l+1)(2l+1)(-1)^{n+3l+1}$$
$$+ (n+l+1)^2(-1)^{n+3l+2} - (n^2 - n - l^2 - l)(-1)^{n+3l+2}\}].$$

If $(n + l)$ is even, then $(n + 3l + 1)$ must be odd and $(n + 3l + 2)$ even; if $(n + l)$ is odd, then $(n + 3l + 1)$ must be even and $(n + 3l + 2)$ odd; in either event, the expression above reduces to,

$$\int_0^\infty x^{2l+2} e^{-x} [L_{n+l}^{2l+1}(x)]^2 \, dx = 2n[(n + l)!]^3 [(n - l - 1)!]^{-1}. \qquad \text{(q.3.106.)}$$

If (q.3.106.) is substituted into (q.3.99.), there is obtained,

$$\int_0^\infty [Y_{n,l}(r)]^2 \, dr = (na_0/2)2n[(n + l)!]^3 [(n - l - 1)!]^{-1}, \qquad \text{(q.3.107.)}$$

which enables the normalizing constant for the function $Y_{n,l}(r)$, given by (q.3.84.) to be obtained. If this normalizing factor is designated $N_{n,l}$, then the normalization condition is that

$$\int_0^\infty [N_{n,l} Y_{n,l}(r)]^2 \, dr = 1,$$

since $Y_{n,l}(r)$ is real, and so

$$N_{n,l} = \left[\int_0^\infty \{Y_{n,l}(r)\}^2 \, dr \right]^{-1/2},$$

or

$$N_{n,l} = \left[\frac{(n - l - 1)!}{n^2 a_0 [(n + l)!]^3} \right]^{1/2} \qquad \text{(q.3.108.)}$$

Therefore the normalized function, $Y_{n,l}(r)$ is given by the expression,

$$Y_{n,l}(r) = \left[\frac{(n - l - 1)!}{n^2 a_0 [(n + l)!]^3} \right]^{1/2} x^{l+1} e^{-x/2} L_{n+l}^{2l+1}(x). \qquad \text{(q.3.109.)}$$

This evaluation of the normalization constant for the function $Y_{n,l}(r)$ follows the method described by Slater.[1] A more elegant and general proof is described by Sommerfeld,[2] based on the use of generating functions. A generating function represents a simple and convenient way of describing various special functions, such as the Hermite polynomials, the Legendre and Laguerre polynomials, the associated Legendre and Laguerre polynomials, and Bessel functions, by means of a function of two arguments, which when expanded in a power series with respect to one argument, contains the functions to be generated as coefficients involving the other argument as parameters.[3]

Thus, the normalized functions are, $N_r R_{n,l}(r) = (1/r) N_{n,l} Y_{n,l}(r)$, with $N_{n,l}$ given by (q.3.108.) and $Y_{n,l}(r)$ given by (q.3.84.). Expressing (q.3.109.) as a function of the radial distance r only, using (q.3.80.), gives

$$Y_{n,l}(r) = \left[\frac{(n - l - 1)!}{n^2 a_0 [(n + l)!]^3} \right]^{1/2} (2r/na_0)^{l+1} e^{-r/na_0} L_{2l+1}^{n+l}(2r/na_0),$$

and therefore the normalized radial function, $N_r R_{n,l}(r)$ is obtained as,

$$N_r R_{n,l}(r) = (1/r) Y_{n,l}(r) = \left[\frac{4(n - l - 1)!}{n^4 a_0^3 [(n + l)!]^3} \right]^{1/2} (2r/na_0)^l e^{-r/na_0} L_{n+l}^{2l+1}(2r/na_0),$$

$$\text{(q.3.110.)}$$

---

[1] Slater, John C., *Quantum Theory of Atomic Structure*, **Vol. I**, 1960, McGraw-Hill Book Co. Inc., New York.
[2] Sommerfeld, A., *Atombau und Spektrallinien*, 1939, Verlag Vieweg, Brunswick, Germany.
[3] Schrödinger, E., *Ann. der Physik*, 1926, **80**, 483.

with $L_{n+l}^{2l+1}(x)$ defined as (q.3.94.) or (q.3.95.). It may be proved in the same manner as that used in finding the normalization constant for these functions, that they are also mutually orthogonal to one another in pairs.

Functions such $U(r)$ and $U(x)$ defined by (q.3.77.) and (q.3.97.), are known as confluent hypergeometric functions and have many interesting properties, only some of which are of importance in quantum mechanical applications.[1] In fact, as a consequence of the property of finiteness, which must be imposed on acceptable wave functions used to describe physico-chemical systems, only a relatively small number of the functions described by (q.3.110.) are of any immediate interest. Thus, from (q.3.74.), $n = 1, 2, 3, \ldots$, and it has previously been deduced from the solution of the $\Theta(\theta)$ equation that $l = 0, 1, 2, \ldots$, but from the definition of the associated Laguerre function given by (q.3.94.), $(n + l)$ must equal or exceed $(2l + 1)$, in order that this function and therefore $Y(r)$ and $R(r)$ should have a non-zero value, as already expressed in (q.3.104.); in other words, n and l cannot independently adopt any integral values, if $R(r)$ is to be finite, but $(n + l) \geq 2l + 1$ or $(n - l - 1) \geq 0$ or

$$l \leq (n - 1). \tag{q.3.111.}$$

The significance of the integers, n, as is apparent from (q.3.75.), is the same in Schrödinger mechanics as in the older Bohr theory; its value determines the total energy of the system describable by such a quantum number, still referred to as the principal or total quantum number. The only allowable combinations of the n and l quantum numbers are then those for which:

$$
\begin{aligned}
n &= 1, & l &= 0; \\
n &= 2, & l &= 0, 1; \\
n &= 3, & l &= 0, 1, 2; \\
n &= 4, & l &= 0, 1, 2, 3; \\
n &= 5, & l &= 0, 1, 2, 3, 4; \\
n &= 6, & l &= 0, 1, (2, 3, 4, 5).
\end{aligned}
\tag{q.3.112.}
$$

In fact, as will become apparent later, all of the normal states of atoms may be described in terms of the combinations of n and l quantum numbers listed above, excepting those l values contained in parentheses for $n = 6$. Then associated Laguerre polynomials, calculated by the use of (q.3.94.) for the above combinations of the n and l quantum numbers, which are those of use in atomic quantum mechanics, are:

---

[1] Slater, L. J., *Confluent Hypergeometric Functions*, 1960, Cambridge Univ. Press, London.

$$L_1{}^1(x) = -1;$$

$$L_2{}^1(x) = -2! \, (2 - x); \quad L_3{}^3(x) = -3!;$$

$$L_3{}^1(x) = -3(6 - 6x + x^2); \quad L_4{}^3(x) = -4! \, (4 - x); \quad L_5{}^5(x) = -5!;$$

$$L_4{}^1(x) = -4(24 - 36x + 12x^2 - x^3); \quad L_5{}^3(x) = -(5!/2)(20 - 10x + x^2);$$

$$L_6{}^5(x) = -6! \, (6 - x); \quad L_7{}^7(x) = -7!; \qquad\qquad\qquad \text{(q.3.113.)}$$

$$L_5{}^1(x) = -5(120 - 240x + 120x^2 - 20x^3 + x^4);$$

$$L_6{}^3(x) = -5! \, (120 - 90x + 18x^2 - x^3);$$

$$L_7{}^5(x) = -(7!/2)(42 - 14x + x^2); \quad L_8{}^7(x) = -8! \, (8 - x); \quad L_9{}^9(x) = -9!;$$

$$L_6{}^1(x) = -6(720 - 1800x + 1200x^2 - 300x^3 + 30x^4 - x^5);$$

$$L_7{}^3(x) = -5! \, (840 - 840x + 252x^2 - 28x^3 + x^4).$$

The normalized radial functions obtained by substituting the appropriate associated Laguerre polynomial into (q.3.110.) are listed below:

$$R_{10}(r) = 2a_0{}^{-3/2}e^{-r/a_0};$$

$$R_{20}(r) = (1/2\sqrt{2})a_0{}^{-3/2}e^{-r/2a_0}[2 - r/a_0];$$

$$R_{21}(r) = (1/2\sqrt{6})a_0{}^{-3/2}e^{-r/2a_0}[r/a_0];$$

$$R_{30}(r) = (2/9\sqrt{3})a_0{}^{-3/2}e^{-r/3a_0}[3 - 2r/a_0 + 2r^2/9a_0{}^2];$$

$$R_{31}(r) = (2/9\sqrt{6})a_0{}^{-3/2}e^{-r/3a_0}(2r/3a_0)[2 - r/3a_0];$$

$$R_{32}(r) = (1/9\sqrt{30})a_0{}^{-3/2}e^{-r/3a_0}(2r/3a_0)^2;$$

$$R_{40}(r) = (1/96)a_0{}^{-3/2}e^{-r/4a_0}[24 - 18r/a_0 + 3r^2/a_0{}^2 - r^3/8a_0{}^3];$$

$$R_{41}(r) = (1/32\sqrt{15})a_0{}^{-3/2}e^{-r/4a_0}(r/2a_0)[20 - 5r/a_0 + r^2/4a_0{}^2];$$

$$R_{42}(r) = (1/96\sqrt{5})a_0{}^{-3/2}e^{-r/4a_0}(r/2a_0)^2[6 - r/2a_0];$$

$$R_{43}(r) = (1/96\sqrt{35})a_0{}^{-3/2}e^{-r/4a_0}(r/2a_0)^3; \qquad\qquad \text{(q.3.114.)}$$

$$R_{50}(r) = (2/75\sqrt{5})a_0{}^{-3/2}e^{-r/5a_0}$$
$$\times \, [15 - 12r/a_0 + 12r^2/5a_0{}^2 - 4r^3/25a_0{}^3 + 2r^4/625a_0{}^4];$$

$$R_{51}(r) = (2/75\sqrt{30})a_0{}^{-3/2}e^{-r/5a_0}(2r/5a_0)$$
$$\times \, [30 - 9r/a_0 + 18r^2/25a_0{}^2 - 2r^3/125a_0{}^3];$$

$$R_{52}(r) = (1/75\sqrt{70})a_0{}^{-3/2}e^{-r/5a_0}(2r/5a_0)^2[21 - 14r/5a_0 + 2r^2/25a_0{}^2];$$

$$R_{53}(r) = (1/150\sqrt{70})a_0{}^{-3/2}e^{-r/5a_0}(2r/5a_0)^3[4 - r/5a_0];$$

$$R_{54}(r) = (1/900\sqrt{70})a_0{}^{-3/2}e^{-r/5a_0}(2r/5a_0)^4;$$

$$R_{60}(r) = (1/2160\sqrt{6})a_0{}^{-3/2}e^{-r/6a_0}$$
$$\times \, [720 - 600r/a_0 + 400r^2/3a_0{}^2 - 100r^3/9a_0{}^3$$
$$+ \, 10r^4/27a_0{}^4 - r^5/243a_0{}^5];$$

$$R_{61}(r) = (1/756\sqrt{210})a_0{}^{-3/2}e^{-r/6a_0}(r/3a_0)$$
$$\times \, [840 - 280r/a_0 + 28r^2/a_0{}^2 - 28r^3/27a_0{}^3 + r^4/81a_0{}^4].$$

The radial functions quoted above have all been multiplied throughout by -1, and are strictly applicable only to the hydrogen atom functions, where the nuclear charge Z is unity. They are equally applicable to all hydrogenic atoms if the parameter x of (q.3.80.) is defined as $x = 2Zr/na_0$; this has the effect of introducing a factor $Z^{3/2}$ into the numerator of (q.3.110.), as well as into each of the other three factors in this expression. For this reason, some authors prefer to define the radial functions in terms of a parameter, $\rho = 2Zr/na_0$, and write (q.3.110.) as [1,2]

$$R_{n,l}(r) = -\left[\frac{4Z^3(n-l-1)!}{n^4 a_0{}^3[(n+l)!]^3}\right]^{1/2} \rho^l e^{-\rho/2} L_{n+l}^{2l+1}(\rho). \qquad \text{(q.3.115.)}$$

In Fig. Q.3.5. the normalized radial functions of (q.3.114.) are shown plotted as a function of r, expressed in Bohr atomic units, a_0, for a few different values of the quantum numbers n and l. The common property of these functions is that they all approach zero asymptotically as the radial distance r increases. For small values of r, the various radial functions behave in a sinusoidal manner, the range of this type of behavior increasing rapidly with increasing value of the principal quantum number n. The total energy of a particle moving in a central field, for each value of the principal quantum number n, is given by (q.3.75.), $-Z^2/n^2$ Rydbergs or $-Z^2/2n^2$ Hartree units. In fact, the range of sinusoidal behavior of the radial functions is within the region where the classical kinetic energy of the particle would be positive, and the radial functions would fall off to zero in the region where the classical kinetic energy would be negative. The infinite number of eigenvalues, given by (q.3.75.), for all integral values of n, all correspond to negative total energies of the particle, described by the radial function corresponding to the same value of n; there are obviously degeneracies, since all states described by the same n value, but different l and m_l values, have the same energy; the degeneracy is given by

$$\sum_{l=0}^{n-1} (2l+1) = 2\sum_{l=0}^{n-1} l + n = 2n(n-1)/2 + n = n^2, \qquad \text{(q.3.116.)}$$

due to the limitations imposed on n, l, and m_l, as already found. As the principal quantum number increases, the total energy adopts smaller and smaller negative values, each negative value of the energy still corresponding to a bound state of the particle concerned; the corresponding radial function for

---

[1] Pauling, L., and Wilson, E. B., *Introduction to Quantum Mechanics*, 1935, McGraw-Hill Book Company, New York.
[2] Pauling, L., *Nature of the Chemical Bond*, 1962, 3rd edition, Cornell University Press, Ithaca, New York.

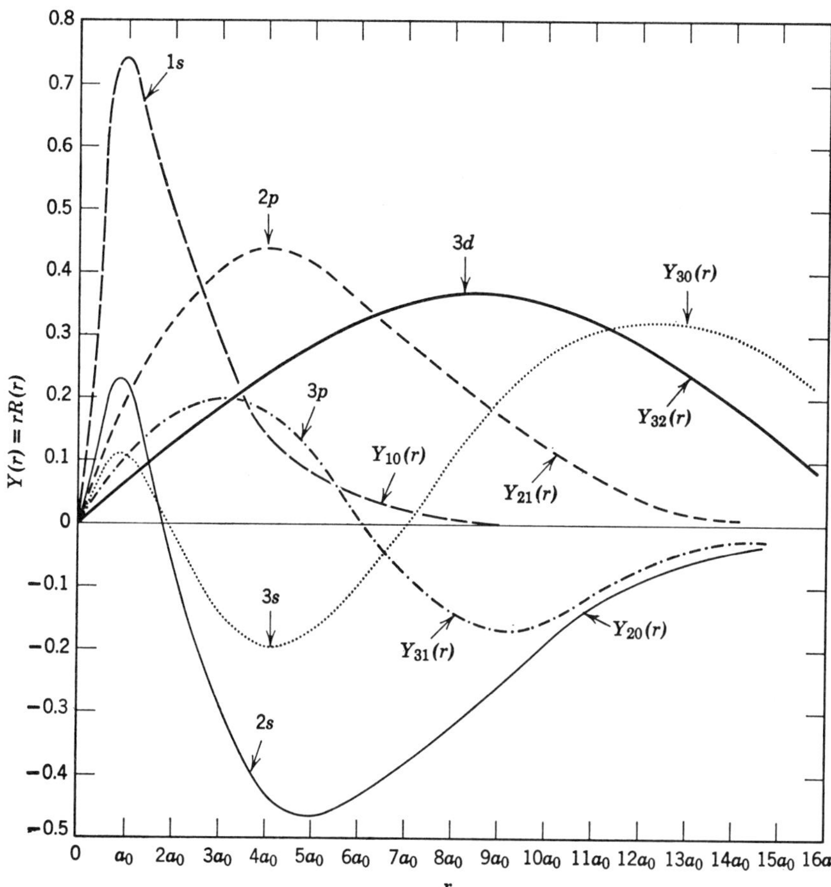

Fig. Q.3.5. The radial wave functions for hydrogen, $Y(r) = rR(r)$, plotted as a function of the radial distance of the electron from the nucleus. The radial distances are expressed in Bohr units.

the particle with large n values extends to an increasingly greater distance in a diagram of the type of Fig. Q.3.5., and for positive energy values extends to infinity. At the same time, the energy levels are no longer discretely different, are no longer quantized, and the energy spectrum becomes continuous. For example, a hydrogen atom may absorb sufficient energy to raise the energy of its constituent electron to the continuous band of energies, the electron in the process becoming free of the nucleus or ionized off; the minimum energy required to accomplish this is referred to as the ionization potential energy. It is still possible to describe unnormalizable radial functions, corresponding to imaginary values of n, but such continuum eigenfunctions while of importance in the theory of scattering[1] are of relatively little importance in atomic quantum mechanics.

For electrons in atoms, described by the radial functions, (q.3.114.), the physically significant quantity is $[R_{n,l}]^2 4\pi r^2\,dr$, which is a measure of the probability of finding the electron within a spherical region of space of thickness dr around a sphere of radius r, and represents the quantum mechanical equivalent of the classical density of negative electrical charge. This quantity is referred to as the radial charge density, and is shown plotted for several stationary states, corresponding to various n and l values, in Fig. Q.3.6. In

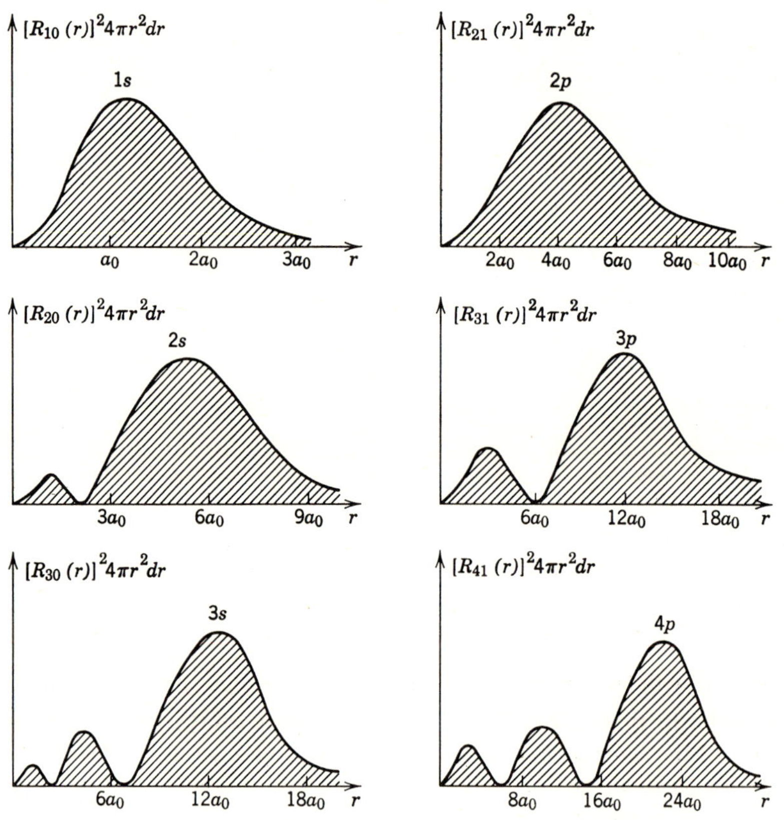

FIG. Q.3.6. (*Continued on facing page.*)

[1] Bethe, H. A., and Salpeter, E. E., *Quantum Mechanics of One- and Two-electron Atoms,* 1957, Springer-Verlag, Berlin.

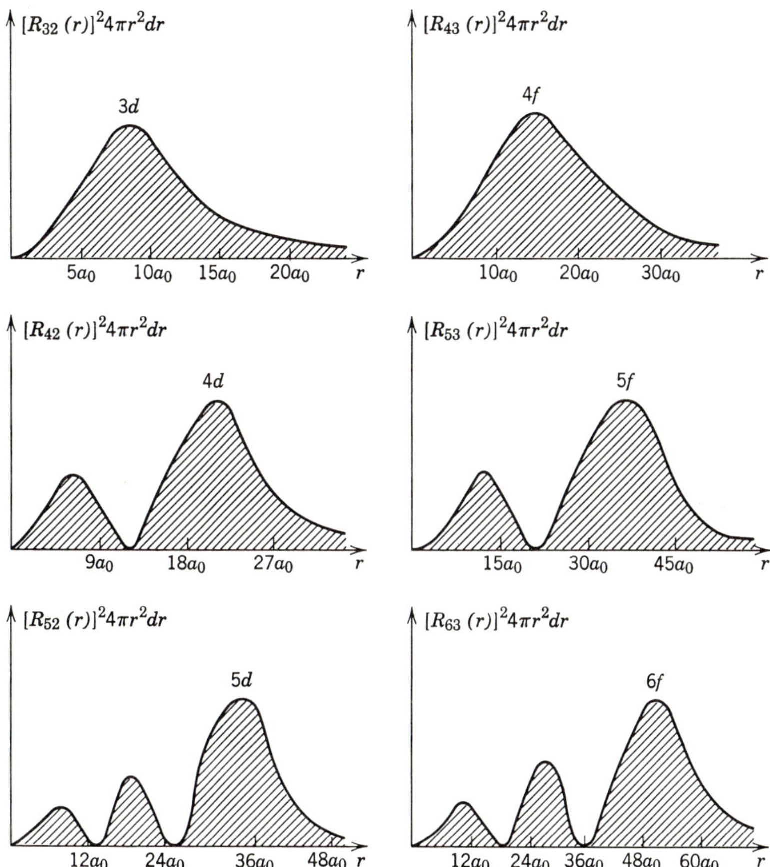

FIG. Q.3.6. Radial probability density functions, $[R_{nl}(r)]^2 4\pi r^2\,dr$, plotted as a function of the radial distance of the electron from the nucleus in hydrogenic atoms in various quantum states. The radial distance in each instance is expressed in Bohr units.

these diagrams, the effect of the normalization process is illustrated by the fact that the area enclosed beneath each curve is unity, assuming that the area up to $r = \infty$ is included. The well defined maxima in these curves representing the radial probability densities, and the nodes corresponding to zero probability density are of course related to the analytical form of the corresponding expressions given by (q.3.114.). For example, the single maxima in the radial probability density curves for $R_{10}(r)$, $R_{21}(r)$, $R_{23}(r)$, and $R_{43}(r)$, alternatively referred to in the terminology described on page 70, as 1s, 2p, 3d, and 4f states respectively, occur for r values of a_0, $4a_0$, $9a_0$, and $16a_0$, respectively. Also, the nodal positions in the radial charge density curves for the 2s, 3p, 4d, and 5f, states are those which correspond to the r values which make the expressions enclosed within square braces in (q.3.114.) for $R_{20}(r)$, $R_{31}(r)$, $R_{42}(r)$, and $R_{53}(r)$, identically zero.

Q.3.7.   The Angular Momentum Operator

The relationship between angular momentum and magnetic moment has already been referred to in (q.1.38.) and the angular momentum about some specific direction in (q.3.23.). It still remains to demonstrate the relationship between the quantum number l and the angular momentum associated with a particle which is described by some particular integral value of l.

In classical mechanics, the angular momentum or the moment of momentum of a particle, L, of mass m, rotating with a linear velocity v or an angular velocity ω, about some fixed point is given by the product, $mrv = mr^2\omega = I\omega$, where $I = mr^2$ is the moment of inertia of the particle; alternatively, $L = rp$, where p is the linear momentum of the particle. Since r and p are both vector quantities, L is also a vector, such that in vector notation

$$\mathbf{L} = \mathbf{r} \times \mathbf{p} = (ix + jy + kz)(ip_x + jp_y + kp_z)$$

$$= \begin{bmatrix} i & j & k \\ x & y & z \\ p_x & p_y & p_z \end{bmatrix} = h/2\pi i \begin{bmatrix} i & j & k \\ x & y & z \\ \partial/\partial x & \partial/\partial y & \partial/\partial z \end{bmatrix}$$

$$= i(yp_z - zp_y) + j(zp_x - xp_z) + k(xp_y - yp_x),$$

$$(q.3.117.)$$

where i, j, and k, are unit vectors in the x, y, and z, Cartesian coordinate directions, and so the components of the angular momentum L, in the three coordinate directions are

$$L_x = yp_x - zp_y = m(yv_x - xv_y),$$

$$L_y = zp_x - xp_z = m(zv_x - xv_z), \qquad (q.3.118.)$$

$$L_z = xp_y - yp_x = m(xv_y - yv_x).$$

If these angular momenta components are transformed into the corresponding expressions in spherical polar coordinates, rather than Cartesian coordinates, by the use of the relationships given on pages 173 and 175, there is obtained, in quantum mechanics,

$$\hat{L}_x = (h/2\pi i)[r \sin\theta \sin\phi(\partial r/\partial z \cdot \partial/\partial r + \partial\theta/\partial z \cdot \partial/\partial\theta + \partial\phi/\partial z \cdot \partial/\partial\phi)$$
$$- r \cos\theta(\partial r/\partial y \cdot \partial/\partial r + \partial\theta/\partial y \cdot \partial/\partial\theta + \partial\phi/\partial y \cdot \partial/\partial\phi)]$$
$$= (h/2\pi i)[-\sin\phi\,\partial/\partial\theta - \cot\theta\cos\phi\,\partial/\partial\phi]. \qquad (q.3.119.)$$

In (q.3.119.), it has to be noted that $i = \sqrt{-1}$, arising of course, from the substitution in the quantum mechanical expression of the dynamical variables p_x and p_y by their appropriate operators (page 108). Similarly,

$$\hat{L}_y = (h/2\pi i)[z(\partial r/\partial x \cdot \partial/\partial r + \partial\theta/\partial x \cdot \partial/\partial\theta + \partial\phi/\partial x \cdot \partial/\partial\phi)$$
$$- x(\partial r/\partial z \cdot \partial/\partial r + \partial\theta/\partial z \cdot \partial/\partial\theta + \partial\phi/\partial z \cdot \partial/\partial\phi)]$$
$$= (h/2\pi i)[\cos\phi\,\partial/\partial\theta - \cot\theta\sin\phi\,\partial/\partial\phi], \qquad (q.3.120.)$$

and

$$\hat{L}_z = (h/2\pi i)[r \sin\theta \cos\phi(\partial r/\partial y \cdot \partial/\partial r + \partial\theta/\partial y \cdot \partial/\partial\theta + \partial\phi/\partial y \cdot \partial/\partial\phi)$$
$$- r \sin\theta \sin\phi(\partial r/\partial x \cdot \partial/\partial r + \partial\theta/\partial x \cdot \partial/\partial\theta + \partial\phi/\partial x \cdot \partial/\partial\phi)]$$
$$= (h/2\pi i)\,\partial/\partial\phi. \qquad (q.3.121.)$$

It is apparent, on comparing equations (q.3.23.) and (q.3.121.) that $p_\phi = \hat{L}_z$, and therefore the z component of the angular momentum is a constant of the motion of the system with eigenvalues $(h/2\pi)m_l$ in the state $\psi(r, \theta, \phi)$; alternatively, it may be said that $\hat{L}_z$ and $\hat{H} = [(h^2/8\pi^2\mu)\nabla^2 + (Z\varepsilon^2/r]$, have the simultaneous eigenfunctions, $\psi(x, y, z) = R(r)\Theta(\theta)\Phi(\phi)$, with $R(r)$, $\Theta(\theta)$, and $\Phi(\phi)$, given by (q.3.110.), (q.3.51.), and (q.3.22.), respectively; the eigenvalues of $\hat{H}$ are the energies $-Z^2/2n^2$ Hartree units of (q.3.75.), and the eigenvalues of $\hat{L}_z$ are $(h/2\pi)m_l$ of (q.3.23.). Since there are $(2l + 1)$ values of m_l for any value of l, there are $(2l + 1)$ values of L_z for any l. $\hat{L}_x$ and $\hat{L}_y$ do not have simultaneous eigenfunctions with $\hat{H}$.

Of particular interest is the scalar product of the total angular momentum,

$$\mathbf{L \cdot L} = L^2 = L_x^2 + L_y^2 + L_z^2$$
$$= (yp_x - zp_y)^2 + (zp_x - xp_y)^2 + (xp_y - yp_x)^2. \qquad (q.3.122.)$$

This is the classical mechanical expression for the square of the total angular momentum expressed in terms of Cartesian coordinates and linear momenta. This equation may be transformed into the corresponding quantum mechanical expression by substituting the appropriate quantum mechanical operators for the dynamical variables, p_x, p_y, and p_z, so that

$$\hat{L}^2 = -(h^2/4\pi^2)[(x\,\partial/\partial y - y\,\partial/\partial x)^2 + (y\,\partial/\partial z - z\,\partial/\partial y)^2$$
$$+ (z\,\partial/\partial x - x\,\partial/\partial z)^2]$$
$$= -(h^2/4\pi^2)[(-\sin\phi\,\partial/\partial\theta - \cot\theta\cos\phi\,\partial/\partial\phi)^2$$
$$+ (\cos\phi\,\partial/\partial\theta - \cot\theta\sin\phi\,\partial/\partial\phi)^2 + \partial^2/\partial\phi^2], \qquad (q.3.123.)$$

on substitution for $\hat{L}_x$, $\hat{L}_y$, and $\hat{L}_z$, from (q.3.119.), (q.3.120.), and (q.3.121.), respectively. Equation (q.3.123.) may be expanded as

$$\hat{L}^2 = (-h^2/4\pi^2)[\sin\phi\,\partial/\partial\theta(\sin\phi\,\partial/\partial\theta) + \cot\theta\cos\phi\,\partial/\partial\phi(\cot\theta\cos\phi\,\partial/\partial\phi)$$
$$+ \sin\phi\,\partial/\partial\theta(\cot\theta\cos\phi\,\partial/\partial\phi) + \cot\theta\cos\phi\,\partial/\partial\phi(\sin\phi\,\partial/\partial\theta)$$
$$+ \cos\phi\,\partial/\partial\theta(\cos\phi\,\partial/\partial\theta) + \cot\theta\sin\phi\,\partial/\partial\phi(\cot\theta\sin\phi\,\partial/\partial\phi)$$
$$- \cos\phi\,\partial/\partial\theta(\cot\theta\sin\phi\,\partial/\partial\phi) - \cot\theta\sin\phi\,\partial/\partial\phi(\cos\phi\,\partial/\partial\theta) + \partial^2/\partial\phi^2],$$

which after performing the indicated operations and collecting together the terms containing the same differential coefficients, simplifies to

$$\hat{L}^2 = -(h^2/4\pi^2)[\partial^2/\partial\theta^2 + (1/\sin^2\theta)\,\partial^2/\partial\phi^2 + \cot\theta\,\partial/\partial\theta]$$
$$= -(h^2/4\pi^2)[(1/\sin^2\theta)\,\partial^2/\partial\phi^2 + (1/\sin\theta)\,\partial/\partial\theta(\sin\theta\,\partial/\partial\theta)]. \quad (q.3.124.)$$

Now an expression has already been obtained for the angular functions relating these functions to their differential coefficients. Thus by substituting (q.3.12.) into (q.3.14.), the result obtained is

$$[R(r)]^{-1}\,d/dr[r^2\,dR(r)/dr] + 8\pi^2\mu h^{-2}[E + Z\varepsilon^2/r]r^2$$
$$= -(1/\sin^2\theta)[\Phi(\phi)]^{-1}\,d^2\Phi(\phi)/d\phi^2 - (1/\Theta(\theta)\sin\theta)\,d/d\theta(\sin\theta\,d\Theta(\theta)/d\theta)$$
$$= \beta = l(l+1), \quad (q.3.125.)$$

making use of equations (q.3.15.) and (q.3.27.). Therefore

$$[\sin^2\theta\Phi(\phi)]^{-1}\,d^2\Phi(\phi)/d\phi^2 + [\Theta(\theta)\sin\theta]^{-1}\,d/d\theta[\sin\theta\,d\Theta(\theta)/d\theta]$$
$$= -l(l+1),$$

and if this equation is multiplied on both sides by $\psi(r,\theta,\phi) = R(r)\Theta(\theta)\Phi(\phi)$, it becomes

$$[\sin^2\theta]^{-1}\,d^2\psi(r,\theta,\phi)/d\phi^2 + (\sin\theta)^{-1}\,d/d\theta[\sin\theta\,d\psi(r,\theta,\phi)/d\theta]$$
$$= -l(l+1)\psi(r,\theta,\phi). \quad (q.3.126.)$$

But the operator, $\hat{L}^2$, defined by (q.3.124.) operating on $\psi(r,\theta,\phi)$ would reproduce equation (q.3.126.), multiplied by $-(h^2/4\pi^2)$, so that

$$\hat{L}^2\psi(r,\theta,\phi) = -(h^2/4\pi^2)[-l(l+1)\psi(r,\theta,\phi)], \quad (q.3.127.)$$

or

$$L = \pm(h/2\pi)\sqrt{l(l+1)}, \quad (q.3.128.)$$

since (q.3.127.) is a typical eigenvalue equation. Thus $\hat{L}^2$, as well as $\hat{L}_z$, has simultaneous eigenfunctions with $\hat{H}$, the corresponding eigenvalues being given by (q.3.128.). The quantum number l is thus deduced to have the significance that its numerical value determines the total angular momentum of the system which it refers to. The angular momentum of a system in the state $\psi(r, \theta, \phi)$ thus has the constant value $(h/2\pi)\sqrt{l(l + 1)}$, and of course this is the reason for referring to l as the angular momentum quantum number. From (q.3.128.), it may be deduced that when $l = 0$, or for s-states, the angular momentum is zero, and so the energy of s-states is due to motion in a radial direction only. The quantum mechanical expression for the total angular momentum of a system given by (q.3.128.) thus differs from the Bohr–Sommerfeld expression, $L = nh/2\pi$ with $n = 1, 2, 3, \ldots$, and from the classical expression given by $L = I\omega = \pm\varepsilon\sqrt{mr}$ (page 214), the latter expression arising from the equivalence between $mv^2/r = \varepsilon^2/r^2 = mr\omega^2 = I\omega^2/r$.

Q.3.8. Atomic Orbitals

It has now been shown how the central field equation of (q.3.7.) may be transformed into spherical polar coordinates and separated into three equations, one in each of the coordinates, r, θ, and ϕ, which have the orthonormal solutions given by (q.3.22.), (q.3.51.), and (q.3.110.), respectively. In the process, it has been necessary, in order to obtain well-behaved solutions and in particular solutions which are finite everywhere for all values of the variables, to introduce the three constants or quantum numbers, n, l, and m_l. The constant n enters into the solution of the equation in $R(r)$; the constant l enters into the solution of the equation in $\Theta(\theta)$; and the constant m_l enters into the solution of the equation in the angular function $\Phi(\phi)$, such that $R(r)$ is a function of n and l, $\Theta(\theta)$ is a function of l and m_l, and $\Phi(\phi)$ is a function of m_l only, and

$$
\begin{aligned}
n &= 1, 2, 3, \ldots \infty, \\
l &= 0, 1, 2, \ldots (n - 1), \\
m_l &= 0, \pm 1, \pm 2, \pm 3, \ldots \pm l.
\end{aligned}
\qquad \text{(q.3.129.)}
$$

The significance of these quantum numbers is essentially the same as that described in Chapter Q.1., but in Schrödinger mechanics they emerge as the necessary conditions required to obtain analytical solutions for the total wave function which are finite and single-valued everywhere. The principal or n quantum number determines the total energy of a hydrogenic atom in accordance with $E_n = -Z^2/n^2$ Rydbergs, the angular momentum quantum number l, determines the total angular momentum of the hydrogenic atom in accordance with (q.3.128.), and the magnetic quantum number m_l determines the number of units of angular momentum of the electron in the atom about some specific direction in accordance with (q.3.23.).

The complete expression for the orthonormalized eigenfunctions for a hydrogen atom may be obtained by combining the three parts of the wave function $\psi(r, \theta, \phi)$ obtained as $R(r)$, $\Theta(\theta)$, and $\Phi(\phi)$, to give

$$\psi_{n,l,m_l}(r, \theta, \phi) = R_{n,l}(r)\Theta_{l,m_l}(\theta)\Phi_{m_l}(\phi)$$

$$= \left[\frac{(2l + 1)[(l - m)!][(n - l - 1)!]}{\pi n^4 a_0^3[(l + m)!][(n + l)!]^3}\right]^{1/2} e^{-r/na_0}\left[\frac{2r}{na_0}\right]^l L_{n+l}^{2l+1}\left[\frac{2r}{na_0}\right]$$

$$+ \sin^m\theta P_l^m(\cos\theta)e^{im\phi}. \tag{q.3.130.}$$

The corresponding expression for any hydrogenic atom, where $Z > 1$ and the factor Z only affects the radial function $R(r)$, if expressed in atomic units with $a_0 = 1$, becomes

$$\psi_{n,l,m}(r, \theta, \phi) = \left[\frac{Z^3(2l + 1)[(l - m)!][(n - l - 1)!]}{\pi n^4[(l + m)!][(n + l)!]^3}\right]^{1/2} e^{-Zr/n}\left[\frac{2Zr}{n}\right]^l L_{n+l}^{2l+1}$$

$$\times \left[\frac{2Zr}{n}\right] \sin^m\theta P_l^m(\cos\theta)e^{im\phi}. \tag{q.3.131.}$$

By substituting into (q.3.130.) or (q.3.131.) the values given on page 192 for the spherical harmonics $\Theta(\theta)\Phi(\phi)$ and the expressions given by (q.3.114.), the one-electron wave function for the various states of any hydrogenic atom may be obtained. These one-electron wave functions are commonly referred to as atomic orbitals, after a suggestion originally made by Mulliken. Since such orbitals or orbital functions are necessarily one-electron wave functions, it is unnecessary to speak of one-electron orbitals and quite incorrect to refer to two-electron orbitals, although such terminology is frequently encountered. It is, of course, quite ridiculous to describe one or two electrons as being "in" a certain orbital, although this manner of speaking and writing is quite prevalent, especially in the chemical literature.

The s-states with l and therefore m_l both zero, are all non-degenerate, so that for hydrogenic atoms, expressed in atomic units, the s-type wave functions are:

$$\psi_{100}(r, \theta, \phi) = \psi_{1s} = Z^{3/2}[1/\sqrt{\pi}]e^{-Zr};$$

$$\psi_{200}(r, \theta, \phi) = \psi_{2s} = Z^{3/2}[1/4\sqrt{2\pi}]e^{-Zr/2}[2 - Zr];$$

$$\psi_{300}(r, \theta, \phi) = \psi_{3s} = Z^{3/2}[1/81\sqrt{3\pi}]e^{-Zr/3}[27 - 18Zr + 2Z^2r^2]; \tag{q.3.132.}$$

$$\psi_{400}(r, \theta, \phi) = \psi_{4s} = Z^{3/2}[1/1536\sqrt{\pi}]e^{-Zr/4}[192 - 144Zr + 24Z^2r^2 - Z^3r^3];$$

$$\psi_{500}(r, \theta, \phi) = \psi_{5s} = Z^{3/2}[1/93750\sqrt{2\pi}]e^{-Zr/5}$$
$$+ [9375 - 7500Zr + 1500Z^2r^2 - 100Z^3r^3 + 2Z^4r^4];$$

$$\psi_{600}(r, \theta, \phi) = \psi_{6s} = Z^{3/2}[1/1049760\sqrt{6\pi}]e^{-Zr/6}$$
$$\times [174960 - 145800Zr + 32400Z^2r^2$$
$$- 2700Z^3r^3 + 90Z^5r^5 - Z^6r^6]$$

The p-states with $l = 1$ and $m_l = -1$, 0, or 1, are three-fold degenerate and the various wave functions for the np states with $n = 2$ to $n = 6$ are:

$$\psi_{211}(r, \theta, \phi) = \psi_{2p_1} = \psi(2p_x) = Z^{5/2}[1/4\sqrt{2\pi}]e^{-Zr/2}r \sin\theta \cos\phi;$$

$$\psi_{210}(r, \theta, \phi) = \psi_{2p} = \psi(2p_x) = Z^{5/2}[1/4\sqrt{2\pi}]e^{-Zr/2}r \cos\theta;$$

$$\psi_{21-1}(r, \theta, \phi) = \psi_{2p_{-1}} = \psi(2p_y) = Z^{5/2}[1/4\sqrt{2\pi}]e^{-Zr/2}r \sin\theta \sin\phi;$$

$$\psi_{311}(r, \theta, \phi) = \psi_{3p_1} = \psi(3p_x) = Z^{5/2}[2/81\sqrt{2\pi}]e^{-Zr/3}[6 - Zr]r \sin\theta \cos\phi;$$

$$\psi_{310}(r, \theta, \phi) = \psi_{3p_0} = \psi(3p_z) = Z^{5/2}[2/81\sqrt{2\pi}]e^{-Zr/3}[6 - Zr]r \cos\theta;$$

$$\psi_{31-1}(r, \theta, \phi) = \psi_{3p_{-1}} = \psi(3p_y) = Z^{5/2}[2/81\sqrt{2\pi}]e^{-Zr/3}[6 - Zr]r \sin\theta \sin\phi;$$

$$\psi_{411}(r, \theta, \phi) = \psi_{4p_1} = \psi(4p_x) = Z^{5/2}[1/512\sqrt{5\pi}]e^{-Zr/4}$$
$$\times\, [80 - 20Zr + Z^2r^2]r \sin\theta \cos\phi;$$

$$\psi_{410}(r, \theta, \phi) = \psi_{4p_0} = \psi(4p_z) = Z^{5/2}[1/512\sqrt{5\pi}]e^{-Zr/4}$$
$$\times\, [80 - 20Zr + Z^2r^2]r \cos\theta;$$

$$\psi_{41-1}(r, \theta, \phi) = \psi_{4p_{-1}} = \psi(4p_y) = Z^{5/2}[1/512\sqrt{5\pi}]e^{-Zr/4}$$
$$\times\, [80 - 20Zr + Z^2r^2]r \sin\theta \sin\phi;$$

$$\psi_{511}(r, \theta, \phi) = \psi_{5p_1} = \psi(5p_x) = Z^{5/2}[2/46875\sqrt{10\pi}]e^{-Zr/5}$$
$$\times\, [3750 - 1125Zr + 90Z^2r^2 - 2Z^3r^3][r \sin\theta \cos\phi];$$

$$\psi_{510}(r, \theta, \phi) = \psi_{5p_0} = \psi(5p_z) = Z^{5/2}[2/46875\sqrt{10\pi}]e^{-Zr/5}$$
$$\times\, [3750 - 1125Zr + 90Z^2r^2 + 2Z^3r^3][r \cos\theta];$$

$$\psi_{51-1}(r, \theta, \phi) = \psi_{5p_{-1}} = \psi(5p_y) = Z^{5/2}[2/46875\sqrt{10\pi}]e^{-Zr/5}$$
$$\times\, [3750 - 1125Zr + 90Z^2r^2 - 2Z^3r^3][r \sin\theta \sin\phi];$$

$$\psi_{611}(r, \theta, \phi) = \psi_{6p_1} = \psi(6p_x) = Z^{5/2}[1/367416\sqrt{70\pi}]e^{-Zr/6}[r \sin\theta \cos\phi]$$
$$\times\, [68040 - 22680Zr + 2268Z^2r^2 - 84Z^3r^3 + Z^4r^4];$$

$$\psi_{610}(r, \theta, \phi) = \psi_{6p_0} = \psi(6p_z) = Z^{5/2}[1/367416\sqrt{70\pi}]e^{-Zr/6}[r \cos\theta]$$
$$\times\, [68040 - 22680Zr + 2268Z^2r^2 - 84Z^3r^3 + z^4r^4];$$

$$\psi_{61-1}(r, \theta, \phi) = \psi_{6p_{-1}} = \psi(6p_y) = Z^{5/2}[1/367416\sqrt{70\pi}]e^{-Zr/6}[r \sin\theta \sin\phi]$$
$$\times\, [68040 - 22680Zr + 2268Z^2r^2 - 84Z^3r^3 + Z^4r^4].$$

$$(q.3.133.)$$

These p-type functions listed in (q.3.133.) have all been expressed as real functions by eliminating the complex part contained in the $\Phi(\phi)$ function using the method described on page 180. It is common practice, as in the third designation given above, to refer to the p-type functions for which $m_l = 1$, 0, and -1, respectively, as p_x, p_z, and p_y, functions, the subscript referring to the Cartesian coordinate equivalent of the corresponding spherical polar expression contained in the wave function.

The d-states with $l = 2$ and $m_l = 2, 1, 0, -1,$ and -2, are five-fold degenerate and the wave functions for the $3d$, $4d$, and $5d$, states are listed below, again expressed as real functions, and the corresponding polynomial description of the angular part is also given:

$$\psi_{322}(r, \theta, \phi) = \psi_{3d_2} = \psi(3d_{x^2-y^2}) = Z^{7/2}[1/81\sqrt{2\pi}]e^{-Zr/3}r^2 \sin^2\theta \cos 2\phi;$$

$$\psi_{32-2}(r, \theta, \phi) = \psi_{3d_{-2}} = \psi(3d_{xy}) = Z^{7/2}[1/81\sqrt{2\pi}]e^{-Zr/3}r^2 \sin^2\theta \sin 2\phi;$$

$$\psi_{321}(r, \theta, \phi) = \psi_{3d_1} = \psi(3d_{xz}) = Z^{7/2}[2/81\sqrt{2\pi}]e^{-Zr/3}r^2 \sin\theta \cos\theta \cos\phi;$$

$$\psi_{32-1}(r, \theta, \phi) = \psi_{3d_{-1}} = \psi(3d_{yz}) = Z^{7/2}[2/81\sqrt{2\pi}]e^{-Zr/3}r^2 \sin\theta \cos\theta \sin\phi;$$

$$\psi_{320}(r, \theta, \phi) = \psi_{3d_0} = \psi(3d_{z^2}) = Z^{7/2}[1/81\sqrt{6\pi}]e^{-Zr/3}r^2(3\cos^2\theta - 1);$$

$$\psi_{422}(r, \theta, \phi) = \psi_{4d_2} = \psi(4d_{x^2-y^2}) = Z^{7/2}[\sqrt{3}/3072\sqrt{\pi}]e^{-Zr/4}$$
$$\times [12 - Zr]r^2 \sin^2\theta \cos 2\phi;$$

$$\psi_{42-2}(r, \theta, \phi) = \psi_{4d_{-2}} = \psi(4d_{xy}) = Z^{7/2}[\sqrt{3}/3072\sqrt{\pi}]e^{-Zr/4}$$
$$\times [12 - Zr]r^2 \sin^2\theta \sin 2\phi;$$

$$\psi_{421}(r, \theta, \phi) = \psi_{4d_1} = \psi(4d_{xz}) = Z^{7/2}[\sqrt{3}/1536\sqrt{\pi}]e^{-Zr/4}$$
$$\times [12 - Zr]r^2 \sin\theta \cos\theta \cos\phi;$$

$$\psi_{42-1}(r, \theta, \phi) = \psi_{4d_{-1}} = \psi(4d_{yz}) = Z^{7/2}[\sqrt{3}/1576\sqrt{\pi}]e^{-Zr/4}$$
$$\times [12 - Zr]r^2 \sin\theta \cos\theta \sin\phi;$$

$$\psi_{420}(r, \theta, \phi) = \psi_{4d_0} = \psi(4d_{z^2}) = Z^{7/2}[1/3072\sqrt{\pi}]e^{-Zr/4}$$
$$\times [12 - Zr]r^2(3\cos^2\theta - 1);$$

$$\psi_{522}(r, \theta, \phi) = \psi_{5d_2} = \psi(5d_{x^2-y^2}) = Z^{7/2}[\sqrt{3}/46875\sqrt{14\pi}]e^{-Zr/5}$$
$$\times [525 - 70Zr + 2Z^2r^2]r^2 \sin^2\theta \cos 2\phi;$$

$$\psi_{52-2}(r, \theta, \phi) = \psi_{5d_{-2}} = \psi(5d_{xy}) = Z^{7/2}[\sqrt{3}/46875\sqrt{14\pi}]e^{-Zr/5}$$
$$\times [525 - 70Zr + 2Z^2r^2]r^2 \sin^2\theta \sin 2\phi;$$

$$\psi_{521}(r, \theta, \phi) = \psi_{5d_1} = \psi(5d_{xz}) = Z^{7/2}[2\sqrt{3}/46875\sqrt{14\pi}]e^{-Zr/5}$$
$$\times [525 - 70Zr + 2Z^2r^2]r^2 \sin\theta \cos\theta \cos\phi;$$

$$\psi_{52-1}(r, \theta, \phi) = \psi_{5d_{-1}} = \psi(5d_{yz}) = Z^{7/2}[2\sqrt{3}/46875\sqrt{14\pi}]e^{-Zr/5}$$
$$\times [525 - 70Zr + 2Z^2r^2]r^2 \sin\theta \cos\theta \sin\phi;$$

$$\psi_{520}(r, \theta, \phi) = \psi_{5d_0} = \psi(5d_{z^2}) = Z^{7/2}[1/46875\sqrt{14\pi}]e^{-Zr/5}$$
$$\times [525 - 70Zr + 2Z^2r^2]r^2(3\cos^2\theta - 1). \qquad \text{(q.3.134.)}$$

For f-states with $l = 3$, and $m_l = 3, 2, 1, 0, -1, -2,$ and -3, the degree of degeneracy is 7. The $4f$ and to a lesser extent the $5f$-states are the only f-states of interest as far as the constitution of chemical compounds are concerned. There are listed below the wave functions for the $4f$- and $5f$-states of a hydrogenic atom, expressed as the simplest set of real functions, together

with the shortened polynomial description characteristic of the angular part
of the total wave function:

$$\psi_{433}(r, \theta, \phi) = \psi_{4f_3} = \psi[4f_{x(x^2-3y^2)}] = Z^{9/2}$$
$$\times [\sqrt{2}/6144\sqrt{\pi}]e^{-Zr/4}r^3 \sin^3 \theta \cos 3\phi;$$

$$\psi_{43-3}(r, \theta, \phi) = \psi_{4f_{-3}} = \psi[4f_{y(3x^2-y^2)}]$$
$$= Z^{9/2}[\sqrt{2}/6144\sqrt{\pi}]e^{-Zr/4}r^3 \sin^3 \theta \sin 3\phi;$$

$$\psi_{432}(r, \theta, \phi) = \psi_{4f_2} = \psi[4f_{x^2-y^2}]$$
$$= Z^{9/2}[\sqrt{3}/3072\sqrt{\pi}]e^{-Zr/4}r^3 \sin^2 \theta \cos \theta \cos 2\phi;$$

$$\psi_{43-2}(r, \theta, \phi) = \psi_{4f_{-2}} = \psi[4f_{xyz}]$$
$$= Z^{9/2}[\sqrt{3}/3072\sqrt{\pi}]e^{-Zr/4}r^3 \sin^2 \theta \cos \theta \sin 2\phi;$$

$$\psi_{431}(r, \theta, \phi) = \psi_{4f_1} = \psi[4f_{xz^2}]$$
$$= Z^{9/2}[\sqrt{6}/6144\sqrt{5\pi}]e^{-Zr/4}r^3 \sin \theta \cos \phi(5 \cos^2 \theta - 1);$$

$$\psi_{43-1}(r, \theta, \phi) = \psi_{4f_{-1}} = \psi[4f_{yz^2}]$$
$$= Z^{9/2}[\sqrt{6}/6144\sqrt{5\pi}]e^{-Zr/4}r^3 \sin \theta \sin \phi(5 \cos^2 \theta - 1);$$

$$\psi_{430}(r, \theta, \phi) = \psi_{4f_0} = \psi[4f_{z^3}]$$
$$= Z^{9/2}[1/3072\sqrt{5\pi}]e^{-Zr/4}r^3(5 \cos^3 \theta - 3 \cos \theta);$$

$$\psi_{533}(r, \theta, \phi) = \psi_{5f_3} = \psi[5f_{x(x^2-3y^2)}]$$
$$= Z^{9/2}[1/93750\sqrt{\pi}]e^{-Zr/5}[20 - Zr]r^3 \sin^3 \theta \cos 3\phi;$$

$$\psi_{53-3}(r, \theta, \phi) = \psi_{5f_{-3}} = \psi[5f_{y(3x^2-y^2)}]$$
$$= Z^{9/2}[1/93750\sqrt{\pi}]e^{-Zr/5}[20 - Zr]r^3 \sin^3 \theta \sin 3\phi;$$

$$\psi_{532}(r, \theta, \phi) = \psi_{5f_2} = \psi[5f_{x^2-y^2}]$$
$$= Z^{9/2}[2\sqrt{3}/93750\sqrt{2\pi}]e^{-Zr/5}[20 - Zr]r^3 \sin^2 \theta \cos \theta \cos 2\phi;$$

$$\psi_{53-2}(r, \theta, \phi) = \psi_{5f_{-2}} = \psi[5f_{xyz}]$$
$$= Z^{9/2}[2\sqrt{3}/93750\sqrt{2\pi}]e^{-Zr/5}[20 - Zr]r^3 \sin^2 \theta \cos \theta \sin 2\phi;$$

$$\psi_{531}(r, \theta, \phi) = \psi_{5f_1} = \psi[5f_{xz^2}] = Z^{9/2}[\sqrt{3}/93750\sqrt{5\pi}]$$
$$\times e^{-Zr/5}[20 - Zr]r^3 \sin \theta \cos \phi(5 \cos^2 \theta - 1);$$

$$\psi_{53-1}(r, \theta, \phi) = \psi_{5f_{-1}} = \psi[5f_{yz^2}] = Z^{9/2}[\sqrt{3}/93750\sqrt{5\pi}]$$
$$\times e^{-Zr/5}[20 - Zr]r^3 \sin \theta \sin \phi(5 \cos^2 \theta - 1);$$

$$\psi_{530}(r, \theta, \phi) = \psi_{5f_0} = \psi[5f_{z^3}]$$
$$= Z^{9/2}[2/93750\sqrt{10\pi}]e^{-Zr/5}[20 - Zr]r^3(5 \cos^3 \theta - 3 \cos \theta).$$
$$(q.3.135.)$$

It should be noted, as already discussed (page 168), that the degree of
degeneracy of any particular energy state is equivalent to the number of
linearly independent wave functions, which correspond to that particular
energy level. It is not the wave functions which are degenerate; thus each of
the seven f-orbitals are distinctively different, although each corresponds to

the same energy value. For hydrogenic atoms, the total degree of degeneracy for any value of n is n^2, as given by (q.3.116.), where the potential energy of such systems is given by (q.3.5.). In general, for other central fields $U(r)$, the degree of degeneracy is less than this, but is at least $(2l + 1)$. The various p, d, and f, orbitals described by (q.3.133.), (q.3.134.), and (q.3.135.), respectively, are the simplest possible real functions which may be written, but there are an infinite number of other sets of three p-type functions, five d-type functions, and seven f-type functions, which may be constructed by taking various linear combinations of the given functions. It is sometimes particularly advantageous to choose some specific linear combination for some particular purpose; thus, in the ligand field theoretical discussion of the constitution of complex compounds containing central atoms or ions in which electrons in f-states are involved in interaction with ligands, it is more convenient to choose a set of seven f-orbitals with symmetry characteristics different from those given by (q.3.135.). This so-called, cubic set of f-orbitals, may be constructed by retaining the three orbitals given as f_{z^2}, f_{xyz}, and $f_{z(x^2-y^2)}$, and forming four new orbitals, f_{x^3}, f_{y^3}, $f_{x(z^2-y^2)}$, and $f_{y(z^2-x^2)}$, as the following particular linear combinations of the remaining four

$$f_{x^3} = -(\sqrt{6}/4)f_{xz^2} + (\sqrt{10}/4)f_{x(x^2-3y^2)},$$

$$f_{y^3} = -(\sqrt{6}/4)f_{yz^2} - (\sqrt{10}/4)f_{y(3x^2-y^2)}, \qquad \text{(q.3.136.)}$$

$$f_{x(z^2-y^2)} = (\sqrt{10}/4)f_{xz^2} + (\sqrt{6}/4)f_{x(x^2-3y^2)},$$

and

$$f_{y(z^2-x^2)} = (\sqrt{10}/4)f_{yz^2} - (\sqrt{6}/4)f_{y(3x^2-y^2)}.$$

All possible shapes of d-type orbitals may be expressed as a linear function of d_{z^2} and $d_{x^2-y^2}$, with a coefficient a ranging from 1 to $\sqrt{3/2}$, or

$$d = ad_{z^2} + (1 - a)^{1/2}\, d_{x^2-y^2}\sqrt{3/2} = 0.866025 \le a \le 1.$$

Alternatively,[1] if a shape parameter $\xi = 4(1 - a^2)$, is so defined then the function d_{z^2} corresponds to $\xi = 0$ and the function $d_{x^2-y^2}$ to $\xi = 1$. There are actually two sets of five equivalent d-type orbitals[2,3] having symmetry axes oriented along five equivalent directions in space, related by a five-fold symmetry axis. These five directions correspond to the body diagonals of a pentagonal antiprism, one set corresponding to an oblate antiprism and the other to a prolate antiprism.

The s-type orbitals, (q.3.132.), with the exception of the $1s$-orbital which only falls off to zero at an infinite distance from the origin (Fig. Q.3.6.), each

[1] Pauling, Linus, and McClure, V., *J. Chem. Ed.*, 1970, **47**, 15.
[2] Kimball, G. E., *J. Chem. Phys.*, 1940, **8**, 188.
[3] Powell, R. E., *J. Chem. Ed.*, 1968, **45**, 45.

have $(n - 1)$ nodal spheres, centered at the origin, but no nodal planes. On the other hand, functions with the angular momentum quantum number l, each have l nodal planes passing through the origin (Fig. Q.3.4.). Thus, the $\psi(3d_{xz})$ orbital of (q.3.134.) is zero when either $x = r \sin \theta \cos \phi$ or $z = r \cos \theta$ is zero, or has nodal planes in the yz plane and the xy plane, corresponding to two nodal planes for $l = 2$. The radial functions, $R_{nl}(r)$, as depicted in Fig. Q.3.5., pass through zero $(n - l - 1)$ times, excluding the origin.

As the one-electron wave functions or orbitals given by (q.3.130.) are normalized, then according to the probability density interpretation of the wave function, the quantity

$$|\psi|^2 \, dr = |R(r)|^2 r^2 [\Theta_l^m(\theta)]^2 \sin \theta [\Phi(\phi)]^2 \, dr \, d\theta \, d\phi, \qquad \text{(q.3.137.)}$$

represents the probability that the electron in a hydrogen atom will be found in the small volume element dr at the position r. It is sometimes convenient to regard the probability density description in another way: in terms of electronic charge density. Since the electronic charge is $-\varepsilon$, the average charge density at position r may be regarded as $-\varepsilon|\psi|^2$. The electron charge cloud densities of all s-states have spherical symmetry, but that of p-, d-, f-, g-, $\ldots$, states consists of a number of regions separated from one another by nodal planes, these regions being distributed in a characteristically symmetrical manner for such states with respect to the axial reference system used. However, it is possible that in atoms which contain more than one electron in different energy states for the electron charge clouds to combine to produce a spherical charge cloud, even when the electrons concerned are not in s-states. In other words, the sum of the squares of the magnitudes of all the angular wave functions for some particular l value, leads to a result which is independent of the values of the polar angles θ and ϕ. This result is known as Unsöld's theorem, and may be proved analytically as follows. If γ is the angle between two directions in spherical polar coordinate space defined by θ_1, ϕ_1 and θ_2, ϕ_2, then by the so-called addition theorem in spherical harmonics (see page 190 and the references given on page 186),

$$P_l(\cos \gamma) = \sum_{m=-l}^{l} [(l - |m|)!][(l + |m|)!]^{-1} P_l^{|m|}(\cos \theta_1) P_l^{|m|}(\cos \theta_2) e^{im(\phi_1 - \phi_2)}.$$

For the special case where $\gamma = 0$ or $\theta_1 = \theta_2$ and $\phi_1 = \phi_2$, $P_l(\cos \gamma)$ becomes

$$P_l(\cos \gamma) = \sum_{m=-l}^{l} [(l - |m|)!][(l + |m|)!]^{-1} [P_l^{|m|}(\cos \theta)]^2,$$

where

$$\theta_1 = \theta_2 = \theta.$$

But from the definition of Legendre functions (see page 182), $P_l(\cos \gamma) = P_l^0(\cos \gamma) = 1$, when $\gamma = 0$ or $\cos \gamma = 1$, so that

$$\sum_{m=-l}^{l} (2l + 1)[(l - |m|)!][(l + |m|)!]^{-1}[P_l^{|m|}(\cos \theta)e^{im\phi}]^*[P_l^{|m|}(\cos \theta)e^{im\phi}]$$
$$= 2l + 1, \quad \text{(q.3.138.)}$$

which is Unsöld's Theorem.[1] The equation (q.3.138.) implies that the sum of the squares of the magnitudes of the angular wave functions of all states with the same l values but different m values, is independent of the polar angles, or since the probability densities are given, when $\Theta(\theta)$ and $\Phi(\phi)$ are normalized, by the quantity $[|\Theta(\theta)\Phi(\phi)|]^2$, which is independent of the polar angles, thus systems which contain electrons in all possible m-states for any particular l-state have spherical charge cloud densities. Unsöld's Theorem therefore has the important consequence that in many-electron atoms if there are $(2l + 1)$ electrons, each describable by a different m_l quantum number in any n–l-set of states, then the total electron charge cloud distribution has spherical symmetry.

Q.3.9.   The Problem of Interacting Particles

While the simplest atom, that of hydrogen, actually consists of two particles, an electron and a proton, and other hydrogenic atoms consist of an electron and a nucleus (regarded as one particle) of charge $Z > 1$, it is apparent from the foregoing that these systems can be reduced to that of a single particle moving in a central field of force. Then by Schrödinger mechanics the energies of this particle may be calculated and from the energy values in different states, the transition frequencies or their associated wavelengths may be obtained by application of the Bohr frequency rule, $h\nu = hc/\lambda = E_m - E_n$, with results in good agreement with experimentally observed spectroscopic data. However, there are certain aspects of the electronic structure of atoms (and molecules) which the Schrödinger theory does not account for, even in the case of the simplest atom of hydrogen. The resolution of the lines in the line spectrum of hydrogen, and other atoms such as the alkali metal atoms, copper, silver, and gold, the splitting of the apparently single lines in atomic spectra when observed in the presence of magnetic and electric fields (the Zeeman and Stark effects), and the other experimentally observed phenomena discussed in Chapter Q.1., have no explanation in Schrödinger mechanics; these phenomena do however find a logical explanation in Dirac relativistic mechanics, where the Schrödinger

---

[1] Unsöld, A., *Ann. Physik*, 1927, **82**, 355.

equation when formulated in the relativistically correct form using not only space coordinates but a time coordinate, leads to the introduction of a fourth quantum number which happens to be independent of the other three and can adopt only two possible values. For this reason, Schrödinger mechanics can be modified relatively easily to include another variable, apart from the three positional variables, namely the electron spin variable, so that the complete state of an electron in an atom is described in terms of a function

$$\Psi(r, \theta, \phi, \zeta) = \psi(r, \theta, \phi)\chi(\zeta),$$

where Ψ is a function of the three positional variables as well as the spin variable ζ, ψ is a function of the positional coordinates only, and χ is a function of the spin variable only, the value of which is independent of $R(r)$, $\Theta(\theta)$, and $\Phi(\phi)$. The question of the angular momentum which an electron possesses in addition to its orbital angular momentum, which may be attributed to what is referred to as electron spin, and the effect of this on atomic spectra will be discussed in Chapter Q.5.

Meanwhile, some of the other difficulties which arise in attempting to describe the behavior of the second simplest atom, helium, in terms of Schrödinger mechanics, will be examined. If the approximation involved in assuming that the nucleus of the hydrogen atom is stationary with respect to the motion of the electron makes only a negligible difference in this case, then for a helium atom where the nuclear mass is four times greater than for hydrogen, this approximation should be even more inconsequential. The helium atom, then, may be reasonably regarded as one in which two electrons, each of charge $-\varepsilon$ move about in a field due to a stationary nucleus of charge $+2\varepsilon$, so that this is a problem involving three interacting particles. If, as in Fig. Q.3.7., the positional vectors of the two electrons in a helium atom are denoted by r_1 and r_2, with respect to the nucleus, and r_{12} is the interelectronic distance, then if the zero of potential energy is assumed to apply when the two electrons are infinitely far apart, the total potential energy of this system will be

$$U(r_1, r_2) = -2\varepsilon^2/r_1 - 2\varepsilon^2/r_2 + \varepsilon^2/r_{12}. \qquad \text{(q.3.139.)}$$

Not only the potential energy, but also the wave function for this system, will necessarily be a function of all of the positional coordinates of both electrons. The classical Hamiltonian for the system will be

$$H = p_1^2/2m + p_2^2/2m + U(r_1, r_2), \qquad \text{where} \qquad p_1^2 = p_{x_1}^2 + p_{y_1}^2 + p_{z_1}^2,$$

and a similar expression applies to p_2^2. Then the Hamiltonian operator for the system will be

$$\hat{H} = [-h^2/8\pi^2 m(\nabla_1^2 + \nabla_2^2) - 2\varepsilon^2/r_1 - 2\varepsilon^2/r_2 + \varepsilon^2/r_{12}],$$

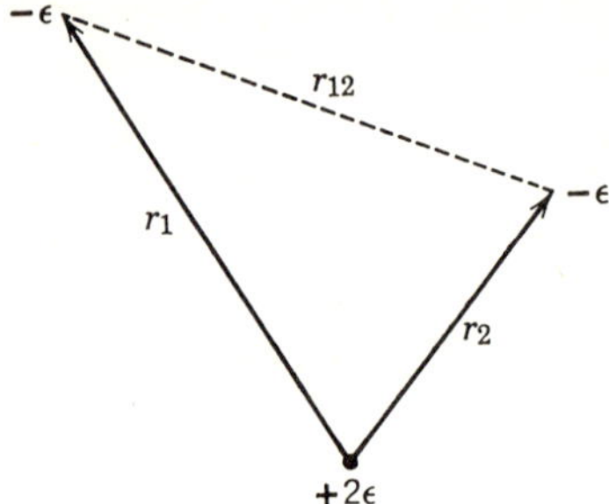

FIG. Q.3.7. The helium atom, represented as a nucleus of charge $+2\varepsilon$, and two electrons each of charge $-\varepsilon$, located by the position vectors r_1 and r_2, with respect to the nucleus.

where m is the electron mass. The Schrödinger equation for the helium atom is then

$$[-h^2/8\pi^2 m(\nabla_1^2 + \nabla_2^2) - 2\varepsilon^2/r_1 - 2\varepsilon^2/r_2 + \varepsilon^2/r_{12}]\psi(r_1, r_2) = E\psi(r_1, r_2),$$
$$(q.3.140.)$$

where E is the total energy of the system. In fact, generally, it is always a simple matter to write down the Schrödinger equation for a system composed of any number of particles; thus for a system composed of n particles such that the ith particle has the mass m_i, Cartesian coordinates $(x_i, y_i\ z_i)$, and momenta coordinates $(p_{x_i}, p_{y_i}, p_{z_i})$, the Schrödinger equation is

$$-\sum_{i=1}^{n} [h^2/8\pi^2 m_i]\nabla_i^2 + U(r_1, r_2, \ldots r_n)]\psi(r_1, r_2, \ldots r_n) = E\psi(r_1, r_2, \ldots r_n),$$

where both ψ and U are functions of all of the positional coordinates of all of the n particles. The problem, of course, is to find solutions for such equations. Even for the helium atom, the equation (q.3.140.) cannot be solved analytically, due to the presence of the term ε^2/r_{12} in the potential energy function. If it is assumed that there is no interaction between the electrons and the term ε^2/r_{12} is entirely neglected, then equation (q.3.140.) may be easily separated into two hydrogenic-type equations. Thus if $\psi_a(r_1)$ represents the wave function of electron 1 moving in the potential field of a nucleus of charge $+2\varepsilon$, then its Schrödinger equation is

$$-h^2/8\pi^2 m\nabla_1^2\psi_a(r_1) - [2\varepsilon^2/r_1]\psi_a(r_1) = E_a\psi_a(r_1), \qquad (q.3.141.)$$

and similarly if $\psi_b(r_2)$ describes the independent behavior of electron 2 moving in the same nuclear potential field, then the equation is

$$-h^2/8\pi^2 m \nabla_2^2 \psi_b(r_2) - [2\varepsilon^2/r_2]\psi_b(r_2) = E_b\psi_b(r_2). \qquad \text{(q.3.142.)}$$

If equation (q.3.141.) is multiplied by $\psi_b(r_2)$ and added to equation (q.3.142.) multiplied by $\psi_a(r_1)$, the resulting equation is

$$[-h^2/8\pi^2 m(\nabla_1^2 + \nabla_2^2) - 2\varepsilon^2/r_1 - 2\varepsilon^2/r_2]\psi_a(r_1)\psi_b(r_2)$$
$$= [E_a + E_b]\psi_a(r_1)\psi_b(r_2). \qquad \text{(q.3.143.)}$$

This is the same equation as (q.3.140.) with the term ε^2/r_{12} omitted and the substitutions

$$\psi(r_1, r_2) = \psi_a(r_1)\psi_b(r_2) \qquad \text{and} \qquad E = E_a + E_b. \qquad \text{(q.3.144.)}$$

It is found, therefore, that if the inter-electronic reaction in a helium atom is neglected that the equation (q.3.140.) may be separated into two hydrogenic-type equations with the total energy of the helium atom being given by the sum of two one-electron energy levels, each of which is given by (q.3.75.). The ground state energy level for the helium atom, with $n = 1$, is then obtained by putting $Z = 2$ in (q.3.75.) as

$$E_g = 2[-8\pi^2 m\varepsilon^4/h^2] = -8 \text{ rydbergs} = -4 \text{ Hartree units of energy.}$$
$$\text{(q.3.145.)}$$

However, the ground state energy of helium measured experimentally is only -5.81 Rydbergs, which is about 40% greater. It is obviously not justifiable to neglect the inter-electronic reaction energy in this manner, but on the other hand if this interaction is included in the helium atom Schrödinger equation, no separation is possible. Various approximation methods for obtaining solutions of the Schrödinger equation of atoms including the electron interaction energy have been proposed, which do not simply neglect this effect entirely, and some of the more important and useful of those methods are discussed in the following chapters.

Q4

Approximation Techniques in Quantum Mechanics

Q.4.1. Introduction

It is only for a relatively small number of systems that an exact solution in closed terms is possible for the Schrödinger equation describing the system. The problem of the electronic interaction energy, as has been indicated for the helium atom, is the principal factor which precludes the separation of the wave equation in any known coordinate system; apparently, this interaction energy, even when there are only two electrons is by no means an insignificant quantity. Furthermore, exact solutions, where they are obtainable, as for the hydrogen atom, are not particularly simple expressions, and if it were possible to obtain exact, formal, solutions for the wave equation of a many-electron atom such as copper, for example, it might reasonably be expected that such solutions would be even more complicated, to such an extent that to assign any physical significance to them would be rather difficult. While it is always possible to write down the Schrödinger equation for any system, as discussed on page 226, the difficulties in obtaining solutions of this equation for even a simple atom such as oxygen, which would involve the manipulation of a partial differential equation in 27 independent variables, are presently insuperable. The application of group theoretical and symmetry arguments, as is possible for many crystallographic problems, might be expected to reduce the magnitude of the problem somewhat, but it is still beyond the reach of any presently known analytical method. The fixed nucleus approximation, to

228

which Born and Oppenheimer[1] have provided a quantum-mechanical justification, is of some help, and it is certainly a considerable aid to know the exact one-electron functions for hydrogenic atoms. The Born–Oppenheimer approximation actually turns out to be a remarkably good one; for example, the calculated energy for the simplest poly-nuclear system, the hydrogen molecule ion, differs by only 0.024% from the best measured value, if it is assumed in the calculation that the two protons are stationary.[2]

Of the various approximation methods which have been proposed and developed to construct approximate wave functions for discrete physical systems, only a few of them are of general use, since in principle, every physical problem differs from every other one in some respect making generalizations difficult. One important approximation method, variously referred to as the WKB or BWK or WBK[3-5] method, is primarily adapted to problems which may be cast into one-dimensional form by separation of variables; it is[6] a semi-classical approximation in which classical behavior of a system is corrected for by the use of a power series in Planck's constant, taking note of the Bohr Correspondence Principle that classical and quantum mechanics converge when h is ignored. This method will not be discussed further.

One of the most useful approximation techniques is to construct wave functions containing a number of parameters, which by suitable variation may be manipulated into a form approximating very closely to the true wave functions of the system. This variational method is based on a scheme devised by Ritz in 1909 and makes use of the variation principle to find approximations to the ground state wave function and energy of atomic systems. In fact, this method has been used to provide the most accurate approximations to solutions of the Schrödinger equation which are known; indeed, the variation principle is an alternative formulation of the Schrödinger equation. Another important approximational method is the perturbation approach, which is used mainly to calculate the change in energy suffered by a system when it is perturbed or acted on by some small external force or perturbation. For computational purposes the original or unperturbed state of a system may be entirely fictitious. Some very powerful techniques involving perturbation theory have been developed. The variation and perturbation methods will be discussed in this chapter.

[1] Born, M., and Oppenheimer, J. R., *Ann. der Phys.*, 1927, **84**, 457.
[2] Johnson, V. A., *Phys. Rev.*, 1941, **60**, 373.
[3] Wentzel, G., *Zeit. f. Physik*, 1926, **38**, 518.
[4] Kramers, H. A., *Zeit. f. Physik*, 1926, **39**, 828.
[5] Brillouin, L., *Comp. rend.*, 1926, **183**, 24; *J. de Physique*, 1926, **7**, 353.
[6] Bates, D. R. (editor), *Quantum Theory*, 1961, **Vol. I**, Academic Press Inc., New York.

Q.4.2. The Variation Method

If it is assumed that a system, which may be composed of any number of particles, has a Hamiltonian operator $\hat{H}$, and an infinite set of discrete energy states E_i, each corresponding to a normalized wave function ψ_i, then the appropriate wave equation may be written as $\hat{H}\psi_i = E_i\psi_i$. If both sides of this equation are multiplied by ψ_i^* and integrated over all of the coordinates involved or over all configuration space of the system, to be denoted simply by $d\tau$, then

$$E_i = \int \psi_i^* \hat{H}\psi_i \, d\tau, \tag{q.4.1.}$$

since if the functions ψ_i are normalized, $\int \psi_i^* \psi_i \, d\tau = 1$. This very important equation enables E_i to be calculated if ψ_i is known. In particular, the ground state energy of the system, E_1, described by the wave function ψ_1 is given by $E_1 = \int \psi_1^* \hat{H}\psi_1 \, d\tau$. It should be noted that unless ψ_i actually happens to be the true wave function for the system, that the function $\hat{H}\psi_i/\psi_i$ will in general be a complicated function of all of the coordinates involved and not a constant such as E_i. Of course, for the actual allowed wave functions, $\hat{H}\psi_i$ is simply a constant multiple of ψ_i, as is implied by the eigenvalue equation, $\hat{H}\psi_i = E_i\psi_i$. For any other single-valued, continuous, normalized function of the coordinates of the system, which is not necessarily an allowed wave function, f, the variation principle, which is the basis of the variation method, indicates that the integral

$$E = \int f^* \hat{H} f \, d\tau = \langle f|\hat{H}|f \rangle = \langle \hat{H} \rangle_{\text{av}}, \tag{q.4.2.}$$

is equal to or greater than the ground state energy of the system. The integral (q.4.2.) is only equal to the ground state energy if the function f is identically the function ψ_1, or, if that function which minimizes the integral (q.4.2.) is the ground state wave function of the system.

One way of proving the variation principle is to expand the function, f, in terms of the orthonormal set of eigenfunctions ψ_i, which are assumed to constitute a complete set, as discussed in Q.2.4. If any of the functions ψ_i correspond to the same degenerate state, in which event they may not be orthogonal, orthogonal linear combinations of them may be constructed as described on page 124. Writing the expansion of f, then, in terms of the functions ψ_i, gives $f = \sum_i a_i \psi_i$, where the coefficients a_i are constants. Since f is assumed to be normalized and the functions ψ_i are assumed to constitute an orthonormal set, then

$$\int f^* f \, d\tau = \sum_i \sum_j a_j^* a_i \int \psi_j^* \psi_i \, d\tau = 1, \quad \text{or} \quad \sum_i a_i^* a_i = 1. \tag{q.4.3.}$$

If $f = \sum a_i \psi_i$ is substituted into (q.4.2.),

$$E = \sum_i \sum_j a_j^* a_i \int \psi_j^* \hat{H} \psi_i \, d\tau = \sum_i \sum_j a_j^* a_i E_i \int \psi_j^* \psi_i \, d\tau = \sum_i a_i^* a_i E_i,$$

$$(q.4.4.)$$

since $\hat{H}\psi_i = E_i\psi_i$, and the functions ψ_i are orthonormal. If the ground-state energy of the system is subtracted from both sides of (q.4.4.) and since $\sum_i a_i^* a_i = 1$, there is obtained

$$E - E_1 = \sum_i a_i^* a_i (E_i - E_1), \qquad (q.4.5).$$

and since E_1 is the lowest possible energy state for the system, $E_i - E_1 \geq 0$ for all possible values of i; also $a_i^* a_i = |a_i|^2$ cannot be negative, and therefore the right-hand side of (q.4.5.) must be zero or positive, or

$$E \geq E_1. \qquad (q.4.6.)$$

E can only be equal to E_1 if all the coefficients a_i are zero with the exception of a_1, which is equivalent to f being identical to ψ_1.

An alternative manner of proving the variation principle and a way which shows that this principle is simply an alternative formulation of the Schrödinger equation is to make use of the method of the calculus of variations, as described on page 101. Suppose that the function f in (q.4.2.) is varied from f to $f + \delta f$, where both f and δf are functions of all of the coordinates involved, then stated in terms of the calculus of variations, the problem of calculating the eigenvalue in (q.4.2.) or the mean value of the quantum Hamiltonian of a system, resolves itself into that of finding the values of the function f which make the integral in (q.4.2.) adopt stationary values. This amounts to finding the function δf, which will not make any first order change in E or $\langle \hat{H} \rangle_{\mathrm{av}}$, but only a second-order change (of the order of magnitude of the square of δf). It has to be noted that if the function f is not normalized, the variation principle applies to the integral

$$E = \langle \hat{H} \rangle_{\mathrm{av}} = \int f^* \hat{H} f \, d\tau \Big/ \int f^* f \, d\tau, \qquad (q.4.7.)$$

which is equivalent to (q.4.2.) if f is normalized. Generally, an arbitrary variation in f will mean that, even if it were initially normalized, the normalization would be destroyed and (q.4.2.) only applies to normalized f's. Assuming that the unvaried function is normalized, then when f is replaced by $(f + \delta f)$, if the corresponding value of the average energy is

$$[\langle \hat{H} \rangle_{\mathrm{av}} + \delta \langle \hat{H} \rangle_{\mathrm{av}}],$$

it is found by substituting the varied function in (q.4.7.) that

$$\langle \hat{H} \rangle_{\mathrm{av}} + \delta\langle \hat{H} \rangle_{\mathrm{av}} = \frac{\int (f^* + \delta f^*)\hat{H}(f + \delta f)\, d\tau}{\int (f^* + \delta f^*)(f + \delta f)\, d\tau}. \tag{q.4.8.}$$

Expanding (q.4.8.) and at the same time neglecting terms in the expansion which contain second and higher order factors of δf, δf^*, and $\delta\langle \hat{H} \rangle_{\mathrm{av}}$, the result is

$$\delta\langle \hat{H} \rangle_{\mathrm{av}} = \int \delta f^* \hat{H} f\, d\tau + \int f^* \hat{H} \, \delta f\, d\tau - \langle \hat{H} \rangle_{\mathrm{av}}\left[\int \delta f^* f\, d\tau + \int f^* \, \delta f\, d\tau\right]$$
$$+ \text{ terms of higher order.} \tag{q.4.9.}$$

Since $\hat{H}$ is a Hermitean operator, then by (q.2.50.) as on page 139,

$$\int f^* \hat{H} \, \delta f\, d\tau = \int \delta f \hat{H}^* f^* \, d\tau, \tag{q.4.10.}$$

and since, by the calculus of variations, the condition that $\langle \hat{H} \rangle_{\mathrm{av}}$ should have a stationary value is that

$$\delta\langle \hat{H} \rangle_{\mathrm{av}} = 0, \tag{q.4.11.}$$

equation (q.4.9.) may be rewritten as

$$\int \delta f^*[\hat{H} - \langle \hat{H} \rangle_{\mathrm{av}}]f\, d\tau + \int \delta f[\hat{H}^* - \langle \hat{H} \rangle_{\mathrm{av}}]f^* \, d\tau = 0, \tag{q.4.12.}$$

since $\langle \hat{H} \rangle_{\mathrm{av}}$ is real. If then, as is implied by (q.4.12.), the sum of a quantity and its complex conjugate is zero, that quantity itself must be zero (and so of course must its complex conjugate). Therefore, the first term in (q.4.12.) must be zero, but this term is the integral of the variation of f^* multiplied by a particular function, $[\hat{H} - \langle \hat{H} \rangle_{\mathrm{av}}]f$, which can only be so if this function is zero, for all values of the coordinates involved. Therefore,

$$[\hat{H} - \langle \hat{H} \rangle_{\mathrm{av}}]f = 0 \qquad \text{or} \qquad \hat{H}f = \langle \hat{H} \rangle_{\mathrm{av}}f = Ef, \tag{q.4.13.}$$

which is simply the Schrödinger equation for the system.

The problem of varying a function such as f so as to make $\langle \hat{H} \rangle_{\mathrm{av}}$ stationary, but subject to the stipulation that the normalization integral $\int f^* f\, d\tau$ should retain a constant value of unity, may also be phrased in terms of Lagrange's method of undetermined multipliers, as used in the derivation of the Boltzmann constant (page 34). In the present case the condition which has to be satisfied is that

$$\delta[\langle \hat{H} \rangle_{\mathrm{av}} + \lambda \int f^* f\, d\tau] = 0, \tag{q.4.14.}$$

where λ is the multiplier to be determined. If both $\delta\langle\hat{H}\rangle_{av}$ and $\delta\int f^*f\,d\tau$ are written in the form of the sum of some quantity and its complex conjugate, as in (q.4.9.) and (q.4.12.), where again this sum must be zero, then by the same argument

$$\int \delta f^*[\hat{H} + \lambda]f\,d\tau + \int \delta f[\hat{H}^* + \lambda]f^*\,d\tau = 0, \qquad (q.4.15.)$$

and, as before, to satisfy the condition that the integral of some arbitrary variation of a function multiplied by another function should be zero, that function must itself be zero, or

$$\hat{H}f + \lambda f = 0 \qquad \text{or} \qquad \hat{H}f = -\lambda f. \qquad (q.4.16.)$$

Again, this is Schrödinger's equation for the system concerned, and furthermore the undetermined multiplier λ must correspond to $-\langle\hat{H}\rangle_{av} = -E$.

It is thus apparent that for any function, f, of the positional coordinates of a system, which is otherwise well-behaved, and is capable of completely arbitrary variation, the functions which give $\langle\hat{H}\rangle_{av}$ a stationary value or make $\delta\langle\hat{H}\rangle_{av} = 0$, in order to satisfy the method of the calculus of variations, constitute solutions of the Schrödinger equation describing the system. The variation principle is of comparable importance in classical as well as in quantum mechanics.[1] In practice, the function, f, referred to as the trial function, is chosen to be of an analytical form which is believed to be similar to the true ground-state wave function of the system, on the basis of whatever information is available, but containing several undetermined parameters; the values of these parameters are then chosen which make the energy calculated by the use of (q.4.2.) a minimum. This minimum value will then constitute the best approximation to the ground-state energy of the system, E_1, for the particular analytical form of the function used. In the selection of a trial function, the analytical form may be chosen quite arbitrarily, but then the calculated energy would probably be a very poor approximation to the correct ground-state energy. Where it is not possible to deduce with sufficient accuracy how closely a trial function approximates to the correct form, a variety of different analytical forms may be chosen, and the one which corresponds to the lowest calculated energy may be taken to be the best available approximation.

Q.4.3. Examples of Variational Calculations

In order to illustrate the applicability of the variational method to the calculation of the ground-state energies of systems, three examples will be used:

---

[1] Temple, G., and Bickley, W. G., *Rayleigh's Principle*, 1933, Oxford University Press.

(a) The ground-state energy of the linear harmonic oscillator, which is an example of a one-dimensional system, where the eigenvalues are known exactly (page 141) and so it is possible to compare the result obtained with the known value.

(b) The calculation of the ground-state energy of the hydrogen atom, as an example of a three-dimensional system, where again the exact eigenvalues are known, and so the effect of variational parameters in the trial function on the calculated value of E_1 may be compared with the exactly known eigenvalue (page 200).

(c) The calculation of the ground-state energy of the helium atom, where it has been found (page 227) that a very poor value for the ground state energy is obtained if the electron interaction energy is neglected.

(a) The quantum Hamiltonian for the linear harmonic oscillator is given on page 141 as $\hat{H} = -h^2[8\pi^2 m]^{-1} d^2/dx^2 + m\omega^2 x^2/2$; the parabolic nature of the potential energy function, which is symmetrical about the origin, has no nodes except at $x = 0$, and tends to infinity for $x = \pm\infty$, suggests a ground-state energy trial function of the form which has no nodes and should fall off to zero for large positive and negative values of x. The simplest type of function of this type is $f(x) = e^{-\alpha x^2}$, where the variable parameter is α, which has to be a positive constant. The integral

$$\int [f(x)]^* f(x)\, dx = \int_{-\infty}^{\infty} e^{-2\alpha x^2}\, dx = \sqrt{\pi/2\alpha}, \qquad \text{(q.4.17.)}$$

as shown on page 44. The quantity $\hat{H}f(x)$ is given by

$$\hat{H}f(x) = -h^2[8\pi^2 m]^{-1}\, d^2/dx^2(e^{-\alpha x^2}) + (1/2)m\omega^2 x^2 e^{-\alpha x^2}$$

$$= -h^2[8\pi^2 m]^{-1}[(4\alpha^2 - h^{-2}m^2\omega^2 4\pi^2)x^2 - 2\alpha]e^{-\alpha x^2}, \qquad \text{(q.4.18.)}$$

making use of which it is found that

$$\int [f(x)]^* \hat{H}f(x)\, dx = -h^2[8\pi^2 m]^{-1} \int_{\infty}^{\infty} [(4\alpha^2 - h^{-2}m\omega^2 4\pi^2)x^2 - 2\alpha]e^{-2\alpha x^2} dx$$

$$= -h^2[8\pi^2 m]^{-1}\{(4\alpha^2 - 4\pi^2 m^2\omega^2 h^{-2})(1/4\alpha) - 2\alpha\}\sqrt{\pi/2\alpha},$$

$$\text{(q.4.19.)}$$

making use of the integrals given on page 44. Now substituting (q.4.19.) and (q.4.17.) into the variational expression for the energy given by (q.4.7.), it is found that

$$E_\alpha = \int [f(x)]^* \hat{H}f(x)\, dx \left[\int \{f(x)\}^* f(x)\, dx\right]^{-1} = [m\omega^2/8\alpha] + [h^2\alpha/8\pi^2 m].$$

$$\text{(q.4.20.)}$$

The analytical condition which makes E_α a minimum is that $dE_\alpha/d\alpha = 0$, and so

$$dE_\alpha/d\alpha = 0 = -[m\omega^2/8\alpha^2] + [h^2/8\pi^2 m], \qquad \text{or} \qquad \alpha = \pi m\omega/h, \qquad \text{(q.4.21.)}$$

and substituting this value for α into (q.4.20.) it emerges that

$$E_{\min} = h\omega/4\pi = h\nu/2. \qquad \text{(q.4.22.)}$$

This is actually the true ground-state energy as given on page 141. However, in this particularly trivial example of the application of the variational method, the wave function of the ground-state is of such a simple analytical form that it may be deduced right away. Seldom is this the case, and in any event, had the ground-state energy of the linear harmonic oscillator not been known otherwise, the variation method would not have indicated that the calculated energy with $\alpha = \pi m\omega/h$ is actually the ground-state energy; it could conceivably be that some other form of trial function would lead to a lower energy value.

(b) The quantum Hamiltonian for the hydrogen atom is,

$$\hat{H} = [-(h^2/8\pi^2 m)\nabla^2 - (\varepsilon^2/r)].$$

Here the spherically symmetrical potential function suggests that the ground-state wave function should have similar properties, and of course for a ground-state function it would not be expected to have any nodes and must fall off to zero at large distances from the nucleus. A plausible function of these characteristics which falls off to zero exponentially for large values of the radial distance of the electron from the nucleus is $f(r) = e^{-r}$, or introducing a variable parameter λ, $f(r) = e^{-\lambda r}$. Expressing the Laplacian in spherical polar coordinates, then

$$\nabla^2 f(r) = \nabla^2[e^{-\lambda r}] = (1/r^2)\, d/dr[r^2\, d/dr(e^{-\lambda r})] = e^{-\lambda r}[\lambda^2 - 2\lambda/r],$$
$$\text{(q.4.23.)}$$

making use of (q.3.8.). Also, since $f(r)$ is real

$$\int [f(r)]^* f(r)\, dr = \int_0^\infty [e^{-\lambda r}]^2 4\pi r^2\, dr = 4\pi \int_0^\infty r^2 e^{-2\lambda r}\, dr$$
$$= 4\pi |e^{-2\lambda r}[-(r^2/2\lambda) - (r/2\lambda^2) - (1/4\lambda^3)]|_0^\infty = \pi/\lambda^3.$$
$$\text{(q.4.24.)}$$

The quantity $\hat{H}f(r)$ becomes

$$\hat{H}f(r) = [-(h^2/8\pi^2 m)\nabla^2 - (\varepsilon^2/r)]f(r)$$
$$= [-(h^2/8\pi^2 m)(\lambda^2 - 2\lambda/r)e^{-\lambda r} - \varepsilon^2 e^{-\lambda r}/r],$$

and

$$\int [f(r)]^*\hat{H}f(r)\,d\tau = \int \{[-(h^2/8\pi^2m)(\lambda^2 - 2\lambda/r)e^{-\lambda r} - \varepsilon^2 e^{-\lambda r}/r]e^{-\lambda r}\}\,d\tau$$

$$= -(h^2\lambda^2/8\pi^2m)\int_0^\infty 4\pi r^2 e^{-2\lambda r}\,dr$$

$$+ (2\lambda h^2/8\pi^2m)\int_0^\infty 4\pi r e^{-2\lambda r}\,dr - \varepsilon^2\int_0^\infty 4\pi r e^{-2\lambda r}\,dr$$

$$= -(h^2\lambda^2/8\pi^2m)(4\pi/4\lambda^3) + (2\lambda h^2/8\pi^2m)(4\pi/4\lambda^2)$$

$$- \varepsilon^2(4\pi/4\lambda^2), \tag{q.4.25.}$$

since $\int_0^\infty r^2 e^{-2\lambda r}\,dr = 1/4\lambda^3$ and $\int_0^\infty r e^{-2\lambda r}\,dr = 1/4\lambda^2$, as shown on page 44. Therefore, substituting (q.4.25.) and (q.4.24.) into the variational expression (q.4.7.) gives

$$E_\lambda = \left[\int \{f(r)\}^*\hat{H}f(r)\,d\tau\right]\left[\int \{f(r)\}^*f(r)\,d\tau\right]^{-1}$$

$$= -(h^2\lambda^2/8\pi^2m) + (h^2\lambda^2/4\pi^2m) - \varepsilon^2\lambda$$

$$= (h^2\lambda^2/8\pi^2m) - \varepsilon^2\lambda. \tag{q.4.26.}$$

The analytical condition for a minimum value of E_λ occurs when

$$dE_\lambda/d\lambda = 0 = 2\lambda h^2/8\pi^2m - \varepsilon^2. \tag{q.4.27.}$$

Thus the optimum value of the parameter λ, to minimize the calculated energy is

$$\lambda = 4\pi^2m\varepsilon^2/h^2 = 1/a_0, \tag{q.4.28.}$$

and substituting this value for λ into (q.4.26.) gives

$$[E_\lambda]_{min} = -(2\pi^2m\varepsilon^4/h^2) = -(\varepsilon^2/2a_0) = -(1/2)\text{ Hartrees.} \tag{q.4.29.}$$

Again, due to the choice of the function of the simple exponential type, it happens that the true ground-state energy of the hydrogen atom is arrived at by the variational method. If a Gaussian type trial function had been selected instead, as in the case of the linear harmonic oscillator, $f(r) = e^{-\lambda r^2}$, then $E_\lambda = [3\lambda h^2/8\pi^2m] - 2\sqrt{2}\,\varepsilon^2\sqrt{\lambda/\pi}$, which adopts a minimum value when $\lambda = (8/9\pi a_0^2)$, giving a value to $[E_\lambda]_{min} = -0.424\varepsilon^2/a_0$, or only 15% greater than the real value. If a two-parameter trial function is used such as $f(r) = e^{-\lambda r^\mu}$, and the above procedure repeated, this time minimizing $[E_{\lambda,\mu}]$ with respect to both λ and μ, it is found that $[E_{\lambda,\mu}]_{min}$ occurs when $\mu = 1$ and $\lambda = 1/a_0$ as found before. Generally, the variational method leads to reasonably good approximations to the true ground-state energies of systems to be found, such that the percentage error in the calculated ground-state energy is usually considerably less than in the trial function chosen, and it is essentially for this reason that it is possible to calculate approximate energies for more complicated atoms and molecules.

(c) For the helium atom the potential energy function is given by (q.3.139.) and the quantum Hamiltonian by the term in square brackets in (q.3.140.). Expressed in atomic units, the quantum Hamiltonian for helium becomes

$$\hat{H} = -(1/2)[\nabla_1^2 + \nabla_2^2] - 2[1r_1 + 1/r_2] + 1/r_{12}, \qquad \text{(q.4.30.)}$$

with the energy expressed in Hartree units and distances expressed in Bohr units. It was found in Q.3.9. that if the Coulomb interaction energy between the two electrons in the helium atom is neglected that the Schrödinger equation for helium was separable into two hydrogenic equations. For helium, with nuclear charge $Z = 2$, the product of two ground-state hydrogenic wave function is, in atomic units, given by $e^{-2(r_1 + r_2)}$. But the use of this approximate wave function for the helium atom leads to a calculated energy of -4 Hartrees, which is about 40% less than the experimentally measured value of -2.905 Hartrees. In order to take account of the fact that there is some interaction between the electrons, and that each of the electrons must reduce the attractive force exerted on the other by the nucleus, due to its "screening" effect, such that the effective charge attracting each electron to the nucleus is less than 2, a trial function of the type

$$f(r_1, r_2) = e^{-\lambda(r_1 + r_2)}, \qquad \text{(q.4.31.)}$$

is suggested, where λ is again a variable parameter, assumed to be real and positive. Since this is a real function, $[f(r_1, r_2)]^* = f(r_1, r_2)$. Then

$$\int [f(r_1, r_2)]^* f(r_1, r_2)\, d\tau = \iint [f(r_1, r_2)]^2\, dr_1\, dr_2 = \iint [e^{-2\lambda(r_1 + r_2)}\, dr_1\, dr_2$$

$$= \int_0^\infty e^{-2\lambda r_1} 4\pi r_1^2\, dr_1 \int_0^\infty e^{-2\lambda r_2} 4\pi r_2^2\, dr_2$$

$$= 16\pi^2 [2/(2\lambda)^3][2/(2\lambda)^3] = \pi^2/\lambda^6, \qquad \text{(q.4.32.)}$$

since $\int_0^\infty r^2 e^{-2\lambda r}\, dr = |e^{-2\lambda r}[-(r^2/2\lambda) - (r/2\lambda^2) - (1/4\lambda^3)]|_0^\infty = [2/(2\lambda)^3]$. In spherical polar coordinates

$$\nabla_1^2 e^{-\lambda r_1} = (1/r^2)\, d/dr[r^2\, d/dr(e^{-\lambda r_1})] = [\lambda^2 - 2\lambda/r_1]e^{-\lambda r_1}, \qquad \text{(q.4.33.)}$$

and

$$\nabla_2^2 e^{-\lambda r_2} = (1/r^2)\, d/dr[r^2\, d/dr(e^{-\lambda r_2})] = [\lambda^2 - 2\lambda/r_2]e^{-\lambda r_2}. \qquad \text{(q.4.34.)}$$

Thus,

$$\hat{H}f(r_1, r_2) = -(1/2)[[\lambda^2 - 2\lambda/r_1]e^{-\lambda(r_1+r_2)} + [\lambda^2 - 2\lambda/r_2]e^{-\lambda(r_1+r_2)}]$$
$$+ [-(2/r_1) - (2/r_2) + (1/r_{12})]e^{-\lambda(r_1+r_2)}$$
$$= [e^{-\lambda(r_1+r_2)}][-\lambda^2 + (\lambda - 2)/r_1 + (\lambda - 2)/r_2 + (1/r_{12})],$$
$$\text{(q.4.35.)}$$

and

$$\iint [f(r_1, r_2)]^* \hat{H}f(r_1, r_2)\, dr_1\, dr_2$$

$$= \iint [-\lambda^2 + (\lambda - 2)/r_1 + (\lambda - 2)/r_2 + (1/r_{12})]e^{-2\lambda(r_1+r_2)}\, dr_1\, dr_2.$$
$$\text{(q.4.36.)}$$

With a little effort‡ it can be shown that

$$\iint (1/r_1)e^{-2\lambda(r_1+r_2)}\, dr_1\, dr_2 = \iint (1/r_2)e^{-2\lambda(r_1+r_2)}\, dr_1\, dr_2$$

$$= \int_0^\infty (1/r_1)e^{-2\lambda r_1}4\pi r_1^2\, dr_1 \int_0^\infty e^{-2\lambda r_2}4\pi r_2^2\, dr_2$$

$$= 16\pi^2(1/4\lambda^2)(1/4\lambda^3) = \pi^2/\lambda^5, \qquad \text{(q.4.37.)}$$

and

$$\iint (1/r_{12})e^{-2\lambda(r_1+r_2)}\, dr_1\, dr_2 = 5\pi^2/8\lambda^5. \qquad \text{(q.4.38.)}$$

Now substituting the integrals (q.4.37.) and (q.4.38.) into (q.4.36.), it is found that

$$\iint [f(r_1, r_2)]^* \hat{H}f(r_1, r_2)\, dr_1\, dr_2 = -\lambda^2(\pi^2/\lambda^6) + 2(\lambda - 2)(\pi^2/\lambda^5) + 5\pi^2/8\lambda^5$$

$$= (\pi^2/\lambda^4) - (27\pi^2/8\lambda^5). \qquad \text{(q.4.39.)}$$

‡ The first integral in the expression (q.4.36.) is given by (q.4.32.). The second and third integrals in (q.4.36.), as indicated in (q.4.37.), both involve the evaluation of integrals of the type, $\int re^{-2\lambda r}\, dr$ and $\int r^2 e^{-2\lambda r}\, dr$, which are readily integrated by parts to give

$$\int re^{-2\lambda r}\, dr = e^{-2\lambda r}[-r/2\lambda - 1/4\lambda^2]$$

and

$$\int r^2 e^{-2\lambda r}\, dr = e^{-2\lambda r}[-r^2/2\lambda - r/2\lambda^2 - 1/4\lambda^3].$$

The last integral in (q.4.36.) is of the type, $\iint (1/r_{12})f_1(r_1)f_2(r_2)\, dr_1\, dr_2$, where in this case, $r_{12} = r_1 - r_2$, $f_1(r_1) = e^{-2\lambda r_1}$ and $f_2(r_2) = e^{-2\lambda r_2}$. Such integrals may often be evaluated exactly by introducing the two sets of polar coordinates (r_1, θ_1, ϕ_1) and (r_2, θ_2, ϕ_2) and then expanding r_{12} as a power series in r_1 and r_2, where the actual form of the series expansion to be used may be derived from the theory of spherical harmonics (see page 353). However, since $e^{-2\lambda r_1}$ and $e^{-2\lambda r_2}$ are both spherically symmetrical functions, it is convenient to think of them as spherically symmetrical charge distributions of density $\rho(r_1)$ and $\rho(r_2)$ respectively, so that the integral may then be regarded such that the integrand represents the product of two charges divided by a distance; that is the integral may be regarded as the potential which exists between two charges separated by some distance. For example, suppose that a spherical charge of desity $\rho(r)$ is accumulated on a sphere of radius r by adding charge from the origin until a radius r has been achieved. Then the potential at the surface of the charged sphere according to the electrostatic definition of potential is, $(1/r) \int_0^r 4\pi r^2 \rho(r)\, dr$. The amount of work which would now be required to be done to bring up a sufficient charge of density $\rho(r)$, from an infinite distance away to form a shell of thickness dr, at the surface of the charged sphere would be

$$4\pi r^2 \rho(r)\, dr[(1/r) \int_0^r 4\pi r^2 \rho(r)\, dr,$$

and since the difference of potential between two points is the work done in transferring charge from one point to the other, the self-potential of the whole charge distribution would be

$$\int_0^\infty 4\pi r \rho(r)\, dr \int_0^r 4\pi r^2 \rho(r)\, dr = 16\pi^2 \int_0^\infty r\rho(r)\, dr \int_0^r r^2 \rho(r)\, dr.$$

But the integral required is of the form $\iint (1/r_{12})\rho(r_1)\rho(r_2)\, dr_1\, dr_2$, which is twice the self-potential energy of a spherically symmetrical charge distribution of density $\rho(r)$, or

$$32\pi^2 \int_0^\infty r\rho(r)\, dr \int_0^r r^2 \rho(r)\, dr.$$

Therefore

$$\iint (1/r_{12})e^{-2\lambda(r_1+r_2)}\, dr_1\, dr_2$$

$$= 32\pi^2 \int_0^\infty re^{-2\lambda r}\, dr \int_0^r r^2 e^{-2\lambda r}\, dr$$

$$= 32\pi^2 \int_0^\infty \{re^{-2\lambda r}[-e^{-2\lambda r}(r^2/2\lambda + r/2\lambda^2 + 1/4\lambda^3) + 1/4\lambda^3]\}\, dr$$

$$= 32\pi^2[(1/4\lambda^3) \int_0^\infty re^{-2\lambda r}\, dr - (1/2\lambda) \int_0^\infty r^3 e^{-4\lambda r}\, dr - (1/2\lambda^2) \int_0^\infty r^2 e^{-4\lambda r}\, dr$$

$$- (1/4\lambda^3) \int_0^\infty re^{-4\lambda r}\, dr]$$

$$= 32\pi^2[(1/4\lambda^3)|e^{-2\lambda r}(-r/2\lambda - 1/4\lambda^2)|_0^\infty$$

$$- (1/2\lambda)|e^{-4\lambda r}(-r^3/4\lambda - 3r^2/16\lambda^2 - 3r^2/32\lambda^3 - 3/128\lambda^4)|_0^\infty$$

$$- (1/2\lambda^2)|e^{-4\lambda r}(-r^2/4\lambda - r/8\lambda^2 - 1/32\lambda^3)|_0^\infty - (1/4\lambda^3)|e^{-4\lambda r}$$

$$\times (-r/4\lambda - 1/16\lambda^2)|_0^\infty]$$

$$= 32\pi^2[5/256\lambda^5] = 5\pi^2/8\lambda^5.$$

If (q.4.39.) is now divided by (q.4.32.) to obtain the variational expression for the energy, there results

$$\iint [f(r_1, r_2)]^* \hat{H} f(r_1, r_2)\, dr_1\, dr_2 \left[\iint [f(r_1, r_2)]^2\, dr_1\, dr_2 \right]^{-1}$$

$$= E_\lambda = [\lambda^2 - 27\lambda/8]. \quad \text{(q.4.40.)}$$

Applying the usual analytical condition for the minimum value of E_λ,

$$d/d\lambda[E_\lambda] = 0 = 2\lambda - 27/8 \quad \text{or} \quad \lambda = 27/16, \qquad \text{(q.4.41.)}$$

so that the minimum value of $E_{27/16}$, for the trial function (q.4.31.) is

$$[E_{27/16}]_{\min} = (27/16) - (27/8)(27/16) = -(729/256) = -2.85 \text{ Hartrees,}$$
$$\text{(q.4.42.)}$$

which differs from the experimentally observed value of -2.905 Hartrees by only about 2%.

A two-parameter trial function may be written for helium as $f(r_1, r_2) = e^{-\lambda_1 r_1}e^{-\lambda_2 r_2}$, the significance of which is that it takes account of the fact that one electron may experience a nuclear attractive force effectively greater than that due to a charge of $+2\varepsilon$, because of the repulsion due to the other electron. Eckart[1] has calculated the ground-state energy of helium using a trial function of the form

$$f(r_1, r_2) = e^{-\lambda_1 r_1}e^{-\lambda_2 r_2} + e^{-\lambda_2 r_1}e^{-\lambda_1 r_2}, \qquad \text{(q.4.43.)}$$

this form of the trial function arising from the indistinguishability of the two electrons, which essentially requires that the combined function should be symmetrical in form. If the two factors λ_1 and λ_2 are regarded as independent variational parameters and the energy minimized with respect to each of them, Eckart found $\lambda_1 = 1.19$ and $\lambda_2 = 2.18$, with an average value as given by (q.4.41.), and a corresponding ground-state energy of -2.8757 Hartrees, which differs by only about 1% from the experimental value. This Eckart-type function describes the tendency of one electron to be further away from the nucleus than the other, or may be regarded as a radial correlation effect. For the helium atom or any two-electron ion, the number of parameters in the trial function may be extended to any number. Hylleraas[2] has devised trial functions of the form

$$f(r_1, r_2) = e^{-\lambda s}[1 + f(s, t, u)], \qquad \text{(q.4.44.)}$$

and of other types, in which $f(s, t, u)$ is a power series in the variables

$$s = r_1 + r_2; \; t = r_1 - r_2; \; u = r_{12}, \qquad \text{(q.4.45.)}$$

which have come to be referred to as Hylleraas variables. A six-parameter trial function of the Hylleraas type is

$$f(r_1, r_2) = e^{-\lambda s}[1 + c_1 s + c_2 s^2 + c_3 t^2 + c_4 u + c_5 u^2], \qquad \text{(q.4.46.)}$$

---

[1] Eckart, C. E., *Phys. Rev.*, 1930, **36**, 878.
[2] Hylleraas, E. A., *Z. Physik*, 1930, **65**, 209; *Rev. Mod. Phys.*, 1963, **35**, 421.

and if the calculated energy is minimized with respect to each of the parameters c_1, c_2, c_3, c_4, and c_5, for several different values of λ, choosing the value of λ which gives the minimum $E = \langle \hat{H} \rangle_{\mathrm{av}}$, the calculated energy is obtained as -2.90324 Hartrees, in essentially complete agreement with the experimental value determined to a comparable number of significant figures of -2.90372 Hartrees. Kinoshita[1] and Pekeris,[2] in particular, the latter using a trial function composed of 1078 terms, have calculated the ground-state energy of the helium atom as -2.903724375 Hartrees, believed to be reliable to within ± 1 in the last digit; there is of course no means of determining experimentally the ground-state energy of helium to a comparable degree of precision, and so it is likely that this is the most accurate value known, neglecting relativistic effects. Others[3-6] have investigated the effect on the calculated ground-state energy of helium of both radial and angular correlation effects by the use of trial functions including $1s^2$, $1s2s$, $2p^2$, $3d^2$, $4f^2$ combinations of hydrogenic wave functions, leading to the conclusion that for helium, angular correlation effects influence the energy to a greater extent than radial correlation effects alone. Angular correlation effects refer to the tendency of two electrons to behave in such a manner that their probability density functions are high in opposing directions with respect to the nucleus.

Accurate calculations have been carried out on other two-electron systems, such as Li^+ and Be^{++} by Weiss[7] and Burke[8] with $Z = 3$ and $Z = 4$, respectively.

Q.4.4. Linear Variation Functions

A particularly useful form of trial function to be used in the variational method, the nature of which has been alluded to above, is one which consists of a linear combination of definite functions, with the linear coefficients regarded as the variational parameters. The definite functions which are chosen as components of the linear combination for any particular system are most usefully chosen from those which form a complete set and which enable parameters to be inserted in such a manner as to provide maximum variation of the function in the region where its value undergoes the most

[1] Kinoshita, T., *Phys. Rev.*, 1959, **115**, 366.
[2] Pekeris, C. L., *Phys. Rev.*, 1958, **112**, 1649; 1959, **115**, 1216.
[3] Taylor, G. R., and Parr, R. G., *Proc. Nat. Acad. Sci. U.S.*, 1952, **38**, 154.
[4] Munschy, G., and Pluvinage, P., *Rev. Mod. Phys.*, 1963, **35**, 494.
[5] Wilson, E. B., Jr., *J. Chem. Phys.*, 1965, **43**, S172.
[6] Shull, H., and Löwdin, P. O., *J. Chem. Phys.*, 1955, **23**, 1362.
[7] Weiss, A. W., *Phys. Rev.*, 1961, **122**, 1826.
[8] Burke, E. A., *Phys. Rev.*, 1963, **130**, 1871.

critical change. It is practically convenient if the linear combination of functions chosen leads to integrals over the functions which are readily evaluated in closed terms. Many of the functions used to describe atomic behavior include combinations of simple exponential functions, Gaussian functions, Legendre functions, Laguerre functions, and hydrogenic-type functions, but in general it does not particularly matter what the component functions correspond to, as when the linear coefficient is treated as a variational parameter and the energy is minimized with respect to that parameter, this coefficient will usually be found to have an insignificantly small value if the function concerned is unimportant in the description of the system concerned. The applicability of this method is most easily illustrated by the use of a linear combination of only two functions, which may then be generalized to any number of components.

Suppose that the function

$$F = c_1 f_1 + c_2 f_2 \qquad\qquad (q.4.47.)$$

is the trial function under consideration, where for convenience it is assumed that both the component functions and the coefficients are real, and the functions f_1 and f_2 are not necessarily normalized or orthogonal to one another. In terms of the variational method, c_1 and c_2 have to be chosen so as to minimize the function

$$E_{c_1,\,c_2} = \int F \hat{H} F \, d\tau \left[\int F^2 \, d\tau \right]^{-1},$$

$$= \frac{c_1{}^2 \int f_1 \hat{H} f_1 \, d\tau + c_1 c_2 \int f_1 \hat{H} f_2 \, d\tau + c_1 c_2 \int f_2 \hat{H} f_1 \, d\tau + c_2{}^2 \int f_2 \hat{H} f_2 \, d\tau}{c_1{}^2 \int f_1{}^2 \, d\tau + 2 c_1 c_2 \int f_1 f_2 \, d\tau + c_2{}^2 \int f_2{}^2 \, d\tau},$$

$$\qquad\qquad (q.4.48.)$$

which may be simplified by making use of the notation

$$\int f_i \hat{H} f_j \, d\tau = \langle f_i | \hat{H} | f_j \rangle = H_{ij} \qquad \text{and} \qquad \int f_i f_j \, d\tau = \langle f_i | f_j \rangle = S_{ij}$$

$$\qquad\qquad (q.4.49.)$$

where i and j may have the value 1 and 2. Obviously $S_{ij} = S_{ji}$. If the quantum Hamiltonian for a single particle system is

$$\hat{H} = -h^2/8\pi^2 m \nabla^2 + U(r),$$

then

$$H_{12} = -h^2/8\pi^2 m \int f_1 \nabla^2 f_2 \, d\tau + \int f_1 U f_2 \, d\tau. \qquad (q.4.50.)$$

Since the potential function $U(r)$ is a function of the coordinates of the system only, it is apparent that $\int f_1 U f_2 \, d\tau = \int f_2 U f_1 \, d\tau$. What is not so apparent, but may be proved by the application of Green's theorem or in a straightforward manner,‡ is that $\int f_1 \nabla^2 f_2 \, d\tau = \int f_2 \nabla^2 f_1 \, d\tau$, and therefore $H_{12} = H_{21}$, for a single particle.

‡ In elementary calculus, the values of a function such as $f(x, y)$ at two points (x_1, y_1) and (x_2, y_2) are obtained by integration of an equation like $df(x, y) = A(x, y) \, dx + B(x, y) \, dy$. The integrals in this equation are referred to as line integrals. A line integral is independent of the path and depends only on the values of x and y at the upper and lower limits of the integral, so that the function $f(x, y)$ is called a point function or a scalar point function. In turn, a scalar field is defined as a region in space, with each point of which there is associated a scalar point function; the temperature of points in the atmosphere, at any time, is an example of a scalar field. One way of stating Green's theorem is that, if $\zeta(r)$ and $\xi(r)$ are scalar point functions, then

$$\int_v [\xi \nabla^2 \xi - \xi \nabla^2 \zeta] \, dr = \int_S [\zeta \nabla \xi - \xi \nabla \zeta] n \, dS,$$

where the left-hand integral is over the volume v within the closed surface S, and the right-hand integral is over the area S, and n is a unit vector along the outward normal to the surface S at the position of the element dS. ∇ is defined as $\partial/\partial x + \partial/\partial y + \partial/\partial z$ and ∇^2 as $\partial^2/\partial x^2 + \partial^2/\partial y^2 + \partial^2/\partial z^2$. If differentiation in the direction of n is denoted by the symbol $\partial/\partial n$, then Green's theorem may be alternatively stated in the form

$$\int_v [\zeta \nabla^2 \xi - \xi \nabla^2 \zeta] \, dr = \int_S [\zeta \, \partial\xi/\partial n - \xi \, \partial\zeta/\partial n] \, dS,$$

which enables a volume integral to be converted into a surface integral. For functions $\zeta(r)$ and $\xi(r)$ which satisfy the same boundary conditions as wave functions, in particular that they fall off to zero as $r \to \infty$, then the volume concerned in the volume integral is all of space and so the boundary surface involved is infinitely large, as far as the surface integral is concerned. Therefore, for wave functions, the surface integral in the statement of Green's theorem is zero, since the integrand is zero. The volume integral must therefore also be zero; that is $\int f_1 \nabla^2 f_2 \, d\tau = \int f_2 \nabla^2 f_1 \, d\tau$, where f_1 and f_2 satisfy the same boundary conditions as wave functions.

Alternatively, it may be proved that $\int f_1 \nabla^2 f_2 \, d\tau = \int f_2 \nabla^2 f_1 \, d\tau$, as follows. Writing $\int f_1 \nabla^2 f_2 \, d\tau = \iiint_{-\infty}^{\infty} f_1 [\partial^2 f_2/\partial x^2 + \partial^2 f_2/\partial y^2 + \partial^2 f_2/\partial z^2] \, dx \, dy \, dz$, and now integrating each of the three members twice by parts, there is obtained

$$\int_{-\infty}^{\infty} f_1 \, \partial^2 f_2/\partial x^2 \, dx = |f_1 \, \partial f_2/\partial x|_{-\infty}^{\infty} - \int_{-\infty}^{\infty} \partial f_1/\partial x \cdot \partial f_2/\partial x \, dx$$

$$= |f_1 \, \partial f_2/\partial x|_{-\infty}^{\infty} - |\partial f_1/\partial x f_2|_{-\infty}^{\infty} + \int_{-\infty}^{\infty} f_2 \, \partial^2 f_1/\partial x^2 \, dx$$

$$= \int_{-\infty}^{\infty} f_2 \, \partial^2 f_1/\partial x^2 \, dx,$$

and similarly

$$\int_{-\infty}^{\infty} f_1 \, \partial^2 f_2/\partial y^2 \, dy = \int_{-\infty}^{\infty} f_2 \, \partial^2 f_1/\partial y^2 \, dy$$

and

$$\int_{-\infty}^{\infty} f_1 \, \partial^2 f_2/\partial z^2 \, dz = \int_{-\infty}^{\infty} f_2 \, \partial^2 f_1/\partial z^2 \, dz.$$

Therefore

$$\int f_1 \nabla^2 f_2 \, d\tau = \left[\int_{-\infty}^{\infty} f_1 \, \partial^2 f_2 / \partial x^2 \, dx + \int_{-\infty}^{\infty} f_1 \, \partial^2 f_2 / \partial y^2 \, dy + \int_{-\infty}^{\infty} f_1 \, \partial^2 f_2 / \partial z^2 \, dz \right]$$

$$= \left[\int_{-\infty}^{\infty} f_2 \, \partial^2 f_1 / \partial x^2 \, dx + \int_{-\infty}^{\infty} f_2 \, \partial^2 f_1 / \partial y^2 \, dy + \int_{-\infty}^{\infty} f_2 \, \partial^2 f_1 / \partial z^2 \, dz \right]$$

$$= \iiint_{-\infty}^{\infty} f_2 [\nabla^2 f_1] \, dx \, dy \, dz = \int f_2 \nabla^2 f_1 \, d\tau.$$

In a similar manner, it may be proved that $\int f_i^* \nabla^2 f_j \, d\tau = \int f_j \nabla^2 f_i^* \, d\tau$, which is required to prove that $H_{ij} = [H_{ji}]^*$.

For a system composed of any number of particles, the same condition is satisfied, in view of the nature of the Hamiltonian. It is thus possible to write (q.4.48.) as

$$E_{c_1,c_2}[c_1^2 S_{11} + 2c_1 c_2 S_{12} + c_2^2 S_{22}] = c_1^2 H_{11} + 2c_1 c_2 H_{12} + c_2^2 H_{22}. \tag{q.4.51.}$$

The analytical condition which must be satisfied in order that E_{c_1,c_2} should have a minimum value is that

$$\partial E / \partial c_1 = \partial E / \partial c_2 = 0. \tag{q.4.52.}$$

On differentiating E in (q.4.51.) with respect to c_1, there is obtained for the partial derivative

$$E[c_1 S_{11} + c_2 S_{12}] = c_1 H_{11} + c_2 H_{22}$$

or

$$c_1[H_{11} - ES_{11}] + c_2[H_{12} - ES_{12}] = 0. \tag{q.4.53.}$$

Similarly, the first partial derivative of E in (q.4.51.) with respect to c_2 is

$$E[c_1 S_{12} + c_2 S_{22}] = c_1 H_{12} + c_2 H_{22},$$

or since $H_{12} = H_{21}$ and $S_{12} = S_{21}$, this equation may be rewritten as

$$E[c_1 S_{21} + c_2 S_{22}] = c_1 H_{21} + c_2 H_{22}$$

or

$$c_1[H_{21} - ES_{21}] + c_2[H_{22} - ES_{22}] = 0. \tag{q.4.54.}$$

Equations (q.4.53.) and (q.4.54.) constitute a pair of simultaneous equations in c_1 and c_2, the variational parameters, and the analytical condition which must be satisfied in order that these two equations should have simultaneous, non-zero, solutions is that the determinant of the coefficients of c_1 and c_2 should vanish (see page 152), or

$$\begin{vmatrix} H_{11} - ES_{11} & H_{12} - ES_{12} \\ H_{21} - ES_{21} & H_{22} - ES_{22} \end{vmatrix} = 0, \tag{q.4.55.}$$

which, of course, is the same result as is obtained by the elimination of c_1 and c_2 from (q.4.53.) and (q.4.54.) by ordinary algebraic methods. This is a quadratic equation in the energy E, and in accordance with the variation principle, its lower root (the one with the negative sign) is an approximation to the lowest energy state of the system concerned, being equal to or greater than E_1. The other root is an approximation to E_2, although not necessarily a particularly good one.

If F, in (q.4.47.) is a linear combination of three functions, then by the same procedure, there is obtained a cubic equation in E, with three roots, the lowest of which gives an approximation to E and the other two poorer approximations to the first two excited states. Generally, for a trial function composed of a linear combination of N functions, such that

$$F = \sum_{i=1}^{N} c_i f_i, \qquad (q.4.56.)$$

this method leads to a set of N linear equations containing N terms each, and the condition which must be satisfied in order that the parameters c_i should have non-zero values is that the determinant of their coefficients should vanish, as before, and a polynomial in E of degree N is obtained of the type

$$\begin{vmatrix} H_{11} - ES_{11} & H_{12} - ES_{12} & \cdots & H_{1N} - ES_{1N} \\ H_{21} - ES_{21} & H_{22} - ES_{22} & \cdots & H_{2N} - ES_{2N} \\ \vdots & \vdots & & \vdots \\ H_{N1} - ES_{N1} & H_{N2} - ES_{N2} & \cdots & H_{NN} - ES_{NN} \end{vmatrix} = 0. \qquad (q.4.57.)$$

The lowest root of this equation gives an approximation to E_1 and the other $(N - 1)$ roots provide, generally poorer, approximations to the next $(N - 1)$ energy states. Determinants of the type (q.4.57.) arise in the solution of a variety of problems involving wave motion and vibratory systems and in the calculation of the periods of such systems; they are therefore commonly referred to as secular determinants and the equations like (q.4.53.) and (q.4.54.) as secular equations. The various eigenvalues of a system described by the linear combination function (q.4.56.) are thus obtained as the roots of the N secular equations, only the lowest eigenvalue, however, being a particularly good approximation. The quantities defined in (q.4.49.) as H_{ij} and S_{ij} are usually referred to as the matrix components of the Hamiltonian operator of the system with respect to the functions f_i and f_j, and the matrix components of unity or overlap integrals, respectively. It should be noted that in the general case where both the functions f_i and the parameters c_i may be complex and where

$$H_{ij} = \int f_i^* \hat{H} f_j \, d\tau \qquad \text{and} \qquad S_{ij} = \int f_i^* f_j \, d\tau, \qquad (q.4.58.)$$

that $S_{ij} \neq S_{ji}$; nor is H_{ij} equal to H_{ji}. Instead, $S_{ji}{}^* = \int f_j f_i{}^* \, d\tau = S_{ij}$, and by the same argument as that used on page 243, $H_{ij} = H_{ji}{}^*$, which is a statement of the Hermitean property of $\hat{H}$ (see page 139); also, as shown on page 139, since the eigenvalues of any Hermitean operator are all real, the energy levels obtained by the application of the linear variation method are necessarily real. Thus generally, the elements $H_{1N} - ES_{1N}$ of (q.4.57.) is not equal to $H_{N1} - ES_{N1}$, although in (q.4.55.) it is the case that

$$H_{12} - ES_{12} = H_{21} - ES_{21}.$$

Q.4.5. Perturbation Theory

The technique which is generally referred to as Rayleigh–Schrödinger perturbation theory provides another means of constructing approximate wave functions and obtaining approximations to the various energy states of a system, in addition to the ground-state wave function and the corresponding energy, from a different point of view. Many of the results which may be obtained by the application of the variational method can also be obtained by perturbation theoretical means, and the method which is referred to as the Brillouin–Wigner perturbation technique can be derived from the variation theorem. These two perturbation theories can be transformed into one another and the distinction between them is more that of the purpose towards which they are applied rather than that of their conceptual basis. Many powerful techniques based on perturbation theory have been proposed which effectively provide the possibility of solving perturbational problems to any desired degree of accuracy.[1] Only first-order perturbation theory and an introduction to generalized theory will be presented here. It is, in fact, largely a matter of preference whether a variational or a perturbational approach is used in the attempt to solve some specific problem. In some respects the variational method allows a more flexible trial function to be used, whereas the perturbation method enables an estimate to be made of the magnitude of the errors in the calculated energy values.

The perturbation method is mainly used to calculate the change in energy of a system when the system is disturbed or perturbed by the action of some external force, referred to as a perturbation. The state of the system pertaining to the action of the external force is called the perturbed state, and in the absence of the external force the state is called the unperturbed state. For example, if the Hamiltonian of a system contains a term which makes it

---

[1] Hirschfelder, J. O., Brown, W. B., and Epstein, S. T., *Recent Developments in Perturbation Theory, in Advances in Quantum Chemistry*, 1964, **Vol. I**, Academic Press, Inc., New York.

difficult or not actually possible to calculate the eigenvalues of a system, but the eigenvalues can be obtained by neglecting this term, then perturbation theory enables this term in the Hamiltonian to be regarded as a perturbation and the additional energy due to its presence to be calculated. This is precisely the case for the helium atom, which in the absence of the electron correlation energy term, ε^2/r_{12}, in the Hamiltonian of this system, is described by a Schrödinger equation which may be separated into two hydrogenic equations for which the solutions are known in closed terms. In this case, the unperturbed state does not correspond to any real system, but perturbation theory may be applied to calculate the energy correction value due to the perturbation.

To illustrate the applicability of perturbation theory to a non-degenerate state of a system, let it be assumed that the Hamiltonian of the system can be expressed as the sum of two parts

$$\hat{H} = \hat{H}^0 + \hat{H}', \qquad\qquad \text{(q.4.59.)}$$

where $\hat{H}$, $\hat{H}^0$, and $\hat{H}'$, denote the Hamiltonian of the perturbed state, the Hamiltonian of the unperturbed state, and the perturbation term in the Hamiltonian of the perturbed state, respectively, and the effect of $\hat{H}'$ on the energy states and eigenfunctions of the unperturbed system is assumed to be small. The Schrödinger equation for the perturbed state of the system is then

$$\hat{H}\psi_n = E_n\psi_n, \qquad\qquad \text{(q.4.60.)}$$

and if it is assumed that the eigenvalues and eigenfunctions of the unperturbed state, satisfying the equation

$$\hat{H}^0 f_n = E_n{}^0 f_n \qquad\qquad \text{(q.4.61.)}$$

are known, subject to the same boundary conditions as in (q.4.60.), then if the perturbation is indeed small, the eigenfunctions ψ_n will not differ substantially from the unperturbed eigenfunctions f_n. It is not necessary to specify how small the effect of $\hat{H}'$ on the eigenvalues $E_n{}^0$ actually is, since in practice if the calculated eigenvalues E_n converge to a finite value, or at least in a satisfactory manner, it may be concluded that this is so; otherwise it is not sufficiently small. In order to simplify the description of the system, which tends to become rather complicated by the use of a variety of subscripts and superscripts, let it be further assumed that only one of the energy states of the system is under consideration; specifically, let it be assumed that the unperturbed state described by the eigenfunction f is converted into the perturbed state described by the eigenfunction ψ by the perturbation $\hat{H}'$. Then, if the perturbation is quite small, ψ will not differ greatly from f, so that an approximation to the energy of the perturbed state may be obtained by substituting f for ψ in the integral expression for the energy, of the form given by (q.4.1.); thus, an approximate value for E is given by

$$E = \int f^*\hat{H}f \, d\tau \left[\int f^*f \, d\tau\right]^{-1} = \int f^*(\hat{H}^0 + \hat{H}')f \, d\tau \left[\int f^*f \, d\tau\right]^{-1}$$

$$= E^0 + \int f^*\hat{H}'f \, d\tau \left[\int f^*f \, d\tau\right]^{-1}, \tag{q.4.62.}$$

making use of (q.4.59.) and (q.4.61.). It is assumed, in the above expressions, that f if not normalized; if f is normalized, the denominators in the above expressions reduce to unity. Equation (q.4.62.) shows that the energy of the perturbed state, in terms of first order perturbation theory, is given by the sum of the energy of the unperturbed state and the energy term obtained by averaging the perturbation term over the unperturbed state of the system. Alternatively stated, the first order perturbation energy is

$$E' = \langle f|\hat{H}'|f\rangle. \tag{q.4.63.}$$

It may conceivably happen, depending on the specific nature of f and $\hat{H}'$, that E' is zero, in which event, second order perturbation theory has to be applied to find a non-zero correction energy to be added to the unperturbed energy to obtain the perturbed energy value.

If the unperturbed state E^0 is degenerate, there will be several eigenfunctions, $f_1, f_2, f_3, \ldots$, each corresponding to the same eigenvalue E^0, so that

$$\begin{aligned}
\hat{H}^0 f_1 &= E^0 f_1; \\
\hat{H}^0 f_2 &= E^0 f_2; \\
&\vdots \qquad \vdots \\
\hat{H}^0 f_\nu &= E^0 f_\nu.
\end{aligned} \tag{q.4.64.}$$

For simplicity of illustration, let it be assumed that E^0 is only doubly degenerate and that f_1 and f_2 are each normalized and orthogonal to one another. If this condition is not satisfied, it may be brought about by the method described on page 123. But, for any such pair of eigenfunctions, any linear combination of them is also an eigenfunction corresponding to the same eigenvalue, E^0; thus, if the first two equations of (q.4.64.) are multiplied by c_1 and c_2, respectively, and added, the following results

$$\hat{H}^0[c_1 f_1 + c_2 f_2] = E^0[c_1 f_1 + c_2 f_2], \tag{q.4.65.}$$

so that the linear combination function $[c_1 f_1 + c_2 f_2]$, is also an eigenfunction of the unperturbed system. If again it is assumed that to a first approximation the wave function of the perturbed state is not appreciably different from the linear combination function describing the unperturbed state, then

$$[\hat{H}^0 + \hat{H}'][c_1 f_1 + c_2 f_2] = [E^0 + E'][c_1 f_1 + c_2 f_2], \tag{q.4.66.}$$

where the perturbation energy is denoted by E'. If (q.4.65.) is substituted into (q.4.66.) it is found that

$$[\hat{H}' - E'][c_1 f_1 + c_2 f_2] = 0. \qquad \text{(q.4.67.)}$$

If equation (q.4.67.) is now multiplied successively by f_1^* and f_2^*, and each of the two resulting expressions integrated over the configuration space of the system, the two equations which are obtained are

$$\int f_1^*[\hat{H}' - E'][c_1 f_1 + c_2 f_2]\, d\tau = 0,$$

and

$$\int f_2^*[\hat{H}' - E'][c_1 f_1 + c_2 f_2]\, d\tau = 0, \qquad \text{(q.4.68.)}$$

which may be alternatively written as

$$c_1 \langle f_1 | \hat{H}' | f_1 \rangle + c_2 \langle f_1 | \hat{H}' | f_2 \rangle - E' c_1 \langle f_1 | f_1 \rangle - E' c_2 \langle f_1 | f_2 \rangle = 0,$$

and

$$c_1 \langle f_2 | \hat{H}' | f_1 \rangle + c_2 \langle f_2 | \hat{H}' | f_2 \rangle - E' c_1 \langle f_2 | f_1 \rangle - E' c_2 \langle f_2 | f_2 \rangle = 0. \qquad \text{(q.4.69.)}$$

Since f_1 and f_2 are assumed to be orthonormal functions, $\langle f_i | f_j \rangle = \delta_{ij}$, and the above equations reduce to

$$c_1[\langle f_1 | \hat{H}' | f_1 \rangle - E'] + c_2 \langle f_1 | \hat{H}' | f_2 \rangle = 0,$$

and

$$c_1 \langle f_2 | \hat{H}' | f_1 \rangle + c_2[\langle f_2 | \hat{H}' | f_2 \rangle - E'] = 0. \qquad \text{(q.4.70.)}$$

This pair of simultaneous equations can only have non-zero solutions for c_1 and c_2, if the determinant of the coefficients of c_1 and c_2 is zero, or

$$\begin{vmatrix} [\langle f_1 | \hat{H}' | f_1 \rangle - E'] & \langle f_1 | \hat{H}' | f_2 \rangle \\ \langle f_2 | \hat{H}' | f_1 \rangle & [\langle f_2 | \hat{H}' | f_2 \rangle - E'] \end{vmatrix} = 0. \qquad \text{(q.4.71.)}$$

This secular determinant for the perturbation corresponds to a quadratic equation in E', so that generally there will be two distinct roots. Thus the perturbation H' will generally remove the degeneracy of the unperturbed state. Equation (q.4.71.) may also be used to calculate the two values of the perturbed energy states directly, since it may be alternatively written in terms of the Hamiltonian and energy levels of the perturbed state, by virtue of the following relationships:

$$\langle f_1 | \hat{H}' | f_1 \rangle - E' = \langle f_1 | \hat{H} - \hat{H}^0 | f_1 \rangle - E'$$
$$= \langle f_1 | \hat{H} | f_1 \rangle - E^0 - E' = \langle f_1 | \hat{H} | f_1 \rangle - E$$

and

$$\langle f_1 | \hat{H}' | f_2 \rangle = \langle f_1 | \hat{H} - \hat{H}^0 | f_2 \rangle = \langle f_1 | \hat{H} | f_2 \rangle - E^0 \langle f_1 | f_2 \rangle = \langle f_1 | \hat{H} | f_2 \rangle,$$

which are a consequence of the orthonormality of f_1 and f_2. In an exactly similar manner, it follows that

$$\langle f_1|\hat{H}'|f_2\rangle = \langle f_1|\hat{H}|f_2\rangle \qquad \text{and} \qquad \langle f_2|\hat{H}'|f_1\rangle = \langle f_2|\hat{H}|f_1\rangle,$$

so that the secular equation (q.4.71.) may equally well be expressed as

$$\begin{vmatrix} \langle f_1|\hat{H}|f_1\rangle - E & \langle f_1|\hat{H}|f_2\rangle \\ \langle f_2|\hat{H}|f_1\rangle & \langle f_2|\hat{H}|f_2\rangle - E \end{vmatrix} = 0. \qquad \text{(q.4.72.)}$$

It is actually quite generally true that the secular determinant is of the same form whether expressed in terms of the Hamiltonian and energy of the perturbed state or in terms of the perturbation H' and the energy correction term E'.

The method described above for a doubly-degenerate unperturbed energy state E^0 is easily extended to apply to any degree of degeneracy. For example, if the unperturbed energy state E^0 is n-fold degenerate, an approximation to the wave function for the perturbed state may be written as

$$F = \sum_{i=1}^{n} c_i f_i, \qquad \text{(q.4.73.)}$$

assuming that the functions f_i constitute an orthonormal set, each corresponding to the energy E^0. Proceeding in the same manner, the secular determinant in this case is found to be an $n \times n$ determinant, such that

$$\begin{vmatrix} \langle f_1|\hat{H}'|f_1\rangle - E' & \langle f_1|\hat{H}'|f_2\rangle & \cdots & \langle f_1|\hat{H}'|f_n\rangle \\ \langle f_2|\hat{H}'|f_1\rangle & \langle f_2|\hat{H}'|f_2\rangle - E' & \cdots & \langle f_2|\hat{H}'|f_n\rangle \\ \vdots & \vdots & & \vdots \\ \langle f_n|\hat{H}'|f_1\rangle & \langle f_n|\hat{H}'|f_2\rangle & \cdots & \langle f_n|\hat{H}'|f_n\rangle - E' \end{vmatrix} = 0. \qquad \text{(q.4.74.)}$$

This constitutes an nth degree equation which may or may not have n distinct roots. In the event that there are not n distinct roots, then the degeneracy is not completely removed to the first order. This secular equation, as in the case of the doubly degenerate unperturbed state, may be expressed alternatively in terms of H and E, so that

$$\begin{vmatrix} \langle f_1|\hat{H}|f_1\rangle - E & \langle f_1|\hat{H}|f_2\rangle & \cdots & \langle f_1|\hat{H}|f_n\rangle \\ \langle f_2|\hat{H}|f_1\rangle & \langle f_2|\hat{H}|f_2\rangle - E & \cdots & \langle f_2|\hat{H}|f_n\rangle \\ \vdots & \vdots & & \vdots \\ \langle f_n|\hat{H}|f_1\rangle & \langle f_n|\hat{H}|f_2\rangle & & \langle f_n|\hat{H}|f_n\rangle - E \end{vmatrix} = 0. \qquad \text{(q.4.75.)}$$

On examination of equations (q.4.75.) and (q.4.57.), it is apparent that, apart from the different terminology, the two expressions are similar. In fact, if the linear combination functions (q.4.56.) had been chosen to be an orthonormal set, the two equations are identical; or alternatively, if the perturbation functions (q.4.73.) had not been chosen to be orthonormal, then (q.4.75.) would become identical with (q.4.57.). There is thus a great similarity between the variational approach and the perturbation method; however, while the variation method applies generally to the lowest energy state of a system and gives considerably poorer approximations to the higher energy states, the perturbation method applies to all the energy states of a system. Also, the variational functions are usually chosen arbitrarily, while in the perturbation method, the known eigenfunctions of the unperturbed state are used to approximate solutions for the perturbed state of a system.

If the determinant given by (q.4.75.) is expanded and written in descending powers of E, a result is obtained which is sometimes referred to as the diagonal-sum rule. This may be stated by saying that the sum of the terms, $\langle f_n|\hat{H}|f_n\rangle$, the diagonal matrix components of energy with respect to the unperturbed wave functions of the system, equals the sum of the eigenvalues of the perturbed system. Effectively, this means that the perturbations have spread the energy levels apart but have not altered their center of gravity. Thus, if the expansion of (q.4.75.) is written in descending powers of E, it is apparent that the highest and next highest terms in E are all derived from the products of terms lying on the principal diagonal of the determinant, so that

$$[-E]^n + [-E]^{n-1} \sum_{i=1}^{n} \langle f_i|\hat{H}|f_i\rangle + \cdots = 0. \tag{A}$$

But since any equation of the nth degree may be written in the form

$$C(E - E_1)(E - E_2)(E - E_3)\cdots(E - E_n) = 0,$$

where C is some constant and $E_1, E_2, \ldots, E_n$ are its roots, this equation may also be expanded and written in descending powers of E, to give

$$CE^n - CE^{n-1} \sum_{i=1}^{n} E_i + \cdots = 0. \tag{B}$$

If $C = (-1)^n$, then the terms in E^n in equations (A) and (B) become identical, in which event all terms in both equations must agree with one another, and specifically the terms in E^{n-1} must be the same. Therefore $\sum_{i=1}^{n} \langle f_i|\hat{H}|f_i\rangle = \sum_{i=1}^{n} E_i$, which is the diagonal-sum rule.

Q.4.6. Examples of First Order Perturbation Calculations

As examples of the application of first order perturbation theory to simple systems, the calculation of the non-degenerate ground-state energy of the helium atom, and the calculation of the effect of an imaginary perturbation on the energy of a multiply-degenerate state for a particle moving in a potential box, will be considered.

For the helium atom, the quantum Hamiltonian, expressed in atomic units, is given by (q.4.30.), and if this term is separated into two parts, as in the perturbation argument, so that the electron interaction term is regarded as a perturbation, then

$$\hat{H} = \hat{H}^0 + \hat{H}', \tag{q.4.76.}$$

where

$$\hat{H}^0 = -(1/2)[\nabla_1^2 + \nabla_2^2] - 2/r_1 - 2/r_2, \tag{q.4.77.}$$

and

$$\hat{H}' = 1/r_{12}. \tag{q.4.78.}$$

As has already been shown in Q.3.9., the Schrödinger equation for the unperturbed state, neglecting the electron interaction term, may be separated into two hydrogenic equations, the eigenfunctions and eigenvalues of which are known exactly. For the ground-state, in terms of this independent particle approximation, the ground-state wave function and ground-state energy are given by $\psi^0 = e^{-2(r_1 + r_2)}$ and $E^0 = -4$ Hartrees, respectively. Applying first order perturbation theory, making use of equation (q.4.62.), the ground-state energy of the perturbed state is

$$E = \iint \psi^{0*} \hat{H} \psi^0 \, d\mathbf{r}_1 \, d\mathbf{r}_2 \left[\iint \psi^{0*} \psi^0 \, d\mathbf{r}_1 \, d\mathbf{r}_2 \right]^{-1}$$

$$= E^0 + \iint e^{-4(r_1 + r_2)}(1/r_{12}) \, d\mathbf{r}_1 \, d\mathbf{r}_2 \left[\iint e^{-4(r_1 + r_2)} \, d\mathbf{r}_1 \, d\mathbf{r}_2 \right]^{-1}, \tag{q.4.79.}$$

since ψ^0 is real. The integrals in this expression for the ground-state energy of the perturbed state have already been evaluated, since they are similar to (q.4.38.) and (q.4.32.), respectively; in fact, if $\lambda = 2$ is substituted into these two latter expressions, the presently required integrals are found to be

$$\iint e^{-4(r_1 + r_2)}(1/r_{12}) \, d\mathbf{r}_1 \, d\mathbf{r}_2 = 5\pi^2/8(2^5)$$

and

$$\iint e^{-4(r_1 + r_2)} \, d\mathbf{r}_1 \, d\mathbf{r}_2 = \pi^2/2^6,$$

so that

$$E = E^0 + (5/4) = -4 + 1.25 = -2.75 \text{ Hartrees.} \tag{q.4.80.}$$

This is not quite such a good value as that calculated by the variation method, being about 5% higher than the experimental value of -2.905 Hartrees. The reason for this is quite apparent; the unperturbed ground-state wave function and the variation function given by (q.4.31.) are of the same type, but in the variation method the optimum value of the parameter λ was used, namely $\lambda = 27/16$, whereas in the first order perturbation calculation above, the effective value of $\lambda = 2$ was used. This simple example serves to illustrate the value of the perturbation method in providing information about the order of magnitude of the perturbation energy; in this case, the perturbation energy is $E' = E - E^0 = -2.75 + 4 = 1.25$ Hartrees. First order perturbation theory thus does not give a particularly good value for the ground state of the helium atom; it requires the application of higher order perturbations to do this. Scherr and Knight[1] have applied nth order perturbation corrections to the ground-state energy of helium and other two-electron systems by expanding the wave functions and energies of the corresponding Schrödinger equations in terms of the nuclear charge Z, finally obtaining a 13th order correction to the ground-state energy, leading to a calculated ground-state energy of -2.90372433 Hartrees, which is comparable to Pekeris's value (page 241), using a 1078-term variational function.

As an example of the application of first order perturbation theory to a system for which the unperturbed energy state is multiply-degenerate, it is convenient to consider the case of a particle in a box, when this system is subjected to a small perturbation. Suppose, specifically, that the triply-degenerate state, for which the unperturbed energy,

$$E^0 = 11h^2/8mL^2, \tag{q.4.81.}$$

is considered, and that this state is subjected to a small perturbing potential field in the y-Cartesian coordinate direction, and proportional to y^2, so that

$$\hat{H}' = ky^2. \tag{q.4.82.}$$

While this is an entirely artificial problem, it serves to illustrate the manner in which first order perturbation theory may be applied to a degenerate energy state of a system. From (q.3.4.), the quantum numbers n_x, n_y, and n_z may have the values $(1, 1, 3)$, $(1, 3, 1)$, and $(3, 1, 1)$, and the three unperturbed wave functions, from (q.3.3.), would be

$$\psi_1 = \sqrt{8/L^3}\, (\sin \pi x/L)(\sin \pi y/L)(\sin 3\pi z/L),$$

$$\psi_2 = \sqrt{8/L^3}\, (\sin \pi x/L)(\sin 3\pi y/L)(\sin \pi z/L), \tag{q.4.83.}$$

and

$$\psi_3 = \sqrt{8/L^3}\, (\sin 3\pi x/L)(\sin \pi y/L)(\sin \pi z/L).$$

---

[1] Scherr, C. W., and Knight, R. E., *Phys. Rev.*, 1962, **128**, 2675; *J. Chem. Phys.*, 1964, **40**, 3034.

These, of course, are normalized functions, and they are also orthogonal to one another since

$$\int \psi_1{}^*\psi_2 \, d\mathbf{r} = \int\!\!\!\int\!\!\!\int\limits_{0\ 0\ 0}^{\ \ L\ L\ L} \psi_1\psi_2 \, dx \, dy \, dz$$

$$= (8/L^3) \int_0^L (\sin^2 \pi x/L) dx \int_0^L (\sin \pi y/L)(\sin 3\pi y/L) dy$$

$$\times \int_0^L (\sin 3\pi z/L)(\sin \pi z/L \, dz).$$

$$= 0,$$

and similarly for the other possible pairs of functions (see pages 122–125). For a particle in a potential box, $U(x, y, z) = 0$, when x, y, or z is equal to 0 or L, and since the potential is infinite outside the box, the wave functions are zero there. The quantum Hamiltonian of the perturbed system is then

$$\hat{H} = \hat{H}^0 + \hat{H}' = (-h^2/8\pi^2 m) \, \nabla^2 + ky^2. \tag{q.4.84.}$$

Now, extending the method described on pages 246–251 to a triply degenerate case, the following secular equation is arrived at:

$$\begin{vmatrix} [\langle\psi_1|\hat{H}'|\psi_1\rangle - E'] & \langle\psi_1|\hat{H}'|\psi_2\rangle & \langle\psi_1|\hat{H}'|\psi_3\rangle \\ \langle\psi_2|\hat{H}'|\psi_1\rangle & [\langle\psi_2|\hat{H}'|\psi_2\rangle - E'] & \langle\psi_2|\hat{H}'|\psi_3\rangle \\ \langle\psi_3|\hat{H}'|\psi_1\rangle & \langle\psi_3|\hat{H}'|\psi_2\rangle & [\langle\psi_3|\hat{H}'|\psi_3\rangle - E'] \end{vmatrix} = 0,$$

where E' is the perturbation energy, to the first order. The various integrals in the above matrix are easily evaluated. Thus

$$\langle\psi_1|\hat{H}'|\psi_2\rangle = 8k/L^3 \int_0^L (\sin^2 \pi x/L) \, dx \int_0^L (y^2 \sin \pi y/L)(\sin 3\pi y/L) dy$$

$$\times \int_0^L (\sin 3\pi z/L)(\sin \pi z/L) \, dz$$

$$= 0,$$

and similarly,

$$\langle\psi_2|\hat{H}'|\psi_1\rangle = \langle\psi_1|\hat{H}'|\psi_3\rangle = \langle\psi_3|\hat{H}'|\psi_1\rangle = \langle\psi_3|\hat{H}'|\psi_2\rangle = \langle\psi_2|\hat{H}'|\psi_3\rangle = 0,$$

since they all contain integrals of the type $\int_0^L (\sin \pi q/L)(\sin 3\pi q/L) \, dq$, which is zero. Thus, all the off-diagonal elements which do not contain E' in the above secular determinant are zero. The secular equation for this system thus reduces to the product

$$[\langle\psi_1|\hat{H}'|\psi_1\rangle - E'][\langle\psi_2|\hat{H}'|\psi_2\rangle - E'][\langle\psi_3|\hat{H}'|\psi_3\rangle - E'] = 0,$$

and so the three roots of this equation give the three possible values for the perturbation energy. But

$$\langle\psi_1|\hat{H}'|\psi_1\rangle = [8k/L^3]\int_0^L\int_0^L\int_0^L \{(\sin^2 \pi x/L)(y^2 \sin^2 \pi y/L)(\sin^2 3\pi z/L)\}\, dx\, dy\, dz$$

$$= [8k/L^3]\int_0^L (\sin^2 \pi x/L)\, dx \int_0^L (y^2 \sin^2 \pi y/L)\, dy \int_0^L (\sin^2 3\pi z/L)\, dz,$$

and

$$\int_0^L (\sin^2 \pi x/L)\, dx = \int_0^L (1/2)[1 - \cos 2\pi x/L]\, dx = L/2 = \int_0^L (\sin^2 3\pi z/L)\, dz;$$

also

$$\int_0^L (y^2 \sin^2 \pi y/L)\, dy = (1/2)\int_0^L y^2[1 - \cos 2\pi y/L]\, dy = L^3/6 - L^3/4\pi^2$$

$$= L^3[1/6 - 1/4\pi^2].$$

Therefore

$$\langle\psi_1|\hat{H}'|\psi_1\rangle = [8k/L^3][L/2][L/2]L^3[1/6 - 1/4\pi^2] = kL^2[1/3 - 1/2\pi^2]. \quad \text{(q.4.85.)}$$

Similarly,

$$\langle\psi_3|\hat{H}'|\psi_3\rangle = [8k/L^3]\int_0^L (\sin^2 3\pi x/L)\, dx \int_0^L (y^2 \sin^2 \pi y/L)\, dy \int_0^L (\sin^2 \pi z/L)\, dz$$

$$= kL^2[1/3 - 1/2\pi^2] = \langle\psi_1|\hat{H}'|\psi_1\rangle. \quad \text{(q.4.86.)}$$

The remaining integral

$$\langle\psi_2|\hat{H}'|\psi_2\rangle = [8k/L^3]\int_0^L (\sin^2 \pi x/L)\, dx \int_0^L (y^2 \sin^2 3\pi y/L)\, dy \int_0^L (\sin^2 \pi z/L)\, dz$$

$$= [8k/L^3][L/2][L^3(1/6 - 1/36\pi^2)][L/2]$$

$$= kL^2[1/3 - 1/18\pi^2]. \quad \text{(q.4.87.)}$$

It is thus found that two of the roots of the secular equation for this system are equal, namely those given by (q.4.85.) and (q.4.86.), while the third root is given by (q.4.87.), indicating that the three-fold degeneracy of the unperturbed state is not completely removed to the first order, one energy level remaining doubly-degenerate. This doubly-degenerate perturbed energy value is given by

$$E_1 = 11h^2/8mL^2 + kL^2[1/3 - 1/2\pi^2],$$

and the second non-degenerate state by

$$E_2 = 11h^2/8mL^2 + kL^2[1/3 - 1/18\pi^2].$$

Q.4.7. General Perturbation Theory

In principle, perturbation theory and the variational method ultimately lead to the same result, although in practice there may be some particular reason for preferring one method rather than the other. General perturbation theory may be developed so that no approximations are involved in the general statement of the theory, but this leads to secular determinants of an infinite order, which therefore require an infinite amount of effort for their evaluation, so that again in practice, it is necessary to be satisfied with an approximation of some type. However, techniques of perturbation theory have been developed, which for all practical purposes, enable perturbation problems to be solved essentially exactly. In addition to the reference given on page 246, some of these methods are described in detail by Byers Brown,[1] Löwdin,[2] Dalgarno and Stewart,[3] and by Sinanoglu.[4] In this section, it will be shown how second order and higher order perturbation corrections to the energy of a system may be obtained from the point of view of the general principles involved.

If it is assumed that all of the eigenfunctions f_i and eigenvalues $E_i{}^0$ of the unperturbed system are known, or that the equation

$$\hat{H}^0 f_i = E_i{}^0 f_i \qquad \text{(q.4.88.)}$$

can be completely solved, then the problem in perturbation theory is to find the eigenfunctions ψ_i and the eigenvalues E_i of the equation

$$\hat{H}\psi_i = E_i\psi_i \qquad \text{where} \qquad \hat{H} = \hat{H}^0 + \hat{H}'. \qquad \text{(q.4.89.)}$$

If the various energy states of the unperturbed system $E_i{}^0$ are all assumed to be discrete and the wave functions f_i to form a complete orthonormal set, then the exact statement can be made that any wave function of the perturbed state can be expressed in the form

$$\psi = \sum_{i=1}^{\infty} c_i f_i, \qquad \text{(q.4.90.)}$$

provided the coefficients c_i are appropriately chosen. If this function is substituted into the wave equation for the perturbed state, given by (q.4.88.), then

$$\sum_{i=1}^{\infty} c_i \hat{H} f_i = E \sum_{i=1}^{\infty} c_i f_i, \qquad \text{(q.4.91.)}$$

---

[1] Brown, W. Byers, *J. Chem. Phys.*, 1966, **44**, 567.
[2] Löwdin, P. O., *Studies in Perturbation Theory, XI, J. Chem. Phys.*, 1965, **S175**, 43; the complete list of this series of publications is given in the Bibliography of John C. Slater's *Quantum Theory of Matter*, 2nd edition, 1968, McGraw-Hill Book Company, New York.
[3] Dalgarno, A., and Stewart, A. L., *Proc. Roy. Soc.*, 1956, **A160**, 94.
[4] Sinanoglu, O., *Phys. Rev.*, 1961, **122**, 493.

which on multiplication through by f_j^* and integration throughout the configuration space of the system leads to

$$\sum_{i=1}^{\infty} c_i \int f_j^* \hat{H} f_i \, d\tau = E \sum_{i=1}^{\infty} c_i \int f_j^* f_i \, d\tau. \qquad \text{(q.4.92.)}$$

Expressing the orthonormality condition as

$$\int f_i^* f_j \, d\tau = \delta_{ij}, \qquad \text{(q.4.93.)}$$

it is then possible to write (q.4.92.) as

$$\sum_{i=1}^{\infty} c_i[\langle f_j | \hat{H} | f_i \rangle - E\delta_{ji}] = 0, \qquad \text{(q.4.94.)}$$

or as the infinite series of equations

$$c_1[\langle f_1 | \hat{H} | f_1 \rangle - E] + c_2\langle f_1 | \hat{H} | f_2 \rangle + c_3\langle f_1 | \hat{H} | f_3 \rangle + \cdots = 0 \ ,$$
$$c_1\langle f_2 | \hat{H} | f_1 \rangle + c_2[\langle f_2 | \hat{H} | f_2 \rangle - E] + c_3\langle f_2 | \hat{H} | f_3 \rangle + \cdots = 0 \ , \qquad \text{(q.4.95.)}$$
$$c_1\langle f_3 | \hat{H} | f_1 \rangle + c_2\langle f_3 | \hat{H} | f_2 \rangle + c_3[\langle f_3 | \hat{H} | f_3 \rangle - E] + \cdots = 0 \ ,$$
$$\cdots\cdots\cdots\cdots\cdots\cdots\cdots\cdots\cdots\cdots\cdots\cdots\cdots\cdots\cdots\cdots\cdots .$$

in which each equation has an infinite number of terms.

The condition that these equations should have solutions corresponding to non-zero values of the coefficients c_i is that the determinant of the coefficients of the c_i's should vanish, or

$$\begin{vmatrix} [\langle f_1 | \hat{H} | f_1 \rangle - E] & \langle f_1 | \hat{H} | f_2 \rangle & \langle f_1 | \hat{H} | f_3 \rangle & \cdots \\ \langle f_2 | \hat{H} | f_1 \rangle & [\langle f_2 | \hat{H} | f_2 \rangle - E] & \langle f_2 | \hat{H} | f_3 \rangle & \cdots \\ \langle f_3 | \hat{H} | f_1 \rangle & \langle f_3 | \hat{H} | f_2 \rangle & [\langle f_3 | \hat{H} | f_3 \rangle - E] & \cdots \\ \vdots & \vdots & \vdots & \end{vmatrix} = 0. \qquad \text{(q.4.96.)}$$

The above expression is identical with that which would result by the use of the linear variation method with a trial function of the form of (q.4.90.). The fact that this is a determinant of an infinite order makes it strictly insolvable with any finite amount of effort. It should also be noted that a similar expression would result, by the use of either the perturbation approach or the variational method, if any complete set of orthonormal functions in (q.4.90.) were to be used, whether or not this set of functions constitutes wave functions of the unperturbed state. In the variation method, however, the set of functions describing the unperturbed state is used. Since the functions f_i are eigenfunctions of the operator $\hat{H}^0$ of the unperturbed system, then

$$\hat{H}_{ij}{}^0 = \int f_i^* \hat{H}^0 f_j \, d\tau = E^0 \int f_i^* f_j \, d\tau = E^0 \, \delta_{ij}, \qquad \text{(q.4.97.)}$$

and since $E = E + E'$, then as on page 250,

$$\hat{H}_{ij} - E\,\delta_{ij} = \hat{H}^0_{ij} + \hat{H}'_{ij} - E^0\,\delta_{ij} - E'\,\delta_{ij} = \hat{H}'_{ij} - E'\,\delta_{ij}, \qquad \text{(q.4.98.)}$$

so that the secular determinant is generally of the same form, whether expressed in terms of H and E or H' and E'.

If the perturbation energy were identically zero, so that the functions f_i were exact eigenfunctions of $\hat{H}$, all of the non-diagonal elements of (q.4.96.) would be zero, for then

$$\hat{H}_{ij} = \int f_i^*\hat{H}f_j\,d\tau = E_i \int f_i^*f_j\,d\tau = 0 \qquad \text{for all} \qquad i \neq j. \quad \text{(q.4.99.)}$$

Actually, (q.4.96.) is true, irrespective of the magnitude of $\hat{H}'$. At any rate, for $E' = 0$, (q.4.96.) would reduce to a diagonal matrix, or become equal to

$$[\langle f_1|\hat{H}|f_1\rangle - E][\langle f_2|\hat{H}|f_2\rangle - E][\langle f_3|\hat{H}|f_3\rangle - E]\cdots = 0, \qquad \text{(q.4.100.)}$$

with roots given by $\langle f_1|\hat{H}|f_1\rangle$, $\langle f_2|\hat{H}|f_2\rangle$, $\langle f_3|\hat{H}|f_3\rangle$,.....

The perturbed energy values given by the roots of equation (q.4.100.), namely,

$$E_1 = \langle f_1|\hat{H}|f_1\rangle;\ E_2 = \langle f_2|\hat{H}|f_2\rangle;\ E_3 = \langle f_3|\hat{H}|f_3\rangle;\ E_4 = \cdots, \qquad \text{(q.4.101.)}$$

although only true if the perturbation were exactly zero, would obviously constitute approximate solutions if the perturbation were small. In other words, for assumed small values of the perturbation energy, the perturbed energy states are given by averaging the Hamiltonian of the perturbed system over the wave functions of the unperturbed states. This, of course, is the same result as already obtained by first order perturbation theory as given by (q.4.63.), since, as already shown, these results may be alternatively expressed in terms of H and E or H' and E'. The foregoing applies to non-degenerate unperturbed states. In the event that any of the unperturbed states are degenerate, (q.4.101.) may not necessarily give particularly good values for the perturbed states, corresponding to these degenerate energy values, due to the fact that there is essentially an infinite choice of orthonormal sets of functions for a degenerate state, and any one of these would lead to an equation of the type of (q.4.96.). But not all of them would make the non-diagonal elements of the secular determinant equally small in value. This means that, even although the unperturbed states are degenerate, (q.4.101.) would still give rather approximate values for the perturbed states but would probably not provide very useful information about the separation or splitting of the degeneracies in the perturbed system. Such a situation may be avoided by retaining the non-diagonal elements which are derived from the degenerate unperturbed state functions, so that all terms of possible consequence in determining the perturbed energy states are included, but setting all other non-diagonal elements equal to zero. For example, suppose that of the unperturbed state functions, f_i, f_1, and f_2, correspond to the same unperturbed energy, so that $E_1{}^0 = E_2{}^0$, and that all of the other $E_i{}^0$ values are non-degenerate. Then setting all non-diagonal elements in the secular determinant equal to zero except those containing f_1 and f_2, there would result instead of (q.4.100.) the expression

$$\begin{vmatrix} [\langle f_1|\hat{H}|f_1\rangle - E] & \langle f_1|\hat{H}|f_2\rangle & 0 & \cdots \\ \langle f_2|\hat{H}|f_1\rangle & [\langle f_2|\hat{H}|f_2\rangle - E] & 0 & \cdots \\ 0 & 0 & [\langle f_3|\hat{H}|f_3\rangle - E] & \cdots \\ \vdots & \vdots & \vdots & \end{vmatrix} = 0,$$

which may be factored into

$$\begin{vmatrix} [\langle f_1|\hat{H}|f_1\rangle - E] & \langle f_1|\hat{H}|f_2\rangle \\ \langle f_2|\hat{H}|f_1\rangle & [\langle f_2|\hat{H}|f_2\rangle - E] \end{vmatrix}$$
$$\times [\langle f_3|\hat{H}|f_3\rangle - E][\langle f_4|\hat{H}|f_4\rangle - E]\cdots = 0. \quad \text{(q.4.102.)}$$

The 2×2 determinant in (q.4.102.), if set equal to zero, provides the perturbed energies corresponding to the degenerate unperturbed states $E_1{}^0 = E_2{}^0$, and is identical to the equation (q.4.72.) already found for two-fold degeneracy in the unperturbed system. The other perturbed energy levels are given by $E_3 = \langle f_3|\hat{H}|f_3\rangle$, $E_4 = \langle f_4|\hat{H}|f_4\rangle$,.... For n-fold degeneracy in any unperturbed state, where $E_1{}^0 = E_2{}^0 = \cdots = E_n{}^0$, it is apparent that (q.4.75.) would appear in (q.4.102.) instead of simply a 2×2 determinant. This result is perfectly general, irrespective of where the degeneracies occur among the set of unperturbed energy states $E_i{}^0$. Also, this same argument would be equally applicable if any subset $E_p{}^0$, of the complete set $E_i{}^0$, of unperturbed energy states were not completely degenerate but of very similar value. Then, a better approximation to the perturbed energy states would be obtained by retaining those non-diagonal elements in the secular determinant which were derived from the "almost" degenerate sub-set of states.

To find the second-order perturbation correction for a non-degenerate unperturbed state, it is convenient to make the substitution of $\langle f_1|\hat{H}|f_1\rangle$ for E, which has already been found to constitute a first approximation to E_1, into every diagonal element of (q.4.96.) except the first one, which may now be written as $\langle f_1|\hat{H}|f_1\rangle - E_1$. Of the non-diagonal elements in (q.4.96.), the only ones which contain f_1 are the elements of the first row and the first column; all the other non-diagonal elements, which only affect E_1 indirectly, may be set equal to zero, to obtain

$$\begin{vmatrix} [\langle f_1|\hat{H}|f_1\rangle - E_1] & \langle f_1|\hat{H}|f_2\rangle & \langle f_1|\hat{H}|f_3\rangle & \langle f_1|\hat{H}|f_4\rangle & \cdots \\ \langle f_2|\hat{H}|f_1\rangle & [\langle f_2|\hat{H}|f_2\rangle - \langle f_1|\hat{H}|f_1\rangle] & 0 & 0 & \cdots \\ \langle f_3|\hat{H}|f_1\rangle & 0 & [\langle f_3|\hat{H}|f_3\rangle - \langle f_1|\hat{H}|f_1\rangle] & 0 & \cdots \\ \langle f_4|\hat{H}|f_1\rangle & 0 & 0 & [\langle f_4|\hat{H}|f_4\rangle - \langle f_1|\hat{H}|f_1\rangle] & \cdots \\ \vdots & \vdots & \vdots & \vdots & \end{vmatrix} = 0,$$

as the equation which has to be solved to obtain the perturbed eigenvalue E_1, corrected for second order perturbation effects. To diagonalize the above expression, advantage is taken of the property of determinants that their value is unchanged, if a constant multiple of any row (or column) is added to (or subtracted from) any row (or column). Thus the element at the intersection of the first row and second column may be eliminated by multiplying the second row by $\langle f_1|\hat{H}|f_2\rangle[\langle f_2|\hat{H}|f_2\rangle - \langle f_2|\hat{H}|f_1\rangle]^{-1}$, and subtracting it from the first row. As a result of this operation, the first diagonal element becomes

$$\langle f_1|\hat{H}|f_1\rangle - E_1 - \frac{[\langle f_1|\hat{H}|f_2\rangle][\langle f_2|\hat{H}|f_1\rangle]}{[\langle f_2|\hat{H}|f_2\rangle - \langle f_1|\hat{H}|f_1\rangle]},$$

the first element in the second column becomes zero, and every other element remains unchanged. The first element in the third column may be similarly eliminated, by multiplying the third row by

$$\frac{\langle f_1|\hat{H}|f_3\rangle}{[\langle f_3|\hat{H}|f_3\rangle - \langle f_1|\hat{H}|f_1\rangle]},$$

and again subtracting from the first row. This time the first element of the third column becomes zero, the first diagonal element becomes

$$\langle f_1|\hat{H}|f_1\rangle - E_1 - \frac{[\langle f_1|\hat{H}|f_2\rangle][\langle f_2|\hat{H}|f_1\rangle]}{[\langle f_2|\hat{H}|f_2\rangle - \langle f_1|\hat{H}|f_1\rangle]} - \frac{[\langle f_1|\hat{H}|f_3\rangle][\langle f_3|\hat{H}|f_1\rangle]}{[\langle f_3|\hat{H}|f_3\rangle - \langle f_1|\hat{H}|f_1\rangle]},$$

and all the other elements remain unchanged. If this procedure is repeated, eliminating the first element of successive columns each time, eventually all the non-diagonal elements of the first row are eliminated, and each time this is done a term of the type

$$- \frac{[\langle f_1|\hat{H}|f_n\rangle][\langle f_n|\hat{H}|f_1\rangle]}{[\langle f_n|\hat{H}|f_n\rangle - \langle f_1|\hat{H}|f_1\rangle]}$$

is added to the first element; thus, for a secular determinant of infinite order, the first element finally becomes

$$\langle f_1|\hat{H}|f_1\rangle - E_1 - \sum_{n=2}^{\infty} \frac{[\langle f_1|\hat{H}|f_n\rangle][\langle f_n|\hat{H}|f_1\rangle]}{[\langle f_n|\hat{H}|f_n\rangle - \langle f_1|\hat{H}|f_1\rangle]}.$$

and the determinant may be expanded as

$$\left[\langle f_1|\hat{H}|f_1\rangle - E_1 - \sum_{n=2}^{\infty} \frac{[\langle f_1|\hat{H}|f_n\rangle][\langle f_n|\hat{H}|f_1\rangle]}{[\langle f_n|\hat{H}|f_n\rangle - \langle f_1|\hat{H}|f_1\rangle]}\right]$$
$$\times [\langle f_2|\hat{H}|f_2\rangle - \langle f_1|\hat{H}|f_1\rangle][\langle f_3|\hat{H}|f_3\rangle - \langle f_1|\hat{H}|f_1\rangle] = 0. \quad \text{(q.4.103.)}$$

Now, since f_1 is assumed to be non-degenerate, none of the factors in (q.4.103.) can be zero, except the first one. Therefore

$$E_1 = \langle f_1|\hat{H}|f_1\rangle - \sum_{n=2}^{\infty} \frac{[\langle f_1|\hat{H}|f_n\rangle][\langle f_n|\hat{H}|f_1\rangle]}{[\langle f_n|\hat{H}|f_n\rangle - \langle f_1|\hat{H}|f_1\rangle]}. \quad \text{(q.4.104.)}$$

This expression gives the perturbed energy state E_1, such that the first term on the right-hand side of equation (q.4.104.) is the first order perturbation correction and the second term, the summation term, gives the second order perturbation correction. In principle, this process may be continued, including additional non-diagonal elements in (q.4.96.), until the whole perturbation series is essentially summed to infinity. However, this procedure is seldom used beyond the second order perturbation correction term; instead, Dalgarno's or one of the methods developed and described in the references on page 256 is used.

Equation (q.4.104.), which is specifically applicable to the first perturbed energy state of a system, may be easily generalized to any perturbed energy state E_i, corresponding to any non-degenerate unperturbed state E_i^0. Thus

$$E_i = \langle f_i|\hat{H}|f_i\rangle - \sum_{j\neq i}^{\infty} \frac{[\langle f_i|\hat{H}|f_j\rangle][\langle f_j|\hat{H}|f_i\rangle]}{[\langle f_j|\hat{H}|f_j\rangle - \langle f_i|\hat{H}|f_i\rangle]}. \quad \text{(q.4.105.)}$$

This expression may be simplified somewhat, by noting that the difference between two first order correction terms, $[\langle f_j|\hat{H}'|f_j\rangle - \langle f_i|\hat{H}'|f_i\rangle]$, would be expected to be very small compared to the difference between the two corresponding unperturbed energy states, $E_j^0 - E_i^0$, and since

$$\langle f_j|\hat{H}|f_j\rangle - \langle f_i|\hat{H}|f_i\rangle = E_j^0 - E_i^0 + \langle f_j|\hat{H}'|f_j\rangle - \langle f_i|\hat{H}'|f_i\rangle \simeq E_j^0 - E_i^0,$$

to a fairly good approximation, (q.4.105.) may be written as

$$E_i = \langle f_i|\hat{H}|f_i\rangle - \sum_{j\neq i}^{\infty} \frac{[\langle f_i|\hat{H}|f_j\rangle][\langle f_j|\hat{H}|f_i\rangle]}{E_j^0 - E_i^0}. \quad \text{(q.4.106.)}$$

Alternatively, since

$$\langle f_i|\hat{H}|f_j\rangle = \langle f_i|\hat{H}^0 + \hat{H}'|f_j\rangle = E_j^0\langle f_i|f_j\rangle + \langle f_i|\hat{H}'|f_j\rangle = E_j^0\,\delta_{ij} + \langle f_i|\hat{H}'|f_j\rangle,$$

and $\hat{H}$ is a Hermitian operator, so that

$$\langle f_i|\hat{H}|f_j\rangle = [\langle f_j|\hat{H}|f_i\rangle]^* \quad \text{or} \quad [\langle f_i|\hat{H}|f_j\rangle][\langle f_j|\hat{H}|f_i\rangle] = |\langle f_i|\hat{H}|f_j\rangle|^2,$$

then by substituting these relationships into (q.4.106.), the value of any perturbed energy state E_i may be expressed directly in terms of the perturbation $\hat{H}'$ as

$$E_i = E_i^0 + \langle f_i|\hat{H}'|f_i\rangle - \sum_{i\neq j}^{\infty} \frac{|\langle f_i|\hat{H}'|f_j\rangle|^2}{E_j^0 - E_i^0}. \qquad \text{(q.4.107.)}$$

which is particularly convenient, since it expresses the perturbed energy as the sum of the unperturbed energy, and the first and second order perturbation energy corrections, each in terms of the perturbation $\hat{H}'$.

Q.4.8.   The Self-Consistent Field Method

For polyelectronic atoms, a fundamental difficulty in obtaining solutions of the Schrödinger equation for such systems is imposed by the fact that each electron interacts with every other electron in the system, so that the motion of any one electron is dependent on the motion of all of the others. When there are only two electrons, as in the helium atom or in the mono-positively charged lithium ion or the di-positively charged beryllium ion, the very elaborate approximations given by the methods of Hylleraas and others (pp. 240, 253) give results for the calculated energies agreeing very closely with experimental observation. But those methods which work for helium are not applicable to systems involving more than two electrons; however, methods have been devised which make use of approximation methods giving good qualitative agreement and fairly good quantitative agreement with experimentally observed results. As indicated on page 226, while it is always a simple matter to write down the Schrödinger equation for any system, the real complexity of the problem involving polyelectronic systems is that both the potential function and the possible eigenfunctions of the system are functions of all of the coordinates of all of the particles composing the system. Of the several methods which have been proposed to simplify this problem, one of the most popular and most successful is a method, usually referred to as the self-consistent field method, originally suggested by D. R. Hartree.[1,2]

---

[1] Hartree, D. R., *Proc. Cambridge Phil. Soc.*, 1928, **24**, 89, 111, 426.
[2] Hartree, D. R., *The Calculation of Atomic Structures*, 1957, John Wiley & Sons, Inc., New York.

The starting point for the Hartree self-consistent field method is the assumption that each electron moves in a central, or spherically symmetrical, force field, produced by the nucleus of an atom and the other electrons. Such an assumption greatly simplifies the problem, as each electron may then be dealt with by ignoring its instantaneous reaction with the other electrons and may then also be considered to move in an effective potential field which is obtained by some suitable average over all positions of all of the other electrons. Each electron is then described by a function which involves only the coordinates of that electron and not of any other. The assumption that the average potential due to the nucleus and other electrons in an atom has spherical symmetry enables the behavior of any specific electron to be described approximately by hydrogenic wave functions. This approximation is actually sufficiently good to account for the general size and energy of atoms and for the periodic variation in the properties of atoms of the various elements as a function of their nuclear charge. The self-consistent field concept has a plausible physical basis in terms of electron charge cloud distributions and, as will be shown below, may be justified by and derived from the variation principle.

In the application of the Hartree self-consistent field method, the difficulty that the wave equation for any particular electron can only be written down if the wave functions for all of the other electrons are known is overcome by the use of an iterative technique; the wave functions of all of the other electrons must be known in order to obtain their electronic charge densities, which then have to be averaged to describe the potential field in which the one electron under consideration moves. In fact, initially, any plausible trial functions for each of the electrons may be selected and any single electron chosen, with respect to which the average field due to the others is found. The wave equation for this particular electron is then solved, assuming that the potential field in which it moves has spherical symmetry, giving a first improved wave function for this electron. This new function is then used to calculate the average field for a second electron in the assembly under consideration, enabling a first improved wave function for this second electron to be obtained. By repeating this procedure, a complete set of first improved wave functions may be obtained, and making use of these a second improved set of functions may be obtained, one by one, and by successive iteration this process may be continued until a set of functions results which is not appreciably altered on going through the process once again. The final set of atomic orbitals are then said to be self-consistent.

The wave functions of electrons in such assumed central fields are very similar to the hydrogenic functions given by (q.3.131.), and in fact only the radial part of the total function, $R(r)$, depends on the particular form of the central field (page 179), the product of the two angular functions being given by (q.3.55.), as in the case of the hydrogen atom. Self-consistent field type calculations have, of course, been programmed for execution on digital computers, and one of the best known of such programs is that of Herman and Skillman,[1] who have tabulated the wave functions of the various electrons in all atoms as well as the eigenvalues of the corresponding Schrödinger equations. Complete self-consistent field type calculations necessarily have to take into account the concept of electron spin and the applicability of the Pauli Exclusion Principle to the description of the electrons in polyelectronic atoms. The quantum mechanical interpretation of the Pauli Principle will be dealt with in the following chapter, after some discussion of the concept of spin wave functions. Meanwhile, in order to give some quantitative expression to the Hartree method, it is necessary to examine what is referred to as the one-electron approximation for the description of many-electron atoms.

Q.4.9.   The One-Electron Approximation Method

All atoms and ions may be regarded as species with a nuclear charge $+Z\varepsilon$ containing N electrons; if the nuclear charge number Z is equal to N, the species is a neutral atom, and if N is greater than or less than Z, then the species is an anion or a cation, respectively. For any atom or ion with a nuclear charge greater than $+2\varepsilon$, the fixed nucleus approximation, in describing the associated electronic motion, is even more justifiable than for helium or hydrogen. If, in an atomic system containing N electrons, the radial distance of the ith electron from the nucleus is r_i and the distance of the ith from the jth electron is denoted by r_{ij}, then the potential energy of the ith electron due to the nuclear field is $-Z\varepsilon^2/r_i$, and the mutual potential energy of the ith and jth electrons, by virtue of their Coulomb, electrostatic interaction is ε^2/r_{ij}. The total potential energy of the N electrons will be given by the sum of all such attractive and repulsive energy terms. The former is given by $-\sum_{i=1}^{N} Z\varepsilon^2/r_i$, and the latter by either of the double summations,

$$(1/2) \sum_{i \neq j}^{N} \sum^{N} \varepsilon^2/r_{ij} \qquad \text{or} \qquad \sum_{i > j}^{N} \sum^{N} \varepsilon^2/r_{ij},$$

without the factor of one-half. Frequently, such double sums and more generally multiple sums are represented by a single summation sign; i and j obviously cannot be the same, since any particular electron does not repel itself, and if the sum is written, as in the first alternative representation, this would include the repulsion between the ith and the jth electron and that

---

[1] Herman, F., and Skillman, S., *Atomic Structure Calculations*, 1963, Prentice-Hall, Inc., Englewood Cliffs, New Jersey.

between the jth and the ith electron as two separate terms of the sum, so that the factor of one-half has to be taken into account. On the other hand, if the double sum is taken over all values of i (or j) and over all those values of j (or i) which are greater than i (or j), then the factor of one-half is unnecessary; this may be easily checked by considering some small finite number of repulsive interactions between electrons, such as three or four. The potential energy of the collection of electrons is generally assumed to be zero when they are infinitely far apart, and if only the electrostatic interaction terms are considered (in any event, the magnetic interaction between electrons is of a relatively small order of magnitude in comparison), then the total potential energy of the assembly of N electrons is

$$U(r_1, r_2, r_3, \ldots, r_n) = - \sum_{i=1}^{N} Z\varepsilon^2/r_i + \sum_{i>j}^{N}\sum^{N} \varepsilon^2/r_{ij}, \qquad (q.4.108.)$$

where $r_1, r_2, \ldots, r_n$ are the positional vectors of the various electrons. The Hamiltonian operator for the system is then, as on page 226,

$$\hat{H} = - \sum_{i=1}^{N} [h^2/8\pi^2 m]\, \nabla_i^2 - \sum_{i=1}^{N} Z\varepsilon^2/r_i + \sum_{i>j}^{N}\sum^{N} \varepsilon^2/r_{ij}$$

$$= - \sum_{i=1}^{N} [(h^2/8\pi^2 m)\, \nabla_i^2 + Z\varepsilon^2/r_i] + \sum_{i>j}^{N}\sum^{N} \varepsilon^2/r_{ij}, \qquad (q.4.109.)$$

and the corresponding Schrödinger equation is

$$\hat{H}\Psi(r_1, r_2, \ldots, r_n) = E\Psi(r_1, r_2, \ldots, r_n), \qquad (q.4.110.)$$

where the wave function Ψ of the system of N electrons is a function of the positional coordinates of all of the N electrons. If it were not for the last term in the Hamiltonian (q.4.109.), that is, if the electron interactions were neglected, then, as in the case of the helium atom equation as given by (q.3.140.), the wave equation (q.4.110.) could be written as

$$- \sum_{i=1}^{N} [(h^2/8\pi^2 m)\, \nabla_i^2 + Z\varepsilon^2/r_i]\psi_1\psi_2\cdots\psi_N = E\psi_1\psi_2\psi_3\cdots\psi_N, \qquad (q.4.111.)$$

with the wave function for the N-electron system written as the product function of N separate one-electron functions, that is,

$$\Psi(r_1, r_2, \ldots, r_N) = \psi_1(r_1)\psi_2(r_2)\cdots\psi_N(r_N). \qquad (q.4.112.)$$

In this event, each of the equations involved in the sum given by (q.4.111.) would be of the hydrogenic type and could be solved analytically. This is easily seen, by dividing equation (q.4.111.) by $\Psi(r_1, r_2, \ldots, r_N)$, when there results

$$- \sum_{i=1}^{N} [1/\psi_i][(h^2/8\pi^2 m)\, \nabla_i^2 + Z\varepsilon^2/r_i]\psi_i = E, \qquad (q.4.113.)$$

in which each term of the sum is dependent on the coordinates of one electron only, so that (q.4.113.) may be separated into the N equations,

$$-[(h^2/8\pi^2 m)\,\nabla_i^2 + Z\varepsilon^2/r_i]\psi_i = E_i\psi_i, \qquad \text{(q.4.114.)}$$

with $i = 1, 2, 3, \ldots, N$, and with $\sum_{i=1}^{N} E_i = E$.

However, as has already been demonstrated, if the electron interaction term is neglected, even for the two-electron system, helium, the ground-state energy calculated in this manner turns out to be about 40% too low, and the error increases with N. The use of the variational method or the perturbation method to calculate the energies of polyelectronic atoms, with reasonable accuracy, is not particularly useful; the variational function for helium, which gives good agreement with experimental observation, is not by any means a simple function, and for N greater than 2 or 3 the variational trial function required would necessarily be even more elaborate. The first order perturbation calculation of the ground state of helium differs from the experimentally observed value by about 5%, and again even poorer approximate values for the ground-state energies of atoms with Z greater than 2 would be expected by the use of this method; second order and higher order perturbation calculations become difficult because of the problem of convergence of the infinite series involved. It has to be admitted that even an approximate solution of equation (q.4.110.), in analytical terms, cannot be realized. Neither is straightforward numerical tabulation of the values of a function of many variables, even with the aid of digital computers, and calculating the values of the function at a reasonable number of points in configuration space, a feasible proposition. Even for the relatively light fluorine atom, this would require tabulating the values of a function of 27 variables. The so-called, one-electron approximation, in terms of which each electron in an atom is described by a separate wave function, provides a relatively simple solution to such a complicated problem.

A one-electron wave function is assumed to be a function of the coordinates of the one electron which it describes, as would be the case if (q.4.114.) were applicable. Such one-electron functions may be calculated if it is assumed that each electron in a polyelectronic system behaves as a single electron moving in the field of a nucleus plus some additional spherically symmetrical potential field, representing the average effect of all of the other electrons, satisfying an equation of the type

$$[-(h^2/8\pi^2 m)\,\nabla_i^2 - Z\varepsilon^2/r_i + U_i(r_i)]\psi_i = E_i\psi_i, \qquad \text{(q.4.115.)}$$

where $U_i(r_i)$ is the potential function which approximately describes the interaction of the ith electron in the system with all of the other electrons. The fundamental assumption of the self-consistent field approximation that $U_i(r_i)$ has spherical symmetry is a very plausible one for atoms, and although (q.4.115.) is not a hydrogenic equation, it is an equation of motion of a particle in a central field, which is soluble, provided a suitable form can be found for $U_i(r_i)$. If $U_i(r_i)$ were identically zero, then of course equation (q.4.115.) would reduce to (q.4.114.), or if $U_i(r_i)$ were proportional to the reciprocal of r_i, (q.4.115.) would correspond to a hydrogenic equation. This, however, is not generally so for polyelectronic atoms, and there would be N equations like (q.4.115.), one for each electron, for a system containing N electrons. In the application of the one-electron approximation method to a polyelectronic atom, there would be N equations like (q.4.115.) to solve and, having solved these, an approximate total wave function for the system could be constructed by taking any one eigenfunction of each to form the product function like (q.4.112.), which could then be used in a variation-type expression like $E = \left[\int \Psi^* \hat{H} \Psi \, d\tau\right]\left[\int \Psi^* \Psi \, d\tau\right]^{-1}$ to calculate the approximate energy of the state described by the function Ψ. The total energy calculated in this manner would not be expected, generally, to be the sum of the various eigen-values E_i, since these values of E_i do not correspond to any quantities which are susceptible to experimental measurement. If the potential function $U(r_i)$ is properly chosen, E_i emerges as being approximately the ionization energy of the electron in the state ψ_i.

The various eigenfunctions of (q.4.115.), of which there are an infinite number, and similarly for each of the other $(N - 1)$ equations, may be expressed in spherical polar coordinates as the product of three functions, as in (q.3.130.), and the equation separated into three ordinary differential equations for $R(r)$, $\Theta(\theta)$, and $\Phi(\phi)$. The angular functions will be exactly the same for all central fields, and are defined by the expressions already found for the hydrogenic case, as given by (q.3.22.) and (q.3.51.), respectively. The radial function will not, in general, be the same as the hydrogenic radial function, given by (q.3.110.), since $U_i(r_i)$ will not usually be proportional to (l/r_i), but will have a similar form. The eigenfunctions, ψ_i, of (q.4.115.) will be atomic orbitals similar to (q.3.131.), but the eigenvalues, E_i, will depend on both the quantum numbers, n and l, and not simply on n, as is the case for hydrogenic atoms. To illustrate the application of these one-electron functions to a specific problem, and to show the manner in which the Hartree concept is used in choosing the potential function, $U_i(r_i)$, the ground state of the helium atom will be considered.

Q.4.10. The Ground State of the Helium Atom and the Hartree Method

In accordance with the general concept inherent in the one-electron approximation, an approximate total wave function for the helium atom may be written as

$$\Psi(r_1, r_2) = \psi_1(r_1)\psi_2(r_2), \tag{q.4.116.}$$

with the one-electron functions, $\psi_1(r_1)$ and $\psi_2(r_2)$, satisfying the equations

$$[-(h^2/8\pi^2 m)\,\nabla_1{}^2 - 2\varepsilon^2/r_1 + U_1(r_1)]\psi_1(r_1) = E_1\psi_1(r_1)$$

and   (q.4.117.)

$$[-(h^2/8\pi^2 m)\,\nabla_2{}^2 - 2\varepsilon^2/r_2 + U_2(r_2)]\psi_2(r_2) = E_2\psi_2(r_2),$$

respectively. Actually, for the ground state of the helium atom, both $\psi_1(r_1)$ and $\psi_2(r_2)$ are $1s$ type orbitals. The Hartree expression for the potential function $U_1(r_1)$ is chosen to be the potential energy of the first electron (electron 1) in the field of the average charge distribution or charge cloud of the second electron (electron 2). The average charge distribution of electron 2 may be regarded (page 223) as a charge cloud with a density at any point r_2, given by $-\varepsilon|\psi_2(r_2)|^2$; the potential energy of electron 1 due to a volume element dr_2 of this charge cloud will then be $\varepsilon^2|\psi_2(r_2)|^2\,dr_2/r_{12}$. Now, considering the whole charge cloud due to electron 2, the potential energy of electron 1 will be given by

$$U_1(r_1) = \varepsilon^2 \int [|\psi_2(r_2)|^2/r_{12}]\,dr_2. \qquad (q.4.118.)$$

This definite integral has to be evaluated throughout all of space, and could be written alternatively as

$$U_1(r_1) = \varepsilon^2 \int [|\psi_2(r)|^2/|r_1 - r|]\,dr \qquad (q.4.119.)$$

to indicate that r_2 in (q.4.118.) is the radius vector from the nucleus to the volume element dr_2 in the charge cloud of electron 2, and that r_{12} is the radius vector from electron 1 to the volume element dr_2 situated within the total charge cloud of electron 2; in other words, the definite integral given in (q.4.118.) is not specifically a function of r_2. Now, substituting (q.4.118.) into the first of equations (q.4.117.), the Hartree equation for $\psi_1(r_1)$ is found to be

$$[-(h^2/8\pi^2 m)\,\nabla_1{}^2 - 2\varepsilon^2/r_1 + \varepsilon^2 \int \{|\psi_2(r_2)|^2/r_{12}\}\,dr_2]\psi_1(r_1) = E_1\psi_1(r_1),$$

$$(q.4.120.)$$

and the equation for $\psi_2(r_2)$ is similar but with the subscripts 1 and 2 interchanged. For the ground state of helium, the functions $\psi_1(r_1)$ and $\psi_2(r_2)$ are both the same and are both spherically symmetrical functions, so that there is indeed no necessity to make any distinction between $\psi_1(r_1)$ and $\psi_2(r_2)$; the single symbol r without any subscript is sufficient provided some symbol other than r is used to denote the variable with respect to which the integration is carried out. The same equation may be used for both electrons, for example,

$$[-(h^2/8\pi^2 m)\,\nabla^2 - 2\varepsilon^2/r + \varepsilon^2 \int \{|\psi(r')|^2/|r - r'|\}\,dr']\psi(r) = E\psi(r). \qquad (q.4.121.)$$

Integro-differential equations of this type, in which in this case the potential energy depends on the function $\psi(r)$, cannot generally be solved either by analytical means or by direct numerical integration. An iterative procedure using successive approximations has to be used. The self-consistent field approach is to choose some initial approximation for $\psi(r)$, normalize it [equation (q.4.121.) is only applicable if $\psi(r)$ is normalized], and then make use of it to evaluate the integral in (q.4.121.), which then becomes an ordinary Schrödinger equation, which may be solved numerically. The value thus found for $\psi(r)$ is used again in the same procedure until successive repetitions of this process do not lead to any substantial change in the function used for calculating the integral, giving the potential energy and the function which results on solving the Schrödinger equation. Convergence to self-consistency is the more rapid the better the initial approximation is to the true ground-state wave function. Of the infinite number of solutions to an equation like (q.4.121.), the ground-state function necessarily has to be a $1s$ type function, with spherical symmetry and no nodes. If the one-electron functions, $\psi(r_1)$ and $\psi(r_2)$ are normalized, then the total wave function given by (q.4.116.) must also be normalized, since

$$\iint \Psi^* \Psi \, dr_1 \, dr_2 = \iint \psi^*(r_1)\psi(r_1)\psi^*(r_2)\psi(r_2) \, dr_1 \, dr_2$$

$$= \int |\psi(r_1)|^2 \, dr_1 \int |\psi(r_2)|^2 \, dr_2 = 1.$$

The problem of finding the ground-state eigenfunction of (q.4.120.), which is the lowest eigenvalue, is found during the calculation of $\psi(r)$, and is the same for both electrons. The approximate ground-state energy of the helium atom is then found, by the use of the variation expression $E_g = \int \Psi^* \hat{H} \Psi \, d\tau$, as

$$E_g = \int \Psi^* \hat{H} \Psi \, d\tau$$

$$= \iint \psi^*(r_1)\psi^*(r_2)[-(h^2/8\pi^2 m)(\nabla_1{}^2 + \nabla_2{}^2) - 2\varepsilon^2/r_1 - 2\varepsilon^2/r_2 + \varepsilon^2/r_{12}]$$

$$\times \psi(r_1)\psi(r_2) \, dr_1 \, dr_2$$

$$= 2E_1 - \iint (\varepsilon^2/r_{12})|\psi(r_1)|^2|\psi(r_2)|^2 \, dr_1 \, dr_2. \ddagger \tag{q.4.122.}$$

$\ddagger$ The Hartree equation for each electron in the helium atom is given by (q.4.121); now writing $\varepsilon^2 \int [|\psi(r')|^2/r - r'] \, dr' = U(r) = \varepsilon^2/r_{12}$, for convenience, and multiplying the equation for the first electron by $\psi_1{}^*(r_1)$ and integrating over all spatial coordinates, the result is

$$\int \psi_1{}^*(r_1)[-(h^2/8\pi^2 m)\nabla_1{}^2 - 2\varepsilon^2/r_1 + \varepsilon^2/r_{12}]\psi_1(r_1) \, dr_1 = \int E_1\psi_1(r_1)\psi_1{}^*(r_1) \, dr_1 = E_1,$$

if $\psi_1(r_1)$ and $\psi_2(r_2)$ are separately normalized. This latter equation may be alternatively written as

$$\int \psi_1{}^*(r_1)[-(h^2/8\pi^2 m)\nabla_1{}^2 - 2\varepsilon^2/r_1]\psi_1(r_1)\,dr_1 = E_1 - \int \psi_1{}^*(r_1)\psi_1(r_1)(\varepsilon^2/r_{12})\,dr_1.$$

Similarly, from the equation for the second electron

$$\int \psi_2{}^*(r_2)[-(h^2/8\pi^2 m)\nabla_2{}^2 - 2\varepsilon^2/r_2]\psi_2(r_2)\,dr_2 = E_1 - \int (\varepsilon^2/r_{12})\psi_2{}^*(r_2)\psi_2(r_2)\,dr_2.$$

If the total wave function for helium is $\Psi(r_1, r_2) \simeq \psi_1(r_1)\psi_2(r_2)$, then

$$E = \iint \Psi^*(r_1, r_2)\hat{H}\Psi\,dr_1\,dr_2$$

$$= \iint \psi_1{}^*(r_1)\psi_2{}^*(r_2)[-(h^2/8\pi^2 m)(\nabla_1{}^2 + \nabla_2{}^2) - 2\varepsilon^2/r_1 - 2\varepsilon^2/r_2$$

$$+ \varepsilon^2/r_{12}]\psi_1(r_1)\psi_2(r_2)\,dr_1\,dr_2$$

$$= \iint |\psi_2(r_2)|^2\psi_1{}^*(r_1)[-(h^2/8\pi^2 m)\nabla_1{}^2 - 2\varepsilon^2/r_1]\psi_1(r_1)\,dr_1\,dr_2$$

$$+ \iint |\psi_1(r_1)|^2\psi_2{}^*(r_2)[-(h^2/8\pi^2 m)\nabla_2{}^2 - 2\varepsilon^2/r_2]\psi_2(r_2)\,dr_1\,dr_2$$

$$+ \iint (\varepsilon^2/r_{12})|\psi_1(r_1)|^2|\psi_2(r_2)|^2\,dr_1\,dr_2$$

$$= 2E_1 - \int \psi_1{}^*(r_1)\psi_1(r_1)(\varepsilon^2/r_{12})\,dr_1 - \int (\varepsilon^2/r_{12})\psi_2{}^*(r_2)\psi_2(r_2)\,dr_2$$

$$+ \iint [\varepsilon^2/r_{12}]|\psi_1(r_1)^2|\,|\psi_2(r_2)|^2\,dr_1\,dr_2$$

$$= 2E_1 - 2\iint |\psi_1(r_1)|^2\psi_2(r_2)|^2[\varepsilon^2/r_{12}]\,dr_1\,dr_2 + \iint [\varepsilon^2/r_{12}]|\psi_1(r_1)|^2|\psi_2(r_2)|^2\,dr_1\,dr_2$$

$$= 2E_1 - \iint [\varepsilon^2/r_{12}]|\psi_1(r_1)|^2|\psi_2(r_2)|^2\,dr_1\,dr_2,$$

since $\int |\psi_1(r_1)|^2\,dr_1 = \int |\psi_2(r_2)|^2\,dr_2 = 1$.

The eigenvalue E_1 in equation (q.4.122.) may be found during the process of calculating $\psi(r)$; a good first approximation to $\psi(r)$ would obviously be $\psi(r) = e^{-27r/16}$, which, as shown on page 240, gave a ground-state energy of -2.85 Hartrees. The integral expression in (q.4.122.), which represents the mean interaction energy of the two electrons, may be evaluated by the use of the expression given on page 239; here, however, since $\psi(r)$ will be obtained as a tabulated function, numerical integrations are necessary. This method was first applied to the helium atom by Wilson,[1] who calculated the value of $E_g = -2.86$ Hartrees, which differs from the experimental value of -2.905 Hartrees by about 1.4%, which is only two-thirds of the error obtained in the variational type calculation. The energy calculated by the

---

[1] Wilson, W. S., *Phys. Rev.*, 1935, **48**, 536.

Hartree self-consistent field method, making use of a wave function expressed as a product of two self-consistent field atomic orbitals, cannot be improved by the method used in the variational type calculation, since the Hartree function of this type is the optimum one of this particular form. The fundamental problem for helium and other polyelectronic systems is that the electron correlation effect, due to the mutual repulsion between the electrons tending to keep them apart, is not properly accounted for in the Hartree method. The energy calculated by the Hartree method is too high, for it assumes that the electrons interact with one another at closer proximities than can actually be so. In the variational type calculations, this correlation effect may be accounted for by including terms containing r_{12} in the trial function. Actually, the Hartree wave function, from the point of view of the variational principle, is the best wave function of the simple product type, given by (q.4.116.).

The Hartree method, as given above for two electrons, may now be generalized to any number of electrons, N.

Q.4.11. The Variation Principle and the Hartree Method in General

For an N-electron atom, each electron is assumed to move in the potential field of the nucleus and that, due to the charge cloud of all the other $(N - 1)$ electrons. The potential energy of the ith electron of the set N is given by the sum of a number of expressions like (q.4.118.), one for each of the $(N - 1)$ other electrons; that is,

$$U_i(r_i) = \varepsilon^2 \sum_{j \neq i}^{N} \int [|\psi_j(r_j)|^2/r_{ij}]\, dr_j. \qquad (q.4.123.)$$

The function $U_i(r_i)$, for atoms, will be spherically symmetrical if $\psi_j(r_j)$ is an s-type function, or if there are sufficient electrons to constitute a set of 3 or 6 p-states, 5 or 10 d-states, 7 or 14 f-states, etc. In the event that a particular atom contains such a total number of electrons that there is an incomplete set of p, d, f, states, an average may be taken over the angular coordinates so as to obtain a potential contribution which is spherically symmetrical. It is only in the case of atoms, with a single center of force present, that the Hartree method of taking an averaged spherically symmetrical field due to all the electrons present in the system is justified, and this process in itself introduces an approximation. The Hartree equations for the system, which result from the substitution of (q.4.123.) into (q.4.115.), are then obtained as

$$\left[-(h^2/8\pi^2 m)\, \nabla_i^2 - Z\varepsilon^2/r_i + \varepsilon^2 \sum_{j \neq i}^{N} \int \{|\psi_j(r_j)|^2/r_{ij}\}\, dr_j \right] \psi_i(r_i) = E_i \psi_i(r_i),$$

$$(q.4.124.)$$

where $i = 1, 2, 3, \ldots, N$.

Since the potential function in each of these N integro-differential equations depends upon the solutions of each of the others, the problem becomes that of finding simultaneous solutions to these N equations. For any particular electronic configuration of the atom concerned, as for example $1s^2 2s^2 2p^6$-$3s^2 3p^6 4s^2 3d^5$ (ground state of the manganese atom), a set of approximate ψ_j, functions are selected which are used to calculate the N potential functions; the Hartree equations are thus converted into ordinary Schrödinger equations which may be solved numerically, and the appropriate eigenfunctions are selected dependent on whether it is required to have the $1s$, $3s$, $3d$, ... functions. The iterative process of using these solutions to recalculate the potential functions, solving again for another set of wave functions and repeating this process until self-consistency is achieved, is then applied. From the self-consistent set of functions thus calculated, the approximate total wave function for the system is constructed as

$$\Psi(r_1, r_2, r_3, \ldots, r_N) = \psi_1(r_1)\psi_2(r_2)\psi_3(r_3)\cdots\psi_N(r_N), \qquad \text{(q.4.125.)}$$

and the approximate total energy calculated by using this expression for the total wave function of the system in the variational type expression,

$$E = \int \Psi^* \hat{H} \Psi \, d\tau.$$

If the total wave function Ψ is to be normalized, then each of the ψ_j's have to be normalized at each stage of the iterative calculations involved in arriving at the self-consistent set of ψ_j functions. Thus

$$E = \int \Psi^* \hat{H} \Psi \, d\tau$$

$$= \int \cdots \int \psi_1^*(r_1)\cdots\psi_N^*(r_N)$$

$$\times \left[-\sum_{i=1}^{N} \{(h^2/8\pi^2 m)\,\nabla_i^2 + Ze^2/r_i\} + (1/2)\sum_{i \neq j}^{N}\sum^{N}(e^2/r_{ij}) \right]$$

$$\times \psi_1(r_1)\cdots\psi_N(r_N)\,dr_1\cdots dr_N$$

$$= -\sum_{i=1}^{N} \int \psi_i^*(r_i)[(h^2/8\pi^2 m)\,\nabla_i^2 + Ze^2/r_i]\psi_i(r_i)\,dr_i$$

$$+ (1/2)\sum_{i \neq j}^{N}\sum^{N} \iint \psi_i^*(r_i)\psi_j^*(r_j)[e^2/r_{ij}]\psi_i(r_i)\psi_j(r_j)\,dr_i\,dr_j, \qquad \text{(q.4.126.)}$$

since all the other factors in the product of the N integrals are equal to unity, if all the ψ_i's are normalized. Now making use of (q.4.124.), as in (q.4.122.), it is found that

$$E = \sum_{i=1}^{N} E_i - (1/2)\sum_{i \neq j}^{N}\sum^{N} \iint [e^2/r_{ij}]|\psi_i(r_i)|^2|\psi_j(r_j)|^2\,dr_i\,dr_j. \qquad \text{(q.4.127.)}$$

The first term in (q.4.127.) is simply the sum of the one-electron energies, while the second term is the average interaction energy of the electrons.

The program of Herman and Skillman (page 264), for the self-consistent field calculation of the atomic structure of various atoms, tabulates the radial functions $rR(r)$ for a great many values of r, and also the radial potential energies. The speed of these computer programs is such that the calculation of the spherically symmetrical potential functions, the numerical integration of the corresponding Schrödinger equation, the calculation of the eigenvalues, and all the iterations necessary for self-consistent results can be performed for any atom in a few minutes. This includes the normalization of the various functions at each stage of the iterative process, where the integration required for this operation is performed in the machine programs by the application of Simpson's rule. The eigenvalues corresponding to E_i of (q.4.115.), calculated by these programs with the best choice of the potential function $U_i(r_i)$, approximate to the ionization energies of electrons in the states ψ_i, which can be determined experimentally. The agreement between the results calculated by the Hartree self-consistent field method and these experimentally measured values is remarkably good. The central field, self-consistent field approximation is sufficiently good to account for the periodic system of the elements and the general size and energies of atoms. According to equation (q.4.125.) and the Pauli Exclusion Principle, the order of increasing energy states of electrons in atoms should follow the hydrogen-like atom order

$$1s < 2s = 2p < 3s = 3p = 3d < 4s = 4p = 4d = 4f \cdots,$$

but, from a great deal of accumulated information about the line spectra of atoms, it is evident that this is not so for polyelectronic atoms; for example, the configuration $1s^2 2s^2$ is found to correspond to a lower energy than $1s^2 2p^2$. In fact, one of the effects of interelectronic repulsions is to remove the degeneracies associated with the l quantum number, so that the modified order of increasing electronic energies in atoms is

$$1s < 2s < 2p < 3s < 3p < 3d < 4s < 4p < 4d < 4f \cdots.$$

There is another effect of electron repulsion in atoms which is not accounted for by the above, namely the apparent reversal of the order of increasing electron energies, such that for example the configuration $1s^2 2s^2 2p^6 3s^2 3p^6 4s$ corresponds to a lower energy than the configuration $1s^2 2s^2 2p^6 3s^2 3p^6 3d$. Herman and Skillman have made calculations for several different configurations, assigning different electrons to various nl states, including all of those which lie particularly close to the ground state. In this manner, Herman and Skillman have calculated absolute values for the one-electron energies for atoms of the elements with nuclear charges from 19 to 30, for example, with the following results (energies expressed in Hartree units; Herman and Skillman's results were actually expressed in Rydberg units):

TABLE 4.1.1. One-electron energies of the first transition group of metals, from calculations of Herman and Skillman.[a]

Element	Configuration	$1s$	$2s$	$2p$	$3s$	$3p$	$3d$	$4s$
K	$1s^22s^22p^63s^23p^64s$	131.045	13.530	11.004	1.47605	0.8664		0.1543
Ca	$1s^22s^22p^63s^23p^64s^2$	146.76	15.8135	13.090	1.9375	1.24115		0.19935
Sc	$1s^22s^22p^63s^23p^63d4s^2$	163.145	17.9865	15.065	2.2154	1.44155	0.2654	0.21545
Ti	$1s^22s^22p^63s^23p^63d^24s^2$	180.385	20.2605	17.1395	2.49065	1.6380	0.31395	0.2289
V	$1s^22s^22p^63s^23p^63d^34s^2$	198.475	22.6445	19.323	2.76855	1.852	0.35915	0.24095
Cr	$1s^22s^22p^63s^23p^63d^54s$	217.225	24.9115	21.3895	2.8557	1.8455	0.23045	0.2156
Mn	$1s^22s^22p^63s^23p^63d^54s^2$	237.235	27.7545	24.025	3.3402	2.2388	0.44295	0.26265
Fe	$1s^22s^22p^63s^23p^63d^64s^2$	257.905	30.4785	26.542	3.63455	2.4456	0.48125	0.27255
Co	$1s^22s^22p^63s^23p^63d^74s^2$	279.445	33.324	29.179	3.93925	2.6600	0.52075	0.28235
Ni	$1s^22s^22p^63s^23p^63d^84s^2$	301.85	36.285	31.93	4.2500	2.878	0.55755	0.29155
Cu	$1s^22s^22p^63s^23p^63d^{10}4s$	324.85	30.075	34.51	4.317	2.8545	0.37155	0.25455
Zn	$1s^22s^22p^63s^23p^63d^{10}4s^2$	349.2	42.56	37.775	4.8965	3.3305	0.6921	0.30925

[a] Energies given in Hartree units. The energies of the electrons in potassium and calcium are included for comparison at the head of the table.

It may be noted from the above table that the $3d$ electron energy for chromium, at -0.23045 Hartree units, is only slightly smaller than the $4s$ electron energy at -0.2156 Hartree units. This is in agreement with spectroscopic observation, although the self-consistent field calculation itself is not actually sufficiently accurate to predict the greater stability of the $3d^5 4s$ configuration compared with the $3d^4 4s^2$ configuration. Herman and Skillman's tables are, of course, much more extensive than the above abstract indicates. The structure of atoms will be discussed in more detail after the discussion in the next chapter of spin orbitals.

The numerical tables of $rR(r)$ as a function of r, which are obtained as solutions of the Hartree self-consistent field integro-differential equations, are relatively unwieldy to use, and Slater[1] has proposed analytical functions which approximate very closely to these self-consistent field atomic orbitals. Such functions have been termed Slater-type orbitals by Mulliken. They are wave functions describing the motion of a single electron in a central field, in which the potential energy is given by

$$U(r) = -[Z_{n,l}/r] + [n^*(n^* - 1)/2r^2], \qquad \text{(q.4.128.)}$$

where n^* is an effective principal quantum number and $Z_{n,l}$ is an effective nuclear charge, both of these quantities being regarded as empirical parameters. Slater has shown that it is possible to choose satisfactory values for n^* and $Z_{n,l}$ by the use of some simple rules. Thus $n^* = n$, the actual principal quantum number, for electrons described by $n = 1, 2,$ and 3 (the K, L, and M quantum groups, respectively). For the principal quantum numbers $4, 5,$ and 6 (the N, O, and P quantum groups, respectively), $n^* = 3.7$, 4.0, and 4.2, respectively. The effective nuclear charge, $Z_{n,l} = Z - S_{n,l}$, where Z is the actual nuclear charge in atomic units and $S_{n,l}$ is a screening constant depending on the atomic orbital and the particular electronic configuration involved. The screening constants are assigned values according to the empirical Slater rules given on page 81 of Chapter Q.1. The Slater rules are not particularly reliable for values of the principal quantum number of 4 and greater. From examination of (q.4.128.), it is apparent that when r is large, the second term becomes small and $U(r)$ approaches $-Z_{n,l}$, which is just the potential experienced by an electron interacting with a nucleus screened by the other electrons. For the special case of the hydrogen atom, $n^* = 1$ and $Z_{n,l} = Z = 1$, and the potential reduces to $-1/r$. The general analytical form the Slater-type orbitals in the unnormalized form is

$$\psi_{n^*,l} = r^{n^* - 1} e^{-Z_{n^*}r/n^*} Y_{l,m}(\theta, \phi), \qquad \text{(q.4.129.)}$$

---

[1] Slater, J. C., *Phys. Rev.*, 1930, **36**, 57.

where $Y_{l,m}(\theta, \phi)$ is spherical harmonic in real form [as in (q.3.55.)], and n^* is an effective principal quantum number. The first six Slater-type orbitals are:

$$\psi_{1s} = \sqrt{Z_{1s}{}^3/\pi}\,[e^{-Z_{1s}r}]; \quad \psi_{2s} = \sqrt{Z_{2s}{}^5/96\pi}\,[re^{-Z_{2s}r/2}];$$

$$\psi_{3s} = \sqrt{2Z_{3s}{}^7/5.39\pi}\,[r^2 e^{-Z_{3s}r/3}]; \quad \psi_{2p_x} = \sqrt{Z_{2p}{}^5/32\pi}\,[xe^{-Z_{2p}r/2}];$$

$$\psi_{3p_x} = \sqrt{2Z_{3p}{}^7/5.38\pi}\,[xre^{-Z_{3p}r/3}]; \quad \psi_{3d_{xy}} = \sqrt{Z_{3d}{}^7/2.39\pi}\,[xye^{-Z_{3d}r/3}].$$

$$\text{(q.4.130.)}$$

It should be noted that while hydrogenic atomic orbitals have $[n - l - 1]$ nodes in the radial part of the expression, both the self-consistent field atomic orbitals and the Slater-type orbitals have no nodes in the radial portion of their expression.

Actually, the application of the variational method to the problem of an N-electron atom, making use of a trial function in the form of a single product, such as (q.4.125.), leads to the same one-electron functions as are obtained by the Hartree self-consistent field method. If no assumptions are made about the nature of the product functions, ψ_i of (q.4.125.), except that they be normalized, then the variational problem is to choose these one-electron functions so as to obtain the best approximation to the ground-state energy of the system, that is, to choose the ψ_i's so as to minimize the quantity E given by (q.4.127.). If from (q.4.126.) there is isolated the terms which contain any particular function ψ_i, there is obtained

$$-\int \psi_i{}^*(r_i)[(h^2/8\pi^2 m)\,\nabla_i{}^2 + Z\varepsilon^2/r_i]\psi_i(r_i)\,dr_i$$

$$+ \sum_{j\neq i}^{N} \iint \psi_i{}^*(r_i)\psi_j{}^*(r_j)[\varepsilon^2/r_{ij}]\psi_i(r_i)\psi_j(r_j)\,dr_i\,dr_j,$$

or if

$$[-(h^2/8\pi^2 m)\,\nabla_i{}^2 - Z\varepsilon^2/r_i] + \varepsilon^2 \sum_{j\neq i}^{N} \int [|\psi_j(r_j)|^2/r_{ij}]\,dr_j = \hat{H}_i,$$

this expression above may be written as

$$\int \psi_i{}^*(r_i)\hat{H}_i\psi_i(r_i)\,dr_i. \qquad \text{(q.4.131.)}$$

Now, if the one-electron functions are to be chosen so as to minimize E, the ψ_i must be chosen so as to minimize the integral in (q.4.131.), since no other term contains ψ_i. But, according to the variation principle, the function ψ_i which minimizes (q.4.131.) is the eigenfunction corresponding to the lowest eigenvalue of the equation $\hat{H}_i\psi_i = E_i\psi_i$, or of the equation

$$[-(h^2/8\pi^2 m)\,\nabla_i{}^2 - Z\varepsilon^2/r_i] + \varepsilon^2 \sum_{j\neq i}^{N} \int [|\psi_j(r_j)|^2/r_{ij}]\,dr_j\}\psi_i(r_i) = E_i\psi_i(r_i).$$

$$\text{(q.4.132.)}$$

Comparing this equation with (q.4.124.), it is apparent that (q.4.132.) is identical with the Hartree equation for the ith electron. Thus, the Hartree functions must be the best one-electron functions which may be used in a single product form of wave function for the ground state. There are, however, additional complications brought about by the application of the Pauli Exclusion Principle. Except for helium, the Pauli Principle leads to the fact that all one-electron functions are not eigenfunctions of the lowest energy of the corresponding Hartree equations.

It is seen that if the ith electron is removed from an N-electron atom, this is equivalent to removing the factor ψ_i from the total wave function Ψ of (q.4.125.), and therefore the terms (q.4.131.) from the total E. Thus, if it is assumed that this operation leaves all the other one-electron functions unaltered, then E is reduced by the amount E_i, which is the value of (q.4.131.) as given by (q.4.132.). This is the justification for saying that E_i is approximately the ionization potential of the ith electron. Unfortunately, this is not quite accurate, as the removal of any electron alters the self-consistent fields and therefore the wave functions of all of the other electrons, but it is very nearly true for those electrons whose charge clouds are concentrated closely around the nucleus of an atom, that is, for those electrons in the lowest energy states.

In summary, the Hartree method enables the best ground-state wave functions of a polyelectronic atom to be deduced of the single product type given by (q.4.125.); also, the Variation Principle using a trial function of the same type leads to the Hartree method. The Hartree method, however, is not the most accurate method of calculating atomic energies, essentially because the single product of one-electron functions is not the best approximation to the total wave function of an atomic system. A much improved total wave function is given as a determinantal function, which in turn leads to the so-called Hartree–Fock equations, taking account of the electron exchange effects. It is necessary, before discussing this development of quantum theory, to examine the concept of electron spin and the application of the Pauli Exclusion Principle to atomic systems in some detail, which is the major topic of the next chapter.

Q5

Spin Wave Functions and the Pauli Exclusion Principle

Q.5.1. Complex Atomic Spectra

The principal features of *the theory of complex atomic spectra* were deduced in the period between 1920 and 1925, before the advent of quantum mechanics.[1] However, there were many discrepancies and difficulties in this area, which were only explained after the development of quantum mechanics. Apart from the spectrum of hydrogen, the only elements which have relatively simple atomic spectra are those of the alkali metals, and of copper, silver, and gold. All of the other elements present complicated spectra, apparently composed of thousands of lines. It was, in fact, the apparent simplicity of the atomic spectra of the alkali metals, copper, silver, and gold, which led to the central field model. The pre-quantum mechanical description of atomic spectra in terms of arbitrary selection principles has already been discussed briefly in Sections Q.1.5. and Q.1.6. The apparent simplicity of the spectra of the alkali metals was early interpreted to be due to the fact that in these atoms, a single electron moved in a central field of force provided by the atomic nucleus and a set of electrons in a complete set of energy states, described by some particular values of the n and l quantum numbers. Thus, the alkali metal atoms, lithium, sodium, potassium, rubidium, cesium, and francium, have one $2s$, $3s$, $4s$, $5s$, $6s$, and $7s$ electron, respectively, moving in a spherically symmetrical potential field provided by a nucleus of charge

---

[1] Hund, F., *Linienspektren und periodisches System der Elemente*, 1927, Springer-Verlag OHG, Berlin.

278

$+3$, $+11$, $+19$, $+37$, $+55$, and $+87$, respectively, and a closed set of $1s^2$, $1s^22s^22p^6$, $1s^22s^22p^63s^23p^6$, $1s^22s^22p^63s^23p^63d^{10}4s^24p^6$, $1s^22s^22p^63s^23p^63d^{10}$-$4s^24p^64d^{10}5s^25p^6$, and $1s^22s^22p^63s^23p^63d^{10}4s^24p^64d^{10}5s^25p^64f^{14}5d^{10}6s^26p^6$ electrons, respectively. Similarly, atoms of the elements copper, silver, and gold have one $4s$, $5s$, and $6s$ electron moving in a field of a nucleus of charge $+29$, $+47$, and $+79$ and a closed set of electrons in states $1s^22s^22p^63s^23p^63d^{10}$, $1s^22s^22p^63s^23p^63d^{10}4s^24p^64d^{10}$, and $1s^22s^22p^63s^23p^63d^{10}$-$4s^24p^64d^{10}5s^25p^64f^{14}5d^{10}$, respectively. The similarity and simplicity of the spectra of these elements and their relationship to the spectrum of hydrogen suggested that complete sets of s and p and s, p, and d, electrons, acted on electrons in higher energy states, which are further from the nuclei, like spherical force fields. The discussion on page 224 provides a partial explanation of this effect in terms of Unsöld's theorem, according to which, if there are $(2l + 1)$ electrons in a polyelectronic atom with each electron describable by a different m_l quantum number in any nl set of states, then the total charge-cloud distribution has spherical symmetry. The Hartree method then constitutes an appropriate description of such atomic systems. This conclusion was supported by the fact that the observed spectra of positive ions were found to be similar to those of neutral atoms with the same number of electrons. Thus the spectra of singly-charged positive ions of the alkaline earth metals, beryllium, magnesium, calcium, strontium, and barium, as well as those of zinc, cadmium, and mercury, are observed to be similar to those of the corresponding neutral alkali metal atoms. This observation, in turn, implies that the complexity of the observed spectra of atoms with two electrons, beyond the number required to form a complete quantum set, was due to this subset of two electrons influencing one another so as to destroy the applicability of the central field effect provided by the electrons in the complete set of energy states.

It was, in fact, the presumption that the spectra of the alkaline earth metal atoms would be the next simplest to interpret (after the alkali metal spectra) and the detailed study of the line spectrum of calcium, in particular, which led Russell and Saunders to suggest the scheme which proved capable of accounting for the complexities of atomic spectra, in terms of the interactions between the angular momenta and the various electrons present in atoms. The Russell–Saunders angular momenta coupling scheme and Hund's rules have already been referred to in section Q.1.9.

Q.5.2. Electron Spin

The motion of an electron with respect to the nucleus of an atom has often been compared to the planetary motion of the earth around the sun. Bohr referred to an electron orbiting the nucleus, Mulliken originally suggested the term orbital, for the one-electron wave functions such as (q.3.131.) partially to maintain some continuity with the older theory, and the terminology (secular equations and determinants) of the variation theory, all make reference to this analogy between atomic systems and the solar system. The extension of this analogue between electronic and terrestrial motion then leads to the speculation as to whether an electron may execute some sort of rotation around an axis comparable to the earth's diurnal rotation. It is then reasonable to enquire whether or not such a rotation could be detected experimentally.

According to the simple classical concept of an electron moving in a small circular path of radius r, with radial velocity v, in the xy Cartesian coordinate plane, the components of the angular momenta of the electron about the origin would be $L_x = L_y = 0$, and $L_z = mrv$, where m is the electron mass. As on page 66, such a state of motion would be equivalent to an electric current $I = -\varepsilon v/2\pi r$, where $-\varepsilon$ is the electronic charge. The direction of this current flow would be the same as that of the motion of the electron. Also, from the Faraday laws of electromagnetism, such an electric current in a closed loop of area A is equivalent to a magnetic dipole of magnitude IA/c, expressed in Gaussian units, and acting perpendicular to the plane of the loop. Thus, the motion of the electron would cause a magnetic moment: its orbital magnetic moment, with a z-component of $M_z = -\varepsilon vr/2c = -\varepsilon L_z/2mc$, the other two components being zero. Now, in the presence of an applied magnetic field of field strength H acting in the $+z$ Cartesian coordinate direction, the electron, by virtue of the interaction between its orbital magnetic moment in this direction of the magnetic field, would acquire a potential energy of

$$-M_z H = [\varepsilon/2mc]L_z H. \qquad \text{(q.5.1.)}$$

This classical mechanical result, which is independent of the shape and orientation of the electron's path of motion, is equally true in quantum mechanics, provided that L_z is quantized. For an electron moving in a central electrostatic force field, its plane of motion would actually precess around the direction of the applied magnetic field, but the energy of this precessional motion would still be given by (q.5.1.). Also, for a particle moving in a central field of force, $L_z = (h/2\pi)m_l$ (page 215), and may have $(2l + 1)$ values corresponding to the various possible values of the magnetic quantum number ranging from $-l, (-l + 1), \ldots, l$. This means that the magnetic moment may have any value given by

$$M_z = -(\varepsilon h/4\pi mc)m_l. \qquad \text{(q.5.2.)}$$

There are $(2l + 1)$ separate energy values, therefore, for the energy of the electron in a magnetic field H in the positive z-direction separated by intervals of $\varepsilon hH/4\pi mc$; the removal of the $(2l + 1)$-fold degeneracy in this way in the presence of a magnetic field is what is referred to as the normal Zeeman effect, whereby certain atomic spectral lines are resolved, generally into three components. Electronic transitions in atoms from one state to another, brought about by the absorption or emission of electromagnetic radiation, generally involve a change of m_l by 0 or ± 1, so that in the normal Zeeman effect a splitting into three components is usually observed rather than the more general $(2l + 1)$ components.

A simple quantum mechanical interpretation of the above leads to the same result. Thus, the Schrödinger equation of an electron moving in a central force field is

$$[-(h^2/8\pi^2 m)\,\nabla^2 + U(r)]\psi = E_0\psi, \qquad (q.5.3.)$$

and the eigenfunctions of this equation are given by (q.3.131.), or more simply as

$$\psi_{n,l,m_l}(r,\,\theta,\,\phi) = NR_{n,l}(r)P_l^{m_l}(\cos\theta)e^{im_l\phi}; \qquad (q.5.4.)$$

there are $(2l + 1)$ such functions corresponding to the m_l values of $-l$, $(-l + 1),\ldots, l$, each leading to the eigenvalue E_0. In the presence of a magnetic field acting in the $+z$ direction,‡ the potential energy of the electron would be increased by an amount given by (q.5.1.), so that the appropriate Schrödinger equation for the electron in the presence of a magnetic field would be obtained by adding this additional potential energy term to the Hamiltonian in (q.5.3.). Since the operator for L_z is $(h/2\pi i)\,\partial/\partial\phi = \hat{L}_z$, this equation becomes

$$[-(h^2/8\pi^2 m)\,\nabla^2 + U(r) + (\varepsilon h H/4\pi mci)\,\partial/\partial\phi]\psi = E_1\psi. \qquad (q.5.5.)$$

‡ The angular momentum is quantized in the direction of the applied magnetic field. In the above discussion, where an electron has a component of its angular momentum in the z-direction, and a zero component of its angular momentum in x- and y-directions, it is therefore L_z which is quantized. The wave functions (q.5.4.) are eigenfunctions of the operators in equations (q.5.3.) and (q.5.5.) only because of the fact that the magnetic field is chosen to act in the z-direction. If the magnetic field had been chosen to act in the x- or y-directions, the eigenfunctions of the corresponding Schrödinger equations would be found to be eigenfunctions simultaneously of L_x or L_y, rather than L_z, and the values of L_x or L_y would happen to be quantized. The magnitude of the energy separation, however, would be the same in each case.

Now, since ϕ only appears in (q.5.4.) as the factor $e^{im_l\phi}$, therefore

$$\partial/\partial\phi[\psi_{n,l,m_l}(r,\,\theta,\,\phi)] = im_l\psi_{n,l,m_l}(r,\,\theta,\,\phi), \qquad (q.5.6.)$$

and so (q.5.4.) is also an eigenfunction of (q.5.5.). On combining equations (q.5.3.), (q.5.5.), and (q.5.6.), it follows that

$$E_1 = E_0 + [\varepsilon h H/4\pi mc]m_l, \qquad (q.5.7.)$$

indicating that the $(2l + 1)$ eigenfunctions of (q.5.5.) each correspond to a different eigenvalue, the separate $(2l + 1)$ eigenvalues being separated by intervals of energy of $(\varepsilon hH/4\pi mc)$, for any given values of n and l. The quantity $[\varepsilon h/4\pi mc]$, which is frequently denoted by μ_B, is referred to as the Bohr magneton; it has the numerical value $9.2732 \times 10^{-24} \, \mathrm{J\,T^{-1}}$. It is apparent from (q.5.7.) that if $l = 0$, then $m_l = 0$, and so $E_1 = E_0$; therefore, for an electron in an s-state, in a magnetic field, there would be no splitting of the corresponding energy level, due to its orbital magnetic moment.

For an N-electron atom, in terms of the one-electron approximation, atoms with a complete group of nl electron states or with $(2l + 1)$ p, d, or f electron states will have z-components of their total orbital angular moment and magnetic moments of zero, since for every electron with a given positive value of m_l there will be an electron with an equal negative value of m_l. Thus, such atoms should also be unaffected in an applied magnetic field; this should apply, for example, to the alkali metal atoms and to atoms of copper, silver, and gold.

The above deductions, however, based on the application of Schrödinger mechanics, are at variance with well-known experimental facts. For example, the H_α-line at 6.5628×10^{-7} m in the Balmer region of the line spectrum of hydrogen, is observed under conditions of moderate resolution to be composed of two lines separated by about 3×10^{-11} m; similarly, the so-called D-line in the line spectrum of sodium is actually observed to consist of two closely spaced lines of wavelength 5.890×10^{-7} m and 5.896×10^{-7} m, and lines in the spectra of helium and of the alkaline earth metals are observed to consist of single lines and triple lines. In general, such groups of spectral lines are referred to as multiplets, one line as a singlet, two lines as a doublet, and a group of three closely spaced lines as a triplet, and so on. For hydrogen and the alkali metal atoms observed in the presence of a magnetic field, the triplets in the observed spectra, which conform with the normal Zeeman effect, are also accompanied by doublets and other multiplets of a higher order than three, which are inexplicable in terms of the above concepts; this is referred to as the anomalous Zeeman effect.

Another such example is provided by the Stern–Gerlach experiment,[1,2] involving the splitting of beams of atoms in an inhomogeneous magnetic field. In terms of classical electromagnetic theory, an atom with a magnetic moment M in a magnetic field H will be acted on by a force

$$[M_x \, \partial/\partial x + M_y \, \partial/\partial y + M_z \, \partial/\partial z]H = [M \cdot \nabla]H,$$

where the derivatives are evaluated at the position of the atom. If the applied magnetic field is assumed to act in the z-direction and vary only with z, the total force would be $M_z \, dH/dz$ in the z-direction. For a uniform field, which

[1] Stern, O., Z. Physik, 1921, 7, 249.
[2] Gerlach, W., and Stern, O., Z. Physik, 1922, 8, 110, 9, 349, 353.

is homogeneous, the derivative dH/dz would be zero, since in such an event the force on each pole of the imaginary elementary magnet would be equal and opposite. So, a beam of atoms would not be anticipated to be deflected in a uniform magnetic field. But for an inhomogeneous magnetic field, acting for example on hydrogenic atoms, each in the same state with the orbital quantum number l, it would be expected from (q.5.1.) that a beam of such atoms would be separated into $(2l + 1)$ beams, corresponding to the $(2l + 1)$ values of M_z. This same effect would be expected if the atoms contain complete nl groups of electrons or $(2l + 1)$ p, d, or f electron groups, and one additional electron in a state with orbital quantum number l; however, if this single electron is in an s-state with $l = 0$, then no splitting would be expected. Nevertheless, Stern and Gerlach found that a beam of silver atoms, which have 46 electrons in states $1s^22s^22p^63s^23p^63d^{10}4s^24p^6$ and one electron in a $5s$-state, was split by an inhomogeneous magnetic field into two separate beams. Similar observations have been made on hydrogen, lithium, and sodium atoms, all of which are similar "one-electron" atoms.

The hydrogenic atom doublets and the splitting of the atomic beam into two components in the Stern–Gerlach experiment, where for both cases the Schrödinger theory indicates an angular momentum of zero, suggest either that the component of the angular momentum in the field direction is actually $(1/2)(h/2\pi)$, or alternatively that $L_z = 0$, but that there is some other kind of angular momentum effect such that its projections on the magnetic field direction are $+h/4\pi$ and $-h/4\pi$. An explanation for all of these observed facts was provided by a suggestion in 1925 by Uhlenbeck and Goudsmit[1] that an electron does indeed behave as if it possessed an intrinsic angular momentum or spin momentum S_z and a corresponding magnetic moment M_s. The electron spin postulate of 1925 is actually quite comparable to the Planck quantum postulate of 1900; in both situations a bold assumption is made which affords a reasonable explanation of otherwise incomprehensible experimental data, but which could not be satisfactorily justified until considerably later.

In terms of the concept of electron spin, an electron may be regarded as a charged particle spinning about an axis, or at least it appears to behave in such a manner. If it is assumed that in the presence of a magnetic field acting in the z-direction, S_z is quantized in the same manner as L_z, then the quantized values may be deduced from the results of the Stern–Gerlach experiment. Thus, from the above discussion, if as a result of the orbital angular momentum of an electron, a beam of atoms is separated into $(2l + 1)$ beams in an applied magnetic field, then the quantized values of L_z were deduced to be $-l(h/2\pi)$, $(-l + 1)(h/2\pi), \ldots, l(h/2\pi)$. In a similar manner, since a beam of "one-electron" atoms is separated into two beams, then if this result is due to a

---

[1] Uhlenbeck, G. E., and Goudsmit, S., *Naturwiss.*, 1925, **13**, 953; *Nature*, 1926, **117**, 264.

spin momentum S, this means that $2S + 1 = 2$ or $S = 1/2$, and the quantized values of S_z must be $-(1/2)(h/2\pi)$ and $(1/2)(h/2\pi)$. The amount of separation of the beams also indicates that the corresponding values of M_s must be $-\varepsilon h/4\pi mc$ and $+\varepsilon h/4\pi mc$. This leads to the conclusion that the ratio of the spin magnetic moment to the spin angular momentum is not the same as for the ratio of the orbital magnetic moment to the orbital angular momentum. From the above, the ratio of $M_s/S_z = [-\varepsilon/mc]$, but from (q.5.1.) $M_z/L_z = [-\varepsilon/2mc]$. The fact that these two ratios are not the same implies that the idea of an electron as a spinning charged particle cannot be true.

This and other difficulties concerning the motion of a spinning electron were removed by the application of relativity theory[1] to the problem. The Schrödinger equation is not invariant with respect to relativistic transformations and therefore cannot be correct from the relativistic point of view; ultimately, the Uhlenbeck and Goudsmit assumption is consistent with electromagnetic theory when relativistic effects are taken into account. In 1928 Dirac[2] expressed the Schrödinger equation for the hydrogen atom in a relativistically correct form, such that the space and time coordinates were included in the equation in the symmetrical manner required by special relativity theory. This Dirac equation was expressed in four-dimensional space, and the solutions involved a fourth quantum number, which emerged to have only two possible values, $(1/2)$ and $(-1/2)$. In the relativistic Dirac equation, no hypothesis concerning electron spin was introduced, although the fourth quantum number may be interpreted only on the basis of a fourth degree of freedom, equivalent to a spinning motion about the center of mass of the system. In this sense, the Dirac theory is not really to be interpreted to mean that the electron actually spins about an axis, but rather that the internal angular momentum of $h/4\pi$ and the associated magnetic moment of one Bohr magneton should be regarded as fundamental properties of the electron, like its charge and mass. In any event, a strictly literal interpretation of an electron as a spinning or rotating particle leads to the absurd conclusion that the electronic magnetic energy is so large that the electronic mass would be greater than the proton mass, or alternatively, if the known electron mass is accepted, an electron would occupy more space than a whole atom. The Dirac equation reduces to the Schrödinger equation when relativistic effects are negligibly small. Electron spin has to be regarded as a quantum mechanical property which is inapplicable to particles obeying classical laws of motion, and which vanishes in the limit.

Relatively few attempts have been made to incorporate the theory of invariant transformations into quantum mechanics in a systematic manner

---

[1] Thomas, L. H., *Nature*, 1926, **117**, 514; *Phil. Mag.*, 1927, **3**, 1.
[2] Dirac, P. A. M., *Proc. Roy. Soc.*, 1928, **A117**, 610; **A118**, 351, 1928.

and to develop the subject from the point of view of concepts such as spin, which have only a quantum mechanical interpretation.[1] However, without making use of relativistic quantum mechanics, it is possible to develop the previously stated theory in such a manner as to include the concept of electron spin, by means of the so-called Pauli spin matrices. It is also possible to develop a consistent theory of particles of spin (1/2) without invoking relativistic invariance.[2]

Q.5.3. Spin Matrices and Spin Wave Functions

To incorporate the idea of electron spin into Schrödinger mechanics, it is necessary to adopt a mathematical formalism describing a spin quantum number and a spin variable, recognizing the necessity, to conform with experimental observation, that the one-electron states of Schrödinger mechanics each corresponds to two states, in one of which the electron has a positive spin and in the other a negative spin. These two one-electron states would be described by the same quantum numbers, n, l, and m_l, but by two different spin quantum numbers of $+1/2$ and $-1/2$, the former corresponding to S_z of $h/4\pi$ and the latter to S_z of $-h/4\pi$. It is convenient to designate the spin quantum number as m_s and the spin variable as ζ, so that the complete description of a one-electron state becomes

$$\Psi_{n,l,m_l,m_s}(x, y, z, \zeta), \qquad \text{(q.5.8.)}$$

where Ψ is a function of both the positional and spin coordinates of an electron, to be distinguished from the orbital function ψ, which is a function of the positional coordinates only. To conform with experimental observations, and the already found characteristics of the magnetic quantum number m_l, a mathematically consistent scheme requires the definition of the spin quantum number m_s such that $S_z = (h/2\pi)m_s$, where the only possible values of m_s are $\pm 1/2$. For this to be so, the spin variable must be allowed to assume only two values, and these ideas are self-consistent if the two values for the spin variable are taken to be ± 1.‡ Thus, corresponding to each of the one-electron states of Schrödinger theory there are two separate state functions

$$\Psi_{n,l,m_l,1/2}(x, y, z, \zeta) \quad \text{and} \quad \Psi_{n,l,m_l,-1/2}(x, y, z, \zeta). \qquad \text{(q.5.9.)}$$

These functions, as in (q.5.8.) and (q.5.9.), are referred to as spin-orbitals.

‡ This is a matter of mathematical convenience. Since, as discussed on page 284, it is not possible to regard an electron as a spinning charged particle, it is therefore of no

[1] See, for example, Kaempffer, F. A., *Concepts in Quantum Mechanics*, 1965, Academic Press.
[2] Galindo, A., and Sanchez del Rio, C., *Am. J. Phys.*, 1961, **29**, 582.

significance to assign an angle of rotation to it. Spin is of purely quantum mechanical significance, and there is no physical analogy of the spin variable, ζ. The spin angular momentum of an electron may only be parallel or opposed to the direction of an applied magnetic field, and if this direction is the z-direction of the Cartesian coordinate system, it may be imagined that the spin axis makes an angle of either 0 or π with the z-direction, the cosines of which angles are $+1$ and -1. There is some justification, then, for imagining that the values of ± 1 for the spin variable ζ, correspond to the cosines of the angle which the imaginary spin axis makes with the direction of the applied magnetic field.

Since for an electron not in an s state there are two sources of magnetic moment, one due to its orbital motion and one due to its spin, there must be some interaction between them. This interaction is what is referred to as spin-orbit coupling, and it has the significance that the orbital motion of an electron and consequently its energy depends on the spin direction, even in the absence of a magnetic field. It may be imagined that an electron behaves as an elementary magnet moving in the magnetic field which results from its own orbital motion, so that the spin magnetic moment of the electron is forced to be either in the same direction as the orbital magnetic moment or in the opposed direction, and the observed fine structure of atomic spectra may be attributed to this interaction. The order of magnitude of this interaction energy may be estimated by noting that the magnetic moment due to the orbital motion of an electron is of the order of magnitude of $[\varepsilon h/2\pi mc]$, from (q.5.2.), equivalent to a magnetic dipole situated at the nuclear center. The spin magnetic moment is of the same order of magnitude, but situated on the orbital path. Two magnetic dipoles situated some distance, r, apart have a mutual potential energy of the order of M^2/r^3. If r is of the order of magnitude of the Bohr radius, $a_0 = [h^2/4\pi^2 m\varepsilon^2]$, then the spin-orbit coupling energy is of the order of magnitude of

$$\Delta E \simeq [\varepsilon h/2\pi mc]^2[4\pi^2 m\varepsilon^2/h^2]^3 \simeq [\varepsilon^2/a_0][2\pi\varepsilon^2/hc]^2. \qquad \text{(q.5.10.)}$$

In this expression, the first factor, $[\varepsilon^2/a_0]$, is the potential energy of an electron in an atom, and the second factor, $[2\pi\varepsilon^2/hc]^2$, is the square of the fine structure constant, $[2\pi\varepsilon^2/hc]$, previously encountered on page 59. Since the fine structure constant is a dimensionless constant with the numerical magnitude of $1/137$, it follows from (q.5.10.) that the energy separation due to the spin orbit coupling is 137^2 times smaller than the ordinary electronic energies in atoms, which is very small. As a consequence, the spin-orbit coupling energy may generally be neglected in calculating the energy of an atom. So, the spin orbital functions given by (q.5.9.) may be approximated by the product of an orbital function and a spin function, the orbital functions being the same,

$$\psi_{n,l,m_l}(x, y, z)\chi_{1/2}(\zeta) \qquad \text{and} \qquad \psi_{n,l,m_l}(x, y, z)\chi_{-1/2}(\zeta).$$

The spin function χ of the spin variable ζ is generally written as α or β, dependent on whether m_s is $(1/2)$ or $(-1/2)$, so that the two spin orbital functions of (q.5.9.) may be written as

$$\Psi_{n,l,m_l,1/2}(x, y, z, \zeta) = \psi_{n,l,m_l}(x, y, z)\alpha(\zeta)$$

and   (q.5.11.)

$$\Psi_{n,l,m_l,-1/2}(x, y, z, \zeta) = \psi_{n,l,m_l}(x, y, z)\beta(\zeta).$$

Defined in this manner, an electron in a positive spin state α will have spin magnetic moment S_z of $(1/2)(h/2\pi)$ and in a magnetic field H applied in the z-direction will increase in energy by an amount $\mu_B H$. Similarly, an electron in a negative spin state β will have a spin magnetic moment S_z of $[-(1/2)(h/2\pi)]$ and will acquire an increase in energy of $-\mu_B H$ when in the presence of a magnetic field H applied in the z-direction.

If the spin variable or spin coordinate may only adopt the values of $+1$ or -1, then the above definitions of spin functions become compatible with the way of interpreting orbital functions, if the spin functions can only have the values 0 or 1. This means that $\alpha(1) = 1$, $\alpha(-1) = 0$, $\beta(1) = 0$, and $\beta(-1) = 1$, so that these spin functions may be normalized and orthogonalized in exactly the same manner as for orbital functions. However, since the spin variable ζ can only have the values of $+1$ or -1, an integration of the spin functions over all values of the spin variable would really amount to a summation over the two values of ζ, so that

$$\int |\alpha(\zeta)|^2 \, d\zeta \equiv \sum_{\zeta=\pm 1} |\alpha(\zeta)|^2 = |\alpha(1)|^2 + |\alpha(-1)|^2 = 1, \quad \text{(q.5.12.)}$$

denoting that the spin functions defined as above are normalized. They must also be orthogonal, since for such real functions

$$\int \alpha(\zeta)\beta(\zeta) \, d\zeta \equiv \sum_{\zeta=\pm 1} \alpha(\zeta)\beta(\zeta) = \alpha(1)\beta(1) + \alpha(-1)\beta(-1) = 0. \quad \text{(q.5.13.)}$$

The probability interpretation of such spin functions is also mathematically consistent with that of the orbital functions, since $|\alpha(1)|^2 = 1$ and $|\alpha(-1)|^2 = 0$. Thus, $|\alpha(\zeta)|^2$, denoting the probability that in the spin state α the spin variable has the value ζ, is unity, or ζ must certainly have the value 1, or alternatively the spin momentum must be in the z-direction and so $S_z = (1/2)(h/2\pi)$.

Since the spin functions $\alpha(\zeta)$ and $\beta(\zeta)$ are normalized, this means that if the orbital functions are normalized, then the spin-orbitals (q.5.11.) must also be normalized. The shortened notation described on page 121 for orbitals, whereby $\psi(x, y, z)$ is conveniently denoted by $\psi(\mathbf{r})$, may be extended to spin-orbitals by making use of vector symbol $\mathbf{x}$ instead of (x, y, z, ζ), and $d\mathbf{x}$ would then denote a volume element of the total configurational space, including spin "space." Thus, if $f(\mathbf{x})$ represents any function of the positional and spin coordinates of an electron, then

$$\int f(\mathbf{x})\, d\mathbf{x} = \int f(\mathbf{r}, \zeta)\, d\mathbf{r}\, d\zeta = \sum_{\zeta = \pm 1} \int f(\mathbf{r}, \zeta)\, d\mathbf{r}.$$

For example, if some spin-orbital $\Psi(\mathbf{x}) = \psi(\mathbf{r})\alpha(\zeta)$ is normalized, this means that

$$\int \Psi^*(\mathbf{x})\Psi(\mathbf{x})\, d\mathbf{x} = 1 = \sum_{\zeta = \pm 1} \int |\psi(\mathbf{r})|^2 |\alpha(\zeta)|^2\, d\mathbf{r} = 1 = \int |\psi(\mathbf{r})|^2\, d\mathbf{r},$$

on substituting for the two values of α, according to (q.5.12.). This expression then indicates that if the orbital function $\psi(\mathbf{r})$ is normalized, the spin-orbital $\Psi(\mathbf{x})$ must also be normalized.

Q.5.4.   The Symmetry of Wave Functions

In any polyelectronic system, the electrons are quite indistinguishable from one another. There is no experimental technique known, by means of which the course or motion of an individual electron can be followed. It is possible to know that a particular number of electrons are in certain energy states, but it is not possible to know which electron is in any given state. As a consequence, this means that any observable quantity which depends on the electronic coordinates must be symmetric with respect to those coordinates. Therefore, any interchange of the coordinates of any pair of electrons, including their spin coordinates, must not alter the quantity which is under observation. The state function of an electron, $\Psi(\mathbf{x})$, is not a physically observable quantity; it is $|\Psi(\mathbf{x})|^2$ which has physical significance. Thus, if $\Psi_N[\mathbf{x}_1, \mathbf{x}_2, \mathbf{x}_3, \ldots, \mathbf{x}_N]$ is the state function of an N-electron system, which is a function of all of the spatial coordinates and all of the spin coordinates of the N particles, then interchanging the coordinates $\mathbf{x}_1$ and $\mathbf{x}_2$ of two of the electrons, which includes the spin coordinates, must lead to

$$|\Psi_N[\mathbf{x}_1, \mathbf{x}_2, \mathbf{x}_3, \ldots, \mathbf{x}_N]|^2 = |\Psi_N[\mathbf{x}_2, \mathbf{x}_1, \mathbf{x}_3, \ldots, \mathbf{x}_N]|^2. \qquad \text{(q.5.14.)}$$

For an N-electron system, there are $N!$ ways of permuting the electronic coordinates and equation (q.5.14.) would have to be equally valid for any other pair of interchanges. Generally, if any one of these $N!$ different permutations of the electronic coordinates is represented by the permutation operator P, then

$$|\Psi_N[\mathbf{x}_1, \mathbf{x}_2, \mathbf{x}_3, \ldots, \mathbf{x}_N]|^2 = |P\Psi_N[\mathbf{x}_1, \mathbf{x}_2, \mathbf{x}_3, \ldots, \mathbf{x}_N]|^2, \qquad \text{(q.5.15.)}$$

If, specifically, P_{12} is the operator which denotes the interchange of $\mathbf{x}_1$ and $\mathbf{x}_2$, then from (q.5.15.)

$$\Psi_N[\mathbf{x}_2, \mathbf{x}_1, \mathbf{x}_3, \ldots, \mathbf{x}_N] = P_{12}\Psi_N[\mathbf{x}_1, \mathbf{x}_2, \mathbf{x}_3, \ldots, \mathbf{x}_N],$$

which may alternatively be written as

$$\Psi_N[\mathbf{x}_2, \mathbf{x}_1, \mathbf{x}_3, \ldots, \mathbf{x}_N] = e^{ic_{12}}\Psi_N[\mathbf{x}_1, \mathbf{x}_2, \mathbf{x}_3, \ldots, \mathbf{x}_N], \qquad \text{(q.5.16.)}$$

where $e^{ic_{12}}$ would represent a real constant, such that $|e^{ic_{12}}| = 1$; thus, by taking the moduli of both sides of (q.5.15.), this equation corresponds to (q.5.16.). If the operator P_{12} is applied again to (q.5.16.), the result is

$$\begin{aligned}
P_{12}\Psi_N[\mathbf{x}_2, \mathbf{x}_1, \mathbf{x}_3, \ldots, \mathbf{x}_N] &= \Psi_N[\mathbf{x}_1, \mathbf{x}_2, \mathbf{x}_3, \ldots, \mathbf{x}_N] \\
&= P_{12}P_{12}\Psi_N[\mathbf{x}_1, \mathbf{x}_2, \mathbf{x}_3, \ldots, \mathbf{x}_N] \\
&= e^{2ic_{12}}\Psi_N[\mathbf{x}_1, \mathbf{x}_2, \mathbf{x}_3, \ldots, \mathbf{x}_N]. \qquad \text{(q.5.17.)}
\end{aligned}$$

Therefore, $e^{2ic_{12}} = 1$ and $e^{ic_{12}} = \pm 1$, which must necessarily mean that

$$\Psi_N[\mathbf{x}_2, \mathbf{x}_1, \mathbf{x}_3, \ldots, \mathbf{x}_N] = \pm\,\Psi_N[\mathbf{x}_1, \mathbf{x}_2, \mathbf{x}_3, \ldots, \mathbf{x}_N]. \qquad \text{(q.5.18.)}$$

An exactly similar result must apply to the interchange of the coordinates of any of the pairs of electrons in the collection of N-electrons. A function which behaves in such a manner that it changes sign on interchange of the coordinates of any pair of electrons is said to be antisymmetric, and if this interchange leaves the sign of the function unchanged, it is said to be symmetric. Equation (q.5.18.) shows that the state function for any polyelectronic system must be either symmetric or antisymmetric with respect to the interchange of the coordinates of any pair of electrons; it cannot be symmetric with respect to the interchange of the coordinates of one pair of electrons and antisymmetric with respect to the interchange of the coordinates of another pair, otherwise an absurd conclusion results. Suppose, for example, that some function $\Psi(\mathbf{x}_1, \mathbf{x}_2, \mathbf{x}_3)$ were symmetric with respect to the interchange of the coordinates $\mathbf{x}_1$ and $\mathbf{x}_2$ but antisymmetric with respect to the interchange of the coordinates $\mathbf{x}_2$ and $\mathbf{x}_3$, then

$$\begin{aligned}
\Psi(\mathbf{x}_1, \mathbf{x}_2, \mathbf{x}_3) &= \Psi(\mathbf{x}_2, \mathbf{x}_1, \mathbf{x}_3) = -\Psi(\mathbf{x}_3, \mathbf{x}_1, \mathbf{x}_2) = -\Psi(\mathbf{x}_3, \mathbf{x}_2, \mathbf{x}_1) \\
&= \Psi(\mathbf{x}_2, \mathbf{x}_3, \mathbf{x}_1) = \Psi(\mathbf{x}_1, \mathbf{x}_3, \mathbf{x}_2) = -\Psi(\mathbf{x}_1, \mathbf{x}_2, \mathbf{x}_3), \qquad \text{(q.5.19.)}
\end{aligned}$$

which cannot be true.

As a consequence of the indistinguishability of electrons, it may be said, from the point of view of equation (q.5.14.), that for any physically measurable property of a system which depends on the coordinates, including the spin, of its constituent electrons, the result of any measurement of that property must be independent of any attempt to identify or label the electrons.

This means that the observable property, and so also its operator, must be a symmetric function of the coordinates of the constituent electrons. This result has to be regarded as a basic symmetry law of quantum mechanics. On the other hand, from equation (q.5.18.), it follows that the state function of any polyelectronic system must be either symmetric in all the coordinates of the electrons or antisymmetric in all the coordinates, but in order to decide which it is necessary to consider the results of experimental measurement. The fundamental postulate, arrived at as the result of much study of many-electron functions,[1-3] that the wave function of polyelectronic systems must be antisymmetric in the electronic coordinates, including spin, has to be regarded as the quantum mechanical statement of the Pauli Principle, or the fundamental postulate of antisymmetry. In fact, as will be discussed below, when this antisymmetry principle is applied to the one-electron approximation for a polyelectronic system, it leads to the Pauli Exclusion Principle in the form previously given on pages 70 and 160. The Pauli Exclusion Principle would not result if it were assumed that the total wave function of a system were symmetric in the electronic coordinates. The antisymmetry principle cannot be derived directly from either Schrödinger's equation nor from Dirac's relativistic equation; it appears to be an independent natural law. It is not generally applicable to all multiple-particle systems, but is found to be applicable to all systems composed of particles of half-integral spin, such as electrons, protons, neutrons, $[^3He_2]^{++}$ ions, and generally all particles with odd mass number. Such particles obey the laws of quantum statistics referred to as Fermi–Dirac statistics and are referred to collectively as fermions. It is found that fermions repel one another to a greater extent than would be expected for otherwise similar particles.

All particles which are describable by symmetric state functions have zero or integral values for their spin and obey Bose–Einstein statistics; they are referred to collectively as bosons. Bosons generally attract one another to a greater extent than would be expected for otherwise similar particles. Photons and deuterons, with spin values of unity, and for example $[^4He_2]^{++}$ ions, and other particles with even mass number, with zero spin, are examples of bosons. The distinction between Fermi–Dirac and Bose–Einstein quantum statistics, and their relationship to the classical Maxwell–Boltzmann statistics (page 40), will be discussed in a subsequent chapter.

The interrelationships between fermions, Fermi–Dirac statistics, the antisymmetric postulate, and the Pauli exclusion principle will be illustrated in the following pages.

---

[1] Heisenberg, W., *Z. Physik*, 1926, **38**, 411; **39**, 499; 1927, **41**, 239.
[2] Dirac, P. A. M., *Proc. Roy. Soc.*, London, 1926, **A112**, 661.
[3] Slater, J. C., *Phys. Rev.*, 1929, **34**, 1293.

Q.5.5. Determinantal Wave Functions

The only approximate total wave functions which have been dealt with so far have been of the simple product type, such as (q.4.112.). Taking proper account of the spin properties of electrons, such a product function would be written as

$$\Psi'_1(\mathbf{x}_1)\Psi'_2(\mathbf{x}_2)\Psi'_3(\mathbf{x}_3)\cdots\Psi'_N(\mathbf{x}_N), \qquad (q.5.20.)$$

where, for example,

$$\Psi'_2(\mathbf{x}_2) = \psi_2(r_2)\beta(\zeta_2). \qquad (q.5.21.)$$

A set of such spin orbitals may be found, for instance, for an N-electron atom by the Hartree method. The various spin orbitals in the product function (q.5.20.) could be rearranged in any order and still correspond to the same calculated energy for the system concerned. But none of these product functions would be antisymmetric. Thus, for a two-electron product function, $\Psi'_1(\mathbf{x}_1)\Psi'_2(\mathbf{x}_2)$, if the coordinates of the two electrons are interchanged, there results $\Psi'_1(\mathbf{x}_2)\Psi'_2(\mathbf{x}_1)$, which is not the negative of the first two-electron product function. However, a linear combination of these two products may be written, for example, $\Psi'_1(\mathbf{x}_1)\Psi'_2(\mathbf{x}_2) - \Psi'_1(\mathbf{x}_2)\Psi'_2(\mathbf{x}_1)$, which does have the antisymmetry property, namely,

$$\Psi'_1(\mathbf{x}_1)\Psi'_2(\mathbf{x}_2) - \Psi'_1(\mathbf{x}_2)\Psi'_2(\mathbf{x}_1) = -[\Psi'_1(\mathbf{x}_2)\Psi'_2(\mathbf{x}_1) - \Psi'_1(\mathbf{x}_1)\Psi'_2(\mathbf{x}_2)].$$

Similarly, a linear combination of all of the possible product functions such as (q.5.20.) may be written as the sum of all of the $N!$ permutations P of the electronic coordinates of the N-electron atom, to give

$$\sum_P c_P P\Psi'_1(\mathbf{x}_1)\Psi'_2(\mathbf{x}_2)\Psi'_3(\mathbf{x}_3)\cdots\Psi'_N(\mathbf{x}_N), \qquad (q.5.22.)$$

where the c_P terms are constants. This linear combination function will be antisymmetric if all of the c_P factors have the same absolute magnitude and are positive or negative according to whether P is an even or odd permutation; P is an even or odd permutation if there is an even or odd number of interchanges of pairs of electronic coordinates, respectively. If the product function (q.5.22.) is not required to be normalized, the constant c_P may be chosen to be unity and in that event

$$F = \sum_P (-1)^p P\Psi'_1(\mathbf{x}_1)\Psi'_2(\mathbf{x}_2)\Psi'_3(\mathbf{x}_3)\cdots\Psi'_N(\mathbf{x}_N), \qquad (q.5.23.)$$

where p is the number of interchanges in the permutation P (see page 152). The permutation, for example, which changes [123] into [321] corresponds to

$p = 1$, for this permutation arises by interchanging 1 and 3. But the summation in (q.5.23.) is simply another way of writing the determinant:

$$F = \begin{vmatrix} \Psi'_1(\mathbf{x}_1) & \Psi'_1(\mathbf{x}_2) & \Psi'_1(\mathbf{x}_3) & \cdots & \Psi'_1(\mathbf{x}_N) \\ \Psi'_2(\mathbf{x}_1) & \Psi'_2(\mathbf{x}_2) & \Psi'_2(\mathbf{x}_3) & \cdots & \Psi'_2(\mathbf{x}_N) \\ \Psi'_3(\mathbf{x}_1) & \Psi'_3(\mathbf{x}_2) & \Psi'_3(\mathbf{x}_3) & \cdots & \Psi'_3(\mathbf{x}_N) \\ \vdots & \vdots & \vdots & & \vdots \\ \Psi'_N(\mathbf{x}_1) & \Psi'_N(\mathbf{x}_2) & \Psi'_N(\mathbf{x}_3) & \cdots & \Psi'_N(\mathbf{x}_N) \end{vmatrix}. \qquad \text{(q.5.24.)}$$

The proof that the determinantal function (q.5.24.) is antisymmetric is a simple consequence of one of the fundamental properties of determinants, namely, that interchanging any two rows or any two columns of a determinant changes its sign; interchanging any pair of electronic coordinates such as $\mathbf{x}_2$ and $\mathbf{x}_3$ obviously involves interchanging the second and third columns of the determinant and hence its sign. What is equally true, but not quite so easy to prove (the actual proof requires the use of group theory, but see below), is that the determinantal function (q.5.24.) is the only antisymmetric function of the type of (q.5.22.).

The antisymmetry principle leads automatically to the Pauli Exclusion Principle. This is so because, if two of the one-electron spin orbitals are the same, then two rows of the determinant will be the same, and any determinant with two equal rows is zero (page 152). A wave function cannot be zero and so no two electrons may be described by the same spin orbital or be in the same one-electron state. Alternatively, since interchanging any two rows of a determinant changes its sign, and if two rows are identical, interchanging them cannot make any difference, and the only number whose sign can be changed without making any difference is zero. This means that no determinantal wave function may be constructed which is in violation of the Pauli Exclusion Principle. In this sense, the Pauli Principle is an immediate consequence of the antisymmetry principle and does not need to be assumed as a separate postulate. Also, a symmetric function of the form of (q.5.22.), which would result by choosing all of the factors c_P to have the same magnitude and sign, does not lead to the Pauli Principle, for if two of the one-electron spin orbitals are the same, then the total wave function would not vanish in this case. On the contrary, for systems of particles described by symmetric wave functions, all of the particles may be in the same one-particle state.

The determinantal function given by (q.5.23.) or (q.5.24.) is not normalized. It is not even normalized, when the one-electron spin orbitals are already normalized. Of course, the spin orbitals may be normalized and also orthogonalized by the methods already described on pages 120 and 124, respectively. The normalizing constant for the total determinantal function (q.5.23.)

depends on the spin orbitals $\Psi_i(\mathbf{x}_i)$, and if these functions are orthogonal, that is, if

$$\int \Psi_i^*(\mathbf{x})\Psi_j(\mathbf{x})\, d\mathbf{x} = \delta_{ij}, \qquad \text{(q.5.25.)}$$

then the value of the normalizing constant for F turns out to be $[N!]^{-1/2}$, and the normalized determinantal function to be

$$F = [N!]^{-1/2} \sum_P (-1)^p P\Psi_1(\mathbf{x}_1)\Psi_2(\mathbf{x}_2)\Psi_3(\mathbf{x}_3)\cdots\Psi_N(\mathbf{x}_N). \qquad \text{(q.5.26.)}$$

The most straightforward proof of (q.5.26.) depends on the properties of symmetric operators. A symmetric operator is one which is symmetric in the electronic coordinates, such that if P is a permutation operator, and $\hat{\Omega}$ is a symmetric operator, then $P\hat{\Omega} = \hat{\Omega}$; the Hamiltonian, for example, is a symmetric operator. Suppose that equation (q.5.26.) is true; that is, suppose that the determinantal function (q.5.23.) is indeed normalized by dividing it by $\sqrt{N!}$ and that $\hat{\Omega}$ is any symmetric operator. Then the expectation or mean value of the symmetric operator, from (q.2.66.), is $\langle \Omega \rangle = \int F^*\hat{\Omega}F\, d\tau' = \langle F|\hat{\Omega}|F \rangle$, where F is given by (q.5.26.) and $d\tau' = d\mathbf{x}_1\, d\mathbf{x}_2\, d\mathbf{x}_3\cdots d\mathbf{x}_N$, denoting that a summation over the values of the spin variables is included in the integration throughout the configuration space of the system of N electrons. The mean value of the symmetric operator, then, or more properly the mean value of the dynamical variable represented by the symmetric operator $\hat{\Omega}$ for the state whose normalized wave function is F given by (q.5.26.), is

$$\langle \Omega \rangle = \int F^*\hat{\Omega}F\, d\tau' = \int F^*\hat{\Omega}[N!]^{-1/2}$$

$$\times \sum_P (-1)^p P\Psi_1(\mathbf{x}_1)\Psi_2(\mathbf{x}_2)\Psi_3(\mathbf{x}_3)\cdots\Psi_N(\mathbf{x}_N)\, d\tau'. \qquad \text{(q.5.27.)}$$

The constant factor $[N!]^{-1/2}$ and the summation over P may be written to the left of the integral sign in (q.5.27.) without altering the mean value expression, so

$$\langle \Omega \rangle = [N!]^{-1/2} \sum_P (-1)^p \int F^*\hat{\Omega}P\Psi_1(\mathbf{x}_1)\Psi_2(\mathbf{x}_2)\Psi_3(\mathbf{x}_3)\cdots\Psi_N(\mathbf{x}_N)\, d\tau'$$

$$= [N!]^{-1/2} \sum_P (-1)^p \int\cdots\int [F^*\hat{\Omega}P\Psi_1(\mathbf{x}_1)\Psi_2(\mathbf{x}_2)\Psi_3(\mathbf{x}_3)\cdots\Psi_N(\mathbf{x}_N)]$$

$$\times\, d\mathbf{x}_1\, d\mathbf{x}_2\cdots d\mathbf{x}_N. \qquad \text{(q.5.28.)}$$

In (q.5.28.), the permutation operator P only acts on the product of the spin orbitals written to the right of P; if the operator P is written to the left of the integral sign, then in this position it would operate on both F^* and $\hat{\Omega}$ as well.

The fact that $\hat{\Omega}$ is a symmetric operator means by definition $P\hat{\Omega} = \hat{\Omega}$, and the effect of operating on F^* by P can be negated by allowing the operator P^{-1} to operate on F^*, written to the right of the integral sign, since $P[P^{-1}F^*] = F^*$. Therefore, (q.5.28.) may be written as

$$\langle\Omega\rangle = [N!]^{-1/2} \sum_P (-1)^p P \int [P^{-1}F^*]\hat{\Omega}\Psi_1(\mathbf{x}_1)\Psi_2(\mathbf{x}_2)\Psi_3(\mathbf{x}_3)\cdots\Psi_N(\mathbf{x}_N)\,d\tau'.$$

$$(q.5.29.)$$

Now, if P^{-1} is the inverse of the operation P, then the number of sign interchanges involved for both operators is the same, p, since the two operators simply affect the sign interchanges in the reverse order, each producing p interchanges of pairs of columns of the determinant (q.5.24.). Therefore, $P^{-1}F^* = (-1)^p F^*$. If this result is substituted in (q.5.29.), and the term $(-1)^p$ thus introduced to the right of the integral sign is factored out and combined with the other term $(-1)^p$ to the left of the integral sign, then since $(-1)^{2p} = 1$, (q.5.29.) becomes

$$\langle\Omega\rangle = [N!]^{-1/2} \sum_P P \int F^*\hat{\Omega}\Psi_1(\mathbf{x}_1)\Psi_2(\mathbf{x}_2)\Psi_3(\mathbf{x}_3)\cdots\Psi_N(\mathbf{x}_N)\,d\tau'. \quad (q.5.30.)$$

For each of the integrations in (q.5.30.) with respect to each of the variables of integration $d\mathbf{x}_i$, the same result is obtained for each of the $N!$ permutations P, since the value of a definite integral is a function only of its limits and not of the variable of integration. Therefore, since there are $N!$ permutations altogether, (q.5.30.) becomes

$$\langle\Omega\rangle = [N!]^{-1/2}N! \int\cdots\int F^*\hat{\Omega}\Psi_1(\mathbf{x}_1)\Psi_2(\mathbf{x}_2)\Psi_3(\mathbf{x}_3)\cdots\Psi_N(\mathbf{x}_N)$$
$$d\mathbf{x}_1\,d\mathbf{x}_2\,d\mathbf{x}_3\cdots d\mathbf{x}_N$$
$$= [N!]^{1/2} \int F^*\hat{\Omega}\Psi_1(\mathbf{x}_1)\Psi_2(\mathbf{x}_2)\Psi_3(\mathbf{x}_3)\cdots\Psi_N(\mathbf{x}_N)\,d\tau'$$
$$= \int F^*\hat{\Omega}F\,d\tau'. \qquad (q.5.31.)$$

Equation (q.5.31.) has been deduced on the assumption that the determinantal function F was normalized and represents a very useful theorem in quantum mechanics. It may be stated as follows: For a symmetric operator $\hat{\Omega}$ and a normalized determinantal wave function F, the mean value of the dynamical variable corresponding to the symmetric operator, $\int F^*\hat{\Omega}F\,d\tau'$, is given by an expression in which F is replaced by $\sqrt{N!}$ times the product of the diagonal elements of the determinant.

Making use of this theorem and considering the determinantal function given by (q.5.26.), it is apparent that the expression $\int F^*F\,d\tau'$ represents the mean value of the operator unity, with respect to the function F, or alternatively this expression may be formally regarded as the matrix component of unity (see page 125). Therefore, replacing F by $\sqrt{N!}$ multiplied by the product of the diagonal elements of the determinant, in accordance with the symmetric operator theorem, since unity is obviously symmetric with respect to the various electronic coordinates as it does not depend on any of them, then

$$\int F^*F\,d\tau' = \sqrt{N!}\int F^*\Psi_1(\mathbf{x}_1)\Psi_2(\mathbf{x}_2)\Psi_3(\mathbf{x}_3)\cdots\Psi_N(\mathbf{x}_N)\,d\tau',$$

and now substituting for the complex conjugate of F, from (q.5.26.), there results

$$\begin{aligned}
\int F^*F\,d\tau' &= \int\left[\left\{\sum_P(-1)^P P\Psi_1^*(\mathbf{x}_1)\Psi_2^*(\mathbf{x}_2)\Psi_3^*(\mathbf{x}_3)\cdots\Psi_N^*(\mathbf{x}_N)\right\}\right.\\
&\qquad\qquad\left.\times\ \Psi_1(\mathbf{x}_1)\Psi_2(\mathbf{x}_2)\cdots\Psi_N(\mathbf{x}_N)\right]d\tau'\\
&= \int\cdots\int|\Psi_1(\mathbf{x}_1)|^2|\Psi_2(\mathbf{x}_2)|^2|\Psi_3(\mathbf{x}_3)|^2\cdots|\Psi_N(\mathbf{x}_N)|^2\\
&\qquad\qquad\times\ d\mathbf{x}_1\,d\mathbf{x}_2\,d\mathbf{x}_3\cdots d\mathbf{x}_N\\
&= 1,\qquad\qquad\qquad\qquad\qquad\qquad\qquad\qquad\text{(q.5.32.)}
\end{aligned}$$

since all the other terms in the sum are equal to zero owing to the orthonormality of the spin orbitals Ψ_i, as given by (q.5.25.). Thus, since $\int F^*F\,d\tau' = 1$, the determinantal functions given by (q.5.26.) must indeed be normalized.

Q.5.6. Two-Electron Spin Orbitals

For polyelectronic atoms, experimental observation indicates that, if two electrons are in the same orbital state, they must have opposed spin moments. An atom of helium in its ground state is observed to have no resultant magnetic moment, so that the two electrons which are both in $1s$-states, must have their spin moments opposed to one another. This, of course, represents one way of stating the Pauli exclusion principle, that if spin is included in describing a one-electron state, not more than two electrons may be in the same energy state.

The ground-state energy of the helium atom has previously been calculated, neglecting the electronic interaction energy (page 224, Q.3.9.), by the variational method (page 237, Q.3.4.c), by the perturbation method (page 251, Q.4.6.), and by the Hartree self-consistent field method (page 267, Q.4.10.).

The helium atom is still the simplest two-electron system to use in illustration of the use of determinantal wave functions, including the spin interaction effect.

The Hamiltonian for the helium atom is given by (q.4.30.), and if the electronic interaction term ε^2/r_{12} is neglected, the Schrödinger equation, (q.3.140.), separates into two hydrogenic equations, (q.3.141.) and (q.3.142.), each of which is of the form

$$[h^2/8\pi^2 m\nabla^2 - 2\varepsilon^2/r]\psi = E\psi. \tag{q.5.33.}$$

Of the various methods referred to above, which have been used previously to calculate the ground-state energy of helium, the Hartree method gave the best calculated result. The Hartree orbitals would therefore constitute the best starting point in constructing determinantal functions; however, they are not orthogonal, which makes it a little more difficult to normalize the total determinantal wave function. On the other hand, the eigenfunctions of (q.5.33.) are orthogonal, so that it is perhaps easier to construct normalized determinantal functions from them, and to make use of the first order perturbation theory to take account of the electronic interaction term. The first order perturbation theory calculation, in any event, leads to quite a reasonable value for the ground-state energy of helium. The general technique of forming properly antisymmetrized functions is essentially the same, no matter what type of orbital functions are used.

For the ground state of helium, it is known from experimental observation that both electrons are in $1s$-states and they must have opposed spins. If it is assumed that the orbitals are eigenfunctions of (q.5.33.), with eigenvalues corresponding to the lowest possible eigenvalue of -2.0 Hartrees, then the two one-electron spin orbitals may be written as

$$\Psi_1(\mathbf{x}_1) = \psi(r_1)\alpha(\zeta) \qquad \text{and} \qquad \Psi_2(\mathbf{x}_2) = \psi(r_2)\beta(\zeta),$$

respectively, and the normalized determinantal function, as given by (q.5.26.), would be

$$F_0(1/\sqrt{2})\begin{vmatrix} \Psi_1(\mathbf{x}_1) & \Psi_1(\mathbf{x}_2) \\ \Psi_2(\mathbf{x}_1) & \Psi_2(\mathbf{x}_2) \end{vmatrix}$$

$$= (1/\sqrt{2})[\Psi_1(\mathbf{x}_1)\Psi_2(\mathbf{x}_2) - \Psi_1(\mathbf{x}_2)\Psi_2(\mathbf{x}_1)]$$

$$= (1/\sqrt{2})[\psi(r_1)\psi(r_2)][\alpha(\zeta_1)\beta(\zeta_2) - \alpha(\zeta_2)\beta(\zeta_1)]. \tag{q.5.34.}$$

The determinantal function (q.5.34.), expressed as a product of an orbital function and a spin function, is such that the orbital factor is symmetric in the positional coordinates of the two electrons, while the spin factor is antisymmetric with respect to the spin coordinates of the two electrons. The ground-state energy calculated by the use of this function is then given by

$$\iint F_0{}^*\hat{H}F\, d\mathbf{x}_1\, d\mathbf{x}_2 = (1/2) \sum_{\zeta_1,\zeta_2 = \pm 1} \iint \psi^*(r_1)\psi^*(r_2)\hat{H}\psi(r_1)\psi(r_2)$$
$$\times\, dr_1\, dr_2[\alpha(\zeta_1)\beta(\zeta_2) - \alpha(\zeta_2)\beta(\zeta_1)]^2$$
$$= \iint [\psi^*(r_1)\psi^*(r_2)\hat{H}\psi(r_1)\psi(r_2)]\, dr_1\, dr_2, \qquad \text{(q.5.35.)}$$

since the spin factor vanishes when ζ_1 and ζ_2 have the same sign and is unity when they have opposite signs. With $\hat{H}$ as given by (q.4.30.), this leads to the same value for the ground-state energy as given by the use of the simple product function $\psi(r_1)\psi(r_2)$ and the first order perturbation calculation, as in Q.4.6.(c). For this reason, it has been possible to neglect explicit account of electron spin in the previous perturbation calculation. While this is not true for excited states, in which the two electrons are not both described by $1s$ orbitals, it is always true, even for excited states, if both electrons are described by the same orbital functions. States for which the two electrons are described by the same orbital function must have opposed spins and may be described, as for the ground state, by a single determinantal function; such states are therefore non-degenerate and are known as singlets or singlet states. It is also common practice to refer to determinantal functions such as (q.5.34.) as Slater determinants.[1]

For the various excited states of a two-electron atom, the two orbitals may be different, in which event the electrons need not have opposing spins. For example, if the two different normalized orbitals are $\psi_a(r)$ and $\psi_b(r)$, and these are taken to be eigenfunctions of (q.5.33.) corresponding to the eigenvalues E_a and E_b, then there are four possible ways of associating the two different spin functions with the two different orbitals, to conform with the Exclusion Principle, and therefore four possible determinantal wave functions, which are:

$$F_1 = (1/\sqrt{2})\begin{vmatrix} \psi_a(r_1)\alpha(\zeta_1) & \psi_a(r_2)\alpha(\zeta_2) \\ \psi_b(r_1)\alpha(\zeta_1) & \psi_b(r_2)\alpha(\zeta_2) \end{vmatrix}$$
$$= (1/\sqrt{2})[\psi_a(r_1)\psi_b(r_2) - \psi_a(r_2)\psi_b(r_1)][\alpha(\zeta_1)\alpha(\zeta_2)];$$

$$F_2 = (1/\sqrt{2})\begin{vmatrix} \psi_a(r_1)\beta(\zeta_1) & \psi_a(r_2)\beta(\zeta_2) \\ \psi_b(r_1)\beta(\zeta_1) & \psi_b(r_2)\beta(\zeta_2) \end{vmatrix}$$
$$= (1/\sqrt{2})[\psi_a(r_1)\psi_b(r_2) - \psi_a(r_2)\psi_b(r_1)][\beta(\zeta_1)\beta(\zeta_2)]; \qquad \text{(q.5.36.)}$$

$$F_3 = (1/\sqrt{2})\begin{vmatrix} \psi_a(r_1)\alpha(\zeta_1) & \psi_a(r_2)\alpha(\zeta_2) \\ \psi_b(r_1)\beta(\zeta_1) & \psi_b(r_2)\beta(\zeta_2) \end{vmatrix}$$
$$= (1/\sqrt{2})[\psi_a(r_1)\psi_b(r_2)\alpha(\zeta_1)\beta(\zeta_2) - \psi_a(r_2)\psi_b(r_1)\alpha(\zeta_2)\beta(\zeta_1)];$$

$$F_4 = (1/\sqrt{2})\begin{vmatrix} \psi_a(r_1)\beta(\zeta_2) & \psi_a(r_2)\beta(\zeta_2) \\ \psi_b(r_1)\alpha(\zeta_1) & \psi_b(r_2)\alpha(\zeta_2) \end{vmatrix}$$
$$= (1/\sqrt{2})[\psi_a(r_1)\psi_b(r_2)\alpha(\zeta_2)\alpha(\zeta_1) - \psi_a(r_2)\psi_b(r_1)\alpha(\zeta_1)\beta(\zeta_2)].$$

---

[1] Slater, J. C. *Phys., Rev.*, 1929, **34**, 1283.

The functions F_1, F_2, F_3, and F_4, since they are constructed from the ortho-normal eigenfunctions $\psi_a(r)$ and $\psi_b(r)$ of (q.5.33.) and the spin functions $\alpha(\zeta)$ and $\beta(\zeta)$, must all be normalized. Also, the unperturbed state, corresponding to the state where the electron interaction term is neglected, is four-fold degenerate, each of these four functions corresponding to the unperturbed energy $E_a + E_b$. Therefore, any linear combination of these four functions also corresponds to the same unperturbed energy. F_1 and F_2 differ from F_3 and F_4 in that the former pair appear as the product of a linear combination of orbital functions and spin functions, whereas the latter pair do not. However, by selecting the linear combinations $F_3 + F_4$ and $F_3 - F_4$, another pair of determinantal functions may be constructed, in which the orbital part and spin part of the total functions are written as separate products. Thus,

$$F_3 + F_4 = (1/\sqrt{2})[\psi_a(r_1)\psi_b(r_2) - \psi_a(r_2)\psi_b(r_1)][\alpha(\zeta_1)\beta(\zeta_2) + \alpha(\zeta_2)\beta(\zeta_1)]$$

and

$$F_3 - F_4 = (1/\sqrt{2})[\psi_a(r_1)\psi_b(r_2) + \psi_a(r_2)\psi_b(r_1)][\alpha(\zeta_1)\beta(\zeta_2) - \alpha(\zeta_2)\beta(\zeta_1)].$$

$$(q.5.37.)$$

These linear combination functions are not now normalized, as written above, but may be normalized by multiplying again by $1/\sqrt{2}$. The four possible determinantal functions corresponding to the unperturbed eigenvalue $E_a + E_b$ may therefore be selected as F_1, F_2, $(1/\sqrt{2})[F_3 + F_4]$, and $(1/\sqrt{2})[F_3 - F_4]$, each of which is normalized and of the form in which the function is written as the product of two factors, one containing orbital functions and one containing spin functions only.

Apart from the advantage conferred by similarity of form, there is a further advantage in using functions of this type when calculating the perturbation energy due to the term ε^2/r_{12}. In applying first order perturbation theory to a four-fold degenerate unperturbed state (see section Q.4.5.), the first order perturbation energies, for the perturbation ε^2/r_{12}, are obtained as the roots of a secular equation of the type given by (q.4.74.), which involves a fourth order secular determinant; as on page 250, this secular determinant may be written such that each element is the matrix component of either the total Hamiltonian or of the perturbation ε^2/r_{12}. The advantage of using the functions F_1, F_2, $(1/\sqrt{2})[F_3 + F_4]$, and $(1/\sqrt{2})[F_3 - F_4]$ rather than simply the set F_1, F_2, F_3, and F_4 in constructing this fourth order secular determinant is that for the former set of functions each of the non-diagonal elements of the determinant vanishes due to the orthogonality of the spin factors; each of the off-diagonal elements is of the form $\langle F_n|\hat{H}|F_m\rangle$ or $\langle F_n|\hat{H}'|F_m\rangle$, where $n \neq m$, and is thus zero. The correct perturbed energy values will result by the use of either set of functions, but it is obviously more convenient to choose the set which makes the non-diagonal elements of the secular determinant vanish, for then the perturbed energy states are immediately given by the four equations

$$E = \langle F|\hat{H}|F\rangle = \int F^*\hat{H}F\, d\tau', \qquad \text{(q.5.38a.)}$$

where F may be either F_1, F_2, $(1/\sqrt{2})[F_3 + F_4]$, or $(1/\sqrt{2})[F_3 - F_4]$.

The values of the perturbed energy states calculated by the use of (q.5.38a.) for each F is independent of the normalized spin factors contained in each of the four functions. For example, for F_1

$$\langle F_1|\hat{H}|F_1\rangle = (1/2) \sum_{\zeta_1,\zeta_2 = \pm 1} \iint [\psi_a{}^*(r_1)\psi_b{}^*(r_2) - \psi_a{}^*(r_2)\psi_b{}^*(r_1)]$$
$$\times \hat{H}[\psi_a(r_1)\psi_b(r_2) - \psi_a(r_2)\psi_b(r_1)]|\alpha(\zeta_1)|^2|\alpha(\zeta_2)|^2\, dr_1\, dr_2$$
$$= (1/2) \iint [\psi_a{}^*(r_1)\psi_b{}^*(r_2) - \psi_a{}^*(r_2)\psi_b{}^*(r_1)]$$
$$\times \hat{H}[\psi_a(r_1)\psi_b(r_2) - \psi_a(r_2)\psi_b(r_1)]\, dr_1\, dr_2.$$

Two of the other determinantal functions, namely F_2 and $(1/\sqrt{2})[F_3 + F_4]$, also contain the same antisymmetric orbital factor as F_1, and since the normalized spin factors contained in these functions do not affect the calculated perturbed energies, one of the perturbed states is triply-degenerate. The remaining function, $(1/\sqrt{2})[F_3 - F_4]$, contains the symmetric orbital factor, $(1/\sqrt{2})[\psi_a(r_1)\psi_b(r_2) + \psi_a(r_2)\psi_b(r_1)]$, and thus leads to the perturbed energy value,

$$E = (1/2) \iint [\psi_a{}^*(r_1)\psi_b{}^*(r_2) + \psi_a{}^*(r_2)\psi_b{}^*(r_1)]$$
$$\times \hat{H}[\psi_a(r_1)\psi_b(r_2) + \psi_a(r_2)\psi_b(r_1)]\, dr_1\, dr_2.$$

There are, therefore, only two perturbed energy states, a triply-degenerate state described by the three functions with antisymmetric orbital factors and a non-degenerate state, corresponding to the function with the symmetric orbital factor. Incidentally, F_1, F_2, and $(1/\sqrt{2})[F_3 + F_4]$, with antisymmetric orbital factors, have spin factors which are symmetric in the spin coordinates, while the function $(1/\sqrt{2})[F_3 - F_4]$, with a symmetric orbital factor, has an antisymmetric spin factor, so that all four of these functions are antisymmetric in all of the electron coordinates as required by the generalized form of the Pauli Exclusion Principle.

It is thus found that for the two-electron system which is the helium atom, the unperturbed excited states, neglecting the electronic interaction term, are four-fold degenerate, corresponding to the four determinantal functions given by F_1, F_2, $(1/\sqrt{2})[F_3 + F_4]$, and $(1/\sqrt{2})[F_3 - F_4]$ of (q.5.36.). Also, treating the electronic interaction term ε^2/r_{12} by first order perturbation theory, the excited states are only two-fold degenerate, corresponding to the symmetric and antisymmetric orbital functions, respectively, so that

$$E = (1/2) \iint [\psi_a^*(r_1)\psi_b^*(r_2) \pm \psi_a^*(r_2)\psi_b^*(r_1)]$$
$$\times \hat{H}[\psi_a(r_1)\psi_b(r_2) \pm \psi_a(r_2)\psi_b(r_1)] \, dr_1 \, dr_2. \quad \text{(q.5.38}b.)$$

Now, making use of the fact that $\psi_a(r)$ and $\psi_b(r)$ are orthonormal eigenfunctions of (q.5.33.) and substituting (q.4.30.) for $\hat{H}$, there is found for E the two possible values of the energy of the perturbed excited states,

$$E = E_a + E_b + (1/2) \iint (\varepsilon^2/r_{12})$$
$$\times [|\psi_a(r_1)|^2|\psi_b(r_2)|^2 + |\psi_a(r_2)|^2|\psi_b(r_1)|^2$$
$$\pm [\psi_a^*(r_1)\psi_b^*(r_2)\psi_a(r_2)\psi_b(r_1) + \psi_a(r_1)\psi_b(r_2)\psi_a^*(r_2)\psi_b^*(r_1)]] \, dr_1 \, dr_2$$
$$= E_a + E_b + \iint [\varepsilon^2/r_{12}][|\psi_a(r_1)|^2|\psi_b(r_2)|^2 \, dr_1 \, dr_2$$
$$\pm \iint [\varepsilon^2/r_{12}][\psi_a^*(r_1)\psi_b^*(r_2)\psi_a(r_2)\psi_b(r_1)] \, dr_1 \, dr_2, \quad \text{(q.5.39.)}$$

since

$$|\psi_a(r_1)|^2|\psi_b(r_2)|^2 + |\psi_a(r_2)|^2|\psi_b(r_1)|^2 = 2[|\psi_a(r_1)|^2|\psi_b(r_2)|^2$$

and

$$[\psi_a^*(r_1)\psi_b^*(r_2)\psi_a(r_2)\psi_b(r_1) + \psi_a(r_1)\psi_b(r_2)\psi_a^*(r_2)\psi_b^*(r_1)]$$
$$= 2[\psi_a^*(r_1)\psi_b^*(r_2)\psi_a(r_2)\psi_b(r_1)],$$

as each pair of terms in these two expressions differ from one another only in that r_1 and r_2 have been interchanged, which does not affect the value of the double integral containing these terms.

The integrals in (q.5.39.) have special significance. The first of the double integrals is referred to as a coulombic integral and, variously written as

$$C_{ab} = \iint [\varepsilon^2/r_{12}]|\psi_a(r_1)|^2|\psi_b(r_2)|^2 \, dr_1 \, dr_2$$
$$= \langle \psi_a(r_1)\psi_b(r_2)|\varepsilon^2/r_{12}|\psi_a(r_1)\psi_b(r_2)\rangle, \quad \text{(q.5.40.)}$$

is of the same form as the electronic repulsion integral of (q.4.37.) and represents the electrostatic energy due to the interaction between the two charge clouds $|\psi_a(r_1)|^2$ and $|\psi_b(r_2)|^2$, and would occur even if a simply product type of wave function were used to describe the system. The second of the double integrals in (q.5.39.) is referred to as the exchange integral,

$$J_{ab} = \iint [\varepsilon^2/r_{12}]\psi_a^*(r_1)\psi_b^*(r_2)\psi_a(r_2)\psi_b(r_1) \, dr_1 \, dr_2$$
$$= \langle \psi_a(r_1)\psi_b(r_2)|\varepsilon^2/r_{12}|\psi_a(r_2)\psi_b(r_1)\rangle, \quad \text{(q.5.41.)}$$

and differs from the coulombic integral in that it has no simple classical interpretation. It appears in the energy expression only because an anti-symmetric total wave function has been used, and the antisymmetry principle has only a quantum mechanical interpretation. The antisymmetry ensures that the total charge density remains unaltered when the two electrons are exchanged, and the value of this integral corresponds to what is referred to as the exchange energy. If the indistinguishability of electrons had not been taken into account by requiring that the total wave function should be anti-symmetric in all of the electronic coordinates, the energy expression of (q.5.39.) would not contain this exchange integral J_{ab}. Coulombic integrals representing electrostatic repulsion energies are always positive, since the integrand of (q.5.40.) must, for example, always be positive. The integrand of J_{ab}, the exchange integral, since it has no classical interpretation, is such that its sign cannot be immediately deduced; it may be positive for some values of r_1 and r_2 and negative for others.

If $\psi_a(r)$ and $\psi_b(r)$, which are eigenfunctions of a hydrogenic equation, are assumed for simplicity to be real, then when r_1 and r_2 are very nearly equal, the integrand of J_{ab} will have almost the same value as the integrand of C_{ab}, which will then be very large and positive due to the small value of r_{12}. On the other hand, for the integrand of J_{ab} to be negative would require that either $\psi_a(r_1)$ and $\psi_a(r_2)$ or $\psi_b(r_1)$ and $\psi_b(r_2)$ should have opposite signs, which in turn would require that r_1 and r_2 should be quite different from one another, so that r_{12} would be large and so the integrand would be small. Thus, the contributions to the integral from regions where the integrand is positive generally exceed those from regions where the integrand is negative, making the exchange integral generally positive. The two energy values given by equation (q.5.39.) may therefore be written as

$$E_t = E_a + E_b + C_{ab} - J_{ab} \quad \text{and} \quad E_s = E_a + E_b + C_{ab} + J_{ab}. \quad \text{(q.5.42.)}$$

If J_{ab} is positive, then E_t will correspond to the lower energy value and E_s to the higher energy value. The lower energy state is triply-degenerate, a triplet state, the three-wave functions, each of which has the same antisymmetric orbital factor but different symmetric spin factors being

$$[1/\sqrt{2}][\psi_a(r_1)\psi_b(r_2) - \psi_a(r_2)\psi_b(r_1)][\alpha(\zeta_1)\alpha(\zeta_2)],$$

$$[1/\sqrt{2}][\psi_a(r_1)\psi_b(r_2) - \psi_a(r_2)\psi_b(r_1)][\beta(\zeta_1)\beta(\zeta_2)],$$

and
$$\text{(q.5.43.)}$$

$$[1/2][\psi_a(r_1)\psi_b(r_2) - \psi_a(r_2)\psi_b(r_1)][\alpha(\zeta_1)\beta(\zeta_2) + \alpha(\zeta_2)\beta(\zeta_1)].$$

The orbital and spin factors in these functions are, of course, separately normalized. The higher energy state E_s is non-degenerate and is therefore a singlet state, like the ground state, and is described by the wave function with an antisymmetrical spin factor and a symmetric orbital factor,

$$[1/2][\psi_a(r_1)\psi_b(r_2) + \psi_a(r_2)\psi_b(r_1)][\alpha(\zeta_1)\beta(\zeta_2) - \alpha(\zeta_2)\beta(\zeta_1)]. \quad \text{(q.5.44.)}$$

Since $\psi_a(r)$ and $\psi_b(r)$ are eigenfunctions of a hydrogenic equation, all the eigenvalues E_a, E_b will be degenerate, except the lowest one, and therefore for any unperturbed energy state $E_a + E_b$, except the ground state, there will generally be a larger number of determinantal wave functions than for the case described above. Also, it should be noted that the state E_t is only triply-degenerate if the spin-orbit coupling effect is disregarded, and it has already been seen that this effect is small. These three states described by the functions given in (q.5.43.) would no longer be degenerate in an external magnetic field. In the first of these three states, the two electrons each have positive spins, in the second each have negative spins, and in the third there is a linear combination of states such that the spins of the two electrons are opposed to one another, so that for a magnetic field applied in the z-Cartesian coordinate direction, the total component of the spin momentum in this direction would be $+h/2\pi$, $-h/2\pi$, and 0, respectively. Alternatively stated, if a magnetic field of field strength H were to be applied in the z-Cartesian coordinate direction, the energies of the three states would be altered by $+2\mu_B H$, $-2\mu_B H$, and 0, respectively, thus removing the degeneracy.

Q.5.7. Polyelectronic Atoms—The Hartree–Fock Equations

It has previously been shown on pages 276–277 that the application of the variational method to the description of an N-electron system, making use of a trial function of a single product type, as given by (q.4.125.), gives rise to the same one-electron functions as those obtained by the application of the Hartree self-consistent field method. It was also concluded that the Hartree functions must be the best one-electron functions, of the single product type, to describe the ground state of a system. However, the requirement that the antisymmetry principle should be obeyed necessitates the use, not of a single product function, but an antisymmetrized product or a determinantal function to adequately describe such an N-electron system. The use of the variational principle, using a determinantal function instead of a single product function composed of the Hartree one-electron functions, enables a much improved many-electron function to be calculated, which in turn leads to an improved calculated energy value for an N-electron system. This is the basis of the Hartree–Fock method,[1,2] which differs essentially from the Hartree method in that an additional integral expression appears in the calculated

[1] Fock, V., *Z. Physik*, 1930, **61**, 126; **62**, 795.
[2] Slater, J. C., *Phys. Rev.*, 1930, **35**, 210.

energy values, of a nature similar to the integral (q.5.41.), which occurs in the expression (q.5.39.); it is this additional term which is referred to as the exchange integral or the exchange term.

It is a rather complicated process to derive the Hartree–Fock equations by the direct application of the variational method. The methods described by Slater and Fock, and also that of Roothaan,[1] all make use of Lagrange's method of undetermined multipliers to deal with the problem of varying the determinantal function for a system, subject to the restriction that the average value of the Hamiltonian of the system has a stationary value and that the various spin orbitals composing the determinantal function remain normalized; this method has been described in Q.4.2., in the proof of the variational method. It is fairly easily possible, however, to give a plausible, if not completely rigorous, derivation of the Hartree–Fock equations for a polyelectronic system, by a method similar to that used in deriving Hartree's equations, as described in Q.4.11.

The Hamiltonian operator for an N-electron atom, of nuclear charge Z, is given by equation (q.4.109.), and the corresponding normalized, antisymmetric total state function by (q.5.26.); also the condition that the various spin orbitals $\Psi_i(\mathbf{x}_i)$ should be orthogonal to one another is given by (q.5.25.). It is assumed that each of the spin orbitals in (q.5.26.) is the product of a function of the electronic coordinates and a function α or β of the spin coordinates; in other words, each spin orbital is assumed to correspond to a definite value of m_s. The Hamiltonian is a symmetric operator, so that it remains unchanged for any permutation of the electronic coordinates. Therefore, the total energy of the system may be found by the application of the theorem given by equation (q.5.31.) or

$$
E = \langle \hat{H} \rangle = \langle F | \hat{H} | F \rangle = [N!]^{1/2} \int F^* \hat{H} \Psi_1(\mathbf{x}_1) \cdots \Psi_N(\mathbf{x}_N)\, d\tau'
$$

$$
= \int \left[\sum_P (-1)^p P \Psi_1{}^*(\mathbf{x}_1) \cdots \Psi_N{}^*(\Psi_N) \right]
$$

$$
\times \left[-\sum_i^N [h^2/8\pi^2 m)\, \nabla_i{}^2 + \sum_i^N Z\varepsilon^2/r_i + (1/2) \sum_{i \neq j}^N \sum^N \varepsilon^2/r_{ij} \right]
$$

$$
\times\ \Psi_1(\mathbf{x}_1) \cdots \Psi_N(\mathbf{x}_N)\, d\tau'. \tag{q.5.45.}
$$

Because of (q.5.25.), the orthogonality and normalization of the $\Psi_i(\mathbf{x}_i)$ terms, the expression (q.5.45.) may be written as

$$
E = \sum_i^N \int \Psi_i{}^*(\mathbf{x}_i) \left[-(h^2/8\pi^2 m)\, \nabla_i{}^2 + \sum_i^N Z\varepsilon^2/r_i \right] \Psi_i(\mathbf{x}_i)\, d\mathbf{x}_i
$$

$$
+ (1/2) \sum_{i \neq j}^N \sum^N \iint [\Psi_i{}^*(\mathbf{x}_i)\Psi_j{}^*(\mathbf{x}_j) - \Psi_i{}^*(\mathbf{x}_j)\Psi_j{}^*(\mathbf{x}_i)]\varepsilon^2/r_{ij}
$$

$$
\times\ [\Psi_i(\mathbf{x}_i)\Psi_j(\mathbf{x}_j)]\, d\mathbf{x}_i\, d\mathbf{x}_j, \tag{q.5.46.}
$$

---

[1] Roothaan, C. C. J., *Rev. Mod. Phys.*, 1951, **23**, 69.

since, as far as the first term is concerned, it is only the identity operator in
the expression for F^*, which leaves the electronic coordinates unchanged,
which has a non-zero value; all the other permutations result in factors like
$\int \Psi_1{}^*(\mathbf{x}_1)\Psi_2(\mathbf{x}_1)\,d\mathbf{x}_1$, which are zero by (q.5.25.). On the other hand, due to
the factor ε^2/r_{ij} in the integrand of the second term containing the double
integral, both the identity operator and that which interchanges $\mathbf{x}_i$ and $\mathbf{x}_j$,
leaving the other coordinates unchanged, give non-zero results, and in the
latter case the operator $p = 1$ leads to the minus sign in front of the second
term in the square brackets immediately to the right of the double integral
sign of (q.5.46.). The equation (q.5.46.) may be alternatively written in the
expanded form,

$$
E = \sum_{i}^{N} \int \Psi_i{}^*(\mathbf{x}_i)\left[-(h^2/8\pi^2 m)\,\nabla_i{}^2 + \sum_{i}^{N} Z\varepsilon^2/r_i \right]\Psi_i(\mathbf{x}_i)\,d\mathbf{x}_i
$$

$$
+ (1/2)\sum_{i \neq j}^{N}\sum^{N} \iint [\varepsilon^2/r_{ij}]|\Psi_i(\mathbf{x}_i)|^2|\Psi_j(\mathbf{x}_j)|^2\,d\mathbf{x}_i\,d\mathbf{x}_j
$$

$$
- (1/2)\sum_{i \neq j}^{N}\sum^{N} \iint [\varepsilon^2/r_{ij}]\Psi_i{}^*(\mathbf{x}_j)\Psi_j{}^*(\mathbf{x}_i)\Psi_i(\mathbf{x}_i)\Psi_j(\mathbf{x}_j)\,d\mathbf{x}_i\,d\mathbf{x}_j. \quad \text{(q.5.47.)}
$$

This expression differs from the expression given by (q.4.126.) obtained in the
application of the Hartree method, using a simple product type function, in
that a third term, the exchange term, is added. This exchange integral arises
in the same manner as in the discussion of the helium atom given in Q.5.6.,
due to the exchange of the electronic coordinates in the various terms of the
expanded determinantal function. The second term in (q.5.47.) has the same
significance as (q.5.40.); it is the Coulomb energy representing the interaction
of the electronic charge clouds of the various electrons.

If the spin-orbit coupling effect is neglected, and this effect is small accord-
ing to the discussion in Q.5.3., then $\Psi_i(\mathbf{x}) = \psi_i(r)\chi_i(\zeta)$, where the spin function $\chi_i(\zeta)$ may be either α or β, and therefore the summation over the spin
variables may be easily carried out. By virtue of (q.5.12.) and (q.5.13.), the
summation over the spin variables of the first two terms in (q.5.47.) simply
gives rise to the result that $\Psi_i(\mathbf{x}_i)$ is replaced by $\psi_i(r_i)$. Furthermore, since the
value of a definite integral is independent of the variable of integration, it is
possible to write r instead of r_i in the first term and r_1 and r_2 instead of r_i and
r_j in the second term.

For the exchange term in (q.5.47.), the situation is different, however; it is
necessary to consider separately the cases where $\Psi_i(\mathbf{x}_i)$ and $\Psi_j(\mathbf{x}_j)$ have the
same spin, either α or β, and where they have different spins. Firstly, if $\Psi_i(\mathbf{x}_i)$
and $\Psi_j(\mathbf{x}_j)$ have the same spin, either α or β, the integral in the exchange

term adopts the value, for example, if they are both α

$$\sum_{\zeta_1,\zeta_2=\pm1}\iint [\varepsilon^2/r_{ij}]\psi_i^*(r_j)\psi_j^*(r_i)\psi_i(r_i)\psi_j(r_j)\,dr_i\,dr_j|\alpha(\zeta_i)|^2|\alpha(\zeta_j)|^2$$

$$=\iint [\varepsilon^2/r_{ij}]\psi_i^*(r_j)\psi_j^*(r_i)\psi_i(r_i)\psi_j(r_j)\,dr_i\,dr_j, \quad \text{(q.5.48.)}$$

with an identical result, if both spins are β. Secondly, the exchange integral vanishes if $\Psi_i(\mathbf{x}_i)$ and $\Psi_j(\mathbf{x}_j)$ have opposite spins for then

$$\sum_{\zeta_1,\zeta_2=\pm1}\iint [\varepsilon^2/r_{ij}]\psi_i^*(r_j)\psi_j^*(r_i)\psi_i(r_i)\psi_j(r_j)\,dr_i\,dr_j[\alpha(\zeta_j)\beta(\zeta_i)\alpha(\zeta_i)\beta(\zeta_j)]=0.$$

Thus, the expression for the exchange energy in (q.5.47.) becomes

$$-(1/2)\sum_{\substack{i\neq j}}^{N}\sum^{N}\iint [\varepsilon^2/r_{ij}]\psi_i^*(r_j)\psi_j^*(r_i)\psi_i(r_i)\psi_j(r_j)\,dr_i\,dr_j$$

$$=-(1/2)\sum_{\substack{i\neq j}}^{N}\sum^{N}\iint [\varepsilon^2/r_{12}]\psi_i^*(r_1)\psi_j^*(r_2)\psi_i(r_2)\psi_j(r_1)\,dr_1\,dr_2, \quad \text{(q.5.49.)}$$

where the double sum is now over those functions ψ_i and ψ_j which are associated with the same spin functions; thus the double sum is to be taken with respect to the electrons which have parallel spins only.

The expression (q.5.45.) for the total energy of an N-electron atom may thus be written in terms of orbital functions as

$$E = \sum_i^N \int \psi_i^*(r)\left[-(h^2/8\pi^2m)\,\nabla^2 + \sum_i^N Z\varepsilon^2/r\right]\psi_i(r)\,dr$$

$$+ (1/2)\sum_{\substack{i\neq j}}^{N}\sum^{N}\iint [\varepsilon^2/r_{12}]|\psi_i(r_1)|^2|\psi_j(r_2)|^2\,dr_1\,dr_2$$

$$- (1/2)\sum_{\substack{i\neq j}}^{N}\sum^{N}\iint [\varepsilon^2/r_{12}]\psi_i^*(r_1)\psi_j^*(r_2)\psi_i(r_2)\psi_j(r_1)\,dr_1\,dr_2. \quad \text{(q.5.50.)}$$

Since the second and third terms in the above expression become equivalent if $i = j$ and they have opposite signs, it is unnecessary to retain the stipulation that $i \neq j$. Since the value of a definite integral is independent of the symbols used for the integration variable, but is a function only of the limits of the integration, (q.5.50.) could equally well be written with the first term expressed in terms of the variable r_1 rather than r. Then, the only terms in (q.5.50.) which would be dependent on the particular function ψ_i would be

$$\int \psi_i^*(r_1)\left[-(h^2/8\pi^2m)\,\nabla_1^2 + \sum_i^N Z\varepsilon^2/r_1\right]\psi_i(r_1)\,dr_1$$

$$+ \sum_j^N\iint [\varepsilon^2/r_{12}]|\psi_i(r_1)|^2|\psi_j(r_2)|^2\,dr_1\,dr_2$$

$$- \sum_j^N\iint [\varepsilon^2/r_{12}]\psi_i^*(r_1)\psi_j^*(r_2)\psi_i(r_2)\psi_j(r_1)\,dr_1\,dr_2, \quad \text{(q.5.51.)}$$

since the factors of (1/2) vanish due to the fact that the number i occurs in both sums of the double summations, and the spin i = spin j. The expression (q.5.51.) is of the form

$$\int \psi_i{}^*(r_1)\hat{H}_i\psi_i(r_1)\, dr_1, \qquad (q.5.52.)$$

with the operator $\hat{H}_i$ written as

$$\hat{H}_i = -(h^2/8\pi^2 m)\,\nabla_1{}^2 + \sum_i^N Z\varepsilon^2/r_1 + \varepsilon^2 \sum_j^N \int [|\psi_j(r_2)|^2/r_{12}]\, dr_2$$

$$- \varepsilon^2 \sum_j^N \int [\psi_j{}^*(r_2)\psi_i(r_2)\psi_j(r_1)]/[r_{12}\psi_i(r_1)]\, dr_2, \quad (q.5.53.)$$

again with spin j = spin i in the summations.

The ground state of an N-electron system, corresponding to the minimum value of E given by (q.5.50.), will be given by suitably choosing the functions ψ_i which minimize this expression. But the only terms contributing to E which contain any particular function ψ_i are those described by (q.5.52.), and in terms of the variation principle, the appropriate functions ψ_i are obtained as eigenfunctions of the equations $\hat{H}_i\psi_i = E_i\psi_i$, with their lowest eigenvalues E_i. These are the equations which are referred to as the Hartree–Fock equations for ψ_i. If the operator $\hat{H}_i$, as given by (q.5.53.), is substituted in these equations, then the Hartree–Fock equations for the functions ψ_i are

$$\left[-(h^2/8\pi^2 m)\,\nabla_1{}^2 + \sum_i^N Z\varepsilon^2/r_1 + \varepsilon^2 \sum_j^N \int [|\psi_j(r_2)|^2]/r_{12}\, dr_2 \right]\psi_i(r_1)$$

$$- \varepsilon^2 \sum_j^N \left[\int [\psi_j{}^*(r_2)\psi_i(r_2)\psi_j(r_1)]/[r_{12}\psi_i(r_1)]\, dr_2 \right]\psi_i(r_1) = E_i\psi_i(r_1). \quad (q.5.54.)$$

The operator enclosed within the first set of large square brackets in (q.5.54.) is the same as the Hartree operator of (q.4.132.), except that here the terms in the sums where $i = j$ are not excluded, for spin j = spin i. It has to be kept in mind that, in terms of the Pauli Exclusion Principle, there may be not more than two electrons in the same orbital state provided that these have opposed spins, so that although the functions ψ_i are eigenfunctions of the equation (q.5.54.) corresponding to its lowest eigenvalue, they must in fact also satisfy the requirement that they be the lowest eigenvalue allowed by the Pauli Principle.

There will be an equation of the same form as (q.5.54.) for each of the one-electron functions ψ_i, and these equations must be solved simultaneously to obtain the energy of the N-electron system. The difficulty in the application of the Hartree–Fock method to atoms is that the operator $\hat{H}_i$ depends upon ψ_i, so that it is not immediately obvious that the variation principle can be applied directly to (q.5.52.). The second term in (q.5.54.), the exchange term, may be alternatively written as

$$-\varepsilon^2 \sum_{j}^{N} \left[\int \{[\psi_j{}^*(r_2)\psi_i(r_2)]/r_{12}\} \, dr_2 \right] \psi_j(r_1). \qquad \text{(q.5.55.)}$$

For small values of N, or for the lighter atoms, the Hartree–Fock equations can be solved directly by the method of successive approximations devised by Hartree, until self-consistency to whatever degree of accuracy is desired has been attained. However, since the functions ψ_j occur in each of the equations, this requires the manipulation of a complicated set of simultaneous equations. The Hartree–Fock method can also be applied to solids, and metals in particular, and also to simple molecules composed of light atoms. For solid materials, the term representing the interaction of the various electrons with the atomic nucleus has to be replaced by a term representing the interaction with all of the nuclei in the solid, and an additional term has also to be added to describe the Coulomb interaction of the nuclei with one another. For the idealized case of metals, regarded as free-electron gases, the appropriate Hartree–Fock equations can be solved analytically, as will be shown in a subsequent chapter. But for solids generally there would be as many spin orbitals as there are electrons in the solid of the order of magnitude of $Z \times 10^{23}$ per cubic centimeter, with each electron moving in a different potential field according to (q.5.54.). The physical impossibility of accomplishing such a calculation has led to many proposals for simplifying the exchange charge density term or the exchange potential energy by some type of average which would be the same for each ψ_i. Much progress has been made in the application of the Hartree–Fock method to atoms,[1-5] molecules,[6-11] and to metals,[12-17] some of which methods will be referred to in subsequent chapters.

[1] Eckart, C., *Phys. Rev.*, 1930, **36**, 878.
[2] Dirac, P. A. M., *Proc. Camb. Phil. Soc.*, 1930, **26**, 376; **27**, 240.
[3] Löwdin, P.-O., *Phys. Rev.*, 1955, **97**, 1474.
[4] Roothaan, C. C. J., *Rev. Mod. Phys.*, 1951, **23**, 69; 1960, **32**, 179.
[5] Froese, C., *J. Chem. Phys.*, 1966, **45**, 1417.
[6] Zener, C., *Phys. Rev.*, 1930, **36**, 51.
[7] Slater, J. C., *Phys. Rev.*, 1930, **36**, 57; 1932, **42**, 33; 1951, **81**, 385; 1953, **91**, 528.
[8] Clementi, E., and Raimondi, D. L., *J. Chem. Phys.*, 1963, **38**, 2686.
[9] Hund, F., *Z. Physik*, 1927, **40**, 742; **42**, 93.
[10] Mulliken, R. S., *Phys. Rev.*, 1928, **32**, 186.
[11] Lennard-Jones, J. E., *Trans. Faraday Soc.*, 1929, **25**, 668.
[12] Bloch, F., *Z. Physik*, 1929, **57**, 545.
[13] Wigner, E., and Seitz, F., *Phys. Rev.*, 1933, **43**, 804.
[14] Lenz, W., *Z. Physik*, 1932, **77**, 713.
[15] Slater, J. C., *Quantum Theory of Molecules and Solids*, McGraw-Hill Book Co., 1967, New York.
[16] Sommerfeld, A., *Z. Physik*, 1928, **47**, 1; **47**, 43.
[17] Pines, D., *Phys. Rev.*, 1953, **92**, 626.

The total energy of the electrons in an N-electron atom is apparently not simply the sum of the Hartree–Fock eigenvalues E_i. Thus, combining equations (q.5.50.) and (q.5.54.), it is found that

$$E = \sum_i^N E_i - (1/2) \sum_i^N \sum_j^N \iint [\varepsilon^2/r_{12}]|\psi_j(r_2)|^2|\psi_i(r_1)|^2 \, dr_1 \, dr_2$$

$$+ (1/2) \sum_i^N \sum_j^N \iint [\varepsilon^2/r_{12}]\psi_i^*(r_1)\psi_j^*(r_2)\psi_i(r_2)\psi_j(r_1) \, dr_1 \, dr_2, \quad \text{(q.5.56.)}$$

where again the double summations in the exchange term are to be taken over those functions ψ_i and ψ_j associated with the same spin functions, that is, over electrons with parallel spins only. As for the Hartree equations (page 277), the eigenvalue E_i may be interpreted as the negative value $-E_i$ of the energy required to remove an electron in the state ψ_i from the atom, assuming that the functions ψ_i remain virtually unchanged after the removal of a single electron; this implies the assumption that the orbital functions of the electrons for the ion are the same as those for the atom. That this is so is easily proved by noting that any given function ψ_i only appears in the expression for the total energy E through the terms described by (q.5.51.), and if an electron in the state ψ_i is removed from the atom, then the total energy will be reduced by the value of the expression given by (q.5.51.), which is E_i. Therefore, the energy required to remove the electron is $-E_i$. This result is usually referred to as Koopmans' theorem.[1]

The Hartree–Fock equations, (q.5.54.), have N solutions representing the N spin orbitals used in writing the determinantal function describing the state of the N-electron atom concerned; there is also an infinite number of other solutions corresponding to states described by spin orbitals representing vacant or excited energy states. All of these solutions may be shown to correspond to functions which are mutually orthogonal to one another by applying the method described on page 122. Also, since these spin orbitals may be normalized, they form a complete set of spin orbitals, in terms of which any arbitrary function of coordinates and spin may be expanded, as described in section Q.2.4.

Q.5.8.   The Fermi Hole Concept

A typical Hartree–Fock equation, as given by (q.5.54.), may be regarded as giving the function ψ_i as the solution of a Schrödinger equation for which the Hamiltonian operator is expressed as the sum of the kinetic energy of an

---

[1] Koopmans, T. C., *Physica*, 1933, **1**, 104.

electron, the potential energy of an electron in the field of the atomic nucleus, plus the potential energy of the electron in the field of all the other electrons in the atom, distributed over the energy states described by a determinantal function, reduced by a correction term involving an exchange integral. The attempt to assign some physical significance to the exchange integral leads to an interesting conclusion.

Since electrons repel one another, they obviously cannot move independently but in such a manner as to preclude the possibility of any two occupying the same region in space. The general effect of the Coulomb force between electrons is, then, to couple or correlate the individual electronic motions such as to reduce the probability of any two electrons from closely approaching one another. In the simple Hartree method, such Coulomb correlation effects (page 271) are completely neglected, and each electron is assumed to move in an average charge distribution provided by all of the other electrons in the system. The total wave function for the system is written as a single product of one-electron functions, so that the probability of any particular configuration depends only upon the one-electron functions and is not directly related to the distances between pairs of electrons.

The Hartree–Fock method does not specifically take into account these Coulomb correlation effects either, but nevertheless does include another type of correlation effect. The Hartree–Fock correlation effect is not due to Coulomb forces between the electrons, but arises as a consequence of the Pauli exclusion principle, as is inherent in the use of a determinantal function, so that this correlation affects the positions of electrons with parallel spins only. This consideration is apparent from the form of a determinantal function, which is given by

$$
F = \begin{vmatrix}
\Psi_1(\mathbf{x}_1) & \Psi_1(\mathbf{x}_2) & \Psi_1(\mathbf{x}_3) & \cdots & \Psi_1(\mathbf{x}_N) \\
\Psi_2(\mathbf{x}_1) & \Psi_2(\mathbf{x}_2) & \Psi_2(\mathbf{x}_3) & \cdots & \Psi_2(\mathbf{x}_N) \\
\Psi_3(\mathbf{x}_1) & \Psi_3(\mathbf{x}_2) & \Psi_3(\mathbf{x}_3) & \cdots & \Psi_3(\mathbf{x}_N) \\
\vdots & \vdots & \vdots & & \vdots \\
\Psi_N(\mathbf{x}_1) & \Psi_N(\mathbf{x}_2) & \Psi_N(\mathbf{x}_3) & \cdots & \Psi_N(\mathbf{x}_N)
\end{vmatrix},
$$

where $\Psi(\mathbf{x}) = \psi(r)\chi(\zeta)$, and the spin function $\chi(\zeta)$ may be either α or β. The use of such a determinantal function actually means that no particular electron may be strictly regarded as being in any specific one-electron state, but rather that each electron may be considered to be partially in all of the states, so that two electrons with the same spin directions or coordinates are not to be regarded as being in specific one-electron states described by the same spin functions. Since a determinant vanishes if any two rows or columns are identical, this means that a total determinantal function becomes identically zero if two electrons have parallel spins; alternatively, it may be said that there is zero probability of two electrons with parallel spins being in the same region of space. Thus, if two electrons have parallel spins, then $\zeta_1 = \zeta_2$, and if they occupy the same region of space so that $r_1 = r_2$, then $\mathbf{x}_1 = \mathbf{x}_2$ and the first two rows of the determinantal function would be identical, so that F, and therefore the probability density $|F|^2$, would be zero. However, if any two electrons have opposed or antiparallel spins, even although the orbital factor of the spin orbitals are the same, the ζ_1 being different from ζ_2 would prevent the total determinantal function from adopting a zero value. In fact, not only does the use of a determinantal function give rise to a zero probability of finding two electrons in the same spatial region if they have parallel spins, but it also means that there is only a small probability of two such electrons adopting a position near to one another. The over-all behavior of electrons appears to be as if each electron, during the course of its motion, were surrounded by a "hole" in the distribution of electrons with parallel spins; this is usually referred to as the Fermi or exchange hole. The concept of the Fermi hole emerges from the attempt to assign some physical significance to the exchange term in the Hartree–Fock equations.

By reference to (q.4.123.), the Hartree potential energy of the electron at position r_1 is

$$\varepsilon^2 \sum_{j}^{N} \int [|\psi_j(r_2)|^2]/r_{12} \, dr_2,$$

and is due to a charge distribution which has a density at r_2 of $-\varepsilon \sum_{j}^{N} |\psi_j(r_2)|^2$. If the electron at position r_1 is assumed to be in the particular one-electron state Ψ'_i, then the exchange term in the Hartree–Fock equation (q.5.54.) may be interpreted in a similar manner to the Hartree case; it has to be noted, however, that as pointed out on page 309, this is not strictly accurate when a determinantal function rather than a simple product function is used. The exchange term in equation (q.5.54.) may be written as equivalent to $U_i(r_1)\psi_i(r_1)$ with

$$-\varepsilon^2 \sum_{j}^{N} [\psi_j{}^*(r_2)\psi_i(r_2)\psi_j(r_1)]/[r_{12}\psi_i(r_1)] \, dr_2 = U_i(r_1)$$

with spin $i =$ spin j, and by analogy with the Hartree case, the function $U_i(r_1)$ may be regarded as the potential energy of the electron at r_1 due to some fictitious exchange charge distribution with a density at r_2 of

$$\varepsilon \sum_{j}^{N} [\psi_j{}^*(r_2)\psi_i(r_2)\psi_j(r_1)]/\psi_i(r_1), \tag{q.5.57.}$$

with the summation again over spin $i =$ spin j. From this expression the total exchange charge would be

$$\varepsilon \sum_{j}^{N} \int \{[\psi_j{}^*(r_2)\psi_i(r_2)\psi_j(r_1)]/\psi_i(r_1)\} \, dr_2, \tag{q.5.58.}$$

which latter expression is equivalent to simply ε, since if the functions ψ_i and ψ_j are orthonormal,

$$\int \psi_j^*(r_2)\psi_i(r_2)\, dr_2 = \delta_{ij}.$$

Thus the total exchange charge is a positive charge of magnitude one electronic charge or one atomic unit of charge. Also, the exchange charge density at the position r_1, which may be obtained by letting $r_2 = r_1$ in (q.5.57.) is $\varepsilon \sum_j^N |\psi_j(r_1)|^2$, with the sum over spin $i =$ spin j. This quantity is equal in magnitude but opposite in sign to the average charge density of electrons with spins parallel to that of the electron at r_1. The electron at position r_1 is here assumed to be in the orbital state ψ_i, associated with some spin state $\chi(\zeta)$, and the summation is taken over all orbital states described by the same spin function $\chi(\zeta)$. Now, since the total charge density is only one atomic unit (ε), the exchange charge density must fall off rapidly from $\varepsilon \sum_j^N |\psi_j(r_1)|^2$ as the distance r_{12} increases. Actually, the exchange charge density given by (q.5.57.) cannot be spherically symmetrical about the position r_1, since it is not the same for different functions ψ_i. However, from the above considerations, it appears that, due to the exchange term in the Hartree–Fock equations, an electron is surrounded by a positive charge cloud of total charge $+\varepsilon$, which has a limited extension in space and effectively negates the average charge distribution of electrons with parallel spins in the vicinity of that particular electron. This situation is what is regarded as giving rise to the Fermi hole. In the simple free electron theory of metals the Fermi hole may be regarded as a spherical region with a radius equal to the atomic radius of the metal atoms concerned.

Q6

The Electronic Structure of Atoms

Q.6.1. Introduction

The concept which proved capable of leading to the interpretation of complex atomic spectra, even before the development of quantum mechanical theory, was that of angular momentum. Much information about spherically symmetrical atoms, in particular, may be derived by assigning angular momentum vectors to the various electrons in atoms and ions and considering how angular momentum is conserved during electronic transitions between different energy states. These considerations have led to the vector model of atoms and the development of multiplet theory. The behavior of angular momenta has continued to play an important part in quantum theory.

The ultimate objective of theoretical chemistry is to obtain and derive information about the structure of substances, to predict the properties of materials, and to arrive at an understanding of the nature of the chemical bond in many-electron systems. As far as quantum mechanical calculations are concerned, the evaluation of the energy of atoms, molecules, and crystals, is the most frequently used means of checking the accuracy of such calculations. The calculation and the determination of other properties of many-electron systems has often been of subsidiary interest, merely complementing the energy criterion. For atomic and ionic states, the energy values which are accessible to experimental observation, largely from spectroscopic data, are term values or differences between energy states. For metals and solids in general, it is usually the difference between the total energy of the electrons and nuclei in the solid and the energy of the system of free atoms which is of importance; this energy difference is referred to as the cohesive energy of the solid. It is a measure of the strength of the forces which bind the atoms

312

together in the solid state and, of course, a stable solid exists as such only if its total energy is less than that of a system of free atoms at the same temperature.

The Hartree–Fock method has been applied not only to atoms, but to molecules and solids, and much progress has been made in designing approximation techniques to handle the Hartree–Fock equations for relatively complicated systems. Exact Hartree–Fock calculations have been made for atomic and molecular systems with relatively few electrons. Considerable progress has also been made in designing improved state functions for systems by considering the influence of smaller terms in the Hamiltonian of the system. Such terms include spin-spin interaction effects, spin-orbital interactions, the influence of electron spin on nuclear spin, quadruple moments, and the effect on all of these of externally applied electric and magnetic fields. The experimental technique which supplies information about the magnitude of the energy effects of these terms is essentially based on microwave measurements, where energy differences are generally minute. However, by making use of such experimental data, it is possible to examine the accuracy of the wave function for a system and the effect on the calculated energies of such interaction terms.

In this chapter, some of these developments will be examined briefly in so far as free atoms and ions are concerned.

Q.6.2. Angular Momenta

It has been shown in section Q.3.7. that the magnitude of an angular momentum vector such as that of a single electron moving in a spherically symmetrical potential field is quantized, its value being given by (q.3.128.), and also that its component along a fixed axial direction is quantized, with the value $m_l(h/2\pi)$. The electronic energy is independent of m_l, which only defines the orientation in space of the angular momentum. Also, the discussion in section Q.4.11. of the Hartree method has shown that it is a reasonably good approximation to assume that each electron in an atom moves in a spherically symmetrical force field. However, for an atom containing two or more electrons not in s-states, each electron would have a charge distribution depending on m_l (on angle), and these charge distributions would exert an electrostatic force on each other. The spherical symmetry of the force field, in which the angular momentum of each electron would be constant, would therefore no longer apply, since as a result of the electrostatic forces brought into play, the angular momentum of each electron would change with time. But since the electrostatic forces are completely internal to the atom, the angular momentum of the whole system would remain constant. For a system of N-electrons, the total orbital angular momentum is given by the vector sum

$$\mathbf{L} = \sum_{j=1}^{N} \mathbf{L}_j = L_x\mathbf{i} + L_y\mathbf{j} + L_z\mathbf{k}, \qquad \text{(q.6.1.)}$$

where $\mathbf{i}$, and $\mathbf{j}$, and $\mathbf{k}$ are unit vectors in the x, y, and z Cartesian coordinate directions. It thus happens that for a system of electrons, the total orbital angular momentum is quantized with a value of $\sqrt{L(L + 1)}(h/2\pi)$, with a component along a fixed axial direction of $M(h/2\pi)$ similarly to the single electron case.

In quantum mechanics, the angular momentum is described by an operator and, since it is a vector, there are three component operators resolved along the x-, y-, and z-coordinate directions. These operators are expressed in Cartesian and in spherical polar coordinates in section Q.3.7.; also the operator $\hat{L}^2$ is given in (q.3.124.). For a single electron moving in a central field of force, it is also proved in section Q.3.7. that $\hat{L}^2$, $\hat{L}_z$, and $\hat{H}$ commute with one another; alternatively expressed, it may be said that $\hat{L}_z$ and $\hat{L}^2$ have diagonal matrices, since each of these operators, operating on the corresponding one-electron function, reproduces the function multiplied by a constant. To find the matrix components of $\hat{L}_x$ and $\hat{L}_y$, given by (q.3.119.) and (q.3.120.), respectively, it is necessary to know the derivative of the Legendre function, $P_l^m(\cos\theta)$, with respect to the polar angle θ. From the analytical nature of the Legendre function as discussed in section Q.3.5., it is found that

$$d/d\theta[P_l^m(\cos\theta)] = -P_l^{m+1}(\cos\theta) + m\cot\theta P_l^m(\cos\theta), \qquad \text{(q.6.2.)}$$

and

$$m\cot\theta P_l^m(\cos\theta) = (1/2)[P_l^{m+1}(\cos\theta) + (l + m)(l - m + 1)P_l^{m-1}(\cos\theta)], \qquad \text{(q.6.3.)}$$

where (q.6.2.) is valid for only positive values of m, but (q.6.3.) is valid for $m = 0$ as well as for any positive value of m. Making use of this derivative, considering the fact that the associated Legendre functions are not normally defined for negative values of m (page 186), then the result of operating with $\hat{L}_x$ and $\hat{L}_y$ on the central field eigenfunctions given by (q.3.130.) may be obtained as

$$\hat{L}_x\psi_{n,l,m} = (1/2)[\sqrt{(l - m)(l + m + 1)}\psi_{n,l,m+1} \\ + \sqrt{(l - m + 1)(l + m)}\psi_{n,l,m-1}]$$

and $\qquad\qquad\qquad\qquad\qquad\qquad\qquad\qquad\qquad\qquad$ (q.6.4.)

$$\hat{L}_y\psi_{n,l,m} = (1/2)[-\sqrt{(l - m)(l + m + 1)}\psi_{n,l,m+1} \\ + \sqrt{(l - m + 1)(l + m)}\psi_{n,l,m-1}],$$

where it is understood that the right-hand side of each of these expressions is to be multiplied by $(h/2\pi)$, the ordinary unit of angular momentum, and for simplicity the subscript l has been omitted from m_l. If the total quantum numbers for a polyelectronic atom are written as L and M_L, where these symbols designate the many-electron equivalents of the one-electron quantum numbers l and m_l, then equations (q.6.4.) are also applicable when l and m are substituted by L and M. If the linear combination operators $[\hat{L}_x + i\hat{L}_y]$ and $[\hat{L}_x - i\hat{L}_y]$ are allowed to operate on the same eigenfunctions as above, then there results

$$[\hat{L}_x + i\hat{L}_y]\psi_{n,l,m} = \sqrt{(l-m)(l+m+1)}\,\psi_{n,l,m+1}$$

and $\hspace{10cm}$ (q.6.5.)

$$[\hat{L}_x - i\hat{L}_y]\psi_{n,l,m} = \sqrt{(l-m+1)(l+m)}\,\psi_{n,l,m-1},$$

which again are equally applicable for polyelectronic atoms if l and m are replaced by L and M. The third component of the total angular momentum of an electron in a central field, L_z, the operator form of which is given by (q.3.121.), leads in a similar manner to the expression

$$\hat{L}_z\psi_{n,l,m} = m\psi_{n,l,m}, \hspace{4cm} \text{(q.6.6.)}$$

again equally applicable to the case where l and m are substituted by L and M. It is interesting to note that the expressions (q.6.4.), (q.6.5.), and (q.6.6.) are valid for both positive and negative values of m and M if the angular function $\Theta_l^m(\theta)$ given by (q.3.51.) is considered to be multiplied by the factor $(-1)^{m+|m|/2}$.

Another significantly important feature of these angular momentum operators is that, when operating on central field eigenfunctions, $\hat{L}_x$ and $\hat{L}_y$ each give rise to a linear combination of just two wave functions with the same n and l (L) values but with m (and M) values different by ± 1 from the original m (or M) value. On the other hand, the linear combination angular momentum operators $[\hat{L}_x \pm i\hat{L}_y]$, when operating on the same central field eigenfunction, each gives rise to a single wave function, the value of the m (or M) quantum number for which is greater or less than the original value by one unit. It is for this reason that the operators $[\hat{L}_x \pm i\hat{L}_y]$ are generally referred to as step-up and step-down operators, respectively, since when they operate on a wave function they increase or decrease m (or M) by one unit, depending on the sign of the second term in the linear combination operator.

The matrix components of each of the three angular momentum component operators $\hat{L}_x$, $\hat{L}_y$, and $\hat{L}_z$ may be derived from equations (q.6.4.), (q.6.5.), and (q.6.6.) by multiplying the left-hand sides of these equations by the complex conjugate of one of the wave functions, integrating over all of the coordinates involved, noting that the wave functions constitute an orthonormal set, when it is found that all of the matrix components lead to the relatively simple result, conveniently expressed in the Dirac notation as (see page 125)

$$\langle n, l, m|\hat{L}_x|n, l, m + 1\rangle$$
$$= \langle n, l, m + 1|\hat{L}_x|n, l, m\rangle = -i\langle n, l, m|\hat{L}_y|n, l, m + 1\rangle$$
$$= i\langle n, l, m + 1|\hat{L}_y|n, l, m\rangle = (1/2)\sqrt{(l - m)(l + m + 1)} \quad \text{(q.6.7.)}$$

and

$$\langle n, l, m|\hat{L}_z|n, l, m\rangle = m;$$

the equations (q.6.7.) are again applicable to polyelectronic atoms if l and m are substituted by L and M. All the matrix components of $\hat{L}_x$, $\hat{L}_y$, and $\hat{L}_z$, which do not conform with (q.6.7.), are found to vanish.

The quantum mechanical expressions of (q.6.7.) have a simple equivalent in classical mechanics. The simplest mechanical equivalent of two electrons with angular momentum interacting with one another is that of a rotating symmetrical top spinning under the influence of the earth's gravitational field. A rotating top possesses an angular momentum, which under the influence of the gravitational force acting on it, changes with time, giving rise to the characteristic precessional motion of the rotating top. In classical mechanics, the rate of change of momentum of a system is a measure of a force, and the rate of change of angular momentum is a measure of the torque acting on a system. The relationship between angular momentum, torque, and the precessional motion of a rotating top is shown in Fig. Q.6.1. Both the angular momentum and the torque are vector quantities, and for a rotating top the direction of the torque is at right angles to the angular momentum vector, such that its action changes the direction of the angular momentum vector while maintaining its magnitude constant. The total angular momentum of the spinning top is represented by the vector which is at right angles to the plane of rotation of the top; the z-component of the angular momentum remains constant at $L \cos \theta$ (see Fig. Q.6.1.), while L_x and L_y, the angular momenta components in the x and y Cartesian coordinate directions, vary as sinusoidal functions of time $90°$ out of phase with one another, and given by $L_x = L \sin \theta \cos \omega t$ and $L_y = L \sin \theta \sin \omega t$, where ω is the angular velocity of the precession of the top. The precession of the rotating top is such that the angular momentum vector moves around the vertical axis on the surface of a cone of semi-vertical angle θ. In the absence of any gravitational field effect, the torque would be zero, and the angular momentum would be independent of time. The quantum mechanical equations of (q.6.7.) are the expression of this same precessional motion, described in terms of quantum mechanics. Thus, from (q.6.7.) the matrix component of $\hat{L}_z$ is the constant M, equivalent to the classical constancy of the z-component

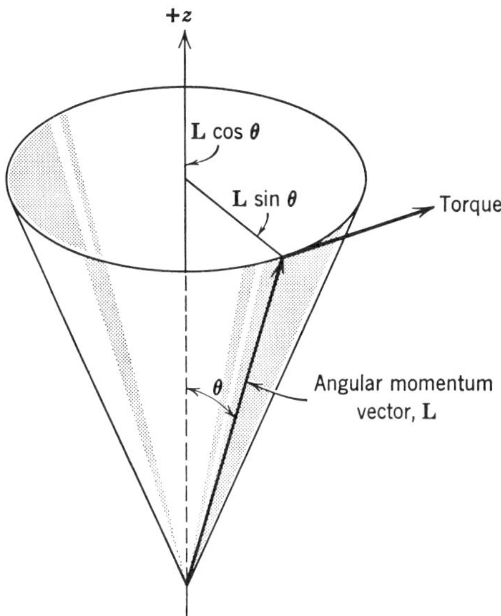

FIG. Q.6.1. Relationship between the angular momentum, torque, and the precessional motion of a symmetrical rotating top.

of the angular momentum of a rotating top in a gravitational field; alternatively expressed, $\hat{L}_z$ has a diagonal matrix. M in quantum mechanics is, of course, expressed in units of $h/2\pi$. The non-diagonal nature of the matrix components of $\hat{L}_x$ and $\hat{L}_y$ in the quantum mechanical expressions, is related to the classical precessional motion, where the angular momenta components L_x and L_y vary sinusoidally with time.

Q.6.3. Matrix Mechanics and Angular Momenta

The matrix components of the angular momentum vector may be derived by another method, which is quite independent of any knowledge of the properties of associated Legendre functions, but dependent only on the commutation behavior of the type of operators which occur in quantum mechanics, as described on page 137. This is the method of matrix mechanics, originally described by Born, Heisenberg, and Jordan,[1,2] before the development of Schrödinger mechanics. As far as the problem of the motion of a

---

[1] Born, M., and Jordan, P., *Z. Physik*, 1925, **34**, 858.
[2] Born, M., Heisenberg, W., and Jordan, P., *Z. Physik*, 1926, **35**, 557.

single electron in a central field is concerned, matrix mechanics does not provide any substantially simpler description than the Schrödinger method; but for polyelectronic atoms, where the total wave function is generally quite complicated, the method of matrix mechanics, which enables the angular momentum of atoms to be described only in terms of commutation relationships, avoids the difficulties associated with the evaluation of the matrix components of the angular momenta operators with respect to such complicated wave functions. In this respect, the Born–Heisenberg–Jordan method of matrix mechanics is a more general one than the Schrödinger method of wave mechanics.

The angular momentum of a particle in a central field of force, for which the potential energy depends only on a radial distance r, is a constant according to classical mechanics, since in this case there would be no torque acting on the particle. In quantum mechanics, the characteristics of quantities which are constants of the motion of a system, is that the operators describing these types of motion commute with one another, as described in section Q.2.6. Thus, considering the kinetic energy of a particle in a central field, where the potential energy is a function of the radial distance r only, the quantum mechanical operator appropriate to the description of the kinetic energy is the Laplacian operator, and this operator may be shown to commute with the operator $\hat{L}_x$, for example, since

$$\nabla^2[y(\partial\psi/\partial z) - z(\partial\psi/\partial y)]$$
$$= y\nabla^2(\partial\psi/\partial z) + 2\,\partial^2/\partial y\,\partial z[\psi] - z\nabla^2(\partial\psi/\partial y) - 2\,\partial^2/\partial y\,\partial z[\psi]$$

and
$$\text{(q.6.8.)}$$

$$[y\,\partial/\partial z - z\,\partial/\partial y]\nabla^2\psi = y\nabla^2(\partial\psi/\partial z) - z\nabla^2(\partial\psi/\partial y),$$

from which it follows that

$$[y\,\partial/\partial z - z\,\partial/\partial y]\nabla^2\psi = \nabla^2[y\,\partial/\partial z - z\,\partial/\partial y]\psi. \qquad \text{(q.6.9.)}$$

Therefore the Laplacian or energy operator commutes with the angular momentum component operator $\hat{L}_x$. In a similar manner, it may be shown that for each component of the angular momentum, the appropriate operator commutes with the Hamiltonian for any spherically symmetrical potential energy $U(r)$. However, although the operators representing each component of the angular momentum commute with the Hamiltonian, they do not commute with one another, for ‡

$$\hat{L}_y\hat{L}_z - \hat{L}_z\hat{L}_y = [\hat{L}_y, \hat{L}_z] = i\hat{L}_x$$

and
$$\text{(q.6.10.)}$$

$$[\hat{L}_z, \hat{L}_x] = i\hat{L}_y \qquad \text{and} \qquad [\hat{L}_x, \hat{L}_y] = i\hat{L}_z,$$

where in equations (q.6.10.) it is assumed that atomic units of angular momentum ($h/2\pi$) are used. Thus, since $\hat{L}_x$, $\hat{L}_y$, and $\hat{L}_z$ do not commute with one another, even though each of them may be diagonalized separately, they cannot be diagonalized simultaneously, as has already been shown by the fact that $\hat{L}_z$ has a diagonal matrix, while $\hat{L}_x$ and $\hat{L}_y$ do not. It should also be noted that $\hat{L}_x$, $\hat{L}_y$, and $\hat{L}_z$ each commutes with $\hat{L}^2 = \hat{L}_x{}^2 + \hat{L}_y{}^2 + \hat{L}_z{}^2$, which also commutes with the Hamiltonian, so that a set of wave functions exist which have diagonal matrices for $\hat{L}^2$ as well as $\hat{L}_z$.

‡ From equations (q.3.18.), by substitution of the operator forms of L_x, L_y, and L_z,

$$\hat{L}_x = (h/2\pi i)[y\, \partial/\partial z - z\, \partial/\partial y],$$
$$\hat{L}_y = (h/2\pi i)[z\, \partial/\partial x - x\, \partial/\partial z],$$

and

$$\hat{L}_z = (h/2\pi i)[x\, \partial/\partial y - y\, \partial/\partial x].$$

Therefore

$$\hat{L}_x\hat{L}_y = -[y\, \partial/\partial z - z\, \partial/\partial y][z\, \partial/\partial x - x\, \partial/\partial z]$$
$$= -[y\, \partial/\partial z\, z\, \partial/\partial x - y\, \partial/\partial z\, x\, \partial/\partial z - z\, \partial/\partial y\, z\, \partial/\partial x + z\, \partial/\partial y\, x\, \partial/\partial z],$$

and similarly

$$\hat{L}_y\hat{L}_x = -[z\, \partial/\partial x\, y\, \partial/\partial z - z\, \partial/\partial x\, z\, \partial/\partial y - x\, \partial/\partial z\, y\, \partial/\partial z + x\, \partial/\partial z\, z\, \partial/\partial y].$$

Therefore

$$\hat{L}_x\hat{L}_y - \hat{L}_y\hat{L}_x = [\hat{L}_y, \hat{L}_z] = -[y\, \partial/\partial x(\partial/\partial z\, z - z\, \partial/\partial z) + x\, \partial/\partial y(z\, \partial/\partial z - \partial/\partial z\, z)]$$
$$= \{y\, \partial/\partial x[\partial/\partial z, z] + x\, \partial/\partial y[z, \partial/\partial z]\}$$
$$= -\{(y\, \partial/\partial x - x\, \partial/\partial y)[\partial/\partial z, z]\} = i\hat{L}_z[\partial/\partial z, z] = i\hat{L}_z.$$

Making use of the cyclical permutation of x, y, and z, the other commutation relationships for the components of $\hat{L}$ may be derived in a similar manner.

From the discussion in sections Q.2.4., Q.2.6., and Q.2.7., it may be shown that any dynamical variable whose operator commutes with the energy or the Hamiltonian of a system, such as the momentum of a particle on which there are no forces acting or the angular momentum of a particle on which there is no torque acting, represents a quantity which is independent of time. Thus, if $\hat{F}$ is some operator which commutes with the Hamiltonian $\hat{H}$, then from (q.2.48.)

$$[\hat{F}, \hat{H}]_{mn} = [\hat{F}\hat{H} - \hat{H}\hat{F}]_{mn} = 0, \qquad (q.6.11.)$$

for any two wave functions Ψ_m and Ψ_n. Since $\hat{H}$ has a diagonal matrix, then expressing (q.6.10.) in terms of the matrix elements of $\hat{F}$ and $\hat{H}$,

$$\hat{F}_{mn}[\hat{H}_{nn} - \hat{H}_{mm}] = 0 \qquad \text{or} \qquad \hat{F}_{mn}[E_n - E_m] = 0. \qquad (q.6.12.)$$

Therefore, the non-diagonal matrix elements F_{mn} must be zero, except when $E_m = E_n$, or when the matrix elements are with respect to two degenerate states. But the average value of the quantity F represented by the operator $\hat{F}$ with respect to the time dependent function

$$\Psi(q, t) = \sum_m A_m \psi_m(q) e^{-2\pi i E_m t/h}, \qquad \text{(q.6.13.)}$$

as given on page 110, is obtained by the application of equation (q.2.66.) as

$$\langle F \rangle \text{ or } \bar{F} = \int \left[\sum_m A_m{}^* \psi_m{}^*(q) e^{+2\pi i E_m t/h} \hat{F} \sum_n A_n \psi_n(q) e^{-2\pi i E_n t/h} \right] d\tau$$

$$= \sum_m \sum_n A_m{}^* A_n \hat{F}_{mn} e^{2\pi i t(E_m - E_n)/h}, \qquad \text{(q.6.14.)}$$

where $\hat{F}_{mn} = \int \psi_m{}^* \hat{F} \psi_n \, d\tau$. Thus, all the non-vanishing terms of $\bar{F}$ when $\hat{F}$ commutes with the energy are either diagonal and therefore independent of time, or have $E_m = E_n$, in which event the term containing time or the exponential quantity $e^{2\pi i(E_m - E_n)t/h}$, must actually be a constant. In either case, if the operator $\hat{F}$ commutes with the Hamiltonian of a system, it represents a quantity whose average value does not depend on time. This statement is equally true the other way round, for if an operator is time-independent, it must commute with the Hamiltonian of a system.‡

‡ This may be seen by differentiating (q.6.14.) with respect to time, when

$$\partial/\partial t[\bar{F}] = \sum_m \sum_n A_m{}^* A_n \hat{F}_{mn} [i(E_m - E_n)2\pi/h] e^{2\pi i(E_m - E_n)/h}$$

$$= \sum_m \sum_n A_m{}^* A_n [\partial \bar{F}/\partial t]_{mn} e^{2\pi i(E_m - E_n)/h},$$

if the matrix elements $[\partial \bar{F}/\partial t]_{mn} = \hat{F}_{mn}[2\pi i(E_m - E_n)/h]$. Making use of the fact that $\hat{H}$ has a diagonal matrix, and applying the normal method of matrix multiplication, this latter expression may be written as

$$(ih/2\pi)[\partial \bar{F}/\partial t] = (\hat{F}\hat{H} - \hat{H}\hat{F})_{mn} = [\hat{F}, \hat{H}]_{mn}.$$

Therefore, the same equation is applicable to each matrix component. But, if two operators have matrix components all of which are equal to one another, with respect to a complete set of functions, the operators must be identical with one another. This means that the last expression above may be written equally well as the operator equation

$$(ih/2\pi)[\partial \bar{F}/\partial t] = (\hat{F}\hat{H} - \hat{H}\hat{F}) = [\hat{F}, \hat{H}].$$

Thus, the rate of change of an operator $\hat{F}$, representing a quantity which is time-independent, is such that it must commute with the Hamiltonian of the system concerned. This result, as well as the technique of operating with matrices was first established by Heisenberg.

Now, making use of equations (q.6.9.) and (q.6.12.) and the fact that while each of the angular momentum component operators commutes with the Hamiltonian but not with each other, then if $\hat{L}_z$, for example, has a diagonal matrix, $\hat{L}_x$ and $\hat{L}_y$, which do not commute with one another, cannot have diagonal matrices, at least for degenerate states. But since $\hat{L}_x$ and $\hat{L}_y$ each commutes with the energy, then by virtue of (q.6.12.) they can have non-diagonal matrix components only between degenerate states. If it is assumed that there are a finite number of such degenerate states between which $\hat{L}_x$ and $\hat{L}_y$ have non-diagonal matrix components, and that the quantum number describing the various wave functions appropriate to this degenerate set of states is M, then there must be $2L + 1$ such states. L must be either integral or integral plus one-half, since the spin angular momentum has the magnitude $(1/2)(h/2\pi)$. Then the range of values of M will be from $-L$ to $+L$ in unit intervals, and $\hat{L}_x$ and $\hat{L}_y$ will have matrix components with respect to states of different M values, which are therefore finite matrices. Now since $\hat{L}_z$ has a diagonal matrix, if the matrix components with respect to M and M' are written down for the operators defined by the commutation relationships of (q.6.10.), there results for the commutator $[\hat{L}_y, \hat{L}_z]$

$$\langle M|\hat{L}_y|M'\rangle\langle M'|\hat{L}_z|M'\rangle - \langle M|\hat{L}_z|M\rangle\langle M|\hat{L}_y|M'\rangle$$
$$\langle M|\hat{L}_y|M'\rangle(\langle M'|\hat{L}_z|M'\rangle - \langle M|\hat{L}_z|M\rangle) = i\langle M|\hat{L}_z|M'\rangle. \quad \text{(q.6.15.)}$$

Similarly, for the commutator $[\hat{L}_z, \hat{L}_x]$ of (q.6.10.), there is found

$$\langle M|\hat{L}_x|M'\rangle(\langle M'|\hat{L}_z|M'\rangle - \langle M|\hat{L}_z|M\rangle) = -i\langle M|\hat{L}_y|M'\rangle. \quad \text{(q.6.16.)}$$

From (q.6.15.), the ratio of the matrix components

$$\langle M|\hat{L}_x|M'\rangle/\langle M|\hat{L}_y|M'\rangle = (1/i)[\langle M'|\hat{L}_z|M'\rangle - \langle M|\hat{L}_z|M\rangle], \quad \text{(q.6.17.)}$$

and from (q.6.16.), the same ratio is found to be

$$\langle M|\hat{L}_x|M'\rangle/\langle M|\hat{L}_y|M'\rangle = -i/[\langle M'|\hat{L}_z|M'\rangle - \langle M|\hat{L}_z|M\rangle], \quad \text{(q.6.18.)}$$

and equating these two expressions given by (q.6.17.) and (q.6.18.) gives an equation for $\hat{L}_z$,

$$[\langle M'|\hat{L}_z|M'\rangle - \langle M|\hat{L}_z|M\rangle]^2 = 1. \quad \text{(q.6.19.)}$$

This equation, (q.6.19.), thus leads to the conclusion that the difference of the diagonal matrix components of $\hat{L}_z$ between two states, between which $\hat{L}_x$ and $\hat{L}_y$ have non-diagonal matrix components, is ± 1. It should be noted that equation (q.6.19.) is only valid for two states such as M and M', between which $\hat{L}_x$ and $\hat{L}_y$ have non-diagonal matrix components, otherwise the ratios given by (q.6.17.) and (q.6.18.) would be indeterminate. If the range of values of M is from $-L$ to L, this implies that the range of values of $\langle M|\hat{L}_z|M\rangle$ must also be from $-L$ to L, so that for every state designated by the quantum number M and with a given value of $\langle M|\hat{L}_z|M\rangle$, there must be another state with the negative of this value. The two important consequences of (q.6.19.) are, then, that the differences between the diagonal matrix components of $\hat{L}_z$ must be integers (± 1) and that $\hat{L}_x$ and $\hat{L}_y$ can adopt non-diagonal components only between states differing by a single unit of angular momentum.

Using the fact that M and $\langle M|\hat{L}_z|M\rangle$ vary by unit intervals from $-L$ to L, it may be deduced from equations (q.6.15.) and (q.6.19.) that

$$\langle M|\hat{L}_x|M \pm 1\rangle/\langle M|\hat{L}_y|M \pm 1\rangle = \mp i, \tag{q.6.20.}$$

which indicates that if the matrix components of $\hat{L}_x$ are real (or imaginary), that those of $\hat{L}_y$ are imaginary (or real), respectively. The same result arises by combining equations (q.6.16.) and (q.6.19.). Furthermore, since $\hat{L}_x$ and $\hat{L}_y$ are Hermitian operators (see page 139), the matrices derived from them are also Hermitian, and so

$$\langle M|\hat{L}_x|M + 1\rangle = \langle M + 1|\hat{L}_x|M\rangle = -i\langle M|\hat{L}_y|M + 1\rangle$$
$$= i\langle M + 1|\hat{L}_y|M\rangle, \tag{q.6.21.}$$

from (q.6.20.). Now equation (q.6.21.) may be combined with the commutation relationship given by the third equation of (q.6.10.), $[\hat{L}_x, \hat{L}_y] = i\hat{L}_z$, to deduce the squares of the matrix components of $\hat{L}_z$. Thus

$$\langle M|\hat{L}_x|M + 1\rangle\langle M + 1|\hat{L}_y|M\rangle + \langle M|\hat{L}_x|M - 1\rangle\langle M - 1|\hat{L}_y|M\rangle$$
$$- \langle M|\hat{L}_y|M + 1\rangle\langle M + 1|\hat{L}_x|M\rangle - \langle M|\hat{L}_y|M - 1\rangle\langle M - 1|\hat{L}_x|M\rangle$$
$$= i\langle M|\hat{L}_z|M\rangle = iM, \tag{q.6.22.}$$

by the application of the matrix product expression to the M, $M \pm 1$ matrix components of the commutator $[\hat{L}_x, \hat{L}_y] = i\hat{L}_z$, which by virtue of the Hermitian relationships of (q.6.21.) may be written as

$$[\langle M|\hat{L}_x|M + 1\rangle]^2 - [\langle M|\hat{L}_x|M - 1\rangle]^2 = -M/2. \tag{q.6.23.}$$

However, the first term in (q.6.23.) must vanish since the maximum value of M is L, so that there cannot be any matrix component of $\hat{L}_x$ with respect to states described by L and $L + 1$. Therefore

$$[\langle L|\hat{L}_x|L - 1\rangle]^2 = L/2. \tag{q.6.24.}$$

In a similar manner, the various successive squares of the matrix components of $\hat{L}_x$ may be obtained from (q.6.23.), using (q.6.24.), for the allowed values of M, as

$$[\langle L - 1|\hat{L}_x|L - 2\rangle]^2 = [\langle L|\hat{L}_x|L - 1\rangle]^2 + (L - 1)/2$$
$$= (1/2)[L + (L - 1)]$$
$$[\langle L - 2|\hat{L}_x|L - 3\rangle]^2 = (1/2)[L + (L - 1) + (L - 2)]$$
$$\cdots\cdots\cdots\cdots\cdots\cdots\cdots\cdots\cdots\cdots\cdots\cdots\cdots\cdots\cdots \tag{q.6.25.}$$
$$[\langle M + 1|\hat{L}_x|M\rangle]^2$$
$$= (1/2)[L + (L - 1) + (L - 2) + \cdots + (M + 1)]$$
$$= (1/4)(L - M)(L + M + 1).$$

If the square root of equation (q.6.25.) is extracted, and noting the Hermitian nature of the matrix components of $\hat{L}_x$ as expressed in (q.6.21.), there is obtained

$$\langle M|\hat{L}_x|M + 1\rangle = \langle M + 1|\hat{L}_x|M\rangle = -i\langle M|\hat{L}_y|M + 1\rangle$$
$$= i\langle M + 1|\hat{L}_y|M\rangle = (1/2)\sqrt{(L - M)(L + M + 1)},$$

$$(q.6.26.)$$

which is the same result as (q.6.7.). Thus the application of the commutation rules for the angular momentum operators leads to the same result for the matrix components of these operators as the Schrödinger method.

The Born–Heisenberg–Jordan matrix method is, however, more general than the Schrödinger wave-function method. Thus the equations (q.6.25.) allow of solutions if L or M are integral and also if L or M are integral plus $1/2$, whereas there is no wave function in the form of an associated Legendre function which could lead to such matrix components of the angular momentum operators, but this is not inconsistent with the commutation rules. It is also worth noting that the equations (q.6.25.) lead to the vanishing of the matrix components between states with $M = -L$ and $M = -L - 1$, otherwise the series involved in these expressions do not terminate to give a finite number of states. This situation arises for the negative values of M because of the fact that for each state described by $+M$ there is one with the corresponding negative value, which in turn is a consequence of the integral or integral plus $1/2$ values of L.

The accumulated spectroscopic evidence referred to in section Q.5.2., which led to the proposal of Uhlenbeck and Goudsmit (page 283) that an electron has an intrinsic angular momentum of $h/4\pi$ or $1/2$ of an angular momentum unit, provides the justification for describing the spin angular momentum in a similar manner to the orbital angular momentum of an electron. It was Pauli[1] who first suggested describing spin angular momentum by the use of two wave functions associated with the spin quantum numbers of $\pm(1/2)$ (see section Q.5.3.). If the spin angular momentum, which would correspond to $L = 1/2$ in the above expressions, is denoted as before in section Q.5.3. as S with spin operator components $\hat{S}_x$, $\hat{S}_y$, and $\hat{S}_z$ with the spin quantum number being $\pm 1/2$, then the matrix components of the spin angular momentum between the functions described by the two possible spin quantum numbers are

$$\langle 1/2|\hat{S}_x| - 1/2\rangle = \langle -1/2|\hat{S}_x|1/2\rangle = -i\langle -1/2|\hat{S}_y|1/2\rangle = i\langle 1/2|\hat{S}_y| - 1/2\rangle$$
$$= 1/2, \qquad\qquad (q.6.27.)$$
$$\langle 1/2|\hat{S}_z|1/2\rangle = 1/2, \quad \text{and} \quad \langle -1/2|\hat{S}_z| - 1/2\rangle = -1/2.$$

These are the so-called Pauli spin matrices, which may also be deduced directly by substituting $L = 1/2$ in (q.6.7.).

---

[1] Pauli, W., *Z. Physik*, 1927, **43**, 601.

For any polyelectronic atom, in which there are two or more electrons each characterized by different angular momentum vectors $L_1, L_2, \ldots$, the total resultant of the angular momentum of the atom will behave in the same manner as the quantity L in the above expressions. The commutation relationships of (q.6.10.) will still apply. There will be states described by an M quantum number with M values ranging from $-L$ to L, and the resultant angular momentum operator components will have matrix elements only between degenerate states which differ only in the value of the M quantum number, these matrix elements being described by equations (q.6.4.), (q.6.5.), and (q.6.6.). The detailed consideration of the way in which two or more angular momenta vectors interact or couple with one another provides the foundation for the multiplet theory of atomic structure.

The Hamiltonian for an N-electron atom is as given by (q.4.109.); properly this is the non-relativistic Hamiltonian. However, in order to interpret and comprehend many of the finer details of the spectra of atoms and ions, it is necessary to add several terms to this expression, some of the more important of which are associated with various types of angular momentum interaction effects. The terms of greatest magnitude, which give rise to effects observable in ordinary optical spectroscopy, are the orbital and spin angular momenta coupling among electrons and the coupling between electronic spin angular momenta alone. The interaction between the orbital magnetic moments of electrons is of lesser magnitude. The coupling of spin angular momenta among nuclei of atoms and of the spin angular momenta of electrons with the spin angular momenta of nuclei are of great importance in the electron-spin resonance and nuclear magnetic resonance spectra of molecules and complex ions, and are generally referred to as spin-spin interactions or couplings.[1] Of lesser magnitude are the effects of the coupling of nuclear spin angular momenta with electronic orbital magnetic moments and nuclear electric-quadrupole moment interactions in molecules. There is still the relativistic mass effect (see page 28), which is of greater consequence for atoms of higher nuclear charge and systems composed of many atoms, although from the point of view of spectroscopic data, where it is only term values or energy differences that are observable, this effect is generally small.

Q.6.4. Angular Momentum Vector Coupling

There is much experimental evidence, such as the study of atomic spectra and their behavior in magnetic (Zeeman effect) and electric fields (Stark effect), the Stern–Gerlach experiment, the spectrum of helium and other

---

[1] Pople, J. A., Schneider, W. G., and Bernstein, H. J., *High-Resolution Nuclear Magnetic Resonance*, 1959, McGraw-Hill Book Co. Inc., New York.

atoms, to indicate that atomic states ostensibly characterized by some given value of L and S are not associated with a single energy level but with two or more closely spaced energy levels. Such a set of closely spaced energy levels described by the same L and S quantum numbers are what are referred to as multiplets. The existence of such multiplets indicates that the total angular momentum of a system is not L but a vector sum of L and some additional angular momentum vector, and since it is found empirically that the total angular momentum quantum number associated with any multiplet is always integral or half-integral, it is plausible to deduce that the additional angular momentum is due to the electron spin. It will be shown later that for light atoms the vector sum of the orbital and spin angular momenta, denoted usually by $J = L + S$. For heavier atoms, there is still a coupling effect between orbital and spin angular momenta, but it is more complicated. The coupling of L and S means that neither L nor S can be true constants of the motion of the atomic system, since both must precess about their resultant J, as is indicated in Fig. Q.6.2. It is perhaps best to examine firstly the coupling between two angular momenta vectors, each due to electronic orbital effects.

The notation commonly used to describe atomic energy states has already been referred to in section Q.1.9., as well as the Russell–Saunders type term symbols. However, the discussion at that point was in terms of the empirical rules governing the quantum numbers used to describe the behavior of electrons in atoms, which had been deduced empirically before the advent of quantum mechanics. It remains in this chapter to show that there is a logical mathematical justification for those empirical rules.

If it is assumed that for an atom containing two electrons, with angular momenta l_1 and l_2, respectively, the operator $\hat{L}_x = \hat{l}_{x_1} + \hat{l}_{x_2}$, and that similar relationships apply to the other operator components of the vector sum, then the commutator $\hat{L}_y\hat{L}_z - \hat{L}_z\hat{L}_y$ contains terms which are double sums over the two electrons. In some of these terms, both operators apply to the same electron and in others both operators refer to different electrons. Since the coordinates and momenta of one electron commute with those of another, then the double sum in which the two operators refer to different electrons must contain operators which commute with one another too. Therefore the only non-zero terms in the double sum arise from the single sum of terms like $\hat{l}_{iy}\hat{l}_{iz} - \hat{l}_{iz}\hat{l}_{iy}$, in which both operators refer to the same electron, and in this case the first equation of (q.6.10.) applies, and therefore

$$\hat{L}_y\hat{L}_z - \hat{L}_z\hat{L}_y = i\hat{L}_x. \qquad \text{(q.6.28.)}$$

Similarly, it may be shown that the other two commutation relationships of (q.6.10.) apply, where $\hat{L}_x$, $\hat{L}_y$, and $\hat{L}_z$ are the vector sums of the one-electron operators. Now for two electrons with angular momentum, where a torque arises due to the electrostatic force between them, and this force depends on

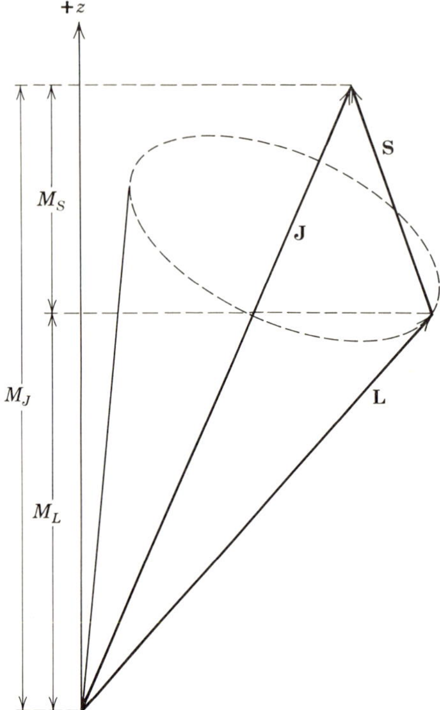

FIG. Q.6.2. The precession of the orbital angular momentum vector *L*, and the spin angular momentum vector *S*, about the total resultant angular momentum vector *J*, as in a light atom.

the interelectronic distance, the energy of their interaction will depend only on the difference $(\phi_1 - \phi_2)$, where ϕ_1 and ϕ_2 are the corresponding spherical polar coordinate angles for the two electrons. For a single electron, $\hat{L}_z = (1/i)\,\partial/\partial\phi = -i\,\partial/\partial\phi$ [equation (q.3.121.) expressed in angular momenta units], and for two electrons, a similar expression applies,

$$\hat{L}_z = -i[\partial/\partial\phi_1 + \partial/\partial\phi_2]. \qquad (q.6.29.)$$

If this operator acts on the Hamiltonian of the two-electron system, wherein the potential energy is related to the polar angles ϕ_1 and ϕ_2 only through $(\phi_1 - \phi_2)$, two terms will result which are equal and opposite, so that $\hat{L}_z\hat{H}\psi = 0$. Therefore

$$\hat{L}_z\hat{H}\psi - \hat{H}\hat{L}_z\psi = 0, \qquad (q.6.30.)$$

or $\hat{L}_z$ commutes with the Hamiltonian. In a similar manner, by choosing the spherical polar coordinate axis to be parallel with the x or y Cartesian coordinate axial direction, instead of parallel with z, it may be shown that $\hat{L}_x$ and $\hat{L}_y$ commute with the Hamiltonian of the two-electron system. Thus, the vector sum of two angular momentum vectors, coupled together by torques depending only on the relative position of the two angular momenta and not on their absolute orientation in space, are quantized in an exactly similar manner to that of the angular momentum of a single electron, and the same commutation rules apply.

The simplest situation involving two electrons with angular momentum is for two p-electrons, where l_1 and l_2 each equal 1, and therefore where L may adopt the values 2, 1, or 0; that is, two p-electrons may couple their angular momenta to give a D-, P-, and an S-state. If the principal quantum numbers of the two electrons are chosen to be dissimilar, such as $2p$ and $3p$, then there is no problem associated with the applicability of the Exclusion Principle. If each electron is assumed to move in a spherically symmetrical potential field $U(r)$ independent of the other, as in the self-consistent field approximation, then the Hamiltonian of the system in atomic units is

$$-(1/2)[\nabla_1{}^2 + \nabla_2{}^2] + U(r_1) + U(r_2), \qquad (q.6.31.)$$

and since ∇_1 only operates on the coordinates of electron 1 and ∇_2 on the coordinates of electron 2 only, the total wave function of the two-electron system may be written as

$$\Psi(r_1, r_2) = \psi(r_1)\psi(r_2). \qquad (q.6.32.)$$

The Schrödinger equation for the system

$$-(1/2)[\nabla_1{}^2 + \nabla_2{}^2]\Psi(r_1, r_2) + [U(r_1) + U(r_2)]\Psi(r_1, r_2) = E\Psi(r_1, r_2)$$

will then have a solution if

$$-(1/2)\nabla_1{}^2\psi_1(r_1) + U(r_1)\psi_1(r_1) = E_1\psi_1(r_1)$$

and $\qquad\qquad\qquad\qquad\qquad\qquad\qquad\qquad\qquad\qquad$ (q.6.33.)

$$-(1/2)\nabla_2{}^2\psi_2(r_2) + U(r_2)\psi_2(r_2) = E_2\psi_2(r_2),$$

where $E = E_1 + E_2$. If the torque between the two p-electrons is regarded as a perturbation to the central field Hamiltonian of the system, then the problem of computing the matrix components of the total Hamiltonian arises. Characterizing the $2p$ and the $3p$ orbitals by their quantum numbers, as in the previous section, the function (q.6.32.) would be

$$\Psi(r_1, r_2) = \psi(2p, m_1; r_1)\psi(3p, m_2; r_2), \qquad (q.6.34.)$$

and there would be nine such functions corresponding to the various ways of combining m_1 and m_2, each with possible values of 0, ± 1, each of these nine combinations corresponding to a different value of the total magnetic quantum number $M = m_1 + m_2$, as shown in Table Q.6.1.

TABLE Q.6.1.

No.	m_1	$+$	m_2	M
1	$+1$		$+1$	$+2$
2	$+1$		0	$+1$
3	0		$+1$	$+1$
4	$+1$		-1	0
5	0		0	0
6	-1		1	0
7	0		-1	-1
8	-1		0	-1
9	-1		-1	-2

$$(q.6.35)$$

In a central field each of the nine functions (q.6.34.) would correspond to the same energy, but because of the torque exerted by the two electrons on one another, there will be different diagonal matrix components of the Hamiltonian with respect to the various functions, and there will generally be non-diagonal matrix components with respect to some of them. It should be noticed that generally, for any two electrons with angular quantum numbers l_1 and l_2, there would be $(2l_1 + 1)(2l_2 + 1)$ combinations. In terms of the linear combination principle, it would be expected that a linear combination of the above nine functions, regarded as the unperturbed basis functions, would give a better approximation to the true eigenfunction of the system than (q.6.34.) themselves. To calculate the energy of the system with such a linear combination function requires the solution of a 9×9 secular equation. However, by making use of the properties of angular momenta, as discussed in the previous section, the situation is considerably simpler.

The angular momentum component operator of the two-electron system in the z-direction, as given by (q.6.29.), operating on the function (q.6.34.), gives simply m_1 times the function plus m_2 times the function, since the angular momentum operator of the first electron operates only on ϕ_1 and that of the second electron operates only on ϕ_2, thus indicating that the operator (q.6.29.) is diagonal. In other words, the functions (q.6.34.) are such that they diagonalize the z-component of the angular momentum and the corresponding quantum number is given as $M = m_1 + m_2$; $\hat{L}_z^2$, however, is not diagonal. Now, since $[\hat{L}_z, \hat{H}] = 0$ from (q.6.30.), the matrix components of this operator equation between two of the unperturbed functions (q.6.34.) corresponding to different M values, say M' and M'', making use of the fact that $\hat{L}_z$ is diagonal with eigenvalues M are

$$[M' - M'']\langle M'|\hat{H}|M''\rangle = 0. \qquad (q.6.36.)$$

$\langle M'|\hat{H}|M''\rangle$ is the non-diagonal matrix component of the Hamiltonian between unperturbed functions corresponding to the quantum numbers M' and M''. Therefore, the non-diagonal matrix components of the Hamiltonian are zero unless $M' = M''$, which greatly reduces the number of non-diagonal matrix components of the total Hamiltonian which have to be calculated. It is only necessary to make linear combinations of those unperturbed functions with some value of M and diagonalize the matrix of the Hamiltonian with respect to the functions of this particular M value. Thus, of the nine functions of (q.6.34.), there is one each with $M = \pm 2$, two each with $M = \pm 1$, and three with $M = 0$, so that the process of solving the 9×9 secular equation resolves itself into that of solving at most a 3×3 secular equation, in addition to a quadratic equation. Also, any linear combination of the unperturbed functions, all of which have the same diagonal matrix components of $\hat{L}_z$ (a diagonal operator), of the same M value will still diagonalize this operator, with the same value of M, and therefore M must be diagonal in the final functions arrived at by diagonalizing the Hamiltonian. Incidentally, the number of unperturbed functions of the type given by (q.6.34.) with the same M value agrees with the number which would be expected for two electrons with angular momenta quantum numbers l_1 and l_2 coupling together to give a resultant quantum number L ranging from $(l_1 + l_2)$ to $|l_1 - l_2|$. Thus, for two p-electrons, $L = 2, 1, 0$, and for $L = 2$, M may adopt the values $\pm 2, \pm 1, 0$ for $L = 1$, M may be $\pm 1, 0$, and for $L = 0$, M can only be 0, in agreement with (q.6.35.), where there is one unperturbed function with $M = \pm 2$, two each with $M = \pm 1$, and three with $M = 0$, a total of nine altogether.

There are various ways in which the properties of angular momenta operators may be used to calculate the energies of atoms, including the interaction energy of electrons with angular momenta. After solving the appropriate Schrödinger equation, since it is shown in the previous section that $\hat{H}, \hat{L}_z,$ and $\hat{L}^2$ all have diagonal matrices, it is sometimes more convenient to find the appropriate eigenfunctions of a system by diagonalizing $\hat{L}^2$, for example, rather than $\hat{H}$, or to make use of a step-down operator (page 315) to find all the functions corresponding to a given L value once one of them is known. For the present case, involving two p-electrons, where for each value of L there are $2L + 1$-degenerate states, with M values ranging from L to $-L$, the state described by the function with $M = 2$, for example, must arise from the case where $L = 2$, and since there is no secular equation involved for this event $\hat{L}^2$ must be diagonal for this unperturbed function with an eigenvalue of $L(L + 1) = 6$. Also, this must be a D-state ($L = 2$), and its energy would be given by the diagonal matrix component of the Hamiltonian for this state. When $M = 1$ (two unperturbed functions), a 2×2 secular equation has to be solved to diagonalize the Hamiltonian, which at the same time diagonalizes $\hat{L}^2$; of the two resulting states, $\hat{L}^2$ must have the eigenvalue of $2(2 + 1)$, corresponding to $L = 2$ and the eigenvalue of the Hamiltonian of this state must be the same as that of the state with $L = 2$, $M = 2$, and the other eigenvalue of $\hat{L}^2$ in this case must be $1(1 + 1)$, corresponding to $L = 1$, for which the eigenvalue of the Hamiltonian must be that of a P-state ($L = 1$). Similarly, for the three states with $M = 0$, which involved the solution of a 3×3 secular equation, and after diagonalizing the Hamiltonian in this case, one of the states would have to correspond to $L = 2$, one to $L = 1$, and one to $L = 0$. Actually, the cubic secular equation does not indeed have to be solved if it is only the eigenvalues which are required, since knowing the eigenvalues of the Hamiltonian for $L = 2$ and $L = 1$, the diagonal sum rule (page 251) may be applied to find that for $L = 0$. The eigenvalues of $\hat{L}^2$ for the states with $M = 0$ are $2(2 + 1)$, $2(1 + 1)$, and $0(1)$.

Q.6.5. Step-down and Projection Operators

The step-down angular momentum operator of (q.6.5.) may be used to find the wave function associated with one value of the quantum number M from that of another value.[1] For the specific example of two p-electrons, described by basis functions like (q.6.34.), which are products of one-electron functions, each a solution of a central field equation, it is possible by the use of a step-down angular momentum operator to find the linear combination of such basis functions which provides the best approximation to the corresponding solution of the appropriate Schrödinger equation.

The step-down operator of (q.6.5.) for the case of two p-electrons, described by the nine basis functions given by (q.6.34.) and characterized by the quantum numbers of (q.6.35.), written in Dirac notation, is

$$(\hat{L}_x - i\hat{L}_y)|L, M\rangle = \sqrt{(L - M + 1)(L + M)}|L, M - 1\rangle,$$

$$(q.6.37.)$$

where $\hat{L}_x$ and $\hat{L}_y$ refer to the two-electron system where each of these operator components is the sum of the two one-electron operators (page 325). These operators operate on wave functions described in Dirac notation as $|L, M\rangle$, $|L, M - 1\rangle$; the n quantum numbers included in (q.6.5.) remain constant and so it is not necessary to include them in this shortened notation. The nine functions of (q.6.34.) are each characterized by different values of the l and m

---

[1] Gray, N. M., and Wills, L. A., *Phys. Rev.*, 1931, **38**, 248.

quantum numbers of the two p-electrons, so that is requires a Dirac symbol of the type $|l_1, m_1; l_2, m_2\rangle$ to describe the product function of the two one-electron functions, where the subscripts 1 and 2 refer to electrons 1 and 2, respectively. Now, keeping in mind that $\hat{L}_x = \hat{l}_{x_1} + \hat{l}_{x_2}$ and that $\hat{l}_{x_1}$ only operates on the coordinates of electron 1 and that $\hat{l}_{x_2}$ similarly only operates on the coordinates of electron 2, then the equation

$$(\hat{L}_x - i\hat{L}_y)|l_1, m_1; l_2, m_2\rangle = \sqrt{(l_1 - m_1 + 1)(l_1 + m_1)}|l_1, m_1 - 1; l_2, m_2\rangle$$
$$+ \sqrt{(l_2 - m_2 + 1)(l_2 + m_2)}|l_1, m_1; l_2, m_2 - 1\rangle$$

(q.6.38.)

describes the effect of the step-down operator of (q.6.37.) operating on one of the nine product functions of (q.6.34.) characterized by the one-electron quantum numbers l_1, m_1 and l_2, m_2. For the first of the nine functions of (q.6.34.) and (q.6.35.), with $m_1 = 1$ and $m_2 = 1$, which in terms of the above would be described by the Dirac ket symbol $|1, 1; 1, 1\rangle$ and would correspond to the resultant $L = 2$, $M = 2$. The result of operating on this function with the step-down operator in (q.6.37.) would be $\sqrt{(2 - 2 + 1)(2 + 2)}$ times the function corresponding to $L = 2$, $M = 1$. But from (q.6.38.) the result of this operation would be $\sqrt{2}\,(|1, 0; 1, 1\rangle + |1, 1; 1, 0\rangle)$, and equating these two results

$$|2, 1\rangle = (1/\sqrt{2})(|1, 0; 1, 1\rangle + |1, 1; 1, 0\rangle), \qquad (q.6.39.)$$

where the symbol on the left-hand side is expressed in terms of L and M and that on the right-hand side is in terms of m_1 and m_2. If the same step-down operator is applied again to the function expressed in terms of L and M on the left-hand side of (q.6.39.), there results

$$(\hat{L}_x - i\hat{L}_y)|2, 1\rangle = \sqrt{(2 - 1 + 1)(2 + 1)}|2, 0\rangle = \sqrt{6}|2, 0\rangle, \quad (q.6.40.)$$

and on operating with the same step-down operator on the function on the left-hand side of (q.6.39.), expressed as the product function of two one-electron functions, gives

$$\begin{aligned}
(\hat{L}_x - i\hat{L}_y)|2, 1\rangle &= (1/\sqrt{2})(\hat{L}_x - i\hat{L}_y)(|1, 0; 1, 1\rangle + |1, 1; 1, 0\rangle) \\
&= (1/\sqrt{2})[\{\sqrt{(1 - 0 + 1)(1 + 0)}|1, -1; 1, 1\rangle \\
&\qquad + \sqrt{(1 - 1 + 1)(1 + 1)}|1, 0; 1, 0\rangle\} \\
&\qquad + \{\sqrt{(1 - 1 + 1)(1 + 1)}|1, 0; 1, 0\rangle \\
&\qquad\qquad + \sqrt{(1 - 0 + 1)(1 + 0)}|1, 1; 1, -1\rangle\}] \\
&= [|1, -1; 1, 1\rangle + 2|1, 0; 1, 0\rangle + |1, 1; 1, -1\rangle] \\
&= \sqrt{6}|2, 0\rangle,
\end{aligned}$$

(q.6.41.)

from (q.6.40.). Thus, it is found that

$$|2, 0\rangle = (1/\sqrt{6})[|1, -1; 1, 1\rangle + 2|1, 0; 1, 0\rangle + |1, 1; 1, -1\rangle].$$

(q.6.42.)

Similarly, by successive operation with the same step-down operator, it is found that

$$|2, -1\rangle = (1/\sqrt{2})(|1, 0; 1, -1\rangle + |1, -1; 1, 0\rangle)$$

and   (q.6.43.)

$$|2, -2\rangle = |1, -1; 1, -1\rangle.$$

In this manner, all five D-functions corresponding to $L = 2$, $M = \pm 2, \pm 1, 0$, may be expressed as linear combinations of the appropriate basis functions of (q.6.35.). The numerical multiplier which appears in equations (q.6.39.), (q.6.42.), and (q.6.43.) is the correct normalizing constant in each case.

The three P-functions corresponding to $L = 1$ and $M = \pm 1, 0$ and the S-function corresponding to $L = 0$, $M = 0$, may be obtained in a similar manner, as suitable linear combinations of the basis functions of (q.6.35.). For one of the P-functions, the one with $M = 1$, since there are two ways of choosing linear combinations from the basis set functions of (q.6.35.) which lead to $M = 1$ and one of these, namely (q.6.39.), represents a D-state, the other combination must describe the P-state; also, since the D-state, given by (q.6.39.), is the normalized sum of $|1, 0; 1, 1\rangle$ and $|1, 1; 1, 0\rangle$, the P-state must be orthogonal to this combination, which means that it must be represented by the difference of these two same functions. In the absence of any means of deciding which function should be assigned a minus sign, it is convenient to write

$$|1, 1\rangle = (1/\sqrt{2})(-|1, 0; 1, 1\rangle + |1, 1; 1, 0\rangle). \qquad (q.6.44.)$$

Now, operating on this function with the same step-down operator as above, the functions with $M = 0$ and $M = -1$ may be found as

$$|1, 0\rangle = (1/\sqrt{2})(-|1, -1; 1, 1\rangle + |1, 1; 1, -1\rangle)$$

and   (q.6.45.)

$$|1, -1\rangle = (1/\sqrt{2})(-1, -1; 1, 0\rangle + |1, 0; 1, -1\rangle),$$

where again these linear combination functions are normalized and orthogonal to the corresponding D-functions.

For the S-state with $M = 0$, it is necessary to form a linear combination of the three basis functions with $M = 0$, orthogonal to $|2, 0\rangle$ of (q.6.42.) and also to $|1, 0\rangle$ of (q.6.45.). Subject to these conditions, but still with the sign indeterminancy, the S-function may be written as

$$|0, 0\rangle = (1/\sqrt{3})(|1, 1; 1, -1\rangle - |1, 0; 1, 0\rangle + |1, -1; 1, 1\rangle). \quad \text{(q.6.46.)}$$

The D, P, and S linear combination functions given by equations (q.6.39.) and (q.6.42.–46.) all diagonalize the operator $\hat{L}^2$. It has already been shown in (q.3.127.) that the eigenvalues of $\hat{L}^2$ with respect to the one-electron functions $\psi(r, \theta, \phi)$ are $l(l + 1)(h^2/4\pi^2)$; rewriting this expression in the shortened terminology used in this chapter with respect to the two-electron functions of the type (q.6.34.) gives

$$\hat{L}^2|L, M\rangle = L(L + 1)|L, M\rangle, \quad \text{(q.6.47.)}$$

where again the angular momenta are expressed in atomic units of $h/2\pi$, and L and M are the resultants of l_1 and l_2 and m_1 and m_2, respectively, in accordance with the previous discussion. It is perhaps simplest to prove that $\hat{L}^2$ is a diagonal operator with respect to the above functions in the case of the function with $L = 2$, namely, $|2, 2\rangle = |1, 1; 1, 1\rangle$; this is in fact the only one of the nine functions referred to which is not a linear combination of the basis functions of (q.6.35.), indeed chosen that way to illustrate the behavior of step-down operators. From the characteristic property of step-up and step-down operators (page 315) that they only give one function when they operate on any particular wave function, whereas $\hat{L}_x{}^2$ and $\hat{L}_y{}^2$ give two, it is particularly convenient for purposes of the present proof to express $\hat{L}^2$ in terms of these step-up and step-down operators rather than as $\hat{L}_x{}^2 + \hat{L}_y{}^2 + \hat{L}_z{}^2$; thus, making use of the commutation relationships of (q.6.10.)‡

$$\hat{L}^2 = (\hat{L}_x + i\hat{L}_y)(\hat{L}_x - i\hat{L}_y) + \hat{L}_z{}^2 - \hat{L}_z, \quad \text{(q.6.48.)}$$

and operating with this operator on the function $|2, 2\rangle$ gives

$$\begin{aligned}
\hat{L}^2|2, 2\rangle &= [(\hat{L}_x + i\hat{L}_y)(\hat{L}_x - i\hat{L}_y) + \hat{L}_z{}^2 - \hat{L}_z]|2, 2\rangle \\
&= (\hat{L}_x + i\hat{L}_y)[\sqrt{2}(|1, 0; 1, 1\rangle + |1, 1; 1, 0\rangle)] \\
&\qquad\qquad + 4|1, 1; 1, 1\rangle - 2|1, 1; 1, 1\rangle \\
&= 4|1, 1; 1, 1\rangle + 4|1, 1; 1, 1\rangle - 2|1, 1; 1, 1\rangle \\
&= 6|1, 1; 1, 1\rangle = 2(2 + 1)|2, 2\rangle, \quad \text{(q.6.49.)}
\end{aligned}$$

indicating that the function $|2, 2\rangle$ diagonalizes $\hat{L}^2$. In a similar manner it may be proved that each of the other eight linear combination functions diagonalize $\hat{L}^2$.

‡ From the relationships

$$\hat{L}^2 = \hat{L}_x{}^2 + \hat{L}_y{}^2 + \hat{L}_z{}^2,$$

and

$$\begin{aligned}
[\hat{L}_x + i\hat{L}_y][\hat{L}_x - i\hat{L}_y] &= \hat{L}_x{}^2 + \hat{L}_y{}^2 + i\hat{L}_y\hat{L}_x - i\hat{L}_x\hat{L}_y = \hat{L}_x{}^2 + \hat{L}_y{}^2 + i[\hat{L}_y, \hat{L}_x] \\
&= \hat{L}_x{}^2 + \hat{L}_y{}^2 + \hat{L}_z,
\end{aligned}$$

and

$$[\hat{L}_x - i\hat{L}_y][\hat{L}_x + i\hat{L}_y] = \hat{L}_x{}^2 + \hat{L}_y{}^2 + i\hat{L}_x\hat{L}_y - i\hat{L}_y\hat{L}_x = \hat{L}_x{}^2 + \hat{L}_y{}^2 + i[\hat{L}_x, \hat{L}_y]$$
$$= \hat{L}_x{}^2 + \hat{L}_y{}^2 - \hat{L}_z,$$

it is apparent that

$$\hat{L}^2 = [\hat{L}_x + i\hat{L}_y][\hat{L}_x - i\hat{L}_y] + \hat{L}_z{}^2 - \hat{L}_z,$$

as in (q.6.48.).

An even more convenient method of diagonalizing $\hat{L}^2$, rather than by the use of step-down operators, as a means of finding the energy of a system composed of two or more electrons with angular momenta components (it has already been seen that $\hat{H}$ and $\hat{L}^2$ have simultaneous eigenfunctions) is a technique proposed by Löwdin,[1] making use of what are referred to as projection or annihilation operators. This method is based on the fact that each of the basis functions of (q.6.35.), for example, may be expressed as a linear combination of functions corresponding to several different L values. For example, the three functions described as $|1, 1; 1, -1\rangle$, $|1, 0; 1, 0\rangle$, and $|1, -1; 1, 1\rangle$ occur in the three linear combination functions (q.6.42.), (q.6.45.), and (q.6.46.), corresponding to $L = 2$, $L = 1$, and $L = 0$, respectively, in each case with $M = 0$, and each of the three original functions used in forming these linear combinations may itself be written as a linear combination of three functions corresponding to $L = 2, L = 1$, and $L = 0$. Thus,

$$|1, 1; 1, -1\rangle = \alpha|2, 0\rangle + \beta|1, 0\rangle + \gamma|0, 0\rangle, \tag{q.6.50.}$$

where α, β, and γ are meantime undetermined coefficients. Now, if the function (q.6.50.) is operated on with $\hat{L}^2$, there results

$$\hat{L}^2[\alpha|2, 0\rangle + \beta|1, 0\rangle + \gamma|0, 0\rangle]$$
$$= \alpha 2(2 + 1)|2, 0\rangle + \beta(1\cdot 2)|1, 0\rangle + \gamma(0\cdot 1)|0, 0\rangle, \tag{q.6.51.}$$

in which the coefficient γ has been eliminated. In this sense, the operator $\hat{L}^2$ has annihilated the S-function $|0, 0\rangle$. Löwdin's method is to find similar annihilation operators to eliminate the other coefficients (β in the above example) until $\hat{L}^2$ is diagonalized. These annihilation operators are of the general form $[\hat{L}^2 - \nu(\nu + 1)]$, where ν is an integer, and such operators annihilate that part of the linear combination function of the type (q.6.50.) which corresponds to a multiplet with L equal to ν. Thus, the operator $[\hat{L}^2 - 1\cdot 2]$ will annihilate the P-function from (q.6.50.), since

$$[\hat{L}^2 - 1\cdot 2][\alpha|2, 0\rangle + \beta|1, 0\rangle + \gamma|0, 0\rangle]$$
$$= \alpha(2\cdot 3 - 1\cdot 2)|2, 0\rangle + \beta(1\cdot 2 - 1\cdot 2)|1, 0\rangle + \gamma(0\cdot 1 - 1\cdot 2)|0, 0\rangle, \tag{q.6.52.}$$

---

[1] Löwdin, P.-O., *Phys. Rev.*, 1955, **97**, 1509.

and if the function (q.6.50.) is operated on in succession with these two annihilation operators, the result is

$$\hat{L}^2[\hat{L}^2 - 1\cdot 2][\alpha|2, 0\rangle + \beta|1, 0\rangle + \gamma|0, 0\rangle]$$
$$= \alpha(2\cdot 3)(2\cdot 3 - 1\cdot 2)|2, 0\rangle + \beta(1\cdot 2)(1\cdot 2 - 1\cdot 2)|1, 0\rangle$$
$$+ \gamma(0\cdot 1)(0\cdot 1 - 1\cdot 2)|0, 0\rangle = 24\alpha|2, 0\rangle. \qquad \text{(q.6.53.)}$$

The elimination of the $|1, 0\rangle$ and $|0, 0\rangle$ components of the function (q.6.50.) thus diagonalizes $\hat{L}^2$ directly, and apart from the fact that the resulting function is not normalized, it is the same one as given in equation (q.6.42.) by the step-down operator method. It is most convenient to use the Löwdin method on functions with $M = 0$, as the result of operating with $\hat{L}_z$ and $\hat{L}_z^2$ on such a function is zero. Thus, with a function such as (q.6.50.), by operating on it with the product of two annihilation operators so as to eliminate the $|1, 0\rangle$ and $|0, 0\rangle$ components, or the $|2, 0\rangle$ and $|0, 0\rangle$ components, or the $|2, 0\rangle$ and $|1, 0\rangle$ components, the functions $|2, 0\rangle$ (as above) and the $|1, 0\rangle$ and $|0, 0\rangle$ may be found, respectively. After the functions thus found have been normalized, then the step-up or step-down type operators may be applied to obtain the functions corresponding to other values of M.

The reason that the above annihilation operators are also referred to as projection operators is associated with the general idea of regarding a wave function as a vector. Thus, equation (q.6.50.) may be regarded as the linear combination of three orthogonal vectors, $|2, 0\rangle$, $|1, 0\rangle$, and $|0, 0\rangle$, with coefficients α, β, and γ, respectively, so that the function $|2, 0\rangle$ may be regarded as the projection of the vector $|1, 1; 1, -1\rangle$ along the direction of the vector $|2, 0\rangle$ giving the result $\alpha|2, 0\rangle$. The actual operation of taking a linear combination of a number of vectors, or basis functions, and finding the component or projection of this vector along a direction corresponding to one of the vectors, is attributed to a projection operator. The projection of an arbitrary vector in ordinary three-dimensional space onto the axes of the unit orthonormal vectors of the space is illustrated in Fig. Q.6.3. If the vector concerned is in the direction along which the projection is carried out, the projection operator involved leaves the initial vector unchanged; for example, if β and γ are zero in (q.6.50.), and a projection was made along the direction of $|2, 0\rangle$, then the initial function $\alpha|2, 0\rangle$ would remain unchanged. A function which behaves in this manner is said to be idempotent. The above-described method of using annihilation operators may be equally well described as due to the result of the action of a projection operator, by suitable choice of linear combinations of operators. For example, the particular operator

$$\frac{[\hat{L}^2 - \mu(\mu + 1)]}{\nu(\nu + 1) - \mu(\mu + 1)} \qquad \text{(q.6.54.)}$$

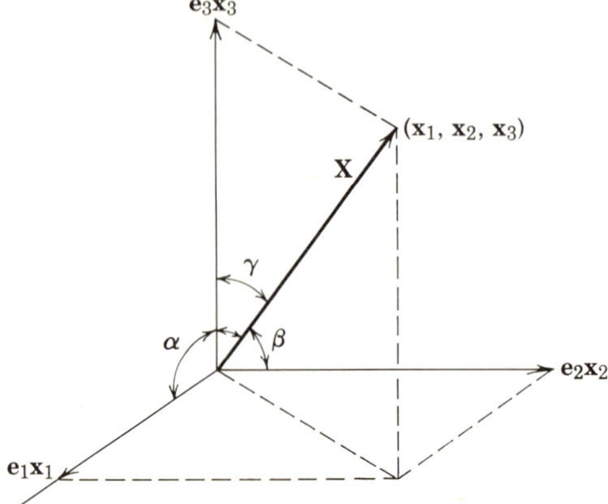

FIG. Q.6.3. The projection of an arbitrary vector, $\chi = x_1 \mathbf{e}_1 + x_2 \mathbf{e}_2 + x_3 \mathbf{e}_3$, in three-dimensional space onto the axes of the three orthonormal unit vectors, $\mathbf{e}_1$, $\mathbf{e}_2$, and $\mathbf{e}_3$, defining the space. The vectors $\mathbf{e}_1$, $\mathbf{e}_2$, and $\mathbf{e}_3$, form an orthonormal basis in three-dimensional space and are represented by the column vectors

$$\mathbf{e}_1 = \begin{vmatrix} 1 \\ 0 \\ 0 \end{vmatrix}, \mathbf{e}_2 = \begin{vmatrix} 0 \\ 1 \\ 0 \end{vmatrix}, \text{ and } \mathbf{e}_3 = \begin{vmatrix} 0 \\ 0 \\ 1 \end{vmatrix}.$$

From the figure it may be deduced that $\mathbf{e}_1^\dagger \chi = x_1 = \chi \cos \alpha$, $\mathbf{e}_2^\dagger \chi = x_2 = \chi \cos \beta$, and $\mathbf{e}_3^\dagger \chi = x_3 = \chi \cos \gamma$, where $\chi = (\chi^\dagger \chi)^{1/2}$ (see page 151).

if allowed to operate on a function of some definite L value, such as $|1, 0\rangle$, will annihilate the function whose L value is equal to μ, as $\hat{L}^2$ operating on this function will provide the same function simply multiplied by $\mu(\mu + 1)$. For any other value of L appearing in a linear combination function of the type (q.6.50.), such as for example L_σ, $\hat{L}^2$ will operate on it so as to multiply it by the factor

$$\frac{\sigma(\sigma + 1) - \mu(\mu + 1)}{\nu(\nu + 1) - \mu(\mu + 1)}. \tag{q.6.55.}$$

Therefore, if a function such as (q.6.50.) is operated on by a product of operator factors such as (q.6.54.) containing all values of μ present as components of the initial function, except for the particular value of ν equal to the L value of interest, all functions except those that contain $L = \nu$ would be annihilated. Each of the factors contained in the product operator would multiply the function for which $L = \nu$ by the value of the quantity defined by (q.6.55.) for which $\sigma = \nu$, which is unity. Thus, the over-all result would be that the function with quantum number L would be left unchanged, or the product operator would be idempotent. The Löwdin projection operator technique is a most useful way of producing a function corresponding to some particular value of L. On the other hand, since the functions produced in this manner are not normalized, there is little advantage in including the denominator of (q.6.54.) in the operator used, and if the numerical value of this denominator is excluded, the method amounts to simply making use of the operator $[\hat{L}^2 - \nu(\nu + 1)]$, as illustrated on page 334.

Much effort has been expended in the attempt to find general expressions for the coefficients, in expressions like (q.6.42.) for example, which enable a function written as $|L, M\rangle$ to be formed by the linear combination of two-electron functions like $|l_1, m_1; l_2, m_2\rangle$. If this general relationship is written as

$$|L, M\rangle = \sum_{m_1} \sum_{m_2} C(l_1 m_1; l_2 m_2 | L, M) |l_1 m_1; l_2 m_2\rangle, \qquad \text{(q.6.56.)}$$

where the coefficients $C(l_1 m_1; l_2 m_2 | L, M)$ are variously referred to in the literature as Clebsch–Gordan, vector addition, or vector-coupling, coefficients, then the problem is that of finding an explicit formula for these coefficients. An expression for these vector-coupling coefficients was first obtained by E. Wigner (1931) and subsequently derived by Racah[1] by the use of step-up and step-down operators in a more symmetrical form. Another derivation has been given by Edmonds.[2] None of these expressions is particularly simple, either to derive or to use, but perhaps Racah's formula is the simplest one, as follows:

$$C(l_1, m_1; l_2, m_2 | L, M)$$

$$= \sqrt{\frac{(2L + 1)(l_1 + l_2 - L)!\,(L + l_1 - l_2)!\,(L - l_1 + l_2)!}{(L + l_1 + l_2 + 1)!}} \times \sum_z (-1)^z$$

$$\times \frac{\sqrt{(l_1 + m_1)!\,(l_1 - m_1)!\,(l_2 + m_2)!\,(l_2 - m_2)!\,(L + M)!\,(L - M)!}}{z!\,(l_1 + l_2 - L - z)!\,(l_1 - m_1 - z)!\,(l_2 + m_2 - z)!}$$

$$\times (L - l_2 + m_1 + z)!\,(L - l_1 - m_2 + z)!$$

$$\text{(q.6.57.)}$$

The summation over z preceding the second factor in Racah's formula is such that z may adopt all integral values which make the factorials in the

---

[1] Racah, G., *Phys. Rev.*, 1942, **62**, 438.
[2] Edmonds, A. R., *Angular Momentum in Quantum Mechanics*, 1957, and 1960, Princeton Univ. Press.

denominator meaningful. For this to be so, the upper limit of z would have to be whichever is the largest of 0, $(-L + l_2 - m_1)$, or $(-L + l_1 + m_2)$, and the lower limit would have to be whichever is the smallest of the terms $(l_1 + l_2 - L)$, $(l_1 - m_1)$, or $(l_2 + m_2)$. C vanishes if $M \neq m_1 + m_2$. As for (q.6.44.), the sign of each eigenfunction is arbitrary, although the signs of the functions with the same L value and different M value are related to the use of the step-up and step-down operators used in the derivation of (q.6.57.). The above expression simplifies for specific values of $L = l_1 + l_2$ and of $M = m_1 + m_2$, and tabulated values of this expression for arbitrary values of l_1 with $l_2 = 1/2, 1, \ldots$, are given by Slater[1] and Condon and Shortley.[2] Other expressions designed to give coupling coefficients for problems involving the coupling of several angular momentum vectors, including spin, into a resultant J, lead to more complicated formulae referred to as 3-j, 6-j, 9-j, and 12-j coefficients.[3–5]

Q.6.6.   Russell–Saunders Coupling

The manner in which atomic spectra were interpreted in terms of empirically deduced selection principles had already been discussed in sections Q.1.5., 6., 7., and 9. It was the definite multiplicities exhibited by atomic spectra, the fact that one-electron atoms displayed doublets, two-electron atoms showed singlets and triplets, three-electron atoms showed doublets and quartets, four-electron atoms, singlets, triplets, and quintets, that led to Sommerfeld's proposal of the so-called inner quantum number J, in terms of which the first partially adequate theory of atomic spectra was developed. The J quantum number proved most useful in accounting for the Zeeman effect, where the number of components into which a spectral line separated in a magnetic field was defined as $2J + 1$, as if there was an angular momentum component characterized by the quantum number J, whose component along an axis was $M(h/2\pi)$ with $M = J, J - 1, \ldots, -J$. The J quantum number could be determined unambiguously from observations on the Zeeman effect and apparently electronic transitions could only take place in atoms between states for which J changed by either ± 1 or 0 and M also by ± 1 or 0. The J

---

[1] Slater, J. C., *Quantum Theory of Atomic Structure*, **Vol. 2**, 1960, McGraw-Hill Book Co., N.Y.

[2] Condon, E. U., and Shortley, G. H., *The Theory of Atomic Spectra*, 1957, Cambridge Univ. Press.

[3] Yutsis, A. P., Levinson, I. B., and Vanagas, V. V., *Mathematical Apparatus of the Theory of Angular Momentum*, 1963, Oldbourne Press, London.

[4] Regge, T., *Il nuovo cimento*, [10], 1959, **11**, 116; 1948, **10**, 544.

[5] Hobson, E. W., *The Theory of Spherical and Ellipsoidal Harmonics*, 1931, Cambridge Univ. Press.

value characteristic of any multiplet was assumed to be the resultant of two vectors which could form resultants with all integrally spaced values between the sum and the difference of the vectors. Thus for an observed triplet state with J values of 4, 3, and 2, it was assumed that there was a vector of magnitude 1, determining the multiplicity and denoted by the symbol S, and another vector denoted by the symbol L, so that the combination of $S = 1$ and $L = 3$ gave $J = L + S, L + S - 1, \ldots, |L - S|$; thus, provided $L \geq S$, there are $2S + 1$ different J values defining the multiplicity. In terms of this notation, a state for which, for example, S was equal to 2, $L = 6$, and $J = L + S = 8$, was denoted as a $^5J_8$-state, as on page 88. This is the basis of the so-called Russell–Saunders coupling scheme.

In the more general application of the Russell–Saunders coupling scheme to the case of polyelectronic atoms, the principle used is to consider the l's of each electron beyond the number forming a closed group and to find the vector sum L, then to do likewise with the spins of the various electrons, finally coupling the vector sums of L and S to form J. As far as the vector sum of the spins are concerned, there is generally no difficulty. Thus for two electrons the vector sum may be either 1 or 0, corresponding to a triplet or singlet state, and with a third electron, S may be either 3/2 or 1/2, corresponding to a doublet or a quartet state, respectively, and so on where the S value appropriate to one number of electrons splits into two values when the number of electrons is increased by unity.‡

‡ Generally for N-electrons, there are 2^N ways of combining their spins. The number of states of different multiplicities is most conveniently determined by the use of what is referred to as the branching diagram, as in Fig. Q.6.4. Such a diagram is easily constructed beginning with a single electron and successively coupling other electron spins to it in all possible algebraic combinations. The number of states of any given multiplicity is indicated in this diagram by a number within a circle whose abscissa is the number of electrons and whose ordinate is the multiplicity. Thus, for four electrons, there would be two singlet states with $S = 0$, three triplet states with $S = -1$, 0, and 1, and one quintet state with $S = -2, -1, 0, 1$, and 2, so that for four independent electrons, not in closed groups, there would, for example, be 16 linearly independent wave functions required to describe them; these functions would not necessarily be orthogonal to one another, but would lead to six different energies.

In coupling the l quantum numbers for individual electrons to form the resultant L, however, the question arises as to which pair of l values to couple together first. From the study of the atomic spectra of both neutral atoms and ions, it was concluded that the lowest energy states should be first dealt with, and then the resultant of this value coupled with the l value of the electron in the next higher energy state. Thus, for an atomic species with three electrons more than the number constituting a closed group, for example a $4d$-, $5s$-, and $6p$-electron, coupling together the $4d$- and $5s$-electrons

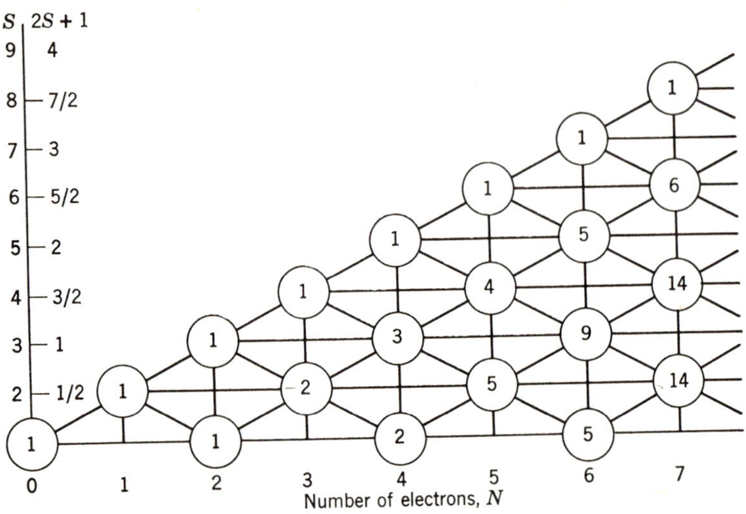

FIG. Q.6.4.   Branching diagram illustrating the number of states of any given multiplicity which results for N independent electrons.

would give $L = 2$, $S = 1, 0$ or to ^3D, ^1D, referred to as the parent multiplets; then coupling the additional electron ($6p$) would give ^4F, ^4D, ^4P, and ^2F states from the ^3D parent multiplet and ^2F, ^2D, and ^2P states from the ^1D parent multiplet. This scheme works all right if the energy of the last coupled electron is considerably higher than that of the others, but if there are two or more electrons in a system with the same n and l quantum numbers, it is not so easily possible to determine the parent multiplets involved. Furthermore, in such cases, it is found that the application of the Pauli Exclusion Principle leads to the result that certain multiplets which would be predicted in terms of the simple vector model are not actually observed in the line spectrum of the system concerned. For example, whereas two p-electrons which are non-equivalent as, for example $2p$ and $3p$, give rise to the multiplets ^3D, ^3P, ^3S, ^1D, ^1P, and ^1S, two $2p$-electrons only give rise to the multiplets ^3P, ^1D, and ^1S.

In the application of the Russell–Saunders coupling scheme to N-electron atoms, it is found that there are certain selection rules or principles which are generally valid and others which are only valid for the lighter atoms. Thus, apart from the restrictions noted above that both J and M may only change by 0 or ± 1, L may only change also by 0 or ± 1 unit, and S may not change, so that transitions between states of different multiplicity cannot take place, but neither of the latter rules are rigorously obeyed. They are valid where the energy separations between the various energy levels of any multiplet are very small compared to the energy differences between multiplets, where the orbital and spin angular momenta, L and S, only interact with one another

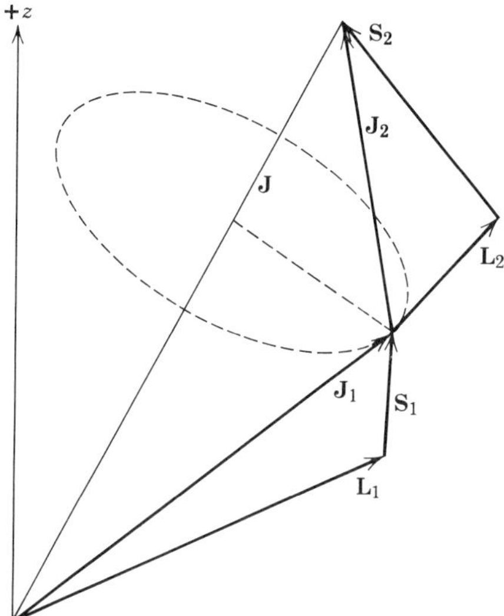

FIG. Q.6.5. Illustration of the spin-orbital interactions involved between a pair of electrons in a heavy atom referred to as *j-j* coupling.

weakly and have approximately constant magnitude. However, in the event that the interaction between L and S is large, resulting in a large separation between the energy levels of differing J values in the same multiplet, implying that there is a large torque effect between L and S, they are no longer of constant value. This turns out to be the case for heavier atoms, where the interaction between L and S increases with the number of electrons, and the Russell–Saunders scheme is no longer applicable. This means that the individual electron orbital angular momentum and spin angular momentum couple rather strongly to form a resultant total angular momentum for a given electron, and that these total angular momenta couple together, more weakly, to form a resultant J, as in Fig. Q.6.5. This type of spin-orbital interaction is referred to as *j-j* coupling. It appears that both *L-S* coupling and *j-j* coupling are approximations to the behavior of real many-electron atoms, representing two extreme types, so that an accurate description of any actual atom would involve interpolation between the two extremes.

Hund's rules are found to apply generally whenever Russell–Saunders coupling is valid (see page 87). In atoms which may be described in terms of the central field approximation, it is found that the only allowed transitions are between states which differ in the l quantum number of only one electron, provided that this l changes by ± 1 unit in the transition involved, but for atomic systems which do not conform with the central field model, this is not so. However, in the latter event, there is a related general principle that transitions may only take place between states of even parity and odd parity, where the parity is said to be even if the sum of the l values of all of the electrons is even and odd if this sum is odd.

Q.6.7. The Anomalous Zeeman Effect—Landé Theory

It has already been shown in section Q.5.2. that the energy of an electron in a homogeneous magnetic field is increased by an amount $[\varepsilon hH/4\pi mc]m_l$, so that any electronic state described by the quantum number l will split (in a magnetic field) into $(2l + 1)$ separate levels, corresponding to the possible values $-l, -l + 1, \ldots, l$ of m_l, these separate levels being spaced at intervals of $\varepsilon hH/4\pi mc$. This energy interval, $\varepsilon hH/4\pi mc = h\nu_L = \mu_B H$, where

$$\nu_L = \varepsilon H/4\pi mc = 1.40 \times 10^6 Hs^{-1}, \tag{q.6.58.}$$

is the so-called Larmor frequency, separating the $2l + 1$ levels, which, in conjunction with the selection rule for m_l that $\Delta m_l = 0, \pm 1$, gives rise to the normal Zeeman effect in atomic spectra, whereby some spectral lines are split into three components in a magnetic field. It should be noted that in the above expressions, m refers to the electron mass and m_l is the magnetic quantum number. Even if the spin of the electron is taken into account, the above expressions would be unaltered if the electron was associated with a magnetic moment with the same ratio to the spin angular momentum as that of the orbital magnetic moment bears to the orbital angular momentum, $M_z/S_z = \varepsilon/mc$ (page 284); for in the latter event the total angular momentum would be j, and the total magnetic moment would be $M_j = [\varepsilon h/4\pi mc]j$, so that j and M_j would have the same direction of quantization in the applied magnetic field and would precess round the field direction together. The one difference, including electron spin, would be that instead of $2l + 1$ possible orientations for m_l, there would be $2j + 1$ possible orientations, and so every undisturbed term would be split up by the magnetic field into $2j + 1$ terms, in such a way that the amount of the splitting would be exactly the same as before. However, the anomalous Zeeman effect, the observed behavior that in a weak magnetic field a spectral line is split up into a number of lines different from three (as in the normal Zeeman effect), can be completely explained by assuming that $M_z/S_z = \varepsilon/mc$, instead of $\varepsilon/2mc$, as for the ratio of the orbital magnetic moment to the orbital angular momentum. The spin quantum number is, of course, always 1/2 and the magnetic moment of an electron is exactly one Bohr magneton, $\mu_B = \varepsilon h/4\pi mc$. This difference in the behavior of electronic spin moments and orbital moments was first put on a theoretical basis by Thomas (1926) and later by Kramers (1935), and follows rigorously from the Dirac relativistic equation, as already mentioned in section Q.5.2.

Much effort has been applied to the measurement and interpretation of Zeeman effects and a theory, originally proposed by Landé,[1] has played an important part in arriving at an understanding of complex atomic spectra. Landé's theory is essentially based on the assumption that the interaction energy between the angular momenta L and S could be described by an expression $A[\mathbf{L}\cdot\mathbf{S}]$, arising from the magnetic coupling between the magnetic moments of the orbital and spin angular momenta. From $\mathbf{J} = \mathbf{L} + \mathbf{S}$, by squaring $\mathbf{J}^2 = \mathbf{L}^2 + 2[\mathbf{L}\cdot\mathbf{S}] + \mathbf{S}^2$, and therefore $A[\mathbf{L}\cdot\mathbf{S}] = (A/2)[\mathbf{J}^2 - \mathbf{L}^2 - \mathbf{S}^2]$. However, in order to obtain a relationship between J, L, and S, compatible with experimental observation, Landé found that this expression had to be modified to read

$$A[\mathbf{L}\cdot\mathbf{S}] = (A/2)[J(J + 1) - L(L + 1) - S(S + 1)], \qquad \text{(q.6.59.)}$$

even although at the time there was no theoretical justification for doing so. However, the quantum mechanical interpretation of (q.6.59.) is quite straightforward. If $[\mathbf{L}\cdot\mathbf{S}]$ is regarded as an operator, then $A[\mathbf{L}\cdot\mathbf{S}] = (A/2)[\mathbf{J}^2 - \mathbf{L}^2 - \mathbf{S}^2]$ represents an operator equation and (q.6.59.) represents the diagonal matrix component of the operator $[\mathbf{L}\cdot\mathbf{S}]$. Now it has already been seen that the operator $\hat{L}^2$ has a diagonal matrix [(q.3.127.)] and it is plausible to assume that the squares of the other angular momentum operators should also have diagonal matrices.

Since all the members of any multiplet have the same L and S values for their quantum numbers, Landé's equation (q.6.59.) may be most usefully applied to finding the energy separation between states with a given J value and another with the value $J + 1$. Thus from (q.6.59.),

$$A[\mathbf{L}\cdot\mathbf{S}] = \Delta E^{LS} = E^{LS}_{J+1} - E^{LS}_{J} = A(J + 1). \qquad \text{(q.6.60.)}$$

In this form, the Landé equation is known as the Landé interval rule, which implies that the multiplet energy separation due to spin-orbital interaction is proportional to the larger of the two J values involved, a relationship which is found to be obeyed for many observed multiplets. The validity of the Landé interval rule is found to depend on the smallness of the $\mathbf{L}\cdot\mathbf{S}$ interaction energy relative to the total energy, and approximately deviations from the Landé interval rule parallel deviations from Russell–Saunders coupling.

By making the further assumption that the energy of an atom in an applied magnetic field could be calculated from the average values of $\mathbf{L}$ and $\mathbf{S}$,

[1] Landé, A., Z. Physik, 1921, **5**, 231; **7**, 398; 1922, **11**, 353; 1923, **15**, 189; **16**, 391; 1924, **24**, 88; **25**, 46.

averaged over the period of their precession about their resultant $\mathbf{J}$, Landé was able to provide a reasonable explanation of the Zeeman effect, which has been of much use in atomic spectroscopy. The average values of $\mathbf{L}$ and $\mathbf{S}$ are equivalent to their components along J, which in vector terminology are given by

$$\langle \mathbf{L} \rangle = \mathbf{J}(\mathbf{L} \cdot \mathbf{J})/J^2 \quad \text{and} \quad \langle \mathbf{S} \rangle = \mathbf{J}(\mathbf{S} \cdot \mathbf{J})/J^2. \qquad \text{(q.6.61.)}$$

In terms of the discussion in section Q.5.2., an electron with orbital angular momentum, $l(h/2\pi)$, will have a magnetic moment of $-(\varepsilon/2mc)(lh/2\pi)$, and in a magnetic field acting in the z-Cartesian coordinate direction, the magnetic energy of interaction will be $-H$ times the component of the magnetic moment along this axis. Since the component of angular momentum along this axial direction is $m_l(h/2\pi)$, the magnetic interaction energy will be $m_l\mu_B H$ (m_l, m, μ_B, and H are the magnetic quantum number, the electron mass, the Bohr magneton, and the magnetic field strength, respectively). If the component of J along the z-direction is M, Landé's assumption is that the energy of the orbital magnetic moment in the external magnetic field is, from (q.6.61.), given by magnetic energy due to

$$L = [M(\mathbf{L} \cdot \mathbf{J})/J^2]\mu_B H. \qquad \text{(q.6.62.)}$$

In order to get agreement with experimentally observed results, Landé, however, found it necessary to assume that the magnetic energy due to the spin angular momentum in the applied field was twice as great as that due to the orbital angular momentum, so that he assumed

$$\text{magnetic energy due to} \quad S = [M(\mathbf{S} \cdot \mathbf{J})/J^2]2\mu_B H. \qquad \text{(q.6.63.)}$$

This assumption is now known to be compatible with the electron's spin angular momentum of $h/4\pi$ and magnetic moment of one Bohr magneton, and is a direct consequence of the Dirac relativistic mechanics (see page 284). From (q.6.62.) and (q.6.63.), the Landé proposal leads to a total magnetic energy of an electron in a magnetic field as

$$E_m = \{[M(\mathbf{L} + 2\mathbf{S}) \cdot \mathbf{J}]/J^2\}[\mu_B H] = gM\mu_B H, \qquad \text{(q.6.64.)}$$

where g is known as Landé's g factor, which is a dimensionless quantity given by

$$g = [\mathbf{L} + 2\mathbf{S}) \cdot \mathbf{J}]/J^2. \qquad \text{(q.6.65.)}$$

Now, making use of the vector properties of $\mathbf{L}$ and $\mathbf{S}$, in particular noting that $\mathbf{L} + 2\mathbf{S} = \mathbf{J} + \mathbf{S}$ and $\mathbf{L}^2 = \mathbf{J}^2 - 2(\mathbf{S} \cdot \mathbf{J}) + \mathbf{S}^2$, the Landé g factor may be written as

$$g = 1 + [\mathbf{S} \cdot \mathbf{J}/J^2] = 1 + [J^2 - L^2 + S^2]/2J^2. \qquad \text{(q.6.66.)}$$

However, just as in the case of the multiplet separation expression of (q.6.59.), it was found by Landé that agreement with experimental observation only resulted if (q.6.66.) was modified to read

$$g = 1 + \frac{J(J + 1) - L(L + 1) + S(S + 1)}{2J(J + 1)}, \qquad \text{(q.6.67.)}$$

which, as for (q.6.59.), represents the average value or the diagonal matrix component of the operator defined by (q.6.66.). For a one-electron atom in a state $^2S_{1/2}$, for which $J = 1/2$, $S = 1/2$, and $L = 0$, g is found to be 2, the value obtained in terms of Dirac's relativistic theory. For singlet states, where $S = 0$, g is equal to unity, so that the energy separation for such states in a magnetic field is independent of J and L, which is in accordance with the normal Zeeman effect. For other multiplicities, g depends on J, L, and S, and this result is in accordance with the anomalous Zeeman effect. It is observed that the magnetic energy of atomic states is directly proportional to the applied magnetic field strength, according to (q.6.64.), provided only that the magnetic separation of an energy level is small compared with the separation of the various levels of the multiplet, corresponding to different J values. In other words, equation (q.6.64.) only applies when the magnetic precession frequency is small compared with the frequency of precession of **L** and **S** about their resultant **J**. For higher values of the magnetic field strength, the expression (q.6.64.) is not found to be applicable, and this is confirmed by experimental observation; such deviations are referred to as the Paschen–Back effect.[1] At these higher field strengths, the anomalous Zeeman effect begins to resemble the normal Zeeman effect, in that three sets of spectral lines appear, each set consisting of very closely spaced lines. This may be attributed to the fact that at these higher magnetic field strengths, the rate of precession of **J** about the field direction becomes higher than the rate of precession of the magnetic moment about **J**, resulting in the uncoupling of **L** and **S**; in fact, if **L** and **S** were completely uncoupled, there would be exactly three lines in the resulting spectrum, but the presence of a small amount of L-S coupling separates the three lines into three groups of closely spaced lines. Spectral lines are also split by an applied electric field, a phenomenon referred to as the Stark effect, whereby the energy splitting is found to depend on $|M|$ rather than on M; this is due to the fact that in an electric field, the energy is unchanged on reversal of the direction of rotation of the angular momenta vectors about the field direction. The Stark effect is of particular value in the discussion of molecular spectra.

From equations (q.6.58.) and (q.6.64.), it may be deduced that

$$E_m = 2\pi\nu_L g M, \qquad \text{(q.6.68.)}$$

---

[1] See, for example, Richtmeyer, F. K., Kennard, E. H., and Lauritsen, T., *Introduction to Modern Physics*, 1955, McGraw-Hill Book Company, New York.

where ν_L is the Larmor frequency. Thus, a multiplet with quantum number J having an energy E_m^0 in the absence of a magnetic field will be split by an applied magnetic field into $2J + 1$ levels having the energies

$$E_m = E_m^0 + 2\pi\nu_L g M \qquad \text{with} \qquad -J \leq M \leq J, \qquad \text{(q.6.69.)}$$

so that the energy splitting is proportional to the field strength, by way of the Larmor frequency. Thus, in the absence of electron spin, when g becomes unity, the angular frequency of precession of the vector $\mathbf{J}$ in a magnetic field is simply the Larmor frequency. But since g is different from unity, the $\mathbf{J}$ vector precesses in a magnetic field with g times the Larmor frequency, giving rise to the anomalous Zeeman effect. The Landé g factor is sometimes also referred to as the gyromagnetic ratio, which is the ratio of magnetic moment to angular momentum.

From the discussion in sections Q.5.6. and Q.5.7., it follows that everything that has been said in this chapter so far is equally applicable, whether simple product functions or determinantal functions, including spin, are used in conjunction with the various angular momenta operators.

Q.6.8. Slater Theory of Multiplets

In section Q.4.8., it has been seen that the central field approximation, in which it is assumed that each electron moves in a spherically symmetrical field provided by the nucleus and the other electrons in an atom, is not a bad approximation. To a next higher approximation, in atoms containing two or more electrons with angular momenta, there are electrostatic torques exerted between such electrons, which effects are neglected in the Hartree approximation. The principle of regarding the torque between electronic angular momenta as a perturbation to the central field energy and the coupling of angular momenta vectors has been described in sections Q.6.4.–7. While the previous discussion in these sections has been carried out with respect to simple product functions of two or more electrons, as mentioned above, due to the fortunate circumstance that the matrix components of the angular momenta operators for electrons have the same values with respect to determinantal functions as they do with respect to simple product functions, the conclusions of the previous sections are generally applicable and valid, even using determinantal functions.

If the Hamiltonian of an atomic system involves the electronic coordinates only, independent of spin effects, which will be the case if all magnetic effects are disregarded, then the Hamiltonian will commute with all spin angular momentum operators of the system. Similarly, if no external torques are involved in the Hamiltonian of an atomic system, it will commute with the orbital angular momentum operator of the whole atom, and the orbital and spin angular momentum operators will commute with one another, and in such an event, where spin-orbital coupling is neglected, the orbital and spin angular momenta may be separately and simultaneously quantized to give the total quantum numbers L, M_L, S, and M_S. There will be, as discussed above, matrix components of the orbital angular momentum operators only between two wave functions of the system concerned which have identical values of the spin quantum numbers of each electron, and also matrix components of the spin angular momentum operators only between two wave functions with identical values of the orbital quantum numbers of each electron. The matrix components will have the same form as given by (q.6.7.), with L and M_L in place of l and m, and the total spin angular momentum of the whole atom will also be of this form. The Hamiltonian of the system will have non-zero matrix components only between two states which have the same M_L value, as in (q.6.36.), and also if the M_S value concerned is the same. It thus works out that if the polyelectronic determinantal wave functions for an atom are examined in terms of both their M_L and M_S values that the secular equation obtained in the application of the perturbation method may be separated into many individual equations, enabling the diagonal sum rule to be applied in finding the energies of the various multiplets involved. Details of how this technique is applied to the calculation of the energies of various polyelectronic atoms are given in the books of Slater,[1] Hartree,[2] and Condon and Shortley,[3] and a simple illustration has already been given in section Q.5.6. in connection with the discussion of the excited states of the helium atom, without actually evaluating the various integrals involved. It should be noted, incidentally, that the first excited state of the helium atom, corresponding to the configuration $1s2s$, leads to the multiplets ${}^1S$ and ${}^3S$, with $S = 0$ and $S = 1$, respectively, and $L = 0$, and that the reason that these two multiplets have different energies is a consequence of the antisymmetry principle, although the Hamiltonian of the system depends only on the electronic coordinates; the four possible unperturbed wave functions, in this case, differ only in their spin functions, with $M_S = 1(m_{s_1} = 1/2, m_{s_2} = 1/2)$, $M_S = 0(m_{s_1} = \pm 1/2, m_{s_2} = \pm 1/2)$, and $M_S = -1(m_{s_1} = -1/2, m_{s_2} = -1/2)$. Here the state with $M_S = 1$ belongs to the triplet state and the diagonal matrix component of the Hamiltonian for this state gives the triplet energy; the two states with $M_S = 0$ leads to the states of $S = 1$ and 0 with $M_S = 0$, and by

---

[1] Slater, J. C., *Quantum Theory of Atomic Structure*, 1960, **Vol. II**, McGraw-Hill Book Co., N.Y.

[2] Hartree, D. R., *The Calculation of Atomic Structures*, 1957, John Wiley & Sons, Inc., N.Y.

[3] Condon, E. U., and Shortley, G. H., *The Theory of Atomic Spectra*, 1957, Cambridge Univ. Press, New York.

the diagonal sum rule the sum of the diagonal matrix components of these states gives the sum of the triplet and singlet energies, from which the singlet energy may be obtained by subtraction. For the next excited state of helium, with the configuration $1s2p$, there are no less than 12 unperturbed states to be considered, since both M_L and M_S have to be taken into account this time, even though M_L arises from only one of the electrons. In terms of the vector model, this gives rise to a ^3P- and a ^1P-multiplet, the former with nine components with $M_L = 1, 0, -1$ and $M_S = 1, 0, -1$, combined in all ways, and the latter with three components with $M_L = 1, 0, -1$ and $M_S = 0$; again the energies of each multiplet may be found by application of the diagonal sum rule. It should also be noted that the operator techniques and properties already discussed for $\hat{L}$, namely the use of step-down operators, $\hat{L}^2$ operators, projection operators, and vector-coupling coefficients, may be equally well applied to $\hat{S}$, with the appropriate substitution of $\hat{S}$ for $\hat{L}$ and $\pm 1/2$ for $l_1, l_2, \ldots$.

In order to calculate the energy values of the multiplets which occur in the observed spectra of polyelectronic atoms, it is necessary to evaluate the matrix components of the Hamiltonian of the system concerned with respect to the unperturbed wave functions used to describe the system. As is evident from the discussion in section Q.5.6., this procedure ultimately involves the evaluation of integrals of the type

$$\int F_i^* \hat{H} F_j \, d\tau = \langle i | \hat{H} | j \rangle, \qquad \text{(q.6.70.)}$$

where $\hat{H}$ is the Hamiltonian of the system and the F's are determinantal functions formed from a set of N spin orbitals according to (q.5.26.). Generally for each complete set of one-electron spin orbitals of the form of (q.5.11.), in which each spin orbital is the product of a function of coordinates like (q.3.130.) combined with either the α or β spin function, it is possible to select N spin orbitals in an infinite number of ways and for each choice of N spin orbitals a separate determinantal function may be formed. The matrix components of the Hamiltonian of the system between these determinantal functions, and the particular linear combination of determinantal functions which make the Hamiltonian diagonal, when found, constitute precise solutions of the Schrödinger equation of the system, consistent with the antisymmetry principle. In actual practice, it is not, of course, possible to use infinite sets of functions, so that the problem is to choose those determinantal functions which give as good an approximation as possible. In the simplest case, only a single configuration is considered, and the determinantal functions arising from this configuration are made up of a set of spin orbitals corresponding to each of the nl values involved, with all possible combinations of m_l and m_s included, which are consistent with the configuration chosen. A still better approximation for any particular system may be obtained by making up linear combinations of determinantal functions arising from different configurations; the detailed consideration of configuration interactions is of special importance in molecular quantum mechanics. Slater has shown that the matrix components of any of the operators of interest in quantum mechanics, whether it be the Hamiltonian, angular momentum, or any of the other operators mentioned in this chapter, with respect to two determinantal functions made up of various spin orbitals, may be reduced to radial integrals expressed in terms of the radial function (q.3.110.). Each spin orbital in the determinantal function describing a polyelectronic atom is of the form given in (q.5.11.), or expressed in spherical polar coordinates

$$\Psi_{n,l,m_l,m_s}(r, \theta, \phi, \zeta) = R_{n,l}(r)\, Y_{l,m}(\theta, \phi)\chi_{m_s}(\zeta), \qquad \text{(q.6.71.)}$$

where $\varUpsilon_{l,m_l}(\theta, \phi)$ is given by (q.3.55.).

If spin-orbital interactions and magnetic effects are neglected, then the central field Hamiltonian for a polyelectronic atom is given, by extension of equations (q.4.76.–78.) to the N-electron case, and expressed in atomic units, as

$$\hat{H} = \sum_i \hat{H}_i^0 + \sum_{i>j}\sum \hat{H}_{ij}', \qquad \text{(q.6.72.)}$$

where

$$\hat{H}_i^0 = -(1/2)\nabla_i^2 - Z/r_i \qquad \text{and} \qquad \hat{H}_{ij}' = 1/r_{ij}, \qquad \text{(q.6.73.)}$$

and r_i is the distance of the ith electron from the origin, r_{ij} is the distance between the ith and jth electron, and ∇_i^2 is the Laplacian operating only on the coordinates of the ith electron. Making use of equations (q.3.8.) and (q.3.24.) [with $\beta = l(l + 1)$], it is found that, with Ψ given by (q.6.71.),

$$-(1/2)\nabla^2\Psi = (1/r)[-(1/2)\, d^2/dr^2 + l(l + 1)(2r^2)^{-1}]P_{n,l}(r)\, Y_{l,m_l}(\theta, \phi)\chi_{m_s}(\zeta),$$
$$\text{(q.6.74.)}$$

where $rR_{n,l}(r)$ is denoted by the Slater symbol $P_{n,l}(r)$, rather than by $Y(r)$ as in (q.3.79.), in order to avoid confusion with the spherical harmonics $Y_{l,m_l}(\theta, \phi)$. If now the set of quantum numbers applicable to the state function Ψ_i', namely n, l, m_l, and m_s, are substituted and equation (q.6.74.) multiplied by the complex conjugate of Ψ_i, thus by Ψ_i^*, and both sides integrated over all space and spin coordinates involved, then by virtue of the orthogonality of the functions $Y_{l,m_l}(\theta, \phi)$, as shown on page 191, the following is obtained:

$$\int_0^\infty \Psi_i^*(-\nabla^2/2)\Psi_i'\, d\tau = \delta(l_i, l_i')\, \delta(m_{l_i}, m_{l_i}')\, \delta(m_{s_i}, m_{s_i}')$$
$$\times \int_0^\infty P_{n_i,l_i}^*(r)[-(1/2)\, d^2/dr^2 + l_i(l_i + 1)(2r^2)^{-1}]$$
$$\times P_{n',l_i}(r)\, dr. \qquad \text{(q.6.75.)}$$

The integral, given by (q.6.75.), by integrating by parts may be expressed in two alternative ways, each of which is more convenient for actual numerical calculation than (q.6.75.) itself and both forms of which express the kinetic energy (the $-\nabla^2/2$ term) as the product of two first derivatives rather than as a second derivative, recast the integral into a more symmetrical form, and exhibit the Hermitian nature of this matrix of the kinetic energy. The first way is due to Slater and the second to Freeman and Löwdin,[1] as follows:

$$\int \Psi_i{}^*(-\nabla^2/2)\Psi_i{}' \, d\tau$$

$$= \delta(l_i, l_i{}') \, \delta(m_{l_i}, m_{l_i}{}') \, \delta(m_{s_i}, m_{s_i}{}')$$

$$\times \int_0^\infty (1/2)[r^2 \, dR^*_{n_i,l_i}/dr \, dR_{n_{i'},l_i}/dr + l_i(l_i + 1)R^*_{n_i,l_i}R_{n_{i'},l_i}] \, dr$$

$$= \delta(l_i, l_i{}') \, \delta(m_{l_i}, m_{l_i}{}') \, \delta(m_{s_i}, m_{s_i}{}')$$

$$\times \int_0^\infty (1/2)r^{2l_i+2} \, d/dr\,(R^*_{n_i,l_i}/r^{l_i}) \, d/dr\,(R_{n_{i'},l_i}/r^{l_i}) \, dr. \qquad \text{(q.6.76.)}$$

The equivalent expressions given by (q.6.75.) and (q.6.76.) for the matrix component of the kinetic energy of an N-electron atom, in the central field approximation, show that these matrix components are zero except when $l_i = l_i{}'$, $m_{l_i} = m_{l_i}{}'$, and $m_{s_i} = m_{s_i}{}'$. In a similar manner, the matrix components of the potential energy of the N-electron atom in the field of the atomic nucleus, corresponding to the second term of $\hat{H}_i$ in equation (q.6.73.), are also zero except when the same conditions are satisfied. Thus,

$$\int_0^\infty \Psi_i^*[-Z/r]\Psi_i{}' \, d\tau = \delta(l_i, l_i{}') \, \delta(m_{l_i}, m_{l_i}{}') \, \delta(m_{s_i}, m_{s_i}{}')\left[-Z\int_0^\infty r R^*_{n_i,l_i}R_{n_{i'},l_i} \, dr\right].$$

$$\text{(q.6.77.)}$$

If now equations (q.6.76.) and (q.6.77.) are combined to obtain the matrix components of $\hat{H}_i{}^0$ of (q.6.73.) with respect to the functions Ψ of (q.6.71.), there results

$$\int \Psi_i^*[-\nabla^2/2 - Z/r]\Psi_i \, d\tau$$

$$= \int_0^\infty [(1/2)r^{2l+2} \, d/dr\,(R^*_{n,l}/r^l) \, d/dr\,(R_{n,l}/r^l) - ZrR^*_{n,l}R_{n,l}] \, dr. \qquad \text{(q.6.78.)}$$

The matrix components of the two-electron operators, $\hat{H}_{ij}{}'$, equivalent to $1/r_{ij}$, involve considerably more complicated expressions, as it is necessary to make use of the expansion of $r_{12}{}^{-1}$ in terms of two sets of polar coordinates, (r_1, θ_1, ϕ_1) and (r_2, θ_2, ϕ_2), as given below.‡

‡ As indicated on page 239, an alternative way of calculating the mutual potential energy of a system of electrical charges is to make use of the theory of spherical harmonics

---

[1] Freeman, A. J., and Löwdin, P.-O., *Phys. Rev.*, 1958, **111**, 1212.

in arriving at the expansion of quantities like $1/r_{12}$, which appear in such calculations. It is also more convenient, in this process, to make use of the generating function approach (see page 207) in defining the spherical harmonics, rather than the simple analytical approach used in Section Q.3.5. Before proceeding with the derivation of (q.6.79.) on this basis, it is necessary to consider the nature of the generating function for Legendre polynomials.

The various Legendre polynomials, as defined by (q.3.45.), may be derived from the generating function, $K(x, h) = [1 - 2xh + h^2]^{-1/2}$; if it is assumed, as may be done without any loss of generality, that the absolute value of the complex variable x is smaller than a finite number $r(b)$, and that h satisfies the condition $2r(b)|h| + |h|^2 < 1$, then $K(x, h)$ may be expanded as a power series in h

$$K(x, h) = \sum_{k=0}^{\infty} P_k(x)h^k,$$

where the Legendre polynomials $P_k(x)$ occur as the coefficients of this expansion. The above-noted restrictions on the values of x and h are necessary in order to apply the binomial expansion theorem. The polynomials, referred to above as $P_k(x)$, are identical to those defined by (q.3.45.), as will be shown below, but the symbol k is used in the present discussion rather than l, in order to make a distinction between this index and the l's used in the Slater and Condon and Shortley expressions for multiplet energies.

Now, if $[1 - z]^{-1/2}$ is expanded by the binomial theorem, if $|z| < 1$, then

$$[1 - z]^{-1/2} = 1 + \frac{1}{2}z + \frac{1}{2}\cdot\frac{3}{2}\cdot\frac{z^2}{2!} + \frac{1}{2}\cdot\frac{3}{2}\cdot\frac{5}{2}\cdot\frac{z^3}{3!} + \frac{1}{2}\cdot\frac{3}{2}\cdot\frac{5}{2}\cdot\frac{7}{2}\cdot\frac{z^4}{4!} + \cdots,$$

and since the product

$$1\cdot 3\cdot 5\cdot 7\cdots(2y - 1) = \frac{1\cdot 2\cdot 3\cdot 4\cdot 5\cdots 2y}{2^y\cdot 1\cdot 2\cdot 3\cdots y} = \frac{(2y)!}{2^y\cdot y!},$$

the above expansion may be written as

$$[1 - z]^{-1/2} = \sum_{y=0}^{\infty} \frac{(2y)!\cdot z^y}{2^{2y}(y!)^2}.$$

Therefore, the generating function $K(x, h)$ may be expanded as

$$K(x, h) = [1 - h(2x - h)]^{-1/2} = \sum_{k=0}^{\infty} \frac{(2k)!\cdot h^k\cdot(2x - h)^k}{2^{2k}\cdot(k!)^2}.$$

If the term, $(2x - h)^k$ in the numerator of this expression is similarly expanded as

$$(2x - h)^k = \sum_{v=0}^{k} \frac{(-1)^v k!\,(2x)^{k-v}h^v}{v!\,(k - v)!},$$

and the result substituted into the expansion for $K(x, h)$, the result is

$$K(x, h) = \sum_{k=0}^{\infty} \frac{(2k)!\,h^k}{k!} \sum_{v=0}^{k} \frac{(-1)^v h^v x^{k-v}}{2^{v+k}v!\,(k - v)!}.$$

The Legendre polynomial of order l may be obtained from the expression immediately above by collecting all terms that contain the factor h^l, which are obtained by taking $k = l$ and $v = 0$, $k = l - 1$ and $v = 1$, $k = l - 2$ and $v = 2$, etc. In other words, $P_l(x)$ is obtained by substituting the index $l - q$ for k and the number q for v into the expression above for $K(x, h)$ and taking the sum over q instead of over k and v; therefore

$$P_l(x) = \sum_{q=0}^{\alpha} \frac{(-1)^q (2l - 2q)! \, x^{l-2q}}{2^l q! \, (l - q)! \, (l - 2q)!}.$$

The maximum value that the summation index α may assume is determined by the condition that the expression above for $K(x, h)v \gtrless k$; if now q is substituted for v and $l - q$ for k, it is found that $2q \gtrless l$, but since q has to be an integer (summation index), it must be so that $\alpha = l/2$ when l is even and $\alpha = (l - 1)/2$ when l is odd. To show that the above expression for $P_l(x)$ is the same as (q.3.45.) is easily accomplished by noting that

$$[x^2 - 1]^l = \sum_{q=0}^{l} \frac{(-1)^q l! \, x^{2l-2q}}{q! \, (l - q)!}.$$

which when differentiated l times gives

$$d^l/dx^l [x^2 - 1]^l = \sum_{q=0}^{\alpha} \frac{(-1)^q l! \, (2l - 2q)! \, x^{l-2q}}{q! \, (l - q)! \, (l - 2q)!}.$$

On comparing this latter expression with the summation form for $P_l(x)$ given above, it is apparent that

$$P_l(x) = [2^l l!]^{-1} \, d^l/dx^l [x^2 - 1]^l,$$

which is the derivative form or Rodriques's expression for the Legendre polynomials, as previously given by (q.3.45.). Thus it has been demonstrated that the Legendre polynomials may be derived from a generating function of the form $K(x, h)$ given by

$$[1 - 2xh + h^2]^{-1/2}.$$

To return now to the derivation of (q.6.79.), it is necessary to consider the expansion of r_{12} as a power series in r_1 and r_2. If $r_{12} = |\mathbf{r}_1 - \mathbf{r}_2|$, and θ_{12} is the angle between $\mathbf{r}_1$ and $\mathbf{r}_2$, then by the cosine rule for any scalene triangle

$$r_{12} = [r_1{}^2 + r_2{}^2 - 2r_1 r_2 \cos \theta_{12}]^{1/2},$$

and if $r(b)$ and $r(a)$ are the greater and the smaller of r_1 and r_2, respectively, then $1/r_{12} = [1/r(b)][1 - 2h \cos \theta_{12} + h^2]^{-1/2}$, where $h = r(a)/r(b)$. But this function is obviously just the generating function for the Legendre polynomials given on page 351, so that $[1/r_{12}]$ may be expanded as

$$[1/r_{12}] = [1/r(b)] \sum_{k=0}^{\infty} [r(a)/r(b)]^k P_k(\cos \theta_{12}).$$

It has been shown in section Q.3.5. that the spherical harmonics $Y_{k,m}(\theta, \phi)$, defined by equation (q.3.55.) form an orthonormal set of functions on the surface of a sphere. Now, with reference to Fig. Q.3.2., showing the polar angles θ and ϕ with respect to the axial reference system xyz, if it is considered that the axial reference system is rotated to $x'y'z'$ as in Fig. Q.6.6., then the direction of the z'-axis in particular is determined by the angles θ' and ϕ'; the other angles α and γ are defined as in Fig. Q.6.6. The spherical harmonics $Y_{k,m}(\theta, \phi) = P_k{}^{|m|}(\cos \theta)e^{im\phi}$, with respect to the reference system xyz, may also be defined with respect to the axial reference system $x'y'z'$ as

$$Y_{k,m}{}'(\theta', \phi') = P_k{}^{|m|}(\cos \theta')e^{im\phi'},$$

and these two sets of functions must be equivalent to one another. Therefore, in conformity with the properties of orthonormal sets of functions, as discussed in section

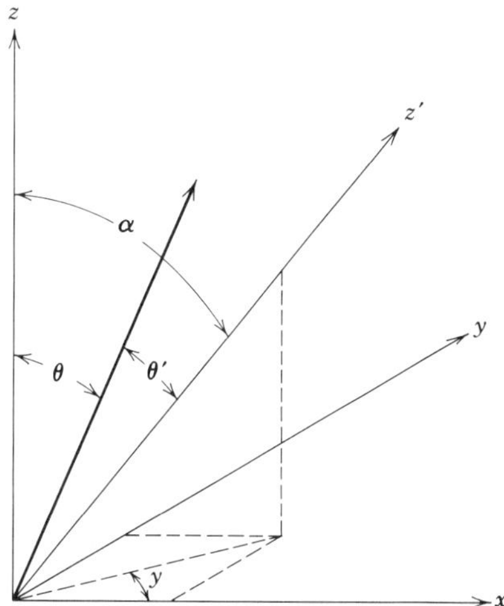

FIG. Q.6.6. Polar coordinates with reference to the Cartesian coordinate system xyz and with reference to the rotated Cartesian coordinate system $x'y'z'$. Only the z' axis is shown, the direction of which with respect to xyz is determined by the angles α and γ.

Q.2.4., each function $Y_{k,m}'(\theta', \phi')$ may be expanded as a linear combination of the functions $Y_{k,m}(\theta, \phi)$, and vice versa. Therefore

$$P_k(\cos \theta') = \sum_{m=-k}^{k} A_m P_k^{|m|}(\cos \theta)e^{im\phi}, \tag{A}$$

and also

$$P_k^{|m|}(\cos \theta)e^{-im\phi} = \sum_{\iota=-k}^{k} B_{m,\iota} P_k^{|\iota|}(\cos \theta')e^{i\iota\phi'}. \tag{B}$$

Since the second expansion (B) is valid for all possible values of θ and ϕ, it is therefore valid for $\theta = \alpha$ and $\phi = \gamma$; in the latter event, however, θ' adopts the value zero and $\cos \theta'$ becomes unity. But, from the definition of associated Legendre polynomials and Legendre polynomials as given in section Q.3.5., $P_k^{|\iota|}(1) = 0$ when $\iota \neq 0$ and $P_k(1) = 1$, so that

$$B_{m,0} = P_k^{|m|}(\cos \alpha)e^{-im\gamma}.$$

Now, with respect to the expansion (A), making use of the orthogonality of spherical harmonics (page 190),

$$2\pi A_m \int_0^\pi [P_k^{|m|}(\cos \theta)]^2 \sin \theta \, d\theta = \int_0^\pi \int_0^{2\pi} P_k(\cos \theta')P_k^{|m|}(\cos \theta)e^{-im\phi} \sin \theta \, d\theta \, d\phi,$$

or making use of (q.3.49.),

$$A_m = \frac{(2k + 1)(k - |m|)!}{4\pi(k + |m|)!} \int_0^{2\pi} e^{-im\phi} \, d\phi \int_0^\pi P_k(\cos \theta')P_k^{|m|}(\cos \theta) \sin \theta \, d\theta.$$

Of course, instead of integrating over θ and ϕ, the integration could also be carried out over θ' and ϕ', and if this is done, at the same time substituting expansion (B) into the expression obtained for A_m above, there results

$$A_m = \frac{(2k+1)(k-|m|)!}{4\pi(k+|m|)!} \sum_{\iota=-k}^{k} B_{m,\iota} e^{i\iota\phi'} \, d\phi' \int_{-1}^{1} P_k(\nu) P_k^{|\iota|}(\nu) \, d\nu,$$

which reduces to

$$A_m = B_{m,0} \left[\frac{(k-m)!}{(k+m)!}\right] = P_k^{|m|}(\cos\alpha) e^{-im\gamma} \left[\frac{(k-|m|)!}{(k+|m|)!}\right].$$

Thus the expansion for $P_k(\cos\theta')$ may be obtained as

$$P_k(\cos\theta') = \sum_{m=-k}^{k} \frac{(k-|m|)!}{(k+|m|)!} P_k^{|m|}(\cos\theta) P_k^{|m|}(\cos\alpha) e^{im(\phi-\gamma)},$$

and if this expression is substituted into the equation for $[1/r_{12}]$ given on page 352, the result finally obtained is

$$[1/r_{12}] = [1/r(b)] \sum_{k=0}^{\infty} [r(a)/r(b)]^k P_k(\cos\theta_{12})$$

$$= \sum_{k=0}^{\infty} \sum_{m=-k}^{k} \left[\frac{(k-|m|)! \, r(a)^k}{(k+|m|)! \, r(b)^{k+1}}\right] P_k^{|m|}(\cos\theta_1) P_k^{|m|}(\cos\theta_2) e^{im(\phi_1-\phi_2)}.$$

In terms of this expansion,

$$1/r_{12} = \sum_{k=0}^{\infty} \sum_{m=-k}^{k} \left[\frac{(k-|m|)! \, r(a)^k}{(k+|m|)! \, r(b)^{k+1}}\right]$$
$$\times P_k^{|m|}(\cos\theta_1) P_k^{|m|}(\cos\theta_2) e^{im(\phi_1-\phi_2)}, \quad \text{(q.6.79.)}$$

where r_{12} is the distance between the two points with the spherical polar coordinates (r_1, θ_1, ϕ_1) and (r_2, θ_2, ϕ_2), and $r(a)$ and $r(b)$ are the smaller and larger, respectively, of r_1 and r_2. Making use of this expansion, it is possible to write the integral

$$\langle ij|\hat{H}_{12}'|rt\rangle = \int \Psi_i^*(1)\Psi_j^*(2)\hat{H}_{12}'\Psi_r(1)\Psi_t(2) \, d\tau_1 \, d\tau_2, \quad \text{(q.6.80.)}$$

as the following very elaborate expression:

$$\langle ij|\hat{H}_{12}'|rt\rangle = \sum_{k=0}^{\infty} \sum_{m=-k}^{k} \left[\frac{(k-|m|)!}{(k+|m|)!}\right] \delta(m_{s_i}, m_{s_r}) \, \delta(m_{s_j}, m_{s_t})$$

$$\times [-1]^{(m_{l_i}+|m_{l_i}|+m_{l_j}+|m_{l_j}|+m_{l_r}+|m_{l_r}|+m_{l_t}+|m_{l_t}|)/2}$$

$$\times \left[\frac{(2l_i+1)(l_i-|m_{l_i}|)!}{(l_i+|m_{l_i}|)!}\right]^{1/2} \left[\frac{(2l_j+1)(l_j-|m_{l_j}|)!}{(l_j+|m_{l_j}|)!}\right]^{1/2}$$

$$\times \left[\frac{(2l_r+1)(l_r-|m_{l_r}|)!}{(l_r+|m_{l_r}|)!}\right]^{1/2} \left[\frac{(2l_t+1)(l_t-|m_{l_t}|)!}{(l_t+|m_{l_t}|)!}\right]^{1/2}$$

$$\times \int_0^{\infty} \int_0^{\infty} R_{n_i,l_i}^*(r_1) R_{n_j,l_j}^*(r_2) \frac{r(a)^k}{r(b)^{k+1}} R_{n_r,l_r}(r_1) R_{n_t,l_t}(r_2)$$

$$\times r_1^2 r_2^2 \, dr_1 \, dr_2$$

$$\times \int_0^{\pi} P_{l_i}^{|m_{l_i}|}(\cos\theta_1) P_{l_r}^{|m_{l_r}|}(\cos\theta_1)(\sin\theta_1/2) P_k^{|m|}(\cos\theta_1) \, d\theta_1$$

$$\times \int_0^{\pi} P_{l_j}^{|m_{l_j}|}(\cos\theta_2) P_{l_t}^{|m_{l_t}|}(\cos\theta_2)(\sin\theta_2/2) P_k^{|m|}(\cos\theta_2) \, d\theta_2$$

$$\times \int_0^{2\pi} (1/2\pi) e^{i(-m_{l_i}+m_{l_r}+m)\phi_1} \, d\phi_1 \int_0^{2\pi} (1/2) e^{i(-m_{l_j}+m_{l_t}-m)\phi_2} \, d\phi_2.$$

$$\text{(q.6.81.)}$$

If the integrations with respect to ϕ_1 and ϕ_2 are considered, it is observed that these integrals are zero, except when the exponent is zero, when the integrals adopt the value of unity; thus, in order that these integrals should have non-zero values

$$m = m_{l_i} - m_{l_r} = -m_{l_j} + m_{l_t} \quad \text{or} \quad m_{l_i} + m_{l_j} = m_{l_r} + m_{l_t}, \quad \text{(q.6.82.)}$$

from which it follows that the matrix component of the Coulomb energy has a non-zero value only between two states for which the sum of the m_l quantum numbers is the same for both the initial and final states. This also leads to a considerable simplification of the expression (q.6.81.), since if m is determined by (q.6.82.), the summation over m in (q.6.81.) reduces to a single term, and no summation is really involved. The integrals with respect to θ in (q.6.81.) involve integration of the product of three associated Legendre functions, which may be evaluated making use of (q.3.41.). A general expression for the latter integrations of triple products of associated Legendre functions with respect to the polar angle θ was worked out by Gaunt[1] and a modified form of this expression was later found by Racah;[2] Gaunt's formula is

$$(1/2) \int_{-1}^{1} P_l^{u}(\mu) P_m^{v}(\mu) P_n^{w}(\mu) \, d\mu$$

$$= (-1)^{s-m-w} \frac{s! \, (m+v)! \, (n+w)! \, (2s-2n)!}{(m-v)! \, (s-l)! \, (s-m)! \, (s-n)! \, (2s+1)!}$$

$$\times \sum_{t} (-1)^t \frac{(l+u+t)! \, (m+n-u-t)!}{t! \, (l-u-t)! \, (m-n+u+t)! \, (n-w-t)!}, \quad \text{(q.6.83.)}$$

where u, v, and w are zero or positive and $[u - v - w] = 0$ and $s = (1/2) \times (l + m + n)$, by definition, is an integer. This means that the integral in (q.6.83.) becomes zero unless $(l + m + n)$ is an even integer or unless $(m + n) \geq l \geq (m - n)$; this condition is equivalent to the assumption that $m \geq n$ and that l, m, and n correspond to the lengths of the sides of a triangle; in Gaunt's formula, as given by (q.6.83.), l, m, and n are simply integers and not quantum numbers. The various factorials in Gaunt's formula only have significance if the summation over t is taken for values of t ranging from the greater of 0, $(-m + n - u)$ to the smallest of $(m + n - u)$, $(l - u)$, and $(n - w)$.

Slater (see page 347) has shown that the matrix components of the Coulomb energy between two states described by the quantum numbers l, m and l', m' as given by (q.6.80.) may be expressed completely in the form of radial integrals, by defining certain coefficients, the so-called Slater coefficients, incorporating the integrals of (q.6.81.) with respect to θ and making use of

[1] Gaunt, J. A., *Phil. Trans. Roy. Soc.* (London), 1929, **A228**, 151.
[2] Racah, G., *Phys. Rev.*, 1942, **61**, 186.

Gaunt's formula given by (q.6.83.), as follows:

$$c^k(lm; l'm') = [-1]^{(m + |m| + m' + |m'| + \{m - m'\} + |m - m'|)/2}$$

$$\times \left[[1/2] \int_{-1}^{1} P_l^{|m|}(\mu) P_{l'}^{|m'|}(\mu) P_k^{|m - m'|}(\mu) \, d\mu \right]$$

$$\times \left[\frac{(k - |m - m'|)! \, (2l + 1)(l - |m|)! \, (2l' + 1)(l' - |m'|)!}{(k + |m - m'|)! \, (l + |m|)! \, (l' + |m'|)!} \right]^{1/2}.$$

$$(q.6.84.)$$

Making use of these Slater coefficients, c^k, the expression given by (q.6.81.) may be written entirely as a radial integral

$$\langle ij|\hat{H}_{12}'|rt \rangle = \delta(m_{s_i}, m_{s_r}) \, \delta(m_{s_j}, m_{s_t}) \, \delta(m_{l_i} + m_{l_j}, m_{l_r} + m_{l_t})$$

$$\times \sum_{k=0}^{\infty} c^k(l_i m_{l_i}; l_r m_{l_r}) c^k(l_t m_{l_t}; l_j m_{l_j}) R^k(ij; rt), \quad (q.6.85.)$$

where

$$R^k(ij; rt) = \int_0^{\infty} \int_0^{\infty} R^*_{n_i l_i}(r_1) R^*_{n_j l_j}(r_2) \frac{r(a)^k}{r(b)^{k+1}}$$

$$\times R_{n_r l_r}(r_1) R_{n_t l_t}(r_2) r_1^2 r_2^2 \, dr_1 \, dr_2. \quad (q.6.86.)$$

According to this definition of the Slater coefficients, c^k is compatible with the notation used in Condon and Shortley's book (see page 347), provided it is noted that

$$c^k(l_j m_{l_j}; l_i m_{l_i}) = [-1]^{(m_{l_i} - m_{l_j})} c^k(l_i m_{l_i}; l_j m_{l_j}), \quad (q.6.87.)$$

or that reversing the order of the subscripts changes the sign. Slater, in his Volume 2 of *Quantum Theory of Atomic Structure*, 1960, McGraw-Hill Book Co., New York, has tabulated the values of c^k for l values up to three (*f*-electrons) in Appendix 20 of that volume, making use of Gaunt's formula for this purpose. For example, for two *f*-electrons, with m_{l_i} and m_{l_j} values of ± 2 and ∓ 1, respectively, when $k = 6$, $c^k = -\sqrt{37800/736164}$. It emerges, in practice, that the summation in (q.6.85.) is not actually over all values of k from 0 to infinity, but only from zero to a k value of $(l + l')$, which is 6 in the above example; in general, the upper limit of k is the smaller of the two quantities, $l_i + l_r$ or $l_j + l_t$.

Slater (*loc. cit.*) has also tabulated the values of c^k, not only for the Coulomb integral, normally referred to as $\langle ij|\hat{H}_{ij}'|ij \rangle$, but also for the exchange integral (see page 300) usually referred to as $\langle ij|\hat{H}_{ij}'|ji \rangle$. It is common practice to write the Coulomb integral

$$\langle ij|\hat{H}_{ij}'|ij \rangle = \sum_{k=0}^{\infty} a^k(l_i m_{l_i}; l_j m_{l_j}) F^k(n_i l_i; n_j l_j), \quad (q.6.88.)$$

where

$$a^k(l_i m_{l_i}; l_j m_{l_j}) = c^k(l_i m_{l_i}; l_i m_{l_i}) c^k(l_j m_{l_j}; l_j m_{l_j}) \quad (q.6.89.)$$

and

$$F^k(n_i l_i; n_j l_j) = R^k(ij; ji) = \int_0^\infty \int_0^\infty R^*_{n_i l_i}(r_1) R^*_{n_j l_j}(r_2) \frac{r(a)^k}{r(b)^{k+1}}$$
$$\times R_{n_i l_i}(r_1) R_{n_j l_j}(r_2) r_1^2 r_2^2 \, dr_1 \, dr_2. \quad \text{(q.6.90.)}$$

Similarly, for the exchange integrals, in Slater's notation,

$$\langle ij | \hat{H}_{ij}' | ji \rangle = \delta(m_{s_i}, m_{s_j}) \sum_{k=0}^\infty b^k(l_i m_{l_i}; l_j m_{l_j}) G^k(n_i l_i; n_j l_j), \quad \text{(q.6.91.)}$$

where

$$b^k(l_i m_{l_i}; l_j m_{l_j}) = [c^k(l_i m_{l_i}; l_j m_{l_j})]^2 \quad \text{(q.6.92.)}$$

and

$$G^k(n_i l_i; n_j l_j) = R^k(ij; ji) = \int_0^\infty \int_0^\infty R^*_{n_i l_i}(r_1) R^*_{n_j l_j}(r_2) \frac{r(a)^k}{r(b)^{k+1}}$$
$$\times R_{n_i l_i}(r_2) R_{n_j l_j}(r_1) r_1^2 r_2^2 \, dr_1 \, dr_2, \quad \text{(q.6.93.)}$$

and he has tabulated separately (*loc. cit.*) the coefficients a^k.

In terms of these Slater coefficients, making use of the initially formidable expression given by (q.6.81.), it is possible to evaluate expressions like (q.6.70.) and hence to calculate the energy of various multiplets.

Q.6.9. Energy Values of Multiplets

The principle involved in calculating the energies of the ground state and the excited states of helium has already been discussed in section Q.5.6., without actually evaluating the integrals concerned. The lowest excited state of helium is represented by the configuration $1s2s$, and this configuration, as already shown, leads to singlet and triplet states, the triplet corresponding to a lower energy value than the singlet. In terms of the discussion in the previous section of this chapter, the energies of these multiplets may be expressed entirely as integrals of radial functions. Thus, for the component of the triplet state with both electrons in the helium atom having $m_s = 1/2$, the diagonal matrix component of the Hamiltonian for this state, firstly with respect to the one electron operators is given by (q.6.78.), with $Z = 2$ and $R_{n,l}$ given by (q.3.114.). If the value of the integral in (q.6.78.) is referred to as $I_{n,l}$, for convenience, then the kinetic energies of the $1s$- and $2s$-electrons and their potential energies in the field of the nucleus of the helium atom is given by $I_{1s} + I_{2s}$. The diagonal matrix element of the two-electron operator, for helium, where in this particular case there is only a single pair of electrons is given by $\langle 12 | 1/r_{12} | 12 \rangle - \langle 12 | 1/r_{12} | 21 \rangle$, which may be evaluated using equations (q.6.88.) and (q.6.91.), making use of the tabulated values of the appropriate Slater coefficients. Thus, in the Slater terminology

$$\langle 12|r_{12}|12\rangle - \langle 12|r_{12}|21\rangle = F^0(1s, 2s) - G^0(1s, 2s), \qquad \text{(q.6.94.)}$$

since the Slater coefficients, applicable here, have non-zero values only for $k = 0$, for the interaction of two s-electrons, and the coefficients for $k = 0$ are equal to unity. The first term in (q.6.94.) represents the Coulomb energy of interaction of an electron in the $1s$-state (represented by 1 in the integrals) and one in the $2s$-state (represented by 2 in the intergals); the second term, with a negative sign is the exchange integral (see page 301). The energy of the state ^3S is then obtained as

$$^3E = I_{1s} + I_{2s} + F^0(1s, 2s) - G^0(1s, 2s). \qquad \text{(q.6.95.)}$$

For the configuration $1s2s$, there are two possible spin combinations which lead to $M_s = 0$, either $m_{s_1} = 1/2$ and $m_{s_2} = -1/2$ or $m_{s_1} = -1/2$ and $m_{s_2} = 1/2$; one of these linear combinations belongs to the singlet state ^1S and one to the triplet state ^3S. Therefore, by the diagonal sum rule, the energies of the ^1S and ^3S multiplets must be equivalent to the sum of the diagonal matrix components of these two states. In this sum, the terms arising from the one-electron operators are the same as in (q.6.95.), also the term arising from the integral $\langle 12|r_{12}|12\rangle$ has the same value as in (q.6.95.), but the term from the integral $\langle 12|r_{12}|21\rangle$ vanishes, since for opposed spins the term $\delta(m_{s_i}, m_{s_j})$ in (q.6.91.) becomes zero. Thus it is found, by the application of equations (q.6.88.) and (q.6.91.) that

$$^1E + {}^3E = 2[I_{1s} + I_{2s} + F^0(1s, 2s)], \qquad \text{(q.6.96.)}$$

which result when combined with (q.6.95.) gives

$$^1E = I_{1s} + I_{2s} + F^0(1s, 2s) + G^0(1s, 2s), \qquad \text{(q.6.97.)}$$

which is the Slater equivalent of equation (q.5.42.). The singlet state ^1S thus corresponds to a higher energy value than the triplet state ^3S for these two multiplets of the configuration $1s2s$, differing by the amount $2G^0(1s, 2s)$, due to the exchange integral. The numerical values of the Coulomb and exchange integrals $F^0(1s, 2s)$ and $G^0(1s, 2s)$ in equations (q.6.95.) and (q.6.97.), when evaluated, using the appropriate radial functions, lead to numerical values of 0.419 and 0.044 Hartrees, respectively; so that

$$^1E = -2.037 \text{ Hartrees} \qquad \text{and} \qquad {}^3E = -2.125 \text{ Hartrees}, \qquad \text{(q.6.98.)}$$

compared with the experimentally observed spectroscopic values of -2.147 and -2.176 Hartrees, respectively. The fact that the triplet state error is lower (0.051 Hartrees) than the singlet state error (0.110 Hartrees) is attributable to the screening effect not being so important for electrons with parallel spins as for electrons with antiparallel spins, which in turn is associated with antisymmetry principle operating to tend to keep electrons with the same spin further apart and electrons with different spins closer together than is actually the case. This tendency for electrons with similar spins to move as far apart from one another as possible is frequently referred to as spin correlation or exchange correlation. The antisymmetry principle thus affords an explanation of the apparent anomaly that although the spin functions do not affect the energy of electronic states directly, they do influence the total energy by determining the form of the spatial part of the total wave function of a system, even though the Hamiltonian depends only on the spatial coordinates. The non-relativistic energies of the ^1S and ^3S states of helium have been calculated by Sharma and Coulson[1] using up to third order perturbation theory (see section Q.4.7.). Their first order perturbation equation was solved by the variational method described in sections Q.4.5. and Q.4.6., using trial functions which were 12-term Hylleraas functions in the variables s, t, and u (see page 240). Sharma and Coulson values to the third order in energy were

$$^1E = -2.14729 \text{ Hartrees} \quad \text{and} \quad ^3E = -2.17400 \text{ Hartrees.} \quad \text{(q.6.99.)}$$

For the ground-state configuration of helium, $1s^2$, the normalized determinantal function describing this state is given by (q.5.34.), the triplet state in this case being disallowed by the Exclusion Principle, and the corresponding energy is obtained in terms of the Slater notation as

$$^1E(1s^2) = 2I_{1s} + F^0(1s, 1s), \quad \text{(q.6.100.)}$$

which does not involve any exchange integral.

By proceeding in the same manner, it is possible to find the energy of the various multiplets corresponding to particular electronic configurations. For many electronic configurations, it is possible to find the energy of the various multiplets involved by the application of the diagonal sum rule alone. However, for many others, this is not so; for those cases where the vector model of atoms leads to several multiplets of the same L and S values, the separate energies of these multiplets cannot be obtained by the diagonal sum rule, but a secular equation of degree equal to the number of such multiplets must be solved. For example, one of the simplest situations where the diagonal sum rule does not work is that involving three non-equivalent s-electrons. Here, two s-electrons give rise to parent multiplets ^3S and ^1S, and with a third s-electron, ^4S and ^2S multiplets arise from the ^3S parent and another ^2S multiplet is generated from the ^1S parent. The state with three parallel spins belongs to the ^4S multiplet, so that its energy may be obtained from the

---

[1] Sharma, C. S., and Coulson, C. A., *Proc. Phys. Soc.* (*London*), 1962, **80**, 81.

corresponding unperturbed function. But for the three states with $M_s = 1/2$, one will necessarily belong to the ^4S and one each to each of the ^2S multiplets. Thus, subtraction of the energy of the ^4S multiplet from the sum of the diagonal matrix components of the three unperturbed states only gives the sum of the energies of the two ^2S multiplets, but does not lead to their separate energies. Instead, it is necessary to formulate the non-diagonal matrix components of the energy and solve the resulting cubic equation. In the present example, however, this cubic secular equation reduces to a quadratic equation, since one of the roots is obtainable by the diagonal sum rule. The most complicated situations arise for atoms containing several electrons each in different nl states.

For atoms containing two or more electrons with different values of their principal quantum number, n, whether or not the value of the l quantum number is the same or different, the number of possible multiplets is generally greater than if the n values are the same. For example, for the configuration $2p3p$, the two orbital angular momentum vectors in terms of the vector model of atoms may combine to give $L = 2, 1,$ and 0 and the two spin values to give $S = 1$ and 0, leading to the multiplets ^3D, ^1D, ^3P, ^1P, ^3S, and ^1S, but for the configuration $2p^2$, namely two electrons with $n = 2$ and $l = 1$, the Pauli Exclusion Principle operates in such a manner that certain multiplets which are appropriate when n is different for the two electrons, are excluded and certain other multiplets are possible; thus, for the configuration $2p^2$, only the multiplets ^3P, ^1S, and ^1D are necessary, the ^1P, ^3S, and ^3D multiplets being ruled out by the Exclusion Principle. For other numbers of p, d, or f, electrons with the same value of their n quantum number, it emerges that the same multiplet structure is applicable to atoms containing N_i equivalent electrons (in terms of their l quantum numbers) and to atoms containing $N - N_i$ electrons, where N is the maximum number of electrons describable by the same nl values, which is, of course, six for p-states, ten for d-states, and 14 for f-states. Thus, a group of five p-electrons and one p-electron, in an atom, exhibit the same spectrum; an atom containing one d-electron or nine d-electrons exhibits the same spectrum; and an atom containing either one or thirteen electrons in f-states exhibits the same spectrum. Hund[1] has tabulated the possible multiplets for various numbers of equivalent of p, d, and f electrons. Thus, the only multiplet corresponding to the configurations p^1 and p^5 is ^2P. The corresponding multiplets to the configurations p^2 and p^4 are ^3P, ^1D, and ^1S, and the corresponding multiplets to the configuration p^3 are ^4S, ^2D, and ^2P. For either of the configurations d^1 or d^9, the only possible multiplet is ^2D; for either of the configurations d^2 or d^8, the possible multi-

---

[1] Hund, F., *Linienspektren und periodisches System der Elemente*, 1927, Springer-Verlag OHG, Berlin.

plets are ^3F, ^3P, ^1S, ^1D, and ^1G; for either of the configurations d^3 or d^7, the appropriate multiplets are ^4F, ^4P, ^2H, ^2G, ^2F, ^2P, and two ^2D's; and for either of the d^4 or d^6 configurations, the corresponding multiplets are ^5D, ^3H, ^3G, ^3D, ^1I, two ^3F's, two ^3P's, two ^1G's, two ^1D's, and two ^1S's, while the d^5 configuration gives rise to the following sixteen multiplets: ^6S, ^4G, ^4F, ^4D, ^4P, ^2I, ^2H, ^2P, ^2S, two ^2G's, two ^2F's, and three ^2D's. For atoms containing equivalent f electrons, the configurations f^1 or f^{13} each correspond to the multiplet ^2F, the configurations f^2 and f^{12} to the multiplets ^3H, ^3F, ^3P, ^1I, ^1G, ^1D, and ^1S, and the relatively large number of multiplets corresponding to the configurations f^3 to f^{11} are listed below:

f^3 or f^{11}: ^4I, ^4G, ^4F, ^4D, ^4S, ^2L, ^2K, ^2I, ^2P, two ^2H's, two ^2G's, two ^2F's, and two ^2D's.

f^4 or f^{10}: ^5I, ^5G, ^5F, ^5D, ^5S, ^3M, ^3L, ^1N, ^1K, ^1F, two ^3K's, two ^3I's, two ^3D's, two ^1L's, two ^1H's, two ^1S's, three ^3G's, three ^3P's, three ^1I's, four ^3H's, four ^3F's, four ^1G's, and four ^1D's.

f^5 or f^9: ^6H, ^6F, ^6P, ^4M, ^4L, ^4S, ^2O, ^2N, two ^4K's, two ^4P's, two ^2M's, three ^4I's, three ^4H's, three ^4D's, three ^2L's, four ^4G's, four ^4F's, four ^2P's, five ^2K's, five ^2I's, five ^2D's, six ^2G's, seven ^2H's, and seven ^2F's.

f^6 or f^8: ^7F, ^5L, ^5K, ^5P, ^5S, ^3O, ^3N, ^1Q, ^1P, two ^5I's, two ^5H's, two ^5F's, two ^1M's, two ^1N's, three ^5G's, three ^5D's, three ^3M's, three ^3L's, three ^1K's, four ^1L's, four ^1H's, four ^1F's, four ^1S's, five ^3D's, six ^3K's, six ^3I's, six ^3P's, six ^1D's, seven ^3G's, seven ^1I's, eight ^1G's, nine ^3H's, and nine ^3F's.

f^7: ^8S, ^6I, ^6H, ^6G, ^6F, ^6D, ^6P, ^4N, ^4M, ^2Q, ^2O, two ^4P's, two ^4S's, two ^2N's, two ^2S's, three ^4L's, three ^4K's, four ^2M's, five ^4I's, five ^4H's, five ^4F's, five ^2L's, five ^2P's, six ^4D's, seven ^4G's, seven ^2K's, seven ^2D's, nine ^2I's, nine $2H$'s, ten ^2G's, and ten ^2F's.

The two most important consequences of the symmetry between the multiplets observed for atoms containing some number of equivalent p-, d-, or f-electrons, and other atoms with N minus this number of electrons, where N is the maximum number of electrons describable by the same n and l quantum numbers, are from the point of view of X-ray spectra and the spectra exhibited by atoms containing a complete quantum group of electrons. In the latter event, since there will be electrons present with all possible values of m_l and m_s appropriate to the nl values concerned, both M_L and M_S will be zero, so that an atom with six p-electrons, ten d-electrons, or 14 f-electrons, as well as two s-electrons, behaves as if it had no orbital or spin angular momentum. It is for this reason that the spectra of atoms, perhaps containing many electrons, may be relatively simple, since a complete quantum group of electrons acts like a spherically symmetrical system, only those electrons in sets of *nl* states which are not completely occupied contributing to the multiplet structure of the atomic state concerned. This result has already been seen to be a consequence of Unsöld's theorem (page 224). In general, then, the completely occupied sets of *nl* energy states in an atom do not contribute to the complexity of atomic spectra, but only affect the absolute positions of the energy terms by virtue of their effect on the one-electron energies discussed in connection with the self-consistent field model of atoms.

As far as X-ray spectra are concerned (section Q.1.7.), where in the process of exciting such high frequency spectra an electron is removed from an inner, complete set of electrons in some *nl* state, the stationary state of this set of electrons with one electron removed, is similar to that of a single electron, giving rise to a spectrum composed of doublets with the electron spin being necessarily aligned either parallel to or opposed to the orbital angular momentum vector. This doubling of the spectral lines in observed X-ray spectra is quite pronounced.

Q.6.10. Electron Correlation Energies

While the energies calculated by means of the Hartree–Fock method, as described in section Q.5.7., are usually within about 1% of the corresponding experimental values of the total energy of the system concerned, the total energy of any system of physical or chemical interest is not generally of much significance. For spectroscopic states, for example, it is normally energy differences which are of consequence, and unfortunately these energy differences are frequently only of the order of 1% of the total energy, so that small absolute errors in total energies may lead to comparatively large errors in their differences. For this reason, there is great interest associated with quantum mechanical calculations which lead to better energy values than those provided by the Hartree–Fock method[1]. Also, as already pointed out in section Q.5.7., although the Hartree–Fock equations may be solved directly for light atoms and for simple molecules composed of light atoms, this is not generally possible for molecules and solids. In the latter case, for example, there would be as many spin orbitals as there are electrons in the solid, and the Hartree–Fock equations such as (q.5.54.) would have to refer to each electron moving in a different potential field, which is virtually an impossible system of equations to solve.

In the Hartree–Fock method, the Coulomb interactions between pairs of electrons, especially between pairs with antiparallel spins, gives rise to the so-called correlation error. The antisymmetry principle leads to electrons

---

[1] Sinanoğlu, O., and Brueckner, K. A., *Three Approaches to Electron Correlation in Atoms*, Yale University Press, 1970.

with parallel spins tending to avoid the occupation of the same region of space (see page 308) and this effect supercedes the Coulomb repulsion, so that electrons of parallel spins are effectively described better than those of anti-parallel spins in the Hartree–Fock equations. Löwdin[1] refers to the electronic correlation energy for a certain state with respect to some specific Hamiltonian as the difference between the exact eigenvalue of the Hamiltonian and its expectation value as given by the Hartree–Fock approximation for the particular state concerned, that is

$$E_{correlation} = \text{exact } \langle \hat{H} \rangle_{average} - \text{Hartree–Fock } \langle \hat{H} \rangle_{average}. \qquad \text{(q.6.101.)}$$

The electronic correlation energy, defined in this manner, is simply a mathematical quantity which serves as a means of distinguishing a good wave function for a system from a Hartree–Fock wave function, where a good wave function is one which accounts for as much as possible of the correlation energy of the system. Since the exact energy of any system must contain contributions from relativistic effects as well as non-relativistic effects, arising from the use of a non-relativistic Hamiltonian, the Löwdin correlation energy is not the difference between the exact experimentally observed energy and the Hartree–Fock energy. Energies calculated by the Hartree–Fock method are generally too high, so that according to the variation principle correlation energies given by (q.6.101.) are negative quantities. In terms of an important quantum mechanical theorem known as the virial theorem,‡ which the Hartree–Fock energies of systems satisfies,

$$K_{correlation} = -(1/2)U_{correlation}, \qquad \text{(q.6.102.)}$$

where K and U are the appropriate kinetic and potential energies, and so the kinetic part of the correlation energy must be greater than zero and the potential part of the correlation energy must be less than zero.

---

‡ The virial theorem, which is valid in both classical as well as quantum mechanics, refers to the fact that in any system composed of atomic nuclei and electrons, that is for any atom, molecule, complex ion, or crystal, which is in its normal state or any possible excited state (any steady state), the average kinetic energy is equal to minus one-half of the average potential energy. Since the total energy is normally the sum of the kinetic and potential energies of a system, the average kinetic energy has the same value as the total energy but with the opposite sign, and the average potential energy is equal to twice the total energy, or $\bar{U} = -2\bar{K}$, $E = -\bar{K}$, and $\bar{U} = 2E$; these relationships are implied, for example, by (q.1.33.), in the discussion of the Bohr theory of the hydrogen atom.

The quantum mechanical form of the virial theorem has actually already been referred to on page 320; in fact, the equation given on page 320,

$$(ih/2\pi)[\partial \bar{F}/\partial t] = (\hat{F}\hat{H} - \hat{H}\hat{F}) = [\hat{F}, \hat{H}] = 0$$

---

[1] Löwdin, P.-O., *Advan. Chem. Phys.*, 1959, **2**, 207.

is usually referred to as the hypovirial theorem,[1,2] which may be alternatively expressed by saying that the expectation values of time-independent operators do not vary with time in stationary states. The relationship between these two forms of the virial theorem may be seen by considering the special case where the operator $\hat{F}$ is given by $\hat{F} = h/2\pi i[x\,\partial/\partial x + y\,\partial/\partial y + z\,\partial/\partial z] = [\hat{r}\cdot\hat{p}]$, with $\hat{r} = ix + jy + kz$ and $\hat{p} = h/2\pi i\nabla$, and $\hat{H}$ is given by $\hat{H} = (-h^2/8\pi^2 m)\nabla^2 + U(r)$. The commutator of $\hat{F}$ and $\hat{H}$ is then $[\hat{H}, \hat{F}] = (h/2\pi i)[-(h^2/4\pi^2 m)\nabla^2 - (r\cdot\widehat{\nabla}U] = (h/2\pi i)[2\hat{T} - (r\cdot\widehat{\nabla}U)]$, and the expectation value of this commutator is $(h/2\pi i)[2\langle\hat{T}\rangle - \langle(r\cdot\widehat{\nabla}U)\rangle] = 0$ or $2\langle\hat{T}\rangle = \langle(r\cdot\widehat{\nabla}U)\rangle$, which is the usual form of the quantum mechanical virial theorem. For a system of particles having position vectors r_1, r_2, r_3, ... and momenta p_1, p_2, p_3, ..., the virial theorem becomes

$$2\langle\hat{T}\rangle = \sum_i \langle r_i\cdot\widehat{\nabla}_i U\rangle.$$

The virial theorem is of particular value in constructing approximate solutions for the Schrödinger equation of systems, and is most useful in discussing the electronic energies of atoms (and molecules) in the form in which the potential energy is given by the relationship, $U(r) = kr^n$, where k is a constant, for the virial theorem then becomes $2\langle\hat{T}\rangle = n\langle\hat{U}\rangle$, where $\hat{U}$ represents the electrostatic interaction between charged particles and $n = -1$; such a potential energy is said to be homogeneous in r and of degree n.

Löwdin,[3] in particular, has shown how the virial theorem may be used to modify an approximation to the exact wave function of a system to find the exact energy of the system, so that the virial equation, $\langle\hat{T}\rangle = -(1/2)\langle\hat{U}\rangle$, is always satisfied, as follows. For any system of N particles, composed of p electrons and q nuclei, having charges ε_1, ε_2, ..., ε_N, masses m_1, m_2, ..., m_N, and positions vectors r_1, r_2, ..., r_N, if it is assumed that the total energy is conserved ($\langle\hat{H}\rangle$), then the Hamiltonian for the system is given by

$$H = -(1/2)\sum_{i=1}^{N}\nabla_i^2/m_i + \sum_{i<j}^{N}\sum^{N}\varepsilon_i\varepsilon_j/r_{ij} = \hat{T} + \hat{U}.$$

For a stationary state of the system described by the Schrödinger equation, $\hat{H}\psi = E\psi$, where E and ψ are the exact energy and wave function, respectively, of the system, if it is assumed that the function $\phi(r) = \phi(r_1, r_2, \ldots, r_N)$ is an approximation to ψ, corresponding to some approximate energy E_1, then the virial theorem is not necessarily satisfied for this E_1. But this function $\phi(r)$ may be modified to satisfy the virial theorem by forming from it a new function in which each position vector r_i is multiplied by an arbitrary scaling factor; if this scaling factor is written as λ^3, then the new scaled wave function becomes $\phi_\lambda = \lambda^{3/2N}\phi(\lambda r)$, which reduces to $\phi(r)$ when $\lambda = 1$, and the factor $\lambda^{3/2N}$ is included to ensure that ϕ_λ is normalized if $\phi(r)$ is normalized. The application of a scaling factor in this manner effectively increases all of the coordinates of the particles with respect to the origin by a uniform amount λ.

If now the expectation value of the kinetic energy operator of the system with respect to the scaled wave function is represented by $\langle\hat{T}^\lambda\rangle$, then $\langle\hat{T}^\lambda\rangle = \langle\phi_\lambda|\hat{T}|\phi_\lambda\rangle$, assuming that ϕ_λ is normalized. Since the configurational space for the system of particles is $d\tau = d\tau_1\,d\tau_2\cdots d\tau_N = dx_1\,dy_1\,dz_1\cdots dx_N\,dy_N\,dz_N$, and representing for convenience

---

[1] Hirschfelder, J. O., *J. Chem. Phys.*, 1960, **33**, 1762.
[2] Hirschfelder, J. O., and Coulson, C. A., *J. Chem. Phys.*, 1962, **36**, 941.
[3] Löwdin, P.-O., *J. Mol. Spectry.*, 1959, **3**, 46.

$\lambda^3 r_i = r_i'$ (meaning that $\lambda x_i = x_i'$, $\lambda y_i = y_i'$, and $\lambda z_i = z_i'$), the quantity $\langle \hat{T}^\lambda \rangle$, may be written as

$$\langle \hat{T}^\lambda \rangle = \langle \phi(r') | -(1/2)\lambda^2 \sum_{i=1}^{N} (\nabla_i')^2 / m_i | \phi(r') \rangle = \lambda^2 \langle \phi(r') | \hat{T}' | \phi(r') \rangle = \lambda^2 \langle \hat{T}^1 \rangle,$$

since all the functions and operators involving r' have exactly the same form as those involving r and so are equivalent. $\langle \hat{T}^1 \rangle$ is the expectation value of the kinetic energy operator with respect to the unscaled function $\phi(r)$. In a similar manner, it may be shown that the expectation value of the potential energy operator for the system is $\langle \hat{U}^\lambda \rangle = \lambda \langle \hat{U}^1 \rangle$. Thus, if the energy associated with the scaled wave function is denoted by E_λ, this quantity may be expressed in terms of the expectation values of the unscaled function, for which $\lambda = 1$, as $\langle \hat{H} \rangle = E_\lambda = \lambda^2 \langle \hat{T}^1 \rangle + \lambda \langle \hat{U}^1 \rangle$. As the exact energy E cannot be dependent on λ, the value of the scaling factor λ may be found, which makes the quantity $|E - E_\lambda|$ a minimum, by the usual procedure for finding the analytical value of a minimum by differentiating E_λ with respect to λ and equating the result to zero. The result thus obtained is $\partial E_\lambda / \partial \lambda = 2\lambda \langle \hat{T}^1 \rangle + \langle \hat{U}^1 \rangle = 0$ or $\lambda = -(1/2)[\langle \hat{U}^1 \rangle / \langle \hat{T}^1 \rangle]$. If this value for λ is now substituted into the expression for the total energy associated with the scaled wave function, there is obtained the Hylleraas relationship,[1] $|E - E_\lambda|_{\min} = |E + [\langle \hat{U}^1 \rangle]^2 / 4 \langle \hat{T}^1 \rangle|$. In the event that $\phi(r)$ happens to be the exact wave function for the system, so that $\lambda = 1$, then the value found for λ above becomes simply a statement of the virial theorem. For any scaled function ϕ_λ, with λ as found above, substitution of this value of λ into the expressions given above for the expectation values of the kinetic and potential energy operators with respect to the scaled wave function gives $\langle T^\lambda \rangle = -(1/2)\langle \hat{U}^\lambda \rangle$, which is an expression of the virial theorem first derived by Fock.[2] In other words, the position vectors of the particles composing a system may always be multiplied by some properly chosen scaling factor so that the average value of the kinetic energy of the system is equal to minus one-half of the average value of the potential energy of the system, or such that the virial theorem is satisfied.

According to Green et al.[3] and to Fröman,[4] the correlation energies for helium and other two-electron ions is of the order of -1.762×10^{-19} J, so that the kinetic part of their correlation energy is about 1.762×10^{-19} J, and the potential energy part of their correlation energies about -3.525×10^{-19} J. The interpretation to be placed on these figures is that the kinetic energy calculated by the Hartree–Fock method is too low or that electrons in atoms move in such a manner as to avoid one another to a greater extent than is implied by the Hartree–Fock equations; similarly, if the potential energy part of the correlation energy of a two-electron system is -3.525×10^{-19} J, this may arise from two electrons with non-parallel spins moving in the same region of space to make their potential energy greater than really so. It has been pointed out by Slater[5] that the simple Hartree self-consistent

[1] Hylleraas, E. A., Z. Physik, 1929, **54**, 347.
[2] Fock, V. A., Z. Physik, 1930, **63**, 855.
[3] Green, L. C., Mulder, M. M., Lewis, M. N., and Wall, J. W., Phys. Rev., 1954, **93**, 757.
[4] Fröman, A., Rev. Mod. Phys., 1960, **32**, 317.
[5] Slater, J. C., Rev. Mod. Phys., 1953, **25**, 199.

field method, which does not take account explicitly of the spin properties of electrons, sometimes leads to better results in accounting for certain properties of atoms, especially magnetic properties, than the Hartree–Fock method, which does include exchange effects. This result may be understood from the point of view that the former method gives an equally poor description of both electron pairs with parallel and those with antiparallel spins, but the errors being of opposite sign, tend to cancel one another. The Hartree–Fock description of electronic behavior, in terms of the antisymmetry principle, gives a better description of the motion of electrons with parallel spins but does not affect the description of the behavior of pairs of electrons with anti-parallel spins, so that no such cancellation of errors takes place.

The magnitude of electronic correlation energies for atoms and ions with more than two electrons is estimated to be quite considerable. Thus, Clementi[1] gives the correlation energy per electron pair with the same n, l, m_l, but different m_s quantum numbers, as -3.204×10^{-19} J, and Fröman (*loc. cit.*) has estimated that for the Al^{3+} ion, the correlation energy is about -1.762×10^{-18} J and for the Al^{11+} ion it is still about -1.442×10^{-18} J, indicating that the greater part of the correlation energy is attributable to electrons in $1s$- and $2s$-states. Apparently the relativistic energies of atoms increases approximately as the fourth power of the nuclear charge, even for atomic species with the same number of electrons (such as H^-, He, Li^+, Be^{++}, B^{3+}, C^{4+}), and on this basis for $Z = 92$, the relativistic energy may account for about 15% of the total energy of atoms. Since bond energies in compounds are only of the order of magnitude $(3 - 8) \times 10^{-19}$ J, it is perhaps surprising that such effects have not been detected experimentally. A good wave function for a system, in order to account for as much as possible of the correlation energy of its constituent electrons, would apparently require to introduce a "Fermi hole" for electron pairs with both parallel and anti-parallel spins. Actually the Eckart type wave functions given by (q.4.43.), which take account of the indistinguishability of electrons, allow for the radial correlation effects by assigning electrons of different spins to different regions in space; also, wave functions of the Hylleraas, Kinoshita, and Pekeris type (see page 241), which enable very accurate values of the energy of helium to be calculated, contain correlation functions of the type given by (q.4.46.).

In the absence of any exact knowledge of the electronic correlation energy in atoms, one of the most successful methods of obtaining an approximate exchange correction in the Hartree–Fock method is that due to Slater,[2] which is based on the concept of density matrices, originally introduced into quan-

[1] Clementi, E., *J. Chem. Phys.*, 1963, **38**, 2248; **39**, 175.
[2] Slater, J. C., *Phys. Rev.*, 1951, **81**, 385; 1953, **91**, 528.

tum mechanics by Landau,[1] and then developed by von Neumann,[2] Dirac,[3] Pauli,[4] Löwdin,[5] Fano,[6] McWeeny,[7] and ter Haar.[8]

Q.6.11.   Slater's Approximation for the Hartree–Fock Exchange Energy

Slater's proposal for finding an average exchange potential, which would be the same for each of the functions ψ_i in equations (q.5.54.), is based on the following considerations. Let it be assumed that the function describing the motion of all of the N electrons in a polyelectronic atom is the antisymmetric function $F(\mathbf{x}_1, \mathbf{x}_2, \ldots, \mathbf{x}_N)$, which could also be a linear combination of determinantal functions, expressed in terms of the notation introduced on page 288. The probability that electron 1 is at the position r_1 while electron 2 is simultaneously at position r_2 is given by the expression

$$\sum_{\zeta_1}^{\zeta_N} \int F^*(\mathbf{x}_1, \mathbf{x}_2, \ldots, \mathbf{x}_N) F(\mathbf{x}_1, \mathbf{x}_2, \ldots, \mathbf{x}_N) \, dr_3 \, dr_4 \cdots dr_N. \qquad \text{(q.6.103.)}$$

On the other hand, the probability that electron 1 is at position r_1 quite independently of how the other electrons are positioned is

$$\sum_{\zeta_1}^{\zeta_N} \int F^*(\mathbf{x}_1, \mathbf{x}_2, \ldots, \mathbf{x}_N) F(\mathbf{x}_1, \mathbf{x}_2, \ldots, \mathbf{x}_N) \, dr_2 \, dr_3 \cdots dr_N. \qquad \text{(q.6.104.)}$$

Therefore, the probability that electron 2 is at position r_2, if electron 1 is at position r_1, is given by the ratio of (q.6.103.) and (q.6.104.); the integrals defined by each of these two expressions above are what are referred to as density matrices.‡

‡ The description of a system by means of its density matrix is actually the most general form of quantum mechanical description of the system and provides a most useful means for the discussion of a variety of the aspects of many-electron and many-body systems in general.[9,10] It enables, for example, a more satisfactory description of the electronic

[1] Landau, L., *Z. Physik*, 1927, **45**, 430.
[2] Neumann, J. von, *Nachr. Akad. Wiss. Göttingen Math.-Phys. Kl.*, 1927, **245**, 273.
[3] Dirac, P. A. M., *Proc. Cambridge Phil. Soc.*, 1929, **25**, 62; 1930, **26**, 376; **27**, 240.
[4] Pauli, W., (S. Flügge, editor), *Encyclopedia of Physics*, **Vol. 5**, Pt. I, 69, 1958, Springer, Berlin.
[5] Löwdin, P.-O., *Phys. Rev.*, 1955, **97**, 1474.
[6] Fano, U., *Rev. Mod. Phys.*, 1957, **29**, 74.
[7] McWeeny, R., *Rev. Mod. Phys.*, 1960, **32**, 335.
[8] Haar, D. ter, *Rept. Prog. Phys.*, 1961, **24**, 304.
[9] *Concepts in Quantum Mechanics* by F. A. Kaempffer, 1965, Academic Press, New York.
[10] *The Mathematics of Physics and Chemistry*, by Henry Margenau and George M. Murphy, Vol. II, 1964, D. Van Nostrand Co. Inc., Princeton, New Jersey.

repulsion in many-electron atoms to be given, as distinguished from the essentially fortuitous description given by the Hartree–Fock self-consistent field method, where the exchange or spin correlation effects arise only from the antisymmetry of the total wave functions used and not from any explicit account of the electronic motion involved.

If $F(\mathbf{x}_1, \mathbf{x}_2, \ldots, \mathbf{x}_N)$ is the antisymmetric function (or it could be a linear combination of determinantal functions) describing the motion of all of the electrons in an N-electron system, where $\mathbf{x}_i$ represents the space and spin coordinates (x_i, y_i, z_i, ζ_i) of the ith electron, then this is a function of $3N$ spatial variables and N spin variables, or a $4N$-dimensional function. There are, however, particular properties of a system which are not dependent on all of these $4N$ variables, such as, for example, the approximate value of the total energy, which may be expressed in terms of averages involving only two electrons at a time. Such properties of a system, which are not dependent on all of the variables required for the description of the total system, but on only a fraction of them, cannot therefore be described by a complete or closed set of functions and thus cannot have a wave function of their own. The mathematical construct, which is referred to as a density matrix, is a function which is more general in nature than a wave function and may be used to describe those parts of a closed system for which no wave function exists. In general, if the set of coordinates of a system which is part of a closed system is denoted by $\mathbf{x}$ and q represents the remaining coordinates of the closed system, the closed system as a whole may be represented by the function $F(q, \mathbf{x})$, and the system which is part of the total closed system will not have its own wave function if $F(q, \mathbf{x})$ is not a product of functions of q and $\mathbf{x}$ alone. If the function $F(q, \mathbf{x})$, under certain circumstances and at a particular time, may be expressed as a product of functions of q and $\mathbf{x}$, this would mean that the measurement as a result of which the state $F(q, \mathbf{x})$ was brought about would completely describe the part of the system under consideration and the remainder of the closed system of which it was a part, separately. Now if p is some physical quantity associated with the part of the total closed system, its quantum mechanical operator would act only on the coordinates $\mathbf{x}$, and not on the remaining coordinates q; the mean value of this quantity for the state concerned would be (see page 148)

$$\bar{p} = \iint F^*(q, \mathbf{x})\hat{p}F(q, \mathbf{x}) \, dq \, d\mathbf{x}.$$

The function $\rho(\mathbf{x}', \mathbf{x}) = \int F^*(q, \mathbf{x}')F(q, \mathbf{x}) \, dq$, with the integration carried out only over the coordinates q, is what is defined as the density matrix of the system. From this definition of a density matrix, it is apparent that $\rho(\mathbf{x}', \mathbf{x})$ is Hermitian (page 139), since $\rho^*(\mathbf{x}, \mathbf{x}') = \rho(\mathbf{x}', \mathbf{x})$, and since the diagonal elements of the density matrix are given by $\rho(\mathbf{x}, \mathbf{x}) = \int |F(q, \mathbf{x})|^2 \, dq$, these quantities define the probability distribution of the co-ordinates of the system. The mean value of the property $\bar{p}$ may thus be written, in terms of the density matrix as $\bar{p} = \int [\hat{p}\rho(\mathbf{x}', \mathbf{x})]]_{\mathbf{x}' = \mathbf{x}} \, d\mathbf{x}$, denoting that $\hat{p}$ operates only on the variables $\mathbf{x}$ in the function $\rho(\mathbf{x}', \mathbf{x})$, and that after obtaining the result of this operation, $\mathbf{x}'$ is set equal to $\mathbf{x}$. In other words, if the density matrix of a system is known, the mean value of any quantity characterizing the system may be calculated. Also as a consequence of knowing $\rho(\mathbf{x}', \mathbf{x})$ for any system, the probabilities of various values of the physical quantities characteristic of the system may be determined. Therefore, the state of any system which does not have a wave function may be described by a density matrix. The description of a system by means of a wave function is actually a particular case of description in terms of a density matrix of the form $\rho(\mathbf{x}', \mathbf{x}) = F^*(\mathbf{x}')F(\mathbf{x})$. States describable by a wave function are sometimes referred to as pure states, in distinction to the mixed states, which are describable by a density matrix. In the former event, the wave functions are eigenfunctions of some operator, but in the latter event, this is not so.

Since the wave function of a system is a special form of density matrix, it is possible, as follows, to obtain another expression for the average value of the property of a system, from which it is possible to deduce whether a pure state or a mixed state is concerned. If, for example, $F_n(\mathbf{x}, t)$ represents the wave functions of the stationary states of a system (the eigenfunctions of its Hamiltonian), then if the system is describable by a wave function its density matrix $\rho(\mathbf{x}', \mathbf{x}, t) = F^*(\mathbf{x}', t)F(\mathbf{x}, t)$, may be expanded in terms of the stationary state wave functions to obtain a double series in the functions $F_n(\mathbf{x}, t)$ and $F(\mathbf{x}', t)$ as (page 127)

$$\rho(\mathbf{x}', \mathbf{x}, t) = \sum_m \sum_n a_{mn} F_n^*(\mathbf{x}', t)F_m(\mathbf{x}, t) = \sum_m \sum_n a_{mn} F'^*(\mathbf{x}')F_m'(\mathbf{x})e^{-2\pi i(E_n - E_m)t/h}. \quad (A)$$

This expression for the density matrix is the analogue of the expansion for the wave function given by equation (B) on page 110; however, the coefficients A_n in the expansion of the wave functions are here replaced by the double set of coefficients a_{mn}, which like the density matrix itself are also Hermitian, as $a_{nm}^* = a_{mn}$. If now, the above expansion for the density matrix is substituted into the expression for the mean value of any property of the system, there is obtained

$$\bar{p} = \int [\hat{p}\rho(\mathbf{x}', \mathbf{x}, t)]_{\mathbf{x}'=\mathbf{x}} \, d\mathbf{x} = \sum_m \sum_n a_{mn} \int F_n^*(\mathbf{x}, t)\hat{p}F_m(\mathbf{x}, t) \, d\mathbf{x}.$$

This expression may be alternatively written in a form equivalent to that given on page 148 for the mean value of any property of a system describable by a wave function, as

$$\bar{p} = \sum_m \sum_n a_{mn} f_{nm}(t) = \sum_m \sum_n a_{mn} f_{nm} e^{-2\pi i t(E_n - E_m)/h},$$

where f_{mn} are the matrix elements of the quantity f. Since the probability distribution of the coordinates of the system, which must be positive quantities, are determined by the diagonal elements of the density matrix $\rho(\mathbf{x}, \mathbf{x})$, the quantities a_{mn} must also necessarily be positive. There are other consequential restrictions which must apply to the quantities. a_{nm} and to the double sums in the above expressions, constructed with these quantities, Thus, in the expansion (A) above, with $\mathbf{x}' = \mathbf{x}$ if ξ_n are any arbitrary complex quantities, then expressions of the type

$$\sum_m \sum_n a_{mn} \xi_n^* \xi_m,$$

must also be positive. Thus, all quantities of the form a_{nn} (diagonal elements) must be positive, and for any three quantities a_{nn}, a_{mm}, and a_{mn}, the inequality $a_{nn}a_{mm} \geq |a_{mn}|^2$, must be satisfied. Now, it becomes apparent that for a pure state, where the density matrix reduces to a product of functions, there will be a corresponding matrix of the form $a_{mn} = a_m a_n^*$. For this pure state

$$[a_{mn}]^2 = \sum_k a_{mk}a_{kn} = \sum_k a_k^* a_m a_n^* a_k = a_m a_n^* \sum_k |a_k|^2 = a_m a_n^* = a_{mn},$$

or the density matrix is equivalent to its own square. Thus, from the form of the matrix a_{mn}, it is possible to decide by the use of this criterion whether a pure or a mixed state is under consideration.

Density matrices satisfy an equation of a similar form to the time-dependent Schröd inger equation, which is satisfied by a wave function. This is most easily seen by again noting that since a wave function is a special case of a density matrix, or since $\rho(\mathbf{x}', \mathbf{x}, t) = F^*(\mathbf{x}', t)F(\mathbf{x}, t)$, then by differentiating this equation with respect to time, there is obtained

$$\partial\rho(\mathbf{x}', \mathbf{x}, t)/\partial t = F^*(\mathbf{x}', t) \, \partial/\partial t[F(\mathbf{x}, t)] + F(\mathbf{x}, t) \, \partial/\partial t[F^*(\mathbf{x}', t)].$$

If this equation is now multiplied throughout by $(-h/2\pi i)$ and making use of the Schrödinger equation (page 109), $(-h/2\pi i)\,\partial/\partial t[F(\mathbf{x}, t)] = \hat{H}F(\mathbf{x}, t)$, it is deduced that

$$(-h/2\pi i)\,\partial/\partial t[\rho(\mathbf{x}', \mathbf{x}, t)] = F^*(\mathbf{x}', t)\hat{H}F(\mathbf{x}, t) - F(\mathbf{x}, t)\hat{H}'^*F^*(\mathbf{x}', t),$$

where $\hat{H}$ is the Hamiltonian of the system acting on a function of χ and $\hat{H}'$ designates the same operator acting on a function of χ'. This latter equation may obviously be rewritten as

$$(-h/2\pi i)\,\partial/\partial t[\rho(\chi', \chi, t)] = [\hat{H} - \hat{H}'^*]\rho(\chi', \chi, t),$$

which is of the same form as the Schrödinger equation satisfied by the wave function for a system.

If this ratio is written, for convenience, as

$$\frac{\displaystyle\sum_{\zeta_1}^{\zeta_N}\int F^*F\,dr_3\,dr_4\cdots dr_N}{\displaystyle\sum_{\zeta_1}^{\zeta_N}\int F^*F\,dr_2\,dr_3\cdots dr_N}, \tag{q.6.105.}$$

then this expression is a function of the coordinates of electron 2, r_2, expressed as a parameter of the coordinates of electron 1. Because of the antisymmetry of the total function F, the same sort of distribution for any one of the other electrons 3, 4, ..., N would result as for electron 2. The total charge distribution of the electrons 2, 3, ..., N, when electron 1 is at the position r_1, would be given by equation (q.6.105.) multiplied by $(N - 1)$. For the special case where the total wave function is a single determinant of spin orbitals, then due to the orthogonality of the spin orbitals as expressed by equation (q.5.25.), there is obtained on carrying out the integrations required by equation (q.6.105.)

$$\sum_{j=1}^{N}\psi_j^*(r_2)\psi_j(r_2) - \frac{\displaystyle\sum_{\substack{i\neq j}}^{N}\sum^{N}\delta(m_{\zeta_i}, m_{\zeta_j})\psi_i^*(r_1)\psi_j^*(r_2)\psi_j(r_1)\psi_i(r_2)}{\displaystyle\sum_{k}^{N}\psi_k^*(r_1)\psi_k(r_1)}, \tag{q.6.106.}$$

which represents the charge density which is $(N - 1)$ times the expression given by (q.6.105.). The first term in (q.6.106.) represents the total charge density of all of the N-electrons and the second term is the correction or exchange term to be subtracted from this, in order to give the net charge due to $(N - 1)$ electrons. This exchange charge density is similar to that in equations (q.5.54.) and (q.5.55.), derived from the Hartree–Fock equation, since

$$\varepsilon^2\sum_{j}^{N}\left[\int \{[\psi_j^*(r_2)\psi_i(r_2)\psi_j(r_1)]/[r_{12}\psi_i(r_1)]\}\,dr_2\right]\psi_i(r_1)$$

$$= \varepsilon^2\sum_{j}^{N}\left[\int \{[\psi_j^*(r_2)\psi_i(r_2)]/[r_{12}]\}\,dr_2\right]\psi_j(r_1)$$

$$= \left[\sum_{j}^{N}\delta(m_{\zeta_i}, m_{\zeta_j})\int \{[\psi_i^*(r_1)\psi_j^*(r_2)\psi_j(r_1)\psi_i(r_2)]/[r_{12}\psi_i^*(r_1)\psi_i(r_1)]\}\,dr_2\right]\psi_i(r_1), \tag{q.6.107.}$$

as the delta function assumes the value unity, in the event where the whole expression has any significance (otherwise it vanishes) and ε^2 is also unity in atomic units. Referring back to equation (q.5.54.), it is obvious that an electron does not exert a Coulomb interaction with itself, which is the meaning to be associated with the third term in this equation, provided that there is an electron in the state described by the function ψ_i. The exchange term in the Hartree–Fock equations, given by the equivalent expressions (q.6.107.) above, provide a correction for this interaction of an electron with itself, and may be interpreted as representing the potential energy, at the position r_1 of the electron concerned, of a charge distribution at the position r_2 of magnitude

$$\sum_{j}^{N} [\psi_i^*(r_1)\psi_j^*(r_2)\psi_j(r_1)\psi_i(r_2)]/[\psi_i^*(r_1)\psi_i(r_1)]$$

$$= \sum_{j}^{N} \delta(m_{\zeta_j}, m_{\zeta_i})[\psi_j^*(r_2)\psi_j(r_1)\psi_i(r_2)]/\psi_i(r_1). \quad \text{(q.6.108.)}$$

Now, the exchange charge in (q.6.106.) is obtainable from the Hartree–Fock exchange charge density of (q.6.108.) by multiplying the latter expression by

$$[\psi_i^*(r_1)\psi_i(r_1)]/\left[\sum_{k}^{N} \psi_k^*(r_1)\psi_k(r_1)\right]. \quad \text{(q.6.109.)}$$

and taking the summation over i. It is apparent, then, that the exchange charge density given by (q.6.106.) is the averaged density of (q.6.108.), for the various electronic states, weighted by the factor (q.6.109.), which determines the probability that if electron 1 is found at position r_1, it will be described by the ith spin orbital of the system of N-electrons. This is, of course, a different situation from the Hartree–Fock case, where a different exchange charge density applies to an electron with coordinates r_1 and ζ_1, depending on which spin orbital Ψ_i describes the electron. In the Hartree–Fock equations, (q.5.54.), the exchange charge densities integrate to one electronic charge, independently of which spin orbital describes the electron, and approach the same limiting value when r_2, ζ_2 approach r_1, ζ_1.

Slater's modification of the Hartree–Fock equations, (q.5.54.), then, is to replace the exchange term in them by the averaged charge density of (q.6.106.) to give instead

$$\left[-(h^2/8\pi^2 m)\nabla_1^2 + Z\varepsilon^2/r_1 + \varepsilon^2 \sum_{j}^{N} \int \{[\psi_j^*(r_2)\psi_j(r_2)]/[r_{12}]\} \, dr_2 \right]\psi_i(r_1)$$

$$- \left[\varepsilon^2 \sum_{i \neq j}^{N}\sum^{N} \int \{[\psi_i^*(r_1)\psi_j^*(r_2)\psi_j(r_1)\psi_i(r_2)]/[r_{12} \sum_{k}^{N} \psi_k^*(r_1)\psi_k(r_1)]\} \, dr_2 \right]\psi_i(r_1)$$

$$= E_i\psi_i(r_1). \quad \text{(q.6.110.)}$$

This Slater equation maintains the essential features of the Hartree–Fock method, but describes all of the N-electrons as moving in the same averaged exchange potential field. Calculations, making use of this equation, of the exchange potential for the various electrons in polyelectronic atoms have been made by Hartree[1] and others.

Q.6.12. Elaborations of the Hartree–Fock Method

The literature of the Hartree–Fock method and its various extensions, modifications, and developments is quite extensive.[2-12] The approximate solutions of the Hartree–Fock equations, such as (q.5.54.), may be carried out by numerical methods or by the expansion method, in which event the orbitals are expressed as $\psi_i = \sum_i c_i \phi_i$, where the ϕ_i terms are analytical functions and the c_i terms are the expansion coefficients. A variety of numerical techniques have been developed and the analytical Hartree–Fock procedure offers advantages in the calculation of other physical properties of systems apart from energies. The direct determination of analytical Hartree–Fock functions may be carried out by two different methods: McWeeny[7] has proposed an iteration process using first order density matrices, whereas Roothaan[4] makes use of a direct diagonalization procedure. A relatively large number of the existing analytical Hartree–Fock functions have been determined by the self-consistent field method of Roothaan. Analytical

[1] Hartree, D. R., *Phys. Rev.*, 1958, **109**, 840.
[2] Yoshizumi, H., *Advan. Chem. Phys.*, 1959, **2**, 323.
[3] Löwdin, P. O., *Advan. Chem. Phys.*, 1959, **2**, 207; *Rev. Mod. Phys.*, 1960, **32**, 328.
[4] Roothaan, C. C. J., and Bagus, P. S., *Atomic Self-Consistent Field Calculations by the Expansion Method*, in *Methods in Computational Physics*, **Vol. II**, 1963, Academic Press, Inc., New York.
[5] Froese, C., *J. Chem. Phys.*, 1966, **45**, 1417.
[6] Cade, P., Greenshields, J., Huo, W., Malli, G., Peyerimhoff, S., Sales, K. D., Wahl, A. C., and others, in the successive *Technical Reports*, beginning in 1964, from the Laboratory of Molecular Structure and Spectra, Department of Physics, University of Chicago.
[7] McWeeny, R., *The Self-Consistent Generalization of Hückel Theory*, in *Molecular Orbitals in Chemistry, Physics, and Biology*, 1964, edited by P. O. Löwdin and B. Pullman, Academic Press, Inc., New York.
[8] Simonetta, M., and Gianinetti, E., *The Self-Consistent Field Molecular Orbital Theory: An Elementary Approach*, in *Molecular Orbitals in Chemistry, Physics, and Biology*, 1964, edited by P. O. Löwdin and B. Pullman, Academic Press, Inc., New York.
[9] Malli, G., and Olive, J. P., *J. Chem. Phys.*, 1965, **43**, 861.
[10] Clementi, E., *Tables of Atomic Functions*, 1965, I.B.M. Corporation, San José, California.
[11] Malli, G., *Can. J. Phys.*, 1966, **44**, 1451.
[12] Pauncz, R., *Alternant Molecular Orbital Method*, 1967, Saunders Publishing Company, Philadelphia.

Hartree–Fock functions for the following atoms and ions have been reported:

(i)   Ground states of Li, F, F^-, Ne, Na^+, and Na.

(ii)   Ground states of K, K^+, and Ca.

(iii)   Ground states of Al, Si, P, S, Cl, Cl^-, Ar, Zn, Ga, Ga^+, Ge, Ge^{++}, As, Se, Br, Br^-, Rb^+, and Kr.

(iv)   Ground and excited states of Fe^{3+}.

(v)   Ground and excited states of Al, Al^+, Cu, Cu^+, Cr, Cr^+, and Mn^{3+}.

(vi)   Excited states of Mn^{3+}.

(vii)   Excited states of Cu and Cu^+.

(viii) Ground and excited states of N and O and of their ions.

(ix)   Ground states of Mo and Mo^+.

(x)   Excited states of neon- and argon-like ions.

(xi)   Ground states of the neutral atoms of the elements with nuclear charges from 1 to 10.

(xii)   Ground and excited states, corresponding to the same configuration as the ground state, of the neutral atoms with nuclear charges from 11 to 18.

(xiii) Ground states of Cr^+, Cr^{++}, Cr^{3+}, Cr^{4+}, Cr^{5+}, Xe, and Rn.

(xiv)   States corresponding to the $3d_n$ configurations in the first transition metal series from Sc to Cu.

(xv)   Ground states and excited states, corresponding to the same configuration as the ground state, of the neutral atoms He to Kr, the negative ions Li^- to Br^-, and the isoelectronic series $1s^2$, ^1S to $K^2L^8M^{18}$, ^1S.

(xvi)   Isoelectronic series $1s^2$, ^1S, $1s^22s^1$, ^2S and $1s^22s^2$, ^1S.

The Hartree–Fock method may be extended to more complicated systems than atoms and ions. Formally, all that is necessary to be done to accomplish this is to substitute the potential energy of the electrons in the field of the single nucleus of charge $+Z\varepsilon$ by the sum of such terms for all the atoms of the system concerned. The mathematical techniques involved in solving the Hartree–Fock equations for molecules and for solids, however, are quite different from those required for atoms. In the first place, the approximation used in the case of atoms and ions of adopting the solutions of the central field problem as the orbital functions is no longer applicable, as there is no question of spherical symmetry for most molecules or for solid crystalline substances. The orbitals used for molecules are referred to as molecular orbitals, and this technique, although originally developed a long time ago,[1–3] is now of predominant applicability. In the case of metals and other solid

---

[1] Hund, F., *Z. Physik*, 1927, **40**, 742; **42**, 93.
[2] Mulliken, R. S., *Phys. Rev.*, 1928, **32**, 186.
[3] Lennard-Jones, J. E., *Trans. Faraday Soc.*, 1929, **25**, 668.

materials, the one-electron energies no longer have discrete values, as for atoms and for certain simple molecules, but occur in continuous energy bands.

Almost all of the calculations on the electronic structure of atoms and molecules have followed a common basic approach. The problem is split or resolved into two parts, the calculation of a molecular orbital wave function and a subsequent calculation or estimation of the correlation effects involved. The wave functions which result from the solution of the Hartree–Fock equations for a system are sufficiently accurate for many purposes; in particular, they give a very accurate account of the charge distribution in atoms and molecules and a certain number of other experimentally observable properties, such as dipole moments, diamagnetic susceptibilities, and also of the energy of the system expressed in terms of percentage accuracy. But, unfortunately, the total energy of most atoms and molecules is so large compared with chemical energies that a few percent inaccuracy of the Hartree–Fock energy is comparable, for example, with the bond energy of the molecule. As a consequence, Hartree–Fock calculations of bond energies are usually only accurate to about 50%, and in some cases stable molecules are predicted to have a negative bond energy; this is so, for example, for the fluorine molecule. Some other atomic and molecular properties calculated by the use of Hartree–Fock wave functions disagree badly with experimentally observed data.

Of the various methods which have been used to introduce electron correlation effects into atomic and molecular wave functions, the two most widely used techniques both date back to the early days of quantum mechanics. These are the introduction of interelectronic distances (r_{ij}) as explicit variables in the wave function, and the method of configuration interaction. In the latter, the wave function is expressed as a linear combination of determinantal wave functions in which the principal term is the ordinary molecular orbital wave function. Both of these methods, when applied to simple systems such as the helium atom or the hydrogen molecule, lead to results of reasonable accuracy. However, for larger systems, both methods become quite unmanageable; the number of r_{ij} coordinates increases with $(N/2)(N-1)$, where N is the number of electrons, while the number of degrees of freedom of the system increases only as $3N$, so that for quite a small value of N the r_{ij} terms cease to be independent and the method fails. Also, in the configuration interaction method, the number of configurations required for a given accuracy increases rapidly with N, and again the problem becomes unmanageable. These difficulties may be overcome by dealing with the correlation effects in small subgroups of electrons independently.[1-3] In principle, this

---

[1] Sinanoğlu, O., *J. Chem. Phys.*, 1962, **36**, 706.
[2] Szasz, L., *Naturforsch.*, 1960, **15a**, 909; *Phys. Rev.*, 1962, **126**, 169.
[3] Tuan, D. F., and Sinanoğlu, O., *J. Chem. Phys.*, 1964, **41**, 2677.

method consists of selecting a group of electrons, such as the pair in the $1s$ energy states, and solving a modified Schrödinger equation, the so-called Bethe–Goldstone equation, for the detailed motion of these electrons in the average potential of the remaining electrons by the variational method, subject to the constraint that the solution must be orthogonal to the Hartree–Fock solution. This method has been applied, with very satisfactory results, to the calculation of the total correlation energies for beryllium and neon atoms.[1,2] Such calculations lend support to the idea that each electron correlates principally with one other, but that the pair correlation energies may not be strictly additive.

The general direction in which much of the work associated with the study of many electron systems is headed is away from the preoccupation with energy alone. Attention is now being focused on those terms which the Hartree–Fock method ignores, such as spin-spin coupling effects, spin-orbital interactions, coupling of electron spin and nuclear spin, quadruple moments, and the various influences which result from the coupling of all of these effects with an applied electric and magnetic field. As a result of the development of many types of microwave experiments, the influence of such smaller energy terms in the total Hamiltonian for a many-electron system makes it possible to test any given wave function for accuracy in one of several chosen respects, rather than in just one respect, that of energy. The starting point for all of these considerations is the selection of the closest approximation possible to the self-consistent Hartree–Fock description of the problem involved.

---

[1] Nesbet, R. K., *Phys. Rev.*, 1967, **155**, 51, 56.
[2] Szasz, L., and Byrne, J. *Phys. Rev.*, 1967, **158**, 34.

PART M

Metals and Intermetallic Compounds

Introduction

Theories about many of the properties of metallic and other crystalline materials were well advanced long before the development and application of quantum mechanical methods to the description of the behavior of metals and alloys. The bulk properties of solid matter, such as electrical and thermal conductivity, thermoelectric effects, optical absorption and dispersion phenomena, symmetry, point and space group representations, elasticity, heat capacity and related thermodynamic quantities, thermal expansion, and such specific properties as the various types of magnetic behavior displayed by solids, had all been extensively studied before the reported work of Schrödinger, Heisenberg, Born, Jordan, and Dirac. In fact, a great deal of the progress in the theory of solids has been associated with continued interest in the specific property of heat capacities, beginning with the discovery of the empirical law of Dulong and Petit (1819) and extending through the theories of isothermal cavity radiation (Jeans, Wien, Rayleigh, and Planck), the theories of the heat capacities of solids (Einstein, Debye, Born, and von Kárman), to solid-state phenomena associated with the diffraction of X-radiation and electrons, and the inelastic scattering of thermal neutrons (1960).

Extensive developments of the theory of metals followed each of the discoveries: of the electron as a universal constituent of all matter, of the quantum postulate, of the diffraction of X-radiation by crystalline water, of the techniques of quantum mechanics, and of the nature of plasmas and superconductors. The first theory of metals, based on the "free" electron concept proposed by Drude and Lorenz, offered plausible explanations of the experimentally discovered law of Wiedemann and Franz, that for metals the thermal conductivity is proportional to the electrical conductivity, of Ohm's law, of the optical properties of solids, of the Hall effect, and of various other thermoelectric properties of matter. The "free" electron theory has continued to serve as a model for the development and evaluation of the more complete energy band and zone theories of metals. The various deficiencies of the "free" electron theory were not removed, however, until after the discovery of the crystal structures of metals and the application of quantum mechanics to the problem. The clear distinction between solid materials classified as metals, insulators, semiconductors, and superconductors was a consequence of the early quantum mechanical description of the behavior of electrons in metals as particles obeying Fermi–Dirac statistics moving in the periodic potential field of a metal crystal, as a result of the efforts of Sommerfeld, Bloch, Brillouin, Jones, and others; the phenomenon of superconductivity, which is not dealt with in any detail in this book, was only qualitatively explained by 1957 (Bardeen), and although there are now over 1400 metallic elements and their alloys which are known superconductors as well as certain semiconductors which become superconducting under pressure, all of this development has taken place since 1962.

The fast and successful application of new conceptual techniques to the description of the behavior and properties of metals and alloys was undoubtedly made more easily possible by the great amount of previously accumulated experimental data in this area. In many cases, also, important experimental discoveries and new technologies related to metals and other solids have followed as a consequence of theoretical predictions. The technology associated with semiconductors and transistors and their various applications is one of the most extensive which has yet been developed. Perhaps an even greater technological revolution is indicated as a result of a theoretical prediction made by Josephson (1962) with respect to another property of metals which has been realized for about 60 years (Onnes, 1911), namely superconductivity. The Josephson prediction that certain effects would occur when electrons are transported through a narrow barrier of insulating material between two superconductors has been confirmed (Anderson and Rowell, 1963) and has led to the development of extremely sensitive instruments for the measurement of magnetic field strengths (down to 10^{-18} T; one tesla is equivalent to 10^4 gauss), electric current, potential difference (down to 10^{-15} V), and temperature (in the range of 10 K to 10^{-3} K). The potential applications of high-field superconducting magnets are quite tremendous; linear accelerators of enormous power, surface and space transportation devices, small-size power transmission systems, computer memory banks of greatly reduced size, cost, and reliability, magnetohydrodynamic propulsion engines, and the production of extremely sensitive instruments for the measurement and detection of radiation and most of the other attributes of matters, are a few possibilities, some of which have already been realized. The superconductivity of mercury at liquid helium temperatures (4.2 K), originally observed by Onnes, was found to be destroyed by externally applied magnetic fields of the order of much less than one tesla, but niobium–germanium–aluminum alloys (for example, $Al_{\sim 0.8}Ge_{\sim 0.2}Nb_3$)

are superconducting up to 20.7 K with a critical field of 20 tesla, which is about 4×10^5 times greater than the earth's magnetic field.

The first chapter of this part contains an outline of some of the characteristics of pure metals and intermetallic phases, including an account of their crystal structures, the Einstein and Debye theories of heat capacities, and the thermal and electrical conductivities of metals, and concludes with an account of the manner in which the electron theories of metals have developed. These are the essential facts about the metallic state which any comprehensive theory of metals has to attempt to explain.

Before embarking on the detailed discussion of the various electron theories, the question of the cohesive energies of metals is taken up in the second chapter, and the manner in which such properties as hardness, melting and boiling points, compressibility, density, expansion coefficients, internuclear distances in the crystalline metals, and magnetic and structural characteristics are related to and dependent on this cohesive energy is outlined. This approach to the nature of the bonding in metals was initiated by Pauling (1938) somewhat later than the purely quantum mechanical electron theories of metals and has only been modified slightly by others. The Pauling theory of metals is not particularly adaptable to quantitative calculations, although it provides a very plausible explanation of the magnetic properties, especially, of the transition metals.

The "free" electron theory of metals, the Sommerfeld extensions of this concept, quantum statistics, and the application of the Hartree and Hartree–Fock methods to the description of the "free" electron gas, phrased in terms of the concept of wave vector space, is taken up in Chapter M.3. There follows in the next chapter some discussion of such properties as soft X-ray spectra, low temperature thermal conductivity, electrical conductivity, surface tension, photoelectric and thermionic emission from metals, paramagnetism, and the Thomas–Fermi method, from the point of view of the free electron approximation.

The development of the zone and band theories of metals is next discussed, and it is shown how the description of the behavior of electrons in metals by means of Bloch functions leads to the concept of Brillouin and Jones zones, and the distinction between metals, insulators, and semiconductors. The relationships between metal crystal structures, reciprocal lattice space, wave vector space, Brillouin zones, electron densities in metals, and electronic energy distributions, are discussed in some detail in this chapter.

In the sixth chapter the electronic structure of metals is discussed in terms of the more elaborate approximation methods which have been developed to calculate the energy bands in metallic crystals. These include the Wigner–Seitz cellular method, the nearly free electron and tight binding approximations, the quantum defect method, the orthogonal plane wave, augmented plane wave, and scattered wave methods. The Bohm and Pines plasma theory of metals is also discussed in this chapter, as well as some of the experimental techniques, such as the Hall effect, the de Haas–van Alphen effect, and related effects involving magnetic fields, which have been found capable of providing experimental information about the nature of the Fermi surfaces in metals and band energy widths.

The final chapter applies the electron and structure theory previously developed to the discussion of the observed behavior and properties of various metals, alloys, and intermetallic compounds.

M1

The Metallic State

M.1.1. Characteristic Properties

The distinction formerly made between substances classified as metals and non-metals is now perhaps of little significance. Of more importance, in terms of the electron theories of metals, are the differences between metals, insulators, and semiconductors, and the realization that the exhibition of what are regarded as metallic or non-metallic properties by a substance, is characteristic of a state of matter rather than a type or class of matter. In any event, such distinctions are only applicable to the solid and liquid states, and completely vanish in the vapor phase. There is no unique property or set of properties which unambiguously differentiates metals from other substances, although certain properties are commonly associated with metallic behavior. The most obvious and important properties of metals are as follows:

(a) The characteristic metallic luster displayed by metals is a consequence of their very strong absorption for visible light of all wavelengths. With the exception of copper and gold and certain of their alloys and compounds, most metals appear grayish-white or grayish-blue in color,‡ and have very low optical transparency so that they are opaque even in thin layers. In the finely divided state all metals, again with the exception of copper and gold and certain of their compounds, are black. Metallic copper and gold, in the form of the pure elements and in certain of their compounds, are red and yellow, respectively. Like most of the other properties listed below, the possession or exhibition of metallic luster by a substance is, by itself, not necessarily an indication that the substance is a metal. For example, many plastic and fibrous substances may present a lustrous appearance.

‡ An exception is the alloy $AuAl_2$, which has a striking purple color. This substance crystallizes with the fluorite structure [Fig. M.1.8.(f)], each gold atom equidistant from eight aluminum atoms, and each aluminum atom equidistant from four gold atoms.

383

(b) The conductivity, both thermal and electrical, of metals is generally high; the study of these properties and their interrelationships has played an important part in the development of the theories of metals. In fact, the first theory of metals, of Drude[1] and Lorenz,[2] to meet with any success was based on the interpretation of the electrical and thermal conductivity of metals in terms of electrons moving relatively freely through the piece of metal. This so-called free electron theory of metals has provided the model, on the basis of which much of the subsequent progress in the development of the quantum mechanical theory has been arrived at, and in many respects the conclusions derived from this simple theory retain their validity in the more detailed quantum mechanical theories.

The classification of substances into electronic or metallic conductors, electrolytic conductors, semiconductors, and insulators actually follows most directly from examination of their conductivities. The ratio of the conductivity, or its reciprocal, the resistivity, of a typical insulator such as sulfur or diamond to that of a typical metallic conductor such as copper or sodium, is of the order of magnitude of 10^{30}, which greatly exceeds the ratio of the maximum and minimum values of any other known property of substances (see page 524). Most pure metals display normal electronic conduction, their electrical resistivity decreasing with decreasing temperature and increasing with increasing temperature and impurity content. Electronic conduction does not involve any material transport, as in the case of electrolytic conduction, which is of little consequence for metals and alloys, although with alloys and intermetallic compounds containing elements of greatly different electrochemical characteristics, electrolytic transport may take place in the solid phase or in melts. Semiconductors, of which several types may be distinguished, differ from metals in that their electrical resistance generally decreases with increasing temperatures and with increasing impurity concentration.

Generally, the metals have high thermal conductivity as well as electrical conductivity, although it has to be noted that diamond, which is an electrical insulator, has a thermal conductivity which is as high as that of most metals at ordinary temperatures.

(c) When exposed to short-wavelength radiation or on being heated to a certain temperature, many metals emit electrons; the former effect referred to as the photoelectric effect (page 45) only occurs with radiation of a certain limiting or minimum frequency, below which no electronic emission occurs, indicating that a certain minimum amount of energy must be supplied before an electron can be expelled from a metal. This type of electronic emission is

---

[1] Drude, P., *Ann. Phys*, 1900, **1**, 566; 1900, **3**, 369.
[2] Lorentz, H. A., *Proc. Amsterdam Acad.*, 1904–1905, **438**, 588, 684.

independent of temperature, up to that point at which thermionic emission takes place, due to the supply of thermal energy.

(d) The majority of the elements exhibit metallic properties; in fact, only hydrogen, the rare gases, the halogens, the light elements boron, carbon, nitrogen, oxygen, silicon, phosphorus, and sulfur, are essentially non-metallic in character. Even among those typically non-metallic elements, some of them exist in polymorphic modifications with at least semimetallic properties, such as boron, silicon, graphite, and black phosphorus. The metallic elements are thus those elements which are typically electropositive in their chemical behavior and of small electronegativity. The cubic modification of the element tin (gray tin) has essentially non-metallic properties in contradistinction to ordinary (white) tin, which crystallizes in the tetragonal system as a recognizable metal. It is common practice to refer to the metallic elements as "pure" metals, to make some distinction between them and alloys, which may contain minor amounts of non-metals in their composition, in addition to other metals or semimetals. Elements such as antimony and gray tin are also frequently referred to a metalloids. There are also many compounds, generally carbides, borides, nitrides, sulfides, and oxides, which display some metallic properties such as high electrical conductivity and optically lustrous appearance, but are brittle without any of the desirable mechanical properties usually associated with metals. It is interesting to note that some of the elements display typically metallic properties only in the liquid state. Liquid silicon and liquid germanium, for example, exhibit metallic properties quite comparable to mercury, but the solid, crystalline elements are semiconductors; arsenic, selenium, and tellurium also display much more pronounced metallic behavior in the liquid state. Mercury is the only metallic element which is liquid under normal conditions of temperature and pressure, but also gallium has a relatively low melting point (302.78 K). Certain other elements, such as hydrogen, boron, carbon, and phosphorus, may exhibit more well-developed metallic properties under conditions of high pressure. All of the normally solid metallic elements, as well as many of their alloys, retain their typically metallic properties in the liquid state. It is only comparatively recently that the distinctions between the liquid metals such as mercury and other metals in the molten state, and "normal" liquids such as gasoline, ether, and aqueous solutions, for instance, have attracted much attention. The published proceedings of an international conference on the properties of liquid metals held at the end of 1966 at the Brookhaven National Laboratory took up, for example, some 600 pages. Very generally, normal liquid properties may be explained in terms of the kinetic molecular theory, but to account for the properties of metallic liquids and superfluids, of which liquid helium is presently the only known example, requires the application of quantum theory. The exhibition of the properties of superfluidity and superconductivity actually represents the illustration of quantum phenomena on a macroscopic scale.

(e) The mechanical properties of malleability and ductility, which enable metals to be rolled, pressed, and hammered into convenient shapes, are indicative of strong bonding forces between the constituent atoms of the metal; the relative magnitude of the bonding force between metal atoms is also deducible from such properties as the hardness, melting point, boiling point, coefficient of expansion, compressibility, and density, from a knowledge of the interatomic distances observed in the metal and its other compounds. A theory of metals based on such information and observations has been formulated by Pauling[1] in the period since 1938 and has had remarkable success in relating the cohesive energy and magnetic properties of metals to electronic descriptions of the bond-type involved between metal atoms. From this point of view, it is perhaps preferable to regard the apparently continuous transition in properties of the substances, for example, fluorine, iodine, sulfur, tellerium, arsenic, tin, silver, and sodium, as a transition of bond type from the essentially atomic extreme in fluorine to the metallic extreme type in sodium; a similar type of transition in bond types may be deduced from the gradual transition in properties of the series of compounds, for example, cesium fluoride, sodium oxide, sodium nitride, sodium phosphide, sodium arsenide, sodium antimonide, sodium bismuthide, and sodium, as indicating a transition in bond type from the electrostatic extreme in cesium fluoride to the metallic extreme in sodium again.

(f) The majority of the metallic elements crystallize with highly symmetrical structural arrangements. Closest packed structures predominate. Thus the alkali metals all adopt body-centered cubic structures under ordinary conditions, although under high pressure cesium adopts a hexagonal close-packed arrangement, and lithium and sodium both transform to a hexagonal close-packed modification at low temperatures. In fact, all of the metallic elements, with the exception of manganese and tin, crystallize with at least one polymorphic modification having either a face-centered cubic, body-centered cubic, or close-packed hexagonal structure.‡ Manganese is known in three different polymorphic modifications, all with complex structures, and tin occurs as gray tin (diamond structure) and as white tin (crystallizing in the tetragonal system). Many other metallic elements exist in several different polymorphic modifications; thus, calcium, strontium, and uranium are all trimorphic.

‡ A close packing is a manner of arranging equidimensional objects in space so that the space is filled as efficiently as possible, and such arrangements arise when each object is

---

[1] Pauling, L., *Nature of the Chemical Bond*, 1962, Cornell University Press, Ithaca, New York.

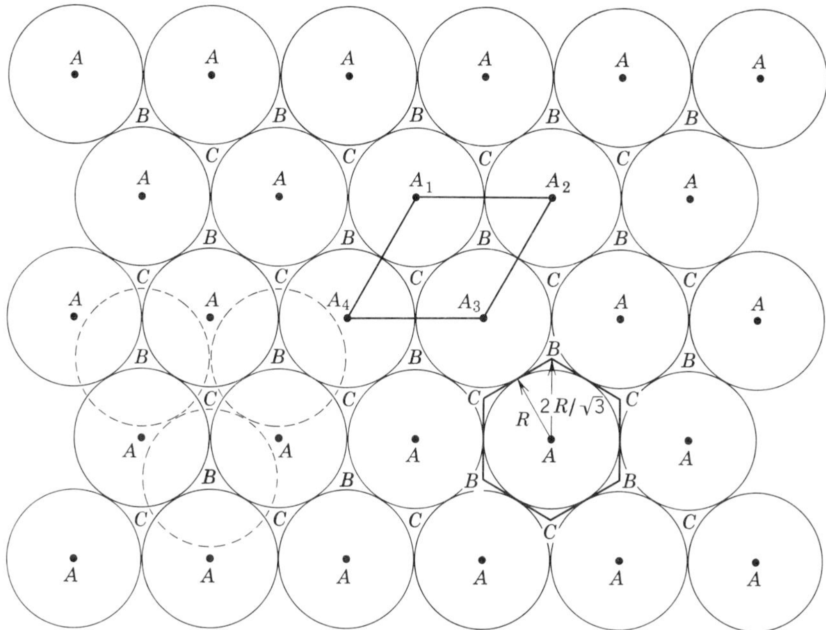

FIG. M.1.1. A close-packed plane array of equidimensional spheres of radius R, illustrating the projection onto the plane through the centers A of the spheres. The quadrilateral $A_1A_2A_3A_4$ represents the unit cell base enclosing one sphere and two holes. The triangular holes between the spheres in the plane can be differentiated into two types, B and C, if a second layer of close packed spheres is centered above them. B represents tetrahedral holes and the set C, octahedral holes.

in contact with as many similar objects as possible. It is found that not only metals but also many crystalline substances composed of ions, complex ions, and atoms, adopt structural arrangements corresponding to rather close packed aggregates, a result which is not too surprising, since the non-directional, electrical, interaction between approximately spherical ions or atoms (Coulomb, van der Waals, and metallic) favors the stability of arrangements with the maximum number of species as close together as possible. Long before the discovery of the diffraction of X-radiation by crystalline materials and the elucidation of the structure of the first compound to be determined in this manner, the question of the most efficient ways of packing spheres to occupy the minimum of space had attracted the attention of crystallographers, and Barlow[1] discovered that there are two ways of close packing equivalent spheres.

If spheres of equal radius are packed together in a plane array, the most efficient way of forming a close packed layer is to have each sphere in contact with six others, such that the center of each sphere corresponds to a point in a plane triangular net, as in Fig. M.1.1. If the radius of each sphere is R, then by comparing the area of the plane projection

---

[1] Barlow, W., *Nature*, 1883, 29, 186, 205, 404.

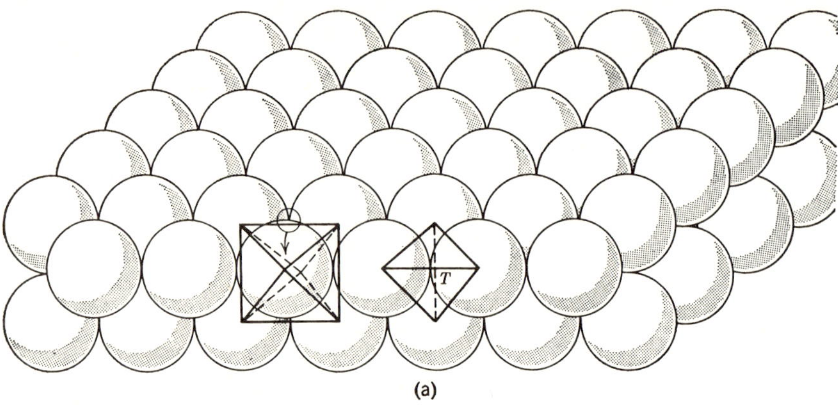

(a)

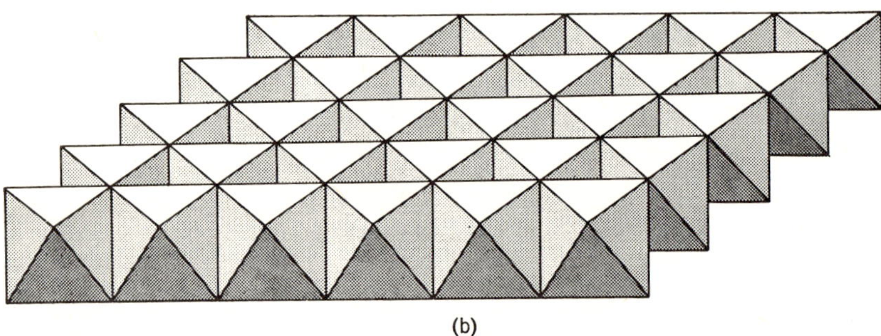

(b)

FIG. M.1.2. Two adjacent layers of close packed spheres, with the centers of the spheres in one layer situated vertically above one half of (the alternate) triangular holes in the layer below. (a) shows the resulting arrangement visualized as spheres and (b) the arrangement visualized by the polyhedra, in this case octahedra, defined by the centers of the spheres in the two adjacent layers.

of a sphere, which is a circle, with the area of the hexagon defined by the six lines tangential to any sphere and the six surrounding spheres, the efficiency with which the available space is utilized in such a layer is given by this formula: efficiency = (area of circle)/(area of hexagon) = $\pi R^2/2\sqrt{3}\,R^2 = \pi/2\sqrt{3} = 90.7\%$. There is no more efficient way of packing equal spheres in a plane array, since not more than six circles can be made to make contact with another circle of the same radius. Such a plane layer is referred to as a hexagonal close packed layer, and each sphere is surrounded by six other spheres and by six spaces and each space is surrounded by three spheres, so that the hexagonal unit cell base (page 581) in a close packed layer contains one sphere and two spaces.

A second similar hexagonal close packed layer may be superimposed on an identical one such that each sphere in either layer is in contact with three spheres in the adjacent layer, or in such a manner that the center of each sphere in either layer forms a regular tetrahedron with respect to the centers of the three nearest spheres in the adjacent close

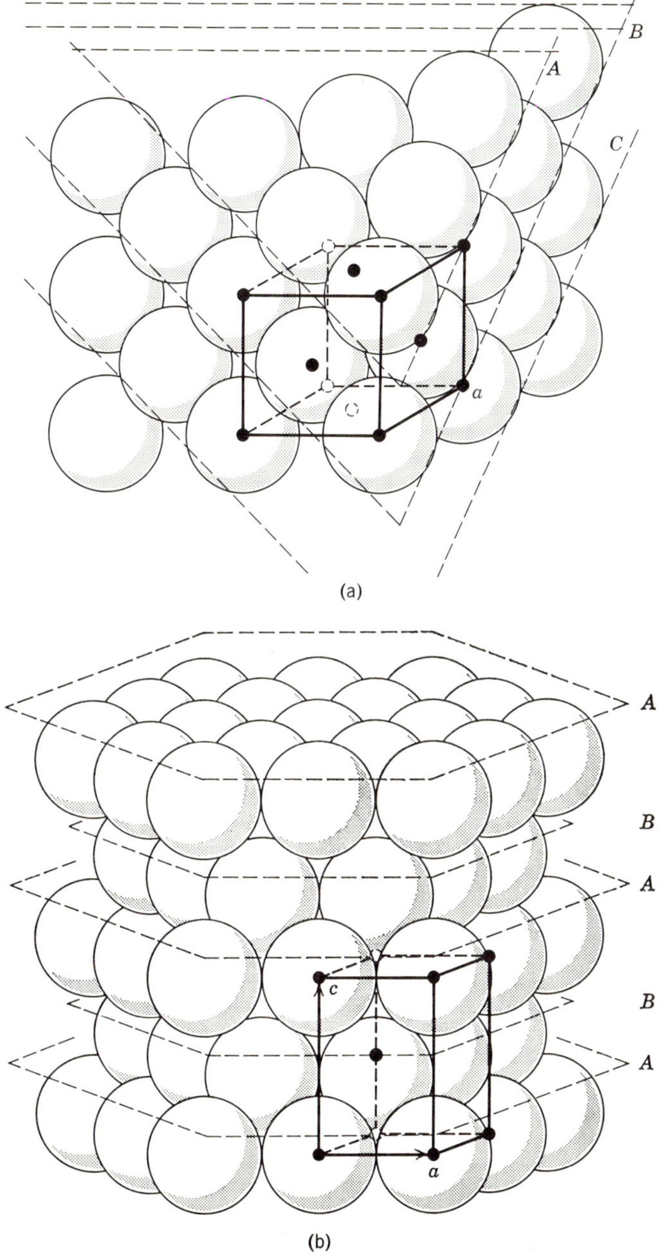

(a)

(b)

Fig. M.1.3. Cubic (a) and hexagonal (b) closest packing of equal spheres. The close packed layers are stacked parallel to the cubic unit cell body diagonal in former case, and parallel to the hexagonal unit cell c axis in latter.

389

(a)

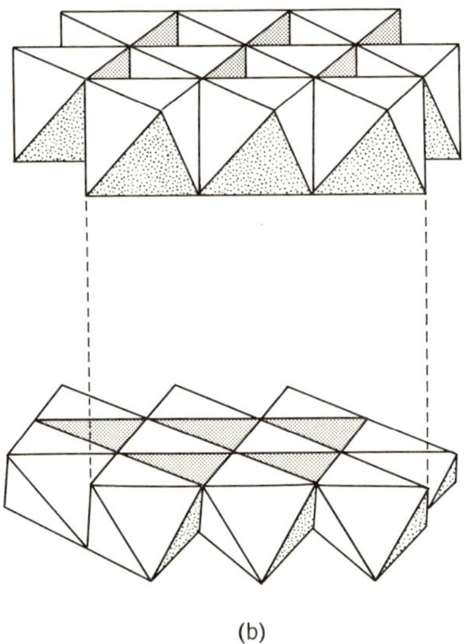

(b)

FIG. M.1.4. Illustration of (a) cubic closest packing and (b) hexagonal closest packing by means of octahedra.

packed layer, as illustrated in Fig. M.1.2. If the centers of the spheres in any one layer are labelled A, then it is immaterial whether the centers of the spheres in the adjacent layer are placed vertically above the set of positions labeled B or the set of positions labeled C in Fig. M.1.1., these two sets of positions being equivalent to one another. It is obviously impossible to place equal sized spheres in one layer vertically above each of the spaces, B and C, in the adjacent layer. Alternatively, as also illustrated in Fig. M.1.2., two such adjacent layers of hexagonal close packed spheres may be visualized as a single layer of close packed octahedra, each octahedral vertex corresponding to the center of a sphere in one or other of the two layers of close packed spheres.

There are, however, two ways of placing a third hexagonal close packed layer of spheres, above or below either of the two layers in Fig. M.1.2., such that each sphere in this third layer is also in contact with three spheres in an adjacent layer. These two ways are such that each sphere in this third layer is either vertically above the set of positions A in the first layer, when the second layer is placed above the positions B in the first layer, or such that each sphere in the third layer is vertically above the set of positions C in the first layer, when the second layer occupies the set of positions B with respect to the first layer. In the first case, the sequence of close packed layers corresponds to the stacking scheme with layers $\cdots ABAB\cdots$, referred to as hexagonal closest packing, and the second case corresponds to the stacking scheme $\cdots ABCABC\cdots$, referred to as cubic closest packing. These two arrangements are illustrated in Fig. M.1.3. Alternatively, the stacking together of hexagonal close packed layers to give rise to the hexagonal and cubic closest packed structures, respectively, may be illustrated, as in Fig. M.1.4., by means of layers of octahedra, the vertices of each octahedron again defining the centers of spheres in two adjacent close packed layers. In the hexagonal closest packed structure, the close packed layers are parallel to the six-fold hexagonal axis and in the cubic closest packed structure, the close packed layers are parallel to the three-fold axis of the unit cube of the structure or normal to the direction of the unit cube body diagonal.

In both cubic and hexagonal closest packing of equal spheres, each sphere is in contact with 12 other spheres, such that the volume occupied per sphere is $4\sqrt{2}\,R^3 = 5.66R^3$, or the efficiency of packing is $\sqrt{2}\,\pi/6 = 74.04\%$. Apart from the small spaces or holes in close packed layers, which are surrounded by three spheres at the vertices of an equilateral triangle, there are larger holes in both cubic and hexatonal closest packed arrangements of two kinds; the smaller of these holes, marked T in Fig. M.1.2., are surrounded by a tetrahedral group of spheres and the number of such holes is equal to twice the number of close packed spheres, while the larger holes which are marked O in Fig. M.1.2. are surrounded by an octahedral group of spheres, and the number of such holes is equal to the number of close packed spheres. These holes are of great importance in determining the composition and structure of many inorganic compounds. For example, many oxides and other complex oxy-compounds consist of oxide ions arranged in either cubic or hexagonal closest packing with the smaller metallic cations occupying one or other or both types of tetrahedral and octahedral holes in a systematic manner; many interstitial alloys contain smaller boron, nitrogen, carbon, or silicon, atoms distributed among the holes in closest packed arrays of metal atoms. In both inorganic compounds and interstitial compounds, the holes in closest packed arrangements of larger atoms are not generally all occupied by other species, but only those related by the operation of some symmetry factor.

In cubic closest packing, also referred to as face-centered cubic packing, the polyhedron which is defined by the centers of the twelve spheres surrounding any one sphere is illustrated in Fig. M.1.5.; it is a cubo-octahedron. This polyhedron is related to the rhombic dodecahedron, also shown in Fig. M.1.5., in the same manner in which the octahedron is related to the cube, or the pentagonal dodecahedron is related to the icosahedron (see Figs. M.5.26.–29.); they are the duals of one another and of much importance in the theory of metals. The rhombic dodecahedron shown in Fig. M.1.5. may be constructed by erecting planes normal to the lines joining the centers of any sphere in a cubic closest packed arrangement to each of the twelve nearest neighboring spheres. The polyhedra which result by similarly considering the twelve nearest neighboring spheres in the hexagonal closest packed structure are also illustrated in Fig. M.1.5. It is worth noting that from Fig. M.1.3., the axial ratio, c/a, for hexagonal closest packing is $c/a = 2\sqrt{2}/\sqrt{3} = 1.633$.

Apart from the stacking sequences of close packed layers of spheres, $\cdots ABABAB\cdots$ and $\cdots ABCABCABC\cdots$, corresponding to hexagonal and cubic closest packing, respectively, there are an infinite number of more complex sequences possible. To describe these more complex sequences, it is convenient to adopt a more compact method of designating the order of the close packed layers, suggested by Pauling.[1] In this method, a layer is designated by h if the two neighboring layers are of the same type, either both A, or both B, or both C, and by c if they are of different types. Thus, hexagonal closest packing is denoted simply by h and cubic closest packing by c. For the more complex sequences containing both c and h layers, since the spheres in those c and h layers have different arrangements of nearest neighboring spheres (refer to the coordination polyhedra of Fig. M.1.5.), then these more complex sequences must have two types of non-equivalent spheres differing in arrangement of nearest neighboring spheres. If there are only two kinds of non-equivalent spheres all of the spheres in both the c and the h layers must also have identical arrangements of more distant neighbors, so that the sequence of h and c layers on each side of an h (or c) layer must always be the same. To conform

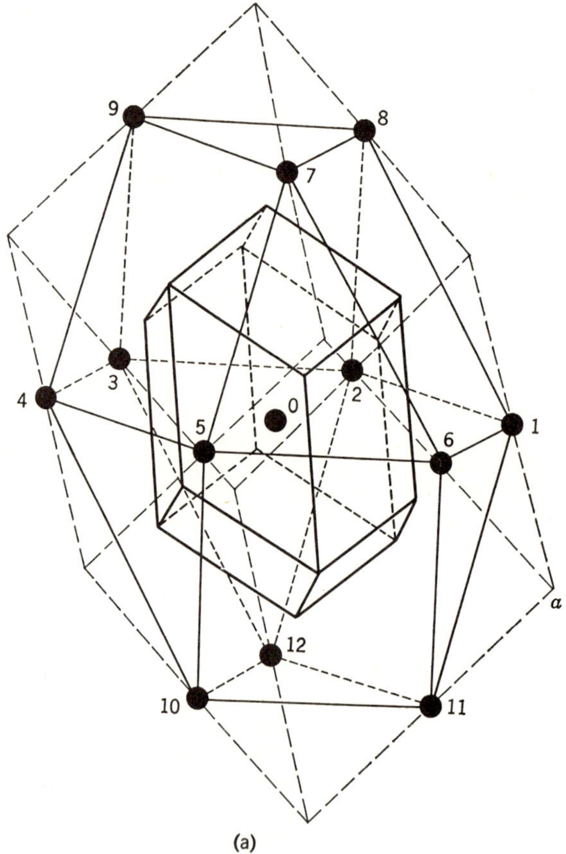

(a)

FIG. M.1.5. (*Continued on facing page*).

---

[1] Pauling, Linus, *The Nature of The Chemical Bond*, 1962, Cornell University Press, Ithaca, New York.

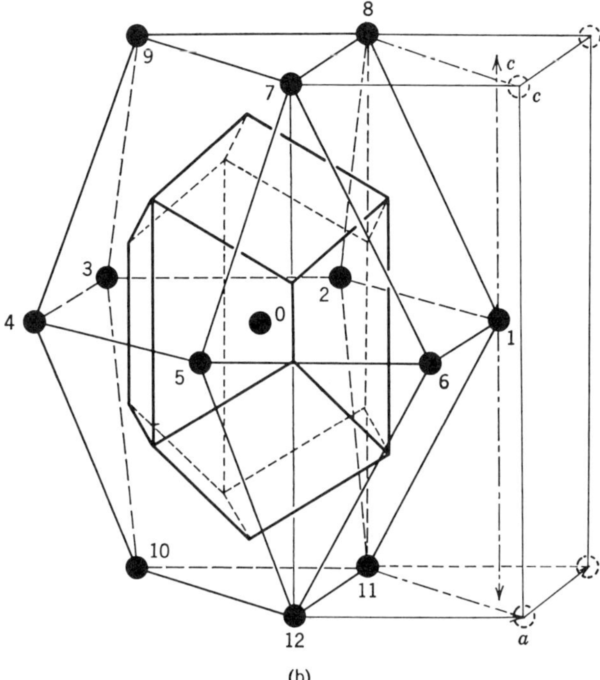

FIG. M.1.5. Coordination polyhedra defined by the 12 nearest neighboring spheres in (a) cubic closest packing and (b) hexagonal closest packing. In (a) this polyhedron is a cuboctahedron and in (b) a truncated hexagonal bipyramid. Also shown are the cubic and hexagonal unit cells of the respective structures and the polyhedra representing the domain of each sphere, which is a rhombic dodecahedron for cubic closest packing and an irregular dodecahedron for hexagonal closest packing. The polyhedra defining these domains are constructed by erecting planes normal to and bisecting the line between each sphere center and its 12 neighboring spheres.

with these requirements, there are only four sequences possible for two types of non-equivalent spheres, namely $\cdots cchh\cdots$ with 12 close packed layers in the repeat unit of the packing arrangement, $\cdots hcc\cdots$ with six layers in the repeat unit, $\cdots hc\cdots$ with four layers in the repeat unit, and $\cdots chh\cdots$ with nine layers in the repeat unit. Closest packing of identical atoms and molecules is confined to metals, the inert gases, and molecules like methane and hydrogen chloride, which become approximately spherical by molecular rotation or random orientation. Some of the lanthanide metals have structures with close packed layers of non-equivalent atoms, such as cerium[1] with the $\cdots hc\cdots$ arrangement and samarium[2] with the $\cdots chh\cdots$ arrangement.

[1] McHargue, C. J., Yakel, H. L., and Jetter, L. K., *Acta Cryst*, 1957, **10**, 832.
[2] Ellinger, F. H., and Zachariasen, W., *J. Am. Chem. Soc.*, 1953, **75**, 5650.

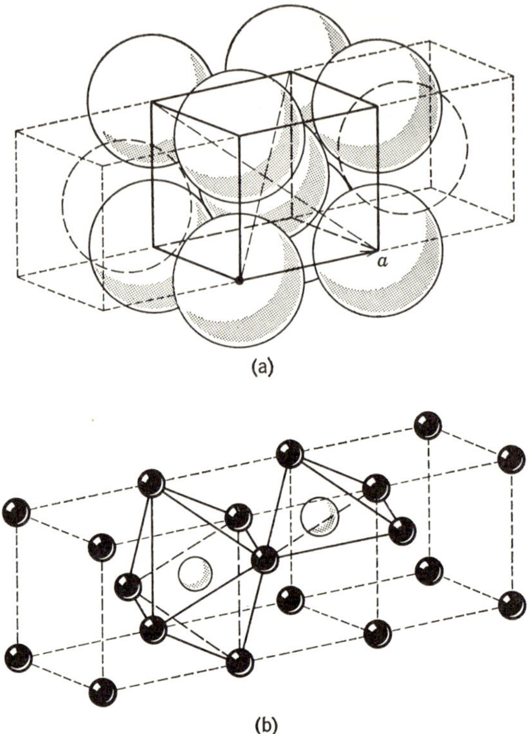

(a)

(b)

FIG. M.1.6. The body-centered cubic structural arrangement, (a), showing a sphere at body center of cubic unit cell surrounded by eight other spheres at vertices of unit cube. The octahedral (irregular) holes and tetrahedral (irregular) holes in this structure are shown in (b).

The most symmetrical manner of arranging equidimensional spherical objects in space so that each has eight equidistant neighbors is to locate them at the lattice sites of a body-centered cubic lattice. This gives rise to one of the commonest types of close packed structures which is not a closest packing. In addition to eight nearest neighbors, each sphere in this arrangement has an additional six neighbors which are only about 15% more distant, as illustrated in Fig. M.1.6. If this structure is therefore compressed along the direction of the four-fold axis of symmetry until the axial ratio c/a becomes 0.82, the coordination number of each sphere becomes 10 and if elongated until c/a becomes 1.41, it becomes identical with the cubic closest packed structure with coordination number 12 for each sphere. There are two important types of holes in the body-centered cubic structure; there are six spheres enclosing irregular octahedral holes at each of the cube faces, which may accommodate spheres of radius $0.154R$, and there are irregular tetrahedral holes located near the cube edges as in Fig. M.1.6., which may accommodate spheres of radius $0.291R$. The ratio of spheres to holes is 1:1 for the former type and 1:2 for the latter. While the tetrahedral holes in this structure are larger than the octahedral holes, they are nevertheless smaller than the regular octahedral holes in closest packed structures.

Of the various intermetallic compounds whose structures have been reported, many of these also adopt structural arrangements, which although apparently of considerable complexity, may be regarded as being derived from the simple cubic closest-packed arrangement. In such compounds containing two or more different kinds of metal atoms, the smaller atoms are found to be closest packed around the larger atoms, such that their centers are located at the vertices of some well-defined polyhedron. Prominent among such polyhedra are the tetrahedron, octahedron, icosahedron, hexagonal antiprism, snub cube, and the Friauf polyhedron,[1] shown in Fig. M.1.7.

(g) The metallic state ceases to exist in the vapor phase. The vapor derived from most metals is monatomic, although diatomic species compose a considerable fraction of the alkali metal vapors. Thus, mercury vapor is colorless, quite transparent, and composed of free atoms. The monatomic nature of the vapor derived from most metals is not attributable to any characteristic of the metallic state; other substances, such as the rare gases, also produce monatomic vapors. It is generally convenient to define the cohesive energy of a metal as the energy required to dissociate a given mass of the solid metal at the absolute zero of temperature, into free atoms. Owing to the relatively large masses of most metallic nuclei, their zero-point vibrational energies may generally be neglected in all except the most precise computational work in the theory of metals, so that the absolute zero of temperature may be regarded as the zero of energy.

(h) It is not generally so easily possible to find solvents which will dissolve metals, as is the case for other major classes of substances. Techniques other than recrystallization from solvents, which are so widely applicable to both organic and other inorganic compounds, have therefore to be used in the preparation of pure metallic phases. Although many metal pairs are mutually soluble, in the liquid phase, in all proportions, it frequently happens that when a solid phase separates from a melt composed of two or more metals, it is either a solid solution or an intermetallic phase of definite composition. Thus, the compound AlB_{12} crystallizes from a solution of boron in aluminum, and the compounds with composition KHg, KHg_2, KHg_3, KHg_5, and KHg_{13} may be crystallized from a solution of potassium in mercury.

(i) When metals react with one another, the heat of reaction is not generally as pronounced as it is in many other chemical reactions. Neither do the

---

[1] Samson, Sten, *"The Structure of Complex Intermetallic Compounds,"* 1968, *Structural Chemistry and Molecular Biology*, W. H. Freeman & Co., San Francisco.

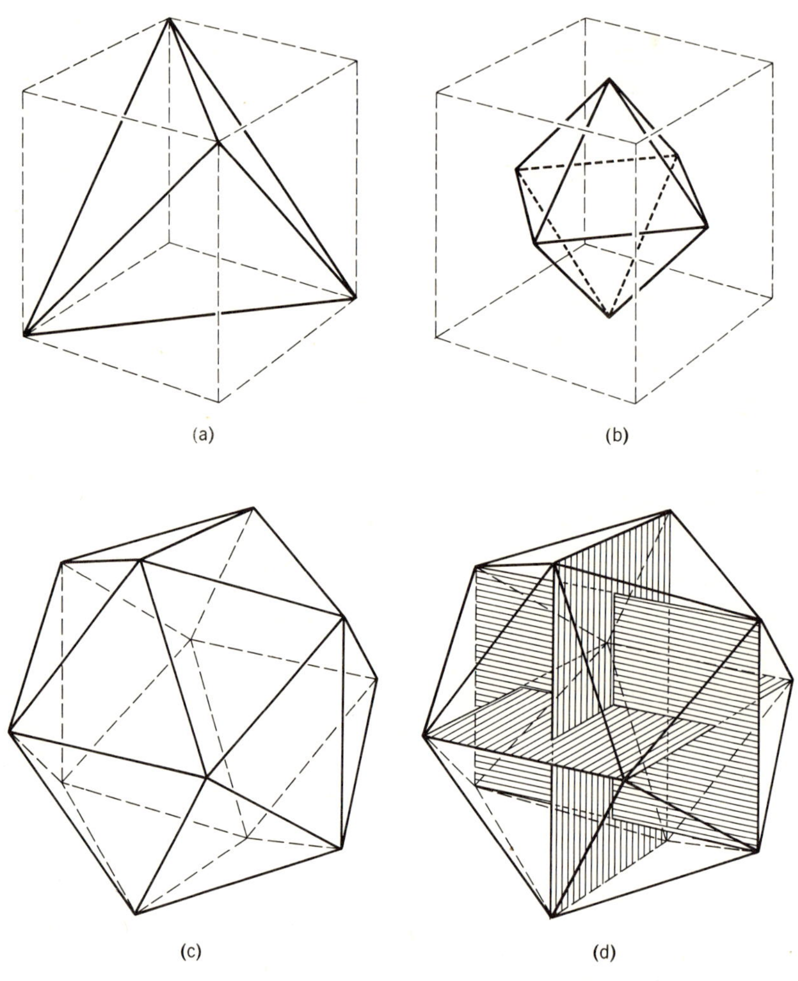

(a)

(b)

(c)

(d)

FIG. M.1.7. (*Continued on facing page*)

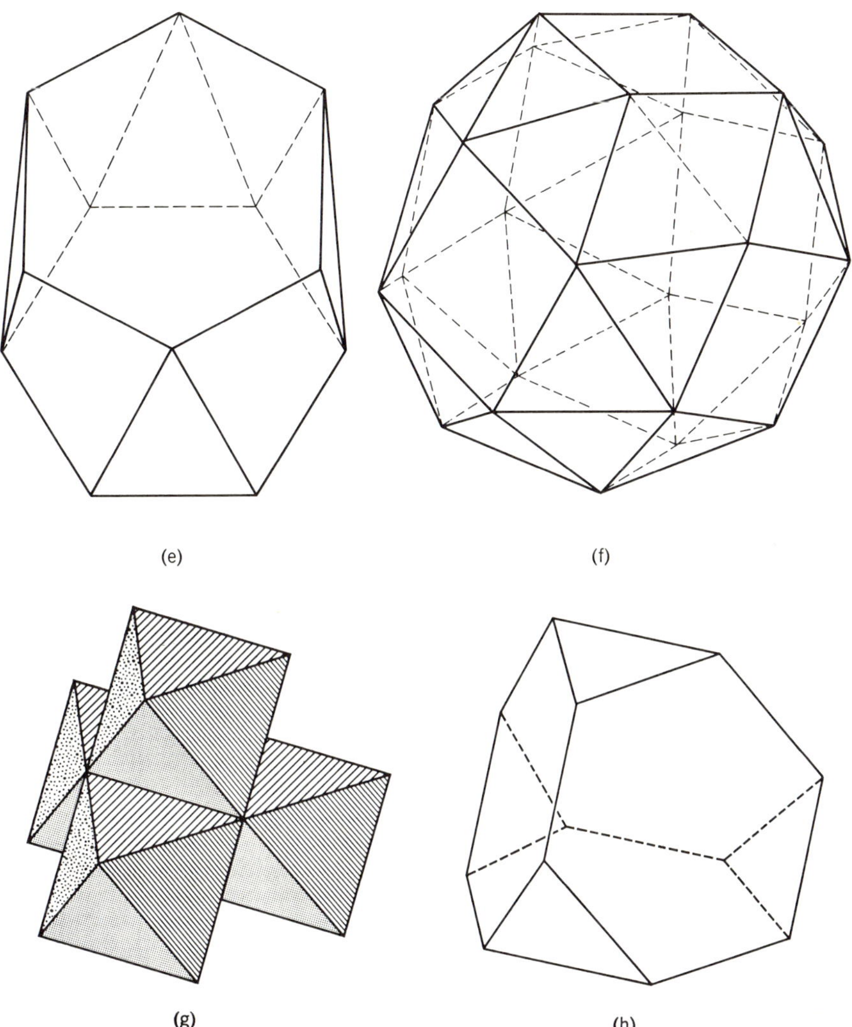

FIG. M.1.7. Some of the polyhedra which are recognizable in the structures of complex intermetallic compounds. These polyhedra are defined by the lines joining the centers of one type of metal atom which are closest to another type of metal atom. The regular polyhedron illustrated in (a) is a tetrahedron, which is a regular polyhedron with 4 equilateral triangular faces, 4 three-fold vertices, and 6 edges; its relationship to the circumscribed cube is also indicated. In (b) is shown an octahedron and its relationship to the corresponding circumscribed cube. This is also a regular polyhedron with 8 equilateral triangular faces, 6 four-fold vertices and 12 edges. The regular icosahedron shown in (c) has 20 equilateral triangular faces, 12 five-fold vertices, and 30 edges. The manner in

(*Caption continued on page 398*)

reaction products exhibit such marked differences in their properties from the reactants as is usual in other chemical phenomena. Discontinuous changes in properties, such as those observed in the formation of so many other chemical compounds, are quite generally much less obvious with intermetallic reactions. In the reaction between copper and zinc, for example, to form brass alloys, there is a continuous change in color from the red, copper-rich brasses containing about 70% copper to yellow brasses containing about 55% copper.

(j) Intermetallic compounds, where these are formed in mixed metal systems, differ most markedly from conventional organic and inorganic compounds in that they generally adopt non-stoichiometric formulas. For example, the following empirical formulas represent the composition of well-defined intermetallic compounds: $CeSn_3$, $PdCu_{13}$, Rh_5Zn_{21}, $NaCd_{11}$, $Mg_{32}(Al, Zn)_{49}$. It would appear that in metals and intermetallic compounds, there are always fewer electrons available in suitable energy states from the constituent metal atoms to form normal electron pair bonds between the various atoms involved; they are electron-deficient compounds in the same sense that the boranes are. Coordination numbers are usually high in both the pure metals and in intermetallic compounds.

(k) All of the alkali and alkaline earth metals, with the exception of beryllium, are soluble to some extent in liquid ammonia, amines, ethers, alcohols, molten salts, hydroxides, amides, and even in oxygen-free water, to form blue paramagnetic solutions. At higher concentrations, such solutions are yellow or bronze colored and the solutions have many of the properties

---

which this polyhedron is related to three mutually intersecting orthogonal planes, the ratio of whose edge lengths are 1.61803:1 (see page 741) is shown in (d). The polyhedron depicted in (e) is a semi-regular one, a hexagonal antiprism, with 14 faces altogether of two different types, 12 equilateral triangular ones and 2 regular hexagonal faces. The hexagonal antiprism has 12 four-fold vertices and 24 edges, conforming to the general description of all regular antiprisms that they possess $(2n + 2)$ faces, $2n$ vertices, and $4n$ edges. Another semi-regular polyhedron is shown in (f), the snub cube, with 38 faces altogether, 6 of which are square and 32 equilateral triangular-shaped, 24 five-fold vertices, and 60 edges. The semi-regular polyhedron depicted in (h), which is a regularly truncated tetrahedron, also variously referred to as a Laves or Friauf polyhedron, may be regarded as being derived from the group of 4 regular octahedra shown in (g), arranged in such a manner that each octahedron has 3 vertices in common with 3 adjacent octahedra and 3 non-common vertices, by joining the 12 non-common vertices. This polyhedron has 8 faces altogether, four of which are regular equilateral triangular-shaped and four regular hexagons, also 12 three-fold vertices, and 18 edges.

of simple metals in the liquid phase. From such solutions metallic solids such as $Ca(NH_3)_6$, have been isolated.

(l) In a magnetic field metals and intermetallic compounds display various types of interaction. The principal types of interaction exhibited are referred to as diamagnetism, paramagnetism, ferromagnetism, antiferromagnetism, and ferrimagnetism. Examination and discussion of the magnetic properties of metals has played an important part in the development of both the detailed electron theories and the Pauling theory of metals.

Very briefly, at this stage, most substances in a magnetic field develop a magnetic moment opposed to the field; such materials are said to be diamagnetic, and those which develop a moment parallel to the field are said to be paramagnetic. Ferromagnetic substances, on the other hand, assume large magnetic moments even in a weak field, which approaches a constant value as the field strength is increased, and many such substances retain their magnetization after the external magnetic field is removed, although above a certain temperature (the Curie point) they become paramagnetic. Antiferromagnetic substances are those which display a pronounced maximum in their paramagnetic behavior at a characteristic temperature (the Néel temperature). Ferrimagnetic substances behave in a manner which is qualitatively similar to that of ferromagnetic materials; however, whereas they display a Curie transition temperature above which they are paramagnetic and below which they are ferromagnetic, the total magnetic moment in the paramagnetic region is very much greater than that corresponding to the saturation moment in the ferromagnetic region. Thus magnetite, Fe_3O_4, the first ferromagnetic substance to be discovered, is in fact ferrimagnetic. Apart from iron and nickel, the two best known ferromagnetic substances, the metals gadolinium and dysprosium, are ferromagnetic, as are uranium hydride (UH_3) and several compounds of manganese ($MnBi$, $MnAs$, $MnSb$, Mn_4N, MnB) and chromium ($CrTe$, CrO_2), even although neither manganese nor chromium are themselves ferromagnetic; the so-called Heussler alloys, of composition Cu_2MnAl and the compound $CoFe_2O_4$, are also ferromagnetic.

(m) Thermodynamic data indicate that nearly all of the metallic elements should be converted spontaneously into their oxides on exposure to the air, since with the exception of gold and platinum, all other metals form oxides with negative free energies of formation from the elements. The rate of oxidation is so slow, however, as for certain metals such as chromium, that this property does not detract from its great commercial value. Some metals form an oxide which occupies a larger volume than the metal from which it is formed, so that while still readily oxidizable, a protective surface layer of oxide soon covers the metal on exposure to an oxidizing atmosphere; this protection of metals by such a covering oxide film or coat is deliberately brought about in the process of anodizing, as of aluminum or titanium.

All of the above, general and specific, properties of metals and intermetallic compounds must, of course, be explicable in terms of any comprehensive theory of metals, and metallic bonding. The discussion of these properties, generally, in terms of the various electron theories, constitutes the subject matter of the present part.

M.1.2.   Intermetallic Phases

Some 84 of the known 103 elements have essentially metallic characteristics, or about 81.5% of them. It might therefore be expected that the chemistry of the metals should constitute the major part of chemistry as a whole, since of all possible binary systems, even if only one binary compound were formed in the interaction of all possible pairs, about 72% of all such compounds would be intermetallic ones. However, while a compilation of binary intermetallic compounds made by Hansen[1] amounts to about 1100 pages, this tabulation obviously comprises only a very small fraction of all possible binary compounds. Some of the reasons for the lack of knowledge about intermetallic compounds are perhaps the fact that there is so little apparent physical distinction between different metallic compounds, the range of composition shown by many such compounds, the deficiency of techniques for the isolation and identification of them, and the general paucity of information about the structure of these substances. The term berthollides was introduced in 1914 to describe solid phases showing a range of composition, in contradistinction to those compounds with definite composition, referred to as daltonides.

The metallic elements may be conveniently classified into four groups:

(1)   The pretransition metals, composing the alkali metals (Li, Na, K, Rb, Cs, Fr), the alkaline earth metals (Be, Mg, Ca, Sr, Ba, Ra) and Al; the chemical analogues of Al, namely Sc, Y, La, Ac, are preferably regarded as members of the other groups below.

(2)   The transition metals, composed of three groups of ten elements each, distinguished as the first transition group (Sc, Ti, V, Cr, Mn, Fe, Co, Ni, Cu, Zn), the second transition group (Y, Zr, Nb, Mo, Tc, Ru, Rh, Pd, Ag, Cd), and the third transition group (La, Hf, Ta, W, Re, Os, Ir, Pt, Au, Hg).

---

[1] Hansen, M., *Der Aufbau der Zweistoffegierungen*, 1936, Berlin.

(3) The lanthanides and actinides, each composed of 14 elements; the lanthanides are: Ce, Pr, Nd, Pm, Sm, Eu, Gd, Tb, Dy, Ho, Er, Tm, Yb, and Lu, and the actinides are: Th, Pa, U, Np, Pu, Am, Cm, Bk, Cf, Es, Fm, Md, No(?), and Lw.

(4) The post-transition metals, composed of the four groups of three elements each; these are Ga, In, Tl; Ge, Sn, Pb; As, Sb, Bi; and Se, Te, Po. Metallic characteristics are not so pronounced among these elements.

The alkali and alkaline earth metals are characterized by the fact that their neutral atoms contain one and two electrons, respectively, in the highest occupied s-states; aluminum atoms in their ground state contain an additional p-electron. The transition metals are composed of atoms, which in their ground states, contain between one and ten $3d$ (first), $4d$ (second), and $5d$ (third) electrons, respectively. Atoms of the various lanthanides and actinides contain in their lowest energy states between one and fourteen $4f$ and $5f$ electrons, respectively. The post-transition elements are composed of atoms containing between one and four p-electrons beyond the number of electrons contained in the ground states of the end member of each of the three respective transition group metal atoms.

Alloys are mixtures of metallic elements which are homogeneous in the liquid state. The term alloy is usually applied to the material which solidifies from the homogeneous melt. In the liquid state, alloys are essentially solutions of metals in one another, although liquid compounds may also be present. Alloys containing mercury are generally referred to as amalgams. Solid alloys vary very widely in composition and types of behavior. They may be homogeneous in the sense that they are composed of a conglomerate of small crystals all of the same kind, as in a pure metal, or they may be quite inhomogeneous, and in either case they may be composed of mixed crystals or intermetallic compounds or of both combined. The metals composing an alloy may become unmixed during the process of solidification resulting in a solid mixture of the component metals. Alternatively, the constituent metals of the alloy may become only partially unmixed on cooling, or not at all, or they may combine partially or completely during cooling of the melt to give rise to compounds which are incapable of existence at the temperature of the melt. These compounds may also form mixed crystals. In all cases, on cooling a liquid metal or mixture of metals, the process of crystallization takes place. In the crystallization of a liquid metal or alloy, however, where complete solidification takes place rather than the growth of individual crystals which occurs on crystallization from solution, the surfaces of the individual crystals are deformed by being constrained by neighboring crystals, so that their forms are not generally identical. Since crystalline materials are anisotropic (different properties in different directions), solid metals or alloys generally appear, on suitable etching and polishing of their surfaces, as conglomerates of crystallites, the surfaces of which are separated from one another by fine lines. Impurities in a metal or intermetallic compound usually appear at such crystallite boundaries.

Alloys may be either of the substitutional or interstitial types. In the former case, dissimilar metal atoms occupy essentially equivalent crystal lattice sites in the solid, while in the latter case the smaller atoms are found to occupy specific cavities or interatomic spaces in the crystal structure adopted by the larger metal atoms.‡ Interstitial compounds are very common among metal carbides, borides, nitrides, and hydrides. The formation of substitutional alloys is determined by essentially the relative sizes of the metallic atoms concerned, their electronegativity difference, and by the electronic structure of the metal atoms, which in turn affects the electron concentration per atom. It has been found that in the formation of alloys of copper, silver, and gold, in particular with other metals, the phase composition limits occur at equal electron concentrations per atom. This generalization was first made by Hume-Rothery[1,2] (Hume-Rothery rules). Examples of intermetallic phases with three different electron to atom ratios are given in Table M.1.1., listed according to the structural type adopted by the phases of these electron concentrations. It has to be noted that the appropriate electron to atom ratio for these Hume-Rothery phases is only applicable if in the case of the compounds containing iron, cobalt, nickel, and the analogues of these elements, ruthenium, rhodium, palladium, and osmium, iridium, and platinum, it is considered that each of these elements do not contribute any electrons towards the formation of the intermetallic phase containing them.

‡ Substitutional solid solutions are preferentially formed between metal pairs which do not differ in metallic radii from one another by more than about 15%, adopt the same crystal structure, are of similar electronic structure, and of similar electro-negativity. For example, as illustrated in Fig. M.1.8., for the copper-gold system, both copper and gold crystallize with face-centered cubic structures and have radii that differ by less than 15% and so form a continuous series of solid solutions. Fig. M.1.8.(a) represents the cubic unit cell of a random solid solution where either copper of gold atoms may be distributed at random over the lattice sites of the face-centered cubic structure. However, at temperatures below 733 K, solid solutions with the compositions containing 50% and 75% of copper, respectively, are more stable if the copper atoms preferentially occupy two-thirds [Fig. M.1.8.(b)] and all [Fig. M.1.8.(c)] of the face centered sites in the cubic unit cell of the structure, respectively, and the gold atoms the remaining positions. This type of behavior is referred to as superlattice or superstructure formation.

Also illustrated in Fig. M.1.8. are a metal with a face-centered cubic unit cell [(d)], an interstitial alloy formed by some small atom occupying one-half of the tetrahedral holes

[1] Hume-Rothery, W., *Electrons, Atoms, Metals, and Alloys*, 1955, Iliffe, London.
[2] Hume-Rothery, W., *J. Inst. Metals*, 1926, **35**, 295.

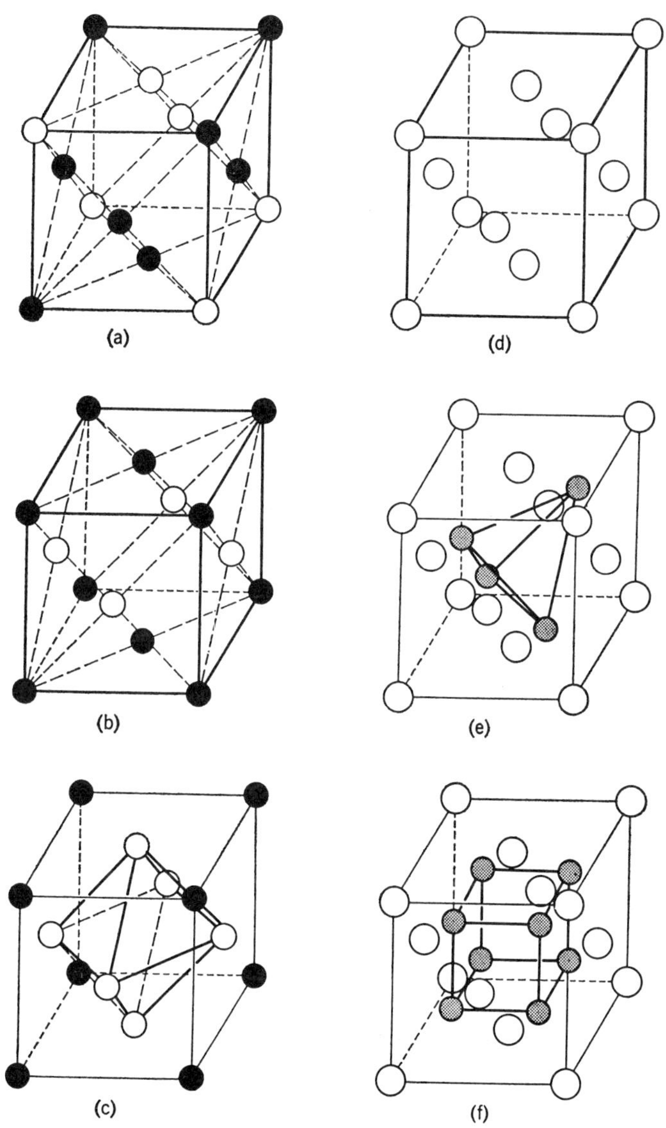

FIG. M.1.8. Substitutional and interstitial alloys.

in this structure [(e)], and an interstitial alloy in which the small atoms occupy all of the tetrahedral holes in the cubic closest packed structure [(f)]. The structure described by (e) was first discovered for the mineral zinc blende, and is found, for example, in ZrH and CrH, and other interstitial hydrides. The structure described by (f) was first discovered for the mineral fluorite, CaF_2, and is also known for interstitual hydrides such as TiH_2.

TABLE M.1.1
Hume-Rothery electron compounds

Electron:atom ratio = 3:2			Electron:atom ratio = 21:13	Electron:atom ratio = 7:4
Body-centered cubic structure[a]	Complex cubic (β-manganese) structure[b]	Closed-packed hexagonal structure[c]	γ-brass structure[d]	Close-packed hexagonal structure[e]
$CuBe$	Cu_5Si	Cu_5Ge	Cu_5Zn_8	$CuZn_3$
$CuZn$	$AgHg$	Cu_3Ga	Cu_5Cd_8	$CuCd_3$
Cu_3Al	Ag_3Al	$AgZn$	Cu_5Hg_8	Cu_3Sn
Cu_3Ga	Au_3Al	$AgCd$	Cu_9Al_4	Cu_3Ge
Cu_3In	$CoZn_3$	Ag_3Al	Cu_9Ga_4	Cu_3Si
Cu_5Si		Ag_3Ga	Cu_9In_4	$AgZn_3$
Cu_5Sn		Ag_3In	$Cu_{31}Si_8$	$AgCd_3$
$AgMg$		Ag_5Sn	$Cu_{31}Sn_8$	Ag_3Sn
$AgZn$		Ag_7Sb	Ag_5Zn_8	Ag_5Al_3
$AgCd$		Au_3In	Ag_5Cd_8	$AuZn_3$
Ag_3Al		Au_5Sn	Ag_5Hg_8	$AuCd_3$
Ag_3In			Ag_9In_4	Au_3Sn
$AuMg$			Au_5Zn_8	Au_5Al_3
$AuZn$			Au_5Cd_8	
$AuCd$			Au_9In_4	
$FeAl$			Mn_5Zn_{21}	
$CoAl$			Fe_5Zn_{21}	
$NiAl$			Co_5Zn_{21}	
$NiIn$			Ni_5Be_{21}	
$PdIn$			Ni_5Zn_{21}	
			$Na_{31}Pb_8$	

[a] In the body-centered cubic structure, as shown in Fig. M.1.6., there are two lattice points per cubic unit cell. If the material situated at the lattice points is composed of atoms, all of which are the same as in the alkali metals, then there are two atoms per unit cell. If atoms of two different kinds are distributed at random over the body-centered cubic lattice points, then compositions such as Cu_3Al and Cu_5Sn may arise. If there are equal numbers of two different kinds of atoms present, then the structure is equivalent to that found for cesium chloride, as in $CuZn$.

[b] The β-manganese structure is a structural arrangement first discovered for one of the polymorphic modifications of manganese, the one stable between about 1073 K and 1373 K. In this complex cubic structure, there are twenty atoms per cubic unit cell, with each atom icosahedrally surrounded by 12 others, at distances varying from $2.36 - 2.67 \times 10^{-10}$ m.

[c] In the hexagonal closest packed structure, as shown in Fig. M.1.3., there are two lattice points per unit cell, such that each lattice point is equidistant from 12 others situated at the vertices of a cubo-octahedron. With random distribution of two different kinds of atom over the lattice sites of the hexagonal closest packed structure compositions such as $AgCd$, Ag_3In, Ag_5Sn, and Au_5Al_3, may result.

The solubility of one metal in another is generally small if the difference in metallic radii is greater than about 15%. The greater the difference in electronegativity between two metals, the greater is their tendency to form a compound rather than simply an alloy, so that in general similar metals dissolve in one another and dissimilar ones form compounds. Thus, among the alkali metals the formation of alloys between them is limited by differences in metallic radii, so that immiscible liquids are formed in the Li–Na, Li–K, Li–Rb, and Li–Cs systems, while continuous solid solutions are formed in the K–Rb, K–Cs, and Rb–Cs systems, where the percentage increase in the atomic radii of Li to Na is 22%, from Na to K it is 24%, from K to Rb it is 6%, from K to Cs it is 14%, and from Rb to Cs it is 7%. The lanthanides form alloys with one another, but the actinides, at least those from thorium to americium, are not particularly soluble in one another. One of the more interesting lanthanide alloys, which is pyrophoric, and finds considerable use in the manufacture of cigarette lighter flints and for the purification of iron since it forms compounds with carbon, oxygen, phosphorus, nitrogen, and sulfur, is "mischmetall" containing 50% cerium, 25% lanthanum, and 25% of neodynium and promethium.

Many binary metal pairs which are completely miscible in the liquid state, but completely immiscible in the solid state, and incapable of forming any chemical compound with one another, give rise to the formation of alloys, which are referred to as eutectics. This type of behavior is characterized by the manner in which each metal of the pair depresses the freezing point or melting point of the other, to an increasing extent with increasing concentration of one metal dissolved in the other. For one particular composition, the melting point of the mixture of the two metals reaches a minimum, and for this composition, the eutectic alloy, the melting point corresponds to the eutectic temperature. Thus, for all possible lead–tin alloys, that containing 26.1 atomic percent of lead and 73.9 atomic percent of tin, melts at 456.5 K. It is characteristic of eutectics that they solidify from the melt as a whole, without any change of composition taking place, to a mixture of crystallites of their components. From any melt richer in one component, that

---

[a] The γ-brass structure is another complex cubic structure first discovered for Cu_5Zn_8. It may be regarded as a defect body-centered cubic arrangement. The cubic unit cell of the body-centered cubic structure contains two atoms per unit cell, and the unit cell of the γ-brass structure, containing 52 atoms per unit cell, may be derived from the simple body-centered unit cell by imagining a cell whose edges are three times as large as this, thus containing 27 simple body-centered cubic unit cells and therefore 54 atoms altogether. If now the central atom of this large cell is imagined to be removed along with each of the eight vertical atoms, thereby removing effectively two atoms from the large cell leaving 52, then after a slight displacement of these remaining 52 atoms the γ-brass unit cell results.

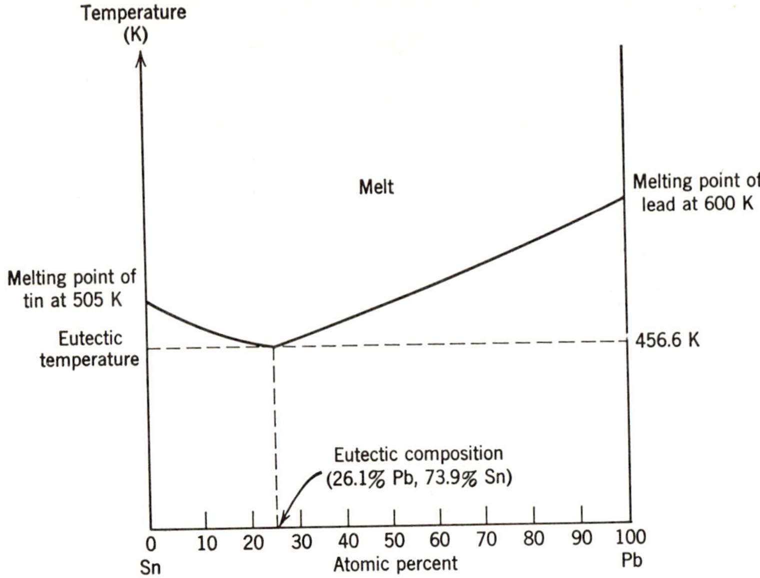

FIG. M.1.9. The formation of a eutectic in the lead-tin system.

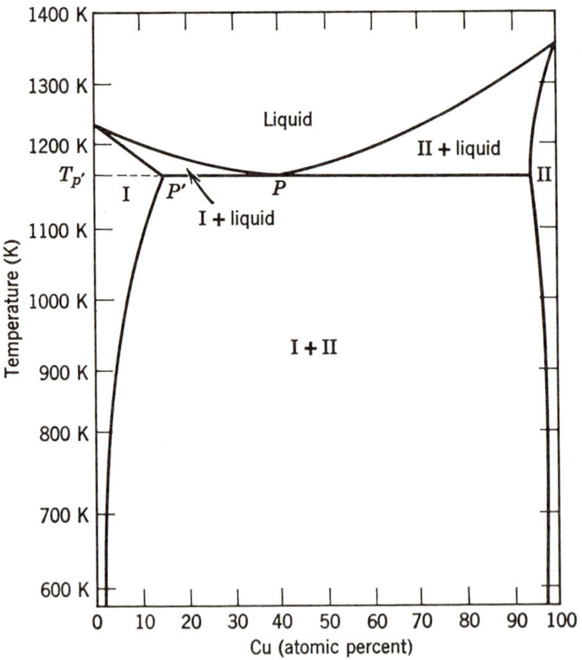

FIG. M.1.10. The phase diagram for the system, copper-silver. There are regions of limited solubility at either end of the diagram, and the eutectic point P is where the three phases, I, II, and the liquid are in equilibrium.

component separates out until, in either case, the composition has changed to that of the eutectic.‡

‡ For example, the melting point of tin (505 K) is lowered by the addition of lead and the melting point of lead (600 K) is lowered by the addition of tin, and the manner in which the melting point of the various possible tin-lead alloys vary with composition is shown in Fig. M.1.9. The two melting point curves for the tin-rich and for the lead-rich compositions intersect at a point corresponding to the composition 73.9 atom percent of tin and 26.1 atomic percent of lead, and the alloy with this composition, the eutectic alloy, melts at 456.6 K, the eutectic temperature for this system.

A slightly more complicated eutectic system is formed by copper and silver, illustrated in Fig. M.1.10, Here there are regions of miscibility at either extreme of the phase diagram. The eutectic point is where the three phases I, II, and the liquid are in equilibrium.

Many metal pairs, especially those differing considerably in their electrochemical characteristics, form stable compounds, many of which have empirical formulas in conformity with the older concepts of valency; for example, compounds such as Mg_3Sb_2 and Mg_2Sn. For the great majority of intermetallic compounds, however, the empirical formulas of these compounds do not agree with these simple ideas of valency. Some intermetallic phases have compositions and structures which are essentially determined by the relative sizes of the atoms involved. For example, the various compounds $MgCu_2$, $NaAu_2$, KBi_2, ZrW_2, $AgBe_2$, $MgZnNi$, $CaAl_2$, and $BiAu_2$ all crystallize with the same structure, and here there is an approximately constant ratio of atomic radii between the various pairs of metal atoms; it is worth noting that the bismuth atom behaves as the smaller atom in KBi_2 and the larger atom in $BiAu_2$. In fact, intermetallic compounds with the general formula AB_2 very frequently crystallize with structures in which each atom of type A is equidistant or nearly so from four other A-type atoms, and each B-type atom from six other atoms of type B, so that each A-type atom is equidistant or nearly so from 12 B-type atoms and each B-type atom from six A-type atoms. Crystalline substances conforming to this structural plan are generally referred to as Laves[1] phases, since they have been principally investigated by Laves in the period from 1934 onwards. In a similar manner the ternary compound $Mg_3Cu_7Al_{10}$ is isostructural with the binary compound Mg_2Zn_{11}; here again the characteristic is that there is an approximately constant ratio of the larger magnesium atoms to the smaller atoms of zinc, aluminum, or copper, since in Mg_2Zn_{11} there are 33 small zinc atoms for each six large magnesium atoms, and in $Mg_3Cu_7Al_{10}$ there are 34 of the smaller copper and aluminum atoms for each six of the larger magnesium atoms. Several groups of intermetallic compounds have been classified in this way, as the result of X-ray crystallographic examination. Three types of Laves phases are actually known, represented by the compounds $MgZn_2$, $MgCu_2$,

---

[1] Laves, F., *Theory of Alloy Phases*, 1956, American Society for Metals, Cleveland, Ohio.

and $MgNi_2$, respectively, the structures of which are derived from hexagonal closest packing, cubic closest packing, and a combination of these two.‡

‡ The Laves phases occur in alloy systems in which the radius ratio of the two component metals is between 1.08 and 1.32. The structure of the three types of Laves phases, of which several hundred examples of intermetallic compounds are known, may be interpreted in several ways, but it is perhaps most easily visualized in terms of the close packed layers of spherical atoms. Two adjacent close packed layers may be represented, as in Fig. M.1.2. by layers of octahedra, enclosing tetrahedral cavities, the vertex of each octahedron corresponding to a sphere center. The group of four octahedra shown in Fig. M.1.12.(a), by comparison with Fig. M.1.4.(a) and Fig. M.1.7.(g), obviously is another way of representing a group of 16 spheres in cubic closest packing, such that there is an occluded tetrahedron; this group of four octahedra may be combined to define the truncated tetrahedron, illustrated in Fig. M.1.7.(h). If the central sphere of each of the four hexagonal faces of this truncated tetrahedron is removed, and this operation is equivalent to removing the four spheres at the vertices of the occluded tetrahedron, then a much larger sphere may be accommodated at the center of the truncated tetrahedron; the radius of this central sphere may be up to 1.35 times that of the surrounding close packed spheres. Similarly, a large sphere (with radius 35% greater than the surrounding spheres) may be located out from the center of each of the hexagonal faces of the truncated tetrahedron, so that the central, large, sphere is surrounded altogether by 16 spheres, 12 small ones and four large ones. It is this group of 16 spheres surrounding a large central one, whose centers on being joined by lines, which is more properly said to define the Friauf polyhedron, although the truncated tetrahedron itself shown in Fig. M.1.7.(h) is also so referred to, as well as being called a Laves polyhedra.

Since it is possible, by removing certain spheres from any two layers in a closest packed arrangement, to transform the arrangement into double layers of Laves poly-

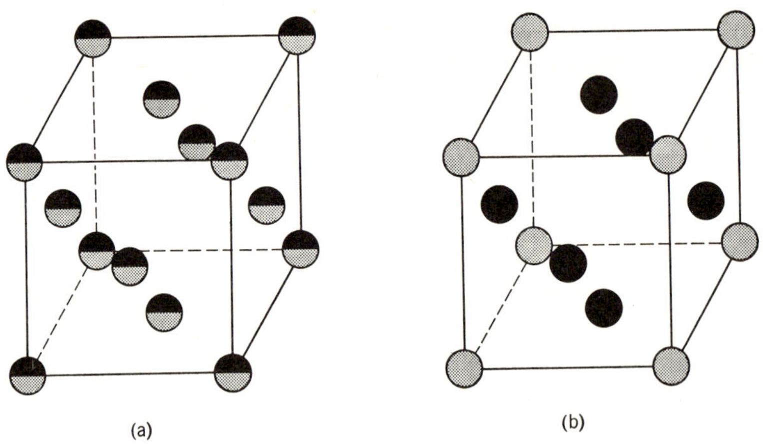

(a)   (b)

FIG. M.1.11. Superstructure formation in the copper-gold system. Above 872 K, the copper and gold atoms are statistically distributed over the lattice sites of a face-centered cubic structure, as in (a). Below 872 K, ordering sets in so that, for example, for the system with composition 75 atomic percent copper, the structure (b) results, corresponding to Cu_3Au.

hedra, then these double layers in turn may be stacked in various ways. Thus, Fig. M.1.12.(*b*) shows a layer of Laves or Friauf polyhedra arranged in such a way as to fill a plane, with each polyhedron having three of its four hexagonal faces in common with three other Laves polyhedra. When packed together in this way, the large sphere which may be located out from the center of the hexagonal faces of the Laves polyhedra is common to two adjacent polyhedra, so that it is immaterial whether one refers to the truncated tetrahedron of Fig. M.1.12.(*a*) as a Laves or Friauf polyhedron. In the close packed, three dimensional arrangement of Laves or Friauf polyhedra, each of the twelve vertices defining the position of the smaller spheres are shared with six other polyhedra, so as to reduce the average number of small spheres per polyhedron to two and the number of large spheres (located at the center and out from the center of each of the hexagonal faces of the polyhedra) to one. The composition which results in this way is thus AB_2, where A designates the large atom and B the smaller atom.

The three Laves structures simply correspond to the three closest packings of Laves polyhedra, differing in the way in which layers of polyhedra are stacked together. In Fig. M.1.12.(*c*), each of the Laves polyhedra in successive layers is rotated 60° with respect to those in adjacent layers. This represents the structure of $MgCu_2$, KBi_2, ZrW_2, $MgZnNi$, and $BiAu_2$, for example. In Fig. M.1.12.(*d*), adjacent layers of Laves polyhedra are stacked together so that one layer is the mirror image of the other. This represents the structure of $MgZn_2$, $TiMn_2$, VBe_2, WFe_2, and $MgAlCu$, for example. In Fig. M.1.12.(*e*), the first and second, as well as the third and fourth, layers are mirror images of one another, but not the second and third; this arrangement of the Laves polyhedra is a combination of the $MgCu_2$ and the $MgZn_2$ structures, and is representative of the $MgNi_2$, $TiCo_2$, and $Mg(Zn, Cu)_2$, structures, for example. The $MgCu_2$ structure is cubic with 24 atoms per unit cell, the $MgZn_2$ structure is hexagonal with 12 atoms per unit cell, and the $MgNi_2$ structure is also hexagonal with 24 atoms per unit cell.

In Fig. M.1.12., where the large (Mg) atoms are imagined located at the centers of the hexagonal faces of the Laves polyhedra, the other vertices define the positions of the centers of the small atoms (Cu, Zn, etc.). The coordination polyhedron around each of these small atoms is an icosahedron [Fig. M.1.7.(*c*)], and in fact each of the three Laves structures may be described in terms of an interpenetrating set of icosahedra and Laves polyhedra, there being twice as many icosahedra as Laves polyhedra to be distinguished. Each icosahedron has six large spheres and six smaller spheres at its vertices, as well as a small sphere at its center.

For metal pairs of large mutual solubility, there is a great tendency for compositions in the range of 25 and 50 atomic percent of the solute species, to undergo an atomic rearrangement on slow cooling or on annealing at a low temperature. This rearrangement involves the transfer of one type of atom from the random positions which they occupy in the substitutional solid solution of the solute in the solvent species, to definite positions in the crystal lattice of the solvent metal, so that the solute atoms are as distant from one another as possible. Such a phenomenon is referred to as superstructure formation. For example, as illustrated in Fig. M.1.11. for the copper–gold system, the alloy with composition $AuCu_3$ at high temperatures, consists of gold atoms randomly distributed over the lattice sites in the face-centered cubic structure of copper, but on slow cooling the gold atoms come to preferentially occupy the sites at the vertices of the unit cell, with the copper atoms

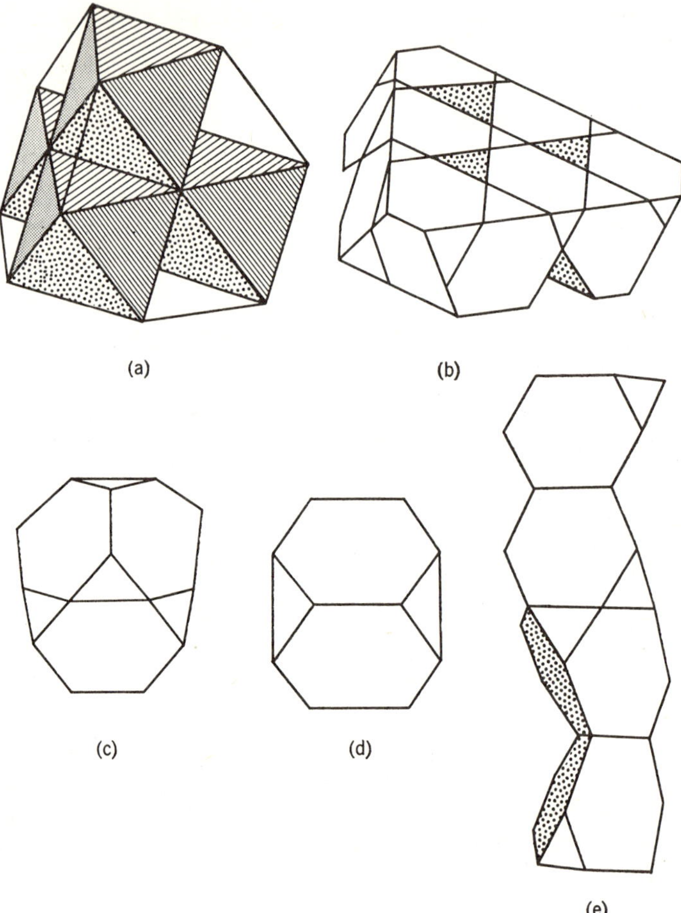

FIG. M.1.12. The structure of the three Laves phases, MgCu$_2$, MgZn$_2$, and MgNi$_2$ described in terms of the packing of truncated tetrahedra. (a) represents the Laves polyhedron, (b) Laves polyhedra close packed in a plane, and (c), (d), and (e), the three different Laves structures.

occupying only the face centers. Superstructure or superlattice formation is accompanied by a sudden and relatively large increase in electrical conductivity, by a small increase in density, and by an increase in ductility.

The ordered or disordered distribution of two or more different kinds of atoms among the lattice sites of cubic or hexagonal closest-packed arrangements of atoms determines the structure of a great many intermetallic compounds. In many of the structures which have been determined for intermetallic compounds, the importance of geometrical factors alone is becoming more evident, especially the manner in which these factors lead to high coordination numbers ranging from 12 to 16. If attention is focused on the polyhedral shapes which are defined by joining the centers of the metal atoms of one kind which are equidistant from the center of other metal atoms of the same or different kind, in the structures of many complex intermetallic compounds, it is possible to understand how these high coordination numbers arise.[1-3] Prominent among the coordination polyhedra which may be distinguished in such structures are cubo-octahedra, icosahedra, Friauf polyhedra, 14-, 15-, and 16-hedra.[4,5] The common occurrence of polyhedra with only equilateral triangular faces ‡ is attributable to the fact that the smallest kind of space between spherical atoms (idealized) when packed together to occupy the minimum space is a tetrahedral cavity, so that in close-packed arrangements coordination polyhedra with those triangular faces predominate. Indeed, the icosahedron, together with the 14-, 15-, and 16-hedra, are the only possible polyhedra bounded only by triangular faces (with five or six faces meeting in a common vertex), and crystal structures containing only the coordination numbers 12, 14, 15, and 16 for their constituent atoms contain only tetrahedral cavities.[2,4]

‡ The smallest holes which occur in the closest packing of spheres are tetrahedral ones, and closest packing results when the number of such holes is the maximum possible. However, in closest packing of equal spheres there are half as many octahedral holes as tetrahedral holes. A localized density of packing greater than ordinary closest packing results if twenty spheres are located out from the centers of the twenty faces of an icosahedron [Fig. M.1.7.(c)] to surround another sphere at the center of this polyhedron. But this type of packing cannot be extended to fill all space, without other holes appearing which reduce the overall packing density to below that of normal closest packing. On the other hand, if there are other spheres between the icosahedral groups, which are larger in radius than the spheres composing the icosahedral groups, a very efficient form of packing arises in which each of the larger spheres is surrounded by 14, 15, or 16, smaller spheres. Consistent with tetrahedral holes being the smallest for close packing, it follows that the cavities to be expected in closest packings would be bounded by triangular faces, and likewise for the coordination polyhedra of the closest packed species. The most nearly equilateral triangular faces on a polyhedron arise when either five or six such triangular faces meet at one vertex. Of the five regular polyhedra (see Fig. M.5.29.), the icosahedron has 12 vertices at each of which five equilateral triangles faces meet, as shown in Fig. M.1.13.(a). In fact, there are only four possible polyhedra, bounded by nearly equilateral triangular faces such that either five or six of these faces have a common vertex and such that no two vertices common to six faces are adjacent to one

[1] Boerdijk, A. H., *Philips Res. Rep.*, 1952, **7**, 303.
[2] Frank, F. C., and Kasper, J. S., *Acta Cryst.*, 1958, **11**, 184.
[3] Frank, F. C., and Kasper, J. S., *Acta Cryst.*, 1959, **12**, 483.
[4] Shoemaker, D. P., and Shoemaker, C. B., *Structural Chemistry and Molecular Biology*, 1968, edited by A. Rich and N. Davidson, W. H., Freeman, San Francisco, 718–730.
[5] Bergman, B., Waugh, J. L. T., and Pauling, L., *Acta Cryst.*, 1957, **10**, 254.

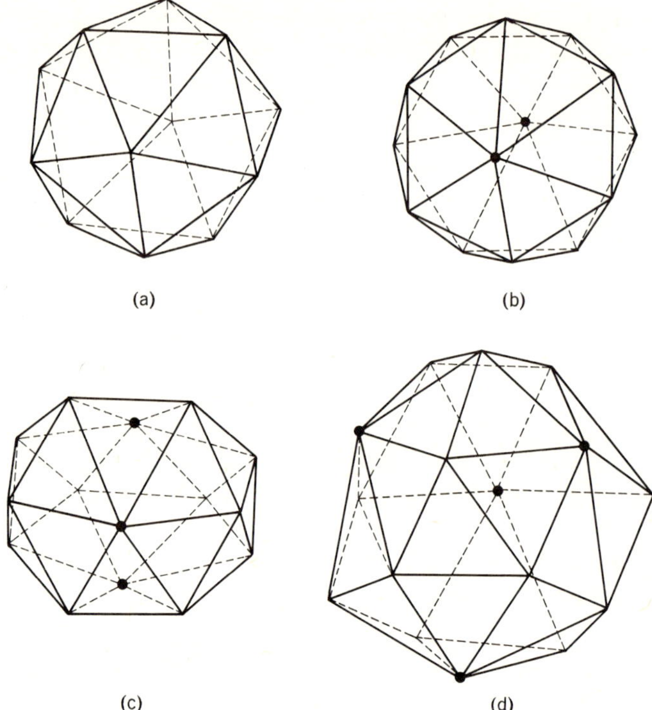

(a)   (b)

(c)   (d)

FIG. M.1.13. The four triangulated polyhedra with nearly equilateral faces and vertices at which either five or six faces meet, such that no two six-fold vertices (indicated by the small black circles) are adjacent.

another. These are shown in Fig. M.1.13.; (a), the icosahedron, has 12 five-fold vertices and 20 faces, (b) has 12 five-fold vertices, two six-fold vertices which are collinear and 24 faces, (c) has 12 five-fold vertices, three six-fold vertices, which are coplanar and triangular with one another, and 26 faces, while (d) has 12 five-fold vertices and four six-fold vertices, which are tetrahedrally arranged with respect to one another, and 28 faces. The icosahedron and the triangulated polyhedron with 16 vertices are recognizable in the Laves phases, as already discussed, and these and the other triangulated polyhedra have been recognized in a number of intermetallic structures.

M.1.3. The Atomic Heat of Metals

One of the fundamental and characteristic properties of the metallic state, the detailed examination of which has led to very major developments in the theory of metals and the solid state in general, is that of their heat capacity.

It had been noted as long ago as 1819 that the product of the specific heat and the atomic weight of many of the then known metals is very nearly independent of the metallic element concerned, this observation being usually referred to as Dulong and Petit's law.[1] The mean value of the product is about 26.4 J(K-at. wt.)$^{-1}$, at constant pressure, making use of normal temperature measurements of specific heats. The atomic heat of a metallic element at constant volume may be calculated from the conventional thermodynamic expression $C_p - C_v = \beta^2 KVT$, where C_p and C_v are the atomic heats at constant pressure and constant volume, respectively, β is the temperature coefficient of volume expansion at constant pressure, K is the isothermal bulk modulus, V is the atomic volume, and T is the Kelvin temperature of the substance. By the use of this expression, from experimentally measured values of C_p, the mean atomic heat of the metallic elements at constant volume, C_v, becomes about 25.1 J(K-at. wt.)$^{-1}$. The first attempted theoretical explanation of the Dulong–Petit law was that of Richarz.[2] From the known fact that the atomic heat of Avogadro's number of ideal gas molecules is $3R/2$, where R is the gas constant (see page 19), and from Maxwell's equipartition of energy principle, the kinetic energy of an atom of a monoatomic gas is $Mc^2/2 = M/2[c_1{}^2 + c_2{}^2 + c_3{}^2]$, then $Mc_1{}^2/2 = Mc_2{}^2/2 = Mc_3{}^2/2 = kT/2 = RT/2N_A$. Here M is the atomic mass, k is Boltzmann's constant, N_A is the Avogadro number, and c_1, c_2, and c_3 are the components of the root-mean-square velocity c. Thus, $C_v = 3R/2$, or the heat capacity per atom is made up of a contribution of $k/2$ from each degree of freedom of the atom. For a solid (metal), where each atom has a certain amount of potential energy by virtue of its position in the solid as well as its kinetic energy of vibration about this position, the mean kinetic energy is only half the total energy. Thus, if $3R/2$ represents the contribution to the atomic heat of a metal associated with the mean kinetic energy, then C_v must be equivalent to $3R$, or $C_v = 3R = 24.94$ J(K^{-1} at. wt.)$^{-1}$, which is approximately the value of C_v given by the Dulong–Petit law.‡

‡ It is proved on pages 42–43 that the average value of the kinetic energy of a free particle such as a gas molecule, in terms of classical Maxwell–Boltzmann statistics is $3kT/2$, or $3N_AkT/2$ for Avogadro's number of particles, or $3RT/2$; hence the heat capacity, $C_v = 3R/2$. For a single particle undergoing simple harmonic oscillations in one dimension, its energy is, $E = p^2/2m + m\omega^2 x^2$ (see pages 140–141). Substituting this expression for energy of a one-dimensional harmonic oscillator into the Maxwell–Boltzmann Distribution Law for the energies of free particles (page 42), gives for the average value of the energy of the oscillator in thermal equilibrium as

$$\bar{E} = (1/2m) \int_{-\infty}^{\infty} p^2 e^{-p^2/2mkT}\, dp \left[\int_{-\infty}^{\infty} e^{-p^2/2mkT}\, dp \right]^{-1}$$
$$+ (m\omega^2/2) \int_{-\infty}^{\infty} x^2 e^{-mx^2\omega^2/2kT}\, dx \left[\int_{-\infty}^{\infty} e^{-m\omega^2 x^2/2kT}\, dx \right]^{-1},$$

[1] Dulong, P. L., and Petit, A. T., *Ann. Chim. et Phys.*, 1819, [2], 10, 395.
[2] Richarz, F., *Ann. Physik*, 1893, **48**, 708; 1899, **67**, 702.

which making use of the integrals (c) on page 43 evaluates to $\bar{E} = kT/2 + kT/2 = kT$. So that for the simple harmonic oscillator in three dimensions, $\bar{E} = 3kT$, or $3N_A kT$ per Avogadro's number of oscillators, or $3RT$, corresponding to the Dulong–Petit value of the specific heat of $3R$.

As a result of much experimental work on atomic heat measurement, it was subsequently observed that certain substances such as diamond, boron, and silicon had atomic heats much lower in value than most metals, and that the atomic heats of most of the metallic elements converged to a low or zero value as temperatures approached 0 K. The first application of quantum theory to this area was Einstein's[1] attempt to account for the low atomic heat of the light elements. Einstein assumed that all solids were composed of atoms, each of which vibrated with the same characteristic frequency, ν. He also assumed that each atom in a solid vibrated about its mean position such that it was equivalent to three mutually perpendicular Planck oscillators (see page 36), which could assume discrete energy values $nh\nu$, where n was an integer and h is Planck's constant (section Q.1.3.). Then, according to the argument on page 36, the mean or expectation value of the energy of an oscillator, $\bar{E}$, is given by

$$\bar{E} = E_0[e^{E_0/kT} - 1]^{-1} = h\nu[e^{h\nu/kT} - 1]^{-1}, \qquad \text{(m.1.1.)}$$

where $E_0 = h\nu$. Then for a three-dimensional oscillator, the mean energy would be three times as much as this, and for an atomic weight of a mon-atomic material containing N_A such oscillators, the total energy would be

$$E = N_A\bar{E} \times 3 = 3N_A h\nu[e^{h\nu/kT} - 1]^{-1}. \qquad \text{(m.1.2.)}$$

On differentiating (m.1.2.) with respect to temperature, to obtain the atomic heat,

$$C_v = \partial E/\partial T = 3R[h\nu/kT]^2 e^{h\nu/kT}[e^{h\nu/kT} - 1]^{-2}. \qquad \text{(m.1.3.)}$$

It was assumed at the time of Einstein's proposal about atomic heats that solids absorbed thermal energy in the infrared region of the spectrum by the vibration of each of their constituent atoms as a whole, and that the observed optical dispersion of solid materials could be accounted for by the additional assumption that the higher frequency vibrations in the ultraviolet region of the electromagnetic spectrum were attributable to the internal vibration of atoms. However, such assumptions, while affording a plausible explanation of the optical dispersion of solids, would imply that each atom in a solid material had additional degrees of freedom which would lead to an atomic heat greatly in excess of the Dulong–Petit value. Equation (m.1.3.) expresses the atomic heat in terms of the infrared vibrational frequency without

---

[1] Einstein, A., *Ann. Physik*, 1907, **22**, 180.

taking into account the internal vibrations of the constituent atoms of the solid. Einstein pointed out that if the ultraviolet frequencies corresponding to the internal vibrations of the atoms were represented by $\nu_1, \nu_2, \nu_3, \ldots$, while retaining ν as the characteristic infrared vibrational frequency, the atomic heat would then be represented by

$$
C_v = 3R[h\nu/kT]^2 e^{h\nu/kT}[e^{h\nu/kT} - 1]^{-2}
$$

$$
+ 3R\left\{\sum_i [h\nu_i/kT]e^{h\nu_i/kT}[e^{h\nu_i/kT}[E^{ho/kT} - 1]^{-2}\right\}, \quad \text{(m.1.4.)}
$$

which expression enables not only the difficulty about the optical dispersion of solids to be accounted for, but also provides a plausible explanation for the low atomic heats of the light elements. Thus, those elements such as diamond, silicon, and boron, which have atomic heats less than $3R$ at ordinary temperatures are deduced to be those for which $kT < h\nu$ at that temperature; when $T = 300$ K, $kT = h\nu$ for a wavelength $\lambda = c/\nu = 4.8 \times 10^{-5}$ m, where c is the velocity of electromagnetic radiation. Hence, most metallic elements have approximately the Dulong–Petit value of C_v at ordinary temperatures because they have a characteristic wavelength greater than 4.8×10^{-5} m, and the light elements with $C_v < 3R$ are associated with an infrared wavelength $\lambda < 4.8 \times 10^{-5}$ m. Also, when $kT < 0.1h\nu_i$ at any particular temperature, the second term in equation (m.1.4.) falls off to a negligible value. For $T = 300$ K, the inequality $kT < 0.1h\nu_i$ is satisfied for $\lambda < 4.8 \times 10^{-6}$ m, so that for internal atomic vibrations in the ultraviolet region, where λ is certainly less than 4.8×10^{-6} m, the second term in (m.1.4.) approximates to zero and (m.1.4.) becomes identical with (m.1.3.), thus accounting for this concept of optical dispersion. However, much careful experimental work on the determination of the heat capacity of metals at low temperatures were subsequently found to be only in rough qualitative agreement with Einstein's formula, and also the decrease in C_v with decreasing temperature was found to be slower at low temperatures than what would be expected from (m.1.3.) as well as the fact that C_v did not fall to zero when $kT < 0.1h\nu$. A modified Einstein formula proposed by Nernst[1] and Lindemann,‡ which arbitrarily

‡ F. A. Lindemann later became Lord Cherwell, the scientific advisor to Winston S. Churchill during World War II.

assumed that half of the atomic oscillators describing the behavior of solids vibrate with the frequency ν and half with the frequency $\nu/2$, was found to give better agreement with the experimental results, leading to the so-called

[1] Nernst, W., and Lindemann, F. A., *Z. Elektrochem.*, 1911, **17**, 817.

Nernst–Lindemann formula:‡

$$C_v = 3R\{(1/2)[h\nu/kT]^2 e^{h\nu/kT}[e^{h\nu/kT} - 1]^{-2}$$
$$+ (1/2)[h\nu/2kT]^2 e^{h\nu/2kT}[e^{h\nu/2kT} - 1]^{-2}\}. \quad \text{(m.1.5.)}$$

While no theoretical basis could be provided for such an allocation of frequencies, the suggestion was made independently by Madelung[1] and Sutherland[2] that the characteristic vibrational frequency of the atoms in solids, in general, is associated not only with infrared or optical type absorption but also with the frequencies characteristic of elastic or sound vibrations. In fact, if it is assumed that the wavelength of the shortest sound wave is of the order of magnitude of twice the interatomic distance between atoms, and if the frequency associated with this wavelength is calculated, making use of the value of the velocity of sound or the elastic constants,[3] it is found that the frequency thus calculated has a value close to the optical or infrared frequency, ν. Madelung‡ proposed that there should be a whole range or spectrum of acoustic waves associated with atomic vibrations in all crystalline solids, and not just a single wavelength or frequency, as in Einstein's theory of atomic heats, with wavelengths ranging from the order of the dimensions of the solid for the longest wavelength to the order of twice the interatomic distance for the shortest.

‡ This proposal of Madelung's about atomic vibrational frequencies provided the foundation for the development of the methods of Debye[4] and Born and von Kármán[5] for the calculation of crystal lattice vibrational frequencies, quantities which did not become accessable to experimental measurement until about 50 years later, with the availability of high-flux nuclear reactors and the technique of inelastic scattering of thermal neutrons.[6]

The single Einstein frequency is replaced in the Debye theory of atomic heats by a frequency spectrum, by considering that the atoms in a solid do not vibrate independently of one another but that they are coupled together and vibrate collectively. Debye assumed that the solid material involved is a continuous medium, which is dispersionless, that is that the velocity of sound through the medium is independent of its frequency. He made use of the number of stationary sound waves which can be accommodated in the solid with frequencies in the range between ν and $(\nu + d\nu)$ to determine the spectrum.

[1] Madelung, E., Ges. Wiss. *Göttingen Nachr. Math. Phys.*, 1909, **Kl 1**, 100.

[2] Sutherland, W., *Phil. Mag.*, 1910, **[6]**, 20, 657.

[3] Huntington, H. B., *The Elastic Constants of Crystals*, 1958, Academic Press, Inc., New York.

[4] Debye, P., *Ann. Physik*, 1912, **39**, 789.

[5] Born, M., and von Kármán, T., *Physik. Z.*, 1912, **13**, 297.

[6] Brockhouse, B. N., *Inelastic Scattering of Neutrons in Solids and Liquids*, 1961, International Atomic Energy Agency, Vienna.

The simplifying assumption made by Debye that a solid is a continuous, dispersionless medium is, of course, not true for any solid, but the discrete nature of solid materials means that the total number of vibrational modes is finite rather than the infinity of modes characterizing a continuum. If a solid contains N atoms, and their displacements were described by a triple Fourier series (see section Q.2.4.), then only the first $3N$ terms of this series would be required to describe the atomic displacements uniquely, and these terms would represent the $3N$ lowest acoustical modes. Thus, if the solid is regarded as being continuous, the total number of acoustical modes would be limited to $3N$; it would, of course, require an infinite series to describe the displacement of all of the points of a continuous medium. Actually, this method of limiting the total number of vibrational modes of a solid implies minimum wavelengths of about twice the interatomic distance, as originally suggested by Madelung and Sutherland. If now, n_1 of the Planck oscillators describing a solid vibrate with the frequency v_1, n_2 with the frequency v_2, and so on, then

$$3N = \sum_1 n_i, \tag{m.1.6.}$$

and if the mean energy of each oscillator is given by (m.1.1.), the total energy of the system will be

$$E = \sum_i n_i h v_i [e^{hv_i} - 1]^{-1}. \tag{m.1.7.}$$

For a frequency spectrum, where the number of frequencies in each frequency range is so large that the quantity n_i may be conveniently replaced by a number density $g(v)$, then

$$n_i = g(v)\, dv, \tag{m.1.8.}$$

where n_i now represents the number of modes or oscillators with frequencies between v and $(v + dv)$, and so the summations of (m.1.6.) and (m.1.7.) may be replaced by the integrals

$$3N = \int_0^{v_D} g(v)\, dv \tag{m.1.9.}$$

and

$$E = \int_0^{v_D} \{[e^{hv/kT} - 1]^{-1} g(v) hv\}\, dv. \tag{m.1.10.}$$

The maximum frequency v_D is determined by the integral (m.1.9.).

In order to evaluate $g(v)$, it is necessary to consider the number of stationary waves which can exist in some volume V of material. As already discussed on pages 30–31, in connection with thermal radiation, let it be considered that the piece of material concerned is cube-shaped, of edge length L, so that $V = L^3$. Now, making use of the expression on page 31,

$$n_x{}^2 + n_y{}^2 + n_z{}^2 = 4L^2k^2, \qquad \text{(m.1.11.)}$$

where n_x, n_y, and n_z are positive integers and k_x, k_y, and k_z are the components of the wave vector $\mathbf{k}$. For each such wave vector, associated with the vibration in three dimensions of an isotropic medium, there is one longitudinal and two transverse modes of vibration, and since by definition $k = 1/\lambda = \nu/U$, with U denoting the velocity of propagation of the wave motion through the material, then the relation between the integral numbers n_x, n_y, and n_z and the frequencies become

$$n_x{}^2 + n_y{}^2 + n_z{}^2 = 4L^2\nu^2/U_l^2, \qquad \text{(m.1.12.)}$$

for the longitudinal vibrational modes, and

$$n_x{}^2 + n_y{}^2 + n_z{}^2 = 4L^2\nu^2/U_t^2, \qquad \text{(m.1.13.)}$$

for each of the two transverse vibrational modes. For elastic solids, the velocity U_l is usually greater than U_t by a factor of two or three.

Equations (m.1.12.) and (m.1.13.) correspond to the equations of spheres, of radii $2L\nu/U_l$ and $2L\nu/U_t$, respectively. However, since n_x, n_y, and n_z can only adopt positive integral values, these equations actually must refer to only one octant of a sphere, whose (radius)^2 is expressible as the sum of the squares of these three integers. Therefore, the frequencies ν must vary discontinuously, since L, U_t, and U_l are all constants, although for large numbers of atoms in a solid, corresponding to large numbers of sets of integers n_x, n_y, and n_z, the frequencies would appear to be continuous. The volume of the octant of the sphere involved, as far as the longitudinal vibrational motion is concerned, is $(1/8)(4\pi/3)(2L\nu/U_l)^3$, and for the two transverse modes the corresponding expressions are $(1/8)(4\pi/3)(2L\nu/U_t)^3$ for each of those modes, so that the total volume of the three octant spheres is

$$(1/8)(4\pi/3)[(2L\nu/U_l)^3 + (2L\nu/U_t)^3 + (2L\nu/U_t)^3]$$
$$= (4\pi/3)V\nu^3[1/U_l^3 + 2/U_t^3]. \qquad \text{(m.1.14.)}$$

For vibrational modes with the maximum frequency ν_D, the corresponding volume would be $(4\pi/3)V\nu_D^3[1/U_l^3 + 2/U_t^3]$, and within this volume all modes with frequency up to ν_D would be accommodated, that is, all $3N$ modes. Therefore,

$$(4\pi/3)V\nu_D^3[1/U_l^3 + 2/U_t^3] = 3N. \qquad \text{(m.1.15.)}$$

It is now possible to find the value of the integral of (m.1.9.), since $g(\nu)\,d\nu$ can be obtained as the difference between the volume of two spherical surfaces of radii $(\nu + d\nu)$ and ν as

$$g(\nu)\,d\nu = 4\pi V\nu^2[1/U_l^3 + 2/U_t^3]\,d\nu, \qquad \text{(m.1.16.)}$$

or making use of (m.1.15.),

$$g(\nu)\, d\nu = [9N/\nu_D{}^3]\nu^2\, d\nu, \qquad \text{(m.1.17.)}$$

which expression when inserted into (m.1.10.) gives the so-called Debye energy of a solid,

$$E = [9Nh/\nu_D{}^3] \int_0^{\nu_D} \{\nu^3[e^{h\nu/kT} - 1]^{-1}\}\, d\nu. \qquad \text{(m.1.18.)}$$

It is common practice to express ν_D in terms of the Debye characteristic temperature, Θ, defined by the relation

$$h\nu_D = k\Theta. \qquad \text{(m.1.19.)}$$

Expressing ν_D in terms of Θ and making the further substitution of $h\nu/kT = \xi$ and $d\xi/d\nu = h/kT$, the Debye energy becomes

$$E = 9R[T^4/\Theta^3] \int_0^{\theta/T} [\xi^3(e^\xi - 1)^{-1}]\, d\xi. \qquad \text{(m.1.20.)}$$

From this expression, it is deduced that for very high values of T, as $T \to \infty$, then $\xi \to 0$ and $(e^\xi - 1) \to \xi$; thus the integral part of (m.1.20.) approaches the value $[\Theta/T]^3/3$, leading to a limiting value of E of $3RT$. Thus, according to the Debye theory of specific heat, the limiting value of C_v at high temperatures is simply the Dulong–Petit value $3R$.

When T becomes very small, or when $h\nu_D/kT = \Theta/T \gg 1$, the wavelengths involved are large. Also, at very low temperatures, where $\Theta/T \to \infty$, the integral in (m.1.20.) becomes effectively the integral over the limits from 0 to ∞, and its value is thus independent of temperature, and E is directly proportional to T^4, or

$$E = 9R[T^4/\Theta^3] \int_0^\infty [\xi^3(e^\xi - 1)^{-1}]\, d\xi. \qquad \text{(m.1.21.)}$$

It is proved on pages 33–40 that

$$\int_0^\infty [\xi^3(e^\xi - 1)^{-1}]\, d\xi = \sum_{n=1}^\infty \int_0^\infty \xi^3 e^{-n\xi}\, d\xi = \sum_{n=1}^\infty [6/n^4] = \pi^4/15, \qquad \text{(m.1.22.)}$$

and making this substitution in (m.1.21.) gives

$$E = [3R\pi^4 T^4]/[5\Theta^3], \qquad \text{(m.1.23.)}$$

which on differentiation with respect to T gives the limit of C_v at low temperatures as

$$C_v = [12R\pi^4/5][T/\Theta]^3 = 1.943 \times 10^3 [T/\Theta]^3 \text{ J K}^{-1}\text{ mol}^{-1}. \qquad \text{(m.1.24.)}$$

This equation is the Debye T^3-law for the low temperature specific heats of substances. The integral in equation (m.1.20.) may actually be evaluated without setting the upper limit equal to infinity, to obtain C_v as an infinite series,

$$C_v/3R = [4\pi^4/5][T/\Theta]^3 - [3\Theta/T][e^{\Theta/T} - 1]^{-1} + 12 \ln [1 - e^{-\Theta/T}]$$

$$- 36[T/\Theta] \sum_{n=1}^{\infty} \{[1 + (2T/n\Theta) + (2/n^2)(T/\Theta)^2][e^{-n\Theta/T}/n^2]\}. \text{ (m.1.25.)}$$

The various values of $C_v/3R$, as a function of T/Θ, have been tabulated by Beattie.[1]

The Debye T^3-law for heat capacities at low temperatures arose as a strictly theoretical prediction; it was not predicated by experimental data. It is, in fact, the acoustical analogue of the Stefan–Boltzmann law for black body radiation (page 33). The order of magnitude of the Debye wavelength at any given temperature may be estimated by comparing the Wien law for black body radiation with the Debye law. Thus, as given by (q.1.19.), the maximum wavelength in the energy density–wavelength distribution curve for thermal radiation is given by $\lambda_{max} = c_2/5T$, with $c_2 = hc/k$ (see page 37). Therefore, $\lambda_{max}T = hc/5k$, where c is the velocity of electromagnetic radiation. To compare this expression with the Debye equation, it is only necessary to regard c as representing the velocity of sound and introduce the Debye characteristic temperature by the relation $\Theta = h\nu_D/k = hc/\lambda_D k$, where λ_D is the wavelength corresponding to the maximum Debye frequency and is about twice the interatomic distance in the solid substance concerned. If the interatomic distance is denoted by a, then

$$\lambda_{max}T = 2a\Theta/5, \qquad\qquad \text{(m.1.26.)}$$

from which it is deduced that when $T = \Theta/5$, $\lambda_{max} = \lambda_D = 2a$, and for $T > \Theta/5$, λ_{max} remains at the same value as λ_D as no shorter waves can exist, unlike the case for isothermal radiation. Therefore, for $T < \Theta/5$, $\lambda_{max}T$ is constant, and for $T > \Theta/5$, $\lambda_{max} = 2a$. By fitting the experimentally measured values of C_v at various temperatures to the Debye formula given by (m.1.24.), the characteristic Debye temperature Θ may be obtained. The experimental data cannot always be fitted to a constant Θ value. For metallic lead, for example, Θ is approximately 90 K, and $a = 3 \times 10^{-10}$ m; therefore, from the above relations, the Debye wavelength of lead from room temperature to 18 K is about 6×10^{-10} m, and at 1.8 K it is about 6×10^{-9} m.

For many metals the Debye formula does not agree with the experimentally observed data. If the Debye theory were to fit the observed data, the plot of Θ as a function of T would be a straight line, $\Theta = $ constant. For most metals it emerges that there is a rapid decrease in Θ with decreasing temperature at low temperatures. The Debye T^3-law should generally be

---

[1] Beattie, J. A., *J. Math. and Phys.*, 1926–27, **6**, 1.

valid in the region of $0 < T/\Theta < 1/12$, where the upper limit is the value of T/Θ for which deviation from the T^3-law is 1%, but for metals at low temperatures there is a contribution to C_v which is directly proportional to the temperature. If C_v/T for a metal is plotted as a function of T^2 in the region where the T^3-law is valid, then the intercept of the resulting curve with the C_v/T axis gives the proportionality constant relating the C_v to T. If this constant, γ, is then used to calculate γT, which quantity is then subtracted from the measured C_v values at each temperature and the resultant corrected value of C_v used in fitting the Debye expression, a more horizontal curve results on the plotting Θ as a function of T. The contribution to the specific heat of metals which is proportional to T has been recognized since 1928, when Sommerfeld[1] showed that it is due to what are referred to as conduction electrons, the remainder of the specific heat being referred to as the lattice specific heat, attributable to the three-dimensional harmonic oscillation of the constituent atoms about their mean positions in the crystal lattice.

As first pointed out by Blackman,[2] a discrete lattice has a much more complicated vibrational spectrum than the continuum of the Debye theory, and the Born–von Kármán[3] approach gives better agreement with experimental results. Lattice dynamics has developed rapidly in the period since 1960, when the direct experimental measurement of lattice vibrational frequencies first became possible by the technique of inelastic neutron scattering.[4,5] From such accurately measured lattice vibrational frequencies, and experimental information about heat capacities, elastic constants and cell dimensions of metallic crystals, it is alternatively possible to assess electronic heat capacities.

It is found experimentally that the heat capacity of typically non-metallic crystals approaches zero as $T \to 0$ as T^3, but that typical metals have heat capacities which approach zero as T, as $T \to 0$. The T^3 behavior is characteristic of lattice vibrations in isotropic materials, while the T approach is characteristic of that part of the heat capacity associated with the conduction electrons in a metal. The Debye characteristic temperature of a substance is of importance with respect to several other aspects of the behavior of solid materials, such as their thermal conductivity, electrical resistivity, and broadening of their X-ray diffraction maxima.

In a crystal containing N atoms or generally for N particles of a crystalline material, the crystal lattice vibrations of the constituent particles will be associated with N different values of a wave vector $\mathbf{k}$, each value in turn

[1] Sommerfeld, A., *Z. Physik*, 1928, **47**, 1.
[2] Blackman, M., *Proc. Roy. Soc.*, 1935, **A148**, 365, 385; ibid. 1935, **A149**, 117, 126.
[3] Born, M., and von Kármán, T., *Physik. Z.*, 1912, **13**, 297.
[4] Wallis, R. F. (editor), *Lattice Dynamics*, 1965, Pergamon Press, Oxford.
[5] Ziman, J. M., *Electrons and Phonons*, 1962, Oxford Univ. Press.

having associated with it three different modes, having different, orthogonal, polarization directions, so that there are a total of $3N$ lattice modes associated with N particles. Just as the term photon is used to describe a quantum of energy in the electromagnetic region of the spectrum, so the word phonon is generally used to describe a quantum of crystal lattice vibrational energy in the acoustical or sound frequency region of the spectrum, associated with the elastic vibration of solid crystalline materials.

There is no experimental evidence that the heat capacity of metals at high temperatures is particularly different from that of other solid substances. In other words, the conduction electrons in a metal, which apparently account for the low temperature heat capacity of metals behaving as if it were proportional to T as $T \rightarrow 0$, instead of the T^3 behavior characteristic of crystal lattice vibrational energy, cannot contribute to the high temperature heat capacity to anything like the extent which would be expected on the basis of classical statistics. Thus, if the conduction electrons in a metal behaved as free classical particles, they would contribute $C_v = 3Nk/2$ to the heat capacity, where N is the number of conduction electrons per unit volume of material. So that for $T \gg 0$, it would be expected that for metals the total heat capacity would be $(3 + 3/2)R = 9R/2$. Actually, as will be discussed in a following chapter, the behavior of electrons has to be described in terms of the laws of quantum statistics (Fermi–Dirac) rather than classical Maxwell–Boltzmann statistics. Generally, for metals, $C_v = C_v(\text{crystal lattice}) + C_v(\text{conduction electrons})$, or $C_v = \alpha T^3 + \gamma T$, and if experimentally determined values of C_v for $T \ll \Theta$ are used to determine Cv/T as a function of T^2, then since $C_v/T = \gamma + \alpha T^2$, the plot of Cv/T as a function of T^2 gives a straight line whose extrapolated intercept to $T = 0$ gives γ and whose slope gives α and therefore the value of the Debye temperature Θ, since α may be defined with reference to T^3 by equation (m.1.24.). At normal temperatures, the electronic contribution to the heat capacity of metals is typically of the order of 0.5%. At low temperatures, generally below 4 K, the electronic contribution to the heat capacity exceeds the crystal lattice vibrational contribution. At temperatures of the order of 1 K, paramagnetic effects may affect the heat capacity of a metal to a tremendous extent.

M.1.4.   The Thermal and Electrical Conductivity of Metals

A great many arguments may be and have been proposed for regarding a metal as a system in which electrons of a number of the order of one electron per metal atom, are free to move throughout the metallic crystal, composed of a three-dimensional array of cations, and thus to behave in much the same manner as gas molecules in a confining container. This assumption forms the basis of the free electron theory of metals, which in turn provides the basis for all of the more elaborate metal theories. This electron–gas concept of metals as a three-dimensional array of cations immersed in a "gas" of freely mobile electrons may, in fact, be used to account for the experimental observations that no material transport is associated with the conduction of electricity in metals, as is the case with electrolytic conduction, and also that electrons may be emitted from metals under the influence of electromagnetic radiation (photoelectric effect) or by heating to high temperatures (the Richardson or thermionic effect), and the Wiedemann–Frantz law, according to which the ratio of the thermal conductivity of any metal to its electrical conductivity is the product of a universal constant and the absolute temperature. It has already been noted that the electron–gas idea, however, leads to the conclusion that the heat capacity of metals should be considerably higher than is actually observed to be the case. This difficulty, as first shown by Sommerfeld (1927), has been removed by the application of quantum theory (Fermi–Dirac statistics) to the description of the behavior of the free electrons, as is discussed later. It emerges that the electron contribution to the heat capacity of a metal is about 30 times as great as the contribution from the lattice vibrational effects.

The classical expression for the thermal conductivity, κ_t, of metals contains two terms which generally are not accessible to experimental observation. These are: the free electron concentration, n_ε, and the diameter of the metal cations, σ_c. If the thermal conductivity of metals is attributable to the motion of free electrons, then it may be deduced by the arguments of the kinetic-molecular theory $\ddagger$ that the coefficient of thermal conductivity is given by

$$\kappa_t = (1/2)n_\varepsilon \bar{c} \lambda_\varepsilon C_v, \qquad \text{(m.1.27.)}$$

where n_ε, $\bar{c}$, λ_ε, and C_v represent the electron concentration, the average free electron velocity, the mean free path length, and the specific heat of the free electrons, respectively. By substituting the equipartition of energy value for C_v of $3k/2$, there results

$$\kappa_t = (3/4)n_\varepsilon \bar{c} \lambda_\varepsilon k, \qquad \text{(m.1.28.)}$$

and making use of the kinetic-molecular theory values of $\bar{c}$ and $\lambda_\varepsilon \ddagger\ddagger$, it is found that

$$\kappa_t = (3/2\pi)(n_\varepsilon/n_a)(2kT/\pi m_\varepsilon)^{1/2}(k/\sigma_a{}^2). \qquad \text{(m.1.29.)}$$

$\ddagger$ The coefficient of thermal conductivity of a substance or system is defined as the amount of heat transported across unit area in unit time under unit temperature gradient, the flow of heat being opposed to the direction of the temperature gradient. The general problem of transport phenomena in non-equilibrium systems where there is a gradient

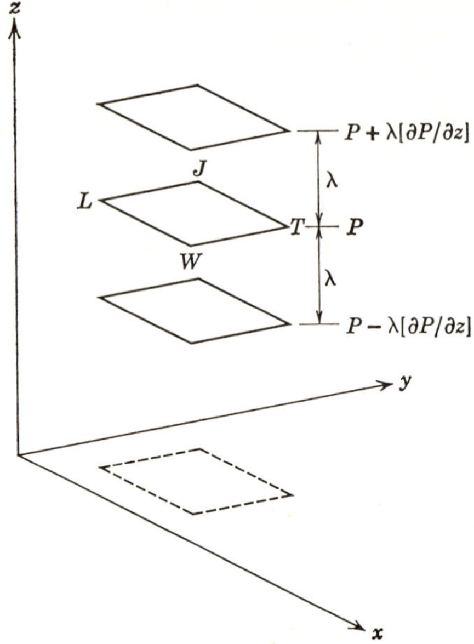

FIG. M.1.14. The rate of transport of some property of an ideal gas which varies with the z-coordinate only, across a plane.

of temperature, pressure, or composition, is rather complicated,[1] but for systems where the free paths of the particles involved terminate by collisions, as for ideal gases, transport of energy or matter in such cases may be regarded in terms of a slight disturbance applied to a Maxwell–Boltzmann distribution of some property among the particles.

Suppose, for example, as in Fig. M.1.14., that $LJTW$ is a plane of area A m², parallel to the plane of the xy-Cartesian coordinate system and perpendicular to the z-axial direction, located in an ideal gas composed of only one kind of molecule, with a concentration of n molecules per cubic meter. For such a system of classical particles at equilibrium, the number of particles crossing the plane in one second in the $+z$ direction is equivalent to the number passing through the plane in the $-z$ direction. If the average velocity of the particles is $\bar{c} = \sqrt{kT/2\pi m}$ (see page 43), then the number of particles crossing the plane from one side in the equilibrium system is $n\bar{c}A$. Now suppose that some property P is distributed among the particles in a non-uniform manner, such that the value of this property is independent of the x and y coordinates of the particles and dependent only on the z-coordinate. In such an event, more of the property P would be transported through the plane from above ($+z$ direction) than through the plane from below ($-z$ direction), so that there would be a net transport of P over the plane. The rate of change of the property P with respect to Z, $\partial P/\partial z$, would not be continuous but would

---

[1] Chapman, G., and Cowling, A., *The Mathematical Theory of Non-Uniform Gases*, 1939, Cambridge University Press.

be dependent on redistributions and exchanges of the property which take place as a consequence of particle collisions. Particles reaching the plane from either above or below, would have on the average only have traveled a distance corresponding to the mean free path length λ, since being in a previous collision with another particle; the mean free path for an ideal gas is $\lambda = 1/\sqrt{2}\,\pi n\sigma^2$ (see page 19). As in Fig. M.1.14., the rate of transport of P from above the plane is $nA\bar{c}[P + \lambda(\partial P/\partial z)]$, and from below is $nA\bar{c}[P - \lambda(\partial P/\partial z)]$, so that the net rate of transport of the property P in the $+z$ direction is

$$Q = -2nA\bar{c}\lambda(\partial P/\partial z).$$

As far as the thermal conductivity of an ideal gas is concerned, the property involved would be the average energy of a gas molecule, $\bar{E} = 3kT/2$ (page 413). The specific heat per molecule would then be, by definition, $\bar{E}/T(C_v = \partial E/\partial T)$. If the temperature of the gas varies in the z-direction, but is uniform in the xy plane, $\partial P/\partial z = C_v[\partial T/\partial Z]$, and the net rate of transport of energy, from the expression for Q above is given as

$$Q = -(n/2)\bar{c}A\lambda C_v(\partial T/\partial z).$$

From the definition of thermal conductivity and the observation that thermal energy flows in the direction opposed to the temperature gradient, it is necessary, to make the thermal conductivity a positive quantity, that Q be defined as $Q = \partial E/\partial t = -\kappa_t A[\partial T/\partial z]$, and therefore the coefficient of thermal conductivity is

$$\kappa_t = (1/2)n\bar{c}\lambda C_v.$$

If the electrons in a metal are regarded as comparable to the molecules of an ideal gas, then the coefficient of thermal conductivity of a metal may be similarly written as $\kappa_t = (1/2)n_\varepsilon\bar{c}_\varepsilon\lambda_\varepsilon C_v$, where the subscript refers to the free electrons and C_v for electrons may be taken as $3k/2$.

‡‡ The average space velocity of an ideal gas molecule is deduced on page 42 as $\bar{c} = [8kT/\pi m]^{1/2}$.

The mean free path of a gas molecule in an ideal gas is determined by the number of collisions with the gas molecules made by some definite molecule in its course of travel through the gas. In any detailed treatment of the frequency of collisions in gas systems, it may be shown that by a change of variable, it is possible to express for example, the total kinetic energy of two different gas molecules in collision as the sum of a component associated with the motion of the combined mass of the two molecules in space, and that associated with the relative motion of one molecule with respect to the other. Thus for two molecules with masses m_1 and m_2, and space velocities c_1 and c_2 (with components u_1, v_1, w_1, u_2, v_2, and w_2, respectively), the kinetic energies of the two molecules may be expressed alternatively as,

$$
\begin{aligned}
E &= (1/2)m_1[u_1{}^2 + v_1{}^2 + w_1{}^2] + (1/2)m_2[u_2{}^2 + v_2{}^2 + w_2{}^2] \\
&= (1/2)[m_1c_1{}^2 + m_2c_2{}^2] \\
&= (1/2)[m_1 + m_2]C^2 + (1/2)[m_1m_2(m_1 + m_2)^{-1}]V^2 \\
&= (1/2)MC^2 + (1/2)\mu V^2,
\end{aligned}
$$

with

$$
\begin{aligned}
(m_1 + m_2) &= M, \ [m_1m_2][m_1 + m_2]^{-1} = \mu, \ u = [u_1m_1 + u_2m_2[M^{-1}. \\
v &= [v_1m_1 + v_2m_2]M^{-1}, \ w = [w_1m_1 + w_2m_2]M^{-1}, \\
\alpha &= u_2 - u_1, \ \beta = v_2 - v_1, \ \gamma = w_2 - w_1, \ C^2 = u^2 + v^2 + w^2,
\end{aligned}
$$

and

$$V^2 = \alpha^2 + \beta^2 + \gamma^2.$$

As a consequence of these considerations, collision problems among ideal gas molecules may be dealt with by considering the motion of one molecule regarded as a sphere of diameter σ traveling through the gas and colliding with stationary spheres. A collision occurs when the center of the moving sphere, in the course of its flight, comes nearer to the center of another sphere of the same diameter at rest than the distance σ. The number of collisions made by the moving sphere is proportional to the number of molecules of gas per unit volume, n, and to their cross-sectional area (kinetic theory), $\pi\sigma^2$; the total number of collisions is than $n\pi\sigma^2$, and the mean free path of a molecule between collisions must be the reciprocal of this number, or the mean free path length, $\lambda = 1/n\pi\sigma^2$.

Although this simple argument leads to the correct value for the mean free path for electrons in "collision" with atoms or cations in a metal, namely, $\lambda_\varepsilon = 1/\pi n_a \sigma_a^2$, this is actually a coincidence as the electronic diameter and mass are negligibly small in comparison with the corresponding atomic quantities, so that the mean free path for electrons is determined by the concentration of the atoms or cations and their diameters. In the detailed theory of binary collisions between two different kinds of particles, the numerical factor of $(1/\sqrt{2})$ appears in the expression for the mean free path length, in the limiting case where all of the particles are the same, thus as $\lambda = 1/\sqrt{2}\,\pi n\sigma^2$.

The electrical conductivity of metals, as distinguished from the conductivity of electrolytic solutions, is attributable to the presence of free electrons. It is, however, not true to imply that the electrons are quite free to move in metals, but that this is not so in other materials. In terms of the quantum mechanical description of electronic behavior, the electrons are in constant motion in all kinds of substances, ionic compounds, insulators, and metals. In ionic compounds, in which the units of structure are oppositely charged anions and cations, it is not possible to identify the motion or interchange of electrons between one ion and another, although the electrons associated with each type of ion are in constant motion around the respective ionic nuclei. Similarly, in substances such as sulfur or diamond, or the large number of other compounds whose existence may be attributed to the formation of electron pair bonds between their constituent atoms, the electrons are in continual motion between the various nuclei. The significant fact is that in metals, under the influence of an externally applied electric field, there is a resultant flow of electrons in the direction of the applied field, while in the absence of such an applied field the electronic motion inside the metal takes place at random so that the resultant flow in any direction is zero. If the applied electric field were the only force acting on the metallic electrons, they would accelerate indefinitely, but experimental observation shows that the flow of electricity through a metal proceeds with a uniform velocity under a constant applied electrical field (Ohm's law), indicating that there must be an opposing frictional force to the electronic motion, both in the absence and in the presence of an applied electrical field. The nature of this opposing force is best interpreted as an interaction between the thermal vibrations of the metallic crystal lattice and the electronic wave motion.

If the retarding force on the motion of an electron of mass m_ε and charge of $(-\varepsilon)$ moving in a metal with some velocity u, under the influence of an applied electric field of strength V, is assumed to be directly proportional to the electronic velocity, then the equation of motion of the electron is

$$m_\varepsilon \, du/dt = -V_\varepsilon - Cu. \tag{m.1.30.}$$

As the impressed velocity on the electron is initially zero, the result obtained on integration of this equation is

$$u = -[V\varepsilon\tau/m_\varepsilon][1 - e^{-t/\tau}], \tag{m.1.31.}$$

where τ, defined as the relaxation time, is written for m_ε/C. The electronic velocity therefore falls from its initial value of zero to a steady, time-independent value of

$$u = -V\varepsilon/C = -V\varepsilon\tau/m_\varepsilon, \tag{m.1.32.}$$

which result follows from (m.1.30.) by letting du/dt become zero. If the concentration of electrons in the metal is n, then since the current density, i, is given by the product of the number of electrons crossing unit area in unit time and the charge carried by each, and the number of electrons crossing unit area in unit time is nu, so that

$$i = -nu\varepsilon = +nV\varepsilon^2\tau/m_\varepsilon. \tag{m.1.33.}$$

The specific electrical conductivity is defined as the quantity i/V, which is given by

$$\kappa_\varepsilon = n\varepsilon^2\tau/m_\varepsilon. \tag{m.1.34.}$$

To calculate the electrical conductivity by the use of this expression requires a knowledge of both n and τ. If it is assumed that each metal atom contributes one free electron, then from the experimentally measured values of the electrical conductivity it is possible to calculate τ; in this manner, it is found that for most metals τ ranges from about 10^{-15} s to about 4.5×10^{-14} s. A more approximate value of τ is obtained from $\tau = \lambda_\varepsilon/2\bar{c}$, and substituting this value into (m.1.34.) gives

$$\kappa_\varepsilon = n_\varepsilon\varepsilon^2\lambda_\varepsilon/2m_\varepsilon\bar{c}. \tag{m.1.35.}$$

From the comparison of (m.1.28.) and (m.1.35.), the ratio of the thermal and electrical conductivities is obtained as

$$\kappa_t/\kappa_\varepsilon = (3/2)[k(\bar{c})^2 m_\varepsilon/2\varepsilon^2], \tag{m.1.36.}$$

or by substitution of the kinetic-molecular theory value for $\bar{c}$ (see page 43),

$$\bar{c} = [8kT/\pi m_\varepsilon]^{1/2}, \tag{m.1.37.}$$

it is found that

$$\kappa_t/\kappa_\varepsilon = [12/\pi][k/\varepsilon]^2 T. \tag{m.1.38.}$$

This equation contains two of the very important generalizations about metals, both of which were discovered empirically before the discovery of electrons. These are the Wiedemann–Franz law, 1853, that the ratio of the thermal and electrical conductivities of all metals is a constant at a given temperature, and Lorenz's law, 1872, that this ratio varies as the absolute temperature.

By fitting experimentally determined values for the thermal conductivity of metals to (m.1.29.), the ratio of the concentration of free electrons to the concentration of atoms (actually the sum of concentrations due to atoms and ions) may be deduced, if values of σ_a are available. Alternatively, if integral values of n_ε/n_a are assumed, corresponding to $1, 2, \ldots$ electrons per atom, then σ may be calculated. This leads to atomic diameters which are considerably less than the values determined for internuclear distances by X-ray diffraction methods.

M.1.5. Development of Theories of Metals

The assumptions involved in the simple free electron theory of metals constitute quite major simplifications of the problem of the interaction of metal atoms in the solid, crystalline, state. Inherent in this treatment is the primary assumption that the structure of the solid is of no consequence, the solid metal simply acting as a container for a number of electrons which move about quite freely as though they constituted an electron gas, between the particles of which there is no electrostatic repulsion. The charge on the electrons is assumed to be neutralized by that of the cations of the metal, uniformly distributed throughout the metal. Nevertheless, these simple ideas, apart from providing at least a qualitative understanding of the thermoelectric effect and enabling the Wiedemann–Franz law to be deduced, provide a reasonable description of the state of affairs in the alkali metals, although constituting a very poor approximation for describing all other metals.

One of the useful aspects of the free electron theory is that it depicts in a simplified manner some of the differences between a "gas" of electrons at the densities at which electrons exist in metals, and ordinary gases obeying Maxwell–Boltzmann statistics. It was Sommerfeld[1] who first applied quantum statistics, making use of the newly formulated Fermi–Dirac[2,3] statistics, at that time, to the description of the behavior of the electrons in a metal and was able to show that the heat capacity per electron should be very much less than the value to be expected in terms of the equipartition of energy principle.

---

[1] Sommerfeld, A., *Zeit. f. Phys.*, 1928, **47**, 1.
[2] Fermi, E., *Zeit. f. Phys.*, 1926, **36**, 902.
[3] Dirac, P. A. M., *Proc. Roy. Soc.*, 1926, **A112**, 661.

In quantum statistics, particles with half-integral spins such as electrons, protons, and neutrons behave quite differently from particles like photons with integral spins, and from spinless particles such as ordinary gas molecules. In Fermi–Dirac statistics, each energy state of a system can accommodate at most two particles of opposed spin, the particles being describable by anti-symmetrical functions and referred to as fermions; the statistical system required for the adequate description of the behavior of photons was proposed at an earlier stage by S. N. Bose and later generalized by Einstein (1924), and such Bose–Einstein statistics applies to particles described by symmetrical functions, referred to as bosons. The interchange of particles in identical states is not significant in quantum statistics, contrary to classical Maxwell–Boltzmann statistics. Briefly, at this stage, if n_i is the number of particles which will be present with energy E_i and statistical weight g_i (the number of states i) in the most probable distribution, the distinction between the three statistical systems is as follows: $g_i/n_i = Be^{E_i/kT} + a$, where $a = 0$ for Maxwell–Boltzmann statistics, $a = +1$ for Fermi–Dirac statistics, and $a = -1$ for Bose–Einstein statistics. If g_i is very large compared with n_i, so that $g_i/n_i \gg 1$, then the difference between the three forms disappears, or classical statistics then applies. The factor B in the above expression is determined by the condition that $\sum n_i = N$, the total number of particles involved. For mono-atomic gases, characterized only by translational energy, the factor $B = [2\pi mkT]^{3/2}h^{-3}[V/N]$, which is always much smaller for electrons than for atoms because $m_\varepsilon \ll m_H$ and also because the particle density N/V is always much higher for electrons than for any gas. Thus the classical laws, such as the law of equipartition of energy will only apply to electrons at very high temperatures; thus, a temperature of 1 K for helium atoms at 1 atmosphere pressure corresponds to a temperature of about 25200 K for electrons, a temperature at which any known metal has long ceased to exist as such.

The application of quantum mechanical methods to the description of electronic behavior in metals, taking into account the periodic field provided by the three-dimensional array of cations, was first developed by Bloch[1] and elaborations of this model were subsequently described by Brillouin,[2] Mott and Jones,[3] Peierls,[4] and others, leading to the so-called band and zone theories. This enabled the deduction from the theory to be made of many of the properties of metals and other phenomena associated with them on a quantitative basis. The thermal, thermoelectric, electrical, and magnetic

---

[1] Bloch, F., *Zeit. f. Phys.*, 1928, **52**, 555.
[2] Brillouin, L., *Die Quantenstatistik*, 1931, Berlin.
[3] Mott, N. F., and Jones, H., *The Theory of the Properties of Metals and Alloys*, 1958, Dover Publications Inc., New York.
[4] Peierls, R., *Elektronentheorie der Metalle*, 1932, *Ergebnisse d. exakt. Naturwiss.*, **11**, 264–322.

properties of metals, important for an understanding of some of the special attributes and chemical behavior of metals, found an explanation in the band theories. There is a major distinction between the discrete energy levels of free atoms and the electron distribution found in compact, solid, crystalline metals.

Of fundamental importance in the application of quantum theory to the calculation of the energy of a metal, is the consideration of the cohesive energy. The cohesive energy of a metal is defined as the difference between the energy of the system of free atoms composing the metal and the total energy of the electrons and nuclei in the metal, and is a measure of the forces which units the atoms in the solid metal. Metals, of course, only exist as stable solids because the energy of the electrons and nuclei is less than that of the free atoms at the same temperature. Alternatively, it is convenient to regard the cohesive energy of a metal as the energy required to dissociate Avogadro's number of atoms of the metal, at the absolute zero of temperature, into free atoms. For any macroscopic sized piece of a metal, there are of the order of 10^{29} atoms per cubic meter. If it is assumed that this large number is represented by N, each of these N atoms containing Z electrons so that there are ZN electrons altogether each with charge $-\varepsilon$, and N nuclei, each with charge $+Z\varepsilon$, then adopting the terminology used in Q.5.7., the Hamiltonian operator for these nuclei and electrons may be expressed as

$$\hat{H} = -\sum_{i=1}^{ZN} [h^2/8\pi^2 m]\nabla_i^2 - \sum_{i=1}^{ZN}\sum_{a=1}^{N} [Z\varepsilon^2/R_{ai}] + (1/2)\sum_{i\neq j}^{ZN}\sum^{ZN} [\varepsilon_2/r_{ij}]$$

$$+ (1/2)\sum_{a\neq b}^{N}\sum^{N} [Z^2\varepsilon^2/R_{ab}], \tag{m.1.39.}$$

where the first term represents the kinetic energy of the electrons, the second is the potential energy of the electrons in the field of the nuclei, the third is the potential energy due to the Coulomb interaction of the electrons, and the last term is the potential energy due to the Coulomb interaction of the nuclei. In this expression, R_{ai} designates the quantity $|\mathbf{R}_a - \mathbf{r}_i|$, r_{ij} the quantity $|\mathbf{r}_i - \mathbf{r}_j|$, and R_{ab} the quantity $|\mathbf{R}_a - \mathbf{R}_b|$, where r_i is the position vector of the ith electron, and R_a that of the ath nucleus.

This type of Hamiltonian is the same for all solid materials composed of only one kind of atom, and does not imply any metallic characteristics. The ground-state energy of the system would be given by the lowest eigenvalue of the equation $\hat{H}\Psi = E\Psi$. Since the Hamiltonian depends on the internuclear distances, so do E and Ψ, but for any known structure where the internuclear distances are determined, Ψ can, at least in principle, be determined as a function of the electronic coordinates (including spin), and the energy found. In principle, it is possible to assume arbitrary positions for the nuclei and show that the minimum ground-state energy results when they are placed at the correct positions in the crystal structural arrangement, but due to the other approximations which have necessarily to be made, differences in the calculated energies are introduced which are of greater magnitude than the difference in energy of different crystal structural arrangements. There is no possibility of solving the Schrödinger equation containing the operator (m.1.39.) directly, and most of the approximation techniques which are fairly successful for free atoms cannot be applied directly either, but nevertheless there are certain simplifying features associated with the wave functions appropriate to the description of the behavior of electrons in metals which enable reasonably accurate calculations to be carried out.

One of these simplifying features is associated with the fact that some of the electrons in each of the atoms in a metal crystal have practically the same energy as they have in the free atom. Thus, for sodium atoms the ground-state configuration is $1s^2 2s^2 2p^6 3s$ with one electron in a $3s$-state and all of the others in closed quantum groups. The radial charge densities of the various electron groups for this atom have been calculated by the Hartree method and are depicted in Fig. M.1.15. The radial charge density of the K and L quantum groups is confined almost entirely to within about 10^{-10} m from the nucleus, while the charge-cloud of the $3s$-electron extends to within several times this distance from the nucleus. The $3s$-electron in the sodium atom, which essentially determines the chemical properties of sodium, is what is referred to as the valency electron, or in metal theory as the conduction electron. Similarly, for atoms of other elements those electrons whose charge clouds extend the greatest distance from the nuclei may be regarded as the valency electrons, the nucleus and the other electrons being regarded as the kernel or ion core of the atom. In solid sodium the internuclear distance is known to be 3.67×10^{-10} m, so that, as illustrated in Fig. M.1.15., the ion cores of the atoms in the solid are well separated but the electron distribution attributable to the valence electrons overlap one another. It is for this reason that it is considered that when sodium atoms condense to form sodium metal, that the K and L quantum groups are not appreciably altered, but that the wave functions of the valence electrons are substantially different, with $3s$ atomic wave functions on adjacent atoms extending throughout the piece of metal and hence associated equally with all the atoms, rather than being associated with one particular sodium atom. Similar considerations apply to the other alkali metals, which because of the small extension in space of their ionic electron cloud densities compared with the internuclear distance observed in the solid metals, are referred to as "open" metals. The small interaction between the ion cores leads to their playing a negligible part in determining the cohesive energies or the crystal lattice constants. On the other hand, the forces resulting from the slight ionic overlap vary rapidly

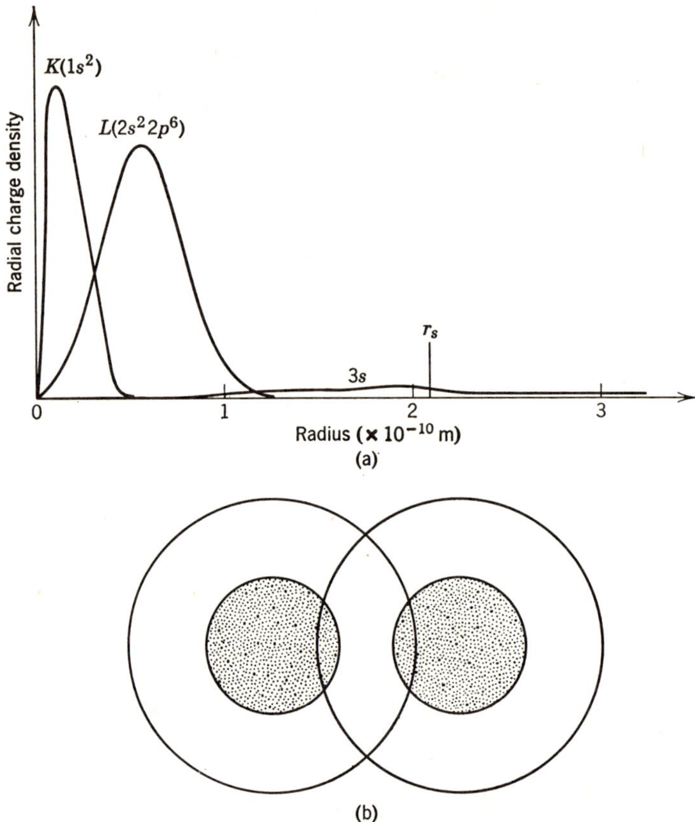

FIG. M.1.15. Radial charge density of the electrons in a free sodium atom, (a). Also shown at (b) a representation of the electron charge clouds of the K and L group electrons denoted by shading (the ion cores), compared with the valency electron charge clouds of two sodium atoms in a piece of sodium metal at their observed internuclear distance, 3.67×10^{-10} m.

with distance, so that these forces only play a very slight part in determining the compressibilities but a major part in determining the elastic constants of the alkali metals. It is thus essentially for the reason that the properties of the alkali metals are so largely determined by the valence electrons that these metals are regarded as the "ideal" metals of the various electron theories and are of such interest, even although they are too reactive to be anything but ideal from the experimental point of view or as the constituents of alloys.

It is generally assumed that if the effective radius of the ion core is small compared with the atomic radius, defined as given by r_s,

$$4\pi r_s^3/3 = V/N, \tag{m.1.40a.}$$

where V is the volume of the metal, that the ion core of an atom is essentially unchanged in the solid state. This condition is satisfied not only by the alkali metals but also more or less by the alkaline earth metals and aluminum, but not by the transition metals. Since the cohesive energy of a metal is represented by the difference between the energy of a collection of free atoms and the collection of electrons and ion cores in the metal, then if the ion cores may be regarded as the same in the metal as in the free atom, the calculation of the choesive energy is slightly simplified by regarding the metal as a collection of valence or conduction electrons moving in a fixed three-dimensional array of ion cores. So, by writing $U_a(r_i)$ for the potential energy of the ith electron in the field of the ath ion core, the Hamiltonian operator of (m.1.39.) becomes

$$\hat{H} = -\sum_i^N [h^2/8\pi^2 m]\nabla_i^2 + \sum_i^N \sum_a^N U_a(r_i) + (1/2)\sum_{i\neq j}^N \sum^N [\varepsilon^2/r_{ij}]$$
$$+ (1/2)\sum_{a\neq b}^N \sum^N [\varepsilon^2/R_{ab}]. \quad (\text{m.}1.40b.)$$

The greatest simplification is in the last term representing the interaction of the ion cores, which for the alkali metals may be regarded as spheres of single positive charge, the remaining charge of the nucleus being screened from the valence electrons by the core electrons. The alkaline earth metals would have charges of $+2\varepsilon$, in which event the last term in (m.1.40b.) would be four times as great.

The detailed electron theories of metals are essentially quantum mechanical in nature and many approximation techniques have been proposed. The theory which has developed from the simple electron "gas" concept due to the efforts of Sommerfeld, Bloch, Brillouin, Wilson, Jones, Pines, and others is the precise theory which lends itself to the more exact and quantitative interpretation of the properties of metals and intermetallic compounds, although in many respects it is applicable only to sodium and the other alkali metals. In fact, even the most successful of the approximation techniques in accounting for the cohesive energies of the alkali metals appear rather inadequate as far as the transition metals are concerned. However great the progress which has been made along this direction, there is an alternative approach. Just as there have been two alternative ways since the first application of quantum theory to the attempted explanation of the nature of the interaction between the atoms of other types of compounds, namely the valence bond method of Heitler–London–Heisenberg–Pauling and the molecular orbital method of Mulliken–Longuet-Higgins and others, so also with metal theory. Pauling has attempted to apply the valence bond method to the problem of bonding in metals with considerable success. Although proposed and developed much later than the Sommerfeld–Bloch–Brillouin theory, it is probably the more comprehensive and simpler approach to the problems involved in metallic bonding. Although very largely qualitative in nature, the Pauling method provides some insight into the relationships between cohesive energy, structure, and magnetic properties of the transition metals especially, and will be considered first before examining the electron theories in any detail.

M2

Metallic Cohesion

M.2.1. Cohesive Energy and Physical Properties

The cohesive energy of a metal, regarded as the difference between the energy of a system of free atoms and the total energy of the electrons and nuclei in the metal, is the energy required to dissociate some finite mass of the metal into free atoms. Many physical properties of metals are related to their cohesive or binding energies. There is a general, though approximate, correlation between cohesive energies and the melting point, boiling point, hardness, compressibility, density, coefficient of expansion, and internuclear distance or metallic radius. It would not be expected that there would be an exact correspondence between cohesive energies and these properties, as other factors must be operative besides the cohesive energy in determining them. Thus for the transition and lanthanide and actinide metals the isolated atom contraction effect with increasing nuclear charge must influence the internuclear distances as well as any change in cohesive energy. The metallic elements of the third transition group (La–Hg) display the greatest regularity in cohesive energy, reaching a pronounced maximum at tungsten.

The Pauling[1] hypothesis about the nature of bonding in metals was originally developed on the basis of the observation of the relationships referred to above and correlation of this information with structural and magnetic data. Thus, as pointed out by Pauling in his 1938 paper, if it is considered that the potassium atoms in potassium metal each contribute one electron to the formation of metal–metal bonds, and that in the metallic element with nuclear charge one unit greater, calcium, that two electrons per atom are involved, then such an assumption is consistent with the observed properties of the two metals. Thus, calcium is harder, denser, with higher melting and boiling points, and stronger, than potassium, consistent with the idea that the bonds holding the calcium atoms together are twice as strong as in

---

[1] Pauling, Linus, *Phys. Rev.*, 1938, **54**, 899; *Proc. Roy. Soc.*, 1949, **A189**, 343; 350; *J. Am. Chem. Soc.*, 1947, **69**, 542; *Proc. Nat. Acad. Soc.*, 1953, **39**, 551; Nature, 1961, **189**, 656.

potassium. In a similar way, there is a further increase in the hardness, melting point, boiling point, density, and certain other properties, with successive increases in the nuclear charges of the atoms of elements corresponding to scandium, titanium, vanadium, and chromium, indicating an increase in the number of electrons per atom involved in metallic bonding from three for scandium, four for titanium, five for vanadium, to six for chromium. These numbers do actually correspond to the maximum values of the oxidation numbers of these elements, as exhibited, for example, in their oxides, K_2O, CaO, Sc_2O_3, TiO_2, V_2O_5, and CrO_3. The properties essentially dependent on cohesive energy of the transition elements with nuclear charges greater than that of chromium in the first transition series, do not however, indicate that there is a further increase in the number of electrons per atom involved in metallic bonding for the elements manganese, iron, cobalt, nickel, copper, and zinc. On the contrary, these properties apparently remain approximately constant for chromium, manganese, iron, cobalt and nickel, then there is a decrease in the magnitude of these properties from nickel to copper, and still a further decrease from copper to zinc. Pauling has concluded that the number of electrons per atom engaged in metallic bonding in the elements copper, zinc, and the succeeding metallic elements gallium, germanium, and arsenic, are approximately 5.56, 4.56, 3.56, 2.56, and 1.56, respectively.

M.2.2. Electronic Energy States in Metals

Generally, the coordination number or the number of nearest neighboring atoms of any atom in metals exceeds the number of electrons per atom available for bond formation between the atoms; this number is twelve for cubic and hexagonal closest packed structures and eight (plus six more neighbors at about 15% father away) for body-centered cubic structures. While this state of affairs obviously precludes the formation of localized bonds between metal atoms, as in simple molecules like hydrogen and chlorine, Pauling nevertheless considers that in metals there are bonds between metal atoms of essentially the same nature as in these molecules. In the development of chemical theory, much data has been accumulated about internuclear distances in crystals of substances, whose existence is best understood in terms of electron pair bonds being responsible for the compound formation; from these observed distances, it has been possible to deduce a set of atomic radii which account for the internuclear distances observed in these molecular compounds. If slight corrections are made for the variation of the coordination numbers in different metallic structures, it is observed that the same radii, approximately, account for the observed internuclear distances in metallic crystals as in many other compounds. The so-called Pauling valence-bond theory of metals is essentially the same as for molecules. For metals, it is necessary to consider the various ways in which the electrons from the metal atoms (the valence electrons) may be paired together, each such arrangement is considered to be a partial description of the metal, and all of these possible arrangements contribute to the complete wave function required to describe the metal. This concept has been described in various ways: there are not enough electrons in metals to form one-electron or two-electron bonds between all of the constituent atoms; in metals electrons are shared between all of the atoms; bond formation between metal atoms is such that wave mechanical exchange or resonance takes place; the metal bonds resonate between all possible atom pairs; and so on. If the coordination number of a metal atom is Z and there are N valence electrons per atom, Pauling refers to the quantity, $n = Z/N$, as the average bond order. Thus, in the vapor phase, where sodium exists as the diatomic molecules Na_2, each sodium atom provides one $3s$-electron to form an electron pair bond between the two atoms, and $n = 1$. In sodium metal, as in the other alkali metals, where each atom is associated with eight nearest neighbors and six slightly more distant neighbors, and there is still only one electron per atom available for interaction between the atoms, the bond order must be $n = 1/8$ or less. From much accumulated information about internuclear distances, particularly organic compounds described as containing multiple electron bonds, the effective radius of an atom is deduced to vary directly with the bond order, and so as a consequence of the low bond order in the alkali metals, the metal–metal bonds are weak, the cohesion is low, and these metals are soft, low melting, and the least dense of the known metals. Proceeding in this manner, Pauling has attempted to deduce the bond order and therefore the distribution of electron density from the observed internuclear distances in metals and intermetallic compounds. However, the corrections which must be applied for the variation of the atomic radius with coordination number and with the bond order are purely empirical.

For sodium, for example, each sodium atom provides one $3s$-valence electron, and Pauling has shown that with respect to the eight nearest neighbors of each sodium atom, for a sodium crystal containing $2N$ atoms there, are 3.14^N ways in which electron pair bonds may be formed between equivalent pairs of sodium atoms. If it is assumed that the valence electrons are accommodated in the $3s$-energy states and thus that no sodium atom in the metal forms more than one bond, then a synchronous exchange of bonds between

$$
\begin{array}{ccc}
\text{Na—Na} & & \text{Na} \quad \text{Na} \\
& \text{and} & | \qquad | \\
\text{Na—Na} & & \text{Na} \quad \text{Na}
\end{array}
$$

throughout the crystal has to be assumed. An approximate estimate of the so-called exchange energy may be obtained from comparison of the observed dissociation energy of Na_2 from spectroscopic data, and the sublimation energy of the metal, both of which are accessible to experimental observation. This value would appear to be considerably larger than can be accounted for by the number of distinguishable configurations which can result from the synchronized bond exchange, although it would be more compatible with a completely random exchange, and Pauling's argument leads to the conclusion that there are 2.32×3.14^N configurations involving exchanges of the type

$$
\begin{array}{ccc}
\text{Na—Na} & & \text{Na} \quad \text{Na}^+ \\
& \text{to} & | \\
\text{Na—Na} & & \text{Na}^-\text{—Na}
\end{array} \quad .
$$

This involves the introduction of Na^+ and Na^- states, and the necessity of some sodium atoms forming two bonds at one time, which therefore requires the availability of some higher energy state than the $3s$-state to accommodate the additional bonding electrons. Extending this idea, Pauling proposes that the availability of an extra, low-lying, energy state for bond formation in this manner is an essential and necessary condition for the formation of a metallic structure, and the conduction of electricity through metals is only possible in this way. For sodium, the next higher available energy states are the three $3p$-states, which may be combined with the $3s$-state to give rise to mixed or hybrid sp-states of lower energy than $3s$ and $3p$ states alone. It is a fundamentally important idea that such mixed energy states play a part in metallic bonding. It provides for example, a connection between the typical facts of the general chemistry of the transition metals and their distinctive metallic properties.

In Pauling's view, for the transition metals, not only the s, but also the three p and also the five d states (for the first transition series of metals, the $4s$, $4p$, and $3d$ states) are involved in metallic bonding, being divisible into three groups or sets:

(a) A group of s, p, and d, states, hybridized into a $d^n sp^3$-set, essentially similar to the $d^2 sp^3$ set of atomic bonds which describe the state of affairs in octahedral molecules and complexes such as in the hexacyanoferrates, $[Fe(CN)_6]^{3-}$.

(b) A group of m pure d-states with non-bonding properties localized within the atom.

(c) Vacant metallic states, $[5 - (m + n)]$ in number, which are necessary for the exhibition of metallic behavior and to provide suitable energy states for the exchange of metallic bonding between atoms.

It is considered that only the m-set of d-states contribute towards the paramagnetic properties of the transition elements, and that this quantity can be deduced from the magnetic properties of the transition metals and their alloys. Both m and n may be fractional quantities, which may be simply interpreted by the assumption that the various metal atoms are present in two or more different oxidation states. In metallic iron, for example, Pauling considers that there are 6.04 electrons per atom involved in forming electron pair bonds between iron atoms, 2.24 electron states occupied by electrons not involved in bond formation (which account for the paramagnetic moment), and 0.72 vacant electron states ("metallic orbitals"), the sum of these quantities being $(6.04 + 2.24 + 0.72) = 9$, corresponding to the one $4s$ + three $4d$ + five $3d$ states. In his 1938 paper Pauling assigned to iron the number 5.78, for the number of electrons per atom involved in metallic bonding, but in his 1953 paper, taking account of the fact that not all the metal–metal bonds may be electron pair bonds, he concluded that it would be more plausible to assume that this number should be six, as also for chromium, manganese, cobalt, and nickel.‡

‡ Pauling argued originally (1938) that for iron with nuclear charge 26, there are eight electrons per atom beyond the rare gas number 18 (argon), and that if iron atoms formed normal electron pair bonds, the spin moments of these eight electrons would be paired with that of other atoms so that the magnetic moment of the metal would be zero. But the saturation magnetic moment of iron is observed to correspond to 2.22 Bohr magnetons per atom, and therefore not more than $(8 - 2.22) = 5.78$ electrons can be involved in bond formation giving rise to electron pair bonds. It was Zener[1] who pointed out, that on the basis of the free electron theory of metals that 0.26 electrons per atom represent conduction electrons in iron, with therefore uncoupled spins and behaving as one-electron bonds. The sum of $(5.78 + 0.26) = 6.04$ electrons must be involved in bond formation; so Pauling's *metallic valence* for iron is 6.04.

For the transition metals copper and zinc, their magnetic properties indicate that their metallic valence is less than six. For example, it is found experimentally that for the ferromagnetic alloys of copper and nickel the saturation magnetic moment falls off linearly with increasing atomic percentage of copper in the alloy, until it reaches zero for 56 atomic percent of copper. Nickel and copper, with nuclear charges 28 and 29, respectively, have 10 and 11 electrons in excess of the rare gas number 18 in argon. If all of these electrons are available for bond formation, then for the alloy with 56 atomic percent of copper, the number of electrons per atom would be $(10 \times 0.44) + (11 \times 0.56) = 10.56$. If the statistical nickel-copper atoms in the alloy utilize six electrons per atom for electron pair bond formation, then 4.56 electrons per atom remain, which would have to occupy 2.28 energy states

---

[1] Zener, C., *Phys. Rev.*, 1951, **81**, 440.

with paired spin moments in the diamagnetic alloy. This would mean that of the nine available energy states beyond the stable argon configuration, $[9 - 6 - 2.28] = 0.72$ energy states per atom would remain vacant, the Pauling *metallic orbital*. If it is assumed that this is a constant quantity for all of the transition metals, then for pure copper with eleven electrons per atom (beyond 18 in the argon closed configuration) only $(9 - 0.72)$ or 8.28 energy states would be available for both bonding electrons and non-bonding electrons. The eleven electrons per atom of copper could be distributed over these 8.28 states in such a manner that 2.72 electron pairs $(11 - 8.28)$, and 5.56 unpaired electrons are present and available for bond formation; then $(2 \times 2.72) + 5.56 = 11$. In this way, Pauling concludes that the *metallic valence* of copper is 5.56, and, extending this argument to other alloys, the metallic valences for the following are: 4.56 for zinc, 3.56 for gallium, 2.56 for germanium, and 1.56 for arsenic.

The number of non-bonding electrons in the first transition group of metals is obtainable from the measured values of their saturation magnetic moments; these are obtained as: 0 for vanadium, 0.22 for chromium, 1.22 for manganese, 2.22 for iron, 1.66 for cobalt, 0.66 for nickel, and 0 for copper. Examination of the magnetic properties of other alloys of iron and vanadium, chromium, nickel, and cobalt, and of these metals with one another, provide additional evidence for the number of unoccupied electron states in the transition metals being 0.72 states per atom. If the saturation magnetic moments of these alloys, in Bohr magnetons, is plotted as a function of nuclear charge of the metals, then instead of the anticipated values of two Bohr magnetons for iron, three Bohr magnetons for cobalt, and two Bohr magnetons for nickel, (on the assumption that the number of bonding electrons per atom is constant at six for chromium, manganese, iron, cobalt, and nickel) the magnetization reaches a maximum at about nuclear charge number 26.28 (Fig. M.2.1.). The maximum is not actually too sharp, but if the two linear legs of the curve are projected to intersection, it lies close to the value of 26.28 electrons per atom. Pauling interprets this value as indicating that 8.28 is the maximum number of electron states that may be occupied by bonding plus non-bonding electrons in the transition metals, leaving $(9 - 8.28) = 0.72$ vacant energy states as the number of "metallic orbitals" per atom, which feature is responsible for the characteristic behavior of metals. It is believed that this vacant energy state permits the unsynchronized exchange of electron pair bonds between metal atoms, giving rise to atoms with formal charges of $+1$, -1, and 0. An approximate derivation of this value of 0.72 has been formulated by Pauling[1] on the basis of a simple statistical argument.

For the alkali metals, for example, crystallizing with body-centered cubic

---

[1] Pauling, L., *Nature*, 1961, **189**, 4765.

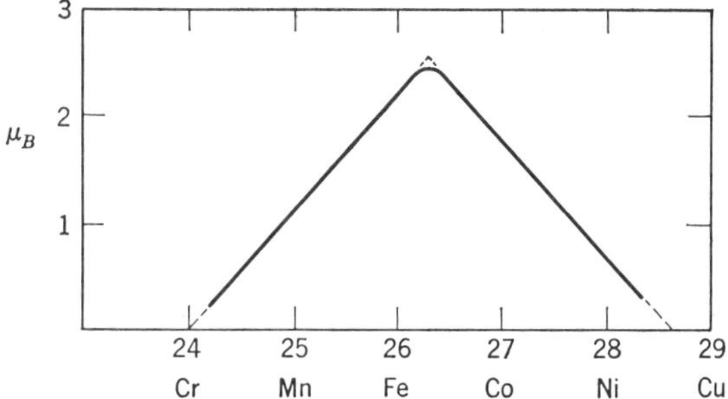

FIG. M.2.1. The saturation magnetic moments per atom of the first transition metal series and their alloys. The maximum occurs at 26.28 electrons per atom.

structures, the metal–metal bonds would be formed principally between any atom and its eight nearest neighbors. Since there is only one valency electron per atom for these metals, and any particular atom may only form two bonds by becoming an M^- atom, at the same time giving rise to an M^+ species by the scheme,

$$\begin{array}{cc} M^-\!\!\!-\!\!\!M \\ | \\ M \quad M^+ \end{array},$$

the relative probabilities of M^+ species being formed forming no bonds, the M species with one bond, and the M^- species with two bonds, are in the ratio of $7:8:4$; in the body-centered cubic crystal the probability of a metal–metal bond existing between any atom and its eight nearest neighbors is $1/8$ and the probability of no bond existing is $7/8$. Since every time an M^- species comes into existence so does an M^+ species, Pauling suggests that the probabilities of M^+ and M^- should be made equal by taking the geometric mean of the two relative probabilities; this gives for the alkali metals the relative probabilities of $M^+:M:M^- = 0.285:0.430:0.285$. The M^- species, already engaged in forming two bonds does not require an additional vacant state, but both M and M^+ require available energy states to form an additional bond. So the number of available energy states per atom required for the alkali metal atoms to switch bonds from one set of equivalent positions to another is given by the sum of the probabilities of M^+ and M, which is 0.715, corresponding to 28.5% of the species M^+ and M^-. For other metals, the similarly calculated value is obtained in the range between 0.715 and 2/3.

It is not possible to offer any explanation of the electrical conductivity of metals in terms of the Pauling valence-bond theory, without the concept of "metallic orbitals." Another major justification for this concept is the experimental observation that the cohesive energy of metals is so much greater than that of diatomic molecules. The dissociation energy of lithium molecules, Li_2, for example, is 5.61×10^4 J (g-atom)$^{-1}$ while the sublimation energy of lithium metal is 1.55×10^5 J (g-atom)$^{-1}$. Also in the alkali metals, the height of excitation of the three p-levels above the energy level of the s-valency electron is not too great (1.78×10^5 J for lithium, 1.95×10^5 J for sodium and 1.61×10^5 J for cesium, for example), so that the formation of mixed or hybrid sp-bonds almost certainly plays an important part in forming stronger metal–metal bonds than would be possible for pure s-states.

For the metals, beginning with the element with nuclear charge 19, potassium, up to nuclear charge 34, selenium, the number of bonding electrons per atom is regarded in terms of the Pauling hypothesis, to increase progressively from 1 in potassium to 5 in vanadium, and the bonds formed in the metals are regarded as essentially electron pair bonds as in simple molecules. With chromium, six of the group of nine available energy states (one $4s$, five $3d$, and three $4p$) would be occupied by bonding electrons and the cohesion is a maximum for this element. The number of bonding electrons per atom then remains constant for manganese, iron, cobalt, and nickel, but with increasing nuclear charge beyond 24 (chromium) paramagnetism appears as electrons occupy non-bonding energy states, reaching a maximum with iron, at which stage all of non-bonding energy states are occupied, while retaining 0.72 states per atom unoccupied. With one electron more in cobalt, pairing begins to occur and the paramagnetic moment falls successively for cobalt and nickel. For copper, the additional electron must be paired with those in the bonding states, so that the metallic valence of copper falls to 5.56. For nuclear charge numbers above that of copper (29), due to the progressive increase of the number of electrons in the bonding energy states, with spin pairing taking place, the metallic valence of these elements decreases to 4.56 in zinc, 3.56 in gallium, 2.56 in germanium, 1.56 in arsenic, 0.56 in selenium, to 0 in bromine.‡ These fractional values are quite compatible with other experimental observations. Magnetic susceptibility measurements indicate, for example, for the lanthanides that about 80% of metallic cerium atoms are present as Ce^{3+}, 20% as Ce^{4+}, similarly for samarium, 20% as Sm^{++}, 80% as Sm^{3+}.

‡ The description of the elements with nuclear charges from 19(K) to 36(Kr), in terms of the Pauling concept of metals is shown in Table M.2.1.

M.2.3.   Internuclear Distances in Metals and Intermetallic Compounds

For simple molecules like H_2, Li_2, Cl_2, and so on, the average number of electrons per bond between the atoms is two. The fractional bond numbers

TABLE M.2.1.[a]

		Available Energy States									Electron pair bonds per atom	Structure	Metallic radius	
Nuclear charge Z	Metal	6 bonding — 4s	3d	3d	3d	3d	3d	2.28 non-bonding — 4p	4p	0.72 conduction			R_{12} $\times 10^{-10}$ m	R_1 $\times 10^{-10}$ m
19	K	↓									1	bcc	2.349	2.025
20	Ca	↓	↓								2	hcp, fcc	1.970	1.736
21	Sc	↓	↓	↓							3	hcp, fcc	1.620	1.439
22	Ti	↓	↓	↓	↓						4	hcp, bcc	1.467	1.324
23	V	↓	↓	↓	↓	↓					5	bcc	1.338	1.224
24	Cr	↓	↓	↓	↓	↓	↓				6 (max)	bcc	1.276	1.186
25	Mn	↓	↓	↓	↓	↓	↓	↓			6	bcc, fcc, complex	1.268	1.178
26	Fe	↓	↓	↓	↓	↓	↓	↓	↓		6	bcc, fcc, hcp	1.260	1.170
27	Co	↓	↓	↓	↓	↓	↓	↓	↓δ	δ	6	hcp, fcc	1.252	1.162
28	Ni	↓	↓	↓	↓	↓	↓	↓↑	↓δ	δ	6	fcc	1.244	1.154
29	Cu	↓δ	↓	↓	↓	↓	↓	↓↑	↓↑	δ	5.56	ccp	1.276	1.176
30	Zn	↓↑	↓δ	↓	↓	↓	↓	↓↑	↓↑	δ	4.56	hcp	1.339	1.213
31	Ga	↓↑	↓↑	↓δ	↓	↓	↓	↓↑	↓↑	δ	3.56	orthorhombic	1.404	1.246
32	Ge	↓↑	↓↑	↓↑	↓δ	↓	↓	↓↑	↓↑	δ	2.56	diamond	1.444	1.242
33	As	↓↑	↓↑	↓↑	↓↑	↓δ	↓	↓↑	↓↑	δ	1.56	rhombohedral	1.476	1.210
34	Se	↓↑	↓↑	↓↑	↓↑	↓↑	↓δ	↓↑	↓↑	δ	0.56	hex. chains	1.40	1.17
35	Br	↓↑	↓↑	↓↑	↓↑	↓↑	↓↑	↓↑	↓↑	↓(nc)	0	orthorhombic (solid)	—	—
36	Kr	↓↑	↓↑	↓↑	↓↑	↓↑	↓↑	↓↑	↓↑	↓↑(nc)	0	ccp (solid)	—	—

[a] ↓—electrons with positive spin moment, α; ↑—electrons with negative spin moment, β; δ—fractional electron density in state concerned; fcc or ccp—face centered cubic; bcc—body centered cubic; hcp—hexagonal closest packed; nc—non-conducting; R_1—single bond metallic radius; R_{12}—metallic radius for coordination number 12.

defined by Pauling as $n = Z/N$, where Z is the number of valency electrons per atom and N is the coordination number of the atom concerned in the crystalline material, are such that $2n$ is the total number of valency electrons associated with the bond. In metallic lithium and the other alkali metals, crystallizing with body-centered cubic structures, there is one valency electron per atom available for forming the bond between any atom and the eight nearest neighboring and six more distant atoms. The two types of bond in these metals will obviously have different n values. If these are referred to as n_8 and n_6, then $8n_8 + 6n_6 = 1$, and the specific values of n_8 and n_6 are determined by the internuclear distances observed in the actual material. These internuclear distances have been tabulated[1] and it is observed that generally the distance between atoms in many compounds varies inversely with the number of electrons involved in bond formation. From the accurately known internuclear distances between carbon atoms in a great many organic compounds, Pauling has noted that the distance observed in bond lengths between two-electron, four-electron, and six-electron bonds ($C\!-\!C$, $C\!=\!C$, $C\!\equiv\!C$, given as 1.537×10^{-10} m, 1.335×10^{-10} m, and 1.202×10^{-10} m, respectively) are related by the equation, $D_n = D_1 - 0.71 \log n$, where $n = 1, 2,$ and 3. For bond orders of less than one, as in most metals, noting that the observed differences in bond lengths between multiple electron bonds are nearly the same for all atoms and that the law of force between atoms bonded in this way is exponential in nature, Pauling has proposed that a similar relationship applies for bonds of fractional order, viz.

$$D_1 - D_n = 0.600 \log n, \qquad \text{or} \qquad R_1 - R_n = 0.300 \log n, \quad \text{(m.2.1.)}$$

where the proportionality factor is 0.600 for metals and intermetallic compounds. The form of this expression, although suggested by theoretical considerations, is nevertheless empirical in nature, and the coefficient 0.300 is probably not really a constant for all kinds of atoms. Thus, for lithium, there are eight other lithium atoms at a distance of 3.039×10^{-10} m from any one and six at a distance of 3.503×10^{-10} m. Substituting these internuclear distance values into (m.2.1.) and solving this equation simultaneously with $8n_8 + 6n_6 = 1$, it is found that $n_8 = 0.111$ and $n_6 = 0.018$ with R_1 for lithium as 1.230×10^{-10} m ($R_1 = D_1/2$), and this value is considerably smaller than R_1 for Li_2, which is observed to be 1.336×10^{-10} m, but for a nearly pure s-type bond. It is apparent from calculations of this type that even a small increase in bond length makes a large difference in the bond order n.

---

[1] *Tables of Interatomic Distances and Configuration in Molecules and Ions*, Special Publication No. 11, 1958, The Chemical Society, London, W.1.; Supplement 1956–59, Special Publication No. 18, 1965, The Chemical Society, London, W.1.

Pauling has shown that the values of R_1 may often be obtained by other means and then made use of to calculate internuclear distances in metals and intermetallic compounds which may then be used to make deductions about the bond type and the total number of valency electrons per atom. However, the corrections that must be applied for the variation of metallic radii with coordination number and with bond order are purely empirical, so that care has to be exercised in assigning significance to the conclusions arrived at in this manner.

As an example of the way in which Pauling's ideas account for the polymorphism, common among metals, and the use of other means of obtaining R_1, tin may be considered. Here, there are two known modifications, gray tin and white tin; the gray modification, stable below 291 K, is non-conducting with none of the typically metallic attributes, and crystallizes with the diamond-type structure‡ with each tin atom tetrahedrally coordinated at a distance of 2.810×10^{-10} m.

‡ This structure, initially discovered by W. H. and W. L. Bragg,[1] is illustrated in Fig. M.2.2. The structure has been re-examined many times[2] since then and the crystal lattice constant has been determined with great precision, as $a_0 = 3.56679 \times 10^{-10}$ m. Each carbon atom is bonded to four equidistant neighbors, so that the tetrahedral environment shown for only four carbon atoms in Fig. M.2.1. extends throughout the whole crystal. The bond angles, 109° 28′ and the C—C bond lengths, $(1.54452 \pm 0.00014) \times 10^{-10}$ m are the same as in simple molecules like $C(CH_3)_4$ and in many saturated hydrocarbons.[3] In the cubic unit cell of the diamond structure carbon atoms are located at each of the lattice points of the face-centered cubic structure as well as occupying one half of the tetrahedral holes in the cubic closest packed arrangement. These four additional carbon atoms are themselves arranged tetrahedrally about the center of the cubic unit cell. Diamond is actually a metastable phase in the carbon system, graphite being the phase normally stable under ordinary conditions, although the value and stability of diamond is associated with the fact that the rate of transition to the stable graphite phase is extremely slow under normal conditions. Much interest has been centered in the high thermal conductivity of diamond, almost as great as that of copper; attempts to synthesise it;[4] and in calculations of the lattice vibrational frequency spectrum of diamond.[5] It has not yet been possible to measure experimentally the lattice vibrational frequencies for diamond by the technique of inelastic scattering of thermal neutrons, although this has been done for silicon and germanium, the other elements apart from gray tin, which crystallize with the diamond structure.[6]

If the four carbon atoms in the diamond structure, occupying the tetrahedral holes, are replaced by a different kind of atom, then the composition becomes MX, where M

[1] Bragg, W. H., and Bragg, W. L., *Nature*, 1913, **91**, 557; *Proc. Roy. Soc.*, 1913, **A89**, 277.
[2] Göttlicher, S., and Wölfel, E., *Z. Elektrochem.*, 1959, **63**, 891.
[3] Lonsdale, K., *Phil. Trans. Roy. Soc.*, 1946, **A240**, 219.
[4] Suits, C. G., *Am. Sci.*, 1964, **52**, 395.
[5] Smith, Helen M. J., *Proc. Roy. Soc.*, 1947, **A241**, 105.
[6] Cochran, W., *Proc. Roy. Soc.*, 1959, **A253**, 260.

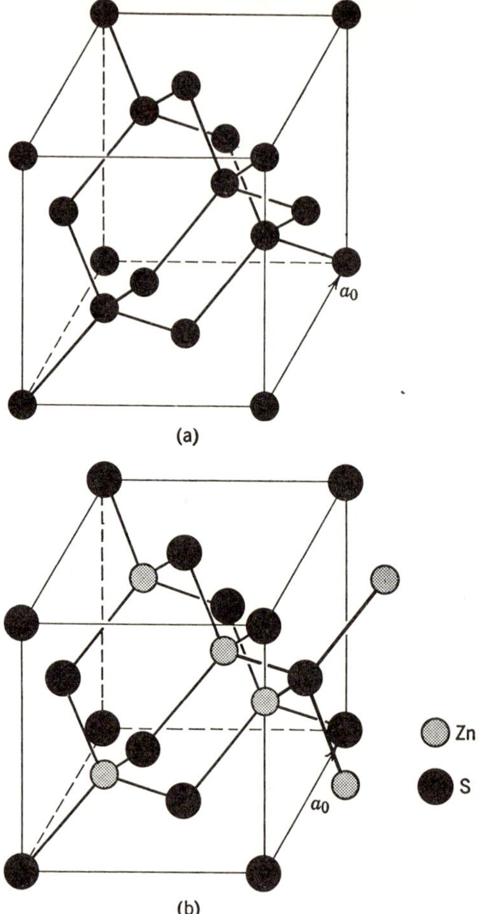

FIG. M.2.2. The diamond (a) and the zinc blende
or sphalerite structures (b).

denotes the atoms occupying the cubic closest packed positions and X the atoms in the tetrahedral holes. This arrangement also was found by Bragg and Bragg[1] for the mineral sphalerite or zinc blende, ZnS, and has since been found to be adopted by a large number of other compounds,[2] such as AgI, AlP, BN, HgSe, GaAs, SiC, and BeS. This structure is shown in Fig. M.2.1.(b). In many of these compounds the bonds are partially ionic in character and they function as semi-conductors. Much interest attaches to the arsenides and antimonides of gallium and indium, all of which crystallize with the zinc blende

[1] Bragg, W. H., and Bragg, W. L., *Proc. Roy. Soc.*, 1913, **A89**, 277.
[2] Wyckoff, R. W. G., *Crystal Structures, 1958*, Second Edition, **Vol., I**, Interscience, New York.

structure, especially since the discovery of the ability of gallium arsenide to function as a solid state laser material (page 148). The lattice vibrational spectra of gallium arsenide and other crystals having the zinc blende structure have attracted a great deal of theoretical interest.[1-6]

In the compound tetramethyltin, $Sn(CH_3)_4$, where quite apparently the tin atoms are tetrahedrally coordinated with the four carbon atoms of four methyl groups and form four two-electron bonds, the tin–carbon distance is observed to be 2.18×10^{-10} m. Subtracting from the Sn–C distance the known radius of a carbon atom (the so-called single bond radius) of 0.77×10^{-10} m, gives the single (electron pair) bond radius of tin as 1.41×10^{-10} m. The observed internuclear distance in gray tin is thus what would be anticipated for an electron pair bond. The nuclear charge of tin atoms is 50, with 14 electrons in excess of the rare gas number for krypton (36). The nine stable energy states available for these 14 electrons would be the five $4d$-states, the $5s$-state, and the three $5p$-states. With 10 of the available 14 electrons from each tin atom distributed among these nine energy states so that 10 electrons with opposed spin moments are in the $4d$-states, and four unpaired electrons in the $5s$ and three $5p$-states, such an atom would be expected to form four tetrahedral (sp^3) bonds, like carbon (in diamond), silicon, and germanium, with no vacant state for metallic conduction electrons. White tin, on the other hand, does exhibit metallic characteristics and crystallizes in the tetragonal system, with each atom having four nearest neighbors at 3.022×10^{-10} m and two others at 3.181×10^{-10} m, as shown in Fig. M.2.3. Assuming that the single bond distance for tin is 2.82×10^{-10} m, the bond orders for the Sn–Sn bonds of length 3.022×10^{-10} m and 3.181×10^{-10} m may be calculated, using (m.2.1.) as 0.46 and 0.25, respectively, corresponding to a value of $[(4 \times 0.46) + (2 \times 0.25)] = 2.34$, for the number of valency electrons per atom. Pauling gives reasons for believing that the single bond radius for tin has the value 2.848×10^{-10} m rather than 2.82×10^{-10} m, which leads to the value 2.64 for the number of valency electrons per atom for tin. This value agrees reasonably well with that deduced on the assumption that a metal has 0.72 states per atom vacant for electronic conduction. Thus, for tin with 14 electrons per atom in excess of the rare gas number to be accommodated among the nine states $4d^5$, $5s$, and $5p^3$, leaving

[1] Dolling, G., *Inelastic Scattering of Neutrons in Solids and Liquids*, 1962, International Atomic Energy Agency, **Vol. II**, 37.

[2] Cowley, R. A., *Proc. Roy. Soc.*, 1962, **A268**, 121.

[3] Demidenko, Z. A., Kucher, T. I., and Tolpygo, K. B., *Fiz. tverd. tela.*, 1961, **3**, 2482; translation, Sov. Phys. Solid-State, 1962, **3**, 1803.

[4] Kaplan, H., and Sullivan, J. J., *Phys. Rev.*, 1963, **130**, 120.

[5] Waugh, J. L. T., and Dolling, G., *Phys. Rev.*, 1963, **132**, 2410.

[6] Dolling, G., and Waugh, J. L. T., *Lattice Dynamics*, 1965, Ed. R. F. Wallis, Pergamon Press, Oxford, 19.

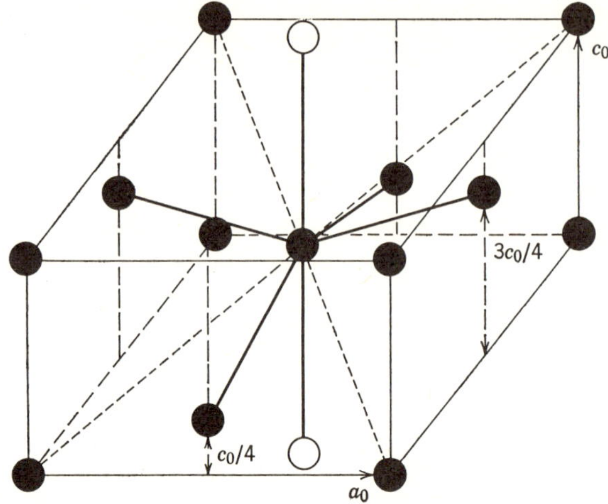

FIG. M.2.3. The tetragonal unit cell of the white tin structure. Each tin is surrounded by a distorted tetrahedron of neighbours and by two other atoms slightly more distant; the coordination polyhedron around the central tin atom is shown, being completed by the four atoms in special positions in the unit cell and the two atoms ○ in adjacent unit cells.

0.72 states vacant, there would be 8.28 states for accommodation of both the bonding and non-bonding electron pairs. This necessitates that $5.72 = (14 - 8.28)$ non-bonding electron pairs be present and leaving 2.56 electrons per atom available for forming metal–metal bonds. It is not clear why in white tin the structure with 72% of its atoms having the configuration $4d^{10}5s^25p^2$ and 28% of the configuration $4d^{10}5s5p^3$ is more stable than the arrangement containing all the tin atoms with the configuration $4d^{10}5s^25p^2$ and one vacant $5p$ per atom.

The occurrence of fractional valencies of all kinds may be interpreted in a similar manner to the above description for tin. Zinc, as another example, crystallizes in the hexagonal system, but not with a closest packed structure. Ideal hexagonal closest packing corresponds to an axial ratio, $c/a = 1.633$ (page 391), but for zinc $c/a = 1.856$. Therefore each zinc atom has six nearest neighboring atoms coplanar with itself (at 2.6649×10^{-10} m) and six others in two planes above and below (at 2.9129×10^{-10} m). In terms of (m.2.1.), these bond lengths correspond to six bonds of order 1/2 and six bonds of order 1/4, equivalent to $4\frac{1}{2}$ electrons per atom for bond formation. This is interpreted as signifying that 75% of the zinc atoms in the metal have the configuration $1s^2 2s^2 2p^6 3s^2 3p^6 3d^9 4s^1 4p^2$ and that 25% have the configuration $1s^1 2s^2 2p^6 3s^2 3p^6 3d^8 4s^1 4p^3$. Similarly, the observed magnetic moment of nickel of 0.61 Bohr magnetons is to be interpreted as due to the metal containing 70% of nickel atoms of configuration $\cdots 3d^7 4s^1 4p^2$ and 30% with configuration $\cdots 3d^6 4s^1 4p^3$.

With the use of equation (m.2.1.) and the observed values of the internuclear distances in metal crystals, applying certain corrections for differing coordination numbers, Pauling[1] has calculated a complete set of metallic radii for coordination number twelve and single bond radii.

M.2.4. Metallic Radii

For metals with ideal closest packed structures, the metallic radius for coordination number 12 is simply one-half of the internuclear distance between any atom and its twelve equidistant neighboring atoms. In almost all hexagonal closest packed metal structures the axial ratio, c/a, is slightly under the ideal value of 1.633, falling between 1.63 and 1.57. This means that atoms in the same basal plane are a little further apart than those immediately above or below them. With the exception of beryllium, the axial ratios of these elements increase with temperature, thus tending to equalize the internuclear distances at higher temperatures. The axial ratios for zinc and cadmium are greater than the ideal value, being 1.8563 and 1.8858, respectively, which effectively makes the internuclear separation in the basal plane in excess of 10% less than that between planes of atoms, and while the axial ratio of both these metals also increases with temperature, the effect of increasing temperature is to further distort the approximation to hexagonal closest packing. For metals which have axial ratios only slightly different from the ideal value, the metallic radius for coordination 12 may be taken as the mean of the two different sets of six internuclear distances; but for zinc and cadmium and other metals which crystallize with structures with lower coordination numbers, the radii for 12-coordination have to be found in some other manner.

The use of an empirical factor, deduced from the examination of the known internuclear distances in metals and alloys, as originally proposed by Goldschmidt, by the use of which the observed internuclear distance for one coordination number could be converted to that for 12-coordination by simple multiplication, provides the basis for an approximate set of metallic radii. Alternatively, from the observed value of the internuclear distances in

---

[1] Pauling, Linus, *The Nature of the Chemical Bond*, 1962, Third Edition, Cornell University Press, Ithaca, New York.

alloys, crystallizing with closest packed structures, and on the assumption that one of the constituents of the alloy maintains the same metallic radius as in the elementary metal, the metallic radius of the other constituent metal may be obtained. For example, the 12-coordination radius of germanium may be obtained knowing the internuclear distance in the hexagonal closest packed structure of Cu_3Ge and the Cu–Cu distance in metallic copper. Another method is to deduce the 12-coordination metallic radius of a metal by examination of the variation of the cell dimensions of the solid solutions formed by that metal with another which is known to crystallize with a closest packed structure, and then to extrapolate to 100% of the metal concerned on the assumption that there is a linear relationship between concentration of the metal concerned, as solute species in the solvent species (known to have a closest packed structure), and the cell dimensions of the solid solutions.

The 12-coordinate radii of the various metallic elements, no matter how deduced, display a general periodic variation with the nuclear charges of the metal atoms, in general agreement with other cohesive-energy-dependent properties. There is however, no general increase in metallic radii with nuclear charge. Thus, uranium ($Z = 92$) has about the same value for its metallic radius as lithium ($Z = 3$). Also, while there is a general increase in magnitude of metallic radii of the metals of the second transition series (Y–Cd) compared with the first transition series (Sc–Zn), there is no similar increase in radii of the third transition series metals (La–Hg) compared with the second transition series; thus, the metallic radius of silver is about 13% greater than that of copper but essentially the same as that of gold.

Pauling has sought to explain the large decrease in the single bond radius of the transition elements from potassium to chromium in the first transition series, and the similar type of behavior for the second and third transition series, in terms of the fraction of s, p, and d states engaged in bond formation. By comparing these single bond metallic radii with tetrahedral and octahedral radii of the same metal atom, as deduced from the observed internuclear distances in substances containing that metal atom where it is tetrahedrally or octahedrally surrounded by atoms of otherwise known radii, he has concluded that while the bond in potassium is mainly of the s-type with about 26% p-type character, that in chromium the bond type is about 40% d-type. In the application of the Pauling equation, (m.2.1.), to the calculation of single bond radii and bond orders, it is found that although bond orders may be generally non-integral quantities, as for example, 0.11 in lithium, bond orders such as 1/2, 1/3, 2/3, and 1/4 are of quite general occurrence, and it has been suggested by Pauling and by Rundle[1] that special stability is associated with bonds of these orders.

---

[1] Rundle, R. E., *J. Am. Chem. Soc.*, 1947, **69**, 1327; *J. Chem. Phys.*, 1949, **17**, 671.

M.2.5.   Extensions and Modifications of the Pauling Concept of Metals

For free copper atoms, the set of $3d$-states are all occupied by electrons and there is only one other electron in a $4s$-state apparently available for bond formation. But there is certainly no comparison between the lattice energy, and cohesive-energy-dependent properties, of copper and potassium, indicating that electrons in d-states play a considerable part in the bonding between copper atoms. The Pauling valence bond description of the nature of the transition metals relates their internuclear distances and lattice energies to well known chemical phenomena, but it does not provide any basis for the independent calculation of the magnitude of these chemical properties. It is for this reason that criticism has been leveled against the Pauling hypothesis[1] as simply a description of the known experimental data in terms of empirical assumptions that cannot be deduced independently. The most satisfactory theoretical explanations of natural phenomena are those which, on the basis of a relatively small number of assumptions, enable the generalization of a large number of facts to be made or which enable the magnitude of the properties of substances to be calculated by techniques which are independent of any assumptions about these properties or their relative values.

Various modifications and minor extensions of the Pauling ideas about metals have been made;[2-5] these are predicated on the basis that closer observation of the magnitudes of the various physical properties of the transition metals indicate that the strength of bonding in chromium is less than in its congeners molybdenum and tungsten, and that in manganese the strength of the metal–metal bonds is exceptionally low. Also, the coefficients of linear expansion, the compressibilities, the heats of sublimation, the melting points, and metallic radii, of iron, cobalt, and nickel in the first transition group of metals, resemble one another more closely than do the respective properties of ruthenium, rhodium, and palladium of the second group of transition metals, or those of osmium, iridium, and platinum of the third transition group. There is at the same time a smaller difference between the physical properties of the elements from iron to copper, than there is in the corresponding elements ruthenium to silver and osmium to gold. Manganese is abnormal among the first transition group metals in that it crystallizes with four different crystallographic modifications, one of which is a

---

[1] Hume-Rothery, W., *Ann. Rep. Chem. Soc.*, 1950, **46**, 42.
[2] Hume-Rothery, W., Irving, H. M., and Williams, R. J. P., *Proc. Roy. Soc.*, 1951, **A208**, 431.
[3] Altmann, S. L., Coulson, C. A., and Hume-Rothery, W., *Proc. Roy. Soc.*, 1957, **A240**, 145.
[4] Griffith, J. S., *J. Inorg. & Nucl. Chem.*, 1956, **3**, 15.
[5] Griffith, J. S., and Orgel, L. E., *Nature*, 1958, **181**, 170.

simple closest packed structure.‡ Hume-Rothery, Irving, and Williams, attribute the increase in metallic radii from chromium to manganese (γ and δ) and the relatively small change from molybdenum to technetium and from tungsten to rhenium, as due to the exceptional stability associated with the dipositive manganese ion, Mn^{++}, with the configuration $\cdots 3d^5$.

‡ All four of the crystallographic modifications of manganese have cubic symmetry, two of them complex and the other two simpler.

α-manganese has a structure derived from the simple body centered cubic arrangement: if a larger unit cube is imagined of edge length three times the edge length of a simple body-centered cubic cell, then such a large unit cell would contain 54 atoms, and the α-manganese structure is derived from this by suitable distortion of all of the atomic positions so that four additional manganese atoms can be accommodated, to give 58 atoms per unit cell. In this complex arrangement each manganese atom has between 12 and 16 neighbors at distances varying from 2.24×10^{-10} m and 3.00×10^{-10} m. By assigning coordinates to the various atoms, choosing one vertex of the unit cell as origin, the positions of all of the atoms in the α–Mn unit cell may be described as:

(a)   two atoms at the body centered positions, 000 and 1/2 1/2 1/2, each atom with coordination number 16;

(b)   eight atoms at www, $w\bar{w}\bar{w}$, $\bar{w}w\bar{w}$, $\bar{w}\bar{w}w$, with $w = 0.317$, and each of these 8 atoms 16-coordinated;

(c)   twenty four atoms, each 13-coordinated, at uuv, vuu, uvu, $u\bar{u}\bar{v}$, $v\bar{u}\bar{u}$, $\bar{u}\bar{u}\bar{u}$, $\bar{u}u\bar{v}$, $\bar{v}u\bar{u}$, $\bar{u}v\bar{u}$, $\bar{u}\bar{u}v$, $\bar{v}\bar{u}u$, and $\bar{u}\bar{v}u$, with $u = 0.317$ and $v = 0.042$;

(d)   another twenty four atoms in the same general set of positions as in (c) above, but with each manganese atom 12-coordinated and with $u' = 0.089$ and $v' = 0.278$.

β-manganese, which is the stable modification between temperatures of about 1073 K and 1373 K, has a structure such that the cubic unit cell contains 20 atoms, such that each atom is 12-coordinated at distances varying between 2.36×10^{-10} m and 2.67×10^{-10} m. The positions of all 20 atoms in the unit cell, again in fractions of the unit cell edge are:

(a)   8 atoms at uuu, $(u + 1/2)(1/2 - u)\bar{u}$, $\bar{u}(u + 1/2)(1/2 - u)$, $(1/2 - u)\bar{u}(u + 1/2)$, $(1/4 - u)(1/4 - u)(1/4 - u)$, $(u + 3/4)(3/4 - u)(u + 1/4)$, $(3/4 - u)(u + 1/4)$ $(u + 3/4)$, $(u + 1/4)(u + 3/4)(3/4 - u)$;

(b)   12 atoms at the positions $(1/4 - v)v1/8$, $(3/4 - v)(1/2 - v)7/8$, $(v + 3/4)(v + 1/2)$ $3/8$, $(v + 1/4)\bar{v}5/8$, $1/8(1/4 - v)v$, $7/8(3/4 - v)(1/2 - v)$, $3/8(3/4 + v)(1/2 + v)$, $5/8(1/4 + v)\bar{v}$, $v1/8(1/4 - v)$, $(1/2 - v)7/8(3/4 - v)$, $(1/2 + v)3/8(3/4 + v)$, $\bar{v}5/8(1/4 + v)$, with $u = 0.061$ and $v = 0.206$.

In γ-manganese, stable above 1373 K, the structure is cubic closest packed, and at 1408 K γ-manganese transforms to δ-manganese with a body centered cubic structure, which is the stable modification up to the melting point at about 1518 K.

In rhenium compounds, displaying an array of higher oxidation states, the dipositive state being apparently much less stable, it is argued by the same authors, i.e., Hume-Rothery, Irving, and Williams, that the high melting point and small metallic radius of rhenium in the metal is attributable to the greater number of electrons per atom involved in metallic bonding than the two per atom, which are of such importance in γ- and δ-manganese. Hume-Rothery, Irving, and Williams, interpret the available data as indicating that the number of bonding electrons per atom for chromium is less than six; that this number decreases further for manganese and then increases for iron; but that for the second and third transition metal series there are six electrons per atom engaged in metal–metal bonding for molybdenum, technetium, ruthenium, and for tungsten, rhenium, and osmium, as in Pauling's scheme; there is a decrease in the number of bonding electrons per atom for ruthenium to palladium and for osmium to platinum, but much less of a decrease in this number for the metals iron to nickel. These postulates are additionally supported by a detailed examination of metallic binding energies.[1]

Consideration of the possible directional nature of the metal–metal bonds in the transition metals and the number of electrons per atom and the energy states available have led others[2,3] to the conclusion that most of the cohesion is due to electrons in d-states with one electron per atom existing in an independent s-state and being responsible for the electrical conductivity. In this view, it is considered that the five d-states are split into a group of three bonding states and two repulsive states. This description, however, is contrary to the valency bond description of the bonding in many chemical compounds in terms of mixed sp and spd types.

Altmann, Coulsen, and Hume-Rothery[4] have attempted to understand the directional character of metallic bonding in the transition metals and to account for the relative incidence of close packed structures in terms of the symmetry of the s, p, and d-type wave functions. In the transition metals electrons described by s, p, and d-type orbitals are present and the electron charge densities of all of these are centrosymmetrical with respect to the atom which contains electrons in these states. The combination of s and p type wave functions gives rise to a non-centrosymmetrical charge distribution and the s and d combination gives rise to a symmetrical charge distribution around the atomic nuclei concerned. It is considered that for the combination of wave functions leading to a non-centrosymmetrical charge distribution, the probability of a strong bond being formed is greater in the direction of greater electron density, thus favoring the directional properties. It is not possible to form eight combinations of s, p, and d type functions, corresponding to

---

[1] Baughan, E. C., *Trans. Faraday Soc.*, 1954, **50**, 322.
[2] Bader, F., Ganzhorn, K., and Dehlinger, U., *Z. Physik*, 1954, **137**, 190.
[3] Dehlinger, U., *Theoretische Metallkunde*, 1955, Springer Verlag, Berlin.
[4] Altmann, S. L., Coulson, C. A., and Hume-Rothery, W., *Proc. Roy. Soc.*, 1957, **A240**, 145.

cubic symmetry for the set, as would be necessary to account for the body-centered cubic arrangement, but a combination with tetrahedral symmetry may be formed by suitably combining an s and three d type functions,[1] and since these combinations are centrosymmetrical they give rise to eight equivalent directions, oriented towards the eight vertices of a cube, of bonding. The same would apply to the combination of four d-type functions, again giving rise to eight equivalent directions of similar charge density. In the view of Altmann, Coulsen, and Hume-Rothery, the combination of three d-type orbitals, again corresponding to centrosymmetrical charge distribution, could account for the formation of six bonds to the six neighboring atoms which are 15% more distant than the eight nearest neighbors in the body-centered cubic structure, so that the total bonding for such an arrangement couldbe written as $[sd^3]^a[d^4]^b[d^3]^c$. It is not necessarily so that $(a + b + c) = 1$, since other possible combinations may occur or be necessary for the description of other aspects of the metallic crystal concerned apart from the directional characteristics of the metal–metal bonds involved.

Proceeding in this manner by considering the directions of the bonds in the closest packed structures, Altmann, Coulson, and Hume-Rothery conclude that for the hexagonal closest packed arrangement, there are six equivalent sd^2 bonds in the basal plane (centrosymmetrical) and six pd^5 and spd^4 mixed, symmetrical and non-centrosymmetrical, bonds to the six neighboring atoms in planes above and below the central one. In the face-centered cubic or cubic closest packed arrangement, they consider that there are four equivalent sets of p^3d^3 mixed centrosymmetrical and non-centrosymmetrical bonds together with twelve sd^5 centrosymmetrical bonds. According to these descriptions, the average concentration of d-type functions involved in bond formation for the hexagonal closest packed structure, ranging from 2/3 to 5/6, is lower than for the body centered cubic structure, where it varies from 3/4 to unity. The mixed p^3d^3 combination has the lowest concentration of d-type functions. Thus, the body centered cubic structure would be expected to arise at the point where the maximum number of electrons in d-states are involved in bond formation with the hexagonal closest packed structure occurring for more than or less than this number, and the cubic closest packed structure occurring for the minimum concentration of d-electron states engaged in bond formation, at the beginning and end of each series of transition metals.

For each of the three transition series of metals the order of the electronic energy levels for the free atoms is $np > (n - 1)d > ns$, but for the terminal members of each of the three series the order of the energy levels is $np > ns > (n - 1)d$, so that for increasing nuclear charge along the transition

---

[1] Ganzhorn, K., *Naturforsch.*, 1953, **A8**, 330.

metal series the d-energy states become of lower energy. Now, since for the face-centered cubic structure requires mixing of p-functions with only three d-functions to form p^3d^3 types but with four or five d-functions in the case of cubic and hexagonal closest packed structures, and this process becomes less probable at the end of the transition series where the d-states adopt successively lower energy values, it appears plausible why the hexagonal closest packed structure is observed as one of the polymorphic modifications of the metals at the beginning of the transition metal series but is of less importance for the terminal members of the series.

The occurrence of hexagonal closest packed structures for technetium, ruthenium, rhenium, and osmium, but not for manganese or iron, may also be interpreted, in Hume-Rothery's view, as evidence substantiating the belief that there are fewer than six electrons per atom available for bond formation in the series chromium through iron in the first transition metal series, but about six electrons per atom available for the metals molybdenum to ruthenium and tungsten to osmium, in the second and third transition metal series, respectively. This arises because, if there are fewer d-electrons available for bond formation, then the hexagonal closest packed structure becomes less probable since it requires four or five d-orbitals to form the mixed bonds necessary to account for the directional nature of the neighboring atoms surrounding any other atom, compared with the three d-orbitals required for forming the mixed types to account for the characteristics of the face centered and body-centered cubic structures.

Neither the Dehlinger, nor Hume-Rothery, nor Griffith (page 451, which also describes the electronic distribution in the transition metals in terms of two groups with up to six electrons per atom in bonding states and the others in non-bonding states) modifications of the Pauling concept of metals lends itself particularly to quantitative calculations. They are all essentially qualitative descriptions. The detailed mathematical theory of metals, which has originated with the Wiedemann–Franz law, and is necessarily quantum-mechanical in nature, has had many spectacular successes, and some of the more important aspects and deductions which have been made from the so-called electron theories of metals will be discussed in the following chapters.

The Free Electron Theory of Metals

M.3.1. The Soft X-Ray Spectra of Metals

As discussed in sections Q.1.4. and Q.1.7., a free atom may exist in a number of definite energy states, each characterized by a definite distribution of its constituent electrons among the possible quantum states. In the normal state of a free atom, the electrons are distributed among the different energy states so as to lead to the lowest possible energy subject to the applicability of the Pauli Exclusion Principle that not more than one electron is in a state defined by any given combination of the four quantum numbers, n, l, m_l, and m_s. The characteristic lines of the optical and the high frequency or X-ray spectra of free atoms are attributable to transitions of the atom from one energy state to another, such that $E_1 - E_2 = h\nu$, where E_1 and E_2 are the energies of the excited and the normal states and ν is the frequency of the emitted radiation. The sharpness of the spectral lines is a consequence of the definite energy associated with each atomic state. For spectra in the visible region with wavelengths of the order of 5×10^{-7} m, since the velocity of electromagnetic radiation is of the order of 3×10^8 m s^{-1}, the frequencies of optical spectra are of the order of 6×10^{14} s^{-1} and thus as h is approximately 6.6×10^{-34} J s, the electronic transitions associated with optical spectra involve energy differences of about 10^{-19} J atom^{-1}. For X-ray spectra, with emission lines of wavelength of the order of 10^{-10} m, energy changes of the order of 10^{-16} J atom^{-1} are involved, which is about 10^3 times greater than for optical spectra. Not all electronic transitions are possible, however, but only those which conform to certain Selection Principles having a high transition probability (pages 54–58, 70, 360–362). It is of course from the detailed study of the frequencies and intensities of the emission lines in the spectra of atoms that it has been possible to deduce the energies of the various electronic states and the probability of transitions taking place between one state and another.

For solid crystalline materials, the conditions pertaining are very different from those of free atoms, for here large numbers of atoms of the order of 10^{23}, at distances from one another of the order of 10^{-9} m to 10^{-10} m are involved. Very approximately, however, the wave functions of the electrons associated with the atoms composing the periodic field of a crystal may be regarded as being derived from the wave functions of the electrons for the free atoms, retaining some of the symmetry characteristics of the s, p, d, f, ... states. In potassium, for example, the valency electrons of the free atoms are in $4s$-states, and in solid potassium metal there are electrons in states derived from the $4s$-states of the free atom, and similarly the $1s$, $2s$, $2p$, $3s$, $3p$, states of the free atom will correspond to the appropriate electronic states in the solid material. In a similar way to that in which the energy characteristics of the electrons in a free atom have been determined from the examination of the emission spectra of free atoms, so the electronic energies in a solid crystalline material may be arrived at by examination of the spectra emitted by solids, under suitable conditions of excitation, most generally brought about by electron bombardment. When a metal is irradiated with high energy electrons, it may happen that an electron is expelled from say the $1s$ or K quantum group as a result of the electron impact, and then another electron may undergo the transition from the L quantum group to the K group, with the emission of a quantum of radiation of frequency given by $E_L - E_K = h\nu$, where the frequency of the observed $K\alpha$ emission corresponds to the difference in energy between the K and L electrons in the solid, as distinguished from their energy difference in the free atom. By selecting the appropriate X-ray emission spectra, it is similarly possible to examine the energy characteristics of electrons in solids. For example, the $K\beta$ emission would provide information about the energy difference between the $3p$ and the $1s$ states, the L_3 emission information about the difference in energy between the $3s$ and the $2p$ states, and so on.

It is observed that the $K\alpha$ emission line in the X-ray spectrum of, for example, nickel is found to have the same frequency whether produced by solid nickel metal, or any other nickel compound, and in fact the $K\alpha$ wavelength for nickel is apparently unaffected by the state of combination of the nickel atoms. Similarly, for the lowest energy states of all but the lightest elements, the observed $K\alpha$ emission line wavelengths are independent of the state of combination of the element. This effect may be understood from the point of view that, while interaction between atoms involves the redistribution of the valency electrons, in states of lower energy and more distant from the nucleus, among the various nuclei, those electrons in the highest energy states nearest to the nucleus are so greatly influenced by the one nucleus that they are effectively undisturbed by any changes in the electron distribution further away from the atomic nuclei, brought about by the influence of other atoms at distances away from them of the order of 10^{-10} m. It may be imagined that as the free atoms in a dilute gas of a monatomic metal vapor are compressed together, until the metal nuclei adopt their equilibrium distance pertaining in the solid material, the sharp energy levels of the free atoms become spread over a range of energy values in the large assembly of atoms in the solid metal, the broadening or spreading of the energy levels being greatest for the electrons in states most distant from the nuclei, and least for the electrons in the highest energy states closest to the various nuclei. The actual extent to which this process takes place will depend on the particular electronic state in the atom concerned, the environment of the atoms in the structural arrangement involved, or on the structure and type of interatomic bond formed by the atoms in the solid. In this manner, it is possible to account for the fact that the observed soft X-ray spectrum of a metal is a band spectrum rather than the line spectrum of free atoms. Among the earliest observations on the soft X-ray spectra of metals were those of Skinner,[1] who by making use of liquid air-cooled targets was able to observe the influence of temperature on the electronic energies in metals. It is a direct conclusion from the experimental work on soft X-ray spectra that the energies of the valency electrons of a metal are distributed over a range or band of energies, in distinct contrast to the sharp energy levels in free atoms.

The soft X-ray band spectrum of sodium, for instance, may be interpreted in the following manner. In the ground-state of the free sodium atoms, the electrons occupy the lowest available energy states with the configuration, $1s^2 2s^2 2p^6 3s^1$, with one electron in the $3s$-state and none in the next higher available $3p$-state. These sharp energy levels may be represented diagramatically as in Fig. M.3.1. The observed X-ray band spectrum of sodium may be interpreted on the assumption that as the assembly of sodium atoms in the metal are brought together until the internuclear distance adopts its equilibrium value of 3.67×10^{-10} m, the sharp energy levels of the electrons in the free atoms become spread out over a range of values, as illustrated in Fig. M.3.1., to an extent dependent on their energy in the free atoms and their distances from the metal nuclei. Thus, at a distance of 3.67×10^{-10} m between the sodium nuclei in metallic sodium, the $1s$, $2s$, and $2p$ energy levels would scarcely be affected, but the $3s$ and $3p$ energy levels would be broadened to such an extent that they overlap. The experimental work of Skinner has shown that the width of the valency electron band in the metals sodium, magnesium, and aluminum is of the order of $(1 - 20) \times 10^{-19}$ J. In the case of metallic sodium, the valency electrons of lowest energy can best be de-

---

[1] Skinner, H. W. B., *Phil. Trans. Roy. Soc.*, 1940, **A239**, 95.

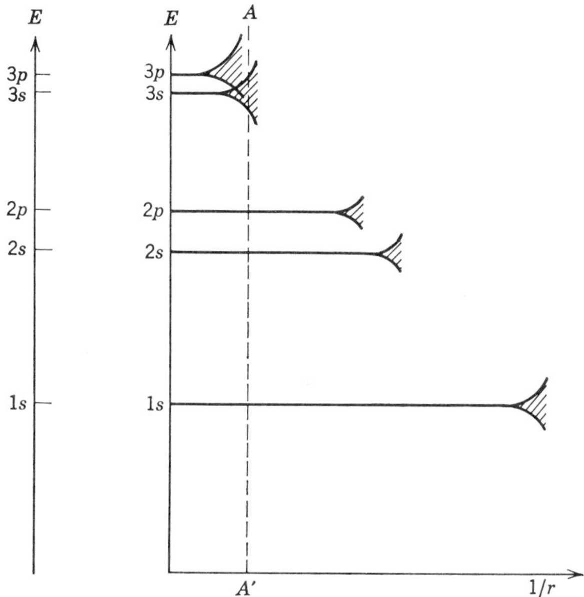

FIG. M.3.1. Diagrammatic representation of the energy levels in free atoms of sodium (at left), and in sodium metal (at right), illustrating the broadening of the free atom energy levels as the atoms are brought together. AA' corresponds to the internuclear distance in sodium metal (3.67×10^{-10} m).

cribed by $3s$ wave functions but modified by the periodic potential field of the crystal lattice, and with increasing energy extending towards the top of the valency or conduction band, $3p$ type characteristics become more important. It is no longer possible to regard the valency electrons in sodium metal as being described by either s or p type functions, but rather by a combination of these. In any transition of an electron from the valency band to the K quantum group in sodium, it is apparent that there cannot be one constant difference in energy between the states before and after the transition. At the internuclear distance applicable in the metal, if the valency electron energies extend over a range of values, there will be a range of possible transition frequencies and the emission will consist of an emission band, instead of a single, well defined, line. Very generally, for any assembly of atoms, the valency electron band width becomes larger for decreasing internuclear distance, and the number of energy states within the band increases with the number of atoms concerned and thus with the size of the piece of crystalline material involved. The band width would be the same for a large crystal as for

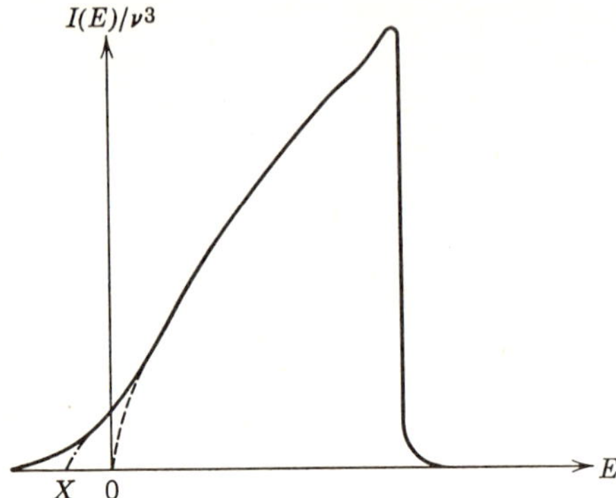

FIG. M.3.2. L_3 soft X-ray emission band for sodium, according to Skinner (1940). $I(E)$ represents the intensity and ν the emission frequency. The point X is obtained by making a linear extrapolation and 0 by making a parabolic extrapolation of the main part of the curve, to define the "empirical" and the "reduced" band widths, respectively, of 4.81×10^{-19} J and 4.01×10^{-19} J.

a small one, but the number of electronic energy states would be greater in the former case.

By measurement of the variation of intensity across the emission band observed for the soft X-ray spectrum of a metal, it is possible to deduce the distribution of the electrons among the various possible energy states. This energy distribution is commonly described by what is referred to as an $N(E)$ curve. As will be apparent later, for any piece of metal of macroscopic dimensions the number of energy levels in each energy range is very large, so that it may be regarded as almost continuous making it convenient to define the density of states in energy $N(E)$ as the number of states with energies falling between E and $(E + dE)$ as $N(E)\,dE$. The curve which results on plotting $N(E)$ as a function of E is the so-called $N(E)$ curve. The intensity of the emitted radiation in the soft X-ray spectrum of a metal will depend on the density of valence electrons in the valency or conduction band, and since there is an almost continuous or quasicontinuous energy distribution of valency electrons, hence the continuously varying intensity of the observed X-ray spectra. The intensity of an observed X-ray spectrum does not precisely reflect the density of states $N(E)$ of the valency electrons, however, due to various complications arising from the fact that the probability of a transition is not constant for all of the valency electrons, but depends on the wave function of the state into which the transition takes place as well as on that of the valency electron. Nevertheless, the soft X-ray emission spectrum of a metal gives a reasonable view of the $N(E)$ curve and in most cases gives the width of the band even in those cases where the shape of the curve is not well defined. The soft X-ray L_3 emission band for sodium is illustrated in Fig. M.3.2. The form of such curves at the low energy end is rather indefinite, only partly due to experimental difficulty.

The shape of the observed emission bands for metals at the high energy end is strongly temperature dependent, falling off more steeply as the temperature is lowered. Also their shape becomes more complicated with increasing number of valency electrons per atom in the metal. The important point to be made at this stage is that the methods of soft X-ray spectroscopy provide direct experimental evidence that the energies of the valency electrons in a metal are distributed over a range of values, and from examination of the intensity distribution across an emission band if it possible to obtain information about the distribution of electrons among the various possible energy states of the valency band. These are additional facts which have to be accounted for by any satisfactory theory of metals (see section M.4.5.).

M.3.2. The Free Electron Approximation

The exact details of the process which takes place when free atoms interact with one another to form a solid in which the sharp energy levels of the electrons in the free atoms become broadened into bands of greater width as the internuclear distance is decreased pose very complicated problems. The total charge distribution of a closed quantum group of electrons in a free atom is spherically symmetrical about the nucleus, and the potential energy of a valency electron in the potential field of the ion-core is also spherically symmetrical. For sodium, for example, the function $V(r)$ very close to the nucleus is essentially $-11\varepsilon^2/r$ (nuclear charge of sodium is 11ε), but at positions beyond the effective radius of the ion-core (see Fig. M.1.15.) $V(r)$ is very approximately $-\varepsilon^2/r$, because the nuclear field is screened from an electron at this distance by the ten electrons in the ion-core so that it behaves effectively as the field due to a single proton. There is no simple way of estimating the function $V(r)$ at intermediate distances. In a solid, even if the ion-core retains the same structure in the solid as in the free atom, the interaction of the valency electrons with one another has also to be considered. This may be done approximately by using the Hartree method (Q.4.8.) to obtain the potential energy of any electron due to the average charge distribution of the valency electrons, or more accurately by means of the Hartree–Fock (Q.5.7.), or some modification of this, method. The general effect of the other electrons in a metal, however, is to screen any given electron from the field of the ion-cores of all atoms except those in its immediate vicinity, so that the potential energy is considerably less than if only the ion-cores were present alone. Because of the crystalline nature of metals, the actual electrostatic field in which a valency electron moves must be periodic and the potential energy must be a periodic function with the periodicity of the metallic crystal lattice.

The first application of quantum mechanics to the description of the electronic behavior in metals was made by Sommerfeld,[1,2] who simply assumed that the potential energy of each electron was constant within the metal, or that the total field of the electrons and ion-cores at any point in the metal was zero. The actual value assumed for the constant potential energy is of course of great concern in any attempt to calculate such quantities as the cohesive energy or the energy required to remove an electron from the metal, but for many purposes only the distribution of energy levels relative to the lowest energy level is of importance and not necessarily their relative positions to an external zero. In its simplest form, the free-electron theory completely disregards the structure of the metal, merely assuming that the metal contains a number of electrons which move about in a region of constant potential as though they were particles of an "electron gas." The repulsion and other interaction between the electrons is ignored, and the metal is regarded as containing a uniformly distributed positive charge which neutralizes the charge on the electrons so that they may be treated as "gas" particles of small mass. From experimental observation the potential barrier at the surface of most metals, which prevents the electrons from freely escaping from the surface, is of the order of magnitude of 1.6×10^{-18} J, and this quantity cannot be independent of either temperature or the volume of the metal although in its simplest form, the free electron theory assumes that it is infinite compared with the zero potential within the metal. On the basis of these assumptions, the energy of electrons within the metal is purely kinetic, and the problem resolves itself into that of essentially non-interacting particles contained in a simple potential box, as in Q.3.2. If the metal is assumed to be in the form of a cube of edge length L, then the normalized wave functions describing the stationary states of the electrons (neglecting spin) are

$$\psi_{n_x,n_y,n_z} = \sqrt{[8/L^3]} \sin(n_x \pi x/L) \sin(n_y \pi y/L) \sin(n_z \pi z/L), \quad \text{(m.3.1.)}$$

inside the metal and zero outside the metal, and the corresponding energy levels are

$$E_{n_x,n_y,n_z} = [h^2/8mL^2][n_x^2 + n_y^2 + n_z^2], \quad \text{(m.3.2.)}$$

where m is the electron mass and n_x, n_y, and n_z, are integers.

[1] Sommerfeld, A., Houston, W. V., and Eckart, C., *Z. Physik*, 1928, **47**, 1.
[2] Sommerfeld, A., and Frank, N. H., *Rev. Mod. Phys.*, 1931, **3**, 1.

The free electron theory is particularly useful in illustrating the manner in which a "gas" of electrons, at the densities at which they exist in metals, differs from an ordinary classical gas of free particles.‡

‡ In classical Maxwell–Boltzmann statistics, the state of a gas molecule can be described by its position, x, y, z, referred to three mutually orthogonal axial directions, and by the three components of the particle's momentum, p_x, p_y, and p_z, parallel to the same three axial directions. For a collection of N particles, the state of the whole aggregate at any time may be represented by a collection of points in this six-dimensional molecular phase space (page 41). For the description of the energy distribution of the particles, alone, it is only necessary to make use of the three-dimensional momentum space, since if the energy of the particles is entirely kinetic, $E = mv^2/2 = p^2/2m$ or $p = \sqrt{2mE}$, and then the length of any vector from the origin to any point P in this momentum space is equivalent to the energy of the particle defined by the position of that point and the direction of the vector gives the direction of motion of the particle concerned. Also, in classical statistics, where the mean energy of a particle is proportional to the absolute temperature, reducing the temperature of the collection of particles results in the collection of points, representing their energies, becoming more concentrated around the origin of the momentum diagram, until at the absolute zero of temperature all of the points become concentrated with infinite density at the origin.

For a free electron, associated with a de Broglie wavelength $\lambda = h/mv = h/p$, the momentum p or the energy $\sqrt{2mE}$ is equivalent to h/λ, so that the energy distribution of a collection of free electrons may equally well be represented by a reciprocal wave space diagram, with respect to three mutually orthogonal axial direction denoting the components of a wave vector $\mathbf{k}$ defined as $k = 1/\lambda = p/h$. This concept of reciprocal wave space or k-space is particularly convenient in interpreting the behavior and properties of metals in terms of the zone and band theories (Chapter M.5.), as it enables the way in which electrons associated with a wavelength λ are scattered by a crystal lattice to be related to the very useful crystallographic concept of describing a crystal in terms of a reciprocal crystal lattice (M.5.6.).

For a collection of free electrons, the distribution of whose energies is described in terms of a momentum diagram, similar to that for a collection of gas molecules, the application of the Pauli and Heisenberg principles, found necessary for the adequate description of the behavior of electrons, immediately leads to results which are entirely different from those of classical statistics; this is the basis of the Fermi–Dirac statistics. If the "electron gas" is imagined to be confined within a cube of edge length L and volume $V = L^3$, then since each particle would then be restricted to a volume V, in terms of the Uncertainty Relationships (pages 132, 157), its momentum can no longer be exactly specified. Under these circumstances, the minimum uncertainty in the momentum of the particle would be of the order of h in each of the dimensions L, and since there would be an uncertainty in the momentum of the particle of the order of h/L in each dimension, the momentum or energy of the particle could no longer be described by the position of a point in momentum space but could only be defined as being associated with a volume $h^3/L^3 = h^3/V$ of momentum space. Now, in terms of the Pauli Exclusion Principle, not more than two electrons may occupy the same energy state, provided these two electrons have opposed spins, or since in momentum space there is an uncertainty in the energy of any electron of the order of h^3/V, each electron in momentum space may be located in a volume element of $h^3/2V$, and so a total collection of N electrons would occupy a volume of $Nh^3/2V$ of momentum space. At the absolute zero of temperature each element of volume of momentum space would contain two electrons, and the lowest energy of the total assembly would apply when the assembly is symmetrically disposed around the origin of the momentum diagram, thus defining a sphere with radius corresponding to the maximum momentum of any electron in the assembly, p_{max}. The volume of this sphere, $[4\pi(p_{max})^3/3]$ must be equivalent to the volume occupied by the N electrons given by $Nh^3/2V$, or $(4/3)\pi(p_{max})^3 = Nh^3/2V$, or $p_{max} = [3Nh^3/8\pi V]^{1/3}$. The maximum kinetic energy of any electron in the assembly is then, $E^* = (p_{max})^2/2m = [3Nh^3/8\pi V]^{2/3}[1/2m] = [3N/\pi V]^{2/3}[h^2/8m]$. This simple approach shows that at the absolute zero, the energies of the electrons in a metal, in terms of the simple free electron theory, are distributed over a range of energy values up to a maximum value given by E^*, which is dependent on N/V and therefore independent of the size of the piece of metal concerned, although since the volume element in momentum space occupied by each electron, $h^3/2V$, is inversely proportional to the volume, the difference in energy between successive energy states must decrease with increasing volume of the metal. This is the same result as that given in (m.3.8.), arrived at in a slightly different manner. The high energy of the "electron gas" at 0 K is a direct consequence of the Pauli Exclusion Principle, which prevents more than two electrons from being present in the same energy state, provided they are described by the appropriate spin quantum numbers; this is equivalent to the description that the maximum density of states is $2V/h^3$ or the maximum density of states in molecular phase space is $2/h^3$.

Electrons are particles which must conform with the restrictions imposed by the operation of the Pauli Exclusion Principle and the Heisenberg Uncertainty Principle. In accordance with the Pauli Principle not more than two electrons may occupy the same state, describable by any particular orbital function, so that at the absolute zero of temperature a collection of N electrons would occupy the $N/2$ lowest energy states; two in the ground-state, two in each of the states of next higher energy, and so on until all of the N electrons are accommodated in the lowest available energy states. Since for any macroscopic-sized piece of metal the number of electrons involved is very large, it thus arises that even at the absolute zero of temperature, some electrons must have energies of considerable magnitude ($n \times 10^{-19}$ J). For a piece of metal equivalent to a one centimeter cube, $n_x = n_y = n_z = 1$ in (m.3.2.) and the ground-state energy would be of the order of 1.6×10^{-34} J, which for most practical purposes may be taken as zero. For other integral values of n_x, n_y, and n_z, the maximum spacing between consecutive energy levels is of the order of 10^{-25} J, so that the distribution of energy may be regarded as essentially continuous, which is the justification for defining $N(E)$ as was done on page 460. If, then, the number of states with energies between E and $(E + dE)$ is $N(E)\,dE$ and dE is about 10^{-22} J, each energy interval will contain about 10^3 energy levels, so that the number of states defined as $\nu(E)$, with energies less than E will be given by $\nu(E) = \int_0^E N(E)\,dE$, so that

$$N(E) = d\nu/dE. \tag{m.3.3.}$$

Each energy state given by (m.3.2.) is represented by a triple set of integers n_x, n_y, and n_z, which set of integers define a point in the positive octant of

"n-space." If the octant is imagined to be divided up into unit cubic cells, then each point representing an energy state will fall at the vertex of such a unit cube, so that each volume unit contains one representative point. If (m.3.2.) is rearranged to read

$$n_x{}^2 + n_y{}^2 + n_z{}^2 = [8mL^2/h^2]E, \qquad \text{(m.3.4.)}$$

then it becomes apparent that this is the equation of a sphere in the "n-space," the square of whose radius is given by the right-hand side of this expression. The number, $\nu(E)$, of states with energy less than E is then the same as the number of representative points in the positive octant of this sphere and with one representative point per unit volume this number $\nu(E)$ is equal to the volume of the octant, or

$$\nu(E) = (1/8)(4\pi/3)[8mL^2E/h^2]^{3/2} = [V/6\pi^2][8\pi^2m/h^2]^{3/2}E^{3/2}, \quad \text{(m.3.5.)}$$

where $V = L^3$. From (m.3.3.) the density of states $N(E)$ may be obtained by differentiating (m.3.5.) with respect to E as

$$N(E) = [V/4\pi^2][8\pi^2m/h^2]^{3/2}E^{1/2} = 2\pi V[2m/h^2]^{3/2}E^{1/2}. \qquad \text{(m.3.6.)}$$

If $N(E)$ is plotted as a function of E the parabolic curve shown in Fig. M.3.3. results, and at the absolute zero of temperature a collection of N electrons would occupy the $N/2$ lowest energy states extending up to $E = E^*$, where E^* is the maximum energy which any electron in the assembly could possess at the absolute zero of temperature. N may be expressed as $2\nu(E)$ or

$$N = 2\int_0^{E^*} N(E)\,dE = [V/2\pi^2][8\pi^2m/h^2]^{3/2}(2/3)(E^*)^{3/2}, \qquad \text{(m.3.7.)}$$

from which it is deduced that the highest occupied energy state has the energy

$$E^* = [3\pi^2N/V]^{2/3}[h^2/8\pi^2m] = [3N/\pi V]^{2/3}[h^2/8m]; \qquad \text{(m.3.8.)}$$

for the alkali metals with one valency electron per atom, N for any given mass of metal can be evaluated knowing the specific gravity and the atomic weight of the metal, and by substituting for the other constants h, π, and m_ε, E^* may be calculated. For sodium E^* has the calculated value 5.06×10^{-19} J, for example. If E^* is the maximum kinetic energy of any electron at 0 K, then since the potential energy is assumed to be zero so that all the electron energy is kinetic, the total energy of all N electrons is

$$2\int_0^{E^*} N(E)E\,dE = [V/4\pi^2][8\pi^2m/h^2]^{3/2}(4/5)(E^*)^{5/2} = (3/5)NE^*. \quad \text{(m.3.9.)}$$

The average kinetic energy per electron, E_F, is then given as

$$E_F = (3/5)E^*. \qquad \text{(m.3.10.)}$$

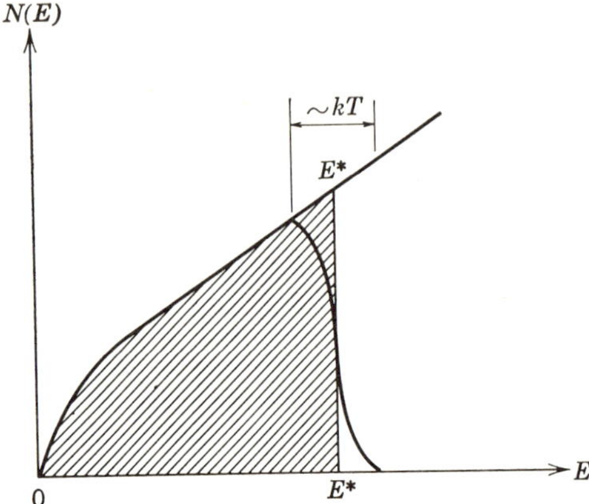

FIG. M.3.3. Density of states as a function of energy in the free electron approximation. At 0 K the $N/2$ lowest energy states are occupied up to states with a maximum energy of E^*. The boundary of the region of occupied states is absolutely sharp at 0 K, but at higher temperatures becomes less definite.

The maximum kinetic energy of an electron on the surface of the sphere of occupied states, referred to as the Fermi surface, is the so-called Fermi energy and the mean Fermi energy is related to this by (m.3.10.).

It is possible, alternatively, to express the mean Fermi energy E_F as a function of the radius r_s defined by (m.1.40.). Thus, substituting for N/V from (m.1.40.) into (m.3.8.) and (m.3.10.) gives

$$E^* = [h^2/8\pi^2 m][9\pi^2/4\pi r_s^3]^{2/3} = [h^2/8\pi^2 m][9\pi/4]^{2/3}[1/r_s^2], \quad \text{(m.3.11.)}$$

and

$$E_F = (3/5)E^* = [3h^2/40\pi^2 m][9\pi/4]^{2/3}[1/r_s^2]. \quad \text{(m.3.12.)}$$

For metals with more than one valency electron per atom, the above results can still be applied by defining another radius, r_ε, as the radius of a sphere whose volume is equal to the average volume per electron. If z is the number of electrons (valency) per atom then N is equation (m.1.40.) has to be substituted by N/z, and for all metals

$$(4\pi/3)r_\varepsilon^3 = V/N = (4\pi/3z)r_s^3 \quad \text{or} \quad r_\varepsilon/r_s = z^{-1/3}, \quad \text{(m.3.13.)}$$

and $r_\varepsilon = r_s$ for the alkali metals, for example, and other metals with one valency electron per atom. Since for all metals the quantity r_s is of the order of 10^{-10} m, it follows from (m.3.11.) and (m.3.12.) that E^* and E_F are of the order of $n \times 10^{-19}$ J, where n is a small integer.

At 0 K the $N/2$ lowest energy states, in the free electron approximation, would be occupied by electrons, each state containing two electrons up to that state with the energy E^*, on the Fermi surface. At any temperature above 0 K, an electron in a low energy state cannot be raised in energy by only a small amount, since all of the states of just slightly greater energy are already occupied, and the probability of a large energy excitation up to the energy of an electron occupying a state near to the Fermi surface is small for ordinary temperatures. Therefore, above 0 K, electrons of low energy are unaffected and only electrons in energy states near to those on the Fermi surface are excited. At higher temperatures still, electrons in lower and lower energy states would become thermally excited and at very high temperatures all of the electrons would be affected, but at the electron densities which occur in most metals, these temperatures are well above the melting point and boiling point of all known metals. At higher temperatures the distribution of occupied states does not end abruptly at E^* but tails off smoothly as in Fig. M.3.3.

This type of behavior of the "electron gas" is considerably different from that of a classical gas. In classical mechanics the general effect of an increase in temperature is to increase the energy of each gas particle by an amount of the order of kT, which at ordinary temperatures is of the order of 3.2×10^{-21} J. But, as has already been seen the valency or conduction electrons in a metal distribute their energy, by virtue of the Pauli Exclusion Principle, over a range of $n \times 10^{-19}$ J, even at the absolute zero of temperature. It is therefore most unlikely that any electron in this distribution will be excited at room temperature if it lies more than about 1.6×10^{-22} J below the value of the Fermi energy, as the states within an energy range of kT of such an electron are essentially filled, and it is not very probable that an electron will gain the $5kT \simeq 1.6 \times 10^{-22}$ J of energy necessary to excite it into an otherwise unoccupied state above the Fermi energy level. Thus it happens that on heating from 0 K every electron does not receive an energy gain of $\sim kT$, as happens classically, but only those electrons already within an energy range of the order of kT of the Fermi energy level may be thermally excited, and these particular electrons gain in energy by about kT.

The simple free electron approximation thus not only gives an immediate result in conformity with experimental observation, namely that even at the absolute zero of temperature the electronic energies in a metal are spread over a range or band of values, but also gives a qualitative solution to the problem of electronic heat capacities. For any temperature T, only a fraction of the order of T/T_F of the total number of N electrons in a metal lie within an energy range of the order of kT of the top of the energy distribution, and are therefore capable of being excited thermally at the temperature T. The Fermi temperature T_F may be defined as E^*/k, and corresponds to about 37000 K for sodium, for example. Now if NT/T_F electrons have each gained a thermal energy of the order of kT, the total electronic thermal energy would be about $[NT/T_F]kT = RT^2/T_F$, and so the electronic heat capacity is

$$C_v = \partial/\partial T[RT^2/T_F] \simeq RT/T_F,$$

which is proportional to T (page 421), in accordance with experimental observation. In general T_F is of the order of 5×10^4 K and C_v is smaller than the classical value of $3R/2$ by a factor of about 0.01.

These results constitute the simplest application to metals of what is referred to as Fermi–Dirac statistics.

M.3.3.   The Fermi–Dirac Distribution Function

The Maxwell–Boltzmann distribution law (pages 34–36, 40–43) describes the behavior of ordinary gas molecules under normal conditions and is the result of classical theory and a logical extension of the kinetic theory of gases. In classical statistical theory any number of particles may have the same energy and momentum, and in particular, in the lowest state of a system describable in terms of classical Maxwell–Boltzmann statistics, all of the constituent particles may have zero energy and momentum.

Electrons have very much smaller masses than molecules of gases and in a metal, the concentration of electrons is higher by a factor of about 10^4 than the concentration of molecules in a gas at standard conditions of temperature and pressure. In quantum statistics, in particular as applied to electrons (photons obey a different type of quantum statistics; see M.1.5.), each electron must be regarded as indistinguishable from another and each energy state of the system may be occupied by at most one electron, characterized by four quantum numbers, n, l, m_l, and m_s. There is no experimental technique known by means of which one particular electron may be distinguished from another. It necessarily follows that if only one electron may occupy each energy state in a system of electrons, then even in the lowest state of such a system many states describable by high quantum numbers will be occupied, up to the highest energy states. At 0 K all available energy states up to some maximum value of the energy will be occupied and the distribution ends abruptly at this value.

At higher temperatures, the distribution of occupied energy states is not sharply terminated but tails off smoothly. To describe the electron distribution at temperatures above 0 K in a metal, it is necessary to make use of what is referred to as the Fermi–Dirac distribution function. The derivation of this function and the general nature of quantum statistics will be discussed in the next section. Meantime, if it is accepted that this Fermi–Dirac distribution function is given by

$$f(E) = [e^{(E-\theta)/kT} + 1]^{-1}, \qquad\qquad \text{(m.3.14.)}$$

where θ is a function of T, which is determined by the relationship

$$2 \int_0^\infty N(E) f(E)\, dE = N, \qquad \text{(m.3.15.)}$$

which simply states that the total number of electrons in the system is N. At any temperature, T, making use of the Fermi–Dirac Distribution Function, the number of electrons in a system with energies lying between E and $E + dE$, may then be expressed as $2N(E)f(E)\,dE$. If $\theta(0)$ represents the value of θ at 0 K, then if $\theta(0)$ is positive and greater than E, the exponential quantity $e^{(E-\theta)/kT}$ tends to zero as T tends to zero, and therefore $f(E) = 1$ at 0 K. Also, if $E > \theta(0)$ and $\theta(0)$ is positive, then $e^{(E-\theta)/kT}$ approaches infinity as $T \to 0$, and $f(E) = 0$ at 0 K. Therefore,

$$f(E) = 1, \quad \text{if } E < \theta(0) \quad \text{and} \quad f(E) = 0, \quad \text{if } E > \theta(0), \text{ at 0 K.}$$
$$\text{(m.3.16.)}$$

Making use of these values for $f(E)$, equation (m.3.15.) becomes

$$\int_0^{\theta(0)} N(E)\, dE = N, \qquad \text{(m.3.17.)}$$

which on comparison with (m.3.7.) shows that $\theta(0) = E^*$, and the density of states, $N(E)f(E)$ with $f(E)$ as given by (m.3.16.) turns out to exactly the same as found previously. At temperatures above 0 K, θ is not equal to $\theta(0) = E^*$, but up to temperatures of the order of 2 to 3×10^3 K, that is up to the temperatures corresponding to about the melting point of many metals, $\theta(0) \simeq E^*$.

The Fermi–Dirac Distribution Function, $f(E) = [e^{(E-\theta)/kT} + 1]^{-1}$, has the value 0.5 for $E = \theta$, so that $f(\theta)$ is always 0.5. For values of E much less than θ, $f(E) \simeq 1$, and for $E \gg \theta$, $f(E) \simeq 0$; it emerges indeed that for temperatures of the order of 300 K, $f(E)$ is only appreciable different from zero or unity in the small region around θ. The width of this region where $f(E)$ differs from 0 or 1 is easily determined. For example,

$$f(\theta - 2kT) = [e^{-2} + 1]^{-1} = 0.88 \quad \text{and} \quad f(\theta + 2kT) = [e^{2} + 1]^{-1} = 0.12.$$

Thus, for $E = \theta \pm 2kT$, $f(E)$ ranges from 0.88 to 0.12, or it decreases approximately from unity to zero in a region of width about $4kT$ around θ. Since kT has the numerical value of about 4×10^{-21} J at $T = 300$ K, and $E^* = \theta$ has a value of $n \times 10^{-19}$ J, this means that at ordinary temperatures, $kT \ll \theta$. Thus the fall in the value of the Fermi–Dirac distribution function near the value of the energy $\theta \simeq E^*$ is so rapid that the distribution of the energy of the electrons is almost the same as at 0 K. For example, for sodium with $E^* \simeq 5.13 \times 10^{-19}$ J, the function $f(E)$ at 0 K and at 10^3 K is shown in Fig. M.3.4.; the corresponding $N(E)$ curve appears as in Fig. M.3.3. These

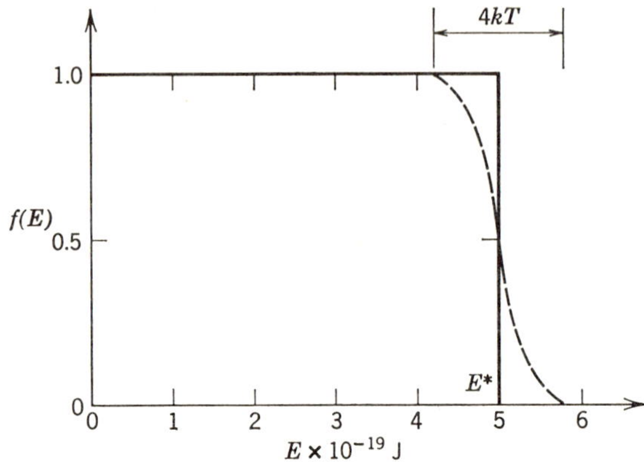

FIG. M.3.4. The Fermi–Dirac distribution function
for sodium at 0 K (———) and at 10³ K (— — —).

curves have the characteristic that they pass through the point $f = 0.5$ and
$E = E^*$, and are symmetrical with respect to change of sign about these
points. The sharp falling-off of $f(E)$ at $E = E^*$ simply means that up to
$E = E^*$ all the energy states are occupied by two electrons and $f(E) = 1$,
while no states are occupied for $E > E^*$, and the Fermi surface is perfectly
sharp. The "gas" of electrons in this condition is said to be degenerate. It is
only at very high temperatures that these curves differ appreciably from the
$f(E)$ at 0 K.

M.3.4. Quantum Statistics

Prior to 1906 it was assumed that the laws of classical mechanics which
apply to macroscopic objects, could be applied equally well to the description
of all systems of particles, and classical statistical mechanics was a soundly
developed area of knowledge, based on the atomic nature of matter, classical
mechanics as successively developed by Newton, Lagrange, and Hamilton,
and the axioms of probability. The formulation of the quantum postulate by
Planck, found necessary to explain such varied phenomena as isothermal
cavity radiation, the photoelectric effect, and the line spectra of atoms,
indicated that not all naturally occurring phenomena could be understood or
described in terms of classical mechanics. Actually, all real gases deviate from
ideal behavior and especially under conditions of low temperatures, high
pressures, and high concentration, where intermolecular forces are high.
Originally, it was thought that this situation was due to some inadequacy in
the Maxwell–Boltzmann statistical laws. However, this problem was solved

by the development of the Bose–Einstein system of quantum statistics, which was applied by Bose to light quanta, enabling the deduction of the Planck radiation law to be made in a straightforward manner, and later by Einstein to gas molecules in such a manner as to account for the phenomenon of gas degeneracy. The other system of quantum statistics, which is based on Pauli's Exclusion Principle, was introduced by Fermi and Dirac. The derivation of the Boltzmann factor has been described on pages 34–36, and the Maxwell–Boltzmann distribution law on pages 40–43. The two systems of quantum statistics arise by the introduction of the principle of indistinguishability into statistical arguments. This may be seen on the basis of the following argument, due to Born.[1]

Quantum mechanics appears to be the fundamental discipline required for the organization and understanding of natural phenomena, classical mechanics simply being a special case applicable to macroscopic events. In quantum mechanics, every particle, whether a gas molecule, a photon, or an electron, is described by specifying its wave function or state function. For a collection of particles, if it is assumed that the state function of the first, second, . . ., particles are respectively, $\psi_k{}^1$, $\psi_l{}^2$, . . ., where $k, l, \ldots$, represent the state of the particle involved, then to a first approximation and in the absence of any inter-particulate reaction, the state of the whole system of particles may be represented by the product of all of these particle state functions, by

$$\Psi_{klm\ldots} = \psi_k{}^1\psi_l{}^2\psi_m{}^3\cdots. \qquad \text{(m.3.18.)}$$

If particle 1 and particle 2 are interchanged, another total state function results,

$$\Psi'_{klm\ldots} = \psi_k{}^2\psi_l{}^1\psi_m{}^3\cdots, \qquad \text{(m.3.19.)}$$

which corresponds to the same energy value for the whole system,

$$E_{klm\ldots} = E_k + E_l + E_m + \cdots.$$

Other state functions, each corresponding to this same energy value may be obtained by writing down arbitrary linear combinations of the one-state functions, by permuting the arguments 1, 2, 3, . . . of the separate functions, as

$$\Psi_{klm\ldots} = \sum_P a_{123\ldots}\psi_k{}^1\psi_l{}^2\psi_m{}^3\cdots, \qquad \text{(m.3.20.)}$$

where $a_{123}\ldots$ represent arbitrary coefficients, and the summation is taken over all permutations of the arguments, 1, 2, 3, . . . (see page 152). There will be as many different state functions as there are linearly independent wave

---

[1] Born, Max, *Atomic Physics, Seventh Edition*, 1962, Blackie & Son, Limited, Glasgow.

functions among them, according to classical statistics. However, if the particles are assumed to be indistinguishable from one another, as is the case for photons and electrons, for example, then those cases which arise from one another by merely permuting the particles belong to the same state. Since it is only the quantity $|\Psi|^2$ which has any physical significance, state functions which arise from one another by permuting the particles do not change when such a permutation is carried out, or at most can only change in sign. The only possible state function which does not change when the particles are permuted is that one in which all the coefficients are identical to unity, and this is the symmetrical state function (see page 288),

$$\Psi_s = \sum_P \psi_k{}^1\psi_l{}^2\psi_m{}^3\cdots. \tag{m.3.21.}$$

The other possible function, with changed sign but not the value of its square is the antisymmetrical form

$$\Psi_a = \sum_P \pm\psi_k{}^1\psi_l{}^2\psi_m{}^3\cdots, \tag{m.3.22.}$$

where the plus sign refers to an even permutation of the particles and the minus sign to an odd permutation. The antisymmetrical form is well known, from the theory of determinants (page 150), and may be expressed as a determinant,

$$\Psi_a = \begin{vmatrix} \psi_k{}^1 & \psi_k{}^2 & \psi_k{}^3 & \cdots \\ \psi_l{}^1 & \psi_l{}^2 & \psi_l{}^3 & \cdots \\ \psi_m{}^1 & \psi_m{}^2 & \psi_m{}^3 & \cdots \\ \cdots & \cdots & \cdots & \end{vmatrix}. \tag{m.3.23.}$$

There is no other possible function which satisfies the condition of indistinguishability of the particles. As already pointed out in Q.2.7., the special property of the antisymmetric function that it vanishes if any two functions ψ_k and ψ_l composing any two rows or columns of the determinental form, are the same, is an expression of the Pauli Principle that two particles cannot be in the same state. The only two possible ways of describing a state by a wave function, either a symmetrical or an antisymmetrical wave function, the latter corresponding to the Pauli Principle, give rise to the two quantum statistical systems, the Bose–Einstein and the Fermi–Dirac systems, respectively.

In classical statistics, the distribution of the energies of a collection of gas molecules is described in terms of the distribution of a collection of points in three-dimensional momentum space. In the absence of any energy restrictions the most probable distribution of point particles in space, is a uniform one,‡ the relative size of the volume units or cells into which momentum space is divided being of no consequence, as far as the density distribution of the molecules is concerned. There is only one limiting consideration in the deduction of this result, namely that the sum of all of the point particles, any number of which may be contained in each of the arbitrary-sized and shaped cells into which the momentum space may be imagined to be divided, remains constant. If the distribution of the velocity, or momentum, or kinetic energy of such a system of point particles is similarly considered, then as shown on pages 34–36, the Boltzmann Law results, and here there are two subsidiary conditions, that the total number of particles remains constant

$$[n_1 + n_2 + \cdots = N]$$

and that the total energy remains constant $[n_1 E_1 + n_2 E_2 + \cdots = E]$.

In this system, the ratio of the size of the cell in momentum space to the total volume of momentum space, $g_i = V_i/V$, is called the statistical weight of the cell. Also, in classical statistics, the momentum space is regarded as being covered discontinuously, the statistical weight of a cell being simply the number of points contained within it. For cells of any arbitrary size, there must then be a smallest cell, among those containing any point at all, and thus a lower limit to the volume element which has any meaning. This is the type of problem which is involved in the statistics of real systems, such as real gases, light quanta as in isothermal cavity radiation, where it becomes necessary to determine the size of the cells into which phase space may be divided or to determine the number of cells in phase space. These real systems, radiation and gases, were indeed the systems, the examination of which by Bose and Einstein, respectively, led to the necessity for the formulation of the first important type of quantum statistics. As a preliminary to, and an illustration of the type of statistical argument involved in, the discussion of Bose–Einstein statistics, it is instructive to consider the most probable distribution in space of molecules or particles of finite size, rather than the point particles of classical statistics.

‡ This result, although apparently self evident, may be deduced by the following argument, again due to Max Born. Let it be imagined that there are N point particles whose most probable distribution in a total volume V is to be found, without any restrictions about energy being involved. Let the total volume V be sub-divided into arbitrary volume elements, V_1, V_2, $V_3, \ldots$, into each of which is inserted N_1, N_2, $N_3, \ldots$ particles. Then the probability of finding one particle in the volume element V_1 is V_1/V (see page 22), and since the particles are assumed to be point particles, the presence of one of them in a certain volume element does not affect the chance that another one should occupy the same volume element. The probability that there will be two particles in the same volume element is therefore $(V_1/V)^2$. Similarly, the probability that there will be

N_1 particles in the same volume element is $(V_1/V)^{N_1}$, and the probability that there will be N_2 particles in the volume element V_2 is $(V_2/V)^{N_2}$. Since there are $N!$ ways in which the N particles may be permuted among themselves, then if the total number N is split up into sub-groups $N_1, N_2, N_3, \ldots$, the total number of permutations $N!$ includes the number of permutations of the groups $N_1, N_2, N_3, \ldots$ within themselves which do not contribute to any additional distributions, so that the total number of ways of distributing N particles so that $N_1, N_2, N_3, \ldots$ fall into separate sub-groups is $N! [N_1! N_2! N_3! \cdots]^{-1}$. The absolute probability that the whole system of N particles can be distributed in space so that N_1 are in volume V_1, N_2 in volume $V_2, \ldots$, is then

$$W = N! [N_1! N_2! N_3! \cdots]^{-1}(V_1/V)^{N_1}(V_2/V)^{N_2}(V_3/V)^{N_3}\cdots = N! \prod_i [N_i!]^{-1} \prod_i (V_i/V)^{N_i}.$$

The sum of all possible probabilities must be unity, that is $V_1/V + V_2/V + V_3/V + \cdots = 1$. The most probable distribution of the N particles among the various volume elements is found by looking for the conditions which make W a maximum, which is obtained by the usual mathematical technique of equating the derivatives of W with respect to each of the variables involved to zero. By taking logarithms of both sides of the probability expression, using Stirling's approximation for the factorials of large numbers (page 35), and differentiating with respect to N_1, there results

$$[\partial \ln W/\partial N_1]_{V, N_2, \ldots} = -1 - \ln N_1 + \ln (V_1/V).$$

The condition for a maximum is that

$$\delta \ln W = \sum_i [-1 - \ln N_i + \ln (V_i/V)] \, dN_i = 0,$$

subject to the condition that the total number of particles remains constant, or that

$$\delta N = \sum_i dN_i = 0,$$

both conditions applying simultaneously. Applying Lagrange's method of undetermined multipliers, and multiplying the first conditional expression by -1 and the second by λ, gives

$$\sum [1 + \ln N_i - \ln (V_i/V) + \lambda] \, dN_i = 0,$$

and since the variations in the numbers are arbitrary, the expression within the square brackets above must itself be zero for all groups N_i, and therefore

$$N_i = (V_i/V)e^{-\lambda - 1}.$$

Since $\sum_i N_i = N$ and $\sum V_i = V$, then by adding up all the equations of the type for N_i, the term $e^{-\lambda - 1} = N$, and so

$$N_i/V_i = N/V.$$

If the concentrations $N_1/V_1, N_2/V_2, N_3/V_3, \ldots$ are denoted by $n_1, n_2, n_3, \ldots$, then obviously $n_1 = n_2 = n_3 = \cdots = N/V$, or the concentrations in each volume element is the same, or the most probable distribution, in the absence of any energy restriction, of a collection of point particles in space is a uniform one.

Suppose again that for a collection of N particles, each occupying a finite volume V_0, they are distributed in space so that $N_1, N_2, N_3, \ldots$ of the particles occupy the volume elements $V_1, V_2, V_3, \ldots$, respectively. Then the probability that any particle will occupy the volume element V_1 is V_1/V, but the probability that a second particle will be found simultaneously in the same volume element is less than this, since if each particle occupies volume, there will be less space available for the second particle. The probability of two particles occupying the volume element V_1 at the same time is then $(V_1 - V_0)/V$. Similarly, the probability that N_1 particles will be found in the volume element V_1 simultaneously is

$$
\begin{aligned}
w_1 &= [V_1/V][(V_1 - V_0)/V][(V_1 - 2V_0)/V][(V_1 - 3V_0)/V] \cdots \text{ to } N_1 \text{ terms} \\
&= [V_1/V]^{N_1}(1 - v)(1 - 2v)(1 - 3v) \cdots \text{ to } N_1 \text{ terms, where } V_0/V = v \\
&= [V_1/V]^{N_1}(v)^{N_1}(1/v - 1)(1/v - 2)(1/v - 3) \cdots \text{ to } N_1 \text{ terms} \\
&= [V_1 v/V]^{N_1}(v^{-1} - 1)(v^{-1} - 2) \cdots [v^{-1} - (N_1 - 1)] \\
&= [V_1 v/V]^{N_1}(v^{-1})! \, [(v^{-1} - N_1)!]^{-1},
\end{aligned}
$$

on multiplying and dividing by $(v^{-1} - N_1)!$.

Similarly for the probabilities $w_2, w_3, w_4, \ldots$, so that for the total system

$$
W = N! \, [N_1! \, N_2! \, N_3! \cdots]^{-1} \prod [V_i v/V]^{N_i}(v^{-1})! \, [(v^{-1} - N_i)!]^{-1}, \quad \text{(m.3.24.)}
$$

where $N! \, [N_1! \, N_2! \, N_3! \cdots]^{-1}$ is the total number of distinguishable ways of arranging N particles into sub-groups $N_1, N_2, N_3 \ldots$. Obviously, there must be an upper limit to the number of particles N_i which may be accommodated in each volume unit V_i, due to the finite size of the particles. The most probable arrangement will be that which makes W a maximum. Proceeding as above, by taking logarithms of both sides of (m.3.24.), applying Stirling's approximation, differentiating the resulting expression with respect to N_1, $N_2, \ldots$, taking the sum of all such expressions, making use of Lagrange's theorem of undetermined multipliers to express the condition that, for arbitrary variations in the magnitudes of the probabilities subject to the constancy of the total number of particles, the distribution of the particles corresponding to the most probable state of the system is found to be given by

$$
N_i = (1/v)[V_i/(e^\lambda + V_i)] = [V/V_0][e^\lambda/V_i + 1]^{-1}, \quad \text{(m.3.25.)}
$$

where λ is a constant. If now the concentration of particles in the ith volume element is represented as $n_i = N_i/V_i$, then

$$
n_i = N_i/V_i = [V/V_0][e^\lambda + V_i]^{-1}, \quad \text{(m.3.26.)}
$$

from which it is apparent that the concentration of particles in any particular volume element i is dependent on the magnitude of this volume element compared with that of the constant volume e^λ, and that when $e^\lambda \gg V_i$ then $e^\lambda = \sum_i [VV_i/N_i V_0] = V^2/NV_0$, and therefore the concentration is again uniform throughout all of the volume elements; but in general, the concentration of particles of finite volume in any volume element V_i increases with decreasing size of the volume element. It therefore becomes pertinent to enquire as to what is the limiting size of the volume elements into which phase space may be divided, or what amounts to the same thing, the number of cells in phase space. When such an enquiry is carried out, the two different quantum statistical systems arise. These are discussed, in turn, below.

(A) BOSE–EINSTEIN STATISTICS.

In classical Maxwell–Boltzmann statistics, the number of volume elements or cells into which it is imagined that phase space is divided is the same as the number of particles in the system. What distinguishes Bose–Einstein statistics from the classical statistical description of a system, is that the number of cells in phase space is greater than the number of particles composing the system, so that some of the cells may be unoccupied by particles. It was on this basis that Bose found it possible to deduce the Planck radiation law (q.1.26.) and Einstein found it possible to understand the problem of gas degeneracy. Thus for N_1 particles, each having the energy E_1, distributed among M_1 cells, where $M_1 > N_1$, the particles in each group N_1, though having the same total energy may have their energies distributed in different ways, which is the type of behavior typical of degeneracy (page 168). Bose–Einstein statistics is appropriate to the description of the behavior of particles describable by symmetrical state functions, such as photons, generally referred to as bosons, but in determining the number of different states in which a system may exist corresponding to the number of linearly independent wave functions, it is necessary to think in terms of particles.

Suppose that the group of M_1 cells are distinguished as M_a, M_b, M_c, ..., and that the group of N_1 particles are distinguished as N_a, N_b, N_c, ..., then any particular distribution of the N_1 particles among the M_1 cells, there being no limitation to the number of particles in each cell, may be described as

$$M_a N_a M_b N_b N_c M_c N_d N_e N_f M_d M_e N_g N_h N_i N_j M_f \cdots,$$

denoting particle N_a in cell M_a, particles N_b and N_c in cell M_b, particles N_d, N_e, and N_f in cell M_c, no particles in cell M_d, particles N_g, N_h, N_i, and N_j in cell M_e, and so on. There are altogether $(M_1 + N_1)!$ ways in which such distributions may be written down, but if the particles and the cells are indistinguishable from one another, then permutations of the particles or of the cells with one another do not contribute to different distributions of the particles, but represent the same state of the system. The number of such permutations is $M_1! \, N_1!$, so the number of distinguishable arrangements of the N_1 particles among the M_1 cells, or the probability of such an arrangement is only

$$w_1 = (M_1 + N_1)!/M_1! \, N_1!. \qquad \text{(m.3.27.)}$$

Similarly for the group N_2 of particles distributed among the group M_2 of cells, and so on, so that the total probability that the whole system of N particles distributed among M cells shall consist of groups of N_i particles distributed among groups of M_i cells is

$$W = \prod_i w_i = \prod_i [(M_i + N_i)!/M_i!\, N_i!]. \qquad \text{(m.3.28.)}$$

Taking logarithms of W, and applying Stirling's theorem that $\ln A! = A \ln A - A$,

$$\ln W = \sum_i [(M_i + N_i) \ln (M_i + N_i) - (M_i + N_i) - (M_i \ln M_i - M_i)$$
$$- (N_i \ln N_i - N_i)]$$
$$= \sum_i [M_i \ln (M_i + N_i) + N_i \ln (M_i + N_i) - M_i \ln M_i - N_i \ln N_i]$$
$$= \sum_i [M_i \ln \{(M_i + N_i)/M_i\} + N_i \ln \{(M_i + N_i)/N_i\}]. \qquad \text{(m.3.29.)}$$

From the third law of thermodynamics, as on page 35, this expression for $\ln W$ must be equivalent to $(S - C)/k$, where k and C are constants and S is the maximum of entropy, so

$$\ln W = (S - C)/k,$$

or

$$S = k \ln W + C = k \sum_i [M_i \ln \{(M_i + N_i)/M_i\} + N_i \ln \{(M_i + N_i)/N_i\}] + C.$$
$$\text{(m.3.30.)}$$

Also

$$\partial S/\partial N_i = k \sum_i [\ln \{(M_i + N_i)/N_i\}] = \partial/\partial N_i[k \ln W + C]. \ddagger \quad \text{(m.3.31.)}$$

$\ddagger \ \partial S/\partial N_i = \partial/\partial N_i[k \ln W + C]$

$$= \partial/\partial N_i\left[\sum_i (k[M_i \ln \{(M_i + N_i)/M_i\} + N_i \ln \{(M_i + N_i)/N_i\}] + C)\right]$$

$$= k \sum_i [M_i \, \partial/\partial N_i[\ln \{(M_i + N_i)/M_i\}] + N_i \, \partial/\partial N_i[\ln \{(M_i + N_i)/N_i\}]$$

$$+ \ln \{(M_i + N_i)/N_i\}]$$

$$= k \sum_i [M_i(M_i + N_i)^{-1} + (N_i)(-M_i)[N_i(M_i + N_i)]^{-1} + \ln \{(M_i + N_i)/N_i\}]$$

$$= k \sum_i [\ln \{(M_i + N_i)/N_i\}],$$

since

$$\partial/\partial N_i[\ln \{(M_i + N_i)/M_i\}] = \partial y/\partial N_i,$$

where

$$y = \ln \{(M_i + N_i)/M_i\} \qquad \text{or} \qquad e^y = (M_i + N_i)/M_i,$$

so

$$de^y/dN_i \cdot dy/de^y = dy/dN_i = (1/M_i e^y) = (1/M_i)[M_i/(M_i + N_i)] = (M_i + N_i)^{-1},$$

and

$$d/dN_i[\ln\{(M_i + N_i)/N_i\}] = dy'/dN_i,$$

where

$$y' = \ln\{(M_i + N_i)/N_i\} \quad \text{or} \quad e^{y'} = (M_i + N_i)/N_i,$$

so

$$dy'/dN_i = de^{y'}/dN_i \cdot dy'/de^{y'} = -(M_i/N_i^2)(1/e^{y'}) = -(M_i/N_i^2)[N_i/(M_i + N_i)]$$
$$= -M_i[N_i(M_i + N_i)]^{-1}.$$

The most probable distribution will apply when the probability, W, of (m.3.28.) has its maximum value, that is when the following condition is satisfied, that

$$\delta \ln W = 0 = \sum_i [\ln\{M_i + N_i)/N_i\}] \, dN_i. \tag{m.3.32.}$$

Since k cannot be zero, this is the same condition that $\delta S = \sum_i dS_i = 0$. The two subsidiary conditions which must be satisfied are that the total number of particles and their total energy should remain constant, that is that

$$N = \sum_i N_i = \text{constant}, \quad \text{or that} \quad \delta N = \sum_i dN_i = 0, \text{ and} \tag{m.3.33.}$$

$$E = \sum_i N_i E_i = \text{constant}, \quad \text{or that} \quad \delta E = \sum_i E_i \, dN_i = 0. \tag{m.3.34.}$$

To obtain a solution to the three equations of condition, (m.3.32.–34.) by the application of the theorem of undetermined multipliers, it is convenient to multiply the equations of numbers (m.3.33.) by λ and the energy condition equation by $-\mu$, to obtain

$$\sum_i [\ln\{(M_i + N_i)/N_i\} + \lambda - \mu E_i] \, dN_i = 0,$$

or

$$\delta \ln W + \lambda \, \delta N - \mu \, \delta E = 0. \tag{m.3.35.}$$

As the variations dN_i are arbitrary, each of the terms in parentheses in the first equation of (m.3.35.) must each be zero, if this equation is to be generally valid. Therefore,

$$\ln\{(M_i + N_i)/N_i\} + \lambda - \mu E_i = 0,$$

or

$$[(M_i + N_i)/N_i] = M_i/N_i + 1 = e^{-\lambda + \mu E_i},$$

or

$$N_i/M_i = [e^{-\lambda + \mu E_i} - 1]^{-1}. \tag{m.3.36.}$$

Now, from equation (m.3.30), it is apparent that $\delta \ln W = \delta[(S - C)/k]$, which when substituted into the second equation of (m.3.35.) gives

$$\delta[(S - C)/k] + \lambda \, \delta N - \mu \, \delta E = 0, \tag{m.3.37.}$$

from which it is found that

$$[\delta E/\delta S]_{N,V} = 1/\mu k. \tag{m.3.38.}$$

But the thermodynamic definition of temperature, from $dE = T\,dS - P\,dV$, is $[dE/dS]_V = T$, so that

$$1/\mu k = T, \qquad \text{or} \qquad \mu = 1/kT, \tag{m.3.39.}$$

exactly as in classical Maxwell–Boltzmann statistics. Substituting this value for μ into (m.3.36.) gives

$$N_i/M_i = [e^{-\lambda}e^{E_i/kT} - 1]^{-1}, \tag{m.3.40.}$$

which is the Bose–Einstein distribution law.

It is still necessary to evaluate the constant λ. From the discussion of classical statistics on page 141, it is known that for particles with kinetic energy lying between p_x and $p_x + dp_x$, p_y and $p_y + dp_y$, p_z and $p_z + dp_z$, and with position lying between x and $x + dx$, y and $y + dy$, and z and $z + dz$, that the volume element in six-dimensional phase space is

$$dp_x\,dp_y\,dp_z\,dx\,dy\,dz = dV2\pi(2m)^{3/2}E^{1/2}\,dE. \tag{m.3.41.}$$

Now for particles described in terms of the quantum concept, since it requires three coordinates to describe the position of a particle (and also to describe its energy), then in terms of the Heisenberg Uncertainty Principle, if the energy (or position) of any particle is defined there will be an uncertainty in each of three coordinate directions of the order of magnitude of h in the position (or energy) of the particle. So that for a quantized system of particles, the minimum extension in space of a particle so described is of the order of h^3. Therefore the number of volume elements or cells per particle, subject to the limitations imposed by (m.3.41.) is

$$dV2\pi(2m)^{3/2}E^{1/2}\,dE/h^3. \tag{m.3.42.}$$

The number of cells per particle, without any restriction as to their positions, obtained by integrating dV over the complete system, gives the same number as that which has been denoted in the above discussion by M_i, so that

$$M_i = 2\pi V(2m)^{3/2}E^{1/2}\,dE/h^3. \tag{m.3.43.}$$

If this value for M_i is substituted into (m.3.40.), the Bose–Einstein distribution law is obtained as

$$N_i = M_i[e^{-\lambda}e^{E_i/kT} - 1]^{-1} = [2\pi V(2m)^{3/2}E^{1/2}\,dE/h^3][e^{-\lambda}e^{E_i/kT} - 1]^{-1}. \tag{m.3.44.}$$

In order to solve this equation for λ, it is necessary to integrate over all particles and over all possible energy values of these particles, in the system.

For large values of N,

$$\sum_i N_i \equiv \int dN = N = [2\pi V(2m)^{3/2}/h^3] \int_0^\infty E^{1/2}[e^{-\lambda}e^{E/kT} - 1]^{-1}\, dE. \quad \text{(m.3.45.)}$$

Unfortunately, there is no general integration of this expression which can be carried out, but if the term, (-1), in the integrand is neglected, then (m.3.45.) becomes

$$N = [2\pi V(2m)^{3/2}h^{-3}e^{\lambda}] \int_0^\infty e^{-E/kT}E^{1/2}\, dE = e^{\lambda}(2\pi mkT)^{3/2}Vh^{-3}, \quad \text{(m.3.46.)}$$

since $\int_0^\infty e^{-E/kT}E^{1/2}\, dE$ is of the type,

$$\int_0^\infty e^{-ax}x^n\, dx = [n!][a^{n+1}]^{-1} \qquad \text{and} \qquad (1/2)! = \sqrt{\pi}/2,$$

as proved on page 43. By rearranging equation (m.3.46.), it is found that

$$e^{\lambda} = [Nh^3/V][2\pi mkT]^{-3/2} = nh^3[2\pi mkT]^{-3/2}, \qquad \text{where} \qquad n = N/V,$$

or

$$\lambda = -\ln\, [(2\pi mkT)^{3/2}(nh^3)^{-1}]. \quad \text{(m.3.47.)}$$

By dividing (m.3.44.) by (m.3.46.), there results

$$N_i/N = dN/N = [2\pi V(2m)^{3/2}E^{1/2}\, dE][h^3(e^{-\lambda}e^{E/kT} - 1)]^{-1}h^3$$
$$\times\, [e^{\lambda}V(2\pi mkT)^{3/2}]^{-1}$$
$$= 2E^{1/2}e^{-E/kT}\, dE[\sqrt{\pi}(kT)^{3/2}]^{-1}, \quad \text{(m.3.48.)}$$

which on comparison with page 42 is seen to be identical to the Maxwell–Boltzmann energy distribution law. Thus, returning to (m.3.45.), if $e^{-\lambda}e^{E/kT} \gg 1$, the Bose–Einstein expression becomes identical with the classical Maxwell–Boltzmann distribution. In terms of λ, this statement may be expressed alternatively by saying that if

$$e^{-\lambda} = 1/e^{\lambda} = [2\pi mkT]^{3/2}[nh^3]^{-1} \gg 1, \qquad \text{or if} \qquad [2\pi mkT]^{3/2}h^{-3} \gg n,$$

then classical behavior describable in terms of Maxwell–Boltzmann statistics is applicable to a system; conversely, if for any system, the particle concentration $n \gg [2\pi mkT]^{3/2}h^{-3}$, then Bose–Einstein statistics is the appopriate means for the description of the system. Incidentally, from equation (m.3.37.), it is seen that $[\delta E/\delta N]_{s,v} = +[\lambda/\mu] = +\lambda kT$, and this quantity is equivalent to the thermodynamic chemical potential, for a system.

The constant $e^{\lambda} = nh^3[2\pi mkT]^{-3/2}$, is called the degeneracy parameter, since when this quantity is small compared with unity, the Bose–Einstein distribution becomes identical with the classical one. On the other hand, when e^{λ} is comparable with unity, deviations from classical properties occur and the gas of particles is said to be degenerate. The degeneracy generally increases with increasing value of n, the concentration of particles, or with increasing density, but it diminishes with increasing temperature and atomic weight of the particles concerned. For example, for hydrogen gas at ordinary temperatures of 300 K, $n \simeq 3 \times 10^{19}$ cm^{-3} and $e^\lambda \simeq 3 \times 10^{-5} \ll 1$, and for heavier gases e^λ is even smaller, so that gases are never degenerate at normal temperatures and pressures, but behave classically. Degeneracy would only become apparent at very low temperatures or very high pressures, where classical theory does not anticipate ideal behavior. If $e^\lambda \gg 1$, then $\lambda > 0$, and the whole theory becomes meaningless, for then the expression $[e^{-\lambda}e^{E/kT} - 1]$ in (m.3.45.) vanishes; this happens when $E = \lambda/\mu = \lambda kT$, and if energy is less than λkT, it would be negative, which is impossible. The energy quantity λkT is the upper limit for energy of particles.

(B) Fermi–Dirac Statistics.

It has been seen that the introduction of the principle of indistinguishability of particles into statistics leads to two, and only two, new systems of statistics (page 473), one of which has been discussed above, the Bose–Einstein statistics. The second possible quantum statistical system, which is based on Pauli's Principle, was introduced by Fermi and Dirac. In section Q.5.4., it has also been seen that the proper description of the behavior of electrons requires the use of an antisymmetrical state function, and particles of this type, referred to as fermions, include not only electrons but protons, neutrons, and particles generally of half-integral spin (page 290).

In order to deduce an analytical expression, it is convenient to use the model of an "electron gas," since it is known from the experimental examination of atomic spectra that electrons obey the Pauli Principle. This procedure is perhaps a little unrealistic, since charged particles such as electrons interact with one another, but approximately this interaction may be neglected, since in a metal the electronic charges are neutralized by the charge on the metal ions. The problem then is to deduce the distribution of electrons over the individual volume elements of phase space, making use of the concept that there are twice as many volume elements or cells as in the case of the quantum statistics (Bose–Einstein) of gas molecules or photons, due to the two possible values of the spin quantum number of each electron. On this basis, no volume element or cell can be occupied by more than one electron. Alternatively, it may be assumed that the number of cells in phase space is only half as great but that there may be two electrons in each cell, corresponding to the two directions of spin.

Suppose that there are N_i electrons in the total collection with energy E_i, and that there are M_i cells available. Because of the two directions of spin possible for each electron, M_i must be greater than N_i, and in the ground state M_0 is twice N_0 or there are two cells for each electron. If N_i cells are singly occupied by electrons, then $(M_i - N_i)$ are unoccupied. To enumerate the number of distinguishable distributions, it is convenient to write down the cells in order and assign the symbol 1 or 0 to each on the basis of whether it is occupied or not; thus, a possible distribution could be

$$M_1 \quad M_2 \quad M_3 \quad M_4 \quad M_5 \quad M_6 \quad M_7 \quad \cdots \quad M_i$$
$$1 \quad\; 1 \quad\; 0 \quad\; 1 \quad\; 0 \quad\; 0 \quad\; 1 \quad \cdots \quad 0,$$

where the first, second, fourth, and seventh cells are occupied, and the third, fifth, sixth, and ith cells are unoccupied. There are $M_i!$ ways in which the M_i cells may be written down, corresponding to the number of permutations of the M_i cells. For all the occupied cells, the occupation number is the same, 1, and since electrons are indistinguishable from one another, then if there are N_i electrons there are $N_i!$ indistinguishable arrangements of them possible. Similarly, there are $(M_i - N_i)!$ indistinguishable arrangements of unoccupied cells, those with occupation number 0. Altogether, there are therefore

$$w_i = M_i! \, [N_i! \, (M_i - N_i)!] \tag{m.3.49.}$$

distinguishable arrangements of N_i electrons among the M_i cells, and this expression replaces equation (m.3.27.), describing the number of possible distributions in Bose–Einstein statistics, corresponding to a smaller value for w_i. The total probability of a system of N electrons being distributed in such a manner over all phase space available to them is then

$$W = \prod_i M_i! \, [N_i! \, (M_i - N_i)!]. \tag{m.3.50.}$$

Proceeding as before, in the case of Bose–Einstein statistics, considering the entropy of a system of electrons in all conceivable energy states, and assuming that the equilibrium distribution is that which corresponds to the maximum probability of such a distribution existing, corresponding to maximum entropy, the system being subject to the constancy of the total number N of electrons, and to the constancy of their total kinetic energy, it is found that that

$$N_i/M_i = [e^{-\lambda}e^{E_i/kT} + 1]^{-1}, \ddagger \tag{m.3.51.}$$

and that, as before, the chemical potential (page 480),

$$[dE/dN]_{s,v} = \lambda kT. \tag{m.3.52.}$$

‡ Taking logarithms of (m.3.50.) gives, after the application of Stirling's theorem,

$$\ln W = \sum_i [M_i \ln M_i - M_i - (N_i \ln N_i - N_i) - \{(M_i - N_i) \ln (M_i - N_i) - (M_i - N_i)\}]$$

$$= \sum_i [M_i \ln \{M_i(M_i - N_i)^{-1}\} + N_i \ln \{(M_i - N_i)(N_i)^{-1}\}]$$

$$= (S - C)/k, \quad \text{as on page 477.}$$

As before,

$$S = k \ln W + C = k \sum_i [M_i \ln \{M_i(M_i - N_i)^{-1}\} + N_i \ln \{(M_i - N_i)(N_i)^{-1}\}] + C,$$

and

$$\partial S/\partial N_i = k \sum_i [M_i \, \partial/\partial N_i \ln \{M_i(M_i - N_i)^{-1}\} + N_i \, \partial/\partial N_i \ln \{(M_i - N_i)(N_i)^{-1}\}$$
$$+ \ln (M_i - N_i)/N_i]$$
$$= k \sum_i [M_i(M_i - N_i)^{-1} - M_i(M_i - N_i)^{-1} + \ln \{(M_i - N_i)(N_i)^{-1}\}]$$
$$= k \sum_i \ln \{(M_i - N_i)(N_i)^{-1}\} = \partial/\partial N_i[k \ln W + C],$$

since

$$\partial/\partial N_i \ln \{M_i(M_i - N_i)^{-1}\} = \partial y/\partial N_i,$$

where

$$y = \ln \{M_i(M_i - N_i)^{-1} \quad \text{and} \quad e^y = M_i(M_i - N_i)^{-1},$$

so that

$$\partial y/\partial N_i = \partial e^y/\partial N_i \cdot \partial y/\partial e^y = M_i(M_i - N_i)^{-2} \cdot (M_i - N_i)(M_i)^{-1} = (M_i - N_i)^{-1},$$

and

$$\partial/\partial N_i \ln \{(M_i - N_i)(N_i)^{-1}\} = \partial y'/\partial N_i,$$

where

$$y' = \ln \{M_i(M_i - N_i)^{-1}\} \quad \text{and} \quad e^{y'} = (M_i - N_i)/N_i,$$

so that

$$\partial y'/\partial N_i = \partial e^{y'}/\partial N_i \cdot \partial y'/\partial e^{y'} = -(M_i)(N_i)^{-2} \cdot (N_i)(M_i - N_i)^{-1} = -M_i\{N_i(M_i - N_i)\}^{-1}.$$

The most probable distribution will apply for the maximum value of the total probability W, which is when

$$\delta \ln W = 0 = \sum_i [\ln \{(M_i - N_i)/N_i\}] \, dN_i = \delta S = \sum_i dS_i.$$

The subsidiary conditions which must be complied with are given by (m.3.33.–34.), and then applying the theorem of undetermined multipliers to these three equations of condition, multiplying the equatins of numbers by the constant λ and the energy equation by $-\mu$, there results

$$\delta \ln W + \lambda \, \delta N - \mu \, \delta E = 0 \quad \text{or} \quad \sum_i [\ln \{(M_i - N_i)/N_i\} + \lambda - \mu E_i] \, dN_i = 0.$$

Again, since the variations dN_i are arbitrary, each of the terms in the above sum must each be identically zero, or

$$\ln \{(M_i - N_i)/N_i\} + \lambda - \mu E_i = 0, \quad \text{or} \quad [(M_i - N_i)/N_i] = M_i/N_i - 1 = e^{-\lambda + \mu E_i},$$

so that $N_i/M_i = [e^{-\lambda}e^{\mu E_i} + 1]^{-1}$, the result given in (m.3.51.), with $\mu = 1/kT$.

Equation (m.3.51.) differs from (m.3.40.) only in the sign preceding unity. However, there is a major distinction from the Bose–Einstein case, since now the expression in square brackets must always be greater than unity and λ can adopt all values from $-\infty \rightarrow \infty$, or $e^{-\lambda}$ may adopt all values from 0 to ∞.

In order to evaluate λ, it is convenient to make use of a previous expression deduced on the basis of the assumptions made in the simple free electron theory of metals. From equation (m.3.7.),

$$N = [V/2\pi^2][8\pi^2 m/h^2]^{3/2}(2/3)(E^*)^{3/2} = [(8\pi V)/(3h^3)][2mE^*]^{3/2}, \quad (m.3.53.)$$

and therefore the fractional number of electrons with energies between E and $E + dE$ is

$$dN = (2m)^{3/2}E^{1/2}[4\pi V/h^3]\, dE, \quad\quad (m.3.54.)$$

which on the present basis of allocating one electron to each cell, enables dN to be identified as M_i, the number of occupied cells. Therefore,

$$M_i = [4\pi V/h^3][2m]^{3/2}E^{1/2}\, dE, \quad\quad (m.3.55.)$$

which is twice the number given in (m.3.43.) for the Bose–Einstein expression. On substituting into (m.3.51.), the Fermi–Dirac distribution law is obtained as

$$N_i = [4\pi V/h^3][2m]^{3/2}[e^{-\lambda}e^{E/kT} + 1]^{-1}E^{1/2}\, dE. \quad\quad (m.3.56.)$$

Integration of the left-hand side of this equation gives the total number of electrons in the system, but because of the factor $+1$ included with the exponential terms it is not possible to carry out any general integration of the left-hand side, so that

$$N = [4\pi V/h^3][2m]^{3/2}\int_0^\infty [e^{-\lambda}e^{E/kT} + 1]^{-1}E^{1/2}\, dE. \quad\quad (m.3.57.)$$

An approximation to the value of the integral in (m.3.57.) may be obtained, however, from the following considerations. From the equation of condition given on page 483, $\delta \ln W + \lambda\, \delta N - \mu\, \delta E = 0$, the thermodynamic chemical potential $[dE/dN]_{S,N}$, is obtained as $[dE/dN]_S = +\lambda/\mu$, and since from (m.3.39.), equally applicable to both Bose–Einstein and Fermi–Dirac statistics, $\mu = 1/kT$, therefore $[dE/dN]_{S,V} = \lambda kT$, which on integration gives

$$E = N\lambda kT. \quad\quad (m.3.58.)$$

If then the chemical potential, $[dE/dN]$ is large, λ must be large, and therefore $e^{-\lambda}$ must be small and $e^{-\lambda}e^{E/kT}$ will be small in comparison with unity. Thus, an approximate value of the integral in (m.3.57.) is simply $\int_0^\infty E^{1/2}\, dE$. Since the upper physical limit for the energy of particles is λkT, rather than the mathematical one of ∞ (page 481), this means that the integral of (m.3.57.) may be approximated by $\int_0^{\lambda kT} E^{1/2}\, dE$ and so

$$N = [4\pi V/h^3][2m]^{3/2}\int_0^{\lambda kT} E^{1/2}\, dE = [4\pi V/h^3][2m]^{3/2}(2/3)(\lambda kT)^{3/2}$$

$$= [8\pi V/3h^3][2m\lambda kT]^{3/2}. \quad\quad (m.3.59.)$$

On comparison of (m.3.59.) and (m.3.53.), it is apparent that $\lambda kT = E^*$, the energy of electrons on the Fermi surface of momentum space, and therefore

$$\lambda = E^*/kT. \quad\quad (m.3.60.)$$

The Fermi–Dirac distribution law then becomes

$$N_i = M_i[e^{E-E^*/kT} + 1]^{-1} = [4\pi V/h^3][2m]^{3/2}[e^{E-E^*/kT} + 1]^{-1}E^{1/2} \, dE.$$

$$(\text{m.3.61.})$$

This expression contains the distribution function used in (m.3.14.), and it should be noted that when $E = E^*$, there are two cells for each electron.

The free energy of the system of electrons is given by

$$A = E - TS = N(\lambda kT) - T(k \ln W)$$

$$= kT[\lambda N - \ln W] = kT\left[\lambda N - \sum_i (M_i \ln \{M_i(M_i - N_i)^{-1}\} \right.$$

$$\left. + N_i \ln \{(M_i - N_i)(N_i)^{-1}\}\right]$$

$$= NE^* - kT\sum_i M_i \ln [1 + e^{(E^* - E_i)/kT}] = NE^* - \sum_i M_i[E^* - E_i]$$

$$= NE^* - [4\pi V/h^3][2m]^{3/2}\int_0^{E^*} [E^* - E]E^{1/2} \, dE$$

$$= NE^* - [8\pi V/3h^3][2m]^{3/2}(E^*)^{3/2}(2/5)E^*$$

$$= NE^* - (2/5)_N E^* = (3/5)_N E^*.\qquad(\text{m.3.62.})$$

This is the value of the mean Fermi energy, as given in (m.3.9.). The total energy of the collection of electrons may be obtained directly as

$$E = \sum_i N_i E_i \simeq [4\pi V/h^3][2m]^{3/2}\int_0^{E^*} E^{3/2} \, dE = [4\pi V/h^3][2m]^{3/2}(2/5)(E^*)^{5/2}$$

$$= (3/5)NE^*,\qquad(\text{m.3.63.})$$

so that the total energy of the electrons is the same as their free energy.

From equation (m.3.62.), it is seen that the quantity, S, the entropy of the electrons, given by $S = -[dA/dT]_V = 0$. The mean Fermi energy and the free energy of the electrons depend only on the concentration, $n = N/V$, and independent of temperature [(m.3.8.) and (m.3.10.)]. Also, the chemical potential of the electrons, $[dA/dN]_{V,T} = E^*$, as in (m.3.58.) and (m.3.60.). The specific heat, $[dE/dT] = [dA/dT] = S = 0$.

If equation (m.3.57.) is integrated by neglecting the term $+1$ in the integrand, there is obtained

$$N = [2V/h^3]e^{\lambda}[2\pi mkT]^{3/2},\qquad(\text{m.3.64.})$$

or

$$e^{\lambda} = [Nh^3/2V][2\pi mkT]^{-3/2} = [nh^3/2][2\pi mkT]^{-3/2},\qquad(\text{m.3.65.})$$

giving

$$\lambda = -\ln ([2\pi mkT]^{3/2}[nh^3/2]^{-1}).\qquad(\text{m.3.66.})$$

Now, dividing equation (m.3.56.) for N_i by (m.3.64.) for N, omitting the term $+1$ in the former equation gives

$$N_i/N = dN/N = [4\pi V/h^3][2m]^{3/2}[e^{-\lambda}e^{E/kT}]^{-1}E^{1/2}\,dE$$
$$\times\ [(2V/h^3)(e^{\lambda})(2\pi mkT)^{3/2}]^{-1}$$

$$= 2E^{1/2}e^{-E/kT}[\sqrt{\pi}(kT)^{3/2}]^{-1}\,dE, \qquad\qquad (m.3.67.)$$

which again is identical with the classical Maxwell–Boltzmann energy distribution law (page 42). Thus, if the term $e^{-\lambda}e^{E/kT} \gg 1$, the Fermi–Dirac distribution law of (m.3.56.) becomes identical with the classical Maxwell–Boltzmann distribution law. This may be alternatively expressed in terms of the constant λ of (m.3.66.), by saying that if $e^{-\lambda} = [2/nh^3][2\pi mkT]^{3/2} \gg 1$, or if $[Nh^3/2V][2\pi mkT]^{-3/2} \ll 1$, which condition applies at very high temperatures or at very low particle densities, then Fermi–Dirac statistics becomes identical with classical statistics. However, for all metals at ordinary temperatures, $[Nh^3/2V][2\pi mkT]^{-3/2}$ is always very much greater than unity, so that under ordinary conditions all metals correspond to a degenerate electron gas, the state of degeneracy being favored by the small mass of electrons and by the large value of the density N/V or the electron concentration. Very approximately, an electron gas at ordinary temperature conditions and at the density concerned in metals is degenerate to the extent that would correspond to a gas such as helium at a temperature close to 0 K, if helium under these temperature conditions obeyed Fermi–Dirac statistics (but of course, helium actually obeys Bose–Einstein statistics at low temperatures and high densities). It may alternatively be regarded that the condition, $e^{-\lambda} = [2/nh^3][2\pi mkT]^{3/2} \gg 1$, applicable to Maxwell–Boltzmann statistics, is an expression of the Heisenberg Uncertainty Principle. If the small numerical factors in the statement of this condition are disregarded, then it becomes

$$[1/nh^3][mkT]^{3/2} \gg 1, \qquad \text{or} \qquad [V/N]^{1/3}[mkT]^{1/2} \gg h. \quad (m.3.68.)$$

But since $(V/N)^{1/3}$ is of the order of magnitude of the average distance between two particles, and in classical mechanics $[mkT]^{1/2}$ represents the average momentum of a particle, the product of these two quantities has the same magnitude as the product of the uncertainties in the position and momentum of a particle, and the condition of (m.3.68.) is simply a statement that this product must be very much greater than h, when classical mechanics and statistics is applicable to a system.

The relationship between the three statistical systems may be generalized by considering the fundamental problem of how N_i particles, each possessing the same energy, and having available to them a greater number M_i of cells in phase space, may be distributed. If $M_i > N_i$ and only one particle may be placed in each cell, then there are $M_i(M_i - 1)(M_i - 2)(M_i - 3) \cdots \times [M_i - (N_i - 1)]$ ways in which the N_i particles may be placed in the M_i cells, since there are M_i ways in which the first particle may be allocated, and $(M_i - 1)$ vacant cells after this has been done, so that there are $M_i(M_i - 1)$ ways of allocating two particles to the M_i cells, and so on until all N_i particles have been distributed, one to each cell. There are thus N_i terms in this expression. If, when one cell is occupied, there is however, the reduced probability by some amount δ of another particle entering that cell, then the number of ways of distributing the N_i particles among the M_i cells is

$$w_i = M_i(M_i - \delta)(M_i - 2\delta)(M_i - 3\delta) \cdots [M_i - (N_i - 1)\delta] \qquad \text{to } N_i \text{ terms}$$

$$= \delta^{N_i}[M_i/\delta][(M_i/\delta) - 1][(M_i/\delta) - 2] \cdots [(M_i/\delta) - (N_i - 1)],$$

on multiplying and dividing each of the N_i terms by δ. On multiplying both the numerator and denominator of this expression again by $[(M_i/\delta) - N_i]!$, the number of ways of distributing the N_i particles among the available M_i cells becomes

$$w_i = \frac{\delta^{N_i}[M_i/\delta][(M_i/\delta) - 1][(M_i/\delta) - 2] \cdots [(M_i/\delta) - (N_i - 1)] \times [(M_i/\delta) - N_i] \cdots 1}{[(M_i/\delta) - N_i]!}$$

$$= \frac{\delta^{N_i}[(M_i/\delta)]!}{[(M_i/\delta) - N_i]!}.$$

The total probability that for a system containing N particles, there would be N_1 particles distributed in this manner, N_2 similarly distributed in another energy group, and so on is

$$W = N! \, [N_1! \, N_2! \, N_3! \, \cdots]^{-1} \prod_i w_i$$

$$= N! \, [N_1! \, N_2! \, \cdots]^{-1} \prod_i ([\delta^{N_i}(M_i/\delta)!][\{(M_i/\delta) - N_i\}!]^{-1}). \qquad \text{(m.3.69.)}$$

In the event that δ is non-integral, when its factorial has no meaning except for the special case of $(1/2)!$, this expression is still valid if the factorials are substituted by gamma functions, noting that $n! = \Gamma(n + 1)$ as shown on page 43, when

$$W = N! \, [N_1! \, N_2! \, N_3! \, \cdots]^{-1} \prod_i ([\delta^{N_i}\Gamma\{(M_i/\delta) + 1\}]$$

$$\times [\Gamma\{(M_i/\delta) - N_i + 1\}]^{-1}). \qquad \text{(m.3.70.)}$$

Thus, it is only when $\delta = 0$, when all of the cells in phase space are equally occupiable, that these expressions simplify to give the probability corresponding to Maxwell–Boltzmann statistics. When $\delta = +1$, Fermi–Dirac statistics arises and when $\delta = -1$ the Bose–Einstein statistics is described.

M.3.5.   Quantum Mechanical Formulation of the Free Electron Theory

In order to adequately describe the properties of metals, such as electrical conductivity, electronic specific heat, electronic emission, and thermal conductivity, the theory must be concerned with the behavior of large numbers or assemblies of electrons. In its simplest aspects, the periodic field provided by the metallic nuclei, the Coulomb and exchange interactions between the electrons, and the perturbations due to the thermal vibrations of the nuclei on the periodic field, are disregarded, and the behavior of an assembly of N electrons in a volume V is assumed to be describable in terms of the sum of the N one-electron states, subject to the restriction that not more than two electrons are in any one state. This simple free electron model actually provides a reasonable description for the state of affairs in the alkali metals, but is otherwise a very crude approximation for all other metals.

The simple electron in a box model has already been seen to lead to the Fermi–Dirac relationship between N, E, and V (pages 462–468). Wave functions, such as (m.3.1.) represent stationary waves and describe the states of an electron moving to and fro between the walls of a box of edge length L, since the average value of the momentum of the electron, so described, is zero for all such functions (see page 149). To deal with problems such as the electrical conductivity of metals, it is however, necessary to describe the electrons by means of traveling waves, rather than stationary waves, in terms of the "cyclical metal" model. It has been seen also in Chapter Q.2. that a simple function corresponding to a traveling wave, in one dimension, is for example $\psi_k = e^{ikx}$, which represents a free particle traveling in the x-direction with constant momentum $hk/2\pi$ and energy $E = h^2k^2/8\pi^2m = p_x{}^2/2m$, and satisfying the equation, $d^2\psi_k/dx^2 + [8\pi^2m/h^2]E\psi_k = 0$. The "cyclical" model is designed to meet the objection that ψ_k does not satisfy the correct boundary conditions, in that it does not vanish at $x = 0$ and at $x = L$. This condition is however satisfied, in the one-dimensional case, by imagining the two ends making contact, thus eliminating the infinite potential barriers at its ends and also the necessity of the wave function having a vanishing value at these points, but leaving the Schrödinger equation inside the metal unchanged. Since now the point x and the point $(x + L)$ coincide, the wave function becomes a periodic function of x with the period L, or $\psi(x) = \psi x) + L)$ for all values of x, and for the function to be smooth and continuous, $\psi(0) = \psi(L)$ and $d\psi(0)/dx = d\psi(L)/dx$. These are referred to as periodic boundary conditions. When applied to the function $\psi_k = e^{ikx}$, the condition that $\psi_k(0) = \psi_k(L)$ results in $e^{ikL} = 1$, so that $k = 2\pi n/L$, where $n = 0, \pm 1, \pm 2, \ldots$, and $E_k = h^2k^2/8\pi^2m = [h^2/8mL^2][2n]^2$. The fact that the quantities n_x of (m.3.2.)

may only have positive integral values, whereas the values E_k include only one-quarter of the energy levels E_n but in this case n may be either positive or negative, or zero, leads to the density of states being identical in both cases. The properties of the whole assembly of electrons are therefore independent of the boundary conditions assumed, although the actual boundary conditions greatly affect the energies and wavelengths of the individual states. Thus, the electron-in-the-box model of (m.3.1.) corresponds to a lowest energy of $3h^2/8mL^2$ and a wavelength of $2L/\sqrt{3}$ ($\lambda = h/p = h/\sqrt{2mE}$), while the periodic boundary conditions lead to a lowest energy of zero and corresponding wavelength of infinity. For real metals, the number of electrons is so large and the spacing between levels is so small that the only quantity of real importance is the density of states, and the individual energy levels are only of consequence in their contribution to the total energy density. Periodic boundary conditions and the use of traveling waves are generally more mathematically convenient than the stationary wave description, although both correspond to the same density of energy states.

In three-dimensional space, the function required to describe traveling waves is of the type

$$\psi_k(\mathbf{r}) = e^{i\mathbf{k}\cdot\mathbf{r}} = e^{i(k_1 x + k_2 y + k_3 z)}, \tag{m.3.71.}$$

which describes the motion of a particle moving with constant momentum $h\mathbf{k}/2\pi$, and is such that it is triply periodic in x, y, and z, with the period L. Thus,

$$\psi_k(x, y, z) = \psi_k(x + L, y, z) = \psi_k(x, y + L, z) = \psi_k(x, y, z + L). \tag{m.3.72.}$$

These periodic boundary conditions lead to the same density of states already given in (m.3.6.).

M.3.6. Wave Vector Space or k-Space

It is assumed in the Sommerfeld free electron gas model of metals that the electrons are completely free and non-interacting, even though they are assumed to be contained in the metal by the neutralization of their collective charges by the uniform distribution of positive charge provided by the metallic cations. If it is assumed that there is, in a cube of metal of edge length L, N electrons, then the positive charge density in this cube must be $N\varepsilon/L^3$, and the appropriate Hartree equation is simply

$$-h^2/8\pi^2 m\nabla^2\psi = E\psi. \tag{m.3.73.}$$

When periodic boundary conditions are applied, describing the electrons by traveling waves, then since

$$\nabla^2 e^{i\mathbf{k}\cdot\mathbf{r}} = -k^2 e^{i\mathbf{k}\cdot\mathbf{r}}, \tag{m.3.74.}$$

where $\mathbf{k} = k_1\mathbf{e}_1 + k_2\mathbf{e}_2 + k_3\mathbf{e}_3$, $\mathbf{r} = x\mathbf{e}_1 + y\mathbf{e}_2 + z\mathbf{e}_3$, and $\mathbf{e}_1$, $\mathbf{e}_2$, and $\mathbf{e}_3$, are a mutually orthogonal set of unit vectors, so that $\mathbf{k} \cdot \mathbf{r} = k_1 x + k_2 y + k_3 z$, a solution of (m.3.73.) is given by

$$\psi_k(\mathbf{r}) = Ce^{i\mathbf{k}\cdot\mathbf{r}}, \tag{m.3.75.}$$

where C is any constant. The eigenvalue $E = E(\mathbf{k})$ must be

$$E(\mathbf{k}) = h^2 k^2/8\pi^2 m = h^2/8\pi^2 m[k_1{}^2 + k_2{}^2 + k_3{}^2].$$

If the periodic boundary conditions are applied to (m.3.75.),

$$e^{ik_1 L} = e^{ik_2 L} = e^{ik_3 L} = 1, \qquad \text{or} \qquad k_1 = 2\pi n_1/L,\ k_2 = 2\pi n_2/L,\ k_3 = 2\pi n_3/L, \tag{m.3.76.}$$

where n_1, n_2, and n_3, may be zero, or positive or negative integers. On this basis, then the vector $\mathbf{k}$, given by

$$\mathbf{k} = [2\pi/L][n_1\mathbf{e}_1 + n_2\mathbf{e}_2 + n_3\mathbf{e}_3], \tag{m.3.77.}$$

is referred to as the wave vector of the state described by the function $\psi_k(\mathbf{r})$, or simply as the state $\mathbf{k}$. If λ is the wavelength associated with ψ_k, the magnitude of k is then $2\pi/\lambda$. The functions ψ_k are normalized if $\int |\psi_k{}^2|\ d\mathbf{r} = 1 = C^2 L^3$, so that the normalized functions are given by

$$\psi_k(\mathbf{r}) = [L]^{-3/2}e^{i\mathbf{k}\cdot\mathbf{r}}, \tag{m.3.78.}$$

and therefore since the charge density is given by $|\psi_k|^2 = L^{-3}$, the average electron charge density is uniform. The average charge density of all the electrons is $-\varepsilon N/L^3$, which offsets the uniform positive charges on the cations leading to the zero Hartree field. The energy levels of (m.3.76.) may be alternatively written as

$$E(\mathbf{k}) = [h^2/8\pi^2 m][4\pi^2/L^2][n_1{}^2 + n_2{}^2 + n_3{}^2] = h^2/2mL^2[n_1{}^2 + n_2{}^2 + n_3{}^2], \tag{m.3.79.}$$

which on comparison with (m.3.2.) shows that only one-quarter of the energy levels there described are now present. However, again since n_x, n_y, and n_z, in (m.3.2.) could only have positive integral values, while now n_1, n_2, and n_3, may have both positive and negative integral values, the degeneracy of each level is now four times as large as before, leading to four times as many states to each energy level, so that the density of states $N(E)$ remains unchanged in both modes of description.

The concept of reciprocal wave space, wave vector space, or $\mathbf{k}$-space, is a very useful one in many aspects of solid state phenomena-metal theory, X-ray, electron, and neutron diffraction theories. Instead of momentum space or $\mathbf{n}$-space, this $\mathbf{k}$-space, referred to the orthogonal set of reference axes k_1, k_2, and k_3, is such that a point with coordinates $2\pi n_1/L$, $2\pi n_2/L$, and

$2\pi n_3/L$, where n_1, n_2, and n_3 are integers, represents an orbital state. Thus a cube with edge length $2\pi/L$ contains exactly one orbital state, and therefore the number of such states per unit volume of **k**-space is $L^3/8\pi^3$ or $V/8\pi^3$, or the number of orbital states in the volume element $d\mathbf{k}$ of **k**-space is $[V/8\pi^3]\,d\mathbf{k}$. Even although $d\mathbf{k}$ may correspond to a very small volume compared with the occupied region of **k**-space for the metal concerned, it may still contain a large number of states, as described on page 464 for $N(E)$.

From (m.3.76.), it is apparent that surfaces of constant energy in **k**-space are spheres, the so-called Fermi spheres. Since each orbital state contains two electrons, the total number of electrons in a metal must be denoted by twice the number of orbital states within the Fermi sphere, corresponding to the highest states present at 0 K,

$$N = 2[V/8\pi^3][4\pi/3]k^3; \qquad \text{(m.3.80.)}$$

the radius of the Fermi sphere, at the absolute zero of temperature, when all states up to the energy of those on the Fermi surface are occupied is then

$$k_0 = [3\pi^2 N/V]^{1/3} = [9\pi/4]^{1/3}r_s^{-1},$$

and the Fermi energy

$$E^* = h^2 k^2/8\pi^2 m = [h^2/8\pi^2 m][9\pi/4]^{2/3}r_s^{-2},$$

as in (m.3.11.). The density of electron states in terms of the wave vector model may be derived from the fact that the number of states per unit unit volume in **k**-space is $V/8\pi^3$. Therefore the number of orbital states with energy less than some value k is $[4\pi k^3/3][V/8\pi^3] = [Vk^3/6\pi^2]$, and the density of states is the derivative with respect to E of this quantity, so that

$$N(E) = d/dE[Vk^3/6\pi^2] = d/dk[Vk^3/6\pi^2][dk/dE] = [Vk^2/2\pi^2][dE/dk]^{-1}$$

$$= [V/2\pi^2][4\pi^2 mk/h^2] = [V/4\pi^2][8\pi^2 m/h^2]^{3/2}E^{1/2} = 2\pi V[2m/h^2]^{3/2}E^{1/2},$$

since $E = h^2 k^2/8\pi^2 m$ and $dE/dk = h^2 k/4\pi^2 m$, and this is the same result as has already been obtained in (m.3.6.).

M.3.7. The Hartree–Fock Method and the Free Electron Approximation

The Fermi–Dirac theory of the electron gas completely neglects exchange effects and assumes that the electrons in a metal move about as if they had no potential energy. Since, actually, the free electrons would exert repulsive forces on one another, it is necessary in order to prevent the "electron gas" from dispersing to think of a uniform distribution of positive charge being present in the metal, of just the right density to balance the average negative charge density of the electrons. This assumption gives rise to the Hartree self-consistent-field description of the electron gas, which is a slightly more sophisticated way of describing a metal than the Sommerfeld model, although it leads to exactly the same results as the Sommerfeld method. In fact, the previous section, M.3.6., constitutes the application of the Hartree method to a free electron gas, with the energies expressed in terms of their distribution in wave-vector or reciprocal wavelength space rather than momentum space, and it has been seen that the resulting expressions for the Fermi energy and the density of electron states are identical to those of the Sommerfeld model, given in section M.3.2.

Actually, the exchange effect for a free electron gas can be calculated exactly, and the free electron approximation to the Hartree–Fock exchange energy was one of the early applications of quantum mechanics to metal theory. These type of calculations were originally carried out by Bloch[1] and Dirac[2] and extended by Wigner and Seitz.[3] The Hartree–Fock equations for a one-electron-per-atom metal, such as the alkali metals, has been found, as discussed in section Q.5.7. to be given by equations of the form (q.5.54.),

$$\left[-(h^2/8\pi^2 m)\nabla_1^2 + Z\varepsilon^2/r_1 + \varepsilon^2 \sum_j^N \int [|\psi_j(\mathbf{r}_2)|^2]/r_{12}\, d\mathbf{r}_2 \right]\psi_i(\mathbf{r}_1)$$

$$- \varepsilon^2 \sum_j^N \left[\int [\psi_j^*(\mathbf{r}_2)\psi_i(\mathbf{r}_2)\psi_j(\mathbf{r}_1)]/[r_{12}\psi_i(\mathbf{r}_1)]\, d\mathbf{r}_2 \right]\psi_i(\mathbf{r}_1) = E_i\psi_i(\mathbf{r}_1). \quad \text{(m.3.81.)}$$

This equation is comparable with Slater's equation approximating the Hartree–Fock exchange energy, given by (q.6.110.). For the free "electron gas" model, since as noted above the Hartree field is zero, the second and third terms in the first large square brackets in (m.3.81.), which represent the Hartree energy, may be neglected. If then equation (m.3.81.) is rewritten in terms of the wave vector description of electron energies, the eigenfunctions there distinguished by the subscripts i and j may be more conveniently be replaced by the subscripts $\mathbf{k}$ and $\mathbf{k}'$, so that

$$-[h^2/8\pi^2 m]\nabla_1^2\psi_k(\mathbf{r}_1) - \varepsilon^2 \sum_{k'}^N \left[\int [\psi_{k'}^*(\mathbf{r}_2)\psi_k(\mathbf{r}_2)\psi_{k'}(\mathbf{r}_1)]/(r_{12}\psi_k(\mathbf{r}_1))\, d\mathbf{r}_2 \right]$$

$$\times \psi_k(\mathbf{r}_1) = E(\mathbf{k})\psi_k(\mathbf{r}_1). \quad \text{(m.3.82.)}$$

In this equation the integrals are applicable throughout the whole volume of the metal concerned, V. At 0 K, when each state contains two electrons with opposed spins, a sum taken over all electrons with one kind of spin is the same as a sum taken over all electrons with the opposite spin, so that the

[1] Bloch, F., *Z. Physik*, 1929, **56**, 545.
[2] Dirac, P. A. M., *Proc. Cambridge Phil. Soc.*, 1930, **26**, 376; **27**, 240.
[3] Wigner, E., and Seitz, F., *Phys. Rev.*, 1933, **43**, 804.

summation above over $\mathbf{k}'$ must necessarily include also summation over $\mathbf{k}$, without it being necessary to explicitly state that this is a double sum.

If it is assumed that the normalized plane wave functions, (m.3.78.), which have been shown to be eigenfunctions of the Hartree equation are also eigenfunctions of the Hartree–Fock equation (m.3.82.), then the exchange term in the latter becomes

$$-[\varepsilon^2/V]\sum_{\mathbf{k}'}\left(\int [e^{i(\mathbf{k}-\mathbf{k}')\cdot(\mathbf{r}_2-\mathbf{r}_1)}]/r_{12}\,d\mathbf{r}_2\right)\psi_k(\mathbf{r}_1). \qquad \text{(m.3.83.)}$$

There are several ways of getting at the value of the integral in (m.3.83.), which is

$$e^{-i(\mathbf{k}-\mathbf{k}')\cdot\mathbf{r}_1}\int [e^{i(\mathbf{k}-\mathbf{k}')\cdot\mathbf{r}_2}]/r_{12}\,d\mathbf{r}_2, \qquad \text{(m.3.84.)}$$

as pointed out by Slater,[1] but an approximate method due to Raimes[2] making use of an analogy based on electrostatics suggests a simple solution. Thus, if $f(\mathbf{r})$ is the potential at some point $\mathbf{r}$ due to an imaginary charge distribution of density $e^{i(\mathbf{k}-\mathbf{k}')\cdot\mathbf{r}}$, then the potential $f(\mathbf{r})$ must satisfy the inhomogeneous partial differential equation known as Poisson's equation,[3]

$$\nabla^2 f(\mathbf{r}) = -4\pi e^{i(\mathbf{k}-\mathbf{k}')\cdot\mathbf{r}}. \qquad \text{(m.3.85.)}$$

A solution of this equation, as may be easily shown by differentiation is

$$f(\mathbf{r}) = 4\pi e^{i(\mathbf{k}-\mathbf{k}')\cdot\mathbf{r}}[|\mathbf{k}-\mathbf{k}'|^2]^{-1}. \qquad \text{(m.3.86.)}$$

The potential at some point $\mathbf{r}_1$ represented by

$$f(\mathbf{r}_1) = \int [e^{i(\mathbf{k}-\mathbf{k}')\cdot\mathbf{r}_2}]/r_{12}\,d\mathbf{r}_2, \qquad \text{(m.3.87.)}$$

is then approximately given by analogy with (m.3.86.) as

$$[4\pi e^{i(\mathbf{k}-\mathbf{k}')\cdot\mathbf{r}_1}][|\mathbf{k}-\mathbf{k}'|^2]^{-1}, \qquad \text{(m.3.88.)}$$

so that the integral (m.3.84.) is obtained as

$$4\pi[|\mathbf{k}-\mathbf{k}'|^2]^{-1}. \qquad \text{(m.3.89.)}$$

The exchange energy term, (m.3.83.) thus becomes

$$E_k(\mathbf{k}) = -[4\pi\varepsilon^2/V]\sum_{k'}[|\mathbf{k}-\mathbf{k}'|^2]^{-1}. \qquad \text{(m.3.90.)}$$

---

[1] Slater, John C., *Quantum Theory of Atomic Structure*, **Vol. II**, 1960, McGraw-Hill, N.Y.

[2] Raimes, S., *The Wave Mechanics of Electrons in Metals*, 1961, North-Holland, Amsterdam.

[3] Margenau, H., and Murphy, G. M., *The Mathematics of Physics and Chemistry*, 2nd Edn., 1961, D. Van Nostrand Co., Inc., Princeton, New Jersey.

The functions (m.3.78.) must therefore be eigenfunctions of the Hartree–Fock equations but corresponding not to the eigenvalues $h^2k^2/8\pi^2m$ of the Hartree equation, but to the different eigenvalues

$$E(\mathbf{k}) = h^2k^2/8\pi^2m + E_k(\mathbf{k}), \qquad \text{(m.3.91.)}$$

where $E_k(\mathbf{k})$ have still to be evaluated.

To evaluate (m.3.90.), it is necessary to convert to spherical polar co-ordinates, (k', θ, ϕ), regarding the vector $\mathbf{k}$ as coincident with the Cartesian $+z$ direction, θ as the angle between $\mathbf{k}$ and $\mathbf{k}'$, and ϕ as the angle in the xy-Cartesian plane between the x-direction and the projection of $\mathbf{k}'$ onto this plane. Then, making use of the fact that the volume element $d\mathbf{k}$ of reciprocal wave space is $V\,d\mathbf{k}/8\pi^3$ (page 491), and replacing the summation in (m.3.90.) by an integration,

$$E_k(\mathbf{k}) = -[4\pi\varepsilon^2/V][V/8\pi^3] \int [|\mathbf{k} - \mathbf{k}'|^2]^{-1}\, d\mathbf{k}'$$

$$= -[\varepsilon^2/2\pi^2] \int_0^{k_0} dk' \int_0^{\pi} d\theta \int_0^{2\pi} [k'^2 \sin\theta][|\mathbf{k} - \mathbf{k}'|^2]^{-1}\, d\phi$$

$$= -[\varepsilon^2/\pi] \int_0^{k_0} dk' \int_{-1}^{1} [k'^2][k^2 + k'^2 - 2kk' \cos\theta]^{-1}\, d(\cos\theta)$$

$$= -[\varepsilon^2/\pi k] \int_0^{k_0} k' \ln\left|\frac{k + k'}{k - k'}\right| dk'. \qquad \text{(m.3.92.)}$$

When $k = k'$, the integrand in this expression becomes infinite, so that for values of $k < k_0$, the integral has to be expressed as the sum of two parts, the integral from 0 to k and the integral from k to k'; both of these converge. Integrating by parts, (m.3.92.) becomes

$$E_k(\mathbf{k}) = -[\varepsilon^2 k_0/2\pi]\left[2 + \frac{k_0^2 - k^2}{kk_0} \ln\left|\frac{k_0 + k}{k_0 - k}\right|\right]. \qquad \text{(m.3.93.)}$$

The eigenvalues of the Hartree–Fock equation, (m.3.91.) then become, on substitution of (m.3.93.)

$$E(\mathbf{k}) = [h^2k^2/8\pi^2m] - [\varepsilon^2 k_0/2\pi]\left[2 + \frac{k_0^2 - k^2}{kk_0} \ln\left|\frac{k_0 + k}{k_0 - k}\right|\right] \equiv E(k). $$
$$\text{(m.9.34.)}$$

The limiting value of the logarithmic function contained in this expression behaves such that

$$\lim_{k \to 0} \frac{k_0^2 - k^2}{kk_0} \ln\left|\frac{k_0 + k}{k_0 - k}\right| = 2 \qquad \text{and} \qquad \lim_{k \to k_0} \frac{k_0^2 - k^2}{kk_0} \ln\left|\frac{k_0 + k}{k_0 - k}\right| = 0,$$
$$\text{(m.3.95.)}$$

and therefore

$$E(0) = -[2\varepsilon^2 k_0/\pi] \quad \text{and} \quad E(k_0) = [h^2 k_0{}^2/8\pi^2 m] - [\varepsilon^2 k_0/\pi].$$
$$\text{(m.3.96.)}$$

The energy difference between the highest and the lowest energy states at 0 K for any electrons is thus

$$E(k_0) - E(0) = [h^2 k_0{}^2/8\pi^2 m] + [\varepsilon^2 k_0/\pi] = E^* + [\varepsilon^2 k_0/\pi], \quad \text{(m.3.97.)}$$

and it is this quantity which is referred to as the band width of the "electron gas." This is quite a different value from that obtained on the basis of the Hartree description, which is simply E^*, the additional band-width in the Hartree–Fock description, $[\varepsilon^2 k_0/\pi]$, amounting to a substantial addition. For example, E^* for sodium is 5.13×10^{-19} J while $[\varepsilon^2 k_0/\pi]$ is 6.57×10^{-19} J. The soft X-ray emission spectrum of sodium indicates a band width close to the Hartree value. Several other physical properties of metals, which are dependent on the band width and thus on the shape of their $N(E)$ curves, are in better agreement with the deductions made from the Hartree theory than from the Hartree–Fock theory. Since the Hartree method completely ignores the Coulomb interactions between electrons, this agreement with experimental observation must be fortuitous. The Hartree–Fock method, as indicated in section Q.5.8., does however, include some account of parallel spin correlation effects but neglects the correlation between electrons with anti-parallel spins. The above results would imply that it is preferable to completely ignore all correlation effects between electrons, rather than to take account of only some of them. Actually, the Hartree–Fock method enables the calculated total energy value to be in better agreement with the experimental value than the Hartree value.

Since the band width corresponding to the region of energy states over which the occupied states are distributed is so different in the Hartree and Hartree–Fock methods, then the $N(E)$ curves present quite a different appearance for the two modes of description. The constant-energy surfaces in wave-vector space are spheres, so that as on page 491, the Hartree–Fock density of states is given by

$$N(E) = [Vk^2/2\pi^2][dE/dk]^{-1}$$
$$= [Vk^2/2\pi^2]\left[h^2 k^2/8\pi^2 m - (\varepsilon^2/2\pi)[2k_0/k - \{1 + k_0{}^2/k^2\} \ln \left|\frac{k_0 + k}{k_0 - k}\right|\right].$$
$$\text{(m.3.98.)}$$

When $k = k_0$ in this expression, dE/dk becomes infinite and therefore $N(E)$ becomes zero, at the surface of the Fermi sphere. In the case of the Hartree description, the relationship between $N(E)$ and E is a simple parabolic one, but for the Hartree–Fock method, where there is no simple functional relationship between k and E, $N(E)$ and E must both be computed for selected values of k in order to find the shape of the $N(E)$ curve. Thus, while the $N(E)$ curve in the Hartree case is unaltered by change of temperature and number of electrons, N, only the density of states being affected, that is the function $N(E)f(E)$, the situation is quite different in the Hartree–Fock method. Here, the exchange energy $E_k(\mathbf{k})$ depends on which states are occupied and so does $N(E)$. The general shape of the $N(E)$ curves at 0 K for sodium, according to the Hartree and to the Hartree–Fock descriptions are shown in Fig. M.3.5. The band width for the Hartree–Fock case is wider than for the Hartree method, and the whole $N(E)$ curve is displaced to the low energy side, so that the total Hartree–Fock energy is lower than the Hartree energy. Actually, the Hartree method leads to a negative cohesive energy, which is ridiculous, but the Hartree–Fock method gives a positive cohesive energy but not large enough. At higher temperatures and for an increase in the total number of electrons, N, the shape of the Hartree–Fock curve would be altered due to the change in $E_k(\mathbf{k})$, and the excited or added electrons would not necessarily be accommodated in the unoccupied states beyond the region where $N(E)$ falls off rapidly to zero.

The total energy of a free electron gas of N electrons, according to the Hartree–Fock scheme is given by equation (q.5.56.). But, since in the free electron model there are assumed to be no cations as such but only a uniform positive charge distributed in such a manner as to cancel out the electron charge distribution, the second and fourth terms in (q.5.56.) are inapplicable, so that the total Hartree–Fock energy becomes

$$E = \sum_i^N E_i + (1/2) \sum_i^N \sum_j^N \iint [\varepsilon^2/r_{12}]\psi_i{}^*(\mathbf{r}_1)\psi_j{}^*(\mathbf{r}_2)\psi_i(\mathbf{r}_2)\psi_j(\mathbf{r}_1)\,d\mathbf{r}_1\,d\mathbf{r}_2. \quad \text{(m.3.99.)}$$

As for (m.3.81.), if this equation is rewritten in wave vector notation, changing the subscripts for the eigenfunctions from i and j to $\mathbf{k}$ and $\mathbf{k}'$, keeping in mind that summations over $\mathbf{k}$ and $\mathbf{k}'$, which refer to orbital states, must be doubled to provide the summations with respect to the number of electrons, the total Hartree–Fock energy for a free electron gas is obtained as

$$N = 2\sum_k E(\mathbf{k}) + \sum_k \sum_k \iint [\varepsilon^2/r_{12}]\psi_k{}^*(\mathbf{r}_1)\psi_{k'}{}^*(\mathbf{r}_2)\psi_k(\mathbf{r}_2)\psi_{k'}(\mathbf{r}_1)\,d\mathbf{r}_1\,d\mathbf{r}_1$$

$$= 2\sum_k [h^2k^2/8\pi^2m + E_k(\mathbf{k})] - \sum_k E_k(\mathbf{k}) = 2\sum_k [h^2k^2/8\pi^2m] + \sum_k E_k(\mathbf{k}).$$

$$\text{(m.3.100.)}$$

The first term in this expression is simply the total kinetic energy of the electrons given by (m.3.12.) in the Sommerfeld and Hartree descriptions, and the second term is the total Hartree–Fock exchange energy. To evaluate the latter, it may be handled in the same manner as $E_k(\mathbf{k})$ in (m.3.92.), with the

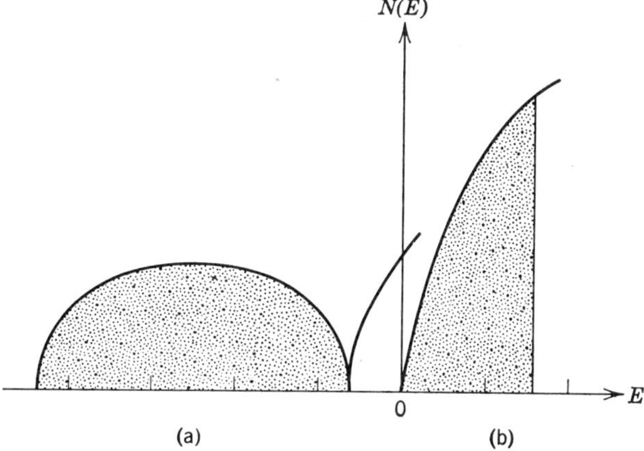

FIG. M.3.5. Representation of the density of states, $N(E)$, for sodium (a) according to the Hartree–Fock description of the free electron system at 0 K, and (b) in accordance with the Hartree description. The occupied states at 0 K are indicated.

result that

$$\sum_k E_k(\mathbf{k}) = -[\varepsilon^2/2\pi] \sum_k [2k_0 + \{(k_0{}^2 - k^2)/k\} \ln \{(k_0 + k)/(k_0 - k)\}]$$

$$= -[\varepsilon^2/2\pi][V/8\pi^3] \int_0^{k_0} [2k_0 + \{(k_0{}^2 - k^2)/k\} \ln$$

$$\times \{(k_0 + k)/(k_0 - k)\}]4\pi k^2 \, dk$$

$$= -[\varepsilon^2 V/4\pi^3] \int_0^{k_0} [2k_0 k^2 + k(k_0{}^2 - k^2) \ln \{(k_0 + k)/(k_0 - k)\}] \, dk$$

$$= -[\varepsilon^2 V/4\pi^3]k_0{}^4 = -[\varepsilon^2 V/4\pi^3][3\pi^2 N/V]^{4/3}$$

$$= -[3\varepsilon^2 N/4][9/4\pi^2]^{1/3}[1/r_s], \tag{m.3.101.}$$

where $k_0 = [3\pi^2 N/V]^{1/3} = [9\pi/4]^{1/3}[1/r_s]$, as shown on page 491. The average energy per electron given by the Hartree–Fock method when applied to the free "electron gas" is then the sum of the Fermi energy (m.3.12.) and $1/N$ of the Hartree–Fock exchange energy of (m.3.101.), or

$$E \text{ (Hartree–Fock)} = E_F + E \text{ (exchange)}/N$$

$$= [3h^2/40\pi^2 m][9\pi/4]^{2/3}[1/r_s{}^2] - [3\varepsilon^2/4][9/4\pi^2]^{1/3}[1/r_s]$$

$$= [3h^2/40\pi^2 m][9\pi/4]^{2/3}(1/r_s{}^2][4\pi^2 m\varepsilon^2/h^2]^2[h^2/4\pi^2 m\varepsilon^4]$$

$$\quad - [3\varepsilon^2/4][9/4\pi^2]^{1/3}[1/r_s][4\pi^2 m\varepsilon^2/h^2][h^2/4\pi^2 m\varepsilon^4]$$

$$\text{Hartree units}$$

$$= ([3/5][9\pi/4]^{2/3}[1/r_s{}^2] - [3/2][9/4\pi^2]^{1/3}[1/r_s]) \text{ Hartrees}$$

$$= [1.105/r_s{}^2] - [0.458/r_s] \text{ Hartrees (see Q.1.10.).}$$

$$\tag{m.3.102.}$$

Since the atomic radii, r_s, for all of the alkali metals are between three and six atomic Bohr units, the Hartree–Fock exchange energy lowers the Fermi energy E_F by a sufficiently large amount to make it negative. In the more detailed electron theories, where proper account is taken of all Coulomb correlations, and not just those with one spin orientation as in the Hartree–Fock scheme, the total energy becomes even lower and the density of states closely approaches that of the Hartree theory to give good agreement with observed energy values.

The concept of the Fermi hole as a spherical region of approximate radius r_s in the distribution of electrons with parallel spins, may also be deduced from the Hartree–Fock equations of (m.3.82.).‡ The idea of the Fermi hole has been discussed in section Q.5.8., and the general expression for the exchange charge density written as a sum is given by (q.5.57.). If this is rewritten in wave-vector notation, it becomes the exchange charge density at the position $\mathbf{r}_2$ for an electron in the skate $\mathbf{k}$ at position $\mathbf{r}_1$, as

$$\varepsilon \sum_{\mathbf{k}'} [\psi_{k'}{}^*(\mathbf{r}_2)\psi_k(\mathbf{r}_2)\psi_{k'}(\mathbf{r}_1)][\psi_k(\mathbf{r}_1)]^{-1}. \qquad \text{(m.3.103.)}$$

‡ An expression for the average charge density at some position $\mathbf{r}_2$ due to all electrons with spins parallel to that of the electron at $\mathbf{r}_1$, may be deduced by means of a probability argument, quite independently of the use of the Hartree–Fock equation. Suppose, for example, that $\mathbf{k}$ and $\mathbf{k}'$ refer to the states of two free electrons with parallel and positive spins, $\alpha(\zeta)$ and $\alpha(\zeta')$ (see section Q.5.3.). Then if the electrons in the states $\mathbf{k}$ and $\mathbf{k}'$ are described by plane wave functions of the type given by (m.3.78.) the determinental wave function (see section Q.5.5.) for this system is

$$\frac{1}{V\sqrt{2}} \begin{vmatrix} e^{i\mathbf{k}\cdot\mathbf{r}_1} & e^{i\mathbf{k}\cdot\mathbf{r}_2} \\ e^{i\mathbf{k}'\cdot\mathbf{r}_1} & e^{i\mathbf{k}'\cdot\mathbf{r}_2} \end{vmatrix} \alpha(\zeta)\alpha(\zeta').$$

From the probability interpretation of wave functions and the general probability laws, the probability that $\mathbf{r}_1$ lies within the volume element $d\mathbf{r}_1$ and that $\mathbf{r}_2$ lies within the volume element $d\mathbf{r}_2$ is

$$[2V^2]^{-1}|e^{i(\mathbf{k}\cdot\mathbf{r}_1+\mathbf{k}'\cdot\mathbf{r}_2)} - e^{i(\mathbf{k}\cdot\mathbf{r}_2-\mathbf{k}'\cdot\mathbf{r}_1)}|^2\, d\mathbf{r}_1\, d\mathbf{r}_2 \quad = [2V^2]^{-1}|1 - e^{i(\mathbf{k}'-\mathbf{k})\cdot(\mathbf{r}_1-\mathbf{r}_2)}|^2\, d\mathbf{r}_1\, d\mathbf{r}_2$$

$$= [1/V^2][1 - F_R e^{i(\mathbf{k}'-\mathbf{k})\cdot(\mathbf{r}_1-\mathbf{r}_2)}]\, d\mathbf{r}_1\, d\mathbf{r}_2,$$

where F_R is used to denote the real part of the exponential function following it. An exactly similar expression would describe the probability applicable to all pairs of states $\mathbf{k}$ and $\mathbf{k}'$ which have the same spin, so the total probability would result from averaging over all states $\mathbf{k}$ and $\mathbf{k}'$ within the Fermi sphere. If $\mathbf{k} = \mathbf{k}'$, the above expression vanishes as two electrons with the same spin cannot occupy the same state. In determining the total probability, for an N-electron system where N is very large, the double sum of the probabilities over all electron pairs in states $\mathbf{k}$ and $\mathbf{k}'$ may be replaced by a double integral, and also since N is large the average is negligibly different whether the total is divided by $N^2/4$ or by $[N/2](N/2 - 1)$; $N/2$ is the number of electrons with parallel spins. Thus, the total probability that electron 1 is in the volume element $d\mathbf{r}_1$ while electron 2 occupies the volume element dr_2 simultaneously, with a parallel spin orientation, is

$$\frac{[V^{-2}] \sum\limits_{k} \sum\limits_{k} [1 - F_R e^{i(\mathbf{k}'-\mathbf{k})\cdot(\mathbf{r}_1 - \mathbf{r}_2)}]\, d\mathbf{r}_1\, d\mathbf{r}_2}{[N^2/4]}$$

$$= [4N^{-2}][8\pi^3]^{-2}\, d\mathbf{r}_1\, d\mathbf{r}_2 \iint [1 - F_R e^{i(\mathbf{k}'-\mathbf{k})\cdot(\mathbf{r}_1 - \mathbf{r}_2)}]\, d\mathbf{k}\, d\mathbf{k}',$$

on replacing the summations by integrals, and noting that the number of states per unit volume of k-space is $V/8\pi^3$:

$$= [4N^{-2}][8\pi^3]^{-2}\, d\mathbf{r}_1\, d\mathbf{r}_2([4\pi k_0^3/3]^2 - F_R \int [e^{i\mathbf{k}'\cdot(\mathbf{r}_1 - \mathbf{r}_2)}]\, d\mathbf{k}' \int [e^{-i\mathbf{k}\cdot(\mathbf{r}_1 - \mathbf{r}_2)}]\, d\mathbf{k}),$$

where k_0, as before, is the radius of the Fermi sphere of occupied states of total volume $[4\pi k_0^3/3]$;

$$= [4N^{-2}][8\pi^3]^{-2}\, d\mathbf{r}_1\, d\mathbf{r}_2([4\pi k_0^3/3]^2 - 16\pi^2[\sin k_0 r_{12} - k_0 r_{12} \cos k_0 r_{12}]^2 [r_{12}^3]^{-2}),$$

where both the integrals are real and equal in value and therefore unaffected by the F_R operator, and may be integrated as in (m.3.105.);

$$= [4N^{-2}][8\pi^3]^{-2}\, d\mathbf{r}_1\, d\mathbf{r}_2[4\pi k_0^3/3]^2(1 - 9[\sin k_0 r_{12} - k_0 r_{12} \cos k_0 r_{12}]^2 [k_0^3 r_{12}^2]^{-2},$$

on factoring out the term $[4\pi k_0^3/3]^2$,

$$= V^{-2}\, d\mathbf{r}_1\, d\mathbf{r}_2(1 - 9[\sin \Lambda - \Lambda \cos \Lambda]^2 [\Lambda^3]^{-2}),$$

on making the substitutions given by (m.3.80.) and (m.3.106.) for N and $k_0 r_{12}$, respectively;

$$= V^{-2}\, d\mathbf{r}_1\, d\mathbf{r}_2 F(\Lambda),$$

making use of the substitution for $F(\Lambda)$, given by (m.3.108.). The probability of electron 2 being in the volume element dr_2, if electron 1 is known to be in the volume element dr_1, with parallel spin, is thus $F(\Lambda)[d\mathbf{r}_2/V]$, since the probability of any particular electron being in the volume element $d\mathbf{r}$, irrespective of the positions occupied by the other electrons is $d\mathbf{r}/V$ (see page 22). Therefore, the average charge density of electron 2 at $\mathbf{r}_2$ is $-[\varepsilon/V]F(\Lambda)$, and the average charge density at $\mathbf{r}_2$ due to all electrons with spins parallel to that of the electron at $\mathbf{r}_1$ is $-[\varepsilon N/2V]F(\Lambda)$, which is the same result as that deduced below from the Hartree–Fock theory, given by (m.3.108.).

For functions of the plane wave type, as given by (m.3.78.), equation (m.3.103.) becomes after replacement of the summation by an integral over the Fermi sphere

$$[\varepsilon/8\pi^3] \int e^{i(\mathbf{k}'-\mathbf{k})\cdot(\mathbf{r}_1 - \mathbf{r}_2)}\, d\mathbf{k}' = [\varepsilon/8\pi^3][e^{-i\mathbf{k}\cdot(\mathbf{r}_1 - \mathbf{r}_2)} \int e^{i\mathbf{k}'\cdot)\mathbf{r}_1 - \mathbf{r}_2)}\, d\mathbf{k}'. \tag{m.3.104.}$$

Again, this integral may be evaluated by transforming to spherical polar coordinates (k', θ, ϕ), where θ this time is the angle between the vectors $\mathbf{r}_1 - \mathbf{r}_2$ and $\mathbf{k}'$. With this transformation, the expression for the exchange charge density now becomes

$$[\varepsilon/8\pi^3][e^{-i\mathbf{k}\cdot(\mathbf{r}_1 - \mathbf{r}_2)}] \int_0^{k_0} 2\pi k'\, dk' \int_{-1}^{1} e^{ik' r_{12}\cos\theta}\, d(\cos\theta)$$

$$= [\varepsilon/2\pi^2][e^{-i\mathbf{k}\cdot(\mathbf{r}_1 - \mathbf{r}_2)}] \int_0^{k_0} [(k' \sin k' r_{12})/r_{12}]\, dk'$$

$$= [\varepsilon/2\pi^2][e^{-i\mathbf{k}\cdot(\mathbf{r}_1 - \mathbf{r}_2)}]([\sin k_0 r_{12} - k_0 r_{12} \cos k_0 r_{12}]/(r_{12})^3). \tag{m.3.105.}$$

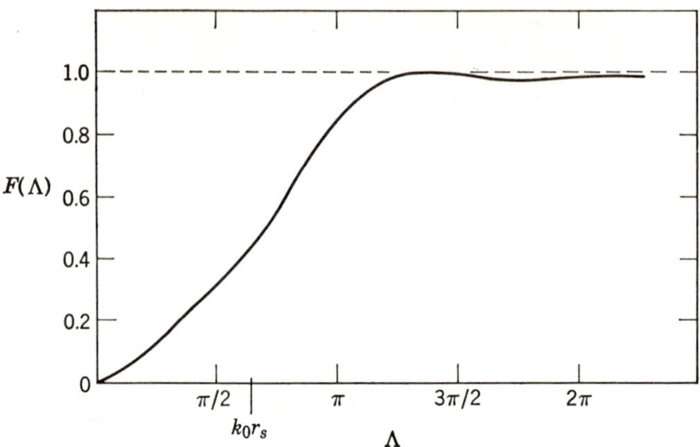

FIG. M.3.6. Shape of the function $F(\Lambda)$. Since the average charge density of electrons with parallel spins at a distance of r_{12} from any given electron is obtained as the product of this function with $-[\varepsilon N/2V]$, the idea of the average Fermi hole as a spherical region of radius of the order of r_s in the distribution of electrons with parallel spins is seen to have plausible justification.

This exchange charge density is thus a complex function, dependent on **k**, and there is no spherical symmetry about the electronic position $\mathbf{r}_1$; it falls off rapidly with increasing value of r_{12}, and in the limit as $r_{12} \to 0$, the exchange charge density becomes $[\varepsilon N/2V]$, which is equivalent to the charge density of electrons with one type of spin (one-half of total electron density) but of opposite sign. To find the average exchange charge density at the point $\mathbf{r}_2$ for an electron at the position $\mathbf{r}_1$, it is necessary to integrate equation (m.3.105.) over all values of **k** within the Fermi sphere, which is equivalent to doing so over all electrons with one kind of spin, and divide the result by $N/2$ (the number of electrons with one kind of spin). Thus, the average exchange charge density obtained in this way, making use of the substitution

$$k_0 r_{12} = [9\pi/4]^{1/3}[r_{12}/r_s] = \Lambda, \qquad \text{(m.3.106.)}$$

where the wave vector for a state on the Fermi surface is as given on page 491, is

$$[\varepsilon/2\pi^2]([\sin \Lambda - \Lambda \cos \Lambda]/(r_{12})^3)[2V/8N\pi^3] \int e^{-i\mathbf{k}\cdot(\mathbf{r}_1 - \mathbf{r}_2)}\, d\mathbf{k}$$

$$= [\varepsilon V(k_0)^6/2\pi^4 N]([\sin \Lambda - \Lambda \cos \Lambda][k_0 r_{12}]^{-3})^2$$

$$= [9N\varepsilon/2V]([\sin \Lambda - \Lambda \cos \Lambda][\Lambda]^{-3})^2. \qquad \text{(m.3.107.)}$$

Now, since the average charge density of electrons with one kind of spin is $-[\varepsilon N/2V]$, then if this quantity is added to the average exchange charge density given by (m.3.107.), there is obtained the average charge density of electrons with parallel spins, distant r_{12} from an electron at r_1, namely

$$-[\varepsilon N/2V] + [9\varepsilon N/2V]([\sin \Lambda - \Lambda \cos \Lambda][\Lambda]^{-3})^2$$

$$= -[1 - 9([\sin \Lambda - \Lambda \cos \Lambda][\Lambda^{-3}])^2][\varepsilon N/2V]$$

$$= -[\varepsilon N/2V]F(\Lambda). \qquad \text{(m.3.108.)}$$

Thus the average charge density of electrons with parallel spins is proportional to the function, $F(\Lambda)$; the asymptotic values of this function are unity for large values of Λ and zero for small values, and the general shape of the function is shown in Fig. M.3.6., from which it may be seen that there is good justification for regarding thinking of the average Fermi hole as a spherical hole of radius about r_s in the electron distribution of electrons with parallel spins. If there were no correlation at all of electronic motion, this function would be unity and the charge density of electrons with parallel spins would simply be $-[\varepsilon N/2V]$.

M4

Some Applications and Consequences of the Free Electron Theory

The metallic crystals for which the free electron theory provides the best approximation are the alkali metals, as already discussed on pages 431–434. In those metals the valency or conduction electrons are moving in the potential field of the ion cores of the metal atoms, which have a rare gas type configuration. For the alkaline earth metals and for the metals with three electrons per atom available as conduction electrons, the free electron approximation becomes progressively poorer, although sufficiently good to provide a valuable guide and background against which to check the results of the more elaborate theories of metals. In fact, the problem provided by the motion of the electrons in the alkali metals is similar to that discovered experimentally by Ramsauer[1] in the early 1920's involving the scattering of a beam of electrons by the rare gases,‡ a quantum mechanical explanation of which was provided later by Faxen, Holtsmark, Allis, and Morse,[2-5] based essentially on the calculations of Ewald[6] for the corresponding problem in X-ray scattering by crystals.

‡ The so-called Ramsauer effect was discovered experimentally by Ramsauer while investigating the way in which the velocity or energy of a beam of electrons was affected

[1] Ramsauer, C., *Ann. Physik*, 1921, **64**, 513; 1921, **66**, 545; 1923, **72**, 345.
[2] Faxen, H., and Holtsmark, J., *Z, Physik*, 1927, **45**, 307.
[3] Holtsmark, J., *Z. Physik*, 1928, **48**, 231.
[4] Allis, W. P., and Morse, P. M., *Z. Physik*, 1931, **70**, 567.
[5] Morse, P. M., *Rev. Mod. Phys.*, 1932, **4**, 577.
[6] Ewald, P. P., *Ann. Physik*, 1916, **49**, **1**, 117; 1917, **54**, 519.

by scattering by a gas. The atoms of most gases behave as if they had quite considerable cross sections, causing the beam of electrons to be reduced in intensity as if the electrons were scattered out of the incident beam by either elastic or inelastic collisions with the gas atom cross sections. Scattering refers to the fact that as a result of the interaction between the electrons in the incident beam and the target atoms, the direction of motion of the incident particle is changed. In elastic scattering, there is merely an exchange of kinetic energy between the projectile particle and the target atoms, both the total kinetic energy and momentum being conserved (see pages 74–76); in inelastic scattering the target atom acquires excitation energy as a result of the collision with the incident particle, so that kinetic energy is not conserved. For very slow or low energy electrons, with energies below the excitation energies of the target atoms, the scattering has to be entirely elastic. Ramsauer observed that the gas atom cross sections varied only slowly with the incident electron energy, being comparable in size with atomic dimensions in most cases. For the rare gases, however, he observed that the cross section apparently decreased rapidly with decreasing electron energies, becoming extremely small for very low energy electrons. This remarkable and non-classical Ramsauer effect was subsequently given a quantum mechanical explanation by Faxen, Holtsmark, Allis, and Morse.

It has already been seen in the previous chapters how the application of the Fermi–Dirac statistics to the free "electron gas" in metals leads immediately to results which are quite different from those of an ordinary gas obeying classical statistics, and provides some understanding of the heat capacity of metals at low temperatures, the soft X-ray spectra of metals, and the energy distribution of the electrons in metals. The application of the free electron theory to some of the other properties of metals, where the explanation of these properties suggested by the free electron approximation is valid, with various modifications, in the more detailed electron theories, is taken up in this chapter.

M.4.1. Electrical Conductivity and the Fermi–Dirac Distribution

The classical expression for the electrical conductivity of a metal is given by (m.1.33.) as $\kappa_e = ne^2\tau/m_e$, where n is the electron concentration and τ is the relaxation time between collisions as defined on page 427. This expression for the conductivity may be derived by an alternative argument, for a classical Maxwell–Boltzmann distribution of the electron velocities, which is unaltered by the application of Fermi–Dirac statistics.

Suppose, as in Fig. M.4.1., that (a) represents the classical distribution of electron velocities in one dimension, when the electrons in the "electron gas" are in thermal equilibrium. Then, when an electric field V is applied for a time τ in the $+x$ direction, the distribution will be moved to the right by an amount $eV\tau/m_e$, as shown in (b), where τ is the relaxation time. This distribution then represents the steady-state distribution in the presence of the electric field. The difference between the steady-state distribution and the thermal

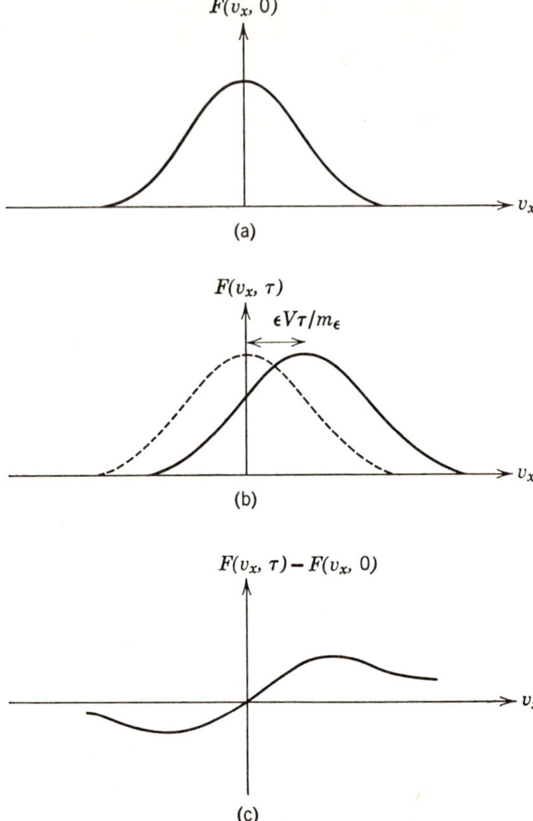

FIG. M.4.1.   Classical Maxwell–Boltzmann distribution of velocity of electrons in "electron gas." Equilibrium distribution, (a), steady-state distribution as shifted by application of an electric field E for time τ, (b), and difference between steady-state distribution and equilibrium distribution, (c). The quantity, $F(v_x)\,dv_x$, represents the fraction of the total number of electrons with velocities between v_x and $v_x + dv_x$.

equilibrium distribution, represented in (c), is given to the first order of the electric field by Taylor's theorem (pages 643–645) as $-[\varepsilon V\tau/m_\varepsilon]x[dF/dv_x]_{t=0}$, where it is assumed that the distribution function F satisfies the condition that $\int F\,dv_x = 1$, that is it is normalized to unity. Then the average velocity taken over the steady-state distribution,

$$F(v_x, \tau) \equiv F(v_x, 0) - [\varepsilon V\tau/m_\varepsilon][dF/dv_x]_{t=0}, \qquad (\text{m.4.1.})$$

is obtained as

$$\bar{v}_x = -[\varepsilon V\tau/m_\varepsilon] \int v_x [dF/dv_x]_0 \, dv_x. \qquad \text{(m.4.2.)}$$

Since the term in $F(v_x, 0)$ is symmetrical in v_x and does not therefore contribute to the average value of v_x, this means that the average or drift velocity of the electrons in thermal equilibrium is zero; also, in evaluating the integral in (m.4.2.), since $[v_x F_0]^\infty_{-\infty} = 0$, it is found on integrating by parts that

$$\int [v_x(dF/dv_x)_0 \, dv_x = -\int F_0 \, dv_x = -1, \qquad \text{(m.4.3.)}$$

where $F(v_x, 0)$ is written as F_0. Therefore, on substituting this value for the integral into (m.4.2.), it is found that

$$\bar{v}_x = \varepsilon V\tau/m_\varepsilon, \qquad \text{(m.4.4.)}$$

and therefore, as on page 427,

$$i = n\varepsilon\bar{v}_x = n\varepsilon^2 V\tau/m_\varepsilon, \qquad \text{(m.4.5.)}$$

or

$$\kappa_\varepsilon = n\varepsilon^2\tau/m_\varepsilon, \qquad \text{(m.4.6.)}$$

as before. On the application of Fermi–Dirac statistics to the electron velocity distribution, since each electron is the assembly undergoes the same velocity change, $\varepsilon V\tau/m_\varepsilon$, under the influence of the applied electric field, the Pauli Exclusion Principle has no effect on the way the assembly of electrons responds to the applied field; there is always an available state, created by the change of state of one electron, to accommodate another electron which is changing its state under the influence of the applied electric field. The corresponding Fermi–Dirac velocity distribution of the assembly of electrons is shown in Fig. M.4.2. The relaxation time, τ, defined on page 427, is related to the mean time of flight of the electrons between collisions and also to the mean free path of the conduction electrons or the average distance traveled between collisions. While it might appear that the application of the Exclusion Principle would disallow many possible electron collisions as the final states are already occupied in the electron distribution, Fig. M.4.2. (c) shows that there are excess electrons in part $+X$ and vacant states in part $-X$ in the steady state distribution, so that the relaxation effects enable electrons, as a result of collisions, to transfer from states $+X$ to states $-X$. The overall effect of the Pauli Exclusion Principle is, then, to prevent many collisions taking place although allowing those to take place which are required to restore equilibrium. For this reason, the above argument in equally applicable to the Fermi–Dirac distribution of the electron velocities, and therefore equation (m.4.6.) is still applicable. This is a surprising result, despite the fact that in

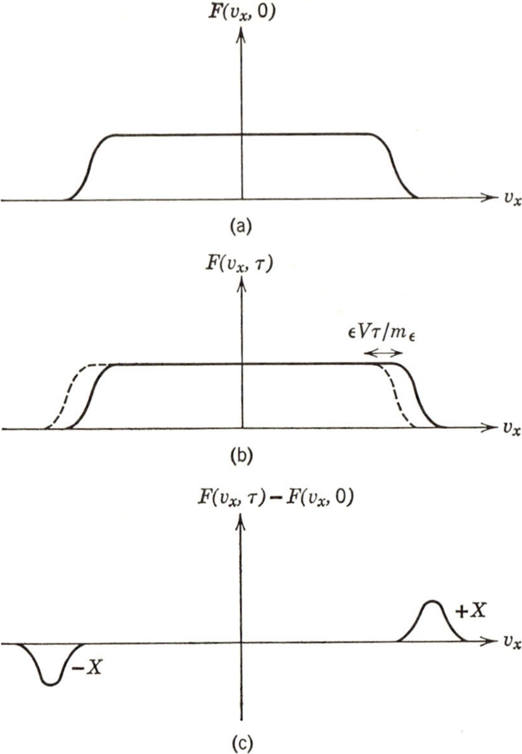

FIG. M.4.2. One-dimensional representation of the Fermi–Dirac velocity distribution of electrons in an "electron gas." The equilibrium distribution is shown at (a), the distribution as shifted by the application of an electric field for a time τ at (b), and the difference between the steady-state and the equilibrium distributions at (c). F has the same significance as in Fig. M.4.1.

the Fermi–Dirac distribution only those electrons near the Fermi surface can take part in collisions, but all collisions either conserve energy or, at most, change it by no more than a quantity of the order of kT, the thermal energy of the metal crystal lattice vibrations. The actual physical process, referred to above as a collision, should actually be more properly described as the interaction between an electron and a phonon, the former referring to a quantum of electromagnetic energy and the latter as a quantum of lattice vibrational energy of the metal crystal. At any rate, the description of electrical conductivity in the free electron theory is that all the electrons are accelerated in the direction of the applied electric field, but only those electrons in states at the edge of the Fermi surface are able to produce a resultant flow in any direction, or are able to undergo collisions which produce a resistance to the resultant flow. It is not possible, in the free electron theory to deal with problems involving the dependence of electrical conductivity on specific directions in metal crystals. The fact that it leads to the correct value for the Wiedemann–Franz ratio (page 428) is only coincidentally due to the cancellation of the terms containing the mean free path length, which emerges to have values of the order of several hundred times the interatomic distances in metals, on substitution of appropriate numerical values, and is therefore quite incompatible with the assumptions involved in the theory of collisions between electrons and atoms.

M.4.2. Heat Capacity at Low Temperatures

The two separate and distinct factors contributing to the heat capacity of metals, the so-called lattice heat capacity attributable to the thermal vibrations of the ion-cores of the metal atoms, and the electronic component, attributable to the thermal motion of the conduction electrons, have already been alluded to in section M.1.3. As on page 422, the total heat capacity at low temperatures is given by, $C_v = \alpha T^3 + \gamma T$, the lattice heat capacity following the Debye T^3-Law and the electronic heat capacity being directly proportional to T. By plotting C_v/T as a function of T^2, the resulting straight line intersects the C_v/T axis in the value of γ, to give the electronic heat capacity.

For very small values of T, at very low temperatures, γT is larger than αT^3, or the electronic heat capacity is larger than the lattice heat capacity. At normal values of T, however, the electronic heat capacity will be negligibly small in comparison with αT^3. The free electron approximation enables this result to be understood, as indicated on page 419; the only electrons which may be excited by energies of the order of kT are those in states near the Fermi surface, and although these electrons have normal heat capacities, they are so few in number that their contribution to the total heat capacity is negligibly small compared to the lattice heat capacity, which at ordinary temperatures is of the order of 3k per atom. This may also be seen from Fig. M.3.3.; the density of states $N(E)$ is a parabolic function of E, and at low temperatures $N(E)$ is not substantially different from $N(E^*)$ in an energy range of the order of kT about E^*. Thus, the number of electrons which may be excited thermally by an amount of the order of kT is proportional to $N(E^*)kT$, and the increase in thermal energy for a temperature rise from 0 K to T K is proportional to $N(E^*)(kT)^2$, which on differentiating with respect to T to obtain C_v, gives the electronic heat capacity at low temperatures as $C_v \propto T$. Thus the Sommerfeld, and the Hartree free electron method lead to the correct temperature dependence, which will generally be valid provided that $N(E)$ does not change greatly near the Fermi surface.

The Hartree–Fock description of the density of electron states, as represented in Fig. M.3.5., shows that the $N(E)$ curve falls very quickly to zero at the Fermi surface and then increases again, so that this method cannot lead to the correct temperature dependence of the electronic heat capacity. In this case, the number of electrons near to the Fermi surface which may be excited thermally by energies of the order of kT, cannot simply be obtained by multiplying kT by $N(E)$ at the Fermi surface, since now $N(E)$ is zero. In fact, according to the Hartree–Fock theory, $C_v \propto T/\ln T,$‡ so that as T tends to zero, C_v/T also tends to zero, which is contrary to the finite value observed experimentally and indicated by the simpler Sommerfeld–Hartree method. It is indeed only when the electron correlation effects are taken into account that the Hartree–Fock method leads to the observed temperature dependence of the electronic heat capacities of metals. This is a major defect in the free electron theory, that the wave functions ascribed to electrons in the assembly of electrons at the high concentration at which they occur in metals (of the order of one per atom), are the same as those used to describe the motion of an electron in a potential box, without making any allowance for the Coulomb and other correlation effects.

‡ This result may be established as follows.[1] Let it be assumed that the Hartree–Fock $N(E)$ curve has essentially the same shape at low temperatures as it has at 0 K and depicted in Fig. M.3.5. Since $N(E)$ falls off to zero at the Fermi surface, it is necessary in order to estimate the number of electrons in states within an energy range of the order of kT from the Fermi surface, to find the value of the wave vector for a state whose energy lies kT below the energy E^* at the Fermi surface, where T is very small. If k_0 denotes the radius of the Fermi sphere and the Boltzmann constant is distinguished from the wave vector symbol by writing it as k_B, then it is necessary to find the value of k for a state, such that

$$E(k_0) - E(k) = k_B T. \qquad \text{(m.4.7.)}$$

For a spherical shell of thickness xk_0 in k-space, at the Fermi surface, $k = k_0 - k_0 x$, and the number of electrons which may be excited by thermal energy of the amount $k_B T$ will be the number contained within the spherical region of thickness $k_0 x$, which is given by

$$2[1/8\pi^3][4\pi k_0{}^2 x k_0] = [k_0{}^3/\pi^2]x, \qquad \text{(m.4.8.)}$$

for each unit volume of metal, since the number of states-per-unit-volume of k-space is given by $V/8\pi^3$ (page 491) and the volume of the spherical shell is of the form $4\pi r^2 \, dr$. Now from equation (m.3.94.),

$$E(k) = [h^2 k^2/8\pi^2 m] - [\varepsilon^2/2\pi]\left[2k_0 + \frac{k_0{}^2 - k^2}{k} \ln \left|\frac{k_0 + k}{k_0 - k}\right|\right]. \qquad \text{(m.4.9.)}$$

---

[1] Raimes, S., *The Wave Mechanics of Electrons in Metals*, 1961, North-Holland Publishing Company, Amsterdam.

and

$$E(k_0) = [h^2 k_0{}^2/8\pi^2 m] - \varepsilon^2 k_0/\pi,$$

$$(m.4.10.)$$

and if the assumed relationship, $k = k_0 - k_0 x = k_0(1 - x)$, is substituted into $(m.4.9.)$, the result is

$$E(k) = [h^2/8\pi^2 m][k_0{}^2(1 + x^2 - 2x)] - [\varepsilon^2/2\pi]\left[2k_0 + \frac{2k_0 x - k_0 x^2}{1 - x}\ln\frac{2 - x}{x}\right]$$

$$= [h^2 k_0{}^2/8\pi^2 m][1 - 2x] - [\varepsilon^2/2\pi][2k_0 + 2k_0 x \ln(2/x)], \qquad (m.4.11.)$$

since if x is very small, then x^2 will be negligible, $2/x$ will be large compared with unity, and unity is large compared with x. Substituting $(m.4.10.)$ and $(m.4.11.)$ into $(m.4.7.)$ gives

$$k_B T = h^2 k_0{}^2 x/m + [\varepsilon^2 k_0/\pi]x \ln(2/x). \qquad (m.4.12.)$$

As $x \to 0$, the first term on the right-hand side of this equation becomes negligibly small compared with the second term, so that to a good approximation

$$k_B T = [\varepsilon^2 k_0/\pi]x \ln(2/x). \qquad (m.4.13.)$$

This equation may be rewritten as

$$[\pi k_B T/\varepsilon^2 k_0] = x \ln(2/x) = \lambda,$$

so that $x = \lambda/\ln(2/x)$, and therefore $2/x = [2/\lambda] \ln(2/x)$. Thus

$$x = \lambda/[\ln(2/x)] = \lambda/[\ln\{2/\lambda \ln(2/x)\}] = \lambda/[\ln(2/\lambda) + \ln \ln\{2/\lambda \ln(2/x)\}]$$

$$= \lambda/[\ln 2/\lambda + \ln \ln 2/\lambda + \ln \ln \ln 2/\lambda + \cdots],$$

and when $\lambda = [\pi k_B T/\varepsilon^2 k_0]$ is very small, x approximates very closely to

$$x = \lambda/\ln 2/\lambda = [\pi k_B T/\varepsilon^2 k_0][\ln(2\varepsilon^2 k_0/\pi k_B T)]^{-1}. \qquad (m.4.14.)$$

If the number of electrons which may be excited by an energy amount of $k_B T$ is given by equation $(m.4.8.)$, then the increase in thermal energy of the assembly of electrons when the temperature is increased from 0 K to T K will be proportional to

$$k_B T[k_0{}^3/\pi^2]x = [k_B{}^2 k_0{}^2/\pi \varepsilon^2]T^2[\ln(2\varepsilon^2 k_0/\pi k_B T)]^{-1}, \qquad (m.4.15.)$$

on substituting $(m.4.14.)$ for x. In order to obtain the low temperature heat capacity, this expression, $(m.4.15.)$, has to be differentiated with respect to T, to give

$$C_v \propto 2T[\ln 2\varepsilon^2 k_0/\pi k_B T]^{-1} + T[\ln 2\varepsilon^2 k_0/\pi k_B T]^{-2} \propto T/\ln T, \qquad (m.4.16.)$$

since as $T \to 0$ the term $T[\ln 2\varepsilon^2 k_0/\pi k_B T]^{-2}$ is negligibly small compared with $2T[\ln 2\varepsilon^2 k_0/\pi k_B T]^{-1}$, and $\ln 2\varepsilon^2 k_0/\pi k_B T = \ln 2\varepsilon^2 k_0/\pi k_B - \ln T$, where again the term $\ln 2\varepsilon^2 k_0/\pi k_B \to 0$ as $T \to 0$.

M.4.3. Photoelectric and Thermionic Emission

Many metals emit electrons when exposed to radiant energy of a suitable frequency, or when heated to a high temperature in a vacuum. In the absence of other influences, the emitted electrons congregate on the surface and prevent the escape of further electrons. If an anode is placed near the surface,

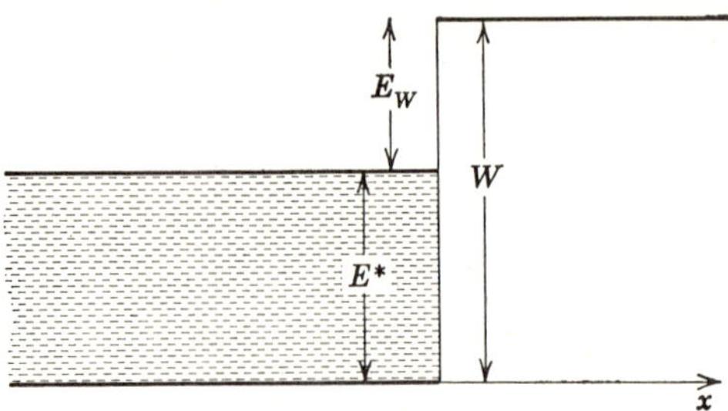

FIG. M.4.3. Representation of the energy terms involved in electronic emission phenomena.

a current of negative electricity flowing from the surface may be measured. The saturation current is found experimentally to depend only on the frequency of the radiant energy, or the temperature, and the nature of the metal. If the potential energy of an electron at rest outside the metal is taken as being zero, and in the free electron approximation the potential inside the metal has the constant value $-W$, then, as in Fig. M.4.3., the minimum amount of energy required to remove an electron to infinity from the lowest free electron state in the metal is

$$E_W = [W - E^*], \qquad\qquad (\text{m.4.17.})$$

where E^* is the energy of an electron on the Fermi surface. E_W is referred to as the work function or work term, and from experimental observation is of the order of magnitude of $3\text{–}8 \times 10^{-19}$ J, for most metals; it is a quantity which is very sensitive to surface conditions, particularly to surface films and non-uniform surfaces.[1]

For photoelectric emission, brought about by the absorption of a quantum of electromagnetic radiation $h\nu$, of frequency ν, it is apparent from Fig. M.4.3. that, at the absolute zero of temperature, emission may only take place if $h\nu > [W - E^*]$, corresponding to the sharply defined photoelectric frequency given by

$$h\nu = [W - E^*]. \qquad\qquad (\text{m.4.18.})$$

Since E^* is very approximately constant at low temperatures, the threshold photoelectric frequency would be expected also to be independent of temperature, as is observed experimentally. At higher temperatures, as the Fermi surface becomes more diffuse, the threshold frequency is less sharply defined,

---

[1] Herring, C., and Nichols, M. H., *Rev. Mod. Phys.*, 1949, **21**, 185.

and the photoelectric process becomes confused with the pure thermionic emission process.

The thermionic effect, on heating metals to high temperatures, is quite independent of any photoelectric effects. Richardson's[1] original discussion of this effect, based on the classical theory of the behavior of electrons in metals has to be modified to take account of the Fermi–Dirac statistics. The process of thermionic emission is essentially a problem in chemical kinetics, involving a system which is tending towards equilibrium but for which equilibrium is not established between the number of electrons emitted from the metal surface per second and the number striking the surface per second. From equation (m.3.61.), the Fermi–Dirac distribution law for electron energies inside a metal, when $E_i - E^* \gg kT$ may be written as

$$N_i = dN = [4\pi V/h^3][2m]^{3/2}e^{-(E-E^*)/(kT}E^{1/2}\,dE, \qquad \text{(m.4.19.)}$$

and if the energies are expressed in terms of momenta (page 41) as

$$2\pi(2m)^{3/2}E^{1/2}\,dE = dp_x\,dp_y\,dp_z, \qquad \text{(m.4.20.)}$$

then

$$dN = [2V/h^3]e^{-[\{(p_x{}^2+p_y{}^2+p_z{}^2)/2m\}-E^*]/kT}\,dp_x\,dp_y\,dp_z. \qquad \text{(m.4.21.)}$$

For electrons escaping from the metal surface in the z-direction, the number of electrons with momenta components between p_z and $p_z + dp_z$, independently of the values of the momenta components in the x- and y-directions, may be obtained by integrating over all values of p_x and p_y from $-\infty$ to ∞ (see page 42), as

$$dN = [2V/h^3][2\pi mkT]e^{-[(p_z{}^2/2m)-E^*]/kT}\,dp_z. \qquad \text{(m.4.22.)}$$

The number of electrons with such restrictions applying to their momenta components, which would cross in unit time, unit area of a plane normal to the z-direction, will be $[dN/V]v_z$, where $v_z = p_z/m$ is the velocity of flow in the z-direction; if this number is designated N_z, then

$$dN_z = [dN/V]v_z = [dN/V][p_z/m] = [4\pi kT/h^3]e^{-[(p_z{}^2/2m)-E^*]/kT}p_z\,dp_z.$$
$$\text{(m.4.23.)}$$

The total number of electrons crossing unit area per second with kinetic energy components which are at least W, in the z-direction, may now be obtained by integrating this expression from W to ∞, to give

$$N_z = [4\pi mk^2T^2h^{-3}]e^{-(W-E^*)/kT}, \qquad \text{(m.4.24.)}$$

and it being assumed that every electron with such energy values escapes from the metal surface, the thermionic current is then given by $N_z(-\varepsilon)$, or

$$i_t = [4\pi m/h^3][kT]^2\varepsilon e^{-(W-E^*)/kT}, \qquad \text{(m.4.25.)}$$

---

[1] Richardson, O. W., *Thermionic Emission*, 1934, Chapman & Hall, London.

which is referred to as the Richardson–Dushman equation. The quantity

$$[4\pi m\varepsilon h^{-3}k^2] = 120 \text{ A cm}^{-2}\text{ K}^{-2}, \qquad \text{(m.4.26.)}$$

on substitution of the appropriate values for the constants, π, ε, m, h, and k, although the experimentally measured values vary widely for different metals, dependent on their surface conditions. Fairly close agreement is usually observed between the numerical values of the photoelectric and the thermionic work functions.

M.4.4. Surface Tension of Metals

It has been noted that, in terms of the simple electron in a cubical box of edge length L, the possible electron energies are given by (m.3.2.),

$$E_{n_x,n_y,n_z} = [h^2/8mL^2][n_x^2 + n_y^2 + n_z^2],$$

where n_x, n_y, and n_z are all positive integers, and each possible combination of this triple set of integers represents a possible energy state, each such energy state containing two electrons. The total energy of an assembly of N electrons results from summing the squares of all possible integral values of n_x, n_y, and n_z, or

$$E(\text{total}) = [h^2/8mL^2] \sum [n_x^2 + n_y^2 + n_z^2], \qquad \text{(m.4.27.)}$$

the summation being applicable over the $N/2$ lowest values of

$$(n_x^2 + n_y^2 + n_z^2).$$

In n-space, each combination of positive values of the integers, n_x, n_y, and n_z, will represent a point, located in the positive octant of this n-space. For large values of N, the $N/2$ points nearest to the origin of n-space may be regarded as lying within the positive octant of a sphere of radius $[(8mL^2/h^2)E_{\max}]^{1/2}$, and corresponding to the integral values of n_x, n_y, and n_z, these points would be located in a cubic array. The number of points in this cubic array of points would then be $(1/8)[4\pi r^3/3]$, or

$$N/2 = (1/8)(4\pi/3)[(8mL^2/h^2)E]^{3/2},$$

leading to $E^* = [3N/\pi V]^{3/2}[h^2/8m]$, as in (m.3.8.). However, this approximation overlooks the facts that since the minimum value of n_x, n_y, or n_z, is 1, there are no points in n-space in the planes $n_x = 0$, $n_y = 0$, or $n_z = 0$, and also that the sphere in n-space cannot be fitted to a cubic array of points. In other words, for the cubic array of points located in the positive octant of n-space, the value of the right-hand side of equation (m.4.27.) does not correspond exactly to the volume of the octant of a sphere of radius

$[(8mL^2/h^2)E]^{1/2}$. This type of discrepancy has been examined by Brager and Schuchowitzky,[1] and they have proposed that the total energy, of (m.4.27.), may be better expressed in the form

$$E\text{(total)} = [\alpha \pi b^5/5\Delta^3]V + [\alpha \pi b^4/16\Delta^2]S + \text{a smaller term}, \quad \text{(m.4.28.)}$$

where $\alpha = [h^2/8m\Delta^2]$, $\Delta = [L/N^{1/3}]$, $b = [3/\pi]^{1/3}$, and N is the number of electrons in a cube of edge length L. The first term in Brager and Schuchowitzky's equation is obviously just the term $(3/5)NE^*$ of (m.3.9.), or the volume–energy–term of the simple theory. The second term of equation (m.4.24.) represents the additional energy of the total assembly of electrons due to the surface, S, and therefore the surface tension is deduced to be given by

$$\sigma = [\alpha \pi b^4/16\Delta^2] = [\pi h^2 b^4/128m]n^{4/3} = [3h^2/128m][3/\pi]^{1/3}n^{4/3}, \quad \text{(m.4.29.)}$$

where n is the electron concentration N/V. The numerical values of the surface tension of metals calculated by the use of this expression are generally of the order of one to three times greater than the experimentally measured values, which in view of the other approximations involved is not too unreasonable.

M.4.5. Soft X-Ray Spectra

The general nature of the soft X-ray emission spectra of metals has been dealt with in M.3.1., and the observed emission band widths are consistent with the free electron approximation. The band width given by the Hartree–Fock theory in (m.3.97.) agrees very poorly with the experimentally observed value, but again with proper account taken of Coulomb correlation effects, the Hartree–Fock values come into good agreement with experiment.

In principle, the method of soft X-ray spectroscopy should allow the determination of the band width and the form of the $N(E)$ curve, from the experimentally observed results, as well as enabling the distinction to be made between the fraction of s, p, and d states contributing to a mixed band. If E_0 and E_m denote the lowest and highest occupied energy states of a metallic crystal at 0 K, $\nu^3 f(E)$ the probability per unit time that an electron of energy E will fall into an inner level, when a vacancy has been created in such an inner level, and $I(E)\,dE$ the intensity or number of quanta per second, of the radiation in the emitted band in the range of energies between E and $E + dE$, then assuming the energy of the inner level to be zero.

$$I(E) \propto \nu^3 f(E) \quad \text{if} \quad E \ll E_\oplus \qquad \text{and} \qquad I(E) = 0 \quad \text{if} \quad E > E_m.$$

---

[1] Brager, A., and Schuchowitzky, A., *Acta Physicochim.*, 1946, U.R.S.S., **21**, 13.

The energy of this transition is $h\nu$ and the ν^3 factor applies to the transition probability description (see pages 54–58). For the corresponding absorption process at 0 K, the energy absorbed must raise the energy of an electron from an inner level to E_m, since all the electron states below E_m are occupied. For a thin layer of a metal, if μ_E denotes the partial absorption coefficient expressed as a function of the quantum energy of the absorbed radiation, then

$$\mu_E = 0 \quad \text{if} \quad E < E_m \quad \text{and} \quad \mu_E \propto f(E)N(E) \quad \text{if} \quad E > E_m.$$

These considerations lead to the conclusion that there should be a sharp edge at the high energy, short wave-length, side of the emission band and also on the low energy long wave-length side of the absorption band, and both these sharp edges should be at the same wave-length. This result is confirmed by experimental observation. Since emission spectra result from electrons making the transition from an occupied region of the corresponding valency band to the inner levels, emission spectra provide information about occupied electron states. On the other hand, absorption spectra provide information about unoccupied electron states, and both types of spectra are required to provide information about the complete $N(E)$ curve for the valency band of a metal; generally, both K and L spectra have to be examined, in order to assess the fraction of s, p, and d states contributing to the mixed states in the valency bands of that metal. The low energy limit of an emission band is not always easily determined from experimental observations. Because of the applicability of certain selection principles in determining the electronic transitions which may take place in the emission of optical and X-ray spectra, each emission spectrum can only provide information about some particular part of a mixed set of valency states. Thus, in a K-emission spectrum no clue as to the s-component of a mixed state is revealed although the p-nature of such a mixed valency state is indicated.

In the case of a solid solution, each kind of atom will have its own characteristic K-level, so that for a binary solid solution, for example, emission spectra may result from electrons making the transition into two different K-levels. From the study of such spectra, it may be possible to distinguish between cases where bonding is due to electrons associated only with the solute atoms, and those types where the two kinds of atoms give rise to a common valency band, the width of which increases with the number of valency electrons per atom.

Another cause of energy broadening in X-ray emission spectra leading to difficulty in estimating band widths is the probability of Auger transitions. In this type of event, an electron may undergo the transition from one energy state to another but the energy associated with the transition may be acquired by another electron instead of being emitted as radiation; even though a transition may be excluded on the basis of the selection principles, it may have a high probability for an Auger transition of this type.

Under certain circumstances, such as subjecting metals to irradiation by high energy electron beams, it may happen that both K-electrons may be ejected from the metal atoms, and such species provide a soft X-ray emission spectrum containing satellite bands of the same shape and essentially the same width as the main bands. The production of such exciton states with energy levels below the normal states is attributable to the greater attractive energy between the nucleus of an atom and the other electrons on which the screening effect of the K-group is reduced, so that in the metal such atoms behave as solute species containing discrete one-electron or exciton states lying below the normal energy levels of the solvent species.

M.4.6. Paramagnetism

The free electron theory implies two types of magnetic behavior for simple metals, such as the alkali metals: a paramagnetic effect attributable to the electron spin, and a diamagnetic effect due to the electronic translational or orbital motion. The simple theory is not capable of accounting for ferromagnetism, nor is it capable of accounting for the type of abnormal paramagnetism displayed by some of the transition metals, such as platinum. Generally, if any substance (including metals) is placed in a magnetic field, of field strength $\mathbf{H}$, it becomes magnetized, the magnetic moment per unit volume, $\mathbf{M}$, being related to the field strength by the equation

$$\mathbf{M} = \chi\mathbf{H}, \qquad \text{(m.4.30.)}$$

where χ is the magnetic susceptibility of the substance, or in this particular case, the volume susceptibility. If χ is negative with an induced magnetic moment opposed to the external magnetic field, the material is diamagnetic; if χ is positive with an induced magnetic moment parallel to the magnetic field, the substance is paramagnetic; if χ is large and positive, the substance is ferromagnetic (see page 399). The magnetic properties of atoms, molecules, and crystals, generally, provide some of the most significant and interesting information about their electronic behavior. The magnetization of an atom, metal, or crystal, is usually derived from a combination of the orbital and spin characteristics of its constituent electrons, and it is not always easy to distinguish between these effects. Most metals are weakly paramagnetic or diamagnetic, the alkali metals, for example, being paramagnetic while copper, silver, and gold are diamagnetic.

Diamagnetism is a phenomenon which arises from an induced magnetic moment being developed in a substance, proportional to and oppositely directed to the external magnetic field. The simplest example is that found for a monoatomic gas whose atoms are in the ^1S-state. The diamagnetism of metals is more complicated and the study of the electronic motions leading to cyclotron resonance was first made by Landau[1,2] in connection with the theory of the diamagnetism of metals. Basically however, diamagnetism is due to the orbital motion of the electrons, while paramagnetism is attributable to electron spin, and except in the case of some transition metals, the contribution to the magnetic susceptibility from the metal ion cores is diamagnetic, so that any induced paramagnetic moment is due to the conduction electrons. The free "electron gas" theory, according to the Sommerfeld and Hartree interpretation, is primarily concerned with the paramagnetic susceptibility of the conduction electrons in a metal. In all cases where theoretical and experimental values of magnetic susceptibilities are being compared, it is of great importance to take care that the observed magnetic properties are not influenced by ferromagnetic impurities.

For a free "electron gas," in the absence of any external magnetic field, all of the electronic states of lowest available energy are occupied by two electrons each, with opposed spins. The nature of the Fermi–Dirac distribution function is such that, the susceptibility of metals is essentially the same at all normal temperatures as it is at 0 K. This state of affairs may be represented diagrammatically as in Fig. M.4.4., where the two $N(E)$ curves, one for electrons with positive spin and one for electrons with negative spin, are each drawn to contain electrons up to the same energy level. In the presence of a magnetic field $\mathbf{H}$, an electron whose spin magnetic moment is parallel to the field direction will have its energy decreased by an amount $\mu_B H$, where μ_B is the Bohr magneton, $\mu_B = \varepsilon h / 4\pi mc$ (see section Q.5.2.). Similarly, an electron with its magnetic moment opposed to the field direction will have its energy increased by a like amount. Therefore, in the presence of an external magnetic field, the densities of states curves for electrons with spins parallel to and opposed to the field become displaced relative to one another by an amount $2\mu_B H$, as shown in Fig. M.4.4. The energy of the system may now, however, be reduced if some of the electrons with spins opposed to the field direction change to states with spins parallel to the field direction. Such a change leads to a decrease in magnetic energy of $2\mu_B H$ for each electron involved, but since any particular state may only contain one electron of a given spin direction, the electrons undergoing such a reversal must enter states lying above those already occupied by electrons with parallel spins, leading to an increase in kinetic energy of the system. This process will continue until an equilibrium

---

[1] Landau, L., *Z. Physik*, 1930, **64**, 629.
[2] Slater, J. C., and Frank, N. H., *Electromagnetism*, 1947, McGraw-Hill Book Co., New York.

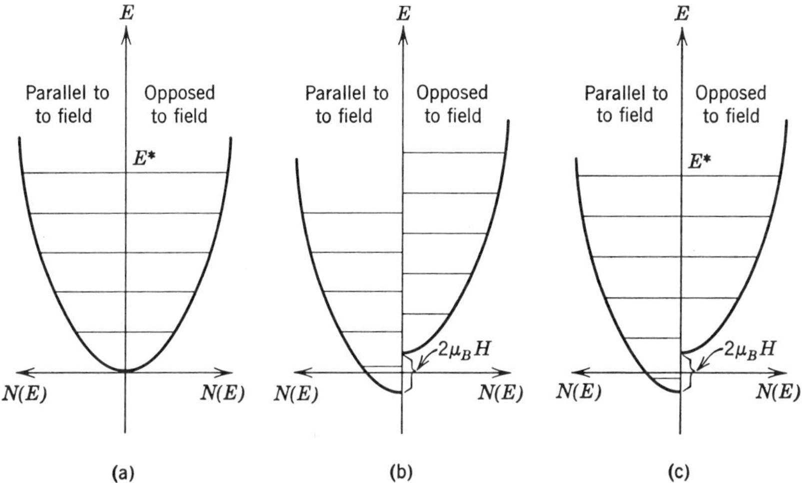

FIG. M.4.4.   Spin paramagnetism (Pauli) at 0 K. The distribution of the free
electrons in a metal with spin moments parallel and opposed to a magnetic
field **H** (a) before, (b) immediately after field is switched on, and (c) after
equilibrium has been established. The levels in the shaded regions are occupied,
and E denotes the total electronic energies, kinetic plus magnetic.

state is arrived at, in which the increase in kinetic energy during a transfer is
equal to the decrease in magnetic energy so that the total energy, kinetic
plus magnetic, is unaffected during a transfer. There will thus arise an excess
of electrons with parallel spins and so a paramagnetic effect. The experimentally observed fact that the paramagnetism is almost independent of
temperature is simply the result of the "electron gas" being essentially
completely degenerate at ordinary temperatures. Classical theory leads to the
conclusion that the magnetic susceptibility varies inversely as the temperature.

Even for very large applied magnetic fields, $\mu_B H$ is only of the order of
magnitude of 2×10^{-22} J, so that for an energy range of width $\mu_B H$ about
E^*, the density of states may be assumed to be $N(E^*)$. The number of electrons undergoing a reversal of spin direction to become parallel to the
external field is then $N(E^*)\mu_B H$, which means that the number of electrons
with magnetic moments parallel to **H** becomes $2N(E^*)\mu_B H$ greater than the
number with magnetic moments opposed to **H**. The magnetic moment per
unit volume is thus $2\mu_B^2 H N(E^*)/V$ parallel to **H**, and from (m.4.26.), the
paramagnetic susceptibility becomes

$$\chi = 2\mu_B^2 N(E^*)/V. \qquad \text{(m.4.31.)}$$

It is seen from equations (m.3.6.) and (m.3.8.) that

$$N(E^*) = [3\pi^2 N/V]^{1/3}[2Vm/h^2], \qquad \text{(m.4.32.)}$$

and writing the electron density $N/V = n$, then

$$\chi = [4m\mu_B^2/h^2][3\pi^2 m]^{1/3}, \qquad \text{(m.4.33.)}$$

which corresponds to the paramagnetic susceptibility of the free "electron gas" in a metal, in accordance with the Sommerfeld–Hartree approximation.

According to equation (m.4.29.), when E^* is very small, the susceptibility should become very large, but according to (m.3.8.) E^* becomes smaller as the concentration of the "electron gas" becomes less, and it is one of the defects of the free electron theory that it would require a very dilute "electron gas" to exhibit ferromagnetic properties. Thus, for cesium, with a large atomic volume, the free electron theory indicates that the conduction electrons should be ferromagnetic, quite contrary to experimental observation. Actually, the values of the paramagnetic susceptibility calculated by the use of (m.4.29.) cannot be easily compared with experimental values, even for sodium metal, where the conduction electrons are very nearly free. This is a consequence of the difficulty of separating contributions to the observed susceptibility arising from the spin paramagnetism of the conduction electrons, the diamagnetic contribution from the conduction electrons, and the diamagnetic contribution from the ion cores of the metal atoms. It is possible, however, to estimate the magnitude of the latter effect, on the assumption that it is the same as in the free atoms, and it has been shown that the diamagnetic contribution, for completely free electrons, is exactly $-(1/3)$ of the value of the paramagnetic contribution.[1] The total susceptibility calculated in this manner is approximately of the right order of magnitude for some simple metals, but apart from variation in experimentally observed values, presumably due to ferromagnetic contamination, the agreement between calculated and observed paramagnetic susceptibilities is not particularly good.

Since the Hartree–Fock density of states curve (Fig. M.3.5.) is so different from the Sommerfeld–Hartree one, it is apparent that the paramagnetic susceptibility calculated by this method would correspond to an entirely different value from that given by (m.4.27.). The Hartree–Fock susceptibility obtained in this way is actually found to be much too great, but again with proper account taken of the Coulomb correlation effects, and also the use of the appropriate effective electron mass, m^* (see page 595), rather than simply the electronic mass, m_ε, much better agreement between calculated and experimental values results. It was Pauli[2] who first developed the quantum theory of spin paramagnetism.

---

[1] Landau, L., *Z. Physik*, 1930, **64**, 629.
[2] Pauli, W., *Z. Physik*, 1927, **41**, 81.

M.4.7.   The Thomas–Fermi Method

An application of the Fermi–Dirac statistics, which provides a very simple and rather good approximation to the potential and charge distribution in an atom, was suggested by Thomas[1] and Fermi.[2] The method applies the ideas of the free electron approximation to polyelectronic atoms in which the potential energy is not constant; it has also been used in calculating the charge distributions in heavy atoms, molecules, and crystals.

From the expression given on page 464 for p_{max} or from equation (m.3.8.) for E^*, it is apparent that the electron concentration in the free "electron gas," the quantity $N/V = n$, may be written as

$$n = N/V = [8\pi p_{max}{}^3][3h^3]^{-1} = [8mE^*]^{3/2}[\pi/3h^3], \qquad \text{(m.4.34.)}$$

and the electronic charge density is $-\varepsilon$ times this amount; in the Hartree approximation, the continuous distribution of positive charge has the same magnitude but opposite sign. The Thomas–Fermi approximation assumes that this equation remains true for the lowest energy state of a system even when the electrons are not free or when the potential field in which they move is not constant. In this event, both the electron density, n, and the maximum momentum possessed by any electron, are functions of the position vector, $\mathbf{r}$, so that

$$n(\mathbf{r}) = [8\pi/3h^3]p^3{}_{max}(\mathbf{r}). \qquad \text{(m.4.35.)}$$

If the electrostatic potential at some point within the system is $V(\mathbf{r})$, a function of $\mathbf{r}$ also, then the potential energy of an electron at the point r will be $-\varepsilon V(\mathbf{r})$. The approximation, stated in (m.4.31.), is only valid if the electron density is so large that there are a large number of electrons in each of the volume elements of real space, over which the potential energy changes very slightly. Now if $-\varepsilon V_0$ represents the total energy (kinetic plus potential) of the fastest electron in the assembly, then

$$-\varepsilon V_0 = p_{max}{}^2(\mathbf{r})/2m - \varepsilon V(\mathbf{r}),$$

and therefore

$$p_{max}(\mathbf{r}) = [2m\varepsilon(V - V_0]^{1/2},$$

which when substituted into (m.4.31.) gives

$$n(\mathbf{r}) = [8\pi/3h^3][2m\varepsilon(V - V_0)]^{3/2}. \qquad \text{(m.4.36.)}$$

---

[1] Thomas, L. H., *Proc. Camb. Phil. Soc.*, 1927, **23**, 542.
[2] Fermi, E., *Z. Physik*, 1928, **48**, 73.

In this expression, $-\varepsilon V_0$ corresponds to the highest occupied energy level of the system, so that V is independent of $\mathbf{r}$. The self-consistency condition is that the electrostatic potential $V(r)$ must be determined by Poisson's equation from the charge density of the electrons, as given by (m.4.32.), and from the nuclear charge (for atomic systems). Since the electronic charge density is $-\varepsilon n(r)$, Poisson's equation is $\nabla^2 V = 4\pi\varepsilon n$, or making use of (m.4.32.)

$$\nabla^2 V(r) = [32\pi^2\varepsilon/3h^3][2m\varepsilon]^{3/2}[V - V_0]^{3/2}, \qquad \text{(m.4.37.)}$$

which is the basic form of the Thomas–Fermi equation. It cannot be solved analytically, but numerical solutions have been given, subject to appropriate boundary conditions. The equations (m.4.32–33.) are only valid as long as $V > V_0$, and in any region where $V < V_0$, or where the potential energy $-\varepsilon V$ is greater than the maximum energy $-\varepsilon V_0$ of any electron, $n(r)$ must be zero. There are many references to the use of the Thomas–Fermi method in the literature. It has been used, in the first place, to calculate the electron density as a function of r, for atoms, and properties such as X-ray scattering of atoms, which depend on the electron density, were calculated by this method, long before the self-consistent field type calculations were carried out. Before the accurate calculations of Hermann and Skillman became available (see pages 264, 273), Latter[1] made calculations of the wave functions of an electron in a Thomas–Fermi potential, which effectively apply the self-consistent field argument to just one cycle. It is indeed found that the Thomas–Fermi method gives a remarkably good approximation to the detailed self-consistent field calculations. It is possible to calculate an exchange correction energy for the Thomas–Fermi model atom, leading to the method usually referred to as the Thomas–Fermi–Dirac method.[2] It has been shown that the Thomas–Fermi–Dirac equation may be derived from a variation method,[3-4] in the same manner that the Hartree–Fock method can be so derived, thus enabling the Thomas–Fermi–Dirac equation to be applied to problems more complicated than those presented by spherical atoms, such as for some molecules and crystals. However, since it is now possible to carry through the self-consistent-field type calculations, which are more accurate, the Thomas–Fermi–Dirac method is of less consequence.

The Thomas–Fermi method is nevertheless of considerable value, in the theory of metals and alloys, in enabling the potential due to another atomic species, as an impurity for example, in a metal to be deduced, and to provide an estimate of the screening effect of all other electrons on the field of any one

---

[1] Latter, R., *Phys. Rev.*, 1955, **99**, 510.
[2] Slater, J. C., *Quantum Theory of Molecules and Solids*, **Vol. 3**, 1967, McGraw-Hill Book Company, New York.
[3] Lenz, W., *Z. Physik*, 1932, **77**, 713.
[4] Jensen, H., *Z. Physik*, 1934, **89**, 713.

of them. These are valuable considerations in dealing with the problem of Coulomb correlation effects. For example, if a metal impurity atom with $(Z + 1)$ valency electrons per atom is dissolved in an alkali metal, these atoms may be regarded as behaving as point positive charges of magnitude $Z\varepsilon$ in a free electron gas, and their potential may be calculated by the Thomas–Fermi method. Thus, suppose that the point charge provided by such an impurity atom is $q = Z\varepsilon$, and that this point-charge is introduced into an electron gas of uniform density. The "electron gas" system, consisting of a uniformly distributed positive charge and the "electron gas" of uniform density, is such that the total charge is zero, and therefore the potential is also zero everywhere, so that $-\varepsilon V_0 = E^*$. The electron density initially, before the introduction of the point-charge, is given by (m.3.40.) as

$$n_0 = [8mE^*]^{3/2}[\pi/3h^3],$$

and it may be reasonably assumed that no change takes place in $-\varepsilon V_0$, due to the introduction of the point charge q. From equation (m.4.32.), the electron density which results, as a consequence of introducing the point charge q at the origin [in defining the resulting potential as $V(r)$] is given by

$$n(r) = [8\pi/3h^3][2m\varepsilon(V - V_0)]^{3/2},$$

$$= [8\pi/3h^3][2m\varepsilon(V + E^*/\varepsilon)]^{3/2}, \qquad \text{on substituting } V_0 = -E^*/\varepsilon,$$

$$= n_0[1 + \varepsilon V/E^*]^{3/2}, \qquad \text{on substituting for } n_0, \text{ as given above.}$$

Both $V(r)$ and $n(r)$ may be regarded as spherically symmetrical about the point-charge q, for a large free "electron gas" system, and since there is a uniform distribution of positive charge, εn_0, Poisson's equation becomes

$$\nabla^2 V = 4\pi\varepsilon(n - n_0) = 4\pi\varepsilon n_0[(1 + \varepsilon V/E^*)^{3/2} - 1]. \qquad \text{(m.4.38.)}$$

This equation again cannot be solved analytically, but on the assumption that V is sufficiently small, so that its square may be neglected in comparison with unity, an approximate analytical solution may be obtained. Thus, the binomial expansion of

$$[1 + \varepsilon V/E^*]^{3/2} \qquad \text{gives} \qquad [1 + 3\varepsilon V/2E^* + \cdots,$$

and so

$$\nabla^2 V = 6\pi n_0 \varepsilon^2 V/E^*. \qquad \text{(m.4.39.)}$$

Since however, $V(r) \to \infty$ as $r \to 0$, this approximation cannot be valid in the immediate vicinity of q, and in any event equation (m.4.34.) does not apply when $\varepsilon V/E^* < -1$. The boundary conditions applicable to $V(r)$ may be deduced from the considerations that, by the introduction of the point charge q into the "electron gas," a redistribution of the electron density in the vicinity of q will take place, such that if q is positive there will be an increase in the electron density around q, and if q is negative, there will be a decrease. So $V(r) \to 0$ as $r \to \infty$, and $V(r) \to q/r$ as $r \to 0$. The general effect is that the "electron gas" will screen the charge q so that the effective field at large distances is zero, but this screening effect will be barely perceptible close to q. Now, since $V(r)$ is spherically symmetrical, $\nabla^2 V$ may be written as $d^2V/dr^2 + (2/r)\,dV/dr$, and thus equation (m.4.34.) becomes

$$d^2V/dr^2 + (2/r)\,dV/dr = 6\pi n_0 \varepsilon^2 V/E^*. \qquad \text{(m.4.40.)}$$

A solution of this second order differential equation is given by

$$V(r) = [q/r]e^{-r/\sqrt{E^*/6\pi n_0 \epsilon^2}} = [q/r]e^{-r/\sigma}, \qquad \text{where} \qquad \sigma = [E^*/6\pi n_0 \varepsilon^2]^{1/2},$$
$$\text{(m.4.41.)}$$

as is apparent by direct substitution. Such a potential function is referred to as a screened potential, and σ is called the screening distance. The approximate value of the effective range of the potential field due to the charge q is given by the screening distance σ; approximately, the screening distance is the distance over which the unscreened potential falls off to about one-third of its value. For the alkali metals, from equation (m.1.40.),

$$n_0 = 1/V = 3/4\pi r_s^3,$$

and from equation (m.3.11.),

$$E^* = [h^2/8\pi^2 m][9\pi/4]^{2/3} r_s^{-2},$$

and substituting these expressions into (m.4.37.) gives

$$\sigma = [1/3][9\pi/4]^{1/3}[h^2 r_s/4\pi^2 m \varepsilon^2]^{1/2};$$

but from page 91, as the Bohr radius $a_0 = h^2/4\pi^2 m\varepsilon^2$, so that

$$\sigma = [1/3][9\pi/4]^{1/3}[a_0 r_s]^{1/2} = 0.642\sqrt{r_s} \text{ atomic units.}$$

For sodium, in particular, it is seen from Fig. M.1.14. that r_s is of the order of four atomic units, so that the screening distance for the "electron gas" from sodium is

$$\sigma = 1.28 \text{ Bohr units} = 6.4 \times 10^{-11} \text{ m.}$$

This is actually a shorter screening distance than is indicated from a more detailed study of the Coulomb correlation effects.

M5

Band and Zone Theories of Metals

M.5.1. Deficiencies of the Free Electron Model

The various approximations involved in the simple free electron theory of metals have been outlined in the previous chapters. Despite these very major approximations, this approach enables a reasonable understanding to be attained of several of the physical properties of metals, such as electronic emission phenomena, electrical conductivity, spin paramagnetic effects, the soft X-ray spectra, and even surface tension effects. The application of Fermi–Dirac statistics to the free "electron gas," made originally by Pauli to account for the paramagnetic susceptibilies of the alkali metals, and the subsequent elaborations of the theory by Sommerfeld[1] and others, affords an understanding and explanation of the essentially zero contribution to the heat capacity of metals made by the "electron gas," despite the high concentration of electrons in most metals. The concept of a degenerate "gas" of electrons, whose energy is almost independent of temperature, represents one of the great advances in the theory of metals, and the general conclusions of the free electron theory are still valid in the more complete zone and band theories.

The major defect of the free electron theory is that no allowance is made for the interaction of the various electrons with one another. It is obviously quite unsatisfactory to assume that the wave functions of the various electrons, at the density at which they are present in metals, are the same as those pertaining to a single electron moving in a field of constant potential energy. Coulomb

---

[1] Sommerfeld, A., and Bethe, H., *Elektronentheorie der Metalle, Handbuch der Physik,* 1933, Zweite Auflage, **Band XXIV**, Zweiter Teil, Berlin.

523

repulsion effects alone, precluding any two electrons from ever being close together, cannot be described by means of simple one-electron wave functions. Other electronic correlation effects, the influence of quantum mechanical exchange forces between electrons and the applicability of the Pauli Exclusion Principle, would be expected to affect the behavior of electrons in a similar manner to the effects in the hydrogen molecule, described elsewhere. The Coulomb interaction between the electrons in a metal actually is so strong and effective over such a long range that its magnitude cannot be calculated by ordinary perturbation theory, even in the case of free electrons. The so-called independent particle approximation, which assumes that the electrons are non-interacting, is an inherent feature of nearly all band structure calculations and also of all applications of the Hartree and Hartree–Fock methods, where only average effects of the electron interactions are considered. A reasonable justification of the independent particle approximation, based on the theory of plasma oscillations in metals developed by Bohm and Pines,[1] has been proposed by Raimes[2] (see section M.6.10.).

The tremendous range of values displayed by the electrical resistivities of metallic and other solids cannot be at all comprehended in terms of free electrons. Apart from the phenomenon of superconductivity, the ratio of the resistivities of an insulator such as diamond and a typical metal such as silver is of the order of 10^{28-30}. Other solids are semiconductors with electrical properties varying significantly with temperature. Perhaps the above range of values of the electrical resistivities of solids is wider than that of any other commonly observed property of matter, with the exception of solubilities; the ratio of the molar solubilities of ammonium nitrate and mercuric sulfide in water at 373 K is also of the order of 10^{29}. It is remarkable, incidentally, that the thermal conductivity of the electrical insulator, diamond, is almost as high as that of most metals at ordinary temperatures. The fundamental nature of the difference between metals and insulators only becomes apparent by extending the free electron model to take account of the interaction of electrons with the periodically varying potential field, such as must exist in a crystalline metal. If a metal behaves as though some of its constituent electrons were free, and the solid metal is crystalline, then these electrons must move about in the field due to the metallic cations, arranged with the three-dimensional periodicity of the appropriate crystal lattice. In order to gain some insight into the effects of the periodic field on the electronic motion, it is valuable, initially, to examine the idealized problem of the motion of a single electron along a row of equispaced cations.

---

[1] Bohm, D., and Pines, D., *Phys. Rev.*, 1953, **92**, 609.
[2] Raimes, S., *Reports on Progress in Physics*, 1957, **20**, 1.

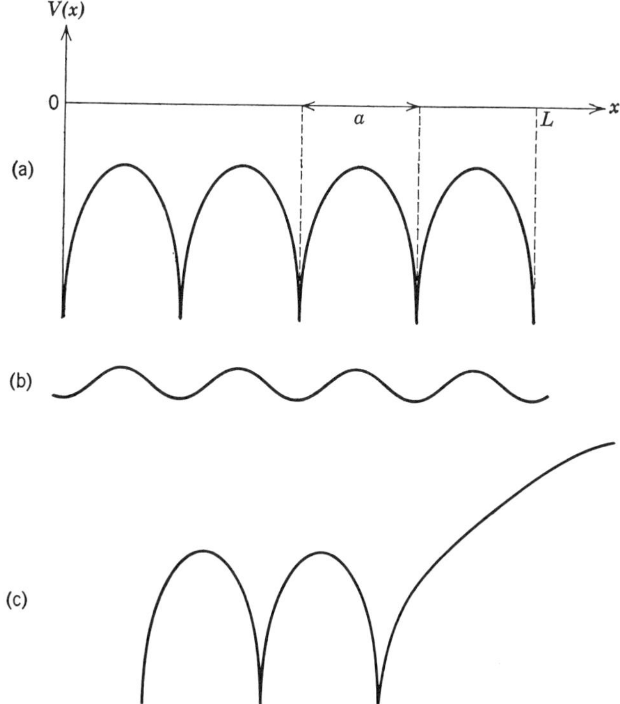

FIG. M.5.1. Potential energy of an electron in a metal: (a) along a line through the centers of the metal nuclei; (b) along a line parallel to (a); (c) at the surface of the metal.

M.5.2. Motion of an Electron in a One-Dimensional Lattice

If a line passing through the centers of the atoms in a metal is considered, the potential energy would vary with the distance along this line, as shown in Fig. M.5.1.(a); if the distance between the atomic nuclei is a, of the order of a few units of 10^{-10} m and the length of the lattice row is $L = Ga$, where G is some integer, then the potential energy, $V(x)$, is a periodic function of x, the distance measured along the row. There will be a singularity in the vicinity of each atomic nucleus, which would be less pronounced along a line not passing through the atomic centers, as shown in Fig. M.5.1.(b) above, while the variation of potential at the surface of the metal might appear as in Fig. M.5.1.(c). Both the maxima and the minima of the potential curve would have the identical period. In accordance with the ideas developed in Part Q, electrons of low energy would have a finite probability of passing from one minimum position to another, even although their energies, according to classical mechanics, would not be great enough to surmount the maxima. If cyclic boundary conditions are applied, it may be imagined that the lattice row is bent around until the two ends are coincident.

When the potential energy is constant, it has been shown in a previous chapter that the wave function of an electron is of the type [equation (m.3.71.)]

$$\psi_k(x) = e^{ikx}, \tag{m.5.1.}$$

where

$$k = \frac{2\pi g}{L} \quad \text{and} \quad g = 0, \pm 1, \pm 2, \ldots.$$

Under the influence of the periodically varying field, it would be expected that the electronic wave function would be affected or modified in some way by the field, and furthermore that the modulating potential function would also have the period of the lattice row, since the assumption of cyclic boundary conditions implies that surface effects or, in the case of a lattice row, end effects, are not taken into account. If this modulating function is designated $\phi_k(x)$, then the expected form of the wave function would be

$$\psi_k(x) = e^{ikx}\phi_k(x), \tag{m.5.2.}$$

where, if $\phi_k(x)$ is periodic in the lattice row spacing and dependent on k, then

$$\phi_k(x + a) = \phi_k(x). \tag{m.5.3.}$$

The theory of the behavior of electrons in a periodic field was first developed by Bloch,[1] and the proof that the wave functions of an electron in a periodic field are of the type designated in (m.5.2.) has come to be known in metal theory as Bloch's theorem, although a similar theorem in the theory of differential equations was first formulated by Floquet in 1883.[2,3]

The problem which is to be solved here is to find the general form of the wave functions which provide acceptable solutions of the one-electron Schrödinger equation,

$$\frac{d^2\psi}{dx^2} + \frac{8\pi^2 m}{h^2}[E - V(x)]\psi = 0, \tag{m.5.4.}$$

where

$$V(x) = V(x + a). \tag{m.5.5.}$$

---

[1] Bloch, F., *Zeit. f. Phys.*, 1928, **52**, 555.
[2] Whittaker, E. T., and Watson, G. N., *A Course of Modern Analysis*, 1940, 4th Ed., Cambridge.
[3] Floquet, G., *Ann. Sci. Ec. Norm. Sup.*, 1883, 12, 47.

The periodic boundary condition implies that

$$\psi(x) = \psi(x + Ga). \qquad \text{(m.5.6.)}$$

If equation (m.5.4.) is rewritten in Hamiltonian form, then

$$\left[-\frac{h^2}{8\pi^2 m} \cdot \frac{d^2}{dx^2} + V(x) \right]\psi = E\psi = \hat{H}\psi,$$

where the expression in square brackets is the Hamiltonian operator. If $\hat{Q}$ is some other operator which, by operating on $\psi(x)$, changes it into $\psi(x + a)$ so that

$$\hat{Q}\psi(x) = \psi(x + a),$$

then

$$\begin{aligned}
\hat{Q}[\hat{H}\psi(x)] &= \hat{Q}\left[\left\{ -\frac{h^2}{8\pi^2 m} \cdot \frac{d^2}{dx^2} + V(x) \right\}\psi(x) \right] \\
&= \left[-\frac{h^2}{8\pi^2 m} \cdot \frac{d^2}{dx^2} + V(x) \right]\psi(x + a) \\
&= \left[-\frac{h^2}{8\pi^2 m} \cdot \frac{d^2}{d(x + a)^2} + V(x + a) \right]\psi(x + a) \\
&= \hat{H}\psi(x + a) = \hat{H}\hat{Q}\psi(x),
\end{aligned}$$

since $V(x) = V(x + a)$ and $d/d(x + a) = d/dx$. Therefore the operators $\hat{Q}$ and $\hat{H}$ have simultaneous eigenfunctions, they commute with one another, or

$$\hat{Q}\psi(x) = \lambda\psi(x) = \psi(x + a). \qquad \text{(m.5.7.)}$$

Alternatively, this result could have been deduced from the probability density, which must be a periodic function of the lattice row repeat unit a. That is,

$$|\psi(x)|^2 = |\psi(x + a)|^2, \qquad \text{(m.5.8.)}$$

which necessarily means that

$$\psi(x + a) = \lambda\psi(x), \qquad \text{(m.5.9.)}$$

where λ is some constant. By repeating this process,

$$\psi(x + 2a) = \lambda\psi(x + a) = \lambda^2\psi(x).$$

The moduli in (m.5.8.) must be unity, otherwise $\psi(x)$ would become infinite after unlimited displacements, each of magnitude a. If g is any integer, then

$$\psi(x + ga) = \lambda^g\psi(x).$$

Since G is any integer, therefore,

$$\psi(x + Ga) = \lambda^G\psi(x) = \psi(x),$$

if the eigenfunction is to be single-valued. Hence,

$$\lambda^G = 1,$$

since it is known that

$$e^{\pm 2\pi i} = 1$$
$$= e^{\pm 2\pi i g} \qquad \text{if} \qquad g = 0, \pm 1, \pm 2, \dots.$$

Therefore, λ is one of the G roots of unity, or alternatively expressed

$$\lambda = e^{2\pi i g/G}$$

and

$$\lambda^G = e^{2\pi i g} \qquad \text{where} \qquad g = 0, \pm 1, \pm 2, \dots.$$

Thus, the following relationship must exist between satisfactory eigenfunctions,

$$\psi(x + a) = e^{2\pi i g/G}\psi(x) = e^{ika}\psi(x), \qquad \text{(m.5.10.)}$$

if k is defined as in (m.5.1.),

$$k = \frac{2\pi g}{L} = \frac{2\pi g}{Ga}, \qquad g = 0, \pm 1, \pm 2, \dots. \qquad \text{(m.5.11.)}$$

For any x, therefore, functions of the type

$$\psi_k(x) = e^{ikx}\phi_k(x) \qquad \text{(m.5.12.)}$$

satisfy (m.5.10.) if

$$\phi_k(x) = \phi_k(x + a). \qquad \text{(m.5.13.)}$$

This result, making no assumptions about the nature of $\phi_k(x)$ apart from its periodicity in the lattice row constant a, is a quite general one, viz. the wave functions of an electron moving in a one-dimensional periodic potential field are of the type (m.5.12.). Such functions are referred to as Bloch-type functions or are said to satisfy the Bloch condition expressed in (m.5.10.).

The expressions (m.5.12.) and (m.5.13.) do not determine k uniquely, since

$$e^{ikx}\phi_k(x) = e^{i(k + 2\pi n/a)x}[e^{-2\pi i n x/a}\phi_k(x)], \qquad \text{(m.5.14.)}$$

where n can be any integer without affecting the fact that the function in the square parentheses is periodic in the lattice row constant a. In other words, k may be replaced by $(k + 2n\pi/a)$ and the functional expression which results will still satisfy the Bloch condition. Therefore, the values of k may be restricted or confined to an interval of length $2\pi/a$, and if this length interval is assumed for convenience to be symmetrical about $k = 0$, then k can be chosen such that

$$-\frac{\pi}{a} \lesssim k \lesssim \frac{\pi}{a}. \qquad \text{(m.5.15.)}$$

This region of k-space, or reciprocal wavelength space, is variously referred to as the basic Brillouin zone, the first Brillouin zone, the reduced zone, or the fundamental domain of k. From the definition of $k = 2\pi g/Ga$, g being an integer, k can only adopt the values $\pm\pi/a$ if G is an even integer. In any macroscopic piece of metal, however, G would correspond to a very large number, such that k would vary almost continuously with g, and it is quite inconsequential as to whether or not the terminal points at $\pm\pi/a$ are included in the range of values of k. Within this range (m.5.15.), there would then be an infinite number of electron states, which may be distinguished from one another by assigning to each an integral label, say $l = 0, \pm 1, \pm 2, \ldots$, corresponding to each value of k. The wave functions for such states would therefore require to be described by two labels, e.g.,

$$\psi_{kl}(x) = e^{ikx}\phi_{kl}(x), \tag{m.5.16.}$$

where

$$\phi_{kl}(x + a) = \phi_{kl}(x). \tag{m.5.17.}$$

If (m.5.16.) is substituted in the one-electron equation (m.5.4.), then

$$i^2 k^2 e^{ikx}\phi_{kl}(x) + \frac{d^2}{dx^2}\phi_{kl}(x)\cdot e^{ikx} + ike^{ikx}\frac{d}{dx}\phi_{kl}(x) + \frac{d}{dx}\phi_{kl}(x)ike^{ikx}$$
$$+ \frac{8\pi^2 m}{h^2}[E_l(k) - V(x)]e^{ikx}\phi_{kl}(x) = 0.$$

On dividing throughout by e^{ikx} and rearranging,

$$\frac{d^2\phi_{kl}(x)}{dx^2} + 2ik\frac{d}{dx}\phi_{kl}(x) + \frac{8\pi^2 m}{h^2}\left[E_l(k) - V(x) - \frac{h^2 k^2}{8\pi^2 m}\right]\phi_{kl}(x) = 0. \tag{m.5.18.}$$

Designating the complex conjugate of $\phi_{kl}(x)$ by $\phi_{kl}{}^*(x)$, then on the assumption that $V(x)$ is a real function, the complex conjugate of this equation is

$$\frac{d^2\phi_{kl}{}^*(x)}{dx^2} - 2ik\frac{d}{dx}\phi_{kl}{}^*(x) + \frac{8\pi^2 m}{h^2}\left[E_l(k) - V(x) - \frac{h^2 k^2}{8\pi^2 m}\right]\phi_{kl}{}^*(x) = 0. \tag{m.5.19.}$$

Also, if k is replaced by $-k$ in equation (m.5.18.), there results

$$\frac{d^2\phi_{-kl}(x)}{dx^2} - 2ik\frac{d\phi_{-kl}(x)}{dx} + \frac{8\pi^2 m}{h^2}\left[E_l(-k) - V(x) - \frac{h^2 k^2}{8\pi^2 m}\right]\phi_{-kl}(x) = 0. \tag{m.5.20.}$$

On comparison of equations (m.5.19.) and (m.5.20.), it becomes apparent that

$$\phi_{kl}{}^*(x) = \phi_{-kl}(x) \tag{m.5.21.}$$

and

$$E_l(k) = E_l(-k). \tag{m.5.22.}$$

The functions $\phi_{kl}(x)$ and $\phi_{-kl}(x)$ are obviously degenerate. Also, from the definition of k, (m.5.1.), neither ϕ_{kl} nor $E_l(k)$ would vary very much for a small change in the value of k, so that for each value of the quantum number l there will correspond a set of electron states, one for each value of k, for which the energy $E_l(k)$ is essentially a continuous function of k. It is such sets of states, each corresponding to a definite value of l, which are usually referred to as Brillouin zones.

To illustrate the relationship between k and l in the simplest possible example, Bloch-type functions could be imagined to be used to describe the electrons in the free electron model, although this would be completely artificial, since here $V(x)$ is zero (however, compare the illustration of the application of the variation principle in Part Q in the case of the hydrogen atom). When $V(x)$ is zero, the one-electron, one-dimensional Schrödinger equation is simply

$$\frac{d^2\phi(x)}{dx^2} + \frac{8\pi^2 m}{h^2} E\phi(x) = 0.$$

If $\phi_{kl} = e^{ikx} e^{2\pi i l x/a}$, which is a Bloch-type function, is substituted in the above equation, then

$$\begin{aligned}
\frac{d^2\phi_{kl}(x)}{dx^2} &= i^2 k^2 e^{ikx} \cdot e^{2\pi i l x/a} + 2\pi i l/a \cdot e^{2\pi i l x/a} \cdot ike^{ikx} \\
&\qquad + 4\pi^2 i^2 l^2/a^2 \cdot e^{2\pi i l x/a} \cdot e^{ikx} + ike^{ikx} \cdot 2\pi i l/a \cdot e^{2\pi i l x/a} \\
&= e^{ikx} \cdot e^{2\pi i l x/a}(-k^2 - 4\pi k l/a - 4\pi^2 l^2/a^2) \\
&= -\frac{8\pi^2 m}{h^2} E_l(k)\phi_{kl}(x).
\end{aligned}$$

Therefore,

$$E_l(k) = \frac{h^2}{8\pi^2 m}[k + 2nl/a]^2. \tag{m.5.23.}$$

Similarly, it may be shown that for

$$\phi_{kl}(x) = e^{ikx} \cdot e^{-2\pi i l x/a},$$

that

$$E_l(k) = \frac{h^2}{8\pi^2 m}[k - 2\pi l/a]^2; \tag{m.5.24.}$$

(m.5.23.) is evidently valid for $0 < k \leq \pi/a$ and (m.5.24.) for $-\pi/a < k \leq 0$, and all possible states are obtained by letting l adopt the values 0, 1, 2, ..., each l value corresponding to one Brillouin zone.

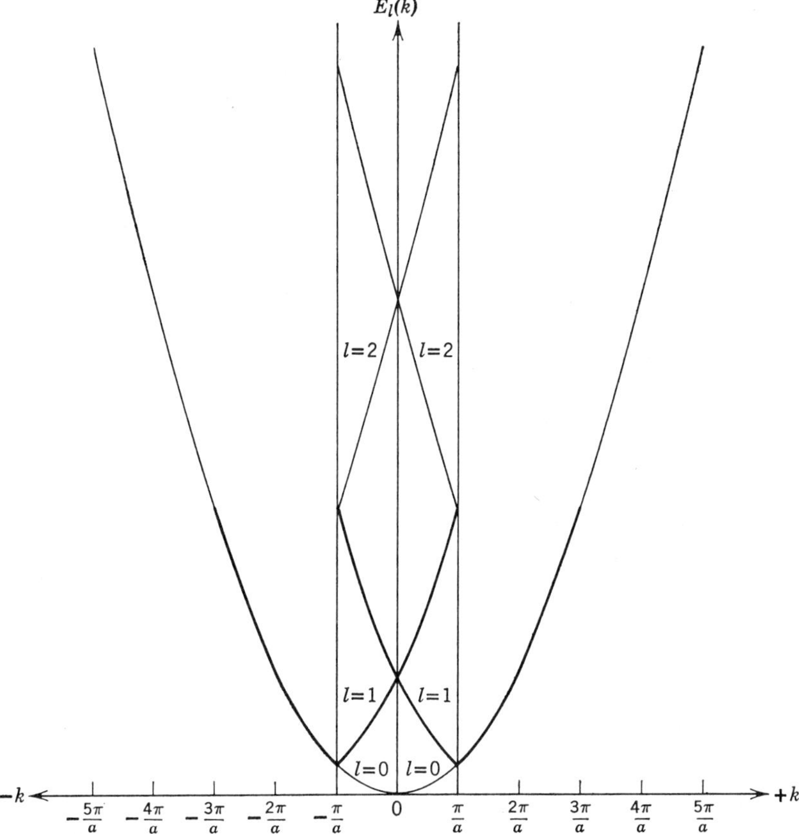

FIG. M.5.2.   Energy of an electron in a one-dimensional lattice as a function of k (energy in units of $h^2/8ma^2$). The parabolic curve extending over the limits of k of $\pm 5\pi/a$ represents the energy of a free electron ($E = h^2k^2/8\pi^2m$) in terms of the extended zone scheme; the curves within the limits of k of $\pm \pi/a$ represents the reduced zone scheme in terms of equations (m.5.23–24). The l index denotes the number of translational units of $\pm 2\pi/a$ necessary to transform from one representation to the other.

If units of energy of magnitude $h^2/8ma^2$ are adopted, then the result of plotting $E_l(k)$ as a function of k is shown in Fig. M.5.2. It is seen that the energy curves of adjacent zones are coincident for values of k corresponding to 0 and $\pm \pi/a$, which is what would be expected for the free electron model, where the energy distribution of the free electrons is continuous up to the energy of electrons on the Fermi surface. This use of Bloch functions to describe the behavior of free electrons, while mathematically quite correct, is of course rather unnecessary, since, where the potential energy is constant, there is no periodicity involved and therefore there is no reason to restrict the range of values which k may adopt. As had already been seen, the use of the simple traveling wave function, $\psi_k(x) = e^{ikx}$, which leads to the one-electron energies, $E(k) = h^2k^2/8\pi^2m$, is quite sufficient in this case, and the one-electron energies are related to k in a simple parabolic manner. Since a Bloch function is such that it is unaltered if k is substituted by $[k + 2\pi n/a]$, where n may be any positive or negative integer, it is apparent that the energy curves shown in Fig. M.5.2. are related to the free-electron parabola by means of translations of parts of these curves through some multiple of the distance π/a parallel to the k abscissa. The result obtained by the use of the Bloch-type function, restricting k to values between $\pm\pi/a$, is referred to as the reduced zone scheme, although as of course in the free-electron model $E(k)$ is a continuous function of k, there are no zones to be distinguished, and the straightforward parabola which results from the use of a simple traveling wave function is referred to as the extended zone scheme. This relationship is also illustrated in Fig. M.5.2.

If $V(x)$ is not zero, the situation is rather different. In particular, if the periodic potential is small, the coincidence of the $E_l(k)$ values for $l = 0$ and $l = 1$, etc., vanishes. While it is possible to obtain solutions of the one-dimensional Schrödinger equation for a single electron in a periodic field to any degree of accuracy using various approximation methods, it is only possible to obtain exact solutions by simple analytical methods when the potential function is either constant (the free electron case) or is approximated by an array of equispaced Dirac delta functions. This latter approximation gives rise to a highly artificial model, although it does show that for an electron moving in a periodic potential field its energy is not a continuous function of its wave vector, k, but that there are regions in k-space for which there is no specific energy solution. This solution was first proposed by Kronig and Penney.[1,2]

M.5.3.   The Kronig–Penney Model

It is initially assumed in this model that the one-dimensional potential function is of the square-well variety, with period $(a + b)$, as depicted in Fig. M.5.3. The Schrödinger equation for this problem is

$$\frac{d^2\psi}{dx^2} + \frac{8\pi^2m}{h^2}\,[E - V(x)]_\psi = 0. \qquad \text{(m.5.25.)}$$

[1] Kronig, R. de L., and Penney, W. G., *Proc. Roy. Soc.*, 1931, **A130**, 499.
[2] Luttinger, J. M., *Philips Research Reports*, 1951, **6**, 303.

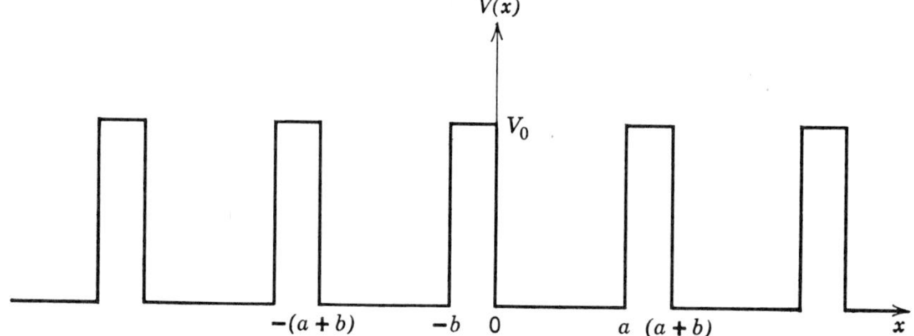

FIG. M.5.3.   Kronig–Penney one-dimensional periodic potential.

By Bloch's Theorem, solutions must be of the form

$$\psi(x) = \phi_k(x) \cdot e^{ikx}, \qquad (m.5.26.)$$

where $\phi_k(x)$ is a periodic function in x with the period $(a + b)$. Substituting (m.5.26.) into (m.5.25.), there results

$$\frac{d^2\psi(x)}{dx^2} = i^2 k^2 e^{ikx}\phi_k(x) + \frac{d\phi_k(x)}{dx} ik e^{ikx} + \frac{d^2\phi_k(x)}{dx^2} e^{ikx} + ik e^{ikx}\frac{d\phi_k(x)}{dx}$$

$$= e^{ikx}\left[-k^2\phi_k(x) + 2ik\frac{d\phi_k(x)}{dx} + \frac{d^2\phi_k(x)}{dx^2}\right]$$

$$= -\frac{8\pi^2 m}{h^2}[E - V(x)]e^{ikx}\phi_k(x).$$

Therefore,

$$\frac{d^2\phi_k(x)}{dx^2} + 2ik\frac{d\phi_k(x)}{dx} + \frac{8\pi^2 m}{h^2}\left[E - V(x) - \frac{h^2 k^2}{8\pi^2 m}\right]\phi_k(x) = 0. \qquad (m.5.27.)$$

Referring to the simple problem of a particle in a rectangular potential well discussed in Part Q, equation (m.5.27.) has the solution

$$\phi_k(x) = A e^{i(\alpha - k)x} + B e^{-i(\alpha + k)x}, \qquad (m.5.28.)$$

where

$$\alpha = \left[\frac{8\pi^2 mE}{h^2}\right]^{1/2}, \qquad (m.5.29.)$$

in the region where $0 < x < a$. A and B are constants.

In the region where $a < x < (a + b)$, the solution is

$$\phi_k(x) = C e^{(\beta - ik)x} + D e^{-(\beta + ik)x}, \qquad (m.5.30.)$$

where

$$\beta = \left[\frac{8\pi^2 m}{h^2}[V_0(x) - E]\right]^{1/2}, \qquad \text{(m.5.31.)}$$

and C and D are constants.

In order that (m.5.28.) and (m.5.30.) should be satisfactory solutions of the equation (m.5.27.), $\phi_k(x)$ should be continuous at $x = 0$ and at $x = a$, and also its first derivative, $d\phi_k(x)/dx$, should be continuous at these points. To satisfy the Bloch periodicity condition, a satisfactory wave function must have the same value at $x = a$ and at $x = -b$. The application of these conditions to $\phi_k(x)$ enable the constants A, B, C, and D to be eliminated. For the function $\phi_k(x)$ to be continuous at $x = 0$ [equivalent to its value at $x = (a + b)$],

$$A + B = C + D. \qquad \text{(m.5.32.)}$$

For continuity of $\phi_k(x)$ at $x = a$ (equivalent to its value at $x = -b$),

$$Ae^{ia(\alpha - k)} + Be^{-ia(\alpha + k)} = Ce^{-(\beta - ik)b} + De^{(\beta + ik)b}. \qquad \text{(m.5.33.)}$$

In order that $d\phi_k/dx$ should be continuous at $x = 0$,

$$i(\alpha - k)A - i(\alpha + k)B = (\beta - ik)C - (\beta + ik)D. \qquad \text{(m.5.34.)}$$

For $d\phi_k/dx$ to be continuous at $x = a$ (equivalent to its value at $x = -b$),

$$i(\alpha - k)Ae^{i(\alpha - k)a} - i(\alpha + k)Be^{-i(\alpha + k)a}$$
$$= (\beta - ik)Ce^{-(\beta - ik)b} - (\beta + ik)De^{(\beta + ik)b}. \qquad \text{(m.5.35.)}$$

These four equations of condition, (m.5.32.–35.), have a simultaneous solution only if the determinant of their coefficients vanish. The condition which must be satisfied for this to occur is given by ‡

$$\frac{(\beta^2 - \alpha^2)}{2\alpha\beta}(\sin a\alpha)(\sinh b\beta) + (\cos a\alpha)(\cosh b\beta)$$

$$= \frac{e^{ik(a + b)} + e^{-ik(a + b)}}{2} = \cos[(k)(a + b)]. \qquad \text{(m.5.36.)}$$

‡ The determinant of the coefficients of equations (m.5.32.–35.) is,

$$\begin{vmatrix} 1 & 1 & -1 & -1 \\ [i(\alpha - k)] & [-i(\alpha + k)] & [-(\beta - ik)] & [\beta + ik] \\ [e^{ia(\alpha - k)}] & [e^{-ia(\alpha + k)}] & [-e^{-b(\beta - ik)}] & [-e^{b(\beta + ik)}] \\ [i(\alpha - k)e^{ia(\alpha - k)}] & [-i(\alpha + k)e^{-i(\alpha + k)a}] & [-(\beta - ik)e^{-b(\beta - ik)}] & [(\beta + ik)e^{b(\beta + ik)}] \end{vmatrix}$$

Selecting the element at the intersection of the first row and first column as the pivotal element, this fourth order determinant reduces to the third order determinant:

$$\begin{vmatrix} [-2i\alpha] & [i\alpha - \beta] & [i\alpha + \beta] \\ [e^{-ia(\alpha + k)} - e^{ia(\alpha - k)}] & [-e^{-b(\beta - ik)} + e^{ia(\alpha - k)}] & [-e^{b(\beta + ik)} + e^{ia(\alpha - k)}] \\ [-i(\alpha + k)e^{-ia(\alpha + k)} - i(\alpha - k)e^{ia(\alpha - k)}] & [(\beta + ik)e^{b(\beta + ik)} + i(\alpha - k)e^{ia(\alpha - k)}] \\ & [-(\beta - ik)e^{-b(\beta - ik)} + i(\alpha - k)e^{ia(\alpha - k)}] \end{vmatrix}$$

Again selecting the element at the intersection of the first row and the first column as pivotal element, this third order determinant reduces to:

$$(-2i\alpha)\begin{vmatrix} T & W \\ L & J \end{vmatrix},$$

where,

$$T = (-e^{-b(\beta - ik)} + e^{ia(\alpha - k)} + \frac{[i\alpha - \beta]}{2i\alpha}\{e^{-ia(\alpha + k)} - e^{ia(\alpha - k)}\}),$$

$$W = (-e^{b(\beta + ik)} + e^{ia(\alpha - k)} + \frac{[i\alpha + \beta]}{2i\alpha}\{e^{-ia(\alpha + k)} - e^{ia(\alpha - k)}\}),$$

$$L = (-(\beta - ik)e^{-b(\beta - ik)} + i(\alpha - k)e^{ia(\alpha - k)}$$
$$+ \frac{[i\alpha - \beta]}{2i\alpha}\{-i[\alpha + k]e^{-ia(\alpha + k)} - i[\alpha - k]e^{ia(\alpha - k)}\}),$$

$$J = ((\beta + ik)e^{b(\beta + ik)} + i(\alpha - k)e^{ia(\alpha - k)}$$
$$+ \frac{[i\alpha + \beta]}{2i\alpha}\{-i[\alpha + k]e^{-ia(\alpha + k)} - i[\alpha - k]e^{ia(\alpha - k)}\}).$$

The product, TJ, is,

$$\left[-e^{-b(\beta - ik)} + e^{ia(\alpha - k)} + \frac{e^{-ia(\alpha + k)}}{2} - \frac{e^{ia(\alpha - k)}}{2} - e^{-ia(\alpha + k)}\frac{\beta}{2i\alpha} + \frac{\beta}{2i\alpha}e^{ia(\alpha - k)}\right]$$

$$\left[\beta e^{b(\beta + ik)} + ike^{b(\beta + ik)} + i\alpha e^{ia(\alpha - k)} - ike^{ia(\alpha - k)} - \frac{i\alpha}{2}e^{-ia(\alpha + k)} - \frac{ik}{2}e^{-ia(\alpha + k)}\right.$$

$$\left. - \frac{i\alpha}{2}e^{ia(\alpha - k)} + \frac{ik}{2}e^{ia(\alpha - k)} - \frac{\beta}{2}e^{-ia(\alpha + k)} - \frac{\beta k}{2\alpha}e^{-ia(\alpha + k)} - \frac{\beta}{2}e^{ia(\alpha - k)} + \frac{\beta k}{2\alpha}e^{ia(\alpha - k)}\right].$$

On multiplication, the expanded product contains the 72 terms,

$$TJ = \left[e^{2ibk}(-\beta - ik) + e^{-b(\beta - ik) + ia(\alpha - k)}[-i\alpha + ik] + e^{-b(\beta - ik) - ia(\alpha + k)}\left(\frac{i\alpha}{2} + \frac{ik}{2}\right)\right.$$

$$+ e^{-b(\beta - ik) + ia(\alpha - k)}\left(\frac{i\alpha}{2} - \frac{ik}{2}\right) + e^{-b(\beta - ik) - ia(\alpha + k)}\left(\frac{\beta}{2} + \frac{\beta k}{2\alpha}\right)$$

$$+ e^{-b(\beta - ik) + ia(\alpha - k)}\left(\frac{\beta}{2} - \frac{\beta k}{2\alpha}\right) + e^{b(\beta + ik) + ia(\alpha - k)}[\beta + ik]$$

$$+ e^{2ia(\alpha - k)}[i\alpha - ik] + e^{-2ika}\left(-\frac{i\alpha}{2} - \frac{ik}{2}\right) + e^{2ia(\alpha - k)}\left(-\frac{i\alpha}{2} + \frac{ik}{2}\right)$$

$$+ e^{-2ika}\left(-\frac{\beta}{2} - \frac{\beta k}{2\alpha}\right) + e^{2ia(\alpha - k)}\left(-\frac{\beta}{2} + \frac{\beta k}{2\alpha}\right) + e^{b(\beta + ik) - ia(\alpha + k)}\left(\frac{\beta}{2} + \frac{ik}{2}\right)$$

$$+ e^{-2ika}\left(\frac{i\alpha}{2} - \frac{ik}{2}\right) + e^{-2ia(\alpha + k)}\left(-\frac{i\alpha}{4} - \frac{ik}{4}\right) + e^{-2ika}\left(-\frac{i\alpha}{4} + \frac{ik}{4}\right)$$

$$+ e^{-2ia(\alpha + k)}\left(-\frac{\beta}{4} - \frac{\beta k}{4\alpha}\right) + e^{-2ika}\left(-\frac{\beta}{4} + \frac{\beta k}{4\alpha}\right) + e^{b(\beta + ik) + ia(\alpha - k)}\left(-\frac{\beta}{2} - \frac{ik}{2}\right)$$

$$+ e^{2ia(\alpha - k)}\left(-\frac{i\alpha}{2} + \frac{ik}{2}\right) + e^{-2ika}\left(\frac{i\alpha}{4} + \frac{ik}{4}\right) + e^{2ia(\alpha - k)}\left(\frac{i\alpha}{4} - \frac{ik}{4}\right)$$

$$+ e^{-2ika}\left(\frac{\beta}{4} + \frac{\beta k}{4\alpha}\right) + e^{2ia(\alpha - k)}\left(\frac{\beta}{4} - \frac{\beta k}{4\alpha}\right) + e^{-2ika}\left(-\frac{\beta}{2} + \frac{\beta k}{2\alpha}\right)$$

$$+ e^{b(\beta + ik) - ia(\alpha + k)}\left(-\frac{\beta^2}{2i\alpha} - \frac{\beta k}{2\alpha}\right) + e^{-2ia(\alpha + k)}\left(\frac{\beta}{4} + \frac{\beta k}{4\alpha}\right) + e^{-2ika}\left(\frac{\beta}{4} - \frac{\beta k}{4\alpha}\right)$$

$$+ e^{-2ia(\alpha + k)}\left(\frac{\beta^2}{4i\alpha} + \frac{\beta^2 k}{4i\alpha^2}\right) + e^{-2ika}\left(\frac{\beta^2}{4i\alpha} - \frac{\beta^2 k}{4i\alpha^2}\right) + e^{b(\beta + ik) + ia(\alpha - k)}\left(\frac{\beta^2}{2i\alpha} + \frac{\beta k}{2\alpha}\right)$$

$$+ e^{2ia(\alpha - k)}\left(\frac{\beta}{2} - \frac{\beta k}{2\alpha}\right) + e^{-2iak}\left(-\frac{\beta}{4} - \frac{\beta k}{4\alpha}\right) + e^{2ia(\alpha - k)}\left(-\frac{\beta}{4} + \frac{\beta k}{4\alpha}\right)$$

$$+ e^{-2iak}\left(-\frac{\beta^2}{4i\alpha} - \frac{\beta^2 k}{4i\alpha^2}\right) + e^{2ia(\alpha - k)}\left(-\frac{\beta^2}{4i\alpha} + \frac{\beta^2 k}{4i\alpha^2}\right)\right].$$

This expression may be simplified somewhat and rearranged to give:

$$TJ = e^{2ibk}(-\beta - ik) + e^{-2iak}\left[-\beta - \frac{ik}{2} - \frac{\beta^2 k}{2i\alpha^2}\right]$$

$$+ e^{-b(\beta - ik) + ia(\alpha - k)}\left[-\frac{i\alpha}{2} + \frac{ik}{2} - \frac{\beta k}{2\alpha} + \frac{\beta}{2}\right]$$

$$+ e^{-b(\beta - ik) - ia(\alpha + k)}\left[\frac{i\alpha}{2} + \frac{ik}{2} + \frac{\beta}{2} + \frac{\beta k}{2\alpha}\right]$$

$$+ e^{b(\beta + ik) + ia(\alpha - k)}\left[\frac{\beta}{2} + \frac{ik}{2} + \frac{\beta^2}{2i\alpha} + \frac{\beta k}{2\alpha}\right] + e^{b(\beta + ik) - ia(\alpha + k)}\left[\frac{\beta}{2} + \frac{ik}{2} - \frac{\beta^2}{2i\alpha} - \frac{\beta k}{2\alpha}\right]$$

$$+ e^{2ia(\alpha - k)}\left[\frac{i\alpha}{4} - \frac{ik}{4} - \frac{\beta^2}{4i\alpha} + \frac{\beta^2 k}{4i\alpha^2}\right] + e^{-2ia(\alpha + k)}\left[-\frac{i\alpha}{4} - \frac{ik}{4} + \frac{\beta^2}{4i\alpha} + \frac{\beta^2 k}{4i\alpha^2}\right].$$

Similarly, the product WL, on expansion and rearrangement, leads to a comparable type of expression.

$$WL = \left[-e^{b(\beta + ik)} + e^{ia(\alpha - k)} + \frac{e^{-ia(\alpha + k)}}{2} - \frac{e^{ia(\alpha - k)}}{2} + e^{-ia(\alpha + k)}\frac{\beta}{2i\alpha} - e^{ia(\alpha - k)}\frac{\beta}{2i\alpha}\right]$$

$$\left[-\beta e^{-b(\beta - ik)} + ik e^{-b(\beta - ik)} + i\alpha e^{ia(\alpha - k)} - ik e^{ia(\alpha - k)} - \frac{i\alpha}{2}e^{-ia(\alpha + k)}\right.$$

$$- \frac{ik}{2}e^{-ia(\alpha + k)} + \frac{\beta}{2}e^{-ia(\alpha + k)} + \frac{\beta k}{2\alpha}e^{-ia(\alpha + k)} + \frac{\beta}{2}e^{ia(\alpha - k)} - \frac{\beta k}{2\alpha}e^{ia(\alpha - k)}$$

$$\left.- \frac{i\alpha}{2}e^{ia(\alpha - k)} + \frac{ik}{2}e^{ia(\alpha - k)}\right]$$

$$= \left[e^{2ibk}(\beta - ik) + e^{b(\beta + ik) + ia(\alpha - k)}(-i\alpha + ik) + e^{b(\beta + ik) - ia(\alpha + k)}\left(\frac{i\alpha}{2} + \frac{ik}{2}\right)\right.$$

$$+ e^{b(\beta + ik) - ia(\alpha + k)}\left(-\frac{\beta}{2} - \frac{\beta k}{2\alpha}\right) + e^{b(\beta + ik) + ia(\alpha - k)}\left(-\frac{\beta}{2} + \frac{\beta k}{2\alpha}\right)$$

$$+ e^{b(\beta + ik) + ia(\alpha - k)}\left(\frac{i\alpha}{2} - \frac{ik}{2}\right) + e^{-b(\beta - ik) + ia(\alpha - k)}(-\beta + ik) + e^{-2ika}\left(-\frac{i\alpha}{2} - \frac{ik}{2}\right)$$

$$+ e^{2ia(\alpha - k)}(i\alpha - ik) + e^{-2iak}\left(\frac{\beta}{2} + \frac{\beta k}{2\alpha}\right) + e^{2ia(\alpha - k)}\left(\frac{\beta}{2} - \frac{\beta k}{2\alpha}\right)$$

$$+ e^{2ia(\alpha - k)}\left(-\frac{i\alpha}{2} + \frac{ik}{2}\right) + e^{-b(\beta - ik) - ia(\alpha + k)}\left(-\frac{\beta}{2} + \frac{ik}{2}\right) + e^{-2iak}\left(\frac{i\alpha}{2} - \frac{ik}{2}\right)$$

$$+ e^{-2ia(\alpha + k)}\left(-\frac{i\alpha}{4} - \frac{ik}{4}\right) + e^{-2ia(\alpha + k)}\left(\frac{\beta}{4} + \frac{\beta k}{4\alpha}\right) + e^{-2iak}\left(\frac{\beta}{4} - \frac{\beta k}{4\alpha}\right)$$

$$+ e^{-2iak}\left(-\frac{i\alpha}{4} + \frac{ik}{4}\right) + e^{2ia(\alpha - k)}\left(-\frac{i\alpha}{2} + \frac{ik}{2}\right) + e^{-b(\beta - ik) + ia(\alpha - k)}\left(\frac{\beta}{2} - \frac{ik}{2}\right)$$

$$+ e^{-2iak}\left(\frac{i\alpha}{4} + \frac{ik}{4}\right) + e^{-2iak}\left(-\frac{\beta}{4} - \frac{\beta k}{4\alpha}\right) + e^{2ia(\alpha - k)}\left(-\frac{\beta}{4} + \frac{\beta k}{4\alpha}\right)$$

$$+ e^{2ia(\alpha - k)}\left(\frac{i\alpha}{4} - \frac{ik}{4}\right) + e^{-b(\beta - ik) - ia(\alpha + k)}\left(-\frac{\beta^2}{2i\alpha} + \frac{\beta k}{2\alpha}\right) + e^{-2iak}\left(\frac{\beta}{2} - \frac{\beta k}{2\alpha}\right)$$

$$+ e^{-2ia(\alpha + k)}\left(-\frac{\beta}{4} - \frac{\beta k}{4\alpha}\right) + e^{-2ia(\alpha + k)}\left(\frac{\beta^2}{4i\alpha} + \frac{\beta^2 k}{4i\alpha^2}\right) + e^{-2iak}\left(\frac{\beta^2}{4i\alpha} - \frac{\beta^2 k}{4i\alpha^2}\right)$$

$$+ e^{-2iak}\left(-\frac{\beta}{4} + \frac{\beta k}{4\alpha}\right) + e^{-b(\beta - ik) + ia(\alpha - k)}\left(\frac{\beta^2}{2i\alpha} - \frac{\beta k}{2\alpha}\right) + e^{2ia(\alpha - k)}\left(-\frac{\beta}{2} + \frac{\beta k}{2\alpha}\right)$$

$$+ e^{-2iak}\left(\frac{\beta}{4} + \frac{\beta k}{4\alpha}\right) + e^{-2iak}\left(-\frac{\beta^2}{4i\alpha} - \frac{\beta^2 k}{4i\alpha^2}\right) + e^{2ia(\alpha - k)}\left(\frac{\beta}{4} - \frac{\beta k}{4\alpha}\right)$$

$$\left.+ e^{2ia(\alpha - k)}\left(-\frac{\beta^2}{4i\alpha} + \frac{\beta^2 k}{4i\alpha^2}\right)\right].$$

$$= e^{2ibk}(\beta - ik) + e^{-2iak}\left[\beta - \frac{ik}{2} - \frac{\beta^2 k}{2i\alpha^2}\right] + e^{-b(\beta - ik) + ia(\alpha - k)}\left[\frac{ik}{2} - \frac{\beta}{2} - \frac{\beta k}{2\alpha} + \frac{\beta^2}{2i\alpha}\right]$$

$$+ e^{-b(\beta - ik) - ia(\alpha + k)}\left[\frac{ik}{2} - \frac{\beta}{2} + \frac{\beta k}{2\alpha} - \frac{\beta^2}{2i\alpha}\right]$$

$$+ e^{b(\beta + ik) + ia(\alpha - k)}\left[-\frac{i\alpha}{2} + \frac{ik}{2} - \frac{\beta}{2} + \frac{\beta k}{2\alpha}\right]$$

$$+ e^{b(\beta + ik) - ia(\alpha + k)}\left[\frac{i\alpha}{2} + \frac{ik}{2} - \frac{\beta}{2} - \frac{\beta k}{2\alpha}\right] + e^{2ia(\alpha - k)}\left[\frac{i\alpha}{4} - \frac{ik}{4} - \frac{\beta^2}{4i\alpha} + \frac{\beta^2 k}{4i\alpha^2}\right]$$

$$+ e^{-2ia(\alpha + k)}\left[-\frac{i\alpha}{4} - \frac{ik}{4} + \frac{\beta^2}{4i\alpha} + \frac{\beta^2 k}{4i\alpha^2}\right].$$

Substituting these expressions for TJ and WL in the determinant

$$(-2i\alpha)\begin{vmatrix} T & W \\ L & J \end{vmatrix},$$

which must vanish, in order that the four equations, (m.5.32.–35.) should have simultaneous solutions, then

$$-2i\alpha(TJ - WL) = 0$$

$$\therefore\ -2i\alpha\left[e^{2ibk}(-2\beta) + e^{-2iak}(-2\beta) + e^{b(\beta + ik) + ia(\alpha - k)}\left(\beta + \frac{i\alpha}{2} + \frac{\beta^2}{2i\alpha}\right)\right.$$

$$+ e^{b(\beta + ik) - ia(\alpha + k)}\left(\beta - \frac{i\alpha}{2} - \frac{\beta^2}{2i\alpha}\right) + e^{-b(\beta - ik) + ia(\alpha - k)}\left(\beta - \frac{i\alpha}{2} - \frac{\beta^2}{2i\alpha}\right)$$

$$\left. + e^{-b(\beta - ik) - ia(\alpha + k)}\left(\beta + \frac{i\alpha}{2} + \frac{\beta^2}{2i\alpha}\right)\right] = 0$$

$$\therefore\ 4i\alpha\beta(e^{2ibk} + e^{-2iak}) + (i\alpha - \beta)^2[e^{b(\beta + ik) - ia(\alpha + k)} + e^{-b(\beta - ik) + ia(\alpha - k)}]$$

$$+ (\alpha - i\beta)^2[e^{b(\beta + ik) + ia(\alpha - k)} + e^{-b(\beta - ik) - ia(\alpha + k)}] = 0.$$

Dividing throughout by $4i\alpha\beta$,

$$\therefore\ (e^{2ibk} + e^{-2iak}) + \frac{\beta^2 - \alpha^2}{4i\alpha\beta}\left[e^{b(\beta + ik) - ia(\alpha + k)} + e^{-b(\beta - ik) + ia(\alpha - k)}\right.$$

$$\left. - e^{b(\beta + ik) + ia(\alpha - k)} - e^{-b(\beta - ik) - ia(\alpha + k)}\right]$$

$$- \tfrac{1}{2}[e^{b(\beta + ik) - ia(\alpha + k)} + e^{-b(\beta - ik) + ia(\alpha - k)} + e^{b(\beta + ik) + ia(\alpha - k)} + e^{-b(\beta - ik) - ia(\alpha + k)}] = 0.$$

This expression may be rewritten as:

$$(e^{2ibk} + e^{-2iak}) + \frac{(\beta^2 - \alpha^2)}{4i\alpha\beta}\left[-e^{b\beta + ibk - ika}(e^{ia\alpha} - e^{-ia\alpha}) + e^{-b\beta + ibk - ika}(e^{ia\alpha} - e^{-ia\alpha})\right]$$

$$- \tfrac{1}{2}[e^{b\beta + ibk - ika}(e^{ia\alpha} + e^{-ia\alpha}) + e^{-b\beta + ibk - ika}(e^{ia\alpha} + e^{-ia\alpha})] = 0.$$

By now transforming the exponential functions, in the inner parentheses, to the equivalent circular functions, i.e., by making the substitutions,

$$\frac{e^{ia\alpha} + e^{-ia\alpha}}{2} = \cos a\alpha \qquad \text{and} \qquad \frac{e^{ia\alpha} - e^{-ia\alpha}}{2i} = \sin a\alpha,$$

then

$$(e^{2ibk} + e^{-2iak}) + \frac{(\beta^2 - \alpha^2)}{2\alpha\beta}\left[\sin a\alpha(-e^{ik(b - a)}\{e^{b\beta} - e^{-b\beta}\})\right]$$

$$- \tfrac{1}{2}[2\cos a\alpha(e^{ik(b - a)}\{e^{b\beta} + e^{-b\beta}\})] = 0.$$

Now, by transforming the exponential functions, in the inner parentheses, to the equivalent hyperbolic functions, i.e., by making the substitutions,

$$\frac{e^{b\beta} + e^{-b\beta}}{2} = \cosh b\beta \qquad \text{and} \qquad \frac{e^{b\beta} - e^{-b\beta}}{2} = \sinh b\beta,$$

then,

$$(e^{2ibk} + e^{-2iak}) + \frac{(\beta^2 - \alpha^2)}{2\alpha\beta} (\sin a\alpha)(2 \sinh b\beta)(-e^{ik(b-a)})$$
$$- \tfrac{1}{2}(2 \cos a\alpha)(2 \cosh b\beta)(e^{ik(b-a)}) = 0.$$

Finally, by dividing this equation throughout by $[-2e^{ibk}e^{-ika}]$, there results,

$$-\frac{(e^{2ibk} + e^{-2iak})}{(2e^{ibk-ika})} + \frac{(\beta^2 - \alpha^2)}{2\alpha\beta} (\sin a\alpha)(\sinh b\beta) + (\cos a\alpha)(\cosh b\beta) = 0,$$

which, on minor rearrangement, becomes,

$$\frac{(\beta^2 - \alpha^2)}{2\alpha\beta} (\sin a\alpha)(\sinh b\beta) + (\cos a\alpha)(\cosh b\beta) = \frac{e^{ik(a+b)} + e^{-ik(a+b)}}{2} = \cos k(a + b),$$

which is the expression given above in equation (m.5.36.).

This equation, (m.5.36.), expresses the condition that must be satisfied if wave functions of form (m.5.26.) are to exist. The term α contains the energy factor, and therefore to obtain a solution for the energy states of the electron, it has to be possible to isolate and evaluate α. Kronig and Penney have shown that this can be done if the square–well potential function is represented by a periodic delta function.[1] A δ-function is a mathematically useful way, introduced by Dirac, of saying that $\delta(x) = 0$ if $x \neq 0$ and that $\int_{-a}^{b} \delta(x)\,dx = 1$, where a and b are any positive integers; in other words, $\delta(x)$ is 0 everywhere except at $x = 0$, where it becomes infinite. A delta function is thus not a real function in the sense that a proper function has a definite value for each point in its domain; it is an improper function in this respect. It should also not be confused with the Kronecker delta symbol, δ_{ij}, which is 1 if $i = j$ and 0 if $i \neq j$. Alternatively, it is convenient to think of the δ-function as the limit, as c tends to 0, of a function which has the value 0 outside the interval defined by $-c$ and c, and the value $c/2$ inside this interval. If it is imagined that the square–well potential barriers in this problem are represented by periodic δ-functions, such that, since the potential must be real and finite, then in the limit as the potential barrier width b tends to 0 and the actual potential V_0 tends to infinity, the product $V_0 b$ has to remain finite. If the first δ-function is symmetrical with respect to the origin, then it would be centered at the point $b/2$. From (m.5.31.), V_0 is some constant multiple of β^2, so that if $V_0 b$ has to remain finite, then $\beta^2 b/2$ has to remain finite, and since a is the lattice row constant, then $\beta^2 ab/2$ will be a finite constant in the limit when

$$b \to 0 \quad \text{and} \quad \left\{ \begin{matrix} V_0 \to \infty \\ \beta \to \infty \end{matrix} \right\}.$$

Let

$$\lim_{\substack{b \to 0 \\ \beta \to \infty}} \frac{\beta^2 ab}{2} = P. \tag{m.5.37.}$$

---

[1] Dirac, P. A. M., *Quantum Mechanics, 3rd Edition*, 1947, Oxford University Press.

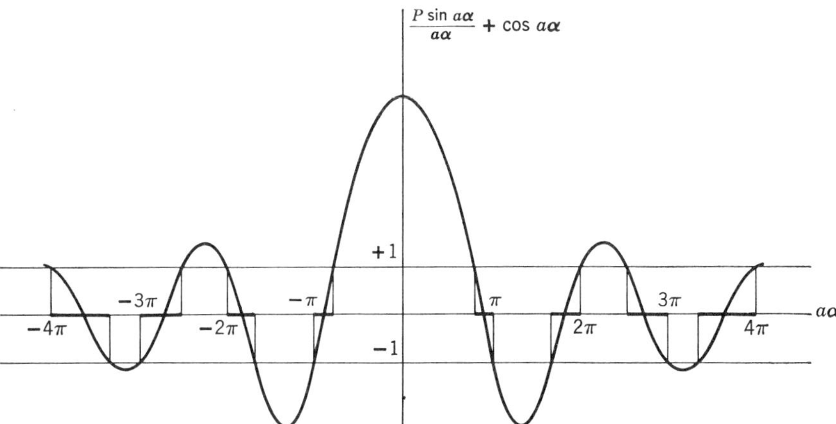

FIG. M.5.4. The left-hand side of equation (m.5.38), the function $(P \sin a\alpha)/a\alpha + \cos a\alpha$ plotted as a function of $a\alpha$, for the arbitrary value of P of $3\pi/2$. Since the values of the function plotted as the ordinate are only allowable in the range between $+1$ and -1, the only acceptable values of a and therefore $E\,(\alpha = \sqrt{[8\pi^2 mE/h^2]})$ are those lying in the limited ranges indicated. (According to Kronig and Penney.)

Since the limit as $b \to 0$ and $V_0 \to \infty$ of $\cos k(a + b)$ is $\cos ka$, and the limit of $\cosh b\beta = (e^{b\beta} + e^{-b\beta})/2$ is 1, and also

$$\lim_{\substack{b \to 0 \\ V_0 \to \infty}} \frac{(\beta^2 - \alpha^2)}{2\alpha\beta} \sinh b\beta = \lim \frac{(\beta^2 - \alpha^2)}{2\alpha\beta} \left[b\beta + \frac{(b\beta)^3}{3!} + \cdots \right]$$

$$= \lim \frac{\beta^2 b}{2\alpha} = \frac{P}{a\alpha},$$

then the condition expressed by (m.5.36.) becomes

$$P\frac{\sin a\alpha}{a\alpha} + \cos a\alpha = \cos ka. \tag{m.5.38.}$$

This equation, (m.5.38.), must have a solution for $\alpha = [8\pi^2 mE/h^2]^{1/2}$ in order that functions of the form $\psi(x) = \phi_k(x)e^{ikx}$ should exist.

The right-hand side of equation (m.5.38.) can only adopt values between $+1$ and -1 corresponding to the limitations on the cosine of an angle, and therefore the left-hand side can only provide values of $a\alpha$ or E which fall within this range. A plot of the left-hand side of equation (m.5.38.) as a function of $a\alpha$ is shown in Fig. M.5.4., using the arbitrary value $P = 3\pi/2$. The permitted ranges of $a\alpha$ are indicated by the heavily lined parts of this curve, and by (m.5.29.) define the allowed ranges of the energy E. The boundaries of the allowed energy ranges are defined by the values of $k = n\pi/a$. If the energy, in units of $8ma^2/h^2$, is plotted as a function of ka, again taking

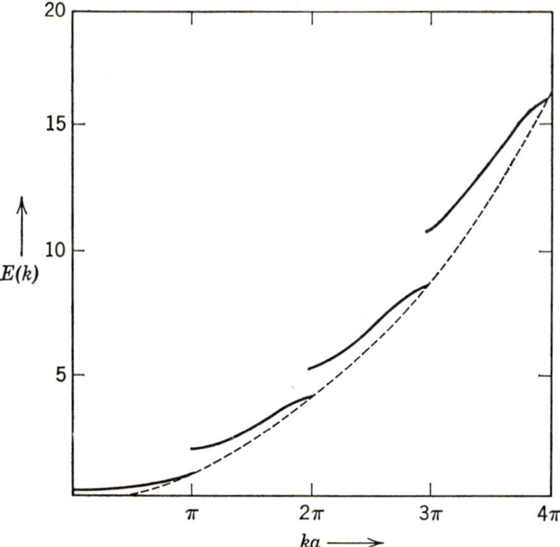

FIG. M.5.5. The energy of an electron as a function of its wave number for the Kronig–Penney potential. The energy is expressed in units of $h^2/8ma^2$ and P in equation (m.5.37) is arbitrarily assumed to be $3\pi/2$. (According to Sommerfeld and Bethe).

P arbitrarily as $3\pi/2$, the curves shown in Fig. M.5.5. result, clearly showing the bands of allowed energy values and the discrete energy gaps, where no energy solution is possible. If P is very small, the disallowed regions vanish, but if P is large and approaches infinity, the allowed ranges of $a\alpha$ shrink to points located at $n\pi$, where $n = \pm 1, \pm 2, \ldots$. In this latter case, the energy spectrum becomes discrete and the appropriate eigenvalues, $E = n^2h^2/8ma^2$, become identical to those of an electron in a box of edge length a.‡

‡ If $ka = n\pi = 2\pi a/\lambda$, by definition of k, then $2a = n\lambda = nh/p = nh/(2mE)^{1/2}$ or $E = n^2h^2/8ma^2$.

M.5.4. Reduced and Extended Zones in One Dimension

It has been shown in equation (m.5.22.) that $E_l(k) = E_l(-k)$, so that the curves shown in Fig. M.5.5. are actually symmetrical about the energy ordinate. If the energy of an electron, described in terms of the Kronig–Penney potential model, is therefore plotted as a function of its wave vector to show its dependence on both positive and negative values of k, as shown in Fig. M.5.6., this is equivalent to the extended zone description of this relation-

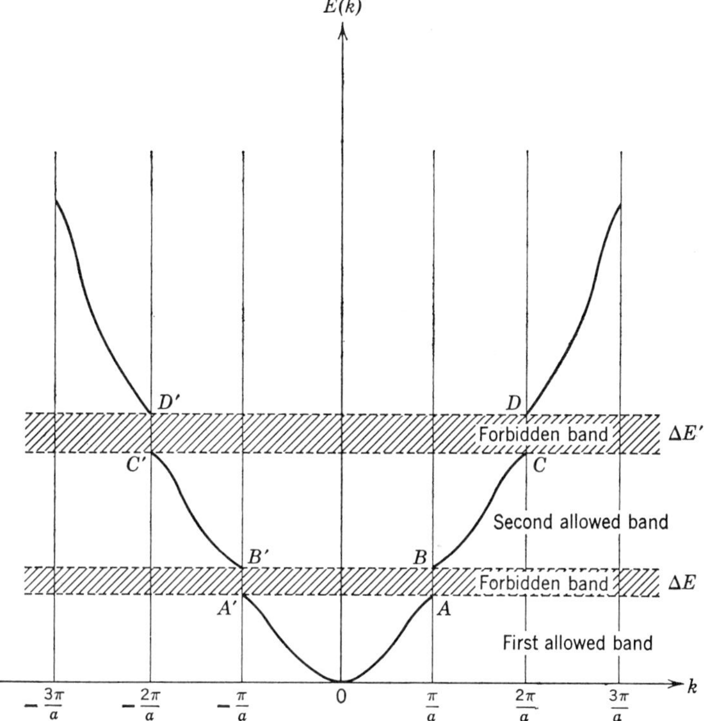

Fig. M.5.6. Energy of an electron as a function of its wave-vector in a one-dimensional lattice of period, a, for a small periodic potential field, according to the extended zone scheme.

ship. In the reduced zone scheme, where k is limited to values lying between $+\pi/a$ and $-\pi/a$, due to the characteristic property of Bloch functions that they are unaltered if k is replaced by $[k + (2\pi n/a)]$, all possible energy states are still included in a single period. The important point is that the energy is a periodic function of k, with a period of $2\pi/a$, and it is simply a matter of convenience to choose this period to be symmetrical about the origin. Any period of length $2\pi/a$ is appropriate and the reduced and extended zone descriptions are equally valid, being related to one another by translations of parts of the $E - k$ curve parallel to the k-axis.

Actually, in solving the Schrödinger equation for an electron in a periodic field, assuming solutions of the form (m.5.12.), $E(k)$ is normally found as a many-valued function of k, the different values of which are distinguished by the second index, l, and if k is unrestricted, the $E - k$ curves resemble those shown in Fig. M.5.7., rather than being immediately recognizable as those of the extended zone scheme.

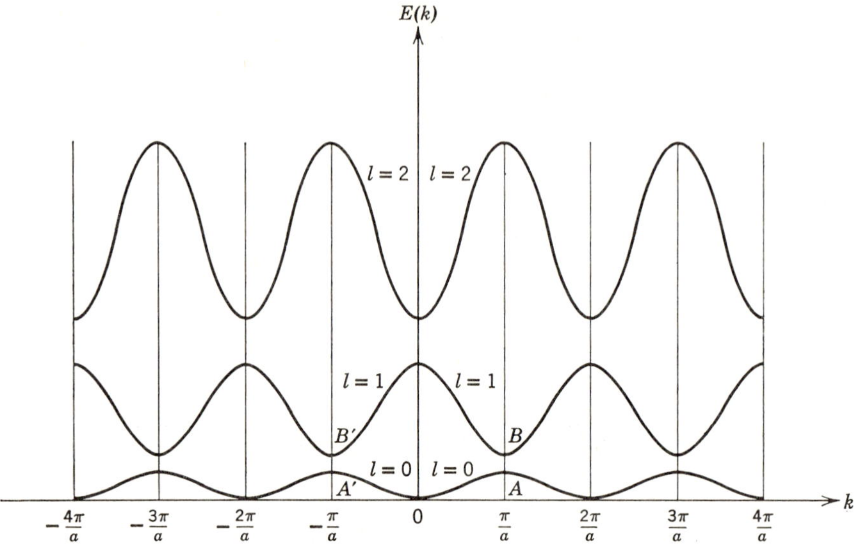

Fig. M.5.7. Energy of an electron as a many-valued function of its wave-vector when the wave-vector is unrestricted, in a one-dimensional lattice for a small periodic potential field.

While the shape of the $E - k$ curves shown in Fig. M.5.6. are very similar to the corresponding free electron curves, it should be pointed out that this is only so if the periodic potential field is small. For a periodic field of large amplitude, there may be very considerable deviations from the free electron curves, although the general characteristics of quasicontinuous bands of allowed energy separated by regions for which there is no energy solution are still observed. In the allowed energy regions, the energy curves are smooth and continuous and as a consequence of their periodicity and symmetry they have zero slope at $k = 0$ and at $k = \pm \pi/a$. Due to the symmetry about $k = 0$,

$$\frac{dE_l(k)}{dk} = -\frac{dE_l(-k)}{dk}, \tag{m.5.39.}$$

and due to the periodicity,

$$dE_l/dk \quad \text{at} \quad \pi/a = dE_l/dk \quad \text{at} \quad -\pi/a. \tag{m.5.40.}$$

Therefore,

$$dE_l/dk \quad \text{at} \quad -\pi/a = -dE_l/dk \quad \text{at} \quad \pi/a = 0,$$

and at the origin

$$dE_l/dk = 0.$$

The occurrence of energy gaps for electrons moving in a one-dimensional lattice array of period a at points $n\pi/a$ is related to the condition for Bragg reflexion or cooperative scattering by the row of metal cations of the electrons traveling along the lattice row. The Bragg law of diffraction, which is described on pages 544–545 for X-radiation, is not limited to the scattering of electromagnetic radiation by crystalline materials, but is equally applicable to the scattering of electrons, neutrons, phonons (quanta of lattice vibrational energy with frequencies corresponding to acoustic frequencies), magnons (quanta of magnetic energy), and other elastic waves by periodic structures, which have a period of magnitude comparable to the wavelength associated with the particular form of energy involved. The Bragg condition for a three-dimensional lattice is ‡

$$\lambda n = 2d_{hkl} \sin\theta, \qquad \text{(m.5.41.)}$$

where λ is the wavelength associated with the energy quanta involved, n is an integer, d_{hkl} is the interplanar spacing of planes with Miller indices hkl (see page 552), and θ is the angle of incidence of the radiation with respect to the plane hkl. For a one-dimensional lattice row of period a, the expression

$$n\lambda = 2a \qquad \text{(m.5.42.)}$$

is the one-dimensional equivalent of the Bragg law, where $\sin\theta = 1$ or $\theta = \pi/2$. Since $k = 2\pi/\lambda$ or conversely $\lambda = 2\pi/k$, the Bragg condition for reflexion through 90° becomes

$$2\pi n/k = 2a \qquad \text{or} \qquad k = n\pi/a. \qquad \text{(m.5.43.)}$$

The fact that this value of the wave vector k, deduced from the Bragg law, agrees with the k-value at which energy gaps occur in the Kronig–Penney model gives rise to the interpretation that electrons with this value for their wave vector cannot travel along the lattice row, but are totally reflected.

---

‡ Following the discovery of X-ray diffraction by crystalline matter, by Friedrich and Knipping,[1] at the instigation of von Laue,[2] the phenomenon was first described in terms of the cooperative scattering of the incident beam of X-radiation from three non-coplanar rows of atoms. W. H. and W. L. Bragg[3] subsequently showed that the problem of diffraction could be described in a very much simpler manner as the reflexion of the incident beam of radiation from planes of atoms in a crystal, or more exactly that diffraction of a beam of X-radiation by a crystal is geometrically equivalent to reflexion from planes of atoms in the crystal. Later still, another approach to X-ray diffraction was given by Ewald,[4] the so-called dynamical theory of X-ray scattering, using the mathematical methods which he had previously used for describing optical diffraction. The Bragg method is by far the simplest and forms the basis of the most widely used interpretation of X-ray diffraction events.

[1] Friedrich, W., Knipping, P., and Laue, M. von, *Ann. Physik*, 1913, **41**, 971.
[2] Laue, M. von, *Sitzber. Math.-Physik, Kl. Bayer. Akad. Wiss.*, 1912, 363.
[3] Bragg, W. H., and Bragg, W. L., *Proc. Roy. Soc.*, 1913, **A88**, 428; **A89**, 246.
[4] Ewald, P. P., *Ann. Physik*, 1916, **49**, 1, 117; 1918, **54**, 519, 557.

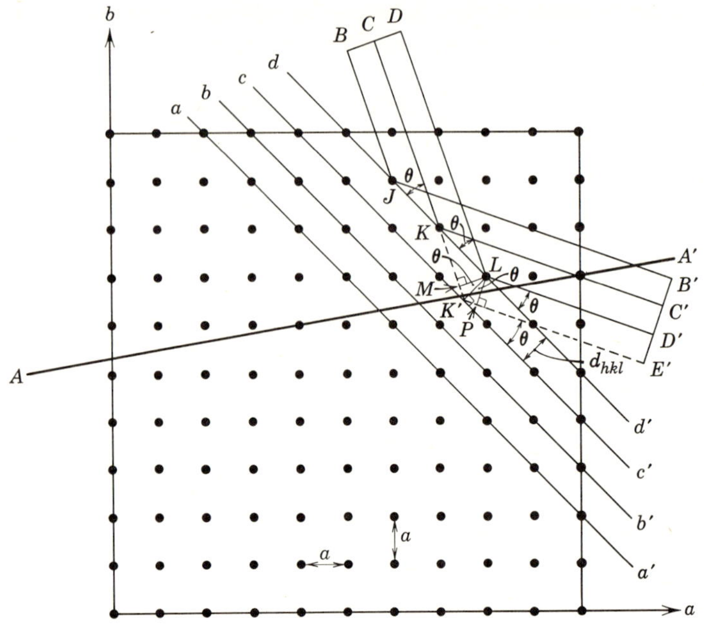

FIG. M.5.8.   Two-dimensional square plane lattice, of period, *a*.

Suppose, as illustrated in Fig. M.5.8., that this represents a plane lattice array of atoms, where some possible parallel planes of atoms are shown at *aa′*, *bb′*, *cc′*, and *dd′*, with indices *hkl*, and spacing d_{hkl}. If the incident beam of X-radiation, incident at some angle θ on these planes of atoms is in phase along the wavefront *BCD*, then the several rays scattered by the atoms at *J*, *K*, and *L*, in particular, can combine after scattering to form a wavefront, provided that they are in phase with one another. For this to occur, the total path lengths *BJB′*, *CKC′*, and *DLD′*, must contain the same number of waves, so that *BJB′* = *CKC′* = *DLD′*. This condition is of course satisfied for the continuation of the direct beam, and the only other way in which it may be satisfied is for reflection of the incident beam by the plane *dd′*. It is not only for the points *J*, *K*, and *L* in the plane *dd′* that reflexion of the incident beam can take place, but every point in this plane, irrespective of the angle θ, reflects the incident radiation. Any point in the plane *dd′* would act as a reflecting point leading to the construction of the wavefront *B′C′D′*, provided only that there were an atom at that point. The ability of a plane of atoms to reflect an incident beam of radiation is quite independent of how the atoms are arranged in the plane, but of course if the atoms in the plane are not in a plane lattice array, then there could not be any other planes of atoms in the assembly. Similarly for any other plane of atoms in the array. For cooperative scattering from the stack of planes *aa′*, *bb′*, *cc′*, *dd′*,..., to take place, the reflected beam from each of the planes in the stack must be in phase with one another. Since each plane further down the stack involves a longer path length for the radiation reflected from it, the condition which must be satisfied for cooperative scattering by the whole stack of planes, which constitutes the whole lattice array of atoms, is that the additional path length traversed by the incident beam on reflexion by each successive plane must be an integral number of wavelengths greater than the path length for the incident and reflected beam from the plane above. Thus, the additional path length for the beam reflected from the plane cc' compared with that of the beam reflected from the adjacent plane dd' is the length $MK'P$ (Fig. M.5.8.). If the wavelength of the incident radiation is λ, and this length has to be an integral number of wavelengths, then $MK'P$ must equal $n\lambda$, where n is any integer. It is apparent from Fig. M.5.8. that the angles $CKJ = LKM = MLK' = K'LP$, and that the path lengths $DL = CM$ and $LD' = PE'$, so that $MK'P = 2d_{hkl} \sin \theta$. Thus

$$n\lambda = 2d_{hkl} \sin \theta \quad \text{or} \quad \theta = \sin^{-1}[n\lambda/2d_{hkl}],$$

which is Bragg's Law as given in (m.5.41.). It is apparent that for cooperative reflexion by all of the planes (hkl) of spacing d_{hkl}, in a three-dimensional lattice, the angle of incidence θ can only have the specific values, $\sin^{-1}[n\lambda/2d_{hkl}]$, with $n = 1, 2, 3, \ldots$, wavelengths' path length difference between adjacent planes in the set, and that co-operative scattering or diffraction by the whole lattice array can only occur as long as $n\lambda < 2d_{hkl}$.

Each zone of allowed energy states extends over an interval of $2\pi/a$. Also, in terms of equation (m.5.11.), the interval between successive k-values is $2\pi/Ga$. Therefore in each zone of k-values there must be

$$(2\pi/a) \div (2\pi/Ga) = G$$

states; for a real metal, where a is of the order of 10^{-10} m, G is a very large number for any macroscopic-sized piece of metal.

Various authors have examined the solution of the Schrödinger equation for an electron in potential fields of special forms and by various approxima-tion methods, some of which will be discussed later for the three-dimensional case. In all cases it is found that only for certain zones of E has the Schrödinger equation a solution of the Bloch type with real values of k; k is found to be complex in the forbidden zones. In particular, Morse[1] has shown that if $V = A \sin x$, the solutions are Mathieu functions.[2] For real metals there are, of course, many electrons present which not only greatly affect the field acting on any one of them but also make the matter of Coulomb correlations of major importance, so that it is not surprising that Bloch functions are not always the most appropriate. However, for the non-transition metals gener-ally, where the conduction electrons originate from incomplete quantum groups and all of the other electrons in the metal cation are distributed among closed quantum groups, the Bloch method is valid.

M.5.5. Brillouin Zones in Two Dimensions

For simplicity, consider a two-dimensional metal which crystallizes with a simple cubic lattice, as in Fig. M.5.8. The distribution of the energies of the conduction electrons moving in such a square lattice can be described in

---

[1] Morse, M., *Phys. Rev.*, 1930, **35**, 1310.
[2] Mathieu Functions, *Handbook of Mathematical Functions*, N.B.S., 1964, 722–750.

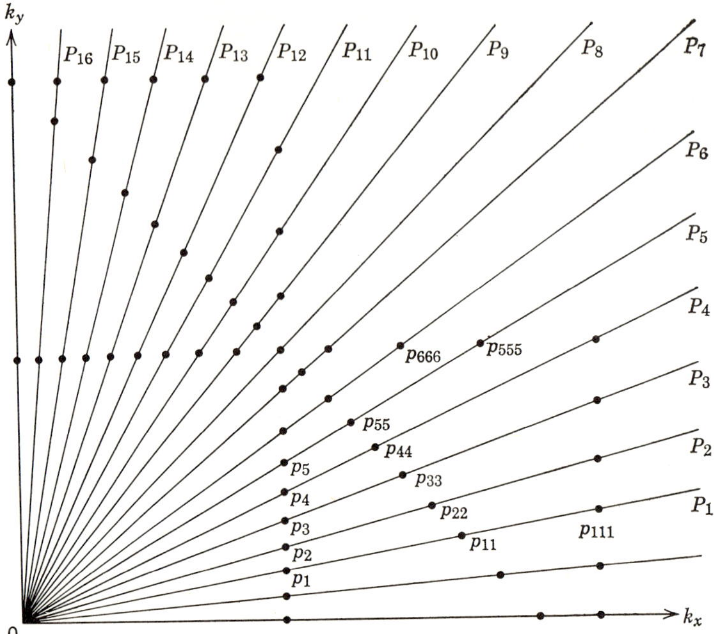

FIG. M.5.9. Two-dimensional wave-vector diagram for the plane, square lattice shown in Fig. M.5.8.

terms of a two-dimensional wave vector, or reciprocal wavelength, or k-space, where again, for convenience, let it be assumed that the k_x and k_y axial directions are parallel to the edges of the square lattice, as in Fig. M.5.9.

For any line OP_1 in the wave vector diagram, a series of points such as $p_1, p_{11}, \ldots$ represents a series of electron states with increasing values of k. For any one such direction as OP_1, the relation between E and k will be given by a curve of the same general form, as in the one-dimensional case shown in Fig. M.5.6., so that as k increases there are energy gaps in the $E - k$ curve, corresponding to the forbidden energy ranges. For other directions, OP_2, $OP_3, \ldots$, in the wave vector diagram, energy gaps will occur at some different intervals, $p_2, p_{22}, \ldots, p_3, p_{33}, \ldots$, corresponding to the different period of the potential in these directions and the dissimilar spacing of the metal cations along these directions of the plane lattice. If, as in Fig. M.5.8., the lines $aa_1, bb_1, cc_1, \ldots$, represent a series of rows of cations in the two-dimensional metal crystal, and AA_1 is a line parallel to OP_1 in the wave vector diagram shown in Fig. M.5.9., then for an electron described by a wave number on OP_1, corresponding to the points $p_1, p_{11}, p_{111}, \ldots$, Bragg reflexion will take place and the electron will not be able to travel through the lattice, since for these particular wave vector values, the relationship $k = n\pi/a_1$ is satisfied,

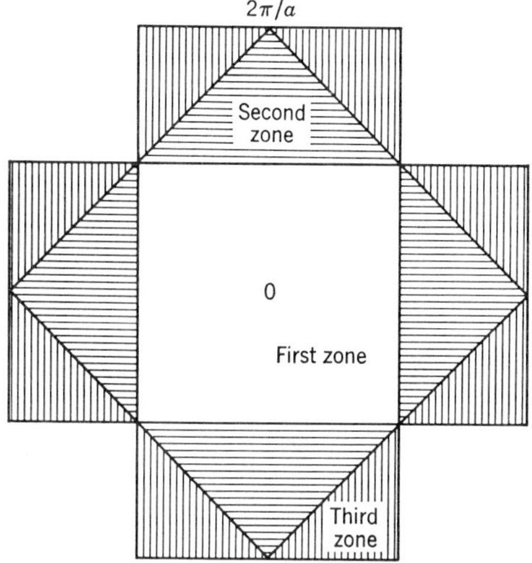

FIG. M.5.10. First three Brillouin zones for
the plane square lattice shown in Fig. M.5.8.

which, as has already been shown, is equivalent to the Bragg condition
$n\lambda = 2d \sin \theta$. In this particular example, d would represent the spacing
between the rows aa_1, bb_1, ..., and the electronic wavelength λ is given by
$2\pi/k$. Similarly, for the points p_2, p_{22}, ..., p_3, p_{33}, ..., where the Bragg condition is satisfied for these particular directions of k. Since the Bragg equation
involves the angle, θ, between the direction of the electron waves and the
lattice lines, therefore for a series of directions in k-space, OP_1, OP_2, OP_3, ...,
the points corresponding to p_1 in Fig. M.5.9. will occur at different distances
from the origin depending on the direction of OP_1, OP_2, ..., at points
designated p_2, p_3, ... In k-space, these points, p_1, p_2, ..., corresponding to
the different directions of the wave vectors for which the Bragg condition is
first satisfied, lie on straight lines in the two-dimensional model, enclosing
zones of equal area. The polygons which result by joining the points p_1, p_2,
p_3, ..., p_{11}, p_{22}, p_{33}, ..., p_{111}, p_{222}, p_{333}, ..., define the first, second, and third
Brillouin zones, respectively. For the hypothetical, plane, square lattice, the
first three such zones are shown in Fig. M.5.10. The electron energy suffers a
discontinuity across the edges of these zones. Such zones were first discussed
in detail for cubic structures by Brillouin.[1] Each metal, with its own characteristic crystal structure and lattice constants, gives rise to its own characteristic Brillouin zones.

In the free electron theory the occupied electron states form a sphere in

---

[1] Brillouin, L., *Die Quantenstatistik*, 1931, Berlin.

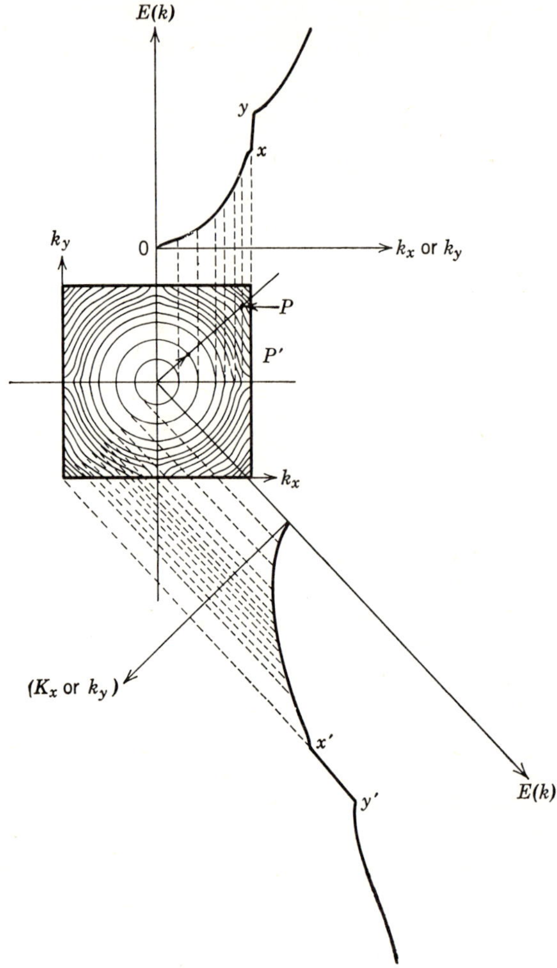

Fig. M.5.11. Derivation of the energy contours inside a zone from the corresponding $E(k)$ curves.

momentum space, or in two dimensions the boundary surface of occupied states would be a circle. Similarly, in wave vector space electron states having momenta, or k-values proportional to $\sqrt{1}, \sqrt{2}, \sqrt{3}, \ldots$, in some arbitrary units, would have energies proportional to $1, 2, 3, \ldots$, since $E = h^2k^2/8\pi^2m$, and these constant energy surfaces or energy contours, in the free electron case, are always circular (in two dimensions). In a periodic potential field, the concept of energy contours in k-space may still be applied and for low energy electrons, for which the $E - k$ curve closely resembles the corresponding free electron curve, the energy contours would again be very nearly circular. The actual relationship between the $E - k$ curve and the energy contour shape for the hypothetical, two-dimensional, square lattice referred to above is shown in Fig. M.5.11., illustrating the derivation of the energy contours inside the first zone, according to Raynor.[1] The central square represents the first Brillouin zone and, as the slope of the $E - k$ curve increases, the energy contours become more closely spaced, as k increases, and then as the $E - k$ curve changes inflection from a positive value of d^2E/dk^2 at the lower end of the zone to a negative d^2E/dk^2 at the upper end of the band, the energy contours become more distorted from circles as the zone boundary is approached.

The foregoing description of the behavior of electrons in hypothetical one- and two-dimensional periodic potential fields, in terms of wave-number space, or reciprocal wavelength space, or k-space, in which generally the increase in wave vector of an electron is equivalent to a continuous increase in energy of the electron within the Brillouin zone, and in which there is an energy discontinuity at the zone boundary, is usually referred to as the extended wave-number space approach. In the case of real, three-dimensional metal crystals, however, it is necessary to consider the motion of electrons through, and their scattering by, planes of atoms rather than rows, and to refer to zones which are polyhedral shapes rather than polygonal figures. Many such three-dimensional problems are considerably simplified by thinking in terms, not of the real crystal lattice itself, but in terms of an idealized lattice which is the reciprocal of the real lattice. This concept has been found extremely useful in the interpretation of the scattering of X-radiation by crystalline materials, and in fact in all situations involving the interaction of a wave motion with a crystal or other periodic structure; it was first introduced into X-ray crystallographic theory by Ewald,[2] based on earlier ideas of Bravais,[3] and considerably extended by Buerger.[4] The reciprocal lattice model is especially appropriate in metal theory, where the energies of the conduction electrons are so conveniently expressed in terms of their k-values, which are the reciprocals of the wavelengths associated with the electronic motion.

M.5.6. Crystal Lattices and Their Reciprocals

All crystalline materials may be regarded, ideally, as substances composed of atoms, ions, molecules, complex ions, or some combination of these arranged in three-dimensional space at the sites of a point space lattice. A

[1] Raynor, G. V., *Inst. Metals Monograph and Rep. Series No. 4*, 1949.
[2] Ewald, P. P., *Z. Krist. (A)*, 1921, **56**, 148.
[3] Bravais, A., No. 90 of Ostwald's *Klassiker der exakten Wissenschaften*, Leipzig, 1897, 112.
[4] Buerger, M. J., *Z. Krist. (A)*, 1935, **91**, 276.

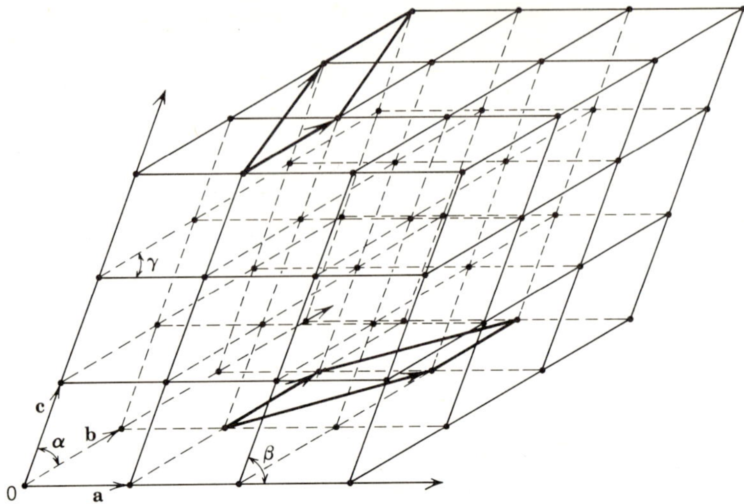

FIG. M.5.12. The general triclinic point space lattice (black points) with one possible line space lattice drawn through these points defining the triclinic unit cell with primitive translations **a, b, c**, and interaxial angles α, β, γ. Two other possible ways of selecting the base of the unit cell are shown.

point space lattice is characterized by three fundamental translation vectors, *a, b, c*, with the property that the arrangement of points appears the same in all respects when viewed from any point *P*, as when viewed from the point *P'*, within the lattice, where

$$P' = P + n_1\mathbf{a} + n_2\mathbf{b} + n_3\mathbf{c}, \qquad \text{(m.5.44.)}$$

and n_1, n_2, n_3, are arbitrary integers. The points in a point space lattice may be connected by a three-dimensional grid of lines, referred to as a line space lattice. Any line space lattice determines a unique point space lattice, but the points of a point space lattice may be joined in various ways to form an infinite number of different line space lattices. The space lattice of a real crystal is a representation of its periodic translation repetition; any point in the space lattice may be selected as the origin of the lattice. The line space lattice can be described by specifying the three non-coplanar translation vectors **a, b, c**, which by repeated action generate the origin point at every point of the space lattice. These are generally referred to as the three unit translations. For any given point space pattern, they may be chosen in innumerable ways, each choice defining a different line lattice, as illustrated in Fig. M.5.12.

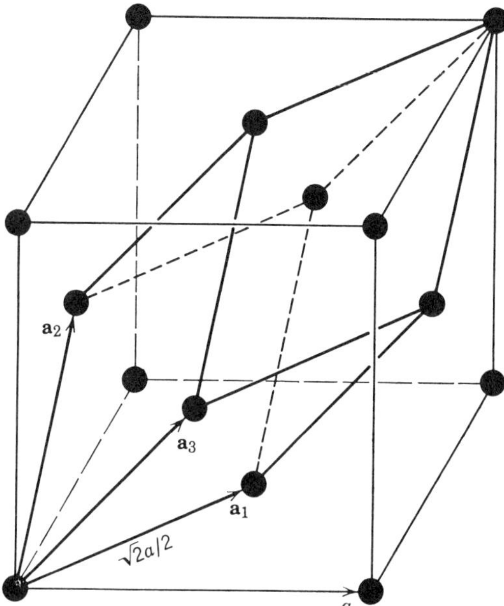

FIG. M.5.13. The primitive Bravais cell and the quadruply primitive cubic cell of the face-centered cubic lattice.

The small parallelepiped constructed with the three unit translations is known as the unit cell of the structure concerned and, though imaginary, has a definite shape and volume. A unit cell does not necessarily have a definite origin or position in space, although it is convenient to make an arbitrary choice for a particular purpose. It does, however, have a definite orientation, defined by the translation vectors. Unit cells are said to be primitive, doubly primitive, ..., if they contain 1, 2, ..., translation equivalent pieces of pattern. The primitive is not necessarily the most convenient way of describing the periodic repetition. For example, the face-centered cubic cell is generally described in terms of the cubic unit cell shown in Fig. M.5.13., which contains four pattern units, whereas the primitive unit cell is rhombohedral with interplanar angles of 60°. The cubic unit cell has the advantages of orthogonality and if the edges of the cell are chosen as the unit translations, the various expressions for interplanar spacings and interatomic distances are fairly simple. The rhombohedral cell is not unique, of course, since generally for oblique cells the base of the cell can be chosen as any one of a large number of ways, as is shown in the plane rectangular pattern in Fig. M.5.14. It is usually most convenient to choose a primitive cell with edges as nearly equal as possible.

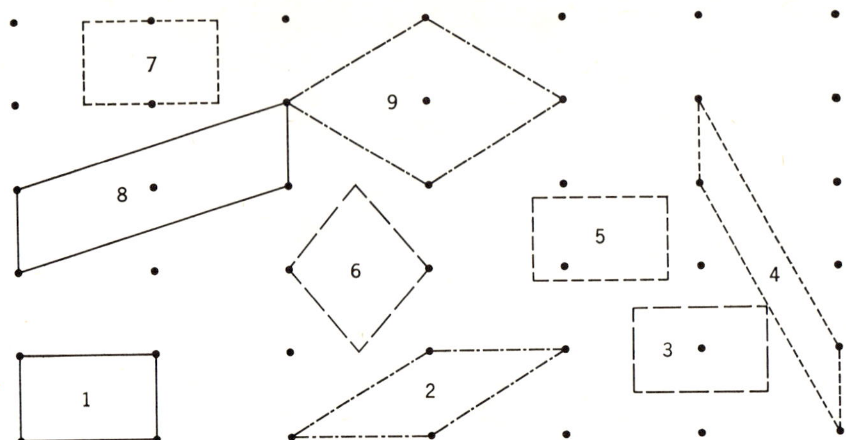

FIG. M.5.14. Various ways of choosing the unit cell in a plane rectangular lattice: cells 1–7 are primitive (contain one lattice point per cell) and cells 8–9 are doubly primitive.

The direction of any plane drawn through a point space lattice is commonly shown, in crystallography, by what are referred to as its Miller indices. In lattice geometry, a plane, unless otherwise specified, means a rational plane, or one which contains lattice points, not necessarily all colinear. In Fig. M.5.15, the rational plane obviously contains lattice points, one A units along the a-axis, a second B units along the b-axis, and a third C units along the c-axis, where a, b, c, are the three fundamental unit cell translations. It is the common practice in crystallography to refer to the directions of the three unit translations as the a, b, c, crystallographic axes (not necessarily orthogonal). They are generally parallel to actual or possible edges of the crystal concerned. For the cubic system, the orthogonal Cartesian coordinate axes are coincident with the crystallographic axes. The equation of the plane in Fig. M.5.15. in intercept form is

$$x/A + y/B + z/C = 1,$$

and every combination of three lattice points, one on each axis, defines a rational plane, but each such plane is repeated by the action of the lattice translations, so that one intercept equation implies a whole set of parallel planes. If A, B, C, do not contain any common multiple, then the equation of the first plane from the origin is

$$BCx + ACy + ABz = 1.$$

However, if A and B have the highest common factor l, B and C the highest common factor m, and A and C the highest common factor n, then the number

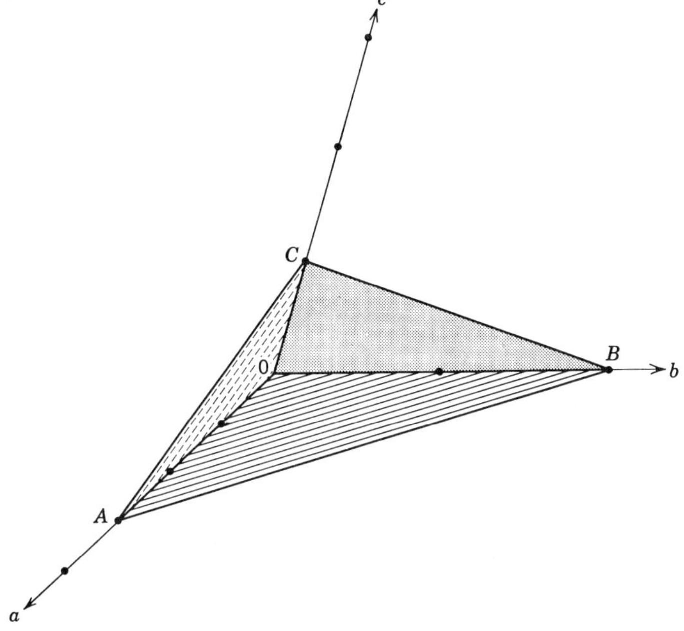

FIG. M.5.15. The rational plane *ABC* intercepts the *a*-axis at 3 translation units of *a*, the *b*-axis at 2 translation units of *b*, and the *c*-axis at 1 translation unit of *c*. Since *A*, *B*, *C* do not contain any common factor, the rational plane is the *ABC*th (sixth) plane from the origin.

of planes from the origin to the first rational plane is *ABC*/*lmn*,[1] and the equation of the first plane is

$$(BC/lmn)x + (AC/lmn)y + (AB/lmn)z = 1.$$

If

$$[BC/lmn = h, \ AC/lmn = k, \ AB/lmn = 1],$$

the equation of the first plane from the origin is

$$hx + ky + lz = 1. \qquad\qquad (m.5.45.)$$

The coefficients *h, k, l* are called the Miller indices of the plane. Thus, the Miller symbol for the rational plane depicted in Fig. M.5.15. is 236. If this plane corresponds to an actual crystal face, it is represented as (236), enclosing the Miller symbol in parentheses. The unbracketed symbol, 236, would denote

---

[1] Buerger, M. J., *X-Ray Crystallography*, 1949, John Wiley & Sons, Inc., N.Y.

the whole family of parallel planes, not all of which are rational. If the symbol is enclosed in braces, {236}, it denotes all the faces of the same form generated from the face (236) by the operation of the various symmetry elements possessed by the crystal containing this plane as a natural face. In the event that the various faces {236} have mutually parallel intersections, this common direction is referred to as a zone axis and represented as [hkl]. Alternatively, in classical crystallography, the Miller indices of a plane are simply defined as the reciprocals of the number of units of the three unit translations which the plane in question makes with the three crystallographic axes, a, b, c. In Fig. M.5.15., where the rational plane intercepts the a-axis at 3 units of a, the b-axis at 2 units of b, and the c-axis at 1 unit of c, the Miller indices would be written

$$1/3,\ 1/2,\ 1/1,$$

which on reduction to the smallest integers gives 236, as before. In a real crystal, the naturally occurring faces necessarily correspond to rational planes; in fact the naturally observed faces on crystals generally correspond to planes containing the maximum amount of material, i.e., the highest density of lattice points possible, consistent with the symmetry displayed by the crystal, its composition, and the conditions under which it was formed. Before the advent of X-ray crystallography, this generalization was referred to as the Law of Rational Indices by classical crystallographers, viz., that the Miller indices of any face of a crystal are always rational, either simple numbers or zero. The practice of distinguishing between rational planes through a point space lattice and the large number of conceivable non-rational planes arose originally in X-ray crystallography, where obviously X-ray scattering can only take place from electron-containing material in a crystal, such as atoms, ions, molecules, or complex ions, and the only Bragg reflexions which may be observed experimentally are those from rational planes, although it is necessary, nevertheless, to distinguish between the different orders of reflexion. In X-ray crystallography the second reflexion from the (111) plane, for example, is referred to as reflexion from the (222) plane. This is essentially a matter of geometrical convenience, in interpreting diffraction effects in terms of the imaginary construct of the reciprocal lattice.

An important consequence of the lattice structure of crystalline materials is the limitation which it imposes on the number and type of axes of rotational symmetry which a crystal may possess. The fundamental property of translational periodicity, defined by (m.5.44.), which is the prime characteristic of all solid, crystalline materials, means that a point space lattice can only have one-fold, two-fold, three-fold, four-fold, six-fold, or some combination of these, axes of rotational symmetry. Suppose that in Fig. M.5.16. AA' is a lattice row, with period t, and there is an n-fold axis of rotational symmetry

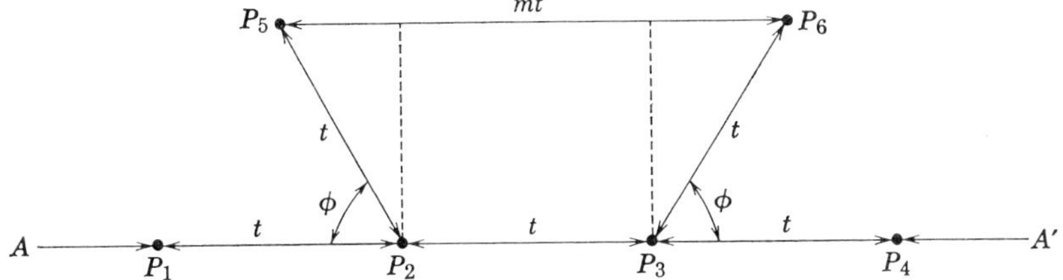

FIG. M.5.16. The limitations imposed by the periodicity of a space lattice on the number and type of axes of rotational symmetry which may be possessed by the lattice.

operative at each lattice point. An n-fold rotation axis is one which, after n rotations of $2\pi/n° = \phi°$, leads to an arrangement which is indistinguishable from the arrangement which existed initially. Since n rotations by an amount ϕ lead to superposition, it does not matter whether the rotations are carried out in a clockwise or counterclockwise manner. Therefore, if it is imagined that at two adjacent lattice points in Fig. M.5.16. rotations of $+\phi°$ and $-\phi°$ are performed, the lattice point at P_1 would be reproduced at P_5 and the lattice point at P_4 would be reproduced at P_6. Also, as a result of these rotations, the points P_5 and P_6 would necessarily be equidistant from the original lattice row, so that the line P_5P_6 would be parallel to the translation t. If the row AA' is part of one lattice row of a plane lattice, which possesses an n-fold axis of rotational symmetry, then P_5 and P_6 must also be lattice points in the plane away, and the line P_5P_6 must have a length equal to some multiple of the translation t. If this is not so, then the line joining P_5 and P_6 cannot be a translation of the plane lattice and the array would not be periodic. Suppose, then, that the length P_5P_6 is of length mt, where m is 0, ± 1, ± 2, $\pm 3, \ldots$, and the $\pm$ values of m depend on whether the translation is measured to the right or left. Then

$$mt = t + 2t \cos \phi$$

or

$$\cos \phi = (m - 1)/2. \qquad \text{(m.5.46.)}$$

Since $(m - 1)$ is an integer if m is an integer and the cosine of an angle is restricted to values between $+1$ and -1, ϕ and therefore $n = 2\pi/\phi$ can only have a limited number of values. The various possibilities consistent with (m.5.46.) are shown in Table M.5.1. No such limitations on axes of rotational symmetry apply to geometrical objects in general, which may exhibit axes of symmetry of all degrees up to the axes of n-fold symmetry displayed by a sphere, where n may be infinite. It is interesting to observe that of the only five possible regular polyhedra, the tetrahedron, the octahedron, the cube, the

TABLE M.5.1

m	$\cos \phi$	ϕ (in deg)	n
$+3$	1	0 or 360	1
-1	-1	180	2
0	$-1/2$	120	3
$+1$	0	90	4
$+2$	$+1/2$	60	6

pentagonal dodecahedron, and the icosahedron, only the pentagonal dodecahedron displays a symmetry axis not possible for a space lattice (a five-fold axis) and that this polyhedron has been recognized as an intermediate polyhedral shape in the structure of certain intermetallic compounds.[1]

Another geometrical consequence of the three-dimensional periodicity of lattice arrays was established by the work of the early crystallographers, Wollaston (1812), Frankenheim (1842), and Bravais (1848). From observation of the external symmetry characteristics displayed by natural crystals, it is found possible to assign any crystal to one or another of seven crystal systems. Crystal symmetry is described in terms of: One or more planes of reflexion symmetry, symbolized m; rotation axes, symbolized in terms of their degree 1, 2, 3, 4, 6; axes of rotary inversion or inversion axes, symbolized also in terms of their degree $\bar{1}, \bar{2}, \bar{3}, \bar{4}, \bar{6}$. An inversion axis is a compound symmetry element; it involves the operation of a rotation followed by an inversion through the center of the crystal or object concerned. The seven crystal systems, described in terms of the interrelationships between the edge lengths and interaxial angles of their unit cells, are indicated in Table M.5.2. α, β, γ, are the angles between b and c, c and a, and a and b, respectively. It was first established unequivocally by Bravais (1848) that there are only 14 point space lattices possible, consistent with the above unit

TABLE M.5.2

Triclinic	a, b, c, α, β, γ
Monoclinic	a, b, c, β ($\alpha = \gamma = 90°$)
Orthorhombic	a, b, c, α ($= \beta = \gamma = 90°$)
Rhombohedral	a ($= b = c$), α ($= \beta = \gamma$)
Hexagonal	a ($= b$), c, $\gamma = 120°$, $\alpha = \beta = 90°$
Tetragonal	a ($= b$), c, α ($= \beta = \gamma = 90°$)
Cubic	a ($= b = c$), α ($= \beta = \gamma = 90°$)

---

[1] Bergman, B. G., Waugh, J. L. T., and Pauling, Linus, *Acta. Cryst.*, 1957, **10**, 254.

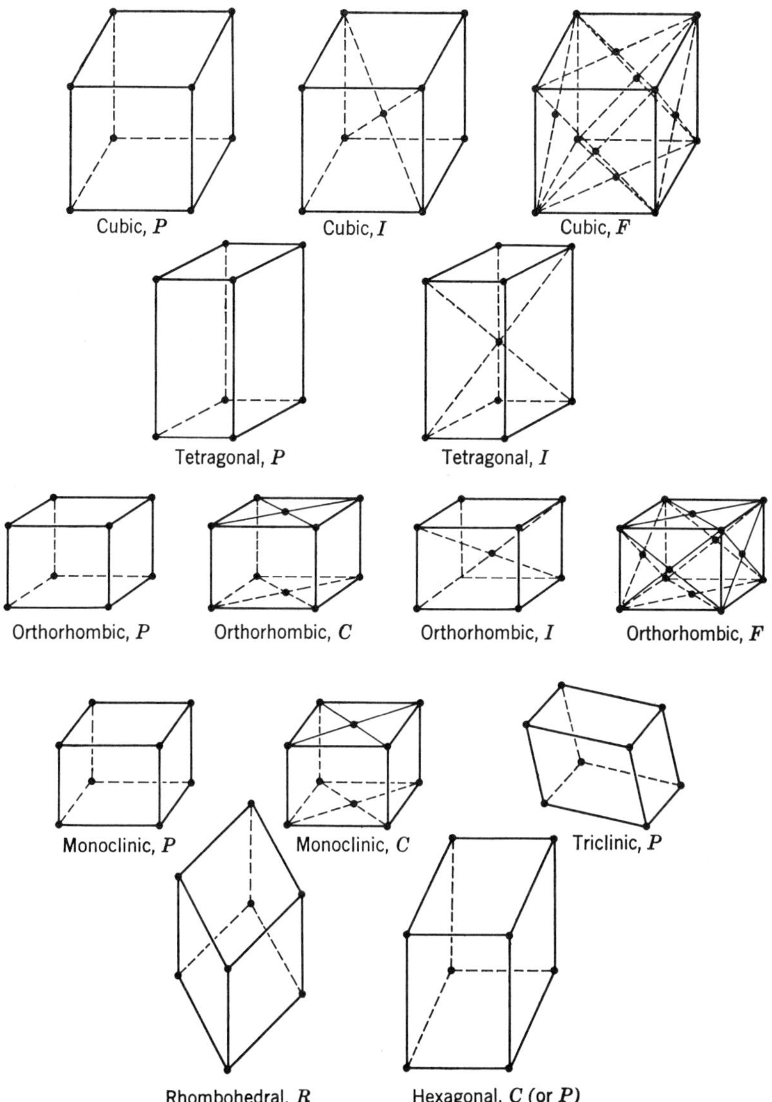

FIG. M.5.17. The 14 Bravais Space Lattices.

cell descriptions. These 14 space lattices are consequently often called Bravais lattices; they are illustrated in Fig. M.5.17. together with their customary identifying symbol. P denotes a primitive lattice with lattice points only at the vertices of the unit cell, I (Innenzentrierte) denotes a unit cell with an additional lattice point at its body center, and F denotes a unit cell with lattice points at the centers of each of its six faces in addition to those at its vertices.

For many purposes, in dealing with three-dimensional space lattices, it is more convenient to think in terms of the normals to planes than in terms of the planes themselves. In optical goniometry, for example, it is easier to measure the angles between the normals to a pair of plane crystal faces rather than the angle between the faces themselves. In X-ray diffraction, where coherent scattering by a crystal lattice occurs in such a manner that it is geometrically equivalent to reflexion by a set of parallel planes (Bragg's law), it is also much more convenient to visualize a collection of lines rather than a collection of planes of different slopes. The slope of a plane is just as adequately defined by the geometry of the plane itself or by the geometry of the normal to the plane, apart from the consideration that the normal has one dimension less than the plane. The reciprocal lattice is a geometrical concept, in which a two-dimensional plane is designated by a point located at the terminus of a vector, the length of which determines the interplanar spacing; it is thus a device which enables both the slopes and the interplanar spacings of the various planes which may be drawn through a space lattice to be conveniently represented by an array of points. The fundamental principle of the reciprocal lattice can be arrived at either by simple geometrical reasoning or in a very compact manner by vector algebraic methods.

Since the geometrical properties of the reciprocal lattice can be completely represented in a plane, it is convenient to examine the properties of the two-dimensional case first. Let $OXYZ$ represent, in Fig. M.5.18, a cell of a plane parallelogram lattice, OX defining the a crystallographic direction and OZ defining the c crystallographic direction, β being the interaxial angle (acute), and AB the trace of the first plane from the origin with indices hl. OP is the normal from O to the plane AB, so if AB is the first plane of the parallel set hl, the length of OP corresponds to the interplanar spacing of this set of planes. The absolute lengths of OA and OB are then a/h and c/l. This is perhaps most easily appreciated if equation (m.5.45.) is rewritten in the form

$$\frac{x}{1/h} + \frac{y}{1/k} + \frac{z}{1/l} = 1,$$

which is recognizable as the equation of a plane in intercept form where the intercepts are $1/h$, i/k, $1/l$. In other words, a series of planes with indices hkl intercepts the lattice axes such that the a-axis is divided into h parts, the b-axis into k parts, and the c-axis into l parts. The length of the intercepted line segment AB is given by the conventional relationship between the sides and angles of a scalene triangle as

$$AB = \sqrt{[(a/h)^2 + (c/l)^2 - 2(ac/hl)\cos\beta]}. \qquad \text{(m.5.47.)}$$

If BM is drawn normal to OX, then in the similar triangles OAP and ABM,

$$OP/OA = BM/BA.$$

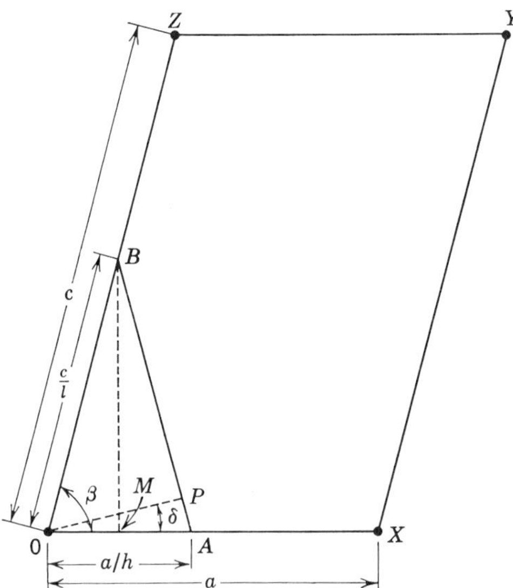

FIG. M.5.18. The unit cell of a plane parallelogram lattice.

Making the appropriate substitutions,

$$\frac{d_{(hl)}}{a/h} = \frac{(c/l)\sin\beta}{\sqrt{[(a/h)^2 + (c/l)^2 - 2(ac/hl)\cos\beta]}},$$

or

$$d_{(hl)} = \frac{(ac/hl)\sin\beta}{\sqrt{[(a/h)^2 + (c/l)^2 - 2(ac/hl)\cos\beta]}}.$$

If the reciprocal of the interplanar spacing of the planes, hl, is denoted by σ_{hl}, then

$$\sigma_{hl} = \frac{1}{d_{(hl)}} = \frac{hl}{ac\sin\beta}\sqrt{[(a/h)^2 + (c/l)^2 - 2(ac/hl)\cos\beta]}$$

$$= \sqrt{[(l/c\sin\beta)^2 + (h/a\sin\beta)^2 - 2(h/a\sin\beta)(l/c\sin\beta)\cos\beta]}. \qquad \text{(m.5.48.)}$$

Equation (m.5.48.) is of the same form as (m.5.47.), and therefore may be interpreted similarly, as shown graphically in Fig. M.5.19. The method of constructing the reciprocal lattice is, starting from the origin of the direct lattice, to draw the line OP normal to the plane hl in the direct lattice and of length equal to the reciprocal of the spacing of the planes hl. This length is given by (m.5.48.). If this operation is repeated for all possible values of h and l, the collection of points which result is referred to as the reciprocal lattice

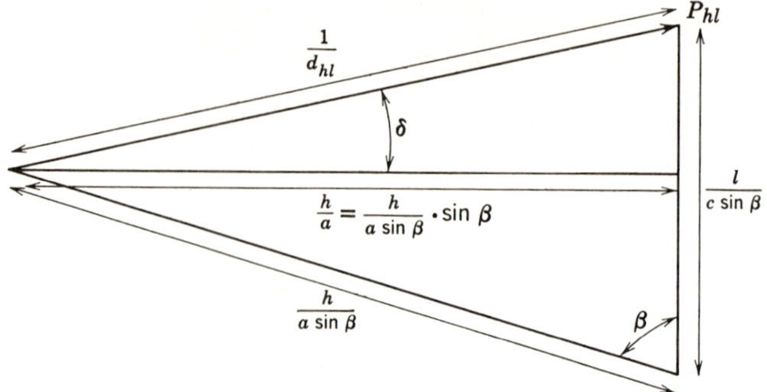

FIG. M.5.19. Reciprocal lattice vectors for plane parallelogram lattice.

of the direct parallelogram lattice. From Fig. M.5.18., the vector σ_{hl} is directed at an angle from the horizontal of

$$\delta = \cos^{-1} \frac{d}{a/h} = \frac{h/a}{1/d}. \qquad \text{(m.5.49.)}$$

So to plot the reciprocal lattice point corresponding to the direct lattice plane hl, it is necessary to lay off the distance $h/a \sin \beta$ along an axial line normal to the direct lattice lines $h0$, and then lay off a distance from this point of $l/c \sin \beta$ parallel with the vertical axis or normal to the direct lattice lines $0l$. These distances are obviously multiples of $1/a \sin \beta$ and $1/c \sin \beta$, respectively, with coefficients h and l. $1/a \sin \beta$ and $1/c \sin \beta$ are, of course, constants for the lattice and represent the lengths of the unit vectors in reciprocal space. If $h = 1$ and $l = 0$, then equations (m.5.48.) and (m.5.49.) become, respectively, $1/a \sin \beta$ and

$$\delta = \cos^{-1} \frac{(h/a)}{(1/d)} = \cos^{-1} \frac{(h/a)}{\sigma} = \cos^{-1} \frac{h/a}{1/a \sin \beta} = \cos^{-1}(\sin \beta) = 90° - \beta.$$

In a similar manner, the length of the other reciprocal lattice unit vector may be obtained by substituting $h = 0$ and $l = 1$ in (m.5.48.) and (m.5.49.), when σ_{01} becomes $1/c \sin \beta$ and δ is given by $\cos^{-1} 0 = 90°$. Thus, having established the positions of the reciprocal points corresponding to the (10) and (01) planes of the direct lattice, thereby defining the unit translations in the reciprocal lattice, it is a simple matter to plot the whole reciprocal lattice. This is shown in Fig. M.5.20. $O'X'$ and $O'Z'$ represent the reciprocal lattice unit translations a^* and c^*, respectively. The points X' and Z' located in reciprocal space correspond to the direct lattice planes (10) and (01). The reciprocal lattice points equivalent to (20), (30), ..., (02), (03), ..., planes of the direct lattice are located 2, 3, ..., units of length OX' along the a^* recip-

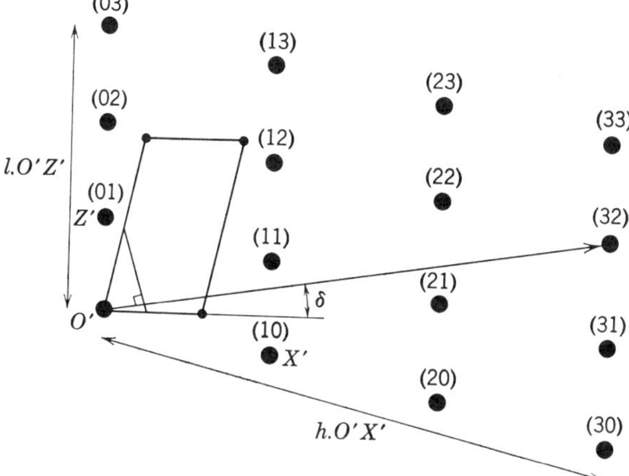

FIG. M.5.20. The reciprocal lattice of the plane parallelogram lattice shown in Fig. M.5.18. The unit cell of the direct lattice is shown in left-hand corner.

rocal lattice axis, and 2, 3, ..., units of length OZ' along the c^* reciprocal lattice axis, respectively. Any point in reciprocal space $(hl)^*$, corresponding to the (hl) plane in the direct lattice, is located h units of length OX' along the a^* axis and l units of length OZ' along the c^* axis, the two lengths being laid off sequentially. If all values of h and l are taken, the series of points arrived at fall at the intersections of a parallelogram, plane line lattice. Accordingly, the collection of points which results from this construction is a true lattice array.

The axial ratio of the reciprocal lattice is

$$\frac{OZ'}{OX'} = \frac{1/c \sin \beta}{1/a \sin \beta} = \frac{a}{c},$$

so that the lattice which is reciprocal to a plane lattice is a similar lattice with an axial ratio reciprocal to that of the plane lattice and with an interaxial angle which is the supplement of that of the direct lattice. Alternatively, it may be said that the lattice reciprocal to a plane lattice is a similar plane lattice with an orientation rotated 90° from the direct one. This relationship is shown in Fig. M.5.20. For square or rectangular lattices the equations (m.5.48.–49.) simplify further by the substitution of $\beta = 90°$ and in the former case there is the further simplification that $a = c$. It is customary to designate the reciprocal lattice constants by starred expressions; thus the reciprocal lattice cell elements are called a^*, b^*, c^*, α^*, β^*, γ^*, in the general case.

To construct a lattice reciprocal to any given three-dimensional space lattice, the construction given above is generalized. For any plane (hkl) in the

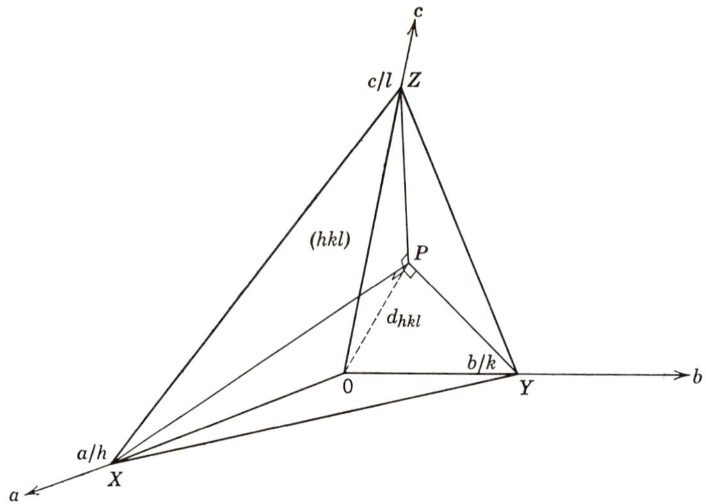

FIG. M.5.21. The interplanar spacing of the series of parallel planes, *hkl*, of which the plane indicated is the first one from the origin, O.

direct lattice, a normal is erected of length equivalent to the interplanar spacing d_{hkl}, and the collection of points thus formed constitutes the reciprocal point space lattice. In X-ray crystallography, it is usually convenient to select

$$\sigma_{hkl} = \lambda \cdot \frac{1}{d_{hkl}},$$

making the proportionality constant equal to the wavelength of the X-radiation used rather than simply unity. The interplanar spacing for any set of planes *hkl*, for the most general case where the unit cell edge lengths *a*, *b*, *c*, are all different and the interaxial angles α, β, γ, are all angles other than 90° may be calculated. For orthogonal systems, for example, as is shown in Fig. M.5.21., if *XYZ* is the first plane from the origin with indices *hkl*, then $OP = d_{hkl}$, where *OP* is normal to *XYZ*, and

$$d_{hkl} = \frac{a}{h} \cos POX = \frac{b}{k} \cos POY = \frac{c}{l} \cos POZ. \qquad \text{(m.5.50.)}$$

But from the law of cosines,

$$\cos^2 POX + \cos^2 POY + \cos^2 POZ = 1. \qquad \text{(m.5.51.)}$$

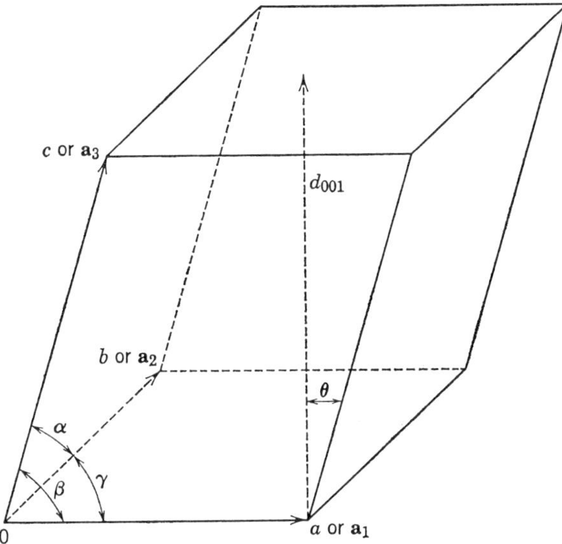

FIG. M.5.22. General triclinic unit cell.

Substituting the values of the cosines of POX, POY, POZ, obtained from (m.5.50.) in (m.5.51.),

$$(hd_{hkl}/a)^2 + (kd_{hkl}/b)^2 + (ld_{hkl}/c)^2 = 1.$$

Then

$$d_{hkl} = \frac{1}{\sqrt{[(h/a)^2 + (k/b)^2 + (l/c)^2]}}. \qquad \text{(m.5.52.)}$$

For non-orthogonal systems, the corresponding expression for the interplanar spacing is more involved.

The crystal lattice and the reciprocal lattice are mutually reciprocal to one another. Just as the vector r^*_{hkl} from the origin of the reciprocal lattice is normal to the (hkl) planes of the direct crystal lattice, so the vector r_{hkl} from the origin of the direct lattice is normal to the corresponding plane in the reciprocal lattice, and the length of the vector r_{hkl} in the real lattice is equal to the reciprocal of the interplanar spacing in the reciprocal lattice. The two statements

$$\sigma_{hkl} = \frac{1}{d_{hkl}} \qquad \text{and} \qquad \sigma_{hkl} = h\mathbf{a}^* + k\mathbf{b}^* + l\mathbf{c}^*$$

are equivalent ways of defining the point hkl in reciprocal space. Since most generally, for the triclinic unit cell shown in Fig. M.5.22. the volume of this cell,

$$V = \text{base} \times \text{altitude} = ab \sin \gamma \times d_{001},$$

therefore

$$d_{001} = \frac{V}{ab \sin \gamma}.$$

But from the foregoing discussion it is evident that d_{001} is related to σ_{001} by the fact that

$$\sigma_{001} = c^* = \frac{1}{d_{001}}.$$

Thus,

$$c^* = \frac{ab \sin \gamma}{V}.$$

Similarly, it may be shown that

$$a^* = \frac{bc \sin \alpha}{V} \quad \text{and} \quad b^* = \frac{ac \sin \beta}{V}.$$

In the reciprocal lattice, the length of the a^* vector is equal to the reciprocal of the a interplanar spacing of the direct unit cell. The length of the vector $2a^*$ in the reciprocal lattice corresponds to the fictitious interplanar spacing of $a/2$, and in the complete reciprocal lattice interplanar spacings of $a/3, a/4, \ldots,$ will also be indicated. This is in contrast with the ideas of classical crystallography, where, for example, the planes (111), (222), (333), would all be regarded as being identical and represented by simply (111). But in lattice crystallography, the plane (222) would indicate a plane of identical slope to plane (111) but with 1/2 of the spacing, or located at one-half of the distance of the (111) plane from the origin. This is particularly advantageous in X-ray crystallography, where the reciprocal lattice point corresponding to a plane with interspacing d_{hkl} is used to indicate the conditions for the nth order Bragg reflexion from the actual direct lattice planes of spacing d_{hkl}.

With reference to two of the crystal structural arrangements which are found to be of predominant importance in metals, it is quite simple to show that the body-centered cubic and the face-centered cubic structural arrangements are mutually reciprocal to one another. Fig. M.5.23. shows one unit cell of a face-centered cubic lattice, of edge length a, and two of the set of 111 planes shaded; also shown in the same figure is one unit cell of a body-centered cubic lattice of edge length $2/a$. The interplanar spacing of the 111 planes of the face-centered cubic lattice is $\sqrt{3}\,a/3 = 1/(\sqrt{3}/a)$, and the interplanar spacing of the 200 planes is $a/2 = 1/(2/a)$. For the body-centered cubic lattice of edge length $2/a$, the length of the vector from a corner to the body center is $(\sqrt{3}/2) \times 2/a = \sqrt{3}/a$, which is the reciprocal of the interplanar spacing of the 111 planes in the face-centered lattice. Also the edge length of

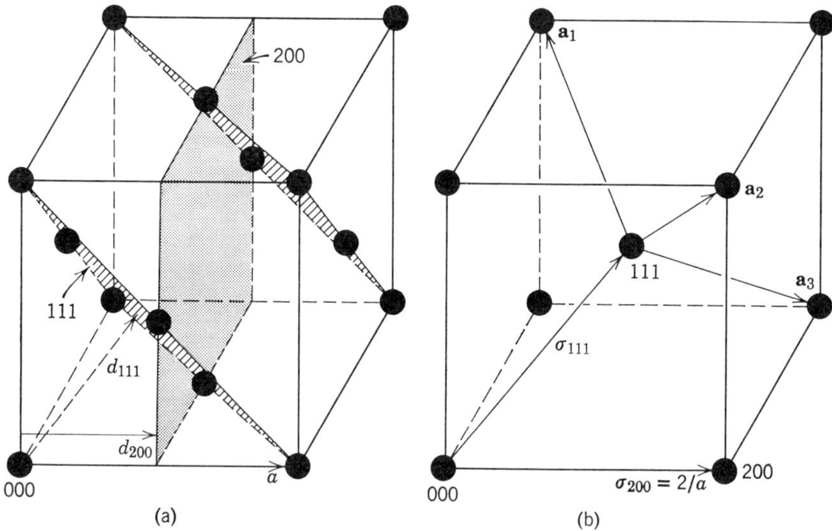

FIG. M.5.23. The mutually reciprocal relationship between the face-centered cubic and body-centered cubic structures. The interplanar spacing vectors for the 111 and 200 planes of the face-centered cubic lattice on the left are indicated; the reciprocals of these are plotted on the right to produce the body-centered cubic lattice. Similarly for all other possible planes—the reciprocal lattice of a face-centered lattice is a body-centered cubic lattice and *vice versa*.

the body-centered cubic cell, $2/a$, is the reciprocal of the interplanar spacing of the 200 planes of the face-centered cell. In a similar manner, the reciprocal lattice point corresponding to any plane in the face-centered cubic lattice is found to lie at a lattice point in the body-centered cubic lattice. So the space lattice reciprocal to a face-centered cubic lattice of edge length a is a body-centered cubic lattice of edge length $2/a$. The converse of this statement is also true, that the lattice reciprocal to a body-centered cubic lattice is a face-centered cubic lattice.

M.5.7. Bloch Functions and Brillouin Zones in a Three-Dimensional Lattice

In this section, the simple geometrical arguments of M.5.6. will be generally rephrased in vector notation, which leads to considerably more compact expressions for the various relationships involved in the three-dimensional problem. In vector language, any Bravais lattice may be described by the collection of points

$$\mathbf{K}_d = n_1\mathbf{a}_1 + n_2\mathbf{a}_2 + n_3\mathbf{a}_3, \qquad \text{(m.5.53.)}$$

where $\mathbf{a}_1$, $\mathbf{a}_2$, $\mathbf{a}_3$, are the three fundamental lattice translational vectors and n_1, n_2, n_3, may adopt all integral values including zero. Then if $\mathbf{b}_1$, $\mathbf{b}_2$, $\mathbf{b}_3$, define the set of vectors normal to the coordinate planes of the direct lattice unit cell (cf. Fig. M.5.22.), the reciprocal lattice of (m.5.53.) may be described as

$$\mathbf{K}_r = m_1\mathbf{b}_1 + m_2\mathbf{b}_2 + m_3\mathbf{b}_3, \qquad \text{(m.5.54.)}$$

where m_1, m_2, m_3, may again take on all integral values including zero. These two expressions may be combined as

$$\mathbf{a}_i \cdot \mathbf{b}_j = \delta_{ij} \qquad (i, j = 1, 2, 3),$$

where δ_{ij} is the Kronecker delta symbol. It is particularly convenient, not only in metal theory where the scattering of electrons by the metal crystal is involved, but also in lattice dynamics where the scattering of thermal neutrons by a crystal is concerned, to define the reciprocal lattice vectors in terms of the fundamental translation vectors of the direct lattice unit cell by

$$\mathbf{a}_i \cdot \mathbf{b}_j = 2\pi \, \delta_{ij}. \qquad \text{(m.5.55.)}$$

By so doing the proportionality constant relating the reciprocal lattice vectors, σ_{hkl}, and the direct lattice spacing vectors, d_{hkl}, is chosen as 2π rather than unity, as was assumed in the previous section, thus making it identical to the proportionality constant relating the wave vector k in the plane wave function e^{ikx} and the wavelength λ associated with the wave motion so described. Substituting the plane wave function e^{ikx} in the one-dimensional case, in the Hamiltonian form of the wave equation with $V(x) = 0$,

$$\frac{h^2 \, d^2}{8\pi^2 m i^2 \, dx^2} (e^{ikx}) = \frac{h^2 i^2 k^2}{8\pi^2 m i^2} \cdot e^{ikx} = E e^{ikx},$$

or

$$E = \frac{h^2 k^2}{8\pi^2 m} \qquad \text{or} \qquad p_x = \frac{hk}{2\pi} \qquad \text{and} \qquad k = \frac{2\pi p_x}{h} = \frac{2\pi}{\lambda},$$

where $\lambda = h/p_x$.

If a crystal lattice has dimensions $G_1 a_1$, $G_2 a_2$, $G_3 a_3$, along $\mathbf{a}_1$, $\mathbf{a}_2$, $\mathbf{a}_3$, respectively, where G_1, G_2, G_3, are integers, then for a simple Bravais lattice containing one pattern unit per unit cell, the number of structural units in the crystal would be $G_1 G_2 G_3$, while for a metal crystal where the pattern or structural units located at the lattice sites are metallic cations, the number of metal atoms in the crystal above would be $G_1 G_2 G_3$. The potential energy, $V(r)$, of an electron moving in this lattice, if the surface effects are neglected, will have the same periodicity as that of the lattice, so that

$$V(\mathbf{r} + \mathbf{R}_n) = V(\mathbf{r}), \qquad \text{(m.5.56.)}$$

where $\mathbf{R}_n$ is a lattice vector, as defined by (m.5.53.). The problem of the possible energies of an electron moving in such a space lattice is then to solve the three-dimensional Schrödinger equation

$$\nabla^2\psi + \frac{8\pi^2 m}{h^2}\,[E - V(\mathbf{r})]\psi = 0. \tag{m.5.57.}$$

If, as in the one-dimensional case, it is assumed that cyclic periodic boundary conditions apply, then

$$\psi(\mathbf{r}) = \psi(\mathbf{r} + G_1\mathbf{a}_1) = \psi(\mathbf{r} + G_2\mathbf{a}_2) = \psi(\mathbf{r} + G_3\mathbf{a}_3). \tag{m.5.58.}$$

The Bloch theorem in three dimensions may be expressed as

$$\psi_k(\mathbf{r} + \mathbf{R}_n) = e^{i\mathbf{k}\cdot\mathbf{R}_n}\psi_k(\mathbf{r}), \tag{m.5.59.}$$

where $\mathbf{k}$ is the wave vector consistent with the boundary conditions. The wave functions must satisfy the Bloch condition and therefore be of the form

$$\psi_k(\mathbf{r}) = e^{i\mathbf{k}\cdot\mathbf{r}}\phi_k(\mathbf{r}), \tag{m.5.60.}$$

where $\phi_k(\mathbf{r})$ has the periodicity of the lattice, or

$$\phi_k(\mathbf{r}) = \phi_k(\mathbf{r} + \mathbf{R}_n). \tag{m.5.61.}$$

From (m.5.58.) and (m.5.61.)

$$\psi_k(\mathbf{r}) = e^{i\mathbf{k}\cdot\mathbf{r}}\phi_k(\mathbf{r}) = \psi_k(\mathbf{r} + G_1\mathbf{a}_1) = e^{i\mathbf{k}\cdot\mathbf{r}}e^{i\mathbf{k}\cdot G_1\mathbf{a}_1}\phi_k(\mathbf{r}).$$

Therefore,

$$e^{iG_1\mathbf{k}\cdot\mathbf{a}_1} = 1.$$

Similarly,

$$e^{i\mathbf{k}G_2\cdot\mathbf{a}_2} = e^{i\mathbf{k}G_3\cdot\mathbf{a}_3} = 1.$$

Since generally if $e^{ix} = \cos x + i\sin x = 1$, then x must be some multiple of 2π; therefore,

$$G_1(\mathbf{k}\cdot\mathbf{a}_1) = 2\pi g_1, \qquad G_2(\mathbf{k}\cdot\mathbf{a}_2) = 2\pi g_2, \qquad G_3(\mathbf{k}\cdot\mathbf{a}_3) = 2\pi g_3, \tag{m.5.62.}$$

where g_1, g_2, g_3, are integers.

From the definition, (m.5.55.), the wave vector may be expressed as

$$\mathbf{k} = g_1\mathbf{b}_1/G_1 + g_2\mathbf{b}_2/G_2 + g_3\mathbf{b}_3/G_3, \tag{m.5.63.}$$

which, as $g_1, g_2, g_3, G_1, G_2, G_3$, are all integers, must be a reciprocal vector of the same form as (m.5.54.) or, in other words, the wave vector $\mathbf{k}$ lies or exists in the reciprocal space of the metal crystal. Also, from equations (m.5.53.–54.),

$$\mathbf{K}_d\cdot\mathbf{K}_r = 2\pi(m_1 n_1 + m_2 n_2 + m_3 n_3),$$

and therefore as m_1, m_2, m_3, n_1, n_2, n_3, are again all integers

$$e^{i\mathbf{K}_d \cdot \mathbf{K}_r} = 1.$$

From the cyclic periodic boundary conditions,

$$e^{i\mathbf{K}_r \cdot (\mathbf{r} + \mathbf{K}_d)} = e^{i\mathbf{K}_r \cdot \mathbf{r}},$$

and thus the function $e^{i\mathbf{K}_r \cdot \mathbf{r}}$ has the periodicity of the lattice. This means that exactly as in the one-dimensional case, the wave vector $\mathbf{k}$ is not uniquely determined by the expressions (m.5.60.) and (m.5.61.), for the wave function in (m.5.60.) can be equally well expressed as

$$\psi_k(\mathbf{r}) = e^{i\mathbf{k} \cdot \mathbf{r}}\phi_k(\mathbf{r}) = e^{i(\mathbf{k} + \mathbf{K}_r) \cdot \mathbf{r}}[e^{-i\mathbf{K}_r \cdot \mathbf{r}}\phi_k(\mathbf{r})]$$

without altering its Bloch characteristics, since the expression in the square brackets still has the periodicity of the lattice.

In view of this fact that the wave vector is undetermined to the extent of an added reciprocal lattice vector, $\mathbf{K}_r$, it is only necessary to consider the $\mathbf{k}$ vectors which exist within a single cell of the reciprocal lattice. This spatial region is usually called the reduced zone of $\mathbf{k}$ or the first Brillouin zone. Solutions of the Schrödinger equation will produce an infinite number of energy states for each $\mathbf{k}$ in the reduced zone, since within the zone the energy is a continuous function of $\mathbf{k}$ but due to the imposition of boundary conditions on the electron wave functions the energy is said to be a quasicontinuous function of $\mathbf{k}$. If each of these infinite sets of states, one for each value of $\mathbf{k}$ within the reduced zone, is described by a quantum label l, just as in the one-dimensional case, then the wave function $\Psi_{kl}(\mathbf{r}) = e^{i\mathbf{k} \cdot \mathbf{r}}\phi_{kl}(\mathbf{r})$ will correspond to the energy level $E_l(\mathbf{k})$.

The reduced zone of $\mathbf{k}$, just as in the case of the unit cell of a space lattice, may be chosen in any number of ways. But as pointed out in the previous section it is most convenient to select a volume unit which is as highly symmetrical as possible and which on infinite repetition reproduces the whole of the reciprocal lattice or the whole region of $\mathbf{k}$. The geometrical construction which meets these criteria is also the identical one which enables a volume unit of a Bravais lattice to be selected, which defines the minimum volume occupied by the structural unit at each lattice site such that the infinite repetition of these volume units occupies all the space of the lattice. If the origin of the reciprocal lattice is joined to each of the other points in the reciprocal lattice and planes are imagined to be erected such that they perpendicularly bisect these vectors, then the smallest volume unit surrounding the origin and bounded by these planes may be chosen as the reduced zone of $\mathbf{k}$. The volume unit which results in this manner will generally be a polyhedron whose symmetry is determined by the symmetry of the direct lattice. If this construction is carried out for a metal crystallizing with a simple

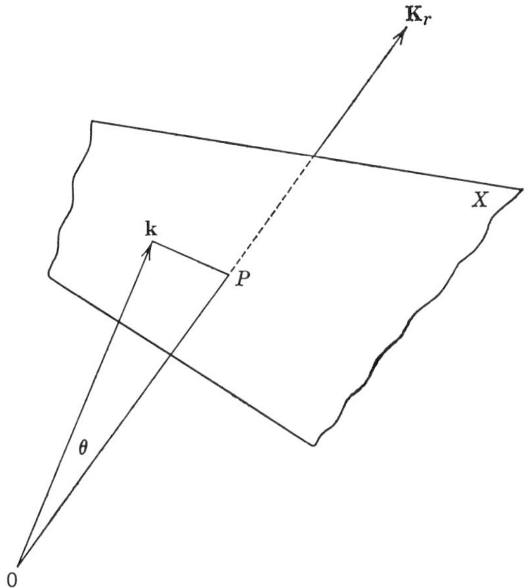

FIG. M.5.24. Geometrical construction of the reduced zone of **k**. The vector $\mathbf{K}_r$ is perpendicularly bisected by the plane X at the point P. **k** is any position of a point in X.

Bravais lattice, the polyhedron is referred to as the Wigner–Seitz[1] cell or the atomic polyhedron of the metal concerned. As will be explained in more detail later, the reduced zone of **k** is thus the Wigner–Seitz cell of the reciprocal lattice, although it is important to note that the Wigner–Seitz cells exist in real crystal space while the Brillouin zones of **k** exist in reciprocal space.

Figure M.5.24. illustrates the mode of defining the polyhedral faces of the reduced zone. X is a plane of presently undefined polygonal shape normal to the reciprocal lattice vector $\overline{OK_r}$, which it bisects at the point P, and **k** is the position vector of any point in the plane X. Then $OP = |K_r|/2 = k\cos\theta$, and from the definition of the scalar product of two vectors as the product of their magnitude multiplied by the cosine of the angle between them, it follows that $\mathbf{k}\cdot\mathbf{K}_r = k|K_r|\cos\theta$. Therefore,

$$\frac{|K_r|}{2} = \frac{\mathbf{k}\cdot\mathbf{K}_r}{|\mathbf{K}_r|} \qquad \text{or} \qquad 2\mathbf{k}\cdot\mathbf{K}_r = |K_r|^2. \qquad \text{(m.5.64.)}$$

---

[1] This method was first used by Wigner and Seitz to calculate the wave functions of the electrons in the alkali metals; E. Wigner and F. Seitz, *Phys. Rev.*, 1933, **43**, 804; *1934*, **46**, 509. See also "*The Modern Theory of Solids*," by F. Seitz, 1940, McGraw-Hill Book Co., New York.

This equation defines quite generally any face of the Brillouin zone. For each face x of each zone there is, by virtue of the construction involved, a parallel face whose equation is $2\mathbf{k}\cdot\mathbf{K}_r + |K_r|^2 = 0$, such that the perpendicular distance between such a pair of faces is K_r.

With reference to the general triclinic direct lattice unit cell shown in Fig. M.5.22., the volume of this cell may be variously expressed in crystallographic notation or in the vector notation of the present section as

$$\begin{aligned} V &= ab \sin \gamma \cdot d_{001} = \mathbf{a}_1 \cdot \mathbf{a}_2 \times \mathbf{a}_3 \\ &= \frac{a \times b}{\sigma_{001}} = \mathbf{a}_1 \cdot \mathbf{a}_2 \times \frac{2\pi}{\mathbf{b}_3} \\ &= \frac{a \times b}{c^*}. \end{aligned}$$

Similarly, V may be expressed in terms of the direct cell interplanar spacing vectors d_{100} and d_{010} as

$$V = \mathbf{a}_2 \cdot \mathbf{a}_3 \times \frac{2\pi}{\mathbf{b}_1} = \mathbf{a}_3 \cdot \mathbf{a}_1 \times \frac{2\pi}{\mathbf{b}_2}. \tag{m.5.65.}$$

Therefore for a simple cubic lattice, where a is the distance between adjacent structural units in the lattice and for which $\mathbf{a}_1 = a\mathbf{e}_1$, $\mathbf{a}_2 = a\mathbf{e}_2$, $\mathbf{a}_3 = a\mathbf{e}_3$ (if $\mathbf{e}_1, \mathbf{e}_2, \mathbf{e}_3$, are unit vectors defining rectangular Cartesian coordinates), the reciprocal lattice vectors are

$$\mathbf{b}_1 = \frac{2\pi}{a} \mathbf{e}_1, \qquad \mathbf{b}_2 = \frac{2\pi}{a} \mathbf{e}_2, \qquad \mathbf{b}_3 = \frac{2\pi}{a} \mathbf{e}_3, \tag{m.5.66.}$$

which means that the reduced zone or the first Brillouin zone of $\mathbf{k}$ for the simple cubic lattice is a cube of edge length $2\pi/a$. Alternatively expressed, if $\mathbf{k}$ is resolved into its three mutually orthogonal components along $\mathbf{e}_1, \mathbf{e}_2, \mathbf{e}_3$, the equations of its six faces are

$$k_1 = \pm\frac{\pi}{a}, \qquad k_2 = \pm\frac{\pi}{a}, \qquad k_3 = \pm\frac{\pi}{a}, \tag{m.5.67.}$$

which is necessarily a logical consequence of equation (m.5.64.).

From the general equation of the wave vector $\mathbf{k}$ given by (m.5.63.), it follows that there must be $G_1G_2G_3$ vectors contained in a unit cell with edge lengths $\mathbf{b}_1/G_1, \mathbf{b}_2/G_2, \mathbf{b}_3/G_3$. This number, $G_1G_2G_3$, of $\mathbf{k}$ vectors contained in the reduced zone or first Brillouin zone must also represent the number of electron energy states in a Brillouin zone, which is also the number of atoms in the crystal, and in the case of a metal crystallizing with a simple Bravais lattice is also numerically equivalent to the number of unit cells in the crystal. In terms of the Pauli Exclusion Principle, since two electrons may exist in the same energy state provided their spin orientations differ or are opposed to one another, a Brillouin zone can therefore accommodate twice as many electrons as there are unit cells in a crystal. Thus, for a metal crystallizing with a simple Bravais lattice, in which each unit cell contains only one metal atom, a Brillouin zone can accommodate two electrons per atom for that metal crystal. In the event that a metal is such that there are more than two electrons per atom available as conduction electrons, the electrons must be distributed over several Brillouin zones. This important conclusion of the simple zone theory of metals enables, as will be discussed in a later section, metals to be differentiated from one another and also from other solid, crystalline materials.

For the most general basic cell of the reciprocal lattice, its volume is

$$\mathbf{b}_1 \cdot \mathbf{b}_2 \times \mathbf{b}_3 = \frac{8\pi^3}{V}(\mathbf{a}_2 \times \mathbf{a}_3) \cdot (\mathbf{a}_3 \times \mathbf{a}_1) \times (\mathbf{a}_1 \times \mathbf{a}_2),$$

making use of (m.5.65.). The vector product

$$(\mathbf{a}_3 \times \mathbf{a}_1) \times (\mathbf{a}_1 \times \mathbf{a}_2) = [\mathbf{a}_1 \cdot (\mathbf{a}_2 \times \mathbf{a}_3)]\mathbf{a}_1 = V\mathbf{a}_1.$$

Therefore,

$$\mathbf{b}_1 \cdot \mathbf{b}_2 \times \mathbf{b}_3 = \frac{8\pi^3}{V^2}\mathbf{a}_1 \cdot \mathbf{a}_2 \times \mathbf{a}_3 = \frac{8\pi^3}{V}. \qquad \text{(m.5.68.)}$$

From this it is seen that the volume of the reduced zone of $\mathbf{k}$ is $8\pi^3$ times as large as the reciprocal of the volume V of the direct unit cell of the crystal concerned. Now, if there are $G_1G_2G_3$ $\mathbf{k}$ vectors in the reduced zone, the number with their termini in a volume $d\mathbf{k}$ of wave vector space must be

$$G_1G_2G_3 V\, d\mathbf{k}/8\pi^3 = v\, d\mathbf{k}/8\pi^3, \qquad \text{(m.5.69.)}$$

where v is the actual volume of the crystal, which is identical to the number of states per volume element $d\mathbf{k}$ deduced for the free electron case (see page 491), where the potential energy is zero. This, of course, must necessarily be so, since the wave vector distribution is independent of the potential energy.

M.5.8. Brillouin Zones in Metal Crystals

Most metals crystallize with simple close-packed structural arrangements, either face-centered cubic, body-centered cubic, or close-packed hexagonal. Many are polymorphic transforming from one close-packed structure to another under appropriate conditions. The major exceptions are: manganese, which adopts complex structures in all three of its polymorphic forms, referred to as α-, β-, and γ-manganese; praseodymium, neodymium, and samarium crystallize with double hexagonal close-packed structures, proto-actinium metal has a body-centered tetragonal structure; and americium, which also adopts a double hexagonal close-packed arrangement. Both body-centered and face-centered cubic structures can be referred to simple Bravais-type lattices having rhombohedral cells, but the close-packed hexagonal structure has a unit cell containing two atoms. However, even in the case of the cubic structures, it is usually more convenient to select a cubic unit cell which is multiply primitive, containing two atoms per unit cell in the body-centered arrangement and four atoms per unit cell in the face-centered cubic structure. These and any more complicated lattices may still be described as Bravais lattices with a basis, where the basis gives the positions of the other atoms in the unit cell, with reference to the one chosen as being situated at the lattice points of the Bravais lattice.

Although no metal is known which crystallizes with a simple cubic structure, this is the simplest possible structural arrangement and is frequently used as a representative model of a three-dimensional lattice. As seen in the previous section, the reciprocal lattice is also a simple cube and Fig. M.5.25. illustrates the Brillouin zone construction which leads to the reduced zone, also a cube. As in the one-dimensional case, regions of the first zone or the reduced zone may be translated through reciprocal lattice vectors to form polyhedra symmetrically surrounding the first zone. The boundary planes of the various polyhedra all bisect the position vectors of reciprocal lattice points and are given by equation (m.5.64.). The space between the first zone and the second smallest polyhedron constitutes the second Brillouin zone, and so on. Each zone has the same volume, and in this extended zone scheme the energy $E(\mathbf{k})$ may be described as a single-valued function of $\mathbf{k}$ ranging over all $\mathbf{k}$-space. This is in contrast to the reduced zone scheme where $\mathbf{k}$ is restricted to the first zone and the energy is an infinitely many-valued function of $\mathbf{k}$. The reduced zone scheme is the more fundamental, as in the extended zone scheme the allocation of energy states to the various zones is not unique. If the primitive unit cell contains n atoms, the first Brillouin zone contains $1/n$ energy states per atom and can contain $2/n$ electrons per atom. Since the various zones are of equal volume, they may all accommodate the same number of electron states. These polyhedral zones, while usually referred to as Brillouin zones in the theory of metals, are variously referred to otherwise as Voronoi polyhedra or Dirichlet regions in a structure.[1,2]

The face-centered cubic unit cell, containing four structural units per unit cell and its relationship to the primitive rhombohedral unit cell, is shown in Fig. M.5.13. Each lattice point is surrounded by twelve others at a distance

[1] Roberts, C. A., *Packaging and Covering*, 1964, Cambridge University Press, New York.
[2] Steinhaus, H., *Mathematical Snapshots*, 1960, Oxford University Press, Toronto.

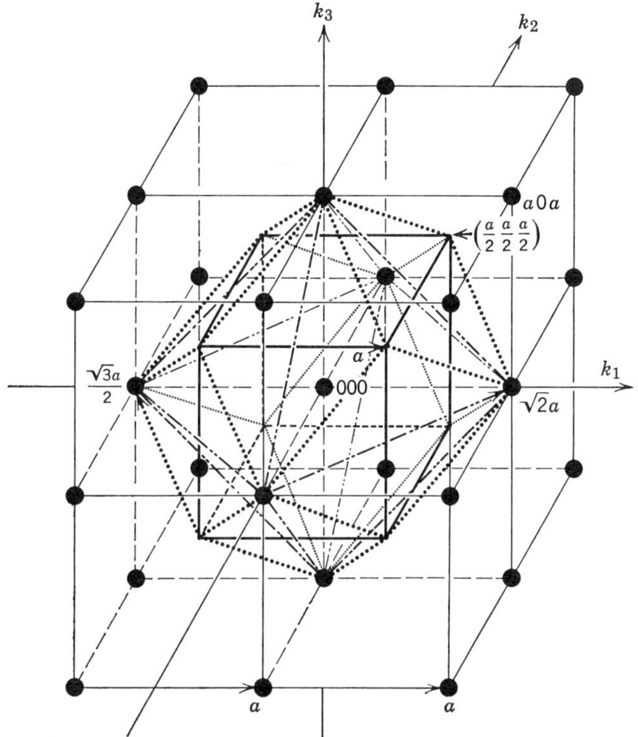

FIG. M.5.25. Construction of Brillouin zones for a simple cubic lattice. The ● denote the reciprocal lattice points of a simple cubic lattice. The reciprocal lattice of a simple cubic lattice is an identical cubic lattice. The first Brillouin zone, ——————— is a cube of edge-length, a, the six faces of which constitute the planes which are the perpendicular bisectors of the lines joining any reciprocal lattice point to its six nearest neighbors. The second Brillouin zone, ... is a rhombic dodecahedron of edge-length, $\sqrt{3}a/2$, and results by translating six parts of the first zone through a reciprocal lattice vector of length a. The six nearest lattice points to any reciprocal lattice point are located at the vertices of a regular octahedron, – – – – – – – of edge-length, $\sqrt{2}a$. The origin of the reciprocal lattice is the common center of the first two Brillouin zones shown and also of the coordination polyhedron surrounding any reciprocal lattice point.

of $a/\sqrt{2}$, where a is the cubic unit cell edge length. The volume of the rhombohedral primitive lattice cell containing one lattice point per unit cell is $a^3/4$. If again $\mathbf{e}_1$, $\mathbf{e}_2$, $\mathbf{e}_3$, denote unit vectors along the cube edges, the rhombohedral cell edge lengths are

$$\mathbf{a}_1 = \frac{a}{2}(\mathbf{e}_1 + \mathbf{e}_2), \qquad \mathbf{a}_2 = \frac{a}{2}(\mathbf{e}_2 + \mathbf{e}_3), \qquad \mathbf{a}_3 = \frac{a}{2}(\mathbf{e}_3 + \mathbf{e}_1), \qquad \text{(m.5.70.)}$$

so that

$$\mathbf{a}_1 \cdot \mathbf{a}_2 \times \mathbf{a}_3 = a^3/4.$$

It has already been shown (Fig. M.5.23.) that the lattice reciprocal to a face-centered cubic lattice is a body-centered cubic lattice. Alternatively, from equations (m.5.65.) and (m.5.70.),

$$\mathbf{b}_1 = \frac{2\pi}{a}(\mathbf{e}_1 + \mathbf{e}_2 - \mathbf{e}_3), \qquad \mathbf{b}_2 = \frac{2\pi}{a}(-\mathbf{e}_1 + \mathbf{e}_2 + \mathbf{e}_3), \qquad \mathbf{b}_3 = \frac{2\pi}{a}(\mathbf{e}_1 - \mathbf{e}_2 + \mathbf{e}_3),$$

which are obviously the reciprocal vectors to those defining the primitive rhombohedral lattice of the body-centered cubic cell shown in Fig. M.5.27., viz.,

$$\mathbf{a}_1 = \frac{a}{2}(-\mathbf{e}_1 + \mathbf{e}_2 + \mathbf{e}_3), \qquad \mathbf{a}_2 = \frac{a}{2}(\mathbf{e}_1 - \mathbf{e}_2 + \mathbf{e}_3), \qquad \mathbf{a}_3 = \frac{a}{2}(\mathbf{e}_1 + \mathbf{e}_2 - \mathbf{e}_3),$$

where a is again the edge length of the body-centered cubic cell and $\mathbf{e}_1$, $\mathbf{e}_2$, $\mathbf{e}_3$, are the unit vectors along the cube edges. Figure M.5.26. depicts a unit cell of a body-centered cubic lattice of edge length a. This is the reciprocal lattice of a face-centered cubic structure. Each lattice point is surrounded by eight lattice points at a distance of $\sqrt{3}a/2$ (numbered 1–8 in the figure) and six other lattice points at a distance a (about 15% more distant and numbered 9–14 in the figure). The first Brillouin zone, constructed by erecting planes normal to the lines joining the origin to these 14 points and positioned halfway along these lines, results in the polyhedron, which is a truncated octahedron, bounded by six square faces and eight hexagonal faces, each of edge length $a/2\sqrt{2}$. The eight octahedral faces result from the planes bisecting the lines joining the origin of the reciprocal lattice to the eight nearest points and the six cube faces are parts of the planes bisecting the lines joining the origin to the six second-nearest points of the reciprocal lattice. The circumscribed polyhedron is a rhombic dodecahedron of edge length $\sqrt{3}a/2$, bounded by 12 rhombus-shaped faces, the acute enclosed angle being $70°32'$ and the obtuse enclosed angle $109°28'$ (the tetrahedral angle); it is the coordination polyhedron for a body-centered cubic lattice and results by drawing planes through the eight nearest plus the six next-nearest lattice points surrounding any one lattice point. From the geometry involved in the con-

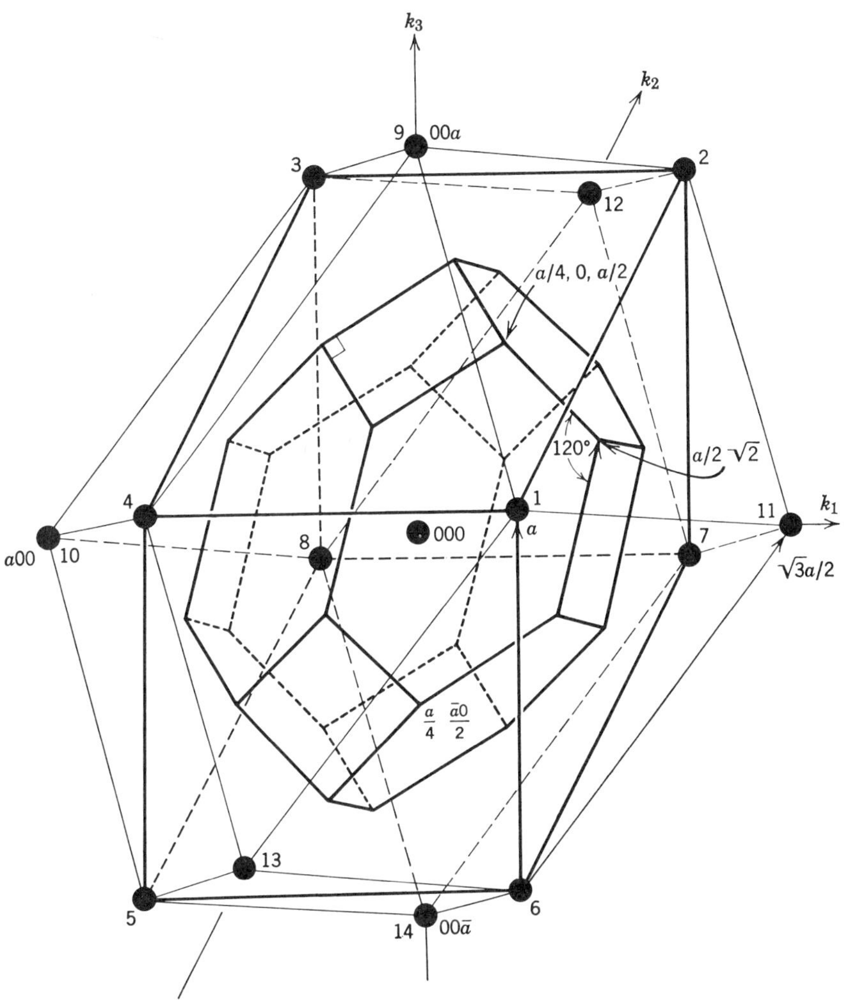

Fig. M.5.26. The first Brillouin zone for a face-centered cubic lattice. The black circles denote the lattice points of a body-centered cubic lattice. The reciprocal lattice of a face-centered cubic array is a body-centered cubic array. The first Brillouin zone, ——————— is a truncated octahedron of edge-length, $a/2\sqrt{2}$, bounded by eight hexagonal faces and six square faces, the eight hexagonal faces constituting the perpendicular planes bisecting the lines joining any reciprocal lattice point with its eight nearest neighoring points, and the six cube faces corresponding to the planes which are the perpendicular bisectors of the lines joining the same reciprocal lattice point to its six next-nearest neighboring points. The group of 14 lattice points consisting of the eight nearest and the six next nearest to any reciprocal lattice point are located at the vertices of a rhombic dodecahedron of edge-length $\sqrt{3}a/2$, which is therefore the coordination polyhedron for a body-centered cubic lattice. This polyhedron is outlined ———————. All three polyhedra shown are concentric.

575

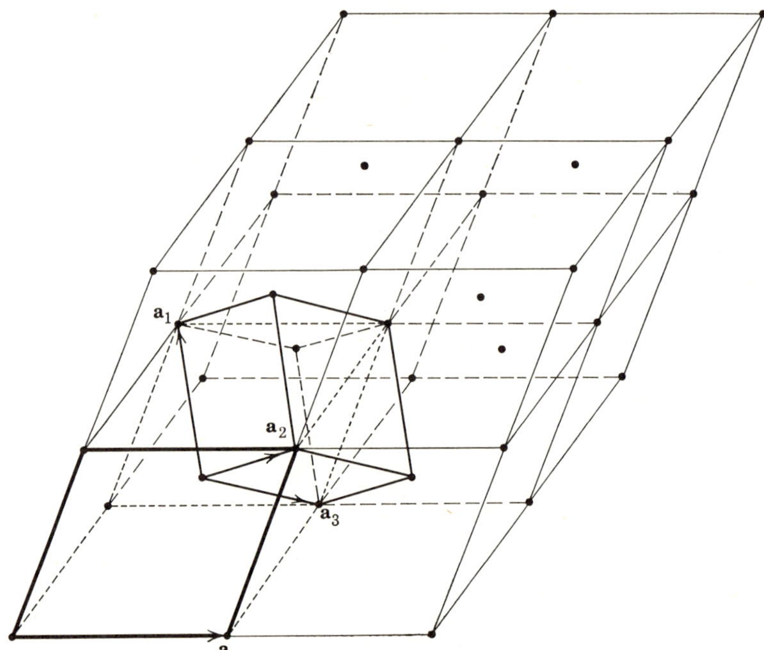

FIG. M.5.27. The body-centered cubic structure. The heavy points denote the lattice sites of the body-centered cubic lattice. A cubic unit cell is shown in lower left front corner and also the rhombohedral Bravais lattice. The lattice vectors of the Bravais cell are $\mathbf{a}_1$, $\mathbf{a}_2$, $\mathbf{a}_3$, and the volume of the Bravais cell is one half that of the cubic unit cell.

struction, the coordinates of the vertices and the edge lengths of the three polyhedra shown in the figure are easily deduced. The eight shortest vectors from the origin of the reciprocal lattice to the centers of the eight hexagonal faces of the Brillouin zone are $(2\pi/a)(\pm\mathbf{e}_1 \pm \mathbf{e}_2 \pm \mathbf{e}_3)$, and the six next-shortest vectors to the centers of the six square faces are $\pm 4\pi\mathbf{e}_1/a$, $\pm 4\pi\mathbf{e}_2/a$, $\pm 4\pi\mathbf{e}_3/a$. The equations of the hexagonal faces which bisect the shortest vectors are $\pm k_1 \pm k_2 \pm k_3 = 3\pi/a$, and the equations of the six square faces which bisect the six next shortest vectors are $k_1 = \pm 2\pi/a$, $k_2 = \pm 2\pi/a$, $k_3 = \pm 2\pi/a$. Since there is one atom per unit cell of the Bravais lattice, the first Brillouin zone of a face-centered cubic lattice, which is a truncated octahedron, can accommodate two electrons per atom of the crystal.

The body-centered cubic unit cell shown in Fig. M.5.27. contains two structural units per cell and the vectors defining the edges of the rhombohedral cell of the Bravais lattice are as shown. Each lattice point is surrounded by eight others at a distance of $\sqrt{3}a/2$. The volume of the rhombohedral cell is

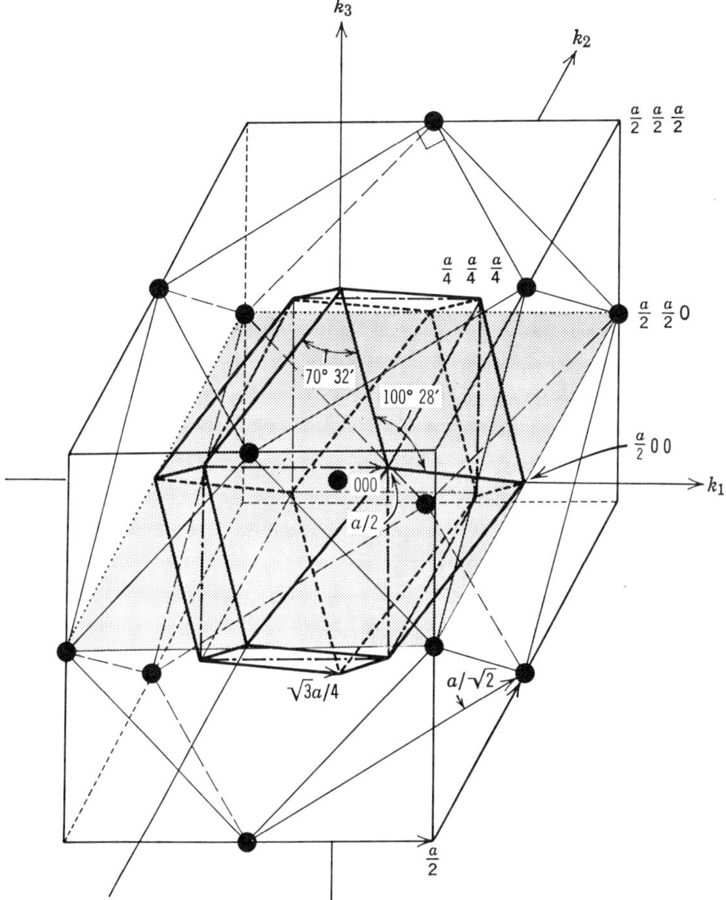

FIG. M.5.28. The first Brillouin zone for a body-centered cubic lattice. The black circles denote the lattice points of a face-centered cubic lattice. The reciprocal lattice of a body-centered cubic lattice is a face-centered cubic lattice. The first Brillouin zone is a rhombic dodecahedron of edge-length, $\sqrt{3}a/4$, bounded by 12 rhombus-shaped faces, each of which corresponds to the plane which is the perpendicular bisector of the line joining any reciprocal lattice point to one of its 12 nearest neighboring points. The 12 reciprocal lattice points, which compose the group of points which are equi-distant from any lattice point are located at the vertices of the polyhedron, which is a cubo-octahedron of edge-length, $a/\sqrt{2}$. The central cube is concentric with the cubic unit cell of the face-centered cubic reciprocal lattice, and is included as an aid in the geometrical construction. All four polyhedra shown have the common center at 000.

$\mathbf{a}_1 \cdot \mathbf{a}_2 \times \mathbf{a}_3 = a^3/2$. The body-centered cubic lattice may be envisaged as two interpenetrating simple cubic lattices, the lattice points of one being at the centers of the unit cells of the other, just as the face-centered cubic lattice may be considered as four interpenetrating simple cubic lattices. Substituting $\mathbf{a}_1$, $\mathbf{a}_2$, $\mathbf{a}_3$, from Fig. M.5.27. into equations (m.5.65.), the reciprocal lattice vectors are $\mathbf{b}_1 = 2\pi(\mathbf{e}_2 + \mathbf{e}_3)/a$, $\mathbf{b}_2 = 2\pi(\mathbf{e}_1 + \mathbf{e}_3)/a$, $\mathbf{b}_3 = 2\pi(\mathbf{e}_1 + \mathbf{e}_2)/a$. It may be seen, by comparing these expressions with (m.5.70.), that the lattice reciprocal to a body-centered lattice is a face-centered cubic lattice, as has already been shown geometrically in Fig. M.5.23. The circled points in Fig. M.5.28. denote the lattice points of a face-centered cubic lattice, the reciprocal lattice of a body-centered cubic array. If the edge length of the face-centered cubic array is again a, each lattice point is equidistant from 12 others at a distance of $a/\sqrt{2}$, six in the same plane and three points in a plane above this central one and three points in a plane at an equivalent distance below the central one. These 12 points define the vertices of a cubo-octahedron of edge length also $a/\sqrt{2}$, and this polyhedron thus constitutes the coordination polyhedron for a face-centered cubic array. The shaded plane in Fig. M.5.28. corresponds to the (001) face of the drawing at the left in Fig. M.5.23. The first Brillouin zone for the body-centered cubic structure is then obtained as before, by constructing the planes which are the perpendicular bisectors of the lines joining any lattice point in the reciprocal lattice (which is the face-centered cubic lattice) to each of its 12 nearest neighboring points. The polyhedron which results is a rhombic dodecahedron of edge length $\sqrt{3}$ $a/4$; each of the 12 faces of this polyhedron is a rhombus with adjacent included angles of $70°32'$ and $109°28'$, respectively. This first Brillouin polyhedron for the body-centered cubic structure is circumscribed about the smaller cube of edge length $a/2$, which is located at the center of the face-centered cubic cell. Again the coordinates of the vertices and the edge lengths of these various polyhedra, as indicated in the figure, are easily deduced from the geometrical construction involved. The 12 shortest vectors from any reciprocal lattice point to the centers of the 12 dodecahedral faces of the first Brillouin polyhedron are given by

$$2\pi(\pm\mathbf{e}_1 \pm \mathbf{e}_2)/a, \qquad 2\pi(\pm\mathbf{e}_2 \pm \mathbf{e}_3)/a, \qquad 2\pi(\pm\mathbf{e}_1 \pm \mathbf{e}_3)/a,$$

and equation (m.5.64.) gives for the equations of the faces of the dodecahedron

$$\pm k_1 \pm k_2 = 2\pi/a, \qquad \pm k_2 \pm k_3 = 2\pi/a, \qquad \pm k_3 \pm k_1 = 2\pi/a.$$

The first Brillouin zone for the body-centered cubic structure, which is a rhombic dodecahedron, therefore is equivalent to the first two zones of the simple cubic lattice, as may be seen by comparison with Fig. M.5.25.

It should be noted that for the first Brillouin zone for the body-centered cubic lattice, the centers of the 12 faces of the zone are equidistant from the reciprocal lattice origin at a distance of $a/2\sqrt{2}$ (Fig. M.5.28.), but six of the vertices of the rhombic dodecahedron (where four faces have a common vertix) are $a/2$ distant from the origin and eight of the dodecahedral vertices are $\sqrt{3}\,a/4$ distant from the origin (where three faces have a common vertix). For the face-centered cubic lattice, the first Brillouin zone has all 24 vertices equidistant from the reciprocal lattice origin at a distance of $\sqrt{5}\,a/4$ (Fig. M.5.26.) but the six square faces of the truncated octahedral zone are $a/2$ distant from the reciprocal lattice origin while the eight hexagonal faces are at a distance of $\sqrt{3}\,a/4$ from the origin. Figures M.5.25., 26., 28. all exist in reciprocal space. The reciprocal lattice and the concept of Brillouin zones are purely mathematical constructs, while a crystal structure described as a collection of atoms, ions, or molecules arranged over the points of a space lattice exists in real space. Similarly, the concept of the Wigner–Seitz cell is a construction in real space. Due to the fact that the reciprocal lattice of a face-centered lattice is a body-centered cubic lattice, the Brillouin polyhedra are therefore the reciprocals of the corresponding Wigner–Seitz cells. Thus, the first Brillouin zone of a face-centered cubic lattice is represented by the same polyhedron as the Wigner–Seitz cell of a body-centered cubic lattice, viz., a truncated octahedron, while the first Brillouin zone of a body-centered cubic lattice is described in terms of the same polyhedron as the Wigner–Seitz cell of a face-centered cubic lattice, which is a rhombic dodecahedron. Both of these polyhedra, the truncated octahedron and the rhombic dodecahedron, are members of the group of semiregular polyhedra known as Archimedean solids. These are distinguished from regular solids, which are both isohedral (having all their faces alike and all regular polygons) and isogonal (having all their vertices surrounded in the same manner), by having polygonal faces of more than one kind (as in the truncated octahedron), or by having dissimilar types of vertices (as in the rhombic dodecahedron); there are only five regular solids, the tetrahedron, the cube, the octahedron, the pentagonal dodecahedron, and the icosahedron,[1] shown in Fig. M.5.29. Of the polyhedra involved here, the cubo-octahedron and the rhombic dodecahedron are conjugate to one another or are the duals of one another, in the sense that where one has f faces and v vertices, the other has v faces and f vertices; the cube and the octahedron and the pentagonal dodecahedron and the icosahedron are related to one another in this same manner, illustrated in Fig. M.5.29. All of these polyhedra referred to above obey the Euler criterion that $v + f = e + 2$, where v, f, e, denote the number of vertices, faces, and edges, respectively, possessed by the polyhedron.

---

[1] Coxeter, H. S. M., *Regular Polytopes*, Methuen, London, 1948, points out that this number is actually 9, if stellated bodies are included; these result by extending outwards the faces of the five regular solids until they meet.

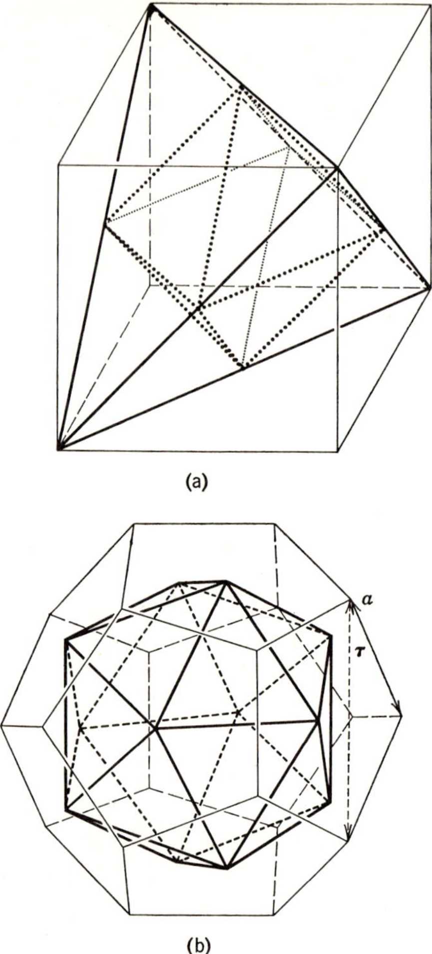

FIG. M.5.29. The five regular polyhedra. The upper figure (a) shows the tetrahedron (four vertices, four faces, outlined in ——————), the cube (eight vertices, six faces, outlined in —————), and the octahedron (six vertices, eight faces, outlined in ....). The cube and the octahedron are conjugate to one another, the tetrahedron is conjugate to itself. The lower figure (b) shows the pentagonal dodecahedron (20 vertices, 12 faces) and the icosahedron (12 vertices and 20 faces, inscribed within the outlines of the dodecahedron); these latter 2 polyhedra are conjugate to one another.

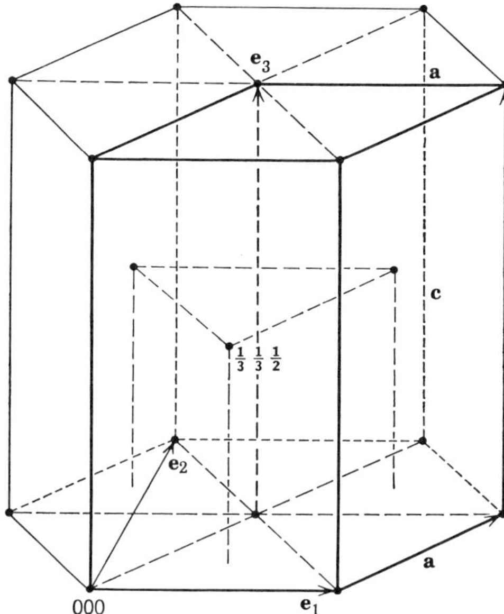

FIG. M.5.30. The close-packed hexagonal structure. The black circles denote the lattice points. The unit cell of the structure is outlined.

In the close-packed hexagonal structure the structural units are arranged in parallel planes, in each of which each of the units of the structure has six equidistant neighbors; alternate planes are stacked vertically above one another but in adjacent planes each unit of structure is arranged vertically above the center of a group of three similar units in the plane below (or vertically below a group of three in the plane above). Unlike the face-centered and body-centered cubic structures, the close-packed hexagonal arrangement cannot be described in terms of a unit cell containing one atom or unit of the structure. The smallest unit cell contains two atoms. The lattice described by taking only alternate planes is a simple hexagonal lattice, so that the close-packed hexagonal structure may be regarded as consisting of two interpenetrating simple hexagonal lattices, as shown in Fig. M.5.30. If c is the distance between alternate planes of the structure and a is the distance between nearest neighboring units in any plane, the lattice may be described as a Bravais lattice with a basis and having unit cell edges

$$\mathbf{a}_1 = a\mathbf{e}_1, \qquad \mathbf{a}_2 = \tfrac{1}{2}a\mathbf{e}_1 + \tfrac{1}{2}\sqrt{3}\,a\mathbf{e}_2, \qquad \mathbf{a}_3 = c\mathbf{e}_3, \qquad \text{(m.5.71.)}$$

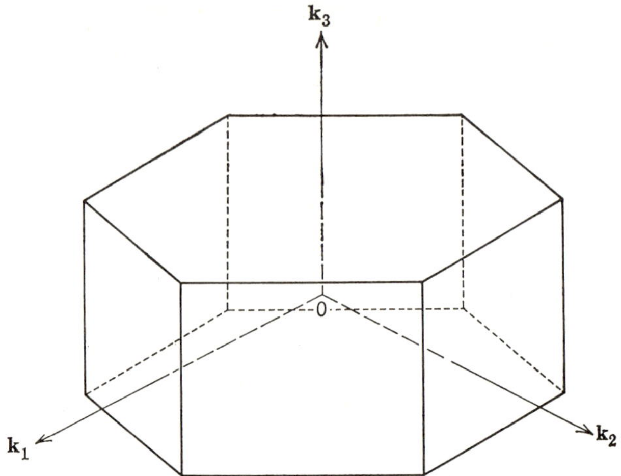

FIG. M.5.31.   The first Brillouin zone for the close-packed hexagonal structure. There is no energy discontinuity across the faces normal to the k_3 axis.

where the orthogonal $\mathbf{e}_1$, $\mathbf{e}_2$, $\mathbf{e}_3$, axes are as indicated. The two structural units are then located at the origin and at the point

$$\tfrac{1}{2}a\mathbf{e}_1 + \tfrac{1}{6}\sqrt{3}\,a\mathbf{e}_2 + \tfrac{1}{2}c\mathbf{e}_3,$$

or, alternatively, in terms of the basic lattice vectors at $(0, 0, 0)$ and $(1/3, 1/3, 1/2)$. The volume of the unit cell is $\tfrac{1}{2}\sqrt{3}\,a^2 c$. Each unit of the structure has six equidistant neighbors at a distance a in the same plane, and six others at a distance of $\sqrt{(\tfrac{1}{3}a^2 + \tfrac{1}{4}c^2)}$, three each in a plane above and below the central plane. In ideal hexagonal closest packing where all 12 neighboring lattice points are equidistant from any one lattice point, $a = \sqrt{(\tfrac{1}{3}a^2 + \tfrac{1}{4}c^2)}$ or $c/a = \sqrt{(8/3)} = 1.633$. For most metals, c/a is considerably different from this ideal; c/a is approximately 1.9 for zinc and cadmium and about 1.6 for magnesium and beryllium. From equations (m.5.65.) and (m.5.71.) the reciprocal vectors of the simple hexagonal Bravais lattice are

$$\mathbf{b}_1 = \frac{2\pi}{a}\left(\mathbf{e}_1 - \frac{1}{\sqrt{3}}\,\mathbf{e}_2\right), \qquad \mathbf{b}_2 = \frac{4\pi}{\sqrt{3}a}\,\mathbf{e}_2, \qquad \mathbf{b}_3 = \frac{2\pi}{c}\,\mathbf{e}_3,$$

which define a simple hexagonal lattice, so that the first Brillouin zone for the hexagonal close-packed structure is a hexagonal prism, shown in Fig. M.5.31. The planes bounding the zone bisect the $\mathbf{b}_1, \mathbf{b}_2, \mathbf{b}_3$, vectors, and since it is customary to refer to four hexagonal axes (three in one plane at 120° and a

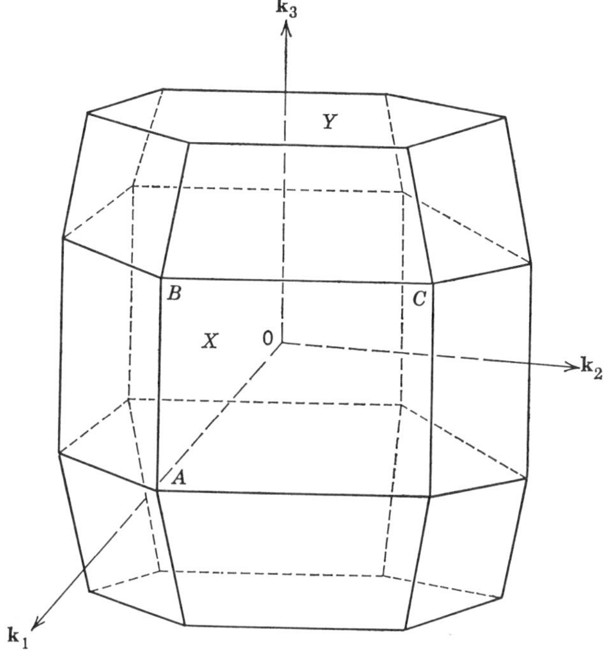

FIG. M.5.32. The first Jones zone for the close-packed hexagonal structure, such that there is an energy discontinuity
across all of the boundary planes of the zone. There is, however, no energy discontinuity along the lines such as AB and
BC.

fourth normal to this plane), also the $(\mathbf{b}_1 + \mathbf{b}_2)$ vector, so that their equations
are

$$\pm k_1 \pm \frac{1}{\sqrt{3}} k_2 = \frac{4\pi}{3a}, \qquad k_2 = \pm (2\pi/\sqrt{3})(1/a), \qquad k_3 = \pm \frac{\pi}{c}.$$

Since the hexagonal closest-packed structure has a unit cell containing two
atoms, the Brillouin zone construction leads to a first Brillouin zone which
contains 1/2 energy states per atom, which can therefore accommodate only
one electron per atom. As will be shown later, the structure factor for the
horizontal planes of the first zone is zero so that there is no energy discontinuity across the hexagonal faces, $k_3 = \pm \pi/c$. This led Jones[1] to propose a

<hr>

[1] Jones, H., *Proc. Roy. Soc.*, 1934, **A144**, 225.

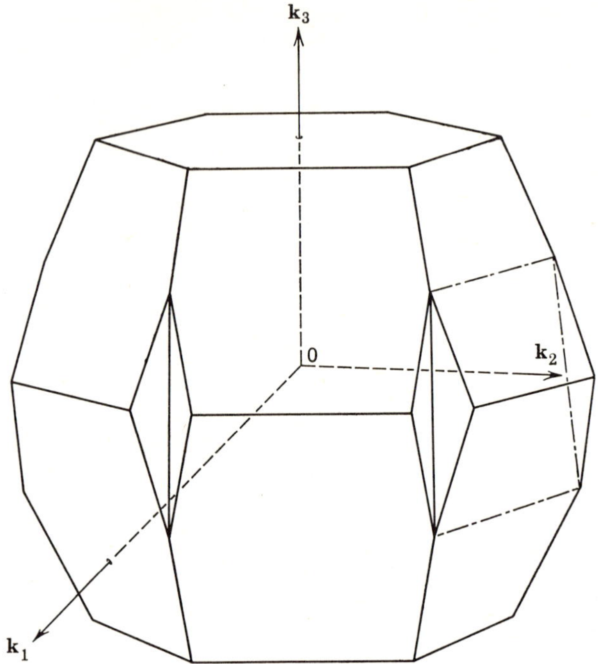

FIG. M.5.33. The smallest region of reciprocal space for the close-packed hexagonal structure surrounded everywhere by an energy discontinuity. It is the Jones zone of Fig. M.5.32. with the addition of the six prisms (dotted) to the six vertical faces.

"first" zone of the form shown in Fig. M.5.32. Within the region of k-space bounded by the planes of the Jones zone, the electronic energy varies continuously with the wave vector for all directions of **k**, but these Jones zones have the peculiar characteristic that the energy discontinuities vanish along edges such as AB and BC, and the smallest volume of k-space which is surrounded everywhere by an energy discontinuity is that of the Jones zone, with the addition on each of the rectangular faces of a small truncated prism, as shown in Fig. M.5.33. The addition of these prismatic faces leads to the second zone, also shown in the figure.

The total volume enclosed within the planes of the second zone contains one energy state, which can accommodate two electrons per atom. This will be so, irrespective of the c/a axial ratio of the direct hexagonal lattice, although the axial ratio of the actual crystal concerned will determine the extension in k-space of the Brillouin zones, such that if c/a for the crystal is greater than the ideal value of 1.633 (page 391), the Brillouin zone will appear compressed along the k_3 axial direction and if c/a for the crystal is less than the ideal value for hexagonal closest packing, the Brillouin zone will appear elongated along the k_3 dimension. Thus, the first zone for the hexagonal closest-packed structure is the hexagonal prism of Fig. M.5.31.; the second zone is that shown in Fig. M.5.33.; while the smallest region in k-space bounded by planes of energy discontinuity is the Jones zone. The first two zones may accommodate two electrons per atom and the Jones zone can accommodate

$$2 - (3/4)(a/c)^2 + (3/16)(a/c)^4 \text{ electrons per atom.}$$

The Brillouin zones constructed as above are obviously related to the interplanar spacings of the metal crystal concerned. It was first pointed out by Jones[1] that the zone faces were equivalent to planes in the direct crystal structure, which had high structure factors.‡ Also, in X-ray and electron diffraction events, it is observed that intense diffracted beams are recorded from crystal planes which have large structure factors. Thus, for face-centered cubic structures, the strongest diffracted beams are observed for scattering by the (111) and (200) planes, and these are the indices of the octahedral and cube faces of the truncated octahedron, which constitutes the first Brillouin zone for such a structure.

‡ An analytical expression for the structure factor, corresponding to any one of the 230 space groups (see page 590) possible for crystalline materials may be deduced on the basis of the following simple geometrical argument. In Fig. M.5.34.(a), there is illustrated a set of ($hk0$) planes, drawn with reference to a plane lattice array for which the repeat unit of the structure is composed of two atoms, so that the complete structural pattern may be thought of as this group of two atoms repeated by the lattice translations. In general, there may be units of any degree of complexity in real crystals, such as complex ions or molecules, which may or may not be related by some additional symmetry operation, but the pattern unit is always repeated by the lattice translations. In Fig. M.5.34.(a), one of the two atoms is chosen as the origin of the unit cell of the structural arrangement, with cell edge-lengths in the plane denoted by a and b. No matter how complex the pattern unit in any crystal may be, a separate lattice grid may be drawn through each constituent atom in this pattern unit, which is identical in dimensions and orientation to the lattice grid which may be drawn through the particular atom in the pattern unit selected as the unit cell origin. Thus, if there are N different atoms in the crystal pattern unit, not related by lattice translations, then N identical lattice arrays of atoms result, each slightly displaced from one another. In Fig. M.5.34.(a), there are only two such lattice arrays; the unit cell base of the structural unit may of course be chosen to correspond with either of these lattice arrays, or any similarly oriented unit of the same dimensions in the plane. The whole crystal would diffract an incident beam of monochromatic X-radiation at angles according to the Bragg condition,

$$\theta = \sin^{-1} (n\lambda/2d_{hk0}),$$

---

[1] Mott, N. F., and Jones, H., *Theory of the Properties of Metals and Alloys*, 1958, Dover Publications, Inc., New York.

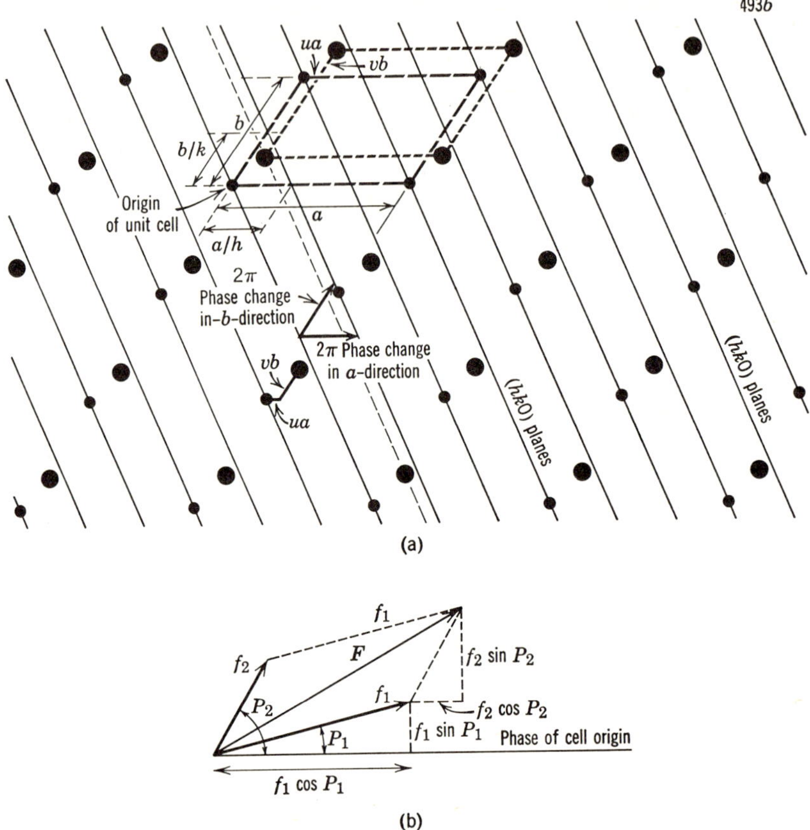

FIG. M.5.34. Phase relations of the reflexions from the (*hk*0) planes of a crystal structure, where the pattern unit of the structure located at each lattice point composed of two different kinds of atoms. The coordinates of the ◓ atoms with respect to the ● atoms, chosen as the origin of the unit cell, are *uv*0.

reflexions occurring simultaneously from each lattice array. However, while reflexions from each lattice array would be in phase with one another, in terms of the Bragg Law, the combined reflexions from the various lattice arrays would not generally be in phase with one another. The amplitudes of the radiation scattered by the whole crystal would be dependent on the phase relations of the reflexions of the separate lattice arrays of the pattern unit. Since these phase relations are determined by the relative displacements of the several lattice arrays from one another, it is apparent that the reflexion intensities of the X-radiation scattered by a crystal are dependent on the shape and symmetry of the pattern unit. The reflexion angles, on the other hand, depend only on the lattice dimensions.

In Bragg scattering, the phase difference for the radiation scattered by adjacent planes is 360° or 2π (page 545), so that if the phase of the radiation scattered by a plane through the origin is $0°$, the phase difference for the adjacent plane is 2π, as indicated in Fig. M.5.34.(a). If the coordinates of the other atom with respect to the origin, in this Figure, are u and v respectively, with reference to the a and b unit cell directions, where u and v are fractions of the respective unit cell translations, the phase difference for radiation scattered by the atoms lying on the dashed planes indicated, with respect to the plane through the unit cell origin, may be deduced. Thus, since the planes $(hk0)$ divide a into h parts, the intecept of the first plane on the a-axis is a/h, and this distance must correspond to a phase change of 2π. If the distance ua corresponds to a phase change of p_a, then $p_a/2\pi = ua/ah^{-1}$, or $p_a = 2\pi hu$. Similarly, if p_b represents the phase change, with respect to the origin, in the b axial direction, then $p_b = 2\pi kv$. The total phase change, attributable to the displacement of the second kind of atom from the kind of atom at the origin will be given by the sum of the phase changes in the a and b unit cell directions, or $P = p_a + p_b = 2\pi(hu + kv)$. More generally, for any set of crystal planes, (hkl), where the coordinates of the second kind of atom with respect to the unit cell origin, are uvw, the three-dimensional phase change would be given by $P = 2\pi(hu + kv + lw)$. Also, if the coordinates of the second kind of atom in the crystal pattern unit are represented by xyz, where $x = ua$, $y = vb$, and $z = wc$, where u, v, and w, are fractions of the lengths of the unit cell translations, a, b, and c, the total phase change may be alternatively expressed as $P = 2\pi(hx/a + ky/b + lz/c)$. This phase angle, P, represents the phase difference between two waves, one scattered by an atom at the origin of the unit cell and the other scattered by an atom with the coordinates uvw.

All atoms, ions, and molecules, scatter X-radiation as a consequence of their constituent electrons oscillating in phase with the changing electric field of the electromagnetic radiation field associated with the X-ray beam. The resulting acceleration and deceleration of the electronic charge gives rise to an electromagnetic disturbance of the same wavelength and frequency as the incident X-ray source, which interaction is referred to as scattering of the electrons by the original source of X-radiation. Any particular atom scatters X-radiation to an extent proportional to the number of electrons it contains, or dependent on its nuclear charge number. However, the scattering efficiency of any atom, varies directly with the angle between the directions of the incident and scattered X-ray beam, and inversely with the wavelength associated with the beam of X-radiation, such that the atomic scattering intensities fall off in proportion to $(\sin\theta)/\lambda$. This decrease in scattering efficiency is attributable to the finite extension of the electronic charge-cloud around the nucleii of atoms, and the fact that each electron does not generally scatter in phase with one another. The net scattering amplitudes of various atoms may be calculated by the Thomas–Fermi (page 519) and other methods, and atomic scattering factors usually designated by f, representing the relative scattering power of an atom in terms of the scattering power of a single electron as unity.

The amplitudes and phases of the radiation scattered by several atoms in the same direction may be compounded vectorially, as indicated in Fig. M.5.34.(b) for the case of two atoms with phases P_1 and P_2 and atomic scattering factors of f_1 and f_2, respectively. Thus, the composite scattering amplitude, $\mathbf{F} = \mathbf{f}_1 + \mathbf{f}_2$, may be written as,

$$|F| = \{[f_1 \cos P_1 + f_2 \cos P_2]^2 + [f_1 \sin P_1 + f_2 \sin P_2]^2\}^{1/2},$$

or in terms of complex quantities, as

$$F = f_1 e^{P_1 i} + f_2 e^{P_2 i}.$$

If the expressions deduced above for the phase angles P_1 and P_2, are substituted in these expressions for F, then

$$|F| = \{[f_1 \cos 2\pi(hx_1/a + ky_1/b + lz_1/c) + f_2 \cos 2\pi(hx_2/a + ky_2/b + lz_2/c)]^2$$
$$+ [f_1 \sin 2\pi(hx_1/a + ky_1/b + lz_1/c) + f_2 \sin 2\pi(hx_2/a + ky_2/b + lz_2/c)]^2\}^{1/2}.$$

Alternatively, in complex notation,

$$F(hkl) = f_1 e^{2\pi i(hx_1/a + ky_1/b + lz_1/c)} + f_2 e^{2\pi i(hx_2/a + ky_2/b - lz_2/c)}.$$

If there are N atoms in the unit cell of the crystal structural arrangement concerned, then the complex resultant, characterized both by an amplitude $|F|$ and a phase α, is known as the structure factor, and may be expressed as the sum,

$$F(hkl) = \sum_0^N f_j e^{2\pi i(hx_j/a + ky_j/b + lz_j/c)},$$

which can be evaluated by means of the expressions, $|F(hkl)| = \sqrt{A^2 + B^2}$ and $\alpha(hkl) = \tan^{-1}(B/A)$, where

$$A = \sum_0^N f_j \cos 2\pi(hx_j/a + ky_j/b + lz_j/c)$$

and

$$B = \sum_0^N f_j \sin 2\pi(hx_j/a + ky_j/b + lz_j/c).$$

The observed diffraction intensity maxima, when a real crystal diffracts a beam of monochromatic X-radiation, are proportional to the square of the structure amplitudes, so that $I \propto |F|^2 = A^2 + B^2$. If the space group for a crystal is known, it is usually more convenient to carry out the summations indicated above over the coordinates of the equivalent point positions, which operation often results in a simplified expression for the structure factor. Particularly, if a center of symmetry is present in the crystal pattern unit, and if the origin of the unit cell is selected as being at this symmetry center, every vector of phase angle $2\pi x_j/a$ must be accompanied by another vector of phase angle $-2\pi x_j/a$, and thus all the sin terms cancel. There are elements of repetition, other than translations, in all space groups except one. For example, for the space group $C_{2h}{}^5 - P2_1/c$, the equivalent points have coordinates

$$x, y, z; \qquad -x, -y, -z; \qquad x, b/2 - y, c/2 + z; \qquad -x, b/2 + y, c/2 - z,$$

so that

$$A = 4 \sum f_j \cos 2\pi[hx/a + lz/c + (k + l)/4] \cdot \cos 2\pi[ky/b - (k + l)/4]$$

and

$$B = 0,$$

where the summations have been carried out over the smaller number of atoms in the asymmetric unit. For certain combinations of indices, such expressions simplify further, so that when $(k + l)$ is odd, $A = -4 \sum f_j \sin 2\pi(hx/a + lz/c) \cdot \sin 2\pi ky/b$ and when $(k + l)$ is even, $A = 4 \sum f_j \cos 2\pi(hx/a + lz/c) \cdot \cos 2\pi ky/b$; the sin expression above is zero when $k = 0$, or when h and l are zero, so that $(h0l)$ reflections are absent when l is odd and $(0k0)$ when k is odd, for this particular space group.

Generally, a large structure factor leads to a large energy gap at a zone boundary and, in fact, in the nearly-free electron approximation method the energy gap is proportional to the magnitude of the structure factor. It was for this reason that Jones proposed constructing zones such that all planes with zero structure factor are omitted and, as far as possible, only planes corresponding to large structure factor values are used as boundary surfaces. The so-called Jones zones, which result from this type of construction, are identical with the Brillouin zones already described for the simpler type of structural arrangements, but for more complicated structures containing symmetry elements such as glide planes and screw axes of symmetry ‡ various difficulties arise, since for such a structure a zero structure factor may not necessarily correspond to a zero energy gap. The property of the Jones zones, that they are bounded by planes across which there is a large energy discontinuity and which correspond to planes in the real crystal structure which have large structure factors, makes them particularly convenient as an aid in interpreting the observed physical properties of metals and intermetallic compounds in terms of the various electron theories.

‡ All crystalline materials are such that their structure may be described in terms of the disposition in a three-dimensional periodic space pattern of either atoms, ions, complex ions, or molecules. The external symmetry displayed by crystals, as finite crystallographic bodies, can be described in terms of the symmetry elements associated with the operations of reflexion, rotation, and inversion. Even if a crystal possesses no other symmetry properties, it must at least possess translational symmetry. A point-space lattice results by imagining that the repetition unit, whether an atom, ion, complex ion, or molecule, of the real crystal, is replaced by a point. The point-space lattice derived in this way from a crystal structure is a collection of points that portrays the translational periodicity of the structure. This fundamental translational periodicity of crystals limits the number of distinct two-dimensional lattices to five, and these plane lattices may be systematically stacked to realize only the 14 Bravais space lattices shown in Fig. M.5.17., which in turn are referable to the six (or seven) coordinate systems listed on page 556. By combining the symmetry elements, which do not involve translation, namely axes of rotation and rotatory inversion, planes of symmetry, and inversion centers, in all possible ways, subject to the restriction imposed by the three-dimensional periodicity of the space lattice, 32 point groups result. These 32 point groups can be arranged into eleven symmetry groups, the so-called *Laue Groups*, each having one centrosymmetrical point group and one or more noncentrosymmetrical point groups, which become identical with it on the addition of a center of symmetry.[1]

In describing the internal symmetry of crystals, as deduced largely from X-ray diffraction studies, which provide information about the detailed arrangement of the constituent atoms, ions, complex ions, or molecules, composing them, it is necessary to add to the above-mentioned translation-free symmetry elements, the additional symmetry elements of screw axes and glide planes. Screw axes of symmetry are associated with the operations of rotation and translation, and glide planes of symmetry with the, operation of reflexion and rotation, but again to be compatible with the finite number of point groups, only a finite number of types of screw axes and glide planes of symmetry are possible. Screw axes are restricted to repetitions at angular intervals of 60°, 90°, 120°, and 180°, while the amount of the translation associated with the screw axis depends on the repeat period of the crystal pattern unit parallel to the screw axis. A two-fold screw axis, designated 2_1, combines a rotation of 180° with a translation of one-half of the

[1] Phillips, F. C., *An Introduction to Crystallography*, 1963, **3rd Edition**, John Wiley & Sons, Inc., New York.

repeat period. Three-fold screw axes combine rotations of 120° counterclockwise (3_1) or clockwise (3_2) with translations of one-third of the repeat period. The operation of a 3-fold screw axis thus produces enantiomorphous configurations. Four-fold screw axes combine rotations of 90° counterclockwise (4_1), or clockwise (4_3) with translations of one-quarter of the repeat unit along the screw axis to produce entiomorphous arrangements, or a rotation of 90° with a translation of one-half of the repeat dimension (4_2). Six-fold screw axes combine rotations of 60° in opposing directions with translations of one-sixth and one-third, respectively, to give two sets of enantiomorphous arrangements, 6_1, 6_5, and 6_2, 6_4; a fifth arrangement produced by a rotation of 60° combined with a translation of one-half of the repeat period along the screw axis gives rise to the 6_3 screw axis.

The translation component of a glide plane is always a simple fraction, one-half or one-quarter, of the repeat period of the crystal pattern. A glide plane is referred to as an axial glide plane, in the a, b, or c, directions, if the glide is parallel to an edge of the unit cell of the structure. If the glide is parallel to the direction of the unit cell diagonal, the plane is described as a diagonal glide (d) if the glide component is one-half the repeat period and as a diamond glide (n) if the glide component is one-quarter of the period. The orientation of a plane limits the direction of the glide, as the translational movement is parallel to the plane involved. Thus, an axial glide plane parallel to the (100) direction cannot have a glide component of $a/2$. For the same kind of reason, the components of diagonal and diamond glide planes are even more restricted; for example, the component of a diagonal glide plane parallel to the (010) direction can only be ($a/2 + c/2$).

If a point group is applied to an infinite lattice, then the point groups become infinite groups, referred to as space groups, and if all possible ways are examined of combining the 32 point groups with the symmetry operations which involve translation components, then 230 space groups result. This number, 230, was deduced independently by three different crystallographers in the period 1890–94, long before the discovery of X-ray diffraction by crystals made it possible to make any use of such knowledge. Diagrams showing the symmetries and the terminology for describing them, of each of the 230 different space groups are listed in the "International Tables for X-ray Crystallography."[1,2]

M.5.9.   The Velocity and the Effective Mass of Electrons in Metals

The quantum mechanical description of the motion of an electron in a varying potential field, or for that matter the motion of any particle in a varying force field, gives rise to phenomena regarding velocity and mass, which are quite absent from the classical mechanical description. The one-dimensional motion of an electron in a varying potential field is comparable to the passage of a beam of light through a medium of varying refractive index; in the optical situation, a certain fraction of the light is transmitted and a certain fraction reflected, and the fraction of the original beam which is reflected becomes smaller as the wavelength of the incident beam becomes

---

[1] *International Tables for X-ray Crystallography*, 1952, **Vol. I**, The Kynoch Press, Birmingham, England.
[2] Buerger, M. J., *Elementary Crystallography*, 1956, John Wiley & Sons, Inc., New York.

less. If the beam of radiation is described in terms of a stream of photons, then it is not possible to identify the motion of individual photons beyond saying that there is a certain probability that the velocity of motion of certain photons will be reversed in direction; similarly, for electrons moving in the varying potential field of a metal crystal, there is a certain probability that the velocity of certain electrons may be reversed, without any change in magnitude. This interpretation of the behavior of electrons in metals has already been arrived at on the basis of the Kronig–Penney model (page 543), where it is deduced that electrons with $k = n\pi/a$ apparently cannot travel along a crystal lattice row, but are totally reflected. In the Brillouin zone theories of electrons in metals, in which the potential may vary very greatly over very small distances, the quantum mechanical description of electronic behavior has to be interpreted especially carefully. An electron, for example, in a state described by the superimposition of two traveling waves of equal amplitude but moving in opposite directions gives rise to a stationary wave of mean velocity zero, although there must be some finite value for the root mean square value of this quantity. It is necessary to distinguish between the electronic velocity, involved in expressions for its momentum and kinetic energy, mv and $mv^2/2$, and its de Broglie wavelength, h/mv, and the velocity of the associated wave motion, λv, where $v = E/h$ as given by the Planck expression.

For a particle like an electron, with a rest mass m_ε, moving with a velocity v, its actual mass is given by the Lorenz expression (page 28), $m = m_\varepsilon[1 - \beta^2]^{-1/2}$, where $\beta = v/c$; its momentum is then $mv = m_\varepsilon v[1 - \beta^2]^{-1/2} = m_\varepsilon \beta c[1 - \beta^2]^{-1/2}$, and its kinetic energy, $T = p^2/2m = [mc^2 - m_\varepsilon c^2] = m_\varepsilon c^2[(1 - \beta^2)^{-1/2} - 1]$. The total energy is given by the Hasenöhrl–Eisenstein equation, $E_r = mc^2$, and for situations where β is small, $E_r = hv \simeq mc^2 + mv^2/2 + U$, where U is the potential energy. The inclusion of the relativity energy, mc^2, in the expression for the total energy of a particle in this way does not affect macroscopic phenomena, since where the potential energy is included in the total energy the addition of a term like this merely alters the arbitrary potential energy zero, although, of course, it makes a tremendous difference in calculating v from the Planck expression whether $E = T + V$ or $E = mc^2 + T + V$ is used; v in fact cannot be uniquely defined since the potential energy may be measured from any arbitrary zero, which amounts to an indeterminate frequency for the Planck quantity, $v = E/h$. If the attempt is made to calculate the so-called phase velocity of particles described in terms of wave mechanics, then if this phase velocity is designated V,

$$V = v\lambda \simeq [h^{-1}(mc^2 + mv^2/2 + U)][h/mv],$$

from $v = E_r/h$ and $\lambda = h/mv$, which since mc^2 is so very much greater than $[mv^2/2 + U]$ gives $V \simeq c^2/v$. Such a velocity, which obviously must always be greater than c, cannot refer to any material system, and even although electrons may be conveniently described in terms of Schrödinger mechanics, the phase velocity of the associated wave motion cannot apparently be related to the electronic velocity. Actually, for free electrons where $v = E_r/h = [mc^2 + mv^2/2 + U]h^{-1}$, the electronic velocity $v = dv/d(1/\lambda) = dv/d(mv/h)$, and this velocity is usually referred to as the group velocity of the wave packet in Schrödinger mechanics, a wave packet being interpreted as a limited spatial region where the probability of locating the electron is substantial. The same expression applies for the electronic velocity, where the electron is described by means of traveling waves in the cyclic model of metals, and since $E = T + V = hv - m_e c^2$, it is apparent that $dE/d(1/\lambda) = d(hv)/d(1/\lambda) = vh$, so that the group velocity of the wave packet or the electronic velocity may be alternatively written as

$$v = dv/d(1/\lambda) = [1/h][dE/d(1/\lambda)] = [2\pi/h][dE/dk], \qquad \text{(m.5.72.)}$$

where $v \ll c$. For free electrons where $E = mv^2/2 = h^2 k^2/8\pi^2 m = h^2/2m\lambda^2$, v is found to be

$$\bar{v} = h/m\lambda = hk/2\pi m, \qquad \text{(m.5.73.)}$$

which is actually the root mean square velocity.

From the form of the one-dimensional $E(k) - k$ curves shown in Fig. M.5.6., it is apparent that $dE(k)/dk$ is zero for electrons at the top and bottom of each band; therefore, from equation (m.5.72.), the electronic velocity is zero at the top and bottom of each energy band. While in the free electron theory an increase in the energy or an increase in the wave number corresponds to an increasing electronic velocity, this is not so for electrons moving in a periodic field. From the bottom of the first energy band, referring to Fig. M.5.6., increasing energy results initially in an increase in velocity, but as the top of the band is approached, $dE(k)/dk$ falls off and eventually becomes zero, where the $E(k) - k$ curve becomes horizontal. The energies and velocities at the top of the first forbidden range of energy values are zero and, with increasing k-value, the electronic velocity firstly increases and then decreases. This relationship between velocity and wave vector of the electrons, in the one-dimensional model, is the same across each allowed energy band, starting at zero at the bottom of the energy band, rising to some maximum value, and then falling to zero again at the top of the energy band. The zero value of the velocity of the electrons, or of the wave packets describing the behavior of the electrons in states near the top and the bottom of the allowed energy ranges, cannot be taken to indicate that the electrons in these states are at rest. In the periodic field, as in the simple case of the electron-in-a-box model, while it is still possible to regard that the electrons in states at the top and bottom of an allowed energy band are represented by stationary waves, which indicate that at these positions in the $E(k) - k$ curve there is an equal probability of finding electrons moving in opposite directions, the fact that the velocity of the representative wave packet as a whole is zero at these points means that the reversals in velocity take place within the length of a wave packet, precluding any detailed visualization of the process which takes place. The occurrence of energy gaps for electrons moving in a one-dimensional lattice of period a at points $n\pi/a$ is, as already discussed on page 542, related to the condition for Bragg reflexion of electrons traveling along the lattice row.

For electrons moving in a three-dimensional periodic field, the relationship between energy and wave number for each particular direction of $\mathbf{k}$ is similar to the one-dimensional curves, and the discontinuities in the different curves occur at the values of $\mathbf{k}$ which lie on the surface of a Brillouin zone, as in Figs. M.5.26. and M.5.28. Again, as for the one-dimensional case, it may be regarded that for electrons in states corresponding to those on the surfaces of the Brillouin zones, the wavelengths of the electrons in such states satisfy the condition for Bragg reflexion within the metallic crystal concerned, so that electrons in those states have a zero component of velocity normal to the zone face concerned. It has to be kept in mind that the Brillouin zones exist in reciprocal wavelength space as imaginary constructs, with the k_x, k_y, and k_z directions regarded as coincident with the x, y, and z Cartesian coordinate directions, so that reference to electrons moving in particular directions with respect to the faces of Brillouin polyhedra may be related to their direction of motion in real space, although it is only for crystals which may be described by reference to mutually orthogonal axial directions that the two directions above are coincident.

As indicated on page 585, there is an intimate relationship between the geometry of the Brillouin zones in a metal crystal and the conditions which must be satisfied for electron (or X-ray or neutron) diffraction to occur within the crystal structure concerned. It is as if for electrons of various $\mathbf{k}$-values moving in a metal crystal, those electrons with wave vectors up to or between certain values are relatively free to move with little scattering, but for those electrons with specific $\mathbf{k}$-values, namely those corresponding to states on the surface of a zone, their $\mathbf{k} = 2\pi/\lambda$ satisfies the condition for Bragg reflexion, $n\lambda = 2d_{hkl} \sin \theta$, and a strong reflexion takes place so that the electrons concerned do not travel further in the lattice. During this reflexion, the component of the velocity normal to the reflecting plane is reversed in direction, while the velocity component parallel to the reflecting plane is unaltered. Thus, for an electron state at the surface of a Brillouin zone, the mean component of velocity normal to the zone face is zero. If $E(\mathbf{k})$ as a function of $\mathbf{k}$ is examined along some particular direction in $\mathbf{k}$-space, such as along the line OP in Fig. M.5.11., then the corresponding $E(k) - k$ curves will be horizontal at the points X and Y if OP is normal to a zone face, and also if OP were to pass through a corner corresponding to two intersecting faces. In general, however, these curves will not be horizontal at such points, since although the velocity component normal to a face is zero, the parallel velocity component is unaffected. The group velocity of the wave packet representing the electron is from (m.5.72.), proportional to dE/dk, and so the direction of the electron velocity is normal to the energy contours in the vicinity of the group of states from which the wave packet is compounded. [In vector analysis, the gradient of $E(\mathbf{k})$ is (see page 243)

$$\operatorname{grad}_k E(k) = \mathbf{e}_1 \, \partial E/\partial k_1 + \mathbf{e}_2 \, \partial E/\partial k_2 + \mathbf{e}_3 \, \partial E/ak_3,$$

which is the derivative of $E(\mathbf{k})$ in a direction normal to the energy surface through the point $\mathbf{k}$.]

As in Fig. M.5.11., the energy contours near the center of a zone are spherical (circular in two dimensions), and for the small group of states in the vicinity of P' the group velocity will be in the same direction as OP', as in the free electron theory. But where the contours are distorted, as near the point P, OP gives the direction of the waves from which the wave packet is compounded, but the group velocity of the wave packet and the velocity of the electrons is in the direction of the arrow indicated, which is normal to the energy contour. This same idea applies in the three-dimensional case, the group velocity being normal to the surface denoted by the energy contours.

In the free electron theory (see page 491), $N(E) \propto \sqrt{E} \propto k$, and so the $N(E) - E$ curve and the $E(k) - k$ curve are both parabolic. From Fig. M.5.6., it is seen that in the Brillouin zone theory, the electrons at the bottom of the first zone behave very like those of the free electron theory. Also, in the various one-electron theories, the same applies to the electrons at the bottom of each of the higher zones, if the energy is determined from the bottom of the zone concerned; thus, if E' is the energy at the bottom of the second allowed band, the lower section of each of the $E(k) - k$ curves in Fig. M.5.6. are also parabolic with $k \propto \sqrt{E - E'} \propto N(E)$, and generally for the few electrons in states at the bottom of the various zones, the relationship between energy and velocity and between energy and $N(E)$ are the same as in the free electron model, always provided that the energy is measured from the bottom of the zone concerned. At the tope of each zone, however, the situation is quite different. For the one-dimensional model, the group velocity v is given by equation (m.5.72.) as $v = [2\pi/h][dE/dk]$. If now the influence on the electrons in a metal of an applied electric field $\mathbf{E}$ is considered, the force exerted on an electron is given by $\mathbf{E}\varepsilon$, and according to classical mechanics this force F is $F = \mathbf{E}\varepsilon = m_\varepsilon(v/t)$, then $t = mv/F = mv/\mathbf{E}\varepsilon$. Under the influence of the applied electric field the wave number $\mathbf{k} = 2\pi/\lambda$ will increase uniformly with time in the direction of the applied field. Since by the de

Broglie relationship, $\lambda = h/p = h/mv$, the variation of the wave vector $\mathbf{k}$ with time may be expressed as

$$dk/dt = d(2\pi/\lambda)/dt = d(2\pi mv/h)/d(mv/\mathbf{E}\varepsilon) = 2\pi\mathbf{E}\varepsilon/h. \quad \text{(m.5.74.)}$$

Therefore, the rate of change of the group velocity, or the acceleration of the electrons in the applied field, normal to the energy contours is

$$\begin{aligned} dv/dt = d/dt\,[(2\pi/h)(dE/dk)] &= [2\pi\mathbf{E}\varepsilon/h]\,d[(2\pi/h)(dE/dk)]/dk \\ &= [4\pi^2\mathbf{E}\varepsilon/h^2]\,d^2E/dk^2. \end{aligned} \quad \text{(m.5.75.)}$$

For comparison, the classical expression for electron acceleration in a field $\mathbf{E}$ is

$$F = m_\varepsilon(dv/dt) = \mathbf{E}\varepsilon \quad \text{or} \quad dv/dt = \mathbf{E}\varepsilon/m, \quad \text{(m.5.76.)}$$

so that for electrons moving in a periodic field under the influence of a field $\mathbf{E}$, the simple mass term, m_ε, has been replaced in equation (m.5.75.) by

$$m^* = h^2[4\pi^2\,d^2E/dk^2]^{-1}. \quad \text{(m.5.77.)}$$

This quantity, m^*, previously referred to on page 518, is what is usually referred to as the effective mass of an electron in a periodic field.

As a consequence of equation (m.5.75.), $dv/dt = \mathbf{E}\varepsilon/m^*$, it follows that when d^2E/dk^2 is negative, the electron acceleration in the field is negative. Now, from Fig. M.5.6., it is apparent that d^2E/dk^2 is negative for states in the upper region of each allowed energy band, so that for electrons in those states the influence of an applied electric field is to reduce the group velocity v, with the result that the effect is similar to that which would result from the electrons having a negative mass. This is a typical quantum mechanical result, which has no analogy in classical mechanics. The resultant effect is that the electric field changes the relative probabilities of an electron moving in opposite directions for electrons in states in the vicinity of the top of an allowed energy zone. Alternatively, it may be regarded that, in the one-dimensional model, the electrons in states at the top of an energy band behave as if they had negative mass or positive charges. For the three-dimensional zone model, those electrons in states at the bottom of a zone behave under the influence of an electric field in such a way as to increase their wave vector $\mathbf{k}$ and the group velocity in the direction of the field, and as long as the three-dimensional surfaces of constant energy are spheres (or ellipsoids), the electrons behave in accordance with the de Broglie relation, $\lambda = h/mv$, with $\mathbf{k} = 2\pi/\lambda = 2\pi mv/h$. For electron states corresponding to those near to the zone boundaries, where the constant energy surfaces become distorted (see page 548), this situation changes so that the electric field alters the group velocity in directions different to that of the field itself, and the applied force and the acceleration acquired by the electrons in those states are not in the same direction as for the wave packet involved. Generally, the Brillouin zone theory of the behavior of electrons in a periodic field indicates then that for only a few electrons in a zone the situation is not substantially different from that indicated by the free electron theory, but for a nearly filled zone or a completely filled zone the electrons in states near the top of the zone behave under the influence of an electric (or magnetic) field as if they were associated with a negative mass or a positive charge. This aspect of the zone theory enables certain experimental observations about the sign of the Hall coefficient (see page 637) of the alkali metals to be understood.

M.5.10.   Energy Density of States

An approximate idea about the shape of the $N(E)$ curves for metals may be obtained from experimental observations of their soft X-ray spectra (M.3.1.) and electronic heat capacities (M.4.2.). In principle, it should be possible to calculate $N(E)$ for a metal if the energy surfaces were known sufficiently accurately; ‡ however, only approximate calculations of this type can be made. Nevertheless, the general shape of typical $N(E)$ curves may be deduced on the basis of some relatively simple considerations.

‡ If the density energy of states is defined in the same manner as in the free electron theory, then for electrons in a periodic field, $N(E)\,dE$ represents the number of orbital states, in each of which there may be two electrons, with energies between E and $E + dE$, where again dE is a small energy interval which however may still contain a large number of energy levels (page 464). Then, as on page 490, $N(E)\,dE = [V/8\pi^3] \int d\mathbf{k}$, where the integral has to be taken over the volume of $\mathbf{k}$-space which is contained between the energy surfaces corresponding to E and $E + dE$. The gradient of $E(\mathbf{k})$, $\mathrm{grad}_\mathbf{k}\, E(\mathbf{k})$ (page 594), is equivalent to the derivative of $E(\mathbf{k})$ in the direction normal to the energy surface through the point $\mathbf{k}$, and since the distance between the surfaces of energy E and $E + dE$ at this particular point is $dE/|\mathrm{grad}_\mathbf{k}\, E(\mathbf{k})|$, $N(E)$ may be obtained as

$$N(E) = [V/8\pi^3] \int [\mathrm{grad}_\mathbf{k}\, E(\mathbf{k})]^{-1}\, dS,$$

where the surface integral is to be taken over the energy surface $E(\mathbf{k}) = E$.

For any particular metallic crystal structure, for electron energy states in any specific direction in $\mathbf{k}$-space, there are critical values of $\mathbf{k}$ for which the Bragg condition is satisfied, and for these values of $\mathbf{k}$ the electrons in the corresponding states are strongly reflected from planes of atoms in the real three-dimensional array of metal atoms. The specific positions at which these $\mathbf{k}$-values occur in $\mathbf{k}$-space will in general depend on the direction considered and will fall on planes which bound the corresponding Brillouin zones. For any one specific direction in $\mathbf{k}$-space, the dependence of E on $\mathbf{k}$ has the form shown in Fig. M.5.6., but the positions and sizes of the energy gaps depend

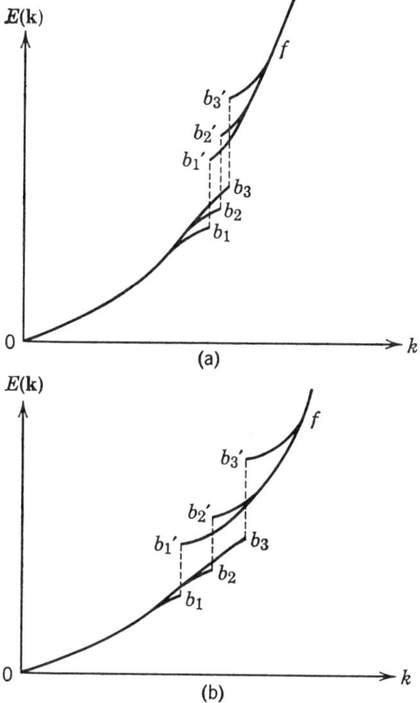

Fig. M.5.35. Diagrammatic representation of the positions and magnitudes of the energy gaps for different directions of the wave vector **k** for a metal crystal. Each of the discontinuous curves, $0b_1b_1'f$, $0b_2b_2'f$, ... corresponds to a different direction in **k**-space. In (a) the energies of states at the top of the first band are less than the energy of any state in the second band, for all directions of **k**. In (b), some of the states at the bottom of the second band have lower energies than the highest states at the top of the first band.

on the direction under consideration. At the top of the first allowed energy band there will be a series of such curves, each referring to a different direction in **k**-space, as shown in Fig. M.5.35. The curves are not generally horizontal at the positions of the energy gaps, only for specific **k**-directions (page 593). It is not possible to show the magnitude of the energy gaps at different positions on the surface of the Brillouin zones, illustrated in Figs. M.5.25.,

M.5.26., and M.5.28., these showing the directional dependence of the points at which gaps occur in the $E(k) - k$ curves.

For any particular Brillouin zone, such as the truncated octahedron which constitutes the first Brillouin zone for face-centered cubic structures and illustrated in Fig. M.5.26., there are two principal possibilities, both shown in Fig. M.5.35. In the first case, (a), the energies for all possible directions of **k** at the top of the first allowed energy band are less than the energy of any state at the bottom of the second band for all directions. In the second case, (b), some of the states at the bottom of the second band have lower energies than the highest states at the top of the first band. If the **k**-value at which the energy gaps occurred were independent of direction, the Brillouin zone concerned would be a sphere. In the free electron theory, not only the complete region of occupied states is spherical, but also all surfaces of constant energy with $E < E^*$. Also in the case of electrons moving in a periodic field, the constant energy surfaces for small energy or **k**-values are approximately spherical, but the energy contours become progressively different from spherical for the constant energy surfaces which correspond to **k**-values closer to the Brillouin zone boundaries. Since in the Brillouin zone theory the geometry of the Brillouin zones is related to the crystal structure of the metal concerned, and the metal crystal structure determines the number of metal atoms per unit cell, the number of electrons per Brillouin zone for any particular metal is determined by the number of conduction electrons per atom of the metal. It is possible to vary the electron concentration, the number of conduction electrons per metal atom, in metals by choosing, for example, two metals with different numbers of electrons per atom and assuming that the metal pair chosen form a continuous series of solid solutions with one another without any change in crystal structure, and then by forming solid solutions of different compositions, each of which will correspond to a different number of electrons per Brillouin zone.

The $N(E)$ curve corresponding to the state of affairs shown in Fig. M.5.35. would, for small electron concentrations, be parabolic as in the free electron case [Fig. M.3.3.]. With increasing electron concentrations, the values of **k** for the highest occupied states approach closer to the surface of the Brillouin zone in specific directions, and when the surface of the zone is reached in those specific directions, it will no longer be possible to accommodate electrons whose states are in these particular directions, since this would necessitate a large increase in energy across an energy gap. After this stage is reached, the $N(E)$ curve will show an abrupt falling off as there will no longer be energy states available to accommodate electrons for some directions of **k**. The corresponding $N(E)$ curve is shown in Fig. M.5.36., the point b_1 being equivalent to the stage at which the surface of the Brillouin zone is first reached by electrons of increasing **k**-value, and corresponding to the point b_1

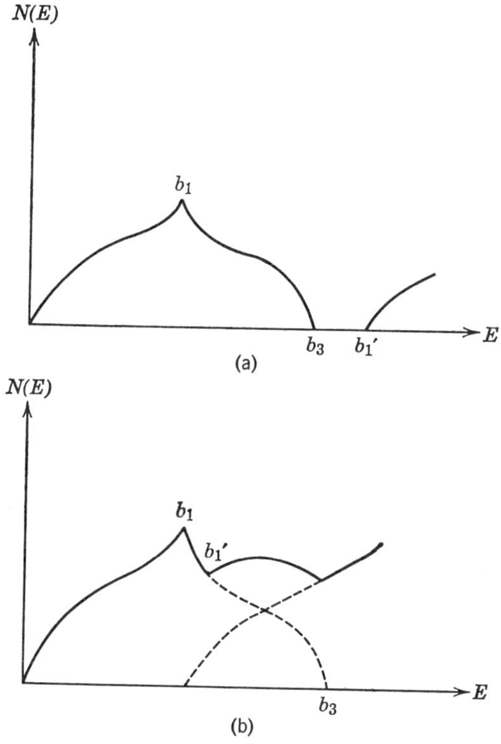

FIG. M.5.36. Representative shapes of the $N(E)$ curves for structures where the Brillouin zones are bounded by only one type of face and these faces intersect in only one type of vertex. In (a), the first and second Brillouin zones contain states of non-overlapping energy values, and in (b) the energies of states in the first two zones overlap.

in Fig. M.5.35. The position of the maximum in the $N(E)$ curve occurs approximately where the energy surfaces of the contained electrons first come in contact with the zone boundary, and the increased slope just before the maximum is reached may be attributed to the bulging of the energy surfaces, which causes an increase in the volume enclosed between two surfaces separated by an energy dE. With a still larger number of electrons, the values of $\mathbf{k}$ increase, and increasingly more directions are affected, until finally all the states in the first Brillouin zone are occupied. As this process occurs, the $N(E)$ curve falls off to zero, where the point b_3 corresponds to the top of the last curve for the first band, as in Fig. M.5.35. If there are more than sufficient electrons to occupy all the states in the first zone, the next electron will be accommodated in a state of considerable higher energy, the increase in energy being the difference between the energy of the point b_1' and the point b_3 in Fig. M.5.35.; these points correspond to the lowest energy of the group of curves for the second allowed energy band and the energy of the top of the highest curve of the first energy band. A similar separation may occur between the second and third Brillouin zones, again with increasing electron concentration, the second zone being completely occupied before any electrons are accommodated in the third zone.

The $N(E)$ curve corresponding to the state depicted in Fig. M.5.35.(b) is shown in Fig. M.5.36.(b). For small electron concentrations, the initial part of this $N(E)$ curve is similar to the first one described above, the falling off at the point b_1 corresponding to the stage at which the surface of occupied states first comes in contact with the first zone surface, and being equivalent to the point b_1 in Fig. M.5.35.(b). The point b_1' in Fig. M.5.36.(b) corresponds to the stage at which electrons begin to occupy the second Brillouin zone and is equivalent to the similarly labeled point in Fig. M.5.35.(b) denoting the lowest curve of the second zone. At this point, the states in the first zone have not been completely filled, since the energy of the point b_3 on the highest of the curves for the first zone is greater than the energy of the point b_1'. With increasing electron concentration, the remaining states of the first zone and the lowest states of the second zone become occupied, so that the corresponding $N(E)$ curve is obtained by summing the ordinates for the two zones where overlap takes place. The availability of a series of new energy states in the second zone which begins at the point b_1' in the $N(E)$ curve gives rise to the sharp rise in the curve at this point. The point b_3 corresponds to the first zone being completely occupied, so that further increases in the electron concentration leads to the filling of the second zone. In a similar manner, a maximum would be reached in the $N(E)$ curve for the second zone, and a second hump would appear in the curve if the second and third zones overlap in energy.

The above descriptions of typical $N(E)$ curves are only applicable if the Brillouin zones involved are bounded by faces of only one type; the point b_1 in each of the two cases referred to represents the stage at which the surface of occupied energy states first makes contact with the Brillouin zone faces, on the assumption that each of the zone faces is equidistant from the origin of k-space. The first Brillouin zones of both the face-centered and body-centered cubic structures are not of this type. For example, the first Brillouin zone for the face-centered cubic structure is bounded by eight hexagonal and six square faces at different distances from the origin of k-space (Fig. M.5.26.), and the first Brillouin zone for the body-centered cubic structure, while bounded by 12

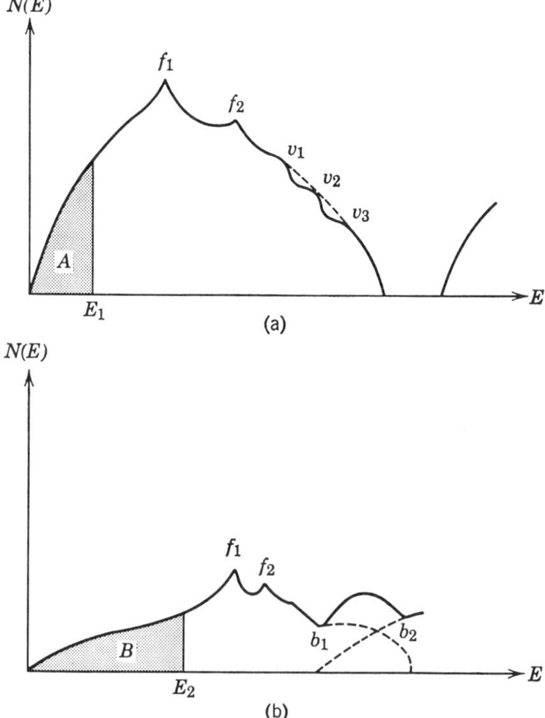

FIG. M.5.37. Representative $N(E)$ curves for structures where (a) the Brillouin zones have two different types of faces bounding the zone, and three different types of vertices, and there are no states of overlapping energy in adjacent zones; in (b), the Brillouin zones have two dissimilar types of bounding faces, only one type of vertex, but there are states of overlapping energy in adjacent zones.

similar rhombus-shaped faces, has two dissimilar types of vertices, so that the 12 face centers are not all equidistant from the origin of k-space. In such cases, there is more than simply one peak in the $N(E)$ curve. For example, for the face-centered cubic structures, where the first Brillouin zone is bounded by octahedral and cube faces at dissimilar distances from the origin, the $N(E)$ curve would appear as in Fig. M.5.37., the first maximum occurring at the stage when the surface of occupied states first comes in contact with the octahedral type faces of the zone, and a second maximum appearing at the point where the surface of occupied states makes contact with the cube faces of the zone. The corresponding $N(E)$ curve falls off quite steeply as the

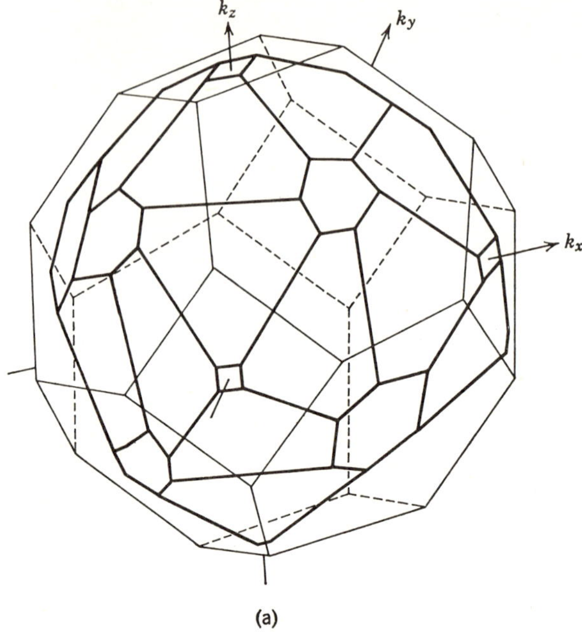

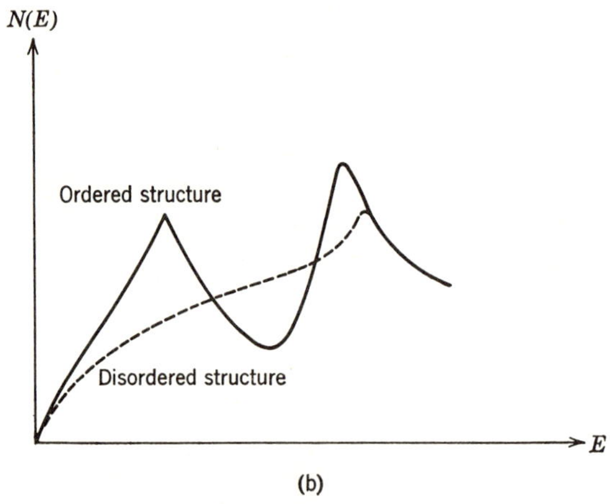

Fig. M.5.38. (a) First Brillouin zone for the superlattice structure, $MgAg_3$, shown with reference to the Brillouin zone for the disordered phase; (b) the corresponding $N(E)$ curves for the random solid solution and the superlattice of composition $MgAg_3$.

surface of occupied states extends to the vertex or vertices of a zone, as the number of states available become very much reduced due to the restrictions imposed by the shape of the zone at these positions; for zones with more than one type of vertex, a rapid falling off of the $N(E)$ curve occurs at the stages where each set of vertices is reached by the surface of occupied states. As illustrated in Fig. M.5.37.(b), it is possible to distinguish between the rounded depressions occurring between the positions f_1 and f_2 where the two types of different zone faces first come in contact with the surface of occupied states, and the sharply defined depressions at b_1 and b_2 due to the zone overlap.

Since the geometry and symmetry of Brillouin zones is determined by the crystal structure of the metal concerned, it is apparent that any change in crystallographic structure of a metal or intermetallic phase, such as that due to phase transformation to another polymorphic modification or superlattice formation due to ordering of a disordered phase (page 409), will give rise to different Brillouin zones, with new planes of energy discontinuity being formed in **k**-space. Such phase transitions or superlattice formation is most easily detected by X-ray diffraction methods. For example, the formation of a superlattice in the Ag–Mg system,[1] involving the ordering transition illustrated in Fig. M.1.11., is depicted in Fig. M.5.38., where the first Brillouin zone for the ordered phase is shown inside the truncated octahedron, which is the first Brillouin zone for the face-centered cubic structure adopted by the disordered phase, and the corresponding change in the $N(E)$ curve is also shown. The increase in stability of a superlattice compared with the disordered phase may be attributed to the two different factors: (a) the reduction in the crystal lattice energy produced by the transfer of the solute atoms in the random solid solution to the ordered positions in the superlattice, and (b) the availability of a larger number of energy states at lower energy values in the Brillouin zones of the superlattice, compared with the smaller number of such states for the face-centered cubic structure of the parent solvent metal species. This latter effect is evidenced by the appearance of a high peak in the $N(E)$ curve of the superlattice structure at a much lower total energy than in the case of the disordered phase, thereby enabling more electrons at low energy to be accommodated, thus reducing the total energy of the system.

M.5.11. Metals, Semiconductors, and Insulators

Metals are most easily differentiated from other solid substances by the fact that they are generally good electrical conductors; under the influence

---

[1] Nicholas, J. F., *Proc. Phys. Soc.*, 1953, **A66**, 201.

of any electric field, an overall flow of electrons in the direction of the applied field must take place, resulting in an electrical current. From the point of view of their electrical conduction, solid materials may be classified into insulators, semiconductors, electrolytic conductors, electronic conductors, and super-conductors. Most pure metals exhibit normal electronic conduction, their resistivity increasing with temperature and impurity content, and approaching zero at 0 K. Electrolytic conductivity is of relatively little importance in metals, although for alloys and intermetallic compounds in which the constituent metals differ greatly in electrochemical properties, it may be of some importance in the liquid state. Some metals exhibit the phenomenon of super-conductivity, where their resistivity remains zero for several degrees above 0 K, and then increases sharply to a finite value at some critical temperature; superconductivity is also destroyed by the application of a magnetic field exceeding a critical temperature-dependent intensity.[1] There is no simple theoretical explanation of these effects, but the Brillouin zone theory enables a satisfactory explanation to be given to account for the behavior of typical metallic conductors, semiconductors, and insulators.

In any solid, there are necessarily a very large number of electrons, which interact with one another in various ways. The approximation of regarding that the same energy band system applies to all the electrons in a solid, with each electron moving in the same self-consistent field, however, enables solid systems to be described in terms of the Brillouin zone theory. At 0 K, all of the electrons within the Fermi surface are contained in the lowest possible set of energy states, two electrons with opposed spins in each orbital state. It has been deduced (page 571) that for substances crystallizing with simple Bravais lattices, each Brillouin zone may accommodate two electrons per atom of the element composing the substance. For many substances there are more than two electrons per atom, necessitating that the occupied states extend over several zones, but important conclusions about the nature of solid materials may be deduced if this number of electrons per atom is even or odd. Thus, for an odd number of electrons per atom, the highest occupied zone, at least, must be only partially occupied, whether or not adjacent zones are of over-lapping energy; for an even number of electrons per atom, the electrons must completely fill the occupied zones, if the highest occupied zone does not over-lap the adjacent one. If the two highest occupied zones contain states of over-lapping energy values, then no matter what the number of electrons per atom may be, the highest zone, at least, must be only partially occupied.

For a solid to exhibit electronic conductivity, it must be possible for an applied electric field to excite electrons to states of slightly higher energy. In the absence of any applied field, by virtue of the symmetry of $E(\mathbf{k})$, the average

---

[1] Bardeen, J., *Advances In Physics*, 1959, **8**, 1, 45.

velocity of an electron in the state **k** is equal and opposed to that of an electron in the state $-\mathbf{k}$, so that at 0 K the symmetrically disposed and fully occupied states within the Fermi surface give rise to no electric current, since for every electron with any particular mean velocity there is one with an equal and opposite average velocity. If the highest occupied zone of a solid is only partially occupied, then a small applied electric field may excite some of the electrons on the Fermi surface of occupied states into slightly higher energy levels, and such a material would exhibit electrical conductivity. If, on the other hand, all of the occupied zones are completely filled such that the occupied region of **k**-space is bounded in all directions by planes of energy discontinuity, then since the electrons within the occupied region of **k**-space cannot be excited and it would require a large field to excite an electron across the energy gap into a vacant state in an unoccupied zone, it is understandable that such a substance would behave as an electrical insulator. This leads to the essential difference between an insulator and a metallic conductor being due to the fact that in the former all the occupied zones are completely filled with electrons and in the latter at least one zone is only partially filled. Of course, for sufficiently high values of an applied electric field, all insulators ultimately break down. For metals the general effect of an electric field on an electron in any particular state within a zone is to change the wave vector **k** continuously in the direction of the applied field until **k** reaches the value of a state on the surface of the Brillouin zone, when a Bragg reflexion takes place within the crystal lattice of the substance concerned, as a result of which the component of the electronic velocity normal to the zone face is reversed. As long as the energy contours within a zone are spherical, the effect of the applied field is to change the electronic velocity and wave vector in the same way, but when **k** approaches the value corresponding to that on the surface of a zone, where the energy contours are no longer spherical (M.5.5.), it is **k** and not v which changes in the direction of the applied field. This process, which takes place within an electronic conductor under the influence of an applied field, means that for any solid with an odd number of electrons per atom, that material must be a conductor. The circumstance that many metals contain an even number of electrons per atom is attributable to the overlapping of energy values of states in adjacent zones. It is only the valency electrons which need to be taken into account, since for any atom the even number of electrons contained in the low-lying energy states cannot contribute to their conductivity. Even though the energy levels occupied by the valence electrons in an atom may extend over several zones, it is usual to refer to this band of allowed energy levels as the conduction band in zone theory. Figure M.5.39. illustrates the distribution of occupied states at 0 K for a one-electron-per-atom metal, like an alkali metal, a two-electron-per-atom metal, such as calcium, and an electrical insulator such as sulfur or diamond, where a zone is completely

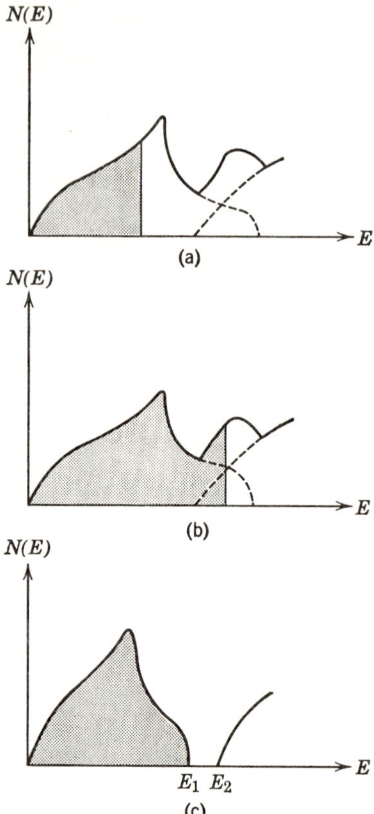

FIG. M.5.39. Representative $N(E)$ curves illustrating the distribution of occupied electron states in (a) a one-electron-per-atom metal, (b) a two-electron-per atom metal, with necessarily overlapping energy values for states in adjacent zones, and (c) an insulator.

filled and separated from an adjacent zone by a large energy gap. For sodium, the zone is only half filled, and it was the natural assumption of all the early theoretical concepts about metals that the Fermi surface of occupied states is spherical and does not reach to the zone surface, which is confirmed by most experimental observations; the electron distribution is essentially parabolic and indeed for sodium the effective electronic mass is not particularly different from the rest mass (see page 595). For calcium and the other alkaline earth metals, with two conduction electrons per atom, there must be overlapping of the energy levels of states in adjacent zones, with two partially filled zones each contributing to their electrical conductivity. These metals crystallize in the close-packed hexagonal structural arrangement, which is not referable to a Bravais lattice, but to the $N(E)$ curve of Fig. M.5.39.(b), which may be regarded as referring to the zone illustrated in Fig. M.5.33. For a typical insulator, the number of valency electrons per atom completely fills a Brillouin zone, with an energy gap across all of its surfaces. For such materials, an applied electric field cannot produce any resultant electron flow in any direction, as electrons are indistinguishable and all of the states in the zone are occupied. The fact that most ionic compounds, such as potassium chlorate or copper sulfate, are insulators is similarly attributable to completely filled energy bands.

At temperatures above 0 K, the distribution of occupied energy states in a metal does not fall off sharply to zero as in Fig. M.5.39., but will fall off more smoothly as illustrated in Fig. M.3.4. for free electrons; according to Fermi–Dirac statistics, however, the range of the thermally disturbed region (page 469) is relatively small at all ordinary temperatures and indeed within the range of the melting points of most metals. For insulators, it is conceivable that a sufficiently large increase in temperature may result in the excitation of some electrons from the highest filled zone to the first available empty zone, with the development of some electrical conductivity. If the energy gap, $(E_2 - E_1)$, is sufficiently large, as it is for diamond (about 10^{-18} J), for example, then negligible conductivity arises in this way. On the other hand, if $(E_2 - E_1)$ is of the order of 1.6×10^{-19} J, as for silicon, germanium, lead sulfide, indium antimonide, and several other substances, which function as insulators at 0 K, then at normal temperatures sufficient numbers of electrons are excited to provide a measurable and detectable conductivity. Such materials are referred to as intrinsic semiconductors.

It is sometimes convenient to describe the conduction properties of a substance in terms of the unoccupied states lying within a zone, rather than with respect to the occupied states. Thus, in Fig. M.5.39.(b), where the density of unoccupied states is very approximately parabolic, and keeping in mind that the current due to a completely filled zone is zero, then the current due to a single unoccupied electron state may be conveniently regarded as the negative of the current due to a zone which is otherwise empty except for a single electron in this state. It is in this sense that in the description of conduction due to an almost completely filled zone, the various electrons may be disregarded and each unoccupied state regarded as if occupied by a positively charged particle, usually referred to as a positive hole, of charge magnitude equal to that of an electron. In the case of a semiconductor of the type referred to above and in Fig. M.5.39., where $(E_2 - E_1)$ is of suitably small magnitude, the conduction process may be regarded as partly due to the electrons which are thermally excited into the conduction band as well as due to the holes left in the so-called, in this case, valency band. It has to be noted that some insulating materials may exhibit the phenomenon of photoconductivity, in which electrons may be excited photoelectrically from the filled zone into a higher energy zone, where they then become free to behave as conduction electrons.

Many compounds with the composition AB_2, crystallizing with the fluorite structure depicted in Fig. M.1.8.(f), behave as intrinsic semiconductors; as the name implies, incidentally, these are semiconductors whose conductivity is a real characteristic of the substance concerned, and is not due to the presence of impurities, as is the case for the most important practical semi-conductors. Calcium fluoride, CaF_2, is usually regarded as a typical ionic compound with the calcium ions occupying the face-centered cubic positions in the unit cell shown in Fig. M.1.8.(f), and the fluoride ions the eight tetra-hedral positions within this unit cell. The compounds Mg_2Pb, Mg_2Sn, Mg_2Ge, and Mg_2Si actually crystallize with the antifluorite structure, with the magnesium atoms occupying the set of positions which are occupied by the fluoride ions in fluorite. These four compounds exhibit some metallic proper-ties, such as having a lustrous appearance and some electrical conductivity, with melting points and ionic character decreasing with increasing electro-positive nature of the elements in the order Si, Ge, Sn, and Pb. Thus, Mg_2Pb behaves as a normal metal[1] with low conductivity and normal temperature coefficient of resistivity, Mg_2Sn is an intrinsic semiconductor with a negative temperature coefficient of resistivity, Mg_2Ge and Mg_2Si having more ionic characteristics. It has been pounted out, alternatively, that the first Brillouin zone for the fluorite structure bounded by (220) faces is capable of accommo-dating 8/3 electrons per atom.[2] This is exactly the number of electrons per atom to be expected if magnesium contributes two electrons per atom and the atoms of silicon, germanium, tin, and lead, each contribute four electrons per atom, and the constitution of these compounds is described from the zone theory point of view.[3] These alternative points of view are further illustrations of the discussion expressed on page 386. In the zone theory description, diamond is an insulator since there is an even number of electrons per atom completely filling a zone with a substantial energy gap across the boundary faces of the zone, and in the Pauling valency bond description, diamond is a strong, high-melting point compound because each carbon atom is bonded by electron-pair bonds to each other carbon atom throughout the crystalline piece of diamond, the electron density being a maximum between pairs of carbon atoms.

---

[1] Robertson, W. D., and Uhlig, H. H., *Trans. Amer. Inst. Min. Met. Eng.*, 1949, **180**, 345.

[2] Mott, N. F., and Jones, H., *The Theory of the Properties of Metals and Alloys*, 1958, Dover.

[3] Hume-Rothery, W., *Atomic Theory for Students of Metallurgy*, 1962, Inst. Metals, London.

It is not necessarily so that the electrical conductivity of a metal increases with the number of electrons per atom available as conduction electrons. Thus, although all of the conduction electrons in a metal take part in the conduction process, the number of electrons actually effective in producing conductivity depends on the nature of the zones involved. In Fig. M.5.39., for example, the metal whose $N(E)$ curve is illustrated in (a) has fewer electrons per atom than that illustrated in (b), but in the latter case the electrical conductivity would be considerably smaller since there are fewer electrons actually available to move into unoccupied states, and indeed the conductivity would be even less if the number of electrons per atom were to be increased to the extent of completely filling the first zone. The electronic conductivity of a metal must also depend on the direction of the applied electric field, under whose influence electron flow takes place into available vacant energy states. It is always only electrons near to the Fermi surface of occupied states which can interact with the thermal vibrations of the metal crystal lattice to affect the velocity distribution of the electrons and therefore create a resistance. The electrical resistivity of pure metals is influenced by the presence of impurity atoms, by lattice defects, dislocations, and other structural imperfections, compared with an "ideal" point-space lattice; it increases with temperature and increasing frequency and amplitude of the lattice vibrations.

For the alkali metals and copper, silver, and gold, the first Brillouin zone is only half filled, and interactions between electrons and thermal vibrations of the metal crystal lattice, electron-phonon interactions, may only transfer an electron into another state in the same zone. With zones of overlapping energy states where the boundary of occupied states is within the region of overlap, electron-phonon interactions may, however, transfer an electron from one zone into another, which is not generally possible under the influence of an electric field alone. The probability of such an electron transition is proportional to the density of states in the final state, so that a metal exhibits a high resistivity if the final state falls on a high part of the $N(E)$ curve for the metal concerned.[1] This is the state of affairs for the transition metals, whose energy band structures are more complicated than for the alkali metals (see page 611), due to the presence of electrons in the free atoms of these elements in d-states, where the density of states is higher but the energies of the electrons in these d-states is not greatly different from those of the valency electrons. The electric current in the transition metals is carried mainly by the s-electrons, which behave approximately like those of the free electron model, but the resistivity is greater because the electron-phonon interactions may cause transitions from states in the s-band to those in the d-band, as well as transitions among the s-states. The metallic elements composing each of the

---

[1] Ziman, J. M., *Electrons and Phonons*, 1962, Oxford University Press, London.

three transition groups of metals each contain in the free atom in its ground state between one and ten electrons in $3d$-, $4d$-, or $5d$-states, respectively. The simplest theory of the transition metals was that first presented by Mott and Jones,[1] who considered in some detail the elements of the first transition group, iron, cobalt, nickel, and copper. In their two-band theory, the $4p$-states were disregarded and the valency electrons were considered to exist in two distinct but overlapping bands derived from the $3d$- and $4s$-states of the free atoms. While, from the foregoing discussion, it is justifiable to regard that there is a Brillouin zone in the metal for each energy state in the free atom, any particular zone can only be considered to be derived from a particular atomic state if there is no overlapping between the various energy bands, or if there is a large energy gap between narrow allowed bands. Where zones overlap, they have to be considered to be derived from the same group of energy states. The general type of correspondence between the energy states of a free atom and the zones in a metal may be thought of as arising from the application of the considerations discussed in connection with the soft X-ray spectra of the metals in M.3.1., and illustrated for sodium in Fig. M.3.1. Metallic copper has been investigated more extensively than any other substance from the point of view of the calculation of the shape of the energy surfaces in the metal and comparison of the results thus obtained with a variety of experimental observations.[2-6] The manner in which the energy bands are believed to widen for copper, and for that matter for the other transition metals with some adjustment of the relative positions of the $3d$ and $4s$ atomic levels, is illustrated diagrammatically in Fig. M.5.40. While it is common practice to refer to the $3d$- and $4s$-bands, these so-referred to bands are not derived in a simple way, such as taking linear combinations of the $3d$- and $4s$-states from the corresponding $3d$ and $4s$ atomic levels.

Calculations of the density of states in copper lead to results in good agreement with experimental observation, and many of the properties of the transition metals may be adequately explained on the basis of the assumption, which is justified by other detailed calculations of Fermi surfaces, densities of states in energy, effective masses, and energy band widths,[7-12] that the $3d$-band

---

[1] Mott, N. F., *Phil. Mag.*, 1953, (**vii**), 44, 187; also reference 2, page 608.

[2] Pippard, A. B., *Phil. Trans. Roy. Soc.*, 1957, **A250**, 325.

[3] Ehrenreich, H., and Philipp, H. R., *Phys. Rev.*, 1962, **128**, 1622.

[4] Shoenberg, D., *Phil. Trans. Roy. Soc.*, 1962, **A255**, 85.

[5] Burdick, G. A., *Phys. Rev.*, 1963, **129**, 138.

[6] Berglund, C. N., and Spicer, W. E., *Phys. Rev.*, 1964, **136**, A1030, A1044.

[7] Joseph, A. S., and Thorsen, A. C., *Phys. Rev.*, 1965, A1159.

[8] Fawcett, E., *The Fermi Surface*, 1960, edited by Harrison, W. A., and Webb, M. B., John Wiley & Sons, Inc., New York.

[9] Wakoh, S., and Yamashita, J., *J. Phys. Soc. Japan*, 1964, **19**, 1342.

[10] Reed, W. A., and Fawcett, E., *Phys. Rev.*, 1964, **136**, A422.

[11] Suffczynski, M., *Phys. Status Solidi*, 1964, **4**, 3.

[12] Coleman, R. V., Funes, A. J., Plaskett, J. S., and Tapp, C. M., *Phys. Rev.*, 1964, **133**, A521.

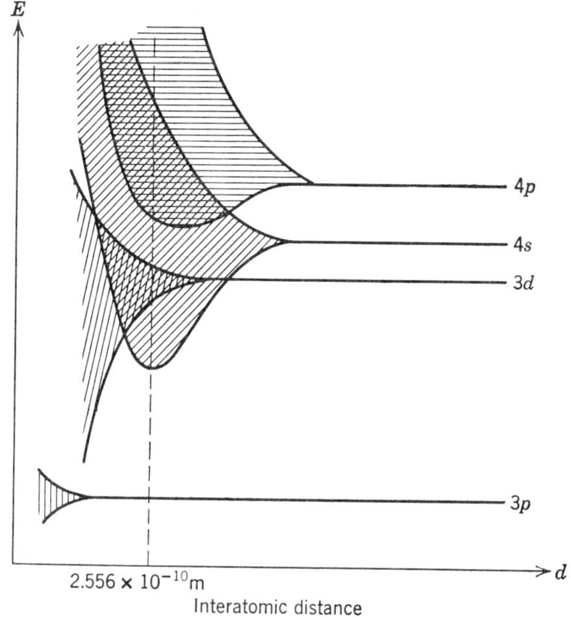

FIG. M.5.40. The energy band widths in metallic copper plotted as a function of the internuclear distance; the dotted line corresponds to the equilibrium internuclear distance in the metal.

is very narrow and overlaps the $4s$-band; the calculated and observed, from photoemission experiments, density of states for copper is shown in Fig. M.5.41. As the $3d$-states are five-fold degenerate in the free metal atoms, the $3d$-band in the metals must range over five Brillouin zones, in spite of its narrow width, so that the density of states in the $3d$-band must be very large compared with that in the $4s$-band, leading to the tall, narrow $3d$-band and the flat, wide $4s$-band shown in Fig. M.5.41. The shape of the Fermi surface in copper was deduced from experimental observation before it was actually calculated, with a result, again in nice agreement with experiment. This is shown in Fig. M.5.42., from which it is seen that the approximately spherical Fermi surface extends along the (111) crystallographic direction cutting the hexagonal faces of the Brillouin zone in such a manner that, due to the periodic repetition of reciprocal space, the Fermi surface in adjacent Brillouin zones is joined up through these necks to form the one surface extending throughout all available space. In copper, with the configuration $\cdots 3d^{10}4s^1$ for the free metal atoms, there is much evidence that in the metal the $3d$-band is filled and the $4s$-band contains one electron per atom. The high electronic specific heats of, e.g., nickel and other transition metals generally, is attributable to

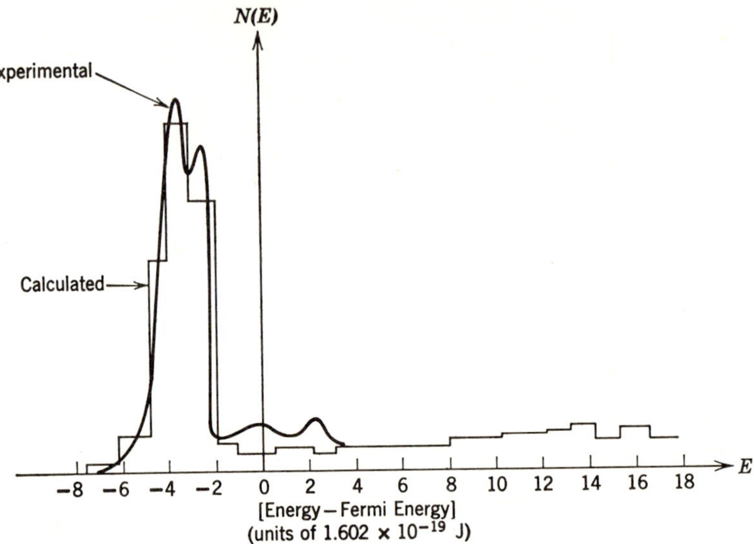

FIG. M.5.41. Comparison of the calculated and experimentally determined density of states in energy for copper. The experimental results are those of Berglund and Spicer (reference 6 on page 610), and the calculated results are those of Burdick (reference 5 on page 610).

the incompletely filled $3d$-band, which also provides an explanation of the ferromagnetism of iron, cobalt, and nickel. At low temperatures, the specific heat of nickel is about ten times as large as that of copper. The zone theories, however, do not offer any explanation as to why ferromagnetism is confined to iron, cobalt, and nickel.

The electrical conductivity of pure metals is very sensitive to the addition of impurity species. Generally, in solid solution formation, even for two metals so similar in many respects as silver and gold, the conductivity decreases with increasing atomic percentage of solute species, and otherwise is greater the greater the electrochemical difference, the atomic radii, the electronic configuration, and the number of valency electrons per atom, between the solvent and solute species. Conversely, any tendency to form super-lattices or ordering in a solid solution leads to increased conductivity in the more ordered phase.

For semiconductors, in contrast to the behavior of normal metals, their electrical conductivity is increased by the presence of impurities and by rise of temperature. For a substance with the $N(E)$ characteristics of Fig. M.5.39.(c) with a fully occupied zone separated from the next zone by a very small energy gap, such a material would behave as an insulator at 0 K. At higher

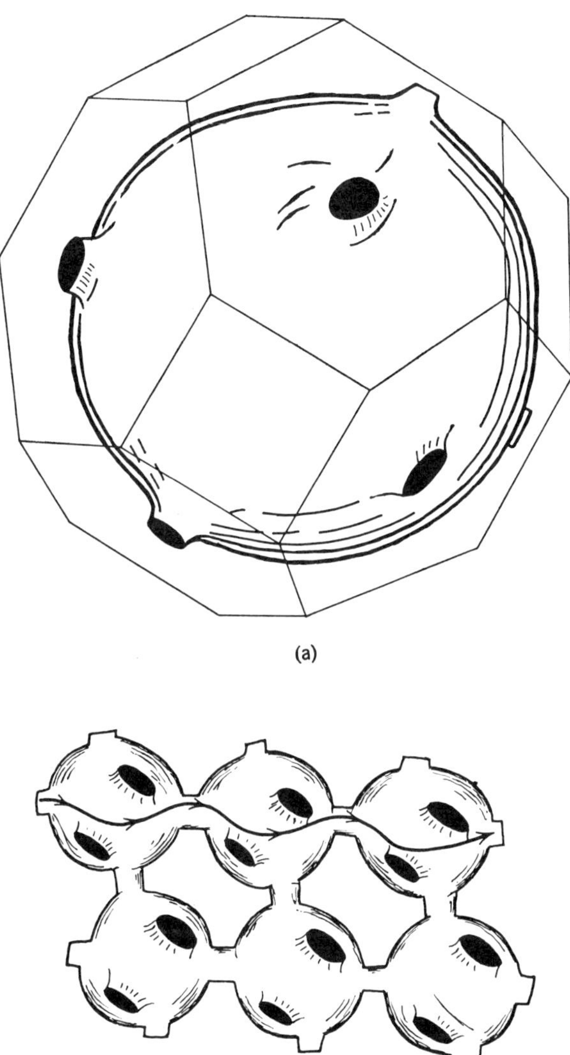

FIG. M.5.42. The Fermi surface of copper. Outlined in (a) is the truncated octahedron, which is the first Brillouin zone for the face-centered cubic structure of copper. Inside this zone is then shown the approximately spherical surface of the occupied region of electron states, with the Fermi surface bulging up to make contact with the octahedral faces of the zone. The way these adjacent zones and the Fermi surfaces join up to form a continuous surface throughout reciprocal space is shown in (b).

TABLE **M.5.3**   Properties of some semiconductors

Material	Crystal Structure	Melting Point (K)	Vapor Pressure at m.p.	Band Gap $\times 10^{-19}$ J	Mobility $cm^2 s^{-1} V^{-1}$
Graphite, C	graphite	incongruent		9.29	1200-p
Silicon, Si	diamond	1695	low	1.76	1900-n
					480-p
Germanium, Ge	diamond	1219	low	1.07	400-n
					1900-p
Silicon carbide, α-SiC	würtzite	incongruent		4.49	400-n
					60-p
Gallium phosphide, GaP	zinc blende	1743	35 atm.	3.52	150-n
					150-p
Gallium arsenide, GaAs	zinc blende	1511	1 atm.	2.24	8500-n
					400-p
Indium antimonide, InSb	zinc blende	798	low	0.27	30000-n
					1000-p
Cadmium sulfide, CdS	würtzite	1743	4 atm.	3.84	200-n
Zinc selenide, ZnSe	zinc blende	1783	1 atm.	4.17	1000-n
Lead sulfide, PbS	rock salt	1393		0.64	800-n
					1000-p
Bismuth telluride, Bi_2Te_3	rhombohedral	859		0.24	1250-n
					515-p

temperatures, however, electrons would be thermally excited from the first to the second zone, and then when so excited into states at the bottom of the second zone, they would behave as free electrons, giving rise to a conductivity which would increase further with temperature, as more electrons would become excited into the second zone; such substances are the intrinsic semiconductors, already referred to on page 607, and it is the thermal excitation and not the influence of an applied electric field which causes them to behave in this manner. The properties of some semiconductors are shown in Table M.5.3.

In dealing with the properties of semiconductors, it is convenient to express equation (m.1.34.) for the conductivity of a metal, or other conductor, in another way, by defining a quantity, referred to as the mobility of a charged particle, under the action of an electric field, as the velocity which it acquires per unit electric field. The terminal velocity given by equation (m.1.31.) is $\varepsilon V/C$, so that the mobility is therefore ε/C, so that the current density i, given by (m.1.33.), may be written in terms of the mobility μ as

$$\kappa_\varepsilon = n\varepsilon^2\tau/m_\varepsilon = n\varepsilon\mu. \qquad \text{(m.5.78.)}$$

Experimental values of the mobility, assuming n to be of the order of magnitude of one electron per atom, as determined by measurement of the Hall effect (page 637), prove to be of the order of magnitude of hundreds of centimeters-per-second-per-volt-per-centimeter of field, for good conductors (see Table M.5.1.). Now for semiconductors, in which the energy gap is small enough so that a substantial number of electrons may be excited into the conduction band, leaving an equal number of holes in the valency band by the application of modest temperatures, the conductivity will be

$$\kappa_\varepsilon = n_h\varepsilon\mu_h + n_\varepsilon\varepsilon\mu_\varepsilon, \qquad \text{(m.5.79.)}$$

where n_h and n_ε are the numbers of positive and negative carriers per unit volume, and μ_h and μ_ε the respective mobilities. The order of magnitude of the mobilities and their temperature variation are not particularly different in semiconductors from what they are in metals, so that the striking differences between semiconductors and metals must be associated with the n's. The number of electrons in the conduction band and the number of holes in the valency band may now be found by the application of the Fermi–Dirac distribution function, given by equation (m.3.14.). At ordinary temperatures, with most ordinary semiconductors, the energy gap is very large compared with the term kT. If the width of the energy gap is ΔE, then from equation (m.3.14.) it is apparent that the exponential quantity, $e^{(E-E^*)/kT}$, is at least as great as $e^{\Delta E/2kT}$ for all energy levels in the conduction band, and therefore unity may be neglected in comparison with it. At ordinary temperatures kT is of the order of 4.8×10^{-21} J. Thus, if E is the energy of a level in the conduction band measured from the bottom of that band, $f(E)$ for one of the levels in the conduction band will be given by the expression

$$f(E) = e^{-\Delta E/2kT}e^{-Ek/T}. \qquad \text{(m.5.80.)}$$

The sum $\sum f(E)$ over all states of both spin types may be converted into an integral, and the integration carried out over the levels of the conduction band, by calculating the number of energy levels per unit range of energy, using equation (m.3.6.) or one of its equivalent expressions given in sections M.3.4.–6. The number of electrons per unit volume in the conduction band is then obtained as

$$f(E) = [2h^{-3}][2\pi m^*kT]^{3/2}e^{-\Delta E/2kT}, \qquad \text{(m.5.81.)}$$

where m^* is the effective mass of the electrons in states near the zone boundaries, assuming that this effective mass is the same for states at the top of the valency band and at the bottom of the conduction band, and is given by equation (m.5.77.). The number of holes in the valency band, which can be obtained by summing $2[1 - N(E)]$ over the levels of that band, has of course the same value as (m.5.81.), as required, in order that the total sum of $f(E)$ over all levels has the correct value. The factor $e^{-\Delta E/kT}$ in the expression for both the numbers of electrons and holes, which decreases very rapidly to zero as $T \rightarrow 0$ K, is the most important term in the expression (m.5.79.) for the conductivity of a conductor. If the logarithm of the conductivity is plotted as a function of the reciprocal of T, $(1/T)$, the result is very nearly a straight line. The dependence of the mobility on temperature, and the factor $T^{3/2}$ in equation (m.5.81.), cause a relatively small departure from linearity with respect to the exponential factor, and their temperature dependencies approximately compensate each other, so that the exponential increase with temperature of the number of electrons in the conduction band overcompensates for the normal increase of resistance with temperature resulting from the increase of the amplitude of the thermal vibrations of the crystal lattice, and therefore the electrical conductivity of a semiconductor increases with temperature. The straight lines obtained by plotting the logarithm of the conductivity of a semiconductor as a function of $1/T$ enable the magnitude of the energy gap ΔE to be found from the slope of the line. It is, however, very difficult to prepare most semiconductors in a pure enough state so that they behave as intrinsic semiconductors. By substitution of the appropriate constants in (m.5.81.), since kT at ordinary temperatures is about 4.8×10^{-21} J, it is seen that for germanium, for instance, the number of conducting electrons is about 10^{14} cm^{-3}, compared with about 10^{22} cm^{-3} for a normal metal.

Another important class of semiconductors owes its properties to the presence of impurities. In these materials, a completely filled zone, which need not be the first one, is again separated from an empty one by a comparatively narrow energy gap. The general effect of introducing impurity atoms is to create new energy levels, and these may lie within the energy gap. At 0 K, the substance would be an insulator, but at higher temperatures, one or both of two alternative processes may take place. First, if electrons are excited thermally from the impurity levels into the vacant zone and enter states at the bottom of this second zone, a limited conductivity will develop. Such semiconductors are known as n-type semiconductors, because the electrons behave as though they had a normal negative charge, denoted by the n, and the impurity atoms are called donor atoms, as they provide electrons for excitation into the second zone. The conductivity of n-type semiconductors increases with temperature, although if the concentration of donor atoms, such as phosphorus atoms with five electrons per atom in silicon with four electrons per atom, is very large, a stage may be reached at which the conductivity may fall, due to the usual effect of thermal vibrations. For substances with a narrow band gap, there may be exhibition of both intrinsic and n-type behavior by such a material, the former being of more significance at high temperatures and the latter at low temperatures. In this event, the conductivity plotted as a function of temperature would show a change of direction at the temperature at which the donor species had given up nearly all its electrons. The other process which may take place is that impurity energy levels to which electrons are thermally excited from the valency band may be added to a substance by dissolving impurity atoms with fewer electrons per atom than the solvent species, such as the addition of boron atoms to silicon. Such a semiconductor is called a p-type, the p denoting the positive charge of the holes in the valency band, which function as the charge carriers in such a case. If the concentration of acceptor atoms in a p-type semiconductor is small, the increase in conductivity with increasing temperature may again show a saturation effect, and providing the energy gap between the zones is sufficiently small, both intrinsic and p-type behavior may occur together.

A tremendous amount of effort has been devoted to the study of the behavior of electrons in extremely pure crystals of semiconducting substances such as germanium and silicon. Much of this technological effort followed the discovery of the transistor in 1948, making use of such substances, and from such work has arisen the new telecommunication systems, computer systems, and the electric power industry generally. It is estimated that perhaps the study of the behavior of electrons in solids has taken up more time than that of any other technological achievement. In 1968 alone, over 10,000 technical papers were published dealing with semiconductors and various electronic devices using them. Scientific and technological interest in this general area is so great that with the realization in 1968 that semiconducting devices may be made from vitreous materials, which are neither crystalline nor ultrapure, sufficient work had been done on this topic to justify holding two international conferences devoted to the study of the behavior of electrons in vitreous materials within ten months.

The vitreous materials of major interest are various oxides of silicon, germanium, phosphorus, vanadium, and tungsten, and glasses composed of two to four of the elements, oxygen, sulfur, selenium, and tellurium, together with either silicon, germanium, zinc, cadmium, or arsenic. Such glasses are essentially intrinsic semiconductors, containing approximately equal numbers of electrons and holes, but without the two current carriers necessarily having the same mobilities. It is believed that the electrons in such glassy or vitreous materials have a band structure somewhat similar to that in metals, despite the lack of order over any long range, on the atomic scale. Perhaps also the disorder in the vitreous state leads to the existence of additional, discontinuous, energy levels, which destroy the sharpness of the band edges and lead to band overlap. One theory of the mode of operation of such vitreous materials as semiconductors is based on the concept that an insulator, such as sulfur, behaves as such not only because there is a deficiency of charge carriers since all the electrons available from the sulfur atoms occupy a zone of states separated from the next higher energy zone by an energy gap, but also because they contain a high concentration of traps for charge carriers. These carrier traps may be due to impurities, disorder in the crystallographic sense, or to some other cause, and the energy levels that are occupied by such trapped charge carriers correspond to the additional, discontinuous, energy levels which exist in the vitreous materials. It is then postulated that in vitreous semiconductors, there are two sets of discontinuous trapping states, one each for carriers and one for holes, and derived from the conduction and valency bands as extensions into the energy gap, such as exists in an insulator. According to this idea, the energy band gap of normal semiconductors is replaced by a mobility gap.

M6

Electron Theories and the Electronic Structure of Metals

M.6.1. Collective Electron Theories Based on Bloch Functions

The simple energy band theory of perfect crystals, with their nuclei assumed to be at rest in an ideal point-space lattice, can still obviously be only an approximation to the actual state of affairs in a real metal crystal. The theoretical interpretation of the properties of solids in general, and metals in particular, involves the study of the interactions between the electronic motion, the crystal lattice vibrations, the effect of spin, and the electro-magnetic field effects, both those imposed externally and those produced by electric and magnetic sources within the material itself. It is a matter of great difficulty with numerous complications to provide a detailed theoretical description of the behavior of electrons in a metal, and it is really only for the alkali metals, for aluminum, and in a few other cases that the various electron theories provide quantitative agreement with experimentally observed properties.

The properties and behavior of metals, like all other chemical compounds, depend essentially on the electrons derived from those in the highest energy states of the constituent atoms, which are ns and np electrons for the light metallic elements; ns, np, and $(n - 1)d$ electrons for the transition metals; and ns, np, $(n - 1)d$, and $(n - 2)f$ electrons for the lanthanide and actinide metals. In the solid metal, however, as indicated in section M.3.1., the sharp energy levels of free atoms become broadened into bands with, in general, overlapping taking place between adjacent bands, so that electron states in the metal have to be regarded as being derived from two or more atomic states. To properly describe the behavior of an electron in a metal would

619

require the use of a state function, which had the characteristics of the metallic ion close to this ion, perhaps the characteristics of a free electron when moving outside the influence of the metal ion cores, and some other characteristic at other positions in the metal crystal. The problem of calculating such functions requires, in turn, some knowledge about the nature of the periodic field in which the electrons move, and how it varies in the metal. Such potential fields cannot usually be calculated accurately, but self-consistent field type calculations provide good approximations. Various proposals and techniques have been proposed and devised to approximate the potential function for particular calculations. In a periodic field, the electron states are described by Bloch functions of the general type given by equation (m.5.60.), $\psi_k(\mathbf{r}) = e^{i\mathbf{k}\cdot\mathbf{r}}\phi_k(\mathbf{r})$, which represents a plane wave of wavelength $2\pi/k$, moving in the direction of the wave vector $\mathbf{k}$, but modulated by the three-dimensional periodicity of the crystal lattice. In the simple band theories, $\phi_k(\mathbf{r})$ has just the periodicity of the crystal lattice, but in the more detailed collective electron theories based on the use of Bloch-type functions, the characteristics of s, p, d, or combinations of these, states is periodically repeated, the complete Bloch function for each electron extending throughout the whole piece of crystalline material. Bloch-type functions are one-electron wave functions, describing each electron in a system singly, assuming that the field produced by all of the others, with respect to the one under consideration, are smoothed out to provide the same periodicity as that of the crystal lattice. They do not allow for correlation effects between electrons and become succeedingly less adequate as volume per electron in a metal decreases. However, energy band calculations of considerable accuracy have been carried out.[1-3] Some of the more important approximation methods which have been described, based on Bloch-type functions, are discussed briefly in the following sections, together with some of the experimental methods which provide direct information about the properties of metals, for comparison with the theoretical calculations.

M.6.2.   The Nearly Free Electron Approximation

In the simple Brillouin zone theory developed in the previous chapter, the metallic valency electrons are regarded as moving in a regular periodic field of period a (Fig. M.5.1.), without making any detailed specification about the

[1] Woodruff, T. O., *Solid State Physics*, 1955, edited by F. Seitz and D. Turnbull, Academic Press, New York.

[2] Slater, John C., *Quantum Theory of Molecules and Solids*, 1965, **Vol. 2**, *Symmetry and Energy Bands in Crystals*, McGraw-Hill Book Company, New York.

[3] Slater, John C., *Insulators, Semiconductors, and Metals*, 1967, **Vol. 3**, McGraw-Hill Book Company, New York.

field. The Brillouin zones arise from the reflexion of electrons by the crystal lattice, so that the faces of the zones are directly related to the planes of the crystal with large structure factors (pages 543 and 585). In the nearly free electron approximation, the periodic potential function is regarded as being of very small amplitude and treated as a small perturbation applied to the free electron system. The nature of the argument involved in the application of this method is most easily appreciated by consideration again of a one-dimensional system.

Suppose that the unperturbed wave function in one dimension is the free electron function, given by equation (m.3.78.), normalized in the length of the lattice row L, and given by

$$\psi_k(x) = L^{-1/2}e^{ikx}, \qquad \text{(m.6.1.)}$$

where k is not restricted to the reduced zone representation, but making use of the extended zone scheme as in Fig. M.5.2. Since this unperturbed function is doubly degenerate, as $\psi_k(x) = \psi_{-k}(x)$, both correspond to the unperturbed energy $h^2k^2/8\pi^2m$ (see section M.3.6.). Now applying first order perturbation theory for a doubly degenerate state, as described in sections Q.4.5. and Q.4.6., since the functions given by (m.6.1.) are orthonormal in the length L, then if the perturbation energy is denoted by E', the appropriate secular equation, from equation (q.4.72.), may be written in matrix notation (page 125) as

$$\begin{vmatrix} V_{k,k} - E' & V_{k,-k} \\ V_{-k,k} & V_{-k,-k} - E' \end{vmatrix} = 0, \qquad \text{(m.6.2.)}$$

where the various matrix elements of the potential energy with respect to the unperturbed functions are

$$V_{k,k} = \int_0^L \psi_k{}^*V(x)\,\psi_k\,dx = L^{-1}\int_0^L V(x)\,dx = V_{-k,-k};$$

$$V_{k,-k} = \int_0^L \psi_k{}^*V(x)\psi_{-k}\,dx = L^{-1}\int_0^L e^{-2ikx}V(x)\,dx; \qquad \text{(m.6.3.)}$$

$$V_{-k,k} = \int_0^L \psi_{-k}{}^*V(x)\psi_k\,dx = L^{-1}\int_0^L e^{2ikx}V(x)\,dx = V_{k,-k}^*.$$

Since $V_{k,k} = V_{-k,-k}$, and these quantities are simply the average value of $V(x)$ over the one-dimensional lattice L, the zero of energy may be conveniently chosen so that the average value is 0, when the secular equation, (m.6.2.), becomes

$$[E']^2 = V_{k,-k}V_{-k,k} = |V_{k,-k}|^2. \qquad \text{(m.6.4.)}$$

If, as on page 525, $L = Ga$, and p is any integer, then the integrals (m.6.3.) may be split up into G intervals of length a, so that

$$\int_0^L e^{ikx}V(x)\,dx = \sum_{p=0}^{G-1}\int_{pa}^{(p+1)a} [e^{ikx}V(x)]\,dx. \qquad \text{(m.6.5.)}$$

Then, by virtue of the periodicity of $V(x)$, if x is expressed as $x = X + pa$,

$$\int_{pa}^{(p+1)a} [e^{ikx}V(x)]\,dx = \int_0^a [e^{ik(X+pa)}V(X+pa)]\,dX = e^{ipka}\int_0^a [e^{ikX}V(x)]\,dX. \tag{m.6.6.}$$

The variable of integration, whether x or X, is of no consequence in a definite integral, so that the same integral is contained as a factor in every term of the sum in (m.6.5.). Therefore,

$$\int_0^L [e^{ikx}V(x)]\,dx = \sum_{p=0}^{G-1}\int_{pa}^{(p+1)a} [e^{ikx}V(x)]\,dx = \sum_{p=0}^{G-1} e^{ipka}\int_0^a [e^{ikx}V(x)]\,dx. \tag{m.6.7.}$$

Now, by writing the exponential expansion of the sum in (m.6.7.),

$$\sum_{p=0}^{G-1} e^{ipka} = 1 + e^{ika} + e^{2kia} + \cdots + e^{i(G-1)ka} = [1 - e^{iGka}][1 - e^{ika}]^{-1}$$

$$= [1 - e^{2\pi ig}][1 - e^{2\pi ig/G}]^{-1} = 0, \tag{m.6.8.}$$

unless $e^{ika} = 1$, which is the case if $k = 2\pi n/a$, where n is zero, or a positive or negative integer. If $k = 2\pi n/a$, then each term in the series is unity and the sum in (m.6.8.) is simply G. Therefore,

$$\int_0^L [e^{ikx}V(x)]\,dx = 0, \tag{m.6.9.}$$

and also since the integrand has the period a,

$$\int_0^L e^{2\pi inx/a}V(x)\,dx = G\int_0^a e^{2\pi inx/a}V(x)\,dx. \tag{m.6.10.}$$

There are the additional consequences of (m.6.9.) that, if $\psi_k(x)$ and $\psi_{k'}(x)$ are Bloch functions of the type $\psi_k(x) = e^{ikx}\phi_k(x)$, as given by (m.5.2.), or if $\psi_{kl}(x)$ and $\psi_{k'l'}(x)$ are Bloch functions defined by (m.5.16.) in the reduced zone scheme, then

$$\int_0^L \psi_k{}^*(x)V(x)\psi_{k'}(x)\,dx = \int_0^L [e^{i(k'-k)x}V(x)\phi_k{}^*(x)\phi_{k'}(x)]\,dx = 0, \tag{m.6.11.}$$

unless $k' - k = 2\pi n/a$, since the integral above is of the same form as (m.6.9.) and $V(x)\phi_k{}^*(x)\phi_{k'}(x)$ has a period a, and

$$\int_0^L [\psi_{kl}{}^*(x)V(x)\psi_{k'l'}(x)]\,dx = 0, \tag{m.6.12.}$$

unless $k = k'$, since when k and k' apply to the reduced zone scheme, the only permissible value of n is zero.

    Making use of (m.6.9.), it is apparent that $V_{k,-k}$ contained in equation (m.6.4.) must be zero, unless $k = n\pi/a$, where $n = 0, \pm1, \pm2, \ldots$, and in this event $V_{k,-k}$ is by the application of (m.6.10.) equivalent to

$$V_n = [L]^{-1} \int_0^L [e^{-2\pi inx/a} V(x)] \, dx = a^{-1} \int_0^a [e^{-2\pi inx/a} V(x)] \, dx. \quad \text{(m.6.13.)}$$

Therefore, first order perturbation theory indicates that no separation of the degenerate energy levels occurs, except when $k = n\pi/a$, and then the unperturbed level splits into the two perturbed levels given by

$$[h^2/8\pi^2 m][n\pi/a]^2 \pm |V_n|. \quad \text{(m.6.14.)}$$

The energy gap at $k = n\pi/a$, between the $|n|$th and the $|n + 1|$th zone, in the reduced zone scheme is thus

$$\Delta E[n\pi/a] = 2|V_n|; \quad \text{(m.6.15.)}$$

the application of first order perturbation theory does not lead to the displacement of the energy curves for other values of k, even although energy gaps only occur for $k = n\pi/a$ and the energy curve must be similar to the one shown in Fig. M.5.6. By reference to the Fourier series expansion given in equation (q.2.32.), it is apparent that the quantities V_n are the Fourier coefficients of the potential energy given by the expansion

$$V(x) = \sum_{n=-\infty}^{\infty} V_n e^{2\pi inx/a}. \quad \text{(m.6.16.)}$$

The fact that the perturbed functions at the zone boundaries, where $k = n\pi/a$, are not traveling waves but stationary waves follows from the consideration that at the zone boundaries the perturbed wave functions are linear combinations of the type $C_1 e^{n\pi ix/a} + C_2 e^{-n\pi ix/a}$, and from their symmetry characteristics or from the perturbation results, $C_1 = \pm C_2$, so that the perturbed functions are proportional to $\sin(n\pi x/a)$ and $\cos(n\pi x/a)$. The reason why such stationary waves correspond to different energies is that their moduli squared have maxima at different positions along the lattice row. If, as in Fig. M.6.1., $V(x)$ has minima at points corresponding to $x = 0, a,$ $2a, \ldots$, then the cosine function must always correspond to the lower energy value, since $\cos^2(n\pi x/a)$ has maxima at the position of minimum potential energy, while $\sin^2(n\pi x/a)$ has minima. For $n = 1$, the cosine function squared would correspond to the points A, A' in Fig. M.5.7. and the $\sin^2(\pi x/a)$ to the points denoted by B or B'. The correspondence between the analytical condition for Bragg reflexion by a one-dimensional lattice of period a through $90°$, as discussed on page 543, namely that $n\lambda = 2a = 2\pi n/k$, so that $k = n\pi/a$, which is equivalent to the wave vector values at which energy gaps occur, gives rise to the interpretation that a beam of electrons with such

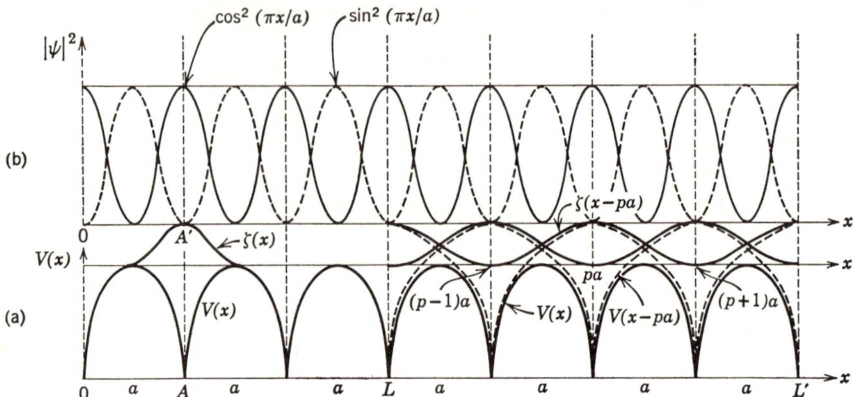

Fig. M.6.1. Representation of the potential energy function, $V(x)$, for an electron moving in a one-dimensional row lattice of period a (a), and the probability densities of electron states at the boundaries of the first Brillouin zone (b).

k-values, or energies, must be totally reflected when incident on a lattice of period a.‡

‡ Reference has been made in several places to the fact that the faces of Brillouin zones correspond usually to planes in the crystal structure of the metal concerned with large values for their structure factors (pages 543, 585, 621), and the simple geometrical interpretation of structure factors has been discussed on page 585. The relationship between the occurrence of energy gaps and the structure factor of the boundary planes of the Brillouin zones emerges from an examination of the direct extension of the perturbation type argument, involved in the nearly free electron approximation method, to real three-dimensional crystals.

In three dimensions, the unperturbed wave functions are the free electron functions

$$\psi_k(\mathbf{r}) = [V]^{-1/2}[e^{i\mathbf{k}\cdot\mathbf{r}}],$$

which are normalized over the whole volume of the crystal. As seen on page 623, for the one-dimensional case, energy gaps occur when the matrix element $V_{k,-k}$ of the potential energy has a non-zero value. In three-dimensional k-space, the degree of degeneracy is very much larger than in one dimension, since all of the unperturbed states located on a sphere of radius k correspond to the same energy value, $h^2k^2/8\pi^2m$. Alternatively, it may be said that the condition for the occurrence of an energy gap is that some of the non-diagonal elements of the secular determinant must have non-zero values, and this situation occurs when the vector $\mathbf{k}$ has such a value that the matrix element $V_{k,k'}$, is not zero, where $|\mathbf{k}| = |\mathbf{k}'|$. The matrix element,

$$V_{\mathbf{k},\mathbf{k}'} = \int \psi_{\mathbf{k}}{}^{*} V_t(\mathbf{r})\psi_{\mathbf{k}'}\, d\mathbf{r} = V^{-1}\int [e^{i(\mathbf{k}'-\mathbf{k})\cdot\mathbf{r}} V_t(\mathbf{r})\, d\mathbf{r},$$

where the integration is carried out over the volume of the crystal, denoted by the non-functionally dependent V. By making use of the periodic nature of the potential function $V_t(r)$, as given by equation (m.5.56.), and writing $\mathbf{r} = \mathbf{r}' + \mathbf{K}_d$, the integral involved in the matrix element above may be reduced to an integration over the unit cell of the crystal structure by means of the technique used in the one-dimensional case (page 622). Thus

$$V_{\mathbf{k},\mathbf{k}'} = [V]^{-1} \sum_d \int [e^{i(\mathbf{k}'-\mathbf{k})\cdot(\mathbf{r}'+\mathbf{K}_d)}V(\mathbf{r}')]\,d\mathbf{r} = V^{-1} \sum_d e^{i(\mathbf{k}'-\mathbf{k})\cdot\mathbf{K}_d} \int [e^{i(\mathbf{k}'-\mathbf{k})\cdot\mathbf{r}}V_t(\mathbf{r})]\,d\mathbf{r},$$

where the integrals apply to a unit cell of the structure and the sums are to be taken over all the cells of the structure. Now, making use of equations (m.5.53.) and (m.5.62.),

$$(\mathbf{k}'-\mathbf{k})\cdot\mathbf{K}_d = 2\pi[(g_1'-g_1)(n_1/G_1) + (g_2'-g_2)(n_2/G_2) + (g_3'-g_3)(n_3/G_3)].$$

Thus, the matrix element $V_{k,k'}$ contains the triple sum

$$\sum_{n_1=0}^{G_1-1} e^{2\pi i(g_1'-g_1)(n_1/G_1)} \sum_{n_2=0}^{G_2-1} e^{2\pi i(g_2'-g_2)(n_2/G_2)} \sum_{n_3=0}^{G_3-1} e^{2\pi i(g_3'-g_3)(n_3/G_3)},$$

which is zero, except when

$$g_1' = g_1 - p_1 G_1, \qquad g_2' = g_2 - p_2 G_2, \qquad g_3' = g_3 - p_3 G_3,$$

where

$$p_1 = p_2 = p_3 = 0, \pm 1, \pm 2, \pm 3, \pm \cdots$$

This may be alternatively expressed, making use of equations (m.5.54.) and (m.5.63.), by saying that $V_{\mathbf{k},\mathbf{k}'}$ will be zero unless

$$\mathbf{k}' = \mathbf{k} - \mathbf{K}_r, \qquad \text{or} \qquad |\mathbf{k}|^2 = |\mathbf{k}-\mathbf{K}_r|^2 \qquad \text{or} \qquad 2\mathbf{k}\cdot\mathbf{K}_r = |\mathbf{K}_r|^2,$$

which is the same equation as (m.5.64.), defining the boundary planes of the Brillouin zone. As on page 568, functions of the type $e^{i\mathbf{K}_r\cdot\mathbf{r}}$ conform with periodic boundary conditions, so that

$$\int_V [e^{i(\mathbf{K}_r-\mathbf{K}_q)\cdot\mathbf{r}}]\,d\mathbf{r} = 0 \quad \text{if} \quad \mathbf{K}_r \neq \mathbf{K}_q \qquad \text{and} \qquad \int_V [e^{i(\mathbf{K}_r-\mathbf{K}_q)\cdot\mathbf{r}}]\,d\mathbf{r} = V \quad \text{if} \quad \mathbf{K}_r = \mathbf{K}_q.$$

As a consequence, it is possible to expand the potential energy function as a triple Fourier series (page 131) of the type

$$V_t(\mathbf{r}) = \sum_{\mathbf{r}} V_{\mathbf{r}} e^{i\mathbf{K}_r\cdot\mathbf{r}},$$

where the summation is applicable over all positive and negative integral values, as well as zero, of the indices r_1, r_2, and r_3. The Fourier coefficient V_r is (pages 128–132)

$$V_{\mathbf{r}} = V^{-1} \int e^{-i\mathbf{K}_r\cdot\mathbf{r}} V_t(\mathbf{r})\,d\mathbf{r},$$

which is exactly what the expression given for the matrix element $V_{\mathbf{k},\mathbf{k}'}$ above becomes when the condition that $\mathbf{k}' = \mathbf{k} - \mathbf{K}_r$ is satisfied, and so there will be an energy gap of finite magnitude on the plane which is defined by $\mathbf{K}_r$, subject to the condition that the Fourier coefficient of the potential energy is itself not zero.

For a compound crystal lattice, which may be described as two or more interpenetrating Bravais lattices, of number corresponding to the number of atoms in a unit cell of the compound lattice, the total potential energy $V_t(\mathbf{r})$ may be conveniently regarded as the sum of the potential energy values due to the separate Bravais Lattices. If the unit cell of the compound lattice contains t atoms with position vectors relative to a vertex of the unit cell of $\mathbf{d}_1, \mathbf{d}_2, \mathbf{d}_3, \ldots, \mathbf{d}_t$, and the potential energy of an electron due to only one of the constituent Bravais lattices is $V'(\mathbf{r})$, then

$$V(\mathbf{r}) = \sum_{j=1}^{t} V'(\mathbf{r} - \mathbf{d}_j),$$

and the Fourier expansion of

$$V'(\mathbf{r}) \quad \text{is} \quad V'(\mathbf{r}) = \sum_r V_r' e^{i\mathbf{K}_r \cdot \mathbf{r}}.$$

The expansion of $V_t(\mathbf{r})$ may therefore be expressed as

$$V_t(\mathbf{r}) = \sum_{j=1}^{t} \sum_r V_r' e^{i\mathbf{K}_r \cdot (\mathbf{r} - \mathbf{d}_j)} = \sum_r V_r' S_r e^{i\mathbf{K}_r \cdot \mathbf{r}}.$$

The quantity S_r in this expression for $V_t(r)$ is what is referred to as the structure factor

$$S_r = \sum_{j=1}^{t} e^{-i\mathbf{K}_r \cdot \mathbf{d}_j},$$

and the Fourier coefficients of $V_t(r)$ may be written as

$$V_r = V_r' S_r.$$

The coefficients V_r are zero and the corresponding energy gap vanishes when the structure factor is zero, which is so for some particular $\mathbf{K}_r$.

The important body centered cubic structure, for example, may be regarded as two interpenetrating simple cubic lattices, or as a simple cubic lattice with a basis at $(0, 0, 0)$ and $(1/2, 1/2, 1/2)$. The position vectors of the two atoms in the body centered cubic structure are $d_1 = 0$ and $d_2 = (a/2)(\mathbf{e}_1 + \mathbf{e}_2 + \mathbf{e}_3)$. The first Brillouin zone is a cube, the faces of which are given by (m.5.67.), but there are no energy gaps on these planes, and from (m.5.66.)

$$K_r = (2\pi/a)(m_1\mathbf{e}_1 + m_2\mathbf{e}_2 + m_3\mathbf{e}_3),$$

so that the structure factor is

$$S_r = 1 + e^{-\pi i(m_1 + m_2 + m_3)}.$$

The values of (m_1, m_2, m_3) are $(\pm 1, 0, 0)$, $(0, \pm 1, 0)$, or $(0, 0, \pm 1)$ at the surface of the first Brillouin zone and for each of these both the structure factor and the energy gap is zero. The first set of planes at which the energy gap has a non-zero value are the rhombus-shaped planes bounding the rhombic dodecahedron shown in Fig. M.5.28, which are the planes bounding the second Brillouin zone for the simple cubic lattice. The first two Brillouin zones for the simple cubic lattice are shown in Fig. M.5.25., which indicate that the second zone in this case is equivalent to the first zone for the body-centered cubic structure, when the latter is constructed from the proper Bravais lattice shown in Fig. M.5.27., with the rhombohedral unit cell defined by the equations on page 574.

There is an important distinction between one-dimensional and actual three-dimensional $E(k) - k$ relationships; while in both cases the energy plotted as a function of k gives rise to curves such as Figs. M.5.6.–7., for any given direction of $\mathbf{k}$ with the energy bands of allowed energy being broken up by energy gaps, in the three-dimensional case the energy values which correspond to energy gaps in one direction may not do so in another, so that there may very well be no energy forbidden to an electron and no gaps in the energy spectrum. The occurrence of overlapping zones or energy bands in this way depends on the magnitude of the energy gaps, as shown in Fig. M.5.35., illustrating (a) large energy gaps with no overlapping and (b) small energy gaps with overlapping energy bands.

For real three-dimensional crystals, the Bloch function is of the form given by equation (m.5.60.), and in the nearly free electron approximation the function $\phi_k(\mathbf{r})$, instead of simply having the periodicity of the lattice, is more

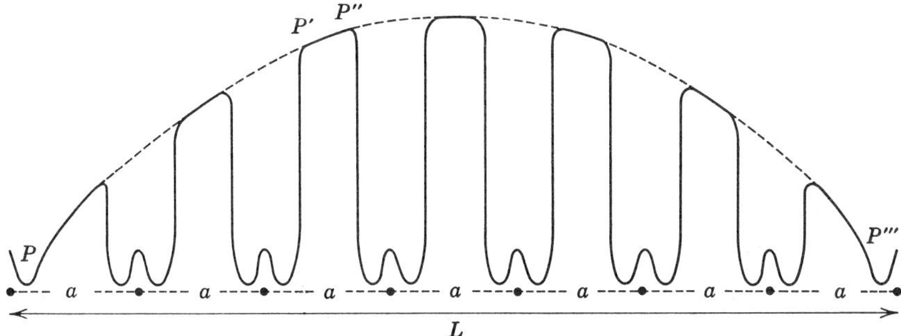

FIG. M.6.2. Representation of the comparison between a free electron function, defined with respect to the lattice row $L = Ga$, the smooth dotted curve, and a Bloch-type function with $3s$ characteristics, the full line curve.

complicated and designed to describe the way in which the s-, p-, d-, . . ., characteristics of the corresponding electron states are repeated at regular intervals. Thus, in Fig. M.6.2., where the smooth dotted curve represents one half wavelength of a free electron function, described by a plane wave with wavelength related to $L = Ga$, the full-lined curve may represent one form of a Bloch function, with singularities at the position of each nucleus along a lattice row, with the characteristics of a $3s$-state (two nodes between each nucleus). However, to regard the deviations at the nuclear positions in Fig. M.6.2. as only a slight perturbation of the smooth dotted curve can only be justified, in the nearly free electron approximation, for electron states of low wave vector and long wavelength. For states of high wave vector and wavelengths of the order of internuclear distances in the crystal, the singularities in the vicinity of the nuclei would extend over such a large part of the internuclear distance as to invalidate the assumption of only a small perturbation being applicable. The nearly free electron approximation is most justifiable for the alkali metals, and of less validity for metals with electron to atom ratios of greater than about 1.8.

M.6.3. The Tight Binding Approximation

The approximation technique, which is referred to as the tight binding method in metal theory, is known as the linear combination of atomic orbitals method in molecular quantum mechanics, so that the molecular orbitals or extended orbitals in the latter are built up from atomic orbitals in the same manner that Bloch orbitals are in the case of metal theory. Basically, while in the nearly free electron approximation, the smooth curve $PP'P''P'''$ of Fig. M.6.2. is regarded as the fundamental function from which the behavior in the vicinity of the atomic nuclei is approximated by the application of perturbation methods, the tight binding approximation regards the singularities in the region of the atomic nuclei as the fundamentally important functions from which the behavior in regions such as $P'P''$ is taken to be a slight perturbation. In this respect, the tight binding method regards the atomic nuclei as being so far apart that the appropriate electron functions resemble those of the free atoms. This method is not particularly appropriate where the electron clouds around adjacent nuclei overlap one another appreciably, as for the transition metals containing d-electrons. The principle involved in the application of this method is again best appreciated by consideration of a one-dimensional lattice.

For an electron moving in the isolated potential well in the vicinity of a single nucleus, as at the position A in Fig. M.6.1., the particular state of the electron described by the function $\zeta(x)$ is such that its wave function satisfies the equation

$$d^2\zeta(x)/dx^2 + [8\pi^2 m/h^2][E_0 - V(x)]\zeta(x) = 0. \qquad \text{(m.6.17.)}$$

If $\zeta(x)$ refers to the ground state of the electron, then it will fall off exponentially to 0 as $x \to \pm\infty$, and have the form illustrated in the vicinity of A' in Fig. M.6.1. There is, however, no good reason why $\zeta(x)$ should not be the wave function of an excited state, displaying nodes, but it is convenient in illustrating the applicability of the tight binding method in one dimension to assume that $\zeta(x)$ is non-degenerate and that the potential function $V(x)$ is an even function, so that $V(x) = V(-x)$. For a one-dimensional lattice row composed of G such one-dimensional atoms, as shown between the points LL' in Fig. M.6.1., where G is a very large number so that the departure from periodicity along the row is negligibly small, the potential energy of an electron in the lattice row is V_t, where

$$V_t(x) = \sum_p V(x - pa), \qquad \text{(m.6.18.)}$$

and the index p ranges over G successive integers. For an electron in this lattice row, the appropriate Schrödinger equation is then

$$d^2\psi(x)/dx^2 + [8\pi^2 m/h^2]\left[E - \sum_p V(x - pa)\right]\psi(x) = 0, \qquad \text{(m.6.19.)}$$

and if a is sufficiently large so that there is only a small overlap between adjacent atomic wave functions, which means that $V(x - pa)\zeta(x - na)$ is small except when $p = n$, then any of the functions $\zeta(x - pa)$ corresponds to an approximate solution of equation (m.6.19.). When $E = E_0$, all the terms in the potential sum of (m.6.18.) fall out except $V(x - pa)$, and the equation which has to be satisfied by the function $\zeta(x - pa)$ is

$$d^2\zeta(x - pa)/dx^2 + [8\pi^2 m/h^2][E_0 - V(x - pa)\zeta(x - pa)] = 0. \qquad \text{(m.6.20.)}$$

When overlap between the atomic wave functions is small enough to be neglected, all of the functions $\zeta(x - pa)$ are equally good approximate solutions of equation (m.6.19.), so that when the overlap is small but finite, a proper solution of equation (m.6.19.) would be expected to be given by very nearly a linear combination of these functions, or

$$\psi(x) = \sum_p C_p \zeta(x - pa), \qquad \text{(m.6.21.)}$$

where the coefficients C_p may, for example, be determined by the variational method (Q.4.2.). This is the approximation known as the tight binding method in metal theory.

The coefficients C_p are also, however, fixed by the requirement that the function (m.6.21.) should satisfy the Bloch condition. Thus, in terms of the wave number k,

$$\psi_k(x) = \sum_p [e^{ikpa}\zeta(x - pa)] \qquad \text{(m.6.22.)}$$

and

$$\psi_k(x + a) = \sum_p [e^{ikpa}\zeta\{x - (p - 1)a\}] = e^{ika} \sum_p [e^{ik(p - 1)a}\zeta\{x - (p - 1)a\}]$$

$$= e^{ika}\psi_k(x) \qquad \text{(m.6.23.)}$$

for a sufficiently long lattice, so that the function (m.6.22.) is a Bloch function and is of the general form of (m.6.21.).

If the atomic orbitals are assumed to be normalized in the length of the lattice row and overlap between adjacent atoms is neglected, then

$$\int \zeta^*(x - na)\zeta(x - pa)\,dx = \delta_{np} \qquad \text{(m.6.24.)}$$

and

$$\int \psi_k^*\psi_k\,dx = \sum_n \sum_p e^{ik(p - n)a} \int \zeta^*(x - na)\zeta(x - pa)\,dx$$

$$= \sum_n \sum_p e^{ik(p - n)a} \delta_{np} = G, \qquad \text{(m.6.25.)}$$

where the integrals are taken over the length of the lattice row. The approximate and normalized wave function in the state k is thus

$$\psi_k(x) = G^{-1/2} \sum_p [e^{ikpa}\zeta(x - pa)], \qquad \text{(m.6.26.)}$$

and the corresponding energy is given by

$$E(k) = \int \psi_k^*(x)\hat{H}\psi_k(x)\,dx, \qquad \text{(m.6.27.)}$$

where

$$H = -[(h^2/8\pi^2 m)] \, d^2/dx^2 + V_t(x). \qquad \text{(m.6.28.)}$$

Making use of equation (m.6.20.),

$$\hat{H}\psi_k = G^{-1/2} \sum_p e^{ikpa} \hat{H}\zeta(x - pa)$$

$$= G^{-1/2} \sum_p e^{ikpa}[-(h^2/8\pi^2 m) \, d^2/dx^2 + V(x - pa)]\zeta(x - pa)$$

$$+ \, G^{-1/2} \sum_p e^{ikpa}[V_t(x) - V(x - pa)]\zeta(x - pa)$$

$$= E_0\psi_k + G^{-1/2} \sum_p e^{ikpa}[V_t(x) - V(x - pa)]\zeta(x - pa), \qquad \text{(m.6.29.)}$$

and so $E(k)$ is obtained as

$$E(k) = \int \psi_k{}^* E_0\psi_k \, dx + G^{-1} \sum_n \sum_p e^{ik(p-n)a}$$

$$\times \int \zeta^*(x - na)[V_t(x) - V(x - pa)]\zeta(x - pa) \, dx$$

$$= \int \psi_k{}^* E_0\psi_k \, dx + G^{-1} \sum_n \sum_p e^{ikqa}$$

$$\times \int \zeta^*(X + qa)[V_t(X + pa) - V(X)]\zeta(X) \, dX$$

$$= E_0 + \sum_q e^{ikqa} \int \zeta^*(x + qa)[V_t(x) - V(x)]\zeta(x) \, dx, \qquad \text{(m.6.30.)}$$

where $X = x - pa$ and $q = p - n$, and it is assumed that for a very long lattice row and consistent with the periodic boundary conditions, all of the integrals are taken over the same range. On the assumption that the overlap between different atomic orbitals is small, it is permissible to neglect all of the integrals in the sum of (m.6.30.) except those for which $q = 0$ or ± 1, which amounts to only taking into account the overlap between nearest neighboring atoms. Thus,

$$E(k) = E_0 + \int \zeta^*(x)[V_t(x) - V(x)]\zeta(x) \, dx$$

$$+ \, e^{ika} \int \zeta^*(x + a)[V_t(x) - V(x)]\zeta(x) \, dx$$

$$+ \, e^{-ika} \int \zeta^*(x - a)[V_t(x) - V(x)]\zeta(x) \, dx$$

$$= E_0 + \int \zeta^*(x)[V_t(x) - V(x)]\zeta(x) \, dx$$

$$+ \left(\int \zeta^*(x + a)[V_t(x) - V(x)]\zeta(x) \, dx \right)(e^{ika} + e^{-ika}), \qquad \text{(m.6.31.)}$$

since $V(x)$ is an even function and $\zeta(x)$ is either even or odd, so that

$$\int \zeta^*(x + a)[V_t(x) - V(x)]\zeta(x)\, dx = \int \zeta^*(x - a)[V_t(x) - V(x)]\zeta(x)\, dx = -C.$$

So the energy $E(k)$ is finally obtained as

$$E(k) = E_0 - A - 2C \cos ka, \qquad (m.6.32.)$$

where

$$-A = \int \zeta^*(x)[V_t(x) - V(x)]\zeta(x)\, dx$$

and the other integrals in (m.6.32.) are designated by $-C$.

It thus emerges that in the tight binding approximation method, the atomic energy level E_0 becomes split up into a band of energy levels in the lattice, and since the cosine of an angle varies between $+1$ and -1, the width of the band is $4C$. Also, all possible states in this band, or Brillouin zone, correspond to values of k between $-\pi/a$ and $+\pi/a$, so that $E(k)$ is a periodic function with period $2\pi/a$ (cf. Fig. M.5.7.). If $V_t(x)$ is negative, $[V_t(x) - V(x)]$ is also negative, and if $\zeta(x)$ is an even function, like that shown in the vicinity of A' in Fig. M.6.1., then $\zeta(x)$ and $\zeta(x + a)$ have the same sign, so that A is positive; in this case, the energy curve appears as shown in Fig. M.6.3. This curve is similar to the curves in Fig. M.5.7. for $l = 0$ and $l = 2$; in the event that there is a node at the origin of the $E(k) - k$ curve, as would be the case for the first excited state where $\zeta(x)$ is an odd function of x, A would be positive but C would be negative, so that the energy curve would have a maximum, rather than a minimum, at the origin of k, and then the energy curve would resemble that for $l = 1$ in Fig. M.5.7. The principal difference between the $E(k) - k$ curve of Fig. M.6.3. and the lowest curve of Fig. M.5.7., which is equivalent to the result obtained by the nearly free electron approximation, is that for the tight binding approximation the energy curve is symmetrical about $k = \pm\pi/2a$, while the latter is essentially parabolic except near to the zone boundaries.

If k is very small, the cosine term in equation (m.6.32.) may be expanded by the normal trigonometric expansion, given by $\cos x = 1 - x^2/2! + x^4/4! - x^6/6! + \cdots$, and for small k retaining only the first two terms of the expansion, the energy becomes

$$E(k) = E_0 - A - 2C + Ca^2k^2, \qquad (m.6.33.)$$

which since the first three terms merely define the zero from which the energy is measured, is of the same form as the free electron energy (section M.3.6.) given by $h^2k^2/8\pi^2m$; the similarity between the two energy expressions is apparent if Ca^2 is written as $h^2/8\pi^2m^*$, where m^* is the effective mass of the electron (page 595) near the value of $k = 0$.

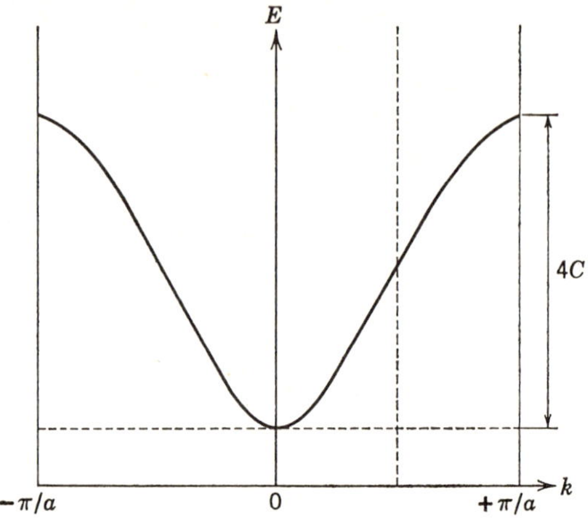

FIG. M.6.3.   The energy of an electron plotted as a function of its wave vector **k**, in the one-dimensional application of the tight binding method.

The perturbation method involved in the nearly free electron approximation has been seen to be appropriate only when the energy bands are wide and the energy gaps small. On the other hand, the approximations involved in the tight binding method, involving the assumptions that there is little overlap between the atomic wave functions, which means that C and therefore the energy bandwidth $4C$ must be small, and the method of forming the Bloch functions from the single atomic function $\zeta(x)$ associated with the energy level E_0, implying not only that this level is non-degenerate but also that it is quite separate from any other atomic energy levels, means that this method is only appropriate when it leads to narrow energy bands with large energy gaps between them. The tight binding method may, of course, be applied equally well to degenerate energy levels. In this event, instead of the single function $\zeta(x)$, a linear combination of the degenerate functions associated with any particular energy level would be used to form Bloch functions similar to (m.6.22.), and the coefficients would be determined by the variational method (Q.4.4.). Generally, every atomic energy level will broaden into such a band as given by (m.6.32.), in the application of the tight binding method, and by taking all of the various atomic levels in turn there would arise an infinite set of energy curves as in Fig. M.5.7.; each of these would be described by the equation (m.6.32.), but with the appropriate values of E_0, A, and C.

The extension of the one-dimensional example to a real three-dimensional metal crystal is fairly straightforward. Let it be assumed that the potential energy of an electron in a free atom is $V(r)$ and that E_0 is the energy of a state with the wave function $\zeta(r)$. If, for simplicity, it is assumed that $\zeta(r)$ is an s-type function, so that it is non-degenerate, then the Schrödinger equation which has to be satisfied by $\zeta(r)$ is

$$\nabla^2\zeta(r) + [8\pi^2 m/h^2][E_0 - V(r)]\zeta(r) = 0. \qquad \text{(m.6.34.)}$$

Now, for an electron in a space lattice, its potential energy is

$$V_t(\mathbf{r}) = \sum_n \mathbf{V}(\mathbf{r} - \mathbf{R}_n), \qquad \text{(m.6.35.)}$$

where the sum is taken over all of the lattice points. The Schrödinger equation becomes

$$\nabla^2\psi(r) + [8\pi^2 m/h^2]\left[E - \sum_n V(\mathbf{r} - \mathbf{R}_n)\right]\psi(r) = 0. \qquad \text{(m.6.36.)}$$

Approximate solutions of equation (m.6.36.) may now be constructed by forming linear combinations, on the assumption that the functions $\zeta(\mathbf{r} - \mathbf{R}_n)$ centered on different lattice sites do not overlap very much, as

$$\psi_k(\mathbf{r}) = \sum_p e^{i\mathbf{k}\cdot\mathbf{R}_p}\zeta(\mathbf{r} - \mathbf{R}_p), \qquad \text{(m.6.37.)}$$

the coefficients of the linear combination functions being selected so that the functions satisfy the Bloch condition as stated in equation (m.5.59.). So

$$\psi_k(\mathbf{r} + \mathbf{R}_n) = \sum_p e^{i\mathbf{k}\cdot\mathbf{R}_p}\zeta(\mathbf{r} - \mathbf{R}_p + \mathbf{R}_n)$$

$$= e^{i\mathbf{k}\cdot\mathbf{R}_n}\sum_p e^{ik\cdot(\mathbf{R}_p - \mathbf{R}_n)}\zeta[\mathbf{r} - (\mathbf{R}_p - \mathbf{R}_n)] = e^{ik\mathbf{R}_n}\psi_k(\mathbf{r}). \qquad \text{(m.6.38.)}$$

If there are N atoms in the whole piece of crystalline metal, and the overlap between the atomic functions centered on different lattice sites is disregarded, then on the assumption that the functions (m.6.38.) are normalized over the volume of the crystal, which approximately means normalization over all space, the normalizing factor for the functions ψ_k is simply $N^{-1/2}$. By the same argument as that used in the one-dimensional case, the approximate energy is found to be given by an expression similar to equation (m.6.30.),

$$E(\mathbf{k}) = E_0 + \sum_n e^{i\mathbf{k}\cdot\mathbf{R}_n}\int \zeta^*(\mathbf{r} + \mathbf{R}_n)[V_t(\mathbf{r}) - V(\mathbf{r})]\zeta(\mathbf{r})\,d\mathbf{r}, \qquad \text{(m.6.39.)}$$

the integral this time being taken throughout the volume of the crystal. Now, neglecting overlap except that between nearest neighboring atoms, and substituting

$$\int \zeta^*(\mathbf{r})[V_t(\mathbf{r}) - V(\mathbf{r})]\zeta(\mathbf{r})\,d\mathbf{r} = -A$$

and $\qquad\qquad\qquad\qquad\qquad\qquad\qquad\qquad\qquad\qquad\qquad\qquad$ (m.6.40.)

$$\int \zeta^*(\mathbf{r} + \mathbf{R}_n)[V_t(\mathbf{r}) - V(\mathbf{r})]\zeta(\mathbf{r})\,d\mathbf{r} = -C,$$

the expression for the energy reduces to

$$E(\mathbf{k}) = E_0 - A - C \sum_n e^{i\mathbf{k}\cdot\mathbf{R}_n}. \qquad \text{(m.6.41.)}$$

Since $\zeta(\mathbf{r})$ is a spherically symmetrical function, the quantity C is the same for all nearest neighbors, so that the sum in (m.6.41.) need only be taken over the position vectors of the nearest neighboring atoms to those at the origin. If $V(\mathbf{r})$ is negative, then $V_t(\mathbf{r}) - V(\mathbf{r})$ is also negative, so that for s-states with small overlap, A and C are both positive.

In the case of a simple cubic lattice (section M.5.8.), the lattice vectors $\mathbf{R}_n$ are given by $\pm a\mathbf{e}_1, \pm a\mathbf{e}_2, \pm a\mathbf{e}_3$ when $E(\mathbf{k})$ becomes

$$E(\mathbf{k}) = E_0 - A - 2C(\cos k_1 a + \cos k_2 a + \cos k_3 a), \qquad \text{(m.6.42.)}$$

so that the energy levels, in this case, are spread over a band of width $12C$. The energy is a triply periodic function of $\mathbf{k}$, and all states are included if $\mathbf{k}$ is restricted to fall within a cube bounded by those planes defined by $k_1 = k_2 = k_3 = \pm\pi/a$. By differentiating (m.6.42.) with respect to the components of k, it is seen that the normal derivative of $E(\mathbf{k})$ vanishes at the bounding planes, as

$$\partial E(\mathbf{k})/\partial k_1 = \partial E(\mathbf{k})/\partial k_2 = \partial E(\mathbf{k})/\partial k_3 = 2Ca \sin k_1 a = 2Ca \sin k_2 a$$
$$= 2Ca \sin k_3 a = 0,$$

when $k_1 = k_2 = k_3 = \pm\pi/a$. By expanding the cosine terms in (m.6.42.) and retaining only the first two terms in the expansion, as is valid for small values of k, then approximately

$$E(\mathbf{k}) = E_0 - A - 6C + Ca^2k^2, \qquad \text{(m.6.43.)}$$

and the effective mass m^* may be defined in the same way as for the one-dimensional case, near the region where $k = 0$, as

$$m^* = h^2/8\pi^2 Ca^2. \qquad \text{(m.6.44.)}$$

It may be deduced from (m.6.43.) that near the region in $\mathbf{k}$-space where $k = 0$, $E(\mathbf{k})$ has the same value independently of direction, so that states with the same energy fall on the surface of a sphere, as for free electrons. For larger values of the energy, since the normal derivative of $E(\mathbf{k})$ vanishes at the bounding planes of the zone, therefore the constant energy surfaces for larger $\mathbf{k}$ intersect the zone boundaries orthogonally. The energy surface for the highest occupied states (the Fermi surface) for copper is shown in Fig. M.5.42. and a cross-section through the energy surfaces in any one of the coordinate planes for a simple cubic lattice is shown in Fig. M.5.11. In the individual k_1, k_2, and k_3 axial directions, the energy is given as a function of k by curves similar to that shown in Fig. M.6.3., with the state of lowest

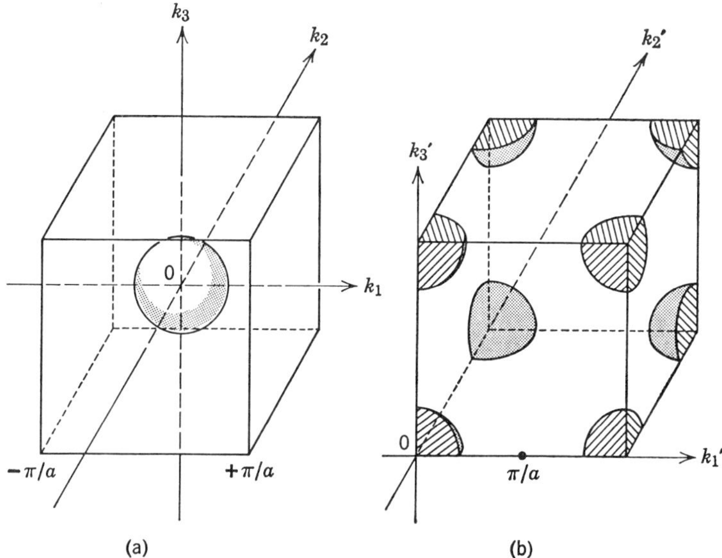

FIG. M.6.4.   Constant energy surfaces for the simple cubic lattice by the tight binding approximation corresponding to (a) energy states near the bottom of the energy band with the cube center chosen as the origin of k-space, or energy states near the top of the band with the cube vertex chosen as origin of k-space; and (b) energy states near the bottom of the energy band or top of the energy band, dependent on whether the cube vertex or the cube center, respectively, is chosen as the origin of k-space.

energy being at the center of a zone and the highest energy states at the vertices. Every state of a free atom will give rise to an energy band, but some of the corresponding energy curves may have maxima either at their centers ($\mathbf{k} = 0$) or at some other point in the zone. Degenerate atomic states will give rise to a number of bands or zones equivalent to the degree of degeneracy of the state involved. The actual three-dimensional shape of the constant energy surfaces in crystals is related to the manner of choice of the origin of the crystallographic unit cell (see Fig. M.5.14.), which in turn determines the origin of reciprocal space in which the k-surfaces are located. For example, Fig. M.6.4. shows two different illustrations of the constant energy surface for a simple cubic lattice, near to the top of the energy band, with the origin of the cubic unit cell and of the reciprocal unit cell chosen to coincide with a lattice point and with the center of each cell coincident with a lattice point, respectively.

From equations (m.6.43.) and (m.6.44.), the energy near $\mathbf{k} = 0$ may be written as

$$E(\mathbf{k}) = [h^2/8\pi^2 m^*]k^2, \qquad (\text{m.6.45.})$$

provided the zero of energy is taken to be the lowest energy in the zone. Therefore, the density of states in the neighborhood of $\mathbf{k} = 0$ is given by the same expression as that for free electrons, but with the effective mass replacing the m,

$$N(E) = [V/4\pi^2][8\pi^2 m^*/h^2]^{3/2}E^{1/2}. \qquad (\text{m.6.46.})$$

With reference to Fig. M.6.4., if k' is one of the vectors given by

$$k_1' = k_1 \pm \pi/a, \qquad k_2' = k_2 \pm \pi/a, \qquad k_3' = k_3 \pm \pi/a, \quad (\text{m.6.47.})$$

then equation (m.6.42.) becomes

$$E(\mathbf{k'}) = E_0 - A + 2C[\cos k_1'a + \cos k_2'a + \cos k_3'a],$$

and if k' is small, which is so in the vicinity of the vertices of Fig. M.6.4.(b), then

$$E(\mathbf{k'}) = E_0 - A + 6C - Ca^2k'^2. \qquad (\text{m.6.48.})$$

Thus, the energy surfaces are again approximately spherical, and the effective mass at the top of the energy band is given as $m_1^* = -h^2/8\pi^2 Ca^2$, which is equal to $-m^*$, the effective mass near $k = 0$. If E_M is the maximum energy in the zone, m^* is the same positive effective mass as in equation (m.6.46.), k' is defined as in (m.6.47.), and the origin of k-space is at the center of the cube as shown in Fig. M.6.4.(a), then

$$E_M - E(\mathbf{k}) = h^2k'^2/8\pi^2 m^*, \qquad (\text{m.6.49.})$$

and for energies in the vicinity of E_M,

$$N(E) = [V/4\pi^2][8\pi^2 m^*/h^2]^{3/2}[E_M - E]^{1/2}. \qquad (\text{m.6.50.})$$

Therefore, from equations (m.6.46.) and (m.6.50.), the $N(E)$ curve for the simple cubic lattice, in the tight binding approximation, is parabolic at both the upper and lower ends, and specifically in the tight binding approximation, it is symmetrical about the mean value of the energy, if the whole curve is calculated making use of equation (m.6.42.) and the surface integral given for $N(E)$ on page 596; the resulting curve is shown in Fig. M.6.5., in comparison with the free electron curve for a single zone (compare Fig. M.3.3.), which is parabolic up to the zone boundary, and the $N(E)$ curve for the valency electrons in a metal (compare with Figs. M.5.35.–36.), which is intermediate between the former two shapes. This latter curve is parabolic at both ends, but with different magnitudes for the effective mass, and is not symmetrical about the mean energy value. The maximum occurs approximately where the constant energy surfaces first approach the zone boundaries (see section

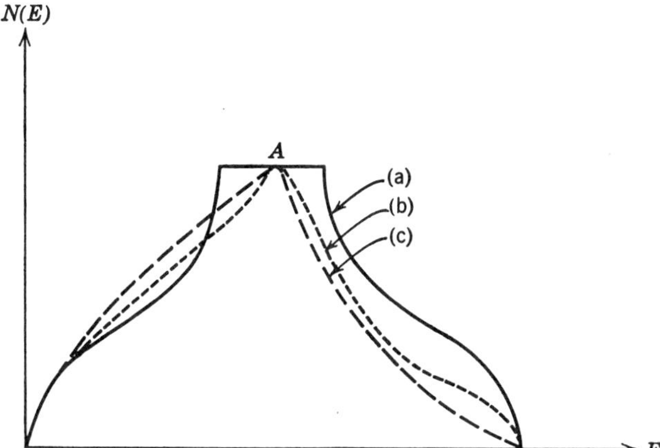

FIG. M.6.5. Comparison of the density of states curves for a single zone of a simple cubic lattice: (a) the symmetrical curve according to the tight binding approximation; (b) the free electron approximation curve, parabolic in the region 0 to A, and then falling off in approximately an exponential manner; (c) most probable curve for a real metal, parabolic at both ends, but asymmetrical.

M.5.10.), and the increase in slope before the maximum is attributable to the bulging of the energy surfaces, which leads to an increase in the volume encompassed between two surfaces separated by an energy dE.

M.6.4. The Hall Effect

It was discovered originally by Hall[1] that if a transverse magnetic field is applied to a conductor carrying an electric current, a difference of potential automatically appears across the conductor, transverse to both the direction of the magnetic field and that of the current flow, and proportional to the magnetic field and to the current, as illustrated in Fig. M.6.6. In terms of the current density, which makes the magnitude of the Hall effect independent of the size of the conductor, the Hall constant is defined as the transverse electric field, which is set up per unit magnetic field and per unit current density. The potential difference which appears across the conductor is referred to as the Hall voltage. Experimental observation of the Hall effect and measurement of the Hall constant provide information about conductors which is not accessible in any other way, indicating the number of current

---

[1] Hall, E. H., *Am. J. Math.*, 1879, **2**, 287.

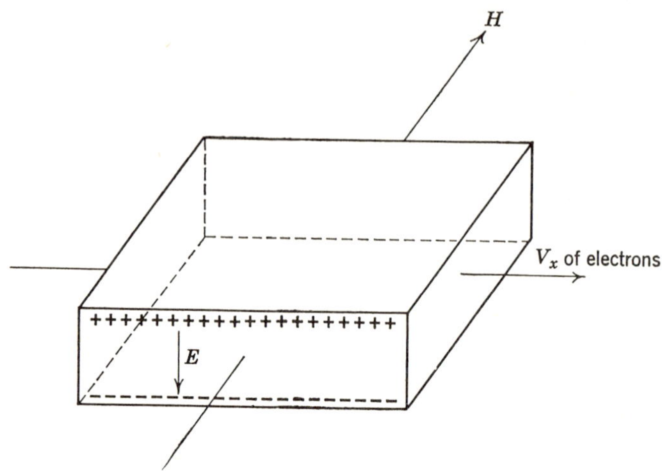

FIG. M.6.6.   Illustration of the Hall effect. The current flow is in the direction of $+x$, the magnetic field in the $+y$ direction, and the Hall potential difference appears in the direction $+Z$.

carriers and the sign of their charge, whether electrons or holes. The Hall effect is an especially important method in the investigation of semiconductors, where there are generally fewer current carriers than in metals, and as a consequence the Hall constant is larger and more easily observable. Conductivity measurements, on the contrary, provide information which is complicated by the effective mass of the charge carriers, and also by the fact that the square of the charge is involved, rather than its first power, so that no distinction may be made between positive and negative charge carriers. For semiconductors, knowing their conductivity and Hall constants, it is possible in many cases to deduce the numbers or concentrations of positive and negative charge carriers and their mobilities. The Hall effect has been widely studied, in particular with respect to semiconductor research.[1,2]

When an electron is in motion in an electric field, it is acted on by a force proportional to the electronic charge and to the electric field, and in the same direction as the field. In a magnetic field, an electron is subjected to a force again proportional to the electronic charge, its velocity, and the magnetic field, in a direction normal to the plane determined by the velocity and the magnetic field. Since these are all vector quantities,

$$\mathbf{F} = q[\mathbf{E} + (\mathbf{v} \times \mathbf{H})],\tag{m.6.51.}$$

[1] Peierls, R. E., *Quantum Theory of Solids*, 1955, Clarendon Press, Oxford.
[2] Ziman, J. M., *Electrons and Phonons*, 1962, Clarendon Press, Oxford.

where $\mathbf{E}$ is the electric field, $(\mathbf{v} \times \mathbf{H})$ is the vector product of the electronic velocity and the magnetic field, and q is the electronic charge. Since in the Hall effect the velocity of the electrons in a direction transverse to the magnetic field, along the conductor, gives rise to a transverse force, this force will tend to cause the electrons to drift to one side of the conductor. The charges which accumulate on one face of the conductor in this manner would in turn give rise to an electric field, which with the accumulation of sufficient charge would grow until large enough to cancel the force exerted by the magnetic field, so that the net transverse force would become zero [as given by (m.6.51.)]. Other electrons would then travel straight down the conductor without being subjected to any transverse acceleration. For such a state to exist, it is necessary that the magnitudes of the transverse $\mathbf{E}$ and the vector product $\mathbf{v} \times \mathbf{H}$ should be equal, or $E = vH$. From equation (m.1.33.), the current density in a conductor is given by $i = -nv\varepsilon$ for electrons or $i = nvq$ for particles of charge q, where n is the number of charge carriers per unit volume. Thus,

$$E = vH = Hi/nq. \tag{m.6.52.}$$

The ratio $E/iH = 1/nq$ is referred to as the Hall constant, and since $q = -\varepsilon$ for electrons, the corresponding Hall constant is negative. The convention regarding signs is usually chosen so that if the current is carried by electrons, the Hall constant is negative, and for positive carriers or holes the Hall constant is positive. Since n is much smaller for semiconductors than for metals, the Hall constant will be much larger.

The Hall constant, designated R_H, sometimes referred to as the Hall coefficient, has values for the alkali metals, copper, silver, and gold, which are negative and the agreement between the observed and calculated values are quite good, in confirmation that these metals conduct electricity by the motion of free electrons present to the extent of about one per atom. For metals like beryllium, zinc, and cadmium, the observed Hall constant is positive, implying that the current is carried by positive charges. For such metals, where the predominant effect is indicated to be conduction by positive holes, and there is an overlap from the first into the second zone, the observed result is to be regarded as the net effect of the negative contribution of the essentially free electrons at the bottom of the second zone and of the positive contribution from the holes at the top of the first zone. For bismuth, the observed Hall constant is about 250 times as large as the free electron value, implying that there are only a small concentration of charge carriers, amounting to a small fraction of the number of atoms, outside nearly filled bands. Ferromagnetic metals and some of their alloys exhibit rather remarkable Hall effects,[1,2] but

---

[1] Pugh, E. M., and Rostoker, N., *Rev. Mod. Phys.*, 1953, **25**, 151.
[2] Karplus, R., and Luttinger, J. M., *Phys. Rev.*, 1954, **95**, 1154.

generally metals with larger numbers of conduction electrons per atom than two exhibit conduction by means of both holes and relatively free electrons and the results are more complicated. The investigation of the variation of the Hall constant of alloys on solid solution formation of one metal with another or larger metallic valency may be used to detect the stage at which Brillouin zone overlap occurs, since at that point the Hall constant-composition curve will be observed to change direction towards the metal with the more negative Hall constant.

From equation (m.1.34.), it is apparent that the conductivity of a metal involves the square of the charge and also the mass, and for states near the top or at the bottom of an energy band, it has been seen that the effective mass may be quite different from the actual electron mass [equations (m.5.77.) and (m.6.44.)]. If only one type of charge carrier is involved in the conductivity of metals, the situation is quite simple, $R_H = \pm 1/n\varepsilon$, but for most intermediate cases where both electrons and holes are involved in the conductivity, the situation is quite complicated. The intermediate case, however, where there is a relatively small number of electrons at the bottom of one band and a relatively small number of holes at the top of another, which is the situation in semiconductors, is somewhat simpler. Here, if n, n_ε, n_h, μ_ε, and μ_h denote the number of atoms per unit volume, the number of electrons, the number of holes, the mobilities of the electrons and holes, respectively, then the Hall constant[1] becomes

$$R_H = [1/n\varepsilon][n_h\mu_h{}^2 - n_\varepsilon\mu_\varepsilon{}^2][n_h\mu_h + n_\varepsilon\mu_\varepsilon]^{-2}, \qquad \text{(m.6.53.)}$$

which reduces to simply $R_H = 1/n\varepsilon$ when either n_ε or n_h is zero, and indicates that some suitable number of electrons and holes together may balance their contributions to the Hall constant, so that it is possible to have intermediate cases between all negative and all positive carriers. The semimetals, arsenic, antimony, and bismuth, also fall into this category. The Hall constant for bismuth is, as already mentioned above, negative and large, but for arsenic and antimony it is small and positive, denoting that the second term in square braces in (m.6.53.) has opposite signs in the two cases, electrons being the more important carriers in bismuth but the holes in arsenic and antimony. The great importance and value of measurements of the Hall constant is that it enables the number of charge carriers to be deduced from direct experimental observation, and combined with direct conductivity measurements, it also enables the mobilities to be deduced, since from (m.5.78.) $\mu = |K_e R_H|$.

---

[1] Ziman, J. M., *Electrons and Phonons*, 1962, Clarendon Press, Oxford, pp. 486–490.

M.6.5. The Orthogonal Plane Wave Approximation

The orthogonalized plane wave approximation method, suggested originally by Herring,[1] combines the best features of the tight binding method and expansion in plane waves, at the same time avoiding the difficulties of each of these other methods.[2] The parts of the wave functions for a metal which are most difficult to expand in plane waves are those which vary greatly from point to point, such as the *s*-functions near the atomic nuclei (Fig. M.6.2.). On the other hand, for a smoothly varying function, the plane wave expansion would be rapidly convergent. Herring's idea was to use the linear combination of atomic orbitals method to build up the rapidly varying part of the wave functions, such that these have been made orthogonal to the remaining part of the wave function, represented by plane waves. This method is referred to as the orthogonalized plane wave method for the following reason. While the solutions of Schrödinger's equation for a system must be orthogonal to one another, a plane wave of the type $e^{i(\mathbf{K}_r + \mathbf{k}) \cdot \mathbf{r}}$ will not be orthogonal to a linear combination of atomic functions or to a Bloch sum. In molecular quantum mechanics, Bloch sums or linear combinations of atomic orbitals may form quite satisfactory solutions of the Schrödinger equation for the corresponding molecular orbital, but a single plane wave cannot constitute a correct solution. But a plane wave can be orthogonalized to all the Bloch sums by forming linear combinations of the plane wave and each of the Bloch sums. The coefficients in the linear sum can be chosen so as to make no overlap between the so-called orthogonalized plane wave and each of the Bloch sums. An orthogonalized plane wave, formed in this way, will vary smoothly between the atoms, but on account of the Bloch sums of atomic orbitals involved in forming it, it will vary in the vicinity of each nucleus more like a true wave function should, and so should constitute a better solution to the appropriate Schrödinger equation than an ordinary plane wave. The Fourier expansion of such an orthogonalized plane wave, involving the Fourier expansions of the Bloch sums, will have the property of having non-zero values for the off-diagonal elements of the corresponding secular determinant out to large $\mathbf{K}_r$ values, so that it might be expected that a linear combination of a fairly small number of such orthogonalized plane waves might approximate quite closely to the true wave function for the system.

The basic principle of the orthogonalized plane wave method is then to use a composite set of basis functions, including a comparatively small number of non-overlapping Bloch sums of atomic orbitals and a few plane waves, with

---

[1] Herring, C., *Phys. Rev.*, 1940, **57**, 1169.
[2] Woodruff, T. O., *Solid State Physics*, Vol. IV, 1957, 367, Academic Press, New York.

fairly small wave vectors or large wavelengths, and then to vary the coefficients of these basis functions to get the best possible approximation to a solution of the corresponding Schrödinger equation. In order to do this, it is necessary to find the matrix components of the one-electron Hamiltonian between these basis functions; there will result matrix components of three different types between two Bloch sums, between two plane waves, and between a Bloch sum and a plane wave. The first type, between two Bloch sums, are the same type as those involved in the application of the tight binding method, and the second type, between two plane waves, are similar to those in the free electron model, involving essentially the expansion of the whole wave function in plane waves. For the third type of matrix component, between a Bloch sum and a plane wave, the Bloch sum is first expanded in plane waves and then the situation is dealt with as in the second type of matrix element. The method works best for crystals in which the metallic atomic orbitals are definitely separated into two types; firstly, those confined to well inside the atomic radii, which may be expanded as linear combinations of atomic orbitals or Bloch sums, and secondly, those which are essentially confined to the outer regions of the atoms and may be expanded in plane waves. The method thus does not work particularly well for the transition metals, although it has been applied to many metals,[1-3] and in some cases it turns out to be one of the most accurate methods for obtaining solutions of Schrödinger's equation in a periodic potential. In the limit, when the number of plane waves is infinitely large, the method would lead to an exact solution, but again, of course, with an infinite amount of computational labor.

M.6.6.   The Cellular (Wigner and Seitz) Method

The Wigner–Seitz polyhedra have been referred to on pages 391, 569, and 579. This method was first applied by Wigner and Seitz to metallic sodium, and has been used successfully in calculating the wave functions in other metals, the more general application in this manner being referred to as the cellular method. Each atom in the metallic structure is regarded as being surrounded by a polyhedron of such a shape that the different polyhedra pack together to fill the whole of space. The polyhedra are constructed by imagining that planes are erected on the perpendicular bisectors of the lines

---

[1] Chatterjee, S., and Sen, S. K., *Proc. Phys. Soc. London*, 1966, **87**, 779.
[2] Herman, F., Kortum, R. L., Kuglin, C. D., and Short, R. A., *Quantum Theory of Atoms, Molecules, and the Solid State*, 1966, Edited by P.-O. Löwdin, Academic Press, New York.
[3] Slater, John C., *Quantum Theory of Molecules and Solids*, 1965, **Vol. II**, McGraw-Hill Book Company, New York.

joining any atom to each of its nearest neighboring atoms. These polyhedra have been variously referred to as the Wigner–Seitz cells, Wigner–Seitz polyhedra, atomic cells or polyhedra, s- or cellular polyhedra, and the appropriate polyhedra for the face-centered cubic, body-centered cubic, and close-packed hexagonal structures, respectively, are the rhombic dodecahedron of Fig. M.5.28., the truncated octahedron of Fig. M.5.26., and the truncated hexagonal bipyramid of Fig. M.1.5.(b). It is important to note the distinction between the Wigner–Seitz cells and the Brillouin polyhedra, as made on page 579. The potential within the polyhedra are regarded as being spherically symmetrical and the Schrödinger equation for an electron in this field is solved, subject to the continuity of the wave function and its normal derivative at the boundary. This boundary condition allows the wave functions to join up smoothly in the region between the atoms.

In the application of this method for calculating the wave functions of the electrons in a metal, it is only necessary to solve the Schrödinger equation, (m.5.57.), within a single unit cell of the crystal structure, because of the periodicity of the function $\phi_k(\mathbf{r})$ contained in the Bloch function, (m.5.60.). Because of the method of constructing the Wigner–Seitz polyhedra, they completely fill all space, and for the cubic structures, corresponding to body-centered and face-centered packing or any other simple Bravais lattice, any one may be brought into coincidence with any other by a translation through a lattice vector $\mathbf{K}_d$, as defined by (m.5.53.). The situation is not quite so simple for the close-packed hexagonal structure, since here not only a translation through a lattice vector but also a rotation is necessary to bring two Wigner–Seitz cells into coincidence. In the face-centered and body-centered cubic structures, for each face of the atomic polyhedra there is a parallel face, and the vector $\mathbf{R}$, which is normal to such a pair of faces, is a translational vector of the crystal lattice. Thus, a vector from the center of the cell to one face of such a pair $\mathbf{r}$, when combined with the lattice translational vector $\mathbf{R}$, defines the vector from the center of the cell to the other parallel face $\mathbf{r}'$, so that $\mathbf{r}' = \mathbf{r} + \mathbf{R}$. If $\delta\mathbf{R}$ is a small displacement in the direction of R, then since any Bloch function must conform to the condition that is expressed in equation (m.5.59.), then

$$\psi_k(\mathbf{r} + \mathbf{R} + \delta\mathbf{R}) = e^{i\mathbf{k}\cdot\mathbf{r}}\psi_k(\mathbf{r} + \delta\mathbf{R}). \qquad \text{(m.6.54.)}$$

By expanding both sides of this equation in Taylor's series,‡ there results

$$\psi_k(\mathbf{r} + \mathbf{R}) + \delta\mathbf{R}\cdot\nabla\psi_k(\mathbf{r} + \mathbf{R}) + \cdots = e^{i\mathbf{k}\cdot\mathbf{r}}[\psi_k(\mathbf{r}) + \delta\mathbf{R}\cdot\nabla\psi_k(\mathbf{r}) + \cdots],$$
$$\text{(m.6.55.)}$$

where all the other terms in the expansion contain higher powers of δ than the first power as factors. The first term on the left-hand side of equation (m.5.55.) is equal to the first term on the right-hand side, since this is simply a statement of the Bloch condition, and dividing through by δ it is found that in the limit as $\delta \to 0$

$$\mathbf{R} \cdot \nabla \psi_k(\mathbf{r} + \mathbf{R}) = e^{i\mathbf{k} \cdot \mathbf{r}} \mathbf{R} \cdot \nabla \psi_k(\mathbf{r}), \qquad \text{(m.6.56.)}$$

and for every pair of faces of each Wigner–Seitz polyhedron, every wave function must satisfy both the Bloch condition, (m.5.59.), as well as equation (m.6.56.).

‡ Taylor's series represents the expansion of a function of the sum or difference of two variables as a series of ascending powers of one of the variables. If u_1 is a function of the sum of the two variables, x and y, such that $u_1 = f(x + y)$, and it is assumed that this function may be expressed as a power series in one of the variables, y, then Taylor's Theorem determines the law for such an expansion. Thus, if

$$u_1 = f(x + y) = A + A'y + A''y^2 + A'''y^3 + \cdots, \qquad \text{(i)}$$

where $A, A', A'', A''', \ldots$, are constants, independent of the variable y, but dependent on the variable x, and also on the constants entering into the original equation, Taylor's theorem enables the values of $A, A', A'', \ldots$, to be found, such that this expansion is true.

If (i) is to be true for all values of x and y, then it will also be true for any value of x, say a. So, if $B, B', B'', B''', \ldots$, are the respective values of $A, A', A'', A''', \ldots$, in (i) when $x = a$, then

$$u_1' = f(a + y) = B + B'y + B''y^2 + B'''y^3 + \cdots. \qquad \text{(ii)}$$

Now substituting $z = a + y$, so that $y = z - a$, the application of Maclaurin's Theorem (see page 114) gives,

$$u_1' = f(z) = B + B'(z - a) + B''(z - a)^2 + B'''(z - a)^3 + \cdots.$$

The successive derivatives of u_1' with respect to x may then be written down as,

$$du_1'/dx = f'(z) = B' + 2B''(z - a) + 3B'''(z - a)^2 + \cdots;$$
$$d^2u_1'/dx^2 = f''(z) = 2B'' + 2 \cdot 3B'''(z - a) + 3 \cdot 4B''''(z - a)^2 + \cdots;$$
$$d^3u_1'/dx^3 = f'''(z) = 2 \cdot 3B''' + 2 \cdot 3 \cdot 4B''''(z - a) + \cdots.$$

Maclaurin's theorem evaluates the series on the assumption that the variable becomes zero, while Taylor's theorem deduces a value for the series when $x = a$. Now, letting $z = a$, then $y = 0$, and so

$$f(a) = B; \quad f'(a) = B'; \quad f''(a) = 2B'' \qquad \text{or} \qquad B'' = f''(a)/2; \quad f'''(a) = 2 \cdot 3B'''$$

or

$$B''' = f'''(a)/3!;$$

If these values for $B, B', B'', B''', \ldots$, are substituted into equation (ii), then

$$u_1' = f(a + y) = f(a) + f'(a)[y/1] + f''(a)[y^2/2!] + f'''(a)[y^3/3!] + \cdots,$$

and this expansion is valid when x assumes some given particular value, a. But since a is any value of x, therefore if $u = f(x)$ and the above values for $B, B', B'', \ldots$, are substituted into the original expression (i), there is obtained

$$u_1 = f(x + y) = u + [du/dx][y/1] + [d^2u/dx^2][y^2/1 \cdot 2] + [d^3u/dx^3][y^3/1 \cdot 2 \cdot 3] + \cdots.$$

The series on the right-hand side of this equation is referred to as Taylor's series. The first term in this series is what the given function becomes when $y = 0$; for example, if $u = f(x)$, which results when $y = 0$, then the Taylor expansion of $(x + y)^5$ gives

$$(x + y)^5 = x^5 + 5x^4y + 10x^3y^2 + 10x^2y^3 + 5xy^4 + y^5,$$

since

$$u = f(x) = x^5; \quad du/dx = f'(x) = 5x^4; \quad d^2u/dx^2 = f''(x) = 4\cdot 5x^3; \cdots.$$

Taylor's series may alternatively be written in these forms:

$$u_1 = f(x + y) = f(x) + f'(x)[y/1] + f''(x)[y^2/2!] + f'''(x)[y^3/3!] + \cdots;$$

or

$$u_1 = f(x + y) = f(y) + f'(y)[x/1] + f''(y)[x^2/2!] + f'''(y)[x^3/3!] + \cdots;$$

and by a similar argument, it may easily be shown that

$$f(x - y) = f(x) - f'(x)[y/1] + f''(x)[y^2/2!] - f'''(x)[y^3/3!] + \cdots.$$

The Maclaurin and Taylor series are related to one another by the fact that the one may be converted into the other by substituting $f(x + y)$ for $f(x)$ in the Maclaurin series or by letting $y = 0$ in the Taylor series. The Maclaurin series may thus be regarded as a special case of the Taylor series, and the latter in turn is a special form of the more general Lagrange series, which in turn is a special form of the Laplace series, which enables an implicit function of several variables to be expanded as a power series.

For the alkali metals in particular, all of which crystallize with the body-centered cubic structure with one valency electron per atom, the lowest energy state corresponding to $\mathbf{k} = 0$ may be regarded as being described by the function $\psi_0(\mathbf{r})$ or $\phi_0(\mathbf{r})$, since each of these functions has the periodicity of the crystal lattice and the exponential part $e^{i\mathbf{k}\cdot\mathbf{r}}$ has the value unity when $\mathbf{k} = 0$. Also, the valency electron in each of these metals is in an s-state, so that $\psi_0(\mathbf{r})$ within the atomic polyhedron would be expected to be similar to an s-type function with no nodal planes. In the metals, $\psi_0(\mathbf{r})$ would not be expected to be completely spherically symmetrical within the Wigner–Seitz polyhedron, but nevertheless to have the full symmetry of the truncated octahedron characteristic of the crystal lattice. Thus, from the symmetry of the ground-state function, $\psi_0(\mathbf{r})$ and its outward normal derivative must have the same values at similarly situated points on opposite faces of the Wigner–Seitz cell, or

$$\psi_0(\mathbf{r} + \mathbf{R}) = \psi_0(\mathbf{r}) \quad \text{and} \quad \mathbf{R}\cdot\nabla\psi_0(\mathbf{r} + \mathbf{R}) = -\mathbf{R}\cdot\nabla\psi_0(\mathbf{r}). \quad \text{(m.6.57.)}$$

For $\mathbf{k} = 0$, the Bloch condition (m.5.59.) is identical with the first equation of (m.6.57.), but the boundary condition (m.6.56.) means that

$$\mathbf{R}\cdot\nabla\psi_0(\mathbf{r} + \mathbf{R}) = \mathbf{R}\cdot\nabla\psi_0(\mathbf{r}), \qquad \text{(m.6.58.)}$$

and from comparison of the second equation of (m.6.57.) with (m.6.58.) it is deduced that $\mathbf{R}\cdot\nabla\psi_0(\mathbf{r})$ must be zero everywhere on the surface of the Wigner–Seitz polyhedron. This means that the normal derivative of $\psi_0(\mathbf{r})$ must indeed vanish at the surface of the atomic polyhedron. In general, the boundary conditions may be modified further, dependent on the actual symmetry of the function $\psi_k(\mathbf{r})$, and this may be determined by the application of group theory. The boundary conditions (m.5.59.) and (m.6.56.) must apply to all wave functions, not only the ground-state function, although for the latter the fact that $\psi_0(\mathbf{r})$ is symmetrical about any atomic nucleus and has the periodicity of the crystal lattice, its normal derivative must be zero at the boundary of the atomic polyhedron, otherwise its gradient would not be continuous at the boundaries without having to apply the general boundary conditions.

For the purpose of calculating $\psi_0(\mathbf{r})$ for sodium in particular, Wigner and Seitz imagined that the atomic polyhedron was replaced by a sphere of equal volume, and denoting the radius of this atomic sphere by r_s, then if there are N metal atoms in the volume V,

$$[4\pi r_s^3/3] = V/N. \qquad \text{(m.6.59.)}$$

The radius r_s is a hypothetical concept and not equivalent to one-half of the internuclear distance in the crystal of the metal. For sodium (pages 431–433), $\psi_0(\mathbf{r})$ is almost constant over a fairly large interval about $r = r_s$, so that the assumption that the ground-state function is spherically symmetrical within the atomic sphere of radius r_s and that its normal derivative vanishes at the surface of this sphere, $[d\psi_0(r)/dr]_{r=r_s} = 0$, means that such a function will very nearly satisfy the true boundary conditions and have a normal derivative which is zero at the surface of the atomic polyhedron.

It is necessary to make some further approximations in order to calculate $\psi_0(\mathbf{r})$, since the potential energy $V(\mathbf{r})$ of the valency electron in the crystal has to be known. To obtain this crystal potential, it is necessary to know the field due to the free ion and also that due to the other valency electrons. For sodium, where the relative extension in space of the electrons in the ion is small compared with the radius of the atomic sphere [Fig. M.1.14.], so that there is no overlapping of the wave functions of the electrons in different metal ions, it is plausible to assume that the potential field of an ion in the metal is the same spherically symmetrical field as in the free atom. The total charge density due to the valency electrons is periodic, with the period of the crystal lattice, as

$$|\psi_k(\mathbf{r})|^2 = |\phi_k(\mathbf{r})|^2, \qquad \text{(m.6.60.)}$$

and therefore each cell is electrically neutral. It is reasonable to assume that since the atomic cells are approximately spherical, the electrostatic field of any one cell is zero at all points outside that cell. So the potential field acting on any valency electron can be taken as that due to the metal ion in the cell containing that particular valency electron plus the self-consistent field of the other electrons in that cell. However, as a consequence of the Pauli Exclusion Principle and the Coulomb correlation effects, the presence of any valency electron in an atomic cell essentially excludes all other electrons from that cell, and so the potential field acting on a valency electron may be assumed to be that of the metal ion in the cell alone. The concept of the Fermi hole has been discussed in section Q.5.8., where it was seen that as a consequence of the Pauli Principle, an electron is always surrounded by a hole in the charge distribution of electrons with parallel spins, approximately equal in radius to r_s; it is believed that a similar hole exists in the charge distribution of electrons with antiparallel spins due to Coulomb correlation effects.

The Schrödinger equation describing the motion of a valency electron within an atomic polyhedron, in spherical polar coordinates, and subject to the boundary condition that $[d\psi_0(r)/dr]_{r=r_s} = 0$, is

$$d^2\psi_0(r)/dr^2 + [2/r][d\psi_0(r)/dr] + [8\pi^2 m/h^2][E_0 - V(r)]\psi_0(r) = 0, \quad \text{(m.6.61.)}$$

where $V(r)$ is thus assumed to be the same spherically symmetrical potential function for the metallic ion as that in a free atom. Expressing equation (m.6.61.) in atomic units and writing $r\psi_0(r) = f(r)$, it becomes

$$d^2f(r)/dr^2 + [E_0 - V(r)]f(r) = 0, \quad \text{(m.6.62.)}$$

where E_0 is the ground-state energy. The boundary condition which must be satisfied, also expressed in terms of $f(r)$, becomes

$$df(r)/dr = f(r)/r, \quad \text{(m.6.63.)}$$

where $r = r_s$. Integration of equation (m.6.61.) should lead to finding $\psi_0(r)$ and E_0. $V(r)$ may be obtained from the Hartree–Fock calculations for the free atom as a tabulated function, and the integration of equation (m.6.61.) has to be carried out numerically. To calculate E_0 as a function of r_s, it is convenient to initially assume some plausible value for E_0 as being of lower energy than the $3s$-electron in the free atom, and to stipulate that $\psi_0(r)$ and $f(r)$ should have two spherical nodes similar to the $3s$ function. Then, with the assumed value of E_0, equation (m.6.62.) is integrated numerically from $f(r) = 0$ to find those points corresponding to a value of r beyond the value of r which corresponds to the second spherical node of $\psi_0(r)$, at which the boundary condition (m.6.63.) is satisfied. For not too low values of E_0, the boundary condition is found, in this way, to be satisfied for two different values of r, both larger than the r value corresponding to the second radial node of $\psi_0(r)$. For the solid metal, E_0 must be negative; by now increasing the magnitude of E_0, which is equivalent to trying another assumed value of E_0 numerically smaller than the initially assumed value, and repeating the integration of (m.6.62.), two new values of r satisfying the boundary condition (m.6.63.) will be found which are closer together, and subsequent repetition of this process will lead to the boundary condition being satisfied for a single inflexion point in $\psi_0(r)$. For assumed values of E_0 smaller and larger numerically than the value which enables the boundary condition to be satisfied at a single inflexion point for $\psi_0(r)$, the boundary condition will not be satisfied at all beyond the r value corresponding to the second node of $\psi_0(r)$, or the two possible values of $r = r_s$ will diverge, until one of them becomes infinite

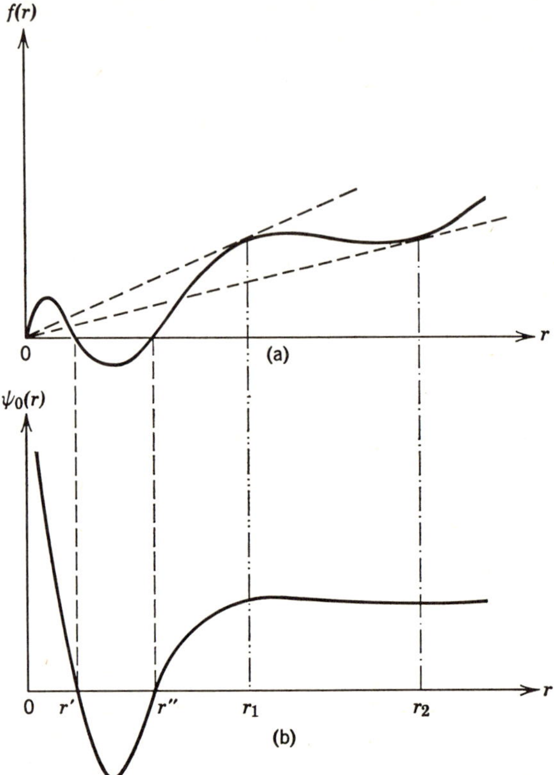

FIG. M.6.7. Approximations to the ground-state wave function for the valency electron in sodium for an energy value near to the ground-state energy. The nodal points in the function are at r' and r''; r_1 and r_2 are the points at which the Wigner–Seitz boundary conditions are satisfied. The function, $f(r) = r\psi_0(r)$ is shown at (a) and $\psi_0(r)$ at (b).

for the E_0 value equal to the negative value of the ionization potential of the free atom. The latter situation arises since $V(r)$ is the potential function of the sodium ion Na^+ in the free atom, and increasing the numerical value of the assumed value for E_0 is equivalent to lowering the magnitude of the energy, which is negative, and the valency electron wave function in the free atom falls off to zero as the distance r tends to infinity. The way in which the calculated values of $r = r_s$ vary as the functions $\psi_0(r)$ and $f(r) = r\psi_0(r)$, for different assumed values of E_0, is shown in Fig. M.6.7. The function $\psi_0(r)$ for sodium has, of course, two nodal points on either side of the nucleus, and

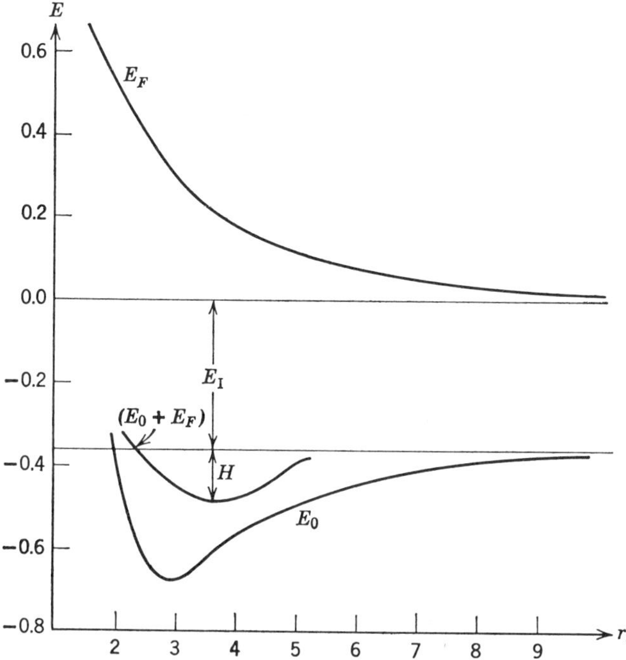

FIG. M.6.8. Energy of sodium as a function of the inter-nuclear distance. The uppermost curve, E_F, describes the variation of the mean Fermi energy with r_s, the lowest one, E_0, the variation of the ground-state energy of the sodium atom containing a sodium cation and the electron in lowest energy state in the metal, with r_s, and the middle curve results from the summation of $E_F + E_0$. The minimum in the latter curve enables the internuclear distance and compressibility to be obtained. H is the approximate cohesive energy or sublimation energy. The energy values are expressed in Rydberg units and the distances in Bohr units.

is almost horizontal for a relatively large part of the distance between two adjacent nuclei; it is this fact which makes the free electron model a satisfactory approximation for the alkali metals. Since the volume is proportional to the cube of the distance, this means that the valency electrons in sodium are essentially free, $\psi_0(r)$ being practically constant in the region of the atomic sphere lying outside the ionic electrons, which is about 90% of the total volume.

The energy associated with the $\psi_0(r)$ from Fig. M.6.7. may then be used to obtain E_0 as a function of r_s, with the result shown in Fig. M.6.8., which enables the approximate cohesive energy and equilibrium atomic radius of the metal to be deduced. The cohesive or binding energy of a crystal, defined as the work required to dissociate the crystal into neutral free atoms, is the same as the heat of sublimation. The other energy quantities in Fig. M.6.8. are the ionization energy and the mean Fermi energy; the ionization energy is the work required to dissociate a neutral atom into a free electron and a free positive ion; and the mean Fermi energy as defined in equation (m.3.10.). With the usual sign convention, E_0 is negative, and E_F and E_I are positive, so that when the atoms in a metal are infinitely far apart, E_F is zero and E_0 and E_I are numerically equal but of opposite sign. For a collection of free atoms, at large distances apart with negligible interaction between the atoms, the work required to remove separately from this assembly one electron and one cation is simply E_I, and if the electron and cation recombine to form a neutral atom again E_I of energy is released, so that there is zero binding energy. On the other hand, in a crystal of sodium metal, the energy of the valency electrons is distributed over a range or band of energies, and the work now required to remove an electron in the lowest state together with a sodium cation is a function of the interatomic distance in the metal and is given by $-E_0$. If this electron and the cation then recombine to form a neutral atom, the energy evolved will be E_I, so that the net work required to remove from the metal crystal one cation and one electron in the lowest energy state, and then to let them recombine into a neutral atom, is $[-E_0 - E_I]$. But the electrons in the metal have energies extending over the complete range of the Fermi distribution, so that the average binding energy of the crystal per atom is $-[E_0 + E_F + E_I]$ instead of $-[E_0 + E_I]$, as it would be if all of the electrons in the metal were in the lowest energy state. Conversely, the energy of the crystal relative to the free atom is $+[E_0 + E_F + E_I]$, so that large values of E_F or E_I make the crystal energy greater than that of the free atoms and therefore less stable. For all metals, E_I is independent of their crystal structure and is a constant of the free atom, while E_F and E_0 are both functions of the internuclear distance in the crystal. E_F and E_0 for sodium can both be calculated as a function of r_s, and by adding the two as in Fig. M.6.8., the curve for $(E_0 + E_F)$ as a function of r_s is found to have a minimum, which is not too different from the minimum in the E_0 curve alone. From the position of the minimum in the $(E_0 + E_F)$ curve, the equilibrium distance may be obtained and hence the atomic volume and lattice spacing of the crystal at 0 K calculated. From calculated curves such as Fig. M.6.8., the binding energy is obtained as $-[E_0 + E_F + E_I]$, so that physical constants such as the compressibility may be calculated, since this is defined as β (compressibility) $= -(1/V)(dV/dp)$, and if the heat effects are small, the change of energy on compression as a consequence of the work done is $dE = -p\,dV$, where E is the total energy, giving $1/\beta = V[d^2E/dV^2]$.

Since $V = 4\pi r_s^3/3$, then

$$1/\beta = [12\pi r_s]^{-1} d^2E/dr_s^2 - [6\pi r_s^2]^{-1} dE/dr_s.$$

When $r_s = r_0$, the equilibrium atomic radius at zero temperature and pressure, then $dE/dr_s = 0$, and

$$1/\beta = [12\pi r_0]^{-1}[d^2E/dr_s^2]_{r_s=r_0}.$$

The compressibilities of the alkali metals calculated by means of this expression agree very closely with the experimentally determined values.

The potential function used by Wigner and Seitz in equation (m.6.61.) was not actually the spherically symmetrical sodium ion potential, such as may be calculated by the Hartree–Fock method, but a semiempirical function constructed by Prokofjew[1] designed to account for the observed spectroscopic transitions of the valency electron in sodium atoms. Several techniques have since been described for obtaining E_0 directly without calculating either $V(r)$ or $\psi_0(r)$ based on the observed spectroscopic values of the atomic energy levels. The device first used by Kuhn and Van Vleck[2] and later improved by Brooks and Ham[3,4] has been successfully used for the study of the alkali metals, and has come to be known as the quantum defect method.

The cohesive and correlation energy of sodium is discussed later (pages 690–697).

M.6.7. The Quantum Defect Method

For sodium in particular, and the other alkali metals generally, the effective radius of the normal cation is small compared with the atomic radius, to such an extent that the potential function $V(r)$ of the alkali metal cations is very approximately of the hydrogenic type, or $V(r) = -\varepsilon^2/r$. As a consequence, in the region outside the ionic radius, where in the Wigner–Seitz method it is necessary to fit the boundary condition that for the ground-state function, $[d\psi_0(r)/dr]_{r=r_s} = 0$, the Schrödinger equation (m.6.61.) becomes identical to the hydrogen atom equation,

$$d^2\psi_0(r)/dr^2 + [2/r][d\psi_0(r)/dr] + [8\pi^2m/h^2][E_0 + \varepsilon^2/r]\psi_0(r) = 0,$$

which may be alternatively expressed in atomic units as

$$d^2\psi_0(r)/dr^2 + [2/r][d\psi_0(r)/dr] + [E_0 + (2/r)]\psi_0(r) = 0. \quad \text{(m.6.64.)}$$

---

[1] Prokofjew, W., *Z. Physik*, 1929, **58**, 255.
[2] Kuhn, T. S., and Van Vleck, J. H., *Phys. Rev.*, 1950, **79**, 382.
[3] Brooks, H., and Ham, F. S., *Phys. Rev.*, 1958, **112**, 344.
[4] Ham, F. S., *Solid State Physics I*, 1955, Academic Press, New York.

The solutions of this equation in closed analytical terms have been discussed in section Q.3.6. The problem of fitting the Wigner–Seitz boundary condition to these known solutions is that there are two independent solutions for each value of E_0, and the general solution is a quite arbitrary linear combination of the two. For the correct linear combination, the wave function inside the ion has to join smoothly with the particular combination function, for the same value of E_0. For some particular eigenvalue, E_0, the general solution of equation (m.6.64.) will be of the form

$$\psi_0(r) = C_1 F_1(r) + C_2 F_2(r), \qquad \text{(m.6.65.)}$$

where $F_1(r)$ and $F_2(r)$ are known independent solutions and C_1 and C_2 are arbitrary constants. The experimentally determined quantum defect enables the coefficients C_1 and C_2 to be chosen so that the linear combination function will join smoothly with the function inside the ion, for the same value of E_0, in the absence of any knowledge about the wave function inside the ion. Of the infinite number of choices of the functions $F_1(r)$ and $F_2(r)$ which are possible, provided that one of them is finite at the origin while the other one is infinite at the same place, then when E_0 is one of the free atom eigenvalues, the correct linear combination of the functions $F_1(r)$ and $F_2(r)$ must approach zero as r tends to infinity; then the function $\psi_0(r)$ defined by (m.6.65.) will describe the free atom wave function outside the ion, and for this particular value of E_0 must join up smoothly with the unknown wave function inside the ion. In the quantum defect method, the ratio of the coefficients C_2/C_1 which leads to such a function is given as[1]

$$C_2/C_1 = -\tan \pi\delta, \qquad \text{(m.6.66.)}$$

where δ, the so-called quantum defect, is a smooth and slowly varying function of E, which may be expressed as low-degree polynomial in E as

$$\delta = \alpha + \beta E + \gamma E^2. \qquad \text{(m.6.67.)}$$

The quantum defect is related to spectroscopic data by the fact that the energy levels of the hydrogen atom are given in atomic units by $E = -(1/n^2)$, and the energy levels of the s-states of the valency electron in the alkali metals by $E = -[1/(n - \delta)^2]$, where n is the principal quantum number for the electron state concerned; the quantum defect has different values for electrons in $p, d, f, \ldots$, states, and is defined for any value of E by (m.6.67.).

For the ground-state of an alkali metal, the energy E_0 will necessarily be lower than the energy of the normal state of a valency electron in a free atom, and this latter energy value is the lowest for which it is possible to obtain an experimental value for the quantum defect. If it is assumed that the ratio

---

[1] Ham, F. S., *Phys. Rev.*, 1962, **128**, 82, 2524.

C_2/C_1 for the function $\psi_0(r)$ outside the metal ion, which joins up smoothly to the true but unknown function inside the ion for all values of E_0, is also given by the same expression as (m.6.66.), then the value of $r = r_s$ outside the ion, for which the normal derivative of $\psi_0(r)$ is zero, may be found from (m.6.65.). For the lowest energy state E_0, the quantum defect is obtained by extrapolation from (m.6.67.). Having thus found the value of $r = r_s$, for which the derivative of $\psi_0(r)$ becomes zero, E_0 may then be obtained as a function of r_s without having to solve the Schrödinger equation inside the metal ion.

The use of the observed atomic spectrum to deduce a general expression for the ratio C_2/C_1 for $s, p, d, f, \ldots$, states, as a function of energy, is believed to lead to calculated values of the cohesive energies of metals, which are perhaps more accurate than the values calculated by numerical integration of some assumed potentials. This is due to the fact that the quantum defect deduced from spectroscopic data automatically contains and thus makes allowance for the polarization effect, for example, of the ion by the valency electron, which is an effect that is completely disregarded in self-consistent field-type calculations.

M.6.8.   The Augmented Plane Wave and Scattered Wave Methods

The type of potential function commonly found in metal crystals in which it may be assumed that there is a region of spherically symmetrical potential surrounding the metallic nuclei and a constant region of potential between these spherically symmetrical regions, referred to by Slater[1] as a "muffin-tin" potential, is such that a solution for the corresponding Schrödinger equation in a periodic potential field may be obtained by the so-called augmented plane wave[2,3] and scattered wave methods.[4-6] For an isolated atom, a spherically symmetrical potential function is a reliable approximation, and for the potential between atoms the smoothly varying potential function in this region may be represented to a good approximation by its average value. Because of the spherical symmetry of the bounding surface between the regions of constant potential, the solution of the appropriate Schrödinger equation in such an assumed "muffin-tin" potential may be obtained in the same manner as for the particle in the one-dimensional potential well (section

---

[1] Slater, J. C., *Phys. Rev.*, 1937, **51**, 846.

[2] Slater, John C., *Quantum Theory of Molecules and Solids*, Vol. 2, 1965, McGraw-Hill, New York.

[3] Loucks, T., *Augmented Plane Wave Method*, 1967, W. A. Benjamin, Inc., New York.

[4] Korringa, J., *Physica*, 1947, **13**, 392.

[5] Kohn, W., and Rostocker, J., *Phys. Rev.*, 1954, **94**, 111.

[6] Morse, P. M., *Proc. Natl. Acad. Sci. Wash.*, 1956, **42**, 276.

Q.2.2.); this, in principle, involves choosing a suitable analytical form for the wave functions in the different regions and then joining them in such a manner that the appropriate function and its derivative are continuous over the bounding surfaces. The appropriate wave function within the region of spherical symmetry surrounding each metallic nucleus will be of the form given by equation (q.3.130.) and the radial part of this function, of the form given by equation (q.3.110.), will constitute a solution of the corresponding radial Schrödinger equation for some assumed value of the spherical potential. Outside the regions of spherically symmetrical potential, the appropriate wave function is represented by a plane wave of the type $e^{i(\mathbf{k}+\mathbf{K}_r)\cdot\mathbf{r}}$, where $\mathbf{k}$ is a reduced wave vector and $\mathbf{K}_r$ is defined by equation (m.5.54.). The plane wave, which when joined to a suitable solution of the spherically symmetrical problem pertaining inside the sphere surrounding the metal nuclei, to produce a continuous function over the surfaces of the spheres, is what is referred to as an augmented plane wave and has many of the characteristics of the orthogonalized plane waves of section M.6.5. Actually, to find suitable augmented plane wave functions for any particular crystal requires expansion of the function within each region of spherical symmetry in spherical harmonics [equation (q.3.55.)], each such expansion multiplied by the appropriate radial function, and the plane wave part expanded over 80–100 different values of $\mathbf{K}_r$, to be fitted to the sum of the spherically symmetrical functions, in order to find a complete wave function which is itself continuous and has a continuous derivative over the surfaces of the spheres. While there is no unique expansion of the wave function of a system in plane waves, it has been proved[1] that the plane wave expansion given by the augmented plane wave method provides the most rapidly convergent plane wave expansion (pages 624–626) which is possible for the wave function in the region between spheres surrounding each nucleus. Details of the method are given in the books of Slater and Loucks, referred to on page 653, who give a description of the various programs required in computing the various quantities involved in arriving at the energy bands of real crystals. In metal crystals of high symmetry, the number of augmented plane waves required to describe the wave function in the region between metal nuclei is actually much less than the 80–100 required for a good approximation, since at symmetry points in the crystal the coefficients of many of the augmented plane waves are equal to one another. The augmented plane wave method of calculating the energy bands in metal crystals indicates that the validity of the free electron and the nearly free electron approximation methods is associated closely with the Ramsauer effect (page 502) and with the phenomena of X-ray and electron scattering by crystals. Thus it emerges that for atoms with small scattering

---

[1] Johnson, K. H., *Phys. Rev.*, 1966, **150**, 429.

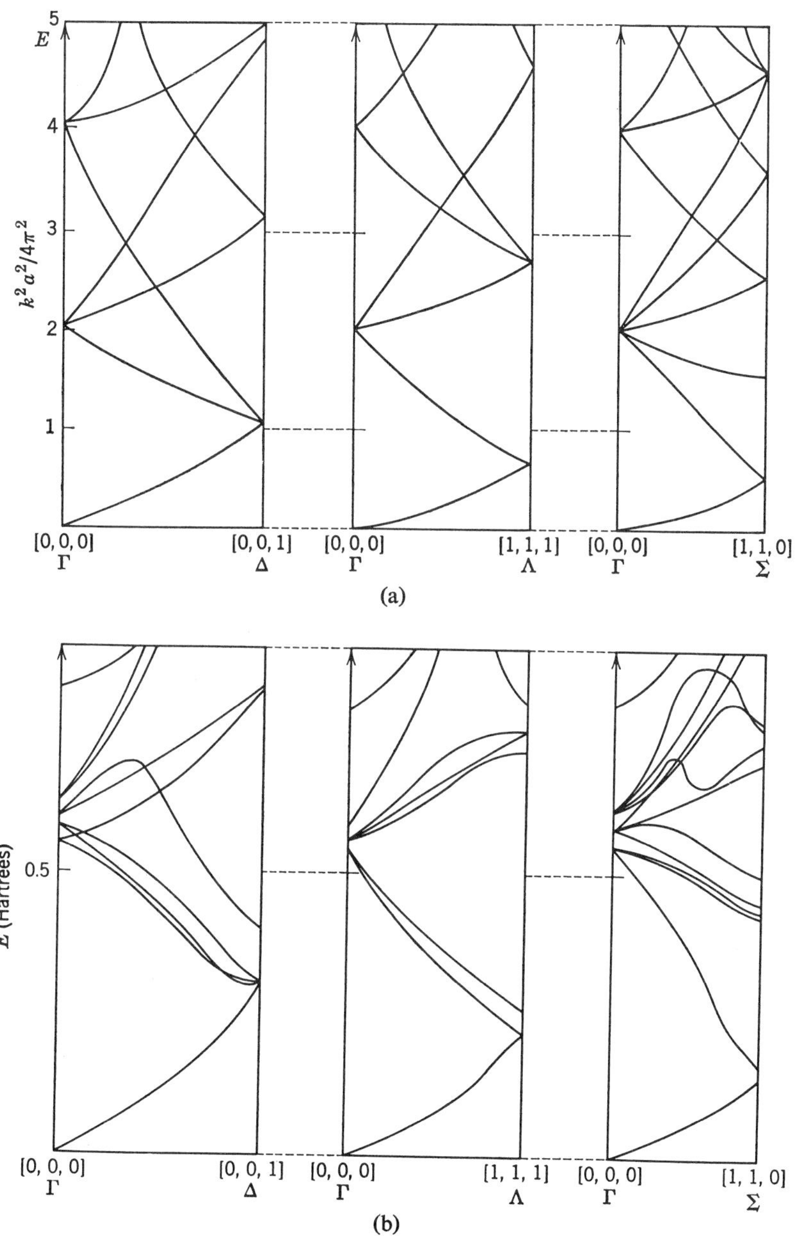

FIG. M.6.9. *Continued on page 656*

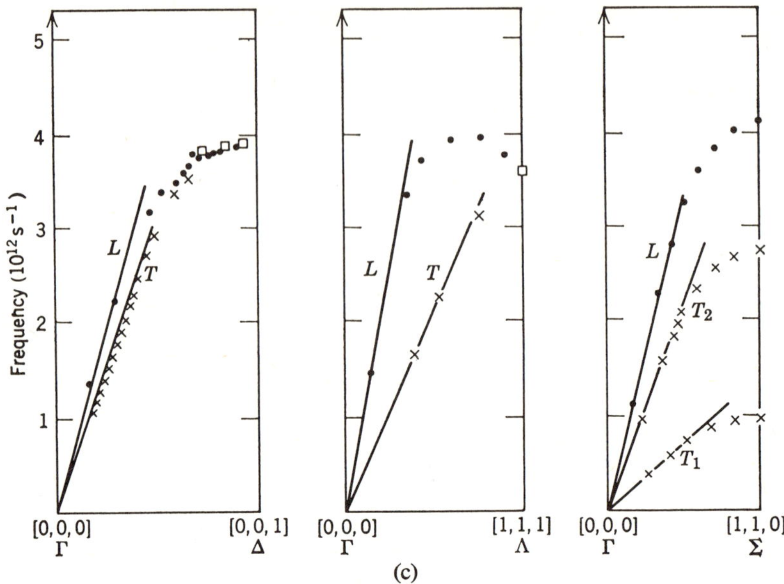

FIG. M.6.9. Comparison of the energy band calculations of sodium according to (a) the free electron approximation; (b) the augmented plane wave method; and (c) the lattice vibrational spectrum of sodium at 90 K obtained experimentally by analysis of the energy distribution of inelastically scattered thermal neutrons. See *Quantum Theory of Molecules and Solids*, Volume 2, 1965, by John C. Slater, McGraw-Hill Book Company, New York, and footnote to page 659.

cross sections, such as the alkali metals, where the Ramsauer effect is applicable, the solution of the Schrödinger equation in the appropriate periodic potential approaches the free electron description of the same problem. In fact, the energy bands calculated by both the augmented plane wave and the scattered wave methods, at least for the alkali metals, not only closely resemble the free electron and nearly free electron approximation results but also the energy wave vector results obtained experimentally by the inelastic scattering of thermal neutrons from the metal concerned (see page 770). In metal theory, the results of energy band calculations are usually shown by plotting the energy as a function of the magnitude of the wave vector **k** along certain directions in **k**-space, while the experimental results obtained by examination of the energy distribution of inelastically scattered neutrons (by the metal crystal) are generally illustrated by plotting the measured lattice vibrational frequencies as a function of the wave vector, also along certain directions, usually of high symmetry, in *k*-space. Figure M.6.9. shows the comparison between the energy bands of sodium according to the free electron approximation, the augmented plane wave method, and the lattice vibrational frequencies of sodium, as determined experimentally, along the three principal symmetry directions 100, 110, and 111, respectively.‡

‡ With reference to Fig. M.6.9., the free electron results shown in (a) have been calculated[1] by setting up the functions $(\mathbf{k} + \mathbf{K}_r)^2 h^2/8\pi^2 m$, where $\mathbf{k}$ is a reduced wave vector and $\mathbf{K}_r$ is one of the reciprocal lattice vectors. Corresponding to the infinite number of vectors $\mathbf{K}_r$, there will be an infinite number of such functions, but in the central Brillouin zone the lowest energy value of these functions will correspond to the lowest energy band, the second lowest to the second energy band, and so on. The results are to be interpreted in terms of the Brillouin zones for the body centered cubic structure, which sodium adopts and shown in Fig. M.5.28., and the various symmetry directions and points for the rhombic dodecahedron which is the first Brillouin zone for this structure. The first Brillouin zone for body-centered cubic structures, showing the various points and lines in the zone such that reduced wave vectors terminating in these points and lines show special symmetry properties, is illustrated in Fig. M.6.10. The terminology used in describing these special symmetry points and directions was originally introduced by Bouchaert, Smoluchowski, and Wigner.[2] In this terminology, the origin of the body-centered cubic Brillouin zone is designated by Γ, the symmetry direction from the center of the zone to a four-fold vertex (the 100 direction) by Δ, the direction from the zone center to a three-fold vertex corresponding to the 111 direction by Λ, and the direction from the center of the zone to the mid-point of a face corresponding to the 110 direction by Σ; the points corresponding to the extremities of the wave vectors in the Δ, Λ, and Σ, directions located on the zone surface are designated by H (four-fold vertex), P (three-fold vertex), and N (mid-point of zone face), respectively. The directions from H to P along a zone edge, from H to N along the line from a four-fold vertex to the center of a face, and from N to P along the line from a face center to a three-fold vertex, are designated F, G, and D, respectively. All of the energy levels in Fig. M.6.9.(a) are segments of parabolas, describing $(\mathbf{k} + \mathbf{K}_r)^2 h^2/8\pi^2 m$ as a function of the magnitude of $\mathbf{k}$ along the directions concerned. The free electron approximation simplifies the problem to the extent that many sets of energy levels which would be separated in the real potential applicable to metallic sodium are degenerate in this description.

In Fig. M.6.9.(b) is shown the result of the augmented plane wave calculation[3] of the energy bands of sodium, showing that the degenerate energy levels which result from the idealized free electron calculations have been separated by the perturbations provided by the real potential in the crystal. The agreement between the two sets of results is only good for electrons of quite small energies, for the energy bands near the bottom of each diagram. The Fermi level for sodium is about half way up the lowest band illustrated and in this region the agreement between the free electron and the calculated energy values is quite good. The agreement between the results of the free electron and the augmented plane wave type calculations becomes worse at higher energies and is also poor at lower energy values. In terms of the free electron approximation, the lowest energy band is the

---

[1] Slater, J. C., *Quantum Theory of Molecules and Solids*, Vol. 2, 1965, McGraw-Hill Book Company, New York.
[2] Bouckaert, L. P., Smoluchowski, R., and Wigner, E., *Phys. Rev.*, 1936, **50**, 58.
[3] Performed by J. F. Kenney at the Massachussetts Institute of Technology; see Slater, J. C., *Quantum Theory of Molecules and Solids*, Vol. 2, 1965, McGraw-Hill Book Company, New York; also Ham, F. S., *Phys. Rev.*, 1962, **128**, 82.

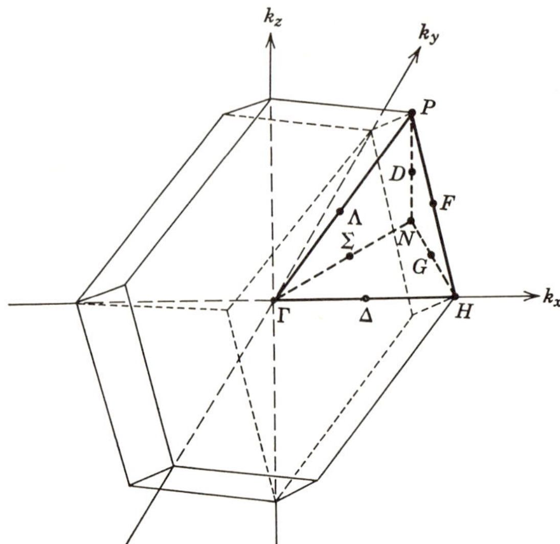

FIG. M.6.10   Brillouin zone for the body-centered cubic structure showing the special symmetry points and directions.

one shown in Fig. M.6.9.(a), but for a real sodium crystal, there are very narrow energy bands far below this, corresponding to the $1s$, $2s$, and $2p$ energy levels of the atom which are scarcely broadened at all in the crystal, because of the minute overlap of the atomic orbitals in these energy bands (compare Fig. M.3.1.).

In Fig. M.6.9.(c) is shown the lattice vibrational frequencies of sodium at 90 K, as measured by the technique of inelastic neutron scattering,[1] illustrating the relationship between the vibrational frequencies of the transverse and longitudinal acoustic modes as functions of the wave vector, along the same symmetry directions as in (a) and (b). In the Δ and Λ directions or the 001 and 111 directions, there is a degenerate transverse mode with two directions of polarization at right angles to the wave vector, and a non-degenerate longitudinal mode, while in the Σ or 110 direction the symmetry does not lead to degeneracy and the three modes are separated.[2-4] In all three directions shown the longitudinal mode has a higher frequency than the transverse mode, which is quite a general phenomenon as the longitudinal oscillations corresponding to compression and expansion of the crystal correspond to a larger restoring force than for transverse oscillations involving a shearing force. There are no optical branches for sodium, as for crystals with only one atom per Bravais unit cell, there is only an acoustic branch to the

---

[1] Woods, A. D. B., Brockhouse, B. N., March, R. H., Stewart, A. T., and Bowers, R., *Phys. Rev.*, 1962, **128**, 1112; *Proc. Phys. Soc.* (London), 1962, **79**, 440.

[2] Brockhouse, B. N., *"Inelastic Scattering of Neutrons in Solids and Liquids,"* 1961, International Atomic Energy Agency, 113.

[3] Wallis, R. F., editor, *Lattice Dynamics*, 1965, Pergamon Press, Oxford.

[4] Banerjee, R., and Varshni, Y. P., *Can. J. Phys.*, 1969, **47**, 451.

lattice vibrational spectrum.[1] The straight lines in Fig. M.6.9.(c) are determined from the measured values of the low frequency elastic constants of sodium metal, at the same temperature at which the frequency measurements were made (90 K), and from these the velocity of sound in the crystal, corresponding to the slope of the curves may be calculated.[1,2] The frequency-wave vector dispersion relations for sodium previously calculated[3] are in good agreement with the experimentally measured values.

The scattered wave method or the Green's function method of calculating the energy bands in metal crystals is based on a method originally evolved by Ewald (page 502) to deal with X-ray scattering by crystalline substances. The starting point for all calculations of this type is the consideration of the scattering of an incident wave by one of the atoms, with a spherically symmetrical potential function, in the crystal concerned; then by superimposing all of the scattered waves, the potential field throughout the crystal is synthesized. The quantum mechanical explanation of the non-classical Ramsauer effect (page 502) indicated that the observed phenomena was just what would be expected on the basis of wave theory. The principle involved in scattered wave method calculations of energy bands in metallic crystals, as in the augmented plane wave method, is to solve the wave equation for the spherical potential in the vicinity of an atom, for some assumed value of the energy, and then to join the function thus found to the solution outside the spherical potential region, so that both the function and its derivative are continuous. If the potential outside the region of spherical symmetry is taken as being constant, the corresponding Schrödinger equation is of the time-independent form and its radial solutions are known to be spherical Bessel and Neumann functions.[4,5] The best solution in the region between the atoms is found to be a spherical Hankel function,[4,5] which is a linear combination of a Bessel and a Neumann function. The analytical expression for the synthesis of the total wave function, according to Ewald's method, from the waves scattered by each atom in the crystal, and the calculation of energies from the resulting wave functions, has been programmed for digital computers,[6,7] and in those cases where both the augmented plane wave and the scattered wave methods have been applied to the calculation of the energy bands of metal crystals for similar potential fields, the results are very similar.

---

[1] Launey, J. de, *Solid State Physics*, Vol. 2, 1956, 219–303, Academic Press, New York.
[2] Born, M., and Huang, K., *Dynamical Theory of Crystal Lattices*, 1956, Oxford University Press, London.
[3] Toya, T., *J. Res. Inst. Catalysis, Hokkaido Univ.*, 1958, **6**, 183.
[4] Margenau, H. M., and Murphy, G. M., *The Mathematics of Physics and Chemistry*, 1961, Vol. I, 2nd Edn., D. Van Nostrand Co. Ltd., London.
[5] Abramowitz, M., and Stegun, I. A., Editors, *Handbook of Mathematical Functions*, 1964, National Bureau of Standards, Washington.
[6] Johnson, K. H., *Phys. Rev.*, 1966, **150**, 429.
[7] Beeby, J. L., *Lectures on the Many-Body Problem*, 1964, Academic Press, Inc., New York.

M.6.9.   Heitler–London–Pauling–Slater Functions

In discussing the energy bands in metals, the various collective electron theories, the free electron, the nearly free electron, the tight binding, the orthogonal and augmented plane wave methods, the cellular method of Wigner and Seitz, and the scattered wave methods focus attention on the electrons which lie outside the closed quantum groups of electrons in the free atoms, and it is assumed that Bloch-type functions are appropriate, extending over large distances on the atomic scale. The concept of a Fermi surface is strictly applicable only to theories of the Bloch type, where the characteristic behavior of the electrons on the Fermi surface is so closely related to many of the magnetic and electrical properties of metals that this surface may be essentially regarded as a physical property of metals. But metals containing, as they do, many electrons, all of which contribute to the field acting on any one of them and giving rise to Coulomb-type correlations, are such that these correlation effects may mean that Bloch functions do not always provide the best description of the electronic motion involved.

In the early development of quantum mechanics, two approaches to the quantum mechanical description of the simplest molecules, the hydrogen molecule ion $H_2{}^+$ and the hydrogen molecule H_2, were adopted: the molecular orbital method[1-3] and the Heitler–London[4,5] method. The latter method initially attracted more attention and led to many methods of providing the solution to quantum mechanical problems, more accurate than the Hartree–Fock method. The Heitler–London method paid scant attention to the directional properties of the chemical bonds in simple molecules, such as the pyramidal structure of the ammonia molecule, the triangular structure of the water molecule, and the tetrahedral structure of the methane molecule, but Pauling[6] and Slater[7] later extended the theory to take account of the geometrical properties of simple molecules. The functions used in section Q.5.6. to describe the excited states of the helium atom are examples of the Heitler–London type, also variously referred to as localized atomic orbitals. The major difference between the Heitler–London and the molecular orbital methods lies in the different ways in which the correlation of the motion between electrons

---

[1] Hund, F., *Z. Physik*, 1927, **40**, 742; **42**, 93.

[2] Mulliken, R. S., *Phys. Rev.*, 1928, **32**, 186.

[3] Lennard-Jones, J. E., *Trans. Faraday Soc.*, 1929, **25**, 668.

[4] Heitler, W., and London, F., *Z. Physik*, 1927, **44**, 455.

[5] Sugiura, Y., *Z. Physik*, 1927, **45**, 484.

[6] Pauling, L., *Proc. Nat. Acad. Sci. Wash.*, *U.S.*, 1928, **14**, 359; 1932, **18**, 293, 414; *Nature of the Chemical Bond*, 1962, Cornell University Press, Ithaca, New York.

[7] Slater, J. C., *Phys. Rev.*, 1931, **37**, 481; **36**, 325, 1109; 1932, **41**, 255.

is dealt with. The Heitler–London method includes a large degree of correlation by assuming that the electrons are associated with separate nuclei, while in the application of the molecular orbital method the only correlation which appears is due to the applicability of the Pauli Exclusion Principle. As a consequence of these considerations, Bloch functions would be expected to be most appropriate for the description of the valency electrons in a metal but not for the very tightly bound electrons in the close vicinity of atomic nuclei. For the closed quantum groups of electrons in the lowest energy states in atoms and molecules, there is no real distinction between the two methods, and both the Heitler–London and Bloch methods lead to the same results. The Pauling development of the Heitler–London method has led to the so-called valency bond theory of atomic bonding and the aspects of electron spin interactions which Heitler and London handled by the application of the methods of group theory was shown by Slater to be more easily dealt with by the determinantal method (Q.5.5.). It is not so easy and there is considerable difference of opinion as to whether atomic or molecular orbitals provide the most appropriate description of the electrons in, for example, the d-states in transition metals.

For insulators, it is generally agreed that the collective electron theories and the Heitler–London–Pauling–Slater method are simply different approaches to the same problem, and that if applied to the limit both methods would lead to the same result. Thus, diamond may be regarded as a substance in which the electrons are described by Bloch functions and that the four electrons per atom completely fill a zone which is surrounded everywhere by an energy gap. Alternatively, it may be described by the valency bond method as a substance in which the electrons are represented by Heitler–London–Pauling–Slater functions derived from mixed sp^3 orbitals of the constituent carbon atoms. The latter type functions, derived from neutral atoms, generally describe an insulator, and electrical conductivity in a substance requires for its explanation the introduction of functions derived from ionized states. Thus, if the valency electrons in sodium were described by Heitler–London-type functions, it would be expected to be an insulator, and in order to allow for the electrical conductivity it is necessary to assume that the descriptive functions which apply are derived from Na^+ and Na^- ions with the electrons and ions well separated from one another. The Pauling theory of metals (Chapter M.2.) is based on the use of such functions. However, even approximate Heitler–London-type functions cannot be calculated for a crystal, and the theories based on the use of such functions for the description of metals are only qualitative or semiquantitative, in contrast to the quantitative methods based on Bloch theory. It has been pointed out[1,2] that there are

---

[1] Mott, N. F., *Progress in Metal Physics*, Vol. III, 1952, Pergamon Press, London.
[2] Mott, N. F., and Stevens, K. W. H., *Phil. Mag.*, 1958, **2**, 1364.

difficulties associated with the use of both Bloch-type and Heitler–London-type functions for the description of the electrons in conductors. Many substances, such as metal oxides, sulfides, selenides, and tellurides, are insulators but crystallize with simple ionic structures, such as the rock salt or fluorite structures. In the case of a transition metal oxide like NiO, which adopts the rock salt structure, then on the assumption that it contains the d^8 dipositive nickel ion, it would be expected to exhibit metallic conductivity as a consequence of the incompletely filled d^{10} band. Since the substance is observed to be an insulator, this would preclude the use of Bloch-type functions for the description of the $3d$ nickel electrons, but the non-conducting Heitler–London-type functions would be appropriate in this case. It is on the basis of this and other considerations that it has been suggested[1,2] that the concept of bonding and non-bonding orbitals adopted by Pauling is better substituted by the idea that the wave functions of electrons in solids should be divided into Bloch (conducting) and Heitler–London (non-conducting) types.

M.6.10. Plasma Theory of Metals

One of the major difficulties in all band theory type calculations is the problem of the correlation energies of the electrons in metals. A theory of metals has been developed in the period since 1951[1-3] which enables the correlation energies to be calculated by the application of ordinary second order perturbation theory, as well as providing a relatively simple means of justifying the independent particle approximation, which is assumed to be valid in nearly all band and zone type calculations, as for example, in equation (m.5.57.), which uses the same potential function for all one-electron states. The concept of a plasma was originally introduced by Langmuir, who introduced this term to describe a gas discharge in which there were free positive ions and free electrons, in just sufficient number to neutralize the ions on the average. Langmuir found that violent electromagnetic oscillations took place in such a gas discharge, which had the effect of accelerating or decelerating any electrons in the plasma so that such electrons came very quickly to equilibrium with a typically Maxwellian distribution of velocities, characteristic of a very high temperature. The explanation of the cause of these oscillations was given by Tonks and Langmuir.[4] Any system composed of a large number of positive ions and virtually free electrons, where the total charge is zero, may be referred to as a plasma. Thus, apart from gas discharge

---

[1] Bohm, D., and Pines, D., *Phys. Rev.*, 1951, **82**, 625; 1952, **85**, 338; 1953, **92**, 609.

[2] Pines, D., *Phys. Rev.*, 1953, **92**, 628; *Solid State Physics*, Vol. I, 367, Academic Press, New York; *Rev. Mod. Phys.*, 1956, **28**, 184; *Nuovo Cimento Suppl.*, 1958, **7**, 329.

[3] Raimes, S., *Reports on Progress in Physics*, 1957, **20**, 1.

[4] Tonks, L., and Langmuir, I., *Phys. Rev.*, 1929, **33**, 195.

tube systems, thermonuclear plasmas, and highly ionized gases generally, metals are also examples of plasmas. In metals, however, the positive ions, apart from their thermal vibrational modes, are to be regarded as stationary.

In the simplest application of the quantum mechanical theory of plasma oscillations in metals, the collection of cations composing the lattice array of metal ions may be imagined to be substituted by a uniform distribution of positive charge, of density equal to the average electronic density but of opposing sign. Since the electrons in a metal are constantly in motion, the electron density must continually vary throughout the plasma. As a consequence of chance fluctuations in the electronic motion, it is conceivable that the electronic charge density in some region of the plasma is reduced below the average value. The positive background density in that region would then no longer be neutralized and the resultant region of positive charge would attract nearby electrons in the tendency to neutralize the charge buildup. As the electrons attracted in this way acquire momentum, they will tend to overshoot the positively charged region, leading to an accumulation of electrons over the number required to neutralize the positive charge. This excess negative charge will then repel electrons giving rise to the setting up of an oscillatory system. The longitudinal oscillations of the electron gas in a metal set up in this manner, which are analogous to sound waves, are referred to as plasma oscillations. The major distinction between gas discharge plasmas and metals is the large difference in the electron densities: in gas discharges, the electron density is of the order of 10^{15} dm^{-3} but in a metal it is of the order of 10^{26} dm^{-3}. Thus, while the electrons in a gas discharge may be suitably described in terms of classical Maxwell–Boltzmann statistics, the motion of the electrons in a metal must be described in terms of Fermi–Dirac statistics.

The frequency of the plasma oscillations may be found most simply, in terms of simple classical theory, by assuming initially that the electrons in the metal plasma are at rest. If the average density of electrons is n, so that the average electronic charge is $-n\varepsilon$ and the average uniform positive charge is $n\varepsilon$, then if it is assumed that the electrons are displaced outwards from some point in the plasma chosen as the origin 0, the displacement may be represented by some function $\xi(r)$ of the radial distance from 0, and it is convenient to suppose that this function and its derivative is everywhere small. As the result of such a displacement, the number of electrons which move from a sphere of radius r, centered at 0, is approximately $4\pi r^2 n\xi(r)$, which is the number initially contained in a spherical shell of thickness $\xi(r)$ at the surface of the sphere. As a consequence of the positive background charge, the sphere would then be left with a charge $4\pi r^2 n\varepsilon\xi(r)$, so that the electric field distant r from the origin 0 would be of magnitude $4\pi n\varepsilon\xi(r)$, directed radially outwards. Each electron therefore is acted upon by a force of $-4\pi n\varepsilon^2\xi(r)$, directed radially outwards, which may be equated to the product of the electron's mass and acceleration, $m\, d^2\xi(r)/dt^2$. Thus, $-4\pi\varepsilon^2 n\xi(r) = m\, d^2\xi(r)/dt^2$ or $d^2\xi(r)/dt^2 + [4\pi n\varepsilon^2 m^{-1}]\xi(r) = 0$, from which, being an equation describing simple harmonic motion, the angular frequency of oscillation is obtained as

$$\omega_p = [4\pi n\varepsilon^2 m^{-1}]^{1/2}, \qquad\qquad \text{(m.6.68.)}$$

referred to as the plasma frequency. ω_p has the magnitude of about 10^{10} rad s^{-1} for gas discharge plasmas and of about 10^{16} rad s^{-1} for metals. The relationship between the concept of plasma oscillations in metals and their properties, especially their optical properties, did not become apparent until attempts were made[1] to account for the discovery that the alkali metals became transparent beyond a certain wavelength in the ultraviolet region of the electromagnetic spectrum, and that this wavelength decreased for the alkali metals of lower nuclear charge number (Cs to Li).[2] If the normal thermal motion of electrons is taken into account, the angular frequency of the plasma oscillations is very similar to ω_p for long wavelengths, but increases with decreasing wavelength. The so-called dispersion relation between frequency and wave vector k $(2\pi/\lambda)$ according to classical theory, where ω is the oscillation frequency $(2\pi\nu)$, k the wave vector, and $\overline{v_\varepsilon^2}$ the average value of the squares of the velocities of the electrons, is: $\omega^2 = \omega_p{}^2 + k^2\overline{v_\varepsilon^2}$. The plasma frequency ω_p corresponds to the point in the electromagnetic spectrum which denotes the boundary between the transparency of a metal, at higher frequency, and its capacity at lower frequencies. The dispersion for metals is small, because of the upper limit corresponding to the wave vector k_c, for wave numbers above which plasma oscillations do not take place. Generally, thermal motion of the electrons disrupts the plasma oscillations and prevents them from taking place with short wavelengths, so that there is a lower limit to the wavelength of plasma oscillation.

Although the electron concentration in metals is so much greater than in gas discharge plasmas, qualitative and semiquantitative deductions about the nature of plasma oscillations in metals have been made by the application of classical methods.[3] Thus, it has been shown in this way that if the electron density in a metal is described by means of a Fourier series, that the coefficients oscillate approximately with the angular frequency ω_p if the wavelength is sufficiently long, and that plasma oscillation will take place provided the wave number k is such that $\omega_p \gg k v_0$, where v_0 is the electron velocity at the Fermi surface, or the maximum electron velocity at 0 K. Alternatively, the upper limit for k, denoted by k_c, may be approximately expressed as $k_c \simeq \omega_p/v_0$. By substituting for the constants in (m.6.68.), ω_p corresponds to a

[1] Zener, C., *Nature*, 1933, **132**, 968.
[2] Wood, R. W., *Phys. Rev.*, 1933, **44**, 353.
[3] Bohm, D., and Gross, E. P., *Phys. Rev.*, 1949, **75**, 1851.

wavelength of about 1.14×10^{-7} m, which lies within the ultraviolet region of the spectrum, and $1/k_c$ has the value of about 10^{-10} m, which is of the order of magnitude of internuclear distances in solids.

The application of quantum theory to plasma oscillations involves quantizing the oscillations and consideration of their excitation by electrons moving through metals. This stage in the development of the theory also followed an experimental discovery[1-3] that electrons of definite energy, after traversing very thin metal foils, emerge with energy losses of the order of 1.6×10^{-18} J. The discrete quanta of energy lost by electrons in interaction with plasma oscillations are referred to as plasmons. The plasma oscillations, whether represented by a finite set of harmonic oscillators with angular frequency ω_p or described by Fourier series, representing the electron density, whose coefficients oscillate with approximately the frequency ω_p, may have their energies quantized quite simply; as in Fig. Q.2.8., the energy of a harmonic oscillator may only adopt the values $(N + 1/2)(h\omega_p/2\pi)$, where N is zero or positive and integral. If $N = 0$, the ground-state energy $h\omega_p/4\pi$ would correspond to the zero-point energy of a plasmon.

By substituting equation (m.3.13.) for the electron concentration into equation (m.6.68.) for the plasma frequency, it is possible to determine the excitation energy, $h\omega_p/4\pi$, required to excite any state into the state of next highest energy. Thus, from (m.6.68.),

$$\omega_p^2 = [4\pi ne^2/m] = [3e^2/mr_s^3], \qquad (m.6.69.)$$

and in atomic units (r_s expressed in Bohr units and energy in Hartrees) there results

$$h\omega_p/2\pi = \sqrt{3}\, r_s^{-3/2} \text{ Hartrees.} \qquad (m.6.70.)$$

On comparison of this energy value with the Fermi energy, some important conclusions may be drawn. In terms of the Sommerfeld–Hartree free electron theory, the maximum energy which may be possessed by any electron in a metal at 0 K is given by equation (m.3.11.), which when expressed in atomic units becomes

$$E^* = 1.84 r_s^{-2} \text{ Hartrees.} \qquad (m.6.71.)$$

Now, combining equations (m.6.70.) and (m.6.71.) leads to

$$h\omega_p/2\pi = 0.47\sqrt{r_s}\, E^*. \qquad (m.6.72.)$$

For example, r_s for sodium is four Bohr units and so $h\omega_p/2\pi = 0.217$ Hartrees, and this quantity must be greater than E^*, since for all metals the

[1] Ruthemann, G., *Ann. Physik.*, 1948, **2**, 113.
[2] Lang, W., *Optik*, 1948, **3**, 233.
[3] Rudberg, E., and Slater, J. C., *Phys. Rev.*, 1936, **50**, 150.

quantity $0.47\sqrt{r_s} > 1$. For all metals $h\omega_p/2\pi$ lies in the range $5 - 40(\times 10^{-19})$ J, so that, generally, for all metals $h\omega_p/2\pi > E^*$. Now, since only a few electrons in states near the Fermi surface may be excited thermally and their thermal energies are of the order of kT (Chapter M.3.), which is very much smaller than E^*, it may be deduced that $h\omega_p/2\pi$ is very much greater than the thermal energy that any electron is possessed of at normal temperatures. As a consequence, the thermal excitation of plasmons may be regarded as negligible, and the plasmons will remain in their ground states unless excited by some other means. This type of excitation has been observed many times by passing beams of electrons through thin metal films of about 10^{-8} m thickness, and determining the energy loss of the scattered electrons; for beryllium, aluminum, and magnesium, electron energy losses are observed at almost exact integral multiples of $h\omega_p/2\pi$, but for other metals, such as copper, silver, and gold, and the transition metals, no such agreement is observed between electron energy loss and the plasmon excitation energy.

The principal use of the plasma oscillation theory actually depends on this simple fact, that the excitation energy of a plasmon is so much greater than the maximum thermal energy of any electron in a metal at normal temperatures, so that plasmons take no part in a great many electronic processes and they may therefore be disregarded. Also, since plasma oscillations do not take place with wavelengths less than a few units of 10^{-10} m, and the plasma oscillations in general involve the cooperative motion of large numbers of electrons, it may be deduced that it is the long-range part of Coulomb interactions forces which is responsible for plasmons; therefore, in those calculations where plasmons may be disregarded, so may the long-range Coulomb forces. In fact, the short-range Coulomb interactions between electrons in metals extends only over about 10^{-10} m, which is so short a distance that these forces also may be reasonably disregarded and the electrons essentially regarded as non-interacting particles. Thus, the simple plasma theory affords some justification for the assumption of the independent particle approximation, which is a common feature of many of the band theories. The short-range Coulomb interaction, in terms of plasma theory, turns out to be a screened potential of the type defined by equation (m.4.37.) in connection with the Thomas–Fermi calculations; however, the screening distance, $\sigma = 6.4 \times 10^{-11}$ m, for sodium, calculated by the Thomas–Fermi method (page 519) is only about one-half of the value of $1/k_c$ required by the plasma theory, so that the screening effect of the electron gas is not as great as indicated by the Thomas–Fermi method.

With respect to the manner in which the plasma oscillation theory enables the electron correlation energy to be calculated, the significant point is that the effects of long-range Coulomb correlations among the positions of all electrons, independently of their spin properties, are associated with the plasmons, which are not normally excited. As discussed in section M.3.7., the Hartree–Fock theory only incidentally allows for the correlation among the positions of electrons with parallel spins, due to the Pauli Exclusion Principle, but disregards entirely Coulomb correlation effects. This gives rise to too wide a calculated density of states curve with an unsatisfactory shape in the region of the Fermi surface. But, if the short-range Coulomb interactions only are used in calculating the Hartree–Fock exchange energy, instead of all of the Coulomb interactions, then the results of the Hartree–Fock calculations should be improved. This is so, since the plasmons are not normally excited, and therefore of no importance, except in so far as they are associated with the long-range Coulomb interactions, which thus may also be disregarded. In the Bohm–Pines plasma theory, the effect of the short-range Coulomb interactions is initially disregarded and then calculated as the first order perturbation to a system of completely non-interacting particles. Thus, the first order perturbation energy gives the exchange energy of the system and the second order perturbation energy corresponds to the effect of short-range Coulomb interactions between the electrons. Second order perturbation energies calculated in this manner[1] are found to be dependent on the ratio of k_c/k_0, with k_c as on page 664 and k_0 as on page 491. The value of this ratio has a substantial effect on the shape of the calculated $N(E)$ curves.

If a typical metal is regarded as a system of n electrons and a uniform distribution of positive charge contained in a space of unit volume, then the plasma theory description of such a system would be approximately that of a set of n particles having a screened Coulomb-type interaction between them and a set of plasmons, one for each value of the wave vector $\mathbf{k}$ with $k < k_c$. Since the number of states, each described by a vector $\mathbf{k}$, contained in the volume element $d\mathbf{k}$ of $\mathbf{k}$-space is $Vd\mathbf{k}/8\pi^3$ (page 491), for a piece of metal of volume V, then the number of $\mathbf{k}$ vectors contained within a sphere of radius k_c, which is equivalent to the number of plasmons, is $k_c^3/6\pi^2$, the density being $1/8\pi^3$ for unit volume. In the ground state of the system, the plasmons are all in their lowest energy states, each having energy equivalent to their zero-point energy of $h\omega_p/4\pi$ and being represented by a product of ground-state harmonic oscillator wave functions. Also, as a first approximation, the screened interaction of the particles which is of such short range that it may be disregarded, means that the particle wave functions may be written as a single determinant of free electron functions including spin, so that the ground-state wave function of the system is approximately given as the product of ground-state harmonic oscillator functions and a single determinant of free electron functions. A difficulty about this description is that it increases the number of degrees of freedom of the system of n particles, which

---

[1] Fletcher, J. G., and Larson, D. C., *Phys. Rev.*, 1958, **111**, 455.

is $3n$, by the number $k_c{}^3/6\pi^2$, the number of plasmons, which is not possible. Therefore, in order to make use of the plasma theory description, it is necessary to impose a number of so-called subsidiary conditions on the system, equal in number to the number of plasmons, in order to reduce the total number of degrees of freedom of the system to $3n$. This, of course, greatly complicates the formal quantum mechanical theory, but again it has been shown[1] that, as far as the ground-state energy of the system is concerned, such subsidiary conditions may be ignored. This leads to an approximate energy for the system being obtained as the sum of the particle kinetic energies, the exchange energy due to the short-range interactions, and the zero-point energies of the plasmons, which is lower than the Hartree–Fock value as calculated by equation (m.3.102.). The principal reason for the lower total energy given by the Bohm–Pines method compared with the Hartree–Fock calculation is that the Bohm–Pines theory takes into account the long-range correlations among the various electronic positions, independently of their spin properties, while the Hartree–Fock method does not. In fact, the difference between the total energies calculated by the two methods gives a measure of the long-range correlation energy per electron in a metal. The short-range correlation energy may be obtained, as already described, from the second order perturbation energy correction to the system of non-interacting particles, and the total correlation energy may be ultimately obtained[1] as a series of terms in r_s, such that the total correlation energy per electron in a metal is given by

$$E_L = - [0.0575 + 0.0155 \log r_s] \text{ Hartrees.} \qquad \text{(m.6.73.)}$$

This expression, calculated by Nozières and Pines,[2] takes into account the interaction between electrons and plasmons in metals, at the densities at which they occur in metals, and is considered to give the total electron correlation energy to within a probable error of less than 15%. For metallic sodium, for example, the correlation energy per electron is thus obtained as -0.036 Hartrees.

M.6.11.   The de Haas–van Alphen and Related Effects

The Hall effect (M.6.4.) is capable of providing much useful information about the electrons in metals and particularly in semiconductors. The energy bands in several metals, insulators, and semiconductors have been calculated with considerable accuracy, mainly by the orthogonal plane wave and the augmented plane wave methods. Thus, accurate energy band calculations are available, for example, for the metals, sodium[3] (body-centered cubic struc-

---

[1] Bohm, D., Huang, K., and Pines, D., *Phys. Rev.*, 1957, **107**, 71.
[2] Nozières, P., and Pines, D., *Phys. Rev.*, 1958, **111**, 442.
[3] Kenney, J. F.; see Slater, John C., *Quantum Theory of Molecules and Solids*, Vol. 2, 1965, McGraw-Hill Book Co., New York.

ture), copper[1,2] (face-centered cubic structure), and beryllium[3] (hexagonal close-packed structure), for the semiconductors germanium[4] (diamond structure) and cuprite[5] Cu_2O (a high symmetry cubic structure with less than four-fold coordination for the copper atoms), and for the insulators diamond[6] (diamond structure) and potassium chloride[7] (rock salt structure).

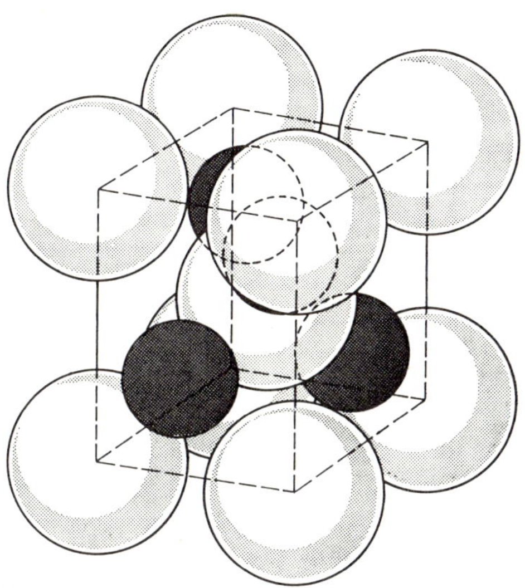

FIG. M.6.11. The crystal structure of cuprite, Cu_2O. This is a cubic structure, with two formula units of Cu_2O in the cubic unit cell. The copper atoms are located at the special positions, 1/4 1/4 1/4, 3/4 3/4 1/4, 3/4 1/4 3/4, 1/4 3/4 3/4, and the oxygen atoms at 000, 1/2 1/2 1/2, so that each copper atom has less than four-fold coordination, having only two close oxygen neighbors, while each oxygen atom is surrounded by a tetrahedron of copper atoms. This structure is also adopted by, for example, Ag_2O, Pb_2O, $Cd(CN)_2$, and $Zn(CN)_2$.

[1] Burdick, G. A., *Phys. Rev.*, 1963, **129**, 138.

[2] Segall, B., *Phys. Rev.*, 1962, **125**, 109.

[3] Loucks, T. L., and Cutler, P. H., *Phys. Rev.*, 1964, **A819**, 135.

[4] Löwdin, P.-O., Editor, *Quantum Theory of Atoms, Molecules, and the Solid State*, 1966, Academic Press, Inc., New York.

[5] Dahl, J. P., and Switendick, A. C., *J. Phys. Chem. Solids*, 1966, **27**, 931.

[6] Herman, F., Kortum, R. L., Kuglin, C. D., and Short, R. A., in *Quantum Theory of Atoms, Molecules, and The Solid State*, P.-O. Löwdin, ed., 1966, Academic Press, New York.

[7] De Cicco, P.; see Slater, John C., *Quantum Theory of Molecules and Solids*, Vol. 3, 1967, McGraw-Hill Book Co., New York.

Apart from the Hall effect, and the deductions which may be made about the shape of the $N(E)$ curves and the nature of the energy bands in solids, based on electronic and lattice heat capacities (sections M.1.4., M.1.5., and M.4.2.) and on soft X-ray spectroscopic data (sections M.3.1. and M.4.5.), a number of other more recently discovered effects involving magnetic fields have emerged which provide useful information about energy bands in solids and experimental data against which the results of theoretical calculations may be checked. The de Haas–van Alphen effect is one of these methods and is related to what is referred to as the cyclotron resonance effect.

Both of these effects refer to the influence of an applied electromagnetic field on the rotational motion of an electron in an impressed constant magnetic field. In terms of classical electromagnetic theory, for an electron of charge $-\varepsilon$ in the presence of a constant external magnetic field $\mathbf{H}$, the force exerted on the electron is $-\varepsilon(\mathbf{v} \times \mathbf{H})$, where $\mathbf{v}$ is the velocity, and this force is exerted at right angles to the magnetic field. The component of velocity parallel to the field direction is not affected by the field but remains constant. In the plane normal to the field direction in which the force acts, the electron travels in a circular path with constant angular velocity ω, referred to as the cyclotron frequency. If the radius of the path of motion of the electron is r, the linear velocity is ωr, and the acceleration of the electron towards the center of the circular path is then $\omega^2 r = v^2/r$, so that the product of this acceleration and the electronic mass gives the force of $m\omega^2 r = \varepsilon\omega r H$, giving $\omega = \varepsilon H/m$. In a cyclotron where positive ions are accelerated in a constant magnetic field, the angular frequency of rotation is given by this same equation, which, of course, is the reason for referring to this effect as cyclotron resonance; the cyclotron frequency for electrons is very much greater than for positive ions, on account of the smaller electronic mass. The energy is related to the radius of the path of motion by the fact that $E = mv^2/2 = m\omega^2 r^2/2$, and in cyclotron resonance experiments with electrons an oscillating high-frequency electric field is applied in a plane normal to $\mathbf{H}$ with the frequency ω. Whereas positive ions in a cyclotron build up energy without change of angular frequency, the radius of the path increasing for electrons as the acceleration is increased by application of the radio frequency electric field, it is found that absorption of high-frequency power occurs when the radio frequency is equal to the cylcotron frequency. Actually, in practice the cyclotron frequency is measured experimentally by changing the magnetic field and tuning the cyclotron frequency until resonance can be detected by the absorption of energy by the circulating electrons; the synchronization between the radio frequency field and the cyclotron frequency of the electrons can be easily detected by electronic methods.

In order to discuss the acceleration of electrons in crystals it is necessary to consider the motion of wave packets. In quantum mechanics, as in electromagnetic theory or any form of wave theory, the concept of a wave packet built up from the superposition of an infinite set of solutions of the Schrödinger equation for some particular system corresponding to different energies and each having some approximate value for a wave vector $\mathbf{k}$ at the position of the packet, constitutes an expression of the Heisenberg Uncertainty Principle. The center of gravity of such a wave packet moves exactly according to the laws of classical Newtonian or Hamiltonian mechanics, when the wave packet or any form of wave function used to describe it moves according to Hamilton's equations, using any hermitian Hamiltonian (see Chapter Q.2.); the group velocity of motion of the wave packet corresponds to the classical velocity of the particle described by the wave packet. The minimum size of a wave packet is limited by the Uncertainty Principle. For an electron in a crystal with momentum or wave vector $hk/2\pi$, if the electron is localized within a small region of a Brillouin zone, the corresponding wave packet must extend over many unit cells of the crystal. If the energy of an electron in the nth energy band of a crystal as a function of $\mathbf{k}$ is known, then from Hamilton's equations the motion of the center of gravity of the appropriate wave packet is given by

$$dx/dt = v_x = \partial E_n(k)/\partial p_x \quad \text{and} \quad dp_x/dt = -dV/dx, \quad \text{(m.6.74.)}$$

with corresponding equations for the y and z components. The first of these equations indicates that the electronic velocity can be found from the slope of the energy bands giving E as a function of $\mathbf{k}$, and the second gives the extra potential or the extra part of the Hamiltonian acting on the electron in addition to the periodic potential in the crystal. For the acceleration of electrons in a crystal under the influence of applied electric and magnetic fields $\mathbf{E}$ and $\mathbf{H}$, respectively, the total external force acting on the wave packet of electrons is given by

$$\mathbf{F} = (h/2\pi)\, d\mathbf{k}/dt = -\varepsilon[\mathbf{E} + (\mathbf{v} \times \mathbf{H})]. \quad \text{(m.6.75.)}$$

In other words, the effect of an external field produces a time rate of change of the wave vector of the center of the wave packet, proportional to the classical force due to the external field; in the absence of an external field, the $\mathbf{k}$ vector of the electron remains constant as in an ordinary periodic potential.

In the free electron theory, the relation between E_n and $\mathbf{k}$ is given by $E_n(k) = h^2k^2/8\pi^2m$ and the velocity is given by the classical mechanical expression p/m, so that (m.6.75.) leads to Newton's second law relating the force acting on the electron and the acceleration of the wave packet. For electrons in states near the top and bottom of energy bands, as discussed in section M.5.9., the electronic mass m has to be replaced by the effective mass m^*, so that $E_n(\mathbf{k}) = h^2\mathbf{k}^2/8\pi^2m^*$.

In the simplest case, then, the measurement of cyclotron frequencies enables effective masses to be measured from $\omega = \varepsilon H/m^*$, where the wave packet describing an electron behaves like a classical particle. But more generally, the energy-wave vector relation is of the form

$$E_n(\mathbf{k}) = [h^2/8\pi^2][k_x{}^2/m_x{}^* + k_y{}^2/m_y{}^* + k_z{}^2/m_z{}^*], \qquad \text{(m.6.76.)}$$

with three effective masses, although two of these may be the same; also, in this event, there is an ellipsoidal relationship between E_n and $\mathbf{k}$, provided k_x, k_y, and k_z are chosen to coincide with the principal axes of the ellipsoid, otherwise the situation is still more complicated with cross terms such as $k_x k_y$ entering into the energy expression. All the different effective masses may be determined by cyclotron resonance experiments, by carrying out the measurements with magnetic fields applied in different directions, and in this way it is possible to establish the form of the energy as a function of the wave vector in the region of maximum and minimum in the energy band of a crystal. The maxima and minima in the energy bands of germanium were first worked out in this way from an analysis of the resonance frequencies found for a magnetic field applied in an arbitrary direction. Generally, reliable values for the effective masses of electrons in crystals may be obtained from such experimental measurements and the general agreement with the calculated energy bands lend support to the approximate validity of the energy band theories.

By application of the requirement that the angular momentum of the electron in its path of motion should be a multiple of $h/2\pi$, quantized energy levels result which differ in energy from one another by $h\omega/2\pi$. For semiconductors such as germanium, there would be quantized levels for the electrons in a magnetic field, both in the valency band and in the conduction band, with different spacings on account of the differing effective masses. Under these circumstances, it is possible to observe optical absorption from the quantized energy levels in the valency band to quantized levels in the conduction band, as sharp absorption lines in the corresponding spectrum, and from their observed frequencies the cyclotron frequencies can be measured; these are referred to as magnetic interband transitions. A very large amount of experimental work has been carried out on the determination of cyclotron frequencies and magnetic interband transition frequencies.[1]

Cyclotron resonance experiments are more difficult to carry out on metals than on semiconductors, because of the attenuation of electromagnetic waves as they penetrate a conductor. The distance in which the amplitude of an electromagnetic wave falls off to $1/e$ of its initial value, as it penetrates a metal, is referred to as the skin depth. The oscillating field used in cyclotron resonance experiments can thus only penetrate a very short distance into a metal, although for semiconductors the skin depth is quite large. A way of

---

[1] Slater, John C., *Quantum Theory of Molecules and Solids*, Vol. 3, 1967, McGraw-Hill Book Co., New York.

overcoming this difficulty with metals is based on the discovery of de Haas and van Alphen,[1] who observed that the diamagnetic susceptibility of bismuth displayed an oscillatory dependence on the applied magnetic field, and that the period of the oscillation was constant when the susceptibility was plotted as a function of $1/H$. This effect has later been observed for copper and for many other metals, although it has not been detected for the alkali metals. In some cases there is only a single periodic variation of the susceptibility with $1/H$ and in others two or more periodic effects may be superimposed. The periodicity is found to be independent of the temperature, but the amplitude falls off rapidly with increasing temperature, vanishing at about 30–40 K, so that such experiments are generally carried out at liquid helium temperatures. The first explanation of the origin of this effect was given by Peierls.[2] In the range of magnetic fields available to experimentation, it is sometimes possible to observe hundreds of oscillations, the spacing between which can be measured with great accuracy, giving much valuable information about the energy bands in the substance under investigation.

For a metal placed in a magnetic field, the circular electronic orbits induced by the field give rise to electric currents and so to a magnetic moment (see section Q.5.2.). The oscillation of these magnetic moments at regularly spaced values of the reciprocal of the applied field, observed in the de Haas–van Alphen effect, is a consequence of the quantized nature of the cyclotron oscillations. For the motion of an electron in an energy band of a crystal, in the presence of a constant magnetic field but in the absence of an electric field, equation (m.6.75.) reduces to

$$[h/2\pi][d\mathbf{k}/dt] = -\varepsilon(\mathbf{v} \times \mathbf{H}), \qquad \text{(m.6.77.)}$$

and for an electron with no velocity component in the direction of the magnetic field, and with a position vector defined by $\mathbf{r}$, its velocity will be $d\mathbf{r}/dt$, so that equation (m.6.77.) becomes simply

$$[h/2\pi][d\mathbf{k}/dt] = -\varepsilon(d\mathbf{r}/dt \times \mathbf{H}). \qquad \text{(m.6.78.)}$$

On integration of this equation with respect to time, it is found that

$$h\mathbf{k}/2\pi = -\varepsilon(\mathbf{r} \times \mathbf{H}), \qquad \text{(m.6.79.)}$$

indicating that the vector $\mathbf{k}$ is normal to $\mathbf{r}$ and $\mathbf{H}$, so that the plane of orbital motion of the electron, normal to the applied magnetic field, is such that the vector $\mathbf{r}$ in this plane is normal to $\mathbf{H}$. Therefore, the magnitudes of $\mathbf{k}$ and $\mathbf{r}$ are related by

$$k = \varepsilon H r 2\pi/h, \qquad \text{(m.6.80.)}$$

---

[1] Haas, W. J. de, and van Alphen, P. M., *Communs. Phys. Lab. Univ. Leiden*, 1930, 212a; 1932, 220d.
[2] Peierls, R., *Z. Physik*, 1933, **80**, 763.

and the vector $\mathbf{k}$ in the plane normal to $\mathbf{H}$ is also normal to the vector $\mathbf{r}$. Thus, the radial orbit in $\mathbf{k}$-space must have the same shape as the orbit in real space, but rotated 90° with respect to it around the magnetic field direction. Therefore, from (m.6.80.), the dimensions of the orbit in $\mathbf{k}$-space, its area, must be related to the area of the orbit in real space by the relation

Area of orbit in $\mathbf{k}$-space $= [4\pi^2\varepsilon^2 H^2/h^2] \times$ area of orbit in real space.

$$(\text{m.6.81.})$$

In order to find the area of the orbit in $\mathbf{k}$-space, it is necessary to find the area of the orbit in real space. In order to do this, it is necessary to make use of the fact that in Maxwell's electromagnetic theory, an external field may be described by means of a scalar potential ϕ and a vector potential $\mathbf{A}$, according to the equations $\mathbf{E} = -\text{grad }\phi - \partial\mathbf{A}/\partial t$; $\mathbf{H} = \text{curl }\mathbf{A}$, where $\mathbf{E}$ is the electric field and $\mathbf{H}$ the magnetic induction,[1] since in the de Haas–van Alphen effect a vector potential is involved. Also, in the application of quantum mechanics to the discussion, it is convenient to use the Bohr–Sommerfeld quantum condition as an approximation (see Chapter Q.1.). The Sommerfeld quantum condition refers to the fact that for a particle with some positional coordinate q, mass m, velocity v, and momentum $p = mv$, the phase integral $\oint p\,dq = nh$, where $\oint$ denotes that the integral has to be taken around a complete cycle. For the orbit in two-dimensional space where the coordinate involved is the radius vector $\mathbf{r}$ and $\mathbf{p}$ is also a vector quantity, the integral concerned becomes $\oint \mathbf{p}\cdot d\mathbf{r} = nh$, and n is a half integral or integral quantum number. In a vector potential field, the momentum has to be represented by $h\mathbf{k}/2\pi = \mathbf{p} + \varepsilon\mathbf{A}$, where $\mathbf{A}$ is the vector potential and $-\varepsilon$ the electronic charge. The momentum vector involved in the Sommerfeld quantum condition is therefore $\mathbf{p} = h\mathbf{k}/2\pi - \varepsilon\mathbf{A}$. How, if $\mathbf{r}$ is the radius vector from the center of the orbit, the vector potential for a constant magnetic field $\mathbf{H}$ is $\mathbf{A} = -[(\mathbf{r} \times \mathbf{H})/2]$, and from (m.6.79.) the momentum $\mathbf{p}$ can now be expressed as

$$\mathbf{p} = h\mathbf{k}/2\pi - \varepsilon\mathbf{A} = -\varepsilon(\mathbf{r} \times \mathbf{H}) + (\varepsilon/2)(\mathbf{r} \times \mathbf{H}) = -(\varepsilon/2)(\mathbf{r} \times \mathbf{H}). \quad (\text{m.6.82.})$$

The Sommerfeld quantum condition is thus expressible as

$$\oint \mathbf{p}\cdot d\mathbf{r} = -(\varepsilon/2) \oint (\mathbf{r} \times \mathbf{H})\cdot d\mathbf{r} = [\varepsilon H/2]\left|\oint (\mathbf{r} \times d\mathbf{r})\right|, \quad (\text{m.6.83.})$$

where H is the magnitude of the magnetic field. From the fact that the quantity $(1/2)\oint (\mathbf{r} \times d\mathbf{r})$ represents the sum of the areas of triangles formed by the vectors $\mathbf{r}$ and $d\mathbf{r}$, the value of this integral obviously gives the area of the

---

[1] Slater, J. C., and Frank, N. H., *Electromagnetism*, 1947, McGraw-Hill Book Company, New York.

orbit; actually, the expression $(1/2) \oint (\mathbf{r} \times d\mathbf{r})$ represents a vector quantity perpendicular to the plane of the orbit and so parallel to $\mathbf{H}$, which is the reason why only the magnitude of $\mathbf{H}$ is involved in equation (m.6.83.). The area of the orbit in real space may therefore be obtained from equation (m.6.83.) as

$$\text{Area of orbit in real space} = \left[\oint \mathbf{p}\, d\mathbf{r}\right]/\varepsilon H = nh/\varepsilon H. \quad \text{(m.6.84.)}$$

Substituting this value into equation (m.6.81.), the area of the orbit in $\mathbf{k}$-space may be obtained as

$$\text{Area of orbit in } \mathbf{k}\text{-space} = 4\pi^2 n\varepsilon H h^{-1}. \quad \text{(m.6.85).}$$

In $\mathbf{k}$-space, the area of the orbit is thus directly proportional to the magnetic field, while in ordinary space it is inversely proportional to the field, being quantized in both cases; in real space, the orbit would thus be very large for small field strengths. The periodic variation of the magnetic susceptibility as a function of the reciprocal of the magnetic field, observed in the de Haas–van Alphen effect, may be understood from equation (m.6.85.), and is more apparent if this equation is rearranged to read

$$H^{-1} = n[4\pi^2\varepsilon h^{-1}]/\text{area of orbit in } \mathbf{k}\text{-space}. \quad \text{(m.6.86.)}$$

Since the electronic energy levels are quantized, there can only be a discrete number of them with energy less than some finite amount, such as the Fermi energy. As the magnetic field is increased, so does the spacing of the energy levels, so that the number of levels with energy less than the Fermi energy decreases. The periodic effect takes place, as when a quantized rotational energy level acquires energy in the applied field in excess of the Fermi energy, it becomes vacant and the remaining electrons redistribute themselves among the remaining levels, and each time the number of energy levels with energy less than the Fermi energy changes by unity this process is repeated. The result is that each time the number of energy levels increases by unity, the quantum number n increases by unity, and $1/H$ increases by the amount $4\pi^2\varepsilon/h$ divided by the area of the orbit in $\mathbf{k}$-space. In principle, by measurement of the period of oscillation of the magnetic field observed, $1/H$, it is possible to determine the area of the electronic orbits in $\mathbf{k}$-space.

Observation of the de Haas–van Alphen effect is greatly complicated by a number of circumstances; the Fermi surface of occupied states is not sharply defined at any temperature above 0 K, the probability of occupation falling off rapidly from unity to zero over an energy range of the order kT, which reduces the amplitude of oscillation of the magnetic susceptibility with $1/H$. Electrons with components of velocity along the direction of the applied field complicate matters further, so that in general quite complicated variations of the de Haas–van Alphen effect may occur for a magnetic field applied in various directions with respect to some particular crystallographic direction in a crystal. However, as is indicated by equation (m.6.86.), it is possible to investigate the area of the Fermi surface of a metal in different directions in **k**-space; the shape of the Fermi surface can, of course, be deduced from detailed energy band calculations, as it is the surface in three-dimensional **k**-space described by all points on an energy band having the Fermi energy. There will be a different value for the observed de Haas–van Alphen periodicity for each plane, normal to the applied magnetic field direction, where this plane intersects the Fermi surface. However, the planes which intersect the Fermi surface near either a maximum or minimum of area of orbit, for any particular orientation of the magnetic field, are more numerous than planes intersecting the Fermi surface corresponding to other areas of orbit, so that the enhanced de Haas–van Alphen effect for these planes is easily detectable. For a Fermi surface such as that of copper[1] shown in Fig. M.5.42., for a plane in the 111 direction intersecting one of the narrow necks connecting the Fermi surfaces between adjacent unit cells in **k**-space, a minimum area would be found in the observation of the de Haas–van Alphen effect corresponding to the largest period of oscillation, as indicated by equation (m.6.86.). Larger areas, detected by shorter periods, would be observed for such planes as intersect the approximately spherical parts of the Fermi surface near to its maximum diameter. The area of the Fermi surface for metals, deduced in this way from measurement of de Haas–van Alphen oscillation periods, are found to agree quite closely with the results of energy band calculations.

Although it is a well known consequence of classical electromagnetic theory that an oscillating electric field cannot penetrate deeply into the bulk of a good conductor such as a metal, but is confined to a thin layer known as the skin depth (page 672), valuable information about the Fermi surface in a metal may be deduced on the basis of observation of surface fields and currents in metals under the influence of radio frequency magnetic fields. Two such methods are the so-called anomalous skin effect[2] and nuclear magnetic resonance phenomena.[3]

The electrical conductivity of a metal, which at ordinary temperatures is given as the proportionality constant relating the current and electric field (Ohm's law), increases greatly as very low temperatures are approached. The plasma frequency remains constant and this depends only on the electron concentration per unit volume of metal, but the increase in conductivity arises from an increase in the time between collisions of the electrons and the lattice vibrations of the crystalline metal; if this relaxation time between collisions

[1] Burdick, G. A., *Phys. Rev.*, 1963, **129**, 138.
[2] London, H., *Proc. Roy. Soc.*, 1940, **A176**, 522.
[3] Rowland, T. J., *Progress in Materials Science*, 1961, Vol. 9, No. 1, p. 1, Pergamon Press, Oxford.

is defined as on page 427, then the reciprocal of this quantity has the dimensions of frequency, and while at normal temperatures this frequency is in the infrared region of the electromagnetic spectrum, at very low temperatures it is in the microwave region. As the relaxation time increases, so does the mean free path of the electrons, and at very low temperatures the mean free path may be quite large compared with the skin depth. The phenomena which are observed under these conditions are referred to as anomalous skin effects. The skin depth in a metal, δ_s, is defined by

$$\delta_s = [2\pi\omega\kappa_\varepsilon]^{-1/2} = R/2\pi\omega, \qquad \text{(m.6.87.)}$$

where ω is the cyclotron frequency, κ_ε is the electrical conductivity, and R is the surface resistance for alternating currents.[1] Most of the current is carried within the thickness of the skin depth and $R \propto \sqrt{1/\kappa_\varepsilon}$. Below 20 K and at microwave frequencies of the order of 10^9–10^{11} s^{-1}, κ_ε increases very rapidly and R does not decrease sufficiently to satisfy the proportionality with $\sqrt{1/\kappa_\varepsilon}$; Ohm's law is no longer applicable; and this anomalous skin effect observed for very pure metals at such low temperatures leads to R becoming independent of κ_ε at what is known as the extreme anomalous limit. Here, unless an electron happens to be moving nearly parallel to the metal surface, it will spend only a small fraction of its time in the region where it will be under the influence of the field. R depends on the orientation of the metal crystal relative to the direction of the electric field, and it is found that in the extreme anomalous limit, the surface resistance depends only on the radius of curvature of the Fermi surface, so that by an analysis of these effects, information may be obtained about the shape of the Fermi surface in a metal. This technique has been applied extensively, especially by Pippard.[2]

Just as the finer details in the line spectra of the elements are attributable to the electron possessing a spin angular momentum and an associated magnetic moment, so the hyperfine structure of certain spectral lines in the presence of a magnetic field may be regarded as arising from atomic nuclei possessing spin angular momenta and corresponding nuclear magnetic moments. The nuclear magneton is, however, only of the order of about one two-thousandths of a Bohr magneton, and nuclear magnetic energy levels are of such a magnitude that transitions between them may be induced by the magnetic field provided by an electron passing close to a nucleus, if the electron can be transferred into a new state. This is only possible for electrons near to the Fermi surface. Electrons responsible for nuclear magnetic effects are mainly those in s-states, which have the highest density of states near to the nuclei in metals. From the observation of nuclear magnetic effects, it is possible to calculate the density of states of the electrons responsible for these effects, which when compared with the total density of states as determined

---

[1] Ziman, J. M., *Electrons and Phonons*, 1962, Clarendon Press, Oxford.
[2] Pippard, A. B., *Proc. Roy. Soc.*, 1954, **A224**, 273; *Phil. Trans.*, 1957, **A250**, 325.

from heat capacity measurements enables the fraction of electrons at the Fermi surface which are in *s*-states to be determined.[1]

The theory of magnetoresistance, which deals with the effect of magnetic fields on the resistance of metals, indicates that there are tremendous differences in metallic resistance depending on the direction of application of the magnetic field. These observed large differences in resistance for small differences in magnetic field direction may be understood in terms of the change from closed to open electronic orbits. For copper, for example, which has been much studied by all of the methods described in this chapter, and for which the Fermi surface is shown in Fig. M.5.42., it is apparent that some planes normal to the field direction will intersect the Fermi surface in closed orbits, but other planes corresponding to special directions of the magnetic field may give rise to orbits which are continuous throughout the metal; such a direction is indicated in Fig. M.5.42.(b) by the arrowed line, passing through several zones of the crystal by means of the extended necks which are normal to the 111 direction in **k**-space. Magnetoresistance measurements of this type can determine whether the Fermi surfaces in a metal are closed, as they would be, for example, in sodium, where the Fermi surfaces are essentially spherically shaped or are open, as in copper, and can also provide information about the direction in which the open orbits are observed.

The way in which ultrasonic frequencies are absorbed by metallic crystals has also been examined,[2] both in the presence and absence of magnetic fields, and used to provide information about the Fermi surface. There are small electric fields associated with ultrasonic vibrations and provided the ultrasonic frequencies are properly selected, attenuation of the ultrasonic oscillations in the crystal takes place with absorption of energy from the field by the electrons. Since only those electrons at the Fermi surface are capable of having collisions and therefore absorbing energy, this provides another way of finding the cross-sectional area of the Fermi surface.[3,4] The conditions which determine when the cyclotron resonance-type absorption of ultrasonic energy takes place are as described in equation (m.6.81.). In many cases where the energy bands have not yet been calculated, experiments of the type involving the cyclotron resonance effect, magnetic interband transitions, the de Haas–van Alphen effect, the anomalous skin effect, and the attenuation of ultrasonic oscillations in metal crystals, have provided information about the nature of the Fermi surface from which the energy band structure may be deduced directly. For copper and the other substances, for which both experimental and theoretical data is available, good agreement prevails.

---

[1] Butterworth, J., *Bull. Ampère*, 1960, p. 417; *Phys. Rev. Letters*, 1960, **5**, 305.

[2] Pippard, A. B., *Phil. Mag.*, 1957, **2**, 1147.

[3] Miller, P. B., *Phys. Rev.*, 1965, **137**, A1937.

[4] Quinn, J. J., and Rodriguez, S., *Phys. Rev.*, 1965, **137**, A889.

M7

Application of Structure and Electron Theories to Metals and Intermetallic Compounds

I. THE ALKALI METALS

M.7.1. The "Ideal" Metals

The alkali metals, lithium, sodium, potassium, rubidium, and cesium, with nuclear charges of 3, 11, 19, 37, and 55, respectively, are all very highly reactive metals but of comparatively little interest in the metallurgical sense.‡ The element with nuclear charge 87, francium, is also an alkali metal; it is, however, naturally radioactive, and is only formed in small amounts by the branching α-decay of actinium. All of the alkali metals, including francium, crystallize with the body-centered cubic type of structure (Fig. M.1.6.), each atom being equidistant from eight others at the vertices of the cubic unit of structure; six other atoms are only 15% more distant, so that this (8 + 6)-type coordination is not too different from the environment of a 12-coordinated atom in the two closest packed structures (Fig. M.1.3.). The lattice constants of the alkali metals have been determined with considerable precision, at various temperatures, as listed in Table M.7.1.

‡ Apart from the use of cesium (and potassium) in photoelectric devices, the metals rubidium and cesium find little use. Metallic sodium, which may be prepared at low cost, is the starting material for the manufacture of many technically important compounds. Sodium and potassium alloys are used as heat-transfer agents in power reactors. Lithium is used quite extensively in alloys with other metals, especially hardened, lead

679

bearing metals; a typical bearing metal is composed of lead alloyed with 0.7% calcium, 0.6% sodium, and 0.04% lithium.

It is remarkable that one of the most complicated crystal structural problems involving intermetallic compounds, concerns a sodium compound with the deceptively simple composition $NaCd_2$. This material crystallizes in the cubic system with the largest unit of structure observed among intermetallic compounds, the unit cell edge length being 3.056×10^{-9} m, so that there are 1152 atoms per unit cell.[1-3] However, this is not the largest crystallographic unit cell observed among inorganic compounds. The hexagonal ferrite with the remarkable stoichiometric formula $Ba_{70}Zn_{66}Fe_{444}O_{802}$, has perhaps this distinction;[4] the hexagonal unit cell of this structure has the dimensions, $a_0 = 5.88 \times 10^{-10}$ m, $c_0 = 1.577 \times 10^{-7}$ m, thus containing over 4100 atoms in the unit cell. Sixty-one members of the hexagonal ferrite series of compounds, so named to distinguish them from the cubic or spinel type ferrites, have been recognized so far, with compositions ranging from $Me_2Fe_4O_8$ to $Ba_{70}Me_{66}Fe_{444}O_{802}$, where Me is usually dipositive iron, nickel, cobalt, or zinc. The a_0 dimension is the same in all the known hexagonal ferrite structures, with the ordered stacking of two basic structural units along the direction of the hexagonal c axial direction parlaying the unit cell dimension in this direction to a known maximum of 1.577×10^{-7} m. The two basic structural units are layers of close-packed oxygen atoms and layers of which every fourth oxygen atom has been substituted by a barium atom, with the Me-type atoms occupying either the interstitial tetrahedral or octahedral cavities. These structural units are assembled into three types of blocks composed of two oxygen layers, five oxygen layers one of which is of the Ba-O type, and six oxygen atom layers with two adjacent ones containing barium atoms, respectively, which when stacked along the c axial direction in a variety of ratios and permutations give rise to the extensive family of ferrites.[5]

Table M.7.1

Element	$a_0 (\times 10^{-10}\,\text{m})$
Lithium, Li	3.5093 (293 K)
	3.491 (78 K)
Lithium, Li^6	3.5107 (293 K)
Lithium, Li^7	3.5092 (293 K)
Sodium, Na	4.2906 (293 K)
	4.225 (5 K)
Potassium, K	5.225 (5 K)
	5.247 (78 K)
Rubidium, Rb	5.585 (5 K)
	5.605 (78 K)
Cesium, Cs	6.045 (5 K)
	6.067 (78 K)

[1] Pauling, L., *J. Am. Chem. Soc.*, 1923, **45**, 2777.
[2] Pauling, L., *Am. Sci.*, 1955, **43**, 285.
[3] Samson, S., *Nature*, 1962, **195**, 259.
[4] Kohn, J. A., Eckart, D. W., and Cook, C. F., Jr., *Mater. Res. Bull.*, 2, 55, 1967; *Sci.*, 172, 519, 1971.
[5] Eckart, D. W., and Kohn, J. A., *Z. Kristallogr.*, 125, 130, 1967.

TABLE M.7.2

| | Binding Energy $[\times 10^5$ J (g-atom)$^{-1}]$ | |
Element	Calc.	Expt.
Lithium	-1.72	-1.53
Sodium	-1.26	-1.09
Potassium	-1.09	-0.96
Rubidium	-1.00	-0.79
Cesium	-0.96	-0.79

The small extension in space of the electron charge cloud in the ions of these metals compared with the internuclear distance in the solids (Fig. M.1.15.) accounts for the fact that most of the properties of these metals are determined by the valency electrons; it is for this reason (page 432) that these metals are of such interest in the development of electron theories of metals. The first Brillouin zone for a body-centered cubic structure is a rhombic dodecahedron (Fig. M.2.28., Fig. M.6.10.), and since each of the alkali metal atoms contributes one valency electron and the zone can accommodate two electrons, the zone is only half filled. The free electron theory assumes, and the other modifications of the electron theories of metals afford confirmation, that the Fermi surface of occupied electron states in these metals is approximately spherical and does not reach to the surface of the zone. The Fermi surface of lithium is believed to be considerably distorted from spherical symmetry, and a distortion occurs with increasing nuclear charge of the alkali metals, from sodium which is indeed close to the "ideal" metal of the electron theories, to potassium, to rubidium, to cesium, where again the Fermi surface of occupied states approaches the zone surface. Lithium is anomalous not only in this respect but in many others;‡ the binding energy of lithium is about 56% greater than the mean value of that of the other alkali metals, as indicated in Table M.7.2. In the liquid phase lithium is quite immiscible with either sodium, potassium, rubidium, or cesium, while sodium, potassium, rubidium, and cesium, are completely miscible in the liquid phase. This type of behavior is quite compatible with the percentage increases in atomic radii of lithium to sodium (22%), sodium to potassium (25%), potassium to rubidium (6%), potassium to cesium (13%), and rubidium to cesium (7%).

‡ Lithium appears to behave abnormally, compared with the other alkali metals, in much of its chemistry, and in many of its interactions with other metals there would seem to be more than one valency electron per atom available. In the silver–lithium system, for example, there occurs the well-defined phase with the approximate composition Ag_3Li_{10}, which crystallizes with the γ-brass structure characteristic of the Hume-Rothery 21:13 electron compounds (page 404). If each lithium atom contributed two electrons

per atom towards compound formation, the composition Ag_3Li_{10} would correspond approximately to that required for the 21:13 electron: atom ratio, although there is no conceivable combination of one-electron-per-atom elements which can lead to this ratio. The lithium–magnesium system exhibits a wide range of solid solutions of one metal in the other, as though both metals had similar valency electron characteristics, although in the sodium–magnesium system, and also the silver–magnesium system, where the radius ratio effects are compatible for solid solution formation, solid solution of sodium, and silver, in magnesium is so limited that in the former case the two metals are essentially immiscible in the liquid phase.

Most of these and the other physical properties of lithium, which differentiate it from the other alkali metals, are attributable to the distribution of s- and p-electron states in the Brillouin zone.[1] In lithium the energy difference between the ns and np energy states is less than would be anticipated on the basis of extrapolation of the $s–p$ energy differences for atoms of the other alkali metals. This gives rise to the ns and np states broadening to such an extent, when the free atoms assemble as in the metallic crystal of lithium, that overlapping takes place (Fig. M.3.1.). The calculations of the energy bands of lithium[1,2] indicate that for states within the first Brillouin zone, those in the Σ direction near to the zone boundary (the center of the rhombus-shaped face designated N in Fig. M.6.10.) are either pure p-type or pure s-type states, but not a mixture of the two. If the energy states just inside the zone, are p-states (or s-states), then the states just outside the point N in the second zone are s-states (or p-states), and it is indicated that in lithium the states inside the first Brillouin zone near the face centers are of the p-type with a large energy gap across the zone face, of the order of 3–5 ($\times\ 10^{-19}$) J. As a result the Fermi surface of occupied states is so distorted from spherical symmetry, although the zone is only half filled, that the Fermi surface reaches to the face centers of the rhombic dodecahedral zone. It is believed that the large energy difference between an s- and a p-state, for lithium, on either side of the surface of the first Brillouin zone, which is responsible for the deformation of the Fermi surface, thus accounts for the behavior of lithium that the zone appears to be filled, in the directions of the face centers, at a lower electron concentration than would be applicable for a spherical Fermi surface. For any specific volume of k-space, a spherical Fermi surface has the minimum surface area, and would so correspond to the lowest value of $N(E)$ at the Fermi surface. The $N(E)$ curve for lithium, as observed from its soft X-ray K-emission spectrum, does rise to a maximum which is greater and at a lower electron concentration than would be found for a spherical Fermi surface. If, in the lithium compounds formed with other metals, which have the γ-brass structure, lithium retains this characteristic of giving rise to a deformed Fermi surface, then the extension of this surface to the zone faces at a lower electron concentration than that found for a spherical Fermi surface, gives rise to a comparable effect to a metal with more than one electron per Brillouin zone.

The apparently anomalous behavior of lithium compounds, compared with the corresponding compounds formed by the other alkali metals, such as differences in solubility, stability, volatility, complex-formation ability, and so on, are attributable to the larger ionic potential of the small lithium ion, as a consequence of the relatively unscreened nuclear charge.

Cesium has the lowest first ionization potential of any known element [0.38×10^6 J (g-atom)$^{-1}$].

---

[1] Schiff, B., *Proc. Phys. Soc.*, 1954, **A67**, 2.
[2] Cohen, M. H., and Heine, V., *Advances in Physics*, 1958, **7**, 395.

M.7.2.   The Effective Masses of the Valency Electrons in the Alkali Metals

The way in which the ground-state wave function and the ground-state energy of the valency electron in sodium may be calculated by the Wigner and Seitz cellular method has already been described in section M.6.6. However, in order to calculate the cohesive energy of the metal, it is necessary to know the energies of all of the occupied states, and not simply the lowest eigenvalue for the valency electron in the metal. Only two electrons with opposed spins may occupy any particular orbital state, according to the Pauli Exclusion Principle, and therefore all but two electrons must be in higher energy states. As discussed on page 646, the potential field $V(\mathbf{r})$ acting on an electron may be assumed to be that only of the metal ion in the cell in which that electron is located, and the corresponding Schrödinger equation for any state $\psi_k(\mathbf{r})$ is

$$\nabla^2\psi_k(\mathbf{r}) + [8\pi^2 m/h^2][E(\mathbf{k}) - V(\mathbf{r})]\psi_k(\mathbf{r}) = 0. \qquad \text{(m.7.1.)}$$

The boundary conditions which must be satisfied at the surface of the cell are given by equations (m.5.59.) and (m.6.56.), and the Schrödinger equation above is valid throughout the piece of metal crystal, although it is only necessary to obtain its solution within a single cell. The boundary conditions for any state $\mathbf{k}$ cannot be simplified, as for the state $\mathbf{k} = 0$, and the atomic sphere approximation is of no particular help; as usual, no analytical solution of equation (m.7.1.) is possible, and suitable approximations must be attempted.

Wigner and Seitz[1] originally made the assumption that

$$\psi_k(\mathbf{r}) = e^{i\mathbf{k}\cdot\mathbf{r}}u_0(\mathbf{r}), \qquad \text{(m.7.2.)}$$

where $u_0(\mathbf{r})$ is written instead of $\psi_0(\mathbf{r})$, used in section M.6.6., but represents the same function continued periodically throughout the crystal; this function is of the Bloch type, since $u_0(\mathbf{r})$ has the period of the crystal lattice, but strictly a function $u_k(\mathbf{r})$ which varies with $\mathbf{k}$ should actually be used rather than the approximation of $u_0(\mathbf{r})$. If $u_0(\mathbf{r})$ were constant, as for free electrons, no approximation would be involved at all, and as already seen the function $\psi_0(\mathbf{r})$ is very nearly constant throughout most of the crystal volume. Also, when $\mathbf{k} = 0$, equation (m.7.2.) reduces to $u_0(\mathbf{r}) = \psi_0(\mathbf{r})$, and since for the alkali metals only the lower half of the Brillouin zone is occupied, it might be expected that $u_k(\mathbf{r})$ would be approximately equal to $u_0(\mathbf{r})$. If it is now assumed that $u_0(\mathbf{r})$ is normalized throughout the volume of the crystal, and since it is

---

[1] Wigner, E., and Seitz, F., *Phys. Rev.*, 1933, **43**, 804.

a real function, the approximate energy $E(\mathbf{k})$ of the state $\psi_k(\mathbf{r})$ is obtained from equations (m.7.1.) and (m.7.2.) as

$$E(\mathbf{k}) = \int [u_0(\mathbf{r})e^{-i\mathbf{k}\cdot\mathbf{r}}[-(h^2/8\pi^2m)\nabla^2 + V(\mathbf{r})]e^{i\mathbf{k}\cdot\mathbf{r}}u_0(\mathbf{r})]\,d\mathbf{r}, \quad \text{(m.7.3.)}$$

where the integral has to be taken throughout the crystal volume. Since

$$e^{i\mathbf{k}\cdot\mathbf{r}}u_0(\mathbf{r}) = e^{i(k_1 x + k_2 y + k_3 z)}u_0(x, y, z),$$

and the first and second derivatives of this function, for example, with respect to x are

$$\partial/\partial x[e^{i\mathbf{k}\cdot\mathbf{r}}u_0(\mathbf{r})] = e^{i\mathbf{k}\cdot\mathbf{r}}\,\partial u_0/\partial x + ik_1 e^{i\mathbf{k}\cdot\mathbf{r}}u_0,$$

and

$$\partial^2/\partial x^2[e^{i\mathbf{k}\cdot\mathbf{r}}u_0(\mathbf{r})] = e^{i\mathbf{k}\cdot\mathbf{r}}\,\partial^2 u_0/\partial x^2 + 2ik_1 e^{i\mathbf{k}\cdot\mathbf{r}}\,\partial u_0/\partial x - k_1{}^2 e^{i\mathbf{k}\cdot\mathbf{r}}u_0,$$

with similar expressions for the differentiations with respect to y and z, the total derivative may be obtained as

$$\nabla^2[e^{i\mathbf{k}\cdot\mathbf{r}}u_0(\mathbf{r})] = e^{i\mathbf{k}\cdot\mathbf{r}}[\nabla^2 + 2i\mathbf{k}\cdot\nabla - k^2]u_0(\mathbf{r}), \quad \text{(m.7.4.)}$$

and $E(\mathbf{k})$ obtained as

$$E(\mathbf{k}) = \int [u_0(\mathbf{r})[-(h^2/8\pi^2m)\nabla^2 + V(\mathbf{r})]u_0(\mathbf{r})\,d\mathbf{r}] + h^2k^2/8\pi^2m$$

$$- (h^2 i/4\pi^2 m)\mathbf{k}\cdot\left[\int u_0(\mathbf{r})\nabla u_0(\mathbf{r})\,d\mathbf{r}\right]. \quad \text{(m.7.5.)}$$

The last integral in this expression vanishes, since $u_0(\mathbf{r}) = u_0(-\mathbf{r})$, an even function, and $\nabla u_0(\mathbf{r}) = -\nabla u_0(-\mathbf{r})$. Since the expression for $u_0(\mathbf{r})$ may be written as

$$[-(h^2/8\pi^2m)\nabla^2 + V(\mathbf{r})]u_0(\mathbf{r}) = E_0 u_0(\mathbf{r}), \quad \text{(m.7.6.)}$$

$E(\mathbf{k})$ may now be obtained as the energy of the ground-state plus the kinetic energy of an electron in the state $\mathbf{k}$, or

$$E(\mathbf{k}) = E_0 + (h^2 k^2/8\pi^2 m). \quad \text{(m.7.7.)}$$

Adopting the same method with respect to the actual wave function for the state $\mathbf{k}$, which is $\psi_k(\mathbf{r}) = e^{i\mathbf{k}\cdot\mathbf{r}}u_k(\mathbf{r})$, as that used in obtaining (m.7.4.), gives

$$\nabla^2\psi_k(\mathbf{r}) = e^{i\mathbf{k}\cdot\mathbf{r}}[\nabla^2 + 2i\mathbf{k}\cdot\nabla - k^2]u_k(\mathbf{r}). \quad \text{(m.7.8.)}$$

If equation (m.7.8.) is substituted into equation (m.7.1.), the following equation for $u_k(\mathbf{r})$ is obtained:

$$\nabla^2 u_k(\mathbf{r}) + 2i\mathbf{k}\cdot\nabla u_k(\mathbf{r}) + (8\pi^2 m/h^2)[E(\mathbf{k}) - h^2 k^2/8\pi^2 m - V(\mathbf{r})]u_k(\mathbf{r}) = 0,$$

$$\text{(m.7.9.)}$$

which reduces to equation (m.7.6.) if $E(\mathbf{k})$ is given by (m.7.7.), and if $2i\mathbf{k}\cdot\nabla u_k(\mathbf{r})$ is sufficiently small to be negligible. If atomic units are used, equation (m.7.9.) may be alternatively, and more simply expressed as

$$\nabla^2 u_k + 2i\mathbf{k}\cdot\nabla u_k + [E(\mathbf{k}) - k^2 - V(\mathbf{r})]u_k = 0, \qquad \text{(m.7.10.)}$$

and the energy $E(\mathbf{k})$ as

$$E(\mathbf{k}) = \left[\int u_k^*\{-\nabla^2 - 2i\mathbf{k}\cdot\nabla + k^2 + V(\mathbf{r})\}u_k\right] d\mathbf{r}\left[\int |u_k|^2 \, d\mathbf{r}\right]^{-1}. \qquad \text{(m.7.11.)}$$

For small values of k, for states near $\mathbf{k} = 0$, the term $2i\mathbf{k}\cdot\nabla u_k$ in the above equations will be small, quite irrespective of the value of ∇u_k, and Wigner and Seitz[1] regarded this term as constituting a small perturbation, but a more general method of obtaining $E(\mathbf{k})$ in terms of the effective mass of the valency electron is given by Bardeen.[2] Bardeen's method leads to a relatively simple analytical expression for $E(\mathbf{k})$, expressed correctly in terms up to the second order in k as

$$E(\mathbf{k}) = E_0 + \alpha k^2, \qquad \text{(m.7.12.)}$$

where $\alpha = m/m^*$, the ratio of the electron mass to the effective mass. This comes about, as will be shown, due to the symmetry of u_0, the first order term in the integrals in equation (m.7.11.) vanishes when $u_k(\mathbf{r})$ is expanded in powers of k up to the first term. That is, if $u_k(\mathbf{r})$ is expanded as

$$u_k(\mathbf{r}) = u_0(\mathbf{r}) + ku_1(\mathbf{r}), \qquad \text{(m.7.13.)}$$

where both $u_0(\mathbf{r})$ and $u_1(\mathbf{r})$ have the lattice periodicity, and substituted into (m.7.11.), the first term, $-2i\mathbf{k}\cdot\left[\int u_0\nabla u_0 \, d\mathbf{r}\right]$, falls out. All of the integrals in (m.7.11.) may be evaluated either over the whole volume of the crystal or over a single cell, again due to the periodicity of u_k.

Thus, substituting equations (m.7.12.) and (m.7.13.) into equation (m.7.10.) gives

$$\nabla^2(u_0 + ku_1) + 2i\mathbf{k}\cdot\nabla(u_0 + ku_1) + [E_0 + \alpha k^2 - k^2 - V(\mathbf{r})](u_0 + ku_1) = 0.$$
$$\text{(m.7.14.)}$$

As $E(\mathbf{k})$ may be regarded as independent of the direction of the wave vector $\mathbf{k}$, $\mathbf{k}$ may be conveniently chosen to be in the z-direction, when it is found that

$$\mathbf{k}\cdot\nabla(u_0 + ku_1) = k \, \partial/\partial z(u_0 + ku_1). \qquad \text{(m.7.15.)}$$

Since it is generally the case that when a solution for a differential equation in the form of a series is substituted into the equation, an identity results, it is possible to equate each of the sets of terms in equation (m.7.14.) which do

---

[1] Wigner, E., and Seitz, F., *Phys. Rev.*, 1934, **46**, 509.
[2] Bardeen, J., *J. Chem. Phys.*, 1938, **6**, 367.

not contain k, and those which contain the first power of k, separately to zero, to give

$$[\nabla^2 + E_0 - V(\mathbf{r})]u_0 = 0 \qquad \text{and} \qquad [\nabla^2 + E_0 - V(\mathbf{r})]u_1 + 2i\,\partial u_0/\partial z = 0.$$
$$\text{(m.7.16.)}$$

The first equation above is similar to (m.6.62.), indicating that $u_0 = \psi_0$. Since u_0 is an even function, $\partial u_0/\partial z$ must be an odd function, and therefore u_1 must be an odd function, if it is to satisfy equation (m.7.16.). Thus, $u_1(\mathbf{r}) = -u_1(-\mathbf{r})$, and if such an odd function, having the lattice periodicity, is to be continuous at the surface of an atomic polyhedron, and yet change sign there, its value must be zero at the surface of an atomic polyhedron. Therefore, the function u_1, in equation (m.7.16.), must have a vanishing value at the surface of an atomic polyhedron, or expressed in terms of the atomic sphere approximation (page 646),

$$u_1(r_s) = 0. \qquad \text{(m.7.17.)}$$

This is the boundary condition, in terms of which a solution u_1 for equation (m.7.16.) has to be found.

It is simple enough to show that a particular solution u_1 for equation (m.7.16.) is $u_1 = -izu_0$, since

$$
\begin{aligned}
[\nabla^2 + E_0 - V(\mathbf{r})](-izu_0) &= \nabla[-iu_0 - iz\nabla u_0] - izu_0[E_0 - V(\mathbf{r})] \\
&= -2i\nabla u_0 - iz\nabla^2 u_0 - izu_0[E_0 - V(\mathbf{r})] \\
&= -2i\nabla u_0 - iz[\nabla^2 + E_0 - V(\mathbf{r})]u_0 \\
&= -2i\nabla u_0 = -2i\,\partial u_0/\partial z,
\end{aligned}
$$

since from the first equation of (m.7.16.),

$$[\nabla^2 + E_0 - V(\mathbf{r})]u_0 = 0.$$

The general solution of the second equation of (m.7.16.) is obtained by adding to the specific solution $u_1 = -izu_0$, the complementary function, which is the general solution of the homogeneous equation,

$$[\nabla^2 + E_0 - V(\mathbf{r})]u_1 = 0.$$

However, for this specific problem, the completely general solution is not actually required, but the solution which conforms with the boundary condition expressed in (m.7.17.). This means that it is required that

$$u_1 = izu_0 \qquad \text{at} \qquad r = r_s. \qquad \text{(m.7.18.)}$$

A function of this type, which is z times a spherically symmetrical function, is a p-type function (see sections Q.3.6. and Q.3.8., and specifically page 219). Suppose that this p-type function is written as

$$u_1(r) = iz[Y(r)/r^2]. \qquad \text{(m.7.19.)}$$

Then

$$\begin{aligned}
\nabla^2 u_1(r) &= \nabla^2[iz\,Y(r)r^{-2}] = \nabla[iY(r)r^{-2} + iz\nabla\{Y(r)r^{-2}\}] \\
&= 2i\,\partial/\partial z[Y(r)r^{-2}] + iz\nabla^2[Y(r)r^{-2}] \\
&= 2izr^{-1}\,d/dr[Y(r)r^{-2}] + iz[d^2/dr^2 + 2r^{-1}\,d/dr][Y(r)r^{-2}]
\end{aligned}$$

(see page 173)

$$\begin{aligned}
&= 2izr^{-1}[-2r^{-3}Y(r) + r^{-2}\,dY(r)/dr] \\
&\quad + iz[d/dr\{-2r^{-3}Y(r) + r^{-2}\,dY(r)/dr\}] \\
&\quad + 2izr^{-1}[-2r^{-3}Y(r) + r^{-2}\,dY(r)/dr] \\
&= -4izr^{-4}Y(r) + 2izr^{-3}\,dY(r)/dr \\
&\quad + iz[2r^{-3}Y(r) - 2r^{-3}\,dY(r)/dr + r^{-2}\,d^2Y(r)/dr^2] \\
&= izr^{-2}[d^2Y(r)/dr^2 - 2r^{-2}Y(r)].
\end{aligned}$$

(m.7.20.)

If equation (m.7.20.) is substituted into the homogeneous equation

$$[\nabla^2 + E_0 - V(\mathbf{r})]u_1 = 0,$$

the result is,

$$d^2Y(r)/dr^2 + [E_0 - 2r^{-2} - V(\mathbf{r})]Y(r) = 0. \qquad \text{(m.7.21.)}$$

The function $Y(r)$ must satisfy the conditions that it is zero at the origin and also such that

$$Y(r_s) = r_s^2 u_0(r_s). \qquad \text{(m.7.22.)}$$

To find $Y(r)$, it is necessary to integrate equation (m.7.21.) outward from the origin, and multiply the solution by the appropriate constant to satisfy the condition (m.7.22.). Then $u_1(r)$ is found as

$$u_1(r) = iz[Y(r)r^{-2} - u_0(r)] = izF(r). \qquad \text{(m.7.23.)}$$

This solution satisfies the condition expressed in (m.7.17.), and is valid in the region where $V(\mathbf{r})$ is spherically symmetrical, within the atomic sphere or cell; it is a purely imaginary function, while u_0 is real. It is not necessary to find the eigenvalue of equation (m.7.21.) since E_0 is known. The expression for $E(\mathbf{k})$ obtained by substituting (m.7.13.) in equation (m.7.11.) is

$$E(\mathbf{k}) = \frac{\int \{[u_0(\mathbf{r}) + ku_1(\mathbf{r})]^*[-\nabla^2 - 2ik\,\partial/\partial z + k^2 + V(r)][u_0(\mathbf{r}) + ku_1(\mathbf{r})]\}\,d\mathbf{r}}{\int \{|u_0(\mathbf{r}) + ku_1(\mathbf{r})|^2\}\,d\mathbf{r}},$$

(m.7.24.)

where the integrals in this expression have to be evaluated throughout an atomic sphere. If the equations of (m.7.16.) are substituted into (m.7.24.), $E(\mathbf{k})$ is obtained as

$$\begin{aligned}
E(\mathbf{k}) &= E_0 + k^2 - \left[2ik^2\int [\{u_0(\mathbf{r}) + ku_1(\mathbf{r})\}^*\,\partial u_1(\mathbf{r})/\partial z]\,d\mathbf{r}\right] \\
&\quad \times \left[\int [\{u_0(\mathbf{r})\}^2]\,d\mathbf{r} + k^2\int |u_1(\mathbf{r})|^2\,d\mathbf{r}\right]^{-1} \\
&= E_0 + k^2 + k^2\left[-2i\int [u_0(\mathbf{r})\,\partial u_1(\mathbf{r})/\partial z]\,d\mathbf{r}\right]\left[\int [u_0(\mathbf{r})]^2\,d\mathbf{r}\right]^{-1}. \quad \text{(m.7.25.)}
\end{aligned}$$

If $u_0(\mathbf{r})$ is normalized within the atomic sphere, so that the denominator in the third term for $E(\mathbf{k})$ in equation (m.7.25.) is unity, that is

$$\int [u_0(\mathbf{r})]^2 \, d\mathbf{r} = 4\pi \int_0^{r_s} [r^2 \{u_0(\mathbf{r})\}^2] \, d\mathbf{r} = 1, \qquad \text{(m.7.26.)}$$

then from equations (m.7.12.) and (m.7.25.),

$$\alpha = 1 - 2i \int [u_0(\mathbf{r}) \, \partial u_1(\mathbf{r})/\partial z] \, d\mathbf{r}$$

$$= 1 + 2 \int [u_0(\mathbf{r})\{F(\mathbf{r}) + z^2 r^{-1} \, dF(\mathbf{r})/dr\}] \, d\mathbf{r}, \quad \text{from equation (m.7.23.),}$$

$$= 1 + 2 \int_0^{r_s} [4\pi r^2 u_0(\mathbf{r}) F(\mathbf{r})] \, dr + 4\pi \int_0^{r_s} r^3 \, dr \int_0^{\pi} u_0(\mathbf{r}) \, dF(\mathbf{r})/dr \, \cos^2 \theta \sin \theta \, d\theta$$

$$= 1 + 8\pi \int_0^{r_s} [r^2 u_0(\mathbf{r}) F(\mathbf{r})] \, dr + [8\pi/3] \int_0^{r_s} [r^3 u_0(\mathbf{r}) \, dF(\mathbf{r})/dr] \, dr$$

$$= 1 + 8\pi \int_0^{r_s} [r^2 u_0(\mathbf{r}) F(\mathbf{r})] \, dr - [8\pi/3] \int_0^{r_s} [F(\mathbf{r}) \, d/dr\{r^3 u_0(\mathbf{r})\}] \, dr,$$

$$\text{(m.7.27)}$$

on integrating by parts, and making use of the fact that $F(r_s) = 0$. So,

$$\alpha = 1 - [8\pi/3] \int_0^{r_s} [r^3 F(\mathbf{r}) \, du_0(\mathbf{r})/dr] \, dr$$

$$= 1 - [8\pi/3] \int_0^{r_s} [r\{Y(r) - r^2 u_0(\mathbf{r})\} \, du_0(\mathbf{r})/dr] \, dr$$

$$= 1 - [8\pi/3] \int_0^{r_s} \{[Y(r) - rf(r)][df(r)/dr - r^{-1}f(r)]\} \, dr, \quad \text{(m.7.28.)}$$

on writing

$$ru_0(\mathbf{r}) = f(r), \qquad \text{(m.7.29.)}$$

as on page 647. Now, to evaluate the integrals in (m.7.28.), use is made of equation (m.7.26.), that

$$4\pi \int_0^{r_s} [f(r)]^2 \, dr = 1 = 4\pi \int_0^{r_s} [ru_0(\mathbf{r})]^2 \, dr,$$

and the fact that

$$\int_0^{r_s} [rf(r)\{df(r)/dr\}] \, dr = |r\{f(r)\}^2 \Big|_0^{r_s} - \int_0^{r_s} [\{f(r)\}^2 + rf(r)\{df(r)/dr\}] \, dr,$$

so that

$$(8\pi/3) \int_0^{r_s} [rf(r)\{df(r)/dr\}] \, dr = (4\pi/3)|r\{f(r)\}^2 \Big|_0^{r_s} - (4\pi/3) \int_0^{r_s} [f(r)]^2 \, dr$$

$$= (4\pi/3)(r_s)^3 [u_0(r_s)]^2 - (1/3), \qquad \text{(m.7.30.)}$$

and

$$\alpha = (4\pi/3)(r_s)^3[u_0(r_s)]^2 - (8\pi/3) \int_0^{r_s} Y(r)[\{df(r)/dr\} - r^{-1}f(r)]\, dr.$$

(m.7.31.)

Another of the integrals in (m.7.28.) is

$$\int_0^{r_s} [\![Y(r)[df(r)/dr - r^{-1}f(r)]]\!]\, dr$$

$$= \int_0^{r_s} [\![[Y(r)r]\, d/dr[f(r)r^{-1}]]\!]\, dr$$

$$= |Y(r)f(r)|_0^{r_s} - \int_0^{r_s} [\![r^{-1}f(r)[r\{dY(r)/dr\} + Y(r)]]\!]\, dr$$

so that

$$2 \int_0^{r_s} [\![Y(r)[df(r)/dr - r^{-1}f(r)]]\!]\, dr$$

$$= |Y(r)f(r)|_0^{r_s} - \int_0^{r_s} [f(r)\{dY(r)/dr\} - Y(r)\, df(r)/dr + 2Y(r)f(r)r^{-1}]\, dr.$$

(m.7.32.)

By combining equations (m.7.21.) and (m.6.62.) for $Y(r)$ and $f(r)$, respectively, there results

$$f(r)[d^2 Y(r)/dr^2] - Y(r)[d^2f(r)/dr^2] = 2r^{-1}Y(r)f(r), \quad \text{(m.7.33.)}$$

so that

$$d/dr[\![r[f(r)\{dY(r)/dr\}] - Y(r)\{df(r)/dr\}]\!]$$
$$= f(r)\{dY(r)/dr\} - Y(r)\{df(r)/dr\} + 2r^{-1}Y(r)f(r). \quad \text{(m.7.34.)}$$

Now, by combining equations (m.7.31.), (m.7.32.), and (m.7.34.), α may be written as a function of r_s, as

$$\alpha = [4\pi/3][r_s]^3[u_0(r_s)]^2 - [4\pi/3] Y(r_s)f(r_s)$$
$$+ [4\pi/3]r_s[\![f(r_s)\{dY(r_s)/dr\} - Y(r_s)\{df(r_s)/dr\}]\!]. \quad \text{(m.7.35.)}$$

This may be simplified further by applying the condition (m.6.63.), which leads to

$$df(r_s)/dr = (r_s)^{-1}f(r_s) = u_0(r_s), \quad \text{(m.7.36.)}$$

and from equation (m.7.22.), it is apparent that

$$Y(r_s) = [r_s]^2 u_0(r_s) = r_s f(r_s), \quad \text{(m.7.37.)}$$

and thus the difference between the two terms enclosed within the open square brackets, $[\![\ ]\!]$, in equation (m.7.35.) may be written as

$$r_s f(r_s)[dY(r_s)/dr] - Y(r_s)[df(r_s)/dr]r_s$$
$$= [r_s]^4[u_0(r_s)]^2[dY(r_s)/dr][Y(r_s)]^{-1} - [r_s]^3[u_0(r_s)]^2. \quad \text{(m.7.38.)}$$

At last, by substitution of equations (m.7.38.) into (m.7.35.), there is obtained

$$\alpha = [4\pi/3][r_s]^3[u_0(r_s)]^2[r\,Y(r)\{dY(r)/dr\} - 1]_{r=r_s}. \qquad \text{(m.7.39.)}$$

For a given value of r_s, $u_0(r_s)$ and E_0 may thus be found and then α can be determined by integrating equation (m.7.21.) once to obtain $dY(r)/dr$. Values of α calculated in this way of Bardeen for lithium and sodium gave 0.65 and 1.07, respectively. The expression (m.7.39.) has also been used in conjunction with the quantum defect method[1] (section M.6.7.), while Bardeen made use of the observed atomic radii, to obtain α for the alkali metals as 0.69 for lithium, 1.02 for sodium, 1.08 for potassium, 1.12 for rubidium, and 1.20 for cesium. Sine $E(\mathbf{k})$ for an electron in the state $\mathbf{k}$, where k is small, is given by (m.7.11.) and it has been shown that this is equivalent to $E(\mathbf{k}) = E_0 + \alpha k^2$ [equation (m.7.12.)], expressed in atomic units, and α may be evaluated as (m.7.39.), then in ordinary units

$$E(\mathbf{k}) = E_0 + \alpha[h^2k^2/8\pi^2m], \quad \text{with} \quad \alpha = m/m^*, \qquad \text{(m.7.40.)}$$

where m^* is the effective mass near $\mathbf{k} = 0$. For sodium, in particular, the effective mass for small k values is very little different from m, and since the first Brillouin zone is only half filled for sodium (and the other alkali metals), the free electron assumption and the spherical nature of the Fermi surface would appear to be reasonably justified.

M.7.3.   The Cohesive and Correlation Energies for Sodium

In the discussion of the Hartree–Fock method in section Q.5.7., an expression was given for the total energy of an N-electron atom, written in terms of orbital functions as equation (q.5.50.). A similar expression may be written for a metal consisting of N atoms, by adding to the Hamiltonian for the electrons the energy of interaction of the metal ions; thus, modifying equation (q.5.50.), by adding a term to account for the ion–ion interaction energy between monopositively charged ions as in the alkali metals, gives

$$
\begin{aligned}
E = {}& \sum_i^N \int \psi_i^*(\mathbf{r}_1)\left[-(h^2/8\pi^2m)\nabla_1^2 + \sum_a^N V_a(\mathbf{r}_1)\right]\psi_i(\mathbf{r}_1)\,d\mathbf{r}_1 \\
& + (1/2)\sum_i^N\sum_j^N \iint (\varepsilon^2/r_{12})|\psi_i(\mathbf{r}_1)|^2|\psi_j(\mathbf{r}_2)|^2\,d\mathbf{r}_1\,d\mathbf{r}_2 \\
& - (1/2)\sum_i^N\sum_j^N \iint (\varepsilon^2/r_{12})\psi_i^*(\mathbf{r}_1)\psi_j^*(\mathbf{r}_2)\psi_i(\mathbf{r}_2)\psi_j(\mathbf{r}_1)\,d\mathbf{r}_1\,d\mathbf{r}_2 \\
& + (1/2)\sum_{a\neq b}^N\sum^N (\varepsilon^2/R_{ab}),
\end{aligned}
\qquad \text{(m.7.41.)}
$$

---

[1] Brooks, H., *Nuovo Cimento Suppl.*, 1958, **7**, 165.

where the first term represents the kinetic energy of the N electrons plus the potential energy of the electrons in the field of the ions, numbered from a to N, the second term is derived from the Coulomb energy of the electrons, the third term is associated with the exchange effects, and the last term is derived from the potential energy due to the Coulomb interactions between the metal ions. If a metal contains dipositively charged or multiply charged ions, of charge Z, with Z valency electrons, then the total number of valency electrons would be ZN and the last term would have to be multiplied by Z^2. In this expression, (m.7.41.), $r_{ij} = |\mathbf{r}_i - \mathbf{r}_j|$, $R_{ab} = |\mathbf{R}_a - \mathbf{R}_b|$, and $V_a(\mathbf{r}_i)$ is the potential energy of the ith electron in the field of ion a, where $\mathbf{r}_i$ is the position vector of the ith electron and $\mathbf{R}_a$ that of nucleus of ion a. The ψ_i's are orbital functions (see pages 303–306) of the one-electron type. Since for a metal at 0 K, there are two electrons with opposed spins in each occupied state, equation (m.7.41.) may be rewritten in terms of the wave vector notation, with two electrons per occupied state $\psi_{\mathbf{k}}$, for a metal such as sodium with one valency electron per atom, in the form

$$NE_m = 2 \sum_k \int \psi_k{}^*(\mathbf{r})[-(h^2/8\pi^2 m)\nabla^2 + \sum_a^N V_a(\mathbf{r})]\psi_k(\mathbf{r})\, d\mathbf{r}$$

$$+ 2 \sum_k \sum_{k'} \iint (\varepsilon^2/r_{12})|\psi_k(\mathbf{r}_1)|^2|\psi_k(\mathbf{r}_2)|^2\, d\mathbf{r}_1\, d\mathbf{r}_2$$

$$- \sum_k \sum_{k'} \iint (\varepsilon^2/r_{12})\psi_k{}^*(\mathbf{r}_1)\psi_{k'}{}^*(\mathbf{r}_2)\psi_k(\mathbf{r}_2)\psi_{k'}(\mathbf{r}_1)\, d\mathbf{r}_1\, d\mathbf{r}_2 \qquad \text{(m.7.42.)}$$

$$+ (1/2) \sum_{a \neq b}^N {}^N (\varepsilon^2/R_{ab}),$$

where the integrals apply throughout the volume of the metal and E_m represents the average energy per electron, for this particular approximation.

Wigner and Seitz originally (1933) postulated that the average energy of a valency electron in sodium metal was simply $E_0 + E_F$, where E_F represents the free electron Fermi energy, given in equation (m.3.102.) as $[1.105/r_s{}^2]$ Hartrees; they assumed that the approximation of (m.7.7.) was valid for $E(\mathbf{k})$ and that the effects of electronic interactions were dealt with by the fact that only one electron was contained in an atomic polyhedron. From the curve showing $E_0 + E_F$ as a function of r_s (Fig. M.6.8.), the value of r_s corresponding to the minimum of $E_0 + E_F$ is assumed to be the equilibrium atomic radius, and the depth of this minimum below the value of E_0 for the infinite value for r_s is taken as the calculated cohesive energy per atom. If the equilibrium atomic radius obtained in this way is designated r_0 and the cohesive energy per atom as E_c, then Wigner and Seitz's first expression for the cohesive energy of sodium was

$$E_c = -I - E_0(r_0) - [1.105/r_0{}^2] \text{ Hartrees}, \qquad \text{(m.7.43.)}$$

where $+I$ is the ionization energy per atom, as in section M.6.6. In their second publication on sodium (1934), Wigner and Seitz improved on the approximate value calculated for the cohesive energy in this way by calculating the total energy by the use of an expression like (m.7.42.), using a wave function consisting of a single determinant of one-electron functions, and subsequently correcting for Coulomb correlation effects by the use of the free electron approximation.

If it is assumed that the functions ψ_k of (m.7.42.) are eigenfunctions of equation (m.7.1.), then in terms of the Wigner–Seitz cells, the potential function in this latter equation, $V(\mathbf{r})$, may be rewritten as

$$V(\mathbf{r}) = \sum_a^N U_a(\mathbf{r}), \qquad \text{(m.7.44.)}$$

with $U_a(\mathbf{r}) = V_a(\mathbf{r})$ in the region within the cell designated a and $U_a(\mathbf{r}) = 0$ in the region outside cell a. The potential of an electron in the field of a metal ion located by the position vector $\mathbf{R}_a$ is spherically symmetrical about the lattice point $\mathbf{R}_a$, so that $V_a(\mathbf{r}) = V(|\mathbf{r} - \mathbf{R}_a|)$. At all points outside the cell a, it is assumed that the potential energy function is hydrogenic in nature, that is that

$$V_a(\mathbf{r}) = -(\varepsilon^2/|\mathbf{r} - \mathbf{R}_a|). \qquad \text{(m.7.45.)}$$

If the validity of the atomic sphere approximation is also assumed, that is that the valency electron charge distribution within any cell a is spherically symmetrical, so that the charge may be regarded as being concentrated at its center, and the density of electrons, with both types of spin, at the position $\mathbf{r}$ is defined by

$$\rho(r) = 2 \sum_k |\psi_k(\mathbf{r})|^2, \qquad \text{(m.7.46.)}$$

with ψ_k normalized throughout the volume of the metal, then the first two terms in the expression for NE_m given by (m.7.42.) may be rewritten in a more tractable form, from the point of view of numerical evaluation.

Thus, the first term in (m.7.42.) may now be expressed as

$$2 \sum_k \int \left(\psi_k{}^*(\mathbf{r}) \left[-(h^2/8\pi^2 m)\nabla^2 + \sum_a^N V_a(\mathbf{r}) \right] \psi_k(\mathbf{r}) \right) d\mathbf{r}$$

$$= 2 \sum_k \int \left(\psi_k{}^*(\mathbf{r}) \left[E(\mathbf{k}) - V(\mathbf{r}) + \sum_a^N V_a(\mathbf{r}) \right] \psi_k(\mathbf{r}) \right) d\mathbf{r}$$

$$= 2 \sum_k E(\mathbf{k}) + \int \left(\rho(\mathbf{r}) \left[\sum_a^N V_a(\mathbf{r}) - V(\mathbf{r}) \right] \right) d\mathbf{r}$$

$$= 2 \sum_k E(\mathbf{k}) + \sum_a^N \int (\rho(\mathbf{r})[V_a(\mathbf{r}) - U_a(\mathbf{r})]) \, d\mathbf{r}$$

$$= 2 \sum_k E(\mathbf{k}) + \sum_a^N \left[\int_a (\rho(\mathbf{r})[V_a(\mathbf{r}) - U_a(\mathbf{r})]) \, d\mathbf{r} \right.$$

$$\left. + \sum_{b \neq a}^N \int_b (\rho(\mathbf{r})[V_a(\mathbf{r}) - U_a(\mathbf{r})]) \, d\mathbf{r} \right],$$

where $\int_a$ and $\int_b$ denote integration throughout the cells a and b, respectively;

$$= 2 \sum_k E(\mathbf{k}) + \sum_{a \neq b}^{N} \sum^{N} \int_b (\rho(\mathbf{r})V_a(\mathbf{r}))\, d\mathbf{r},$$

since $U_a(\mathbf{r}) = V_a(\mathbf{r})$ inside the cell a and $U_a(\mathbf{r})$ is zero outside the cell a;

$$= 2 \sum_k E(\mathbf{k}) - \sum_{a \neq b}^{N} \sum^{N} [\varepsilon^2/\mathbf{R}_{ab}], \qquad \text{(m.7.47.)}$$

since from (m.7.45.),

$$\int_b (\rho(\mathbf{r})V_a(\mathbf{r}))\, d\mathbf{r} = -\int_b ([\varepsilon^2/|\mathbf{r} - \mathbf{R}_a|]\rho(\mathbf{r}))\, d\mathbf{r} = -[\varepsilon^2/\mathbf{R}_{ab}],$$

by (m.7.46.).

The Coulomb interaction term in (m.7.42.), the second term as written there, may similarly be re-expressed as

$$(1/2) \iint ([\varepsilon^2/r_{12}]\rho(\mathbf{r}_1)\rho(\mathbf{r}_2))\, d\mathbf{r}_1\, d\mathbf{r}_2$$

$$= (1/2) \sum_a^N \sum_b^N \int_a \int_b ([\varepsilon^2/r_{12}]\rho(\mathbf{r}_1)\rho(\mathbf{r}_2))\, d\mathbf{r}_1\, d\mathbf{r}_2$$

$$= (1/2) \sum_{a \neq b}^N \sum^N \int_a \int_b ([\varepsilon^2/r_{12}]\rho(\mathbf{r}_1)\rho(\mathbf{r}_2))\, d\mathbf{r}_1\, d\mathbf{r}_2$$

$$\quad + (N/2) \int_a \int_b ([\varepsilon^2/r_{12}]\rho(\mathbf{r}_1)\rho(\mathbf{r}_2))\, d\mathbf{r}_1\, d\mathbf{r}_2$$

$$= (1/2) \sum_{a \neq b}^N \sum^N [\varepsilon^2/R_{ab}] + (N/2) \int_a \int_b ([\varepsilon^2/r_{12}]\rho(\mathbf{r}_1)\rho(\mathbf{r}_2))\, d\mathbf{r}_1\, d\mathbf{r}_2. \quad \text{(m.7.48.)}$$

If now these two expressions given by (m.7.47.) and (m.7.48.) are added and substituted for the first two terms in (m.7.42.), there results

$$NE_m = 2 \sum_k E(\mathbf{k}) + (N/2) \int_a \int_b ([\varepsilon^2/r_{12}]\rho(\mathbf{r}_1)\rho(\mathbf{r}_2))\, d\mathbf{r}_1\, d\mathbf{r}_2$$

$$\quad - \sum_k \sum_{k'} \iint ([\varepsilon^2/r_{12}]\psi_k{}^*(\mathbf{r}_1)\psi_{k'}{}^*(\mathbf{r}_2)\psi_k(\mathbf{r}_2)\psi_{k'}(\mathbf{r}_1))\, d\mathbf{r}_1\, d\mathbf{r}_2, \quad \text{(m.7.49.)}$$

the ionic Coulomb interaction term canceling out.

Numerical values may be assigned to the various terms in (m.7.49.), in the following way. To a good approximation the first term, by making use of (m.7.40.), is given by

$$2 \sum_k E(\mathbf{k}) = N[E_0 + \alpha E_F], \qquad \text{(m.7.50.)}$$

with $E_F = [1.105/r_s{}^2]$ Hartrees, as calculated in (m.3.102.). The Coulomb energy term may be evaluated, using the technique described on pages 238–239, since the expression $(1/2) \int_a \int_b ([\varepsilon^2/r_{12}]\rho(\mathbf{r}_1)\rho(\mathbf{r}_2))\, d\mathbf{r}_1\, d\mathbf{r}_2$ is the self-potential energy of a charge distribution of density $\varepsilon\rho(r)$ within each atomic polyhedron, and in terms of the free electron approximation

$$\rho(r) = N/V = 3/4\pi r_s^3, \qquad\qquad (\text{m.7.51.})$$

so that

$$(1/2) \int_a \int_b ([\varepsilon^2/r_{12}]\rho(\mathbf{r}_1)\rho(\mathbf{r}_2))\, d\mathbf{r}_1\, d\mathbf{r}_2 = [9\varepsilon^2/2][4\pi r_s^3]^{-1} \int_a \int_b ([1/r_{12}])\, d\mathbf{r}_1\, d\mathbf{r}_2$$

$$= [9\varepsilon^2/2][4\pi r_s^3]^{-2} 32\pi^2 \int_0^{r_s} r\, dr \int_0^{r} r^2\, dr$$

$$= [9\varepsilon^2/2][4\pi r_s^3]^{-2}[32\pi^2 r_s^5/15]$$

$$= [3\varepsilon^2 N/5 r_s], \qquad\qquad (\text{m.7.52.})$$

and expressed in atomic units, this becomes

$$= 0.6N/r_s \text{ Hartrees.}$$

The exchange term in (m.7.49.) has already been evaluated on page 497, in the free electron approximation, as $-0.458N/r_s$ Hartrees, and if this term value is added to (m.7.50.) and (m.7.52.), all three terms divided by N, then the average energy per electron in sodium is obtained as

$$E_m(r_s) = E_0(r_s) + \alpha E_F(r_s) + 0.6/r_s - 0.458/r_s \text{ Hartrees}$$
$$= E_0(r_s) + 1.105\alpha/r_s^2 + 0.142/r_s \text{ Hartrees.} \qquad (\text{m.7.53.})$$

The value of the cohesive energy of sodium calculated in this way, by subtracting the minimum value of E_m from the ionization energy, $-I$ per atom, comes to 2.94×10^4 J mol^{-1}, compared with the experimentally observed value of 10.88×10^4 J mol^{-1}, so that this calculated value is much too small. Actually, E_m is equivalent to the average energy per electron given by a total wave function consisting of a single determinant of one-electron functions, approximating to the Hartree–Fock average energy (section M.3.7.); if a total wave function composed of a single product of one-electron functions were used, corresponding to the Hartree method (section Q.4.8.), the only difference would be that no exchange energy term would appear in (m.7.53.). Since the exchange term is apparently large and negative, the Hartree mean energy would be so high that there would be no cohesive energy at all. It has already been seen that the principal value of the Hartree–Fock method is that it enables an improved cohesive energy to be calculated, but there is obviously something else missing from the calculated E_m. Actually, the first two terms of (m.7.53.) alone enable quite a good calculated value of the cohesive energy of sodium to be obtained, as expressed in (m.7.43.); incidentally, it makes little difference to this latter value if α is included or excluded, since α is so close to unity. Also, the only approximations used in obtaining (m.7.53.), apart from the substitution of the atomic polyhedra by spheres, lies in the use of the free electron approximation in the estimation of the Coulomb and exchange energy terms. The disagreement between the calculated E_m and the experimental cohesive energy is then attributable to the exchange energy term, and more precisely to the lack of properly accounting for the Coulomb correlations, which is inherent in the Hartree–Fock method (page 495). As already seen, the Hartree–Fock method accounts partially for correlation effects between electrons with parallel spins, essentially due to the operation of the Pauli Exclusion Principle, but takes no account of correlations between electrons with antiparallel spins.

The simplest definition of the correlation energy, without making any reference to electron spins, is to say that it is the total energy, calculated in such a manner as to make proper allowance for Coulomb correlations, minus the calculated Hartree–Fock energy, and it is convenient to designate the average correlation energy per electron by E_L. An approximate estimate of the magnitude of the correlation energy for sodium may be obtained by consideration of the fact that (m.7.43.) gives a good approximation to the cohesive energy, and this expression only makes use of the first two terms in (m.7.53.), namely $E_0 + E_F$, so that if any calculated value of the cohesive energy is going to be approximately in agreement with experiment, then the correlation energy must be comparable to the sum of the Coulomb and exchange energy terms in (m.7.53.); so, approximately, $E_L \simeq -[0.142/r_s]$ Hartrees, which is of the same order of magnitude as the cohesive energy itself and thus not at all a negligible amount.

The calculation of correlation energies in metals is a matter of great difficulty, but it is relatively easy to calculate the correlation energy of a free electron gas at very low densities, which provides a valuable guide in more detailed correlation energy calculations. For very low electron gas densities, equivalent to large values of r_s, the kinetic energy of the electrons in the electron gas tend to become negligible compared to their potential energies; this is so, since the kinetic energy is inversely proportional to r_s^2, whereas the potential energy is inversely proportional to r_s. Thus, for large r_s values, the electrons in the dilute electron gas will tend to behave as if each electron is at the center of a uniform positive charge of radius r_s and total positive charge ε, so that the atomic sphere approximation may be used to calculate the total energy. The positive charge density is then $[3\varepsilon/4\pi r_s^3]$, and if an electron is at the center of a sphere of radius r, of positive charge of this density, the total charge inside the sphere would be $[-\varepsilon + [4\pi r^3/3][3\varepsilon/4\pi r_s^3]] = [-\varepsilon + \varepsilon r^3/r_s^3]$, and the surface potential would be $[-1 + r^3/r_s^3][\varepsilon/r]$. The energy required to bring sufficient charge from infinity to form a shell of thickness dr and density $[3\varepsilon/4\pi r_s^3]$ is (see pages 238–9) $[(r^3/r_s^3) - 1][\varepsilon/r][3\varepsilon/4\pi r_s^3]4\pi r^2 \, dr =$

$[3\varepsilon^2/r_s{}^3][r^3/r_s{}^3 - 1]r\,dr$, and so at very low electron densities, the energy per electron is given by

$$[3\varepsilon^2/r_s{}^3] \int^{r_s} ([r^3/r_s{}^3 - 1]r)\,dr = -0.9\varepsilon^2/r_s \equiv -0.9/r_s \text{ Hartrees.} \quad \text{(m.7.54.)}$$

According to the Hartree–Fock calculation, the energy per electron, if the kinetic energy is again disregarded, is equivalent to the exchange energy, -0.458 Hartrees [equation (m.3.102.), page 497]. Therefore the average correlation energy per electron at very low electron densities is

$$E_L = -0.9/r_s + 0.458/r_s = -0.44/r_s. \quad \text{(m.7.55.)}$$

This expression has no direct applicability, since the valency electrons in a metal only approximate to a free electron gas at the densities which normally apply to solid metals, and so not at very low densities; at very low densities, when r_s is large and the metallic ions are so widely separated as to form a collection of free atoms, then the valency electrons are no longer describable by free electron theory.

Since all of the correlation effect between electrons in a metal are included in a term like,

$$(1/2) \sum_{\substack{i \neq j}}^{N} \sum^{N} [\varepsilon^2/r_{ij}],$$

and if this term is disregarded then the eigenfunctions of the more complete Hamiltonian,

$$-[h^2/8\pi^2m] \sum_{i=1}^{N} \nabla_i{}^2 + (1/2) \sum_{\substack{i \neq j}}^{N} \sum^{N} [\varepsilon^2/r_{ij}],$$

would be simply determinants of free electron functions, it is apparent that correlation energies can, in principle, be calculated as perturbation effects to the free electron energies. However, first order perturbation theory gives the Coulomb energy and the exchange energy as a first order correction, but the second order perturbation correction leads to a divergent sum (see page 261); a convergent result may only be obtained by including higher order perturbation corrections. Although perturbation calculations of the correlation energy of metals have been carried out, these are only valid at small values of $r_s < 1$ Bohr unit, at high electron densities, so that to estimate the correlation energy at the actual electron densities in metals, it is necessary to interpolate between the high density and low density limiting values. An interpolation expression given by Pines[1] is of the form,

$$E_L = -[0.44/(r_s + 7.8)] \quad \text{(m.7.56.)}$$

---

[1] Pines, D., *Solid State Physics I*, 1955, 367, Academic Press, New York.

which, for sodium with $r_s = 4$ Bohr units, gives $E_L = -0.037$ Hartrees, a very similar value to that obtained by the use of the approximate expression given on page 696, $E_L = -[0.142/r_s]$, of -0.036 Hartrees; it is also close to the value given by equation (m.6.73.), based on Bohm and Pines plasma theory of metals (see section M.6.10.). Another reported calculation of the correlation energy of a free electron gas, claimed to be completely accurate in the high density limit[1] gives

$$E_L = 0.0311 \log r_s - 0.048 + \Omega(r_s) \text{ Hartrees,}$$

where $\Omega(r_s)$ is small and tends to zero as $r_s \to 0$, from which for $r_s = 1$ Bohr unit, disregarding the term $\Omega(r_s)$, $E_L = -0.048$ Hartrees, compared with the value obtained from (m.7.56.) of $E_L = -0.05$ Hartrees.

If either (m.6.73.) or (m.7.56.) is used to give the correlation energy E_L, then the final expression for the mean energy per electron of an alkali metal obtained by adding E_L to equation (m.7.53.) is obtained as

$$E_m(r_s) = E_0(r_s) + 1.105\alpha/r_s^2 + 0.142/r_s + E_L \text{ Hartrees} \quad \text{(m.7.57.)}$$

As before, if r_0 describes the atomic radius corresponding to the minimum value of E, then the cohesive energy per electron is

$$E_c = -I - E_m(r_0). \quad \text{(m.7.58.)}$$

The calculated cohesive energy of the alkali metals, their equilibrium radii, and their compressibilities (page 650), obtained by the use of equations (m.7.57.) and (m.7.58.) agree within a few percent with the experimentally observed values. These expressions, however, do not apply to copper, silver, and gold, the other one-electron-per-atom metals, even although the valency electron wave functions for these metals are equally well described by free electron functions. The principal reason for this is that their ions are so large, and in particular the overlapping of the d-state charge density clouds between adjacent ions in copper, silver, and gold, is so extensive that it is not valid to regard the interaction between the ions as describable in terms of point charges.

II. COPPER, SILVER, AND GOLD

M.7.4. General Characteristics

Neutral atoms of copper, silver, and gold, with nuclear charges 29, 47, and 79, respectively, have ground state electron configurations of $1s^2 2s^2 2p^6 3s^2 3p^6 3d^{10} 4s^1$, $1s^2 2s^2 2p^6 3s^2 3p^6 3d^{10} 4s^2 4p^6 4d^{10} 5s^1$, and $1s^2 2s^2 2p^6 3s^2 3p^6 3d^{10} 4s^2 4p^6 4d^{10} 5s^2 5p^6 4f^{14} 5d^{10} 6s^1$, sharing the common feature of having one

---

[1] Gell-Mann, M., and Brueckner, K. A., *Phys. Rev.*, 1957, **106**, 364.

s-state valency electron beyond a closed quantum group. They all crystallize with the face-centered cubic or cubic closest packed structure (Fig. M.1.3.(a)), with lattice constants of 3.61496×10^{-10} m (copper, at 291 K), 4.0862×10^{-10} m (silver, at 298 K), and 4.07825×10^{-10} m (gold, at 298 K).

The red color of copper and the yellow color of gold are characteristic properties which distinguish them from all of the other known metals. All metals absorb and reflect visible light strongly, these processes being associated with energy changes in the electron bands, although generally the appearance of metals is greatly influenced by the thin film of oxide with which they are usually covered under normal conditions. The $N(E) - E$ curve for copper is shown in Fig. M.5.41. and it is believed that the comparable curves for silver and gold are not substantially different, apart from the energy difference between the top of the filled d-band and the Fermi surface. From X-ray data, it is indicated that the top of the d-band for copper is about 3.7×10^{-19} J below the Fermi surface, the s–p band, and about 6.4×10^{-19} J below the Fermi surface for silver. Electrons from the d-band in copper can therefore be excited to the s–p band by green, blue, or violet light, but not by wavelengths at the red end of the electromagnetic spectrum. For silver, ultraviolet radiation is required to similarly excite electrons from the d-band. This interpretation is supported by calculations of the binding energy of the $3d$ electrons in copper and of the $4d$ electrons in silver; the $4d$ electrons in silver are more firmly bound than the $3d$ electrons in copper, which is in agreement with the general chemistry of the two elements. The $(n - 1)\,d$ sub-group of electrons in copper is more easily broken to give rise to Cu^{++} compounds, but the chemistry of silver is predominantly that of Ag^+ ions. In copper–zinc solid solutions, increasing concentration of zinc increases the electron to atom ratio, so that the s–p band is occupied to higher levels. The absorption peak then moves to lower wavelengths and the color of the brass alloys changes from the red copper color of the low zinc content species to the yellow of the zinc rich alloys. For gold, the absorption peak occurs at a slightly shorter wavelength than for copper, but at longer wavelengths than for silver, the yellow color of gold again being associated with the excitation of electrons from the $(n - 1)\,d$ band to the s–p band. Gold, however, does not exhibit the properties which would be expected by an extrapolation of those of copper and silver, but in many respects is more similar to copper. This is a consequence of the relative differences in ionization energy of atoms and ions containing electrons in $3d$, $4d$, and $5d$, states. From spectroscopic data, it is observed that the binding energy of the electrons in d-states increases steadily with increasing nuclear charge for the metallic elements of each of the three transition metal series, with a distinct break, however, after the half filled d sub-group when electrons are being ionized off orbital states containing two electrons. This is shown in Fig. M.7.1. The repulsion between

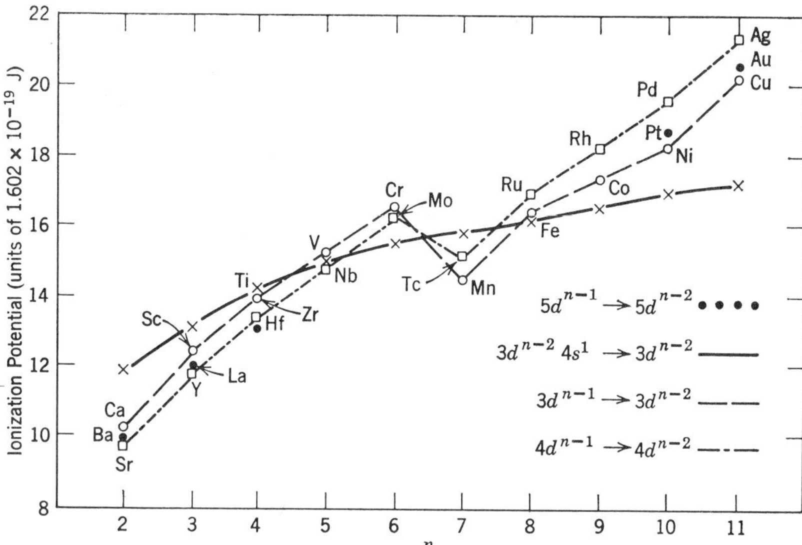

FIG. M.7.1. Ionization energies for the three transition metal series, corresponding to the electron configurational changes indicated at the lower right of the diagram. The discontinuity between $n = 6$ and $n = 7$, for the first two transition metal series, corresponds to the difference in spin-pairing energy.

two d-electrons in the same orbital state is especially notable as a result of the highly directional properties of the d-orbital states (see Fig. Q.3.4.). The discontinuity is less pronounced for the larger atoms containing electrons in $4d$ and $5d$ states, so that the pairing of electrons in the same orbital state is less unfavorable for the second and third transition metal series than it is for the first series of transition metals. This is indeed an important factor in determining the formation and the stability of complex compounds containing these metals. In the second half of the second transition metal series the $4d$-electrons are more firmly bound than the $3d$-electrons in the corresponding members of the first transition series; the reverse applies to the s-electrons, which accounts for the relative stability of the dispositive oxidation state of copper as compared to that of silver. There is less information available for the third transition metal series, but it is apparent that the binding energy of $5d$-electrons is less than that of $4d$-electrons, as a consequence of the interposition of the $4f^{14}$-electron group, corresponding to the 14 lanthanide elements; thus, the d-electrons are less stable in gold than in silver, and this effect is confirmed by the study of the X-ray spectra of these metallic elements, and by the observation that the normal stable oxidation state of gold is the tripositive state, corresponding to Au^{3+} ions.

The existence of d-electron states in copper, silver, and gold, gives rise to a much greater overlap between the ions in these metal crystals, compared with the situation for the alkali metals. The repulsive forces which arise between the metal ions in crystals of these metals, which vary very rapidly with the internuclear distance, account for the fact that the compressibility is so very much smaller than for the alkali metals. Generally, the compressibility of isoelectronic metals increases with atomic volume, but the compressibilities of the alkali metals is so much greater than that of copper, silver, and gold, that considerations other than the greater atomic volume alone must be involved. For the alkali metals, where the ionic overlap is so small that it accounts for only a small fraction of the repulsive force between the ions, the compressibilities are exceptionally large. If the compressibility of copper is calculated by the method of Wigner and Seitz without allowing for the effects of ionic overlap, the calculated value is about six times the experimentally observed value, but if the ionic overlap is allowed for, good agreement results between the calculated and observed values,[1] confirming that the compressibility of copper is essentially due to ionic repulsive effects. The cubic closest packed structure adopted by copper is consistent with the greater stability of this structural arrangement compared to the body-centered cubic arrangement, when the ionic repulsive effects are taken into account, although all of the alkali metals crystallize with the latter structure even although the calculated energies are essentially equal in this case.

The first Brillouin zone for the face-centered cubic structure is shown in Figs. M.5.26. and M.5.42.(a). It is a truncated octahedron bounded by six cube faces (200) and eight octahedral faces (111), and it is from these planes that the first two strongest diffraction maxima are observed in X-ray and electron scattering events.

The considerable evidence, both computational and experimental, on which basis it is concluded that the approximately spherical Fermi surface in copper, and in silver and gold, bulges out to contact the 111 faces of the Brillouin zone, is reviewed on pages 610–4, and the suggested Fermi surface for copper is shown in Fig. M.5.42. The results of energy band calculations for copper are very sensitive to the exact potential field assumed, but indicate that the conduction electrons behave very much like free electrons and range over energy states spanning about 10^{-18} J. The calculated binding energy of copper indicates that the cohesive energy of the metal is substantially greater than can be accounted for by the valency electrons alone, and it is generally believed that the electrons in adjacent ions must be involved in bond formation between neighboring atoms, even as many as 5.56 electrons per atom, as suggested by the Pauling theory of metals (section M.2.2.).

---

[1] Fuchs, K., *Proc. Roy. Soc.*, 1935, **A151**, 585; 1936, **A153**, 622.

The chemistry of these metals, in so far as their interactions with mainly non-metallic elements is concerned, and the closer similarity between the behavior of copper and gold compounds, rather than the observation of any regular gradation of properties in the order of increasing atomic mass and nuclear charge, is largely determined by the greater stability of the $4d^{10}$ group of electrons in silver compared with the $3d^{10}$ and $5d^{10}$ groups in copper and gold. Thus, the only stable oxidation state of silver exhibited in its compounds is $+1$, while for copper the $+1$ and $+2$ states are common, and for gold $+1$ and $+3$ states. Similarly, with such physical properties as compressibility, melting point, coefficient of expansion, and cohesive energy; silver has a higher compressibility and coefficient of expansion, and a lower melting point and binding energy, than either copper or gold (see Figs. M.7.14.–15.). Since 1960, the melting points of silver and gold, 1233.95 K and 1336.15 K respectively, have been selected to define two of the six reference points on the International Practical Temperature Scale (page 15).

M.7.5. Intermetallic Phases Containing Copper, Silver, or Gold

Much important information applicable to the principles determining alloy formation has been deduced from the study of the alloys formed by copper, silver, and gold, especially the study of those intermetallic systems containing these metals and other metals which contribute more than one electron per atom towards the system. Silver and gold, which have very similar lattice constants and therefore similar metallic radii, form a continuous range of solid solutions with one another. Copper and silver have only a very restricted range of solid solution formation and only a simple eutectic is formed in this system, containing about 60 atomic percent of silver. In the copper–gold system, although the relative metallic radius ratio is almost the same as for copper and silver, there is complete miscibility in the solid state at temperatures above 1172 K, but at lower temperatures three well defined superlattices (pages 402–3) exist, with composition Cu_3Au, $CuAu$, and $CuAu_3$; the structures of these phases are illustrated in Fig. M.1.8. In intermetallic systems where the "sizes" of the constituent atoms are sufficiently similar to permit the random substitution of atoms of one element by the other in their mutual solid solution, it is observed that superlattice formation tends to increase with increasing difference in "size" and electrochemical character between the elements involved.[1] As to why superlattices are formed in the copper–gold system but only a eutectic in the copper–silver system, it has been proposed[2] that the explanation lies in the greater polarizability and

---

[1] Hume-Rothery, W., and Powell, H. M., *Z. Krist.*, 1935, **91**, 23.
[2] Hume-Rothery, W., and Raynor, G. V., *Phil. Mag.*, 1938, **26**, 129.

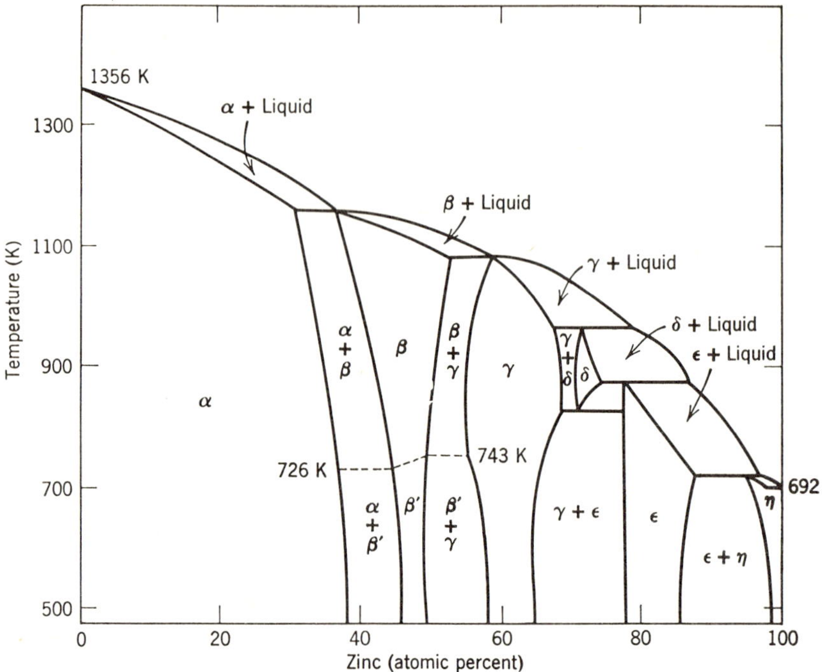

FIG. M.7.2. Equilibrium phase diagram for the copper–zinc system, showing succession of phases which result with increasing concentration of zinc.

deformability of gold compared to silver, again as a consequence of the greater stability of the $4d^{10}$-electron group in the latter.

A much-studied system from the point of view of the theories of alloy formation[1-4] is the copper–zinc one. The phase diagram for this system is shown in Fig. M.7.2.

There is a wide range of composition over which solid solution of zinc in copper takes place, but only a very narrow range of composition for solid solution of copper in zinc. The succession of phases which results with increasing concentrations of zinc being added to the face-centered cubic α-phase of copper, in the order β or β', γ, ε, and η phases, is observed in other systems also. The structurally analogous phases occur at essentially the same electron concentration in different systems. A prime indicates that the structure is

[1] Jones, H., *Proc. Roy. Soc.*, 1934, **A147**, 396; 1937, **A147**, 396.
[2] Andrews, K. W., and Hume-Rothery, W., *Proc. Roy. Soc.*, 1941, **A178**, 464.
[3] Hume-Rothery, W., Reynolds, P. W., and Raynor, G. V., *J. Inst. Metals*, 1940, **66**, 191.
[4] Cohen, M. H., and Heine, V., *Advances in Physics*, 1958, **7**, 395.

ordered, so that the β' structure in the copper–zinc system, in equilibrium with the face-centered cubic α-phase, becomes ordered below a temperature of 726 K.

The first theory of the dependence of equilibrium structures on electron concentration was proposed by Jones (reference 1 on page 702), and involves the assumption that as the electron concentration in one phase of an intermetallic system is increased, the structure of the solid phase will alter to some other structure, which can accommodate the electrons with a lower Fermi energy. The electron concentration in a system may be effectively increased, as by the solution of zinc in copper, by forming a solid solution of a metal of higher valency in a solvent species of lower valency. For example, for the two $N(E)$ curves shown in Fig. M.5.37., if any given number of electrons are included in the two different structures to which (a) and (b) refer, then the number of occupied states would be the same, indicated by the shaded areas A and B in (a) and (b) respectively. However, whereas in Fig. M.5.37.(a) the highest energy of any electron in the region of occupied states indicated is E_1, the region of occupied states in Fig. M.5.37.(b) extends to the higher energy value, E_2. Therefore, a structure with a high $N(E)$ curve may accommodate a certain number of electrons with a lower total energy than a structure with a low $N(E)$ curve; for the electron concentration indicated in Fig. M.5.37., the structure (a) would be more stable than the structure corresponding to (b). The dotted curve in Fig. M.5.37.(a) corresponds to that for the face-centered cubic structure, where the point f_1 corresponds to the state where the Fermi surface, on expansion with increasing electron concentration, first reaches the octahedral (111) faces of the zone (Fig. M.5.42.(a)), and the point f_2 corresponds to the stage where a continuously expanding Fermi surface would reach to the cube faces. The $N(E)$ curve falls off rapidly after the points f_1 and f_2, as the number of electrons per unit energy range decreases and additional electrons can only be accommodated in higher energy states. For a structure, which corresponds to an $N(E)$ curve which remains high beyond, for example, the point f_1 in Fig. M.5.37.(a), compared with another structure for which this is not so, then the latter structure would be less stable than the former, with increasing electron concentration. For any particular electron concentration, the most stable and preferred of several possible structures will be that one which enables the available electrons to be accommodated with the lowest total energy.

The interpretation of the electron concentration per atom values at which phase transitions occur in the formation of copper, silver, and gold alloys, has undergone some change since the recognition of the now well established data that the Fermi surface of copper, with one electron per atom and only a half-filled zone, is quite distorted from spherical symmetry. Many of the earlier interpretations were based on the assumption of spherical Fermi surfaces. The phase transformation in the copper–zinc system, for example, from the α face-centered cubic phase to the β body-centered cubic phase, was originally considered to be accounted for by the fact, that on the basis of a simple calculation assuming the existence of spherical Fermi surfaces in both these phase structures, the first peak on the $N(E)$ curve for the face-centered cubic structure occurred at an electron concentration of 1.36, whereas for the body-centered cubic structure, the corresponding first peak occurred at an electron to atom ratio of 1.48.‡

‡ The basis of this type of calculation is as follows: If it is assumed that the energy-wave vector relations of the free electron approximation are valid, and that the Fermi surface of occupied states is a sphere, then the first peak on the $N(E)$ curve for a face-centered cubic structure, for which the first Brillouin zone is a truncated octahedron bounded by (111) and (200) faces [Fig. M.5.41.(a)], occurs when an expanding sphere centered at the center of the zone reaches the (111) plane faces of the zone. From the free electron approximation expressions given on page 464, the momentum associated with electron states on the spherical Fermi surface is given by $[4\pi(p_{\max})^3/3] = [Nh^3/2V]$, and since for free electrons $p = h/\lambda$, then $[4\pi h^3/3\lambda^3] = [Nh^3/2V]$, or $\lambda = [8\pi V/3N]^{1/3}$. This expression relates the wavelength associated with electron states corresponding to the first peak on the $N(E)$ curve for a face-centered cubic structure with the number of electrons N in the volume V of metal. However, another expression relating λ with the number of metal atoms in the volume V may be obtained by considering that for electron states on the surface of a Brillouin zone, the associated wavelength must satisfy the Bragg condition that $n\lambda = 2d \sin \theta$ (page 593). At the point where the spherical Fermi surface first reaches to the surface of the Brillouin zone, the first order Bragg reflexion occurs for $n = 1$ and at $\theta = 90°$; also, for the set of (111) planes, referred to an orthogonal reference axial system, $d_{111} = a/\sqrt{3}$ [equation (m.5.52.)]. Therefore, from the Bragg relation, on substitution for n, d, and θ, there results, $\lambda = 2a/\sqrt{3}$, where a is the unit cell edge length of the face-centered cubic unit cell of the structure. But a face-centered cubic unit cell [Fig. M.1.3.(a)] contains four atoms per unit cell, or $N_a/V = 4/a^3$, or $a = [4V/N_a]^{1/3}$, and therefore considering a volume equivalent to one unit cell of the structure, $\lambda = 2a/\sqrt{3} = [2/\sqrt{3}][4V/N_a]^{1/3}$. These two expressions relating λ, first to V and N (number of electrons in the volume V) and second to V and N_a (number of atoms in volume V), must be equivalent to one another, and so $\lambda = [8\pi V/3N]^{1/3} = [2/\sqrt{3}][4V/N_a]^{1/3}$, or $N/N_a = \sqrt{3}\,\pi/4 = 1.36$. On the basis of the assumptions noted, this corresponds to the electron concentration per atom at which the first peak occurs on the $N(E)$ curve for a face-centered cubic structure.

For the body-centered cubic structure, containing two atoms per unit cell [Fig. M.1.6.], $N_a/N = 2/a^3$, and since the first peak on the $N(E)$ curve in this case would correspond to the assumed spherical Fermi surface reaching to the (110) faces of the rhombic dodecahedral Brillouin zone [Fig. M.6.10.], for which $d_{110} = a/\sqrt{2}$, then the corresponding Bragg condition which has to be satisfied is that $\lambda = 2a/\sqrt{2}$. Equating this expression for λ, with the one relating N and V as obtained above for the face-centered cubic structure, there is obtained,

$$\lambda = 2a/\sqrt{2} = \sqrt{2}[2V/N_a]^{1/3} = [8\pi V/3N]^{1/3}, \qquad \text{or} \qquad N/N_a = \pi\sqrt{2}/3 = 1.48.$$

The transformation from the α to the β (body-centered cubic) phase observed empirically to take place at the electron to atom ratio of 3:2 (Hume–

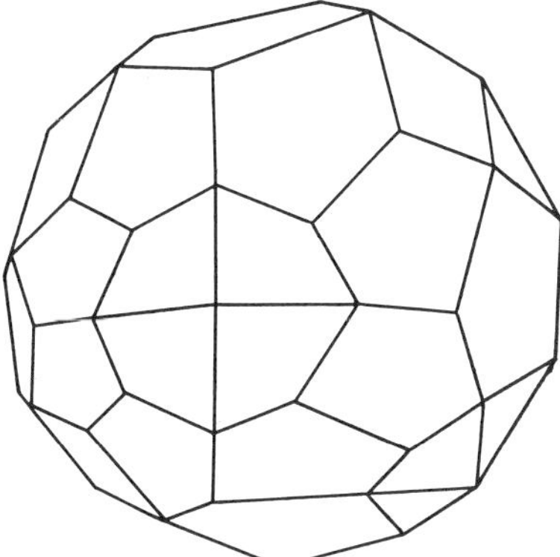

FIG. M.7.3. First Brillouin zone for the complex β-manganese structure. The zone is bounded by 48 planes of the type (221) and (310) and may contain 1.62 electrons per atom.

Rothery rules, page 402), appears on this basis to be nicely explained. However, if the surface of the occupied states in pure copper already reaches to the zone faces, in the (111) direction, at an electron concentration of 1, this agreement may be entirely coincidental. Also, there is not only one β phase structure observed, but several. There may be, for example, instead of either the body-centered cubic β or β' phases, the β-manganese structure (page 404), or close-packed hexagonal ζ or ζ' phases. The β-manganese structure is the complex cubic arrangement (page 404) stable between 1073 K and 1373 K, and obtained by quenching from this temperature range, for metallic manganese. There are 20 atoms in the cubic unit cell of edge length 6.30×10^{-10} m, distributed among two sets of equivalent points, so that each atom has an approximately icosahedral environment. From X-ray diffraction observations, the planes of largest structure factor for the β-manganese structure are the (221) and (310) planes, and the first zone for this structure is a polyhedron with 48 faces bounded by these two types of planes, as shown in Fig. M.7.3. The volume of this zone is $971/60a^3$, and since the volume per atom is $a^3/20$, the zone must contain $9.71/6 = 1.62$ states per atom. A spherical surface first reaches the zone surfaces for this structure at $9\pi/20 = 1.41$ electrons per atom concentration. The Brillouin zone for this β-manganese structure is much more nearly spherical than for the body-centered cubic phases, and since the zone can only contain 1.62 electrons per atom, compared with two for the body-centered cubic structure, the fall in the $N(E)$ would be expected to be more precipitous after the first peak than for the body-centered cubic phase. It is observed that the range of compositions for alloys which adopt the β-manganese structure seldom exceeds an electron concentration of 1.5, while for alloys which crystallize with the body-centered cubic structure the composition may frequently extend to higher electron concentrations.

For close packed hexagonal structures, the first zone may contain a number of electrons which varies with the axial ratio of the hexagonal unit cell, c/a. For ideal hexagonal closest packing, where $c/a = 1.633$, the number of electrons per atom per zone is 1.745. Since fewer than two electrons per atom may be accommodated in each zone, this means that for metals with two valency electrons per atom, in hexagonal close packing, some states in the next highest zone must be occupied by electrons, or the zones must overlap. However, because of the anisotropic nature of this structure, the energy surfaces in the two zones can overlap normal to either the Y or the X faces in Fig. M.5.32., depending on which face has the lower energy. Table M.7.3. shows the number of available states per atom for metals crystallizing with close packed hexagonal structures of varying c/a ratios, and the ratio of the free electron energy values at the points at the center of the X and Y type faces. For metals with $c/a > 1.633$, the ideal value for hexagonal closest packing, $E_Y < E_X$, while the reverse is true for metals with $c/a < 1.633$. This indicates that overlap parallel to the c hexagonal axis distorts the hexagonal closest packing to a greater extent than does overlap in the direction of the a hexagonal axis; the c/a ratio is much nearer to the ideal value in beryllium and magnesium than in zinc or cadmium. The difference in the electron distribution in k-space of the two types of hexagonal close packing with c/a greater and less than, the ideal value of 1.633 respectively, is closely related to differences in the electrical and magnetic properties exhibited by metals and intermetallic compounds crystallizing with structures conforming to these axial ratios. Close packed hexagonal structures may also exhibit lattice dimensional changes for certain electron concentration values. From the observed shape of typical $N(E)$ curves (page 606), there is an upwards inflection just prior to the Fermi surface of occupied states reaching the zone boundary face. For a Fermi surface approaching a zone face, then,

TABLE M.7.3

Metal	Number of Available States per Atom	E_Y/E_X	c/a
Beryllium	1.731	1.196	1.585
Magnesium	1.743	1.137	1.625
Zinc	1.799	0.863	1.861
Cadmium	1.805	0.859	1.890

corresponding to an increase in electron concentration, a lower energy results if the crystal expands in the particular direction involved, since the expansion will lead to a reduction of the zone size in conformity with the reciprocal relation between crystal space and reciprocal space, and this in turn will raise the $N(E)$ curve and so accommodate the electrons with a lower energy. This will be so if the crystal contracts in some other direction, so as to maintain the volume approximately constant.[1] In binary alloys crystallizing with the ζ close packed hexagonal structure, the hexagonal a axial length always increases with increasing electron concentration, while the c axial length may either increase or decrease, so that the overall effect is to decrease the value of the axial ratio c/a.

In the copper–zinc system, the β-phase, containing between 45–50 atomic percent of zinc, has the disordered body-centered cubic structure above about 727 K and the ordered cesium chloride structure below that temperature. The normal vibrational modes of both the ordered and disordered β-brass phase have been studied by the method of inelastic scattering of thermal neutrons,[2] along certain high symmetry directions in the crystal. From such studies it emerges that interatomic forces between copper and zinc atoms extends as far as the seventh nearest neighboring atoms, and that there is a large difference between the second nearest neighbor Cu–Cu and Zn–Zn forces. The frequency-wave vector relations, as indicated by the observed dispersion curves, are found to be substantially unchanged in the disordered phase, the frequencies decreasing and the energy widths increasing, generally smoothly, with increasing temperature, although certain anomalous increases in energy widths of the observed dispersion curves corresponding to specific vibrational frequencies and associated with the order-disorder transition have been noted.

Solid phases in intermetallic systems having such a range of composition as the β-brass phase are sometimes referred to as intermediate phases. At a composition of 61.5 atomic percent of zinc, the γ-brass phase appears, which is only stable over a more limited compositional range [Fig. M.7.2.], and has the complicated cubic structure referred to on page 405. The succeeding ε-phase is of the intermediate type again containing between 82–88 atomic percent zinc, and crystallizing with a hexagonal structure based on hexagonal closest packing. The terminal η-phase containing up to 2.5 atomic percent of copper, crystallizes with the hexagonal closest packed structure of zinc. The electron to atom ratio in these various phases is 3:2 for the β-phase (CuZn), 21:13 in the γ-phase (Cu$_5$Zn$_8$), and 7:4 in the ε phase (CuZn$_3$) [see Table M.1.1.].

---

[1] Goodenough, J. B., *Phys. Rev.*, 1953, **89**, 282.
[2] Gilat, G., and Dolling, G., *Phys. Rev.*, 1965, **A1053**, 138.

Hume-Rothery[1] has deduced some general principles which appear to govern the choice of the body-centered cubic β-phase, the β-manganese phase, or the close packed hexagonal ζ-phase, in intermetallic systems, although there is no strictly quantitative theory which affords an explanation as to why one structure rather than the other is preferred for any particular system. Thus, from the observation that the copper–zinc system has the β-phase only at an electron concentration of 3:2, the copper–gallium system the β-phase stable only at high temperatures, and the copper–germanium system only the ζ-phase, he concludes that increasing valency of the solute species leads to increasing stability of the ζ- or β-manganese structures, at the expense of the β-phase; this conclusion is apparently justified from the study of other systems, apart from the three mentioned above. The greater the difference in electrochemical character between the metals in an intermetallic system, the more probable is the occurrence of ordered structures, since this effect tends to the formation of oppositely charged species which minimize the consequent attractive-repulsive forces between them by adopting an ordered arrangement. For example, β-brass adopts an ordered structure only at lower temperatures, while in the silver–magnesium and gold–magnesium systems where the electrochemical difference between the constituent metals is greater than for the copper–zinc system, no disordered phase is observed but the body-centered cubic β-phase is stable up to the melting point. A ternary intermetallic phase which is completely ordered is found in the magnesium–silicon–copper system, although copper and silicon are known to form binary phases with disordered crystal structures, as for example, Cu_5Si. The intermetallic compound $Mg_6Si_7Cu_{16}$, whose crystal structure has been reported,[2] is completely ordered with 116 atoms (four formula units) per face-centered cubic unit cell. Efficient packing of the constituent atoms is achieved in this arrangement by the magnesium atoms being packed in octahedral groups at the vertices of the cubic unit cell, each magnesium atom being coordinated with four silicon atoms, four other magnesium atoms, and eight copper atoms, at the vertices of the polyhedron shown in Fig. M.1.13.(d). The silicon atoms, in two different crystallographic sets of positions, are coordinated with eight copper atoms located at the vertices of a cube in one set, and with four magnesium atoms, and eight copper atoms, located at the vertices of a slightly distorted icosahedron in the other set. The copper atoms, likewise, are in two different crystallographic sets, with one set having six copper atoms, three magnesium atoms, and three silicon atoms, located at the vertices of a slightly distorted icosahedron, and the other set with co-

---

[1] Hume-Rothery, W., Reynolds, P. W., and Raynor, G. V., *J. Inst. Metals*, 1940, **66**, 191.
[2] Bergman, G., and Waugh, J. L. T., *Acta Cryst.*, 1953, **6**, 93; 1956, **9**, 214.

ordination number 13 composed of three magnesium atoms, four silicon atoms, and six copper atoms located at the vertices of a polyhedron related to the icosahedron. The structure reported for Th_6Mn_{23} is a special case of the more general structure $A_6B_7C_{16}$, as found for $Mg_6Si_7Cu_{16}$, with the thorium atoms in the positions of the magnesium atoms and the manganese atoms substituted for the copper and silicon atoms[1] (see section M.7.10.).

The increased stability of the β-phase at higher temperatures compared with the β-manganese or ζ-phase, originally observed by Hume-Rothery, was later attributed by Zener[2] to the larger entropy associated with the elastic constants[3] of the body-centered cubic structure, which gives rise to a decrease in free energy with increasing temperature $[G = H - TS]$. The fact that the body-centered cubic phase may accommodate larger atoms than the close packed ζ-phase, affords some explanation of the greater stability observed for the β-phase with increasing differences in metallic radii of the constituent metal atoms in intermetallic systems. It is generally observed that as the difference between the atomic radii of the two constituents of binary alloy systems increases, the composition limits over which the β-phases exist is moved in the direction of lower electron concentration.

The equilibrium phase diagram for the copper–zinc system (Fig. M.7.2.) displays an effect which is characteristic of this type of alloy system, involving the equilibrium between the face-centered cubic α-phase and the body-centered cubic β-phase. Generally, the solubility of one metal in another is determined by three factors: size, electrochemical differences and the number of electrons per atom, and mutual solubility decreases with decreasing temperatures. However, in the copper–zinc system, and in many other binary systems composed of copper or silver and other metals, where the relative atomic radii are not too disproportionate, the solubility curves delineating the boundary between the α-regions and the $(\alpha + \beta)$-regions and the $(\alpha + \beta)$-regions and β-regions of the corresponding phase diagrams, exhibit the characteristic that the solubility of the metal concerned in the α-phase increases as the temperature decreases. The negative slope of these phase boundary solubility curves is apparent in Fig. M.7.2., and Hume-Rothery has shown that if the corresponding solubility curves for the metal pairs Cu–Al, Cu–Ga, Cu–Si, Ag–Cd, and Ag–Al, are plotted as a function of electron concentration rather than atomic percent composition, as shown in Fig. M.7.4., they almost coincide with one another. For the Cu–In and Cu–Sn systems, the slope of the corresponding solubility curve is practically the same as the others, but displaced towards the direction of lower electron concentration, corresponding to the greater stability of the β-phase compared with the close packed

---

[1] Florio, J. V., Rundle, R. E., and Snow, A. I., *Acta Cryst.*, 1952, **5**, 449.
[2] Zener, C., *Elasticity and Anelasticity of Metals*, 1948, Chicago University Press.
[3] Huntington, H. B., *The Elastic Constants of Crystals*, 1958, Academic Press, New York.

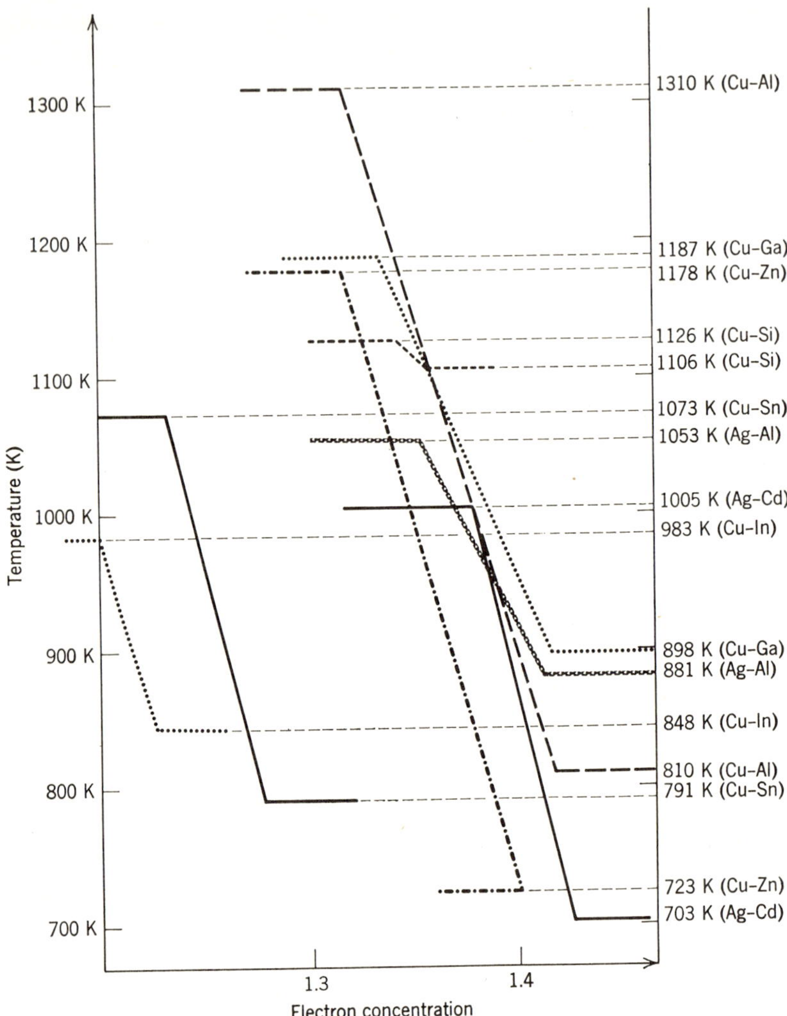

FIG. M.7.4. Solubility of various metals in copper and silver as a function of temperature (K). The temperature range between the horizontal lines for each system shows the temperature limits over which the system can be studied. The abscissa corresponds to electron concentration per atom (after Hume-Rothery).

face-centered cubic α-phase. The transition from the α-phase to the β-phase at an approximately constant electron concentration of 1.4 appears to have a relatively simple explanation in terms of the assumption of free electrons and spherical Fermi surfaces, as indicated on page 703. The occurrence of the first maxima in the $N(E)$ curves for the α-phases and β-phases at electron concentrations of 1.36 and 1.48, respectively, in terms of the assumptions noted, means that there is a region between the two maxima where with increasing electron concentration, the $N(E)$ curve for the α-phase is falling off while that for the β-phase is still increasing, so that in this region the β-phase becomes more stable than the α-phase, at an electron concentration of about 1.4, between 1.36 and 1.48. Hume-Rothery[1] had actually observed much earlier that, where there was no great disparity between the corresponding metallic radii, the maximum solubilities of other metals in copper and silver corresponded to this electron concentration of 1.4. Thus, for metals with two, three, four, and five electrons per atom, respectively, the expected solubilities would be 40%, 20%, 13.3%, and 10%. The observed values are about 40% for the Cu–Zn and Ag–Cd systems, 20% for the Cu–Ga and Cu–Al systems, 12% for the Cu–Ge and Ag–Sn systems, and 7% for the Ag–Sb system.

The equilibrium between the α-phases and β-phases in brass and related intermetallic systems is determined by electron concentration effects, but with increasing differences in radii of the two metals involved in these binary systems, crystal lattice distortions apparently give rise to energy effects which complicate the question of the free energy of formation of the phases. Ideas based on the relative Fermi energies of the two phases have not led to any understanding of the slope of the solubility curve bounding the α-regions and $(\alpha + \beta)$-regions in the phase diagrams of such alloys. The apparent success of the simple Jones theory, based on spherical Fermi surfaces, in accounting for the observed phase change transition at about an electron concentration of 1.4 electrons per atom, is however at variance with the well-established information that the Fermi surface in copper extends to the zone boundaries in the (111) direction at an electron concentration of one electron per atom, so that the first maximum in the $N(E)$ curve for the copper α-phase should actually occur at an electron concentration of less than 1.0, and not at 1.36, as indicated by the simple theory. There are difficulties and objections associated with the attempted explanation of these observations on the basis that the Fermi surface in copper, while initially quite distorted in the pure metal, becomes more nearly spherical on alloying with zinc, for example, and then, with increasing electron concentration, expands as an approximate sphere to reach the zone boundary at the concentration required by the simple theory. For assumed spherical Fermi surfaces, the first maximum in the $N(E)$ curve for the α-phase occurs at an electron concentration of 1.36, as the Fermi surface reaches the (111) faces of the zone, and the second maximum at 1.88 electrons per atom, as the Fermi surface reaches to the (200) zone boundary

---

[1] Hume-Rothery, W., *Phil. Trans. Roy. Soc.*, 1934, **A233**, 1.

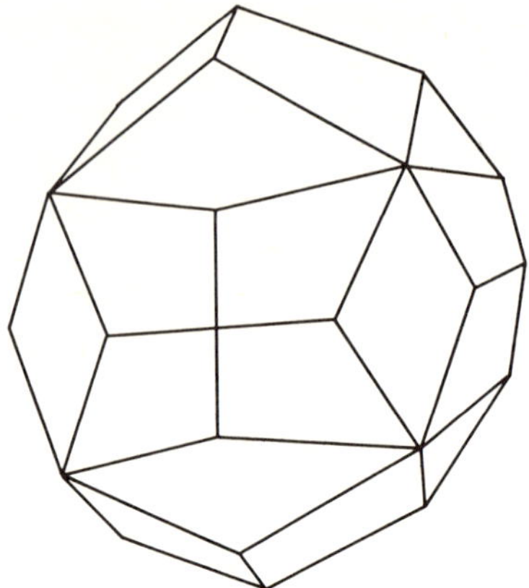

FIG. M.7.5.   First zone for the γ brass structure. The zone is bounded by the planes (330)
(411) and has 36 faces.

faces. Hume-Rothery[1] suggests that if distortion of the Fermi surface from
spherical symmetry in copper reduces the electron concentration at which
the first maximum in the $N(E)$ curve occurs to the value of less than 1.0, then
if the Fermi surface remained distorted on alloy formation, the second maximum could be reduced to the point corresponding to the electron concentration of about 1.4, so that it would become important to regard this second
maximum as being responsible for the α-phase solubility limit.

The γ-brass structure occurs at an electron concentration of 21/13 in the
copper–zinc system. This complex cubic structure, first reported for Cu_5Zn_8,
has 52 atoms per unit cell (see page 405). The planes designated by small
integral values for their indices and observed from X-ray diffraction examination to correspond to large structure factor values are (330) and (411); the
zone described by these planes is shown in Fig. M.7.5., and is bounded by 36
faces. The completely filled zone contains 90 electrons per unit cell and therefore corresponds to an electron concentration of $90/52 = 1.731$ electrons per
atom. The fact that this phase invariably occurs when the electron to atom
ratio is $21/13 = 1.615$, means that the zone is almost completely full. A
centered spherical surface with increasing radius would first reach the zone
faces at an electron concentration of $9\pi/13\sqrt{2} = 1.54$, corresponding to
88.4% of the total possible for the cell. The zone is thus very approximately
spherical. The γ-phase extends to higher electron concentrations in the
Cu–Ga and Cu–Al systems than in brass, but there is no γ-phase observed in
the Ag–Al system, where the metallic radii are similar. It appears that no

---

[1] Hume-Rothery, W., and Roaf, D. J., *Phil. Mag.*, 1961, **6**, 55.

γ-phases exist in systems where the metallic radii are either similar in magnitude or differ from one another by about 20%. These observations can be understood on the basis that the γ-structure is a defect body-centered structure; for atoms of the same size, the normal body-centered cubic structure would apply, but for dissimilar atoms differing slightly in radii, the γ-structure derived from the body-centered cubic arrangement by the elimination of atoms at certain lattice sites and a slight rearrangement of the remainder, is apparently more stable. Examination of these and other defect structures indicates that the number of electrons per unit cell, even though certain lattice sites in the structure are void, is the important criterion in Brillouin zone theory. Increasing differences in metallic radii, for binary systems containing copper, silver, and gold, change the composition limits in the direction of lower electron concentration, and increasing electrochemical differences move the composition limits towards higher electron concentrations.

M.7.6. Ordered and Disordered Structures

Any discussion of the nature of the most stable state formed as a result of the interaction of two pure phases X and Y, whether they are mutually soluble in one another and to what extent, whether they form a solid solution of variable composition or a compound of stoichiometric composition, or an alloy with a range of composition, has to be concerned with thermodynamic considerations.‡

‡ The great accumulation of experimental data which exists describing and relating the various transformations which are possible in solid systems, in the absence of accurate theories describing them, is most conveniently interpreted in terms of thermodynamics, or more exactly, in terms of statistical thermodynamics. The exact treatment of most problems involving metals and solids, and for that matter all of the problems of physics and chemistry, require the application of quantum mechanics, since the various energies that atoms and molecules may exhibit are quantized. It is apparent from the discussion in this and the previous chapters that the quantum mechanical calculation of such a relatively simple problem as that of finding the cohesive energy of sodium, presents considerable difficulty and can only be handled approximately. It is not surprising, therefore, that most of the problems involving the transformations which may occur in metal systems, such as structural transformations (polymorphism), order-disorder transformations, transformations of bond type, deformation transformations, and all of the other observed phase transformations, are not accessible to accurate quantum mechanical calculation.

There is, however, much accumulated information contained in what are referred to as phase diagrams or equilibrium diagrams, which describe transformations which actually take place. Such phase diagrams show the relation between the different phases, the structure of which may be deduced from X-ray crystallographic examination, that exist in a system, and in particular they usually show the most stable phases that occur under equilibrium conditions. For a two-component system, where the components are the minimum number of chemical entities required to completely specify the composition of the system, a two-coordinate diagram is sufficient to indicate the variation of composition with one thermodynamic variable, usually temperature. A three component system requires a three-dimensional diagram to represent the variation of composition with, for example, temperature. Such diagrams, however, are difficult to draw and even more difficult to interpret, so that it is more customary, for three component systems, to represent the phase relationships on a number of isothermal sections corresponding to a series of different temperatures.

The phase diagrams of even binary metal systems may be of all degrees of complexity, ranging from simple solid solution formation to dozens of phases being present and several intermetallic compounds appearing, with structures of great complexity. Thus, a continuous range of solid solutions exist in the potassium–rubidium system, a simple eutectic in the tin–lead system, the single intermetallic compound $PbMg_2$ in the lead–magnesium system, four different intermetallic compounds exist in the sodium–lead system with compositions Na_4Pb, Na_2Pb, $NaPb$, and Na_2Pb_5. The structures found for intermetallic phases, likewise range from the simple rock salt type found for BaTe and the zinc blende structure found for GaAs to the complex structures found for $Mg_{32}(Al, Zn)_{49}$ in the Mg–Zn–Al system,[1] for the intermetallic compound Cu_4Cd_3 in the Cu–Cd system,[2] and the M-phase in the Nb–Ni–Al system.[3]

The change in free energy when two substances react with one another, or when two metals crystallizing with the same crystal structure substitute for one another in the crystal lattice of the other, is in terms of the usual thermodynamical convention, given by the Gibbs free energy function, $G = H - TS = U - TS + PV$, where U represents the change in internal energy of the system under consideration, H is the enthalpy change, PV represents the work done on or by the surroundings, and S is the entropy change. Changes in PV are generally quite small for reactions among condensed phases, so that it is normally quite sufficient to consider changes only in the Helmholtz function, $A = U - TS$.

If two pure phases X and Y do not mix with one another, that is if atoms of X (or Y) do not dissolve in Y (or X), then for any composition defined by the mole fraction x giving X per gram mole of $X_{1-x}Y_x$, then the free energy of the two pure phases X and Y is simply a linear function of the mole fraction x. On the other hand if atoms of X may be randomly distributed over the lattice sites of the crystal lattice of Y, or the other way around, so that mixing of X and Y takes place, then the entropy change due to this mixing may be calculated from the expression

$$S_{\text{mix}} = -R[x \ln x + (1 - x) \ln (1 - x)] \text{ J K}^{-1} \text{ mole}^{-1},$$

(see pages 34–36, Boltzmann law),

---

[1] Bergman, G., Waugh, J. L. T., and Pauling, L., *Acta Cryst.*, 1957, **10**, 254.

[2] Samson, S., *Acta Cryst.*, 1965, **19**, 401.

[3] Shoemaker, C. B., and Shoemaker, D. P., *Acta Cryst.*, 1967, **23**, 331.

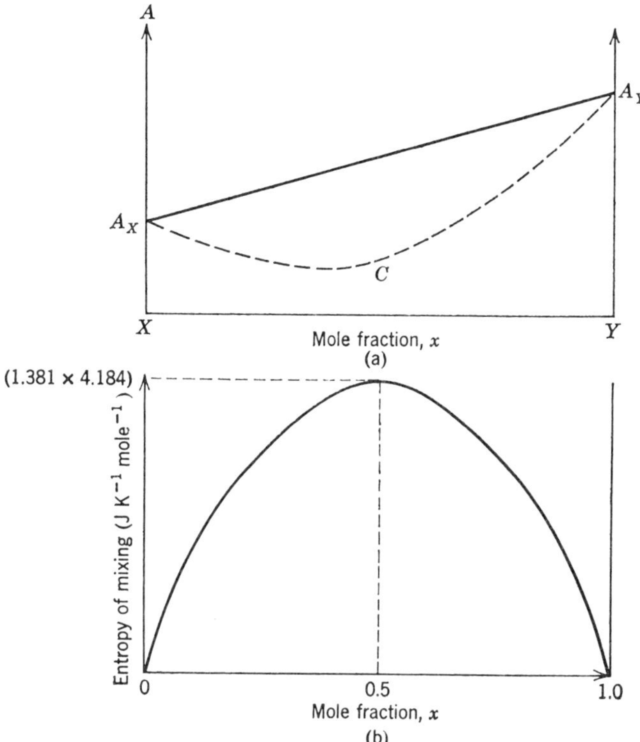

FIG. M.7.6. The Helmholtz free energy and entropy of mixing of two species X and Y as a function of mole fraction x. (a) The line $A_X A_Y$ shows the Helmholtz free energy A as a function of mole fraction x, where no mixing takes place; the curve $A_X C A_Y$ shows the relation between A and x, where there is no energy of mixing U. In (b) the entropy of mixing is a maximum for the composition $X_{0.5} Y_{0.5}$.

where x now represents the atomic concentration of Y in a lattice of Avogadro's number of atoms. These relations are shown in Fig. M.7.6., from which it may be seen that the entropy of mixing increases quite sharply at $x = 0$ and at $x = 1$. This rapid increase in the entropy when a few atoms of Y (or X) are introduced into the crystal lattice of X (or Y), arises because of the large number of ways in which an atom of one species may be arranged among the lattice sites of another species. The entropy of mixing for the composition $X_{0.5} Y_{0.5}$ is 1.381×4.184 J K^{-1} mole^{-1}, which is equivalent to a factor TS or free energy contribution of $(1.8 \times 10^3 \times 4.184)$ J mole^{-1} at 1273 K and 1.67×10^3 J mole^{-1} at normal temperatures. This is in agreement with the common observation that the tendency to mix of two components increases with temperature. Where there is no energy change on mixing, the free energy of the mixed phases will always be lower than that represented by the line $A_X A_Y$ in Fig. M.7.6.(a), so that the mixed phase is always more stable than the mixture of separate X- and separate Y-phases, for all compositions, as indicated by the curve in Fig. M.7.6.(a).

The energy changes U which occur on mixing two species together, depend on whether the bond strength of X—Y is greater or less than that of X—X or Y—Y. If, for example, the attraction X—Y is greater than the mean value of that of X—X and Y—Y, over all possible ranges of composition, then the entropy and the energy effects enhance one another to give rise to the formation of a solution in preference to either X or Y alone; this situation is illustrated in Fig. M.7.7. If any line is drawn in this figure connecting the free energy curve at either pure X or pure Y with any point on the $A - x$ curve and this line intersects the curve at $x = x_1$, then this composition may be regarded as being made up from any combination of $X_{1-x}Y_x$ and Y, for $x < x_1$ and of $X_{1-x}Y_x$ and X for $x > x_1$. Every such combination or any similar combination of $X_{1-x}Y_x$ (for $x < x_1$) with $X_{1-x}Y_x$ (for $x > x_1$) is less stable than the composition x_1 itself, so that the single phase at x_1 is at equilibrium. In fact, the general characteristic of free energy (A) − mole fraction (x) curves which give rise to only single phases is that the free energy at composition x, A_x, is always less than $(1 - x)A_X + xA_Y$, and the tangents to this curve do not intersect it again, as in Fig. M.7.6. Alternatively expressed, d^2A/dx^2 is always positive. In the simple case illustrated in Fig. M.7.6., where $U_{XY} = (1/2)[U_{XX} + U_{YY}]$, the entropy is the predominant factor and d^2A/dx^2 is always positive and no phase separation takes place. If the X—Y attractions are less than the mean of X—Y and Y—Y attractions, or

$$U_{XY} > (1/2)[U_{XX} + Y_{YY}],$$

then the energy of mixing will be proportional to the number of X—Y interactions which are substituted for X—X and Y—Y interactions, again assuming that there is no change in structure from the random arrangement; in this event, the energy of mixing will depend on $x(1 - x)$. The energy will increase linearly with x for small values of x, or with $(1 - x)$ for large values of x, so that dU/dx remains constant. This is equivalent to the situation where, for example, X atoms are introduced into a Y lattice under circumstances where it is very improbable that they will have similar X atoms for nearest neighbors; the energy increase of the entropy under the same circumstances will, on the other hand, be a logarithmic one. The net effect of the energy and entropy effects, in this event, acting in opposition to one another, is such that the entropy term is of predominant importance at the extreme ends of the composition range, and the energy term of predominant importance in the

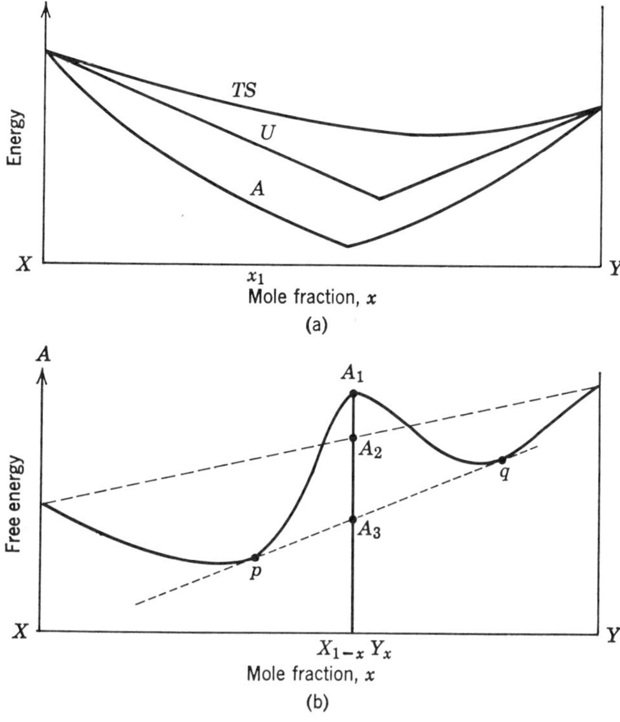

FIG. M.7.7. Illustration of the dependence of the Helmholtz free energy, A, as a function of the composition of a mixture of species X and Y, expressed in terms of the mole fraction, x, where the energy U is linearly dependent on x. In (a), the entropy of mixing S and the energy U enhance one another, and in (b) they oppose one another.

mid-composition range, as illustrated in Fig. M.7.7.(b). For a composition $X_{1-x}Y_x$, assumed to have a free energy of random solution given by A_1, which may or may not be lower than the free energy of the mixture of the pure phases X and Y, given by A_2, dependent on the magnitude of the interactions X—Y, X—X, and Y—Y, the minimum free energy A_3 may be obtained by a mixture of an X-rich phase, p, and a Y-rich phase, q, where the line joining p and q is twice tangential to the curve (Fig. M.7.7.(b)). The phase diagram for such a system would be given by the one like Fig. M.1.10. for the copper–silver system. Similar considerations apply to the case where atoms X are inserted into vacancies in lattice of Y, so that these general considerations of phase equilibra are applicable to all structures, such as those referred to in pages 386–395. If atoms of X differ in size from those of

Y by about 15%, then the strain energy introduced in the crystal lattice of Y by about a 1% molar solubility of X in Y is about equivalent to the entropy term at 1000 K, so that the minimum free energy occurs about this composition, limiting the solubility of X in Y to about 1%. This small primary solubility of one metal in another, when the atomic radii differ from one another by more than about 15%, was discovered empirically by Hume-Rothery as long ago as 1934.

Where two or more phases, a phase being defined as that part of a system which is microscopically and macroscopically homogeneous and distinguished from its environment by distinct physical boundaries, coexist in a system, their equilibrium relationships are conveniently described in terms of the Phase Rule. It was discovered by J. W. Gibbs (1876) that the number of components C and phases P present in a system is related to the number of degrees of freedom F by the rule that $F + P = C + 2$, where the degrees of freedom F represent all of those conditions influencing the system that are independently variable, especially the temperature, pressure, and composition. If $F = 0$, the phase concerned is stable only for a fixed temperature, pressure, and composition, with no degrees of freedom. If $F = 1$, either one of these parameters may be varied without affecting the stability of the phase, and if $F = 2$, two of these parameters may be arbitrarily specified, and so on. Where the pressure is assumed to be constant, as for most experimental work on equilibrium systems, one of the degrees of freedom is thus fixed, so that there remain $F' = F - 1 = C - P + 1$ degrees of freedom. It is interesting to note the similarity between the phase rule expression, $P + F = C + 2$, and the relationship discovered much earlier by Euler between the number of edges, faces, and vertices, of regular and semi-regular polyhedra, $v + f = e + 2$ (page 579). As a simple application of the phase rule, consider the point P in Fig. M.1.10., where three phases are in equilibrium, namely a liquid phase saturated with I and II type atoms, and the two solid phases I and II, which may have the same or different crystal structures; here, $F' = C - P + 1 = 2 - 3 + 1 = 0$. Thus, at this point solidification can only occur at one temperature, and in fact at the eutectic temperature in a system. This is the lowest temperature at which an intermetallic phase can exist in the liquid state. A similar situation, that $F' = 0$, may arise in another way in intermetallic systems, for example, as at the point P' in Fig. M.1.10. Here, phase I reacts with the liquid phase to form crystals of phase II, the three phases being in equilibrium at the temperature $T_{P'}$, referred to as the peritectic temperature. If, in some particular phase diagram, all three of the phases which are in equilibrium with one another are solid, the temperature at which they coexist is referred to as the eutectoid temperature and the corresponding composition as a eutectoid.

It is possible for several different reactions to occur in any one system. An especially important type is where a phase of some intermediate composition, such as XY is so stable that a compound is formed; an intermediate phase, generally, is any new phase whose composition and stability range are limited to intermediate regions of the corresponding phase diagram. When such an intermediate phase has an ordered structure, it usually differs from ordinary compounds, in that its structure may be stable over a range of compositions, whereas most chemical compounds have to be nearly stoichiometric to exist at all. Compound formation will occur when there is a strong chemical bonding between X and Y, which corresponds to $U_{XY} < (1/2)[U_{XX} + U_{YY}]$, and compounds of limited composition ranges will occur when this difference is dependent on composition. Phase relationships in solids are related to the function d^2A/dx^2, which is not necessarily dependent on the magnitude of A. In fact, the distinction between solid solutions, non-stoichiometric phases stable over wide ranges of composition with disordered or ordered structures, and conventional compounds stable over narrow limits of composition and crystallizing with definite structures and corresponding to stoichiometric formulae, depends on changes of A with composition. The three types of behavior corresponding to the free energy of formation varying with the composition in such a way as to give rise to a very narrow, a limited, and a wide region of miscibility of the two components X and Y, are represented in Fig. M.7.8.(a) and the corresponding phase diagram in Fig. M.7.8.(b). The range of composition of the intermediate phase $X_p Y_q$ may be shown by constructing the tangents to the free energy-composition $(A - x)$ curves. The composition range is determined by both the value of d^2A/dx^2 in the vicinity of the composition $X_p Y_q$ and by the relative stability of the X-rich and Y-rich phases with which it is in equilibrium; the latter depends on the magnitude by which $U_{X_p Y_q}$ is less than $(1/2)(pU_{XX} + qU_{YY})$. This is necessarily so since the equilibrium combination of phases for any composition x must be given by a point on a line which is tangential to two points on the free energy-composition curve. Also d^2A/dx^2 can only be very large if X—Y not only has a larger binding energy than the mean of X—X and Y—Y but also is such that the binding energy of X—Y increases as X and Y interact. For such cooperative type interactions to take place, the species XY need to adopt an ordered arrangement in some structure, where this ordered structure is only stabilized by U. The changes which take place successively as Y is added to X at some particular temperature may be observed, as in Fig. M.7.8.(b), by beginning at the point n, corresponding to pure phase X, and tracking across the ordinate for constant temperature. The X-phase remains homogeneous as Y-atoms are included in its structure until composition o is reached, when the solid phase γ begins to appear; at the point p, the α-phase is no longer present and a homogeneous phase exists over the range of composition from p to q. From point q to point r, the phases γ and β coexist

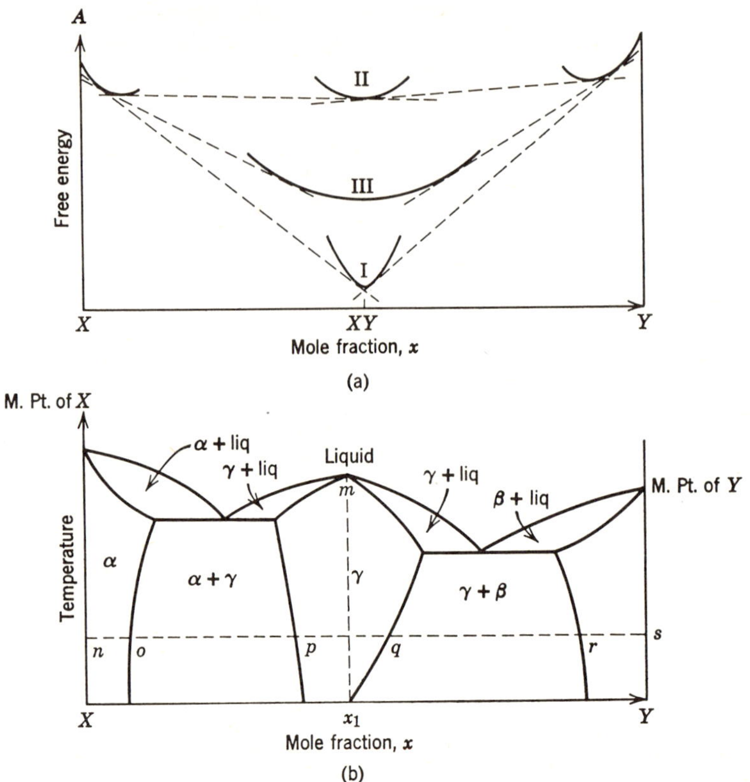

FIG. M.7.8.   Representation of (a) the manner in which the free energy may vary with the composition of mixtures of X and Y so as to give rise to very small (I), small (II), and wide regions of miscibility; and (b) a possible phase diagram corresponding to (a).

in equilibrium, but after point r the γ-phase is no longer present, and finally at the point s, the β phase corresponds to pure Y.

An important feature of phase diagrams depicting the formation of intermetallic compounds is that the liquidus and solidus lines come to a common maximum at the point m corresponding to the composition x_1. The liquidus curve in a phase diagram is such that it denotes a phase boundary, every point on which gives the temperature that solidification takes place for the composition indicated by the abscissa for that point; only the liquid phase exists for all temperatures greater than that corresponding to any point on the liquidus curve. Similarly, for temperatures lower than those corresponding to all points on the solidus curve in a phase diagram, no liquid phase can exist but only solids. Actual phase diagrams, even although quite complicated in appearance may be sub-divided usually into parts describing only one type of reaction, as in Fig. M.7.8.(b) which may be divided by the vertical line mx_1 into two eutectic diagrams.[1–3]

III. THE ALKALINE EARTH METALS AND ZINC, CADMIUM, AND MERCURY

M.7.7. Structure and Brillouin Zones

Neutral atoms of the alkaline earth metals, beryllium, magnesium, calcium, strontium, barium, and radium, and the metals zinc, cadmium, and mercury, have the common characteristic that they have two s-state electrons beyond a closed group of electrons, with the following electronic configurations in the ground state:

Be $1s^2 2s^2$

Mg $1s^2 2s^2 2p^6 3s^2$

Ca $1s^2 2s^2 2p^6 3s^2 3p^6 4s^2$

Sr $1s^2 2s^2 2p^6 3s^2 3p^6 3d^{10} 4s^2 4p^6 5s^2$

Ba $1s^2 2s^2 2p^6 3s^2 3p^6 3d^{10} 4s^2 4p^6 4d^{10} 5s^2 5p^6 6s^2$

Ra $1s^2 2s^2 2p^6 3s^2 3p^6 3d^{10} 4s^2 4p^6 4d^{10} 5s^2 5p^6 4f^{14} 5d^{10} 6s^2 6p^6 7s^2$

Zn $1s^2 2s^2 2p^6 3s^2 3p^6 3d^{10} 4s^2$

Cd $1s^2 2s^2 2p^6 3s^2 3p^6 3d^{10} 4s^2 4p^6 4d^{10} 5s^2$

Hg $1s^2 2s^2 2p^6 3s^2 3p^6 3d^{10} 4s^2 4p^6 4d^{10} 5s^2 5p^6 4f^{14} 5d^{10} 6s^2$

The crystal structure adopted by these metals and their lattice constants are shown in Table M.7.4. Thus mercury is the only one of these metals which does not crystallize with one or another of the simple face-centered cubic, body-centered cubic or hexagonal close packed structures. Calcium and strontium are trimorphic. The α-form of mercury, with one atom per rhombohedral primitive unit cell, if transformed to the corresponding hexagonal system, would contain three atoms per hexagonal unit cell with $a_0' = 3.457 \times 10^{-10}$ m and $c_0' = 6.663 \times 10^{-10}$ m at 5 K and $a_0' = 3.464 \times 10^{-10}$ m and $c_0' = 6.667 \times 10^{-10}$ m at 78 K. A unit cell of this shape is equivalent to a cubic close packed structure deformed by compression along a body diagonal. For most metals crystallizing with hexagonal close packed structures, the axial ratio is slightly

---

[1] Findlay, A., Campbell, A. N., and Smith, N. O., *The Phase Rule and its Applications*, 1951, 9th Edition, Dover Publications, New York.

[2] Smoluchowski, R., Mayer, J. E., and Weyl, W. A., *Phase Transformations in Solids*, 1951, John Wiley & Sons, Inc., New York.

[3] Cottrell, A. H., *Theoretical Structural Metallurgy*, 1957, St. Martin's Press, Inc., New York.

TABLE M.7.4

| Metal | Structure | Lattice Constants ($\times 10^{-10}$ m) | | Axial Ratio, c/a |
		a_0	c_0	
Beryllium, Be	h.c.p.	2.2866 (298 K)	3.5833 (298 K)	1.5671
Magnesium, Mg	h.c.p.	3.20927 (298 K)	5.21033 (298 K)	1.6235
Calcium, α-Ca	c.c.p.	5.576		
β-Ca	h.c.p.	3.98 (723 K)	6.52 (723 K)	1.64
γ-Ca	b.c.c.	4.38 (773 K)		
Strontium, α-Sr	c.c.p.	6.0847 (298 K)		
β-Sr	h.c.p.	4.32 (521 K)	7.06 (521 K)	1.63
γ-Sr	b.c.c.	4.85 (887 K)		
Barium, Ba	b.c.c.	5.025 (299 K)		
Radium, Ra	no structure reported	5.000 (5 K)		
Zinc, Zn	h.c.p.	2.6648 (298 K)	4.9467 (298 K)	1.8563
Cadmium, Cd	h.c.p.	2.97887 (299 K)	5.61765 (299 K)	1.8858
Mercury, α-Hg	rhombo-	2.9863 (5 K)	$\alpha = 70°44.6'$ (5 K)	
β-Hg	hedral	2.9925 (78 K)	$\alpha = 70°44.6'$ (78 K)	

less than the ideal value of 1.633, usually falling between 1.63 and 1.57. This leads to atoms in the same basal plane being a little further apart than those immediately above and below. The only known metal crystallizing with a hexagonal close packed structure and with an axial ratio less than 1.57 is beryllium, with $c/a = 1.5671$. The axial ratios increase with temperature, thus tending to equalize the interatomic distances. Zinc and cadmium, with axial ratios greater than the ideal value, are exceptional, and the principal interest in these metals from the point of view of the zone theories of metals, is associated with their abnormal axial ratios. With axial ratios greater than the ideal value, the atomic separation in the basal planes is about 10% less than the distance between planes; again, their axial ratios (zinc and cadmium) increase with temperature, but this effect exaggerates even more the deviation from close packing of spherical atoms. The compressive distortion exhibited by the α-mercury structure, on the other hand, brings atoms in neighboring layers especially close together; the separation between mercury atoms in a rhombohedral (111) plane is 3.47×10^{-10} m, whereas between the planes of close packed atoms it is only 3.00×10^{-10} m, there being six neighbors at each of these distances. The energy associated with the two types of close packed structures is not substantially different, and there is little tendency to change from one to the other. The hexagonal modification of β-calcium is only obtained above 823 K if the metal is especially pure. The β-modification of mercury, crystallizing with a tetragonal body-centered structure at temperatures below 79 K contains two atoms per tetragonal unit cell, and like the α modification may be considered as a distorted cubic close packing; the distortion in this case corresponds to compression along a cube axis rather than along a three-fold axis, leading to the creation of strings of atoms along the c_0 axis at 2.825×10^{-10} m apart and a next longer interatomic distance at 3.16×10^{-10} m, applying to eight neighboring atoms.

Radium, with atoms whose ground-state electronic configuration contain two s-state electrons beyond the closed group as in the radon atom, is also a congener of the alkaline earth metals, but of little interest from the point of view of the properties and theories of metals; a structure does not seem to have been reported for radium.

All of these metals are electrical conductors, so that those crystallizing with cubic structures must possess overlapping Brillouin zones, since for all simple translational lattices each Brillouin zone contains $2N$ electron states, where there are N atoms in the crystal. With two valency electrons per atom, these metals would be electrical insulators if the zones were completely filled and no zone overlap occurred.

The Brillouin zones of the two close packed structures and of the body-centered cubic structure have been described in section M.5.8. and earlier in this chapter. For the hexagonal close packed structure, special interest is associated with the Jones zone of Fig. M.5.33. As a consequence of the reciprocal relationship between real crystal space and the reciprocal space in terms of which the Brillouin and Jones zones are described, the ratio of the height of the zone to the length of the side of the base varies with the axial ratio of the crystal concerned. Thus, for zinc and cadmium, where c/a is approximately 1.9, corresponding to the crystal being extended in the c axial direction compared with ideal hexagonal closest packing, the corresponding Brillouin zone is short and compressed compared with the zones for beryllium and magnesium, for which c/a is approximately 1.6. The energy of an electron in a-state at the point Y, in Fig. M.5.32., will in general be different from that of an electron at the point X. For zinc and cadmium, with short compressed zones, E_Y will usually be less than E_X, but for beryllium and magnesium the opposite will apply with $E_X < E_Y$. Since the complete zone may contain only two electrons, and all four of these metals are electrical conductors, there must be an overlap into the second zone. However, different types of overlap, in different directions are possible, and in principle these may be of three types. Firstly, overlap from the incomplete first zone across the faces X (of Fig. M.5.32.) into the small truncated prisms of Fig. M.5.33., conveniently referred to as X overlap. Secondly, overlap may occur from the complete first zone into a second outer zone in either one of two ways; it may be that overlap occurs either at Y (in Fig. M.5.32.) or at the re-

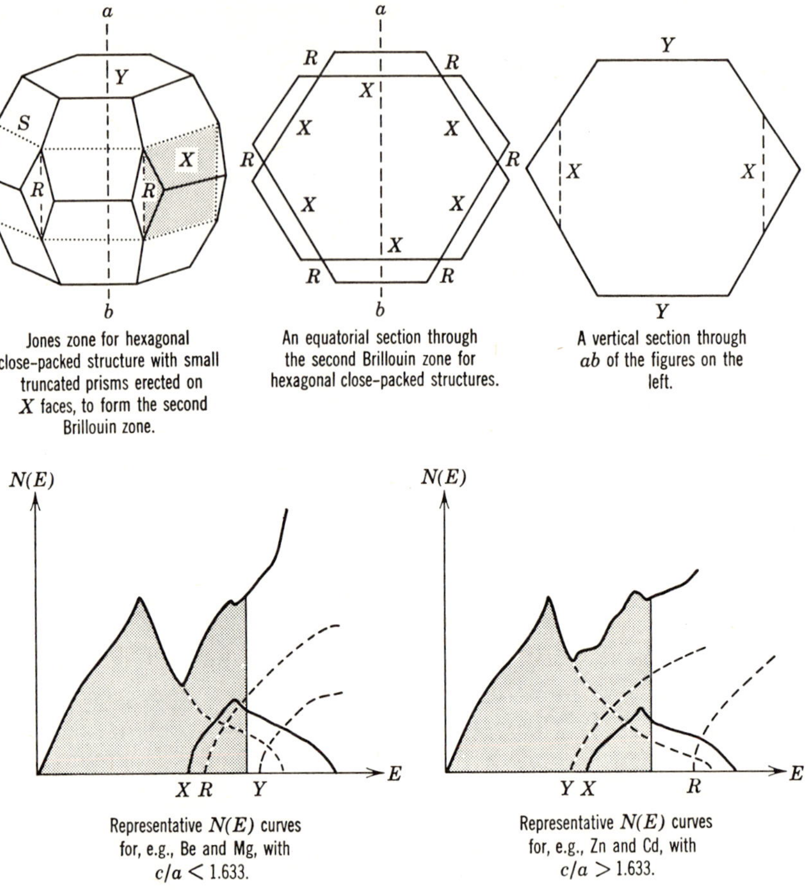

FIG. M.7.9. The nature of the Brillouin zones and $N(E)$ curves for metals crystallizing with hexagonal close packed structures.

entrant corners R, shown in Fig. M.7.9. along with the other directions mentioned; these latter types of overlap may be conveniently designated Y and R overlaps, and the major factor in determining which occurs first is the axial ratio, c/a, of the crystal concerned.

It would appear[1] that for beryllium and magnesium and for other metals with axial ratios less than the ideal value of 1.633, the X-overlap occurs first and then the R-overlap, but for metals with two electrons per atom an overlap of the Y type does not occur. For zinc and cadmium, with c/a of the order of

---

[1] Mott, N. F., and Jones, H., *Theory of the Properties of Metals and Alloys*, 1958, Dover Publications, New York.

1.9, the Y overlap occurs first and then the X-overlap, but in this case, for metals with two electrons per atom, no overlap occurs of the R type. For metals with electron concentrations of about 2.0, the total number of electrons outside the first zone is not greatly affected by the axial ratio, although the order of the overlaps is affected. As a consequence of the anisotropic nature of hexagonal crystals, the shape of the $N(E)$ curves for crystals with overlap occurring into the outer zone at either Y or R is quite different. Overlap in the Y direction, parallel to the c-hexagonal crystal axis, affects electrons with wave vectors in only two directions, whereas overlap at R affects electrons with wave vectors in six directions, normal to the crystal c hexagonal axis. The principal features of the $N(E)$ curves for metals like beryllium and magnesium and for those like zinc and cadmium was first deduced approximately on the basis of such considerations, and are shown in Fig. M.7.9. The exact form of the appropriate $N(E)$ curves must depend on the nature of the potential field in the crystal and on the shape of the Fermi surface of occupied states. Generally, since there are more electron states available, for any particular increase in energy, in the case of R-overlaps than for Y-overlaps, the $N(E)$ curve for the R-overlaps will be expected to climb more steeply away from the energy abscissa than for the Y-overlaps, after the initial sharply rising part has been passed. The small X zone in the $N(E)$ curves of Fig. M.7.9. corresponds to the six small truncated prisms superimposed on the X faces of Fig. M.5.32. The soft X-ray L-emission spectrum of magnesium and more elaborate calculations of the $N(E)$ curves, appear to substantiate the approximate correctness of the shape of the $N(E)$ curves shown in Fig. M.7.9. The effect of pressure on the lattice parameters of magnesium has been reported[1] and there are many studies relating to the investigation of the Fermi surfaces in zinc and cadmium,[2] calcium,[3] and of the de Haas–van Alphen effect in zinc.[4]

The expression given by equation (m.7.57.) for the cohesive energy of the alkali metals, with small ion cores, while strictly applicable only to the alkali metals, can be generalized to metals with any number of valency electrons. If $E_0(r_s)$ is again regarded as the eigenvalue of equation (m.6.61.), subject to the same boundary condition, that $[d\psi_0/dr]_{r=r_s} = 0$, but with $V(r)$ now corresponding to the potential function of the ion M^{+z}, the average energy per electron is then obtained as,

$$E_m(r_s) = E(r_s) + 1.105\alpha Z^{2/3}(r_s)^{-2} + 0.6Z r_s^{-1} - 0.458Z^{1/3} r_s^{-1} + E_L Z^{-1/3} r_s$$

$$(m.7.59.)$$

[1] Clendenen, G. L., and Drickamer, H. G., *Phys. Rev.*, 1964, **A1643**, 135.
[2] Deaton, B. C., and Gavenda, J. D., *Phys. Rev.*, 1964, **A1096**, 136.
[3] Altman, S. L., and Cracknell, A. P., *Proc. Phys. Soc.*, 1964, **84**, 761.
[4] Higgins, R. J., Marcus, J. A., and Whitmore, D. H., *Phys. Rev.*, 1965, **A1172**, 137.

Actually, the justification for using the function $V(r)$ in equation (m.6.61.), for metal atoms of valency Z, as being equivalent to that of an atom which has been ionized Z times, is that the wave function ψ_0 is found to be essentially the same whether or not the self-consistent field of the electrons is included in $V(r)$ (see section M.7.3.), even although the use of this function is not justified by the Coulomb correlation effects, since for a Z-valent metal there have to be on the average Z electrons in each atomic polyhedron. The kinetic energy, the exchange energy, and the correlation energy, parts of equation (m.7.59.), are simply obtained, in the free electron approximation, from those of a one-valency-electron-per-atom metal by substituting $Z^{-1/3}r_s$ in place of r_s, whereas the Coulomb energy per atom is still given by equation (m.7.48.), although now, in the free electron approximation, $\rho(r)$ becomes $\rho(r) = 3Z[4\pi r_s^3]^{-1}$, so that the total Coulomb energy becomes that given by expression (m.7.52.), namely $0.6N/r_s$, multiplied by Z^2, which corresponds to the term $0.6Z/r_s$ in equation (m.7.59.), per electron. Because the electrons in a Z-valent metal occupy states in more than one Brillouin zone, and equation (m.7.40.) for $E(\mathbf{k})$ is strictly only applicable near to the center of the first zone, it would be expected that equation (m.7.59.) would represent a much poorer approximation for the average energy of an electron in a Z-valent metal than for an alkali metal, irrespective of the size of the ion core. Nevertheless, despite the fact that also the free electron approximation would not be expected to describe the electronic interaction terms as well as for the alkali metals, reasonably good values have been calculated for the average energy per electron of metals with small ion cores, such as magnesium and aluminum, by this method.[1] This indicates that, no matter what the effects of the zone boundaries may be, the valency electrons in such metals are still very nearly free electrons, at least in some respects.

M.7.8. The Nature of the Zone Overlap

It is indicated in Fig. M.7.9. that in magnesium metal the zone overlaps occur in the order XRY, while in cadmium the overlaps occur in the order YXR. It is not possible to deduce unambiguously from experimental data on soft X-ray scattering or from the observation of other physical phenomena that the zone overlaps do indeed occur in this order. However, if it is assumed that the overlaps do occur in this particular order, then a plausible explanation may be afforded for the relationships observed between the lattice constants and the composition of various alloys of magnesium.

If, as is indicated on page 706, the effect of a Brillouin zone overlap in any direction in $\mathbf{k}$-space is to increase the lattice parameter in the corresponding

---

[1] Brooks, H., *Nuovo Cimento Suppl.*, 1958, **7**, 165.

direction in real space, then when magnesium is alloyed with a metal of higher valency, with which it forms a solid solution, there would be expected to be a uniform change in the a-parameter with increasing amount of the alloying metal, if the Brillouin zone overlap has already taken place in the pure metallic magnesium. On the other hand, the c-parameter would be expected to show a sudden change, with increasing percentage of an alloying element of higher valency, at the point where the electron concentration becomes sufficiently large to cause the Y-overlap to take place. Such changes have been observed,[1] for solid solution of aluminum, gallium, indium, and thallium, in magnesium, as well as of lead and tin. For solid solutions of cadmium in magnesium, although here both metals contribute two electrons per atom as valency electrons, a similar type of change in lattice parameters with composition would be expected, if the zone overlap in the two metals is as depicted in Fig. M.7.9. Magnesium and cadmium are observed to form a series of solid solutions in one another, disordered at high temperatures and forming superlattices at lower temperatures. If no zone overlaps take place on solution of cadmium in magnesium, then since atoms of cadmium are smaller than those of magnesium, it would be expected that both the c- and a-parameters should decrease. However, since there are only two electrons per atom for the whole range of compositions possible, then if the Y-overlap does not occur in magnesium metal and the R-overlap does not occur in cadmium, successive substitution of cadmium atoms for magnesium atoms in the solid solutions should lead to a stage at which the Y-overlap begins and the R-overlap stops. From such a point onwards, there would be expected an increase in the axial ratio, as a result of the expanding c-parameter compared with the a-parameter. These deductions are borne out completely by the observed variation of the lattice parameters and the axial ratio of magnesium alloys, as increasing amounts of cadmium are put into solid solution,[2] as illustrated in Fig. M.7.10. For the solid solution containing about 65 atomic percent of cadmium, there is evidently a marked change from the slight decrease in both the c- and a-parameters which are caused by the initial solution of cadmium in magnesium, and which barely affects the axial ratio. The rapid increase in the magnitude of the c-parameter at this point, is quite what would be expected from the onset of the Y-overlap and the falling off of the R-overlap. For magnesium, with two valency electrons per atom, Y type overlap may be brought about by solid solution in the metal of smaller concentrations of metals of higher valency.

The general validity of the correctness of the $N(E)$ curves for magnesium shown in Fig. M.7.9., is also indicated by measurements of the Hall constant

---

[1] Raynor, G. V., *The Physical Metallurgy of Magnesium and its Alloys*, 1959, Pergamon Press, London.
[2] Hume-Rothery, W., and Raynor, G. V., *Proc. Roy. Soc.*, 1949, **A27**, 177.

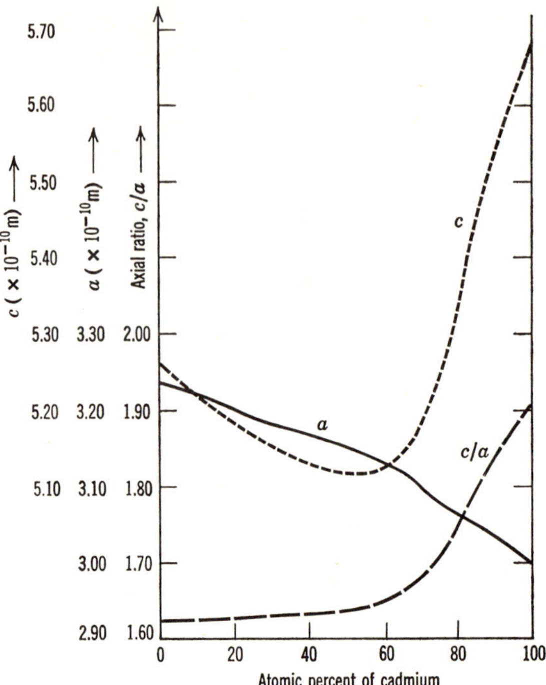

FIG. M.7.10. The variation of the lattice constant
and axial ratio with composition, of magnesium-
cadmium alloys at 583 K.

of magnesium alloys. The negative Hall constant of pure magnesium indicates
an excess of negative charge carriers outside the regions of the X- and R-
overlapping zones, compared with positive holes. Although the exact nature
of the effect of increasing electron concentration on the value of the Hall
constant must depend on the relative mobilities of the two kinds of charge
carrier (see section M.6.4., page 637), it would be expected if the $N(E)$ curve
for magnesium in Fig. M.7.9. is correct, that increasing electron concentration
would lead to an even more negative value for the Hall constant, as the
Y-type overlap begins; this is confirmed by experimental observation[1,2] of
the Hall effect of solid solutions of aluminum and silver in magnesium. It is
concluded from such measurements that if the electron concentration in
magnesium is increased by about 0.05% that overlap takes place in the Y-
direction. It has been suggested that an understanding of the manner in which

[1] Schindler, A. L., and Salkovitz, E. I., *Phys. Rev.*, 1953, **91**, 1320.
[2] Borovik, E. S., *Zh. Eksperim. i Teor. Fiz.*, 1954, **27**, 355.

some of the mechanical properties of metals are altered by compression and the application of tensile stresses, may be attributable to them having such a composition that the region of occupied states extends up to an energy value corresponding to that of a zone boundary, in the direction in which the mechanical stresses are applied.

The phenomenon referred to above is quite different from the alteration in the mechanical properties of metals and alloys brought about by mechanical working, such as hammering, rolling, pressing, and wire-drawing, carried out at low or moderately high temperatures. The properties of metals are not directly related to their chemical nature, nor are those of alloys related to their composition alone, but to a large degree are dependent on their physical texture and the size and arrangement of the crystallites from which the metal is made up. It has long been known that a considerable increase in the hardness and mechanical strength of certain alloys can be brought about by suitable heat treatment, such as quenching from high temperatures, annealing at low temperatures, or the age hardening or precipitation hardening brought about by standing at ordinary temperatures. Thus, the tensile strength of copper may be increased by a factor of 14 times its original value by wire-drawing,[1] and perhaps the best known of the age hardening alloys is duralumin (93–95% aluminum, 2.5–5.5% copper, 0.5–2% magnesium, 0.5–1.2% manganese, and 0.2–1% of silicon). The age-hardening process, originally discovered for copper–aluminum alloys (Wilm, 1911), is quite distinct from the well known hardening of steel, which is based on the fact that iron–carbon alloys with a carbon content below 1.7%, may be converted to austentite (a solid solution of carbon in the face-centered cubic modification of γ-iron) by heating them, which when suddenly cooled or quenched transforms into the very hard material, martensite. It is known that mechanical working does not change the hardness or structure of the individual crystallites in a metal or alloy, but only their shape. Generally, fine grained structures such as rapidly cooled alloys or eutectics, have a greater mechanical strength than materials containing larger crystallites. Alloys may always be age hardened if they consist of mixed crystals forming a mutually, almost saturated solution at high temperatures, which can be brought into a condition of supersaturation by quenching; the age hardening depends on the fact that decomposition of such mixed crystals sets in on aging, with the result that another crystalline phase separates out in a highly dispersed state. The conditions for age hardening are also satisfied if an alloy of some particular composition consists of an intermetallic phase, stable only at high temperatures, which decomposes into two other phases on cooling, provided that this can be obtained

---

[1] Kochendörfer, A., *Plastische Eigenschaften von Kristallen und metallischen Werkstoffen*, 1941, Berlin.

in a metastable state at ordinary temperatures by quenching, for then the decomposition products appear in a highly dispersed state. The tensile strength of alloys, as well as their hardness is increased by age hardening; thus, aluminum alloys containing about 5 weight percent of copper, exhibit a tensile strength of 25 kg mm^{-2}, after cold working, up to 40 kg mm^{-2}, after age hardening, over 50 kg mm^{-2}, and the latter after cold working to 60–70 kg mm^{-2}, and the tensile strength of an age hardened copper–beryllium alloy with 2.5 weight percent of beryllium is about 130–150 kg mm^{-2}, comparable to the best steels.

IV. ALUMINUM AND ITS ALLOYS

M.7.9. Aluminum

Aluminum, with nuclear charge one unit greater than magnesium (13), is a congener boron, gallium, indium, and thallium. The ground-state configuration of aluminum atoms is $1s^2 2s^2 2p^6 3s^2 3p^1$. It crystallizes with the cubic closest packed structure, with $a_0 = 4.04958$ at 298 K. It occurs naturally to a greater abundance than any other metal. The thermal conductivity of aluminum is about half that of copper, and its specific electrical conductivity about 60% that of copper; it has a relatively high heat capacity and latent heat of fusion, compared with other metallic elements. The decrease in compressibility, interatomic distance, and coefficient of expansion, and the increase in melting point, of aluminum compared with the corresponding properties of sodium and magnesium, indicate that there are three electrons per atom responsible for the metallic cohesion. Although the metal is very ductile, its tensile strength is about one-quarter that of copper, but is very much increased by alloying with a few percent of copper.

The first Brillouin zone for aluminum is the truncated octahedron of Fig. M.5.41.(a). The accumulated information derived from the calculation of the energy bands,[1] and conclusions based on the determination of the heat capacity, the observation of the de Haas–van Alphen effect[2] and the anomalous skin effect,[3] indicate that this zone is very nearly filled, with a large overlap across the hexagonal faces and a smaller overlap across the square faces of the zone. Except in the immediate vicinity of the zone boundaries, it is indicated that the relationship between the energy and wave vector of the electrons is very nearly that of the free electron theory, so that to a good approximation, the occupied states must nearly fill the first zone,[4,5] bulging

---

[1] Spong, F. W., and Kip, A. F., *Phys. Rev.*, 1965, **137**, A431.
[2] Gunnerson, E. M., *Phil. Trans. Roy. Soc.*, 1957, **A249**, 299.
[3] Försvoll, K., and Holwech, I., *Phil. Mag.*, 1964, **9**, 435.
[4] Heine, V., *Proc. Roy. Soc.*, 1957, **A240**, 340, 354, 361.
[5] Jones, B. K., *Phil. Mag.*, 1964, **9**, 217.

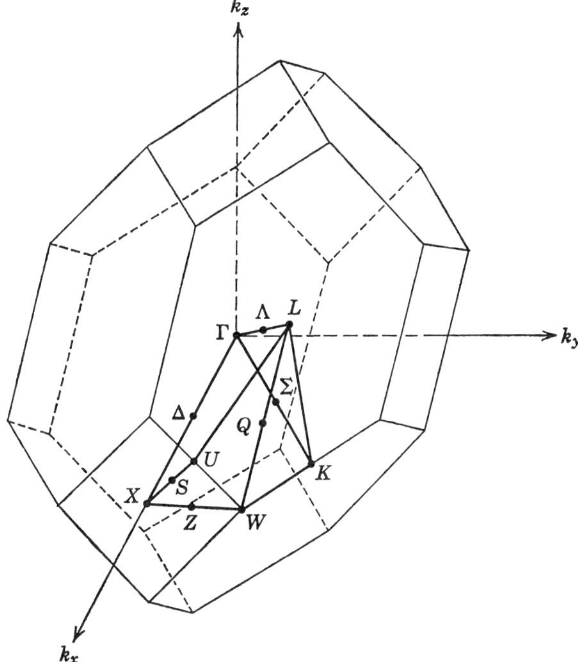

FIG. M.7.11. Special symmetry points and directions in the first Brillouin zone for the face-centered cubic structure.

against the hexagonal faces and to a lesser extent against the square faces of the zone.

The various points and lines in the Brillouin zone for the face-centered cubic structure, such that the reduced wave vectors terminating in these points show special symmetry properties, are shown in Fig. M.7.11. The notation used in this figure is the same as that described on page 658 for the body-centered cubic structure. In aluminum, the overlaps across the hexagonal and square faces of the zone, have spread out from points such as L and X, respectively, to such an extent that points such as K and U at the mid-points of the edges, in the second zone, are below the Fermi surface although this surface has not reached points such as W in the second zone. The de Haas–van Alphen effect studies indicate that the electron states in the first zone are not all occupied, especially in the region of the points W, and that there is some overlap into the third zone together with a large overlap across the hexagonal and square faces of the zone. These zone characteristics of aluminum, when interpreted in terms of the argument given on page 706, that when overlap takes place across a zone face the tendency is to compress the zone or expand the crystallographic parameter in the direction of the overlap, enable an understanding to be attained of some of the otherwise apparently anomalous types of behavior of the metal. Thus, the relatively large coefficient of expansion, could be attributed to the thermal excitation of electrons from the lower to higher zones, and then the expansion effect attributable to overlapping zones would enhance the expansion, compared to the situation where no zone overlaps were involved. Similarly, this expansion effect on heating, leading to a reduced attraction between the aluminum atoms, could account for the abnormally low melting point of the metal (only 10 K higher than that of magnesium).

When a primary solid solution is formed by the dissolution of one metal in another, it is generally found that for dilute solutions, the observed lattice spacing is a linear function of the composition, and if extrapolated to 100% of the solute species the resulting lattice spacing may be regarded as giving the effective radius of the solute species. Quite apart from the difficulty and danger of attempting to represent the size of an atom or ion by a fixed constant, it is found that in aluminum the metallic radius estimated from the observed interatomic distance in the metal (2.8635×10^{-10} m) is considerably larger than the value deduced from its behavior in alloys. Thus, when aluminum is dissolved in copper or silver, it expands the lattice constant of these metals as though the aluminum had a radius of 1.35×10^{-10} m, no zone overlap being involved in this case; when dissolved in magnesium, where zone overlaps have taken place, the crystal lattice constant is reduced as though the aluminum atom had a radius of 1.42×10^{-10} m. The apparently reduced radius of aluminum in alloys is undoubtedly due to the contraction effects associated with overlapping zones, although of course in some cases, differences in electrochemical characteristics between the constituent metal species, leading to charge transfer and ion formation, also accounts for some of the contraction.

M.7.10. Intermetallic Compounds Containing Aluminum

There are numerous intermetallic compounds which have structures involving either the ordered or disordered distribution of two or more kinds of metal atoms among the atomic positions for one or other of the close packed structures,[1] or one of the other simple structural types. Structures based on cubic or hexagonal closest packing are favored for intermetallic compounds composed of atoms differing only slightly from one another in radius. There are a great many intermetallic compounds containing aluminum

---

[1] See Wyckoff, R. W. G., *Crystal Structures*, Vol. I, 1963, 2nd edition, Interscience Publishers, John Wiley & Sons, New York.

known, with structures ranging from the simplest type to those of the greatest complexity.[2,3]

The simple binary compounds, AlCo, AlFe, AlNd, AlNi, crystallize with the body-centered cubic structure, each atom having eight nearest neighbors at the vertices of a cube, all of the opposite type (Fig. M.1.6.). In LiAl, also based on the body-centered cubic structure, each atom however has four nearest neighboring atoms of one type and four of another.

The structure found for the intermetallic compounds Fe_3Al and Cu_2AlMn is also based on the body-centered cubic arrangement. In the disordered phase, for the former compound, the cubic unit cell may be regarded as being derived from eight simple body-centered cubic units, with the iron atoms occupying all of the sites at the vertices of the eight smaller cubes and the aluminum atoms being statistically distributed over one-half of the eight body-centered positions. In the ordered arrangement, the aluminum atoms occupy exactly one-half of the body-centered positions, arranged tetrahedrally about the center of the whole unit cell, as shown in Fig. M.7.12. The intermetallic compound Cu_2AlMn has the same structure, but with the copper and manganese atoms statistically distributed over the positions occupied by iron atoms in Fe_3Al. The crystal structures of the compounds Al_3Ti and Al_3Zr are related to the face-centered cubic structure of metallic aluminum in the same manner as that in which the structure of Fe_3Al is related to that of body-centered cubic iron; in Fe_3Al, one-quarter of the iron atoms in the metal have been replaced in a systematic way to form the superstructure of Fe_3Al, and similarly if one-quarter of the aluminum atoms in aluminum are substituted in a systematic manner by titanium or zirconium, superstructures with tetragonal symmetry arise.[1]

In AlAs, AlP, and AlSb, the zinc blende structure is adopted (Fig. M.1.8.(e)), and in AlN, the würtzite structure (Fig. M.7.12.); in this latter arrangement both types of atoms are hexagonal closest packed with respect to one another, although each atom is still tetrahedrally coordinated with four atoms of the other type, as in the zinc blende structure, involving cubic closest packing of each type of atom. Substances crystallizing with the würtzite structure always have an axial ratio, c/a, close to the ideal value of 1.633. The compound $AuAl_2$ crystallizes with the fluorite structure [Fig. M.1.8.(f)] and Li_3AlN_2 with the antifluorite arrangement, in which the lithium and aluminum atoms are statistically distributed among the tetrahedral holes in the fluorite unit cell, with the nitrogen atoms occupying the vertices and face centers of the

[1] Brauer, G., *Z. anorg. Chem.*, 1939, **242**, 1.

[2] Samson, S., in *Structural Chemistry and Molecular Biology*, 1968, pp. 687–717, W. H. Freeman & Co., San Francisco.

[3] Shoemaker, D. P., and Shoemaker, C. B., in *Structural Chemistry and Molecular Biology*, 1968, pp. 718–730, W. H. Freeman & Co., San Francisco.

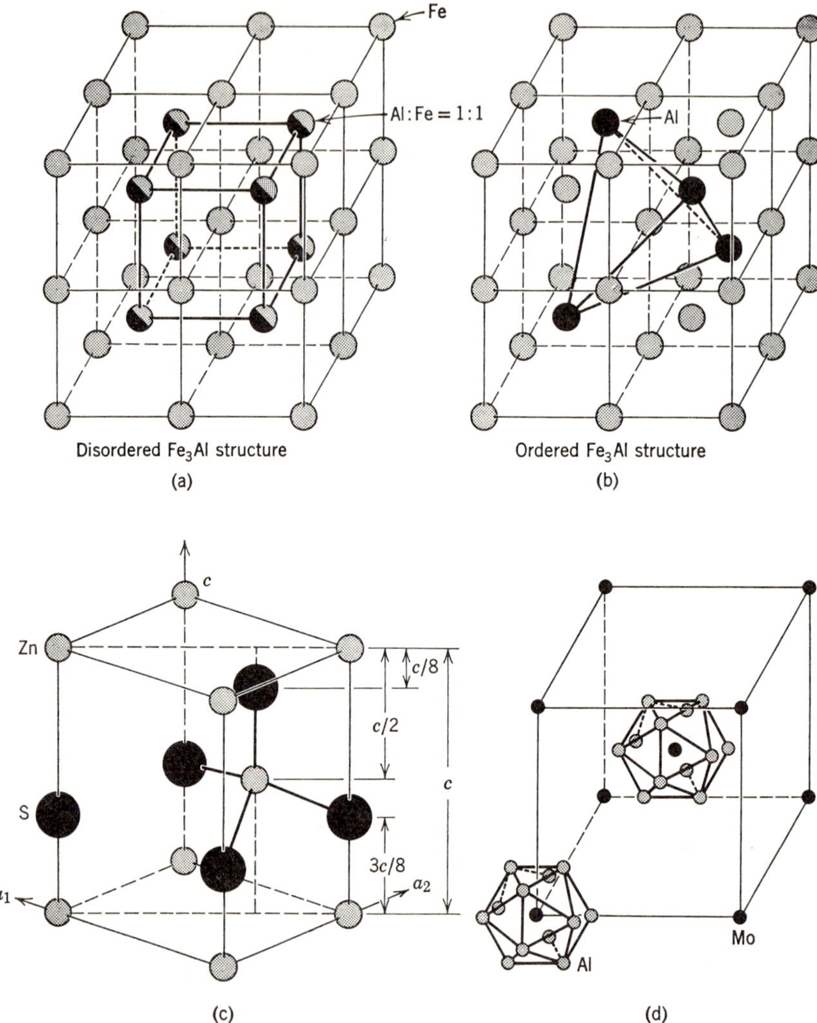

FIG. M.7.12. The crystal structure of some intermetallic compounds containing aluminum. The disordered (a) and ordered (b) structure, based on body-centered cubic packing adopted by Fe_3Al. The würtzite structure, involving hexagonal close packing of two different kinds of atoms is shown in (c). The structure shown at (d) is also based on the body-centered cubic arrangement, found for $MoAl_{12}$; each molybdenum atom at the body-centered cubic lattice sites is icosahedrally surrounded by aluminum atoms. Only two such icosahedra are illustrated.

cubic unit cell. The structure found for Ni_2Al_3 and Pd_2Al_3 is based on the simple body-centered cubic arrangement, but with one-third of the body-centered positions being vacant.[1] A slightly more complicated structure, based on hexagonal closest packing, is found for Al_3Pu; this structure is built up from the sequence of hexagonal close packed layers $\cdots cchcchcch\cdots$ (see page 392), each layer containing plutonium and aluminum atoms in the ratio of one to three.[2]

The Hume-Rothery electron compounds, with closely related structures but apparently unrelated stoichiometric composition, are determined by electron to atom ratios (see page 402). Thus, aluminum compounds with the electron:atom ratio of 3:2 crystallize with either the simple body-centered cubic structure (Cu_3Al, Ag_3Al), the complex β-manganese structure (Ag_3Al, Au_3Al), or the close packed hexagonal structure (Ag_3Al); those with the electron:atom ratio of 21:13 adopt the complex γ-brass structure (Cu_9Al_4), and those with the electron:atom ratio of 7:4 crystallize with the close packed hexagonal structure (Ag_5Al_3, Au_5Al_3). These structures are referred to on page 404. The first Brillouin zone for the γ-brass structure (Fig. M.7.5.) may contain 22.5 electrons per 13 atoms, but Pauling[1] has pointed out that another Brillouin polyhedron for crystals with the γ-brass structure, bounded by the (600) and (442) planes, the only other planes of high structure factor, may contain 63.90 electrons per 13 atoms, where the number 63.90 is exactly that which corresponds to his metallic valencies (Chapter M.2.) of 5.53 for copper and 4.53 for zinc in the case of Cu_5Zn_8, which also adopts the γ-brass structure.

The Laves phases, with compositions and structures which are largely determined by the relative sizes of the atoms involved have been discussed on pages 407, 408–9. Among aluminum intermetallic compounds, the three different Laves phases with structures corresponding to the three closest packings of Laves polyhedra [Fig. M.1.12.(a)], are represented by $CaAl_2$, $CeAl_2$, $LaAl_2$, and UAl_2, with the $MgCu_2$ type structure [Fig. M.1.12.(c)], $MgAlCu$ with the $MgZn_2$ structure [Fig. M.1.12.(d)], and $MgAlCu$ also with the $MgNi_2$ structure [Fig. M.1.12.(e)].

The coordination number 12, characteristic of both cubic and hexagonal closest packing, is also achievable in another way. It is possible to retain this coordination number with a central atom as much as 10% smaller than the 12 surrounding atoms if these latter are arranged at the vertices of a regular icosahedron (page 396). Structures involving icosahedral coordination are usually complex, with 20, 52, 58, 162, or 184, or more atoms per cubic unit

[1] Bradley, A. J., and Taylor, A., *Phil. Mag.*, 1937, (7), 23, 1049.
[2] Larson, A. C., Cromer, D. T., and Stanbaugh, C. K., *Acta Cryst.*, 1957, **10**, 443.
[3] Pauling, L., and Ewing, F. J., *Rev. Mod. Phys.*, 1948, **20**, 112.

cell. A crystal with cubic symmetry cannot retain any more than a maximum of four three-fold symmetry axes (see page 554) and of course no five-fold axes, whereas a regular icosahedron possesses six five-fold, 10 three-fold, and 15 two-fold, axes of symmetry. However, crystals of WAl_{12}, $MoAl_{12}$, and $(Mn,Cr)Al_{12}$ have structures, based on the body-centered cubic arrangement, in which there is a nearly regular icosahedron of 12 aluminum atoms co-ordinated about the smaller tungsten, molybdenum, manganese or chromium, atoms, distributed over the lattice sites of the body-centered cubic structure.[1] This structure is shown in Fig. M.7.12. The intermetallic compounds $Mg_{34}Al_{24}$ and Ag_3Al crystallize with the α-manganese and the β-manganese structures, respectively, in each of which there are icosahedrally coordinated atoms. Of the four reported polymorphic modifications of manganese, α-manganese, with 58 atoms in its cubic unit cell, and β-manganese, also cubic with 20 atoms per unit cell, each contain one kind of manganese atom with an effective radius less than that of the others which displays icosahedral coordination; 24 of the 58 manganese atoms in the α modification and 8 of the 20 manganese atoms in the β-modification behave in this manner. Icosahedral groupings are also to be distinguished in the cubic Laves phases, as for example in the structure of $CaAl_2$; each aluminum atom is coordinated with an icosahedral group composed of six other aluminum atoms and six calcium atoms.

As discussed in Chapter M.1., the structure of the most complex structures which have been reported for intermetallic compounds are most conveniently interpreted in terms of the polyhedral groups which are described by the centers of atoms of one kind which surround those of another or the same type. The most important of these polyhedra are shown in Figs. M.1.7. and M.1.13. The truncated tetrahedron of Figs. M.1.7.(e) and M.1.12.(a), which as indicated in these figures may be regarded as a tetrahedral array of four octahedra, appears to be a particularly convenient intermediate polyhedron for the purposes of describing the more complicated cubic structures. The structure found for the compound, $Mg_3Cr_2Al_{18}$, with 184 atoms in its cubic unit cell,[2,3] for example, can be described as an array of the Friauf polyhedra of Fig. M.1.12.(a), each containing a magnesium atom at its center and an aluminum atom at its vertices, each sharing its four hexagonal faces with four other Friauf polyhedra through the agency of four hexagonal prisms, at the center of which either aluminum or magnesium atoms are randomly dis-tributed; the chromium atoms are accommodated at the centers of distorted icosahedra and the remaining aluminum atoms are located at the centers of

[1] Adam, J., and Rich, J. B., *Acta Cryst.*, 1954, **7**, 813.
[2] Erdmann-Jesnitzer, F., *Aluminum Archiv.*, 1940, 29.
[3] Samson, S., *Acta Cryst.*, 1958, **11**, 851.

pentagonal prisms which are described by the vertices of such an arrangement of Friauf polyhedra. The compound α-VAl_{10} is found to be isostructural[1] with $Mg_3Cr_2Al_{18}$, except that there are no atoms at the centers of the Friauf polyhedra, and the atoms located at the centers of the hexagonal prisms are entirely aluminum atoms.

Many other more complicated structures found for intermetallic compounds can be described in terms of the way in which the Friauf polyhedra are arranged in space, with common hexagonal or triangular faces, to form infinite three dimensional networks. The structure already referred to for α-Mn and for $Mg_{34}Al_{24}$, for example, is neatly described in terms of aggregates of five Friauf polyhedra located at the lattice sites of a body-centered cubic unit cell; each group of five Friauf polyhedra is composed of a central truncated tetrahedron, with four other similar polyhedra each with one hexagonal face in common with the central Friauf polyhedron. When such complexes are positioned at the lattice sites of a body-centered cubic unit cell, each complex has four triangular faces in common with four other such complexes, so that the total number of atoms per unit cell is 58. On comparison of the α-Mn and the $Mg_{34}Al_{24}$ structures, it appears that there are three different kinds of manganese atom in the former structure, of different sizes, located at the centers of the 10 Friauf polyhedra in each unit cell, at the vertices of the Friauf polyhedra in the positions which are common to adjacent complexes of five, and at the centers of icosahedral groups which are to be distinguished in the structure; α-Mn could thus be described as a ternary intermetallic compound, $Mn_{10}{}'Mn_{24}{}''Mn_{24}{}'''$, containing atoms of manganese exhibiting three different valencies. The so-called ε-phase, derived from $Mg_{34}Al_{24}$ by a solid phase transformation, with the composition $Mg_{23}Al_{30}$, and found to be isostructural with the R-phase in the Mo–Co–Cr system, is found to have a structure[2] consisting of an infinite framework of Friauf polyhedra arranged along the edges of a rhombohedral unit cell, containing 53 atoms; along the three-fold axis of the rhombohedral cell, there occurs a string of five more polyhedra with triangular faces in common. In the R-phase structure, the five polyhedra involved are three icosahedra with a Friauf polyhedron at either end, and in the $Mg_{23}Al_{30}$ structure, the three middle polyhedra correspond to those with 13, 14, and 15 vertices, respectively, the polyhedra at either end of the group still being Friauf polyhedra.

A more complex structure, is that of $Mg_{32}(Al, Zn)_{49}$, in which not only Friauf polyhedra but also the four triangulated polyhedra of Fig. M.1.13. may be distinguished.[3] The phases, Mg_4CuAl_6 in the mangesium–copper–

[1] Brown, P. J., *Acta Cryst.*, 1957, **10**, 133.
[2] Komura, Y., Sly, W. G., and Shoemaker, D. P., *Acta Cryst.*, 1960, **13**, 575.
[3] Bergman, B. G., Waugh, J. L. T., and Pauling, L., *Nature*, 1952, **169**, 1057: *Acta Cryst.*, 1957, **10**, 254.

aluminum system,[1] Li_3CuAl_5 in the lithium–copper–aluminum system,[2] $Li_{32}(Zn, Al)_{49}$ in the lithium–zinc–aluminum system,[2] and the quaternary phase in the mangesium–aluminum–zinc–copper system,[3] all appear to be isostructural with $Mg_{32}(Al,Zn)_{49}$, with 162 atoms per body-centered cubic unit cell. Although no well-defined Brillouin polyhedron could be recognized in this structure to account for its stability, geometrical factors associated with the symmetry of the icosahedron and the Friauf polyhedron are apparently of major importance in determining this structural arrangement. The most important geometrical feature about the icosahedron is that the distance from its center to its vertices is about 5% less than the edge lengths, so that a smaller atom at its center may be coordinated with twelve larger atoms at its vertices, and as a consequence of the dual relationship of the icosahedron and the pentagonal dodecahedron as shown in Fig. M.5.29., icosahedral packing may be retained in a structure by the continued addition of atoms of increasing radius centered above the triangular faces of a primary icosahedron. Thus, the $Mg_{32}(Al, Zn)_{49}$ structure is built up from a central small atom of aluminum or zinc surrounded icosahedrally by 12 others, the center of an icosahedron being located at each of the lattice sites of the body-centered cubic unit cell. This group of 13 atoms is then surrounded by a group of 20 atoms, centered above each of the 20 faces of the primary icosahedron, and such that their centers define the vertices of a pentagonal dodecahedron. Another group of 12 atoms, located centrally above the 12 faces of the pentagonal dodecahedron, define by their centers another icosahedron, and effectively giving rise to five triangular faces from each of the twelve dodecahedral faces, or 60 triangular faces in all. The resulting complex of 45 atoms contains an inner group of 13, and 32 in the outer region. A further group of 60 atoms, located at the centers of the 60 faces referred to above, are now found to lie at the vertices of a truncated icosahedron, with 12 pentagonal faces and 20 hexagonal faces. Alternatively, this truncated icosahedron may be regarded as being derived from 20 Friauf polyhedra arranged with their centers at the vertices of a pentagonal dodecahedron,[4] and enclosing a central icosahedron, so that such an aggregate, including an atom at the center of the enclosed icosahedron, would amount to a complex of 113 atoms. If an additional 12 atoms are now located out from the centers of 12 of the hexagonal faces of the truncated icosahedron, which do not lie in the (111) directions, then this outermost group of 72 atoms (60 + 12) is such that their centers lie on the surface of a truncated octahedron, which latter polyhedra may be packed

---

[1] Laves, F., Löhberg, K., and Witte, H., *Metallwirt.*, 1935, **14**, 793.
[2] Cherkashin, E. E., Kripakevich, P. I., and Oleksiv, G. I., *Kristallografiya*, 1964, **8**, 681.
[3] Strawbridge, D. J., Hume-Rothery, W., and Little, A. T., *J. Inst. Met.*, 1947, **74**, 191.
[4] Samson, S., in *Structural Chemistry and Molecular Biology*, 1968, pp. 687–717, Edited by A. Rich and N. Davidson, W. H. Freeman & Co., San Francisco.

together at the sites of a body-centered cubic lattice so as to occupy all available space, each face common to two truncated octahedra. Thus, since each truncated octahedron contains a nucleus of 45 atoms and each of the other 72 is common to two such polyhedra, the number of atoms located at each lattice point in the structure is [45 + 72/2] = 81, and therefore there are 162 atoms per unit cell. In the resulting structure, all 98 zinc and aluminum atoms have icosahedral coordination, and 40 of the 64 magnesium atoms are 16-coordinated, 12 have 15-coordinated environments, and the other 12 are 14-coordinated. The nature of the various polyhedra used to describe this structure are illustrated in Fig. M.7.13.

A still more complicated structure is that reported for β-Mg$_2$Al$_3$, which crystallizes in the cubic system with 1168 atoms per smallest unit cell.[1,2] The basic unit of the structure of this compound is a group of five Friauf polyhedra arranged about an approximate five-fold symmetry axis, and containing 47 atoms. If six of these primary units are arranged at the vertices of an octahedron, then four additional Friauf polyhedra are produced, and the second structural sub-unit composed of 34 Friauf polyhedra and containing 234 atoms, results. Each of these 234-atom complexes is stacked in the crystal so that each one is tetrahedrally surrounded by four others, so that the total number of atoms per 234-atom complex is reduced from 234 to 144 by atom sharing between adjacent faces of the complexes. The crystallographic unit cell of the structure contains eight of the 144-atom complexes, making 1152 atoms per cell; with a further group of eight and another of 32 atoms, added at the centers and out from the centers of the faces, respectively, of eight interstitial polyhedra which arise from the mode of packing, a total of 1192 atoms per unit cell for the ordered structure of this phase is arrived at. The disordered phase contains only 1168 atoms per unit cell, and is more stable, the increase in stability compared with the ordered phase, apparently being associated with the increase in the number of icosahedrally coordinated atoms which arise as a result of the disordering.

In all of these complex structures, a remarkable feature is the tendency towards the formation of five-fold symmetry axes, as in the icosahedron, groups of five icosahedra arranged approximately around a five-fold axis, and groups of five Friauf polyhedra similarly arranged. Another consequence of the formation of such five-fold symmetry axes is the concurrent creation of triangular faces, and for approximately spherical atoms, a triangular net enables other atoms to be inserted above the center of a triangle to create a tetrahedral hole. From the fact that the smallest kind of cavity between spheres in contact is a tetrahedral one, the closest packing or the most

[1] Perlitz, H., *Nature*, 1944, **154**, 607.
[2] Samson, S., *Acta Cryst.*, 1965, **19**, 401.

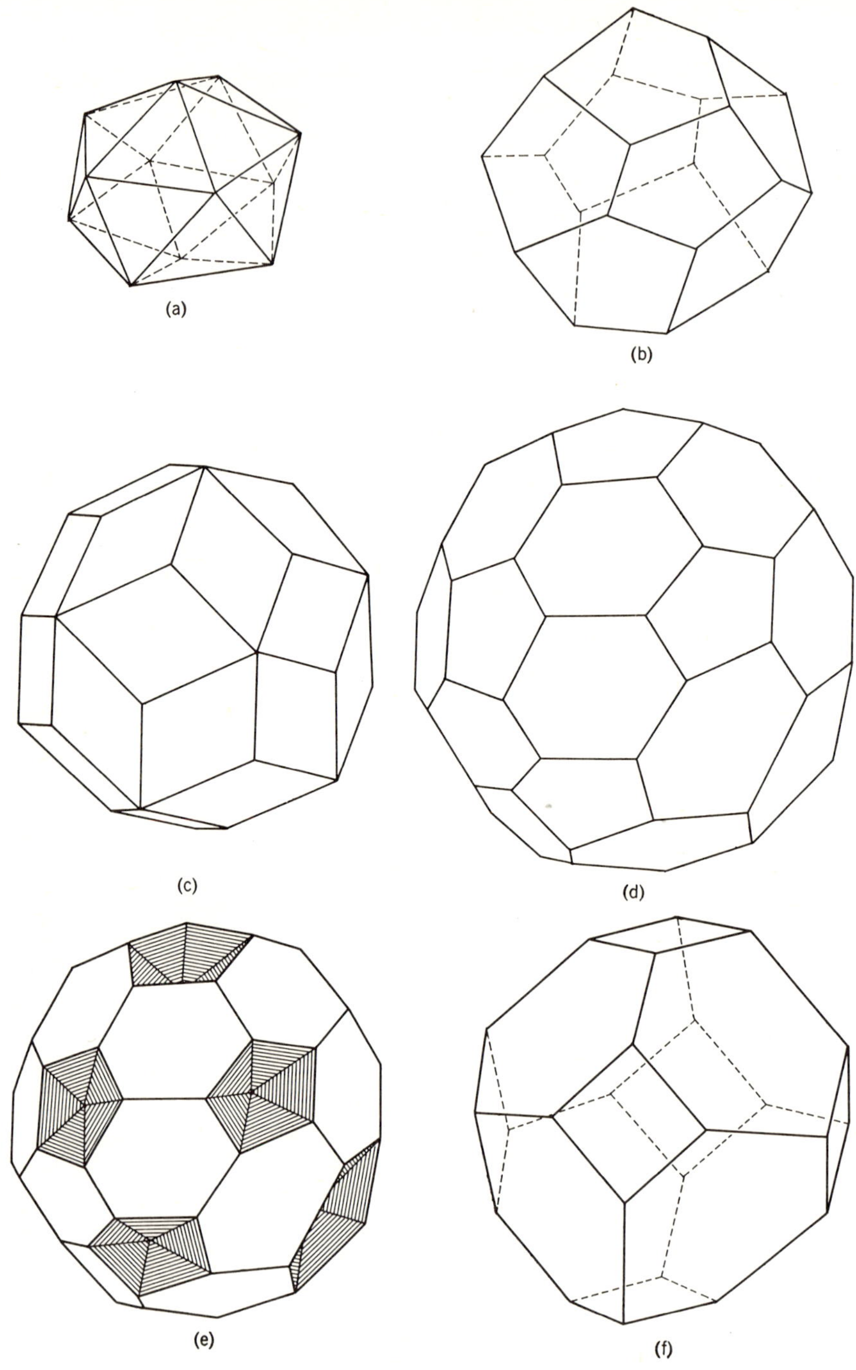

(a)
(b)
(c)
(d)
(e)
(f)

efficient way of occupying space is to make the number of such cavities the maximum possible. Consistent with these considerations, it would be expected that the most efficient and stable arrangement of spherical atoms or ions would be those containing coordination polyhedra with triangular faces. In cubic closest packing, for example, where there are square faces as well as triangular faces, there are also repulsion effects between non-adjacent species, but in a triangular arrangement there are no unattached neighbors to produce repulsion effects. Also, in a pentagonal net, the next nearest neighbor distance is $a \cos 18°/\cos 54° \simeq 1.62a$, where a is the nearest neighbor distance around the periphery of the pentagonal array.‡ The repulsive force would therefore

‡ The irrational numbers, $\pi = 3.14159\cdots$, representing the ratio of the length of the circumference to the diameter of a circle, and $\tau = 1.61803\cdots$, representing the ratio of the length of the diagonal to the edge-length of a plane pentagon, or at least reasonable approximations to these quantities, were known to the ancient Babylonian, Egyptian, and Greek, mathematicians (from about 2000 B.C.). From the magnitude of the angles in a regular pentagon [see Fig. M.5.29.(b)], it is evident that $\tau = [a \cos 18°/\cos 54°] = 2 \cos \pi/5 = (1 + \sqrt{5})/2$, the latter value constituting the positive root of the equation, $\tau^2 - \tau - 1 = 0$. Euclid's (365–275 B.C.) construction for dividing a line in extreme and mean section, the so-called golden section, involves sub-division of the line in the ratio of $\tau/1$, and from the fact that τ is the positive root of the equation, $\tau^2 - \tau - 1 = 0$, this necessarily means that $\tau/1 = 1/(\tau - 1) = (\tau - 1)/(2 - \tau) = (2 - \tau)/(2\tau - 3) = (2\tau - 3)/(5 - 3\tau)$, etc. The 12 vertices of a regular icosahedron, as in Fig. M.1.7.(c) and (d), are located at the vertices of three mutually orthogonal planes, whose edge-lengths are in the ratio of $\tau/1$. τ is also equivalent to the ratio of the distance from the center to any vertex to the edge-length, of a regular decagon, and represents the limiting value of the ratios of successive numbers in the so-called Fibonacci (c. 1175–1250) series of numbers. The latter series is the recurring series defined by $u_{n+2} = u_{n+1} + u_n$,

---

FIG. M.7.13. Illustration of the crystal structure of $Mg_{32}(Al, Zn)_4$ and related phases. The body-centered cubic unit cell of the arrangement contains the complex group of 117 atoms lying within and on the surface of the truncated octahedron (cubo-octahedron) shown at (f). There are 45 atoms within this cubo-octahedron, and 72 lying on its surface, all of the latter being shared with neighboring cubo-octahedra when packed in the crystal. The successive polyhedra from which this complex group is built up, as described in the text, are (a) the icosahedron (12 vertices and 20 triangular faces) surrounding a central zinc or aluminum atom, (b) the pentagonal dodecahedron (20 vertices and 12 pentagonal faces), (c) the rhombic triacontahedron (20 three-fold vertices, 12 five-fold vertices, and 30 rhombus-shaped faces), (d) the truncated icosahedron (60 three-fold vertices and 12 pentagonal and 20 hexagonal faces), (e) the alternative way of regarding the truncated icosahedron of (d) as being built up from 20 of the Friauf polyhedra shown in Fig. M.1.12. (a) arranged with their centers at the vertices of the pentagonal dodecahedron shown in (b), and (f) the final complex group of 117 atoms obtained from (e) by the addition of 12 further atoms out from the centers of 12 of the hexagonal faces of (d), as described in the text.

where $n = 0, 1, 2, \ldots$, and $u_0 = 0$ and $u_1 = 1$, which gives rise to the series, 0, 1, 1, 2, 3, 5, 8, 13, 21, 34, 55, 89, ... The ratios of alternate pairs of numbers in this series approach τ, the upper series of ratios increasing towards the value $1.60803\cdots$, and the lower series decreasing towards the same value, as

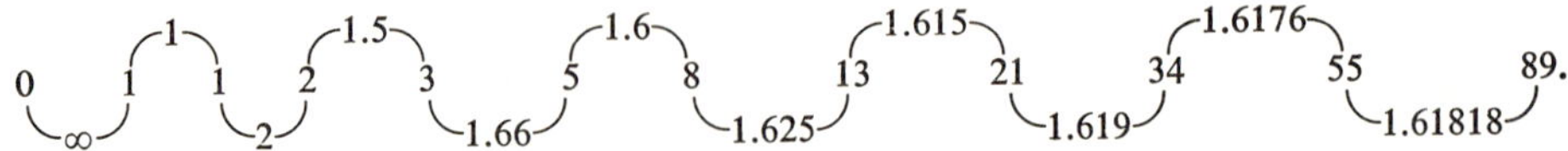

be less in a pentagonal array than in a square array, for example, where the next nearest neighbor distance is $\sqrt{2}\,a = 1.414\,a$. The combination of these circumstances, together with the reduced center-to-vertex distance compared with the edge length of the icosahedron, make it the preferred coordination polyhedron in complex structures involving approximately spherical atoms or ions.‡

‡ Other intermetallic phases containing aluminum, in which icosahedra are the predominant coordination polyhedra observed, usually with the other triangulated polyhedra of Fig. M.1.13., are $Mg_2Cu_6Al_5$ in the magnesium–copper–aluminum system,[1] $Nb_{48}Ni_{39}Al_{13}$, the M-phase in the niobium–nickel–aluminum system,[2] $Zr_{57}Al_{43}$ as representative of the so-called Zr_4Al_3 phase in the zirconium–aluminum system,[3] and the σ-Nb_2Al phase.[4]

V.   THE TRANSITION METALS

M.7.11. The Problem of the Transition Metals

There is not yet any quantitative theory of the transition metals, in spite of much work, and perhaps most of the so-called theories are really attempts to interpret and form generalizations from the experimentally observed data in terms of either: many simplified assumptions about the electronic configurations of atoms of these elements; or, simply, plausible assumptions, which are not in direct contradiction with observed data. A general interpretation of the electronic structures of the transition metals is possible, in terms of the band theories. The s- and d-states of the isolated atoms (as well as the p-states) broaden out into overlapping bands in the metallic crystals, as indicated in Figs. M.5.39.–40., where the most direct experimental evidence for such broadening is available from the study of the soft X-ray spectra of the metals. For most purposes, it is convenient to distinguish between an s-band, which is associated with some p-type character, and a predominantly

[1] Samson, S., *Acta Chem. Scand.*, 1949, **3**, 809.
[2] Shoemaker, C. B., and Shoemaker, D. P., *Acta Cryst.*, 1967, **23**, 231.
[3] Wilson, C. G., Thomas, D. K., and Spooner, F. J., *Acta Cryst.*, 1960, **13**, 56.
[4] Brown, P. J., and Forsyth, J. B., *Acta Cryst.*, 1961, **14**, 362.

d-band. In the transition metal crystals, where there is considerable overlap between the s-states of the metal atoms, the s-band is quite broad, with a low density of states, as it is derived from atomic s-states which can only accommodate two electrons per atom. The d-band will be broad at the beginning of each transition metal series, contracting with increasing nuclear charge along each series, until it is full with ten electrons per atom (see Fig. M.5.40.), and exhibiting a much greater density of states than the s-band. The band theories deal with the collective nature of the valency electrons. The alternative approach, due to Pauling, has been discussed in Chapter Q.2., in which the d-electrons are regarded as being sub-divided into three groups, on the basis of magnetic moment data and interatomic distance measurements, of bonding electrons, non-bonding electrons, and metallic electrons. The two different approaches to electron theories of the transition metals are similar in that they both attribute a strong bonding contribution from d-electrons at the beginning of each transition metal series, where the s and d electron energies are similar, and a weaker contribution from the d-electrons in the later members of each series. The band theories deal with this weakening of d-bonding in terms of the narrowing of the d-band, and the Pauling theory in terms of an increasing number of electrons being involved in non-bonding states. While the band theories provide an explanation of the magnetic properties of the transition metals, they fail to indicate why ferromagnetism should be confined to iron, cobalt, and nickel, of the first transition metal series. It has been proposed that two of the d-orbitals, the d_{z^2} and $d_{x^2-y^2}$ (page 220), would be non-bonding in body-centered cubic metal structures, as they have zero amplitudes along all of the nearest neighboring atom directions.[1] Thus, body-centered cubic iron is ferromagnetic with a saturation magnetic moment corresponding to two electrons with parallel spins, which may be interpreted as one in each of the non-bonding d-orbitals. Iron is unique in having the body-centered cubic structure as the stable form at both high and low temperatures, with the face-centered cubic modification stable in the intermediate temperature range of 1183 K–1462 K; for all of the other transition metals, where the body-centered cubic modification is known to exist, it is the form stable at the highest temperature. The entropy effect (sections M.7.5.–6.) which tends to stabilize the β-phases in the copper alloys, is probably also responsible for the stability of the body-centered cubic structure in iron.

There is little doubt that, as pointed out by Pauling (see Chapter M.2.) and Hume-Rothery and Coles,[2] the cohesive or binding energies of each of the three series of transition metals, as indicated by the measured values of

---

[1] Orgel, L. E., *An Introduction to Transition Metal Chemistry*, 1960, John Wiley & Sons, Inc., New York.

[2] Hume-Rothery, W., and Coles, B. R., *Advances in Physics*, 1954, **3**, 149.

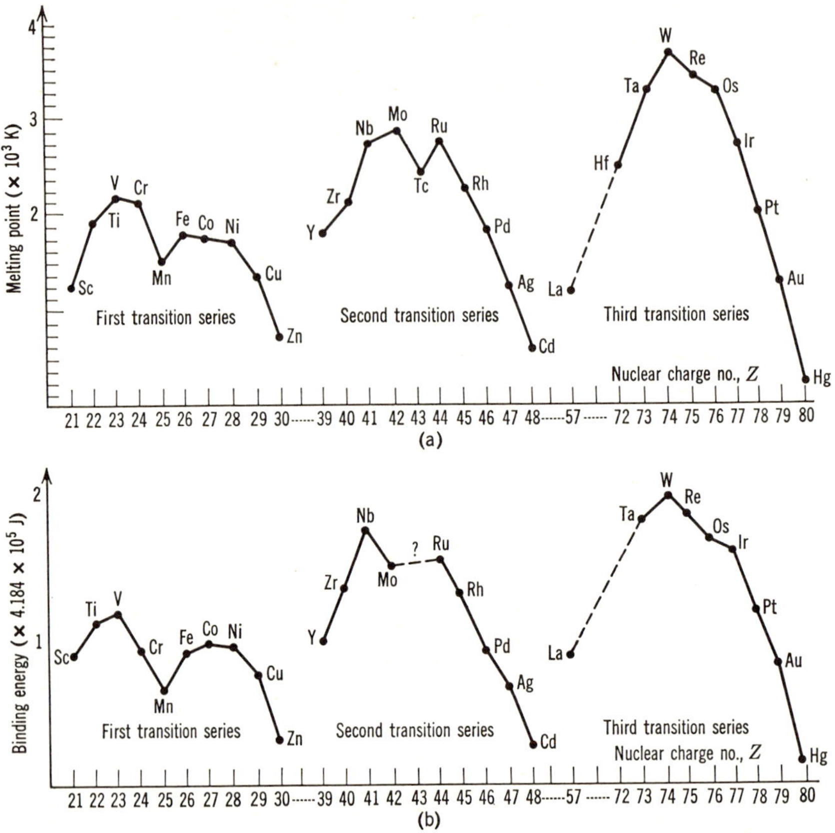

FIG. M.7.14. The melting points of the transition metals (K) shown at (a) and the binding energies of the transition metals shown at (b); the binding energies are obtained from the heats of sublimation of the metals and are expressed in J mole^{-1}.

their heats of sublimation, as well as certain physical properties such as the melting point, compressibility, coefficient of thermal expansion, and atomic diameter (closest distance of approach of the atoms in crystals of the pure metals), which are related to the binding energy, vary in such a way as to indicate that the maximum number of electrons per atom involved in bond formation between the metal atoms is about six. Some of these relationships are shown in Figs. M.7.14.–15. It would not be expected that there would be an exact correspondence between the binding energies and the various physical properties, as some of these must be dependent on factors other than the overall binding energy. The low compressibility of copper and of gold, for example, compared with titanium and hafnium, must be attributable to some extent to the stronger repulsion between the closed *d*-electron groups in

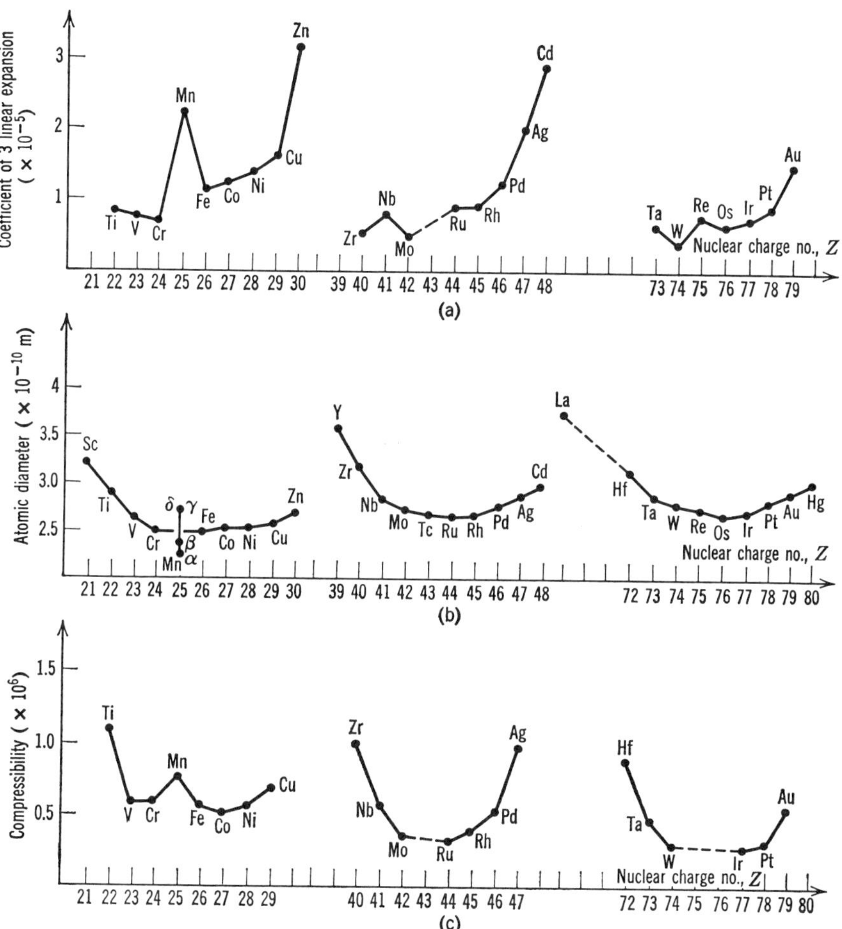

FIG. M.7.15. The coefficients of thermal expansion (a), the atomic diameters (b), and the compressibilities of the transition metals, (c).

these metals, on any attempt to compress them. Also, the contraction of the atomic radii with increasing nuclear charge, for each of the transition metal series, will be superimposed on the changes due to the binding energy. Large binding energies generally correspond to low compressibility, but there is no simple relationship between these two properties, since the binding energy is determined by the amount of work required to remove an atom from the solid state to the vapor state, whereas the compressibility is a measure of the energy required to bring about a small displacement of the atoms from their

TABLE M.7.5 Structure of the transition metals [a]

Metal	Nuclear Charge Number	Lattice Constants ($\times 10^{-10}$ m)				
		c.c.p. a_0	b.c.c. a_0	h.c.p. a_0	c_0	Other
Scandium, Sc	21	4.541		3.3090 (293 K)	5.2733 (293 K)	
Titanium, Ti	22		3.3065 (1173 K)	2.950 (298 K)	4.686 (298 K)	
Vanadium, V	23		3.0240 (298 K)			
Chromium, Cr	24	3.68	2.8839 (298 K)	2.722	4.427	
Manganese, Mn	25	3.868 (1373 K)	3.081 (1413 K)			α,β [b]
Iron, Fe	26	3.5910 (295 K)	2.8665 (298 K) α			
			2.91 (1073 K) β			
			2.94 (1698 K) δ			
Cobalt, Co	27	3.548		2.5071 (293 K)	4.0686 (293 K)	
Nickel, Ni	28	3.52387 (298 K)		2.65	4.33	
Copper, Cu	29	3.61496 (291 K)				
Zinc, Zn	30			2.6648 (298 K)	4.9467 (298 K)	
Yttrium, Y	39			3.6474 (293 K)	5.7306 (293 K)	
Zirconium, Zr	40		3.62 (1123 K)	3.232 (298 K)	5.147 (298 K)	

				?	?
Niobium, Nb	41		3.3004 (291 K)		
Molybdenum, Mo	42	4.16	3.1473 (298 K)		
Technetium, Tc	43				
Ruthenium, Ru	44			2.70389 (293 K)	4.28168 (293 K)
Rhodium, Rh	45	3.8031 (298 K)			
Palladium, Pd	46	3.8898 (298 K)			
Silver, Ag	47	4.0862			
Cadmium, Cd	48			2.97887 (299 K)	5.61765 (299 K)
Lanthanum, La	57	5.296		3.75	6.07
Hafnium, Hf	72			3.1967 (299 K)	5.0578 (299 K)
Tantalum, Ta	73		3.3058 (298 K)		
Tungsten, W	74		3.16469 (298 K)		
Rhenium, Re	75			2.7608 (298 K)	4.4582 (298 K)
Osmium, Os	76			2.7352 (293 K)	4.3190 (293 K)
Iridium, Ir	77	3.8394 (299 K)			
Platinum, Pt	78	3.9231 (298 K)			
Gold, Au	79	4.07285 (298 K)			
Mercury, Hg	80			3.457 (278 K)	6.663 (278 K)[c]
				3.464 (351 K)	6.667 (351 K)

[a] Lattice constants are those listed in *Crystal Structures*, Vol. I, 2nd edition, by Ralph W. G. Wyckoff, Interscience Publishers, John Wiley & Sons, New York, 1963.
[b] See page 452.
[c] See page 721.

equilibrium position in the solid. The coefficient of thermal expansion, the compressibility, and the atomic radius, of a metal are properties of the solid state alone, and although a high melting point denotes resistance to thermal disturbance and strong atomic bonding, this property involves equilibrium between the solid and the liquid state. There is a distinct difference between the metals of the first transition series compared with the other two series, with respect to each of the above properties, and the non-sequential behavior generally occurs in the vicinity of the element in each series where the set of d-states is half filled.

The transition metals are generally polymorphic, with the body-centered cubic modification apparently the most stable near the middle of each of three series. Manganese is exceptional (see page 452), crystallizing with four different structures, two of them complex, and iron is also unique, as noted above. The crystal structures of the various transition metals are listed in Table M.7.5. For structures with each atom having 12 nearest neighbors, as in the face-centered cubic and hexagonal close packed arrangements, if each atom contributes b-electrons towards bond formation, then the number of ways in which these b-electrons can be distributed among the 12 positions is $n = [12!][(12 - b)!\,b!]^{-1}$, which is a maximum for $b = 6$. For the body-centered cubic modification, with eight nearest neighbors and six more neighboring atoms at only a 15% greater distance away, the maximum value of n in the above expression occurs at $b = 4$ and $b = 7$ for coordination numbers 8 and 14, respectively. If the actual coordination number is somewhere between 8 and 14, then the maximum would again be at about $b = 6$. This is consistent with Pauling's theory of metals and indicates maximum cohesion for about six bonding electrons per atom, corresponding to the limited number of atom pairs in any structure between which bonds may be formed.

At the beginning of each transition metal series the order of the electronic energy levels of the free atoms is $ns < (n - 1)\,d < np$, but at the end of each series the order is $(n - 1)\,d < ns < np$. In the solid crystalline metals which result when the individual metal atoms assemble and distribute themselves at the observed internuclear distances over the lattice sites of the particular crystal structure concerned, each sharp atomic energy level broadens out into a band of energies. It is generally agreed that up to the point at which the discontinuities in the sequence of observed physical properties occur, all of the electrons beyond the numbers 18 and 36 for the first two transition metal series, respectively, and beyond the number 54 for the third transition metal series, with the insertion of the 14 $4f$-electrons corresponding to the 14 lanthanide elements between lanthanum and hafnium, are involved in the metallic cohesion process, and that mixed (spd)-states are involved. Again, it is generally agreed that for the elements with nuclear charges greater than

30, 48, and 80, corresponding to the end of each of the transition metal series, the $(nd)^{10}$ electron groups play no part in metallic cohesion. The Pauling theory and some of the modifications which have been proposed along this direction, has already been discussed in Chapter M.2.

M.7.12. Transition Metal Alloys

The qualitative generalizations and empirical principles which appear to apply to the alloys formed by the various transition metals have been deduced from observations of a large number of intermetallic systems, and from the crystal structures determined for a limited number of intermetallic phases. From Fig. M.7.15.(b), it is apparent that among the transition metals there are a considerable number of metal atoms of comparable "size," so that there are many pairs of these metals which can form binary substitutional solid solutions with one another; this situation is especially noticeable among the middle members of the second and third transition metal series, where the contraction effect due to the formation of the lanthanide elements brings the metals in these groups even closer together in "size." Substitutional solid solutions among these metals are limited, as for other metal pairs, by the consideration that extensive solid solution between metal pairs only occurs if the respective atomic radii differ by less than 15%, by electron to atom ratios, and by the tendency to form compounds for metal pairs differing greatly in electronegativity. In considering the "size" factor, it has to be kept in mind that, for example, the effective radius of aluminum varies from 1.35×10^{-10} m to 1.40×10^{-10} m in solid solutions compared with 1.43×10^{-10} m in the pure metal, and that similar variations may well apply to the transition metals, quite apart from the influence of the coordination number on metallic radius in different structural arrangements.

For continuous ranges of solid solution to be possible between metal pairs, the two metals must have identical structures, and the dimensions of the crystallographic unit cell are determined by the relative radii. Thus continuous solid solutions are formed in the face-centered cubic copper–nickel and silver–gold systems, and in the body-centered cubic molybdenum–tungsten system, but not in the body-centered cubic molybdenum and face-centered cubic silver systems, even although the radii in this latter case differ by less than 6%. Solid solutions in the close packed hexagonal modifications of transition metals are much more restricted than in either the face-centered cubic or body-centered cubic forms. The formation of superlattices or superstructures in the transition metal alloys occurs more frequently than would be anticipated from differences in electronegativity and metallic radii, as observed, for example, in the iron–nickel and iron–cobalt systems, but in general ordered phases become more stable with increasing differences in metallic radii.

Continuous solid solution formation is also favored by the mutual metal pairs if each type of atom contributes the same number of valency electrons. Elements providing a larger number of valency electrons per atom are generally more soluble in metals providing fewer valency electrons per atom, than the reverse of this. For example, where the metallic radii, structure, and electronegativity differences, are otherwise favorable to solid solution formation, up to 42.5% of cadmium dissolves in silver, but only 6% of silver dissolves in cadmium; similarly, in the copper–silicon system, up to 11.5% of silicon dissolves in copper but only 2% of copper in silicon. The so-called Vegard's Law, that where complete solid solubility occurs, the unit cell dimensions of the solid solution are a linear function of the composition, is however more commonly applicable to isomorphous inorganic compounds than to intermetallic systems.‡

‡ With increasing electronegativity differences between the two constituents of a binary pair, alloys which may be stable phases over either a wide or narrow range of composition and immiscible with either pure component, may be formed; many such phases are formed with large enthalpies of formation and large volume changes. According to the Pauling concept of electronegativity,[1] this quantity is based on the difference in enthalpy between $[(H_X + H_Y)/2]$ and H_{XY}, where the compound XY is formed from X and Y. The volume change effect on compound formation, which may be correlated with the electronegativity, is the percentage volume decrease, $100[(V_X + Y_Y) - V_{XY}]/[V_X + V_Y]$. If the difference in the heat of formation is plotted as a function of the volume change, the curve shown in Fig. M.7.16. results.[2] There is a continuous relationship between alloys such as AgZn and ionic compounds like MgO; both the rock salt and the cesium chloride structures occur along the whole series of compounds, but changes in electrical conductivity and other physical properties along the series indicate that there is a change from metals to salts. With compound formation, the electrons from the reactant species are on the average exposed to an increased nuclear charge resulting in a contraction in internuclear distances, which is accompanied by a charge transfer, so that the relative sizes of the reactant atoms or ions, are functions of the electronegativity difference.

Apart from the order–disorder transformations in metal systems, which are temperature dependent, as for example, in the copper–gold system (page 403, section M.7.6.), there is another way in which order–disorder transitions may take place; this is a magnetic ordering effect, detectable by neutron diffraction techniques. Thus, when small amounts of copper are alloyed with the γ-phase of manganese, the resultant copper–manganese alloy crystallizes with the face-centered cubic structure of γ-Mn. In this structure, it is found by neutron diffraction examination that each atom has four nearest neighbors in the same plane (001) with parallel spin orientations and eight nearest neighbors, four in the plane above and four in the plane below, with antiparallel spins

[1] Pauling, Linus, *The Nature of the Chemical Bond*, 3rd edition, 1962, Cornell University Press, Ithaca, New York.
[2] Kubaschewski, O., and Evans, E. L., *Metallurgical Thermochemistry*, 1956, 2nd edition, Pergamon Press, London.

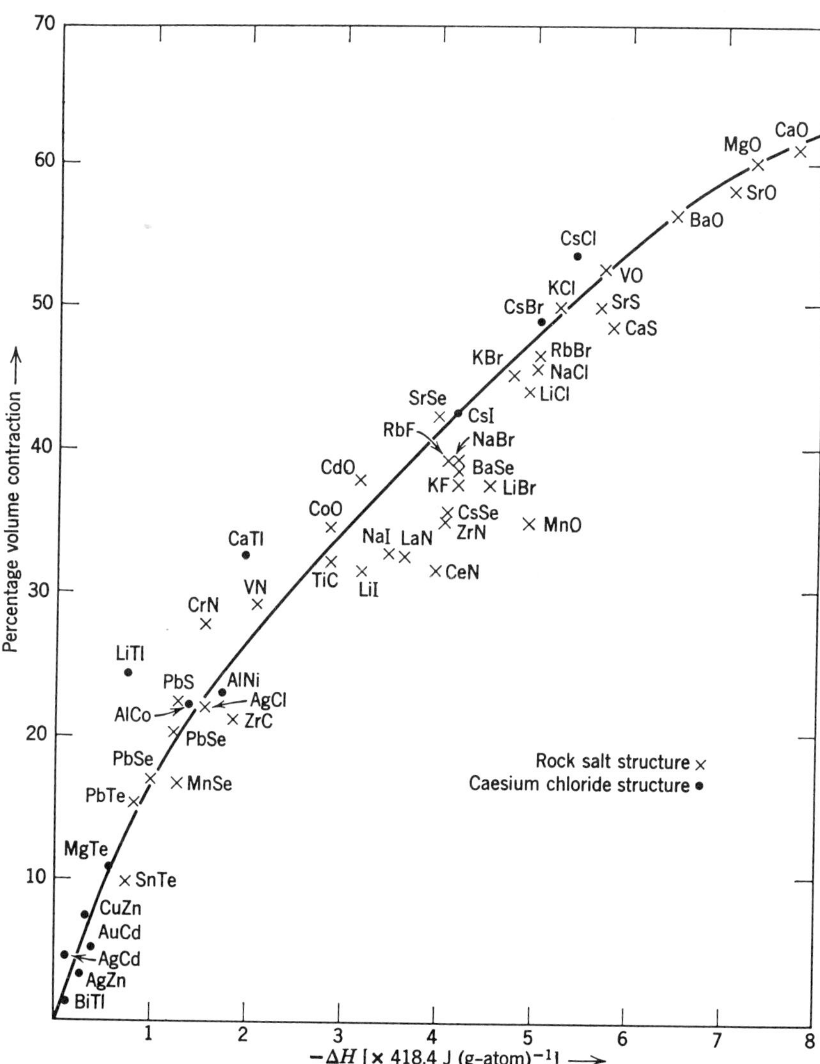

FIG. M.7.16. The heat of formation of various binary compounds plotted as a function of the volume contraction on formation of the compound from the solid elements.

(Fig. M.7.17.). This type of experimental observation is possible since neutrons interact with the magnetic moments of atoms. The spin arrangement in the unit cell of this structure has only tetragonal symmetry. It was Zener[1] who originally, in an attempt to account for the structures adopted by the various

[1] Zener, C., *Trans. A.I.M.E.*, Vol. 203, 1955, 619.

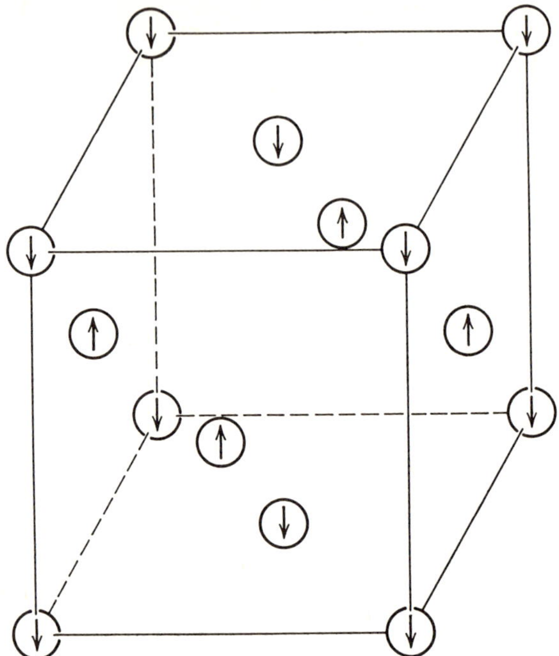

FIG. M.7.17. Ordered magnetic structure for Mn–Cu alloys obtained by alloying a small amount of copper with γ Mn, which crystallizes with the cubic closest packed structure. Alternating planes (001) contain atoms with opposed spin moments.

transition metals, ranging from cubic closest packed to body-centered cubic to hexagonal closest packed, proposed that magnetic interactions between the unpaired d-electrons of adjacent atoms may account for some of the structures found. For chromium, for example, the individual atoms with the configuration $1s^2 2s^2 2p^6 3s^2 3p^6 4s^1 3d^5$, have a completely half filled set of d-states in which the electrons all have parallel spins according to Hund's rules (page 87). By analogy with the hydrogen molecule, and it can be shown that, the energy of the chromium crystal is lowered if the spin orientation of the electrons in overlapping atomic orbitals is such that they are opposed. This would imply that the most stable structure for chromium would be that in which each atom is completely surrounded by atoms having opposed spin moments. But such an arrangement is not possible in a close packing, in which all the atoms must be identical, but it is possible for a body-centered cubic structure. Thus chromium crystallizes with a body-centered cubic structure in which each chromium atom has eight nearest neighbors with opposed spin moments. It has been shown otherwise that the exchange integral favoring antiparallel spin orientations in overlapping $3d$-orbitals lowers the crystal energy and increases its cohesion. Now, if the face-centered tetragonal cell in the Cu–Mn alloy is selected to describe this structure, it emerges that the axial ratio c/a is less than unity, in agreement with Zener's postulate that,, because of the exchange integral favoring antiparallel spins, the energy of the crystal is lowered by this distortion of the unit cell, and this is so because the bonding forces normal to the (001) planes are greater than those within the (001) planes. At high temperatures, the thermal energy is greater than that due to the magnetic interaction, so that the spin moments of nearest neighboring atoms are randomly oriented, but below 523 K, the ordering effect of the exchange forces exceeds the disordering effect of the thermal vibrations, so that the spin moments assume the ordered array shown in Fig. M.7.16. In this particular case, the cubic to the tetragonal disorder and the transformation back to the ordered spin arrangement does not involve any atomic positional rearrangement, apart from a small contraction parallel to the c direction. Similar types of magnetic transformations from paramagnetic to ferromagnetic states in iron, cobalt, nickel and their alloys take place, and many similar types of magnetic transformations have been detected in non-metallic phases.

The type of interaction between binary pairs of transition metals can be classified approximately into three groups:[1]

(a) Where the metal pairs are similar in electronegativity, radii, and electron to atom ratios, there is complete miscibility in the solid state, with superlattice formation at lower temperatures. This occurs, for example in the Ti–V, Ti–Cr, V–Cr, Fe–Cr, V–Fe, Fe–Mn, Fe–Co, Fe–Ni, Ni–Mn, Ni–Co, Ni–Cu, Cr–Mo, Cr–W, Co–Pd, Co–Pt, Ni–Pd, Ni–Pt, Zr–Nb, Mo–W, Ru–Pt, Os–Pt, Rh–Pt, Ir–Pt, Pd–Pt, Pd–Au, Pd–Ag, and Pt–Au, systems.

(b) Where the region of mutual solubility does not extend beyond about 20% of one metal in the other, the metals differ considerably from one another, and a eutectic is formed, and there may be a variety of solid phases between the two regions of mutual solubility. This type of behavior occurs, for example, in the Ti–Mn, Ti–Fe, Ti–Co, Ti–Ni, Ti–Cu, Ti–Zn, Cr–Ni, Cr–Cu, Cr–Zn, Ni–V, Co–W, Cr–Pd, Cr–Pt, Zr–Mo, Zr–W, Zr–Ag, systems.

(c) Where there is more extensive solid solution formation and the behavior between the metal pairs is more or less intermediate between the cases (a) and (b) above. This situation arises in the Co–Cr, Ni–Zn, Ni–W, Au–Zr, Pt–W, and Ag–Pt, systems.

---

[1] Hume-Rothery, W., *J. Less-Common Metals*, 1964, **7**, 152.

For ternary systems, there may be a wide variation in the composition of certain intermediate phases in the system, provided that the electron to atom ratio is not affected by the compositional change. This happens particularly in alloy systems containing the transition metals with aluminum and silicon. In the Fe–Ni–Al system, for example, an intermediate phase exists having the cesium chloride structure which can accommodate a continuously variable amount of iron or nickel. Such intermediate phases may therefore be classified as Hume-Rothery electron compounds (page 404). The type of bond between the metal atoms involved, as well as the relative radii, appear to be important in determining the stability of such phases. The Laves phases (408) in which the electron to atom ratio varies from about 5:4 to 2:1, appear to have their stability determined to a large extent by the electronic structure of the various atoms involved, but a small electronegativity difference between atoms of similar numbers of valency electrons apparently favors the formation of Laves phases.

An intermediate phase formed by alloys of the transition metals, Mn, Fe, Co, and Ni, with Cr, V, Mo, and W, is especially important in the manufacture of special alloy steels, because this so-called σ-phase is very hard, brittle, and non-magnetic at ordinary temperatures. This σ-phase occurs at approximately the 50:50% composition, for example, in the iron–chromium system, as shown in Fig. M.7.18. The composition and the stability range of this phase varies from one system to another; in the Mn–Cr system, the σ-phase is stable up to its melting point over the composition range of 17–28 atomic percent of chromium, but it is formed only in the solid state in the Cr–Fe and Cr–Co systems with compositional ranges of 43–50 atomic percent and 53–58 atomic percent of chromium, respectively. The σ-phase appears to be stabilized by the addition of a third component, and occurs in ternary systems even when the three possible binary pairs which may be formed from the three components do not contain it. The crystal structure of the σ-phase has been determined.[1] For the Fe–Cr system, it crystallizes with a complex tetragonal structure with 30 atoms per unit cell and $a_0 = 8.7990 \times 10^{-10}$ m, and $c_0 = 4.559 \times 10^{-10}$ m. In this structure, only slightly distorted hexagonal layers of atoms are stacked in the tetragonal unit cell, with orientations differing by 90°, and spaced 2.28×10^{-10} m apart, which is a much shorter distance than either the Fe–Fe distance in α-Fe (2.48×10^{-10} m) or the Cr–Cr distance in chromium metal (2.49×10^{-10} m). The proposed Brillouin zone for this structure may contain up to 6.97 electrons per atom and the mean electron to atom ratio for most of the known binary σ-phases is 6.93, although based on the oxidation states deduced from the observed internuclear distances, it may be as few as 5.76 electrons per atom. For the

---

[1] Shoemaker, D. P., and Bergman, B. G., *J. Am. Chem. Soc.*, 1950, **72**, 5793.

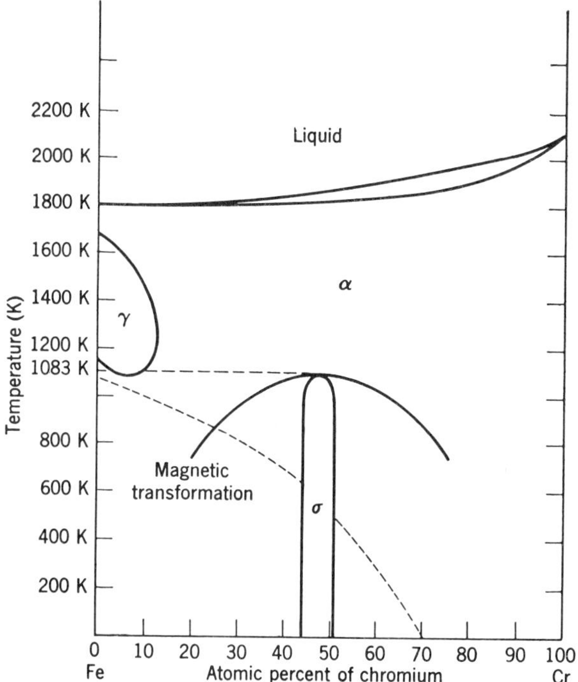

FIG. M.7.18. Equilibrium phase diagram for the iron-chromium system, showing the region of stability of the σ-phase.

Fe–W and Co–W systems the component metal radii differ by about 12% which is the most extreme (high) radius ratio for any σ-phase, and in these systems the phase is only stable at high temperatures. The stability of the σ-phase decreases generally with increasing radius ratio of the component metal atoms. It was suggested by Frank and Kasper[1-3] that the structure of the σ-phase and other related phases can be described in terms of spheres packed together in such a way as to give rise to coordination polyhedra with 12, 14, 15, and 16, vertices (Fig. M.1.13.). Crystal structural arrangements containing only these polyhedra can possess only tetrahedral cavities, and these polyhedra arise by systematically omitting certain spheres in hexagonal close packed layers to form so-called Kagomé nets; these nets may be stacked in several ways to form large voids surrounded by either 13, 14, 15, or 16,

[1] Bergman, B. G., and Shoemaker, D. P., *J. Chem. Phys.*, 1951, **19**, 515; *Acta Cryst.*, 1954, **7**, 857.

[2] Kasper, J. S., *Am. Soc. of Metals Symposium, Theory of Alloy Phases*, 1956, 264.

[3] Frank, F. C., and Kasper, J. S., *Acta Cryst.*, 1958, **11**, 184.

spheres. The analysis of the structures of alloy phases in terms of the different ways of stacking double layers, similar to the stacking of layers of octahedra (Fig. M.1.2.) and of Laves polyhedra (Fig. M.1.12.), has been carried out[1] with reference to the σ-phase, the Laves phases, and certain other intermediate phases observed in alloys of the transition metals.[2] Included among such phases are β-uranium, found to be isostructural with the σ-phase,[3] the μ-phase[4] as for example $Mo_{46}Co_{54}$, the δ-phase[5] as in $Mo_{50}Ni_{50}$, the P-phase[6] as in $Mo_{42}Cr_{18}Ni_{40}$, the R-phase[7] as in $Mo_{31}Cr_{18}Co_{51}$, the χ-phase[8] as in $Mo_{17}Cr_{21}Fe_{62}$, the D-phase[9] as in $V_{26}Fe_{44}Si_{30}$, the Zr_4Al_3 alloys,[10] the β-W phase[11] as in W_3O and Cr_3Si, and the α-Mn arrangement.[12]

M.7.13.   Interstitial Alloys

In closest packed arrangements of atoms (pages 386–395), the octahedral cavities have a radius $r_o = 0.414R$, where R is the radius of the closest packed atoms, and the tetrahedral cavities a radius $r_t = 0.225R$, so that small atoms such as those of hydrogen, boron, carbon, and nitrogen, may enter into interstitial solid solution. Among the transition metals, titanium, vanadium, zirconium, hafnium, and tantalum, in particular, form especially stable carbides and nitrides, with all of the octahedral holes occupied by carbon or nitrogen atoms. The smaller hydrogen and boron atoms usually tend to occupy the smaller tetrahedral holes in such interstitial compounds. The physical properties of interstitial alloys indicate that if the host species is one of the transition metals, the bonding between the interstitial species and the solvent species is essentially metallic, but becomes more similar to normal electron pair bonds with more electropositive metals; stable solid solution tend to be formed if the radius ratio for octahedrally substituted atoms is less than about 0.59 and electron pair bonds tend to be formed between the constituent species for radius ratio values exceeding this amount. In the latter event, the compounds formed are generally semiconductors rather than

[1] Frank, F. C., and Kasper, J. S., *Acta Cryst.*, 1959, 12, 483.

[2] Shoemaker, D. P., and Shoemaker, C. B., in *Structural Chemistry and Molecular Biology*, 1968, Edited by A. Rich and N. Davidson, W. H. Freeman & Co., San Francisco, 718–730.

[3] Tucker, C. W., and Senio, P., *Acta Cryst.*, 1953, **6**, 753.

[4] Forsyth, J. B., and da Veiga, *L.M.d'Alte*, *Acta Cryst.*, 1962, **15**, 543.

[5] Shoemaker, C. B., and Shoemaker, D. P., *Acta Cryst.*, 1963, **16**, 997.

[6] Shoemaker, D. P., Shoemaker, C. B., and Wilson, F. C., *Acta Cryst.*, 1957, **10**, 1.

[7] Komura, Y., Sly, W. G., and Shoemaker, D. P., *Acta Cryst.*, 1960, **13**, 575.

[8] Kasper, J. S., *Acta Metallurgica*, 1954, **2**, 456.

[9] Shoemaker, D. P., and Shoemaker, C. B., *Structural Chemistry and Molecular Biology*, 1968, Edited by A. Rich and N. Davidson, W. H. Freeman & Co., San Francisco.

[10] Wilson, C. G., Thomas, D. K., and Spooner, F. J., *Acta Cryst.*, 1960, **13**, 56.

[11] Hägg, G., and Schönberg, N., *Acta Cryst.*, 1954, **7**, 351.

[12] Bradley, A. J., and Thewlis, J., *Proc. Roy. Soc.*, 1927, **A115**, 456.

metallic substances, and many have the diamond, zinc blende, or würtzite structure.

If all of the octahedral holes in a cubic closest packed structure are occupied by interstitial atoms, the rock salt structure results, as in many nitrides and carbides of the transition metals; these substances generally form continuous series of substitutional solid solutions with one another. The system ZrC–VC is exceptional in this respect, since zirconium atoms are about 21% larger than vanadium. Tungsten and molybdenum form nitrides in which only one-half of the octahedral holes are occupied by nitrogen atoms, giving rise to the compositions W_2N and Mo_2N. If only those holes at the body centers of the face-centered cubic structure are occupied, that is only one quarter of the total number of octahedral holes, then composition of the type Mn_4N and Fe_4N result. It is rather remarkable that in the carbides with the rock salt structure, where the metal atoms are cubic closest packed, the pure metals do not crystallize with this structure. Such refractory carbides, with melting points in the range 3300–4300 K, do not contain purely metallic bonding. If carbon or nitrogen atoms are distributed among the octahedral holes in a hexagonal closest packed arrangement, similar structures arise; thus, in Fe_3N one-third of the available holes between each layer is occupied such that no two nitrogen atoms are nearest neighbors, and in Fe_2N alternate layers have one-third and two-thirds of the available octahedral holes occupied by nitrogen atoms. Similarly, in V_2C and Mo_2C, one-half of the octahedral holes in a hexagonal closest packed array of metal atoms is occupied by carbon atoms. More complicated structural arrangements are possible by combining the two types of layers in different ways, but in all cases the interstitial atoms tend to be as far apart as possible.

The interstitial solid solutions of carbon in iron are especially important, because of the industrial importance of such alloys. Plain carbon steels are iron–carbon alloys to which no other metal has been added. The other transition metals which are widely used in special steels are vanadium, chromium, manganese, nickel, molybdenum, and tungsten. The stable modification of iron at temperatures below 1183 K is body-centered cubic α-Fe; the face centered cubic γ-Fe is stable in the range from 1183 K to 1673 K, and the body centered cubic phase δ-Fe exists up to the melting point of the metal at 1807 K. The γ-modification is the only form which dissolves carbon interstitially to any extent, as due to the octahedral holes in the face-centered cubic structure it can take up to about 2% by weight of carbon. The interstitial solid solution in which carbon is distributed at random among the octahedral holes is referred to as austinite. The α-Fe is ferromagnetic up to 1041 K, above which temperature it becomes paramagnetic without any structural transformation. The body centered cubic α-Fe can only contain up to a maximum of 0.025% by weight of carbon in interstitial solution, and this phase is usually referred to as ferrite. An intermediate phase containing 6.67% by weight of carbon, called cementite, corresponds to the composition Fe_3C, which appears to be stable only when the composition is very nearly stoichiometric, in contradistinction to most intermediate metallic phases. Cementite has a structure which is more complicated than that of a simple interstitial solid solution. It crystallizes in the orthorhombic system with the carbon atoms located at the centers of slightly distorted trigonal prisms and the iron atoms with either 11 or 12 nearest neighbors at distances ranging from 2.49–2.68×10^{-10} m. When the austenite phase is cooled quickly, it undergoes what is referred to as the martensite transformation, to give rise to a body centered tetragonal structure, which is only slightly distorted from the rock salt structure by the carbon content. If austenite is cooled slowly from the melt, it transforms into pearlite, a crystalline intergrowth of characteristic appearance of ferrite and cementite.

If all of the tetrahedral holes in a cubic closest packed structure are occupied by interstitial atoms, the fluorite structure results, and if only half of the tetrahedral holes are occupied, the zinc blende structure is obtained (Fig. M.1.8.). For example, TiH_2 and CrH_2 adopt the fluorite structure, and ZrH, TiH, and CrH, adopt the zinc blende arrangement. If a smaller fraction of the tetrahedral holes are occupied in this way, then compositions such as Pd_2H, with one-quarter of the tetrahedral holes, and Zr_4H, with one-eighth of the tetrahedral holes, occupied by hydrogen atoms, may be obtained. In TiB_2, VB_2, ZrB_2, and NbB_2, hexagonal structures arise, in which the metal atoms form hexagonal layers separated from one another by layers of boron atoms, and such that each boron atom is coordinated with a trigonal prism of metal atoms. The crystal structure of the metal borides are more complicated than those of the interstitial hydrides, nitrides, or carbides, indicating that bond formation takes place between adjacent boron atoms in these structures. Nevertheless, the electrical conductivity of several borides is greater than that of the parent element, as in TiB_2 and ZrB_2. In FeB and CoB, the boron atoms appear to be bonded together in an infinite chain of trigonal prisms. In the boride, UB_{12}, there are four UB_{12} groups per face centered cubic unit cell, such that each uranium atom is coordinated with 24 boron atoms located at the vertices of a cubo-octahedron, and each boron atom has five boron atoms and two uranium atoms as nearest neighbors.[1]

M.7.14. Magnetic Properties

The magnetic properties of metals have already been briefly referred to on page 399 and in section M.4.6. The ferromagnetic properties of the mineral magnetite and of other iron oxides have been known for a very long time,

---

[1] Bertaut, F., and Blum, P., *Compt. rend.*, 1949, **330**, 666.

and it was not until much later that it was realized that other substances exhibited magnetic properties. The origin of all types of magnetization in atoms, molecules, and crystals, is associated with the orbital and spin angular momenta of the electrons in their constituent atoms. Diamagnetism is present in all substances, even in atoms in ^1S-states, although it is usually obscured when other types of magnetic behavior are exhibited, such as paramagnetism, ferromagnetism, anti-ferromagnetism, and ferimagnetism (page 399). The rare gases and the elements with immediately preceeding nuclear charges are generally weakly diamagnetic, with gram atomic susceptibilities (page 515) of the order of magnitude of 0–50×10^{-6}, although antimony and bismuth are more strongly diamagnetic. The alkali and alkaline earth metals are either weakly paramagnetic or weakly diamagnetic, with gram atomic susceptibilities of the order of $\pm 5 \times 10^{-5}$. The transition metals are more strongly paramagnetic, and the lanthanide metals even more so, with gram atomic susceptibilities in the latter group being about ten times as large as for typical transition metals. Ferromagnetism is confined to iron, cobalt, and nickel of the first transition metal series, although the ferromagnetism of nickel is destroyed at a fairly low temperature (626 K), and to a few of the lanthanide metals such as gadolinium, dysprosium, erbium, and holmium, which are ferromagnetic at low temperatures; all ferromagnetic substances become paramagnetic at sufficiently high temperatures. The paramagnetism of barium is abnormally high, and among the three transition metal groups, those of odd nuclear charge (Sc, V, Mn, Y, Nb, Ta, and Re, with nuclear charges, 21, 23, 25, 39, 41, 73, and 75, respectively) have high paramagnetic susceptibilities compared with the intervening metals with even nuclear charge numbers (Ti, Cr, Zr, Mo, Hf, and W, with nuclear charge numbers of 22, 24, 40, 42, 72, and 74, respectively) which have considerably lower susceptibilities. The congeners of the ferromagnetic metals iron, cobalt, and nickel, of the first transition metal series, namely ruthenium, rhodium, and palladium, in the second transition metal series, exhibit a marked increase in paramagnetic susceptibility in this order (Ru, Rh, and Pd), the magnitudes of which are systematically greater than those of osmium, iridium, and platinum, of the third transition metal series.

The simplest example of diamagnetism is that of a monatomic gas whose atoms are in the ^1S-state, and is a purely classical effect,[1] in which an induced magnetic moment, proportional to and oppositely directed to an applied external magnetic field arises due to the induced currents set up by the magnetic field in the process of building up the field from zero. According to Faraday's law of electromagnetic induction, the rate of change of the magnetic field

---

[1] Kittel, C., *Introduction to Solid State Physics*, 2nd Edition, 1957, John Wiley & Sons, Inc., New York.

produces an induced electromagnetic force, which accelerates the electrons and their resulting currents produce the diamagnetic moments of the atoms. The diamagnetism of metals is more complicated (see page 515 and the reference on page 516). Diamagnetic susceptibilities, deduced classically,[1,2] are found to differ from unity by the very small factor of 10^{-6}, as observed experimentally, indicating that diamagnetism is not a quantum phenomenon, provided that such effects as the de Haas–van Alphen effect are disregarded.

For atoms not in a ^1S-state, there is a net magnetic moment, and such atoms, in addition to the diamagnetic effect exhibit a magnetism proportional to and in the direction of an externally applied magnetic field, resulting from the orientation of the permanent magnetic moments in the applied field. This paramagnetism is closely related to the Zeeman effect (sections Q.6.4. and Q.6.7.). In the Zeeman effect, an atom with resultant angular momentum $J(h/2\pi)$, where J is the resultant quantum number, may be oriented in $2J + 1$ directions in which the component of the angular momentum along the axis of the magnetic field is $M(h/2\pi)$, with M ranging from a maximum of $+J$ to a minimum of $-J$. Since the magnetic moment of an atom is proportional to its angular momentum, these $2J + 1$ orientations correspond to different components of magnetic moment along the axis and therefore to different magnetic energies in the externally applied magnetic field. The magnetic energy in the field direction is $gM\mu_B B$ and the magnetic moment along the direction of the field H, is $-gM\mu_B$, where g is the Landé factor (page 344) and μ_B is the Bohr magneton. For any particular temperature, as a consequence of the effect of the Boltzmann factor, $e^{-E/kT}$, in the Boltzmann distribution function, which is applicable in this case, it is more probable that atoms will exist in states of lower energy, and the average magnetic moment in the field direction is

$$\text{Mean magnetic moment} = \sum_{M=-J}^{J} [-gM\mu_B]e^{-gM\mu_B B/kT} \left[\sum_{M=-J}^{J} e^{-gM\mu_B B/kT} \right]^{-1}.$$

$$\text{(m.7.60.)}$$

When J is very large, the summation in the above expression may be substituted by an integration, and the magnetic moment evaluated in terms of what is referred to as the Langevin function, $L(y) = \coth y - 1/y$, where $y = Jg\mu_B B/kT$, as

$$\text{Mean magnetic moment} = Jg\mu_B L(y). \qquad \text{(m.7.61.)}$$

The analytical nature of the Langevin function is such that the induced magnetic moment is proportional to the field strength for small values of H

---

[1] Slater, J. C., and Frank, N. H., *Electromagnetism*, 1947, McGraw-Hill Book Co., New York.

[2] Martius, U. M., *Progress in Metal Physics*, Vol. III, 1952, Pergamon Press, London.

but approaches a saturation value at very high field strengths, the latter corresponding to the situation where the magnetic moment is aligned parallel with the field direction; the magnitude of the magnetic moment of the atom is $Jg\mu_B$. For normal magnetic fields, the magnitude of H is sufficiently small to make the magnetic moment proportional to H, and in this region if the Langevin function $L(y)$ is expanded in power series, it is found that $L(y)$ in this region behaves as $y/3$. Therefore, in this linear range, the average magnetic moment is given by $[Jg\mu_B]^2 B/3kT$; thus, the extent of the magnetization is proportional to the magnetic field strength but inversely proportional to the absolute temperature. It is possible to find the mean magnetic moment by summing (m.7.60.) exactly, instead of assuming that J is very large, by expanding in power series, to give

Average magnetic moment

$$
\begin{aligned}
&= \sum_{-J}^{J} [-gM\mu_B][1 - gM\mu_B B/kT + \cdots]\left[\sum_{-J}^{J} [1 - gM\mu_B B/kT + \cdots]\right]^{-1} \\
&= [(g\mu_B)^2 B][kT(2J + 1)]^{-1} \sum M^2, \qquad \text{since } \sum M = 0, \\
&= J(J + 1)[g\mu_B]^2 B[3kT]^{-1}, \qquad\qquad\qquad\qquad\qquad \text{(m.7.62.)}
\end{aligned}
$$

since $\sum M^2 = (2J + 1)J(J + 1)/3$. It thus emerges that by making use of the exact expression for the summation in (m.7.60.) instead of the integral approximation, the term J^2 is replaced by $J(J + 1)$ in the derived equation for the average magnetic moment.[1] In the above expressions, the quantity B corresponds to the total magnetic flux in the material composed of the atoms concerned when in the presence of a magnetic field of field strength H gauss; it should be noted that the unit of magnetic flux density for magnetic induction referred to as the Gauss is not strictly an *SI* unit (pages 12–15), but is actually 10^{-4} T, where T, the tesla is the *SI* unit, which has the dimensions of $kg\,s^{-2}\,A^{-1} = V\,s\,m^{-2}$ (see, for example, the Notice to Authors, printed on the back inside cover of the *Journal of the Chemical Society*, No. 2, 1970). The quantity B is equivalent to $H + 4\pi I$, where I is the intensity of the induced magnetization. The ratio $B/H = \kappa_m$, is referred to as the magnetic permeability of the material and the quantity I/H is the volume susceptibility of the material, χ. If the expression for the total magnetic flux is divided throughout by H, the equation, $\kappa_m = 1 + 4\pi\chi$ is obtained. In a vacuum $B = H$, and there is no induced magnetization, so that $\chi = 0$. In any other medium, χ is positive for a paramagnetic medium, or negative for a diamagnetic medium. In most chemical problems and applications dealing with paramagnetic salts, it is more usual to deal with the magnetic susceptibility per gram mole rather than the volume susceptibility; the molar susceptibility is defined by $\chi_M = M\chi/\rho$, where ρ is the density. Returning to

---

[1] Fleck, J. H. van, *Electric and Magnetic Susceptibilities*, 1932, Oxford University Press.

equation (m.7.62.), the magnetic permeability may be written as

$$\kappa_m = 1 + 4\pi\chi = 1 + 4\pi I/H = 4\pi N(g\mu_B)^2 J(J+1)/3kT + 1, \quad \text{(m.7.63.)}$$

where N is the number of dipoles per unit volume. This expression is strictly applicable to paramagnetic monatomic gases, but there are few if any such gases on which magnetic measurements can be carried out, although there are some paramagnetic molecules such as oxygen; in the latter case, however, there are in addition to the Zeemann components of the ground-state other multiplet levels lying at low energy values, so that the calculation of the magnetic permeability in this case is more complicated. The major use of equation (m.7.63.) actually lies in its application to inorganic paramagnetic salts, which behave in many ways like isolated magnetic species separated by non-magnetic species, so that there is no direct interaction between the magnetic species; the situation is thus similar to that of a paramagnetic gas. In a typical compound, there are transition or lanthanide metal ions surrounded by an octahedral ligand field, or a ligand field of some other symmetry, of oxide, fluoride, cyanide, or some other simple or complex ionic types. The paramagnetism originates from the partially filled 3d, 4d, 5d, or 4f, sets of states in the metal ion. The lanthanide metal salts behave most nearly like paramagnetic gas atoms, since the 4f-electron states are so far inside the metal atom or ion that they are essentially unaffected by even the nearest neighboring atoms or ions, and the paramagnetic permeabilities are such that the magnetic moment is inversely proportional to temperature, as indicated by (m.7.63.). In the iron group transition metals, however, the 3d-states in such metal ions are not so protected from interaction with adjacent species, and there is considerable interaction between neighboring ions in crystals of these metal salts; in terms of band theory, there is considerable broadening of the 3d-energy bands in the case of the paramagnetic compounds of the first transition metal salts. Where there is appreciable interaction between adjacent magnetic ions, the phenomenon of ferromagnetism arises, in which all the ions in a crystal tend to line up parallel to one another. The temperature dependence of paramagnetic susceptibility is expressed by the empirical Curie law, $\chi = C/T$, where C is a constant, or by the Curie–Weiss law, $\chi = C/(T - \Theta)$, where Θ may be interpreted as representing the mutual interaction between the magnetic ions. The term Θ has the dimensions of temperature and may be positive or negative. When it is positive, there will be a certain value of T at which the susceptibility becomes very large; this is known as the Curie point and may be interpreted as the temperature at which the magnetic orientation is maintained against the thermal agitation. Thus, if a paramagnetic salt is cooled below its Curie point and then subjected to an external magnetic field, the entropy will be reduced by the magnetic ordering which takes place and for any particular starting temperature, the temperature will be reduced as a consequence of the reduction in entropy. This is the principle involved in achieving low temperatures. The ultimate temperature which may be reached depends on the mutual interaction of the elementary magnets; it is impossible to cool to a temperature lower than that at which spontaneous ordering of the elementary magnets occurs. For electronic events the limiting temperature is about 0.003 K. In fact, the basic principle of both electronic and nuclear magnetic cooling is the same, but since nuclear elementary magnets are about 2000 times smaller than electronic magnets, the starting temperature for nuclear cooling must be much lower, and also in practice much higher magnetic fields must be applied. To achieve low temperatures by the use of a paramagnetic salt, the material must first be cooled to about 1 K, and for nuclear magnetic cooling the temperature must first be reduced to about 0.01 K. By using a cerium–magnesium nitrate salt as the paramagnetic material in a so-called adiabatic demagnetization cryostat, temperatures in the region of a few millidegrees above the absolute zero have been reached. Many metals such as lead, tin, zinc, copper, and indium, become superconducting at such low temperatures, and in the superconducting state are much poorer thermal conductors than in the normal state; also, superconductivity can be destroyed by the application of a modest magnetic field, returning a metal to its normal state even although it is at a low temperature. By reducing the temperature to about 0.01 K by means of electronic demagnetization, and then beginning a nuclear demagnetization state by the application of an external field of the order of 6 Tesla, using copper or indium as the nuclear magnetic material, temperatures as low as 10^{-4} K have been achieved, and it is estimated that temperatures of the order of 5×10^{-7} K should be achievable in this manner.[1]

For the ferromagnetic metals, iron, cobalt, and nickel, the Curie temperature is higher than normal room temperatures, and at this point the magnetic susceptibility becomes infinite; below the Curie temperature, magnetization is spontaneous and increases at lower temperatures. Permanent magnets exhibit this spontaneous magnetization, although the situation is complicated in permanent magnets by the fact that such materials become magnetized in separate small domains of microscopic size. Also, ferromagnetic materials are often found in an apparently unmagnetized state, even large single crystals of iron existing in such a state. In such a single crystal, a number of domains each magnetized in one direction may be detected, and in the non-magnetic state the resultant magnetic moment is zero as a consequence of the effects in different directions neutralizing one another. The application of an external field causes the different domains to become magnetized in the same direction with the production of a resultant magnetic moment which may

---

[1] Lounasmaa, O. V., *Sci. Am.*, 1969, **12**, 26.

remain as a permanent moment when the field is withdrawn. When a ferromagnetic substance is magnetized, the saturation moment at the absolute zero represents the maximum degree of magnetization that can be realized, which would be expected to correspond to the situation where all the moments are parallel. The measured values for iron, cobalt, and nickel, expressed in Bohr magnetons are 2.22, 1.71, and 0.61, respectively. There is good reason to believe that essentially all of the magnetic moment of ordinary ferromagnetic substances is due to spin magnetization, the orbital magnetization effect being absent. Thus, the above numbers correspond to the number of unpaired electron spin moments in the metal crystals, and the fact that they are not integral numbers means that these ferromagnetic substances cannot be regarded as consisting of an array of identical atoms or irons, each having the same number of electrons absent from the $3d^{10}$-group of electrons. The energy band theory is however capable of explaining these moments. Alternatively, in the Pauling theory (Chapter M.2.), it is regarded that the ferromagnetic metals consist of atoms in more than one electronic state. The quantity measured experimentally for the saturation magnetic moments is $g\mu_B\sqrt{J(J + 1)}$, where J is the total angular momentum; if the magnetic moment is entirely due to spin momentum, then $g = 2$, and the quantity measured experimentally, the effective magnetic moment is $\sqrt{4S(S + 1)}\mu_B$, where S is the vector sum of the spin moments. As each electron may have a spin of $\pm 1/2$, the effective magnetic moment may be alternatively written as $\sqrt{n(n + 2)}$, where n is the number of electrons with uncompensated spin. Thus, for $n = 1, 2, 3, 4$, and 5, parallel spins, the effective magnetic moments are 1.73, 2.8, 3.9, 4.9, and 5.9, respectively, multiples of the Bohr magneton. In cases where the orbital angular momentum also contributes to the effective magnetic moment, the latter value may be very much greater than the number of unpaired electrons.

The earliest attempt to formulate a theory of ferromagnetism, which could account for the experimentally observed phenomena was due to Weiss,[1] which he was able to do on the basis of a relatively simple assumption. It was assumed that the magnetization of individual magnetic atoms was not due simply to the magnetic induction B, but that it was determined by a quantity $B + \alpha M$, where αM was proportional to the magnetization itself, and that this additional "inner field" was due to some sort of cooperative effect between already magnetized atoms influencing the others to align themselves parallel to the applied field. If the concentration of atoms per unit volume is N and the magnetic moment of an atom is then M/N, so that equation (m.7.61.) becomes

$$M/N = Jg\mu_B L(y), \qquad (\text{m.7.64.})$$

---

[1] Weiss, P., *Ann. Phys.*, Paris, 1932, **17**, 97.

Weiss showed that the correct experimental results are obtained by simply substituting $B + \alpha M$ for B. Making this substitution, the argument y in the Langevin function becomes

$$y = Jg\mu_B/kT[B + \alpha M], \tag{m.7.65.}$$

and for spontaneous magnetization when $B = 0$, this becomes

$$y = Jg\mu_B \alpha M/kT. \tag{m.7.66.}$$

If M is plotted as a function of y, then from (m.7.64.), again as a consequence of the analytical nature of the Langevin function, a curve something like the dotted curve in Fig. M.3.4. results but without the exponential tail on approaching the abscissa. On the other hand, from equation (m.7.66.), by again plotting M as a function of y, a series of straight lines is obtained, each passing through the origin, and with a slope M/y given by $kT/\alpha Jg\mu_B$. Each of these straight lines will intersect the curve given by (m.7.64.) to indicate a magnetization value greater than zero, provided that the slope $kT/\alpha Jg\mu_B$ is less than the initial slope of equation (m.7.64.); this initial slope, determined by the nature of $L(y)$ is $y/3$ (page 760), so that the initial slope of (m.7.64.) is $M/Ny = Jg\mu_B/3$, and if this value is equated with $kT/\alpha Jg\mu_B$, the critical temperature, the Curie temperature, above which spontaneous magnetization cannot occur, is found, for

$$k\Theta = (N\alpha/3)(Jg\mu_B)^2, \qquad \text{where } \Theta \text{ represents the Curie temperature.} \tag{m.7.67.}$$

The Weiss theory is thus capable of accounting for the observed spontaneous magnetization, since the saturation magnetization calculated as a function of temperature by the above method leads to the result where the magnetic moment is a maximum at the absolute zero and falls off with temperature up to the Curie temperature Θ. Above the Curie point, the approximation $L(y) = y/3$ is no longer applicable. By combining equations (m.7.64.) and (m.7.65.), and eliminating y, there is obtained

$$M/N = [(Jg\mu_B)^2/3kT][B + \alpha M], \tag{m.7.68.}$$

from which, on multiplication throughout by T and the elimination of the constant α by the use of (m.7.67.), it is found that

$$(M/N)[T - \Theta] = [Jg\mu_B]^2 B[3k]^{-1} \qquad \text{or} \qquad M/N = [Jg\mu_B]^2 B[3k(T - \Theta)]^{-1}. \tag{m.7.69.}$$

This is the so-called Curie–Weiss law, that the magnetic moment is proportional to the reciprocal of $(T - \Theta)$, and if the susceptibility is calculated from this expression, it is found to be very large compared with unity, as the Curie point is approached. In this latter event, unity may be disregarded in comparison with the term in $(T - \Theta)^{-1}$, and so the Curie law is obeyed.

The difficulty encountered in understanding the Weiss assumption of an "inner field," that causes the electrons to adopt parallel spin orientations with other electrons, did not find any satisfactory answer until the application of quantum mechanics to this problem was made by Heisenberg.[1] Even although it is recognized that there should be a magnetic interaction between electrons of different spin orientations, and that this spin–spin type interaction would be of the correct sign to account for Weiss's "inner field" effect, the resulting values of the constant α are several orders of magnitude too small to account for the observed values of the Curie temperatures. Heisenberg's suggested explanation is the same one that afforded an understanding of the large energy differences between singlet and triplet states and accounts for the origin of Hund's rules, observed in atomic spectra, and also accounts for the increased stability for states with the electron spins parallel to one another, as a consequence of the exchange effects arising from the antisymmetric nature of wave functions leading to the introduction of electrostatic terms in the energy expression, which lower the energy.

Many of the early applications of Heisenberg's idea that the stability of the ferromagnetic state was associated with a positive value of the exchange integral in these cases, compared with the negative exchange integral for the hydrogen molecule, for example, which leads to the electrons having opposed spin orientations and the formation of a normal electron pair bond, were based on a separated-atom interpretation of the magnetic interaction between them. Thus, the dependence of ferromagnetism on some critical value of the ratio of the internuclear distance in the crystal to a distance D (defined below), was believed to be indicated by the fact that although manganese is not ferromagnetic itself, some manganese compounds such as one of the manganese nitrides are ferromagnetic. On the assumption that ferromagnetism is associated with an incomplete group of electrons, which group has a low electron density near to the atomic nuclei, as for d- and f-electron groups, and that the sign of the exchange integral for ferromagnetic substances is positive, it was deduced that the exchange integral had a large positive value for the strongly ferromagnetic metals, and a small positive value for the lanthanide metals,[2] and that if the internuclear distance fell below some critical value the metal atoms formed normal electron pair bonds with one another, with electron pairs of opposed spin orientations. The distance D, referred to above, was taken to be twice the distance from a metal nucleus to the point in the electron density distribution around the nucleus, at which this electron density was a maximum. Thus, the ferromagnetism of man-

---

[1] Heisenberg, W., *Z. Physik.*, 1928, **49**, 619.
[2] Slater, J. C., *Revs. Mod. Phys.*, 1953, **25**, 199.

ganese nitride was explained as being due to the increased separation of the manganese atoms brought about by the interstitial solid solution of nitrogen in manganese. A similar explanation was offered for the ferromagnetism of the Heusler alloys of composition Cu_2AlMn and for some alloys of chromium and tellurium, where again neither copper, manganese, aluminum, chromium, or tellurium, are themselves ferromagnetic.[1] On this basis, it should be possible to synthesize new ferromagnetic alloys of chromium, tungsten, molybdenum, manganese, technicium, rhenium, iron, cobalt, nickel, ruthenium, rhodium, platinum, osmium, iridium, and platinum, by forming suitable alloys with the appropriate internuclear distances. Also, on the bases of these ideas, compression of a strongly ferromagnetic substance should decrease the ratio of internuclear distance to the distance D and thus decrease the magnetization. On the other hand, destruction of the magnetization should lead to a contraction, and it is believed that the low coefficient of expansion of alloys such as Invar (an alloy steel with 36% of nickel) is due to its composition being such that the normal thermal expansion effects are just compensated by a contraction due to the destruction of the magnetization.

The energy band description of ferromagnetism is, however, perhaps more appropriate, if as is apparently the case, ferromagnetism is attributable to the overlapping between the atomic orbitals of adjacent atoms. The $E–k$ energy band curves for a number of the first transition metal series have been calculated by the augmented plane wave method[2] (section M.6.8.). These curves clearly show that the $3d$-band, for these metals, becomes of progressively lower energy after the stage where the $3d$-electron group is complete (for zinc), and also that the Fermi energy increases steadily with increasing number of $3d$-electrons, from scandium to zinc. The Fermi energy level is found to occur within the d-band in all these elements up to copper, at which point it lies above the top of the d-band range of energies. It is indeed this fact, that for the transition elements the Fermi energy level lies within the partially filled d-bands, that provides the explanation of the magnetic behavior of these metals, and not the assumption of a positive exchange integral, for which no experimental or theoretical proof has ever been produced.[3] From the fact that iron, cobalt, and nickel, exhibit a permanent magnetic moment at 0 K, arising from electron spin effects, and consideration of the proper quantum mechanical description of a system composed of atoms with unbalanced spin orientations, an explanation of ferromagnetism in terms of the band theories is available.

---

1 Bates, L. F., *Modern Magnetism*, 1951, Cambridge University Press.
2 Mattheiss, L. F., *Phys. Rev.*, 1964, **A920**, 134.
3 Zener, C., *Phys. Rev.*, 1951, **81**, 440; **82**, 403; **83**, 299.

In the discussion of the Fermi hole concept in section Q.5.8., it was pointed out there and also in section M.3.7. that the Hartree–Fock method, which leads to a negative exchange energy term in the calculated energy of a system, takes no account of antiparallel spin correlation effects although the improvement afforded by the Hartree–Fock method compared with the Hartree method is due to the allowance made for correlation effects associated with electrons of parallel spin orientations. It would be expected, however, that a more complete theory would take account of a similar type of correlation hole in the distribution of all electrons, including those with both antiparallel as well as those with parallel spin orientations. This requires the use of different Hartree–Fock equations and different radial functions for the electrons of each spin type. In turn, this requires the calculation of a separate averaged exchange correction for electrons of each spin type and the calculation of a separate averaged potential energy for each type of electron spin. In the electron hole concept, the radius of the hole is inversely proportional to the one-third power of the charge density, since the volume of the hole is $4\pi R^3/3$, where R is the radius of the atom concerned, and the product of this volume and the density of the electrons of the spin type concerned must be equivalent to one electronic charge. In a ferromagnetic substance, with an excess of electrons of one spin type, say α, there would be a larger density of electrons with spin α than with the other spin type β. Therefore, the exchange correction will be larger for the electrons of spin α, and since the exchange correction is subtracted from the Coulomb energy term to obtain the potential energy in which these electrons move, the electrons with spin α will be moving in a region of lower potential energy than those with spins β. Their energy bands will therefore occur at a lower energy than those for electrons of spin β, so that there will be more occupied states for the electrons with spin α below the Fermi level than for electrons with spin β. By the use of the Hartree–Fock method, with different orbital functions for electrons of different spin types, and finding the separate densities of electrons of each spin type, which lead through the two separate exchange corrections, to potentials which give rise to energy bands producing the initially assumed values of the charge density, it should be possible by the application of a sufficient number of iterations of the self-consistent field type of calculation to calculate the energy bands of a ferromagnetic substance. This type of calculation has been carried out by Connolly[1] for the ferromagnetic state of nickel, and for iron and nickel by Wakoh,[2] by the augmented plane wave method (section M.6.8.). These calculations show that the two sets of energy bands corresponding to the two spin orientations, are separated from one

---

[1] Connolly, J. W. D., *Phys. Rev.*, 1967, **159**, 415.
[2] Wakoh, S., *J. Phys. Soc. Japan*, 1965, **20**, 1894.

another, and that the Fermi level occurs at such a height that the d-type energy bands for the electrons of spin α are entirely filled while those for the other spin type are partly unoccupied, leading to the net magnetization, which is observed experimentally, and in addition the calculated magnetization deduced from these calculations is very close to the experimental value of 0.6 Bohr magnetons. The calculated values for iron also agree well with the experimentally observed value of 2.22 Bohr magnetons, so that the energy band theory is capable of accounting for the fractional number of Bohr magnetons observed for the saturation magnetic moments of these metals. These calculated energy bands also agree well with the high observed values for the electronic specific heats of iron and nickel, consistent with their high density of states, and with the experimental observations on the de Haas–van Alphen effects for these metals, as well as indicating that the correlation effect in reducing the energy, is more important for electrons of opposed spin moments than for those of the same spin orientation. No information about the Curie temperature of ferromagnetic substances can be obtained, however, from the energy separation between the bands corresponding to opposed spin directions.

A class of substances which are characterized by a critical temperature at which their magnetic susceptibilities attain a maximum value, are referred to as antiferromagnetic materials. Long before it was possible to obtain any experimental support for the idea, Néel[1] proposed that in these substances the atomic moments are arranged in definite alternating arrays, rather analogous to superlattices in binary alloy systems (see Fig. M.7.16.). Antiferromagnetic substances are regarded as paramagnetic materials which display a susceptibility above their characteristic or Néel temperature, which depends on temperature in accordance with the Curie–Weiss law with a negative value for the Curie temperature. Below the Néel temperature the susceptibility decreases towards zero with decreasing temperature. Among the transition metals, α-manganese is antiferromagnetic at 4.2 K, and contains three kinds of atom (page 737); slight antiferromagnetism has been detected in chromium, with a Néel temperature of 475 K, and a magnetic moment of 0.4 Bohr magnetons, in reasonable agreement with Pauling's view that the atomic orbitals are just beginning to be of importance at chromium. No indication of antiferromagnetic ordering can be detected for tungsten, molybdenum, vanadium, or niobium.[2] Many crystalline materials have been studied, for which the lowest energy corresponds to the antiferromagnetic state rather than the ferromagnetic state, such as MnF_2, $FeCl_2$, NiO, $TiCl_3$, $MnSe$, FeF_2, Cr_2O_3, MnO, $FeCO_3$, MnS, $MnTe$, FeO,

[1] Néel, L., *Ann. Phys.*, 1932, **18**, 5; 1936, **5**, 232.
[2] Shull, C. G., and Wollan, E. O., *Solid State Physics*, 1962, **Vol. 2**, 137–217, Academic Press, New York.

$CoCl_2$, CoO, $NiCl_2$, some europium compounds (EuS is however, ferromagnetic, and a semiconductor[1]), and $MnBr_2$. In some cases the transition from ferromagnetic to antiferromagnetic behavior takes place with a temperature change of the order of magnitude of a few degrees Kelvin, indicating that the energy differences between the two states are very small. From equation (m.7.67.), it can be seen that the quantity $k\Theta$ may be regarded as the product of one-third of the saturation magnetic moment $Jg\mu_B$ and the Weiss "inner field" $N\alpha Jg\mu_B$, resulting from the saturation magnetization; this means that the energy involved in reversing the direction of the magnetization of a magnetic species (atom or ion), which would be twice the product of the magnetic moment and the "inner field," is about six times the value of $k\Theta$, which for a Curie temperature of 25 K would correspond to an energy of 4.75×10^{-4} Hartrees. This is indeed quite a small quantity, being about 630 times smaller than the energy separation calculated for the two spin types in EuS, for example.[1]

The investigation of the magnetic structure of compounds, like the investigation of the lattice dynamics of crystals, only became possible with the availability of high flux nuclear reactors. Thermal neutrons have wavelengths in the range of X-ray wavelengths and so are diffracted or scattered by crystalline materials in the same manner as X-radiation, but in addition the uncharged neutrons undergo an additional scattering owing to the interaction between the neutron spin moment and the magnetic moment of paramagnetic species, so that an antiferromagnetic substance, which is essentially a magnetic superlattice, will produce extra diffraction lines in the same way that an ordered intermetallic phase produces additional X-ray diffraction lines in X-ray crystallography. Furthermore, the study of the diffuse scattering of beams of thermal neutrons by magnetic materials enables the magnetic moments to be estimated.[2,3] Neither X-radiation nor electron beams behave in this manner. Although electrons have a magnetic moment, the Coulomb repulsion between charged electrons and the electrons in atoms and ions prevent the spin–spin type interaction to be observed. In a crystalline solid the nuclei vibrate about a set of equilibrium positions which constitute a three-dimensional array. On the assumption that the interatomic forces are harmonic in nature, the nuclear motions may be represented by a family of non-interacting plane waves. The coherent part of the scattering process can be split up into terms involving the creation and annihilation of 0, 1, 2, ..., quanta (phonons) of these waves. The one-phonon term is principally of

---

[1] Cho, S. J., *Phys. Rev.*, 1966, **157**, 632.

[2] Bacon, G. E., *Neutron Diffraction*, 1955, Oxford University Press, London.

[3] Shull, C. G., and Wollan, E. O., *Solid State Physics*, 1962, **Vol. 2**, 137, Academic Press, New York.

importance for the study of the frequency-wave vector dispersion relations for solids. If the energy and momentum of the incident and scattered neutrons are E_0, $h\mathbf{k}_0/2\pi$ and E' and $h\mathbf{k}'/2\pi$, respectively, the conditions governing these coherent one-phonon processes may be expressed in terms of the conservation of "crystal momentum" and the conservation of energy, as

$$\mathbf{Q} = \mathbf{k}_0 - \mathbf{k}' = 2\pi\boldsymbol{\tau} + \mathbf{q} \quad \text{and} \quad E_0 - E' = \pm h\nu, \quad \text{(m.7.70.)}$$

where ν is the frequency of the normal vibrational mode whose reduced wave vector is $\mathbf{q}$, $\mathbf{Q}$ is the momentum transfer vector, and $\boldsymbol{\tau}$ is a vector of the reciprocal lattice of the crystal. These two equations are satisfied only when ν and $\mathbf{q}$ satisfy the "dispersion relation" for the normal modes of the crystal, $\nu = \nu_j(\mathbf{q})$, where j denotes the polarization of the normal mode. The energy distribution, therefore, of initially monoenergetic neutrons, after coherent one-phonon scattering at some particular angle from a specimen of a single crystal of a material, consists ideally of a small number of δ-function peaks, expressed by (m.7.70.). From the location of each peak, the frequency and wave vector of the phonon concerned may be deduced by means of equations (m.7.70.). In this manner, the lattice dynamics of a number of materials have been examined.[1-4] The magnetic inelastic scattering method may be employed to study the dynamics of systems of coupled magnetic moments, analogously to the study of crystal and liquid dynamics. The dynamics of any system of ordered magnetic moments can be described in terms of spin waves, the "dispersion relation" for which can be determined in the same way as the "dispersion relation" for crystal lattice vibrations, by use of the two equations of conservation of momentum and energy given in (m.7.70.). In this case, the scattered neutrons are observed to fall into groups in the energy distribution corresponding to the creation or annihilation of a single quantum of spin-wave energy, referred to as a magnon. Neutron groups arising from interaction with magnons can be identified unambiguously from the response of their intensities to an applied magnetic field. Several magnetic substances have been studied in this way.[5,6]

In the antiferromagnetic substances, there are interpenetrating sub-lattices whose opposed magnetic spin orientations cancel each other at very low

---

[1] Brockhouse, B. N., in *Proc. Symp. Inelastic Scattering of Neutrons in Solids and Liquids*, 1961, International Atomic Energy Agency, Vienna, 113–157.

[2] Brockhouse, B. N., Woods, A. D. B., Dolling, G., and Thorson, I. M., *Third U.N. International Conference on the Peaceful Uses of Atomic Energy*, 1964, I.A.E.A., Geneva.

[3] *Lattice Dynamics*, edited by R. F. Wallis, 1965, Pergamon Press, Oxford.

[4] *Inelastic Scattering of Neutrons*, 1965, International Atomic Energy Agency, Vienna.

[5] Sinclair, R. N., and Brockhouse, B. N., *Phys. Rev.*, 1960, **120**, 1638.

[6] Brockhouse, B. N., and Watanabe, H., in *Inelastic Scattering of Neutrons in Solids and Liquids*, I.A.E.A., 1963, 297–308, Vienna.

temperatures where the magnetic ordering approaches perfection. Another type of cooperative magnetic phenomenon, also suggested by Néel, is referred to as ferrimagnetism.[1] In these materials, there is an interaction between atomic magnetic moments such as to cause them to align themselves in antiparallel orientations, as in antiferromagnetic substances, but with a difference in the total moments in the total moments in the two directions, so that the resultant moment is not zero.[2] The properties of ferrimagnetic substances are qualitatively similar to those of ferromagnetic substances; there is a Curie transition temperature, above which the substance concerned is paramagnetic and below which it is ferromagnetic. However, the total magnetic moment indicated in the paramagnetic region is much greater than that given by saturation in the ferromagnetic region. The curves obtained by plotting the reciprocal of the susceptibility as a function of temperature, for ferrimagnetic materials, are hyperbolae instead of the straight lines characteristic of paramagnetic substances. In some ferrimagnetic materials, the two magnetic sublattices may be geometrically identical but may be occupied by atoms with different spin orientations, or in other cases there may be different types of crystallographic sites in the two sub-lattices. The most studied type of ferrimagnetic materials are the ferrites, which are compounds of the general composition XY_2O_4 with the spinel type structure, so-called after the structure found for the mineral spinel $MgAl_2O_4$. The crystallographic unit cell of this structure contains 32 approximately cubic close packed oxygen atoms, with equivalent positions within this unit cell for 8 tetrahedrally coordinated (with oxygen atoms) and 16 octahedrally coordinated atoms. There are altogether 64 and 32 tetrahedral and octahedral holes respectively, in the unit cell. Magnetite, Fe_3O_4, is indeed ferrimagnetic, with 8 iron atoms in one set of equivalent positions and 16 iron atoms in the other; the observed susceptibilities above the Curie temperature are compatible with a magnetic moment of 5.2 Bohr magnetons for Fe^{++} and 5.9 Bohr magnetons for Fe^{3+}, and Néel has interpreted the observed value of the saturation moment of magnetite as indicating that 8 Fe^{++} and 8 Fe^{3+} have parallel spin alignments and that the other 8 Fe^{3+} ions in the unit cubic cell have antiparallel orientations. The ferrites, which may be regarded as being derived from magnetite by substitution of the Fe^{++} ions by one or more of the dipositive metal ions of magnesium, manganese, cobalt, nickel, copper, zinc, or cadmium; these substances are of important practical use in making magnetic tape and in other magnetic applications at high frequencies requiring material of high electrical resistivity. There are, in addition to the ferrites, other structures

---

[1] Néel, L., *Ann. Phys.*, 1948, **3**, 137.
[2] Kittel, C., *Introduction to Solid State Physics*, 1957, John Wiley & Sons, Inc., New York.

which display ferrimagnetic properties; actually, any ordered magnetic system containing non-equivalent lattice sites for the magnetic ions may be expected to be of this type. An intermetallic compound of this type, is Mn_2Sb, which exhibits a ferromagnetic Curie temperature of 550 K and a saturation magnetic moment of 0.936 Bohr magnetons per manganese atom.

Appendix

Table of Physical Constants[a]

Quantity	Symbol	Value in SI Units	Error (ppm)
Avogadro Number	N_A	$6.022169(40) \times 10^{26}$ kmole^{-1}	6.6
Planck Constant	h	$6.626196(50) \times 10^{-34}$ J s	7.6
	$\hbar = h/2\pi$	$1.0545919(80) \times 10^{-34}$ J s	7.6
Electron charge	ε	$1.6021917(70) \times 10^{-19}$ C	4.4
Velocity of light	c	$2.9979250(10) \times 10^{8}$ m s^{-1}	0.33
Atomic mass unit (Dalton)	amu	$1.660531(11) \times 10^{-27}$ kg	6.6
Electron rest mass	m_ε	$9.109558(54) \times 10^{-31}$ kg	6.0
Proton rest mass	m_p	$1.672614(11) \times 10^{-27}$ kg	6.6
Neutron rest mass	m_n	$1.674920(11) \times 10^{-27}$ kg	6.6
Gas Constant	R_0	$8.31434(35) \times 10^{3}$ J kmole^{-1} K^{-1}	42
Boltzmann Constant (R_0/N_A)	k	$1.380622(59) \times 10^{-23}$ J K^{-1}	43
Faraday Constant ($N_A\varepsilon$)	F	$9.648670(54) \times 10^{7}$ C kmole^{-1}	5.5
Electron charge to mass ratio	$\varepsilon/m_\varepsilon$	$1.7588028(54) \times 10^{11}$ C kg^{-1}	3.1
First radiation constant ($8\pi hc$)	C_1	$4.992579(38) \times 10^{-24}$ J m	7.6
Second radiation constant (hc/k)	C_2	$1.438833(61) \times 10^{-2}$ m K	43

[a] The numerical values of the physical constants are those recommended by B. N. Taylor, W. H. Parker, and D. N. Langenberg, on the basis of their final least squares adjusted values reported in the *Reviews of Modern Physics* Monograph, "The Fundamental Constants and Quantum Electrodynamics," published by Pergamon Press, New York, 1969. The numbers in parentheses are the standard deviation uncertainties in the last digits of the quoted value, computed on the basis of internal consistency among the various constants.

Quantity	Symbol	Value in SI Units	Error (ppm)
Bohr radius ($h^2/4\pi^2 m_\varepsilon \varepsilon^2$)	a_0	$5.2917715(81) \times 10^{-11}$ m	1.5
Rydberg Constant ($2\pi^2 m_\varepsilon \varepsilon^4/h^3 c$)	R_∞	$1.09737312(11) \times 10^7$ m^{-1}	0.10
Stefan–Boltzmann Constant ($2\pi^5 k^4/15c^2 h^3$)	σ	$5.66961(96) \times 10^{-8}$ W m^{-2} K^4	170
Fine Structure Constant ($2\pi\varepsilon^2/ch$)	α α^{-1}	$7.297351(11) \times 10^{-3}$ $137.03602(21)$	1.5 1.5
Classical electron radius ($\alpha^3/4\pi R_\infty$)	r_0	$2.817939(13) \times 10^{-15}$ m	4.6
Bohr magneton ($\varepsilon h/4\pi m_\varepsilon c$)	μ_B	$9.274096(65) \times 10^{-24}$ J T^{-1}	7.0
Nuclear magneton ($\varepsilon h/4\pi m_p c$)	μ_n	$5.050951(50) \times 10^{-27}$ J T^{-1}	10
Proton magnetic moment	μ_p	$1.4106203(99) \times 10^{-26}$ J T^{-1}	7.0
Proton magnetic moment in Bohr magnetons	μ_p/μ_B	$1.52103264(46) \times 10^{-3}$	0.30
Compton wavelength of the electron ($h/m_\varepsilon c$)	λ_C $\lambda_C/2\pi$	$2.4263096(74) \times 10^{-12}$ m $3.861592(12) \times 10^{-13}$ m	3.1 3.1
Compton wavelength of the neutron ($h/m_n c$)	$\lambda_{C,n}$ $\lambda_{C,n}/2\pi$	$1.3196217(90) \times 10^{-15}$ m $2.100243(14) \times 10^{-16}$ m	6.8 6.8
Compton wavelength of the proton ($h/m_p c$)	$\lambda_{C,p}$ $\lambda_{C,p}/2\pi$	$1.3214409(90) \times 10^{-15}$ m $2.103139(14) \times 10^{-16}$ m	6.8 6.8
Ratio of proton mass to electron mass (m_p/m_ε)		$1836.109(11)$	6.2
Quantum of circulation	$h/2m_\varepsilon$	$3.636947(11) \times 10^{-4}$ J s kg^{-1}	3.1
Magnetic flux quantum	h/ε	$4.135708(14) \times 10^{-15}$ J s C^{-1}	3.3
Gravitational Constant	G	$6.6732(31) \times 10^{-11}$ N m^2 kg^{-2}	460
Standard molar gas volume	V_0	22.4136 m^3 kmole^{-1}	

Name Index

Subject Index